Area A; circumference C; volume V; curved surface area S.

RIGHT TRIANGLE

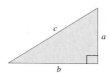

Pythagorean Theorem: $c^2 = a^2 + b^2$

TRIANGLE

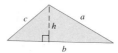

$A = \frac{1}{2}bh$ $C = a + b + c$

EQUILATERAL TRIANGLE

$h = \frac{\sqrt{3}}{2}s$ $A = \frac{\sqrt{3}}{4}s^2$

RECTANGLE

$A = lw$ $C = 2l + 2w$

PARALLELOGRAM

$A = bh$

TRAPEZOID

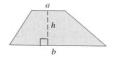

$A = \frac{1}{2}(a+b)h$

CIRCLE

$A = \pi r^2$ $C = 2\pi r$

CIRCULAR SECTOR

$A = \frac{1}{2}r^2\theta$ $s = r\theta$

CIRCULAR RING

$A = \pi(R^2 - r^2)$

RECTANGULAR BOX

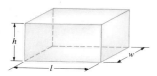

$V = lwh$ $S = 2(hl + lw + hw)$

SPHERE

$V = \frac{4}{3}\pi r^3$ $S = 4\pi r^2$

RIGHT CIRCULAR CYLINDER

$V = \pi r^2 h$ $S = 2\pi rh$

RIGHT CIRCULAR CONE

$V = \frac{1}{3}\pi r^2 h$ $S = \pi r\sqrt{r^2 + h^2}$

FRUSTUM OF A CONE

$V = \frac{1}{3}\pi h(r^2 + rR + R^2)$

PRISM

$V = Bh$ with B the area of the base

THE PRINDLE, WEBER & SCHMIDT SERIES IN MATHEMATICS

Althoen and Bumcrot, *Introduction to Discrete Mathematics*

Brown and Sherbert, *Introductory Linear Algebra with Applications*

Buchthal and Cameron, *Modern Abstract Algebra*

Burden and Faires, *Numerical Analysis,* Third Edition

Cullen, *Linear Algebra and Differential Equations*

Cullen, *Mathematics for the Biosciences*

Eves, *In Mathematical Circles*

Eves, *Mathematical Circles Adieu*

Eves, *Mathematical Circles Revisted*

Eves, *Mathematical Circles Squared*

Eves, *Return to Mathematical Circles*

Fletcher and Patty, *Foundations of Higher Mathematics*

Geltner and Peterson, *Geometry for College Students*

Gilbert and Gilbert, *Elements of Modern Algebra,* Second Edition

Gobran, *Beginning Algebra,* Fourth Edition

Gobran, *College Algebra*

Gobran, *Intermediate Algebra,* Fourth Edition

Gordon, *Calculus and the Computer*

Hall, *Algebra for College Students*

Hall and Bennett, *College Algebra with Applications*

Hartfiel and Hobbs, *Elementary Linear Algebra*

Hunkins and Mugridge, *Applied Finite Mathematics,* Second Edition

Kaufmann, *Algebra for College Students,* Second Edition

Kaufmann, *Algebra with Trigonometry for College Students*

Kaufmann, *College Algebra*

Kaufmann, *College Algebra and Trigonometry*

Kaufmann, *Elementary Algebra for College Students,* Second Edition

Kaufmann, *Intermediate Algebra for College Students,* Second Edition

Kaufmann, *Precalculus*

Kaufmann, *Trigonometry*

Keisler, *Elementary Calculus: An Infinitesimal Approach,* Second Edition

Konvisser, *Elementary Linear Algebra with Applications*

Laufer, *Discrete Mathematics and Applied Modern Algebra*

Nicholson, *Linear Algebra with Applications*

Pasahow, *Mathematics for Electronics*

Powers, *Elementary Differential Equations*

Powers, *Elementary Differential Equations*

Powers, *Elementary Differential Equations with Boundary Value Problems*

Powers, *Elementary Differential Equations with Linear Algebra*

Proga, *Arithmetic and Algebra*

Proga, *Basic Mathematics*, Second Edition

Radford, Vavra, and Rychlicki, *Introduction to Technical Mathematics*

Radford, Vavra, and Rychlicki, *Technical Mathematics with Calculus*

Rice and Strange, *Calculus and Analytical Geometry for Engineering Technology*

Rice and Strange, *College Algebra*, Third Edition

Rice and Strange, *Plane Trigonometry*, Fourth Edition

Rice and Strange, *Technical Mathematics*

Rice and Strange, *Technical Mathematics and Calculus*

Schelin and Bange, *Mathematical Analysis for Business and Economics*, Second Edition

Strnad, *Introductory Algebra*

Swokowski, *Algebra and Trigonometry with Analytic Geometry*, Sixth Edition

Swokowski, *Calculus with Analytic Geometry*, Second Alternate Edition

Swokowski, *Calculus with Analytic Geometry*, Fourth Edition

Swokowski, *Fundamentals of Algebra and Trigonometry*, Sixth Edition

Swokowski, *Fundamentals of College Algebra*, Sixth Edition

Swokowski, *Fundamentals of Trigonometry*, Sixth Edition

Swokowski, *Precalculus: Functions and Graphs*, Fifth Edition

Tan, *Applied Calculus*

Tan, *Applied Finite Mathematics*, Second Edition

Tan, *Calculus for the Managerial, Life, and Social Sciences*

Tan, *College Mathematics*, Second Edition

Venit and Bishop, *Elementary Linear Algebra*, Second Edition

Venit and Bishop, *Elementary Linear Algebra*, Alternate Second Edition

Willard, *Calculus and Its Applications*, Second Edition

Willerding, *A First Course in College Mathematics*, Fourth Edition

Wood and Capell, *Arithmetic*

Wood, Capell, and Hall, *Developmental Mathematics*, Third Edition

Wood, Capell, and Hall, *Intermediate Algebra*

Wood, Capell, and Hall, *Introductory Algebra*

Zill, *A First Course in Differential Equations with Applications*, Third Edition

Zill, *Calculus with Analytic Geometry*, Second Edition

Zill, *Differential Equations with Boundary-Value Problems*

CALCULUS
• A •
FIRST
COURSE

SECOND
ALTERNATE
EDITION

EARL W. SWOKOWSKI
MARQUETTE UNIVERSITY

PWS·KENT PUBLISHING COMPANY
BOSTON

NELSON CANADA
TORONTO · CANADA

PWS–KENT
Publishing Company

20 Park Plaza
Boston, Massachusetts 02116

Dedicated to the memory of

my mother and father,

Sophia and John Swokowski

PWS-KENT Publishing Company is a division of Wadsworth, Inc.

92 91 90 89 88 / 10 9 8 7 6 5 4 3 2 1

LIBRARY OF CONGRESS CATALOGING-IN-PUBLICATION DATA

Swokowski, Earl W.
    Calculus, a first course / Earl W. Swokowski.
  —2nd alternate edition.
      p.   cm.
    Includes index.
    ISBN 0-534-91559-0
    1. Calculus.   2. Geometry, Analytic.   I. Title
QA303.S94   1988c.
515—dc19   88-12404
                                            CIP

COVER IMAGE · Copyright © 1987 Nicholas DeSciose
SPONSORING EDITOR · David Geggis
PRODUCTION AND DESIGN · Katherine Townes
TEXT COMPOSITION · Syntax International Pte. Ltd.
TECHNICAL ARTWORK · J & R Art Services; Deborah Schneck
COVER MANUFACTURING · New England Book Components, Inc.
TEXT MANUFACTURING · Arcata Graphics/Hawkins

Printed in the United States of America

PREFACE

This book is a major revision of the Alternate Edition of *Calculus, A First Course*. One of my goals was to maintain the mathematical soundness of the previous edition, while making discussions somewhat less formal by rewriting and by placing more emphasis on graphs and figures. Another objective was to stress the usefulness of calculus through a variety of new applied examples and exercises from many different disciplines. Finally, suggestions for improvements from instructors led me to change the order of presentation of certain topics.

A great deal of rewriting, reorganization, and new material has gone into this edition, and to list the changes in detail would make the preface excessively long. The following remarks merely highlight the principal changes from the previous edition.

HIGHLIGHTS OF THIS EDITION

- The review of graphs of functions in Chapter 1 includes vertical and horizontal shifts, stretching, and reflections. Many exercises that involve applications are designed to prepare students for later work with extrema and related rates. Graphs of the trigonometric functions are included in the review of trigonometry.
- In Chapter 2 the limit concept is motivated informally, prior to the rigorous approach considered in Section 2.3. To help raise students' level of interest at this early stage of calculus, examples and exercises that involve unusual applications — such as compressed gases, optics, G-forces

experienced by astronauts, drug dosage levels, and the theory of relativity — are included throughout the chapter.
- Derivatives of all six trigonometric functions are considered in Chapter 3. The concept of rate of change (previously in Chapter 4) is introduced in Section 3.3 to provide a greater variety of applications of the derivative early in the text. Related rates are discussed in Section 3.9.
- Chapter 4 consists of concepts pertaining to extrema, graphing, and antiderivatives. Applications to economics (which formerly constituted a separate section) are included, where appropriate, in this and other chapters.
- Properties of the definite integral and the definition of average value of a function are discussed in one section of Chapter 5. Numerical integration involving the use of approximate data is considered at the end of the chapter and in the applications discussed in Chapter 6.
- The concepts of arc length and surfaces of revolution are introduced in Section 6.5 so that mathematical applications of the definite integral are considered in the first half of Chapter 6. The physical applications in the second half of the chapter are independent of one another and may be studied in any order (or omitted), depending on class objectives. Moments and the center of mass of a lamina are discussed in Section 6.8. Nontraditional applications of the definite integral are given in the final section.
- Chapters 7 and 8 include a large number of examples and exercises on applications of the natural logarithmic function, the natural exponential function, and other transcendental functions to diverse fields.

- The discussion in Chapter 9 is limited to techniques of integration. Applications from earlier chapters are reconsidered in exercises that require advanced methods of integration.
- Chapter 10 contains numerous new examples and exercises that involve applications of indeterminate forms and improper integrals.
- The approach to infinite sequences in Chapter 11 provides a geometric motivation for the concepts of convergence and divergence. The ratio test for positive-term series is introduced early, and alternating series and absolute convergence are discussed in one section. A new table summarizes all the tests discussed in the chapter.
- In Chapter 12 the notion of eccentricity of conic sections is prominent. Applications include LORAN navigation and orbits of planets and comets.
- The topics of tangent lines, arc length, and surfaces of revolution that are associated with parametrized curves are consolidated in one section in Chapter 13. Polar equations of conics are considered in the last section.
- Chapter 14 is an introduction to partial differentiation and differential equations. Partial differentiation is an extension of the concept of derivatives to functions of more than one variable. First- and second-order linear differential equations are discussed with applications. The final section applies infinite series to solutions of differential equations.

FEATURES OF THE TEXT

APPLICATIONS Every calculus book has applied problems from engineering, physics, chemistry, biology, and economics. This revision also includes exercises from specialized fields such as physiology, sociology, psychology, ecology, oceanography, meteorology, radiotherapy, astronautics, and transportation.

EXAMPLES Each section contains carefully chosen examples to help students understand and assimilate new concepts. Whenever feasible, applications are included to demonstrate the usefulness of a topic.

EXERCISES Exercise sets begin with routine drill problems and progress gradually to more difficult types. Applied problems generally appear near the end of a set to allow students to gain confidence in manipulations and new ideas before attempting questions that require analyses of practical situations.

Many new exercises involving applications are included to stress the diversity and power of calculus. Many applications are novel, and differ greatly from standard applications that have been used traditionally in calculus books.

A review section at the end of each chapter consists of a list of important topics and pertinent exercises.

Answers to the odd-numbered exercises are given at the end of the text.

CALCULATORS Calculators are referred to where appropriate. It is possible to work most of the exercises without a calculator; however, instructors may wish to encourage its use for computations involving approximate data.

TEXT DESIGN A new use of color makes discussions easier to follow and highlights major concepts. All figures have been redrawn for this edition and, wherever possible, are placed in the margin next to the discussion. Graphs are usually labeled and color-coded to clarify complex figures. Drawings have been added to many exercise sets to help students visualize applied problems.

FLEXIBILITY Syllabi from schools that used the previous edition attest to the flexibility of the text. Sections and chapters can be rearranged in different ways, depending on the objectives and length of the course.

SUPPLEMENTS FOR THE INSTRUCTOR

The following teaching aids may be obtained from the publisher:

> *Complete Solutions Manual,* Volumes I and II
>> by Jeff Cole, Anoka-Ramsey Community College, and Gary Rockswold, Mankato State University
>
> Even-numbered answer booklet
> Computerized test generator (for IBM-PCs and compatibles)
> Printed test bank
> PWS-KENT GradeDisk

SUPPLEMENTS FOR THE STUDENT

The following supplements are available:

> *Student Supplement,* Volumes I and II
>> by Thomas A. Bronikowski, Marquette University, which contains solutions for every third problem in the text
>
> *Programmed Study Guide*
>> by Roy A. Dobyns, Carson-Newman College, which is keyed to the first nine chapters of the text
>
> True BASIC™ Calculus Software by True BASIC, Inc. for IBM-PCs and compatibles and the Apple MacIntosh
>
> *Calculus and the Computer*
>> by Sheldon Gordon, Suffolk Community College

ACKNOWLEDGMENTS

I wish to thank Michael R. Cullen of Loyola Marymount University for supplying most of the new exercises dealing with applications. This large assortment of problems provides strong motivation for the mathematical concepts introduced in the text. Professor Cullen also furnished the computer graphics that accompany some of these exercises. Christian C. Braunschweiger of Marquette University contributed many suggestions that improved the exposition.

I also wish to thank the individuals who reviewed the manuscript and offered suggestions, and the following mathematics educators, who met with me and representatives from PWS-KENT for several days in the summer of 1986, and later reviewed portions of the manuscript:

Cliff Clarridge, *Santa Monica College*
Jeff Cole, *Anoka-Ramsey Community College*
Michael Cullen, *Loyola Marymount University*
Bruce Edwards, *University of Florida*
Michael Schneider, *Belleville Area College*

Their comments on pedagogy and their specific recommendations about the content of calculus courses helped me to improve the book.

I am grateful for the excellent cooperation of the staff at PWS-KENT. Two people in the company deserve special mention. My production editor Kathi Townes did a truly exceptional job in designing the book and taking care of an enormous number of details associated with the production. I cannot thank her enough for her assistance. Managing Editor Dave Geggis supervised the project, contacted many reviewers and users of my texts, and was a constant source of information and advice.

In addition to all the persons named here, I express my sincere appreciation to the many unnamed students and teachers who have helped shape my views on how calculus should be presented in the classroom.

EARL W. SWOKOWSKI

CONTENTS

Calculus was invented in the seventeenth century as a tool for investigating problems that involve motion. Algebra and trigonometry may be used to study objects that move at constant speeds along linear or circular paths; but calculus is needed if the speed varies or if the path is irregular. An accurate description of motion requires precise definitions of *velocity* (the rate at which distance changes per unit time) and *acceleration* (the rate at which velocity changes). These definitions may be obtained by using one of the fundamental concepts of calculus—the *derivative*.

Although calculus was developed to solve problems in physics, its power and versatility have led to uses in many diverse fields of study. Modern-day applications of the derivative include investigating the rate of growth of bacteria in a culture, predicting the outcome of a chemical reaction, measuring instantaneous changes in electrical current, describing the behavior of atomic particles, estimating tumor shrinkage in radiation therapy, forecasting economic profits and losses, and analyzing vibrations in mechanical systems.

The derivative is also useful in solving problems that involve maximum or minimum values, such as manufacturing the least expensive rectangular box that has a given volume, calculating the greatest distance a rocket will travel, obtaining the maximum safe flow of traffic across a long bridge, determining the number of wells to drill in an oil field for the most efficient production, finding the point between two light sources at which illumination will be greatest, and maximizing corporate revenue for a particular product. Mathematicians often employ derivatives to find tangent lines to curves and to help analyze graphs of complicated functions.

Another fundamental concept of calculus—the *definite integral*—is motivated by the problem of finding areas of regions that have curved boundaries. Definite integrals are employed as extensively as derivatives and in as many different fields. Some applications are finding the center of mass or moment of inertia of a solid, determining the work required to send a space probe to another planet, calculating the blood flow through an arteriole, estimating depreciation of equipment in a manufacturing plant, and interpreting the amount of dye dilution in physiological tests that involve tracer methods. We can also use definite integrals to investigate mathematical concepts such as area of a curved surface, volume of a geometric solid, or length of a curve.

The concepts of derivative and definite integral are defined by limiting processes. The notion of *limit* is the initial idea that separates calculus from elementary mathematics. Sir Isaac Newton (1642–1727) and Gottfried Wilhelm Leibniz (1646–1716) independently discovered the connection between derivatives and integrals and are both credited with the invention of calculus. Many other mathematicians have added greatly to its development in the last 300 years.

The applications of calculus mentioned here represent just a few of the many considered in this book. We can't possibly discuss all the uses of calculus, and more are being developed with every advance in technology. Whatever your field of interest, calculus is probably used in some pure or applied investigations. Perhaps *you* will discover a new application for this branch of science.

This chapter contains a review of topics required for the study of calculus. After a brief discussion of real numbers, coordinate systems, and graphs in two dimensions, we will turn our attention to one of the most important concepts in mathematics—the notion of *function*.

FUNCTIONS AND GRAPHS

1.1 REAL NUMBERS

Calculus is based on properties of **real numbers.** If we add the real number 1 successively to itself we obtain the **positive integers** 1, 2, 3, 4, ... The **integers** consists of all positive and negative integers together with the real number 0. We sometimes list the integers as follows:

$$\ldots, \quad -4, \quad -3, \quad -2, \quad -1, \quad 0, \quad 1, \quad 2, \quad 3, \quad 4, \quad \ldots$$

A **rational number** is a real number that can be expressed as a quotient a/b, for integers a and b with $b \neq 0$. Real numbers that are not rational are **irrational.** For example, the ratio of the circumference of a circle to its diameter is irrational. This real number is denoted by π and the notation $\pi \approx 3.1416$ is used to indicate that π is *approximately equal* to 3.1416. Another example of an irrational number is $\sqrt{2}$.

Real numbers may be represented by *nonterminating decimals*. For example, the decimal representation for the rational number 177/55 is found by division to be 3.2181818 . . . , where the digits 1 and 8 repeat indefinitely. Rational numbers may always be represented by *repeating* decimals. Irrational numbers may be represented by nonterminating and *nonrepeating* decimals.

A *one-to-one correspondence* exists between the real numbers and the points on a line l, in the sense that to each real number a there corresponds one and only one point P on l and, conversely, to each point P there corresponds one real number. Such a correspondence is illustrated in Figure 1.1, where we have indicated several points corresponding to real numbers. The

FIGURE 1.1

point O that corresponds to the real number 0 is the **origin.**

The number a that is associated with a point A on l is the **coordinate** of A. An assignment of coordinates to points on l is a **coordinate system** for l, and l is called a **coordinate line,** or a **real line.** A direction can be assigned to l by taking the **positive direction** to the right and the **negative direction** to the left. The positive direction is noted by placing an arrowhead on l as shown in Figure 1.1.

Real numbers that correspond to points to the right of O in Figure 1.1 are **positive real numbers,** whereas those that correspond to points to the left of O are **negative real numbers.** The real number 0 is neither positive nor negative.

If a and b are real numbers, and $a - b$ is positive, we say that **a is greater than b** and write $a > b$. An equivalent statement is **b is less than a** $(b < a)$. The symbols $>$ and $<$ are **inequality signs,** and expressions such as $a > b$ and $b < a$ are **inequalities.** Referring to Figure 1.1, if A and B are points with coordinates a and b, respectively, then $b > a$ (or $a < b$) *if and only if A lies to the left of B.* Since $a - 0 = a$, it follows that $a > 0$ if and only if a is positive. Similarly, $a < 0$ means that a is negative. We can prove the following properties.

<table>
<tr><td>PROPERTIES OF **(1.1)**
INEQUALITIES</td><td>(i) If $a > b$ and $b > c$, then $a > c$.

(ii) If $a > b$, then $a + c > b + c$.

(iii) If $a > b$, then $a - c > b - c$.

(iv) If $a > b$ and c is positive, then $ac > bc$.

(v) If $a > b$ and c is negative, then $ac < bc$.</td></tr>
</table>

Analogous results are true if the inequality signs are reversed. Thus, if $a < b$ and $b < c$, then $a < c$; if $a < b$, then $a + c < b + c$, and so on.

The symbol $a \geq b$, which is read **a is greater than or equal to b,** means that either $a > b$ or $a = b$. The symbol $a < b < c$ means that $a < b$ and $b < c$, in which case we say that **b is *between* a and c.** The notations $a \leq b$, $a < b \leq c$, $a \leq b < c$, $a \leq b \leq c$, and so on, can be interpreted from the preceding definitions.

If a real number a is the coordinate of a point A on a coordinate line l, the symbol $|a|$ is used to denote the number of units (or distance) between A and the origin, without regard to direction. The nonnegative number $|a|$ is the *absolute value* of a. Referring to Figure 1.2, we see that for the point with coordinate -4, we have $|-4| = 4$. Similarly, $|4| = 4$. In general, *if a is negative we change its sign to find $|a|$. If a is nonnegative, then $|a| = a$.* The next definition summarizes this discussion.

FIGURE 1.2

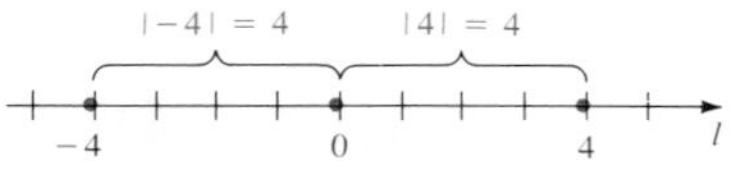

$|-4| = 4$ $|4| = 4$

$-4 \qquad 0 \qquad 4 \qquad l$

<table>
<tr><td>DEFINITION **(1.2)**</td><td>Let a be a real number. The **absolute value** of a, denoted by $|a|$, is

$$|a| = \begin{cases} a & \text{if } a \geq 0 \\ -a & \text{if } a < 0 \end{cases}$$</td></tr>
</table>

EXAMPLE 1 Find $|3|, |-3|, |0|, |\sqrt{2} - 2|$, and $|2 - \sqrt{2}|$.

SOLUTION Since $3, 2 - \sqrt{2}$, and 0 are nonnegative,

$$|3| = 3, \quad |2 - \sqrt{2}| = 2 - \sqrt{2}, \quad \text{and} \quad |0| = 0.$$

Since -3 and $\sqrt{2} - 2$ are negative, we use the formula $|a| = -a$ to obtain

$$|-3| = -(-3) = 3 \quad \text{and} \quad |\sqrt{2} - 2| = -(\sqrt{2} - 2) = 2 - \sqrt{2}. \quad \bullet$$

We can show that for all real numbers a and b,

$$|a| = |-a|, \qquad |ab| = |a||b|, \qquad -|a| \le a \le |a|.$$

The following properties can also be proved.

PROPERTIES OF (1.3)
ABSOLUTE VALUES
$(b > 0)$

(i) $|a| < b$ if and only if $-b < a < b$.
(ii) $|a| > b$ if and only if either $a > b$ or $a < -b$.
(iii) $|a| = b$ if and only if $a = b$ or $a = -b$.

Properties (ii) and (iii) are also true if $b = 0$. Thus, if $b \ge 0$, then

$$|a| \le b \quad \text{if and only if} \quad -b \le a \le b$$

and

$$|a| \ge b \quad \text{if and only if} \quad a \ge b \quad \text{or} \quad a \le -b.$$

THE TRIANGLE INEQUALITY (1.4)

$$|a + b| \le |a| \pm |b|$$

PROOF Consider $-|a| \le a \le |a|$ and $-|b| \le b \le |b|$. Adding corresponding sides, we obtain

$$-(|a| + |b|) \le a + b \le |a| + |b|.$$

Using the remark preceding this theorem gives us the Triangle Inequality. $\quad \bullet\bullet$

We shall use absolute values to define the distance between any two points on a coordinate line. Note that the distance between the points with coordinates 2 and 7 shown in Figure 1.3 equals 5 units on l. This distance is the difference, $7 - 2$, obtained by subtracting the smaller coordinate from the larger. If we employ absolute values, then, since $|7 - 2| = |2 - 7|$, the order of subtraction is irrelevant.

FIGURE 1.3

$5 = |7 - 2| = |2 - 7|$

DEFINITION (1.5)

Let a and b be the coordinates of two points A and B, respectively, on a coordinate line l. The **distance between A and B,** denoted by $d(A, B)$, is

$$d(A, B) = |b - a|$$

The number $d(A, B)$ denotes the **length of the line segment AB.** Observe that, since $d(B, A) = |a - b|$ and $|b - a| = |a - b|$,

$$d(A, B) = d(B, A).$$

Also note that the distance between the origin O and the point A is

$$d(O, A) = |a - 0| = |a|.$$

which agrees with the geometric interpretation of absolute value illustrated in Figure 1.2. The formula $d(A, B) = |b - a|$ is true regardless of the signs of a and b, as illustrated in the next example.

EXAMPLE 2 If $A, B, C,$ and D have coordinates $-5, -3, 1,$ and 6, respectively, find $d(A, B), d(C, B), d(O, A),$ and $d(C, D)$.

SOLUTION The points are sketched in Figure 1.4. By Definition (1.5):

$$d(A, B) = |-3 - (-5)| = |-3 + 5| = |2| = 2$$
$$d(C, B) = |-3 - 1| = |-4| = 4$$
$$d(O, A) = |-5 - 0| = |-5| = 5$$
$$d(C, D) = |6 - 1| = |5| = 5$$

FIGURE 1.4

For some topics, such as inequalities, it is convenient to use the notation and terminology of *sets*. We may think of a **set** as a collection of objects of some type. The objects are **elements** of the set. In our work, $\mathbb{R}$ will denote the set of real numbers. If S is a set, then $a \in S$ means that a is an element of S, whereas $a \notin S$ signifies that a is not an element of S. If every element of a set S is also an element of a set T, then S is a **subset** of T. Two sets S and T are **equal,** and we write $S = T$, if S and T contain precisely the same elements. The notation $S \neq T$ means that S and T are not equal. If S and T are sets, their **union** $S \cup T$ consists of the elements that are either in S, in T, or in *both* S and T. The **intersection** $S \cap T$ consists of the elements that the sets have in common.

We frequently use letters to represent arbitrary elements of a set. For example, we may use x to denote a real number when we do not wish to specify a *particular* real number. A letter that is used to represent *any* element of a given set is sometimes called a **variable.** A symbol that represents a *specific* element is a **constant.** In most of our work, letters near the end of the alphabet, such as x, y, and z, will be used for variables. Letters such as a, b, and c will denote constants. Throughout this text, unless otherwise specified, variables represent real numbers.

The **domain of a variable** is the set of real numbers represented by the variable. To illustrate, $\sqrt{x}$ is a real number if and only if $x \geq 0$, and hence the domain of x is the set of nonnegative real numbers. Similarly, given the expression $1/(x - 2)$, we must exclude $x = 2$ in order to avoid division by zero; consequently, in this case the domain is the set of all real numbers different from 2.

If the elements of a set S have a certain property, we sometimes write $S = \{x: \quad\}$ and state the property describing the variable x in the space after the colon. For example, $\{x: x > 3\}$ denotes the set of all real numbers greater than 3. Finite sets are sometimes identified by listing all the elements within braces. Thus, if the set T consists of the first five positive integers, we may write $T = \{1, 2, 3, 4, 5\}$.

Of major importance in calculus are certain subsets of $\mathbb{R}$ called **intervals.** If $a < b$, the set of all real numbers between a and b is an **open interval** and is denoted by (a, b), as in the following definition.

$$(a, b) = \{x: a < x < b\}$$

The numbers a and b are the **endpoints** of the interval.

The **graph** of a set of real numbers is defined as the points on a coordinate line that correspond to the numbers in the set. In particular, the graph of the open interval (a, b) consists of all points between the points corresponding to a and b. In Figure 1.5 we have sketched the graphs of a general open interval (a, b) and two specific open intervals $(-1, 3)$ and $(2, 4)$. The parentheses on each graph indicate that the endpoints of the interval are not included. For convenience, we shall use the terms *interval* and *graph of an interval* interchangeably.

To denote the inclusion of an endpoint in an interval, we use a bracket instead of a parenthesis. If $a < b$, then a **closed interval,** denoted by $[a, b]$, and a **half-open interval,** denoted by $[a, b)$ or $(a, b]$, are defined as follows.

FIGURE 1.5

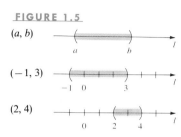

CLOSED AND HALF-OPEN INTERVALS **(1.7)**

$$[a, b] = \{x: a \le x \le b\}$$
$$[a, b) = \{x: a \le x < b\}$$
$$(a, b] = \{x: a < x \le b\}$$

FIGURE 1.6

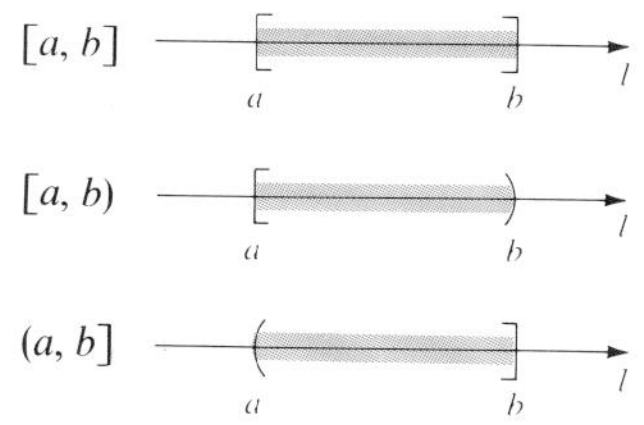

Typical graphs are sketched in Figure 1.6, where a bracket indicates that the corresponding endpoint is part of the graph.

In future discussions of intervals, whenever the numbers a and b are not stated explicitly we will always assume that $a < b$. If an interval is a subset of another interval I, it is a **subinterval** of I. For example, the closed interval $[2, 3]$ is a subinterval of $[0, 5]$.

We use the following notation for **infinite intervals.**

INFINITE INTERVALS **(1.8)**

$$(a, \infty) = \{x: x > a\}$$
$$[a, \infty) = \{x: x \ge a\}$$
$$(-\infty, a) = \{x: x < a\}$$
$$(-\infty, a] = \{x: x \le a\}$$

FIGURE 1.7

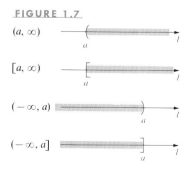

For example, $(1, \infty)$ represents all real numbers greater than 1. The symbol ∞ is read *infinity* and is merely a notational device. It does not represent a real number. Typical graphs of infinite intervals for an arbitrary real number a are sketched in Figure 1.7. The absence of a parenthesis or bracket on the right for (a, ∞) or $[a, \infty)$ and on the left for $(-\infty, a)$ or $(-\infty, a]$ indicates that the graph, shown as the colored portion, extends indefinitely. The set $\mathbb{R}$ of real numbers is sometimes denoted by $(-\infty, \infty)$.

We will often consider inequalities that involve variables, such as

$$x^2 - 3 < 2x + 4.$$

If numbers, such as 4 or 5, are substituted for x in $x^2 - 3 < 2x + 4$, we obtain false statements, such as $13 < 12$ or $22 < 14$, respectively. Other numbers, such as 1 or 2, produce true statements: $-2 < 6$ or $1 < 8$, respectively. If a true statement is obtained when x is replaced by a real number a, then a is a **solution** of the inequality. Thus 1 and 2 are solutions of the inequality $x^2 - 3 < 2x + 4$, but 4 and 5 are not solutions. To **solve** an inequality means to find all solutions. Two inequalities are **equivalent** if they have exactly the same solutions.

To solve an inequality, we usually replace it with a list of equivalent inequalities, terminating in one for which the solutions are obvious. The main tools used in applying this method are properties of inequalities and absolute value. For example, if x represents a real number, then adding the same expression in x to both sides of an inequality leads to an equivalent inequality. We may multiply both sides by an expression containing x if we are certain that the expression is positive for all values of x under consideration. If we multiply both sides of an inequality by an expression that is always negative, such as $-7 - x^2$, then we reverse the inequality sign in the equivalent inequality.

EXAMPLE 3 Solve the inequality $4x + 3 > 2x - 5$ and represent the solutions graphically.

SOLUTION The following inequalities are equivalent (supply reasons):

$$4x + 3 > 2x - 5$$
$$4x > 2x - 8$$
$$2x > -8$$
$$x > -4$$

FIGURE 1.8

Hence the solutions consist of all real numbers greater than -4, that is, the numbers in the infinite interval $(-4, \infty)$. The graph is sketched in Figure 1.8. •

EXAMPLE 4 Solve the inequality $-5 \le \dfrac{4 - 3x}{2} < 1$ and sketch the graph corresponding to the solutions.

SOLUTION We may proceed as follows:

$$-5 \le \frac{4 - 3x}{2} < 1$$
$$-10 \le 4 - 3x < 2$$
$$-14 \le -3x < -2$$
$$\frac{14}{3} \ge x > \frac{2}{3}$$
$$\frac{2}{3} < x \le \frac{14}{3}$$

FIGURE 1.9

Hence the solutions are the numbers in the half-open interval $(\frac{2}{3}, \frac{14}{3}]$. The graph is sketched in Figure 1.9.

FIGURE 1.10

SIGN OF FACTOR:

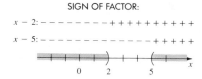

EXAMPLE 5 Solve $x^2 - 7x + 10 > 0$ and represent the solutions graphically.

SOLUTION Since the inequality may be written

$$(x - 5)(x - 2) > 0,$$

it follows that x is a solution if and only if both factors $x - 5$ and $x - 2$ are positive or both are negative. The diagram in Figure 1.10 indicates the signs of these factors for various real numbers. Evidently, both factors are positive if x is in the interval $(5, \infty)$ and both are negative if x is in $(-\infty, 2)$. Hence the solutions consists of all real numbers in the union $(-\infty, 2) \cup (5, \infty)$, as illustrated by the sketch in Figure 1.10. •

EXAMPLE 6 Solve the inequality $|x - 3| < 0.1$.

SOLUTION Using (1.3) (i) and (1.1) (ii), we find that the inequality is equivalent to each of the following:

$$-0.1 < x - 3 < 0.1$$

$$-0.1 + 3 < (x - 3) + 3 < 0.1 + 3$$

$$2.9 < x < 3.1$$

Thus the solutions are the real numbers in the open interval $(2.9, 3.1)$. •

Inequalities similar to the one in Example 6 will be used to define the concept of *limit* in Section 2.3. Specifically, we shall consider inequalities of the type given in the next example. The Greek letter δ (delta) in Example 7 is used frequently in calculus to denote a small positive real number.

EXAMPLE 7 Let a and δ denote real numbers, with $\delta > 0$. Solve the inequality

$$0 < |x - a| < \delta$$

and represent the solutions graphically.

SOLUTION The inequality $0 < |x - a|$ is true if and only if $x \neq a$. The solutions of $|x - a| < \delta$ may be found using (1.3) (i) and (1.1) (ii) as follows:

$$|x - a| < \delta$$

$$-\delta < x - a < \delta$$

$$a - \delta < x < a + \delta$$

Thus, the solutions of $0 < |x - a| < \delta$ consist of all real numbers in the open interval $(a - \delta, a + \delta)$ *except* the number a. This is the union

$$(a - \delta, a) \cup (a, a + \delta)$$

of two open intervals. The graph is sketched in Figure 1.11. •

FIGURE 1.11

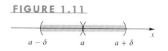

EXAMPLE 8 Solve $|2x - 7| > 3$.

SOLUTION By (1.3) (ii), x is a solution of $|2x - 7| > 3$ if and only if either

$$2x - 7 > 3 \quad \text{or} \quad 2x - 7 < -3.$$

The first of these two inequalities is equivalent to $2x > 10$, or $x > 5$. The second is equivalent to $2x < 4$, or $x < 2$. Hence the solutions of $|2x - 7| > 3$ are the numbers in the union $(-\infty, 2) \cup (5, \infty)$. The graph is identical to the one in Figure 1.10. •

Exer. 1–2: Replace the symbol □ with either <, >, or =.

1 (a) $-2 \ \square \ -5$ (b) $-2 \ \square \ 5$ (c) $6 - 1 \ \square \ 2 + 3$
   (d) $\frac{2}{3} \ \square \ 0.66$ (e) $2 \ \square \ \sqrt{4}$ (f) $\pi \ \square \ \frac{22}{7}$

2 (a) $-3 \ \square \ 0$ (b) $-8 \ \square \ -3$ (c) $8 \ \square \ -3$
   (d) $\frac{3}{4} - \frac{2}{3} \ \square \ \frac{1}{15}$ (e) $\sqrt{2} \ \square \ 1.4$ (f) $\frac{4053}{1110} \ \square \ 3.6513$

Exer. 3–4: Rewrite the expression without using the symbol for absolute value.

3 (a) $|2 - 5|$ (b) $|-5| + |-2|$
   (c) $|5| + |-2|$ (d) $|-5| - |-2|$
   (e) $\left|\pi - \frac{22}{7}\right|$ (f) $(-2)/|-2|$
   (g) $\left|\frac{1}{2} - 0.5\right|$ (h) $|(-3)^2|$
   (i) $|5 - x|$ if $x > 5$ (j) $|a - b|$ if $a < b$

4 (a) $|4 - 8|$ (b) $|3 - \pi|$
   (c) $|-4| - |-8|$ (d) $|-4 + 8|$
   (e) $|-3|^2$ (f) $|2 - \sqrt{4}|$
   (g) $|-0.67|$ (h) $-|-3|$
   (i) $|x^2 + 1|$ (j) $|-4 - x^2|$

5 If A, B, and C are points on a coordinate line with coordinates -5, -1, and 7, respectively, find the following distances:

   (a) $d(A, B)$ (b) $d(B, C)$ (c) $d(C, B)$ (d) $d(A, C)$

6 Rework Exercise 5 if A, B, and C have coordinates 2, -8, and -3, respectively.

Exer. 7–34: Solve the inequality and express the solution in terms of intervals.

7 $5x - 6 > 11$ 8 $3x - 5 < 10$

9 $2 - 7x \leq 16$ 10 $7 - 2x \geq -3$

11 $|2x + 1| > 5$ 12 $|x + 2| < 1$

13 $3x + 2 < 5x - 8$ 14 $2 + 7x < 3x - 10$

15 $12 \geq 5x - 3 > -7$ 16 $5 > 2 - 9x > -4$

17 $-1 < \dfrac{3 - 7x}{4} \leq 6$ 18 $0 \leq 4x - 1 \leq 2$

19 $\dfrac{5}{7 - 2x} > 0$ 20 $\dfrac{4}{x^2 + 9} > 0$

21 $|x - 10| < 0.3$ 22 $\left|\dfrac{2x + 3}{5}\right| < 2$

23 $\left|\dfrac{7 - 3x}{2}\right| \leq 1$ 24 $|3 - 11x| \geq 41$

25 $|25x - 8| > 7$ 26 $|2x + 1| < 0$

27 $3x^2 + 5x - 2 < 0$ 28 $2x^2 - 9x + 7 < 0$

29 $2x^2 + 9x + 4 \geq 0$ 30 $x^2 - 10x \leq 200$

31 $\dfrac{1}{x^2} < 100$ 32 $5 + \sqrt{x} < 1$

33 $\dfrac{3x + 2}{2x - 7} \leq 0$ 34 $\dfrac{3}{x - 9} > \dfrac{2}{x + 2}$

35 The relationship between the Fahrenheit and Celsius temperature scales is given by $C = \frac{5}{9}(F - 32)$. If $60 \leq F \leq 80$, express the corresponding values of C in terms of an inequality.

36 For the electrical circuit shown in the figure, Ohm's law states that $I = V/R$, where R is the resistance (in ohms), V is the potential difference (in volts), and I is the current (in amperes). If the voltage is 110, what values of the resistance will result in a current that does not exceed 10 amperes?

EXERCISE 36

37 According to Hooke's law, the force F (in pounds) required to stretch a certain spring x inches beyond its natural length is given by the formula $F = (4.5)x$ (see figure). If $10 \leq F \leq 18$, what are the corresponding values of x?

EXERCISE 37

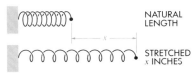

NATURAL LENGTH

STRETCHED x INCHES

38 If two resistors R_1 and R_2 are connected in parallel in an electrical circuit, the net resistance R is given by $1/R = (1/R_1) + (1/R_2)$ (see figure). If $R_1 = 10$ ohms, what values of R_2 will result in a net resistance of less than 5 ohms?

EXERCISE 38

39 A convex lens has focal length $f = 5$ cm. If an object is placed a distance p cm from the lens, the distance q cm of the image from the lens is related to p and f by the *lens equation* $(1/p) + (1/q) = 1/f$ (see figure). How close must the object be to the lens for the image to be more than 12 cm from the lens?

EXERCISE 39

OBJECT   IMAGE

40 Boyle's law for a certain gas states that $pv = 200$, where p denotes the pressure (lb/in.^{2}) and v denotes the volume (in.^{3}). If $25 \leq v \leq 50$, what are the corresponding values of p?

FIGURE 1.12

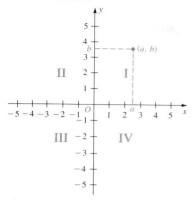

*The term *Cartesian* is used in honor of the French mathematician and philosopher René Descartes (1596–1650), who was one of the first to employ such coordinate systems.

FIGURE 1.13

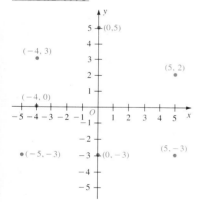

In Section 1.1 we discussed a method of assigning coordinates to points on a line. A coordinate system can also be introduced in a plane by means of *ordered pairs.* The term **ordered pair** refers to two real numbers, where one is designated as the "first" number and the other as the "second." The symbol (a, b) denotes the ordered pair consisting of the real numbers a and b, where a is first and b is second. Ordered pairs have many uses. We used them in Section 1.1 to denote open intervals. In this section they will represent points in a plane. Although ordered pairs are employed in different situations, there is little chance that we will confuse their uses, since it should always be clear from our discussion whether the symbol (a, b) represents an interval, a point, or some other mathematical concept. Two ordered pairs (a, b) and (c, d) are **equal,** and we write

$$(a, b) = (c, d) \quad \text{if and only if} \quad a = c \quad \text{and} \quad b = d.$$

This implies, in particular, that $(a, b) \neq (b, a)$ if $a \neq b$. The set of all ordered pairs will be denoted by $\mathbb{R} \times \mathbb{R}$.

A **rectangular,** or **Cartesian*** **coordinate system** may be introduced in a plane by considering two perpendicular coordinate lines in the plane that intersect in the origin O on each line. Unless specified otherwise, the same unit of length is chosen on each line. Usually one of the lines is horizontal with positive direction to the right, and the other line is vertical with positive direction upward, as indicated by the arrowheads in Figure 1.12. The two lines are **coordinate axes,** and the point O is the **origin.** The horizontal line is usually referred to as the **x-axis** and the vertical line as the **y-axis,** and they are labeled x and y, respectively. The plane is then a **coordinate plane,** or an **xy-plane.** In certain applications different labels, such as s or t, are used for the coordinate axes, and we refer to the system by its labels, such as an *st*-plane. The coordinate axes divide the plane into four parts called the **first, second, third,** and **fourth quadrants** and labeled I, II, III, and IV, respectively (see Figure 1.12).

Each point P in an xy-plane may be assigned a unique ordered pair (a, b) as shown in Figure 1.12. The number a is the **x-coordinate** (or **abscissa**) of P, and b is the **y-coordinate** (or **ordinate**). We say that P has coordinates (a, b). Conversely, every ordered pair (a, b) determines a point P in the xy-plane with coordinates a and b. We often refer to the *point* (a, b), or $P(a, b)$, meaning the point P with x-coordinate a and y-coordinate b. To **plot a point** $P(a, b)$, we locate P in a coordinate plane and represent it by a dot, as illustrated by some points plotted in Figure 1.13.

The next statement provides a formula for finding the distance between two points in a coordinate plane.

DISTANCE FORMULA (1.9)

> The distance $d(P_1, P_2)$ between any two points $P_1(x_1, y_1)$ and $P_2(x_2, y_2)$ in a coordinate plane is
>
> $$d(P_1, P_2) = \sqrt{(x_2 - x_1)^2 + (y_2 - y_1)^2}$$

PROOF If $x_1 \neq x_2$ and $y_1 \neq y_2$, then, as illustrated in Figure 1.14, the points P_1, P_2, and $P_3(x_2, y_1)$ are vertices of a right triangle. By the Pythagorean

FIGURE 1.14

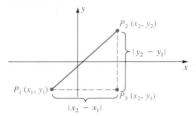

Theorem,

$$[d(P_1, P_2)]^2 = [d(P_1, P_3)]^2 + [d(P_3, P_2)]^2.$$

From the figure we see that

$$d(P_1, P_3) = |x_2 - x_1| \quad \text{and} \quad d(P_3, P_2) = |y_2 - y_1|.$$

Since $|a|^2 = a^2$ for every real number a, we may write

$$[d(P_1, P_2)]^2 = (x_2 - x_1)^2 + (y_2 - y_1)^2.$$

Taking the square root of each side of the last equation gives us the Distance Formula.

If $y_1 = y_2$, the points P_1 and P_2 lie on the same horizontal line and

$$d(P_1, P_2) = |x_2 - x_1| = \sqrt{(x_2 - x_1)^2}.$$

Similarly, if $x_1 = x_2$, the points are on the same vertical line and

$$d(P_1, P_2) = |y_2 - y_1| = \sqrt{(y_2 - y_1)^2}.$$

These are special cases of the Distance Formula.

Although we referred to the points shown in Figure 1.14, the argument used in our proof is independent of the positions of P_1 and P_2. • •

When applying the Distance Formula, we should note that $d(P_1, P_2) = d(P_2, P_1)$ and, hence, the order in which we subtract the x-coordinates and the y-coordinates of the points is immaterial.

EXAMPLE 1 Plot the points $A(-1, -3)$, $B(6, 1)$, and $C(2, -5)$. Prove that the triangle with vertices A, B, and C is a right triangle, and find its area.

FIGURE 1.15

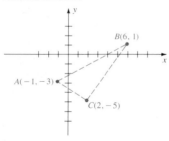

SOLUTION The points and the triangle are shown in Figure 1.15. From plane geometry, a triangle is a right triangle if and only if the sum of the squares of two of its sides is equal to the square of the remaining side. Using the Distance Formula,

$$d(A, B) = \sqrt{(-1 - 6)^2 + (-3 - 1)^2} = \sqrt{49 + 16} = \sqrt{65}$$

$$d(B, C) = \sqrt{(6 - 2)^2 + (1 + 5)^2} = \sqrt{16 + 36} = \sqrt{52}$$

$$d(A, C) = \sqrt{(-1 - 2)^2 + (-3 + 5)^2} = \sqrt{9 + 4} = \sqrt{13}$$

Since $[d(A, B)]^2 = [d(B, C)]^2 + [d(A, C)]^2$, the triangle is a right triangle with hypotenuse AB. The area is $\frac{1}{2}\sqrt{52}\sqrt{13} = \frac{1}{2} \cdot 2\sqrt{13}\sqrt{13}$, or 13 square units. •

It is easy to obtain a formula for the midpoint of a line segment. Let $P_1(x_1, y_1)$ and $P_2(x_2, y_2)$ be two points in a coordinate plane and let M be the midpoint of the segment P_1P_2. The lines through P_1 and P_2 parallel to the y-axis intersect the x-axis at $A_1(x_1, 0)$ and $A_2(x_2, 0)$ and, from plane geometry, the line through M parallel to the y-axis bisects the segment A_1A_2 (see Figure 1.16). If $x_1 < x_2$, then $x_2 - x_1 > 0$, and hence $d(A_1, A_2) = x_2 - x_1$. Since M_1 is halfway from A_1 to A_2, the x-coordinate of M_1 is

FIGURE 1.16

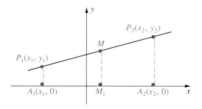

$$x_1 + \tfrac{1}{2}(x_2 - x_1) = x_1 + \tfrac{1}{2}x_2 - \tfrac{1}{2}x_1 = \tfrac{1}{2}x_1 + \tfrac{1}{2}x_2$$

$$= \frac{x_1 + x_2}{2}.$$

It follows that the x-coordinate of M is also $(x_1 + x_2)/2$. Similarly, the

y-coordinate of M is $(y_1 + y_2)/2$. Moreover, these formulas hold for all positions of P_1 and P_2. This gives us the following result.

MIDPOINT FORMULA (1.10)

> The midpoint of the line segment from $P_1(x_1, y_1)$ to $P_2(x_2, y_2)$ is
> $$\left(\frac{x_1 + x_2}{2}, \frac{y_1 + y_2}{2}\right)$$

EXAMPLE 2 Find the midpoint M of the line segment from $P_1(-2, 3)$ to $P_2(4, -2)$. Plot the points P_1, P_2, and M and verify that

$$d(P_1, M) = d(P_2, M).$$

SOLUTION By the Midpoint Formula, the coordinates of M are

$$\left(\frac{-2 + 4}{2}, \frac{3 + (-2)}{2}\right) \quad \text{or} \quad \left(1, \frac{1}{2}\right).$$

FIGURE 1.17

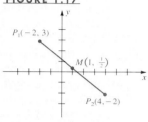

The three points P_1, P_2, and M are plotted in Figure 1.17. Using the Distance Formula,

$$d(P_1, M) = \sqrt{(-2 - 1)^2 + (3 - \tfrac{1}{2})^2} = \sqrt{9 + \tfrac{25}{4}}$$
$$d(P_2, M) = \sqrt{(4 - 1)^2 + (-2 - \tfrac{1}{2})^2} = \sqrt{9 + \tfrac{25}{4}}$$

Hence, $d(P_1, M) = d(P_2, M)$. •

If W is a set of ordered pairs, we may consider the point $P(x, y)$ in a coordinate plane that corresponds to the ordered pair (x, y) in W. The **graph** of W is the set of all such points. To *sketch the graph of* W, we illustrate the significant features of the graph geometrically on a coordinate plane.

FIGURE 1.18

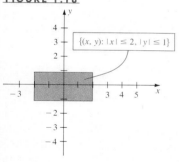

EXAMPLE 3 Sketch the graph of $W = \{(x, y): |x| \le 2, |y| \le 1\}$.

SOLUTION The inequalities are equivalent to $-2 \le x \le 2$ and $-1 \le y \le 1$. Thus, the graph of W consists of all points within and on the boundary of the rectangular region shown in Figure 1.18. •

EXAMPLE 4 Sketch the graph of $W = \{(x, y): y = 2x - 1\}$.

SOLUTION We wish to find the points (x, y) that correspond to the ordered pairs (x, y) in W. It is convenient to list coordinates of several such points in tabular form, where for each x, we obtain the value for y from $y = 2x - 1$:

x	-2	-1	0	1	2	3
y	-5	-3	-1	1	3	5

After plotting the points with these coordinates, we conclude that they appear to lie on a line and we sketch the graph (see Figure 1.19). Ordinarily, the few points we have plotted would not be enough to illustrate the graph; however, in this elementary case we can be reasonably sure that the graph is a line. In Section 1.3 we will prove this fact. •

FIGURE 1.19

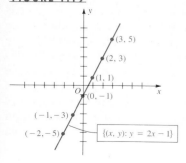

The x-coordinates of points at which a graph intersects the x-axis are the **x-intercepts** of the graph. The y-coordinates of points at which a graph intersects the y-axis are the **y-intercepts.** The graph in Figure 1.19 has one x-intercept, $\frac{1}{2}$, and one y-intercept, -1.

It is impossible to sketch the entire graph in Example 4, since x may be assigned values that are numerically as large as desired. Nevertheless, we call the drawing in Figure 1.19 *the graph of W* or *a sketch of the graph*. It is understood that the drawing is only a device for visualizing the actual graph and the line does not terminate as shown in the figure. In general, the sketch of a graph should illustrate enough of the graph so that the remaining parts are evident.

Given an equation in x and y, an ordered pair (a, b) is a **solution** of the equation if equality is obtained when a is substituted for x and b for y. For example, $(2, 3)$ is a solution of $y = 2x - 1$, since substitution of 2 for x and 3 for y leads to $3 = 4 - 1$, or $3 = 3$. Two equations in x and y are **equivalent** if they have exactly the same solutions. The solutions of an equation in x and y determine a set W of ordered pairs, and we define the **graph of the equation in x and y** as the graph of W. Note that the graph of the equation $y = 2x - 1$ is the same as the graph of the set W in Example 4 (see Figure 1.19).

For some of the equations we shall consider in this chapter, the technique we will use for sketching the graph consists of plotting a sufficient number of points until some pattern emerges, and then sketching the graph accordingly. This is obviously a crude (and often inaccurate) way to arrive at the graph; however, it is a method often employed in elementary courses. As we progress through this text, techniques will be introduced that will enable us to sketch accurate graphs without plotting many points.

EXAMPLE 5 Sketch the graph of the equation $y = x^2$.

SOLUTION To sketch the graph, we must plot more points than in the previous example. Increasing successive x-coordinates by $\frac{1}{2}$, we obtain a table of coordinates:

x	-3	$-\frac{5}{2}$	-2	$-\frac{3}{2}$	-1	$-\frac{1}{2}$	0	$\frac{1}{2}$	1	$\frac{3}{2}$	2	$\frac{5}{2}$	3
y	9	$\frac{25}{4}$	4	$\frac{9}{4}$	1	$\frac{1}{4}$	0	$\frac{1}{4}$	1	$\frac{9}{4}$	4	$\frac{25}{4}$	9

Larger values of $|x|$ produce larger values of y. For example, the points $(4, 16)$, $(5, 25)$, and $(6, 36)$ are on the graph, as are $(-4, 16)$, $(-5, 25)$, and $(-6, 36)$. Plotting the points given by the table and drawing a smooth curve through these points gives us the sketch in Figure 1.20, in which several points are labeled. •

The graph in Example 5 is a **parabola.** In this case, the y-axis is the **axis of the parabola.** The lowest point $(0, 0)$ is the **vertex** of the parabola and we say that the parabola **opens upward.** If we invert the graph, as would be the case for $y = -x^2$, then the parabola **opens downward** and the vertex $(0, 0)$ is the highest point on the graph. In general, the graph of *any* equation of the form $y = ax^2$ for $a \neq 0$ is a parabola with vertex $(0, 0)$. Parabolas may also open to the right or to the left (see Example 6). Parabolas and their properties will be discussed in detail in Chapter 12, where we will show that the graph of every equation of the form $y = ax^2 + bx + c$, with $a \neq 0$, is a parabola whose axis is *parallel* to the y-axis.

FIGURE 1.20

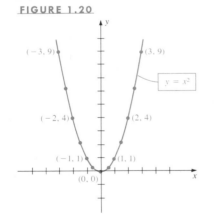

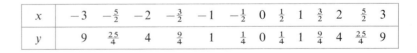

FIGURE 1.21
Symmetries

(i) y-axis

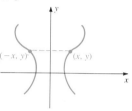

(ii) x-axis

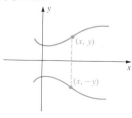

(iii) Origin

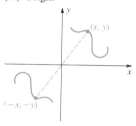

If the coordinate plane in Figure 1.20 is folded along the y-axis, then the graph that lies in the left half of the plane coincides with that in the right half. We say that the **graph is symmetric with respect to the y-axis.** As in Figure 1.21(i), a graph is symmetric with respect to the y-axis provided that the point $(-x, y)$ is on the graph whenever (x, y) is on the graph. As in Figure 1.21(ii), a **graph is symmetric with respect to the x-axis** if whenever a point (x, y) is on the graph, then $(x, -y)$ is also on the graph. Certain graphs possess a **symmetry with respect to the origin.** In this situation, whenever a point (x, y) is on the graph, then $(-x, -y)$ is also on the graph, as illustrated in Figure 1.21(iii).

The following tests are useful for investigating the three types of symmetry for graphs of equations in x and y.

TESTS FOR **(1.11)**
SYMMETRY

> (i) The graph of an equation is symmetric with respect to the y-axis if substitution of $-x$ for x leads to an equivalent equation.
>
> (ii) The graph of an equation is symmetric with respect to the x-axis if substitution of $-y$ for y leads to an equivalent equation.
>
> (iii) The graph of an equation is symmetric with respect to the origin if the simultaneous substitution of $-x$ for x and $-y$ for y leads to an equivalent equation.

If, in the equation of Example 5, we substitute $-x$ for x, we obtain $y = (-x)^2$, which is equivalent to $y = x^2$. Hence, by Symmetry Test (1.11) (i), the graph is symmetric with respect to the y-axis.

If a graph is symmetric with respect to an axis, it is sufficient to determine the graph in half of the coordinate plane, since we may sketch the remainder by taking a mirror image, or reflection, through the axis of symmetry.

EXAMPLE 6 Sketch the graph of the equation $y^2 = x$.

SOLUTION Since substitution of $-y$ for y does not change the equation, the graph is symmetric with respect to the x-axis (see Symmetry Test (ii)). Thus, it is sufficient to plot points with nonnegative y-coordinates and then reflect through the x-axis. Since $y^2 = x$, the y-coordinates of points above the x-axis are given by $y = \sqrt{x}$. Coordinates of some points on the graph are listed in the following table:

FIGURE 1.22

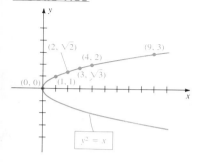

x	0	1	2	3	4	9
y	0	1	$\sqrt{2} \approx 1.4$	$\sqrt{3} \approx 1.7$	2	3

A portion of the graph is sketched in Figure 1.22. The graph is a parabola that opens to the right, with its vertex at the origin. In this case the x-axis is the axis of the parabola. •

EXAMPLE 7 Sketch the graph of the equation $4y = x^3$.

SOLUTION If we substitute $-x$ for x and $-y$ for y, then

$$4(-y) = (-x)^3 \quad \text{or} \quad -4y = -x^3.$$

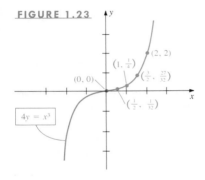

FIGURE 1.23

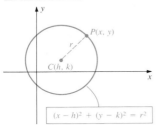

FIGURE 1.24

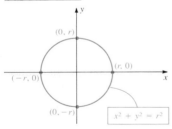

FIGURE 1.25

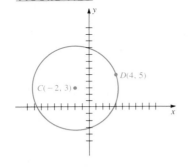

FIGURE 1.26

Multiplying both sides by -1, we see that the last equation has the same solutions as the equation $4y = x^3$. Hence, from Symmetry Test (iii), the graph is symmetric with respect to the origin. The following table lists some points on the graph:

x	0	$\frac{1}{2}$	1	$\frac{3}{2}$	2	$\frac{5}{2}$
y	0	$\frac{1}{32}$	$\frac{1}{4}$	$\frac{27}{32}$	2	$\frac{125}{32}$

By symmetry (or by substitution) we see that the points $(-1, -\frac{1}{4})$, $(-2, -2)$, and so on, are on the graph. Plotting points leads to the sketch in Figure 1.23. •

If $C(h, k)$ is a point in a coordinate plane, then a circle with center C and radius $r > 0$ consists of all points in the plane that are r units from C. As shown in Figure 1.24, a point $P(x, y)$ is on the circle if and only if $d(C, P) = r$ or, by the Distance Formula, if and only if

$$\sqrt{(x - h)^2 + (y - k)^2} = r.$$

The following equivalent equation is called the **equation of a circle of radius r and center (h, k).**

EQUATION (1.12)
OF A CIRCLE

$$(x - h)^2 + (y - k)^2 = r^2, \qquad r > 0$$

If $h = 0$ and $k = 0$, this equation reduces to $x^2 + y^2 = r^2$, which is an equation of a circle of radius r with center at the origin (see Figure 1.25). If $r = 1$, the graph is called a **unit circle.**

EXAMPLE 8 Find an equation of the circle that has center $C(-2, 3)$ and contains the point $D(4, 5)$.

SOLUTION The circle is illustrated in Figure 1.26. Since D is on the circle, the radius r is $d(C, D)$. By the Distance Formula,

$$r = \sqrt{(-2 - 4)^2 + (3 - 5)^2} = \sqrt{36 + 4} = \sqrt{40}.$$

Using the equation of a circle with $h = -2$, $k = 3$, and $r = \sqrt{40}$, we obtain

$$(x + 2)^2 + (y - 3)^2 = 40$$

or

$$x^2 + y^2 + 4x - 6y - 27 = 0. \quad •$$

Squaring terms of $(x - h)^2 + (y - k)^2 = r^2$ and simplifying leads to an equation of the form

$$x^2 + y^2 + ax + by + c = 0$$

for some real numbers a, b, and c. Conversely, if we begin with the last equation, it is always possible, by *completing squares*, to obtain an equation of the form

$$(x - h)^2 + (y - k)^2 = d.$$

The method will be illustrated in Example 9. If $d > 0$ the graph is a circle with center (h, k) and radius $r = \sqrt{d}$. If $d = 0$ the graph consists of only one point (h, k). Finally, if $d < 0$ the equation has no real solutions and hence there is no graph.

EXAMPLE 9 Find the center and radius of the circle with equation

$$x^2 + y^2 - 4x + 6y - 3 = 0.$$

SOLUTION We begin by arranging the equation as follows:

$$(x^2 - 4x) + (y^2 + 6y) = 3.$$

Next we complete the squares for the expressions within parentheses. Of course, to obtain an equivalent equation, we must add the numbers to *both* sides of the equation. To complete the square for an expression of the form $x^2 + ax$, we add the square of half the coefficient of x, that is, $(a/2)^2$, to both sides of the equation. Similarly, for $y^2 + by$, we add $(b/2)^2$ to both sides. In this example $a = -4$, $b = 6$, $(a/2)^2 = (-2)^2 = 4$, and $(b/2)^2 = 3^2 = 9$. This leads to

$$(x^2 - 4x + 4) + (y^2 + 6y + 9) = 3 + 4 + 9$$

or

$$(x - 2)^2 - (y + 3)^2 = 16.$$

Hence, by (1.12) the center is $(2, -3)$ and the radius is 4. •

EXERCISES 1.2

Exer. 1–6: Find (a) the distance $d(A, B)$ between the points A and B, and (b) the midpoint of the segment AB.

1 $A(6, -2)$, $B(2, 1)$

2 $A(-4, -1)$, $B(2, 3)$

3 $A(0, -7)$, $B(-1, -2)$

4 $A(4, 5)$, $B(4, -4)$

5 $A(-3, -2)$, $B(-8, -2)$

6 $A(11, -7)$, $B(-9, 0)$

Exer. 7–8: Prove that the triangle with vectices A, B, and C is a right triangle and find its area.

7 $A(-3, 4)$, $B(2, -1)$, $C(9, 6)$

8 $A(7, 2)$, $B(-4, 0)$, $C(4, 6)$

Exer. 9–14: Sketch the graph of the set W.

9 $W = \{(x, y): x = 4\}$

10 $W = \{(x, y): y = -3\}$

11 $W = \{(x, y): xy < 0\}$

12 $W = \{(x, y): xy = 0\}$

13 $W = \{(x, y): |x| < 2, |y| > 1\}$

14 $W = \{(x, y): |x| > 1, |y| \le 2\}$

Exer. 15–36: Sketch the graph of the equation and use (1.11) to test for symmetry.

15 $y = 3x + 1$

16 $y = 4x - 3$

17 $y = -2x + 3$

18 $y = 2 - 3x$

19 $y = 2x^2 - 1$

20 $y = -x^2 + 2$

21 $4y = x^2$

22 $3y + x^2 = 0$

23 $y = -\frac{1}{2}x^3$

24 $y = \frac{1}{2}x^3$

25 $y = x^3 - 2$

26 $y = 2 - x^3$

27 $y = \sqrt{x}$

28 $y = \sqrt{x} - 1$

29 $y = \sqrt{-x}$

30 $y = \sqrt{x - 1}$

31 $x^2 + y^2 = 16$

32 $4x^2 + 4y^2 = 25$

33 $y = -\sqrt{4 - x^2}$

34 $x = \sqrt{4 - y^2}$

35 $x = \sqrt{9 - y^2}$

36 $y = \sqrt{9 - x^2}$

Exer. 37–44: Find an equation of a circle satisfying the stated conditions.

37 Center $C(3, -2)$, radius 4

38 Center $C(-5, 2)$, radius 5

39 Center at the origin, passing through $P(-3, 5)$

40 Center $C(-4, 6)$, passing through $P(1, 2)$

41 Center $C(-4, 2)$, tangent to the x-axis

42 Center $C(3, -5)$, tangent to the y-axis

43 Endpoints of a diameter $A(4, -3)$ and $B(-2, 7)$

44 Tangent to both axes, center in the first quadrant, radius 2

Exer. 45–50: Find the center and radius of the circle with the given equation.

45 $x^2 + y^2 + 4x - 6y + 4 = 0$

46 $x^2 + y^2 - 10x + 2y + 22 = 0$

47 $x^2 + y^2 + 6x = 0$

48 $x^2 + y^2 + x + y - 1 = 0$

49 $2x^2 + 2y^2 - x + y - 3 = 0$

50 $9x^2 + 9y^2 - 6x + 12y - 31 = 0$

The following concept is fundamental to the study of lines. All lines referred to are considered to be in a coordinate plane.

DEFINITION (1.13)

> Let l be a line that is not parallel to the y-axis, and let $P_1(x_1, y_1)$ and $P_2(x_2, y_2)$ be distinct points on l. The **slope m** of l is
>
> $$m = \frac{y_2 - y_1}{x_2 - x_1}$$
>
> If l is parallel to the y-axis, then the slope is not defined.

FIGURE 1.27
Positive slope

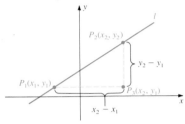

Typical points P_1 and P_2 on a line l are shown in Figures 1.27 and 1.28. The numerator $y_2 - y_1$ in the formula for m is the vertical change in direction from P_1 to P_2 and may be positive, negative, or zero. The denominator $x_2 - x_1$ is the horizontal change from P_1 to P_2, and it may be positive or negative, but never zero, because l is not parallel to the y-axis if the slope exists.

In finding the slope of a line it is immaterial which point we label as P_1 or as P_2, since

$$\frac{y_2 - y_1}{x_2 - x_1} = \frac{y_1 - y_2}{x_1 - x_2}.$$

FIGURE 1.28
Negative slope

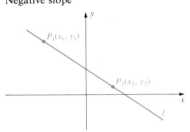

Consequently, we may assume that the points are labeled so that $x_1 < x_2$, as in Figures 1.27 and 1.28. In this situation, $x_2 - x_1 > 0$, and hence the slope is positive, negative, or zero, depending on whether $y_2 > y_1$, $y_2 < y_1$, or $y_2 = y_1$. The slope of the line shown in Figure 1.27 is positive. The slope of the line shown in Figure 1.28 is negative.

A **horizontal line** is a line parallel to the x-axis. Note that *a line is horizontal if and only if its slope is 0*. A **vertical line** is a line parallel to the y-axis. The slope of a vertical line is undefined.

The definition of slope is independent of the two points that are chosen on l. If other points $P_1'(x_1', y_1')$ and $P_2'(x_2', y_2')$ are used, then as in Figure 1.29, the triangle with vertices P_1', P_2', and $P_3'(x_2', y_1')$ is similar to the triangle with vertices P_1, P_2, and $P_3(x_2, y_1)$. Since the ratios of corresponding sides of similar triangles are equal,

FIGURE 1.29

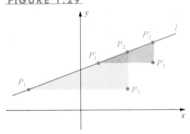

$$\frac{y_2 - y_1}{x_2 - x_1} = \frac{y_2' - y_1'}{x_2' - x_1'}.$$

EXAMPLE 1 Sketch the line through the each pair of points and find its slope.

(a) $A(-1, 4)$ and $B(3, 2)$ (b) $A(2, 5)$ and $B(-2, -1)$

(c) $A(4, 3)$ and $B(-2, 3)$ (d) $A(4, -1)$ and $B(4, 4)$.

SOLUTION The lines are sketched in Figure 1.30. Using Definition (1.13),

(a) $m = \dfrac{2 - 4}{3 - (-1)} = \dfrac{-2}{4} = -\dfrac{1}{2}$

FIGURE 1.30

(a) $m = -\frac{1}{2}$

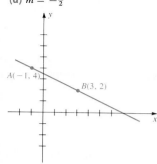

(b) $m = \frac{3}{2}$

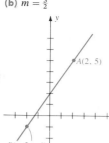

(c) $m = 0$

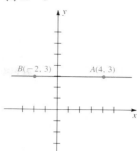

(d) m undefined

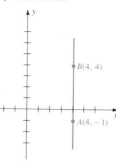

(b) $m = \dfrac{5 - (-1)}{2 - (-2)} = \dfrac{6}{4} = \dfrac{3}{2}$

(c) $m = \dfrac{3 - 3}{-2 - 4} = \dfrac{0}{-6} = 0$

(d) The slope is undefined because the line is vertical. We can also see this by noting that if the formula for m is used, the denominator is zero. •

It is not difficult to obtain an equation whose graph is a given line. We shall begin with the simplest case, in which the line is either vertical or horizontal.

THEOREM (1.14)

> (i) The graph of the equation $x = a$ is a vertical line with x-intercept a.
>
> (ii) The graph of the equation $y = b$ is a horizontal line with y-intercept b.

FIGURE 1.31

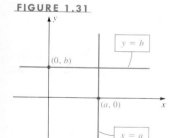

PROOF The equation $x = a$ may be written in the form $x + (0)y = a$. Some typical solutions of this equation are $(a, -2)$, $(a, 1)$, and $(a, 0)$. Evidently, every solution has the form (a, y), where y may have any value and a is fixed. It follows that the graph of $x = a$ is a line with x-intercept a and parallel to the y-axis, as illustrated in Figure 1.31. This proves (i). Part (ii) is proved in similar fashion. • •

Let us next find an equation of a line l through a point $P_1(x_1, y_1)$ with slope m (only one such line exists). If $P(x, y)$ is any point with $x \neq x_1$ (see Figure 1.32), then P is on l if and only if the slope of the line through P_1 and P is m, that is,

FIGURE 1.32

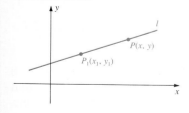

$$\frac{y - y_1}{x - x_1} = m.$$

This equation may be written in the form

$$y - y_1 = m(x - x_1).$$

Note that (x_1, y_1) is a solution of the last equation and hence the points on l are precisely the points that correspond to the solutions. This equation for l is referred to as the **Point-Slope Form.**

POINT-SLOPE FORM (1.15)

An equation for the line through the point (x_1, y_1) with slope m is

$$y - y_1 = m(x - x_1)$$

EXAMPLE 2 Find an equation of the line through the points $A(1, 7)$ and $B(-3, 2)$.

SOLUTION   The slope m of the line is

$$m = \frac{7 - 2}{1 - (-3)} = \frac{5}{4}.$$

We may use the coordinates of either A or B for (x_1, y_1) in the Point-Slope Form (1.15). Using $A(1, 7)$ gives us

$$y - 7 = \tfrac{5}{4}(x - 1)$$

which is equivalent to

$$4y - 28 = 5x - 5 \quad \text{or} \quad 5x - 4y + 23 = 0 \bullet$$

The Point-Slope Form may be rewritten as $y = mx - mx_1 + y_1$, which is of the form

$$y = mx + b$$

where $b = -mx_1 + y_1$. The real number b is the y-intercept of the graph, as we may see by setting $x = 0$. Since the equation $y = mx + b$ displays the slope m and y-intercept b of l, it is called the **Slope-Intercept Form** for the equation of a line. Conversely, if we start with $y = mx + b$, we may write

$$y - b = m(x - 0).$$

Comparing this equation with the Point-Slope Form, we see that the graph is a line with slope m and passing through the point $(0, b)$. This gives us the next result.

SLOPE-INTERCEPT FORM (1.16)

The graph of the equation $y = mx + b$ is a line having slope m and y-intercept b.

We have shown that every line is the graph of an equation of the form

$$ax + by + c = 0$$

for real numbers a, b, and c such that a and b are not both zero. We call such an equation a **linear equation** in x and y. Let us show, conversely, that the graph of $ax + by + c = 0$, with a and b not both zero, is always a line.

If $b \neq 0$, we may solve for y, obtaining

$$y = \left(-\frac{a}{b}\right)x + \left(-\frac{c}{b}\right)$$

which, by the Slope-Intercept Form, is an equation of a line with slope $-a/b$ and y-intercept $-c/b$. If $b = 0$ but $a \neq 0$, we may solve for x, obtaining $x = -c/a$, which is the equation of a vertical line with x-intercept $-c/a$. This establishes the following theorem.

THEOREM (1.17)

> The graph of a linear equation $ax + by + c = 0$ is a line and, conversely, every line is the graph of a linear equation.

For simplicity, we shall use the terminology *the line* $ax + by + c = 0$ instead of the more accurate phrase *the line with equation* $ax + by + c = 0$.

EXAMPLE 3 Sketch the graph of $2x - 5y = 8$.

SOLUTION From Theorem (1.17), the graph is a line, and hence it is sufficient to find two points on the graph. Let us find the x- and y-intercepts. Substituting $y = 0$ in the given equation, we obtain the x-intercept 4. Substituting $x = 0$, we see that the y-intercept is $-\frac{8}{5}$. This leads to the graph in Figure 1.33.

Another method of solution is to express the given equation in Slope-Intercept Form. We begin by isolating the term involving y on one side of the equal sign, obtaining

$$5y = 2x - 8.$$

Next, dividing both sides by 5 gives us

$$y = \frac{2}{5}x + \left(\frac{-8}{5}\right)$$

which is in the form $y = mx + b$. Hence, the slope is $m = \frac{2}{5}$ and the y-intercept is $b = -\frac{8}{5}$. We may then sketch a line through $(0, -\frac{8}{5})$ with slope $\frac{2}{5}$. •

The next theorem specifies the relationship between parallel lines and slope.

THEOREM (1.18)

> Two nonvertical lines are parallel if and only if they have the same slope.

PROOF Let l_1 and l_2 be distinct lines of slopes m_1 and m_2, respectively. By the Slope-Intercept Form (1.16), the lines have equations

$$y = m_1 x + b_1, \qquad y = m_2 x + b_2$$

where b_1 and b_2 are the y-intercepts. The lines intersect at some point (x, y)

FIGURE 1.33

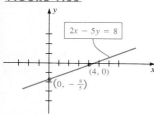

if and only if

$$m_1 x + b_1 = m_2 x + b_2$$

or

$$(m_1 - m_2)x = b_2 - b_1.$$

Since $l_1 \neq l_2$, the last equation can be solved for x if and only if $m_1 - m_2 \neq 0$. We have shown that the lines l_1 and l_2 intersect if and only if $m_1 \neq m_2$. Hence they do *not* intersect (are parallel) if and only if $m_1 = m_2$. • •

EXAMPLE 4 Find an equation of the line through the point $(5, -7)$ that is parallel to the line $6x + 3y - 4 = 0$.

SOLUTION Let us express the given equation in Slope-Intercept Form. We begin by writing

$$3y = -6x + 4$$

and then dividing both sides by 3, obtaining

$$y = -2x + \tfrac{4}{3}.$$

The last equation is in Slope-Intercept Form with slope $m = -2$. Since parallel lines have the same slope, the required line also has slope -2. Applying the Point-Slope Form gives us

$$y + 7 = -2(x - 5).$$

This is equivalent to

$$y + 7 = -2x + 10 \quad \text{or} \quad 2x + y - 3 = 0 \bullet$$

The next result specifies conditions for perpendicular lines.

THEOREM (1.19)

> Two lines with slope m_1 and m_2 are perpendicular if and only if
> $$m_1 m_2 = -1$$

FIGURE 1.34

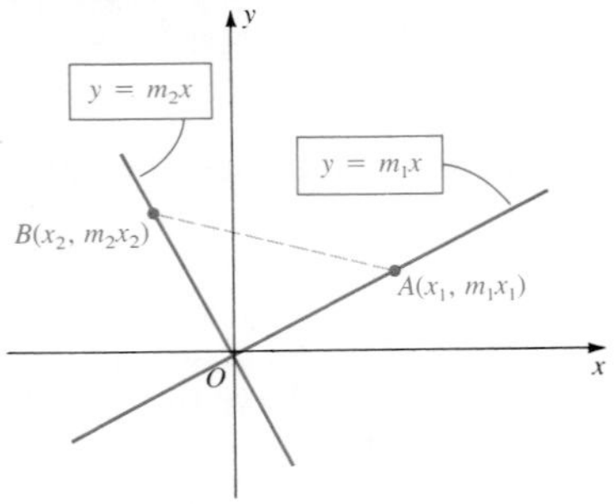

PROOF For simplicity, let us consider the special case of two lines that intersect at the origin O, as illustrated in Figure 1.34. In this case equations of the lines are $y = m_1 x$ and $y = m_2 x$. If, as in the figure, we choose points $A(x_1, m_1 x_1)$ and $B(x_2, m_2 x_2)$ different from O on the lines, then the lines are perpendicular if and only if angle AOB is a right angle. Applying the Pythagorean Theorem, angle AOB is a right angle if and only if

$$[d(A, B)]^2 = [d(O, B)]^2 + [d(O, A)]^2$$

or, by the Distance Formula,

$$(m_2 x_2 - m_1 x_1)^2 + (x_2 - x_1)^2 = (m_2 x_2)^2 + x_2^2 + (m_1 x_1)^2 + x_1^2.$$

Squaring terms and simplifying gives us

$$-2m_1 m_2 x_1 x_2 - 2x_1 x_2 = 0.$$

Dividing both sides by $-2x_1 x_2$, we see that $m_1 m_2 + 1 = 0$. Thus, the lines are perpendicular if and only if $m_1 m_2 = -1$.

The same type of proof may be given if the lines intersect at *any* point (a, b). • •

A convenient way to remember the conditions for perpendicularity is to note that m_1 and m_2 must be *negative reciprocals* of one another, that is, $m_1 = -1/m_2$ and $m_2 = -1/m_1$.

EXAMPLE 5 Find an equation for the perpendicular bisector of the line segment from $A(1, 7)$ to $B(-3, 2)$.

SOLUTION By the Midpoint Formula (1.10), the midpoint M of the segment AB is $(-1, \frac{9}{2})$. Since the slope of AB is $\frac{5}{4}$ (see Example 2), it follows from Theorem (1.19) that the slope of the perpendicular bisector is $-\frac{4}{5}$. Applying the Point-Slope Form,

$$y - \frac{9}{2} = -\frac{4}{5}(x + 1).$$

Multiplying both sides by 10 and simplifying leads to $8x + 10y - 37 = 0$. •

Two variables x and y are **linearly related** if $y = ax + b$ for constants a and b with $a \neq 0$. Linear relationships between variables occur frequently in applications. The following example gives one illustration. For other applications see Exercises 35–40.

EXAMPLE 6 The relationship between the air temperature T (in °F) and the altitude h (in feet above sea level) is approximately linear. When the temperature at sea level is 60°, an increase of 5000 feet in altitude lowers the air temperature about 18°.

(a) Express T in terms of h.

(b) Approximate the air temperature at an altitude of 15,000 feet.

SOLUTION

(a) If T is linearly related to h, then

$$T = ah + b$$

for some constants a and b. Since $T = 60$ when $h = 0$,

$$60 = a(0) + b \quad \text{or} \quad b = 60.$$

Thus, $$T = ah + 60.$$

In addition, if $h = 5000$, then $T = 60 - 18 = 42$. Substituting these values into the formula $T = ah + 60$,

$$42 = a(5000) + 60 \quad \text{or} \quad 5000a = -18.$$

Hence, $$a = -\frac{18}{5000} = -\frac{9}{2500}$$

and the (approximate) formula for T is

$$T = -\frac{9}{2500} h + 60.$$

(b) Using the formula for T obtained in part (a), the (approximate) temperature when $h = 15,000$ is

$$T = -\frac{9}{2500}(15,000) + 60 = -54 + 60 = 6 \,°F \quad •$$

Exer. 1–4: Plot the points A and B and find the slope of the line through A and B.

1 $A(-4, 6)$, $B(-1, 18)$

2 $A(6, -2)$, $B(-3, 5)$

3 $A(-1, -3)$, $B(-1, 2)$

4 $A(-3, 4)$, $B(2, 4)$

5 Show that $A(-3, 1)$, $B(5, 3)$, $C(3, 0)$, and $D(-5, -2)$ are vertices of a parallelogram.

6 Show that $A(2, 3)$, $B(5, -1)$, $C(0, -6)$, and $D(-6, 2)$ are vertices of a trapezoid.

7 Prove that the points $A(6, 15)$, $B(11, 12)$, $C(-1, -8)$, and $D(-6, -5)$ are vertices of a rectangle.

8 Prove that the points $A(1, 4)$, $B(6, -4)$, and $C(-15, -6)$ are vertices of a right triangle.

9 If three consecutive vertices of a parallelogram are $A(-1, -3)$, $B(4, 2)$, and $C(-7, 5)$, find the fourth vertex.

10 Let $A(x_1, y_1)$, $B(x_2, y_2)$, $C(x_3, y_3)$, and $D(x_4, y_4)$ denote the vertices of an arbitrary quadrilateral. Prove that the line segments joining midpoints of adjacent sides form a parallelogram.

Exer. 11–20: Find an equation for the line satisfying the given conditions.

11 Through $A(2, -6)$, slope $\frac{1}{2}$

12 Slope -3, y-intercept 5

13 Through $A(-5, -7)$ and $B(3, -4)$

14 x-intercept -4, y-intercept 8

15 Through $A(8, -2)$, y-intercept -3

16 Slope 6, x-intercept -2

17 Through $A(10, -6)$, parallel to (a) the y-axis; (b) the x-axis.

18 Through $A(-5, 1)$, perpendicular to (a) the y-axis; (b) the x-axis.

19 Through $A(7, -3)$, perpendicular to the line $2x - 5y = 8$.

20 Through $(-\frac{3}{4}, -\frac{1}{2})$, parallel to the line $x + 3y = 1$.

21 Given $A(3, -1)$ and $B(-2, 6)$, find an equation for the perpendicular bisector of the line segment AB.

22 Find an equation for the line that bisects the second and fourth quadrants.

Exer. 23–30: Use the Slope-Intercept Form (1.16) to find the slope and y-intercept of the line with the given equation, and sketch the graph.

23 $3x - 4y + 8 = 0$

24 $2y - 5x = 1$

25 $x + 2y = 0$

26 $8x = 1 - 4y$

27 $5x + 4y = 20$

28 $x + 2 = \frac{1}{2}y$

29 $x = 3y + 7$

30 $x - y = 0$

31 Find a real number k such that the point $P(-1, 2)$ is on the line $kx + 2y - 7 = 0$.

32 Find all values of r such that the slope of the line through the points $(r, 4)$ and $(1, 3 - 2r)$ is less than 5.

33 If a line l has nonzero x- and y-intercepts a and b, respectively, prove that an equation for l is $(x/a) + (y/b) = 1$. (This is called the *intercept form* for the equation of a line.) Express the equation $4x - 2y = 6$ in intercept form.

34 Prove that an equation of the line through $P_1(x_1, y_1)$ and $P_2(x_2, y_2)$ is
$$(y - y_1)(x_2 - x_1) = (y_2 - y_1)(x - x_1).$$
(This is called the *two-point form* for the equation of a line.) Use the two-point form to find an equation of the line through $A(7, -1)$ and $B(4, 6)$.

35 New pharmacological products must specify recommended dosages for adults and children. Two formulas suggested for modification of adult dosage levels for young children are:

$$\text{Cowling's Rule:} \quad y = \frac{t + 1}{24} a$$

$$\text{Friend's Rule:} \quad y = \frac{2}{25} ta$$

where a denotes adult dose (in mg) and t denotes the age of the child (in years).

(a) If $a = 100$, graph the two linear equations on the same axes for $0 \le t \le 12$.

(b) For what age do the two formulas specify the same dosage?

36 Charles' law for gases states that if the pressure remains constant, then the relationship between the volume V (in cm^3) that a gas occupies and its temperature T (in °C) is given by $V = V_0(1 + \frac{1}{273}T)$.

(a) What is the significance of V_0?

(b) What increase in temperature corresponds to an increase in volume from V_0 to $2V_0$?

(c) Sketch the graph of the equation on a TV-plane for the case $V_0 = 100$ and $T \ge -273$.

37 The electrical resistance R (in ohms) for a pure metal wire is linearly related to its temperature T (in °C) by the formula
$$R = R_0(1 + aT)$$
for some constant a and $R_0 > 0$.

(a) What is the significance of R_0?

(b) At *absolute zero* ($T = -273$ °C), $R = 0$. Find a.

(c) At 0 °C, silver wire has a resistance of 1.25 ohms. At what temperature is the resistance doubled?

38 The freezing point of water is 0 °C or 32 °F. The boiling point is 100 °C or 212 °F. Using this information, find a linear relationship between temperature in °F and temperature in °C. What temperature increase in °F corresponds to an increase in temperature of 1 °C?

39 Newborn blue whales measure approximately 24 feet long and weigh 3 tons. When weaned at 7 months, young whales

measure an amazing 53 feet and weigh 23 tons. Let L and W denote the length (in feet) and the weight (in tons), respectively, of a whale that is t months of age.

(a) If L and t are linearly related, what is the daily increase in length? (Use 1 month = 30 days.)

(b) If W and t are linearly related, what is the daily increase in weight?

40 A hammer thrower is working on his form in a small practice area. As the thrower spins, the hammer generates a circle with a radius of 5 feet. When thrown, the hammer hits a tall screen that is 50 feet from the center of the throwing area. Let coordinate axes be introduced as shown in the figure (not to scale).

(a) If the hammer is released at $(-4, -3)$ and travels in the tangent direction, where will it hit the screen?

(b) If the hammer is to hit the screen at $(0, -50)$, where on the circle should it be released?

EXERCISE 40

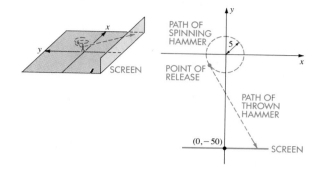

1.4 DEFINITION OF FUNCTION

The notion of **correspondence** occurs frequently in everyday life. As examples,

to each book in a library there corresponds the number of pages in the book;

to each human being there corresponds a birth date;

if the temperature of the air is recorded throughout a day, then at each instant of time there corresponds a temperature.

These examples of correspondence involve two sets, D and E. In our first example D denotes the set of books in a library and E the set of positive integers. For each book x in D there corresponds a positive integer y in E—namely, the number of pages in the book.

We sometimes depict correspondences by diagrams of the type shown in Figure 1.35, where the sets D and E are represented by points within regions (shown in color) in a plane. The curved arrow indicates that the element y of E corresponds to the element x of D. The two sets may have elements in common. As a matter of fact, we often have $D = E$.

Our examples indicate that *to each x in D there corresponds one and only one y in E*; that is, *y is unique* for a given x. However, the same element of E may correspond to different elements of D. For example, two books may have the same number of pages, two people may have the same birthday, and so on.

In most of our work D and E will be sets of numbers. To illustrate, let both D and E denote the set $\mathbb{R}$ of real numbers, and to each real number x let us assign its square x^2. Thus, to 3 we assign 9, to -5 we assign 25, and to $\sqrt{2}$, the number 2. This gives us a correspondence from $\mathbb{R}$ to $\mathbb{R}$.

Each of the preceding examples of a correspondence is a *function*, which we define as follows.

FIGURE 1.35

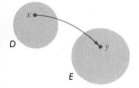

DEFINITION (1.20)

> A **function** f from a set D to a set E is a correspondence that assigns to each element x of D a unique element y of E.

FIGURE 1.36

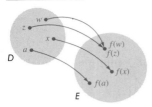

The element y of E is the **value** of f at x and is denoted by $f(x)$ (read "f of x"). The set D is the **domain** of the function. The **range** of f is the subset of E consisting of all possible values $f(x)$ for x in D.

We may now consider the diagram in Figure 1.36. The curved arrows indicate that the elements $f(x)$, $f(w)$, $f(z)$, and $f(a)$ of E correspond to the elements x, w, z, and a of D. It is important to remember that *to each x in D there is assigned precisely one value $f(x)$ in E*; however, different elements of D, such as w and z in Figure 1.36, may have the same value in E.

The symbols

$$D \xrightarrow{f} E, \qquad f: D \to E, \qquad \text{or}$$

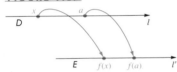

signify that f is a function from D to E. Students are sometimes initially confused by the notations f and $f(x)$. Remember that f is used to represent the function. It is neither in D nor in E. However, $f(x)$ is an element of E, namely the element that f assigns to x.

If the sets D and E of Definition (1.20) are intervals or other sets of real numbers, then instead of using points within regions to represent elements we may use two coordinate lines l and l', as illustrated in Figure 1.37.

FIGURE 1.37

Two functions f and g from D to E are **equal,** and we write

$$f = g \quad \text{provided} \quad f(x) = g(x) \quad \text{for every } x \text{ in } D.$$

For example, if $g(x) = \frac{1}{2}(2x^2 - 6) + 3$ and $f(x) = x^2$ for every x in $\mathbb{R}$, then $g = f$.

EXAMPLE 1 Let f be the function with domain $\mathbb{R}$ such that $f(x) = x^2$ for every x in $\mathbb{R}$.

(a) Find $f(-6)$, $f(\sqrt{3})$, and $f(a + b)$, for real numbers a and b.

(b) What is the range of f?

SOLUTION

(a) We may find values of f by substituting for x in the equation $f(x) = x^2$. Thus

$$f(-6) = (-6)^2 = 36, \qquad f(\sqrt{3}) = (\sqrt{3})^2 = 3,$$

and

$$f(a + b) = (a + b)^2 = a^2 + 2ab + b^2.$$

(b) By definition, the range of f consists of all numbers of the form $f(x) = x^2$, for x in $\mathbb{R}$. Since the square of every real number is nonnegative, the range is contained in the set of all nonnegative real numbers. Moreover, every nonnegative real number c is a value of f since $f(\sqrt{c}) = (\sqrt{c})^2 = c$. Hence, the range of f is the set of all nonnegative real numbers. •

If a function is defined as in Example 1, the symbols used for the function and variable are immaterial; that is, expressions such as $f(x) = x^2$, $f(s) = s^2$, $g(t) = t^2$, and $k(r) = r^2$ all define the same function. This is true because if a is any number in the domain, then the same value a^2 is obtained regardless of which expression is employed.

In the remainder of our work the phrase *f is a function* will mean that the domain and range are sets of real numbers. If a function is defined by means of an expression, as in Example 1, and the domain D is not stated explicitly, then we will consider D to be the totality of real numbers x such that $f(x)$ is real. To illustrate, if $f(x) = \sqrt{x-2}$, then the domain is assumed to be the set of real numbers x such that $\sqrt{x-2}$ is real, that is, $x - 2 \geq 0$, or $x \geq 2$. Thus, the domain is the infinite interval $[2, \infty)$. If x is in the domain, we say that f **is defined at** x, or that $f(x)$ **exists.** If a set S is contained in the domain, f **is defined on** S. The terminology f **is undefined at** x means that x is not in the domain of f.

Many formulas that occur in mathematics and the sciences determine functions. For instance, the formula $A = \pi r^2$ for the area A of a circle of radius r assigns to each positive real number r a unique value of A. This determines a function f such that $f(r) = \pi r^2$, and we may write $A = f(r)$. The letter r, which represents an arbitrary number from the domain of f, is an **independent variable.** The letter A, which represents a number from the range of f, is a **dependent variable,** since its value depends on the number assigned to r. If two variables r and A are related in this manner, we say that "A is a function of r." As another example, if an automobile travels at a uniform rate of 50 miles per hour, then the distance d (miles) traveled in time t (hours) is given by $d = 50t$, and hence, the distance d is a function of time t.

EXAMPLE 2 A steel storage tank for propane gas is to be constructed in the shape of a right circular cylinder of altitude 10 feet with a hemisphere attached to each end. The radius r is yet to be determined. Express the volume V of the tank as a function of r.

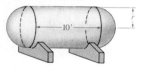

FIGURE 1.38

SOLUTION The tank is sketched in Figure 1.38. The volume of the cylindrical part of the tank may be found by multiplying the altitude 10 by the area πr^2 of the base of the cylinder. This gives us

$$\text{Volume of cylinder} = 10(\pi r^2) = 10\pi r^2.$$

The two hemispherical ends, taken together, form a sphere of radius r. Using the formula for the volume of a sphere, we obtain

$$\text{Volume of the two ends} = \tfrac{4}{3}\pi r^3.$$

Thus, the volume V of the tank is

$$V = \tfrac{4}{3}\pi r^3 + 10\pi r^2.$$

This formula expresses V as a function of r. In factored form:

$$V = \tfrac{1}{3}\pi r^2(4r + 30) = \tfrac{2}{3}\pi r^2(2r + 15) \quad \bullet$$

EXAMPLE 3 Two ships leave port at the same time, one sailing west at a rate of 17 mi/hr and the other sailing south at 12 mi/hr. If t is the time (in hours) after their departure, express the distance d between the ships as a function of t.

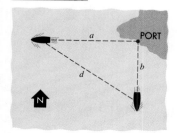

FIGURE 1.39

SOLUTION To help visualize the problem we begin by drawing a picture and labeling it, as in Figure 1.39. Using the Pythagorean Theorem,

$$d^2 = a^2 + b^2 \quad \text{or} \quad d = \sqrt{a^2 + b^2}.$$

Since distance = (rate)(time) and the rates are 17 and 12, respectively,

$$a = 17t \quad \text{and} \quad b = 12t.$$

Substitution in $d = \sqrt{a^2 + b^2}$ gives us

$$d = \sqrt{(17t)^2 + (12t)^2} = \sqrt{289t^2 + 144t^2} = \sqrt{433t^2} \quad \text{or} \quad d = \sqrt{433}t.$$

An *approximate* formula for expressing d as a function of t is $d \approx (20.8)t$. •

If $f(x) = x$ for every x in the domain D of f, then f is the **identity function** on D. A function f is a **constant function** if there is some (fixed) element c in the range such that $f(x) = c$ for every x in the domain. If a constant function is represented by a diagram of the type shown in Figure 1.35, *every* arrow from D terminates at the same point in E.

The types of functions described in the next definition occur frequently.

DEFINITION (1.21)

> Let f be a function such that $-x$ is in the domain D whenever x is in D.
>
> (i) f is **even** if $f(-x) = f(x)$ for every x in D,
>
> (ii) f is **odd** if $f(-x) = -f(x)$ for every x in D.

EXAMPLE 4

(a) If $f(x) = 3x^4 - 2x^2 + 5$, show that f is an even function.

(b) If $g(x) = 2x^5 - 7x^3 + 4x$, show that g is an odd function.

SOLUTION If x is any real number, then

(a)
$$\begin{aligned} f(-x) &= 3(-x)^4 - 2(-x)^2 + 5 \\ &= 3x^4 - 2x^2 + 5 = f(x) \end{aligned}$$

and hence f is even.

(b)
$$\begin{aligned} g(-x) &= 2(-x)^5 - 7(-x)^3 + 4(-x) \\ &= -2x^5 + 7x^3 - 4x \\ &= -(2x^5 - 7x^3 + 4x) = -g(x) \end{aligned}$$

Thus, g is odd. •

A function f may have the same value for different numbers in its domain. For example, if $f(x) = x^2$, then $f(2) = 4$ and $f(-2) = 4$, but $2 \neq -2$. If values are always different, then the function is *one-to-one*.

DEFINITION (1.22)

> A function f with domain D and range E is a **one-to-one function** if whenever $a \neq b$ in D, then $f(a) \neq f(b)$ in E.

EXAMPLE 5

(a) If $f(x) = 3x + 2$, prove that f is one-to-one.

(b) If $g(x) = x^4 + 2x^2$, prove that g is not one-to-one.

SOLUTION

(a) If $a \neq b$, then $3a \neq 3b$ and hence $3a + 2 \neq 3b + 2$, or $f(a) \neq f(b)$. Thus, f is one-to-one.

FIGURE 1.40

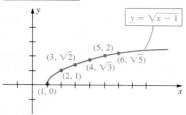

(b) The function g is not one-to-one, since different numbers in the domain may have the same value. For example, although $-1 \neq 1$, both $g(-1)$ and $g(1)$ are equal to 3. •

If f is a function, we may use a graph to show the change in $f(x)$ as x varies through the domain of f. By definition, the **graph of a function** f is the graph of the equation $y = f(x)$ for x in the domain of f. As shown in Figure 1.40, we often attach the label $y = f(x)$ to a sketch of the graph. Note that if $P(a, b)$ is a point on the graph, then the y-coordinate b is the function value $f(a)$. The figure exhibits the domain of f (the set of possible values of x) and the range of f (the corresponding values of y). Although we have pictured the domain and range as closed intervals, they may be infinite intervals or other sets of real numbers.

It is important to note that since there is a unique value $f(a)$ for each a in the domain, only *one* point on the graph has x-coordinate a. Thus, *every vertical line intersects the graph of a function in at most one point*. Consequently, the graph of a function cannot be a figure such as a circle, in which a vertical line may intersect the graph in more than one point.

The x-intercepts of the graph of a function f are the solutions of the equation $f(x) = 0$. These numbers are the **zeros** of the function. The y-intercept of the graph is $f(0)$, if it exists.

EXAMPLE 6 Sketch the graph of f if $f(x) = \sqrt{x - 1}$. What are the domain and range of f?

SOLUTION By definition, the graph of f is the graph of the equation $y = \sqrt{x - 1}$. The following table lists coordinates of several points on the graph.

x	1	2	3	4	5	6
y	0	1	$\sqrt{2}$	$\sqrt{3}$	2	$\sqrt{5}$

FIGURE 1.41

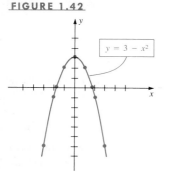

Plotting points, we obtain the sketch shown in Figure 1.41. Note that the x-intercept is 1, and there is no y-intercept.

The domain of f consists of all real numbers x such that $x \geq 1$, or equivalently, the interval $[1, \infty)$. The range of f is the set of all real numbers y such that $y \geq 0$, or equivalently, $[0, \infty)$. •

FIGURE 1.42

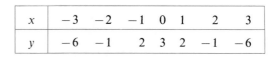

EXAMPLE 7 Sketch the graph of f if $f(x) = 3 - x^2$. What are the domain and range of f?

SOLUTION The following table lists some points (x, y) on the graph.

x	-3	-2	-1	0	1	2	3
y	-6	-1	2	3	2	-1	-6

The x-intercepts are the solutions of the equation $f(x) = 0$, that is, of $3 - x^2 = 0$. These are $\pm\sqrt{3}$. The y-intercept is $f(0) = 3$. Plotting points leads to the parabola sketched in Figure 1.42.

Since x may be assigned any value, the domain of f is $\mathbb{R}$. Referring to the graph we see that the range of f is $(-\infty, 3]$. •

The solution to Example 7 could have been simplified by observing that since $3 - (-x)^2 = 3 - x^2$, the graph of $y = 3 - x^2$ is symmetric with respect to the y-axis. This fact also follows from (i) of the next theorem.

THEOREM (1.23)

(i) The graph of an even function is symmetric with respect to the y-axis.

(ii) The graph of an odd function is symmetric with respect to the origin.

PROOF If f is even, then $f(-x) = f(x)$, and hence the equation $y = f(x)$ is not changed if $-x$ is substituted for x. Statement (i) now follows from Symmetry Test (1.11) (i). The proof of (ii) is left to the reader. • •

EXAMPLE 8 Sketch the graph of f if $f(x) = |x|$ and find the domain and range of f.

FIGURE 1.43

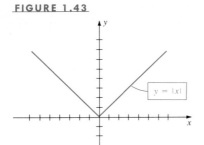

$y = |x|$

SOLUTION If $x \geq 0$, then $f(x) = x$ and hence the part of the graph to the right of the y-axis is identical to the graph of $y = x$, which is a line through the origin with slope 1. If $x < 0$, then by Definition (1.2), $f(x) = |x| = -x$, and hence the part of the graph to the left of the y-axis is the same as the graph of $y = -x$. The graph is sketched in Figure 1.43.

Note that f is an even function, and hence by Theorem (1.23) (i), the graph is symmetric with respect to the y-axis, as indicated in the figure.

Referring to the graph, we see that the domain of f is $\mathbb{R}$, and the range is $[0, \infty)$. •

FIGURE 1.44

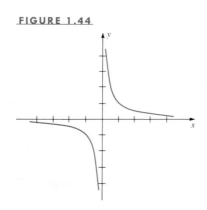

EXAMPLE 9 Sketch the graph of f if $f(x) = \dfrac{1}{x}$.

SOLUTION The domain of f is the set of all nonzero real numbers. If x is positive, then so is $f(x)$, and hence no part of the graph lies in quadrant IV. Quadrant II is also excluded, since if $x < 0$, then $f(x) < 0$. If x is close to zero, then $|1/x|$ is large. As x increases through positive values, $1/x$ decreases and is close to zero when x is large. Similarly, if x is negative and $|x|$ is large, then $1/x$ is close to zero. These remarks and plotting several points gives us the sketch in Figure 1.44.

The graph of f, or equivalently, of the equation $y = 1/x$, is symmetric with respect to the origin. This may be verified by using either Theorem (1.23) (ii) or Symmetry Test (1.11) (iii). •

EXAMPLE 10 Describe the graph of a constant function.

SOLUTION If $f(x) = c$ for a real number c, then the graph of f is the same as the graph of the equation $y = c$, and hence is a horizontal line with y-intercept c. •

Functions are sometimes described in terms of more than one expression, as in the next examples. We call such functions **piecewise-defined functions.**

EXAMPLE 11 Sketch the graph of the function f that is defined as follows:

$$f(x) = \begin{cases} 2x + 3 & \text{if } x < 0 \\ x^2 & \text{if } 0 \leq x < 2 \\ 1 & \text{if } x \geq 2 \end{cases}$$

SOLUTION If $x < 0$, then $f(x) = 2x + 3$. This means that if x is negative, the expression $2x + 3$ should be used to find function values. Consequently, if $x < 0$, then the graph of f coincides with the line $y = 2x + 3$, and we sketch that portion of the graph to the left of the y-axis, as indicated in Figure 1.45.

If $0 \leq x < 2$, we use x^2 to find values of f, and therefore this part of the graph of f coincides with the graph of the parabola $y = x^2$. We then sketch the part of the graph of f between $x = 0$ and $x = 2$, as indicated in the figure.

Finally, if $x \geq 2$, the values of f are always 1. The graph of f for $x \geq 2$ is the horizontal half-line illustrated in Figure 1.45. •

FIGURE 1.45

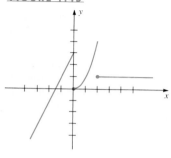

EXAMPLE 12 If x is any real number, then there exist consecutive integers n and $n + 1$ such that $n \leq x < n + 1$. Let f be the function defined as follows: If $n \leq x < n + 1$, then $f(x) = n$. Sketch the graph of f.

SOLUTION The x- and y-coordinates of some points on the graph may be listed as follows:

Values of x	$f(x)$
$\cdots$	$\cdots$
$-2 \leq x < -1$	-2
$-1 \leq x < 0$	-1
$0 \leq x < 1$	0
$1 \leq x < 2$	1
$2 \leq x < 3$	2
$\cdots$	$\cdots$

FIGURE 1.46

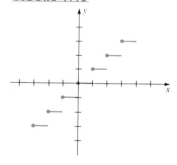

Since f is a constant function whenever x is between successive integers, the corresponding part of the graph is a segment of a horizontal line. Part of the graph is sketched in Figure 1.46. The graph continues indefinitely to the right and to the left. •

The symbol $[\![x]\!]$ is often used to denote the largest integer n such that $n \leq x$. For example $[\![1.6]\!] = 1$, $[\![\sqrt{5}]\!] = 2$, $[\![\pi]\!] = 3$, and $[\![-3.5]\!] = -4$. Using this notation, the function f of Example 12 may be defined by $f(x) = [\![x]\!]$. We refer to f as the **greatest integer function**.

EXERCISES 1.4

1 If $f(x) = x^3 + 4x - 3$, find $f(1)$, $f(-1)$, $f(0)$, and $f(\sqrt{2})$.

2 If $f(x) = \sqrt{x - 1} + 2x$, find $f(1)$, $f(3)$, $f(5)$, and $f(10)$.

Exer. 3–4: If f is the given function, and if a and h are real numbers, find the following:

(a) $f(a)$

(b) $f(-a)$

(c) $-f(a)$

(d) $f(a + h)$

(e) $f(a) + f(h)$

(f) $\dfrac{f(a + h) - f(a)}{h}$ provided $h \neq 0$

3 $f(x) = 3x^2 - x + 2$

4 $f(x) = \dfrac{1}{x^2 + 1}$

Exer. 5–6: If g is the given function, find the following:

(a) $g(1/a)$

(b) $\dfrac{1}{g(a)}$

(c) $g(a^2)$

(d) $[g(a)]^2$

(e) $g(\sqrt{a})$

(f) $\sqrt{g(a)}$

5 $g(x) = \dfrac{1}{x^2 + 4}$

6 $g(x) = \dfrac{1}{x}$

Exer. 7–12: Find the domain of the function f.

7 $f(x) = \sqrt{3x - 5}$

8 $f(x) = \sqrt{7 - 2x}$

9 $f(x) = \sqrt{4 - x^2}$

10 $f(x) = \sqrt{x^2 - 9}$

11 $f(x) = \dfrac{x + 1}{x^3 - 9x}$

12 $f(x) = \dfrac{4x + 7}{6x^2 + 13x - 5}$

Exer. 13–20: Determine if the function f is one-to-one.

13 $f(x) = 2x + 9$

14 $f(x) = \dfrac{1}{7x + 9}$

15 $f(x) = 5 - 3x^2$

16 $f(x) = 2x^2 - x - 3$

17 $f(x) = \sqrt{x}$

18 $f(x) = x^3$

19 $f(x) = |x|$

20 $f(x) = 4$

Exer. 21–30: Determine if f is even, odd, or neither even nor odd.

21 $f(x) = 3x^3 - 4x$

22 $f(x) = 7x^4 - x^2 + 7$

23 $f(x) = 9 - 5x^2$

24 $f(x) = 2x^5 - 4x^3$

25 $f(x) = 2$

26 $f(x) = 2x^3 + x^2$

27 $f(x) = 2x^2 - 3x + 4$

28 $f(x) = \sqrt{x^2 + 1}$

29 $f(x) = \sqrt[3]{x^3 - 4}$

30 $f(x) = |x| + 5$

Exer. 31–44: Sketch the graph and determine the domain and range of f.

31 $f(x) = -4x + 3$

32 $f(x) = 4x - 3$

33 $f(x) = -3$

34 $f(x) = 3$

35 $f(x) = 4 - x^2$

36 $f(x) = -(4 + x^2)$

37 $f(x) = \sqrt{4 - x^2}$

38 $f(x) = \sqrt{x^2 - 4}$

39 $f(x) = \dfrac{1}{x - 4}$

40 $f(x) = -\dfrac{1}{(x - 4)^2}$

41 $f(x) = \dfrac{x}{|x|}$

42 $f(x) = x + |x|$

43 $f(x) = \sqrt{4 - x}$

44 $f(x) = 2 - \sqrt{x}$

Exer. 45–50: Sketch the graph of the piecewise-defined function f.

45 $f(x) = \begin{cases} 2 & \text{if } x < 0 \\ -1 & \text{if } x \geq 0 \end{cases}$

46 $f(x) = \begin{cases} 3 & \text{if } x < -3 \\ -x & \text{if } -3 \leq x \leq 3 \\ -3 & \text{if } x > 3 \end{cases}$

47 $f(x) = \begin{cases} x & \text{if } x < 0 \\ -2 & \text{if } 0 \leq x < 1 \\ x^2 & \text{if } x \geq 1 \end{cases}$

48 $f(x) = \begin{cases} x & \text{if } x \leq 1 \\ -x^2 & \text{if } 1 < x < 2 \\ x & \text{if } x \geq 2 \end{cases}$

49 $f(x) = \begin{cases} \dfrac{x^2 - 4}{x - 2} & \text{if } x \neq 2 \\ 3 & \text{if } x = 2 \end{cases}$

50 $f(x) = \begin{cases} \dfrac{x^2 - 1}{1 - x} & \text{if } x \neq 1 \\ 2 & \text{if } x = 1 \end{cases}$

51 Explain why the graph of the equation $x^2 + y^2 = 1$ is not the graph of a function.

52 Prove that a function f is one-to-one if and only if every horizontal line intersects the graph of f in at most one point.

53 An open box is to be made from a rectangular piece of cardboard having dimensions 20 inches × 30 inches by cutting out identical squares of area x^2 from each corner and turning up the resulting sides (see figure). Express the volume V of the box as a function of x.

EXERCISE 53

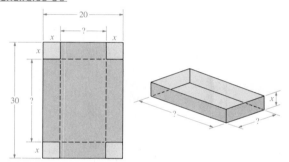

54 A small office building is designed with 500 ft² of floor space. The simple floor plans are shown in the figure.

(a) Express the length y of the building as a function of the width x.

(b) If the walls cost \$100 per running foot, express the cost C of the walls as a function of the width x. (Disregard the wall space above the doors.)

EXERCISE 54

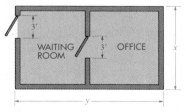

55 A hot-air balloon is released at 1:00 P.M. and rises vertically at a rate of 2 m/sec. An observation point is situated 100 meters from a point on the ground directly below the balloon (see figure). If t denotes the time (in seconds) after 1:00 P.M., express the distance d between the balloon and the observation point as a function of t.

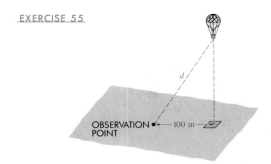

the distance. Suppose he can row at a rate of 3 mi/hr and
can walk at a rate of 5 mi/hr. If T is the total time required
to reach the house, express T as a function of x.

EXERCISE 58

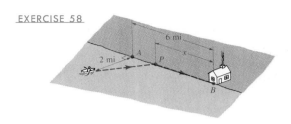

56 An acrobat's tightrope set-up is shown in the figure. Two poles are set 50 feet apart, but the point of attachment P for the rope is yet to be determined.

(a) Express the length L of the rope as a function of the distance x from point P to the ground.

(b) If the rope is be 75 feet, determine the height of the point of attachment P.

59 The relative positions of an aircraft runway and a 20-foot-tall control tower are shown in the figure. The beginning of the runway is at a perpendicular distance of 300 feet from the base of the tower. If x denotes the distance a jet has moved down the runway, express the distance d between the jet and the control booth as a function of x.

EXERCISE 56

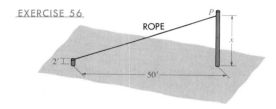

EXERCISE 59

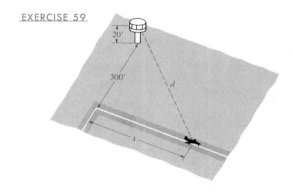

57 From an exterior point P that is h units from a circle of radius r, a tangent line is drawn to the circle (see figure). Let y denote the distance from the point P to the point of tangency T.

(a) Express y as a function of h. (*Hint:* If C is the center of the circle, then PT is perpendicular to CT.)

(b) If r is the radius of the earth, and h is the altitude of a space shuttle, then we can derive a formula for the maximum distance y (to the earth) that an astronaut can see from the shuttle. In particular, if $h = 200$ miles and $r \approx 4000$ miles, approximate y.

60 A right circular cylinder of radius r and height h is inscribed in a cone of altitude 12 and base radius 4, as illustrated in the figure.

(a) Express h as a function of r. (*Hint:* Use similar triangles.)

(b) Express the volume V of the cylinder as a function of r.

EXERCISE 57

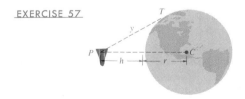

58 A man, who is in a rowboat 2 miles from the nearest point A on a straight shoreline, wishes to reach a house located at a point B, 6 miles further downriver (see figure). He plans to row to a point P that is between A and B and is x miles from the house, and then walk the remainder of

EXERCISE 60

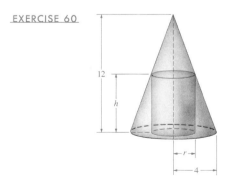

In this section we shall discuss how to obtain a new function by applying algebraic operations, such as addition, multiplication, and so on, to other functions. This will often allow us to obtain properties of the new function in an easy manner. We shall begin by considering the case of adding a constant c to all the values of a function f.

Let f be a function, let c be a constant, and let g be the function defined by

$$g(x) = f(x) + c$$

for every x in the domain of f. We sometimes say that g and f *differ by a constant.* If $c > 0$, the graph of g can be obtained by raising the graph of f a distance c; whereas if $c < 0$, we lower the graph of f a distance $|c|$. This technique is illustrated in the next example.

EXAMPLE 1 Given $f(x) = x^2 + c$, sketch the graph of f if $c = 4$ and if $c = -2$.

SOLUTION We shall sketch both graphs on the same coordinate axes. The graph of $y = x^2$ was sketched in Figure 1.20, and, for reference, is represented in gray in Figure 1.47. To find the graph of $y = x^2 + 4$ we may simply add 4 to the y-coordinate of each point on the graph of $y = x^2$. This amounts to *shifting* the graph of $y = x^2$ *upward* 4 units as shown in the figure. For $c = -2$ we decrease y-coordinates by 2 and, hence, the graph of $y = x^2 - 2$ may be obtained by shifting the graph of $y = x^2$ *downward* 2 units. Each graph is a parabola symmetric with respect to the y-axis. To verify the correct position of each graph, we usually plot several points. •

FIGURE 1.47

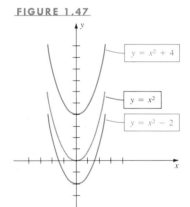

$y = x^2 + 4$

$y = x^2$

$y = x^2 - 2$

The graphs in the preceding example illustrate **vertical shifts** of the graph of $y = x^2$ and are special cases of the following general rules.

VERTICAL SHIFTS OF GRAPHS $(c > 0)$

To obtain the graph of:	shift the graph of $y = f(x)$:
$y = f(x) - c$	c units downward
$y = f(x) + c$	c units upward

Similar rules can be stated for **horizontal shifts.** Specifically, if $c > 0$, consider the graphs of $y = f(x)$ and $y = f(x - c)$ sketched on the same coordinate axes, as illustrated in Figure 1.48. Since $f(a) = f(a + c - c)$, we see that the point with x-coordinate a on the graph of $y = f(x)$ has the same y-coordinate as the point with x-coordinate $a + c$ on the graph of $y = f(x - c)$. This implies that the graph of $y = f(x - c)$ can be obtained by shifting the graph of $y = f(x)$ to the right c units. Similarly, the graph of $y = f(x + c)$ can be obtained by shifting the graph of f to the left c units. These rules are listed for reference in the next box.

FIGURE 1.48

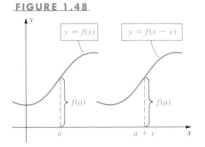

$y = f(x)$ $y = f(x - c)$

$f(a)$ $f(a)$

a $a + c$

To obtain the graph of:	shift the graph of $y = f(x)$:
$y = f(x - c)$	c units to the right
$y = f(x + c)$	c units to the left

FIGURE 1.49

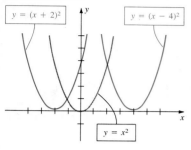

EXAMPLE 2 Sketch the graph of f if $f(x) = (x - 4)^2$ and if $f(x) = (x + 2)^2$.

SOLUTION The graph of $y = x^2$ is sketched in gray in Figure 1.49. According to the rules for horizontal shifts, shifting this graph to the right 4 units gives us the graph of $y = (x - 4)^2$. Shifting to the left 2 units leads to the graph of $y = (x + 2)^2$. Students who are not convinced of the validity of this technique are urged to plot several points on each graph. •

To obtain the graph of $y = cf(x)$ for some real number c, we may *multiply* the y-coordinates of points on the graph of $y = f(x)$ by c. For example, if $y = 2f(x)$, we double y-coordinates, or if $y = \frac{1}{2}f(x)$, we multiply each y-coordinate by $\frac{1}{2}$. If $c > 0$ (and $c \neq 1$) we shall refer to this procedure as **stretching** the graph of $y = f(x)$.

FIGURE 1.50

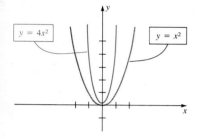

EXAMPLE 3 Sketch the graphs of (a) $y = 4x^2$ and (b) $y = \frac{1}{4}x^2$.

SOLUTION

(a) To sketch the graph of $y = 4x^2$ we may refer to the graph of $y = x^2$ (shown in gray in Figure 1.50) and multiply the y-coordinate of each point by 4. This gives us a narrower parabola that is sharper at the vertex, as illustrated in the figure. To obtain the correct shape, we should plot several points, such as $(0, 0)$, $(\frac{1}{2}, 1)$, and $(1, 4)$.

(b) The graph of $y = \frac{1}{4}x^2$ may be sketched by multiplying y-coordinates of points on the graph of $y = x^2$ by $\frac{1}{4}$. The graph is a wider parabola that is flatter at the vertex, as shown in Figure 1.51. •

FIGURE 1.51

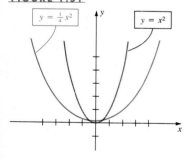

The graph of $y = -f(x)$ may be obtained by multiplying the y-coordinate of each point on the graph of $y = f(x)$ by -1. Thus, every point (a, b) on the graph of $y = f(x)$ that lies above the x-axis determines a point $(a, -b)$ on the graph of $y = -f(x)$ that lies below the x-axis. Similarly, if (c, d) lies below the x-axis (that is, $d < 0$), then $(c, -d)$ lies above the x-axis. The graph of $y = -f(x)$ is a **reflection** of the graph of $y = f(x)$ through the x-axis.

EXAMPLE 4 Sketch the graph of $y = -x^2$.

SOLUTION The graph may be found by plotting points; however, since the graph of $y = x^2$ is well known, we sketch it in gray, as in Figure 1.52, and then multiply y-coordinates of points by -1. This gives us the reflection through the x-axis indicated in the figure. •

Functions are often defined in terms of sums, differences, products, and quotients of various expressions. For example, if

$$h(x) = x^2 + \sqrt{5x + 1},$$

FIGURE 1.52

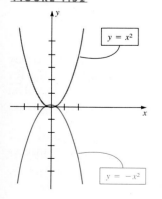

we may regard $h(x)$ as a sum of values of the simpler functions f and g defined by

$$f(x) = x^2 \quad \text{and} \quad g(x) = \sqrt{5x + 1}.$$

We refer to the function h as the *sum* of f and g.

In general, suppose f and g are *any* functions. Let I be *the intersection of their domains*, that is, the numbers *common* to both domains. The **sum** of f and g is the function h defined by

$$h(x) = f(x) + g(x)$$

for every x in I.

It is convenient to denote h by the symbol $f + g$. Since f and g are functions, not numbers, the $+$ used between f and g does not signify addition of real numbers. It is used to indicate that the value of $f + g$ at x is $f(x) + g(x)$; that is,

$$(f + g)(x) = f(x) + g(x).$$

The **difference** $f - g$ and the **product** fg of f and g are defined by

$$(f - g)(x) = f(x) - g(x) \quad \text{and} \quad (fg)(x) = f(x)g(x)$$

for x in I. The **quotient** f/g of f by g is given by

$$\left(\frac{f}{g}\right)(x) = \frac{f(x)}{g(x)}$$

for x in I and $g(x) \neq 0$.

EXAMPLE 5 If $f(x) = \sqrt{4 - x^2}$ and $g(x) = 3x + 1$, find the sum, difference, and product of f and g, and the quotient of f by g.

SOLUTION The domain of f is the closed interval $[-2, 2]$ and the domain of g is $\mathbb{R}$. Consequently, the intersection of their domains is $[-2, 2]$, and the required functions are given by

$$(f + g)(x) = \sqrt{4 - x^2} + (3x + 1), \qquad -2 \leq x \leq 2$$
$$(f - g)(x) = \sqrt{4 - x^2} - (3x + 1), \qquad -2 \leq x \leq 2$$
$$(fg)(x) = \sqrt{4 - x^2}(3x + 1), \qquad -2 \leq x \leq 2$$
$$\left(\frac{f}{g}\right)(x) = \frac{\sqrt{4 - x^2}}{3x + 1}, \qquad -2 \leq x \leq 2, \quad x \neq -\tfrac{1}{3} \bullet$$

If g is a constant function such that $g(x) = c$ for every x, and if f is any function, then cf will denote the product of g and f; that is, $(cf)(x) = cf(x)$ for every x in the domain of f. To illustrate, if f is the function of Example 5, then $(cf)(x) = c\sqrt{4 - x^2}$, $-2 \leq x \leq 2$. Geometrically, the function cf stretches and/or reflects the graph of f (see Figures 1.50–1.52).

Among the most important functions in mathematics are those discussed in the remainder of this section—polynomial, rational, algebraic, transcendental, and composite functions.

DEFINITION (1.24)

A function f is a **polynomial function** if
$$f(x) = a_n x^n + a_{n-1} x^{n-1} + \cdots + a_1 x + a_0$$
where the coefficients $a_0, a_1, \ldots, a_n$ are real numbers and the exponents are nonnegative integers.

A polynomial function may be thought of as a sum of functions whose values are of the form $a_k x^k$ for a real number a_k and a nonnegative integer k.

The expression to the right of the equal sign in Definition (1.24) is a **polynomial in x** (with real coefficients) and each $a_k x^k$ is a **term** of the polynomial. The number a_0 is the **constant term**. We often use the phrase *the polynomial $f(x)$* when referring to an expression of this type. If $a_n \neq 0$, then a_n is the **leading coefficient** of $f(x)$ and we say that f (or $f(x)$) has **degree n.**

If a polynomial function f has degree 0, then $f(x) = c$ for $c \neq 0$, and hence f is a constant function. If a coefficient a_k is zero, we abbreviate (1.24) by deleting the term $a_k x^k$. If *all* the coefficients of a polynomial are zero, it is called the **zero polynomial** and is denoted by 0. The zero polynomial is not assigned a degree.

If some of the coefficients are negative, then for convenience we often use minus signs between appropriate terms. To illustrate, instead of writing $3x^2 + (-5)x + (-7)$, we write $3x^2 - 5x - 7$ for this polynomial of degree 2. Polynomials in other variables may also be considered. For example, $\frac{2}{5}z^2 - 3z^7 + 8 - \sqrt{5}z^4$ is polynomial in z of degree 7. We ordinarily arrange the terms in order of decreasing powers: $-3z^7 - \sqrt{5}z^4 + \frac{2}{5}z^2 + 8$.

According to the definition of degree, if c is a nonzero real number, then c is a polynomial of degree 0. Such polynomials (together with the zero polynomial) are called **constant polynomials.**

If $f(x)$ is a polynomial of degree 1, then $f(x) = ax + b$ for $a \neq 0$. From Section 1.3, the graph of f is a line and, accordingly, f is called a **linear function.**

Any polynomial $f(x)$ of degree 2 may be written

$$f(x) = ax^2 + bx + c$$

for $a \neq 0$. In this case f is called a **quadratic function.** The graph of f or, equivalently, of the equation $y = ax^2 + bx + c$, is a parabola.

In Chapter 4 we will use methods of calculus to investigate graphs of polynomial functions of degree greater than 2.

A **rational function** is a quotient of two polynomial functions. Thus q is rational if, for every x in its domain,

$$q(x) = \frac{f(x)}{h(x)}$$

for polynomials $f(x)$ and $h(x)$. The domain of a polynomial function is $\mathbb{R}$; however the domain of a rational function consists of all real numbers except the zeros of the polynomial in the denominator.

An **algebraic function** is a function that can be expressed in terms of sums, differences, products, quotients, or roots of polynomial functions. For example, if

$$f(x) = 5x^4 - 2\sqrt[3]{x} + \frac{x(x^2 + 5)}{\sqrt{x^3 + \sqrt{x}}},$$

then f is an algebraic function. Functions that are not algebraic are termed **transcendental.** The trigonometric, exponential, and logarithmic functions considered later in this book are examples of transcendental functions.

We conclude this section by describing an important method of using two functions f and g to obtain a third function. Suppose D, E, and K are sets of real numbers. Let f be a function from D to E, and let g be a function from E to K. Using arrow notation we have

$$D \xrightarrow{f} E \xrightarrow{g} K.$$

We shall use f and g to define a function from D to K.

For every x in D, the number $f(x)$ is in E. Since the domain of g is E, we may then find the number $g(f(x))$ in K. By associating $g(f(x))$ with x, we obtain a function from D to K called the *composite function* of g by f. This is illustrated in Figure 1.53, where the gray arrow indicates the correspondence we have defined from D to K.

We sometimes use an operational symbol $\circ$ and denote a composite function as $g \circ f$. The following definition summarizes our discussion.

FIGURE 1.53

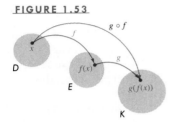

(1.25)

> Let f be a function from D to E and let g be a function from E to K. The **composite function** $g \circ f$ is the function from D to K defined by
>
> $$(g \circ f)(x) = g(f(x))$$
>
> for every x in D.

If the domain of g is a *subset* E' of E, then the domain of $g \circ f$ consists of every x in D such that $f(x)$ is in E'.

EXAMPLE 6 If f and g are defined by $f(x) = x - 2$ and $g(x) = 5x + \sqrt{x}$, find $(g \circ f)(x)$ and the domain of $g \circ f$.

SOLUTION Formal substitutions give us the following:

$$(g \circ f)(x) = g(f(x))$$
$$= g(x - 2)$$
$$= 5(x - 2) + \sqrt{x - 2}$$
$$= 5x - 10 + \sqrt{x - 2}$$

The domain of f is $\mathbb{R}$; however, the last equality implies that $(g \circ f)(x)$ is a real number only if $x \geq 2$. Thus, the domain of the composite function $g \circ f$ is the interval $[2, \infty)$. •

Given f and g, it may also be possible to find $(f \circ g)(x) = f(g(x))$, as illustrated in the next example.

EXAMPLE 7 If $f(x) = x^2 - 1$ and $g(x) = 3x + 5$, find $(f \circ g)(x)$ and $(g \circ f)(x)$.

SOLUTION We may proceed as follows:

$$(f \circ g)(x) = f(g(x)) = f(3x + 5)$$
$$= (3x + 5)^2 - 1$$
$$= 9x^2 + 30x + 24$$

Similarly,

$$(g \circ f)(x) = g(f(x)) = g(x^2 - 1)$$
$$= 3(x^2 - 1) + 5$$
$$= 3x^2 + 2$$

Note that in Example 7, $f(g(x))$ and $g(f(x))$ are not the same, that is, $f \circ g \neq g \circ f$.

In some applications, we must express a quantity y as a function of time t. As the following example illustrates, it is often easier to first introduce a variable x, and express x as a function of t—that is, $x = g(t)$. Next express y as a function of x—that is, $y = f(x)$. Finally form the composite function given by $y = f(x) = f(g(t))$.

EXAMPLE 8 A spherical toy balloon is being inflated with helium gas. If the radius of the balloon is changing at a rate of 1.5 cm/sec, express the volume V of the balloon as a function of time t (in seconds).

SOLUTION Let x denote the radius of the balloon. If we assume that the radius is 0 initially, then after t seconds

$$x = 1.5t \qquad \text{(radius of balloon after } t \text{ seconds).}$$

To illustrate, after 1 second the radius is 1.5 cm; after 2 seconds it is 3.0 cm; after 3 seconds it is 4.5 cm; and so on.

Next we write

$$V = \tfrac{4}{3}\pi x^3 \qquad \text{(volume of a sphere of radius } x\text{).}$$

This gives us a composite-function relationship in which V is a function of x, and x is a function of t. By substitution, we obtain

$$V = \tfrac{4}{3}\pi x^3 = \tfrac{4}{3}\pi(1.5t)^3 = \tfrac{4}{3}\pi(\tfrac{3}{2}t)^3 = \tfrac{4}{3}\pi(\tfrac{27}{8}t^3).$$

Simplifying, we obtain the following formula for V as a function of t:

$$V = \tfrac{9}{2}\pi t^3 \quad \bullet$$

EXERCISES 1.5

Exer. 1–10: On one coordinate system sketch the graphs of f for the three values of c. (Make use of vertical shifts, horizontal shifts, stretching, or reflecting).

1 $f(x) = 3x + c$; $c = 0, c = 2, c = -1$

2 $f(x) = -2x + c$; $c = 0, c = 1, c = -3$

3 $f(x) = x^3 + c$; $c = 0, c = 1, c = -2$

4 $f(x) = -x^3 + c$; $c = 0, c = 2, c = -1$

5 $f(x) = \sqrt{4 - x^2} + c$; $c = 0, c = 4, c = -3$

6 $f(x) = c - |x|$; $c = 0, c = 5, c = -2$

7 $f(x) = 3(x - c)$; $c = 0, c = 2, c = 3$

8 $f(x) = -2(x - c)^2$; $c = 0, c = 3, c = 1$

9 $f(x) = (x + c)^3$; $c = 0, c = 2, c = -2$

10 $f(x) = c\sqrt{9 - x^2}$; $c = 0, c = 2, c = 3$

11 The graph of a function f with domain $0 \le x \le 4$ is shown in the figure. Sketch the graph of each of the following:

(a) $y = f(x + 2)$ (b) $y = f(x - 2)$

(c) $y = f(x) + 2$ (d) $y = f(x) - 2$

(e) $y = 2f(x)$ (f) $y = \frac{1}{2}f(x)$

(g) $y = -2f(x)$ (h) $y = f(x - 3) + 1$

EXERCISE 11

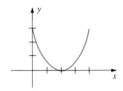

12 The graph of a function f with domain $0 \le x \le 4$ is shown in the figure. Sketch the graph of each of the following:

(a) $y = f(x - 1)$ (b) $y = f(x + 3)$

(c) $y = f(x) - 2$ (d) $y = f(x) + 1$

(e) $y = 3f(x)$ (f) $y = \frac{1}{2}f(x)$

(g) $y = f(x + 2) - 2$ (h) $y = f(2x)$

EXERCISE 12

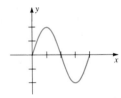

Exer. 13–18: Find the sum, difference, and product of f and g, and the quotient of f by g.

13 $f(x) = 3x^2, \ g(x) = 1/(2x - 3)$

14 $f(x) = \sqrt{x + 3}, \ g(x) = \sqrt{x + 3}$

15 $f(x) = x + (1/x), \ g(x) = x - (1/x)$

16 $f(x) = x^3 + 3x, \ g(x) = 3x^2 + 1$

17 $f(x) = 2x^3 - x + 5, \ g(x) = x^2 + x + 2$

18 $f(x) = 7x^4 + x^2 - 1, \ g(x) = 7x^4 - x^3 + 4x$

Exer. 19–32: Find $(f \circ g)(x)$ and $(g \circ f)(x)$.

19 $f(x) = 2x^2 + 5, \ g(x) = 4 - 7x$

20 $f(x) = 1/(3x + 1), \ g(x) = 2/x^2$

21 $f(x) = x^3, \ g(x) = x + 1$

22 $f(x) = \sqrt{x^2 + 4}, \ g(x) = 7x^2 + 1$

23 $f(x) = 3x^2 + 2, \ g(x) = 1/(3x^2 + 2)$

24 $f(x) = 7, \ g(x) = 4$

25 $f(x) = \sqrt{2x + 1}, \ g(x) = x^2 + 3$

26 $f(x) = 6x - 12, \ g(x) = \frac{1}{6}x + 2$

27 $f(x) = |x|, \ g(x) = -5$

28 $f(x) = \sqrt[3]{x^2 + 1}, \ g(x) = x^3 + 1$

29 $f(x) = x^2, \ g(x) = 1/x^2$

30 $f(x) = 1/(x + 1), \ g(x) = x + 1$

31 $f(x) = 2x - 3, \ g(x) = (x + 3)/2$

32 $f(x) = x^3 - 1, \ g(x) = \sqrt[3]{x + 1}$

33 If f is a linear function and g is a quadratic function, show that $f \circ g$ and $g \circ f$ are quadratic functions.

34 If f and g are polynomial functions of degrees m and n, respectively, show that $f \circ g$ is a polynomial function of degree mn.

Exer. 35–40: Use the method of Example 8 to solve.

35 A fire starts in a dry, open field and spreads in the form of a circle. If the radius of this circle increases at the rate of 6 ft/min, express the total fire area as a function of time t.

36 A 100-foot-long cable of diameter 4 inches is submerged in seawater. Due to corrosion, the surface area of the cable decreases at the rate of 750 in.^{2} per year. Express the diameter of the cable as a function of time. (Ignore corrosion at the ends of the cable.)

37 A hot-air balloon rises vertically as a rope attached to the base of the balloon is released at the rate of 5 ft/sec. The pulley that releases the rope is 20 feet from a platform where passengers board the balloon (see figure). Express the height of the balloon as a function of time.

EXERCISE 37

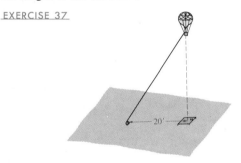

38 The diameter d of a cube is the distance between two opposite vertices. Express d as a function of the edge x of the cube. (*Hint:* First express the diagonal y of a face as a function of x.)

39 Refer to Exercise 59 of Section 1.4. When the jet is 500 feet down the runway, it has reached and will maintain a speed of 150 ft/sec (or about 100 mi/hr) until takeoff. Express the distance d of the jet from the control tower as a function of time t (in seconds). (*Hint:* In Exercise 59, Section 1.4, first write x as a function of t.)

40 Refer to Exercise 56 of Section 1.4. The acrobat moves up the tightrope at a steady rate of 1 ft/sec. If the rope is attached 30 feet up the pole, express the height h of the acrobat above the ground as a function of time t. (*Hint:* Let d denote the total distance traveled along the rope. First express d as a function of t, and then h as a function of d.)

FIGURE 1.54

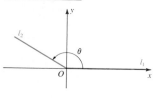

In geometry we say that an angle is determined by two rays, or half-lines, l_1 and l_2, having the same initial point O. If A and B are points on l_1 and l_2, respectively (see Figure 1.54), then we refer to **angle AOB.** An angle can also be considered as two finite line segments with a common endpoint.

In trigonometry we may interpret angles as rotations of rays. Start with a fixed ray l_1 having endpoint O, and rotate it about O, in a plane, to a position specified by ray l_2. We call l_1 the **initial side,** l_2 the **terminal side,** and O the **vertex** of angle AOB. The amount or direction of rotation is not restricted in any way. We might let l_1 make several revolutions in either direction about O before coming to the position l_2. Thus, many angles may have the same initial and terminal sides.

FIGURE 1.55

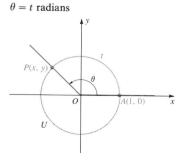

If we introduce a rectangular coordinate system, then the **standard position** of an angle is obtained by taking the vertex at the origin and letting the initial side l_1 coincide with the positive x-axis. If l_1 is rotated in a counterclockwise direction to the terminal position l_2, then the angle is considered **positive,** whereas if l_1 is rotated in a clockwise direction, the angle is **negative.** We often denote angles by lowercase Greek letters and specify the direction of rotation by means of a circular arc or spiral with an arrowhead (see Figure 1.55).

FIGURE 1.56

$\theta = t$ radians

The magnitude of an angle may be expressed in terms of either degrees or radians. An angle of degree measure $1°$ is obtained by $1/360$ of a complete revolution in the counterclockwise direction. In calculus, the most important unit of angular measure is the *radian*. To define radian measure, consider the unit circle U with center at the origin of a rectangular coordinate system, and let θ be an angle in standard position (see Figure 1.56). The angle θ is generated by rotating the positive x-axis about O. As the x-axis rotates to the terminal side of θ, its point of intersection with U travels a certain distance t before arriving at its final position $P(x, y)$.* If t is considered positive for a counterclockwise rotation and negative for a clockwise rotation, then a natural way of assigning a measure to θ is to use the number t. We say that θ **is an angle of t radians** and we write $\theta = t$ or $\theta = t$ radians. Thus θ denotes either an angle or its measure. In Figure 1.56, t is the length of the arc $\overset{\frown}{AP}$ that subtends θ.

*Distances along curves, or *arc length,* will be defined, using concepts of calculus, in Section 6.5.

If $\theta = 1$ (that is, if θ is an angle of 1 radian), then θ is subtended by an arc of unit length on the unit circle U (see Figure 1.57). The notation $\theta = -7.5$ means that θ is the angle generated by a clockwise rotation in which the point of intersection of the x-axis with the unit circle U has traveled 7.5 units. Since the circumference of U is 2π, we see that if $\theta = \pi/2$, then θ is obtained by $\frac{1}{4}$ of a complete revolution in the counterclockwise direction. Similarly, if $\theta = -\pi/4$, then θ is generated by $\frac{1}{8}$ of a revolution in the clockwise direction. These angles, measured in radians, are sketched in Figure 1.57.

FIGURE 1.57

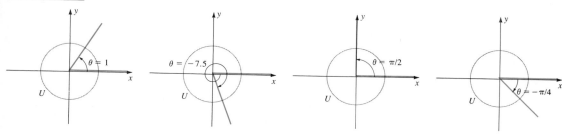

If an angle in standard position is generated by one-half of a complete counterclockwise rotation, then the degree measure is 180° and the radian measure is π. This gives us the following relationships.

RELATIONSHIPS BETWEEN (1.26)
DEGREES AND RADIANS

$$180° = \pi \text{ radians}, \qquad 1° = \frac{\pi}{180} \text{ radian}, \qquad 1 \text{ radian} = \left(\frac{180}{\pi}\right)°$$

If we use a calculator to approximate $\pi/180$ and $180/\pi$, we obtain

$$1° \approx 0.0174533 \text{ radian} \quad \text{and} \quad 1 \text{ radian} \approx 57.29578°.$$

The following theorem is a consequence of the preceding formulas.

THEOREM (1.27)

(i) To change radian measure to degrees, multiply by $180/\pi$.

(ii) To change degree measure to radians, multiply by $\pi/180$.

When radian measure of an angle is used, no units will be indicated. Thus, if an angle has radian measure 5, we write $\theta = 5$ instead of $\theta = 5$ *radians*. There should be no confusion as to whether radian or degree measure is being used, since if θ has degree measure 5°, we write $\theta = 5°$, *not* $\theta = 5$.

EXAMPLE 1

(a) Find the radian measure of θ if $\theta = -150°$ and if $\theta = 225°$.

(b) Find the degree measure of θ if $\theta = 7\pi/4$ and if $\theta = -\pi/3$.

SOLUTION

(a) By Theorem (1.27) (i), we can find the number of radians in $-150°$ by multiplying -150 by $\pi/180$. Thus,

$$-150° = -150\left(\frac{\pi}{180}\right) = -\frac{5\pi}{6}.$$

Similarly,

$$225° = 225\left(\frac{\pi}{180}\right) = \frac{5\pi}{4}.$$

(b) By Theorem (1.27) (ii), we find the number of degrees in $7\pi/4$ radians by multiplying by $180/\pi$, obtaining

$$\frac{7\pi}{4} = \frac{7\pi}{4}\left(\frac{180}{\pi}\right) = 315°.$$

Similarly,

$$-\frac{\pi}{3} = -\frac{\pi}{3}\left(\frac{180}{\pi}\right) = -60°. \quad \bullet$$

The following table displays the relationship between radian and degree measure of several common angles. The entries may be checked by using Theorem (1.27).

Radians	0	$\pi/6$	$\pi/4$	$\pi/3$	$\pi/2$	$2\pi/3$	$3\pi/4$	$5\pi/6$	π
Degrees	0°	30°	45°	60°	90°	120°	135°	150°	180°

FIGURE 1.58

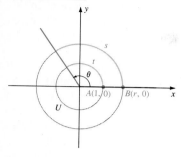

The radian measure of an angle can be found by using a circle of *any* radius. In the following discussion, the terminology **central angle** of a circle refers to an angle whose vertex is at the center of the circle. Suppose that θ is a central angle of a circle of radius r, and that θ is subtended by an arc of length s for $0 \leq s < 2\pi r$. To find the radian measure of θ, we place θ in standard position on a rectangular coordinate system and superimpose a unit circle U, as shown in Figure 1.58. If θ is subtended by an arc of length t on U, then by definition we may write $\theta = t$. From plane geometry, the ratio of the arcs in Figure 1.58 is the same as the ratio of the radii; that is,

$$\frac{t}{s} = \frac{1}{r} \quad \text{or} \quad t = \frac{s}{r}.$$

Substituting θ for t gives us the following result.

THEOREM (1.28)

> If a central angle θ of a circle of radius r is subtended by an arc of length s, then the radian measure of θ is
>
> $$\theta = \frac{s}{r}$$

The formula $\theta = s/r$ for radian measure of an angle is independent of the size of the circle. For example, if the radius of the circle is $r = 4$ cm and an arc of length 8 cm subtends a central angle θ, then the radian measure of θ is

$$\theta = \frac{8 \text{ cm}}{4 \text{ cm}} = 2.$$

If the radius of the circle is 5 km and the arc is 10 km, then

$$\theta = \frac{10 \text{ km}}{5 \text{ km}} = 2.$$

These calculations indicate that the radian measure of an angle is dimensionless and hence may be regarded as a real number. It is for this reason that we employ the notation $\theta = t$ instead of $\theta = t$ radians.

The formula $\theta = s/r$ can be used to find the length of the arc that subtends a central angle θ of a circle. For problems of this type it is convenient to use the equivalent formula

$$s = r\theta.$$

FIGURE 1.59

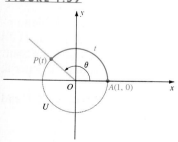

Two standard techniques are often used to introduce the trigonometric functions—one through the use of a unit circle, and the other by means of right triangles. We shall begin with the unit circle approach. Descriptions of trigonometric functions in terms of right triangles are stated in (1.34).

Let U be the unit circle, that is, the circle of radius 1 with center at the origin O of a rectangular coordinate system. Thus U is the graph of the equation $x^2 + y^2 = 1$. Given any real number t, let θ denote the angle (in standard position) of radian measure t. One possibility, with $0 < \theta < 2\pi$, is illustrated in Figure 1.59, where $P(t)$ denotes the point of intersection of the terminal side of θ and the unit circle U. Using the formula $s = r\theta$ with $r = 1$, we see that the arc $\overset{\frown}{AP}$ that subtends θ has length $s = t$. Thus, *the real number t* can be regarded as either the radian measure of the angle θ or as the length of arc $\widehat{AP}$ on *U*.

For any $t > 0$ we may think of the angle θ as having been generated by rotating the positive *x*-axis about *O* in the counterclockwise direction. In this case, *t* is the distance along *U* that the point $P(t)$ travels before arriving at its final position. If $t < 0$, then $|t|$ is the distance traveled by $P(t)$ in the clockwise direction about *U*.

The preceding discussion indicates how we may associate with each real number *t*, a unique point $P(t)$ on *U*. We shall call $P(t)$ **the point on the unit circle *U* that corresponds to *t*.** The rectangular coordinates (x, y) of $P(t)$ may be used to define the six **trigonometric functions.** These functions are referred to as the **sine, cosine, tangent, cotangent, secant,** and **cosecant functions,** and are designated by the symbols **sin, cos, tan, cot, sec,** and **csc,** respectively. If *t* is a real number, then the real number that the sine function associates with *t* will be denoted by either sin (*t*) or sin *t*. Similar notation is used for the other five functions.

When we wish to emphasize the rectangular coordinates of $P(t)$ we will use the notation $P(x, y)$ in place of $P(t)$ and refer to $P(x, y)$ as the point on the unit circle that corresponds to *t*. We do this in the following definition.

TRIGONOMETRIC FUNCTIONS (1.29)
IN TERMS OF A UNIT CIRCLE

If *t* is a real number and $P(x, y)$ is the point on the unit circle *U* that corresponds to *t*, then

$$\sin t = y \qquad\qquad \csc t = \frac{1}{y} \quad (\text{if } y \neq 0)$$

$$\cos t = x \qquad\qquad \sec t = \frac{1}{x} \quad (\text{if } x \neq 0)$$

$$\tan t = \frac{y}{x} \quad (\text{if } x \neq 0) \qquad \cot t = \frac{x}{y} \quad (\text{if } y \neq 0)$$

Since these formulas express values in terms of the coordinates of a point on a unit circle, the trigonometric functions are sometimes referred to as the **circular functions.**

The domain of both the sine and cosine functions is $\mathbb{R}$, because $\sin t = x$ and $\cos t = y$ exist for every real number *t*.

In the definitions of the tangent and secant functions, *x* appears in the denominator, and hence we must exclude values of *t* for which *x* is 0; that is, values of *t* that give us the points $(0, 1)$ and $(0, -1)$ on the *y*-axis. It follows that the domain of the tangent and secant functions consists of all real numbers *except* $(\pi/2) + n\pi$ for every integer *n*. In particular, we exclude $\pm\pi/2$, $\pm 3\pi/2$, $\pm 5\pi/2$, and so on.

For $\cot t = x/y$ and $\csc t = 1/y$, the number *y* appears in the denominator, and hence we must exclude values of *t* that give us the points $(1, 0)$ and $(-1, 0)$ on the *x*-axis. Thus the domain of the cotangent and cosecant functions consist of all real numbers *except* 0, $\pm\pi$, $\pm 2\pi$, $\pm 3\pi$ and, in general, $n\pi$ for every integer *n*.

Note that $P(x, y)$ is a point on the unit circle *U*, and hence $|x| \leq 1$ and $|y| \leq 1$. This implies that

$$|\sin t| \leq 1, \qquad |\cos t| \leq 1, \qquad |\csc t| \geq 1, \qquad |\sec t| \geq 1$$

for every t in the domains of these functions. It will follow from our work in Chapter 2 that $\sin t$ and $\cos t$ take on *every* value between -1 and 1. We can also show that the range of the tangent and cotangent functions is $\mathbb{R}$, and the range of the cosecant and secant functions is $(-\infty, -1] \cup [1, \infty)$.

EXAMPLE 2 Find the values of the trigonometric functions at

(a) $t = 0$ (b) $t = \pi/4$ (c) $t = \pi/2$

SOLUTION The points $P(x, y)$ on the unit circle U that correspond to the given values of t are plotted in Figure 1.60.

(a) For $t = 0$ we let $x = 1$ and $y = 0$ in Definition (1.29), obtaining the values in the first line of the table below. Note that since $y = 0$, $\csc 0$ and $\cot 0$ are undefined, as indicated by the dashes in the table.

(b) If $t = \pi/4$, the line through O and P bisects the first quadrant, and P has coordinates of the form (x, x). Using the equation $x^2 + y^2 = 1$ for U gives us $P(\sqrt{2}/2, \sqrt{2}/2)$. Using the coordinates of P in Definition (1.29) we obtain the second line of the table.

(c) Finally, let $x = 0$ and $y = 1$ in the definition. The results for $t = \pi/2$ are given in the last line of the table.

FIGURE 1.60

(a) $t = 0$

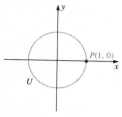

(b) $t = \pi/4$

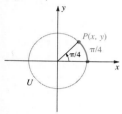

(c) $t = \pi/2$

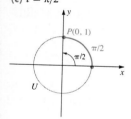

t	(x, y)	$\sin t$	$\cos t$	$\tan t$	$\csc t$	$\sec t$	$\cot t$
0	$(1, 0)$	0	1	0	—	1	—
$\dfrac{\pi}{4}$	$\left(\dfrac{\sqrt{2}}{2}, \dfrac{\sqrt{2}}{2}\right)$	$\dfrac{\sqrt{2}}{2}$	$\dfrac{\sqrt{2}}{2}$	1	$\sqrt{2}$	$\sqrt{2}$	1
$\dfrac{\pi}{2}$	$(0, 1)$	1	0	—	1	—	0

Values corresponding to $t = \pi/6$ and $t = \pi/3$ will be determined in Example 3. By methods developed later in this text, values for every real number t may be approximated to any degree of accuracy. It will be assumed that the reader knows how to use trigonometric tables (see Appendix III) or a calculator to approximate values of the trigonometric functions.

If, in Definition (1.29), $P(x, y)$ is in quadrant I, then x and y are both positive, and hence all values of the trigonometric functions are positive. If $P(x, y)$ is in quadrant II, then x is negative, y is positive, and hence $\sin t$ and $\csc t$ are positive, whereas the other four functions are negative. Similar remarks can be made for the remaining quadrants.

Since the circumference of the unit circle U is 2π, the same point $P(x, y)$ is obtained for $t + 2\pi n$ for every integer n. Hence the values of the trigonometric functions repeat in successive intervals of length 2π. A function f with domain D is **periodic** if there exists a positive real number k such that $t + k$ is in D and $f(t + k) = f(t)$ for every t in D. Geometrically, this means that the graph of f repeats itself as x-coordinates of points vary over successive intervals of length k. If a least such positive real number k exists, it is called the **period** of f. We can show that the sine, cosine, cosecant, and secant functions have period 2π, and the tangent and cotangent functions have period π.

The graphs of the trigonometric functions shown in Figure 1.61 may be obtained by analyzing what happens to the coordinates of the point $P(x, y)$

as P moves around the unit circle U in Figure 1.59, and by plotting several points.

FIGURE 1.61

(i) $y = \sin x$

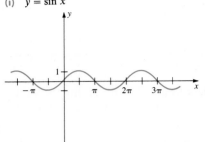

(ii) $y = \cos x$

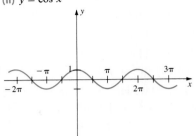

(iii) $y = \tan x$

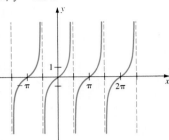

(iv) $y = \csc x$

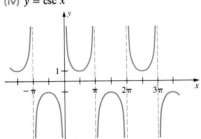

(v) $y = \sec x$

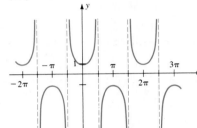

(vi) $y = \cot x$

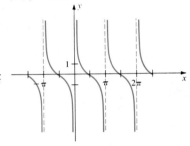

If $P(x, y)$ is the point on the unit circle U corresponding to t, then as illustrated in Figure 1.62, $P(x - y)$ corresponds to $-t$.

FIGURE 1.62

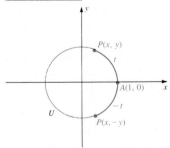

Consequently, $\sin(-t) = -y = -\sin t$ and $\cos(-t) = x = \cos t$. Similarly, $\tan(-t) = -\tan t$. This gives us the following formulas for negatives:

$$\sin(-t) = -\sin t$$
$$\cos(-t) = \cos t$$
$$\tan(-t) = -\tan t$$

If $f(t) = \cos t$, then from (1.30),

$$f(-t) = \cos(-t) = \cos t = f(t).$$

Hence the cosine function is even, and by Theorem (1.23), its graph is symmetric with respect to the y-axis (see Figure 1.61). Similarly, the sine and tangent functions are odd, and their graphs are symmetric with respect to the origin.

Many relationships exist among the trigonometric functions. The formulas listed in the next box are, without doubt, the most important identities in trigonometry, because they may be used to simplify and unify many different aspects of the subject. Since the formulas are true for every allowable value of t and are part of the foundation for work in trigonometry, they are called the *Fundamental Identities*.

Three of the fundamental identities involve squares such as $(\sin t)^2$ and $(\cos t)^2$. In general, if n is an integer different from -1, then powers such as $(\cos t)^n$ are written in the form $\cos^n t$. The symbols $\sin^{-1} t$ and $\cos^{-1} t$ are reserved for inverse trigonometric functions, which we will discuss in Chapter 8.

THE FUNDAMENTAL IDENTITIES (1.31)

$$\csc t = \frac{1}{\sin t} \qquad \tan t = \frac{\sin t}{\cos t} \qquad \sin^2 t + \cos^2 t = 1$$

$$\sec t = \frac{1}{\cos t} \qquad \cot t = \frac{\cos t}{\sin t} \qquad 1 + \tan^2 t = \sec^2 t$$

$$\cot t = \frac{1}{\tan t} \qquad\qquad\qquad\qquad 1 + \cot^2 t = \csc^2 t$$

PROOF The proofs follow from the definition of the trigonometric functions. Thus,

$$\csc t = \frac{1}{y} = \frac{1}{\sin t}, \qquad \sec t = \frac{1}{x} = \frac{1}{\cos t}, \qquad \cot t = \frac{x}{y} = \frac{1}{y/x} = \frac{1}{\tan t},$$

$$\tan t = \frac{y}{x} = \frac{\sin t}{\cos t}, \qquad \cot t = \frac{x}{y} = \frac{\cos t}{\sin t},$$

provided that no denominator is zero.

If (x, y) is a point on the unit circle U, then

$$y^2 + x^2 = 1.$$

Since $y = \sin t$ and $x = \cos t$, this gives us

$$(\sin t)^2 + (\cos t)^2 = 1$$

or equivalently,

$$\sin^2 t + \cos^2 t = 1.$$

If $\cos t \neq 0$, then, dividing both sides of the last equation by $\cos^2 t$, we obtain

$$\frac{\sin^2 t}{\cos^2 t} + 1 = \frac{1}{\cos^2 t}$$

or

$$\left(\frac{\sin t}{\cos t}\right)^2 + 1 = \left(\frac{1}{\cos t}\right)^2.$$

Since $\tan t = \sin t/\cos t$ and $\sec t = 1/\cos t$ we see that

$$\tan^2 t + 1 = \sec^2 t.$$

The proof of the final fundamental identity is left as an exercise. • •

In certain applications it is convenient to change the domain of a trigonometric function from a subset of $\mathbb{R}$ to a set of angles. We may accomplish this by means of the following definition.

DEFINITION (1.32)

> Let θ be an angle with radian measure t. **The value of a trigonometric function at θ is its value at the real number t.**

FIGURE 1.63

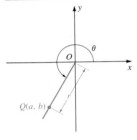

It follows from Definition (1.32) that $\sin \theta = \sin t$, $\cos \theta = \cos t$, and so on, where t is the radian measure of θ. To make the unit of angular measure clear, we shall use the degree symbol, such as $\sin 65°$ or $\tan 150°$, whenever the angle is measured in degrees. Numerals without any symbol attached, such as $\cos 3$ and $\csc (\pi/6)$, will indicate radian measure. This is not in conflict with our previous work, where, for example, $\cos 3$ meant the value of the cosine function at the real number 3, since by definition the cosine of an angle of measure 3 radians is identical with the cosine of the real number 3.

Let θ be an angle in standard position and let $Q(a, b)$ be an arbitrary point on the terminal side of θ, as illustrated in Figure 1.63. The next theorem specifies how the coordinates of the point Q may be used to determine the values of the trigonometric functions of θ.*

TRIGONOMETRIC FUNCTION (1.33)
OF ANGLES

> Let θ be an angle in standard position on a rectangular coordinate system and let $Q(a, b)$ be any point other than O on the terminal side of θ. If $d(O, Q) = r$, then
>
> $$\sin \theta = \frac{b}{r} \qquad\qquad \csc \theta = \frac{r}{b} \quad (\text{if } b \neq 0)$$
>
> $$\cos \theta = \frac{a}{r} \qquad\qquad \sec \theta = \frac{r}{a} \quad (\text{if } a \neq 0)$$
>
> $$\tan \theta = \frac{b}{a} \quad (\text{if } a \neq 0) \qquad \cot \theta = \frac{a}{b} \quad (\text{if } b \neq 0).$$

Note that if $r = 1$, then Theorem (1.33) reduces to (1.29), with $a = x$, $b = y$, and $\theta = t$.

For acute angles, values of the trigonometric functions can be interpreted as ratios of the lengths of the sides of a right triangle. Recall that a triangle is a **right triangle** if one of its angles is a right angle. If θ is an acute angle, then it can be regarded as an angle of a right triangle and we may refer to the lengths of the **hypotenuse**, the **opposite side**, and the **adjacent side** in the usual way. For convenience, we shall use **hyp**, **opp**, and **adj**, respectively, to denote these numbers.

Let us introduce a rectangular coordinate system as in Figure 1.64. The lengths of the adjacent side and the opposite side for θ are the x-coordinate

FIGURE 1.64

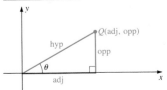

and y-coordinate, respectively, of a point Q on the terminal side of θ. By Theorem (1.33) we have the following.

RIGHT TRIANGLE (1.34)
TRIGONOMETRY

$$\sin \theta = \frac{\text{opp}}{\text{hyp}} \qquad \csc \theta = \frac{\text{hyp}}{\text{opp}}$$

$$\cos \theta = \frac{\text{adj}}{\text{hyp}} \qquad \sec \theta = \frac{\text{hyp}}{\text{adj}}$$

$$\tan \theta = \frac{\text{opp}}{\text{adj}} \qquad \cot \theta = \frac{\text{adj}}{\text{opp}}$$

These formulas are very important in our work with right triangles. The next example illustrates how they may be used.

EXAMPLE 3 Find $\sin \theta$, $\cos \theta$, and $\tan \theta$ for the following values of θ:

(a) $\theta = 60°$ (b) $\theta = 30°$ (c) $\theta = 45°$

SOLUTION Let us consider an equilateral triangle having sides of length 2. The median from one vertex to the opposite side bisects the angle at that vertex, as illustrated in Figure 1.65. By the Pythagorean Theorem, the length of this median is $\sqrt{3}$. Using the colored triangle and (1.34), we obtain the following values:

FIGURE 1.65

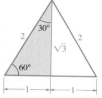

(a) $\sin 60° = \dfrac{\sqrt{3}}{2}$, $\cos 60° = \dfrac{1}{2}$, $\tan 60° = \dfrac{\sqrt{3}}{1} = \sqrt{3}$

(b) $\sin 30° = \dfrac{1}{2}$, $\cos 30° = \dfrac{\sqrt{3}}{2}$, $\tan 30° = \dfrac{1}{\sqrt{3}} = \dfrac{\sqrt{3}}{3}$

FIGURE 1.66

(c) To find the function values for $\theta = 45°$, let us consider an isosceles right triangle whose two equal sides have length 1, as illustrated in Figure 1.66. Thus,

$$\sin 45° = \frac{1}{\sqrt{2}} = \frac{\sqrt{2}}{2} = \cos 45°, \qquad \tan 45° = \frac{1}{1} = 1. \;\; \bullet$$

Some of the following trigonometric identities will be useful later in this text. Proofs may be found in books on trigonometry.

ADDITION AND (1.35)
SUBTRACTION FORMULAS

$$\sin (u + v) = \sin u \cos v + \cos u \sin v$$

$$\cos (u + v) = \cos u \cos v - \sin u \sin v$$

$$\tan (u + v) = \frac{\tan u + \tan v}{1 - \tan u \tan v}$$

$$\sin (u - v) = \sin u \cos v - \cos u \sin v$$

$$\cos (u - v) = \cos u \cos v + \sin u \sin v$$

$$\tan (u - v) = \frac{\tan u - \tan v}{1 + \tan u \tan v}$$

DOUBLE-ANGLE FORMULAS (1.36)

$$\sin 2u = 2 \sin u \cos u$$
$$\cos 2u = \cos^2 u - \sin^2 u = 1 - 2 \sin^2 u = 2 \cos^2 u - 1$$
$$\tan 2u = \frac{2 \tan u}{1 - \tan^2 u}$$

HALF-ANGLE FORMULAS (1.37)

$$\sin^2 \frac{u}{2} = \frac{1 - \cos u}{2} \qquad \cos^2 \frac{u}{2} = \frac{1 + \cos u}{2}$$
$$\tan \frac{u}{2} = \frac{1 - \cos u}{\sin u} = \frac{\sin u}{1 + \cos u}$$

PRODUCT FORMULAS (1.38)

$$\sin u \cos v = \tfrac{1}{2}[\sin(u + v) + \sin(u - v)]$$
$$\cos u \sin v = \tfrac{1}{2}[\sin(u + v) - \sin(u - v)]$$
$$\cos u \cos v = \tfrac{1}{2}[\cos(u + v) + \cos(u - v)]$$
$$\sin u \sin v = \tfrac{1}{2}[\cos(u - v) - \cos(u + v)]$$

FACTORING FORMULAS (1.39)

$$\sin u + \sin v = 2 \sin \frac{u + v}{2} \cos \frac{u - v}{2}$$
$$\sin u - \sin v = 2 \cos \frac{u + v}{2} \sin \frac{u - v}{2}$$
$$\cos u + \cos v = 2 \cos \frac{u + v}{2} \cos \frac{u - v}{2}$$
$$\cos u - \cos v = -2 \sin \frac{u + v}{2} \sin \frac{u - v}{2}$$

EXERCISES 1.6

1 Verify the entries in the table of radians and degrees on page 40.

2 Prove that $1 + \cot^2 t = \csc^2 t$.

3 Find the quadrant containing θ if

(a) $\sec \theta < 0$ and $\sin \theta > 0$.

(b) $\cot \theta > 0$ and $\csc \theta < 0$.

(c) $\cos \theta > 0$ and $\tan \theta < 0$.

4 Find the values of the remaining trigonometric functions if

(a) $\sin t = -\frac{4}{5}$ and $\cos t = \frac{3}{5}$.

(b) $\csc t = \sqrt{13}/2$ and $\cot t = -\frac{3}{2}$.

5 Without the use of a calculator or table, find the values of the trigonometric functions corresponding to each of the following real numbers:

(a) $9\pi/2$ (b) $-5\pi/4$ (c) 0   (d) $11\pi/6$

6 Find, without using a calculator, the radian measure that corresponds to the degree measure:

$$330°, \quad 405°, \quad -150°, \quad 240°, \quad 36°$$

7 Find, without using a calculator, the degree measure that corresponds to the radian measure:

$$9\pi/2, \quad -2\pi/3, \quad 7\pi/4, \quad 5\pi, \quad \pi/5$$

8 A central angle θ is subtended by an arc 20 cm long on a circle of radius 2 meters. What is the radian measure of θ?

9 Find the values of the six trigonometric functions of θ if θ is in standard position and satisfies the stated conditions.

(a) The point $(30, -40)$ is on the terminal side of θ.

(b) The terminal side of θ is in quadrant II and is parallel to the line $2x + 3y + 6 = 0$.

(c) $\theta = -90°$.

10 Find each without the use of a calculator or table.

(a) $\cos 225°$ (b) $\tan 150°$ (c) $\sin(-\pi/6)$

(d) $\sec(4\pi/3)$ (e) $\cot(7\pi/4)$ (f) $\csc(300°)$

Exer. 11–30: Verify the identity.

11 $\cos\theta\sec\theta = 1$

12 $\tan\alpha\cot\alpha = 1$

13 $\sin\theta\sec\theta = \tan\theta$

14 $\sin\alpha\cot\alpha = \cos\alpha$

15 $\dfrac{\csc x}{\sec x} = \cot x$

16 $\cot\beta\sec\beta = \csc\beta$

17 $(1 + \cos\alpha)(1 - \cos\alpha) = \sin^2\alpha$

18 $\cos^2 x\,(\sec^2 x - 1) = \sin^2 x$

19 $\cos^2 t - \sin^2 t = 2\cos^2 t - 1$

20 $(\tan\theta + \cot\theta)\tan\theta = \sec^2\theta$

21 $\dfrac{\sin t}{\csc t} + \dfrac{\cos t}{\sec t} = 1$

22 $1 - 2\sin^2 x = 2\cos^2 x - 1$

23 $(1 + \sin\alpha)(1 - \sin\alpha) = 1/\sec^2\alpha$

24 $(1 - \sin^2 t)(1 + \tan^2 t) = 1$

25 $\sec\beta - \cos\beta = \tan\beta\sin\beta$

26 $\dfrac{\sin w + \cos w}{\cos w} = 1 + \tan w$

27 $\dfrac{\csc^2\theta}{1 + \tan^2\theta} = \cot^2\theta$

28 $\sin x + \cos x\cot x = \csc x$

29 $\sin t\,(\csc t - \sin t) = \cos^2 t$

30 $\cot t + \tan t = \csc t\sec t$

Exer. 31–42: (a) Find the solutions of the given equation that are in the interval $[0, 2\pi)$, and (b) find the degree measure of each solution. (In Exercises 39–42 refer to identities (1.36) and (1.37).)

31 $2\cos^3\theta - \cos\theta = 0$

32 $2\cos\alpha + \tan\alpha = \sec\alpha$

33 $\sin\theta = \tan\theta$

34 $\csc^5\theta - 4\csc\theta = 0$

35 $2\cos^3 t + \cos^2 t - 2\cos t - 1 = 0$

36 $\cos x\cot^2 x = \cos x$

37 $\sin\beta + 2\cos^2\beta = 1$

38 $2\sec u\sin u + 2 = 4\sin u + \sec u$

39 $\cos 2x + 3\cos x + 2 = 0$

40 $\sin 2u = \sin u$

41 $2\cos^2\tfrac{1}{2}\theta - 3\cos\theta = 0$

42 $\sec 2x\csc 2x = 2\csc 2x$

Exer. 43–51: If θ and φ are acute angles such that $\csc\theta = \tfrac{5}{3}$ and $\cos\varphi = \tfrac{8}{17}$, use identities (1.35)–(1.37) to find the numbers.

43 $\sin(\theta + \varphi)$

44 $\cos(\theta + \varphi)$

45 $\tan(\theta - \varphi)$

46 $\sin(\varphi - \theta)$

47 $\sin 2\varphi$

48 $\cos 2\varphi$

49 $\tan 2\theta$

50 $\sin(\theta/2)$

51 $\tan(\theta/2)$

52 Express $\cos(\alpha + \beta + \gamma)$ in terms of functions of α, β, and γ.

53 Is there a real number t such that $7\sin t = 9$? Explain.

54 Is there a real number t such that $3\csc t = 1$? Explain.

Exer. 55–56: If f and g are the given functions, find (a) $f(g(\pi))$ and (b) $g(f(\pi))$.

55 $f(t) = \cos t$, $g(t) = t/4$

56 $f(t) = \tan t$, $g(t) = t/4$

Exer. 57–60: Use Figure 1.61 and results on vertical shifts, horizontal shifts, stretching, and reflecting to sketch the graph of f for $0 \le x \le 2\pi$.

57 (a) $f(x) = 4\sin x$

(b) $f(x) = -\tfrac{1}{4}\sin x$

58 (a) $f(x) = \tfrac{1}{3}\cos x$

(b) $f(x) = -3\cos x$

59 (a) $f(x) = \sin(x - \tfrac{1}{2}\pi)$

(b) $f(x) = \sin x - \tfrac{1}{2}\pi$

60 (a) $f(x) = 2\cos(x + \pi)$

(b) $f(x) = 2\cos x + \pi$

1.7 REVIEW

Define or discuss each of the following.

1 Rational and irrational numbers

2 Coordinate line

3 A real number a is greater than a real number b

4 Inequalities

5 Absolute value of a real number

6 The Triangle Inequality

7 Notation for sets

8 Variable

9 Intervals (open, closed, half-open, infinite)

10 Ordered pair

11 Rectangular coordinate system in a plane

12 The x- and y-coordinates of a point

13 The Distance Formula

14 The Midpoint Formula

EXERCISES 1.7

Exer. 1–8: Solve the inequality and express the solution in terms of intervals.

1 $4 - 3x > 7 + 2x$

2 $\dfrac{7}{2} > \dfrac{1 - 4x}{5} > \dfrac{3}{2}$

3 $|2x - 7| \le 0.01$

4 $|6x - 7| > 1$

5 $2x^2 < 5x - 3$

6 $\dfrac{2x^2 - 3x - 20}{x + 3} < 0$

7 $\dfrac{1}{3x - 1} < \dfrac{2}{x + 5}$

8 $x^2 + 4 \ge 4x$

9 Given the points $A(2, 1)$, $B(-1, 4)$, and $C(-2, -3)$,
   (a) prove that A, B, and C are vertices of a right triangle and find its area.
   (b) find the coordinates of the midpoint of AB.
   (c) find the slope of the line through B and C.

Exer. 10–13: Sketch the graph of the equation and discuss symmetry with respect to the x-axis, y-axis, or origin.

10 $3x - 5y = 10$

11 $x^2 + y = 4$

12 $x = y^3$

13 $|x + y| = 1$

Exer. 14–17: Sketch the graph of the set W.

14 $W = \{(x, y): x > 0\}$

15 $W = \{(x, y): y > x\}$

16 $W = \{(x, y): x^2 + y^2 < 1\}$

17 $W = \{(x, y): |x - 4| < 1, |y + 3| < 2\}$

Exer. 18–20: Find an equation of the circle that satisfies the given conditions.

18 Center $C(4, -7)$ and passing through the origin

19 Center $C(-4, -3)$ and tangent to the line with equation $x = 5$

20 Passing through the points $A(-2, 3)$, $B(4, 3)$, and $C(-2, -1)$

21 Find the center and radius of the circle that has equation
$$x^2 + y^2 - 10x + 14y - 7 = 0.$$

Exer. 22–26: Given the points $A(-4, 2)$, $B(3, 6)$, and $C(2, -5)$, solve each problem.

22 Find an equation for the line through B that is parallel to the line through A and C.

23 Find an equation for the line through B that is perpendicular to the line through A and C.

24 Find an equation for the line through C and the midpoint of the line segment AB.

25 Find an equation for the line through A that is parallel to the y-axis.

26 Find an equation for the line through C that is perpendicular to the line with equation $3x - 10y + 7 = 0$.

Exer. 27–30: Find the domain of f.

27 $f(x) = \dfrac{2x - 3}{x^2 - x}$

28 $f(x) = \dfrac{x}{\sqrt{16 - x^2}}$

29 $f(x) = \dfrac{1}{\sqrt{x - 5}\sqrt{7 - x}}$

30 $f(x) = \dfrac{1}{\sqrt{x}(x - 2)}$

31 If $f(x) = 1/\sqrt{x + 1}$ find each of the following.
   (a) $f(1)$ (b) $f(3)$ (c) $f(0)$
   (d) $f(\sqrt{2} - 1)$ (e) $f(-x)$ (f) $-f(x)$
   (g) $f(x^2)$ (h) $(f(x))^2$

Exer. 32–36: Sketch the graph of f.

32 $f(x) = 1 - 4x^2$

33 $f(x) = 100$

34 $f(x) = -1/(x + 1)$

35 $f(x) = |x + 5|$

36 $f(x) = \begin{cases} x^2 & \text{if } x < 0 \\ 3x & \text{if } 0 \le x < 2 \\ 6 & \text{if } x \ge 2 \end{cases}$

37 Sketch the graph of each equation, making use of shifting, stretching, or reflecting:

(a) $y = \sqrt{x}$ (b) $y = \sqrt{x + 4}$ (c) $y = \sqrt{x} + 4$

(d) $y = 4\sqrt{x}$ (e) $y = \frac{1}{4}\sqrt{x}$ (f) $y = -\sqrt{x}$

38 Determine if f is even, odd, or neither even nor odd:

(a) $f(x) = \sqrt[3]{x^3 + 4x}$ (b) $f(x) = \sqrt[3]{3x^2 - x^3}$

(c) $f(x) = \sqrt[3]{x^4 + 3x^2 + 5}$ (d) $f(x) = 0$

39 If $f(x) = 5 - 7x$, prove that f is one-to-one.

Exer. 40–42: Find $(f + g)(x)$, $(f - g)(x)$, $(fg)(x)$, $(f/g)(x)$, $(f \circ g)(x)$, **and** $(g \circ f)(x)$.

40 $f(x) = x^2 + 3x + 1$, $g(x) = 2x - 1$

41 $f(x) = x^2 + 4$, $g(x) = \sqrt{2x + 5}$

42 $f(x) = 5x + 2$, $g(x) = 1/x^2$

43 What arc on a circle of diameter 8 feet subtends a central angle of radian measure 2?

44 Find the exact values of the trigonometric functions corresponding to the following numbers:

(a) $-3\pi/4$ (b) $7\pi/6$ (c) $-\pi/3$ (d) $5\pi/2$

45 Find the following without using a calculator or tables:

(a) $\sin 225°$ (b) $\cos(-60°)$

(c) $\tan(210°)$ (d) $\sec(5\pi)$

46 If $\sin t = \frac{2}{5}$ and $\cos t < 0$, use fundamental identities to find the values of the other five trigonometric functions.

47 Verify the following identities:

(a) $\cos \theta + \sin \theta \tan \theta = \sec \theta$

(b) $(\sin t)/(1 - \cos t) = \csc t + \cot t$

(c) $(\sec u - \tan u)(\csc u + 1) = \cot u$

48 For each equation find the solutions that are in the interval $[0, 2\pi)$, and the degree measure of each solution:

(a) $4 \sin^2 t - 3 = 0$

(b) $\tan \theta + \sec \theta = 1$

(c) $\tan^2 x \sin x = \sin x$

Exer. 49–50: Sketch the graph of f **for** $0 \le x \le 2\pi$.

49 $f(x) = 4 \sin(x + \frac{1}{2}\pi)$

50 $f(x) = -3 \cos(x - \pi)$

CHAPTER 2

LIMITS OF FUNCTIONS

The concept of *limit of a function* is one of the fundamental ideas that distinguishes calculus from areas of mathematics such as algebra or trigonometry. We can easily develop an intuitive feeling for limits. With this in mind, our discussion in the first two sections is informal. The mathematically precise definition of limit is stated in Section 2.3. The remainder of the chapter deals with important properties of limits.

In calculus and its applications we investigate the manner in which quantities vary, and whether they approach specific values under certain conditions. The quantities will usually involve function values. Our analysis will employ the concept of the *derivative* (Chapter 3) or the *definite integral* (Chapter 5).

The definition of derivative depends on the notion of the *limit of a function*. We shall begin in an intuitive manner. The formal definition of limit is stated in Section 2.3.

Let a function f be defined throughout an open interval containing a real number a, except possibly at a itself. We are often interested in the function value $f(x)$ *when x is very close to a, but not necessarily equal to a.* As a matter of fact, in many instances the number a is not in the domain of f; that is, $f(a)$ is undefined. Roughly speaking, we may ask the question: As x gets closer and closer to a (but $x \neq a$), does $f(x)$ get closer and closer to some number L? If the answer is *yes*, we say that $f(x)$ approaches L as x approaches a, or that $f(x)$ *has the limit L as x approaches a*, and we use the following notation:

LIMIT NOTATION (2.1)

$$\lim_{x \to a} f(x) = L$$

If $f(x)$ approaches some number as x approaches a, but we do not know what that number is, then we use the phrase $\lim_{x \to a} f(x)$ *exists.*

Let us represent the domain and range of the function f by points on two real lines l and l', as illustrated in Figure 2.1 (see also Figure 1.37). If $\lim_{x \to a} f(x) = L$, then, as x approaches a, $f(x)$ approaches L. The manner in

FIGURE 2.1

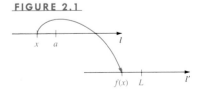

FIGURE 2.2

which x approaches a is irrelevant. Thus in Figure 2.1, x may approach a from the left (denoted by $x \to a^-$), or from the right (denoted by $x \to a^+$), or by oscillating from one side of a to the other. Similarly, the function value $f(x)$ may approach L in a variety of ways, depending on the nature of f.

The notion of limit is fundamental to the investigation of many important mathematical and physical concepts. To illustrate, we shall examine two problems:

(i) Find the tangent line to a curve at a specified point P.

(ii) Find the velocity, at any instant, of an object that is moving along a linear path.

FIGURE 2.3

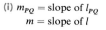

In plane geometry, the tangent line l at a point P on a circle may be defined as the line that has only the point P in common with the circle, as illustrated in Figure 2.2. This definition cannot be extended to arbitrary graphs, since a tangent line may intersect a graph several times, as shown in Figure 2.3 (see also Exercise 11).

To identify the tangent line l at a point P on the graph of a function, it is sufficient to state the slope m of l, since this completely determines the line. To define m we begin by choosing another point Q on the graph and considering the line l_{PQ} through P and Q, as in Figure 2.4(i). The line l_{PQ} is called a **secant line** for the graph.

FIGURE 2.4

(i) m_{PQ} = slope of l_{PQ}
   m = slope of l

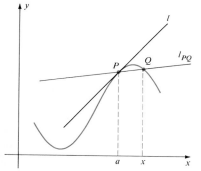

(ii) l_{PQ} approaches l
    m_{PQ} approaches m

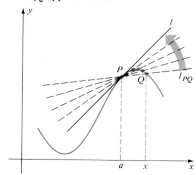

(iii) l_{PQ} approaches l
    m_{PQ} approaches m

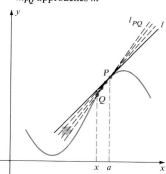

Let m_{PQ} denote the slope of l_{PQ} and consider the variation of m_{PQ} as Q gets closer and closer to P. If Q approaches P from the right we have the situation illustrated in Figure 2.4(ii), where dashed lines indicate possible positions of the secant line l_{PQ} corresponding to several positions of Q. It appears that if Q is close to P, then the slope m_{PQ} should be close to the slope m of l. In Figure 2.4(iii), Q approaches P from the left and, again, m_{PQ} appears to get close to m. These observations suggest that if m_{PQ} approaches some fixed value as Q approaches P, then this value should be *defined* as the slope of the tangent line l at P.

If the function f is defined throughout an open interval containing a, then we may label the coordinates of P and Q as in Figure 2.5. Using the slope formula (1.13), the slope of the secant line l_{PQ} is

FIGURE 2.5

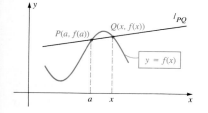

$$m_{PQ} = \frac{f(x) - f(a)}{x - a}.$$

Evidently, the phrase *Q approaches P* may be replaced by *x approaches a*. This leads to the following definition.

DEFINITION (2.2)

> Let f be a function that is defined on an open interval containing a. The **slope m of the tangent line** to the graph of f at the point $(a, f(a))$ is
>
> $$m = \lim_{x \to a} \frac{f(x) - f(a)}{x - a}$$
>
> provided the limit exists.

Note that the formula in Definition (2.2) does not specify the limit of $f(x)$, but rather of the expression $[f(x) - f(a)]/(x - a)$. Also observe that when we are investigating this limit, $x \neq a$. Indeed, if we let $x = a$, then $P = Q$ and m_{PQ} does not exist.

EXAMPLE 1 If $f(x) = x^2$ and if a is any real number, find

(a) the slope of the tangent line to the graph of f at the point $P(a, a^2)$.

(b) an equation of the tangent line to the graph at the point $(\frac{3}{2}, \frac{9}{4})$.

SOLUTION

(a) The graph of $y = x^2$ and typical points $P(a, a^2)$ and $Q(x, x^2)$ are illustrated in Figure 2.6. The slope m_{PQ} of the secant line l_{PQ} is

$$m_{PQ} = \frac{x^2 - a^2}{x - a}.$$

Applying Definition (2.2), the slope m of the tangent line at P is

$$m = \lim_{x \to a} \frac{x^2 - a^2}{x - a}.$$

To find the limit, it is necessary to change the form of the fraction. Since $x \neq a$ in the limiting process, it follows that $x - a \neq 0$, and hence we may divide numerator and denominator by $x - a$, that is, we may *cancel $x - a$*, as follows:

$$m = \lim_{x \to a} \frac{x^2 - a^2}{x - a} = \lim_{x \to a} \frac{(x + a)(x - a)}{(x - a)} = \lim_{x \to a} (x + a).$$

As x gets closer and closer to a, the expression $x + a$ gets close to $a + a$, or $2a$. Consequently, $m = 2a$.

(b) Since the slope of the tangent line l at the point $(\frac{3}{2}, \frac{9}{4})$ is the special case in which $a = \frac{3}{2}$, we have $m = 2a = 2(\frac{3}{2}) = 3$, as illustrated in Figure 2.7. Using the Point-Slope Form (1.15), an equation of l is

$$y - \tfrac{9}{4} = 3(x - \tfrac{3}{2}).$$

This equation simplifies to

$$12x - 4y - 9 = 0. \quad \bullet$$

Let us next consider problem (ii), defining the velocity, at any instant, of an object that is moving on a line. Motion on a line is called **rectilinear**

FIGURE 2.6

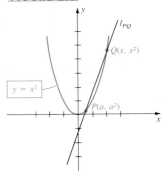

FIGURE 2.7

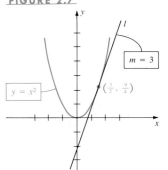

motion. It is easy to find the *average velocity* v_{av}, during any interval of time. We merely use the formula $d = rt$, where r is the rate (average velocity), t is the length of the time interval, and d is the net distance traveled. Solving for r gives us:

AVERAGE VELOCITY (2.3)

$$v_{av} = \frac{d}{t}$$

FIGURE 2.8

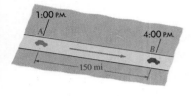

To illustrate, suppose an automobile leaves city A at 1:00 P.M. and travels along a straight highway, arriving at city B, 150 miles from A, at 4:00 P.M., as illustrated in Figure 2.8. Employing (2.3), with $d = 150$ and $t = 3$ (hours), the average velocity during the time interval is

$$v_{av} = \frac{150}{3} = 50 \text{ mi/hr.}$$

This is the velocity that, if maintained for 3 hours, would enable the automobile to travel the 150 miles from A to B.

The average velocity gives no information whatsoever about the velocity at any instant. For example, at 2:30 P.M. the automobile's speedometer may have registered 40, or 30, or the automobile may have been standing still. If we wish to determine the rate at which the automobile is traveling at 2:30 P.M., information is needed about its motion or position *near* this time. For example, suppose at 2:30 P.M. the automobile is 80 miles from A and at 2:35 P.M. it is 84 miles from A, as illustrated in Figure 2.9. The length of time interval from 2:30 P.M. to 2:35 P.M. is 5 minutes, or $\frac{1}{12}$ hour, and the distance d is 4 miles. Substituting in (2.3), the average velocity during this time interval is

FIGURE 2.9

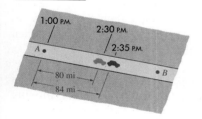

$$v_{av} = \frac{4}{\frac{1}{12}} = 48 \text{ mi/hr.}$$

However, this is still not an accurate indication of the velocity at 2:30 P.M. since, for example, the automobile may have been traveling very slowly at 2:30 P.M. and then increased speed considerably, arriving at the point 84 miles from A at 2:35 P.M. Evidently, we obtain a better approximation by using the average velocity during a smaller time interval, say from 2:30 P.M. to 2:31 P.M. It appears that the best procedure would be to take smaller and smaller time intervals near 2:30 P.M. and study the average velocity in each time interval. This leads us into a limiting process similar to that discussed for tangent lines.

In order to base our discussion on mathematical concepts, let us assume that the position of an object moving rectilinearly may be represented by a point P on a coordinate line l. We sometimes refer to the motion of the *point P* on l, or the motion of an object on l whose position is specified by P. We further assume that we know the position of P at every instant in a given interval of time. If $s(t)$ denotes the coordinate of P at time t, then the function s determined in this way is the **position function** for P. If we keep track of time by means of a clock, then for each t the point P is $s(t)$ units from the origin, as illustrated in Figure 2.10.

FIGURE 2.10

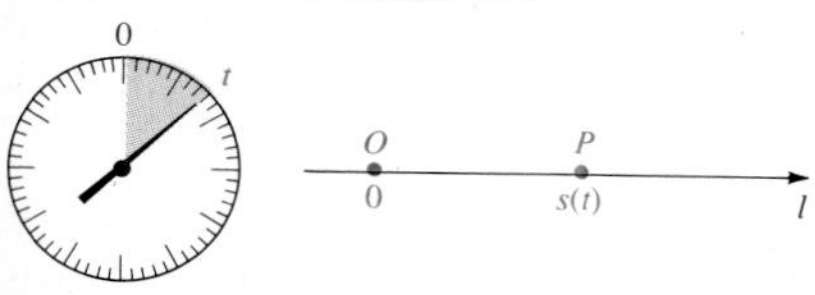

Time   Position of P

FIGURE 2.11

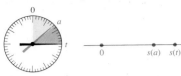

Change in   Change in
time   position of P

To define the velocity of P at time a, we begin by investigating the average velocity in a (small) time interval near a. Thus we consider times a and t, where t is close to a, but $t \neq a$. The corresponding positions of P are $s(a)$ and $s(t)$, as illustrated in Figure 2.11, and hence the amount of change in the position of P is $s(t) - s(a)$. This number may be positive, negative, or zero, depending on whether the position of P at time t is to the right, to the left, or the same as its position at time a. The number $s(t) - s(a)$ is not necessarily the distance traveled by P between times a and t, since, for example, P may have moved beyond the point corresponding to $s(t)$ and then returned to that point at time t.

By (2.3), the average velocity of P between times t and a is

(2.4)
$$v_{av} = \frac{s(t) - s(a)}{t - a}$$

As in our previous discussion, the closer t is to a, the closer v_{av} should approximate the velocity of P at time a. Thus, we *define* the velocity as the limit, as t approaches a of v_{av}. This limit is also called the *instantaneous velocity* of P at time a. Summarizing, we have the following definition.

DEFINITION (2.5)

> If a point P moves on a coordinate line l such that its coordinate at time t is $s(t)$, then the **velocity** $v(a)$ of P at time a is
> $$v(a) = \lim_{t \to a} \frac{s(t) - s(a)}{t - a}$$
> provided the limit exists.

If $s(t)$ is measured in centimeters and t in seconds, then the unit of velocity is centimeters per second (cm/sec). If $s(t)$ is in miles and t in hours, then velocity is in miles per hour. Other units of measurement may, of course, be used.

We shall return to the velocity concept in Chapter 4, where we will show that if the velocity is positive in a given time interval, then the point is moving in the positive direction on l, whereas if the velocity is negative, the point is moving in the negative direction. Although these facts have not been proved, we shall use them in the following example.

EXAMPLE 2 A sandbag is dropped from a hot-air balloon that is hovering at a height of 512 feet above the ground. If air resistance is disregarded, then the distance $s(t)$ from the ground to the sandbag after t seconds can be given by $s(t) = -16t^2 + 512$. Find the velocity of the sandbag at

(a) $t = a$ sec   (b) $t = 2$ sec   (c) the instant it strikes the ground.

FIGURE 2.12

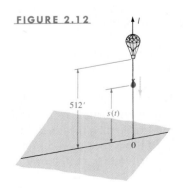

512'

$s(t)$

SOLUTION

(a) As shown in Figure 2.12, we may consider the sandbag to be moving along a vertical coordinate line l with origin at ground level. Note that at the instant it is dropped, $t = 0$ and

$$s(0) = -16(0) + 512 = 512 \text{ ft.}$$

We shall use Definition (2.5) to find the velocity of the sandbag at $t = a$.

First we consider the average velocity, and simplify it:

$$\frac{s(t) - s(a)}{t - a} = \frac{(-16t^2 + 512) - (-16a^2 + 512)}{t - a}$$

$$= \frac{-16t^2 + 16a^2}{t - a} = \frac{-16(t^2 - a^2)}{t - a}.$$

Using Definition (2.5), the velocity $v(a)$ at $t = a$ is

$$v(a) = \lim_{t \to a} \frac{s(t) - s(a)}{t - a} = \lim_{t \to a} \frac{-16(t^2 - a^2)}{t - a}$$

$$= \lim_{t \to a} \frac{-16(t - a)(t + a)}{t - a}$$

$$= \lim_{t \to a} \left[-16(t + a) \right]$$

where (as in Example 1(a)), we have canceled the factor $t - a$ because $t \neq a$ in the limiting process, and hence $t - a \neq 0$. As t approaches a the expression $t + a$ gets closer and closer to $a + a$, or $2a$. Thus

$$v(a) = -16(2a) = -32a \text{ ft/sec}.$$

The negative sign indicates that the motion of the sandbag is in the negative direction (downward) on l.

(b) To find the velocity at $t = 2$ we substitute 2 for a in the formula $v(a) = -32a$, obtaining

$$v(2) = -32(2) = -64 \text{ ft/sec}$$

(c) The sandbag strikes the ground when

$$s(t) = -16t^2 + 512 = 0 \quad \text{or} \quad t^2 = \frac{512}{16} = 32.$$

This gives us $t = \sqrt{32} = 4\sqrt{2} \approx 5.7$ sec. Using the formula $v(a) = -32a$ from part (a) with $a = 4\sqrt{2}$, the impact velocity is

$$v(4\sqrt{2}) = -32(4\sqrt{2}) = -128\sqrt{2} \approx 181 \text{ ft/sec} \quad \bullet$$

You may have noticed the similarity of the formulas in (2.2) and (2.5). Many mathematical and physical applications lead to precisely this same limit. In Chapter 3 we shall call this limit the *derivative* of the function f. The study of properties and applications of derivatives will constitute a major portion of our work in this text.

EXERCISES 2.1

Exer. 1–4: (a) Use (2.2) to find the slope of the tangent line to the graph of f at the point $P(a, f(a))$, and (b) find an equation of the tangent line at the point $P(2, f(2))$.

1 $f(x) = 5x^2 - 4$

2 $f(x) = 3 - 2x^2$

3 $f(x) = x^3$

4 $f(x) = x^4$

Exer. 5–8: (a) Use (2.2) to find the slope of the tangent line at the point with x-coordinate a on the graph of the equation; (b) find an equation of the tangent line at point P; (c) sketch the graph and the tangent line at P.

5 $y = 3x + 2, \ P(1, 5)$

6 $y = \sqrt{x}, \ P(4, 2)$

7 $y = 1/x, \ P(2, \frac{1}{2})$

8 $y = x^{-2}, \ P(2, \frac{1}{4})$

9 Give a geometric argument to show that the graph of $y = |x|$ has no tangent line at the point $(0, 0)$.

10 Give a geometric argument to show that the graph of the greatest integer function (see Figure 1.46) has no tangent line at the point $P(1, 1)$.

11 Refer to Exercise 3. Show that the tangent line to the graph of $y = x^3$ at the point $P(-1, -1)$ intersects the graph at $(-1, -1)$ and $(2, 8)$.

12 Refer to Example 1. Sketch the graph of $y = x^2$ together with tangent lines at the points having x-coordinates -3, -2, -1, 0, 1, 2, and 3. At what point on the graph is the slope of the tangent line equal to 6?

13 In a children's video game, airplanes fly from left to right along the path $y = 1 + (1/x)$ and can shoot their bullets in the tangent direction at creatures placed along the x-axis at $x = 1, 2, 3, 4$, and 5. If a player shoots when the plane is at $P(1, 2)$ will a creature be hit? What if the player shoots at $(\frac{3}{2}, \frac{5}{3})$?

EXERCISE 13

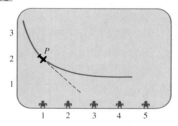

14 If $f(x) = ax + b$, prove that the tangent line to the graph of f at any point coincides with the graph of f.

Exer. 15–16: The position of a point P moving on a coordinate line l is given by $s(t)$ for t in seconds and $s(t)$ in centimeters.

(a) Find the average velocity of P in the following time intervals: $[1, 1.2]$; $[1, 1.1]$; $[1, 1.01]$; $[1, 1.001]$.

(b) Find the velocity of P at $t = 1$.

(c) Determine the time intervals in which P moves in the positive direction.

(d) Determine the time intervals in which P moves in the negative direction.

15 $s(t) = 4t^2 + 3t$ **16** $s(t) = t^3$

17 A balloonist drops a sand bag from a balloon 160 ft above the ground. After t seconds the sand bag is $160 - 16t^2$ feet above the ground.

(a) Find the velocity of the sand bag at $t = 1$.

(b) With what velocity does the sand bag strike the ground?

18 A projectile is fired directly upward from the ground with an initial velocity of 112 ft/sec. Its distance above the ground after t seconds is $112t - 16t^2$ ft. What is the velocity of the projectile at $t = 2$, $t = 3$, and $t = 4$? At what time does it reach its maximum height? When does it strike the ground? What is its velocity at the moment of impact?

19 A man runs the hundred-meter dash in such a way that the distance $s(t)$ run after t seconds is given by $s(t) = \frac{1}{5}t^2 + 8t$ meters (see figure). Find the runner's velocity (a) out of the starting blocks ($t = 0$), (b) at $t = 5$ sec, and (c) when he crosses the finishing line.

EXERCISE 19

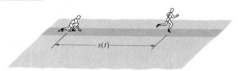

20 (a) If the position of an object moving on a coordinate line is given by a polynomial function of degree 1, prove that the velocity is constant.

(b) If the position function of a rectilinearly moving object is a constant function, prove that the velocity is 0 at all times. Describe the motion of the particle.

2.2 INFORMAL DEFINITION OF LIMIT

In the preceding section we saw how the definitions of tangent line and velocity depend on the notion of *limit* of a function. In the remainder of this chapter we consider limits in more detail. We shall begin with an informal description of the limit concept. A mathematically rigorous discussion is given in the next section.

In Section 2.1 we stated that $\lim_{x \to a} f(x) = L$ means that *the function value $f(x)$ approaches L as x approaches a*. A more precise, but still informal, description is the following.

<table>
<tr><td>**INFORMAL DEFINITION (2.6)**
OF LIMIT</td><td>Let a function f be defined on an open interval containing a, except possibly at a itself, and let L be a real number. The statement

$$\lim_{x \to a} f(x) = L$$

means that $f(x)$ can be made arbitrarily close to L by choosing x sufficiently close to a (but $x \neq a$).</td></tr>
</table>

The phrase $f(x)$ *can be made arbitrarily close to L* in (2.6) means that $|f(x) - L|$ *can be made as small as desired* by choosing x sufficiently close to a (but $x \neq a$). For example, by choosing values of x close enough to a (but $x \neq a$) we could make $|f(x) - L| < 0.0001$, or $|f(x) - L| < 0.00001$, and so on.

As in the last section, the terminology $\lim_{x \to a} f(x)$ *exists* is used if (2.6) is true for some (possibly unknown) number L.

In Section 2.5 we shall discuss limits of trigonometric functions and prove that

$$\lim_{x \to 0} \frac{\sin x}{x} = 1$$

where x denotes a real number or the *radian measure* of an angle. The following table, obtained with a calculator, lends support to this important result.

x	$\dfrac{\sin x}{x}$
± 0.1	0.998334166
± 0.01	0.999983333
± 0.001	0.999999833
± 0.0001	0.999999995
± 0.00001	1
± 0.000001	1

This table is misleading in several respects. First, because a calculator rounds off answers, it appears that $(\sin x)/x$ actually *equals* 1 if x is within 0.00001 of 0. This is *false*. It will follow from the discussion in Section 2.5 that $(\sin x)/x < 1$ for every x. Second, although the table indicates that $(\sin x)/x$ gets closer and closer to 1 as x approaches 0, we cannot be absolutely sure of this fact, since we have merely substituted several numbers for x. Conceivably $(\sin x)/x$ could deviate from 1 in some manner for values of x that are closer to 0 than those listed in the table.

Although a calculator may help give us some idea of whether a limit exists, it cannot be used for *proofs*. It is necessary to have available a precise mathematical theory of limits that does not depend on mechanical devices or guessing. We shall develop this theory in the remainder of this chapter.

The graph of the function f in Figure 2.13 illustrates one way in which we might have $\lim_{x \to a} f(x) = L$. We have not shown a point corresponding to $x = a$, because in the limit process *the value* $f(a)$ *is completely irrelevant*. As we shall see in examples, $f(a)$ may be different from L, may equal L, or may not exist, depending on the nature of the function f.

FIGURE 2.13

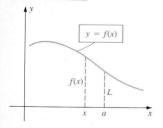

EXAMPLE 1 Let $f(x) = \dfrac{x - 9}{\sqrt{x} - 3}$.

(a) Find $\lim\limits_{x \to 9} f(x)$.

(b) Sketch the graph of f and verify the limit in part (a) graphically.

SOLUTION

(a) Note that the number 9 is not in the domain of f, since substitution of 9 for x leads to the meaningless expression $0/0$. To find the limit we shall change the form of $f(x)$ by rationalizing the denominator as follows:

$$\lim_{x \to 9} f(x) = \lim_{x \to 9} \frac{x - 9}{\sqrt{x} - 3}$$

$$= \lim_{x \to 9} \left(\frac{x - 9}{\sqrt{x} - 3} \cdot \frac{\sqrt{x} + 3}{\sqrt{x} + 3} \right)$$

$$= \lim_{x \to 9} \frac{(x - 9)(\sqrt{x} + 3)}{x - 9}$$

By definition (2.6), when finding the limit of $f(x)$ as $x \to 9$, we can assume that $x \neq 9$. Hence $x - 9 \neq 0$ and we can divide numerator and denominator by $x - 9$; that is, we can *cancel* the expression $x - 9$. This gives us

$$\lim_{x \to 9} f(x) = \lim_{x \to 9} (\sqrt{x} + 3) = \sqrt{9} + 3 = 6$$

(b) If we rationalize the denominator of $f(x)$ as in part (a) we see that the graph of f is the same as the graph of the equation $y = \sqrt{x} + 3$, *except for the point* (9, 6). The fact that (9, 6) is not on the graph of f is illustrated by the small circle in Figure 2.14. As x gets closer and closer to 9, the corresponding y-coordinate $f(x)$ on the graph of f gets closer to the number 6. Note that $f(x)$ never actually attains the value 6; however, it can be made arbitrarily close to 6 by choosing x sufficiently close to 9. •

EXAMPLE 2 If $f(x) = \dfrac{2x^2 - 5x + 2}{5x^2 - 7x - 6}$, find $\lim\limits_{x \to 2} f(x)$.

SOLUTION The number 2 is not in the domain of f since the meaningless expression $0/0$ is obtained if 2 is substituted for x. Factoring the numerator and denominator gives us

$$f(x) = \frac{(x - 2)(2x - 1)}{(x - 2)(5x + 3)}.$$

We cannot cancel the factor $x - 2$ at this stage; however, if we take the *limit* of $f(x)$ as $x \to 2$ this cancellation *is* allowed, because by Definition (2.6), $x \neq 2$ and hence $x - 2 \neq 0$. Thus

$$\lim_{x \to 2} f(x) = \lim_{x \to 2} \frac{2x^2 - 5x + 2}{5x^2 - 7x - 6}$$

$$= \lim_{x \to 2} \frac{(x - 2)(2x - 1)}{(x - 2)(5x + 3)}$$

$$= \lim_{x \to 2} \frac{2x - 1}{5x + 3} = \frac{3}{13} \quad •$$

FIGURE 2.14

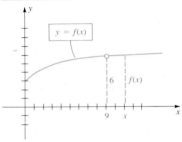

$y = f(x)$

$f(x)$

FIGURE 2.15

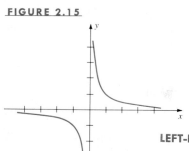

The rational function f defined by $f(x) = 1/x$ provides an illustration in which no limit exists as x approaches 0. Referring to the graph of f sketched in Figure 2.15 (see also Figure 1.44), we see that if x is assigned values closer and closer to 0 (but $x \neq 0$), $|f(x)|$ increases without bound. Properties of rational functions will be discussed in detail in Section 4.6.

We shall also be interested in the following *one-sided* limits.

LEFT-HAND LIMIT (2.7)

Let a function f be defined on an open interval (c, a). The statement

$$\lim_{x \to a^-} f(x) = L_1$$

means that $f(x)$ can be made arbitrarily close to L_1 by choosing x sufficiently close to a, with $x < a$.

RIGHT-HAND LIMIT (2.8)

Let a function f be defined on an open interval (a, c). The statement

$$\lim_{x \to a^+} f(x) = L_2$$

means that $f(x)$ can be made arbitrarily close to L_2 by choosing x sufficiently close to a, with $x > a$.

Graphical illustrations of one-sided limits are shown below. In Figure 2.16 we say that *x approaches a from the left*. In Figure 2.17 *x approaches a from the right*.

FIGURE 2.16
Left-hand limit: $\lim_{x \to a^-} f(x) = L_1$

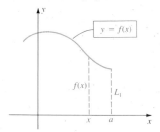

FIGURE 2.17
Right-hand limit: $\lim_{x \to a^+} f(x) = L_2$

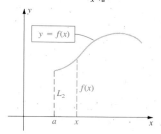

The sketch in Figure 2.16 is not intended to give the impression that $f(x)$ is undefined if $x \geq a$, but rather that for $x \to a^-$ we need only be concerned with values of x that are *less* than a. Similarly, for $x \to a^+$ we only consider values of x that are *greater* than a.

The relationship between one-sided limits and limits is described in the next theorem.

THEOREM (2.9)

If f is defined throughout an open interval containing a, except possibly at a itself, then $\lim_{x \to a} f(x) = L$ if and only if both $\lim_{x \to a^-} f(x) = L$ and $\lim_{x \to a^+} f(x) = L$.

The preceding theorem (which can be proved by using Definition (2.10) of Section 2.3) tells us that *the limit of $f(x)$ as x approaches a exists if and only if both the right-hand and left-hand limits exist and are equal.*

EXAMPLE 3 If $f(x) = \dfrac{|x|}{x}$, find $\lim\limits_{x \to 0^+} f(x)$, $\lim\limits_{x \to 0^-} f(x)$, and $\lim\limits_{x \to 0} f(x)$.

SOLUTION The graph of f is sketched in Figure 2.18. Note that f is undefined at $x = 0$.

If $x > 0$, then $|x| = x$ and $f(x) = x/x = 1$. Consequently,

$$\lim_{x \to 0^+} f(x) = \lim_{x \to 0^+} 1 = 1.$$

If $x < 0$, then $|x| = -x$ and $f(x) = -x/x = -1$. Therefore,

$$\lim_{x \to 0^-} f(x) = \lim_{x \to 0^-} (-1) = -1.$$

Since the right- and left-hand limits are unequal, it follows from Theorem (2.9) that $\lim_{x \to 0} f(x)$ does not exist. •

EXAMPLE 4 Let f be the piecewise-defined function such that

$$f(x) = \begin{cases} 2 - x & \text{for } x < 1 \\ x^2 + 1 & \text{for } x > 1 \end{cases}$$

Find $\lim\limits_{x \to 1^-} f(x)$, $\lim\limits_{x \to 1^+} f(x)$, and $\lim\limits_{x \to 1} f(x)$.

SOLUTION The graph of f is sketched in Figure 2.19. Note that there is no point on the graph with x-coordinate 1. Evidently,

$$\lim_{x \to 1^-} f(x) = \lim_{x \to 1^-} (2 - x) = 1$$

$$\lim_{x \to 1^+} f(x) = \lim_{x \to 1^+} (x^2 + 1) = 2$$

Since the right-hand and left-hand limits are not equal, $\lim_{x \to 1} f(x)$ does not exist by Theorem (2.9). •

EXAMPLE 5 Sketch the graph of f if

$$f(x) = \begin{cases} 3 - x & \text{for } x < 1 \\ 4 & \text{for } x = 1 \\ x^2 + 1 & \text{for } x > 1 \end{cases}$$

Find $\lim\limits_{x \to 1^-} f(x)$, $\lim\limits_{x \to 1^+} f(x)$, and $\lim\limits_{x \to 1} f(x)$.

SOLUTION The graph is sketched in Figure 2.20. We see that

$$\lim_{x \to 1^-} f(x) = \lim_{x \to 1^-} (3 - x) = 2$$

and

$$\lim_{x \to 1^+} f(x) = \lim_{x \to 1^+} (x^2 + 1) = 2$$

Since the right-hand and left-hand limits are equal, it follows from Theorem (2.9) that

$$\lim_{x \to 1} f(x) = 2.$$

Note that the function value $f(1) = 4$ is irrelevant in finding the limit. •

FIGURE 2.18

FIGURE 2.19

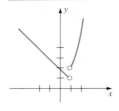

FIGURE 2.20

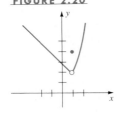

FIGURE 2.21

FIGURE 2.22

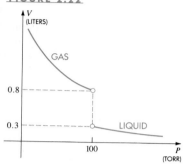

The following application involves one-sided limits.

EXAMPLE 6 A gas (such as water vapor or oxygen) is held at a constant temperature in the piston shown in Figure 2.21. As the gas is compressed, the volume decreases until a certain critical pressure is reached. Beyond this pressure, the gas will assume liquid form. Use the graph in Figure 2.22 to find and interpret

$$\lim_{P \to 100^-} V \quad \text{and} \quad \lim_{P \to 100^+} V$$

SOLUTION We see from Figure 2.22 that when the pressure P (in torrs) is low the substance is a gas and the volume V (in liters) is large. (The definition of the unit of pressure, the *torr*, may be found in textbooks on physics.) If P approaches 100 through values less than 100, V decreases and approaches 0.8, that is,

$$\lim_{P \to 100^-} V = 0.8$$

If P approaches 100 through values greater than 100 the substance is a liquid, and V increases very slowly (since liquids are nearly incompressible), with

$$\lim_{P \to 100^+} V = 0.3$$

At $P = 100$ the gas and liquid forms exist together in equilibrium, and the substance cannot be classified as either a gas or a liquid. •

EXERCISES 2.2

Exer. 1–6: Refer to the graphs to find each limit (a)–(f), if it exists.

(a) $\lim_{x \to 2^-} f(x)$ (b) $\lim_{x \to 2^+} f(x)$ (c) $\lim_{x \to 2} f(x)$

(d) $\lim_{x \to 0^-} f(x)$ (e) $\lim_{x \to 0^+} f(x)$ (f) $\lim_{x \to 0} f(x)$

1

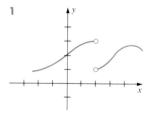

2

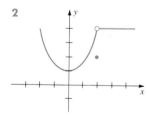

3

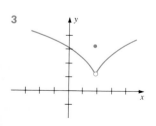

4

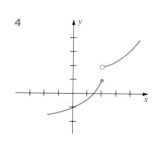

5

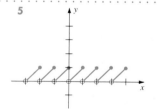

6

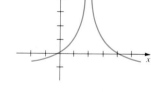

Exer. 7–12: Sketch the graph of the piecewise-defined function f and find each limit (a)–(c), if it exists.

(a) $\lim_{x \to 1^-} f(x)$ (b) $\lim_{x \to 1^+} f(x)$ (c) $\lim_{x \to 1} f(x)$

7 $f(x) = \begin{cases} x^2 + 1 & \text{if } x < 1 \\ 1 & \text{if } x = 1 \\ x + 1 & \text{if } x > 1 \end{cases}$

8 $f(x) = \begin{cases} x^3 & \text{if } x \leq 1 \\ 3 - x & \text{if } x > 1 \end{cases}$

9 $f(x) = \begin{cases} 3x - 1 & \text{if } x \leq 1 \\ 3 - x & \text{if } x > 1 \end{cases}$

10 $f(x) = \begin{cases} |x - 1| & \text{if } x \neq 1 \\ 1 & \text{if } x = 1 \end{cases}$

11 $f(x) = \begin{cases} \dfrac{1}{x - 1} & \text{if } x < 1 \\ 1 & \text{if } x \geq 1 \end{cases}$

12 $f(x) = \begin{cases} \dfrac{|x - 1|}{x - 1} x^2 & \text{if } x \neq 1 \\ 0 & \text{if } x = 1 \end{cases}$

Exer. 13–18: Find each limit, if it exists.

13 (a) $\lim\limits_{x \to 4^-} \dfrac{|x - 4|}{x - 4}$

(b) $\lim\limits_{x \to 4^+} \dfrac{|x - 4|}{x - 4}$

(c) $\lim\limits_{x \to 4} \dfrac{|x - 4|}{x - 4}$

14 (a) $\lim\limits_{x \to -5^+} \dfrac{x + 5}{|x + 5|}$

(b) $\lim\limits_{x \to -5^-} \dfrac{x + 5}{|x + 5|}$

(c) $\lim\limits_{x \to -5} \dfrac{x + 5}{|x + 5|}$

15 $\lim\limits_{x \to -6^+} (\sqrt{x + 6} + x)$

16 $\lim\limits_{x \to 5/2^-} (\sqrt{5 - 2x} - x^2)$

17 $\lim\limits_{x \to 0^+} \dfrac{1}{x^3}$

18 $\lim\limits_{x \to 8^-} \dfrac{1}{x - 8}$

Exer. 19–30: Use algebraic simplifications to help find the limit, if it exists.

19 $\lim\limits_{x \to 2} \dfrac{x^2 - 4}{x - 2}$

20 $\lim\limits_{x \to 3} \dfrac{2x^3 - 6x^2 + x - 3}{x - 3}$

21 $\lim\limits_{x \to 1} \dfrac{x^2 - x}{2x^2 + 5x - 7}$

22 $\lim\limits_{r \to -3} \dfrac{r^2 + 2r - 3}{r^2 + 7r + 12}$

23 $\lim\limits_{x \to 5} \dfrac{3x^2 - 13x - 10}{2x^2 - 7x - 15}$

24 $\lim\limits_{x \to 25} \dfrac{\sqrt{x} - 5}{x - 25}$

25 $\lim\limits_{k \to 4} \dfrac{k^2 - 16}{\sqrt{k} - 2}$

26 $\lim\limits_{h \to 0} \dfrac{(x + h)^3 - x^3}{h}$

27 $\lim\limits_{h \to 0} \dfrac{(x + h)^2 - x^2}{h}$

28 $\lim\limits_{h \to 2} \dfrac{h^3 - 8}{h^2 - 4}$

29 $\lim\limits_{h \to -2} \dfrac{h^3 + 8}{h + 2}$

30 $\lim\limits_{z \to 10} \dfrac{1}{z - 10}$

31 Shown in the figure is a typical graph of the G-forces experienced by astronauts during the takeoff of a spacecraft with two rocket boosters. (A force of 2G's is twice that of gravity, 3G's is three times that of gravity, etc.) If $F(t)$ denotes the G-force t minutes into the flight, find and interpret:

(a) $\lim\limits_{t \to 0^+} F(t)$

(b) $\lim\limits_{t \to 3.5^-} F(t)$ and $\lim\limits_{t \to 3.5^+} F(t)$

(c) $\lim\limits_{t \to 5^-} F(t)$ and $\lim\limits_{t \to 5^+} F(t)$

EXERCISE 31

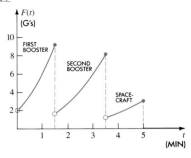

32 A hospital patient receives an initial 200-mg dose of a drug. Additional doses of 100 mg each are then administered every 4 hours. The amount $y(t)$ of the drug present in the bloodstream after t hours is shown in the figure. Find and interpret $\lim\limits_{t \to 8^-} y(t)$ and $\lim\limits_{t \to 8^+} y(t)$.

EXERCISE 32

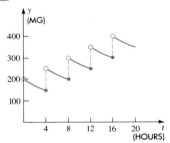

Exer. 33–38: The stated result may be verified by using methods developed later in the text. At this stage of our work, use a calculator to lend support to this result by substituting various real numbers for x. Discuss why this use of a calculator fails to *prove* that the limit exists.

33 $\lim\limits_{x \to 0} (1 + x)^{1/x} \approx 2.72$

34 $\lim\limits_{x \to 0} (1 + 2x)^{3/x} \approx 403.4$

35 $\lim\limits_{x \to 2} \dfrac{3^x - 9}{x - 2} \approx 9.89$

36 $\lim\limits_{x \to 1} \dfrac{2^x - 2}{x - 1} \approx 1.39$

37 $\lim\limits_{x \to 0} \left(\dfrac{4^{|x|} + 9^{|x|}}{2} \right)^{1/|x|} = 6$

38 $\lim\limits_{x \to 0} |x|^x = 1$

2.3 FORMAL DEFINITION OF LIMIT

In Section 2.2 we defined $\lim\limits_{x \to a} f(x) = L$ informally, by stating that $f(x)$ can be made *arbitrarily close* to L by choosing x *sufficiently close* to a (but $x \neq a$). This is a very good description of a limit; however, it lacks mathematical precision because of the phrases *arbitrarily close* and *sufficiently close*. In this section we shall state a formal definition that can be used to give rigorous proofs of results about limits.

The key to arriving at a satisfactory definition is to realize that we must be able to make $|f(x) - L|$ as *small as desired* by choosing x sufficiently close to a (and $x \neq a$); that is, by choosing x such that $|x - a|$ is *sufficiently small* (and $x - a \neq 0$). In calculus it is traditional to use the Greek letters ε (epsilon) and δ (delta) to denote very small positive real numbers. To state that $|f(x) - L|$ can be made as small as desired means that for *every* $\varepsilon > 0$, we can find values of x such that

$$|f(x) - L| < \varepsilon.$$

By (1.3) (i), the last inequality is equivalent to

$$-\varepsilon < f(x) - L < \varepsilon \quad \text{or} \quad L - \varepsilon < f(x) < L + \varepsilon.$$

Similarly, to state that $|x - a|$ is sufficiently small (and $x - a \neq 0$) we may use the inequality

$$0 < |x - a| < \delta \qquad \text{for some } \delta > 0.$$

Again using (1.3) (i), we see that this is equivalent to

$$a - \delta < x < a + \delta \quad \text{and} \quad x \neq a.$$

Graphs of these inequalities are sketched on coordinate lines l and l', respectively, in Figure 2.23, where ε and δ should be regarded as very small numbers, such as 0.0001 or 0.0000001. The small circle in Figure 2.23 (i) indicates that $x \neq a$.

We use the preceding notation in the following definition of limit.

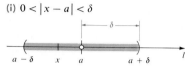
DEFINITION OF LIMIT (2.10)

> Let a function f be defined on an open interval containing a, except possibly at a itself, and let L be a real number. The statement
>
> $$\lim_{x \to a} f(x) = L$$
>
> means that for every $\varepsilon > 0$, there exists a $\delta > 0$, such that
>
> $$\text{if} \quad 0 < |x - a| < \delta, \quad \text{then} \quad |f(x) - L| < \varepsilon$$

It is sometimes convenient to use the following form of Definition (2.10), where the two inequalities involving absolute values have been stated in terms of open intervals (see Figure 2.23).

ALTERNATIVE DEFINITION OF LIMIT (2.11)

> The statement
>
> $$\lim_{x \to a} f(x) = L$$
>
> means that for every $\varepsilon > 0$, there exists a $\delta > 0$, such that if x is in the open interval $(a - \delta, a + \delta)$, and $x \neq a$, then $f(x)$ is in the open interval $(L - \varepsilon, L + \varepsilon)$.

If $f(x)$ *has a limit as x approaches a, then that limit is unique*. A proof of this important theorem is given at the beginning of Appendix II.

To better understand the relationship between the positive numbers ε and δ in Definitions (2.10) and (2.11), let us consider a geometric interpretation similar to that in Figure 1.37. The domain of f is represented by certain points on a coordinate line l, and the range by other points on a coordinate line l'. The limit process may be outlined as follows.

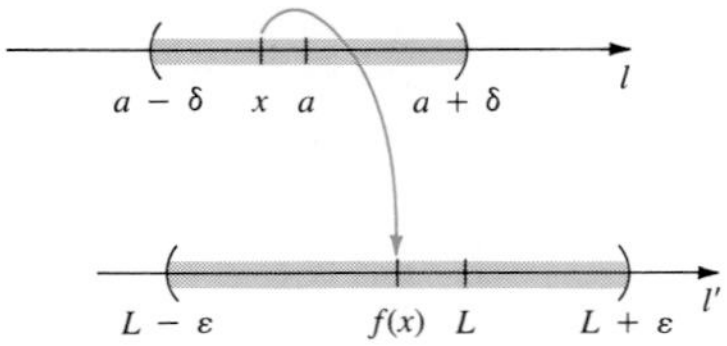

To prove that $\lim\limits_{x \to a} f(x) = L$:

Step 1. For any $\varepsilon > 0$ consider the open interval $(L - \varepsilon, L + \varepsilon)$ in the range of f (see Figure 2.24).

Step 2. Show that there exists an open interval $(a - \delta, a + \delta)$ in the domain of f such that Definition (2.11) is true (see Figure 2.25).

It is extremely important to remember that *first* we consider the interval $(L - \varepsilon, L + \varepsilon)$ and then, *second*, we show that an interval $(a - \delta, a + \delta)$ of the required type exists in the domain of f. One scheme for remembering the proper sequence of events is to think of the function f as a cannon that shoots a cannonball from the point on l with coordinate x to the point on l' with coordinate $f(x)$, as illustrated by the curved arrow in Figure 2.25. Step 1 may then be regarded as setting up a target of radius ε with bull's eye at L. To apply Step 2 we must find an open interval containing a in which to place the cannon such that the cannonball hits the target. Incidentally, there is no guarantee that it will hit the bull's eye; however, if $\lim_{x \to a} f(x) = L$ we can make the cannonball land as close as we please to the bull's eye.

The number δ in the limit definition is not unique, for if a specific δ can be found, then any *smaller* positive number δ' will also satisfy the requirements.

In the following example we verify a limit statement by means of Definition (2.10).

FIGURE 2.24

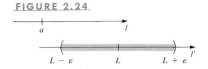

FIGURE 2.25

EXAMPLE 1 Verify that $\lim\limits_{x \to 4} \frac{1}{2}(3x - 1) = \frac{11}{2}$.

SOLUTION Let $f(x) = \frac{1}{2}(3x - 1)$, $a = 4$, and $L = \frac{11}{2}$. According to Definition (2.10), we must show that for every $\varepsilon > 0$, there exists a $\delta > 0$ such that

$$\text{if} \quad 0 < |x - 4| < \delta, \quad \text{then} \quad \left|\tfrac{1}{2}(3x - 1) - \tfrac{11}{2}\right| < \varepsilon.$$

A clue to the choice of δ can be found by examining the inequality involving ε. The following is a list of equivalent inequalities:

$$\left|\tfrac{1}{2}(3x - 1) - \tfrac{11}{2}\right| < \varepsilon$$
$$\tfrac{1}{2}\left|(3x - 1) - 11\right| < \varepsilon$$
$$\left|3x - 1 - 11\right| < 2\varepsilon$$
$$\left|3x - 12\right| < 2\varepsilon$$
$$3\left|x - 4\right| < 2\varepsilon$$
$$\left|x - 4\right| < \tfrac{2}{3}\varepsilon$$

The final inequality gives us the needed clue. If we let $\delta = \frac{2}{3}\varepsilon$, then if $0 < |x - 4| < \delta$, the last inequality in the list is true and, since all inequalities in the list are equivalent, the first is also true. Hence by Definition (2.10), $\lim_{x \to 4} \frac{1}{2}(3x - 1) = \frac{11}{2}$. •

It was relatively easy to use the definition of limit (2.10) in Example 1 because $f(x)$ was a simple expression involving x. Limits of more complicated functions may also be verified by direct applications of the definition; however, the task of showing that for every $\varepsilon > 0$ there exists a suitable $\delta > 0$ often requires a great deal of ingenuity.

Let us next interpret Definitions (2.10) and (2.11) using the graph of f, as illustrated in Figure 2.26. Given any $\varepsilon > 0$, we first consider the open interval $(L - \varepsilon, L + \varepsilon)$ on the y-axis, and the horizontal lines $y = L \pm \varepsilon$. If there exists an open interval $(a - \delta, a + \delta)$ such that for every x in $(a - \delta, a + \delta)$, with the possible exception of $x = a$, the point $P(x, f(x))$ lies between the horizontal lines, that is, within the shaded rectangle shown in Figure 2.26, then

$$L - \varepsilon < f(x) < L + \varepsilon$$

and hence $\lim_{x \to a} f(x) = L$.

The next example illustrates how the geometric process pictured in Figure 2.26 may be applied to a specific function.

FIGURE 2.26

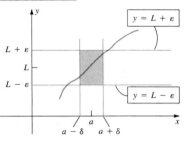

EXAMPLE 2 Prove that $\lim_{x \to a} x^2 = a^2$.

SOLUTION Let us consider the case $a > 0$. We shall apply the Alternative Definition (2.11) with $f(x) = x^2$ and $L = a^2$. The graph of f is sketched in Figure 2.27, together with typical points on the x- and y-axes corresponding to a and a^2, respectively.

For any positive number ε, consider the horizontal lines $y = a^2 - \varepsilon$ and $y = a^2 + \varepsilon$. These lines intersect the graph of f at points with x-coordinates $\sqrt{a^2 - \varepsilon}$ and $\sqrt{a^2 + \varepsilon}$, as illustrated in the figure. If x is in the open interval $(\sqrt{a^2 - \varepsilon}, \sqrt{a^2 + \varepsilon})$, then

$$\sqrt{a^2 - \varepsilon} < x < \sqrt{a^2 + \varepsilon}.$$

Consequently, $\qquad a^2 - \varepsilon < x^2 < a^2 + \varepsilon$,

that is, $f(x) = x^2$ is in the open interval $(a^2 - \varepsilon, a^2 + \varepsilon)$. Geometrically, this means that the point (x, x^2) on the graph of f lies between the horizontal lines.

Choose a number δ smaller than both $\sqrt{a^2 + \varepsilon} - a$ and $a - \sqrt{a^2 - \varepsilon}$ as illustrated in Figure 2.27. It follows that if x is in the interval $(a - \delta, a + \delta)$, then x is also in $(\sqrt{a^2 - \varepsilon}, \sqrt{a^2 + \varepsilon})$, and, therefore, $f(x)$ is in the interval $(a^2 - \varepsilon, a^2 + \varepsilon)$. Hence, by Definition (2.11), $\lim_{x \to a} x^2 = a^2$. Although we have considered only $a > 0$, a similar argument applies if $a \leq 0$. •

FIGURE 2.27

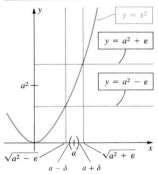

The next two examples, which were also discussed in Section 2.2, indicate how the geometric interpretation illustrated in Figure 2.27 may be used to show that certain limits do not exist.

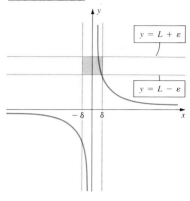

FIGURE 2.28

EXAMPLE 3 Show that $\lim_{x \to 0} \dfrac{1}{x}$ does not exist.

SOLUTION Let us proceed in an indirect manner. Thus, suppose it *were* true that

$$\lim_{x \to 0} \frac{1}{x} = L$$

for some number L. Let us consider any pair of horizontal lines $y = L \pm \varepsilon$, as illustrated in Figure 2.28. Since we are assuming that the limit exists, it should be possible to find an open interval $(0 - \delta, 0 + \delta)$ or equivalently, $(-\delta, \delta)$, containing 0, such that whenever $-\delta < x < \delta$ and $x \neq 0$, the point $(x, 1/x)$ on the graph lies between the horizontal lines. However, since $1/x$ can be made as large as desired by choosing x close to 0, not every point $(x, 1/x)$ with nonzero x-coordinate in $(-\delta, \delta)$ has this property. Consequently our supposition is false; that is, the limit does not exist. •

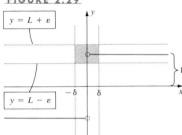

FIGURE 2.29

EXAMPLE 4 If $f(x) = \dfrac{|x|}{x}$, show that $\lim_{x \to 0} f(x)$ does not exist.

SOLUTION If $x > 0$, then $|x|/x = x/x = 1$ and hence, to the right of the y-axis, the graph of f coincides with the line $y = 1$. If $x < 0$, then $|x|/x = -x/x = -1$, which means that to the left of the y-axis the graph of f coincides with the line $y = -1$. If it were true that $\lim_{x \to 0} f(x) = L$ for some L, then the preceding remarks imply that $-1 \leq L \leq 1$. As shown in Figure 2.29, if we consider any pair of horizontal lines $y = L \pm \varepsilon$, where $0 < \varepsilon < 1$, then there exist points on the graph that are not between these lines for some nonzero x in *every* interval $(-\delta, \delta)$ containing 0. It follows that the limit does not exist. •

The following theorem states that *if a function f has a positive limit as x approaches a, then $f(x)$ is positive throughout some open interval containing a, with the possible exception of a.*

THEOREM (2.12)

> If $\lim_{x \to a} f(x) = L$ and $L > 0$, then there exists an open interval $(a - \delta, a + \delta)$ containing a such that $f(x) > 0$ for every x in $(a - \delta, a + \delta)$, except possibly $x = a$.

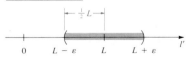

FIGURE 2.30

PROOF Consider the point on a coordinate line that corresponds to the positive number L. If we choose $\varepsilon = \frac{1}{2}L$, then the interval $(L - \varepsilon, L + \varepsilon)$ contains only positive numbers, as illustrated in Figure 2.30. By Definition (2.11), there exists a $\delta > 0$ such that if x is in the open interval $(a - \delta, a + \delta)$ and $x \neq a$, then $f(x)$ is in $(L - \varepsilon, L + \varepsilon)$, and hence $f(x) > 0$. • •

We can show that *if f has a negative limit as x approaches a, then there is an open interval I containing a such that $f(x) < 0$ for every x in I, with the possible exception of $x = a$.*

Formal definitions can also be given for one-sided limits. For the right-hand limit $x \to a^+$, we simply replace the condition $0 < |x - a| < \delta$ in Definition (2.10) by $a < x < a + \delta$. In terms of the Alternative Definition (2.11), we restrict x to the *right* half $(a, a + \delta)$ of the interval $(a - \delta, a + \delta)$.

Similarly, for the left-hand limit $x \to a^-$ we replace $0 < |x - a| < \delta$ in (2.10) by $a - \delta < x < a$. This is equivalent to restricting x to the *left* half $(a - \delta, a)$ of the interval $(a - \delta, a + \delta)$ in (2.11).

EXERCISES 2.3

Exer. 1–12: Prove that the limit exists by means of Definition (2.10).

1 $\lim_{x \to 4} 3x = 12$

2 $\lim_{x \to 5} (-4x) = -20$

3 $\lim_{x \to 2} (5x - 3) = 7$

4 $\lim_{x \to -3} (2x + 1) = -5$

5 $\lim_{x \to -6} (10 - 9x) = 64$

6 $\lim_{x \to 4} (8x - 15) = 17$

7 $\lim_{x \to 3} 5 = 5$

8 $\lim_{x \to 5} 3 = 3$

9 $\lim_{x \to a} c = c$ for all real numbers a and c

10 $\lim_{x \to 6} \left(9 - \frac{x}{6} \right) = 8$

11 $\lim_{x \to a} x = a$ for every real number a

12 $\lim_{x \to a} (mx + b) = ma + b$ for all real numbers m, b, and a

Exer. 13–18: Use the graphical technique illustrated in Example 2 to verify the limit if $a > 0$.

13 $\lim_{x \to -a} x^2 = a^2$

14 $\lim_{x \to a} (x^2 + 1) = a^2 + 1$

15 $\lim_{x \to a} x^3 = a^3$

16 $\lim_{x \to a} x^4 = a^4$

17 $\lim_{x \to a} \sqrt{x} = \sqrt{a}$

18 $\lim_{x \to a} \sqrt[3]{x} = \sqrt[3]{a}$

Exer. 19–26: Use the method illustrated in Examples 3 and 4 to show that the limit does not exist.

19 $\lim_{x \to 3} \frac{|x - 3|}{x - 3}$

20 $\lim_{x \to -2} \frac{x + 2}{|x + 2|}$

21 $\lim_{x \to -1} \frac{3x + 3}{|x + 1|}$

22 $\lim_{x \to 5} \frac{2x - 10}{|x - 5|}$

23 $\lim_{x \to 0} \frac{1}{x^2}$

24 $\lim_{x \to 4} \frac{7}{x - 4}$

25 $\lim_{x \to -5} \frac{1}{x + 5}$

26 $\lim_{x \to 1} \frac{1}{(x - 1)^2}$

27 Give an example of a function f that is defined at a such that $\lim_{x \to a} f(x)$ exists and $\lim_{x \to a} f(x) \neq f(a)$.

28 If f is the greatest integer function and a is any integer, show that $\lim_{x \to a} f(x)$ does not exist.

29 Let f be defined by the following conditions: $f(x) = 0$ if x is rational and $f(x) = 1$ if x is irrational. Prove that for every real number a, $\lim_{x \to a} f(x)$ does not exist.

30 Why is it impossible to investigate $\lim_{x \to 0} \sqrt{x}$ by means of Definition (2.10)?

2.4 TECHNIQUES FOR FINDING LIMITS

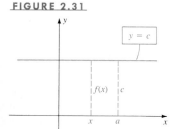

FIGURE 2.31

If f is a constant function, then there is a real number c such that $f(x) = c$ for every x. The graph of f is the horizontal line $y = c$ illustrated in Figure 2.31. To say that $f(x)$ *approaches* c, or that $f(x)$ can be made *arbitrarily close* to c is an understatement, since $f(x)$ has the value c for every x. Thus

(2.13)
$$\lim_{x \to a} c = c$$

This also follows readily from Definition (2.10) (see Exercise 7 of Section 2.3).

The limit in (2.13) is often described by the phrase *the limit of a constant is the constant*.

EXAMPLE 1 Find $\lim_{x \to 3} 8$, $\lim_{x \to 8} 3$, and $\lim_{x \to a} 0$.

SOLUTION Applying (2.13),

$$\lim_{x \to 3} 8 = 8, \quad \lim_{x \to 8} 3 = 3 \quad \text{and} \quad \lim_{x \to a} 0 = 0 \quad \bullet$$

FIGURE 2.32

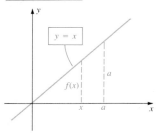

Let us next consider the linear function f such that $f(x) = x$ for every x. The graph of f is the line $y = x$ illustrated in Figure 2.32. Since $f(x) = x$, it is trivial that $f(x)$ approaches a as x approaches a. Using limit notation:

(2.14)

$$\lim_{x \to a} x = a$$

It is easy to prove (2.14) by means of Definition (2.10).

EXAMPLE 2 Find $\lim_{x \to \sqrt{2}} x$ and $\lim_{x \to -4} x$.

SOLUTION By (2.14),

$$\lim_{x \to \sqrt{2}} x = \sqrt{2} \quad \text{and} \quad \lim_{x \to -4} x = -4 \quad \bullet$$

As we shall see, the formulas in (2.13) and (2.14) can be used as building blocks for finding limits of very complicated expressions.

Many functions can be expressed as sums, differences, products, and quotients of other functions. In particular, suppose a function s is a sum of two functions f and g, so that $s(x) = f(x) + g(x)$ for every x in the domain of s. If $f(x)$ and $g(x)$ have limits L and M, respectively, as x approaches a, we would expect $s(x)$ to have the limit $L + M$ as x approaches a. The fact that this and analogous statements are true for products and quotients are consequences of the next theorem.

THEOREM (2.15)

If $\lim_{x \to a} f(x) = L$ and $\lim_{x \to a} g(x) = M$, then

(i) $\lim_{x \to a} [f(x) + g(x)] = L + M$

(ii) $\lim_{x \to a} [f(x) \cdot g(x)] = L \cdot M$

(iii) $\lim_{x \to a} \left[\dfrac{f(x)}{g(x)}\right] = \dfrac{L}{M}$, provided $M \neq 0$

(iv) $\lim_{x \to a} [cf(x)] = cL$, for every real number c

(v) $\lim_{x \to a} [f(x) - g(x)] = L - M$

Although the conclusions of Theorem (2.15) appear to be intuitively evident, the proofs are rather technical. Proofs for (i)–(iii), based on Definition (2.10), may be found in Appendix II. Part (iv) of the theorem follows readily from part (ii) and (2.13) as follows:

$$\lim_{x \to a} [cf(x)] = \left[\lim_{x \to a} c\right]\left[\lim_{x \to a} f(x)\right] = cL.$$

Finally, to prove (v) we may write

$$f(x) - g(x) = f(x) + (-1)g(x)$$

and then use (i) and (iv) (with $c = -1$).

Theorem (2.15) is often written as follows, provided the limits of $f(x)$ and $g(x)$ exist.

(i) $\lim\limits_{x \to a} [f(x) + g(x)] = \lim\limits_{x \to a} f(x) + \lim\limits_{x \to a} g(x)$

(ii) $\lim\limits_{x \to a} [f(x) \cdot g(x)] = \lim\limits_{x \to a} f(x) \cdot \lim\limits_{x \to a} g(x)$

(iii) $\lim\limits_{x \to a} \left[\dfrac{f(x)}{g(x)}\right] = \dfrac{\lim\limits_{x \to a} f(x)}{\lim\limits_{x \to a} g(x)}$, if $\lim\limits_{x \to a} g(x) \neq 0$

(iv) $\lim\limits_{x \to a} [cf(x)] = c\left[\lim\limits_{x \to a} f(x)\right]$

(v) $\lim\limits_{x \to a} [f(x) - g(x)] = \lim\limits_{x \to a} f(x) - \lim\limits_{x \to a} g(x)$

We shall use Theorem (2.15) to establish the following.

THEOREM (2.16)

> If m, b, and a are any real numbers, then
> $$\lim_{x \to a} (mx + b) = ma + b$$

PROOF We know from (2.13) and (2.14) that

$$\lim_{x \to a} m = m, \quad \lim_{x \to a} x = a, \quad \text{and} \quad \lim_{x \to a} b = b.$$

Hence by (i) and (iv) of Theorem (2.15),

$$\lim_{x \to a} (mx + b) = \lim_{x \to a} (mx) + \lim_{x \to a} b$$

$$= m\left(\lim_{x \to a} x\right) + b$$

$$= ma + b.$$

This result can also be proved directly from Definition (2.10) (see Exercise 12 of Section 2.3). • •

If f is a linear function, then according to Theorem (2.16), $\lim_{x \to a} f(x)$ can be found by merely substituting a for x. In Section 2.6 we will discuss many functions that have this property.

EXAMPLE 3 Find $\lim\limits_{x \to 2} \dfrac{3x + 4}{5x + 7}$.

SOLUTION The numerator and denominator of the quotient define linear functions whose limits exist by Theorem (2.16). Moreover, the limit of the denominator is not 0. Hence by Theorem (2.15) (iii) and Theorem (2.16),

$$\lim_{x \to 2} \frac{3x + 4}{5x + 7} = \frac{\lim\limits_{x \to 2} (3x + 4)}{\lim\limits_{x \to 2} (5x + 7)} = \frac{3(2) + 4}{5(2) + 7} = \frac{10}{17} \quad \bullet$$

Theorem (2.15) may be extended to limits of sums, differences, products, and quotients that involve any number of functions.

EXAMPLE 4 Prove that for every real number a, $\lim\limits_{x \to a} x^3 = a^3$.

SOLUTION Since $\lim\limits_{x \to a} x = a$,

$$\lim_{x \to a} x^3 = \lim_{x \to a} (x \cdot x \cdot x)$$

$$= \left(\lim_{x \to a} x \right) \cdot \left(\lim_{x \to a} x \right) \cdot \left(\lim_{x \to a} x \right)$$

$$= a \cdot a \cdot a = a^3 \qquad \bullet$$

The technique used in Example 4 can be extended to x^n for any positive integer n. We merely write x^n as a product $x \cdot x \cdot \cdots \cdot x$ of n factors and then take the limit of each factor. This gives us part (i) of the next theorem. Part (ii) may be proved in similar fashion by using Theorem (2.15) (ii).

THEOREM (2.17)

If n is a positive integer, then

(i) $\lim\limits_{x \to a} x^n = a^n$

(ii) $\lim\limits_{x \to a} [f(x)]^n = \left[\lim\limits_{x \to a} f(x) \right]^n$, provided $\lim\limits_{x \to a} f(x)$ exists.

EXAMPLE 5 Find $\lim\limits_{x \to 2} (3x + 4)^5$.

SOLUTION Applying (2.17) (ii) and Theorem (2.16),

$$\lim_{x \to 2} (3x + 4)^5 = \left[\lim_{x \to 2} (3x + 4) \right]^5$$

$$= [3(2) + 4]^5$$

$$= 10^5 = 100{,}000 \qquad \bullet$$

EXAMPLE 6 Find $\lim\limits_{x \to -2} (5x^3 + 3x^2 - 6)$.

SOLUTION We may proceed as follows (supply reasons):

$$\lim_{x \to -2} (5x^3 + 3x^2 - 6) = \lim_{x \to -2} (5x^3) + \lim_{x \to -2} (3x^2) - \lim_{x \to -2} (6)$$

$$= 5 \lim_{x \to -2} (x^3) + 3 \lim_{x \to -2} (x^2) - 6$$

$$= 5(-2)^3 + 3(-2)^2 - 6$$

$$= 5(-8) + 3(4) - 6 = -34 \qquad \bullet$$

The limit in Example 6 is the number obtained by substituting -2 for x in $5x^3 + 3x^2 - 6$. The next theorem states that the same is true for the limit of *every* polynomial.

THEOREM (2.18)

If f is a polynomial function and a is a real number, then

$$\lim_{x \to a} f(x) = f(a)$$

PROOF We may write $f(x)$ in the form

$$f(x) = b_n x^n + b_{n-1} x^{n-1} + \cdots + b_0$$

for real numbers $b_n, b_{n-1}, \ldots, b_0$. As in Example 6,

$$\lim_{x \to a} f(x) = \lim_{x \to a} (b_n x^n) + \lim_{x \to a} (b_{n-1} x^{n-1}) + \cdots + \lim_{x \to a} b_0$$

$$= b_n \lim_{x \to a} (x^n) + b_{n-1} \lim_{x \to a} (x^{n-1}) + \cdots + \lim_{x \to a} b_0$$

$$= b_n a^n + b_{n-1} a^{n-1} + \cdots + b_0 = f(a) \quad \bullet \bullet$$

COROLLARY (2.19)

> If q is a rational function and a is in the domain of q, then
> $$\lim_{x \to a} q(x) = q(a)$$

PROOF We may write $q(x) = f(x)/h(x)$ where f and h are polynomial functions. If a is in the domain of q, then $h(a) \neq 0$. Using Theorem (2.15) (iii) and (2.18),

$$\lim_{x \to a} q(x) = \frac{\lim_{x \to a} f(x)}{\lim_{x \to a} h(x)} = \frac{f(a)}{h(a)} = q(a) \quad \bullet \bullet$$

EXAMPLE 7 Find $\lim_{x \to 3} \dfrac{5x^2 - 2x + 1}{4x^3 - 7}$.

SOLUTION Applying Corollary (2.19),

$$\lim_{x \to 3} \frac{5x^2 - 2x + 1}{4x^3 - 7} = \frac{5(3)^2 - 2(3) + 1}{4(3)^3 - 7}$$

$$= \frac{45 - 6 + 1}{108 - 7} = \frac{40}{101} \quad \bullet$$

The next theorem states that for positive integral roots of x, we may determine a limit by substitution. The proof makes use of the formal definition of limit (2.10) and may be found in Appendix II.

THEOREM (2.20)

> If $a > 0$ and n is a positive integer, or if $a \leq 0$ and n is an odd positive integer, then
> $$\lim_{x \to a} \sqrt[n]{x} = \sqrt[n]{a}$$

If m and n are positive integers and $a > 0$, then using Theorem (2.17) (ii) and Theorem (2.20),

$$\lim_{x \to a} (\sqrt[n]{x})^m = \left(\lim_{x \to a} \sqrt[n]{x} \right)^m = (\sqrt[n]{a})^m.$$

In terms of rational exponents,

$$\lim_{x \to a} x^{m/n} = a^{m/n}.$$

This limit formula may be extended to negative exponents by writing $x^{-r} = 1/x^r$ and then using Theorem (2.15)(iii).

EXAMPLE 8 Find $\lim\limits_{x \to 8} \dfrac{x^{2/3} + 3\sqrt{x}}{4 - (16/x)}$.

SOLUTION We may proceed as follows (supply reasons):

$$\lim_{x \to 8} \frac{x^{2/3} + 3\sqrt{x}}{4 - (16/x)} = \frac{\lim\limits_{x \to 8} (x^{2/3} + 3\sqrt{x})}{\lim\limits_{x \to 8} [4 - (16/x)]}$$

$$= \frac{\lim\limits_{x \to 8} x^{2/3} + \lim\limits_{x \to 8} 3\sqrt{x}}{\lim\limits_{x \to 8} 4 - \lim\limits_{x \to 8} (16/x)}$$

$$= \frac{8^{2/3} + 3\sqrt{8}}{4 - (16/8)}$$

$$= \frac{4 + 6\sqrt{2}}{4 - 2} = 2 + 3\sqrt{2} \quad \bullet$$

THEOREM (2.21)

> If a function f has a limit as x approaches a, then
> $$\lim_{x \to a} \sqrt[n]{f(x)} = \sqrt[n]{\lim_{x \to a} f(x)}$$
> provided either n is an odd positive integer or n is an even positive integer and $\lim_{x \to a} f(x) > 0$.

The preceding theorem will be proved in Section 2.6. In the meantime we shall use it whenever applicable to gain experience in finding limits that involve roots of algebraic expressions.

EXAMPLE 9 Find $\lim\limits_{x \to 5} \sqrt[3]{3x^2 - 4x + 9}$.

SOLUTION Using Theorems (2.21) and (2.18),

$$\lim_{x \to 5} \sqrt[3]{3x^2 - 4x + 9} = \sqrt[3]{\lim_{x \to 5} (3x^2 - 4x + 9)}$$

$$= \sqrt[3]{75 - 20 + 9} = \sqrt[3]{64} = 4 \quad \bullet$$

We should not be misled by the preceding examples: It is not always possible to find limits merely by substitution. Sometimes other devices must be employed. The next theorem concerns three functions f, h, and g, where $h(x)$ is always "sandwiched" between $f(x)$ and $g(x)$. If f and g have a common limit L as x approaches a, then as stated below, h must have the same limit.

THE SANDWICH THEOREM (2.22)

> Suppose $f(x) \le h(x) \le g(x)$ for every x in an open interval containing a, except possibly at a.
> If $\lim\limits_{x \to a} f(x) = L = \lim\limits_{x \to a} g(x)$, then $\lim\limits_{x \to a} h(x) = L$.

FIGURE 2.33

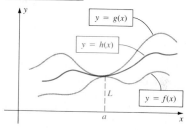

A proof of the Sandwich Theorem based on the definition of limit may be found in Appendix II. The result is also clear from geometric considerations. Specifically, if $f(x) \le h(x) \le g(x)$ for every x in an open interval containing x, then the graph of h lies "between" the graphs of f and g in that interval, as illustrated in Figure 2.33. If f and g have the same limit L as x approaches a, then evidently, h also has the limit L.

EXAMPLE 10 Use the Sandwich Theorem to prove that

$$\lim_{x \to 0} x^2 \sin \frac{1}{x} = 0$$

SOLUTION Since $-1 \le \sin t \le 1$ for every real number t, we may write

$$-1 \le \sin \frac{1}{x} \le 1$$

for every $x \ne 0$. Multiplying by x^2 (which is positive if $x \ne 0$), we obtain

$$-x^2 \le x^2 \sin \frac{1}{x} \le x^2.$$

Since $\lim_{x \to 0} (-x^2) = 0$ and $\lim_{x \to 0} x^2 = 0,$

it follows from the Sandwich Theorem (2.22), with $f(x) = -x^2$ and $g(x) = x^2$, that

$$\lim_{x \to 0} x^2 \sin \frac{1}{x} = 0 \quad \bullet$$

Theorems similar to all of the limit theorems stated in this section can be proved for one-sided limits. For example,

$$\lim_{x \to a^+} [f(x) + g(x)] = \lim_{x \to a^+} f(x) + \lim_{x \to a^+} g(x)$$

and

$$\lim_{x \to a^+} \sqrt[n]{f(x)} = \sqrt[n]{\lim_{x \to a^+} f(x)}$$

with the usual restrictions on the existence of limits and nth roots. Analogous results are true for left-hand limits.

EXAMPLE 11 Find $\lim_{x \to 2^+} (1 + \sqrt{x - 2})$

SOLUTION Using (one-sided) limit theorems,

$$\lim_{x \to 2^+} (1 + \sqrt{x - 2}) = \lim_{x \to 2^+} 1 + \lim_{x \to 2^+} \sqrt{x - 2}$$

$$= 1 + \sqrt{\lim_{x \to 2^+} (x - 2)}$$

$$= 1 + 0 = 1$$

FIGURE 2.34

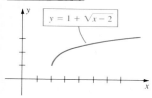

The graph of $f(x) = 1 + \sqrt{x - 2}$ is sketched in Figure 2.34. Note that there is no left-hand limit, since $\sqrt{x - 2}$ is not a real number if $x < 2$. $\bullet$

EXAMPLE 12 In the Theory of Relativity, the *Lorentz contraction formula*

$$L = L_0 \sqrt{1 - \frac{v^2}{c^2}}$$

gives the relationship between the length L of an object that is moving at a velocity v with respect to an observer, and its length L_0 at rest, where c is the speed of light. The formula implies that the measured length of the object is shorter when it is moving than when it is at rest. Find and interpret $\lim_{v \to c^-} L$, and explain why a left-hand limit is necessary.

SOLUTION Using (one-sided) limit theorems,

$$\lim_{v \to c^-} L = \lim_{v \to c^-} L_0 \sqrt{1 - \frac{v^2}{c^2}}$$

$$= L_0 \lim_{v \to c^-} \sqrt{1 - \frac{v^2}{c^2}}$$

$$= L_0 \sqrt{\lim_{v \to c^-} \left(1 - \frac{v^2}{c^2}\right)}$$

$$= L_0 \sqrt{0} = 0$$

Thus, if the velocity of an object could approach the speed of light, then its length, as measured by an observer at rest, would approach zero. This result is sometimes used to justify the theory that the speed of light (approximately 3.0×10^8 m/sec, or 186,000 mi/sec) is the ultimate speed in the universe; that is, no object can have a velocity that is greater than or equal to c.

A left-hand limit is necessary because if $v > c$, then $\sqrt{1 - (v^2/c^2)}$ is not a real number. •

EXERCISES 2.4

Exer. 1–40: Find the limit, if it exists.

1 $\lim_{x \to -2} (3x^3 - 2x + 7)$

2 $\lim_{x \to 4} (5x^2 - 9x - 8)$

3 $\lim_{x \to \sqrt{2}} (x^2 + 3)(x - 4)$

4 $\lim_{t \to -3} (3t + 4)(7t - 9)$

5 $\lim_{x \to 4} \sqrt[3]{x^2 - 5x - 4}$

6 $\lim_{x \to -2} \sqrt{x^4 - 4x + 1}$

7 $\lim_{x \to 7} 0$

8 $\lim_{x \to 1/2} \dfrac{4x^2 - 6x + 3}{16x^3 + 8x - 7}$

9 $\lim_{x \to \sqrt{2}} 15$

10 $\lim_{x \to 15} \sqrt{2}$

11 $\lim_{x \to 1/2} \dfrac{2x^2 + 5x - 3}{6x^2 - 7x + 2}$

12 $\lim_{x \to -3} \dfrac{x + 3}{(1/x) + (1/3)}$

13 $\lim_{x \to 2} \dfrac{x - 2}{x^3 - 8}$

14 $\lim_{x \to 2} \dfrac{x^2 - x - 2}{(x - 2)^2}$

15 $\lim_{x \to 16} \dfrac{x - 16}{\sqrt{x} - 4}$

16 $\lim_{x \to -2} \dfrac{x^3 + 8}{x^4 - 16}$

17 $\lim_{s \to 4} \dfrac{6s - 1}{2s - 9}$

18 $\lim_{x \to \pi} (x - 3.1416)$

19 $\lim_{x \to 1} \left(\dfrac{x^2}{x - 1} - \dfrac{1}{x - 1} \right)$

20 $\lim_{x \to 1} \left(\sqrt{x} + \dfrac{1}{\sqrt{x}} \right)^6$

21 $\lim_{x \to 16} \dfrac{2\sqrt{x} + x^{3/2}}{\sqrt[4]{x} + 5}$

22 $\lim_{x \to -8} \dfrac{16x^{2/3}}{4 - x^{4/3}}$

23 $\lim_{x \to 3} \sqrt[3]{\dfrac{2 + 5x - 3x^3}{x^2 - 1}}$

24 $\lim_{x \to \pi} \sqrt[5]{\dfrac{x - \pi}{x + \pi}}$

25 $\lim_{h \to 0} \dfrac{4 - \sqrt{16 + h}}{h}$

26 $\lim_{h \to 0} \left(\dfrac{1}{h} \right) \left(\dfrac{1}{\sqrt{1 + h}} - 1 \right)$

27 $\lim_{x \to 1} \dfrac{(x - 1)^5}{x^5 - 1}$

28 $\lim_{x \to 6} (x + 4)^3 (x - 6)^2$

29 $\lim_{v \to 3} v^2 (3v - 4)(9 - v^3)$

30 $\lim_{k \to 2} \sqrt{3k^2 + 4} \sqrt[3]{3k + 2}$

31 (a) $\lim_{x \to 5^-} \sqrt{5 - x}$

(b) $\lim_{x \to 5^+} \sqrt{5 - x}$

(c) $\lim_{x \to 5} \sqrt{5 - x}$

32 (a) $\lim_{x \to 2^-} \sqrt{8 - x^3}$

(b) $\lim_{x \to 2^+} \sqrt{8 - x^3}$

(c) $\lim_{x \to 2} \sqrt{8 - x^3}$

33 (a) $\lim\limits_{x \to 1^-} \sqrt[3]{x^3 - 1}$

(b) $\lim\limits_{x \to 1^+} \sqrt[3]{x^3 - 1}$

(c) $\lim\limits_{x \to 1} \sqrt[3]{x^3 - 1}$

34 (a) $\lim\limits_{x \to -8^+} x^{2/3}$

(b) $\lim\limits_{x \to -8^-} x^{2/3}$

(c) $\lim\limits_{x \to -8} x^{2/3}$

35 $\lim\limits_{x \to 5^+} (\sqrt{x^2 - 25} + 3)$

36 $\lim\limits_{x \to 3^-} x\sqrt{9 - x^2}$

37 $\lim\limits_{x \to 3^+} \dfrac{\sqrt{(x - 3)^2}}{x - 3}$

38 $\lim\limits_{x \to -10^-} \dfrac{x + 10}{\sqrt{(x + 10)^2}}$

39 $\lim\limits_{x \to 5^+} \dfrac{1 + \sqrt{2x - 10}}{x + 3}$

40 $\lim\limits_{x \to 4^+} \dfrac{\sqrt[4]{x^2 - 16}}{x + 4}$

Exer. 41–43: n denotes an arbitrary integer. For each function f, sketch the graph of f and find $\lim\limits_{x \to n^-} f(x)$ and $\lim\limits_{x \to n^+} f(x)$.

41 $f(x) = (-1)^n$ if $n \le x < n + 1$

42 $f(x) = \begin{cases} 0 & \text{if } x = n \\ 1 & \text{if } x \ne n \end{cases}$

43 $f(x) = \begin{cases} x & \text{if } x = n \\ 0 & \text{if } x \ne n \end{cases}$

Exer. 44–45: $[\![\,]\!]$ denotes the greatest integer function and n is an arbitrary integer.

44 Find $\lim\limits_{x \to n^-} [\![x]\!]$ and $\lim\limits_{x \to n^+} [\![x]\!]$.

45 If $f(x) = x - [\![x]\!]$, find $\lim\limits_{x \to n^-} f(x)$ and $\lim\limits_{x \to n^+} f(x)$.

46 If $\lim\limits_{x \to a} f(x) = L \ne 0$ and $\lim\limits_{x \to a} g(x) = 0$, prove that $\lim\limits_{x \to a} [f(x)/g(x)]$ does not exist. (*Hint:* Assume there is a number M such that $\lim\limits_{x \to a} [f(x)/g(x)] = M$ and consider

$$\lim\limits_{x \to a} f(x) = \lim\limits_{x \to a} \left[g(x) \cdot \dfrac{f(x)}{g(x)} \right]$$

47 Use the Sandwich Theorem and the fact that $\lim\limits_{x \to 0} (|x| + 1) = 1$ to prove that $\lim\limits_{x \to 0} (x^2 + 1) = 1$.

48 Use the Sandwich Theorem with $f(x) = 0$ and $g(x) = |x|$ to prove that

$$\lim\limits_{x \to 0} \dfrac{|x|}{\sqrt{x^4 + 4x^2 + 7}} = 0$$

49 If c is a nonnegative real number and $0 \le f(x) \le c$ for every x, prove that $\lim\limits_{x \to 0} x^2 f(x) = 0$.

50 Prove that $\lim\limits_{x \to 0} x^4 \sin(1/\sqrt[3]{x}) = 0$. (*Hint:* See Example 10.)

51 Charles' law for gases states that if the pressure remains constant, then the relationship between the volume V that a gas occupies and its temperature T (in °C) is given

by $V = V_0(1 + \frac{1}{273}T)$. The temperature $T = -273$ °C is *absolute zero.*

(a) Find $\lim\limits_{T \to -273^+} V$.

(b) Why is a right-hand limit necessary?

52 According to the Theory of Relativity, the length of an object depends on its velocity v (see Example 12). Einstein also proved that the mass m of an object is related to v by the formula $m = m_0/\sqrt{1 - (v^2/c^2)}$, where m_0 is the mass of the object at rest. Investigate $\lim\limits_{v \to c^-} m$ and use the result to justify that c is the ultimate speed in the universe.

53 A convex lens has focal length f cm. If an object is placed a distance p cm from the lens, the distance q cm of the image from the lens is related to p and f by the *lens equation* $(1/p) + (1/q) = 1/f$. As shown in the figure, p must be greater than f for the rays to converge.

(a) Examine $\lim\limits_{p \to f^+} q$.

(b) What is happening physically to the image as $p \to f^+$?

EXERCISE 53

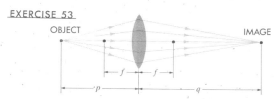

54 Refer to Exercise 53. Shown in the figure for Exercise 54 is a simple magnifier consisting of a convex lens. The object to be magnified is positioned so that its distance p from the lens is less than the focal length f. The linear magnification M is the ratio of the image size to the object size. Using similar triangles, $M = q/p$ for the distance q of the image from the lens.

(a) Find $\lim\limits_{p \to 0^+} M$. Why is a right-hand limit necessary?

(b) Examine $\lim\limits_{p \to f^-} M$. Explain what is happening to the image size.

EXERCISE 54

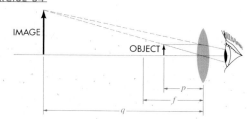

2.5 LIMITS OF TRIGONOMETRIC FUNCTIONS

Whenever we discuss limits of trigonometric expressions involving $\sin t$, $\cos x$, $\tan \theta$, and so on, *we shall assume that each variable represents a real number or the radian measure of an angle.* The following result is important for future developments.

THEOREM (2.23)

$$\lim_{t \to 0} \sin t = 0$$

PROOF Let us first prove that $\lim_{t \to 0^+} \sin t = 0$. Since we are only interested in positive values of t near zero, we may assume that $0 < t < \pi/2$. Let U be the circle of radius 1 with center at the origin of a rectangular coordinate system, and let A be the point $(1, 0)$. If, as illustrated in Figure 2.35, $P(x, y)$ is the point on U such that $\overset{\frown}{AP} = t$, then the radian measure of angle AOP is t. Referring to the figure we see that

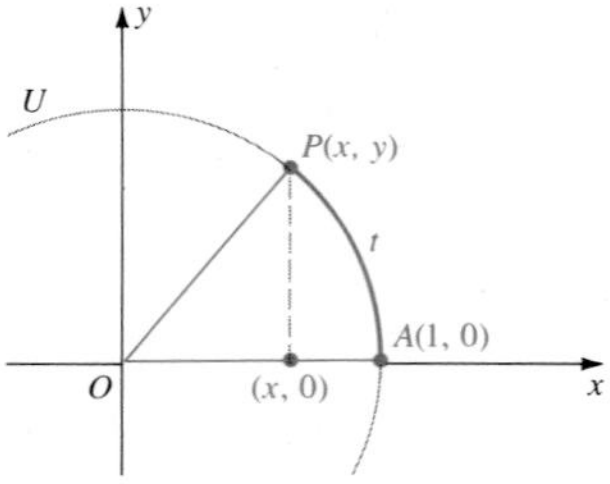

$$0 < y < t$$

or, since $y = \sin t$,

$$0 < \sin t < t.$$

Since $\lim_{t \to 0^+} t = 0$, it follows from the Sandwich Theorem (2.22) that $\lim_{t \to 0^+} \sin t = 0$.

To complete the proof it is sufficient to show that $\lim_{t \to 0^-} \sin t = 0$. If $-\pi/2 < t < 0$, then $0 < -t < \pi/2$ and hence, from the first part of the proof,

$$0 < \sin(-t) < -t.$$

Multiplying the last inequality by -1 and using the trigonometric identity $\sin(-t) = -\sin t$ gives us

$$t < \sin t < 0.$$

Since $\lim_{t \to 0^-} t = 0$, it follows from the Sandwich Theorem that $\lim_{t \to 0^-} \sin t = 0$. • •

FIGURE 2.35

COROLLARY (2.24)

$$\lim_{t \to 0} \cos t = 1$$

PROOF Since $\sin^2 t + \cos^2 t = 1$, it follows that $\cos t = \pm\sqrt{1 - \sin^2 t}$. If $-\pi/2 < t < \pi/2$, then $\cos t$ is positive, and hence $\cos t = \sqrt{1 - \sin^2 t}$. Consequently,

$$\lim_{t \to 0} \cos t = \lim_{t \to 0} \sqrt{1 - \sin^2 t} = \sqrt{\lim_{t \to 0}(1 - \sin^2 t)}$$
$$= \sqrt{1 - 0} = 1$$

• •

For our work in Section 3.4 it will be essential to know the limits of $(\sin t)/t$ and $(1 - \cos t)/t$ as t approaches 0. These are established in Theorems (2.26) and (2.27). In the proof of (2.26) we shall make use of the following.

THEOREM (2.25)

If θ is the radian measure of a central angle of a circle of radius r, then the area A of the sector determined by θ is

$$A = \tfrac{1}{2}r^2\theta$$

FIGURE 2.36

PROOF A typical central angle θ and the sector it determines are shown in Figure 2.36. The area of the sector is directly proportional to θ, that is,

$$A = k\theta$$

for some real number k. For example, the area determined by an angle of 2 radians is twice the area determined by an angle of 1 radian. In particular, if $\theta = 2\pi$, then the sector is the entire circle, and $A = \pi r^2$. Thus

$$\pi r^2 = k(2\pi) \quad \text{or} \quad k = \tfrac{1}{2}r^2$$

and therefore

$$A = \tfrac{1}{2}r^2\theta \qquad \bullet\ \bullet$$

THEOREM (2.26)

$$\lim_{t \to 0} \frac{\sin t}{t} = 1$$

FIGURE 2.37

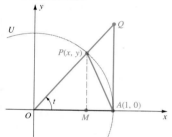

PROOF If $0 < t < \pi/2$ we have the situation illustrated in Figure 2.37, where $P(x, y)$ is the point on the unit circle U such that $AP = t$ and M is the point $(x, 0)$. If A_1 is the area of triangle AOP, A_2 the area of circular sector AOP, and A_3 the area of triangle AOQ, then

$$A_1 < A_2 < A_3.$$

Using the formula for the area of a triangle and Theorem (2.25), we obtain

$$A_1 = \tfrac{1}{2}(1)d(M, P) = \tfrac{1}{2}y = \tfrac{1}{2}\sin t$$

$$A_2 = \tfrac{1}{2}(1)^2 t = \tfrac{1}{2}t$$

$$A_3 = \tfrac{1}{2}(1)d(A, Q) = \tfrac{1}{2}\tan t.$$

Thus

$$\tfrac{1}{2}\sin t < \tfrac{1}{2}t < \tfrac{1}{2}\tan t.$$

Using the identity $\tan t = (\sin t)/(\cos t)$ and then dividing by $\tfrac{1}{2}\sin t$ gives us

$$1 < \frac{t}{\sin t} < \frac{1}{\cos t}$$

or, equivalently,

(*)
$$1 > \frac{\sin t}{t} > \cos t.$$

If $-\pi/2 < t < 0$, then $0 < -t < \pi/2$, and from the result just established,

$$1 > \frac{\sin(-t)}{-t} > \cos(-t).$$

Since $\sin(-t) = -\sin t$ and $\cos(-t) = \cos t$, this inequality reduces to (*). This shows that (*) is also true if $-\pi/2 < t < 0$, and hence is true for every t in the open interval $(-\pi/2, \pi/2)$ except $t = 0$. Since $\lim_{t \to 0} \cos t = 1$, and $(\sin t)/t$ is always between $\cos t$ and 1, it follows from the Sandwich Theorem that

$$\lim_{t \to 0} \frac{\sin t}{t} = 1 \qquad \bullet\ \bullet$$

Roughly speaking, Theorem (2.26) implies that if t is close to 0, then $(\sin t)/t$ is close to 1. Another way of stating this is to write $\sin t \approx t$ for small values of t. It is important to remember that if t denotes an angle, then *radian measure must be used in Theorem (2.26)* and in the approximation formula $\sin t \approx t$. To illustrate, trigonometric tables or a calculator show that to five decimal places:

$$\sin (0.06) \approx 0.05996$$

$$\sin (0.05) \approx 0.04998$$

$$\sin (0.04) \approx 0.03999$$

$$\sin (0.03) \approx 0.03000$$

THEOREM (2.27)

$$\lim_{t \to 0} \frac{1 - \cos t}{t} = 0$$

PROOF We may change the form of $(1 - \cos t)/t$ as follows:

$$\frac{1 - \cos t}{t} = \frac{1 - \cos t}{t} \cdot \frac{1 + \cos t}{1 + \cos t}$$

$$= \frac{1 - \cos^2 t}{t(1 + \cos t)}$$

$$= \frac{\sin^2 t}{t(1 + \cos t)} = \frac{\sin t}{t} \cdot \frac{\sin t}{1 + \cos t}$$

Consequently,

$$\lim_{t \to 0} \frac{1 - \cos t}{t} = \lim_{t \to 0} \left(\frac{\sin t}{t} \cdot \frac{\sin t}{1 + \cos t} \right)$$

$$= \left(\lim_{t \to 0} \frac{\sin t}{t} \right) \left(\lim_{t \to 0} \frac{\sin t}{1 + \cos t} \right)$$

$$= 1 \cdot \left(\frac{0}{1 + 1} \right) = 1 \cdot 0 = 0 \qquad \bullet \ \bullet$$

EXAMPLE 1 Find $\lim\limits_{x \to 0} \dfrac{\sin 5x}{2x}$.

SOLUTION We cannot apply Theorem (2.26) directly, since the given expression is not in the form $(\sin t)/t$. However, we may introduce this form (with $t = 5x$) by using the following algebraic manipulation:

$$\lim_{x \to 0} \frac{\sin 5x}{2x} = \lim_{x \to 0} \frac{1}{2} \frac{\sin 5x}{x}$$

$$= \lim_{x \to 0} \frac{5}{2} \frac{\sin 5x}{5x}$$

$$= \frac{5}{2} \lim_{x \to 0} \frac{\sin 5x}{5x}$$

It follows from the definition of limit that $x \to 0$ may be replaced by $5x \to 0$. Hence, by Theorem (2.26), with $t = 5x$, we see that

$$\lim_{x \to 0} \frac{\sin 5x}{2x} = \frac{5}{2}(1) = \frac{5}{2} \quad \bullet$$

EXAMPLE 2 Find $\lim\limits_{t \to 0} \dfrac{\tan t}{2t}$.

SOLUTION Using the fact that $\tan t = \sin t/\cos t$,

$$\lim_{t \to 0} \frac{\tan t}{2t} = \lim_{t \to 0} \left(\frac{1}{2} \cdot \frac{\sin t}{t} \cdot \frac{1}{\cos t} \right)$$

$$= \frac{1}{2} \cdot 1 \cdot 1 = \frac{1}{2} \quad \bullet$$

EXAMPLE 3 Find $\lim\limits_{x \to 0} \dfrac{2x + 1 - \cos x}{3x}$.

SOLUTION We plan to use Theorem (2.27). With this in mind we begin by isolating the part of the quotient that involves $(1 - \cos x)/x$ and then proceed as follows.

$$\lim_{x \to 0} \frac{2x + 1 - \cos x}{3x} = \lim_{x \to 0} \left(\frac{2x}{3x} + \frac{1 - \cos x}{3x} \right)$$

$$= \lim_{x \to 0} \left(\frac{2x}{3x} \right) + \lim_{x \to 0} \frac{1}{3} \left(\frac{1 - \cos x}{x} \right)$$

$$= \lim_{x \to 0} \frac{2}{3} + \frac{1}{3} \lim_{x \to 0} \frac{1 - \cos x}{x}$$

$$= \frac{2}{3} + \frac{1}{3} \cdot 0 = \frac{2}{3} \quad \bullet$$

EXERCISES 2.5

Exer. 1–26: Find the limit, if it exists.

1 $\lim\limits_{x \to 0} \dfrac{x}{\sin x}$

2 $\lim\limits_{x \to 0} \dfrac{\sin x}{\sqrt[3]{x}}$

3 $\lim\limits_{t \to 0} \dfrac{\sin^3 t}{(2t)^3}$

4 $\lim\limits_{\theta \to 0} \dfrac{30 + \sin \theta}{\theta}$

5 $\lim\limits_{x \to 0} \dfrac{2 + \sin x}{3 + x}$

6 $\lim\limits_{t \to 0} \dfrac{1 - \cos 3t}{t}$

7 $\lim\limits_{\theta \to 0} \dfrac{2 \cos \theta - 2}{30}$

8 $\lim\limits_{x \to 0} \dfrac{x^2 + 1}{x + \cos x}$

9 $\lim\limits_{x \to 0} \dfrac{\sin (-3x)}{4x}$

10 $\lim\limits_{x \to 0} \dfrac{x \sin x}{x^2 + 1}$

11 $\lim\limits_{x \to 0} \dfrac{1 - \cos x}{x^{2/3}}$

12 $\lim\limits_{x \to 0} \dfrac{1 - 2x^2 - 2 \cos x + \cos^2 x}{x^2}$

13 $\lim\limits_{t \to 0} \dfrac{4t^2 + 3t \sin t}{t^2}$

14 $\lim\limits_{x \to 0} \dfrac{x \cos x - x^2}{2x}$

15 $\lim\limits_{t \to 0} \dfrac{\cos t}{1 - \sin t}$

16 $\lim\limits_{t \to 0} \dfrac{\sin t}{1 + \cos t}$

17 $\lim\limits_{t \to 0} \dfrac{1 - \cos t}{\sin t}$

18 $\lim\limits_{x \to 0} \dfrac{\sin \frac{1}{2}x}{x}$

19 $\lim\limits_{x \to 0} \dfrac{x + \tan x}{\sin x}$

20 $\lim\limits_{t \to 0} \dfrac{\sin^2 2t}{t^2}$

21 $\lim\limits_{x \to 0} x \cot x$

22 $\lim\limits_{x \to 0} \dfrac{\csc 2x}{\cot x}$

23 $\lim\limits_{\alpha \to 0} \alpha^2 \csc^2 \alpha$

24 $\lim\limits_{x \to 0} \dfrac{\sin 3x}{\sin 5x}$

25 $\lim\limits_{v \to 0} \dfrac{\cos (v + \frac{1}{2}\pi)}{v}$

26 $\lim\limits_{x \to 0} \dfrac{\sin^2 \frac{1}{2}x}{\sin x}$

Exer. 27–30: Establish the limit for all nonzero real numbers a and b.

27 $\displaystyle\lim_{x\to 0} \frac{\sin ax}{bx} = \frac{a}{b}$

28 $\displaystyle\lim_{x\to 0} \frac{1 - \cos ax}{bx} = 0$

29 $\displaystyle\lim_{x\to 0} \frac{\sin ax}{\sin bx} = \frac{a}{b}$

30 $\displaystyle\lim_{x\to 0} \frac{\cos ax}{\cos bx} = 1$

2.6 CONTINUOUS FUNCTIONS

In the definition of $\lim_{x\to a} f(x)$ we emphasized the restriction $x \neq a$. Several examples in preceding sections have brought out the fact that $\lim_{x\to a} f(x)$ may exist even though f is undefined at a. Let us now turn our attention to the case in which a is in the domain of f. If f is defined at a and $\lim_{x\to a} f(x)$ exists, then this limit may, or may not, equal $f(a)$. If $\lim_{x\to a} f(x) = f(a)$ then f is *continuous* at a according to the next definition.

DEFINITION (2.28)

> A function f is **continuous** at a number a if the following three conditions are satisfied:
>
> (i) f is defined on an open interval containing a.
>
> (ii) $\lim_{x\to a} f(x)$ exists.
>
> (iii) $\lim_{x\to a} f(x) = f(a)$.

If f is not continuous at a, then we say it is **discontinuous** at a, or has a **discontinuity** at a.

If f is continuous at a, then by Definition (2.28) (i) there is a point $(a, f(a))$ on the graph of f. Moreover, since $\lim_{x\to a} f(x) = f(a)$, the closer x is to a, the closer $f(x)$ is to $f(a)$ or, in geometric terms, the closer the point $(x, f(x))$ on the graph of f is to the point $(a, f(a))$ (see Figure 2.38).

Functions that are continuous at every number in a given interval are sometimes thought of as functions whose graphs can be sketched without lifting the pencil from the paper; that is, there are no breaks in the graph. Another interpretation of a continuous function f is that a small change in x produces only a small change in the function value $f(x)$. These are not accurate descriptions, but rather devices to help develop an intuitive feeling for continuous functions.

FIGURE 2.38

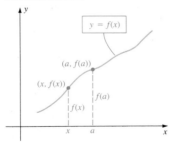

EXAMPLE 1

(a) Prove that the sine function is continuous at 0.

(b) Prove that a polynomial function is continuous at every real number a.

(c) Prove that a rational function is continuous at every real number in its domain.

SOLUTION

(a) We must verify the three conditions of Definition (2.28). Since $\sin x$ is defined for every x, it is defined in an open interval containing 0, and consequently (i) of (2.28) is fulfilled. Moreover, by Theorem (2.23),

$$\lim_{x\to 0} \sin x = 0$$

and hence condition (ii) of (2.28) holds. Finally, since $\sin 0 = 0$, we have

$$\lim_{x \to 0} \sin x = \sin 0$$

and therefore condition (iii) is true. This completes the proof.

(b) A polynomial function f is defined throughout $\mathbb{R}$. Moreover, by Theorem (2.18), $\lim_{x \to a} f(x) = f(a)$ for every real number a. Thus f satisfies conditions (i)–(iii) of Definition (2.28) and hence is continuous at a.

(c) If q is a rational function, then $q = f/h$ for polynomial functions f and h. Consequently q is defined for all real numbers *except* the zeros of h. It follows that if $h(a) \neq 0$, then q is defined throughout an open interval containing a. Moreover, by (2.19), $\lim_{x \to a} q(x) = q(a)$. Applying Definition (2.28), q is continuous at a. •

Graphs of several functions that are *not* continuous at a real number a are sketched in Figure 2.39, where we have also indicated the special names that are associated with these discontinuities.

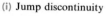
FIGURE 2.39

(i) Jump discontinuity

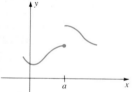

(ii) Infinite discontinuity

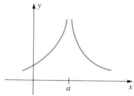

(iii) Removable discontinuity

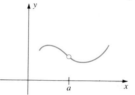

(iv) Removable discontinuity

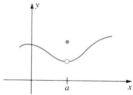

In the case of the jump discontinuity (i), the right-hand and left-hand limits as x approaches a exist, but are unequal, and hence, $\lim_{x \to a} f(x)$ does not exist, as required by (ii) of Definition (2.28).

For the infinite discontinuity (ii), *none* of the three conditions in Definition (2.28) holds. We will have more to say about infinite discontinuities in Section 4.6.

The removable discontinuities in (iii) and (iv) are similar, since $\lim_{x \to a} f(x)$ exists in each case. In (iii), $\lim_{x \to a} f(x) \neq f(a)$, since $f(a)$ does not exist; however, in (iv), $\lim_{x \to a} f(x) \neq f(a)$ even though $f(a)$ *does* exist. If, for a function f with a removable discontinuity at a, we *define* (or *redefine*) $f(a)$ as the number $\lim_{x \to a} f(x)$, then the resulting (new) function *is* continuous at a. To illustrate, if, in Example 1 of Section 2.2 (see Figure 2.14) we *define* the function f such that

$$f(x) = \begin{cases} \dfrac{x - 9}{\sqrt{x} - 3} & \text{if } x \neq 9 \\ 6 & \text{if } x = 9 \end{cases}$$

then the resulting function f is continuous at $a = 9$. In this sense we have "removed" the discontinuity of the original function at $a = 9$.

The functions whose graphs are sketched in Figure 2.39 appear to be continuous at numbers other than a. Most functions considered in calculus are of this type; that is, they may be discontinuous at certain numbers of their domains and continuous elsewhere.

If a function f is continuous at every number in an open interval (a, b), we say that **f is continuous on the interval (a, b).** Similarly, a function is continuous on an infinite interval of the form (a, ∞) or $(-\infty, b)$ if it is continuous at every number in the interval. The next definition covers the case of a closed interval.

DEFINITION (2.29)

> Let a function f be defined on a closed interval $[a, b]$. The **function f is continuous on $[a, b]$** if it is continuous on (a, b) and if, in addition,
> $$\lim_{x \to a^+} f(x) = f(a) \quad \text{and} \quad \lim_{x \to b^-} f(x) = f(b)$$

If a function f has either a right-hand or left-hand limit of the type indicated in Definition (2.29), we say that **f is continuous from the right at a** or that **f is continuous from the left at b,** respectively.

EXAMPLE 2 If $f(x) = \sqrt{9 - x^2}$, sketch the graph of f and prove that f is continuous on the closed interval $[-3, 3]$.

SOLUTION From (1.12), the graph of $x^2 + y^2 = 9$, or equivalently, of $y^2 = 9 - x^2$, is a circle with center at the origin and radius 3. It follows that the graph of $y = \sqrt{9 - x^2}$, and therefore the graph of f, is the upper half of that circle (see Figure 2.40).

If $-3 < c < 3$ then, using Theorem (2.21),

$$\lim_{x \to c} f(x) = \lim_{x \to c} \sqrt{9 - x^2} = \sqrt{9 - c^2} = f(c).$$

Hence, by Definition (2.28), f is continuous at c.

According to Definition (2.29), all that remains is to check the endpoints of the interval using one-sided limits. Since

$$\lim_{x \to -3^+} f(x) = \lim_{x \to -3^+} \sqrt{9 - x^2} = \sqrt{9 - 9} = 0 = f(-3),$$

f is continuous from the right at -3. We also have

$$\lim_{x \to 3^-} f(x) = \lim_{x \to 3^-} \sqrt{9 - x^2} = \sqrt{9 - 9} = 0 = f(3)$$

and hence f is continuous from the left at 3. This completes the proof that f is continuous on $[-3, 3]$. •

FIGURE 2.40

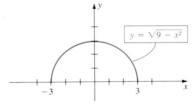
$y = \sqrt{9 - x^2}$

It should be evident how to define continuity on other types of intervals. For example, a function f is continuous on $[a, b)$ or $[a, \infty)$ if it is continuous at every number greater than a in the interval and if, in addition, f is continuous from the right at a. For intervals of the form $(a, b]$ or $(-\infty, b]$ we require continuity at every number less than b in the interval and also continuity from the left at b.

As illustrated in the next example, when asked to discuss the continuity of a function f, we shall list the largest intervals on which f is continuous. Of course, f will also be continuous on any subinterval of those intervals.

EXAMPLE 3 Discuss the continuity of f if $f(x) = \dfrac{\sqrt{x^2 - 9}}{x - 4}$.

SOLUTION The function is undefined if the denominator $x - 4$ is zero (that is, if $x = 4$) or if the radicand $x^2 - 9$ is negative (that is, if $-3 < x < 3$). Any other real number is in one of the intervals $(-\infty, -3]$, $[3, 4)$, or $(4, \infty)$. The proof that f is continuous on each of these intervals is similar to that given in the solution of Example 2. Thus to prove continuity on $[3, 4)$ it is necessary to show that

$$\lim_{x \to c} f(x) = f(c) \quad \text{if } 3 < c < 4$$

and also that

$$\lim_{x \to 3^+} f(x) = f(3).$$

We shall leave the details of the proof for this, and the other intervals, to the reader. •

Limit theorems discussed in Section 2.4 may be used to establish the following theorem.

THEOREM (2.30)

> If the functions f and g are continuous at a, then so are the sum $f + g$, the difference $f - g$, the product fg, and, if $g(a) \neq 0$, the quotient f/g.

PROOF If f and g are continuous at a, then $\lim_{x \to a} f(x) = f(a)$ and $\lim_{x \to a} g(x) = g(a)$. By the definition of sum, $(f + g)(x) = f(x) + g(x)$. Consequently,

$$\lim_{x \to a} (f + g)(x) = \lim_{x \to a} \left[f(x) + g(x) \right]$$

$$= \lim_{x \to a} f(x) + \lim_{x \to a} g(x)$$

$$= f(a) + g(a)$$

$$= (f + g)(a)$$

This proves that $f + g$ is continuous at a. The remainder of the theorem is proved in similar fashion. • •

If f and g are continuous on an interval I it follows that $f + g$, $f - g$, and fg are continuous on I. If, in addition, $g(a) \neq 0$ for every a in I, then f/g is continuous on I. These results may be extended to more than two functions; that is, sums, differences, products, or quotients involving any number of continuous functions are continuous (provided zero denominators do not occur).

A proof of the next result on limits of composite functions is given in Appendix II.

THEOREM (2.31)

> If f and g are functions such that $\lim_{x \to a} g(x) = b$, and if f is continuous at b, then
>
> $$\lim_{x \to a} f(g(x)) = f(b) = f\left(\lim_{x \to a} g(x) \right)$$

The principal use of Theorem (2.31) is to establish other theorems. To illustrate, if n is a positive integer and $f(x) = \sqrt[n]{x}$, then

$$f(g(x)) = \sqrt[n]{g(x)}$$

and

$$f\left(\lim_{x \to a} g(x)\right) = \sqrt[n]{\lim_{x \to a} g(x)}.$$

If we now use the fact that

$$\lim_{x \to a} f(g(x)) = f\left(\lim_{x \to a} g(x)\right),$$

the result stated in Theorem (2.21) is obtained, that is,

$$\lim_{x \to a} \sqrt[n]{g(x)} = \sqrt[n]{\lim_{x \to a} g(x)},$$

provided the indicated nth roots exist.

The next result follows directly from Theorem (2.31).

THEOREM (2.32)

> If g is continuous at a and f is continuous at $b = g(a)$, then
>
> $$\lim_{x \to a} f(g(x)) = f\left(\lim_{x \to a} g(x)\right) = f(g(a))$$

The preceding theorem states that the composite function of f by g is continuous at a. This result may be extended to functions that are continuous on intervals. Sometimes this is expressed by the statement: *The composite function of a continuous function by a continuous function is continuous.*

EXAMPLE 4 If $f(x) = |x|$, prove that f is continuous at every real number a.

SOLUTION Since $|x| = \sqrt{x^2}$, we have, by (2.21) and (2.17),

$$\lim_{x \to a} f(x) = \lim_{x \to a} |x| = \lim_{x \to a} \sqrt{x^2}$$

$$= \sqrt{\lim_{x \to a} x^2} = \sqrt{a^2} = |a| = f(a).$$

Hence, from Definition (2.28), f is continuous at a. •

A proof of the following property of continuous functions may be found in more advanced texts on calculus.

THE INTERMEDIATE VALUE THEOREM (2.33)

> If f is continuous on a closed interval $[a, b]$, and if w is any number between $f(a)$ and $f(b)$, then there is at least one number c in $[a, b]$ such that $f(c) = w$.

Theorem (2.33) states that *as x varies from a to b, the continuous function f takes on every value between $f(a)$ and $f(b)$.* If the graph of the continuous function f is regarded as extending in an unbroken manner from the point

FIGURE 2.41

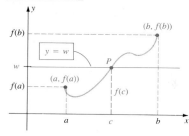

$(a, f(a))$ to the point $(b, f(b))$, as illustrated in Figure 2.41, then for any number w between $f(a)$ and $f(b)$ it appears that a horizontal line with y-intercept w should intersect the graph in at least one point P. The x-coordinate c of P is a number such that $f(c) = w$.

EXAMPLE 5 Verify the Intermediate Value Theorem (2.33) if $f(x) = \sqrt{x + 1}$ and the interval is $[3, 24]$.

SOLUTION The function f is continuous on $[3, 24]$ and $f(3) = 2$, $f(24) = 5$. If w is any real number between 2 and 5 we must find a number c in the interval $[3, 24]$ such that $f(c) = w$, that is, $\sqrt{c + 1} = w$. Squaring both sides of the last equation and solving for c, we obtain $c = w^2 - 1$. This number c is in the interval $[3, 24]$, since if $2 < w < 5$, then

$$4 < w^2 < 25, \quad \text{or} \quad 3 < w^2 - 1 < 24.$$

To check our work we see that

$$f(c) = f(w^2 - 1) = \sqrt{(w^2 - 1) + 1} = w \cdot$$

A corollary of Theorem (2.33) is that *if $f(a)$ and $f(b)$ have opposite signs, then there is a number c between a and b such that $f(c) = 0$; that is, f has a zero at c.* Geometrically, this implies that if the point $(a, f(a))$ on the graph of a continuous function lies below the x-axis, and the point $(b, f(b))$ lies above the x-axis, or vice versa, then the graph crosses the x-axis at some point $(c, 0)$ for $a < c < b$.

A useful consequence of the Intermediate Value Theorem is the following. The interval referred to may be either closed, open, half-open, or infinite.

THEOREM (2.34)

> If a function f is continuous and has no zeros on an interval, then either $f(x) > 0$ or $f(x) < 0$ for every x in the interval.

PROOF The conclusion of the theorem states that under the given hypothesis, $f(x)$ has the same sign throughout the interval. If this conclusion were *false*, then there would exist numbers x_1 and x_2 in the interval such that $f(x_1) > 0$ and $f(x_2) < 0$. By our preceding remarks this, in turn, would imply that $f(c) = 0$ for some number c between x_1 and x_2, contrary to hypothesis. Thus, the conclusion must be true. • •

In Chapter 4 we shall apply Theorem (2.34) to the derivative of a function f to help determine the manner in which $f(x)$ varies on various intervals. As a special case, the following corollary is very useful when considering polynomials. In the statement of the corollary, the phrase *successive solutions* c and d means that there are no other solutions between c and d.

COROLLARY (2.35)

> Let $P(x) = a_n x^n + \cdots + a_1 x + a_0$ be a polynomial. If the real numbers c and d are successive solutions of the equation
>
> $$a_n x^n + \cdots + a_1 x + a_0 = 0,$$
>
> then when x is in the open interval (c, d) either all values of $P(x)$ are positive or all values of $P(x)$ are negative.

The corollary implies that if we choose any number k, such that $c < k < d$, and if the value $P(k)$ of the polynomial is positive, then $P(x)$ is positive for *every* x in (c, d). If $P(k)$ is negative, then $P(x)$ is negative throughout (c, d). We call $P(k)$ a **test value** for the interval (c, d). Test values may also be used on infinite intervals of the form $(-\infty, a)$ or (a, ∞), provided the equation $P(x) = 0$ has no solutions on these intervals. The use of test values is demonstrated in the following example.

EXAMPLE 6 The third-degree Legendre polynomial $P(x) = \frac{1}{2}(5x^3 - 3x)$ occurs in the solution of heat-transfer problems in physics and engineering. Determine where $P(x) > 0$ and where $P(x) < 0$.

SOLUTION Let us begin by finding the solutions of the equation $P(x) = 0$. Writing

$$P(x) = \tfrac{1}{2}x(5x^2 - 3) = 0,$$

we obtain $x = 0$ and $x = \pm\sqrt{3/5} = \pm\sqrt{15}/5 \approx 0.77$. Thus, the successive solutions of $P(x) = 0$ are, in increasing order, $-\sqrt{15}/5$, 0, and $\sqrt{15}/5$. These determine the following four intervals that contain no solutions of $P(x) = 0$:

$$(-\infty, -\sqrt{15}/5), \qquad (-\sqrt{15}/5, 0), \qquad (0, \sqrt{15}/5), \qquad (\sqrt{15}/5, \infty)$$

We next choose a number k in each interval and apply Corollary (2.35). If we choose the number -1 in $(-\infty, -\sqrt{15}/5)$, then the test value is

$$P(-1) = \tfrac{1}{2}(-1)[5(-1)^2 - 3] = -1.$$

Since $P(-1) = -1 < 0$, it follows from the corollary that $P(x) < 0$ for every x in the interval $(-\infty, -\sqrt{15}/5)$.

If we choose $-\frac{1}{2}$ in the interval $(-\sqrt{15}/5, 0)$, then the test value is

$$P(-\tfrac{1}{2}) = \tfrac{1}{2}(-\tfrac{1}{2})[5(\tfrac{1}{4}) - 3] = \tfrac{7}{16} > 0$$

and hence $P(x) > 0$ for every x in $(-\sqrt{15}/5, 0)$.

The remaining intervals are treated in similar fashion. It is convenient to arrange our work in tabular form as follows (check all the entries):

Interval	$(-\infty, -\sqrt{15}/5)$	$(-\sqrt{15}/5, 0)$	$(0, \sqrt{15}/5)$	$(\sqrt{15}/5), \infty)$
k	-1	$-\frac{1}{2}$	$\frac{1}{2}$	1
Test value $P(k)$	-1	$\frac{7}{16}$	$-\frac{7}{16}$	1
Sign of $P(x)$	$-$	$+$	$-$	$+$

Thus $P(x) > 0$ for the intervals $(-\sqrt{15}/5, 0)$ and $(\sqrt{15}/5, \infty)$ and $P(x) < 0$ for the intervals $(-\infty, -\sqrt{15}/5)$ and $(0, \sqrt{15}/5)$. •

EXERCISES 2.6

Exer. 1–12: Show that the function f is continuous at the given number a.

1 $f(x) = \sqrt{2x - 5} + 3x$, $a = 4$

2 $f(x) = 3x^2 + 7 - \dfrac{1}{\sqrt{-x}}$, $a = -2$

3 $f(x) = \dfrac{x}{x^2 - 4}$, $a = 3$

4 $f(x) = 1/x$, $a = 10^{-6}$

5 $f(x) = \sqrt[3]{x^2 + 2}$, $a = -5$

6 $f(x) = \dfrac{\sqrt[3]{x}}{2x + 1}$, $a = 8$

7 $f(x) = \cos x$, $a = 0$ (*Hint:* See Corollary (2.24).)

8 $f(x) = \dfrac{\sin x + \cos x}{x^2 + 3x + 1}$, $a = 0$

(*Hint:* See Example 1 and Exercise 7.)

9 $f(x) = \sec x$, $a = 0$

10 $f(x) = \tan x$, $a = 0$

11 $f(x) = (\cos x)/(1 + \sin x)$, $a = 0$

12 $f(x) = \sqrt{\dfrac{1 - \sin x}{1 + \cos x}}$, $a = 0$

Exer. 13–16: Show that f is continuous on the indicated interval.

13 $f(x) = \sqrt{x - 4}$; $[4, 8]$

14 $f(x) = \sqrt{16 - x}$; $(-\infty, 16]$

15 $f(x) = \dfrac{1}{x^2}$; $(0, \infty)$

16 $f(x) = \dfrac{1}{x - 1}$; $(1, 3)$

Exer. 17–28: Find all numbers for which the function f is continuous.

17 $f(x) = \dfrac{3x - 5}{2x^2 - x - 3}$

18 $f(x) = \dfrac{x^2 - 9}{x - 3}$

19 $f(x) = \sqrt{2x - 3} + x^2$

20 $f(x) = \dfrac{x}{\sqrt[3]{x - 4}}$

21 $f(x) = \dfrac{x - 1}{\sqrt{x^2 - 1}}$

22 $f(x) = \dfrac{x}{\sqrt{1 - x^2}}$

23 $f(x) = \dfrac{|x + 9|}{x + 9}$

24 $f(x) = \dfrac{x}{x^2 + 1}$

25 $f(x) = \dfrac{5}{x^3 - x^2}$

26 $f(x) = \dfrac{4x - 7}{(x + 3)(x^2 + 2x - 8)}$

27 $f(x) = \dfrac{\sqrt{x^2 - 9}\sqrt{25 - x^2}}{x - 4}$

28 $f(x) = \dfrac{\sqrt{9 - x}}{\sqrt{x - 6}}$

29–34 Discuss the discontinuities of the functions defined in Exercises 7–12 of Section 2.2.

35 Suppose $f(x) = \begin{cases} cx^2 - 3 & \text{if } x \le 2 \\ cx + 2 & \text{if } x > 2. \end{cases}$

Find a value of c such that f is continuous on $\mathbb{R}$.

36 Suppose $f(x) = \begin{cases} c^2 x & \text{if } x < 1 \\ 3cx - 2 & \text{if } x \ge 1. \end{cases}$

Determine all values of c such that f is continuous on $\mathbb{R}$.

37 Suppose $f(x) = \begin{cases} c & \text{if } x = -3 \\ \dfrac{9 - x^2}{4 - \sqrt{x^2 + 7}} & \text{if } |x| < 3 \\ d & \text{if } x = 3. \end{cases}$

Find values of c and d such that f is continuous on $[-3, 3]$.

38 Suppose $f(x) = \begin{cases} 4x & \text{if } x \le -1 \\ cx + d & \text{if } -1 < x < 2 \\ -5x & \text{if } x \ge 2. \end{cases}$

Find values of c and d such that f is continuous on $\mathbb{R}$.

39 Suppose

$$f(x) = x^2 \quad \text{and} \quad g(x) = \begin{cases} -4 & \text{if } x \le 0 \\ |x - 4| & \text{if } x > 0. \end{cases}$$

Determine if the composite functions $f \circ g$ and $g \circ f$ are continuous at 0.

40 Let $f(x) = (x - [\![x]\!])^2$ where $[\![\ ]\!]$ denotes the greatest integer function. If n is any integer, show that (a) f is continuous on the interval $[n, n + 1)$, and (b) f is not continuous on $[n, n + 1]$. Sketch the graph of f.

41 Prove that if $f(x) = 1/x$, then f is continuous on every open interval that does not contain the origin. What is true for open intervals containing the origin?

Exer. 42–43: Is f continuous at 3? Explain.

42 $f(x) = \begin{cases} 1 & \text{if } x \ne 3 \\ 0 & \text{if } x = 3 \end{cases}$

43 $f(x) = \begin{cases} \dfrac{|x - 3|}{x - 3} & \text{if } x \ne 3 \\ 1 & \text{if } x = 3 \end{cases}$

Exer. 44–45: Is f continuous at 0? Explain.

44 $f(x) = \begin{cases} \dfrac{\sin x}{x} & \text{if } x \ne 0 \\ 1 & \text{if } x = 0 \end{cases}$

45 $f(x) = \begin{cases} \dfrac{1 - \cos x}{x} & \text{if } x \ne 0 \\ 0 & \text{if } x = 0 \end{cases}$

46 Suppose $f(x) = 0$ if x is rational and $f(x) = 1$ if x is irrational. Prove that f is discontinuous at every real number a.

47 A salesperson receives a base salary of $12,000 plus a $1,000 commission for each $50,000 of sales beyond $100,000. Sketch a graph that shows earnings as a function of sales. Discuss the continuity of the function.

48 The fee charged per car in a parking lot is $1.00 for the first half-hour and $0.50 for each additional half-hour or fraction thereof up to a maximum of $5.00. Find a function f that relates the parking fee to the length of time a car is left in the lot. Sketch the graph of f and discuss the continuity of f.

Exer. 49–52: Verify the Intermediate Value Theorem (2.33) for f on the stated interval $[a, b]$ by showing that if w is any number between $f(a)$ and $f(b)$, then there is a number c in $[a, b]$ such that $f(c) = w$.

49 $f(x) = x^3 + 1$; $[-1, 2]$

50 $f(x) = -x^3$; $[0, 2]$

51 $f(x) = x^2 + 4x + 4$; $[0, 1]$

52 $f(x) = x^2 - x$; $[-1, 3]$

53 If $f(x) = x^3 - 5x^2 + 7x - 9$, use the Intermediate Value Theorem (2.33) to prove that there is a real number a such that $f(a) = 100$.

54 Prove that the equation $x^5 - 3x^4 - 2x^3 - x + 1 = 0$ has a solution between 0 and 1.

55 In models for free-fall, it is usually assumed that the gravitational acceleration g is the constant 9.8 m/sec^2 (or 32 ft/sec^2). Actually, g varies with latitude. If θ is the latitude (in degrees), then a formula that approximates g is

$g = 9.78049(1 + 0.005264 \sin^2 \theta + 0.000024 \sin^4 \theta)$.

Use the Intermediate Value Theorem to show that $g = 9.8$ somewhere between latitudes 35° and 40°.

56 The temperature T (in °C) at which water boils may be approximated by the formula

$$T = 100.862 - 0.0415\sqrt{h + 431.03}$$

where h is the elevation (in meters above sea level). Use the Intermediate Value Theorem to show that water boils at 98 °C at an elevation somewhere between 4000 and 4500 meters.

57 A meteorologist determines that the temperature T (in °F) on a certain cold winter day was given by

$$T = 0.05t(t - 12)(t - 24)$$

for time t (in hours) such that $t = 0$ corresponds to 6 A.M.

(a) Use Corollary (2.35) to determine when T was above 0° and when T was below 0°.

(b) Show that the temperature was 32 °F at some time between 12 noon and 1 P.M. (*Hint:* Use the Intermediate Value Theorem.)

58 The Chebyshev polynomial $T(x) = 8x^4 - 8x^2 + 1$ occurs in statistical studies. Use Corollary (2.35) to determine where $T(x) > 0$. (*Hint:* Let $z = x^2$ and use the Quadratic Formula.)

Exer. 59–60: Use Corollary (2.35) to find all values of x such that (a) $f(x) > 0$; (b) $f(x) < 0$.

59 $f(x) = x^4 - 4x^3 + 3x^2$

60 $f(x) = x(x + 1)^2(x - 3)(x - 5)$

2.7 REVIEW

Define or discuss each of the following.

1 Tangent line to a graph

2 Velocity for rectilinear motion

3 Definition of limit of a function

4 Geometric interpretations of $\lim\limits_{x \to a} f(x) = L$

5 Right-hand and left-hand limits

6 Theorems on limits

7 Limits of polynomial and rational functions

8 The Sandwich Theorem

9 Limits of trigonometric functions

10 Continuous function

11 Types of discontinuities of a function

12 The Intermediate Value Theorem

13 Continuity on an interval

EXERCISES 2.7

Exer. 1–26: Find the limit, if it exists.

1 $\lim\limits_{x \to 3} \dfrac{5x + 11}{\sqrt{x + 1}}$

2 $\lim\limits_{x \to -2} \dfrac{6 - 7x}{(3 + 2x)^4}$

3 $\lim\limits_{x \to -2} (2x - \sqrt{4x^2 + x})$

4 $\lim\limits_{x \to 4^-} (x - \sqrt{16 - x^2})$

5 $\lim\limits_{x \to 3/2} \dfrac{2x^2 + x - 6}{4x^2 - 4x - 3}$

6 $\lim\limits_{x \to 2} \dfrac{3x^2 - x - 10}{x^2 - x - 2}$

7 $\lim\limits_{x \to 2} \dfrac{x^4 - 16}{x^2 - x - 2}$

8 $\lim\limits_{x \to 3^+} \dfrac{1}{x - 3}$

9 $\lim\limits_{x \to 0^+} \dfrac{1}{\sqrt{x}}$

10 $\lim\limits_{x \to 5} \dfrac{(1/x) - (1/5)}{x - 5}$

11 $\lim\limits_{x \to 1/2} \dfrac{8x^3 - 1}{2x - 1}$

12 $\lim\limits_{x \to 2} 5$

13 $\lim\limits_{x \to 3^+} \dfrac{3 - x}{|3 - x|}$

14 $\lim\limits_{x \to 2} \dfrac{\sqrt{x} - \sqrt{2}}{x - 2}$

15 $\lim\limits_{h \to 0} \dfrac{(a + h)^4 - a^4}{h}$

16 $\lim\limits_{x \to -3} \sqrt[3]{\dfrac{x + 3}{x^3 + 27}}$

17 $\lim\limits_{h \to 0} \dfrac{(2 + h)^{-3} - 2^{-3}}{h}$

18 $\lim\limits_{x \to 5/2^-} (\sqrt{5 - 2x} - x^2)$

19 $\lim\limits_{x \to 0} \dfrac{x^2}{\sin x}$

20 $\lim\limits_{x \to 0} \dfrac{x^2 + \sin^2 x}{4x^2}$

21 $\lim\limits_{x \to 0} \dfrac{\sin^2 x + \sin 2x}{3x}$

22 $\lim\limits_{x \to 0} \dfrac{2 - \cos x}{1 + \sin x}$

23 $\lim\limits_{x \to 0} \dfrac{2 \cos x + 3x - 2}{5x}$

24 $\lim\limits_{x \to 0} \dfrac{3x + 1 - \cos^2 x}{\sin x}$

25 $\lim\limits_{x \to 0} \dfrac{x \sin x}{1 - \cos x}$

26 $\lim\limits_{x \to 0} \dfrac{\cos x - 1}{2x}$

Exer. 27–32: Sketch the graph of the piecewise-defined function f and, for the indicated value of a, find

(a) $\lim\limits_{x \to a^-} f(x)$ **(b)** $\lim\limits_{x \to a^+} f(x)$ **and** **(c)** $\lim\limits_{x \to a} f(x)$

provided the limit exists.

27 $a = 2;$ $f(x) = \begin{cases} 3x & \text{if } x \le 2 \\ x^2 & \text{if } x > 2 \end{cases}$

28 $a = 2;$ $f(x) = \begin{cases} x^3 & \text{if } x \le 2 \\ 4 - 2x & \text{if } x > 2 \end{cases}$

29 $a = -3;$ $f(x) = \begin{cases} 1/(2 - 3x) & \text{if } x < -3 \\ \sqrt[3]{x + 2} & \text{if } x \ge -3 \end{cases}$

30 $a = -3;$ $f(x) = \begin{cases} 9/x^2 & \text{if } x \le -3 \\ 4 + x & \text{if } x > -3 \end{cases}$

31 $a = 1;$ $f(x) = \begin{cases} x^2 & \text{if } x < 1 \\ 2 & \text{if } x = 1 \\ 4 - x^2 & \text{if } x > 1 \end{cases}$

32 $a = 0;$ $f(x) = \begin{cases} (x^4 + x)/x & \text{if } x \ne 0 \\ 2 & \text{if } x = 0 \end{cases}$

Exer. 33–34: Find the limit if $[\![\ ]\!]$ denotes the greatest integer function.

33 $\lim\limits_{x \to 3^+} ([\![x]\!] - x^2)$

34 $\lim\limits_{x \to 3^-} ([\![x]\!] - x^2)$

35 Prove, directly from the definition of limit (2.10), that
$$\lim\limits_{x \to 6} (5x - 21) = 9.$$

36 Suppose $f(x) = 1$ if x is rational and $f(x) = -1$ if x is irrational. Prove that $\lim\limits_{x \to a} f(x)$ does not exist for any real number a.

Exer. 37–40: Find all numbers for which f is continuous.

37 $f(x) = 2x^4 - \sqrt[3]{x} + 1$

38 $f(x) = \sqrt{(2 + x)(3 - x)}$

39 $f(x) = \dfrac{\sqrt{9 - x^2}}{x^4 - 16}$

40 $f(x) = \dfrac{\sqrt{x}}{x^2 - 1}$

Exer. 41–46: Find the discontinuities of f.

41 $f(x) = \dfrac{|x^2 - 16|}{x^2 - 16}$

42 $f(x) = \dfrac{1}{x^2 - 16}$

43 $f(x) = \dfrac{x^2 - x - 2}{x^2 - 2x}$

44 $f(x) = \dfrac{x + 2}{x^3 - 8}$

Exer. 45–46: Prove that the function f is continuous at 0.

45 $f(x) = \dfrac{4 + \sin x}{x^2 + 2x + 5}$

46 $f(x) = (2x^2 + 3) \cos x$

47 If $f(x) = 1/x^2$, verify the Intermediate Value Theorem (2.33) for f on the interval $[2, 3]$.

48 Use the Sandwich Theorem to prove that
$$\lim\limits_{x \to 0} x^2 \cos \dfrac{1}{x} = 0.$$

49 If $f(x) = 3x^2 - 2x$, find

(a) the slope of the tangent line to the graph of f at the point $(a, f(a))$.

(b) an equation of the tangent line at the point $(3, 21)$.

50 A toy rocket is projected directly upward with a velocity of 160 ft/sec from a launching pad that is 3 feet above ground level. The distance from the ground to the rocket after t seconds is $-16t^2 + 160t + 3$ feet.

(a) Find the velocity v of the rocket after a seconds.

(b) What is the maximum altitude of the rocket above the ground? (*Hint:* The maximum altitude occurs when $v = 0$.)

CHAPTER 3

THE DERIVATIVE

The *derivative of a function* is one of the most powerful tools in mathematics and the applied sciences. In this chapter we shall define the derivative and discuss many properties associated with this important concept.

3.1 DEFINITION OF DERIVATIVE

Suppose a function f is defined on an open interval containing the real number a. The graph of f and a secant line l_{PQ} through $P(a, f(a))$ and $Q(x, f(x))$ are illustrated in Figure 3.1. The dashed line l represents a possible tangent line at P.

FIGURE 3.1   FIGURE 3.2

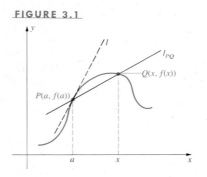

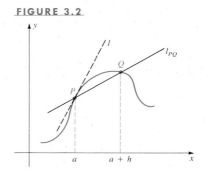

In Section 2.1 we defined the slope m of l as the limiting value of the slope of l_{PQ} as Q approaches P. Thus, by Definition 2.2,

$$m = \lim_{x \to a} \frac{f(x) - f(a)}{x - a}$$

provided the limit exists. If, as illustrated in Figure 3.2, we introduce a new variable h by letting $x = a + h$ (or $h = x - a$), then we obtain the following equivalent formula for m:

$$m = \lim_{h \to 0} \frac{f(a + h) - f(a)}{h}$$

A proof of the equivalence of these two limits may be found in Appendix II. The preceding limit is one of the fundamental concepts of calculus. It is called the *derivative of the function f at a*.

DEFINITION (3.1)

> Let f be a function that is defined on an open interval containing a. The **derivative of f at a**, written $f'(a)$, is
>
> $$f'(a) = \lim_{h \to 0} \frac{f(a + h) - f(a)}{h}$$
>
> provided the limit exists.

As we have observed, the formula for $f'(a)$ may also be written as follows.

ALTERNATIVE DEFINITION (3.1′)

> $$f'(a) = \lim_{x \to a} \frac{f(x) - f(a)}{x - a}$$

The symbol $f'(a)$ is read *f prime of a*. The terminology $f'(a)$ *exists* means that the limit in Definitions (3.1) and (3.1′) exists. If $f'(a)$ exists, we say that the function f **is differentiable at a** or f **has a derivative at a.**

We may use the derivative to restate Definitions (2.2) and (2.5) of Section 2.1 as follows, provided the functions f and s are differentiable at a.

APPLICATIONS OF THE DERIVATIVE (3.2)

> (i) **Tangent line:** The slope of the tangent line to the graph of f at the point $(a, f(a))$ is $f'(a)$.
>
> (ii) **Velocity:** If a point P moves on a coordinate line such that its coordinate at time t is $s(t)$, then the velocity at time a is $s'(a)$.

Additional applications of derivatives will be discussed as we proceed through this text.

A function f **is differentiable on an open interval (a, b)** if it is differentiable at every number c in (a, b). We shall also consider functions that are differentiable on an infinite interval (a, ∞), $(-\infty, a)$, or $(-\infty, \infty)$. For closed intervals we use the following convention, which is analogous to the definition of continuity on a closed interval given in (2.29).

DEFINITION (3.3)

> A function f **is differentiable on a closed interval $[a, b]$** if it is differentiable on the open interval (a, b) and if the following limits exist:
>
> $$\lim_{h \to 0^+} \frac{f(a + h) - f(a)}{h} \quad \text{and} \quad \lim_{h \to 0^-} \frac{f(b + h) - f(b)}{h}$$

FIGURE 3.3

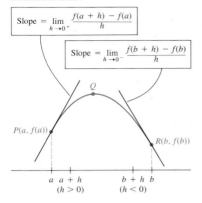

Slope $= \lim\limits_{h \to 0^+} \dfrac{f(a + h) - f(a)}{h}$

Slope $= \lim\limits_{h \to 0^-} \dfrac{f(b + h) - f(b)}{h}$

The one-sided limits in Definition (3.3) are sometimes referred to as the **right-hand** and **left-hand derivatives** of f at a and b, respectively. Note that for the right-hand derivative we have $h \to 0^+$ and $a + h$ approaches a *from the right*. For the left-hand derivative, we have $h \to 0^-$ and $b + h$ approaches b *from the left*.

If f is defined on a closed interval $[a, b]$ and is undefined elsewhere, then the right-hand and left-hand derivatives allow us to define the slopes of the tangent lines at the points $P(a, f(a))$ and $R(b, f(b))$, respectively, as illustrated in Figure 3.3. Thus, for the slope of the tangent line at P we take the limiting value of the slope of the secant line through P and Q as Q approaches P from the right. For the tangent line at R, the point Q approaches R from the left.

Differentiability on an interval of the form $[a, b), [a, \infty), (a, b]$, or $(-\infty, b]$ is defined in the obvious way, using a one-sided limit at an endpoint.

If f is defined on an open interval containing a, then $f'(a)$ *exists if and only if both the right-hand and left-hand derivatives exist at a and are equal.* The functions whose graphs are sketched in Figure 3.4 have right-hand and left-hand derivatives at a that give the slopes of the indicated lines l_1 and l_2, respectively. However, since the slopes of l_1 and l_2 are unequal, $f'(a)$ does not exist. Generally, if the graph of f has a *corner* at the point $P(a, f(a))$, then f is not differentiable at a.

FIGURE 3.4

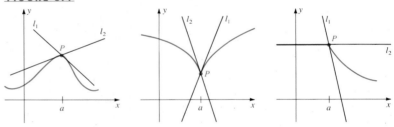

If f is differentiable for every x in some interval, then by associating the number $f'(x)$ with x we obtain a function f' called the **derivative of f**. The value of f' at x is given by the following (or an appropriate one-sided) limit.

THE DERIVATIVE AS A FUNCTION (3.4)

$$f'(x) = \lim\limits_{h \to 0} \frac{f(x + h) - f(x)}{h}$$

Note that in finding $f'(x)$ by means of (3.4), the number x is fixed, but arbitrary, and we take the limit as h approaches zero. The terminology *differentiate $f(x)$* or *find the derivative of $f(x)$* means to find $f'(x)$.

EXAMPLE 1 If $f(x) = 3x^2 - 5x + 4$, find:

(a) $f'(x)$;   (b) the domain of f';   (c) $f'(2)$, $f'(-\sqrt{2})$, and $f'(a)$;

(d) an equation of the tangent line to the graph of f at the point $P(2, f(2))$.

SOLUTION

(a) By (3.4),

$$f'(x) = \lim_{h \to 0} \frac{f(x + h) - f(x)}{h}$$

$$= \lim_{h \to 0} \frac{[3(x + h)^2 - 5(x + h) + 4] - (3x^2 - 5x + 4)}{h}$$

$$= \lim_{h \to 0} \frac{(3x^2 + 6xh + 3h^2 - 5x - 5h + 4) - (3x^2 - 5x + 4)}{h}$$

$$= \lim_{h \to 0} \frac{6xh + 3h^2 - 5h}{h} = \lim_{h \to 0} (6x + 3h - 5)$$

$$= 6x - 5$$

(b) Since $f'(x) = 6x - 5$, the derivative exists for every real number x. Hence the domain of f' is $\mathbb{R}$.

(c) Substituting for x in $f'(x) = 6x - 5$,

$$f'(2) = 6(2) - 5 = 7$$

$$f'(-\sqrt{2}) = 6(-\sqrt{2}) - 5 = -(6\sqrt{2} + 5)$$

$$f'(a) = 6a - 5$$

(d) Since $f(2) = 3(2)^2 - 5(2) + 4 = 12 - 10 + 4 = 6$, the point $P(2, f(2))$ on the graph of f has coordinates $(2, 6)$. By (3.2) (i), the slope of the tangent line at P is $f'(2) = 7$ (see part (c)). Using the Point-Slope Form (1.15), an equation of the tangent line at P is

$$y - 6 = 7(x - 2), \quad \text{or equivalently,} \quad 7x - y - 8 = 0 \quad \bullet$$

EXAMPLE 2 Find $f'(x)$ if $f(x) = \sqrt{x}$. What is the domain of f'?

SOLUTION The domain of f consists of all nonnegative real numbers. We shall examine the cases $x > 0$ and $x = 0$ separately. If $x > 0$, then by (3.4),

$$f'(x) = \lim_{h \to 0} \frac{\sqrt{x + h} - \sqrt{x}}{h}.$$

To find the limit we first rationalize the numerator of the quotient and then simplify:

$$f'(x) = \lim_{h \to 0} \frac{\sqrt{x + h} - \sqrt{x}}{h} \cdot \frac{\sqrt{x + h} + \sqrt{x}}{\sqrt{x + h} + \sqrt{x}}$$

$$= \lim_{h \to 0} \frac{(x + h) - x}{h(\sqrt{x + h} + \sqrt{x})}$$

$$= \lim_{h \to 0} \frac{1}{\sqrt{x + h} + \sqrt{x}}$$

$$= \frac{1}{\sqrt{x} + \sqrt{x}} = \frac{1}{2\sqrt{x}}$$

Since $x = 0$ is an endpoint of the domain of f, we must use a one-sided limit to determine if $f'(0)$ exists. Specifically, if f is differentiable at 0, then using Definition (3.4) with $x = 0$,

$$f'(0) = \lim_{h \to 0^+} \frac{\sqrt{0 + h} - \sqrt{0}}{h}$$

$$= \lim_{h \to 0^+} \frac{\sqrt{h}}{h} = \lim_{h \to 0^+} \frac{1}{\sqrt{h}}.$$

Since the last limit does not exist (see Exercise 46 of Section 2.4), $f'(0)$ does not exist. Hence the domain of f' is the set of positive real numbers.

EXAMPLE 3 If $f(x) = |x|$, show that f is not differentiable at 0.

SOLUTION The graph of f was discussed in Example 8 of Section 1.4 and is resketched in Figure 3.5. It is geometrically evident that f has no derivative at 0, since the graph has a corner at the origin. We can prove that $f'(0)$ does not exist by showing that the right-hand and left-hand derivatives of f at 0 are not equal. Using the limits in Definition (3.3) with $a = 0$ and $b = 0$:

$$\lim_{h \to 0^+} \frac{f(0 + h) - f(0)}{h} = \lim_{h \to 0^+} \frac{|0 + h| - |0|}{h} = \lim_{h \to 0^+} \frac{|h|}{h} = 1$$

$$\lim_{h \to 0^-} \frac{f(0 + h) - f(0)}{h} = \lim_{h \to 0^-} \frac{|0 + h| - |0|}{h} = \lim_{h \to 0^-} \frac{|h|}{h} = -1$$

Thus $f'(0)$ does not exist. •

It follows from Example 3 that the graph of $y = |x|$ does not have a tangent line at the point $P(0, 0)$.

The next example illustrates the use of Alternative Definition (3.1') in finding $f'(a)$.

EXAMPLE 4 If $f(x) = x^{1/3}$ and $a \neq 0$, find $f'(a)$.

SOLUTION Using (3.1'),

$$f'(a) = \lim_{x \to a} \frac{f(x) - f(a)}{x - a}$$

$$= \lim_{x \to a} \frac{x^{1/3} - a^{1/3}}{x - a}$$

provided the limit exists. To investigate the limit it is necessary to change the form of the quotient. One method is to first write

$$\lim_{x \to a} \frac{x^{1/3} - a^{1/3}}{x - a} = \lim_{x \to a} \frac{x^{1/3} - a^{1/3}}{(x^{1/3})^3 - (a^{1/3})^3}.$$

We may factor the denominator by using the formula

$$p^3 - q^3 = (p - q)(p^2 + pq + q^2)$$

with $p = x^{1/3}$ and $q = a^{1/3}$. This gives us

$$\lim_{x \to a} \frac{x^{1/3} - a^{1/3}}{(x^{1/3} - a^{1/3})(x^{2/3} + x^{1/3}a^{1/3} + a^{2/3})}.$$

FIGURE 3.5

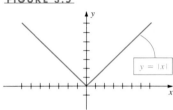

$y = |x|$

Dividing numerator and denominator by $x^{1/3} - a^{1/3}$ and taking the limit, we obtain

$$f'(a) = \lim_{x \to a} \frac{1}{x^{2/3} + x^{1/3}a^{1/3} + a^{2/3}}$$

$$= \frac{1}{a^{2/3} + a^{2/3} + a^{2/3}} = \frac{1}{3a^{2/3}} \, \cdot$$

THEOREM (3.5)

> If a function f is differentiable at a, then f is continuous at a.

PROOF If x is in the domain of f and $x \neq a$, then $f(x)$ may be written as follows:

$$f(x) = f(a) + \frac{f(x) - f(a)}{x - a}(x - a).$$

Employing limit theorems and Definition (3.2),

$$\lim_{x \to a} f(x) = \lim_{x \to a} f(a) + \lim_{x \to a} \frac{f(x) - f(a)}{x - a} \cdot \lim_{x \to a} (x - a)$$

$$= f(a) + f'(a) \cdot 0 = f(a).$$

Thus, by Definition (2.28), f is continuous at a. • •

By using one-sided limits, Theorem (3.5) can be extended to functions that are differentiable on a closed interval.

The converse of Theorem (3.5) is false because *there exist continuous functions that are not differentiable*. To illustrate, if $f(x) = |x|$, then f is continuous at 0; however, as we proved in Example 3, f is not differentiable at 0 (see Figure 3.5).

The following notations for derivatives are used if $y = f(x)$.

NOTATIONS FOR DERIVATIVES (3.6)

$$f'(x) = D_x [f(x)] = D_x y = y' = \frac{dy}{dx} = \frac{d}{dx}[f(x)]$$

All of the above notations are used in mathematics and applications, and it is advisable for you to become familiar with the different forms.

The subscript x in the symbol D_x is employed to designate the independent variable. For example, if the independent variable is t we shall write $f'(t) = D_t [f(t)]$. The symbol D_x or D_t is referred to as a **differential operator.** Standing alone, D_x has no practical significance; however, placing an expression in x to the right of it denotes the derivative. To illustrate, using Example 1,

$$D_x (3x^2 - 5x + 4) = 6x - 5.$$

We say that D_x *operates* on the expression $3x^2 - 5x + 4$. Sometimes $D_x y$ is referred to as **the derivative of y with respect to x.** The symbol d/dx is used in similar fashion, that is,

$$\frac{d}{dx}(3x^2 - 5x + 4) = 6x - 5.$$

As indicated in (3.6), the notations y' and dy/dx also denote the derivative of y with respect to x. We shall justify the notation dy/dx in Section 3.5, where the concept of *differential* is defined.

We shall conclude this section with a specialized application of the derivative.

EXAMPLE 5 In optics, a function f such that $f'(x) > 1$ for every x may be considered as a transformation that magnifies objects. As illustrated in Figure 3.6, an object that extends over the x-interval $[a, a + h]$ is transformed, by f, into an object that extends over the y-interval $[f(a), f(a + h)]$. (Think of a light source to the left of the x-axis that projects a film strip located on the x-axis onto a screen located on the y-axis.) The *magnification* M of f for $[a, a + h]$ is defined as the ratio of the image size to the object size. The value of M may vary, depending on the interval $[a, a + h]$. The *magnification* M_a at $x = a$ is defined as $\lim_{h \to 0} M$.

(a) Express M and M_a in terms of f.

(b) If $f(x) = x^2$, find M_1 and M_2.

FIGURE 3.6

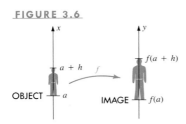

SOLUTION

(a) The size of the object is $(a + h) - a = h$ and the size of the image is $f(a + h) - f(a)$. Hence

$$M = \frac{f(a + h) - f(a)}{h} \quad \text{and} \quad M_a = \lim_{h \to 0} M = f'(a)$$

(b) If $f(x) = x^2$, then from Example 1 of Section 2.1, $f'(a) = 2a$, and hence $M_1 = f'(1) = 2$ and $M_2 = f'(2) = 4$. Note that the magnification at $x = 2$ is twice the magnification at $x = 1$. •

EXERCISES 3.1

Exer. 1–10: (a) Use (3.4) to find $f'(x)$. (b) Find the domain of f'. (c) Find an equation of the tangent line to the graph of f at the point $P(1, f(1))$.

1 $f(x) = 37$

2 $f(x) = 17 - 6x$

3 $f(x) = 9x - 2$

4 $f(x) = 7x^2 - 5$

5 $f(x) = 2 + 8x - 5x^2$

6 $f(x) = x^3 + x$

7 $f(x) = 1/(x - 2)$

8 $f(x) = (1 + \sqrt{3})^2$

9 $f(x) = \sqrt{3x + 1}$

10 $f(x) = 1/(2x)$

Exer. 11–14: Find $D_x y$.

11 $y = 7/\sqrt{x}$

12 $y = (2x + 3)^2$

13 $y = 2x^3 - 4x + 1$

14 $y = x/(3x + 4)$

Exer. 15–20: Find $f'(a)$ by means of (3.1').

15 $f(x) = x^2$

16 $f(x) = \sqrt{2x}$

17 $f(x) = 6/x^2$

18 $f(x) = 8 - x^3$

19 $f(x) = 1/(x + 5)$

20 $f(x) = \sqrt{x}$

Exer. 21–22: Use right-hand and left-hand derivatives to prove that f is not differentiable at $x = 5$.

21 $f(x) = |x - 5|$

22 $f(x) = [\![x]\!]$ (f is the greatest integer function)

Exer. 23–26: Sketch the graph of f and use the graph to find the domain of f'.

23 $f(x) = \begin{cases} 2x & \text{if } x \le 0 \\ x^2 & \text{if } x > 0 \end{cases}$

24 $f(x) = \begin{cases} 2x - 1 & \text{if } x \le 1 \\ x^2 & \text{if } x > 1 \end{cases}$

25 $f(x) = \begin{cases} |x| & \text{if } |x| \le 1 \\ 2 - |x| & \text{if } |x| > 1 \end{cases}$

26 $f(x) = \begin{cases} x - [\![x]\!] & \text{if } n \le x < n + 1 \text{ and} \\ & n \text{ is an even integer} \\ 1 - x + [\![x]\!] & \text{if } n \le x < n + 1 \text{ and} \\ & n \text{ is an odd integer} \end{cases}$

($[\![\ ]\!]$ denotes the greatest integer function)

Exer. 27–28: Each figure is the graph of a function f. Sketch the graph of f'. Where is f *not* differentiable?

27

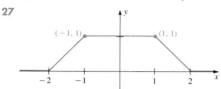

28

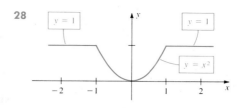

29 Let $f(x) = |x|$. Prove that $f'(x) = 1$ if $x > 0$ and that $f'(x) = -1$ if $x < 0$.

30 Let $f(x) = |x|/x$. Find (a) the domain of f'; (b) $f'(x)$ for every x in the domain of f'.

31 If $f(x)$ is a polynomial of degree 1, prove that $f'(x)$ is a polynomial of degree 0. What if $f(x)$ is a polynomial of degree 2 or 3?

32 Prove that $D_x c = 0$ for every real number c.

3.2 RULES FOR FINDING DERIVATIVES

This section contains some general rules that simplify the task of finding derivatives. In statements of theorems we shall use the differential operator symbol D_x to denote derivatives (see (3.6)). The first result is sometimes phrased: *the derivative of a constant is zero.*

THEOREM (3.7)
$$D_x (c) = 0$$

PROOF Let f be the constant function such that $f(x) = c$ for every x. We wish to show that $f'(x) = 0$. Since every value of f is c, it follows that $f(x + h) = c$ for every h. Applying (3.4),

$$f'(x) = \lim_{h \to 0} \frac{f(x + h) - f(x)}{h} = \lim_{h \to 0} \frac{c - c}{h} = \lim_{h \to 0} 0 = 0$$

The last step follows from Theorem (2.13). • •

THEOREM (3.8)
$$D_x (x) = 1$$

PROOF If we let $f(x) = x$, then

$$f'(x) = \lim_{h \to 0} \frac{f(x + h) - f(x)}{h}$$

$$= \lim_{h \to 0} \frac{(x + h) - x}{h}$$

$$= \lim_{h \to 0} \frac{h}{h} = \lim_{h \to 0} 1 = 1 \quad \bullet \, \bullet$$

In the proof of Theorem (3.10) we shall make use of the following formula.

BINOMIAL THEOREM (3.9)
$$(a + b)^n = a^n + na^{n-1}b + \frac{n(n-1)}{2!}a^{n-2}b^2$$
$$+ \cdots + \binom{n}{r}a^{n-r}b^r + \cdots + nab^{n-1} + b^n$$

for real numbers a and b, any positive integer n, and
$$\binom{n}{r} = \frac{n(n-1)(n-2)\cdots(n-r+1)}{r(r-1)(r-2)\cdots 1} = \frac{n!}{(n-r)!r!}.$$

The Binomial Theorem may be proved by mathematical induction. The special cases $n = 2$, $n = 3$, and $n = 4$ are:
$$(a + b)^2 = a^2 + 2ab + b^2$$
$$(a + b)^3 = a^3 + 3a^2b + 3ab^2 + b^3$$
$$(a + b)^4 = a^4 + 4a^3b + 6a^2b^2 + 4ab^3 + b^4$$

THE POWER RULE (3.10)

> If n is a positive integer, then $D_x(x^n) = nx^{n-1}$.

PROOF Let $f(x) = x^n$. We wish to show that $f'(x) = nx^{n-1}$. By (3.4),
$$f'(x) = \lim_{h \to 0} \frac{f(x+h) - f(x)}{h}$$
$$= \lim_{h \to 0} \frac{(x+h)^n - x^n}{h}$$

provided the limit exists. Employing the Binomial Theorem (3.9) with $a = x$ and $b = h$,
$$(x + h)^n = x^n + nx^{n-1}h + \frac{n(n-1)}{2!}x^{n-2}h^2 + \cdots + nxh^{n-1} + h^n.$$

Every term after the first contains a factor h to some positive integral power. If we subtract x^n and then divide by h we obtain
$$f'(x) = \lim_{h \to 0} \left[nx^{n-1} + \frac{n(n-2)}{2!}x^{n-2}h + \cdots + nxh^{n-2} + h^{n-1} \right].$$

Since each term within the brackets, except the first, contains a power of h, we see that $f'(x) = nx^{n-1}$.

If $x \neq 0$, then (3.10) is also true if $n = 0$, for in this case $f(x) = x^0 = 1$ and by Theorem (3.7), $f'(x) = 0 = 0 \cdot x^{0-1}$. • •

EXAMPLE 1

(a) Find $D_x(x^3)$ and $\dfrac{d}{dx}(x^8)$.   (b) Find $\dfrac{dy}{dx}$ if $y = x^{100}$.

SOLUTION Applying the Power Rule (3.10),

(a) $D_x(x^3) = 3x^2$ and $\dfrac{d}{dx}(x^8) = D_x(x^8) = 8x^7$

(b) $\dfrac{dy}{dx} = D_x\,y = D_x(x^{100}) = 100x^{99}$ •

If symbols other than x are used for the independent variable, then the Power Rule is written in terms of the variable:

$$D_t(t^n) = nt^{n-1}, \qquad D_z(z^n) = nz^{n-1}, \qquad D_v(v^n) = nv^{n-1}.$$

In Section 3.7 we shall prove that the Power Rule is valid for every *rational* number n.

In the statements of Theorems (3.11)–(3.14) we assume that f and g are differentiable at x.

THEOREM (3.11)

$$D_x[cf(x)] = cD_x[f(x)]$$

PROOF If we let $g(x) = cf(x)$, then

$$D_x[cf(x)] = D_x[g(x)]$$

$$= \lim_{h \to 0} \frac{g(x+h) - g(x)}{h}$$

$$= \lim_{h \to 0} \frac{cf(x+h) - cf(x)}{h}$$

$$= \lim_{h \to 0} c\left[\frac{f(x+h) - f(x)}{h}\right]$$

$$= c \lim_{h \to 0} \frac{f(x+h) - f(x)}{h}$$

$$= cf'(x) = cD_x[f(x)] \qquad \bullet\ \bullet$$

For the special case $f(x) = x^n$, Theorems (3.11) and (3.10) give us the next formula, which is true for every real number c and every positive integer n:

$$D_x(cx^n) = (cn)x^{n-1}$$

Thus, *to differentiable cx^n we multiply the coefficient c by the exponent n, and then reduce the exponent by* 1.

EXAMPLE 2

(a) Find $D_x(7x^4)$.   (b) Find $F'(z)$ if $F(z) = -3z^{15}$.

SOLUTION By the remarks preceding this example we have

(a) $D_x(7x^4) = (7 \cdot 4)x^3 = 28x^3$

(b) $F'(z) = D_z(-3z^{15}) = (-3)(15)z^{14} = -45z^{14}$ •

THEOREM (3.12)

$$
\begin{array}{l}
\text{(i)} \ \ D_x \left[f(x) + g(x) \right] = D_x \left[f(x) \right] + D_x \left[g(x) \right] \\[2mm]
\text{(ii)} \ \ D_x \left[f(x) - g(x) \right] = D_x \left[f(x) \right] - D_x \left[g(x) \right]
\end{array}
$$

PROOF To prove (i), let $k(x) = f(x) + g(x)$. We wish to show that $k'(x) = f'(x) + g'(x)$. This may be done as follows:

$$
\begin{aligned}
k'(x) &= \lim_{h \to 0} \frac{k(x + h) - k(x)}{h} \\[2mm]
&= \lim_{h \to 0} \frac{\left[f(x + h) + g(x + h) \right] - \left[f(x) + g(x) \right]}{h} \\[2mm]
&= \lim_{h \to 0} \left[\frac{f(x + h) - f(x)}{h} + \frac{g(x + h) - g(x)}{h} \right] \\[2mm]
&= \lim_{h \to 0} \frac{f(x + h) - f(x)}{h} + \lim_{h \to 0} \frac{g(x + h) - g(x)}{h} \\[2mm]
&= f'(x) + g'(x)
\end{aligned}
$$

A similar argument may be used to prove part (ii). • •

Theorem (3.12) (i), which states that *the derivative of a sum is the sum of the derivatives*, can be extended to sums of any number of functions.

Since a polynomial is a sum of terms of the form cx^n for a real number c and a nonnegative integer n, we may use results on sums and differences to obtain the derivative, as illustrated in the next example.

EXAMPLE 3 Find $f'(x)$ if $f(x) = 2x^4 - 5x^3 + x^2 - 4x + 1$.

SOLUTION

$$
\begin{aligned}
f'(x) &= D_x \left(2x^4 - 5x^3 + x^2 - 4x + 1 \right) \\
&= D_x \left(2x^4 \right) - D_x \left(5x^3 \right) + D_x \left(x^2 \right) - D_x \left(4x \right) + D_x \left(1 \right) \\
&= 8x^3 - 15x^2 + 2x - 4
\end{aligned}
$$

•

THE PRODUCT RULE (3.13)

$$
D_x \left[f(x)g(x) \right] = f(x) D_x \left[g(x) \right] + g(x) D_x \left[f(x) \right]
$$

PROOF Let $k(x) = f(x)g(x)$. We wish to show that

$$
k'(x) = f(x)g'(x) + g(x)f'(x).
$$

If $k'(x)$ exists, then

$$
\begin{aligned}
k'(x) &= \lim_{h \to 0} \frac{k(x + h) - k(x)}{h} \\[2mm]
&= \lim_{h \to 0} \frac{f(x + h)g(x + h) - f(x)g(x)}{h}.
\end{aligned}
$$

To change the form of the quotient so that the limit may be evaluated, we subtract and add the expression $f(x + h)g(x)$ in the numerator. Thus

$$
k'(x) = \lim_{h \to 0} \frac{f(x + h)g(x + h) - f(x + h)g(x) + f(x + h)g(x) - f(x)g(x)}{h}
$$

which may be written

$$k'(x) = \lim_{h \to 0} \left[f(x + h) \cdot \frac{g(x + h) - g(x)}{h} + g(x) \cdot \frac{f(x + h) - f(x)}{h} \right]$$

$$= \lim_{h \to 0} f(x + h) \cdot \lim_{h \to 0} \frac{g(x + h) - g(x)}{h} + \lim_{h \to 0} g(x) \cdot \lim_{h \to 0} \frac{f(x + h) - f(x)}{h}.$$

Since f is differentiable at x, it is continuous at x (see Theorem (3.5)). Hence $\lim_{h \to 0} f(x + h) = f(x)$. Also, $\lim_{h \to 0} g(x) = g(x)$, since x is fixed in this limiting process. Finally, applying the definition of derivative to $f(x)$ and $g(x)$ we obtain

$$k'(x) = f(x)g'(x) + g(x)f'(x) \quad \bullet \ \bullet$$

The Product Rule may be phrased as follows: *The derivative of a product equals the first factor times the derivative of the second factor, plus the second times the derivative of the first.*

EXAMPLE 4 Find $f'(x)$ if $f(x) = (x^3 + 1)(2x^2 + 8x - 5)$.

SOLUTION Using the Product Rule (3.13),

$$f'(x) = (x^3 + 1) D_x (2x^2 + 8x - 5) + (2x^2 + 8x - 5) D_x (x^3 + 1)$$
$$= (x^3 + 1)(4x + 8) + (2x^2 + 8x - 5)(3x^2)$$
$$= (4x^4 + 8x^3 + 4x + 8) + (6x^4 + 24x^3 - 15x^2)$$
$$= 10x^4 + 32x^3 - 15x^2 + 4x + 8$$

We could also find $f'(x)$ in Example 4 by first multiplying the two factors $x^3 + 1$ and $2x^2 + 8x - 5$ and then differentiating the resulting polynomial.

THE QUOTIENT RULE (3.14)

$$D_x \left[\frac{f(x)}{g(x)} \right] = \frac{g(x) D_x [f(x)] - f(x) D_x [g(x)]}{[g(x)]^2}$$

provided $g(x) \neq 0$.

PROOF Let $k(x) = f(x)/g(x)$. We wish to show that

$$k'(x) = \frac{g(x)f'(x) - f(x)g'(x)}{[g(x)]^2}.$$

Using the definitions of $k'(x)$ and $k(x)$,

$$k'(x) = \lim_{h \to 0} \frac{k(x + h) - k(x)}{h}$$

$$= \lim_{h \to 0} \frac{\dfrac{f(x + h)}{g(x + h)} - \dfrac{f(x)}{g(x)}}{h}$$

$$= \lim_{h \to 0} \frac{g(x)f(x + h) - f(x)g(x + h)}{hg(x + h)g(x)}.$$

Subtracting and adding $g(x)f(x)$ in the numerator of the last quotient,

$$k'(x) = \lim_{h \to 0} \frac{g(x)f(x + h) - g(x)f(x) + g(x)f(x) - f(x)g(x + h)}{hg(x + h)g(x)}$$

or, equivalently,

$$k'(x) = \lim_{h \to 0} \frac{g(x)\left[\dfrac{f(x + h) - f(x)}{h}\right] - f(x)\left[\dfrac{g(x + h) - g(x)}{h}\right]}{g(x + h)g(x)}$$

Taking the limit of the numerator and denominator gives us the Quotient Rule. • •

The Quotient Rule may be stated as follows: *The derivative of a quotient equals the denominator times the derivative of the numerator minus the numerator times the derivative of the denominator, divided by the square of the denominator.*

EXAMPLE 5 Find $\dfrac{dy}{dx}$ if $y = \dfrac{3x^2 - x + 2}{4x^2 + 5}$.

SOLUTION By the Quotient Rule (3.14),

$$\frac{dy}{dx} = \frac{(4x^2 + 5) D_x (3x^2 - x + 2) - (3x^2 - x + 2) D_x (4x^2 + 5)}{(4x^2 + 5)^2}$$

$$= \frac{(4x^2 + 5)(6x - 1) - (3x^2 - x + 2)(8x)}{(4x^2 + 5)^2}$$

$$= \frac{(24x^3 - 4x^2 + 30x - 5) - (24x^3 - 8x^2 + 16x)}{(4x^2 + 5)^2}$$

$$= \frac{4x^2 + 14x - 5}{(4x^2 + 5)^2}$$

It is now a simple matter to extend the Power Rule (3.10) to the case in which the exponent is a negative integer.

THEOREM (3.15)

> If n is a positive integer, then $D_x (x^{-n}) = -nx^{-n-1}$.

PROOF Using the definition of x^{-n} and the Quotient Rule (3.14),

$$D_x (x^{-n}) = D_x \left(\frac{1}{x^n}\right) = \frac{x^n D_x (1) - 1 D_x (x^n)}{(x^n)^2}$$

$$= \frac{x^n(0) - 1(nx^{n-1})}{(x^n)^2} = \frac{-nx^{n-1}}{x^{2n}}$$

$$= -nx^{(n-1)-2n} = -nx^{-n-1} \quad \bullet \bullet$$

EXAMPLE 6 Differentiate the following: (a) $g(x) = 1/w^4$ (b) $H(s) = 3/s$

FIGURE 3.7

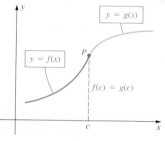

FIGURE 3.8

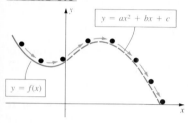

SOLUTION

(a) Writing $g(w) = w^{-4}$ and using Theorem (3.15) (with w as the independent variable),

$$g'(w) = D_w \left(w^{-4} \right) = -4w^{-5} = -\frac{4}{w^5}$$

(b) Since $H(s) = 3s^{-1}$,

$$H'(s) = D_s \left(3s^{-1} \right) = 3(-1)s^{-2} = -\frac{3}{s^2} \, \cdot$$

Suppose f and g are differentiable functions such that $f(c) = g(c)$ for some number c. For certain applications a particle may move from left to right along the graph of f and then make a transition to the graph of g at the point $P(c, f(c))$, as illustrated in Figure 3.7. The transition is **smooth** if $f'(c) = g'(c)$ (or if the left-hand derivative of f at c equals the right-hand derivative of g at c). Thus the tangent lines at $P(c, f(c))$ have the same slope. This concept is used in the next example (also see Exercise 53).

EXAMPLE 7 A ball rolls down a swimming pool slide. After leaving the end of the slide, it makes a smooth transition to a parabolic path until it finally hits the water. The motion of the ball is illustrated in Figure 3.8, where the graph of $y = f(x)$ represents the shape of the slide, and the parabolic path is the graph of $y = ax^2 + bx + c$. The end of the slide is on the y-axis, and the water level lies along the x-axis.

(a) Show that $c = f(0)$ and $b = f'(0)$. (The constant a depends on the momentum of the ball at $x = 0$.)

(b) If $f(x) = 2 + [x/(1 + x^2)]$ and if the ball hits the water at the point $(6, 0)$, find a and estimate the maximum height during flight.

SOLUTION

(a) The number c is the y-intercept of both the parabola and the graph of f, and hence $c = f(0)$.

The slope of the tangent line to the parabola at the point (x, y) is $y' = 2ax + b$. In particular, at $(0, c)$ the slope is $2a(0) + b = b$. Since the slope of the tangent line to the graph of f is $f'(0)$ and the transition is smooth, we must have $b = f'(0)$.

(b) By Theorems (3.12) and (3.7) and the Quotient Rule (3.14),

$$f'(x) = 0 + \frac{(1 + x^2)(1) - x(2x)}{(1 + x^2)} = \frac{1 - x^2}{1 + x^2} \cdot$$

From the results of part (a), $b = f'(0) = 1$ and $c = f(0) = 2$. Thus an equation for the parabola is

$$y = ax^2 + x + 2.$$

Since $(6, 0)$ is a point on the parabola,

$$0 = a(36) + 6 + 2 \quad \text{and} \quad a = -\tfrac{8}{36} = -\tfrac{2}{9}.$$

Hence an equation for the parabola is

$$y = -\tfrac{2}{9}x^2 + x + 2.$$

The maximum height during flight occurs at the vertex of the parabola, which is also the point where the slope of the tangent line is zero, that is, $y' = 0$. We leave it to the reader to verify that this point is $(\tfrac{9}{4}, \tfrac{25}{8})$. Thus the maximum height during flight is $\tfrac{25}{8}$. •

The differentiation formulas (3.12), (3.13), and (3.14) are stated in terms of the function values $f(x)$, $g(x)$, $f'(x)$, and $g'(x)$. If we wish to state these rules without referring to the variable x, we may use the following forms, provided f and g are differentiable.

(3.16)

$$(f + g)' = f' + g' \qquad (f - g)' = f' - g'$$

$$(fg)' = fg' + gf' \qquad \left(\frac{f}{g}\right)' = \frac{gf' - fg'}{g^2}$$

EXERCISES 3.2

Exer. 1–32: Differentiate the function.

1 $f(x) = 10x^2 + 9x - 4$

2 $f(x) = 6x^3 - 5x^2 + x + 9$

3 $f(s) = 15 - s + 4s^2 - 5s^4$

4 $f(t) = 12 - 3t^4 + 4t^6$

5 $g(x) = (x^3 - 7)(2x^2 + 3)$

6 $k(x) = (2x^2 - 4x + 1)(6x - 5)$

7 $h(r) = r^2(3r^4 - 7r + 2)$

8 $g(s) = (s^3 - 5s + 9)(2s + 1)$

9 $f(x) = \dfrac{4x - 5}{3x + 2}$

10 $h(x) = \dfrac{8x^2 - 6x + 11}{x - 1}$

11 $h(z) = \dfrac{8 - z + 3z^2}{2 - 9z}$

12 $f(w) = \dfrac{2w}{w^3 - 7}$

13 $f(x) = 3x^3 - 2x^2 + 4x - 7$

14 $g(z) = 5z^4 - 8z^2 + z$

15 $F(t) = t^2 + (1/t^2)$

16 $s(x) = 2x + (2x)^{-1}$

17 $g(x) = (8x^2 - 5x)(13x^2 + 4)$

18 $H(y) = (y^5 - 2y^3)(7y^2 + y - 8)$

19 $G(v) = (v^3 - 1)/(v^3 + 1)$

20 $f(t) = (8t + 15)/(t^2 - 2t + 3)$

21 $f(x) = \dfrac{1}{1 + x + x^2 + x^3}$

22 $p(x) = 1 + \dfrac{1}{x} + \dfrac{1}{x^2} + \dfrac{1}{x^3}$

23 $g(z) = z(2z^3 - 5z - 1)(6z^2 + 7)$

24 $N(v) = 4v(v - 1)(2v - 3)$

25 $K(s) = (3s)^{-4}$

26 $W(s) = (3s)^4$

27 $h(x) = (5x - 4)^2$

28 $g(r) = (5r - 4)^{-2}$

29 $f(t) = \dfrac{3/(5t) - 1}{(2/t^2) + 7}$

30 $S(w) = (2w + 1)^3$

31 $M(x) = (2x^3 - 7x^2 + 4x + 3)/x^2$

32 $f(x) = (3x^2 - 5x + 8)/7$

Exer. 33–34: Find dy/dx by means of (a) the Quotient Rule (3.14), (b) the Product Rule (3.13), and (c) simplifying algebraically and using (3.12) and (3.10).

33 $y = (3x - 1)/x^2$

34 $y = (x^2 + 1)/x^4$

35 Find an equation of the tangent line to the graph of $y = 5/(1 + x^2)$ at each point.

(a) $P(0, 5)$ (b) $P(1, \frac{5}{2})$ (c) $P(-2, 1)$

36 Find an equation of the tangent line to the graph of $y = 2x^3 + 4x^2 - 5x - 3$ at each point.

(a) $P(0, -3)$ (b) $P(-1, 4)$ (c) $P(1, -2)$

37 Find the x-coordinates of all points on the graph of $y = x^3 + 2x^2 - 4x + 5$ at which the tangent line is (a) horizontal; (b) parallel to the line $2y + 8x - 5 = 0$.

38 Find the point P on the graph of $y = x^3$ such that the tangent line at P has x-intercept 4.

Exer. 39–40: An equation of a classical curve and its graph is given, for positive constants a and b. (Consult books on analytic geometry for further information.) Find the slope of the tangent line at the point P.

39 *Witch of Agnesi:* $y = \dfrac{a^3}{a^2 + x^2}$; $P(a, a/2)$

40 *Serpentine curve:* $y = \dfrac{abx}{a^2 + x^2}$; $P(a, b/2)$

EXERCISE 39

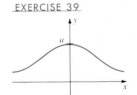

EXERCISE 40

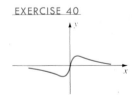

41 Find an equation of the line through the point $P(5, 9)$ that is tangent to the graph of $y = x^2$.

42 Find equations of the lines through $P(3, 1)$ that are tangent to the graph of $xy = 4$.

Exer. 43–46: If f and g are differentiable functions such that $f(2) = 3$, $f'(2) = -1$, $g(2) = -5$, and $g'(2) = 2$, find the indicated numbers.

43 (a) $(f + g)'(2)$ (b) $(f - g)'(2)$
   (c) $(4f)'(2)$ (d) $(fg)'(2)$
   (e) $(f/g)'(2)$

44 (a) $(g - f)'(2)$ (b) $(g/f)'(2)$
   (c) $(4g)'(2)$ (d) $(ff)'(2)$

45 (a) $(2f - g)'(2)$ (b) $(5f + 3g)'(2)$
   (c) $(gg)'(2)$ (d) $\left(\dfrac{1}{f + g}\right)'(2)$

46 (a) $(3f - 2g)'(2)$ (b) $(5/g)'(2)$
   (c) $(6f)'(2)$ (d) $\left(\dfrac{f}{f + g}\right)'(2)$

47 If f, g, and h are differentiable, use the Product Rule to prove that
$$D_x\left[f(x)g(x)h(x)\right] = f(x)g(x)h'(x) + f(x)h(x)g'(x) + h(x)g(x)f'(x).$$
As a corollary, let $f = g = h$ to prove that
$$D_x\left[f(x)\right]^3 = 3\left[f(x)\right]^2 f'(x).$$

48 Extend Exercise 47 to the derivative of a product of four functions, and then find a formula for $D_x\left[f(x)\right]^4$.

Exer. 49–50: Use Exercise 47 to find dy/dx.

49 $y = (8x - 1)(x^2 + 4x + 7)(x^3 - 5)$

50 $y = (3x^4 - 10x^2 + 8)(2x^2 - 10)(6x + 7)$

51 In the 1940s the human cannonball stunt was performed regularly by Emmanuel Zacchini for the Ringling Brothers and Barnum & Bailey Circus. The tip of the cannon, rising 15 feet off the ground, was aimed at a 45°-angle, and Zacchini traveled along a parabolic path, covering a horizontal distance of 175 feet (see figure).

(a) Find an equation $y = ax^2 + bx + c$ that specifies the parabolic flight.

(b) Approximate the maximum height attained by the human cannonball.

EXERCISE 51

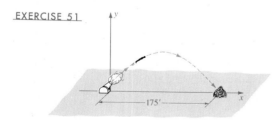

52 A rocket is fired up a hillside following a path given by $y = -0.016x^2 + 1.6x$. The hillside has slope $\frac{1}{5}$ as shown in the figure.

(a) At what slope does the rocket take off?

(b) At what slope does the rocket land?

(c) Find the maximum height of the rocket *above the ground*.

EXERCISE 52

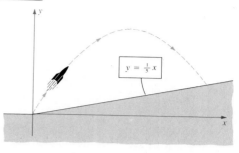

53 Traffic engineers are designing a stretch of highway that will connect a horizontal highway with one having a 20% grade (i.e., slope $\frac{1}{5}$) as illustrated in the figure. The smooth transition is to take place over a horizontal distance of 800 feet using a parabolic path to connect points A and B. Find an equation $y = ax^2 + bx + c$ of a suitable parabola, and find the coordinates of B.

EXERCISE 53

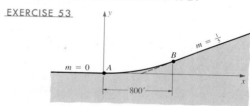

54 Shown in the figure is a computer-generated graph of $f(x) = -x^3 + 9x^2 - 18x + 6$, for $0 \le x \le 3$.

(a) Find the x-coordinate of the point P at which the tangent line is horizontal.

(b) Are there any other points on the *complete graph* such that $f(x) = 0$?

EXERCISE 54

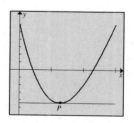

55 A weather balloon is released and rises vertically such that its distance $s(t)$ above the ground during the first 10 seconds of flight is given by $s(t) = 6 + 2t + t^2$ for $s(t)$ in feet and t in seconds.

(a) Find the velocity of the balloon at $t = 1$, $t = 4$, and $t = 8$.

(b) Find the velocity of the balloon at the instant that it is 50 feet above the ground.

56 A ball rolls down an inclined plane such that the distance (in cm) it rolls in t seconds is given by $s(t) = 2t^3 + 3t^2 + 4$

for $0 \leq t \leq 3$ (see figure).

(a) What is the velocity of the ball at $t = 2$?

(b) At what time is the velocity 30 cm/sec?

EXERCISE 56

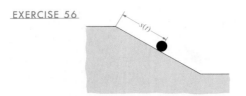

57 Two track men are set to run the hundred-meter dash. The distances $s_1(t)$ and $s_2(t)$ that each will run after t seconds

are given by $s_1(t) = \frac{1}{5}t^2 + 8t$ and $s_2(t) = 1100t/(t + 100)$ for $t \geq 0$. Determine which runner (a) is faster out of the starting blocks; (b) wins the race; and (c) is faster at the finish line.

58 When a particular basketball player leaps straight up for a dunk, his distance in feet off the ground is given by $s(t) = -\frac{1}{2}gt^2 + 16t$.

(a) If $g = 32$, find the player's *hang time*; that is, the total number of seconds that the player is in the air.

(b) Find the initial velocity and, determine the player's *vertical leap*, that is, the maximum distance of his feet off the floor.

(c) On the moon, $g \approx \frac{32}{6}$. Rework parts (a) and (b) for conditions on the moon.

3.3 THE DERIVATIVE AS A RATE OF CHANGE

We have already discussed two very important applications of the derivative: tangent lines to graphs, and the velocity of an object moving on a line. The derivative is useful in many other practical situations. It was used in Example 5 of Section 3.1 to interpret magnifications. In this section we shall discuss several other applications that illustrate the versatility of this powerful concept.

Most quantities encountered in everyday life change with time. This is especially evident in scientific investigations. For example, a chemist may be interested in the rate at which a certain substance dissolves in water. An electrical engineer may wish to know the rate of change of current in part of an electrical circuit. A biologist may be concerned with the rate at which the bacteria in a culture increase or decrease. We could give numerous other examples, including many from fields other than the natural sciences. Let us consider the following general situation, which can be applied to all of the preceding examples.

Suppose a variable w is a function of time such that at time t, $w = g(t)$ for a differentiable function g. The difference between the initial and final values of w in the time interval $[t, t + h]$ is $g(t + h) - g(t)$. As in our development of the velocity concept in Chapter 2, we formulate the following definition.

DEFINITION (3.17)

Let $w = g(t)$ such that g is differentiable and t represents time.

(i) The **average rate of change** of $w = g(t)$ in the interval $[t, t + h]$ is

$$\frac{g(t + h) - g(t)}{h}$$

(ii) The **rate of change** of $w = g(t)$ with respect to t is

$$\frac{dw}{dt} = g'(t) = \lim_{h \to 0} \frac{g(t + h) - g(t)}{h}$$

The units to be used in Definition (3.17) depend on the nature of the quantity represented by w. Sometimes dw/dt is referred to as the **instantaneous rate of change** of w with respect to t.

EXAMPLE 1 A scientist discovers that if a certain substance is heated, the Celsius temperature after t minutes is given by $g(t) = 30t + 6\sqrt{t} + 8$ for $0 \le t \le 5$.

(a) Find the average rate of change of $g(t)$ during the time interval $[4, 4.41]$.

(b) Find the rate of change of $g(t)$ at $t = 4$.

SOLUTION

(a) Letting $t = 4$ and $h = 0.41$ in Definition (3.17) (i), the average rate of change of g in $[4, 4.41]$ is

$$\frac{g(4.41) - g(4)}{0.41} = \frac{[30(4.41) + 6\sqrt{4.41} + 8] - [120 + 6\sqrt{4} + 8]}{0.41}$$

$$= \frac{12.9}{0.41} \approx 31.46 \, {}^\circ C/min$$

(b) By Definition (3.17) (ii), the rate of change of $g(t)$ at time t is $g'(t)$. In Example 2 of Section 3.1 we proved that $D_x(\sqrt{x}) = 1/(2\sqrt{x})$. Hence

$$g'(t) = D_t(30t + 6\sqrt{t} + 8)$$

$$= 30 + 6\left(\frac{1}{2\sqrt{t}}\right) + 0$$

$$= 30 + \frac{3}{\sqrt{t}}.$$

In particular, the rate of change of $g(t)$ at $t = 4$ is

$$g'(4) = 30 + \frac{3}{\sqrt{4}} = 31.5 \, {}^\circ C/min \quad \bullet$$

If a point P moves on a coordinate line l such that its coordinate at time t is $s(t)$, as illustrated in Figure 3.9, then s is the position function of P. By (3.2) (ii), the velocity $v(t)$ of P at time t is $s'(t)$. In terms of Definition (3.17), the velocity is the rate of change of $s(t)$ with respect to time.

The *acceleration* $a(t)$ of P at time t is defined as the rate of change of velocity with respect to time, that is, $a(t) = v'(t)$. Thus the acceleration is the derivative $D_t[s'(t)]$ of $s'(t)$. In Section 3.8, we will call $D_t[s'(t)]$ the *second derivative* of s with respect to t, denoted by $s''(t)$. The next definition summarizes this discussion and also introduces the notion of the *speed* of P.

FIGURE 3.9

Time   Position of P

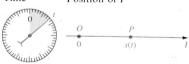

DEFINITION (3.18)

> Let the position of a point P on a coordinate line l at time t be given by $s(t)$ for a differentiable function s.
>
> (i) The **velocity** $v(t)$ of P at time t is $v(t) = s'(t)$.
>
> (ii) The **speed** of P at time t is $|v(t)|$.
>
> (iii) The **acceleration** $a(t)$ of $P(t)$ at time t is $a(t) = v'(t) = s''(t)$.

We shall call v the **velocity function** of P and a the **acceleration function** of P. We sometimes use the notation

$$v = \frac{ds}{dt} \quad \text{and} \quad a = \frac{dv}{dt}.$$

If t is in seconds and $s(t)$ is in centimeters, then $v(t)$ is in cm/sec and $a(t)$ is in cm/sec^2 (centimeters per second per second). If t is in hours and $s(t)$ is in miles, then $v(t)$ is in mi/hr and $a(t)$ is in mi/hr^2 (miles per hour per hour).

We noted in Chapter 2 that if $v(t)$ is positive in a time interval, then the point P is moving in the positive direction on l. If $v(t)$ is negative, the motion is in the negative direction. The velocity is zero at a point where P changes direction. This will be proved in Chapter 4, together with the fact that if the acceleration $a(t)$ is positive, then the velocity is increasing. If $a(t)$ is negative, the velocity is decreasing.

EXAMPLE 2 The position function s of a point P on a coordinate line is given by

$$s(t) = t^3 - 12t^2 + 36t - 20$$

for t in seconds and $s(t)$ in centimeters. Describe the motion of P during the time interval $[-1, 9]$.

SOLUTION Differentiating,

$$v(t) = s'(t) = 3t^2 - 24t + 36 = 3(t - 2)(t - 6),$$

and $\qquad a(t) = v'(t) = 6t - 24 = 6(t - 4).$

Let us determine when $v(t) > 0$ and when $v(t) > 0$, since this will tell us if P is moving to the right or to the left, respectively. We see that $v(t) = 0$ at $t = 2$ and $t = 6$. This suggests that we examine the following time subintervals of $[-1, 9]$:

$$(-1, 2) \quad (2, 6) \quad (6, 9)$$

Since $v(t)$ is a polynomial, it is either positive or negative throughout each of these subintervals (see Corollary (2.35)). As in Example 6 of Section 2.6, we may determine the sign of $v(t)$ by using test values, as indicated in the table (check each entry):

Time interval	$(-1, 2)$	$(2, 6)$	$(6, 9)$
k	0	3	7
Test value $v(k)$	36	-9	15
Sign of $v(t)$	$+$	$-$	$+$
Direction of motion	right	left	right

The next table lists the values of the position, velocity, and acceleration functions at the endpoints of the time interval $[-1, 9]$, and the times at which the velocity or acceleration is zero.

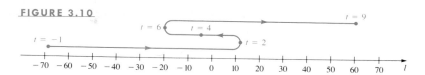

t	-1	2	4	6	9
$s(t)$	-69	12	-4	-20	61
$v(t)$	63	0	-12	0	63
$a(t)$	-30	-12	0	12	30

It is convenient to represent the motion of P schematically as in Figure 3.10. The curve above the coordinate line is not the path of the point, but is intended to show the manner in which P moves on l.

FIGURE 3.10

As indicated by the tables and Figure 3.10, at $t = -1$ the point is 69 cm to the left of the origin and is moving to the right with a velocity of 63 cm/sec. The negative acceleration -30 cm/sec² indicates that the velocity is decreasing at a rate of 30 cm per second, each second. The point continues to move to the right, slowing down until it has zero velocity at $t = 2$, 12 cm to the right of the origin. The point P then reverses direction and moves until, at $t = 6$, it is 20 cm to the left of the origin. It then again reverses direction and moves to the right for the remainder of the time interval, with increasing velocity. The direction of motion is indicated by the arrows on the curve in Figure 3.10. •

EXAMPLE 3 A projectile is fired straight upward with a velocity of 400 ft/sec. Its distance above the ground after t seconds is $s(t) = -16t^2 + 400t$. Find the time and the velocity at which the projectile hits the ground. What is the maximum altitude achieved by the projectile? What is the acceleration at any time t?

FIGURE 3.11

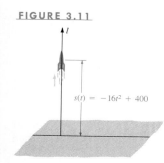

$s(t) = -16t^2 + 400$

SOLUTION The path of the projectile is on a vertical coordinate line with origin at ground level and positive direction upward, as illustrated in Figure 3.11. The projectile will be on the ground when $-16t^2 + 400t = 0$, that is, when $-16t(t - 25) = 0$. This gives us $t = 0$ and $t = 25$. Hence the projectile hits the ground after 25 sec. The velocity at time t is

$$v(t) = s'(t) = -32t + 400.$$

In particular, at $t = 25$, we obtain the *impact velocity:*

$$v(25) = -32(25) + 400 = -400 \text{ ft/sec.}$$

The negative velocity indicates that the projectile is moving in the negative direction on l (downward) at the instant that it strikes the ground. Note that the *speed* at this time is

$$|v(25)| = |-400| = 400 \text{ ft/sec.}$$

The maximum altitude occurs when the velocity is zero, that is, when $s'(t) = -32t + 400 = 0$. Solving for t gives us $t = \frac{400}{32} = \frac{25}{2}$ and hence the maximum altitude is

$$s(\tfrac{25}{2}) = -16(\tfrac{25}{2})^2 + 400(\tfrac{25}{2}) = 2500 \text{ ft.}$$

Finally, the acceleration at any time t is

$$a(t) = v'(t) = -32 \text{ ft/sec}^2.$$

This constant acceleration is caused by the force of gravity. •

We may investigate rates of change with respect to variables other than time, as indicated by the next definition.

DEFINITION (3.19)

> Let $y = f(x)$ for any variable x.
> (i) The **average rate of change of y with respect to x in the interval $[x, x + h]$** is the quotient
> $$\frac{f(x + h) - f(x)}{h}$$
> (ii) The **rate of change of y with respect to x** is the limit of the average rate of change as $h \to 0$, that is, dy/dx.

By Definition (3.19) (ii), if the variable x changes, then y changes at a rate of dy/dx units per unit change in x. To illustrate, suppose a quantity of gas is enclosed in a spherical balloon. If the gas is heated or cooled and if the pressure remains constant, then the balloon expands or contracts and its volume V is a function of the temperature T. The derivative dV/dT gives us the rate of change of volume with respect to temperature.

EXAMPLE 4 The current I (in amperes) in an electrical circuit is given by $I = 100/R$ for resistance R (in ohms). Find the rate of change of I with respect to R when the resistance is 20 ohms.

SOLUTION Writing $I = 100R^{-1}$ and applying the Power Rule (3.10),

$$\frac{dI}{dR} = 100(-1)R^{-2} = -\frac{100}{R^2}.$$

Substituting 20 for R,

$$\frac{dI}{dR} = -\frac{100}{400} = -\frac{1}{4}.$$

Thus, if R is increasing, then when $R = 20$, the current I is decreasing at a rate of $\frac{1}{4}$ amperes per ohm. •

Calculus has become an important tool for solving problems that occur in economics. If a function f is used to describe some economic entity, then the adjective *marginal* is employed to specify the derivative f'.

If x is the number of units of a commodity, then economists often use the functions C, c, R, and P defined as follows:

Cost function: $C(x) = $ Cost of producing x units

Average cost function: $c(x) = \dfrac{C(x)}{x} = $ Average cost of producing one unit

Revenue function: $R(x) = $ Revenue received for selling x units

Profit function: $P(x) = R(x) - C(x) = $ Profit in selling x units

To use the techniques of calculus, we regard x as a real number, even though this variable may take on only integer values. We always assume that $x \geq 0$, since the production of a negative number of units has no practical significance.

EXAMPLE 5 A manufacturer of miniature tape decks has a monthly fixed cost of $10,000, a production cost of $12 per unit, and a selling price of $20 per unit.

(a) Find $C(x)$, $c(x)$, $R(x)$, and $P(x)$.

(b) Find the function values in part (a) if $x = 1000$.

(c) How many units must be manufactured in order to break even?

SOLUTION

(a) The production costs of manufacturing x units is $12x$. Since there is also a fixed monthly cost of $10,000, the total monthly cost of manufacturing x units is

$$C(x) = 12x + 10,000$$

Consequently,
$$c(x) = \frac{C(x)}{x} = 12 + \frac{10,000}{x}.$$

We also see that

$$R(x) = 20x$$

and
$$P(x) = R(x) - C(x) = 8x - 10,000.$$

(b) Substituting $x = 1000$ in part (a),

$$C(1000) = 22,000 \quad \text{(cost of manufacturing 1000 units)}$$
$$c(1000) = 22 \quad \text{(average cost of manufacturing one unit)}$$
$$R(1000) = 20,000 \quad \text{(total revenue received for 1000 units)}$$
$$P(1000) = -2000 \quad \text{(profit in manufacturing 1000 units)}$$

Note that the manufacturer incurs a loss of $2000 per month if only 1000 units are produced and sold.

(c) The break-even point corresponds to zero profit, that is, $8x - 10,000 = 0$. This gives us

$$8x = 10,000 \quad \text{or} \quad x = 1500.$$

Thus, to break even it is necessary to produce and sell 1500 units per month. •

The derivatives C', c', R', and P' are called the **marginal cost function,** the **marginal average cost function,** the **marginal revenue function,** and the **marginal profit function,** respectively. The number $C'(x)$ is referred to as the **marginal cost** associated with the production of x units. If we interpret the derivative as a rate of change, then $C'(x)$ is the rate at which the cost changes with respect to the number x of units produced. Similar statements can be made for $c'(x)$, $R'(x)$, and $P'(x)$.

If C is a cost function and n is a positive integer, then by Definition (3.1),

$$C'(n) = \lim_{h \to 0} \frac{C(n + h) - C(n)}{h}.$$

Hence, if h is small, then

$$C'(n) \approx \frac{C(n + h) - C(n)}{h}.$$

If the number n of units produced is large, economists often let $h = 1$ in the last formula to approximate the marginal cost, obtaining

$$C'(n) \approx C(n + 1) - C(n).$$

In this context, *the marginal cost associated with the production of n units is (approximately) the cost of producing one more unit.*

Some companies find that the cost $C(x)$ of producing x units of a commodity is given by a formula such as

$$C(x) = a + bx + dx^2 + kx^3.$$

The constant a represents a fixed overhead charge for items, such as rent, heat, and light, that are independent of the number of units produced. If the cost of producing one unit were b dollars and no other factors were involved, then the second term bx in the formula would represent the cost of producing x units. If x becomes very large, then the terms dx^2 and kx^3 may significantly affect production costs.

EXAMPLE 6 An electronics company estimates that the cost (in dollars) of producing x components used in electronic toys is given by

$$C(x) = 200 + 0.05x + 0.0001x^2$$

(a) Find the cost, the average cost, and the marginal cost of producing 500 units; 1000 units; and 5000 units.

(b) Compare the marginal cost of producing 1000 units with the cost of producing the 1001st unit.

SOLUTION

(a) The average cost of producing x components is

$$c(x) = \frac{C(x)}{x} = \frac{200}{x} + 0.05 + 0.0001x$$

The marginal cost is

$$C'(x) = 0.05 + 0.0002x.$$

We leave it to the reader to verify the entries in the following table, where numbers in the last three columns represent dollars, rounded off to the nearest cent.

Units x	Cost $C(x)$	Average cost $c(x) = \dfrac{C(x)}{x}$	Marginal cost $C'(x)$
500	250.00	0.50	0.15
1000	350.00	0.35	0.25
5000	2950.00	0.59	1.05

(b) Using the cost function,

$$C(1001) = 200 + 0.05(1001) + (0.0001)(1000)^2 = 350.25$$

Hence the cost of producing the 1001st unit is

$$C(1001) - C(1000) = 350.25 - 350.00 = 0.25$$

which is the same as the marginal cost $C'(1000)$. •

A company must consider many factors in order to determine a selling price for each product. In addition to the cost of production and the profit desired, the company should be aware of the manner in which consumer demand will vary if the price increases. For some products there is a constant demand, and changes in price have little effect on sales. For items that are not necessities of life, a price increase will probably lead to a decrease in the number of units sold. Suppose a company knows from past experience that it can sell x units when the price per unit is given by $p(x)$ for some function p. We sometimes say that $p(x)$ is the price per unit when there is a **demand** for x units, and we refer to p as the **demand function** for the commodity. The total income, or revenue, is the number of units sold times the price per unit, that is, $x \cdot p(x)$. Thus

$$R(x) = xp(x).$$

The derivative p' is called the **marginal demand function.** We will give examples of demand functions later in the text.

EXERCISES 3.3

1 As a spherical balloon is being inflated, its radius (in centimeters) after t minutes is given by $r(t) = 3\sqrt[3]{t}$ where $0 \le t \le 10$. For each of the following, find the rate of change with respect to t at $t = 8$.
(a) $r(t)$ (*Hint:* See Example 4 of Section 3.1.)
(b) The volume of the balloon
(c) The surface area of the balloon (*Hint:* Use the Product Rule.)

2 The volume V (in ft^3) in a small reservoir during spring runoff is given by $V = 5000(t + 1)^2$ for t in months and $0 \le t \le 3$. The rate of change of volume with respect to time is the instantaneous *flow rate* into the reservoir. Find the flow rate at times $t = 0$ and $t = 2$. What is the flow rate when the volume is 11,250 ft^3?

3 Suppose that t seconds after starting to run, an individual's pulse rate (in beats/min) is given by $P(t) = 56 + 2t^2 - t$, for $0 \le t \le 7$. Find the rate of change of $P(t)$ with respect to t at (a) $t = 2$, (b) $t = 4$, and (c) $t = 6$.

4 The temperature T (in °C) of a solution at time t (minutes) is given by $T(t) = 10 + 4t + [3/(t + 1)]$ for $1 \le t \le 10$. Find the rate of change of $T(t)$ with respect to t at (a) $t = 2$, (b) $t = 5$, and (c) $t = 9$.

5 A stone is dropped into a pond, causing water waves that form concentric circles. If, after t seconds, the radius of one of the waves is $40t$ cm, find the rate of change with respect to t of the area of the circle caused by the wave at (a) $t = 1$, (b) $t = 2$, and (c) $t = 3$.

6 Boyle's law for gases states that $pv = c$ for the pressure p, the volume v, and a constant c. Suppose that at time t (in minutes) the pressure is $20 + 2t$ cm/Hg for $0 \le t \le 10$. If the volume is 60 cm^3 at $t = 0$, find the rate at which the volume is changing with respect to t at $t = 5$.

7 A population of flies is growing in a large container. The number of flies P (in hundreds) after t weeks is given by $P = 12t^2 - t^4 + 5$. When does the population stop growing? Over what time intervals is the rate of population growth positive? negative?

8 As a jar containing 10 moles of a gas A is heated, the velocity of the gas molecules increases and a second gas B is formed. When two molecules of gas A collide, two molecules of gas B are formed. If the number y of moles of gas B after t minutes is given by $y = 10t/(t + 4)$, find the reaction rate (in moles/min) when the number of moles of gas A equals the number of moles of gas B.

Exer. 9–14: For the position function s of a point moving rectilinearly, find the velocity and acceleration at time t, and describe the motion of the point during the indicated time interval. Illustrate the motion by means of a diagram of the type shown in Figure 3.10.

9 $s(t) = 3t^2 - 12t + 1$, $[0, 5]$

10 $s(t) = t^2 + 3t - 6$, $[-2, 2]$

11 $s(t) = t^3 - 9t + 1$, $[-3, 3]$

12 $s(t) = 24 + 6t - t^3$, $[-2, 3]$

13 $s(t) = 2t^4 - 6t^2$, $[-2, 2]$

14 $s(t) = 2t^3 - 6t^5$, $[-1, 1]$

15 A projectile is fired directly upward with an initial velocity 144 ft/sec. Its height $s(t)$ in feet above the ground after t seconds is given by $s(t) = 144t - 16t^2$. What are the velocity and acceleration after t sec? After 3 sec? What is the maximum height? When does it strike the ground?

16 As an automobile rolls down an incline (see figure), the number of feet $s(t)$ that it rolls in t seconds is given by $s(t) = 5t^2 + 2$. What is the velocity after 1 sec? 2 sec? When will the velocity be 28 ft/sec?

EXERCISE 16

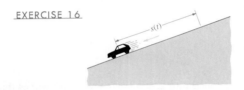

17 The illumination from a light source is directly proportional to the strength of the source and inversely proportional to the square of the distance s from the source. At a distance of 2 feet from a glowing fire, a photographer's light meter registers 120 units. If the photographer backs away from the fire, find the rate of change of the meter reading with respect to s when the photographer is 20 feet from the fire.

18 Show that the rate of change of the volume of a sphere with respect to its radius is numerically equal to the surface area of the sphere.

19 Show that the rate of change of the radius of a circle with respect to its circumference is independent of the size of the circle.

20 The relationship between the temperature F on the Fahrenheit scale and the temperature C on the Celsius scale is given by $C = \frac{5}{9}(F - 32)$. What is the rate of change of F with respect to C?

21 The electrical resistance R of a copper wire of fixed length is inversely proportional to the square of its diameter d. What is the rate of change of R with respect to d?

22 The lens equation $1/f = (1/p) + (1/q)$ is used in optics, where f is the focal length of a convex lens, and p and q are the distances from the lens to the object and image, respectively. (See figure for Exercise 53, Section 2.4.) If f is fixed, find a general formula for the rate of change of q with respect to p.

Exer. 23–26: C is the cost function for a particular product. For each case find (a) the cost of producing 100 units and (b) the average and marginal cost functions, and their values at $x = 100$.

23 $C(x) = 800 + 0.04x + 0.0002x^2$

24 $C(x) = 6400 + 6.5x + 0.003x^2$

25 $C(x) = 250 + 100x + 0.001x^3$

26 $C(x) = 200 + (100/x) + (x/100)$

27 A manufacturer of small motors estimates that the cost (in dollars) of producing x motors per day is given by $C(x) = 100 + 50x + (100/x)$. Compare the marginal cost of producing five motors with the cost of producing the sixth motor.

28 A company conducts a pilot test for the production of a new industrial solvent and finds that the cost of producing x liters of each pilot run is given by $C(x) = 3 + x + (10/x)$. Compare the marginal cost of producing ten liters with the cost of producing the eleventh liter.

3.4 DERIVATIVES OF THE TRIGONOMETRIC FUNCTIONS

Formulas for the derivatives of the trigonometric functions are stated in the next theorem.

THEOREM (3.20)

$$D_x \sin x = \cos x \qquad D_x \cos x = -\sin x$$
$$D_x \tan x = \sec^2 x \qquad D_x \cot x = -\csc^2 x$$
$$D_x \sec x = \sec x \tan x \qquad D_x \csc x = -\csc x \cot x$$

PROOF To find the derivative of the sine function, we let $f(x) = \sin x$. If $f'(x)$ exists, then

$$f'(x) = \lim_{h \to 0} \frac{f(x + h) - f(x)}{h}$$

$$= \lim_{h \to 0} \frac{\sin (x + h) - \sin x}{h}.$$

Using the addition formula for the sine function (see (1.35)),

$$f'(x) = \lim_{h \to 0} \frac{\sin x \cos h + \cos x \sin h - \sin x}{h}$$

$$= \lim_{h \to 0} \frac{\sin x (\cos h - 1) + \cos x \sin h}{h}$$

$$= \lim_{h \to 0} \left[\sin x \left(\frac{\cos h - 1}{h} \right) + \cos x \left(\frac{\sin h}{h} \right) \right].$$

By Theorems (2.27) and (2.26),

$$\lim_{h \to 0} \left(\frac{\cos h - 1}{h} \right) = 0 \quad \text{and} \quad \lim_{h \to 0} \frac{\sin h}{h} = 1.$$

Hence,

$$f'(x) = (\sin x)(0) + (\cos x)(1) = \cos x,$$

that is,

$$D_x \sin x = \cos x.$$

We have shown that the derivative of the sine function is the cosine function.

To obtain the derivative of the cosine function, we let $f(x) = \cos x$. In this case

$$f'(x) = \lim_{h \to 0} \frac{\cos (x + h) - \cos x}{h}.$$

Applying the addition formula for the cosine function,

$$f'(x) = \lim_{h \to 0} \frac{\cos x \cos h - \sin x \sin h - \cos x}{h}$$

$$= \lim_{h \to 0} \frac{\cos x (\cos h - 1) - \sin x \sin h}{h}$$

$$= \lim_{h \to 0} \left[\cos x \left(\frac{\cos h - 1}{h} \right) - \sin x \left(\frac{\sin h}{h} \right) \right].$$

Using Theorems (2.27) and (2.26),

$$f'(x) = (\cos x)(0) - (\sin x)(1) = -\sin x,$$

that is,

$$D_x \cos x = -\sin x.$$

Thus the derivative of the cosine function is the *negative* of the sine function.

To find the derivative of the tangent function, we begin with the fundamental identity $\tan x = \sin x / \cos x$ and then apply the Quotient Rule as follows:

$$D_x \tan x = D_x \left(\frac{\sin x}{\cos x} \right)$$

$$= \frac{\cos x \, (D_x \sin x) - \sin x \, (D_x \cos x)}{\cos^2 x}$$

$$= \frac{\cos x \, (\cos x) - \sin x \, (-\sin x)}{\cos^2 x}$$

$$= \frac{\cos^2 x + \sin^2 x}{\cos^2 x} = \frac{1}{\cos^2 x} = \sec^2 x$$

For the secant function, we first write $\sec x = 1/\cos x$ and use the Quotient Rule:

$$D_x \sec x = D_x \left(\frac{1}{\cos x} \right)$$

$$= \frac{(\cos x)(D_x \, 1) - (1) \, D_x \, (\cos x)}{\cos^2 x}$$

$$= \frac{(\cos x)(0) - (-\sin x)}{\cos^2 x}$$

$$= \frac{\sin x}{\cos^2 x} = \frac{1}{\cos x} \frac{\sin x}{\cos x}$$

$$= \sec x \tan x$$

Proofs of the formulas for $D_x \cot x$ and $D_x \csc x$ are left as exercises. • •

We can use Theorem (3.20) to obtain information about the continuity of the trigonometric functions. For example, since the sine and cosine functions are differentiable at every real number, it follows from Theorem (3.5) that these functions are continuous throughout $\mathbb{R}$. Similarly, the tangent function is continuous on the open intervals $(-\pi/2, \pi/2)$, $(\pi/2, 3\pi/2)$, and so on, since it is differentiable at each number in these intervals.

EXAMPLE 1 Find y' if $y = \dfrac{\sin x}{1 + \cos x}$.

SOLUTION By the Quotient Rule and Theorem (3.20),

$$y' = \frac{(1 + \cos x) \, D_x \, (\sin x) - (\sin x) \, D_x \, (1 + \cos x)}{(1 + \cos x)^2}$$

$$= \frac{(1 + \cos x)(\cos x) - (\sin x)(0 - \sin x)}{(1 + \cos x)^2}$$

$$= \frac{\cos x + \cos^2 x + \sin^2 x}{(1 + \cos x)^2}$$

$$= \frac{\cos x + 1}{(1 + \cos x)^2}$$

$$= \frac{1}{1 + \cos x}$$

In the solution to Example 1 we have used the fundamental identity $\cos^2 x + \sin^2 x = 1$. This and other trigonometric identities are often used to simplify problems that involve derivatives of trigonometric functions.

EXAMPLE 2 Find $g'(x)$ if $g(x) = \sec x \tan x$.

SOLUTION   Using the Product Rule and Theorem (3.20),

$$g'(x) = (\sec x)(D_x \tan x) + (\tan x)(D_x \sec x)$$
$$= (\sec x)(\sec^2 x) + (\tan x)(\sec x \tan x)$$
$$= \sec^3 x + \sec x \tan^2 x$$
$$= \sec x (\sec^2 x + \tan^2 x).$$

The formula for $g'(x)$ can be written in many other ways. For example, since

$$\sec^2 x = \tan^2 x + 1 \quad \text{or} \quad \tan^2 x = \sec^2 x - 1$$

we can write

$$g'(x) = \sec x (2 \tan^2 x + 1) \quad \text{or} \quad g'(x) = \sec x (2 \sec^2 x - 1) \quad \bullet$$

EXAMPLE 3 Find $k'(\theta)$ if $k(\theta) = \sec \theta \cot \theta$.

SOLUTION   We could use the Product Rule as in Example 2; however, it is simpler to first change the form of $k(\theta)$ by using Fundamental Identities as follows:

$$k(\theta) = \sec \theta \cot \theta = \frac{1}{\cos \theta} \frac{\cos \theta}{\sin \theta} = \frac{1}{\sin \theta} = \csc \theta.$$

Applying Theorem (3.20),

$$k'(\theta) = D_\theta (\csc \theta) = -\csc \theta \cot \theta \quad \bullet$$

EXAMPLE 4

(a) Find the slopes of the tangent lines to the graph of $y = \sin x$ at the points with x-coordinates 0, $\pi/3$, $\pi/2$, $2\pi/3$, and π.

(b) Sketch the tangent lines of part (a) on the graph of $y = \sin x$.

(c) For what values of x is the tangent line horizontal?

SOLUTION

(a) The slope of the tangent line at the point (x, y) on the graph of $y = \sin x$ is given by the derivative $y' = \cos x$. In particular, the slopes at the desired points are listed in the following table.

x	0	$\pi/3$	$\pi/2$	$2\pi/3$	π
$y' = \cos x$	1	$1/2$	0	$-1/2$	-1

(b) A portion of the graph of $y = \sin x$ and the tangent lines of part (a) are sketched in Figure 3.12 (on page 120).

(c) A tangent line is horizontal if its slope is zero. Since the slope of the tangent line at the point $P(x, f(x))$ is $f'(x)$, we must solve the equation

$$f'(x) = 0, \quad \text{that is,} \quad \cos x = 0.$$

Thus the tangent line is horizontal if $x = \pm\pi/2$, $x = \pm 3\pi/2$, and, in general, if $x = (\pi/2) + n\pi$ for any integer n.

FIGURE 3.12

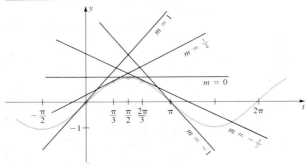

If f is a differentiable function, then the **normal line** at a point $P(a, f(a))$ on the graph of f is the line through P that is perpendicular to the tangent line, as illustrated in Figure 3.13. If $f'(a) \neq 0$, then by Theorem 1.19, the slope of the normal line is $-1/f'(a)$. If $f'(a) = 0$, then the tangent line is horizontal, and in this case, the normal line is vertical and has the equation $x = a$.

FIGURE 3.13

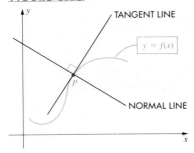

TANGENT LINE

$y = f(x)$

NORMAL LINE

EXAMPLE 5 Find an equation of the normal line to the graph of $y = \tan x$ at the point $P(\pi/4, 1)$. Illustrate graphically.

SOLUTION Since $y' = \sec^2 x$, the slope m of the tangent line at P is

$$m = \sec^2(\pi/4) = (\sqrt{2})^2 = 2$$

and hence the slope of the normal line is $-1/m = -1/2$.

Using the Point-Slope form, an equation for the normal line is

$$y - 1 = -\frac{1}{2}\left(x - \frac{\pi}{4}\right) \quad \text{or} \quad 2y - 2 = -x + \frac{\pi}{4}.$$

FIGURE 3.14

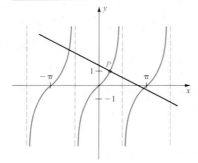

This may also be written

$$x + 2y = 2 + \frac{\pi}{4}.$$

The graph of $y = \tan x$ for $-3\pi/2 < x < 3\pi/2$, together with the normal line at P, is sketched in Figure 3.14. •

Exer. 1–20: Differentiate the function.

1 $f(x) = 4 \cos x$

2 $f(x) = 3x \sin x$

3 $g(t) = t^3 \sin t$

4 $k(t) = t - t^2 \cos t$

5 $f(\theta) = (\sin \theta)/\theta$

6 $T(r) = r^2 \sec r$

7 $f(x) = 2x \cot x + x^2 \tan x$

8 $f(x) = 3x^2 \sec x - x^3 \tan x$

9 $h(z) = (1 - \cos z)/(1 + \cos z)$

10 $R(w) = (\cos w)/(1 - \sin w)$

11 $g(x) = (x + \csc x) \cot x$

12 $K(\theta) = (\sin \theta + \cos \theta)^2$

13 $p(x) = \sin x \cot x$

14 $g(t) = \csc t \sin t$

15 $f(x) = (\tan x)/(1 + x^2)$

16 $h(\theta) = (1 + \sec \theta)/(1 - \sec \theta)$

17 $k(v) = \csc v/\sec v$

18 $g(x) = \sin(-x) + \cos(-x)$

19 $H(\varphi) = (\cot \varphi + \csc \varphi)(\tan \varphi - \sin \varphi)$

20 $f(x) = (1 + \sec x)/(\tan x + \sin x)$

Exer. 21–22: Find equations of the tangent line and the normal line to the graph of f at the point with x-coordinate $\pi/4$.

21 $f(x) = \sec x$ **22** $f(x) = \csc x + \cot x$

Exer. 23–26: Shown is a computer-generated graph of the function f with domain restricted as indicated. Find the points at which horizontal tangent lines occur.

23 $f(x) = \cos x + \sin x;\ 0 \le x \le 2\pi$

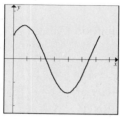

24 $f(x) = \cos x - \sin x;\ 0 \le x \le 2\pi$

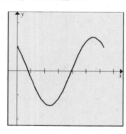

25 $f(x) = \csc x + \sec x;\ 0 < x < \pi/2$

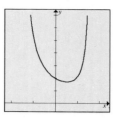

26 $f(x) = 2 \sec x - \tan x;\ -\pi/2 < x < \pi/2$

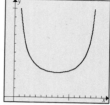

27 If $y = x + 2 \cos x$, find

  (a) the x-coordinates of all points on the graph at which the tangent line is horizontal.

  (b) an equation of the tangent line to the graph at the point where the graph crosses the y-axis.

28 If $y = 3 + 2 \sin x$, find

  (a) the x-coordinates of all points on the graph where the tangent line is parallel to the line $y = \sqrt{2}x - 5$.

  (b) an equation of the tangent line to the graph at the point on the graph with x-coordinate $\pi/6$.

Exer. 29–30: Prove each formula.

29 (a) $D_x \cot x = -\csc^2 x$ (b) $D_x \csc x = -\csc x \cot x$

30 (a) $D_x \sin 2x = 2 \cos 2x$ (*Hint:* $\sin 2x = 2 \sin x \cos x$).

  (b) $D_x \cos 2x = -2 \sin 2x$ (*Hint:* $\cos 2x = 1 - 2 \sin^2 x$).

3.5 INCREMENTS AND DIFFERENTIALS

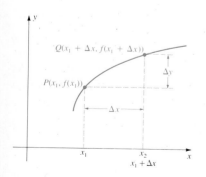

Let us consider the equation $y = f(x)$ for some function f. In many applications the independent variable x may undergo a small change, and it is necessary to find the corresponding change in the dependent variable y. A change in x is often denoted by the symbol Δx (read *delta x*). For example, if x changes from x_1 to x_2, then

$$\Delta x = x_2 - x_1.$$

The number Δx is called an **increment** of x. Note that $x_2 = x_1 + \Delta x$; that is, the new value x_2 equals the initial value x_1 plus the change Δx. The symbol Δy is used to denote the change in the dependent variable y that corresponds to Δx. Thus

$$\Delta y = f(x_2) - f(x_1) = f(x_1 + \Delta x) - f(x_1).$$

The increments Δx and Δy are illustrated graphically in Figure 3.15.

Sometimes we may let x represent the initial value of the independent variable. In this case we use the phrase x *is given an increment* Δx to describe a (small) change in this variable. Our definition of Δy then has the following form.

DEFINITION (3.21)

> Let $y = f(x)$ and let x be given an increment Δx. The **increment Δy of y** is
> $$\Delta y = f(x + \Delta x) - f(x)$$

EXAMPLE 1 Suppose $y = 3x^2 - 5$.

(a) If x is given an increment Δx, find Δy.

(b) Calculate Δy if x changes from 2 to 2.1.

SOLUTION

(a) Let $f(x) = 3x^2 - 5$. Applying Definition (3.21),

$$\begin{aligned}
\Delta y &= f(x + \Delta x) - f(x) \\
&= [3(x + \Delta x)^2 - 5] - [3x^2 - 5] \\
&= [3(x^2 + 2x(\Delta x) + (\Delta x)^2) - 5] - [3x^2 - 5] \\
&= 3x^2 + 6x(\Delta x) + 3(\Delta x)^2 - 5 - 3x^2 + 5 \\
&= 6x(\Delta x) + 3(\Delta x)^2
\end{aligned}$$

(b) We wish to find Δy if $x = 2$ and $\Delta x = 0.1$. Substituting in the formula for Δy,

$$\begin{aligned}
\Delta y &= 6(2)(0.1) + 3(0.1)^2 \\
&= 12(0.1) + 3(0.01) = 1.2 + 0.03 = 1.23
\end{aligned}$$

Thus, y changes by an amount 1.23 if x changes from 2 to 2.1. We could also find Δy directly as follows:

$$\begin{aligned}
\Delta y &= f(x + \Delta x) - f(x) = f(2.1) - f(2) \\
&= [3(2.1)^2 - 5] - [3(2)^2 - 5] = 1.23 \quad \bullet
\end{aligned}$$

The increment notation may be used in the definition of the derivative of a function. All that is necessary is to substitute Δx for h in (3.4) as follows:

(3.22)
$$f'(x) = \lim_{\Delta x \to 0} \frac{f(x + \Delta x) - f(x)}{\Delta x} = \lim_{\Delta x \to 0} \frac{\Delta y}{\Delta x}$$

This may be phrased as follows: *The derivative of f is the limit of the ratio of the increment Δy of the dependent variable to the increment Δx of the independent variable as Δx approaches zero.* In Figure 3.15, note that $\Delta y / \Delta x$ is the slope of the secant line through P and Q. It follows from (3.22) that if $f'(x)$ exists, then

$$\frac{\Delta y}{\Delta x} \approx f'(x) \quad \text{if } \Delta x \approx 0 \qquad \text{or} \qquad \Delta y \approx f'(x)\,\Delta x \quad \text{if } \Delta x \approx 0.$$

We give $f'(x)\,\Delta x$ a special name in the next definition.

DEFINITION (3.23)

> Let $y = f(x)$ for a differentiable function f, and let Δx be an increment of x.
>
> (i) The **differential** dx of the independent variable x is $dx = \Delta x$.
>
> (ii) The **differential** dy of the dependent variable y is
> $$dy = f'(x)\, \Delta x = f'(x)\, dx$$

In (3.23) (i) we see that for the *independent* variable x, there is no difference between the increment Δx and the differential dx. However, for the *dependent* variable y in (ii), the value of dy depends on *both* x and dx.

By Definition (3.23) (ii), $dy = f'(x)\, dx$. If both sides of this equation are divided by dx (assuming that $dx \neq 0$) we obtain the following.

(3.24)

$$\frac{dy}{dx} = f'(x) = \lim_{\Delta x \to 0} \frac{\Delta y}{\Delta x}$$

Thus, the derivative $f'(x)$ may be expressed as a quotient of two differentials. This justifies the notation dy/dx introduced in (3.6) for the derivative of $y = f(x)$ with respect to x.

The discussion preceding Definition (3.23) gives us the following.

(3.25)

> If $\Delta x \approx 0$, then $\Delta y \approx dy = f'(x)\, dx = (D_x\, y)\, dx$.

Consequently, if $y = f(x)$, then dy can be used as an approximation to the exact change Δy of the dependent variable corresponding to a small change Δx in x. This observation is useful in applications where only a rough estimate of the change in y is desired.

EXAMPLE 2 If $y = 3x^2 - 5$, use dy to approximate Δy if x changes from 2 to 2.1.

SOLUTION Let $f(x) = 3x^2 - 5$. We saw in Example 1 that $\Delta y = 1.23$. Using Definition (3.23),

$$dy = f'(x)\, dx = 6x\, dx.$$

In the present example, $x = 2$, $\Delta x = dx = 0.1$, and

$$dy = (6)(2)(0.1) = 1.2$$

Note that the approximation 1.2 is correct to the nearest tenth. •

EXAMPLE 3 If $y = x^3$ and Δx is an increment of x, find (a) Δy, (b) dy, (c) $\Delta y - dy$, and (d) the value of $\Delta y - dy$ if $x = 1$ and $\Delta x = 0.02$.

SOLUTION

Using (3.21) with $f(x) = x^3$,

$$\Delta y = f(x + \Delta x) - f(x) = (x + \Delta x)^3 - x^3.$$

Applying the Binomial Theorem (3.9) with $n = 3$,

$$\Delta y = [x^3 + 3x^2(\Delta x) + 3x(\Delta x)^2 + (\Delta x)^3] - x^3$$
$$= 3x^2(\Delta x) + 3x(\Delta x)^2 + (\Delta x)^3$$

(b) By Definition (3.23) (ii),

$$dy = f'(x)\,dx = 3x^2\,dx = 3x^2(\Delta x)$$

(c) From parts (a) and (b),

$$\Delta y - dy = [3x^2(\Delta x) + 3x(\Delta x)^2 + (\Delta x)^3] - 3x^2(\Delta x)$$
$$= 3x(\Delta x)^2 + (\Delta x)^3$$

(d) Substituting $x = 1$ and $\Delta x = 0.02$ in part (c),

$$\Delta y - dy = 3(1)(0.02)^2 + (0.02)^3$$
$$= 3(0.0004) + (0.000008) = 0.001208 \approx 0.001$$

This shows that if dy is used to approximate Δy when x changes from 1 to 1.02, then the error involved is approximately 0.001. •

EXAMPLE 4 Use differentials to approximate the change in $\sin \theta$ if θ changes from $60°$ to $61°$.

SOLUTION If $y = \sin \theta = f(\theta)$, then

$$dy = f'(\theta)\,d\theta = \cos \theta\,d\theta.$$

When we use derivatives or differentials of trigonometric functions of angles it is essential to employ radian measure. Thus we let $\theta = 60° = \pi/3$ and $\Delta\theta = 1° = \pi/180$ in the formula for dy, obtaining

$$dy = \left(\cos \frac{\pi}{3}\right)\left(\frac{\pi}{180}\right) = \left(\frac{1}{2}\right)\left(\frac{\pi}{180}\right) = \frac{\pi}{360} \approx 0.0087$$

Using a calculator we find that to four decimal places,

$$\Delta y = \sin 61° - \sin 60° \approx 0.8746 - 0.8660 = 0.0086$$

Hence the error involved in using dy to approximate Δy is roughly 0.0001.

•

FIGURE 3.16

(i)

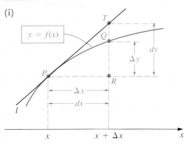

(ii)

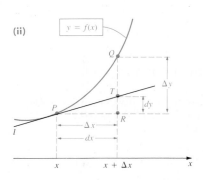

The graphs in Figure 3.16 illustrate geometric interpretations of dx and dy with the tangent line l at the point P. We have pictured both Δx and Δy as positive; however, either may take on negative values. We see from triangle PRT that the slope $f'(x)$ of the tangent line l is the ratio $RT/\Delta x$ with RT the length of the line segment between R and T. Hence

$$\frac{RT}{\Delta x} = f'(x) \quad \text{or} \quad RT = f'(x)\,\Delta x = dy.$$

A similar result holds if increments are negative. In general, if x is given an increment Δx, then dy is the amount that the *tangent line* at P rises (or falls) when the independent variable changes from x to $x + \Delta x$. This is in contrast to the amount Δy that the *graph* rises (or falls) between P and Q. This geometric interpretation also emphasizes the fact that $dy \approx \Delta y$ if Δx is small.

The following example illustrates how differentials may be used to find certain approximation formulas.

EXAMPLE 5

(a) Use differentials to obtain a formula for approximating the volume of a thin cylindrical shell of altitude h, inner radius r, and thickness T.

(b) What error is involved in using the formula?

SOLUTION

(a) Consider the cylindrical shell in Figure 3.17, with the thickness T denoted by Δr. We wish to find the volume contained between the *inner* cylinder of radius r and the *outer* cylinder of radius $r + \Delta r$. The volume V of the inner cylinder is

$$V = \pi r^2 h.$$

If we increase r by an amount Δr, then the volume of the shell is the change ΔV in V. Regarding V as a function of r and applying (3.25),

$$\Delta V \approx dV = (D_r V)\, dr = (2\pi rh)\, dr$$

with $dr = \Delta r$, since r is the independent variable. Hence a formula for approximating the volume of the shell is

$$\Delta V \approx (2\pi rh)\, dr = (2\pi rh)T$$

In words,

$$\text{Volume} \approx (\text{Area of inner cylinder wall}) \times (\text{Thickness}).$$

(b) The exact volume of the shell is

$$\Delta V = \pi(r + \Delta r)^2 h - \pi r^2 h$$

which simplifies to $\quad \Delta V = (2\pi rh)\, \Delta r + \pi h(\Delta r)^2.$

The error involved in using dV to approximate ΔV is

$$\Delta V - dV = \pi h(\Delta r)^2.$$

This indicates that the approximation dV is quite accurate if Δr is small in comparison to h. •

The next example illustrates the use of differentials in estimating errors that may arise because of approximate measurements. As indicated in the solution, *it is important to first consider general formulas involving the variables* that are being considered. Specific values should not be substituted for variables until the final steps of the solutions.

EXAMPLE 6 The radius of a spherical balloon is measured as 12 inches, with a maximum error in measurement of 0.06 inch. Approximate the maximum error in the calculated volume of the sphere.

SOLUTION We begin by considering *general* formulas involving the radius and the volume. Thus we let

$$x = \text{Measured value of the radius,}$$

and $\quad\quad\quad\quad dx = \Delta x = \text{Maximum error in } x.$

FIGURE 3.17

Assuming that Δx is positive, we have

$$x - \Delta x \le \text{exact radius} \le x + \Delta x.$$

FIGURE 3.18

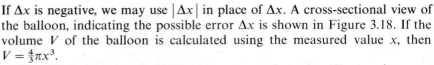

If Δx is negative, we may use $|\Delta x|$ in place of Δx. A cross-sectional view of the balloon, indicating the possible error Δx is shown in Figure 3.18. If the volume V of the balloon is calculated using the measured value x, then $V = \frac{4}{3}\pi x^3$.

Let ΔV be the change in V that corresponds to Δx. We may interpret ΔV as *the error in the calculated volume* caused by the error Δx. We may approximate ΔV by means of dV as follows:

$$\Delta V \approx dV = (D_x V)\,dx = 4\pi x^2\,dx.$$

Finally, we substitute specific values for x and dx. If $x = 12$ and $\Delta x = dx = \pm 0.06$, then

$$dV = 4\pi(12^2)(\pm 0.06) = \pm(34.56)\pi \approx \pm 109$$

Thus the maximum error in the calculated volume due to the error in measurement of the radius is approximately ± 109 in.^{3} •

The radius of the balloon in Example 6 was measured as 12 inches, with a maximum error of 0.06 inch. The ratio of 0.06 to 12 is called the *average error* in the measurement of the radius. Thus

$$\text{Average error} = \frac{0.06}{12} = 0.005$$

The significance of this number is that the error in measurement of the radius is, *on the average*, 0.005 inch per inch. The *percentage error* is defined as the average error multiplied by 100%. In this illustration,

$$\text{Percentage error} = (0.005)(100\%) = 0.5\%$$

The general definition of these concepts follows.

DEFINITION (3.26)

Let x denote a measurement with a maximum error Δx. By definition,

(i) **Average error** $= \dfrac{\Delta x}{x}$

(ii) **Percentage error** $= (\text{Average error}) \times (100\%)$

In terms of differentials, if x represents a measurement with a possible error of dx, then the average error is $(dx)/x$. Of course, if dx is an *approximation* to the error in x, then $(dx)/x$ is an *approximation* to the average error. These remarks are illustrated in the next example.

EXAMPLE 7 The radius of a spherical balloon is measured as 12 inches with a maximum error in measurement of 0.06 inch. Approximate the average error and the percentage error for the calculated value of the volume.

SOLUTION This is a continuation of Example 6, where we found that if the maximum error in the measurement of the radius is 0.06 inch, then the maximum error ΔV in the calculated volume V of the balloon may be approximated by

$$dV = \pm(34.56)\pi \approx \pm 109 \text{ in.}^3$$

The calculated volume is

$$V = \tfrac{4}{3}\pi(12)^3 = 2304\pi \text{ in.}^3$$

Applying Definition (3.26) to the variable V:

$$\text{Average error} = \frac{\Delta V}{V} \approx \frac{dV}{V} = \frac{\pm(34.56)\pi}{2304\pi} \approx \pm 0.015$$

$$\text{Percentage error} \approx \pm(0.015)(100\%) = \pm 1.5\%$$

Thus, *on the average*, there is an error of ± 0.015 in.^{3} per in.^{3} of calculated volume. Following Example 6, we found that the percentage error for the radius was 0.5%. Note that this leads to a percentage error of 1.5% for the volume. •

EXERCISES 3.5

Exer. 1–4: (a) Use Definition (3.21) to find a general formula for Δy; (b) use Δy to calculate the change in y corresponding to the stated values of x and Δx.

1 $y = 2x^2 - 4x + 5$, $x = 2$, $\Delta x = -0.2$

2 $y = x^3 - 4$, $x = -1$, $\Delta x = 0.1$

3 $y = 1/x^2$, $x = 3$, $\Delta x = 0.3$

4 $y = 1/(2 + x)$, $x = 0$, $\Delta x = -0.03$

Exer. 5–12: Find (a) Δy, (b) dy, and (c) $dy - \Delta y$.

5 $y = 3x^2 + 5x - 2$ 6 $y = 4 - 7x - 2x^2$

7 $y = 1/x$ 8 $y = 7x + 12$

9 $y = 4 - 9x$ 10 $y = 8$

11 $y = x^4$ 12 $y = x^{-2}$

Exer. 13–14: Use differentials to approximate the change in $f(x)$ if x changes from a to b.

13 $f(x) = 4x^5 - 6x^4 + 3x^2 - 5$; $a = 1$, $b = 1.03$

14 $f(x) = -3x^3 + 8x - 7$; $a = 4$, $b = 3.96$

15 If $y = 2 \sin \theta + \cos \theta$, find dy. Use dy to approximate the change in y if θ changes from 30° to 27°.

16 If $w = \csc \varphi + \cot \varphi$, find dw. Use dw to approximate the change in w if φ changes from 45° to 46°.

17 The radius of a circular manhole cover is estimated to be 16 inches with a maximum error in measurement of 0.06 inch. Use differentials to estimate the maximum error in the calculated area of one side of the cover. Approximate the average error and the percentage error.

18 The length of a side of a square floor tile is estimated as 1 foot with a maximum error in measurement of $\tfrac{1}{16}$ inch. Use differentials to estimate the maximum error in the calculated area. Approximate the average error and the percentage error.

19 Use differentials to approximate the increase in volume of a cube if the length of each edge changes from 10 inches to 10.1 inches. What is the exact change in volume?

20 A spherical balloon is being inflated with gas. Use differentials to approximate the increase in surface area of the balloon if the diameter changes from 2 feet to 2.02 feet.

21 One side of a house has the shape of a square surmounted by an equilateral triangle. If the length of the base is measured as 48 feet with a maximum error in measurement of 1 inch, calculate the area of the side and use differentials to estimate the maximum error in the calculation. Approximate the average error and the percentage error.

22 Small errors in measurements of dimensions of large containers can have a marked effect on calculated volumes. A silo has the shape of a right circular cylinder surmounted by a hemisphere (see figure). The altitude of the cylinder is exactly 50 feet. The circumference of the base is measured as 30 feet, with a maximum error in measurement of 6 inches. Calculate the volume of the silo from these measurements and use differentials to estimate the maximum error in the calculation. Approximate the average error and the percentage error.

EXERCISE 22

23 As sand leaks out of a container, it forms a conical pile whose altitude is always the same as the radius. If, at a certain instant, the radius is 10 cm, use differentials to approximate the change in radius that will increase the volume of the pile by 2 cm³.

24 Use the technique illustrated in Example 5 of this section to find a formula that can be used to approximate the volume of a thin shell-shaped solid whose surface is (a) spherical; (b) cubical.

25 Newton's law of gravitation states that the force F of attraction between two particles having masses m_1 and m_2 is given by $F = gm_1m_2/s^2$ for a constant g and the distance s between the particles. If $s = 20$ cm, use differentials to approximate the change in s that will increase F by 10%.

26 Boyle's law states that the pressure p and volume v of a confined gas are related by the formula $pv = c$ for a constant c, or equivalently, by $p = c/v$ for $v \neq 0$. Show that dp and dv are related by means of the formula $p\,dv + v\,dp = 0$.

27 Use differentials to approximate $(0.98)^4$. (*Hint:* Let $y = f(x) = x^4$ and consider $f(x + \Delta x) = f(x) + \Delta y$, with $x = 1$ and $\Delta x = -0.02$.) What is the exact value of $(0.98)^4$?

28 Use differentials to approximate

$$N = (2.01)^4 - 3(2.01)^3 + 4(2.01)^2 - 5.$$

What is the exact value of N?

29 The area A of a square of side s is given by $A = s^2$. If s increases by an amount Δs, give geometric illustrations of dA and $\Delta A - dA$.

30 The volume V of a cube of edge x is given by $V = x^3$. If x increases by an amount Δx, give geometric illustrations of dV and $\Delta V - dV$.

31 Constriction of arterioles is a cause of high blood pressure. It has been verified experimentally that as blood flows through an arteriole of fixed length, the pressure difference between the two ends of the arteriole is inversely proportional to the fourth power of the radius. If the radius of an arteriole decreases by 10%, use differentials to find the percentage change in the pressure difference.

32 The electrical resistance R of a wire is directly proportional to its length and inversely proportional to the square of its diameter. If the length is fixed, how accurately must the diameter be measured (in terms of percentage error) to keep the percentage error in R between -3% and 3%?

33 Use differentials to approximate the change in $\tan \theta$ if θ changes from $30°$ to $28°$.

34 An isosceles triangle has equal sides of length 12 inches. If the angle θ between these sides is increased from $30°$ to $33°$, use differentials to approximate the change in the area of the triangle.

35 If an object of weight W pounds is pulled along a horizontal plane by a force applied to a rope that is attached to the object, and if the rope makes an angle θ with the horizontal, then the magnitude of the force is given by

$$F = \frac{\mu W}{\mu \sin \theta + \cos \theta}$$

for a constant μ, called the *coefficient of friction*. Suppose that a man is pulling a 100-pound box along a floor, and that $\mu = 0.2$ (see figure). If θ is changed from $45°$ to $46°$, use differentials to approximate the change in the force that must be applied.

EXERCISE 35

36 It will be shown in Chapter 15 that if a projectile is fired from a cannon with an initial velocity v_0 and at an angle α to the horizontal, then its maximum height h and range R are given by

$$h = \frac{v_0^2 \sin^2 \alpha}{2g} \quad \text{and} \quad R = \frac{2v_0^2 \sin \alpha \cos \alpha}{g}.$$

Suppose $v_0 = 100$ ft/sec and $g = 32$ ft/sec^2. If α is increased from $30°$ to $30°30'$, use differentials to estimate the changes in h and R.

37 At a point 20 feet from the base of a flagpole, the angle of elevation of the top of the pole is measured as $60°$, with a possible error of $15'$. Use differentials to approximate the error in the calculated height of the pole.

38 A spacelab circles the earth at an altitude of 380 miles. When an astronaut views the horizon, the angle θ shown in the figure is $65.8°$ with a possible maximum error of $0.5°$. Use differentials to approximate the error in the astronaut's calculation of the radius of the earth.

EXERCISE 38

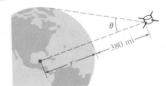

39 The Great Pyramid of Egypt has a square base of 230 meters (see figure). To estimate the height h of this massive structure, an observer stands at the midpoint of one of the sides and views the apex of the pyramid. The angle of elevation φ is found to be $52°$. How accurate must this measurement be to keep the error in h between -1 meter and 1 meter?

EXERCISE 39

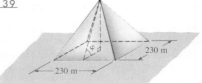

40 For light emanating from a point source P to a surface, the illuminance E on the surface is inversely proportional to the square of the distance s from the source, and is directly proportional to the cosine of the angle θ between the direction of flow and the normal to the surface (see figure). If θ is decreased from $21°$ to $20°$, use differentials to approximate the percentage increase in illuminance.

EXERCISE 40

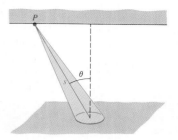

3.6 THE CHAIN RULE

The rules for derivatives obtained in previous sections are limited in scope because they can only be used for sums, differences, products, and quotients that involve x^n, $\sin x$, $\cos x$, $\tan x$, and so on. There is no rule that can be applied *directly* to an expression such as $\sin 2x$. Clearly,

$$D_x \sin 2x \neq \cos 2x,$$

for if we let

$$y = \sin 2x = 2 \sin x \cos x$$

and apply the Product Rule,

$$
\begin{aligned}
D_x y &= 2\, D_x (\sin x \cos x) \\
&= 2[\sin x\, (D_x \cos x) + \cos x\, (D_x \sin x)] \\
&= 2[\sin x\, (-\sin x) + \cos x\, (\cos x)] \\
&= 2[-\sin^2 x + \cos^2 x] = 2 \cos 2x.
\end{aligned}
$$

Hence, $\qquad\qquad\qquad D_x \sin 2x = 2 \cos 2x.$

Since these manipulations are rather cumbersome, let us seek a more direct method of finding the derivative of $y = \sin 2x$. The key is to regard y as a *composite* function of x. Recall that if f and g are functions such that

$$y = f(u) \quad \text{and} \quad u = g(x),$$

and if $g(x)$ is in the domain of f, then we may write

$$y = f(u) = f(g(x)),$$

that is, y is a function of x. This new function is the composite function $f \circ g$ defined in Section 1.5. Note that $y = \sin 2x$ may be expressed in this way by letting

$$y = \sin u \quad \text{and} \quad u = 2x.$$

If we can find a general rule for differentiating $f(g(x))$, then, as a special case, we may apply it to $y = \sin 2x$ and, in fact, to any expression of the form $\sin g(x)$.

To get some idea of the type of rule to expect, let us return to the equations

$$y = f(u) \quad \text{and} \quad u = g(x).$$

Our goal is to find a formula for the derivative dy/dx of the composite function given by $y = f(g(x))$. If f and g are differentiable, then using differential notation,

$$\frac{dy}{du} = f'(u) \quad \text{and} \quad \frac{du}{dx} = g'(x).$$

If we consider the product

$$\frac{dy}{du}\frac{du}{dx}$$

and treat the derivatives as quotients of differentials, then the product *suggests* the following rule:

$$\frac{dy}{dx} = \frac{dy}{du}\frac{du}{dx} = f'(u)g'(x).$$

Note that this rule does, in fact, lead to the correct derivative of $y = \sin 2x$, for if we write

$$y = \sin u \quad \text{and} \quad u = 2x$$

and use the rule, we obtain

$$\frac{dy}{dx} = \frac{dy}{du}\frac{du}{dx} = (\cos u)(2) = 2\cos u = 2\cos 2x.$$

Although we have not *proved* that this rule is valid, it makes the next theorem plausible. We assume that variables are chosen such that the composite function $f \circ g$ is defined, and that if g has a derivative at x, then f has a derivative at $g(x)$.

THE CHAIN RULE (3.27)

> If $y = f(u)$, $u = g(x)$, and the derivatives $\dfrac{dy}{du}$ and $\dfrac{du}{dx}$ both exist, then the composite function defined by $y = f(g(x))$ has a derivative given by
> $$\frac{dy}{dx} = \frac{dy}{du}\frac{du}{dx} = f'(u)g'(x) = f'(g(x))g'(x)$$

PARTIAL PROOF Let Δx be an increment such that both x and $x + \Delta x$ are in the domain of the composite function. Since $y = f(g(x))$, the corresponding increment of y is given by

$$\Delta y = f(g(x + \Delta x)) - f(g(x)).$$

If the composite function has a derivative at x, then by (3.24),

$$\frac{dy}{dx} = \lim_{\Delta x \to 0} \frac{\Delta y}{\Delta x}.$$

Next consider $u = g(x)$ and let Δu be the increment of u that corresponds to Δx, that is,

$$\Delta u = g(x + \Delta x) - g(x).$$

Since

$$g(x + \Delta x) = g(x) + \Delta u = u + \Delta u$$

we may express the formula $\Delta y = f(g(x + \Delta x)) - f(g(x))$ as

$$\Delta y = f(u + \Delta u) - f(u).$$

If $y = f(u)$ is differentiable at u, then

$$\frac{dy}{du} = f'(u) = \lim_{\Delta u \to 0} \frac{\Delta y}{\Delta u}.$$

Similarly, if $u = g(x)$ is differentiable at x, then

$$\frac{du}{dx} = g'(x) = \lim_{\Delta x \to 0} \frac{\Delta u}{\Delta x}.$$

Let us assume that there exists an open interval I containing x such that whenever $x + \Delta x$ is in I and $\Delta x \neq 0$, then $\Delta u \neq 0$. In this case we may write

$$\frac{dy}{dx} = \lim_{\Delta x \to 0} \frac{\Delta y}{\Delta x} = \lim_{\Delta x \to 0} \left(\frac{\Delta y}{\Delta u} \frac{\Delta u}{\Delta x} \right) = \left(\lim_{\Delta x \to 0} \frac{\Delta y}{\Delta u} \right)\left(\lim_{\Delta x \to 0} \frac{\Delta u}{\Delta x} \right)$$

provided the limits exist. Since g is differentiable at x, it is continuous at x. Hence if $\Delta x \to 0$, then $g(x + \Delta x)$ approaches $g(x)$ and, therefore, $\Delta u \to 0$. It follows that the last limit formula displayed above may be written

$$\frac{dy}{dx} = \left(\lim_{\Delta u \to 0} \frac{\Delta y}{\Delta u} \right)\left(\lim_{\Delta x \to 0} \frac{\Delta u}{\Delta x} \right)$$

$$= \left(\frac{dy}{du} \right)\left(\frac{du}{dx} \right) = f'(u)g'(x) = f'(g(x))g'(x)$$

which is what we wished to prove.

In most applications of the Chain Rule, $u = g(x)$ has the property assumed at the beginning of the preceding paragraph. If g does not satisfy this property, then every open interval containing x contains a number $x + \Delta x$, with $\Delta x \neq 0$, such that $\Delta u = 0$. In this case our proof is invalid, since Δu occurs in a denominator. To construct a proof that takes functions of this type into account, it is necessary to introduce additional techniques. *A complete proof of the Chain Rule is given in Appendix II.* • •

EXAMPLE 1 If $y = u^3$ and $u = x^2 + 1$, find dy/dx.

SOLUTION The problem could be worked by substituting $x^2 + 1$ for u in the first equation, obtaining $y = (x^2 + 1)^3 = x^6 + 3x^4 + 3x^2 + 1$, and then calculating dy/dx. Another approach is to use the Chain Rule (3.27):

$$\frac{dy}{dx} = \frac{dy}{du}\frac{du}{dx} = (3u^2)(2x).$$

Since $u = x^2 + 1$,

$$\frac{dy}{dx} = 3(x^2 + 1)^2(2x) = 6x(x^2 + 1)^2 \quad \bullet$$

One of the main uses for the Chain Rule is to establish other differentiation formulas. As a first illustration we shall obtain a formula for the derivative of a power of a function.

THE POWER RULE (3.28)
FOR FUNCTIONS

If g is a differentiable function and n is an integer, then

$$D_x [g(x)]^n = n[g(x)]^{n-1} D_x [g(x)]$$

PROOF If we let $y = u^n$ and $u = g(x)$, then $y = [g(x)]^n$ and, by the Chain Rule,

$$\frac{dy}{dx} = \frac{dy}{du} \frac{du}{dx} = nu^{n-1} D_x u = n[g(x)]^{n-1} D_x [g(x)].$$

This completes the proof. • •

The following alternative form of the Power Rule for Functions is easier to remember than (3.28).

POWER RULE (3.29)
(ALTERNATIVE FORM)

$$D_x(u^n) = nu^{n-1} D_x u, \quad \text{for } u = g(x)$$

Note that if $u = x$, then $D_x u = 1$ and (3.29) reduces to (3.10).

EXAMPLE 2 Find $f'(x)$ if $f(x) = (x^5 - 4x + 8)^7$.

SOLUTION Using the Power Rule (3.29) with $u = x^5 - 4x + 8$ and $n = 7$,

$$f'(x) = D_x (x^5 - 4x + 8)^7 = 7(x^5 - 4x + 8)^6 D_x (x^5 - 4x + 8)$$
$$= 7(x^5 - 4x + 8)^6(5x^4 - 4) \qquad \bullet$$

EXAMPLE 3 Find $\dfrac{dy}{dx}$ if $y = \dfrac{1}{(4x^2 + 6x - 7)^3}$.

SOLUTION Writing $y = (4x^2 + 6x - 7)^{-3}$ and applying the Power Rule (3.29) with $u = 4x^2 + 6x - 7$ and $n = -3$,

$$\frac{dy}{dx} = D_x (4x^2 + 6x - 7)^{-3}$$

$$= -3(4x^2 + 6x - 7)^{-4} D_x (4x^2 + 6x - 7)$$
$$= -3(4x^2 + 6x - 7)^{-4}(8x + 6)$$

$$= \frac{-6(4x + 3)}{(4x^2 + 6x - 7)^4} \qquad \bullet$$

EXAMPLE 4 If $F(z) = (2z + 5)^3(3z - 1)^4$, find $F'(z)$.

SOLUTION Using first the Product Rule, second the Power Rule, and then factoring the result,

$$F'(z) = (2z + 5)^3 D_z (3z - 1)^4 + (3z - 1)^4 D_z (2z + 5)^3$$
$$= (2z + 5)^3 \cdot 4(3z - 1)^3(3) + (3z - 1)^4 \cdot 3(2z + 5)^2(2)$$
$$= 6(2z + 5)^2(3z - 1)^3[2(2z + 5) + (3z - 1)]$$
$$= 6(2z + 5)^2(3z - 1)^3(7z + 9) \qquad \bullet$$

If g is a differentiable function and $y = g(x)$, then (3.29) may also be written in any of the following forms.

$$(3.30) \quad \boxed{D_x(y^n) = ny^{n-1} D_x y = ny^{n-1} y' = ny^{n-1} \frac{dy}{dx}}$$

The dependent variable y represents the expression $g(x)$ and consequently, when differentiating y^n it is *essential* to multiply ny^{n-1} by the derivative y'. Thus, in general, $D_x y^r \neq ry^{r-1}$. Formula (3.30) will be extremely important in our work with implicit functions in the next section.

As another application of the Chain Rule we may prove the following generalization of Theorem 3.20.

THEOREM (3.31)

If $u = g(x)$ and g is a differentiable function, then

$$D_x \sin u = (\cos u) D_x u \qquad\qquad D_x \cos u = (-\sin u) D_x u$$

$$D_x \tan u = (\sec^2 u) D_x u \qquad\qquad D_x \cot u = (-\csc^2 u) D_x u$$

$$D_x \sec u = (\sec u \tan u) D_x u \qquad D_x \csc u = (-\csc u \cot u) D_x u$$

PROOF If we let $y = \sin u$, then by (3.20),

$$\frac{dy}{du} = \cos u.$$

Applying the Chain Rule (3.27),

$$\frac{dy}{dx} = \frac{dy}{du}\frac{du}{dx} = (\cos u) D_x u.$$

The remaining formulas may be obtained in similar fashion. • •

Note that Theorem 3.20 is the special case of Theorem 3.31 in which $u = x$.

EXAMPLE 5 Find $\dfrac{dy}{dx}$ if $y = \cos(5x^3)$.

SOLUTION Using the formula for $D_x \cos u$ in Theorem (3.31) with $u = 5x^3$,

$$\begin{aligned}
\frac{dy}{dx} = D_x y = D_x \cos(5x^3) &= [-\sin(5x^3)] D_x(5x^3)\\
&= [-\sin(5x^3)](15x^2)\\
&= -15x^2 \sin(5x^3) \qquad\qquad •
\end{aligned}$$

EXAMPLE 6 Find $f'(x)$ if $f(x) = \tan^3 4x$.

SOLUTION First note that $f(x) = \tan^3 4x = (\tan 4x)^3$. Applying the Power Rule for Functions (3.29) with $u = \tan 4x$ and $n = 3$,

$$f'(x) = 3(\tan 4x)^2 D_x \tan 4x = (3 \tan^2 4x) D_x \tan 4x.$$

Next, by Theorem (3.31),

$$D_x \tan 4x = (\sec^2 4x) D_x(4x) = (\sec^2 4x)(4) = 4 \sec^2 4x.$$

Thus $\qquad f'(x) = (3 \tan^2 4x)(4 \sec^2 4x) = 12 \tan^2 4x \sec^2 4x$ •

FIGURE 3.19

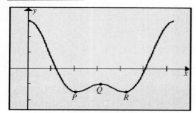

EXAMPLE 7 A computer-generated graph of $y = \cos 2x + 2 \cos x$ for $0 \leq x \leq 2\pi$ is shown in Figure 3.19. Find the points at which the tangent line is horizontal.

SOLUTION Differentiating,

$$D_x y = (-\sin 2x) D_x (2x) + 2(-\sin x)$$
$$= -2 \sin 2x - 2 \sin x.$$

The tangent line is horizontal if its slope $D_x y$ is 0, that is, if

$$-2 \sin 2x - 2 \sin x = 0 \quad \text{or} \quad 2 \sin 2x + 2 \sin x = 0.$$

Using the double-angle formula $\sin 2x = 2 \sin x \cos x$ gives us

$$4 \sin x \cos x + 2 \sin x = 0$$

or equivalently,

$$2 \sin x (2 \cos x + 1) = 0.$$

Thus, either

$$\sin x = 0 \quad \text{or} \quad 2 \cos x + 1 = 0,$$

that is,

$$\sin x = 0 \quad \text{or} \quad \cos x = -\tfrac{1}{2}.$$

The solutions of these equations for $0 \leq x \leq 2\pi$ are

$$0, \quad \pi, \quad 2\pi, \quad 2\pi/3, \quad 4\pi/3.$$

The two solutions $x = 0$ and $x = 2\pi$ tell us that there are horizontal tangent lines at the endpoints of the interval $[0, 2\pi]$. The remaining solutions $2\pi/3$, π, and $4\pi/3$ are the x-coordinates of the points P, Q, and R shown in Figure 3.19. Using $y = \cos 2x + 2 \cos x$ we see that horizontal tangent lines occur at the points

$$(0, 3), \quad (2\pi/3, -1.5), \quad (\pi, -1), \quad (4\pi/3, -1.5), \quad (2\pi, 3).$$

If only approximate solutions are desired, then using $\pi \approx 3.1$, we obtain

$$(0, 3), \quad (2.1, -1.5), \quad (3.1, -1), \quad (4.2, -1.5), \quad (6.3, 3) \quad \bullet$$

In our previous work involving motion on a coordinate line, the position function s was usually algebraic. The following type of motion involves trigonometric functions.

DEFINITION (3.32)

> A point P that moves on a coordinate line l such that its distance $s(t)$ from the origin at time t is given by either
>
> $$s(t) = a \cos (\omega t + b) \quad \text{or} \quad s(t) = a \sin (\omega t + b)$$
>
> for real numbers a, b, and $\omega > 0$, is said to be in **simple harmonic motion.**

Simple harmonic motion may also be defined by requiring that the acceleration $a(t)$ satisfy the condition

$$a(t) = -\omega^2 s(t)$$

for every t. We can show (see Section 19.5) that this condition is equivalent to Definition (3.32).

In simple harmonic motion the point P oscillates between the points on l with coordinates $-a$ and a. The **amplitude** of the motion is the maximum displacement $|a|$ of the point from the origin. The **period** is the time $2\pi/\omega$ required for one complete oscillation. The **frequency** $\omega/2\pi$ is the number of oscillations per unit of time.

Simple harmonic motion takes place in many different types of wave motion, such as water waves, sound waves, radio waves, light waves, and distortional waves that are present in vibrating bodies. Functions of the type defined in (3.32) also occur in the analysis of electrical circuits that contain an alternating electromotive force and current.

As another example of simple harmonic motion, consider a spring with an attached weight, which is oscillating vertically relative to a coordinate line, as illustrated in Figure 3.20. The number $s(t)$ represents the coordinate of a fixed point P in the weight, and we assume that the amplitude $|a|$ of the motion is constant. In this case, there is no frictional force retarding the motion. If friction is present, then the amplitude decreases with time, and the motion is *damped*.

FIGURE 3.20

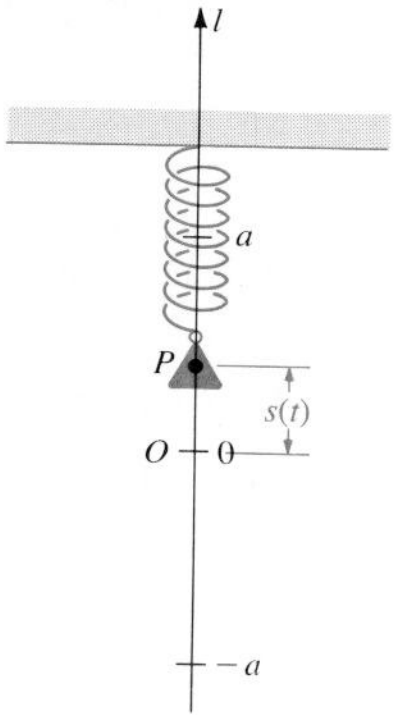

EXAMPLE 8 Suppose the weight shown in Figure 3.20 is oscillating and

$$s(t) = 10 \cos \frac{\pi}{6} t$$

for t in seconds and $s(t)$ in centimeters. Discuss the motion of the weight.

SOLUTION By Definition (3.32), the motion is simple harmonic with $a = 10$, $b = 0$, and $\omega = \pi/6$. The amplitude a is 10 cm. The period is $2\pi/\omega = 2\pi/(\pi/6) = 12$. Thus, one complete oscillation takes place every 12 seconds. The frequency is $\omega/2\pi = (\pi/6)/2\pi = \frac{1}{12}$; that is, $\frac{1}{12}$ of an oscillation takes place each second.

Let us examine the motion during the time interval $[0, 12]$. The velocity and acceleration functions are given by

$$v(t) = s'(t) = 10 \left(-\sin \frac{\pi}{6} t \right) \cdot \left(\frac{\pi}{6} \right) = -\frac{5\pi}{3} \sin \frac{\pi}{6} t,$$

$$a(t) = v'(t) = -\frac{5\pi}{3} \left(\cos \frac{\pi}{6} t \right) \cdot \left(\frac{\pi}{6} \right) = -\frac{5\pi^2}{18} \cos \frac{\pi}{6} t.$$

The velocity is 0 at $t = 0$, $t = 6$, and $t = 12$, since $\sin [(\pi/6)t] = 0$ for these values of t. The acceleration is 0 at $t = 3$ and $t = 9$, since in these cases $\cos [(\pi/6)t] = 0$. The times at which the velocity and acceleration are zero lead us to examine the time intervals $(0, 3)$, $(3, 6)$, $(6, 9)$, and $(9, 12)$. The following table displays the main characteristics of the motion. The signs of $v(t)$ and $a(t)$ in the intervals can be determined using test values (verify each entry).

Time interval	Sign of $v(t)$	Direction of motion	Sign of $a(t)$	Variation of $v(t)$	Speed $|v(t)|$
$(0, 3)$	$-$	Downward	$-$	Decreasing	Increasing
$(3, 6)$	$-$	Downward	$+$	Increasing	Decreasing
$(6, 9)$	$+$	Upward	$+$	Increasing	Increasing
$(9, 12)$	$+$	Upward	$-$	Decreasing	Decreasing

Note that if $0 < t < 3$, the velocity $v(t)$ is negative and decreasing; that is, $v(t)$ becomes *more* negative. This implies that the speed $|v(t)|$ is *increasing*. If $3 < t < 6$, the velocity is negative and increasing ($v(t)$ becomes *less* negative). In this case the speed of P is *decreasing* in the time interval $(3, 6)$. Similar remarks can be made for the intervals $(6, 9)$ and $(9, 12)$.

We may summarize the motion of P as follows: At $t = 0$, $s(0) = 10$ and the point P is 10 cm above the origin O. It then moves downward, gaining speed until it reaches the origin O at $t = 3$. It then slows down until it reaches a point 10 cm below O at the end of 6 sec. The direction of motion is then reversed, and the weight moves upward, gaining speed until it reaches O at $t = 9$, after which it slows down until it returns to its original position at the end of 12 sec. The direction of motion is then reversed again, and the same pattern is repeated indefinitely. •

EXERCISES 3.6

Exer. 1–42: Differentiate the function.

1 $f(x) = (x^2 - 3x + 8)^3$

2 $f(x) = (4x^3 + 2x^2 - x - 3)^2$

3 $g(x) = (8x - 7)^{-5}$

4 $k(x) = (5x^2 - 2x + 1)^{-3}$

5 $f(x) = \dfrac{x}{(x^2 - 1)^4}$

6 $g(x) = \dfrac{x^4 - 3x^2 + 1}{(2x + 3)^4}$

7 $f(x) = (8x^3 - 2x^2 + x - 7)^5$

8 $g(w) = (w^4 - 8w^2 + 15)^4$

9 $F(v) = (17v - 5)^{1000}$

10 $s(t) = (4t^5 - 3t^3 + 2t)^{-2}$

11 $k(x) = \sin(x^2 + 2)$

12 $f(t) = \cos(4 - 3t)$

13 $H(\theta) = \cos^5 3\theta$

14 $g(x) = \sin^4(x^3)$

15 $G(x) = x^3 \cos(1/x)$

16 $K(\theta) = (\sin 2\theta)/(1 + \cos 2\theta)$

17 $h(v) = \tan(8v + 3)$

18 $F(r) = \cot(r/2)$

19 $g(z) = \sec(2z + 1)^2$

20 $k(z) = \csc(z^2 + 4)$

21 $H(s) = \cot(s^3 - 2s)$

22 $f(x) = \tan(2x^2 + 3)$

23 $f(x) = \cos(3x^2) + \cos^2 3x$

24 $g(w) = \tan^3 6w$

25 $F(\varphi) = \csc^2 2\varphi$

26 $M(x) = \sec(1/x^2)$

27 $K(z) = z^2 \cot 5z$

28 $G(s) = s \csc(s^2)$

29 $h(\theta) = \tan^2 \theta \sec^3 \theta$

30 $H(u) = u^2 \sec^3 4u$

31 $N(x) = (6x - 7)^3(8x^2 + 9)^2$

32 $f(w) = (2w^2 - 3w + 1)(3w + 2)^4$

33 $g(z) = \left(z^2 - \dfrac{1}{z^2}\right)^6$

34 $S(t) = \left(\dfrac{3t + 4}{6t - 7}\right)^3$

35 $N(x) = (\sin 5x - \cos 5x)^5$

36 $p(v) = \sin 4v \csc 4v$

37 $T(w) = \cot^3(3w + 1)$

38 $g(r) = \sin(2r + 3)^4$

39 $h(w) = (\cos 4w)/(1 - \sin 4w)$

40 $f(x) = (\sec 2x)/(1 + \tan 2x)$

41 $f(x) = \tan^3 2x - \sec^3 2x$

42 $h(\varphi) = (\tan 2\varphi - \sec 2\varphi)^3$

Exer. 43–46: (a) Find equations of the tangent line and the normal line to the graph of the equation at point P. (b) Find the points on the graph at which the tangent line is horizontal.

43 $y = (4x^2 - 8x + 3)^4$; $P(2, 81)$

44 $y = \left(x + \dfrac{1}{x}\right)^5$; $P(1, 32)$

45 $y = (2x - 1)^{10}$; $P(1, 1)$

46 $y = x + \cos 2x$; $P(0, 1)$

Exer. 47–48: For the function indicated by the computer-generated graph, find the x-coordinates of all points on the graph at which the tangent line is horizontal.

47 $y = \sin 4x - 4 \sin x$; $-2\pi \le x \le 2\pi$ (*Hint:* Use the factoring formulas for trigonometric functions.)

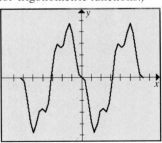

48 $y = \cos 3x - 3 \cos x$; $-2\pi \le x \le -2\pi$

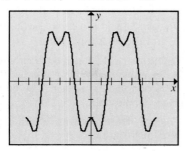

Exer. 49–52: *s is the position function of a particle moving in simple harmonic motion, and t is the time in seconds. Find the amplitude, period, and frequency, and describe the motion of the particle during one complete oscillation.*

49 $s(t) = 5 \cos (\pi/4)t$

50 $s(t) = 4 \sin \pi t$

51 $s(t) = 6 \sin (2\pi/3)t$

52 $s(t) = 3 \cos 2t$

53 The electromotive force V and current I in an alternating-current circuit are given by

$$V = 220 \sin 360\pi t,$$
$$I = 20 \sin (360\pi t - \tfrac{1}{4}\pi)$$

Find the rates of change of V and I with respect to time at $t = 1$.

54 The annual variation in temperature T (in °C) in Vancouver, B.C., is well described by the formula

$$T = 10 + 14.8 \sin \frac{\pi}{6}(t - 3)$$

with t in months and $t = 0$ corresponding to January 1. Approximate the rate at which the temperature is changing at time $t = 3$ (March 1) and at time $t = 10$ (November 1). At what time of the year is the temperature changing most rapidly?

55 The graph in the figure shows the rise and fall of the water level in Boston Harbor during a particular 24-hour period.

(a) Approximate the water level y by means of an expression of the form $y = a \sin (bt + c) + d$, with $t = 0$ corresponding to midnight.

(b) Approximately how fast is the water level rising at 12 noon?

EXERCISE 55

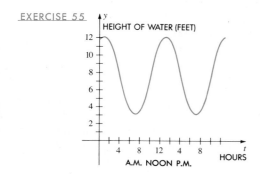

56 A *tsunami wave* is a tidal wave caused by an earthquake beneath the sea. These waves can be more than a hundred feet in height and can travel at great speeds. Engineers represent tsunami waves by trigonometric expressions of the form $y = a \cos bt$ and use these representations to estimate the effectiveness of sea walls. Suppose that a wave has height $h = 25$ feet, period 30 minutes, and is traveling at the rate of 180 ft/sec.

(a) Let (x, y) be a point on the wave represented in the figure. Express y as a function of t if $y = 25$ feet when $t = 0$.

(b) How fast is the wave rising (or falling) when $y = 10$ feet?

EXERCISE 56

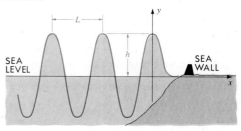

57 A cork bobs up and down in a pail of water such that the distance from the bottom of the pail to the center of the cork at time $t \ge 0$ is given by $s(t) = 12 + \cos (\pi t)$ for $s(t)$ in inches and t in seconds (see figure).

(a) Find the velocity of the cork at $t = 0, t = \tfrac{1}{2}, t = 1, t = \tfrac{3}{2}$, and $t = 2$.

(b) During what time intervals is the cork rising? When is it falling?

EXERCISE 57

58 A particle in a vibrating spring is moving vertically such that its distance $s(t)$ from a fixed point on the line of vibration is given by $s(t) = 4 + \tfrac{1}{25} \sin (100\pi t)$ for $s(t)$ in cm and t in seconds.

(a) How long does it take the particle to make one complete vibration?

(b) Find the velocity of the particle at $t = 1$, $t = 1.005$, $t = 1.01$, and $t = 1.015$.

59 If a body of mass m has velocity v, then its *kinetic energy* K is given by $K = \tfrac{1}{2}mv^2$. If v is a function of time t, use the Chain Rule to find a formula for dK/dt.

60 As a spherical weather balloon is being inflated, its radius r is a function of time t. If V is the volume of the balloon, use the Chain Rule to find a formula for dV/dt.

61 When launched into space, an astronaut's body weight decreases until a state of weightlessness is achieved. The weight W of a 150-pound astronaut at an altitude of x km above sea level is given by

$$W = 150 \left[\frac{6400}{6400 + x} \right]^2.$$

If the space vehicle is moving away from the earth's surface at the rate of 6 km/sec, at what rate is the astronaut's body weight decreasing when $x = 1000$ km?

62 The length-weight relationship for Pacific halibut is well described by the formula $W = 10.375L^3$, for length L in meters and weight W in kg. The rate of growth in length dL/dt is given by $0.18(2 - L)$, for t in years.

(a) Find a formula for the rate of growth in weight dW/dt in terms of L.

(b) Use the formula in part (a) to estimate the rate of growth in weight of a halibut weighing 20 kg.

63 If $k(x) = f(g(x))$, and if $f(2) = -4$, $g(2) = 2$, $f'(2) = 3$, and $g'(2) = 5$, find $k(2)$ and $k'(2)$.

64 Let p, q, and r be functions such that $p(z) = q(r(z))$. If $r(3) = 3$, $q(3) = -2$, $r'(3) = 4$, and $q'(3) = 6$, find $p(3)$ and $p'(3)$.

65 If $f(t) = g(h(t))$, and if $f(4) = 3$, $g(4) = 3$, $h(4) = 4$, $f'(4) = 2$, and $g'(4) = -5$, find $h'(4)$.

66 If $u(x) = v(w(x))$, and if $v(0) = -1$, $w(0) = 0$, $u(0) = -1$, $v'(0) = -3$, and $u'(0) = 2$, find $w'(0)$.

67 Refer to Definition (1.21). Let f be differentiable. Use the Chain Rule to prove that (a) if f is even, then f' is odd, and (b) if f is odd, then f' is even. Use polynomial functions to give examples of (a) and (b).

68 Use the Chain Rule, the derivative formula for $D_x \sin u$, together with the identities

$$\cos x = \sin\left(\frac{\pi}{2} - x\right) \quad \text{and} \quad \sin x = \cos\left(\frac{\pi}{2} - x\right)$$

to obtain the formula for $D_x \cos x$.

69 A point $P(x, y)$ is moving at a constant rate around a circle that has the equation $x^2 + y^2 = a^2$. Prove that the projection $Q(x, 0)$ of P on the x-axis is in simple harmonic motion.

70 If a point P moves on a coordinate line such that

$$s(t) = a \cos \omega t + b \sin \omega t,$$

show that P is in simple harmonic motion by

(a) using the remark following Definition (3.32).

(b) Using only trigonometric methods. (*Hint:* Show that $s(t) = A \cos(\omega t - c)$ for some constants A and c.)

3.7 IMPLICIT DIFFERENTIATION

Given an equation of the form

$$y = 2x^2 - 3$$

we say that y *is a function of* x, since

$$y = f(x) \quad \text{with} \quad f(x) = 2x^2 - 3.$$

The equation

$$4x^2 - 2y = 6$$

defines the same function f, since solving for y gives us

$$-2y = -4x^2 + 6 \quad \text{or} \quad y = 2x^2 - 3.$$

In the case $4x^2 - 2y = 6$ we say that y (or f) is an **implicit function** of x, or that f is determined *implicitly* by the equation. Substituting $f(x)$ for y in $4x^2 - 2y = 6$, we obtain

$$4x^2 - 2f(x) = 6$$

$$4x^2 - 2(2x^2 - 3) = 6$$

$$4x^2 - 4x^2 + 6 = 6$$

The last equation is an identity, since it is true for every x in the domain of f. This fact is characteristic of every function f determined implicitly by an equation in x and y; that is, f is implicit if and only if substitution of $f(x)$ for y *leads to an identity.* Since $(x, f(x))$ is a point on the graph of f, the last statement implies that *the graph of the implicit function f coincides with a* *portion (or all) of the graph of the equation.*

In the next example, we show that an equation in x and y may determine more than one implicit function.

EXAMPLE 1 How many different functions are determined implicitly by the equation $x^2 + y^2 = 1$?

SOLUTION The graph of $x^2 + y^2 = 1$ is the unit circle with center at the origin. Solving the equation for y in terms of x we obtain

$$y = \pm\sqrt{1 - x^2}.$$

Two obvious implicit functions are given by

$$f(x) = \sqrt{1 - x^2} \quad \text{and} \quad g(x) = -\sqrt{1 - x^2}.$$

The graphs of f and g are the upper and lower halves, respectively, of the unit circle (see Figure 3.21(i) and (ii)). To find other implicit functions we may let a be any number between -1 and 1 and then define the function k by

$$k(x) = \begin{cases} \sqrt{1 - x^2} & \text{if } -1 \le x \le a \\ -\sqrt{1 - x^2} & \text{if } a < x \le 1. \end{cases}$$

The graph of k is sketched in Figure 3.21(iii). Note that there is a jump discontinuity at $x = a$. The function k is determined implicitly by the equation $x^2 + y^2 = 1$, since

$$x^2 + [k(x)]^2 = 1$$

for every x in the domain of k. By letting a take on different values, it is possible to obtain as many implicit functions as desired. Many other functions are determined implicitly by $x^2 + y^2 = 1$, and the graph of each is a portion of the graph of the equation. •

It is not obvious that the equation

$$y^4 + 3y - 4x^3 = 5x + 1$$

determines an implicit function f in the sense that if $f(x)$ is substituted for y, then

$$[f(x)]^4 + 3[f(x)] - 4x^3 = 5x + 1$$

is an identity for every x in the domain of f. It is possible to state conditions under which an implicit function exists and is differentiable at numbers in its domain; however, the proof requires advanced methods and hence is omitted. In the examples that follow we will assume that a given equation in x and y determines a differentiable function f such that if $f(x)$ is substituted for y, the equation is an identity for every x in the domain of f. The derivative of f may then be found by the method of **implicit differentiation.**

EXAMPLE 2 Assuming that the equation $y^4 + 3y - 4x^3 = 5x + 1$ determines, implicitly, a differentiable function f such that $y = f(x)$, find its derivative.

SOLUTION We think of y as a symbol that denotes $f(x)$ and regard the equation as an identity for every x in the domain of f. It follows that the derivatives of both sides are equal, that is,

$$D_x(y^4 + 3y - 4x^3) = D_x(5x + 1), \quad \text{or}$$

(*) $$D_x(y^4) + D_x(3y) - D_x(4x^3) = D_x(5x) + D_x(1).$$

FIGURE 3.21

(i)

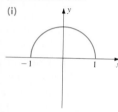

(ii)

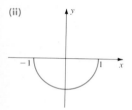

(iii)

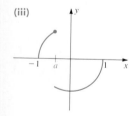

It is very important to remember that, in general, $D_x y^4 \neq 4y^3$. Since $y = f(x)$, we use (3.30), obtaining $D_x (y^4) = 4y^3 y'$. Similarly, $D_x (3y) = 3 D_x y = 3y'$. Substituting in (*),

$$4y^3 y' + 3y' - 12x^2 = 5 + 0.$$

We now solve for y':

$$(4y^3 + 3)y' = 12x^2 + 5$$

and

$$y' = \frac{12x^2 + 5}{4y^3 + 3}$$

provided $4y^3 + 3 \neq 0$. In terms of f,

$$f'(x) = \frac{12x^2 + 5}{4(f(x))^3 + 3} \cdot$$

The last two equations in the solution of Example 2 expose a disadvantage in using the method of implicit differentiation: the formula for y' (or $f'(x)$) may contain the expression y (or $f(x)$). However, these formulas can still be extremely useful in analyzing f and its graph.

In the next example, implicit differentiation is used to find the slope of the tangent line at a point $P(a, b)$ on the graph of an equation. In problems of this type we shall assume that the equation determines an implicit function f whose graph coincides with the graph of the equation for every x in some open interval containing a. Note that, since $P(a, b)$ is a point on the graph, the ordered pair (a, b) must be a solution of the equation.

EXAMPLE 3 Find the slope of the tangent line to the graph of

$$y^4 + 3y - 4x^3 = 5x + 1$$

at the point $P(1, -2)$.

SOLUTION Note that $P(1, -2)$ is on the graph, since

$$(-2)^4 + 3(-2) - 4(1)^3 = 5(1) + 1.$$

The slope m of the tangent line at $P(1, -2)$ is the value of the derivative y' when $x = 1$ and $y = -2$. The given equation is the same as that in Example 2, where we found that $y' = (12x^2 + 5)/(4y^3 + 3)$. Substituting 1 for x and -2 for y gives us

$$m = \frac{12(1)^2 + 5}{4(-2)^3 + 3} = -\frac{17}{29} \cdot$$

EXAMPLE 4 Assume that $x^2 + y^2 = 1$ determines a function f such that $y = f(x)$. Find y'.

SOLUTION Implicit functions determined by $x^2 + y^2 = 1$ were discussed in Example 1. As in Example 2, we differentiate both sides of the equation with respect to x, obtaining

$$D_x (x^2) + D_x (y^2) = D_x (1).$$

Consequently,

$$2x + 2yy' = 0 \quad \text{or} \quad yy' = -x$$

and

$$y' = -\frac{x}{y} \quad \text{if } y \neq 0. \quad \bullet$$

The method of implicit differentiation provides the derivative of *any* differentiable function determined by an equation in two variables. For example, the equation $x^2 + y^2 = 1$ determines many implicit functions (see Example 1). The slope of the tangent line at the point (x, y) on any of the graphs in Figure 3.21 is given by $y' = -x/y$, provided the derivative exists.

EXAMPLE 5 Find y' if $4xy^3 - x^2y + x^3 - 5x + 6 = 0$.

SOLUTION Differentiating both sides of the equation with respect to x,

$$D_x(4xy^3) - D_x(x^2y) + D_x(x^3) - D_x(5x) + D_x(6) = D_x(0).$$

Since y denotes $f(x)$ for some function f, the Product Rule must be applied to $D_x(4xy^3)$ and $D_x(x^2y)$. Thus

$$\begin{aligned}
D_x(4xy^3) &= 4x\, D_x(y^3) + y^3\, D_x(4x) \\
&= 4x(3y^2y') + y^3(4) \\
&= 12xy^2y' + 4y^3
\end{aligned}$$

and

$$\begin{aligned}
D_x(x^2y) &= x^2\, D_x\, y + y\, D_x(x^2) \\
&= x^2y' + y(2x).
\end{aligned}$$

Substituting these in the first equation of the solution and differentiating the other terms leads to

$$(12xy^2y' + 4y^3) - (x^2y' + 2xy) + 3x^2 - 5 = 0.$$

Collecting the terms containing y' and transposing the remaining terms to the right side of the equation gives us

$$(12xy^2 - x^2)y' = 5 - 3x^2 + 2xy - 4y^3.$$

Consequently,

$$y' = \frac{5 - 3x^2 + 2xy - 4y^3}{12xy^2 - x^2}$$

provided $12xy^2 - x^2 \neq 0$. •

EXAMPLE 6 Find y' if $y = x^2 \sin y$.

SOLUTION Differentiating both sides of the equation with respect to x and using the Product Rule, we obtain

$$D_x\, y = x^2\, D_x(\sin y) + (\sin y)\, D_x(x^2).$$

Since $y = f(x)$ for some (implicit) function f, we have, by Theorem (3.31),

$$D_x \sin y = (\cos y)\, D_x\, y.$$

Using this, and the fact that $D_x(x^2) = 2x$, the first equation of our solution may be written

$$D_x\, y = (x^2 \cos y)\, D_x\, y + (\sin y)(2x)$$

or equivalently,

$$y' = (x^2 \cos y)y' + 2x \sin y.$$

Finally, we solve for y' as follows:

$$y' - (x^2 \cos y)y' = 2x \sin y$$

$$(1 - x^2 \cos y)y' = 2x \sin y$$

$$y' = \frac{2x \sin y}{1 - x^2 \cos y}$$

provided $1 - x^2 \cos y \neq 0$. •

EXERCISES 3.7

Exer. 1–8: Find at least one function f determined implicitly by the equation, and state the domain of f.

1 $3x - 2y + 4 = 2x^2 + 3y - 7x$

2 $x^3 - xy + 4y = 1$

3 $x^2 + y^2 - 16 = 0$

4 $3x^2 - 4y^2 = 12$

5 $x^2 - 2xy + y^2 = x$

6 $3x^2 - 4xy + y^2 = 0$

7 $\sqrt{x} + \sqrt{y} = 1$

8 $|x - y| = 2$

Exer. 9–20: Assuming that the equation determines a differentiable function f such that $y = f(x)$, find y'.

9 $8x^2 + y^2 = 10$

10 $4x^3 - 2y^3 = x$

11 $2x^3 + x^2y + y^3 = 1$

12 $5x^2 + 2x^2y + y^2 = 8$

13 $5x^2 - xy - 4y^2 = 0$

14 $x^4 + 4x^2y^2 - 3xy^3 + 2x = 0$

15 $\sin^2 3y = x + y - 1$

16 $x = \sin xy$

17 $x^2y^3 + 4xy + x - 6y = 2$

18 $y^2 + 1 = x^2 \sec y$

19 $y^2 = x \cos y$

20 $xy = \tan xy$

Exer. 21–24: The equation of a classical curve and its graph is given for positive constants a and b. (Consult books on analytic geometry for further information.) Find the slope of the tangent line at the point P for the stated values of a and b.

21 *Ovals of Cassini:* $(x^2 + y^2 + a^2)^2 - 4a^2x^2 = b^4$;
   $a = 2$, $b = \sqrt{6}$, $P(2, \sqrt{2})$

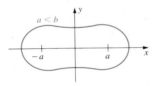

22 *Folium of Descartes:* $x^3 + y^3 - 3axy = 0$; $a = 4$, $P(6, 6)$

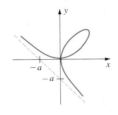

23 *Lemniscate of Bernoulli:* $(x^2 + y^2)^2 = 2a^2xy$;
   $a = \sqrt{2}$, $P(1, 1)$

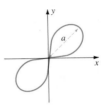

24 *Conchoid of Nicomedes:* $(y - a)^2(x^2 + y^2) = b^2y^2$;
   $a = 2$, $b = 4$, $P(\sqrt{15}, 1)$

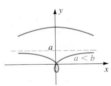

Exer. 25–28: Find an equation of the tangent line to the graph of the given equation at the point P.

25 $xy + 16 = 0$, $P(-2, 8)$

26 $y^2 - 4x^2 = 5$, $P(-1, 3)$

27 $2x^3 - x^2y + y^3 - 1 = 0$, $P(2, -3)$

28 $3y^4 + 4x - x^2 \sin y - 4 = 0$, $P(1, 0)$

29 Show that the equation $x^2 + y^2 + 1 = 0$ does not determine a function f such that $y = f(x)$.

30 Show that the equation $y^2 = x$ determines an infinite number of implicit functions. Sketch the graphs of four such functions.

31 How many implicit functions are determined by each equation?
   (a) $x^4 + y^4 - 1 = 0$
   (b) $x^4 + y^4 = 0$
   (c) $\cos x + \sin y = 3$

32 Use implicit differentiation to prove that if P is any point on the circle $x^2 + y^2 = a^2$, then the tangent line at P is perpendicular to OP.

33 Suppose that $3x^2 - x^2y^3 + 4y = 12$ determines a differentiable function f such that $y = f(x)$. If $f(2) = 0$, use differentials to approximate the change in $f(x)$ if x changes from 2 to 1.97.

34 Suppose that $x^3 + xy + y^4 = 19$ determines a differentiable function f such that $y = f(x)$. If $P(1, 2)$ is a point on the graph of f, use differentials to approximate the y-coordinate b of the point $Q(1.10, b)$ on the graph.

3.8 POWERS AND HIGHER DERIVATIVES

The Power Rule (3.10) can be extended to *rational* exponents. If

$$y = x^{m/n}$$

for integers m and n and $n \neq 0$, then

$$y^n = x^m.$$

Assuming that y' exists and no zero denominators occur we have, by implicit differentiation,

$$D_x\, y^n = D_x\, x^m$$

or

$$ny^{n-1}\, D_x\, y = mx^{m-1}.$$

Thus

$$D_x\, y = \frac{mx^{m-1}}{ny^{n-1}} = \frac{m}{n}\, x^{m-1} y^{1-n} = \frac{m}{n}\, x^{m-1}(x^{m/n})^{1-n}.$$

Simplifying the right side by means of laws of exponents gives us the following **Power Rule for Rational Exponents:** *If m and n are integers and $n \neq 0$, then*

$$D_x\, (x^{m/n}) = \frac{m}{n}\, x^{(m/n)-1}.$$

Our proof of this formula is incomplete in the sense that we have assumed the existence of y'. To give a complete proof the definition of derivative could be used.

EXAMPLE 1 Find y' if $y = 6\sqrt[3]{x^4} + \dfrac{4}{\sqrt{x}}$.

SOLUTION Introducing rational exponents, $y = 6x^{4/3} + 4x^{-1/2}$. Using the Power Rule,

$$y' = 6(\tfrac{4}{3})x^{(4/3)-1} + 4(-\tfrac{1}{2})x^{-(1/2)-1}$$
$$= 8x^{1/3} - 2x^{-3/2}$$

$$= 8x^{1/3} - \frac{2}{x^{3/2}}$$

or, in terms of radicals,

$$y' = 8\sqrt[3]{x} - \frac{2}{\sqrt{x^3}}. \quad \bullet$$

The Power Rule for Functions (3.28) is true for every *rational* exponent. The following examples illustrate uses of this rule.

EXAMPLE 2 Find $f'(x)$ if $f(x) = \sqrt[3]{5x^2 - x + 4}$.

SOLUTION Writing $f(x) = (5x^2 - x + 4)^{1/3}$ and using the Power Rule for Functions with $n = \frac{1}{3}$,

$$f'(x) = \tfrac{1}{3}(5x^2 - x + 4)^{-2/3} D_x (5x^2 - x + 4)$$

$$= \left(\frac{1}{3}\right) \frac{1}{(5x^2 - x + 4)^{2/3}} (10x - 1)$$

$$= \frac{10x - 1}{3\sqrt[3]{(5x^2 - x + 4)^2}}$$

EXAMPLE 3 Find dy/dx if $y = (3x + 1)^6 \sqrt{2x - 5}$.

SOLUTION Since $y = (3x + 1)^6 (2x - 5)^{1/2}$, we have, by the Product and Power Rules,

$$\frac{dy}{dx} = (3x + 1)^6 \tfrac{1}{2}(2x - 5)^{-1/2}(2) + (2x - 5)^{1/2} 6(3x + 1)^5(3)$$

$$= \frac{(3x + 1)^6}{\sqrt{2x - 5}} + 18(3x + 1)^5 \sqrt{2x - 5}$$

$$= \frac{(3x + 1)^6 + 18(3x + 1)^5(2x - 5)}{\sqrt{2x - 5}}$$

$$= \frac{(3x + 1)^5(39x - 89)}{\sqrt{2x - 5}}$$

EXAMPLE 4 Find y' if $y = \sqrt{\sin 6x}$.

SOLUTION Writing $y = (\sin 6x)^{1/2}$ and using the Power Rule for Functions, we obtain

$$y' = \tfrac{1}{2}(\sin 6x)^{-1/2} D_x \sin 6x.$$

Next, by Theorem (3.31),

$$D_x \sin 6x = (\cos 6x) D_x (6x) = (\cos 6x)(6) = 6 \cos 6x.$$

Consequently,

$$y' = \tfrac{1}{2}(\sin 6x)^{-1/2}(6 \cos 6x)$$

$$= \frac{3 \cos 6x}{(\sin 6x)^{1/2}} = \frac{3 \cos 6x}{\sqrt{\sin 6x}}$$

The next example is of interest because it illustrates the fact that after the Power Rule is applied to $[g(x)]^r$, it may be necessary to apply it again in order to find $g'(x)$.

EXAMPLE 5 Find $f'(x)$ if $f(x) = (7x + \sqrt{x^2 + 6})^4$.

SOLUTION Applying the Power Rule,

$$f'(x) = 4(7x + \sqrt{x^2 + 6})^3 D_x (7x + \sqrt{x^2 + 6})$$
$$= 4(7x + \sqrt{x^2 + 6})^3 [D_x (7x) + D_x \sqrt{x^2 + 6}].$$

Again applying the Power Rule,

$$D_x \sqrt{x^2 + 6} = D_x (x^2 + 6)^{1/2} = \frac{1}{2}(x^2 + 6)^{-1/2} D_x (x^2 + 6)$$

$$= \frac{1}{2\sqrt{x^2 + 6}} (2x) = \frac{x}{\sqrt{x^2 + 6}}.$$

Therefore,

$$f'(x) = 4(7x + \sqrt{x^2 + 6})^3 \left[7 + \frac{x}{\sqrt{x^2 + 6}} \right] \cdot$$

EXAMPLE 6 A sperm whale is spotted by a merchant ship, and crew members estimate its length L to be 32 feet with a possible error of ± 2 feet. Whale-research experience has shown that the weight W (in metric tons) is related to L by means of the formula $W = 0.000137 L^{3.18}$. Use differentials to approximate the error in estimating the weight of the whale to the nearest tenth of a metric ton. What is the approximate percentage error?

SOLUTION In order to use calculus methods, we must regard W as a differentiable function of L. If we use the formula to estimate W with $L = 32$, we obtain

$$W = (0.000137)32^{3.18} \approx (0.000137)(61,147.25) \approx 8.4 \text{ metric tons.}$$

Let ΔL denote the error in the estimation of L and let ΔW be the corresponding error in the calculated value of W. As in Section 3.5, these errors may be approximated by dL and dW. Applying Definition (3.23) (ii) and the Power Rule for rational exponents,

$$dW = \left(\frac{dW}{dL} \right) dL = (0.000137)(3.18)L^{2.18} \, dL$$

$$= 0.00043566 L^{2.18} \, dL.$$

Substituting $L = 32$ and $dL = \pm 2$, and using a calculator, we find that the error in estimating W is approximately

$$dW = (0.00043566)32^{2.18}(\pm 2)$$

$$\approx (0.00043566)(1910.85)(\pm 2) \approx \pm 1.7 \text{ metric tons.}$$

By Definition (3.26) we have

$$\text{Average error} = \frac{\Delta W}{W} \approx \frac{dW}{W} \approx \pm \frac{1.7}{8.4} \approx \pm 0.20$$

and

$$\text{Percentage error} \approx \pm (0.20)(100\%) = \pm 20\% \cdot$$

If we differentiate a function f, we obtain another function f'. If f' has a derivative, it is denoted by f'', and is called the **second derivative** of f. Thus

$$f''(x) = D_x [f'(x)] = D_x [D_x (f(x))] = D_x^2 f(x).$$

As we have indicated, the operator symbol D_x^2 is used for second derivatives. The **third derivative** f''' of f is the derivative of the second derivative. Specifically,

$$f'''(x) = D_x [f''(x)] = D_x [D_x^2 f(x)] = D_x^3 f(x).$$

In general, if n is a positive integer, then $f^{(n)}$ denotes the **nth derivative** of f and is found by starting with f and differentiating, successively, n times. Using operator notation, $f^{(n)}(x) = D_x^n f(x)$. The integer n is called the **order** of the derivative $f^{(n)}(x)$. The following summarizes various notations that are used for these **higher derivatives,** with $y = f(x)$.

NOTATIONS FOR **(3.33)**
HIGHER DERIVATIVES

$$f'(x), \quad f''(x), \quad f'''(x), \quad f^{(4)}(x), \quad \ldots, \quad f^{(n)}(x)$$

$$D_x y, \quad D_x^2 y, \quad D_x^3 y, \quad D_x^4 y, \quad \ldots, \quad D_x^n y$$

$$y', \quad y'', \quad y''', \quad y^{(4)}, \quad \ldots, \quad y^{(n)}$$

$$\frac{dy}{dx}, \quad \frac{d^2 y}{dx^2}, \quad \frac{d^3 y}{dx^3}, \quad \frac{d^4 y}{dx^4}, \quad \ldots, \quad \frac{d^n y}{dx^n}$$

The differential notation $d^n y/dx^n$ used for higher derivatives should *not* be interpreted as a quotient.

EXAMPLE 7 If $f(x) = 4x^2 - 5x + 8 - (3/x)$, find the first four derivatives of $f(x)$.

SOLUTION Since $f(x) = 4x^2 - 5x + 8 - 3x^{-1}$,

$$f'(x) = 8x - 5 + 3x^{-2} = 8x - 5 + \frac{3}{x^2}$$

$$f''(x) = 8 - 6x^{-3} = 8 - \frac{6}{x^3}$$

$$f'''(x) = 18x^{-4} = \frac{18}{x^4}$$

$$f^{(4)}(x) = -72x^{-5} = -\frac{72}{x^5} \qquad \bullet$$

EXAMPLE 8 Find the first eight derivatives of $f(x) = \sin x$.

SOLUTION Applying Theorem (3.20),

$$f'(x) = D_x \sin x = \cos x$$
$$f''(x) = D_x \cos x = -\sin x$$
$$f'''(x) = D_x (-\sin x) = -D_x (\sin x) = -\cos x$$
$$f^{(4)}(x) = D_x (-\cos x) = -D_x (\cos x) = -(-\sin x) = \sin x$$

Since $f^{(4)}(x) = \sin x$, it follows that if we continue differentiating, the same pattern repeats, that is,

$$f^{(5)}(x) = \cos x, \qquad f^{(6)}(x) = -\sin x,$$
$$f^{(7)}(x) = -\cos x \qquad f^{(8)}(x) = \sin x. \qquad \bullet$$

The next example illustrates how to find higher derivatives of implicit functions.

EXAMPLE 9 Find y'' if $y^4 + 3y - 4x^3 = 5x + 1$.

SOLUTION The equation was investigated in Example 2 of Section 3.7, where we found that

$$y' = \frac{12x^2 + 5}{4y^3 + 3}.$$

Hence,

$$y'' = D_x(y') = D_x\left(\frac{12x^2 + 5}{4y^3 + 3}\right).$$

We now use the quotient rule, differentiating implicitly as follows:

$$y'' = \frac{(4y^3 + 3)D_x(12x^2 + 5) - (12x^2 + 5)D_x(4y^3 + 3)}{(4y^3 + 3)^2}$$

$$= \frac{(4y^3 + 3)(24x) - (12x^2 + 5)(12y^2y')}{(4y^3 + 3)^2}.$$

Substituting for y' yields

$$y'' = \frac{(4y^3 + 3)(24x) - (12x^2 + 5)\cdot 12y^2\left(\dfrac{12x^2 + 5}{4y^3 + 3}\right)}{(4y^3 + 3)^2}$$

$$= \frac{(4y^3 + 3)^2(24x) - 12y^2(12x^2 + 5)^2}{(4y^3 + 3)^3}$$

●

EXERCISES 3.8

Exer. 1–24: Differentiate the function.

1 $f(x) = \sqrt[3]{x^2} + 4\sqrt{x^3}$

2 $f(x) = 10\sqrt[5]{x^3} + \sqrt[3]{x^5}$

3 $k(r) = \sqrt[3]{8r^3 + 27}$

4 $h(z) = (2z^2 - 9z + 8)^{-2/3}$

5 $F(v) = 5/\sqrt[5]{v^5 - 32}$

6 $k(s) = 1/\sqrt{3s - 4}$

7 $f(x) = \sqrt{2x}$

8 $g(x) = \sqrt[5]{1/x}$

9 $f(z) = 10\sqrt{z^3} + (3/\sqrt[3]{z})$

10 $f(t) = \sqrt[3]{t^2} - (1/\sqrt{t^3})$

11 $g(w) = (w^2 - 4w + 3)/w^{3/2}$

12 $K(x) = \sqrt{4x^2 + 2x + 3}$

13 $f(x) = \sin\sqrt{x} + \sqrt{\sin x}$

14 $f(x) = \tan\sqrt[3]{5 - 6x}$

15 $k(\theta) = \cos^2\sqrt{3 - 8\theta}$

16 $r(t) = \sqrt{\sin 2t - \cos 2t}$

17 $g(x) = \sqrt{x^2 + 1}\,\tan\sqrt{x^2 + 1}$

18 $h(\varphi) = \dfrac{\cot 4\varphi}{\sqrt{\varphi^2 + 4}}$

19 $M(x) = \sec\sqrt{4x + 1}$

20 $F(s) = \sqrt{\csc 2s}$

21 $h(x) = \sqrt{4 + \csc^2 3x}$

22 $f(t) = \sin^2 2t\sqrt{\cos 2t}$

23 $H(x) = (2x + 3)/\sqrt{4x^2 + 9}$

24 $f(x) = (7x + \sqrt{x^2 + 3})^6$

Exer. 25–26: Find an equation of the tangent line to the graph of the equation at the point P.

25 $y = \sqrt{2x^2 + 1}$, $P(-1, \sqrt{3})$

26 $y = (5x - 8)^{1/3}$, $P(7, 3)$

27 Find the point P on the graph of $y = \sqrt{2x - 4}$ such that the tangent line at P passes through the origin.

28 Find the points on the graph of $y = x^{5/3} + x^{1/3}$ at which the tangent line is perpendicular to the line $2y + x = 7$.

Exer. 29–32: Assuming that the equation determines a function f such that $y = f(x)$, find y'.

29 $\sqrt{x} + \sqrt{y} = 100$

30 $x^{2/3} + y^{2/3} = 4$

31 $x^2 + \sqrt{\sin y} - y^2 = 1$

32 $2x - \sqrt{xy} + y^3 = 16$

33 Pinnipeds are a suborder of aquatic carnivorous mammals, such as seals and walruses, whose limbs are modified into flippers. The length-weight relationship during fetal growth is well described by $W = (6 \times 10^{-5})L^{2.74}$ for length L in cm and weight W in kg. Use the Chain Rule to find a formula for the rate of growth in weight with respect to time t. If the weight of a seal is 0.5 kg and is changing at a rate of 0.4 kg per month, how fast is the length changing?

34 The formula for the adiabatic expansion of air is $pv^{1.4} = c$, for pressure p, volume v, and a constant c. Find a formula for the rate of change of pressure with respect to volume.

35 The curved surface area S of a right circular cone having altitude h and base radius r is given by $S = \pi r \sqrt{r^2 + h^2}$. For a certain cone, $r = 6$ cm. The altitude is measured as 8 cm, with a maximum error in measurement of 0.1 cm. Calculate S from these measurements and use differentials to estimate the maximum error in the calculation. Approximate the percentage error.

36 The period T of a simple pendulum of length l may be calculated by means of the formula $T = 2\pi\sqrt{l/g}$ for a constant g. Use differentials to approximate the change in l that will increase T by 1%.

Exer. 37–40: Use the fact that $|a| = \sqrt{a^2}$ to find $f'(x)$. In addition, find the domain of f' and sketch the graph of f.

37 $f(x) = |1 - x|$

38 $f(x) = \sin|\frac{1}{2}\pi - x|$

39 $f(x) = |1 - x^2|$

40 $f(x) = x/|x|$

Exer. 41–46: Find the first and second derivatives.

41 $g(z) = \sqrt{3z + 1}$

42 $k(s) = (s^2 + 4)^{2/3}$

43 $k(r) = (4r + 7)^5$

44 $f(x) = \sqrt[3]{10x + 7}$

45 $f(x) = \sin^3 x$

46 $G(t) = \sec^2 4t$

Exer. 47–50: Find $D_x^3 y$.

47 $y = 2x^5 + 3x^3 - 4x + 1$

48 $y = \cos 3x$

49 $y = \sin(x^2)$

50 $y = (3x + 1)^4$

Exer. 51–54: Assuming that the equation determines a function f such that $y = f(x)$, find y'', if it exists.

51 $x^3 - y^3 = 1$

52 $x^2 y^3 = 1$

53 $x^2 - 3xy + y^2 = 4$

54 $\sqrt{xy} - y + x = 0$

55 If $f(x) = 1/x$, find a formula for $f^{(n)}(x)$ for any positive integer n. Find $f^{(n)}(1)$.

56 If $f(x) = \sqrt{x}$, find a formula for $f^{(n)}(x)$ for any positive integer n.

57 If $f(x)$ is a polynomial of degree n, show that $f^{(k)}(x) = 0$ if $k > n$.

58 Suppose $u = f(x)$, $v = g(x)$, and f and g have derivatives of all orders. If $y = uv$, prove that

$$y'' = u''v + 2u'v' + uv''$$
$$y''' = u'''v + 3u''v' + 3u'v'' + uv'''$$
$$y^{(4)} = u^{(4)}v + 4u'''v' + 6u''v'' + 4u'v''' + uv^{(4)}.$$

Suggest a reasonable formula for $y^{(n)}$. (*Hint:* Use the Binomial Theorem (3.9).)

59 If $y = f(g(x))$ and f'' and g'' exist, use the Chain Rule to express $D_x^2 y$ in terms of the first and second derivatives of f and g.

60 Suppose f has a second derivative. Prove that (a) if f is an even function, then f'' is even, and (b) if f is an odd function, then f'' is odd. Illustrate these facts by using polynomial functions and the sine and cosine functions.

3.9 RELATED RATES

In applications it is not unusual for two variables x and y to be differentiable functions of time t, say $x = f(t)$ and $y = g(t)$. In addition, x and y may be related by means of some equation, such as

$$x^2 - y^3 - 2x + 7y^2 - 2 = 0.$$

Differentiating this equation with respect to t and using the Chain Rule produces an equation involving the rates of change dx/dt and dy/dt:

$$2x\frac{dx}{dt} - 3y^2\frac{dy}{dt} - 2\frac{dx}{dt} + 14y\frac{dy}{dt} = 0.$$

The derivatives dx/dt and dy/dt are called **related rates,** since they are related by means of an equation. Such an equation can be used to find one of the rates when the other is known. This has many practical applications. The following examples give several illustrations.

EXAMPLE 1 A ladder 20 feet long leans against a vertical building. If the bottom of the ladder slides away from the building horizontally at a rate of 2 ft/sec, how fast is the ladder sliding down the building when the top of the ladder is 12 feet above the ground?

FIGURE 3.22

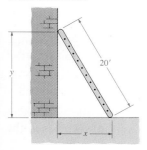

SOLUTION We begin by representing a general position of the ladder schematically, as in Figure 3.22, introducing a variable x to denote the distance from the base of the building to the bottom of the ladder, and a variable y to denote the distance from the ground to the top of the ladder.

Since x is changing at a rate of 2 ft/sec,

$$\frac{dx}{dt} = 2 \text{ ft/sec.}$$

Our objective is to find dy/dt, the rate at which the top of the ladder is sliding down the building, at the instant that $y = 12$ feet.

The relationship between x and y may be determined by applying the Pythagorean Theorem to the right triangle formed by the building, the ground, and the ladder (see Figure 3.22). This gives us

$$x^2 + y^2 = 400.$$

Differentiating both sides of this equation with respect to t,

$$2x \frac{dx}{dt} + 2y \frac{dy}{dt} = 0$$

or, if $y \neq 0$,

$$\frac{dy}{dt} = -\frac{x}{y} \frac{dx}{dt}.$$

The last equation is a *general* formula relating the two rates of change dx/dt and dy/dt. Let us now consider the special case $y = 12$. The corresponding value of x may be determined from

$$x^2 + 144 = 400 \quad \text{or} \quad x^2 = 400 - 144 = 256.$$

Thus $x = \sqrt{256} = 16$ when $y = 12$. Substituting these values into the general formula for dy/dt we obtain

$$\frac{dy}{dt} = -\frac{16}{12}(2) = -\frac{8}{3} \text{ ft/sec} \quad \bullet$$

The following guidelines may be helpful for solving related rate problems of the type illustrated in Example 1.

GUIDELINES FOR SOLVING (3.34)
RELATED RATE PROBLEMS

1 Read the problem carefully several times and think about the given facts, together with the unknown quantities that are to be found.

2 Sketch a picture or diagram and label it appropriately, introducing variables for unknown quantities.

3 Write down all the known facts, expressing the given and unknown rates as derivatives of the variables introduced in Guideline 2.

4 Formulate a *general* equation that relates the variables.

5 Differentiate both sides of the equation formulated in Guideline 4 with respect to t, obtaining a *general* relationship between the rates.

6 Substitute the *known* values and rates, and then find the unknown rate of change.

A common error is introducing specific values for the rates and variable quantities too early in the solution. Always remember to obtain a *general* formula that involves the rates of change at *any* time t. Specific values should not be substituted for variables until the final steps of the solution.

EXAMPLE 2 A water tank has the shape of an inverted right circular cone of altitude 12 feet and base radius 6 feet. If water is being pumped into the tank at a rate of 10 gal/min, approximate the rate at which the water level is rising when it is 3 feet deep (1 gal ≈ 0.1337 ft³).

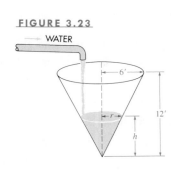

FIGURE 3.23

SOLUTION Applying Guideline 2, we begin by sketching the tank as in Figure 3.23, letting r denote the radius of the surface of the water when the depth is h. Note that r and h are functions of time t.

Following Guideline 3, we write down the known facts and relationships between V, r, and h. We know that $dV/dt = 10$ gal/min, and we wish to find dh/dt when $h = 3$ feet. The volume V of water in the tank corresponding to depth h is

$$V = \tfrac{1}{3}\pi r^2 h.$$

This formula for V relates V, r, and h (see Guideline 4). Before differentiating with respect to t, let us express V in terms of one variable. Referring to Figure 3.23 and using similar triangles, we obtain

$$\frac{r}{h} = \frac{6}{12} \quad \text{or} \quad r = \frac{h}{2}.$$

Consequently, at depth h,

$$V = \frac{1}{3}\,\pi\left(\frac{h}{2}\right)^2 h = \frac{1}{12}\,\pi h^3.$$

Differentiating the last equation with respect to t (see Guideline 5) gives us the following general relationship between the rates of change of V and h at any time:

$$\frac{dV}{dt} = \frac{1}{4}\,\pi h^2\,\frac{dh}{dt}.$$

If $h \neq 0$, an equivalent formula is

$$\frac{dh}{dt} = \frac{4}{\pi h^2}\,\frac{dV}{dt}.$$

Finally (see Guideline 6), we substitute $h = 3$ and $dV/dt = 10$ gal/min ≈ 1.337 ft³/min, obtaining

$$\frac{dh}{dt} \approx \frac{4}{\pi(9)}\,(1.337) \approx 0.189 \text{ ft/min} \quad \bullet$$

In the remaining examples we shall not point out the guidelines that are employed. You should be able to determine specific guidelines by studying the solutions.

EXAMPLE 3 At 1:00 P.M., ship A is 25 miles due south of ship B. If ship A is sailing west at a rate of 16 mi/hr and ship B is sailing south at a rate of 20 mi/hr, find the rate at which the distance between the ships is changing at 1:30 P.M.

FIGURE 3.24

SOLUTION The problem is sketched in Figure 3.24, where x and y denote the miles covered by ships A and B, respectively, in t hours after 1:00 P.M. Points P and Q are their positions at 1:00 P.M., and z is the distance between the ships at time t.

We know that

$$\frac{dx}{dt} = 16 \text{ mi/hr} \quad \text{and} \quad \frac{dy}{dt} = 20 \text{ mi/hr}.$$

Our objective is to find dz/dt. By the Pythagorean Theorem,

$$z^2 = x^2 + (25 - y)^2.$$

This gives us a *general* equation relating the variables x, y, and z. Differentiating both sides with respect to t,

$$2z\frac{dz}{dt} = 2x\frac{dx}{dt} + 2(25 - y)\left(-\frac{dy}{dt}\right)$$

or

$$z\frac{dz}{dt} = x\frac{dx}{dt} + (y - 25)\frac{dy}{dt}.$$

At 1:30 P.M. the ships have traveled for half an hour and we have

$$x = 8, \quad y = 10, \quad \text{and} \quad 25 - y = 15.$$

Consequently,

$$z^2 = 64 + 225 = 289 \quad \text{or} \quad z = 17.$$

Substituting these values into the equation involving dz/dt gives us

$$17\frac{dz}{dt} = 8(16) + (-15)(20)$$

or

$$\frac{dz}{dt} = -\frac{172}{17} \approx -10.12 \text{ mi/hr}.$$

The negative sign indicates that the distance between the ships is decreasing at 1:30 P.M.

Another method of solution is to write $x = 16t$, $y = 20t$, and

$$z = [x^2 + (25 - y)^2]^{1/2} = [256t^2 + (25 - 20t)^2]^{1/2}.$$

The derivative dz/dt may then be found, and substitution of $\frac{1}{2}$ for t produces the desired rate of change. •

EXAMPLE 4 In a lighthouse that is 200 feet from the nearest point P on a straight shoreline, a revolving beacon makes one revolution every 15 seconds. Find the rate at which a ray from the light moves along the shore at a point 400 feet from P.

FIGURE 3.25

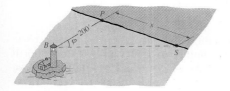

SOLUTION The problem is diagrammed in Figure 3.25, where B denotes the position of the beacon and φ is the angle between BP and a light ray to a point S on the shore x units from P.

Since the light revolves four times per minute, the angle φ changes at a rate of 8π radians per minute, that is, $d\varphi/dt = 8\pi$. Using triangle PBS we see that $$\tan \varphi = \frac{x}{200} \quad \text{or} \quad x = 200 \tan \varphi.$$

Consequently, the rate at which the ray of light moves along the shore is

$$\frac{dx}{dt} = 200 \sec^2 \varphi \, \frac{d\varphi}{dt} = (200 \sec^2 \varphi)(8\pi) = 1600\pi \sec^2 \varphi.$$

If $x = 400$, then

$$\tan \varphi = 400/200 = 2 \quad \text{and} \quad \sec \varphi = \sqrt{1 + \tan^2 \varphi} = \sqrt{1 + 4} = \sqrt{5}.$$

Hence, $$\frac{dx}{dt} = 1600\pi(\sqrt{5})^2 = 8000\pi \approx 25,133 \text{ ft/min} \quad \bullet$$

EXAMPLE 5 Shown in Figure 3.26 is a solar panel that is 10 feet in width and is equipped with a hydraulic lift. As the sun rises, the panel is adjusted so that the sun's rays hit directly on the panel's surface.

(a) Find the relationship between the rate dy/dt at which the panel should be lowered and the rate $d\theta/dt$ at which the angle of inclination of the sun increases.

(b) If, when $\theta = 30°$, $d\theta/dt = 15°/\text{hr}$, find dy/dt.

SOLUTION

(a) If we let φ denote angle BAC in Figure 3.26, then from plane geometry, $\varphi = 90° - \theta = \frac{1}{2}\pi - \theta$. Thus φ decreases at the same rate that θ increases, since $d\varphi/dt = -d\theta/dt$.

Referring to right triangle BAC, we see that

$$\sin \varphi = \frac{y}{10} \quad \text{or} \quad y = 10 \sin \varphi = 10 \sin \left(\tfrac{1}{2}\pi - \theta\right).$$

Differentiating with respect to t, and using the identity $\cos \left(\tfrac{1}{2}\pi - \theta\right) = \sin \theta$,

$$\frac{dy}{dt} = 10 \cos \left(\tfrac{1}{2}\pi - \theta\right)\left(-\frac{d\theta}{dt}\right) = -10 \sin \theta \, \frac{d\theta}{dt}$$

(b) We must use radian measure for $d\theta/dt$. Since $15° = 15(\pi/180) = \pi/12$ radians, we substitute $d\theta/dt = \pi/2$ rad/hr and $\theta = 30° = \pi/6$ in the formula for dy/dt, obtaining

$$\frac{dy}{dt} = -10 \left(\frac{1}{2}\right)\left(\frac{\pi}{12}\right) = -\frac{5\pi}{12} \approx -1.3 \text{ ft/hr} \quad \bullet$$

FIGURE 3.26

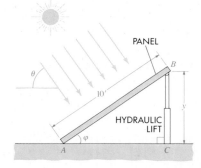

PANEL

B

10'

HYDRAULIC LIFT

θ

φ

y

A C

1 A ladder 20 feet long leans against a vertical building. If the bottom of the ladder slides away from the building horizontally at a rate of 3 ft/sec, how fast is the ladder sliding down the building when the top of the ladder is 8 feet from the ground?

2 As a circular metal griddle is being heated, its diameter changes at a rate of 0.01 cm/min. When the diameter is 30 cm, at what rate is the area of one side changing?

3 Gas is being pumped into a spherical balloon at a rate of 5 ft³/min. Find the rate at which the radius is changing when the diameter is 18 inches.

4 A girl starts at a point A and runs east at a rate of 10 ft/sec. One minute later, another girl starts at A and runs north at a rate of 8 ft/sec. At what rate is the distance between them changing 1 minute after the second girl starts?

5 A light is at the top of a 16-foot pole. A boy 5 feet tall walks away from the pole at a rate of 4 ft/sec (see figure). At what rate is the tip of his shadow moving when he is 18 feet from the pole? At what rate is the length of his shadow increasing?

EXERCISE 5

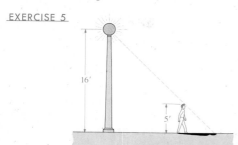

6 A man on a dock is pulling in a boat by means of a rope attached to the bow of the boat 1 foot above water level and passing through a simple pulley located on the dock 8 feet above water level (see figure). If he pulls in the rope at a rate of 2 ft/sec, how fast is the boat approaching the dock when the bow of the boat is 25 feet from a point that is directly below the pulley?

EXERCISE 6

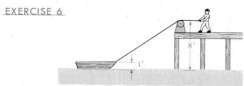

7 The top of a silo has the shape of a hemisphere of diameter 20 feet. If it is coated uniformly with a layer of ice, and if the thickness is decreasing at a rate of $\frac{1}{4}$ in./hr, how fast is the volume of ice changing when the ice is 2 inches thick?

8 As sand leaks out of a hole in a container, it forms a conical pile whose altitude is always the same as its radius. If the height of the pile is increasing at a rate of 6 in./min, find the rate at which the sand is leaking out when the altitude is 10 inches.

9 A boy flying a kite holds the string 5 feet above ground level, and the string is payed out at a rate of 2 ft/sec as the kite moves horizontally at an altitude of 105 feet (see figure). Assuming there is no sag in the string, find the rate at which the kite is moving when 125 feet of string has been payed out.

EXERCISE 9

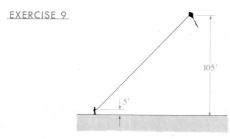

10 A hot-air balloon rises vertically as a rope attached to the base of the balloon is released at the rate of 5 ft/sec. The pulley that releases the rope is 20 feet from the platform where passengers board. At what rate is the balloon rising when 500 feet of rope has been payed out?

EXERCISE 10

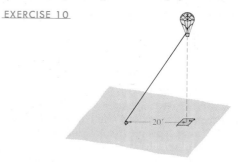

11 Boyle's law for gases states that $pv = c$, for pressure p, volume v, and a constant c. At a certain instant the volume is 75 in³, the pressure is 30 lb/in², and the pressure is decreasing at a rate of 2 lb/in² every minute. At what rate is the volume changing at this instant?

12 Suppose a spherical snowball is melting and the radius is decreasing at a constant rate, changing from 12 inches to 8 inches in 45 minutes. How fast was the volume changing when the radius was 10 inches?

13 The ends of a water trough 8 feet long are equilateral triangles whose sides are 2 feet long (see figure). If water is being pumped into the trough at a rate of 5 ft³/min, find the rate at which the water level is rising when the depth is 8 inches.

EXERCISE 13

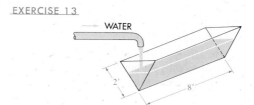

14 Rework Exercise 13 if the ends of the trough have the shape of the graph of $y = 2|x|$ between the points $(-1, 2)$ and $(1, 2)$.

15 A 100-foot-long cable of diameter 4 inches is submerged in seawater. Due to corrosion, the surface area of the cable decreases at a rate of 750 in²/year. Ignoring the corrosion at the ends of the cable, find the rate at which the diameter is decreasing.

16 A fire has started in a dry, open field and spreads in the form of a circle. The radius of the circle increases at the rate of 6 ft/min. Find the rate at which the fire area is increasing when the radius is 150 feet.

17 The area of an equilateral triangle is decreasing at a rate of 4 cm²/min. Find the rate at which the length of a side is changing when the area of the triangle is 200 cm².

18 Gas is escaping from a spherical balloon at a rate of $10 \text{ ft}^3/\text{hr}$. At what rate is the radius changing when the volume is 400 ft^3?

19 A stone is dropped into a lake, causing circular waves whose radii increase at a constant rate of 0.5 m/sec. At what rate is the circumference of a wave changing when its radius is 4 meters?

20 A softball diamond has the shape of a square with sides 60 feet long. If a player is running from second base to third at a speed of 24 ft/sec, at what rate is her distance from home plate changing when she is 20 feet from third?

21 When two resistors R_1 and R_2 are connected in parallel (see figure), the total resistance R is $1/R = (1/R_1) + (1/R_2)$. If R_1 and R_2 are increasing at rates of 0.01 ohm/sec and 0.02 ohm/sec, respectively, at what rate is R changing at the instant that $R_1 = 30$ ohms and $R_2 = 90$ ohms?

EXERCISE 21

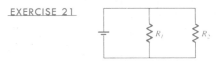

22 The formula for the adiabatic expansion of air is $pv^{1.4} = c$, for pressure p, volume v, and a constant c. At a certain instant the pressure is 40 dyn/cm^2 and is increasing at a rate of 3 dyn/cm^2 per second. If, at that same instant, the volume is 60 cm^3, find the rate at which the volume is changing.

23 If a spherical tank of radius a contains water that has a maximum depth h, then the volume V of water in the tank is given by $V = \frac{1}{3}\pi h^2(3a - h)$. Suppose a spherical tank of radius 16 feet is being filled at a rate of 100 gal/min. Approximate the rate at which the water level is rising when $h = 4$ feet (1 gal $\approx 0.1337 \text{ ft}^3$).

24 A spherical water storage tank for a small community is coated uniformly with a 2-inch layer of ice. If the volume of ice is melting at a rate that is directly proportional to its surface area, show that the outside diameter is decreasing at a constant rate.

25 From the edge of a cliff that overlooks a lake 200 feet below, a boy drops a stone and then, two seconds later, drops another stone from exactly the same position. Discuss the rate at which the distance between the two stones is changing during the next second. (Assume that the distance an object falls in t seconds is $16t^2$ feet.)

26 A metal rod has the shape of a right circular cylinder. As it is being heated, its length is increasing at a rate of 0.005 cm/min and its diameter is increasing at 0.002 cm/min. At what rate is the volume changing when the rod has length 40 cm and diameter 3 cm?

27 An airplane is flying at a constant speed of 360 mi/hr and climbing at an angle of $45°$. At the moment the plane's altitude is 10,560 feet, it passes directly over an air traffic control tower on the ground. Find the rate at which the

airplane's distance from the tower is changing one minute later (neglect the height of the tower).

28 A North–South highway A and an East–West highway B intersect at a point P. At 10:00 A.M. an automobile crosses P traveling north on highway A at a speed of 50 mi/hr. At that same instant, an airplane flying east at a speed of 200 mi/hr and an altitude of 26,400 feet is directly above the point on highway B that is 100 miles west of P. If they maintain the same speed and direction, at what rate is the distance between the airplane and the automobile changing at 10:15 A.M.?

29 A paper cup containing water has the shape of a frustum of a right circular cone of altitude 6 inches and lower and upper base radii 1 inch and 2 inches, respectively. If water is leaking out of the cup at a rate of 3 in.^{3}/hr, at what rate is the water level decreasing when its depth is 4 inches? (*Note:* The volume V of a frustum of a right circular cone of altitude h and base radii a and b is given by $V = \frac{1}{3}\pi h(a^2 + b^2 + ab)$.)

30 The top part of a swimming pool is a rectangle of length 60 feet and width 30 feet. The depth of the pool varies uniformly from 4 feet to 9 feet through a horizontal distance of 40 feet and then is level for the remaining 20 feet, as illustrated by the cross-sectional view in the figure. If the pool is being filled with water at a rate of 500 gal/min, approximate the rate at which the water level is rising when the depth at the deep end is 4 feet (1 gal $\approx 0.1337 \text{ ft}^3$).

EXERCISE 30

31 An airplane at an altitude of 10,000 feet is flying at a constant speed on a line that will take it directly over an observer on the ground. If, at a given instant, the observer notes that the angle of elevation of the airplane is $60°$ and is increasing at a rate of $1°$ per second, find the speed of the airplane.

32 In Exercise 6, let θ be the angle that the rope makes with the horizontal. Find the rate at which θ is changing at the instant that $\theta = 30°$.

33 An isosceles triangle has equal sides 6 inches long. If the angle θ between the equal sides is changing at a rate of $2°$ per minute, how fast is the area of the triangle changing when $\theta = 30°$?

34 A ladder 20 feet long leans against a vertical building. If the bottom of the ladder slides away from the building horizontally at a rate of 2 ft/sec, at what rate is the angle between the ladder and the ground changing when the top of the ladder is 12 feet above the ground?

35 The relative positions of an airport runway and a 20-foot-tall control tower are shown in the figure. The beginning of the runway is at a perpendicular distance of 300 feet from the base of the tower. If an airplane reaches a speed of 100 mi/hr after traveling 300 feet down the runway, at approximately what rate is the distance between the airplane and the control booth increasing at this time?

EXERCISE 35

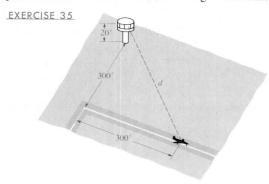

36 The speed of sound in air at 0 °C (or 273 °K) is 1087 ft/sec, but this speed increases as the temperature rises. If T is temperature in °K, the speed of sound v at this temperature is given by $v = 1087\sqrt{T/273}$. If the temperature increases at the rate of 3°C per hour, approximate the rate at which the speed of sound is increasing when $T = 30°C$ (or 303 °K).

37 A jet plane is flying at a constant speed and altitude, on a line that will take it directly over a radar station located on the ground. At the instant that the jet is 60,000 feet from the station, an observer in the station notes that its angle of elevation is 30° and is increasing at a rate of 0.5° per second. Find the speed of the jet.

38 A missile is fired vertically from a point which is 5 miles from a tracking station and at the same elevation (see figure). For the first 20 seconds of flight its angle of elevation θ changes at a constant rate of 2° per second. Find the velocity of the missile when the angle of elevation is 30°.

EXERCISE 38

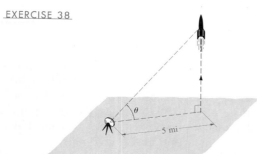

39 The sprocket assembly for a 28-inch bicycle is shown in the figure. Find the relationship between the angular velocity $d\theta/dt$ (in radians/sec) of the pedal assembly and the ground speed of the bicycle (in mi/hr).

EXERCISE 39

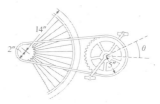

40 A 100-candle spotlight is perched 12 feet above the stage as shown in the figure. The illuminance E (in foot-candles) in the small lighted area of the stage is given by $E = (I \cos \theta)/s^2$ for the intensity I of the light, the distance s that the light must travel, and the indicated angle θ. As the spotlight is rotated through φ degrees, find the relationship between the rate of change in illumination dE/dt and the rate of rotation $d\varphi/dt$.

EXERCISE 40

41 A *conical pendulum* consists of a mass m, attached to a string of fixed length l, which travels around a circle of radius R at a fixed velocity v (see figure). As the velocity of the mass is increased, both the radius and the angle θ increase. Given that $v^2 = Rg \tan \theta$ for a gravitational constant g, find the relationship between the related rates (a) dv/dt and $d\theta/dt$, and (b) dv/dt and dR/dt.

EXERCISE 41

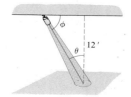

42 Water in a paper conical filter drips into a cup as shown in the figure. Let x denote the height of the water in the filter and y the height of the water in the cup. If 10 in³ of water is poured into the filter, find the relationship between dy/dt and dx/dt.

EXERCISE 42

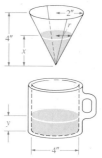

3.10 NEWTON'S METHOD

FIGURE 3.27

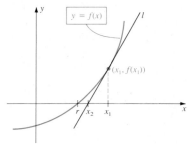

In this section we shall describe a method for approximating a real zero of a differentiable function f, that is, a real number r such that $f(r) = 0$. To use the method we begin by making a first approximation x_1 to the zero r. Since r is an x-intercept of the graph of f, the approximation x_1 can usually be made by referring to a rough sketch of the graph. If we consider the tangent line l to the graph of f at the point $(x_1, f(x_1))$, and if x_1 is sufficiently close to r, then as illustrated in Figure 3.27, the x-intercept x_2 of l should be a better approximation to r.

Since the slope of l is $f'(x_1)$, an equation of the tangent line is

$$y - f(x_1) = f'(x_1)(x - x_1).$$

Since the x-intercept x_2 corresponds to the point $(x_2, 0)$ on l, we obtain

$$0 - f(x_1) = f'(x_1)(x_2 - x_1).$$

If $f'(x_1) \neq 0$, the preceding equation is equivalent to

$$x_2 = x_1 - \frac{f(x_1)}{f'(x_1)}.$$

If we take x_2 as a second approximation to r, then the process may be repeated by using the tangent line at $(x_2, f(x_2))$. If $f'(x_2) \neq 0$, this leads to a third approximation x_3, given by

$$x_3 = x_2 - \frac{f(x_2)}{f'(x_2)}.$$

The process is continued until the desired degree of accuracy is obtained. This technique of successive approximations of real zeros is referred to as **Newton's Method,** which we state as follows.

NEWTON'S METHOD (3.35)

> Let f be a differentiable function and suppose r is a real zero of f. If x_n is an approximation to r, then the next approximation x_{n+1} is given by
>
> $$x_{n+1} = x_n - \frac{f(x_n)}{f'(x_n)}$$
>
> provided $f'(x_n) \neq 0$.

FIGURE 3.28

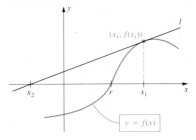

Newton's Method does not guarantee that x_{n+1} is a better approximation to r than x_n for every n. In particular, we must be careful in choosing the first approximation x_1. If x_1 is not sufficiently close to r, it is possible for the second approximation x_2 to be worse that x_1, as illustrated in Figure 3.28. It is evident that we should not choose a number x_n such that $f'(x_n)$ is close to 0, for then the tangent line l is almost horizontal.

We shall use the following rule when applying Newton's Method: If an approximation to k decimal places is required, we shall approximate each of the numbers $x_2, x_3, \ldots$ to k decimal places, continuing the process until two consecutive approximations are the same. This is illustrated in the following examples.

FIGURE 3.29

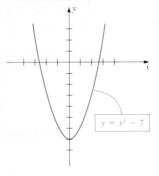

$y = x^2 - 7$

EXAMPLE 1 Use Newton's Method to approximate $\sqrt{7}$ to five decimal places.

SOLUTION The stated problem is equivalent to that of approximating the positive real zero r of $f(x) = x^2 - 7$. The graph of f is sketched in Figure 3.29. Since $f(2) = -3$ and $f(3) = 2$, it follows from the continuity of f that $2 < r < 3$. Moreover, since f is increasing, there can be only one zero in the open interval $(2, 3)$.

If x_n is any approximation to r, then by (3.35) the next approximation x_{n+1} is given by

$$x_{n+1} = x_n - \frac{f(x_n)}{f'(x_n)} = x_n - \frac{x_n^2 - 7}{2x_n}.$$

Let us choose $x_1 = 2.5$ as a first approximation. Using the formula for x_{n+1} with $n = 1$ gives us

$$x_2 = 2.5 - \frac{(2.5)^2 - 7}{2(2.5)} = 2.65000$$

Again using the formula (with $n = 2$), we obtain the next approximation

$$x_3 = 2.65000 - \frac{(2.65000)^2 - 7}{2(2.65000)} \approx 2.64575$$

Repeating the procedure (with $n = 3$),

$$x_4 = 2.64575 - \frac{(2.64575)^2 - 7}{2(2.64575)} \approx 2.64575$$

Since two consecutive values of x_n are the same (to the desired degree of accuracy), we have $\sqrt{7} \approx 2.64575$. It can be shown that to nine decimal places, $\sqrt{7} \approx 2.645751311$. •

EXAMPLE 2 Find the largest positive real root of $x^3 - 3x + 1 = 0$ to four decimal places.

SOLUTION If we let $f(x) = x^3 - 3x + 1$, then the problem is equivalent to finding the largest positive real *zero* of f. The graph of f is sketched in Figure 3.30. Note that f has three real zeros. We wish to find the zero that lies between 1 and 2. Since $f'(x) = 3x^2 - 3$, the formula for x_{n+1} in Newton's Method is

$$x_{n+1} = x_n - \frac{x_n^3 - 3x_n + 1}{3x_n^2 - 3}.$$

FIGURE 3.30

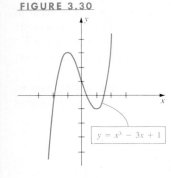

$y = x^3 - 3x + 1$

Referring to the graph we take, as a first approximation, $x_1 = 1.5$ and proceed as follows:

$$x_2 = 1.5 - \frac{(1.5)^3 - 3(1.5) + 1}{3(1.5)^2 - 3} \approx 1.5333$$

$$x_3 = 1.5333 - \frac{(1.5333)^3 - 3(1.5333) + 1}{3(1.5333)^2 - 3} \approx 1.5321$$

$$x_4 = 1.5321 - \frac{(1.5321)^3 - 3(1.5321) + 1}{3(1.5321)^2 - 3} \approx 1.5321$$

Thus the desired approximation is 1.5321. The remaining two real roots can be approximated in similar fashion (see Exercise 15). •

EXAMPLE 3 Approximate the real root of $x - \cos x = 0$ to three decimal places.

SOLUTION We wish to find a value of x such that $\cos x = x$. This coincides with the x-coordinate of the point of intersection of the graphs of $y = \cos x$ and $y = x$. It appears, from Figure 3.31, that $x_1 = 0.8$ is a reasonable first approximation. (Note that the figure also indicates that there is only one real root of the given equation.)

If we let $f(x) = x - \cos x$, then $f'(x) = 1 + \sin x$ and the formula in Newton's Method is

$$x_{n+1} = x_n - \frac{x_n - \cos x_n}{1 + \sin x_n}.$$

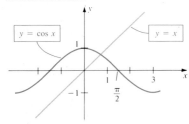

FIGURE 3.31

Following the usual procedure, and using a calculator to evaluate the sine and cosine functions, we obtain:

$$x_2 = 0.8 - \frac{0.8 - \cos 0.8}{1 + \sin 0.8} \approx 0.740$$

$$x_3 = 0.740 - \frac{0.740 - \cos 0.740}{1 + \sin 0.740} \approx 0.739$$

$$x_4 = 0.739 - \frac{0.739 - \cos 0.739}{1 + \sin 0.739} \approx 0.739$$

Hence 0.739 is the desired approximation. •

EXERCISES 3.10

Exer. 1–2: Use Newton's Method to approximate the number to four decimal places.

1 $\sqrt[3]{2}$ 2 $\sqrt[5]{3}$

Exer. 3–10: Use Newton's Method to approximate the real root to four decimal places.

3 The positive root of $x^3 + 5x - 3 = 0$

4 The largest root of $2x^3 - 4x^2 - 3x + 1 = 0$

5 The root of $x^4 + 2x^3 - 5x^2 + 1 = 0$ between 1 and 2

6 The root of $x^4 - 5x^2 + 2x - 5 = 0$ between 2 and 3

7 The root of $x^5 + x^2 - 9x - 3 = 0$ between -2 and -1

8 The root of $\cos x + x = 2$

9 The positive root of $2x - 3 \sin x = 0$

10 The root of $\sin x + x \cos x = \cos x$ between 0 and 1

Exer. 11–18: Use Newton's Method to find all real roots to two decimal places.

11 $x^4 = 125$ 12 $10x^2 - 1 = 0$

13 $x^4 - x - 2 = 0$ 14 $x^5 - 2x^2 + 4 = 0$

15 $x^3 - 3x + 1 = 0$ 16 $x^3 + 2x^2 - 8x - 3 = 0$

17 $2x - 5 - \sin x = 0$ 18 $x^2 - \cos 2x = 0$

19 Approximations to π may be generated by applying Newton's Method to $f(x) = \sin x$ and letting $x_1 = 3$.
(a) Find the first five approximations to π.
(b) What happens to the approximations if $x_1 = 6$?

20 A dramatic example of the phenomenon of *resonance* occurs when a singer adjusts the pitch of her voice to shatter a wine glass. Functions given by $f(x) = ax \cos bx$ occur in the mathematical analysis of such vibrations. Shown in the figure is a computer-generated graph of $f(x) = x \cos 2x$.
(a) Show that horizontal tangents occur at values of x for which $\cos 2x - 2x \sin 2x = 0$.
(b) For the equation in part (a), use Newton's Method to approximate the root between 1 and 2 to three decimal places.

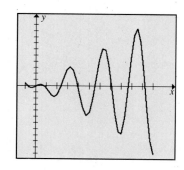

3.11 REVIEW

Define or discuss each of the following.

1 The derivative of a function
2 Differentiable function
3 Differentiability on an interval
4 Right-hand derivative of a function
5 Left-hand derivative of a function
6 The relationship between continuity and differentiability of a function
7 The Power Rule
8 The Product Rule
9 The Quotient Rule
10 The derivative as a rate of change
11 Acceleration in rectilinear motion
12 Applications of derivatives to economics
13 Derivatives of the trigonometric functions
14 Normal line to a graph
15 Increment notation
16 Differentials
17 Average error in measurement
18 Percentage error
19 The Chain Rule
20 The Power Rule for Functions
21 Simple harmonic motion
22 Implicit differentiation
23 Higher derivatives
24 Related rates
25 Newton's Method

EXERCISES 3.11

Exer. 1–2: Find $f'(x)$ directly from the definition of derivative.

1 $f(x) = 4/(3x^2 + 2)$
2 $f(x) = \sqrt{5 - 7x}$

Exer. 3–42: Find the first derivative.

3 $f(x) = 2x^3 - 7x + 2$

4 $k(x) = \dfrac{1}{x^4 - x^2 + 1}$

5 $g(t) = \sqrt{6t + 5}$

6 $h(t) = \dfrac{1}{\sqrt{6t + 5}}$

7 $F(z) = \sqrt[3]{7z^2 - 4z + 3}$

8 $f(w) = \sqrt[5]{3w^2}$

9 $G(x) = \dfrac{6}{(3x^2 - 1)^4}$

10 $H(x) = \dfrac{(3x^2 - 1)^4}{6}$

11 $F(y) = (y^2 - y^{-2})^{-2}$

12 $h(z) = [(z^2 - 1)^5 - 1]^5$

13 $g(x) = \sqrt[5]{(3x + 2)^4}$

14 $P(x) = (x + x^{-1})^2$

15 $r(s) = \left(\dfrac{8s^2 - 4}{1 - 9s^3}\right)^4$

16 $g(w) = \dfrac{(w - 1)(w - 3)}{(w + 1)(w + 3)}$

17 $F(x) = (x^6 + 1)^5(3x + 2)^3$

18 $k(z) = [z^2 + (z^2 + 9)^{1/2}]^{1/2}$

19 $g(y) = \sqrt{1 + \cos 2y}$

20 $p(x) = \dfrac{2x^4 + 3x^2 - 1}{x^2}$

21 $f(x) = \sin^2 (4x^3)$

22 $H(t) = (1 + \sin 3t)^3$

23 $h(x) = (\sec x + \tan x)^5$

24 $K(r) = \sqrt[3]{r^3 + \csc 6r}$

25 $f(x) = x^2 \cot 2x$

26 $f(x) = 6x^2 - \dfrac{5}{x} + \dfrac{2}{\sqrt[3]{x^2}}$

27 $g(z) = \csc\left(\dfrac{1}{z}\right) + \dfrac{1}{\sec z}$

28 $F(t) = \dfrac{5t^2 - 7}{t^2 + 2}$

29 $k(s) = (2s^2 - 3s + 1)(9s - 1)^4$

30 $H(x) = |\cos x|$

31 $f(w) = \sqrt{(2w + 5)/(7w - 9)}$

32 $S(t) = \sqrt{t^2 + t + 1}\,\sqrt[3]{4t - 9}$

33 $P(\theta) = \theta^2 \tan^2 (\theta^2)$

34 $g(v) = \dfrac{1}{1 + \cos^2 2v}$

35 $g(x) = (\cos \sqrt[3]{x} - \sin \sqrt[3]{x})^3$

36 $f(x) = \dfrac{x}{2x + \sec^2 x}$

37 $G(u) = \dfrac{\csc u + 1}{\cot u + 1}$

38 $k(\varphi) = \dfrac{\sin \varphi}{\cos \varphi - \sin \varphi}$

39 $F(x) = \sec 5x \tan 5x \sin 5x$

40 $H(z) = \sqrt{\sin \sqrt{z}}$

41 $g(\theta) = \tan^4 (\sqrt[4]{\theta})$

42 $f(x) = \csc^3 3x \cot^2 3x$

Exer. 43–48: Assuming that the equation determines a differentiable function f such that $y = f(x)$, find y'.

43 $5x^3 - 2x^2y^2 + 4y^3 - 7 = 0$

44 $3x^2 - xy^2 + y^{-1} = 1$

45 $\dfrac{\sqrt{x} + 1}{\sqrt{y} + 1} = y$

46 $y^2 - \sqrt{xy} + 3x = 2$

47 $xy^2 = \sin (x + 2y)$

48 $y = \cot (xy)$

Exer. 49–50: Find equations of the tangent line and the normal line to the graph of f at point P.

49 $y = 2x - \dfrac{4}{\sqrt{x}}$, $P(4, 6)$

50 $x^2y - y^3 = 8$, $P(-3, 1)$

51 Find the x-coordinates of all points on the graph of $y = 3x - \cos 2x$ at which the tangent line is perpendicular to the line $2x + 4y = 5$.

52 Given $y = \sin 2x - \cos 2x$ for $0 \le x \le 2\pi$, find the x-coordinates of all points on the graph at which the tangent line is horizontal.

Exer. 53–54: Find y', y'', and y'''.

53 $y = 5x^3 + 4\sqrt{x}$ **54** $y = 2x^2 - 3x - \cos 5x$

55 If $x^2 + 4xy - y^2 = 8$, find y'' by implicit differentiation.

56 If $f(x) = x^3 - x^2 - 5x + 2$,

  (a) find the x-coordinates of all points on the graph of f at which the tangent line is parallel to the line through $A(-3, 2)$ and $B(1, 14)$.

  (b) find the value of f'' at each zero of f'.

57 If $f(x) = 1/(1 - x)$, find a formula for $f^{(n)}(x)$ for any positive integer n.

58 If $y = 5x/(x^2 + 1)$, find dy and use it to approximate the change in y if x changes from 2 to 1.98. What is the exact change in y?

59 The side of an equilateral triangle is estimated to be 4 inches with a maximum error of 0.03 inch. Use differentials to estimate the maximum error in the calculated area of the triangle. Approximate the percentage error.

60 If $s = 3r^2 - 2\sqrt{r + 1}$ and $r = t^3 + t^2 + 1$, use the Chain Rule to find the value of ds/dt at $t = 1$.

61 If $f(x) = 2x^3 + x^2 - x + 1$ and $g(x) = x^5 + 4x^3 + 2x$, use differentials to approximate the change in $g(f(x))$ if x changes from -1 to -1.01.

62 Use differentials to approximate $\sqrt[3]{64.2}$. (*Hint:* Consider $y = \sqrt[3]{x}$.)

63 Suppose f and g are functions such that $f(2) = -1$, $f'(2) = 4$, $f''(2) = -2$, $g(2) = -3$, $g'(2) = 2$ and $g''(2) = 1$. Find the values of each of the following at $x = 2$.

  (a) $(2f - 3g)'$ (b) $(2f - 3g)''$ (c) $(fg)'$

  (d) $(fg)''$ (e) $\left(\dfrac{f}{g}\right)'$ (f) $\left(\dfrac{f}{g}\right)''$

64 The cost function for producing a microprocessor component is given by $C(x) = 1000 + 2x + 0.005x^2$. If 2000 units are produced, find the cost, the average cost, the marginal cost, and the marginal average cost.

65 A manufacturer of microwave ovens determines that the cost of producing x units is given by

$$C(x) = 4000 + 100x + 0.05x^2 + 0.0002x^3$$

Compare the marginal cost of producing 100 ovens with the cost of producing the 101st oven.

66 Let V and S denote the volume and surface area, respectively, of a spherical balloon. If the diameter is 8 cm and the volume increases by 12 cm^3, use differentials to approximate the change in S.

67 According to Stefan's law, the radiant energy emitted from the surface of a body is given by $R = kT^4$ for the rate of emission R per unit area, the temperature T (in °K), and

a constant k. If the error in the measurement of T is 0.5%, find the resulting percentage error in the calculated value of R.

68 The intensity of illumination from a source of light is inversely proportional to the square of the distance from the source. If a student works at a desk that is a certain distance from a lamp, use differentials to find the percentage change in distance that will increase the intensity by 10%.

69 The position function of a point moving on a coordinate line is given by $s(t) = (t^2 + 3t + 1)/(t^2 + 1)$. Find the velocity and acceleration at time t and describe the motion of the point during the time interval $[-2, 2]$.

70 The position of a moving point on a coordinate line is given by

$$s(t) = a \sin (kt + m) + b \cos (kt + m)$$

for constants a, b, k, and m. Prove that the magnitude of the acceleration is directly proportional to the distance from the origin.

71 The ends of a horizontal water trough 10 feet long are isosceles trapezoids with lower base 3 feet, upper base 5 feet, and altitude 2 feet. If the water level is rising at a rate of $\frac{1}{4}$ in./min when the depth is 1 foot, how fast is water entering the trough?

72 Two cars are approaching the same intersection along roads that run at right angles to each other. Car A is traveling at 20 mi/hr and car B is traveling at 40 mi/hr. If, at a certain instant, A is $\frac{1}{4}$ mile from the intersection and B is $\frac{1}{2}$ mile from the intersection, find the rate at which they are approaching one another at that instant.

73 Boyle's law states that $pv = c$ for pressure p, volume v, and a constant c. Find a formula for the rate of change of p with respect to v.

74 A railroad bridge is 20 feet above, and at right angles to, a river. A man in a train traveling 60 mi/hr passes over the center of the bridge at the same instant that a man in a motor boat traveling 20 mi/hr passes under the center of the bridge (see figure). How fast are the two men moving away from each other 10 seconds later?

EXERCISE 74

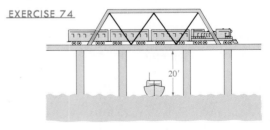

75 A large ferris wheel is 100 feet in diameter and rises 110 feet off the ground as illustrated in the figure. Each revolution of the wheel takes 30 seconds.

  (a) Express the distance h of a seat from the ground as a function of time t (in seconds) if $t = 0$ corresponds to a time when the seat is at the bottom.

(b) If a seat is rising, how fast is the distance changing when $h = 55$ feet?

EXERCISE 75

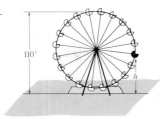

76 A piston is attached to a crankshaft as shown in the figure. The connecting rod AB has length 6 inches and the radius of the crankshaft is 2 inches.

(a) If the crankshaft rotates counterclockwise 2 times per second, find formulas for the position of point A at t seconds after A has coordinates $(2, 0)$.

(b) Find a formula for the position of point B at time t.

(c) How fast is B moving when A has coordinates $(0, 2)$?

EXERCISE 76

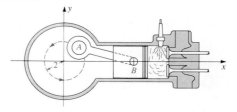

77 Use Newton's Method to approximate the root of the equation $\sin x - x \cos x = 0$ between π and $3\pi/2$ to three decimal places.

78 Use Newton's Method to approximate $\sqrt[4]{5}$ to three decimal places.

CHAPTER 4

EXTREMA AND ANTIDERIVATIVES

In this chapter we use the derivative to investigate maximum and minimum values of functions, applications of extrema, and the graphical concepts of *concavity* and *points of inflection*. Limits involving infinity and rational functions are discussed in Section 4.6. The concept of *antiderivative* is introduced in the last section and applied to problems on motion.

4.1 LOCAL EXTREMA OF FUNCTIONS

FIGURE 4.1

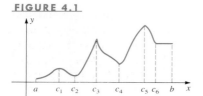

Suppose that the graph in Figure 4.1 was made by a recording instrument that measures the variation of a physical quantity. The x-axis represents time and the y-values represent measurements such as temperature, resistance in an electrical circuit, blood pressure of an individual, the amount of chemical in a solution, or the bacteria count in a culture.

The graph indicates that the quantity increased in the time interval $[a, c_1]$, decreased in $[c_1, c_2]$, increased in $[c_2, c_3]$, and so on. If we restrict our attention to the interval $[c_1, c_4]$, the quantity had its largest (or maximum) value at c_3 and its smallest (or minimum) value at c_2. In other intervals there were different largest or smallest values. For example, over the entire interval $[a, b]$, the maximum value occurred at c_5 and the minimum value at a.

The terminology for describing the variation of physical quantities is also used for functions.

DEFINITION (4.1)

Let a function f be defined on an interval I and let x_1, x_2 denote numbers in I.

(i) f is **increasing** on I if $f(x_1) < f(x_2)$ whenever $x_1 < x_2$.

(ii) f is **decreasing** on I if $f(x_1) > f(x_2)$ whenever $x_1 < x_2$.

(iii) f is **constant** on I if $f(x_1) = f(x_2)$ for every x_1 and x_2.

Graphical illustrations of Definition (4.1) are shown in Figure 4.2.

We shall use the phrases *f is increasing* and *f(x) is increasing* interchangeably. This will also be done for the term *decreasing*. If a function is increasing, then the graph rises as x increases. If a function is decreasing, then the graph falls as x increases. If Figure 4.1 represents the graph of a function f, then f is increasing on the intervals $[a, c_1]$, $[c_2, c_3]$, and $[c_4, c_5]$. It is decreasing on $[c_1, c_2]$, $[c_3, c_4]$, and $[c_5, c_6]$. The function is constant on the interval $[c_6, b]$.

FIGURE 4.2

(i) Increasing function

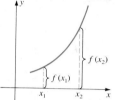

(ii) Decreasing function

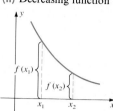

(iii) Constant function

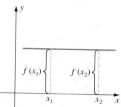

The next definition introduces terminology we shall use for largest and smallest values of functions on an interval.

DEFINITION (4.2)

> Let a function f be defined on an interval I and let c be a number in I.
>
> (i) $f(c)$ is the **maximum value** of f on I if $f(x) \leq f(c)$ for every x in I.
>
> (ii) $f(c)$ is the **minimum value** of f on I if $f(x) \geq f(c)$ for every x in I.

Maximum and minimum values are illustrated in Figures 4.3 and 4.4. We have pictured I as a closed interval $[a, b]$; however, Definition (4.2) may be applied to any interval. Although the graphs shown have horizontal tangent lines at the point $(c, f(c))$, maximum and minimum values can also occur at points where graphs have corners, or at breaks, or at endpoints of domains.

FIGURE 4.3
Maximum value $f(c)$

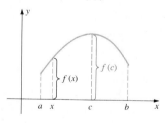

FIGURE 4.4
Minimum value $f(c)$

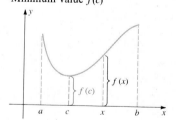

If $f(c)$ is the maximum value of f on I, we say that f *takes on* its maximum value at c, and the point $(c, f(c))$ is a highest point on the graph. If $f(c)$ is the minimum value of f on I we say that f *takes on its* minimum value at c, and $(c, f(c))$ is a lowest point on the graph. Maximum and minimum values are sometimes called **extreme values** or **extrema** of f. A function can take on a maximum or minimum value more than once. If f is a constant function, then $f(c)$ is both a maximum and a minimum value of f for *every* real number c.

EXAMPLE 1 Let $f(x) = 1/x^2$. Determine if f is increasing or decreasing on the following intervals, and find the extrema of f on each interval:

$$[1, 2] \quad (1, 2] \quad (1, 2) \quad (-2, -1] \quad [-1, 2]$$

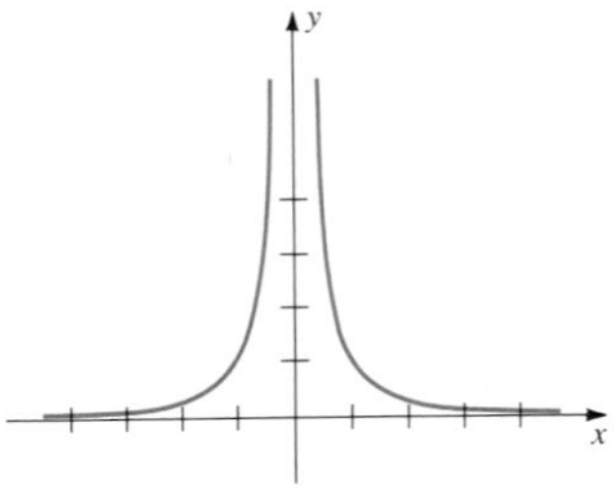

SOLUTION Referring to the graph of f sketched in Figure 4.5, we obtain the following table.

FIGURE 4.5

Interval	f	Maximum value of f	Minimum value of f
$[1, 2]$	decreasing	$f(1) = 1$	$f(2) = \frac{1}{4}$
$(1, 2]$	decreasing	none	$f(2) = \frac{1}{4}$
$(1, 2)$	decreasing	none	none
$(-2, -1]$	increasing	$f(-1) = 1$	none
$[-1, 2]$	neither	none	$f(2) = \frac{1}{4}$

Note that f does not have a maximum value on $(1, 2]$, for if a is any number in $(1, 2]$ and if $1 < c < a$, then $f(c) > f(a)$. On the open interval $(1, 2)$, f takes on *neither* a maximum nor minimum value.

The function is not continuous on $[-1, 2]$. Although f is increasing on $[-1, 0)$ and decreasing on $(0, 2]$, we cannot attach one of these labels to the interval $[-1, 2]$. •

The preceding example indicates that the existence of maximum or minimum values may depend on the type of interval and on the continuity of the function. The next theorem states conditions for which a function takes on a maximum value and a minimum value on an interval. More advanced texts on calculus may be consulted for the proof.

THEOREM (4.3)

> If a function f is continuous on a closed interval $[a, b]$, then f takes on a minimum value and a maximum value at least once in $[a, b]$.

The extrema are also called the **absolute minimum** and **absolute maximum values** for f on an interval. We shall also be interested in *local extrema* of f, defined as follows.

DEFINITION (4.4)

> Let c be a number in the domain of a function f.
>
> (i) $f(c)$ is a **local maximum** of f if there exists an open interval (a, b) containing c such that $f(x) \leq f(c)$ for every x in (a, b).
>
> (ii) $f(c)$ is a **local minimum** of f if there exists an open interval (a, b) containing c such that $f(x) \geq f(c)$ for every x in (a, b).

The term *local* is used because we *localize* our attention to a sufficiently small *open* interval containing c such that f takes on a largest (or smallest) value at c. Outside of that open interval f may take on larger (or smaller) values. The word *relative* may be used in place of *local*. Each local maximum or minimum is called a **local extremum** of f and the totality of such numbers are the **local extrema** of f. Several examples of local extrema are illustrated in Figure 4.6. As indicated in the figure, it is possible for a local minimum to be *larger* than a local maximum.

FIGURE 4.6

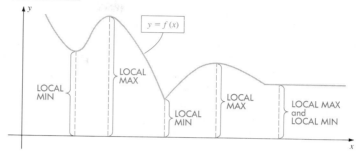

The local extrema may not include the absolute minimum or maximum values of f. For example, in Figure 4.1, $f(a)$ is the minimum value of f on $[a, b]$; however, it is not a local minimum, since there is no *open* interval I contained in $[a, b]$ such that $f(a)$ is the least value of f on I.

At a point corresponding to a local extremum of the function graphed in Figure 4.6, the tangent line is horizontal or the graph has a corner. The x-coordinates of these points are numbers at which the derivative is zero or does not exist. The next theorem specifies that this is generally true. A proof is given at the end of this section.

THEOREM (4.5)

> If a function f has a local extremum at a number c in an open interval, then either $f'(c) = 0$ or $f'(c)$ does not exist.

The following is an immediate consequence of Theorem (4.5).

COROLLARY (4.6)

> If $f'(c)$ exists and $f'(c) \neq 0$, then $f(c)$ is not a local extremum of the function f.

A result similar to Theorem (4.5) is true for the *absolute* maximum and minimum values of a function that is continuous on a closed interval $[a, b]$, provided the extrema occur on the *open* interval (a, b). The theorem may be stated as follows.

THEOREM (4.7)

> If a function f is continuous on a closed interval $[a, b]$ and has its maximum or minimum value at a number c in the open interval (a, b), then either $f'(c) = 0$ or $f'(c)$ does not exist.

The proof of Theorem (4.7) is exactly the same as that of (4.5) with the word *local* deleted.

It follows from Theorems (4.5) and (4.7) that the numbers at which the derivative either is zero or does not exist play a crucial role in the search for extrema of a function. Because of this, we give these numbers a special name.

DEFINITION (4.8) | A number c in the domain of a function f is a **critical number** of f if either $f'(c) = 0$ or $f'(c)$ does not exist.

Referring to Theorem (4.7) we see that if f is continuous on a closed interval $[a, b]$, then the absolute maximum and minimum values occur either at a critical number of f or at the endpoints a or b of the interval. If either $f(a)$ or $f(b)$ is an absolute extremum of f on $[a, b]$ it is called an **endpoint extremum.** The sketches in Figure 4.7 illustrate this concept.

FIGURE 4.7

Endpoint extrema of f on $[a, b]$

(i) Absolute minimum $f(a)$ (ii) Absolute maximum $f(a)$ (iii) Absolute minimum $f(b)$ (iv) Absolute maximum $f(b)$

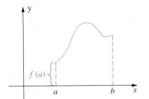 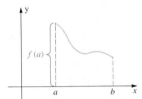 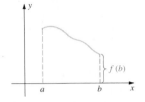 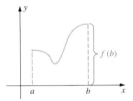

The preceding discussion gives us the following:

GUIDELINES FOR FINDING (4.9)
THE ABSOLUTE EXTREMA
OF A CONTINUOUS FUNCTION f
ON A CLOSED INTERVAL $[a, b]$

1 Find all the critical numbers of f.

2 Calculate $f(c)$ for each critical number c.

3 Calculate $f(a)$ and $f(b)$.

4 The absolute maximum and minimum of f on $[a, b]$ are the largest and smallest function values calculated in Guidelines 2 and 3.

EXAMPLE 2 If $f(x) = x^3 - 12x$, find the absolute maximum and minimum values of f on the closed interval $[-3, 5]$. Sketch the graph of f.

SOLUTION Using Guidelines (4.9), we begin by finding the critical numbers of f. Differentiating,

$$f'(x) = 3x^2 - 12 = 3(x^2 - 4) = 3(x + 2)(x - 2).$$

Since the derivative exists for every x, the only critical numbers are those for which the derivative is zero, that is, -2 and 2. Since f is continuous on $[-3, 5]$, it follows from our discussion that the absolute maximum and minimum are among the numbers $f(-2)$, $f(2)$, $f(-3)$, and $f(5)$. Calculating these values (see Guidelines 2 and 3), we obtain

$$f(-2) = (-2)^3 - 12(-2) = -8 + 24 = 16$$
$$f(2) = 2^3 - 12(2) = 8 - 24 = -16$$
$$f(-3) = (-3)^3 - 12(-3) = -27 + 36 = 9$$
$$f(5) = 5^3 - 12(5) = 125 - 60 = 65$$

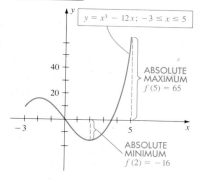

FIGURE 4.8

Thus, by Guideline 4, the minimum value of f on $[-3, 5]$ is $f(2) = -16$ and the maximum value is the endpoint extremum $f(5) = 65$.

Using the function values we have calculated and plotting several more points leads to the sketch in Figure 4.8, with different scales on the x- and y-axes. The tangent line is horizontal at the point corresponding to each of the critical numbers -2 and 2. It will follow from our work in Section 4.4 that $f(-2) = 16$ is a *local* maximum for f, as indicated by the graph. •

Maximum or minimum values are sometimes referred to incorrectly. Note that in Example 2 the minimum occurs *at* $x = 2$, and the minimum *is* $f(2) = -16$. Thus, when asked to find a minimum (or maximum) we do not merely find the value of x at which it occurs: *Be sure to complete the problem by calculating the function value.*

We see from Theorem (4.5) that if a function has a *local* extremum, then it *must* occur at a critical number; however, not every critical number leads to a local extremum, as illustrated by the following example.

EXAMPLE 3 If $f(x) = x^3$, prove that f has no local extremum.

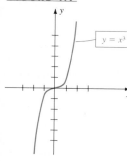

FIGURE 4.9

SOLUTION The graph of f is sketched in Figure 4.9. The derivative is $f'(x) = 3x^2$, which exists for every x and is zero only if $x = 0$. Consequently, 0 is the only critical number. However, if $x < 0$, then $f(x)$ is negative, and if $x > 0$, then $f(x)$ is positive. Thus $f(0)$ is neither a local maximum nor a local minimum. Since a local extremum *must* occur at a critical number (see Theorem (4.5)), it follows that f has no local extrema. Note that the tangent line is horizontal and crosses the graph at the point $(0, 0)$. •

EXAMPLE 4 Find the critical numbers of f if $f(x) = (x + 5)^2 \sqrt[3]{x - 4}$.

SOLUTION Differentiating $f(x) = (x + 5)^2(x - 4)^{1/3}$, we obtain

$$f'(x) = (x + 5)^2 \tfrac{1}{3}(x - 4)^{-2/3} + 2(x + 5)(x - 4)^{1/3}.$$

To find the critical numbers we simplify $f'(x)$ as follows:

$$f'(x) = \frac{(x + 5)^2}{3(x - 4)^{2/3}} + 2(x + 5)(x - 4)^{1/3}$$

$$= \frac{(x + 5)^2 + 6(x + 5)(x - 4)}{3(x - 4)^{2/3}}$$

$$= \frac{(x + 5)[(x + 5) + 6(x - 4)]}{3(x - 4)^{2/3}}$$

$$= \frac{(x + 5)(7x - 19)}{3(x - 4)^{2/3}}$$

Consequently, $f'(x) = 0$ if $x = -5$ or $x = \frac{19}{7}$. The derivative $f'(x)$ does not exist at $x = 4$. Thus f has the three critical numbers -5, $\frac{19}{7}$, and 4. •

EXAMPLE 5 If $f(x) = 2 \sin x + \cos 2x$, find the critical numbers of f that are in the interval $[0, 2\pi]$.

SOLUTION Differentiating,

$$f'(x) = 2 \cos x - 2 \sin 2x.$$

Since $\sin 2x = 2 \sin x \cos x$, this may be rewritten

$$f'(x) = 2 \cos x - 4 \sin x \cos x$$
$$= 2 \cos x (1 - 2 \sin x).$$

The derivative exists for every x, and $f'(x) = 0$ if either $\sin x = \frac{1}{2}$ or $\cos x = 0$. Hence the critical numbers of f in the interval $[0, 2\pi]$ are $\pi/6$, $5\pi/6$, $\pi/2$, and $3\pi/2$. We shall return to this function and sketch the graph of f in Example 5 of Section 4.4. •

PROOF OF THEOREM (4.5) Suppose f has a local extremum at c. If $f'(c)$ does not exist, there is nothing more to prove. If $f'(c)$ exists, then precisely one of the following occurs: (i) $f'(c) > 0$, (ii) $f'(c) < 0$, or (iii) $f'(c) = 0$. We shall arrive at (iii) by proving that neither (i) nor (ii) can occur. Thus, suppose $f'(c) > 0$. Employing Definition (3.1′),

$$f'(c) = \lim_{x \to c} \frac{f(x) - f(c)}{x - c} > 0$$

and hence by Theorem (2.12), there exists an open interval (a, b) containing c such that

$$\frac{f(x) - f(c)}{x - c} > 0$$

for every x in (a, b) different from c. The last inequality implies that if $a < x < b$ and $x \neq c$, then $f(x) - f(c)$ and $x - c$ are either both positive or both negative; that is,

$$\begin{cases} f(x) - f(c) < 0 & \text{whenever } x - c < 0, \quad \text{and} \\ f(x) - f(c) > 0 & \text{whenever } x - c > 0. \end{cases}$$

Another way of stating these facts is: If x is in (a, b) and $x \neq c$, then

$$\begin{cases} f(x) < f(c) & \text{whenever } x < c, \quad \text{and} \\ f(x) > f(c) & \text{whenever } x > c. \end{cases}$$

It follows that $f(c)$ is neither a local maximum nor a local minimum for f, contrary to hypothesis. Consequently (i) cannot occur. Similarly, the assumption that $f'(c) < 0$ leads to a contradiction. Hence (iii) must hold and the theorem is proved. • •

EXERCISES 4.1

Exer. 1–4: Find the absolute maximum and minimum of f on the indicated interval.

1 $f(x) = 5 - 6x^2 - 2x^3$, $[-3, 1]$

2 $f(x) = 3x^2 - 10x + 7$, $[-1, 3]$

3 $f(x) = 1 - x^{2/3}$, $[-1, 8]$

4 $f(x) = x^4 - 5x^2 + 4$, $[0, 2]$

5 (a) If $f(x) = x^{1/3}$, prove that 0 is the only critical number of f, and that $f(0)$ is not a local extremum.

(b) If $f(x) = x^{2/3}$, prove that 0 is the only critical number of f, and that $f(0)$ is a local minimum of f.

6 If $f(x) = |x|$, prove that 0 is the only critical number of f, that $f(0)$ is a local minimum of f, and that the graph of f has no tangent line at the point $(0, 0)$.

Exer. 7–8: Prove that f has no local extrema. Sketch the graph of f. Prove that f is continuous on the interval $(0, 1)$, but f has neither a maximum nor a minimum value on $(0, 1)$. Why doesn't this contradict Theorem (4.3)?

7 $f(x) = x^3 + 1$ 8 $f(x) = 1/x^2$

Exer. 9–32: Find the critical numbers of the function.

9 $f(x) = 4x^2 - 3x + 2$ 10 $g(x) = 2x + 5$

11 $s(t) = 2t^3 + t^2 - 20t + 4$

12 $K(z) = 4z^3 + 5z^2 - 42z + 7$

13 $F(w) = w^4 - 32w$ 14 $k(r) = r^5 - 2r^3 + r - 12$

15 $f(z) = \sqrt{z^2 - 16}$ 16 $M(x) = \sqrt[3]{x^2 - x - 2}$

17 $g(t) = t^2 \sqrt[3]{2t - 5}$ 18 $T(v) = (4v + 1)\sqrt{v^2 - 16}$

19 $G(x) = \dfrac{2x - 3}{x^2 - 9}$ 20 $f(s) = \dfrac{s^2}{5s + 4}$

21 $f(t) = \sin^2 t - \cos t$

22 $g(t) = 4 \sin^3 t + 3\sqrt{2} \cos^2 t$

23 $K(\theta) = \sin 2\theta + 2 \cos \theta$

24 $f(x) = 8 \cos^3 x - 3 \sin 2x - 6x$

25 $f(x) = \dfrac{1 + \sin x}{1 - \sin x}$

26 $g(\theta) = 2\sqrt{3}\theta + \sin 4\theta$

27 $k(u) = u - \tan u$

28 $p(z) = 3 \tan z - 4z$

29 $H(\varphi) = \cot \varphi + \csc \varphi$

30 $g(x) = 2x + \cot x$

31 $f(x) = \sec (x^2 + 1)$

32 $s(t) = \dfrac{\sec t + 1}{\sec t - 1}$

33 Prove that a polynomial function of degree 1 has no local or absolute extrema on the interval $(-\infty, \infty)$. Discuss extrema on a closed interval $[a, b]$.

34 If f is a constant function and (a, b) is any open interval, prove that $f(c)$ is both a local and an absolute extremum of f for every number c in (a, b).

35 If f is the greatest integer function, prove that every number is a critical number of f.

36 Let f be defined by the following conditions: $f(x) = 0$ if x is rational and $f(x) = 1$ if x is irrational. Prove that every number is a critical number of f.

37 Prove that a quadratic function has exactly one critical number on $(-\infty, \infty)$.

38 Prove that a polynomial function of degree 3 has either two, one, or no critical numbers on $(-\infty, \infty)$. Sketch graphs that illustrate how each of these possibilities can occur.

39 Let $f(x) = x^n$ for a positive integer n. Prove that f has either one or no local extremum on $(-\infty, \infty)$ if n is even or odd, respectively. Sketch a typical graph illustrating each case.

40 Prove that a polynomial function of degree n can have at most $n - 1$ local extrema on $(-\infty, \infty)$.

4.2 ROLLE'S THEOREM AND THE MEAN VALUE THEOREM

Finding the critical numbers of a function can sometimes be extremely difficult. As a matter of fact, there is no guarantee that critical numbers even exist. The next theorem, credited to the French mathematician Michel Rolle (1652–1719), provides sufficient conditions for the existence of a critical number. The theorem is stated for a function f that is continuous on a closed interval $[a, b]$ and is differentiable on the open interval (a, b) such that $f(a) = f(b)$. Some typical graphs of functions of this type are sketched in Figure 4.10.

FIGURE 4.10

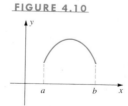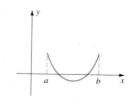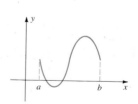

Referring to the sketches in Figure 4.10, we can reasonably expect that there is at least one number c between a and b such that the tangent line at the point $(c, f(c))$ is horizontal, or equivalently, such that $f'(c) = 0$. This is precisely the conclusion of the following theorem.

> If a function f is continuous on a closed interval $[a, b]$ and is differentiable on the open interval (a, b), and if $f(a) = f(b)$, then $f'(c) = 0$ for at least one number c in (a, b).

PROOF The function f must fall into at least one of the following three categories:

(i) $f(x) = f(a)$ *for every* x *in* (a, b). In this case f is a constant function and hence $f'(x) = 0$ for every x. Consequently, *every* number c in (a, b) is a critical number.

(ii) $f(x) > f(a)$ *for some* x *in* (a, b). In this case the maximum value of f in $[a, b]$ is greater than $f(a)$ or $f(b)$ and, therefore, must occur at some number c in the *open* interval (a, b). Since the derivative exists throughout (a, b), we conclude from Theorem (4.7) that $f'(c) = 0$.

(iii) $f(x) < f(a)$ *for some* x *in* (a, b). In this case the minimum value of f in $[a, b]$ is less than $f(a)$ or $f(b)$ and must occur at some number c in (a, b). As in (ii), $f'(c) = 0$. • •

COROLLARY **(4.11)**

> If f is continuous on a closed interval $[a, b]$ and if $f(a) = f(b)$, then f has at least one critical number in the open interval (a, b).

PROOF If f' does not exist at some number c in (a, b), then by Definition (4.8), c is a critical number. Alternatively, if f' exists throughout (a, b), then a critical number exists by Rolle's Theorem. • •

EXAMPLE 1 Let $f(x) = 4x^2 - 20x + 29$. Show that f satisfies the hypotheses of Rolle's Theorem on the interval $[1, 4]$ and find all real numbers c in the open interval $(1, 4)$ such that $f'(c) = 0$.

SOLUTION Since f is a polynomial function, it is continuous and differentiable for every x. In particular, it is continuous on $[1, 4]$ and differentiable on $(1, 4)$. Moreover,

$$f(1) = 4 - 20 + 29 = 13$$

$$f(4) = 64 - 80 + 29 = 13$$

and hence $f(1) = f(4)$. Thus, f satisfies the hypotheses of Rolle's Theorem on $[1, 4]$.

Differentiating,

$$f'(x) = 8x - 20.$$

Setting $f'(x) = 0$ gives us $8x = 20$, or $x = \frac{5}{2}$. Hence

$$f'(\tfrac{5}{2}) = 0 \quad \text{and} \quad 1 < \tfrac{5}{2} < 4.$$

The graph of f (a parabola) is sketched in Figure 4.11. Since $f'(\frac{5}{2}) = 0$, the tangent line is horizontal at the vertex $(\frac{5}{2}, 4)$. •

We can generalize Rolle's Theorem to the case in which $f(a) \neq f(b)$. Let us consider the points $P(a, f(a))$ and $Q(b, f(b))$ on the graph of f as illustrated by any of the sketches in Figure 4.12.

FIGURE 4.11

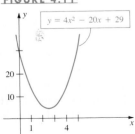

$y = 4x^2 - 20x + 29$

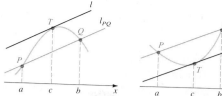

FIGURE 4.12

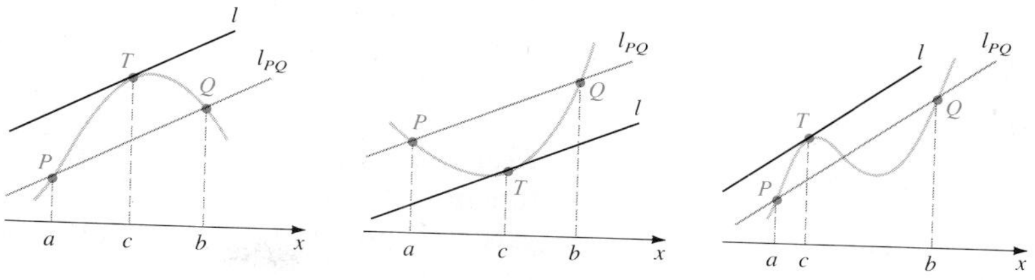

If f' exists throughout the open interval (a, b), then there appears to be at least one point $T(c, f(c))$ on the graph at which the tangent line l is parallel to the secant line l_{PQ} through P and Q. In terms of slopes:

$$\text{slope of } l = \text{slope of } l_{PQ}$$

or

$$f'(c) = \frac{f(b) - f(a)}{b - a}.$$

If we multiply both sides of the last equation by $b - a$, we obtain the formula stated in the next theorem.

THE MEAN VALUE THEOREM (4.12)

> If a function f is continuous on a closed interval $[a, b]$ and is differentiable on the open interval (a, b), then there exists a number c in (a, b) such that
>
> $$f(b) - f(a) = f'(c)(b - a)$$

FIGURE 4.13

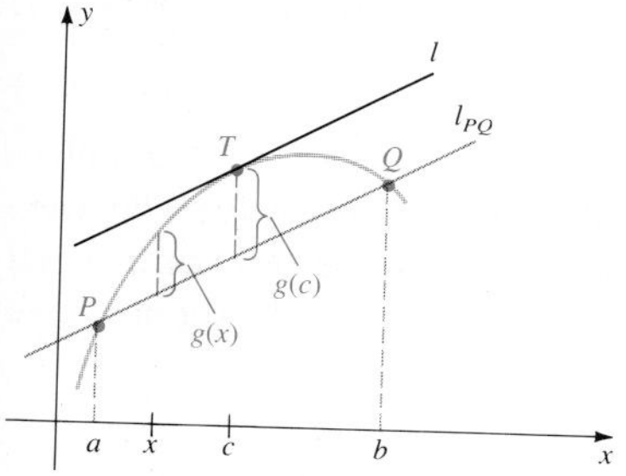

PROOF For any x in the interval $[a, b]$, let $g(x)$ be the vertical (signed) distance from the secant line l_{PQ} to the graph of f, as illustrated in Figure 4.13. (The term *signed distance* means that $g(x)$ is positive or negative if the graph of f lies above or below l_{PQ}, respectively.)

It appears that if $T(c, f(c))$ is a point at which the tangent line l is parallel to l_{PQ}, then the distance $g(c)$ is a relative extremum of g. This suggests that we find the critical numbers of the function g.

We may obtain a formula for $g(x)$ as follows. First, by the Point-Slope Form (1.15), an equation of the secant line l_{PQ} is

$$y - f(a) = \frac{f(b) - f(a)}{b - a}(x - a)$$

or, equivalently,

$$y = f(a) + \frac{f(b) - f(a)}{b - a}(x - a).$$

As illustrated in Figure 4.13, $g(x)$ is the difference of the distances from the x-axis to the graph of f and to the line l_{PQ}, that is,

$$g(x) = f(x) - \left[f(a) + \frac{f(b) - f(a)}{b - a}(x - a) \right].$$

We shall use Rolle's Theorem to find a critical number of g. We first observe that since f is continuous on $[a, b]$ and differentiable on (a, b), the same is true for the function g. Differentiating, we obtain

$$g'(x) = f'(x) - \frac{f(b) - f(a)}{b - a}.$$

Moreover, by direct substitution we see that $g(a) = g(b) = 0$, and hence the function g satisfies the hypotheses of Rolle's Theorem. Consequently, there exists a number c in the open interval (a, b) such that $g'(c) = 0$, or equivalently,

$$f'(c) - \frac{f(b) - f(a)}{b - a} = 0.$$

The last equation may be written in the form stated in the conclusion of the theorem. • •

The Mean Value Theorem is also called **The Theorem of the Mean.** We will use it later to help establish several very important results.

EXAMPLE 2 If $f(x) = x^3 - 8x - 5$, prove that f satisfies the hypotheses of the Mean Value Theorem on the interval $[1, 4]$, and find a number c in the open interval $(1, 4)$ that satisfies the conclusion of the theorem.

SOLUTION Since f is a polynomial function, it is continuous and differentiable for all real numbers. In particular, it is continuous on $[1, 4]$ and differentiable on the open interval $(1, 4)$. According to the Mean Value Theorem, there exists a number c in $(1, 4)$ such that

$$f(4) - f(1) = f'(c)(4 - 1).$$

Since $f(4) = 27$, $f(1) = -12$, and $f'(x) = 3x^2 - 8$, this is equivalent to

$$27 - (-12) = (3c^2 - 8)(3) \quad \text{or} \quad 39 = 3(3c^2 - 8).$$

We leave it to the reader to show that the last equation implies that $c = \pm\sqrt{7}$. Hence the desired number in the interval $(1, 4)$ is $c = \sqrt{7}$. •

In the study of statistics the term *mean* is used for the *average* of a collection of numbers. The word has a similar connotation in *Mean Value Theorem*. As an illustration, if a point P is moving on a coordinate line, and if $s(t)$ denotes the coordinate of P at time t, then by Definition (2.4), the average velocity of P during the time interval $[a, b]$ is

$$v_{av} = \frac{s(b) - s(a)}{b - a}.$$

According to the Mean Value Theorem, this average (mean) velocity is equal to the velocity $v'(c)$ at some time c between a and b.

EXAMPLE 3 The speedometer of an automobile registers 50 mi/hr as it passes a mileage marker along a highway. Four minutes later, as the automobile passes a marker that is five miles from the first, the speedometer registers 55 mi/hr. Use the Mean Value Theorem to prove that the velocity of the automobile exceeded 70 mi/hr at some time while traveling between the two markers.

SOLUTION We may assume that the automobile is moving along a straight highway with the two mileage markers at A and B, as illustrated in Figure 4.14. Let t denote the elapsed time (in hours) after the automobile passes

FIGURE 4.14

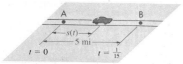

A and let $s(t)$ denote its distance from A (in miles) at time t. Since the automobile passes B at $t = \frac{4}{60} = \frac{1}{15}$ hour, the average velocity during the trip from A to B is

$$v_{av} = \frac{s(1/15) - s(0)}{(1/15) - 0} = \frac{5}{1/15} = 75 \text{ mi/hr}.$$

If we assume that s is a differentiable function, then applying the Mean Value Theorem (4.12) to s, with $a = 0$ and $b = \frac{1}{15}$, there is a time c in the time interval $[0, \frac{1}{15}]$ at which the velocity $s'(c)$ of the automobile is 75 mi/hr. Note that the velocity at A or B is irrelevant. •

EXERCISES 4.2

1 If $f(x) = |x|$, show that $f(-1) = f(1)$ but $f'(c) \neq 0$ for every number c in the open interval $(-1, 1)$. Why doesn't this contradict Rolle's Theorem?

2 If $f(x) = 5 + 3(x - 1)^{2/3}$, show that $f(0) = f(2)$, but $f'(c) \neq 0$ for every number c in the open interval $(0, 2)$. Why doesn't this contradict Rolle's Theorem?

3 If $f(x) = 4/x$, prove that no number c exists such that $f(4) - f(-1) = f'(c)[4 - (-1)]$. Why doesn't this contradict the Mean Value Theorem applied to the interval $[-1, 4]$?

4 If f is the greatest integer function and if a and b are real numbers such that $b - a \geq 1$, prove that no number c exists such that $f(b) - f(a) = f'(c)(b - a)$. Why doesn't this contradict the Mean Value Theorem?

Exer. 5–8: Show that f satisfies the hypotheses of Rolle's Theorem on the interval $[a, b]$, and find all numbers c in (a, b) such that $f'(c) = 0$.

5 $f(x) = 3x^2 - 12x + 11$, $[0, 4]$

6 $f(x) = 5 - 12x - 2x^2$, $[-7, 1]$

7 $f(x) = x^4 + 4x^2 + 1$, $[-3, 3]$

8 $f(x) = x^3 - x$, $[-1, 1]$

Exer. 9–18: Determine if the function f satisfies the hypotheses of the Mean Value Theorem on the interval $[a, b]$. If so, find all numbers c in (a, b) such that $f(b) - f(a) = f'(c)(b - a)$.

9 $f(x) = x^3 + 1$, $[-2, 4]$

10 $f(x) = 5x^2 - 3x + 1$, $[1, 3]$

11 $f(x) = x + (4/x)$, $[1, 4]$

12 $f(x) = 3x^5 + 5x^3 + 15x$, $[-1, 1]$

13 $f(x) = x^{2/3}$, $[-8, 8]$

14 $f(x) = 1/(x - 1)^2$, $[0, 2]$

15 $f(x) = 4 + \sqrt{x - 1}$, $[1, 5]$

16 $f(x) = 1 - 3x^{1/3}$, $[-8, -1]$

17 $f(x) = x^3 - 2x^2 + x + 3$, $[-1, 1]$

18 $f(x) = |x - 3|$, $[-1, 4]$

19 Prove that if f is a linear function, then f satisfies the hypotheses of the Mean Value Theorem on every closed interval $[a, b]$, and that *every* number c satisfies the conclusion of the theorem.

20 If f is a quadratic function and $[a, b]$ is any closed interval, prove that precisely one number c in the interval (a, b) satisfies the conclusion of the Mean Value Theorem.

21 If f is a polynomial function of degree 3 and $[a, b]$ is any closed interval, prove that at most two numbers in (a, b) satisfy the conclusion of the Mean Value Theorem. Sketch graphs that illustrate the various possibilities. Extend this result to a polynomial function of degree 4, and illustrate with sketches. Generalize to polynomial functions of degree n, for any positive integer n.

22 If $f(x)$ is a polynomial of degree 3, use Rolle's Theorem to prove that f has at most three real zeros. Extend this result to polynomials of degree n.

Exer. 23–30: Use the Mean Value Theorem for each proof.

23 If f is continuous on $[a, b]$ and if $f'(x) = 0$ for every x in (a, b), prove that $f(x) = k$ for some real number k.

24 If f is continuous on $[a, b]$ and if $f'(x) = c$ for every x in (a, b), prove that $f(x) = cx + d$ for some real number d.

25 A straight highway 50 miles long connects two cities A and B. Prove that it is impossible to travel from A to B by automobile in exactly one hour without the speedometer reading 50 mi/hr at least once.

26 Let T denote the temperature (in °F) at time t (in hours). If the temperature is decreasing, then dT/dt is the *rate of cooling*. The greatest temperature variation during a twelve-hour period occurred in Montana in 1916, when the temperature dropped from 44 °F to a chilling −56 °F. Show that the rate of cooling exceeded −8 °F/hr at some time during the period of change.

27 If W denotes the weight (in pounds) of an individual, and t denotes time (in months), then dW/dt is the rate of weight gain or loss (in lb/mo). The current speed record for weight loss is a drop in weight from 487 lb to 130 lb over an eight-month period. Show that the rate of weight loss exceeded 44 lb/mo at some time during the eight-month period.

28 The electrical charge Q on a capacitor increases from 2 millicoulombs to 10 millicoulombs in 15 milliseconds.

Show that the current $I = dQ/dt$ exceeded $\frac{1}{2}$ ampere at some instant during this short time interval. (*Note:* 1 ampere = 1 coulomb/sec.)

29 Prove that if u and v are any real numbers, then
$$|\sin u - \sin v| \le |u - v|.$$

30 Prove that $\sqrt{1 + h} < 1 + \frac{1}{2}h$ for $h > 0$.

4.3 THE FIRST DERIVATIVE TEST

The following theorem indicates how the derivative may be used to determine intervals on which a function is increasing or decreasing.

THEOREM (4.13)

Let f be a function that is continuous on a closed interval $[a, b]$ and differentiable on the open interval (a, b).

(i) If $f'(x) > 0$ for every x in (a, b), then f is increasing on $[a, b]$.

(ii) If $f'(x) < 0$ for every x in (a, b), then f is decreasing on $[a, b]$.

PROOF To prove (i), suppose that $f'(x) > 0$ for every x in (a, b), and consider any numbers x_1, x_2 in $[a, b]$ such that $x_1 < x_2$. We wish to show that $f(x_1) < f(x_2)$. Applying the Mean Value Theorem (4.12) to the interval $[x_1, x_2]$,
$$f(x_2) - f(x_1) = f'(w)(x_2 - x_1)$$
for w in the open interval (x_1, x_2). Since $x_2 - x_1 > 0$ and since, by hypothesis, $f'(w) > 0$, the right-hand side of the previous equation is positive, that is, $f(x_2) - f(x_1) > 0$. Hence $f(x_2) > f(x_1)$, which is what we wished to show. The proof of (ii) is similar and is left as an exercise. • •

Theorem (4.13) is illustrated in the graphs shown in Figure 4.15. If $f'(x) > 0$ the tangent line l rises, and so does the graph of f. If $f'(x) < 0$ both the tangent line and the graph fall.

FIGURE 4.15

(i) $f'(x) > 0$; f increasing on $[a, b]$

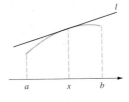 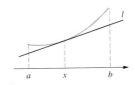 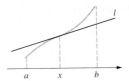

(ii) $f'(x) < 0$; f decreasing on $[a, b]$

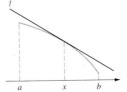 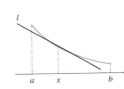 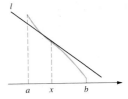

We may also show that if $f'(x) > 0$ throughout an infinite interval $(-\infty, a)$ or (b, ∞), then f is increasing on $(-\infty, a]$ or $[b, \infty)$, respectively, provided f is continuous on those intervals. An analogous result holds for decreasing functions if $f'(x) < 0$.

To apply Theorem (4.13) we must determine intervals in which $f'(x)$ is either always positive or always negative. Theorem (2.34) is useful in this respect. Specifically, if the *derivative* f' is continuous and has no zeros on an interval, then either $f'(x) > 0$ or $f'(x) < 0$ for every x in the interval. Thus, if we choose *any* number k in the interval, and if $f'(k) > 0$, then $f'(x) > 0$ for *every* x in the interval. Similarly, if $f'(k) < 0$, then $f'(k) < 0$ throughout the interval. Recall that $f'(k)$ is referred to as a *test value* of $f'(x)$ for the interval.

EXAMPLE 1 If $f(x) = x^3 + x^2 - 5x - 5$, find the intervals on which f is increasing and the intervals on which f is decreasing. Sketch the graph of f.

SOLUTION Differentiating,

$$f'(x) = 3x^2 + 2x - 5 = (3x + 5)(x - 1).$$

By Theorem (4.13) it is sufficient to find the intervals in which $f'(x) > 0$ and those in which $f'(x) < 0$. The factored form of $f'(x)$ and the critical numbers $-\frac{5}{3}$ and 1 suggest the open intervals $(-\infty, -\frac{5}{3})$, $(-\frac{5}{3}, 1)$, and $(1, \infty)$. On each of these intervals, f' is continuous and has no zeros and, therefore, $f'(x)$ has the same sign throughout the interval. This sign can be determined by choosing a suitable test value for the interval.

If we choose -2 in the interval $(-\infty, -\frac{5}{3})$, we obtain the test value

$$f'(-2) = 3(-2)^2 + 2(-2) - 5 = 12 - 4 - 5 = 3.$$

Since 3 is positive, $f'(x)$ is positive throughout $(-\infty, -\frac{5}{3})$.

If we choose 0 in $(-\frac{5}{3}, 1)$, the test value is

$$f'(0) = 3(0)^2 + 2(0) - 5 = -5.$$

Since -5 is negative, $f'(x) < 0$ throughout $(-\frac{5}{3}, 1)$.

Finally, choosing 2 in $(1, \infty)$ we obtain

$$f'(2) = 3(2)^2 + 2(2) - 5 = 12 + 4 - 5 = 11.$$

Since 11 is positive, $f'(x) > 0$ throughout $(1, \infty)$.

In future examples we will display our work in tabular form as follows. The last row is a consequence of Theorem (4.13).

Interval	$(-\infty, -\frac{5}{3})$	$(-\frac{5}{3}, 1)$	$(1, \infty)$
k	-2	0	2
Test value $f'(k)$	3	-5	11
Sign of $f'(x)$	$+$	$-$	$+$
Variation of f	increasing on $(-\infty, -\frac{5}{3}]$	decreasing on $[-\frac{5}{3}, 1]$	increasing on $[1, \infty)$

Note that

$$f(x) = x^3 + x^2 - 5x - 5$$
$$= x^2(x + 1) - 5(x + 1)$$
$$= (x^2 - 5)(x + 1)$$

FIGURE 4.16

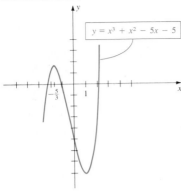

$$y = x^3 + x^2 - 5x - 5$$

and hence the x-intercepts of the graph are $\sqrt{5}$, $-\sqrt{5}$, and -1. The y-intercept is $f(0) = -5$. The points corresponding to the critical numbers are $\left(-\frac{5}{3}, \frac{40}{27}\right)$ and $(1, -8)$. Plotting these six points and using the information in the table gives us the sketch in Figure 4.16. •

Theorem (4.13) may be used to justify the results about the direction of motion that we have used previously (see Example 2 of Section 3.3). Thus, if $s(t)$ is the coordinate of a point P on a coordinate line at time t, and if the velocity $v(t)$ is positive in a time interval I, then $s'(t) > 0$ and hence, by Theorem (4.13), $s(t)$ is increasing; that is, P is moving in the positive direction. Similarly, if $v(t)$ is negative, then P is moving in the negative direction.

Since $v(t) = s'(t)$, the times at which the velocity is zero are critical numbers for the position function s and hence lead to possible local maxima or minima for s. If a local maximum or minimum for s occurs at t_1, then t_1 is usually a time at which P reverses direction.

We saw in Section 4.1 that if a function has a local extremum, then it must occur at a critical number; however, not every critical number leads to a local extremum (see Example 3 of Section 4.1). To find the local extrema, we begin by locating all the critical numbers of the function. Next, each critical number is tested to determine whether or not a local extremum occurs. There are several methods for conducting this test. The following theorem is based on the sign of the first derivative of f. In the statement of the theorem, the terminology f' *changes from positive to negative at c* means that there is an open interval (a, b) containing c such that $f'(x) > 0$ if $a < x < c$, and $f'(x) < 0$ if $c < x < b$. An analogous meaning is applied to the phrase f' *changes from negative to positive at c*.

THE FIRST DERIVATIVE TEST (4.14)

Suppose f is continuous at a critical number c and differentiable on an open interval I containing c, except possibly at c itself.

(i) If f' changes from positive to negative at c, then $f(c)$ is a local maximum of f.

(ii) If f' changes from negative to positive at c, then $f(c)$ is a local minimum of f.

(iii) If $f'(x) > 0$ or if $f'(x) < 0$ for every x in I except $x = c$, then $f(c)$ is not a local extremum of f.

FIGURE 4.17
(i) Local maximum $f(c)$

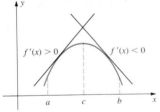

(ii) Local minimum $f(c)$

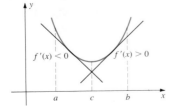

PROOF If f' changes from positive to negative at c, then there is an open interval (a, b) containing c such that $f'(x) > 0$ for $a < x < c$ and $f'(x) < 0$ for $c < x < b$. Furthermore, we may choose (a, b) such that f is continuous on $[a, b]$. Hence by Theorem (4.13), f is increasing on $[a, c]$ and decreasing on $[c, b]$. Thus, $f(x) < f(c)$ for every x in (a, b) different from c; that is, $f(c)$ is a local maximum for f. This proves (i). Similar proofs may be given for (ii) and (iii). • •

To remember the first derivative test, think of the graphs in Figure 4.17. For a local maximum, the slope $f'(x)$ of the tangent line at $P(x, f(x))$ is positive if $x < c$ and negative if $x > c$. For a local minimum the opposite situation occurs. Similar drawings can be sketched if the graph has a corner at the point $(c, f(c))$.

EXAMPLE 2 Find the local extrema of f if $f(x) = x^3 + x^2 - 5x - 5$.

SOLUTION This is the function considered in Example 1. The critical numbers are $-\frac{5}{3}$ and 1. We see from the table in Example 1 that the sign of $f'(x)$ changes from positive to negative as x increases through $-\frac{5}{3}$. Hence, by the First Derivative Test, f has a local maximum at $-\frac{5}{3}$. This maximum value is $f(-\frac{5}{3}) = \frac{40}{27}$ (see Figure 4.16).

A local minimum occurs at 1, since the sign of $f'(x)$ changes from negative to positive as x increases through 1. This minimum value is $f(1) = -8$. •

EXAMPLE 3 Find the local maxima and minima of f if $f(x) = x^{1/3}(8 - x)$. Sketch the graph of f.

SOLUTION By the Product Rule,

$$f'(x) = x^{1/3}(-1) + (8 - x)\tfrac{1}{3}x^{-2/3}$$

$$= \frac{-3x + (8 - x)}{3x^{2/3}} = \frac{4(2 - x)}{3x^{2/3}}$$

and hence the critical numbers of f are 0 and 2. As in Example 1, this suggests that we consider the signs of $f'(x)$ corresponding to the intervals $(-\infty, 0)$, $(0, 2)$, and $(2, \infty)$. Since f' is continuous and has no zeros on each interval, we may determine the sign of $f'(x)$ by using a suitable test value $f'(k)$. The following table summarizes this work (check each entry).

Interval	$(-\infty, 0)$	$(0, 2)$	$(2, \infty)$
k	-1	1	8
Test value $f'(k)$	4	$\frac{4}{3}$	-2
Sign of $f'(x)$	$+$	$+$	$-$
Variation of f	increasing on $(-\infty, 0]$	increasing on $[0, 2]$	decreasing on $[2, \infty)$

The number 8 was chosen in $(2, \infty)$ since $8^{2/3}$ is an integer. We may, of course, choose *any* number k in $(2, \infty)$ and use a calculator to find $f'(k)$.

By the First Derivative Test, f has a local maximum at 2 since f' changes from positive to negative at 2. Thus, we have

Local max: $f(2) = 2^{1/3}(8 - 2) = 6\sqrt[3]{2} \approx 7.6$

The function does not have an extremum at 0, since f' does not change sign at 0.

To sketch the graph we first plot points corresponding to the critical numbers. From the formula for $f(x)$ it is evident that the x-intercepts of the graph are 0 and 8. The graph is sketched in Figure 4.18. •

EXAMPLE 4 Find the local extrema of f if $f(x) = x^{2/3}(x^2 - 8)$. Sketch the graph of f.

SOLUTION By the Product Rule,

$$f'(x) = x^{2/3}(2x) + (x^2 - 8)(\tfrac{2}{3}x^{-1/3}) = \frac{6x^2 + 2(x^2 - 8)}{3x^{1/3}} = \frac{8(x^2 - 2)}{3x^{1/3}}.$$

FIGURE 4.18

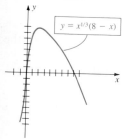

$y = x^{1/3}(8 - x)$

The critical numbers are the solutions of $x^2 - 2 = 0$ and $x^{1/3} = 0$, that is, $-\sqrt{2}$, 0, and $\sqrt{2}$. This suggests that we find the sign of $f'(x)$ in each of the intervals $(-\infty, -\sqrt{2})$, $(-\sqrt{2}, 0)$, $(0, \sqrt{2})$, and $(\sqrt{2}, \infty)$. Arranging our work in tabular form as in previous examples, we obtain the following.

Interval	$(-\infty, -\sqrt{2})$	$(-\sqrt{2}, 0)$	$(0, \sqrt{2})$	$(\sqrt{2}, \infty)$
k	-8	-1	1	8
Test value $f'(k)$	$-\frac{248}{3}$	$\frac{8}{3}$	$-\frac{8}{3}$	$\frac{248}{3}$
Sign of $f'(x)$	$-$	$+$	$-$	$+$
Variation of f	decreasing on $(-\infty, -\sqrt{2}]$	increasing on $[-\sqrt{2}, 0]$	decreasing on $[0, \sqrt{2}]$	increasing on $[\sqrt{2}, \infty)$

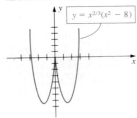

FIGURE 4.19

$y = x^{2/3}(x^2 - 8)$

By the First Derivative Test, f has local minima at $-\sqrt{2}$ and $\sqrt{2}$ and a local maximum at 0. The corresponding function values give us the following:

Local max: $f(0) = 0$

Local min: $f(\sqrt{2}) = f(-\sqrt{2}) = -6\sqrt[3]{2} \approx -7.56$

Note that $f'(0)$ does not exist. The graph is sketched in Figure 4.19. •

As we saw in Section 4.1, an *absolute* maximum or minimum of a function may not be included among the local extrema. Recall from (4.9) that if a function f is continuous on a closed interval $[a, b]$ and we wish to determine the absolute maximum and minimum values, then in addition to finding all the local extrema, we should calculate the values $f(a)$ and $f(b)$ of f at the endpoints a and b of the interval $[a, b]$. The largest number among the local extrema and the values $f(a)$ and $f(b)$ is the absolute maximum of f on $[a, b]$. The smallest of these numbers is the absolute minimum of f on $[a, b]$. To illustrate these remarks, consider the function discussed in the previous example, but let us restrict our attention to certain intervals.

EXAMPLE 5 If $f(x) = x^{2/3}(x^2 - 8)$, find the absolute maximum and minimum values of f on each of the following intervals:

(a) $\left[-1, \frac{1}{2}\right]$ (b) $[-1, 3]$ (c) $[-3, -2]$

SOLUTION   The graph in Figure 4.19 indicates the local extrema and the intervals on which f is increasing or decreasing. Figure 4.20 illustrates the part of the graph of f that corresponds to each of the intervals (a), (b), and (c).

FIGURE 4.20

(a) $\left[-1, \frac{1}{2}\right]$

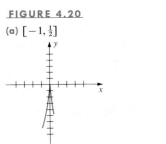

(b) $[-1, 3]$

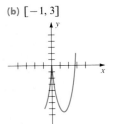

(c) $[-3, -2]$

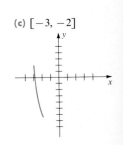

Referring to these sketches, we obtain the following table (check each entry).

Interval	Absolute minimum	Absolute maximum
$[-1, \frac{1}{2}]$	$f(-1) = -7$	$f(0) = 0$
$[-1, 3]$	$f(\sqrt{2}) = -6\sqrt[3]{2}$	$f(3) = \sqrt[3]{9}$
$[-3, -2]$	$f(-2) = -4\sqrt[3]{4}$	$f(-3) = \sqrt[3]{9}$

Note that on some intervals the maximum or minimum value of f is also a local extremum; however, on other intervals this is not the case. •

EXAMPLE 6 Shown in Figure 4.21 is a computer-generated graph of the equation $y = \frac{1}{2}x + \sin x$ for $-2\pi \le x \le 2\pi$. Determine the coordinates of the points A, B, C, and D that correspond to relative extrema.

SOLUTION Letting $f(x) = \frac{1}{2}x + \sin x$ and differentiating, we obtain

$$f'(x) = \tfrac{1}{2} + \cos x.$$

The relative extrema occur if $f'(x) = 0$, that is, if

$$\tfrac{1}{2} + \cos x = 0 \quad \text{or} \quad \cos x = -\tfrac{1}{2}.$$

Using some basic trigonometry (or a calculator) we see that points C and D have x-coordinates

$$2\pi/3 \approx 2.1 \quad \text{and} \quad 4\pi/3 \approx 4.2$$

By symmetry (with respect to the origin) the x-coordinates of A and B are

$$-4\pi/3 \approx -4.2 \quad \text{and} \quad -2\pi/3 \approx -2.1$$

We could now apply the First Derivative Test; however, it is evident from the graph that points A and C correspond to relative maxima and points B and D correspond to relative minima. The following table lists the coordinates of these points.

Point	x-coordinate	$y = \frac{1}{2}x + \sin x$
A	$-4\pi/3 \approx -4.2$	$(-2\pi/3) + (\sqrt{3}/2) \approx -1.2$
B	$-2\pi/3 \approx -2.1$	$(-\pi/3) + (-\sqrt{3}/2) \approx -1.9$
C	$2\pi/3 \approx 2.1$	$(\pi/3) + (\sqrt{3}/2) \approx 1.9$
D	$4\pi/3 \approx 4.2$	$(2\pi/3) + (-\sqrt{3}/2) \approx 1.2$

•

In applications, a physical quantity Q is often described in terms of a formula $Q = f(x)$ for a differentiable function f. (Of course, other symbols may be used for the variables.) We may then use the derivative $D_x Q$ to help find maximum or minimum values of Q. Sometimes it may be necessary to determine an appropriate formula for Q, as in the next example (see also Exercises 45 and 46). We shall consider many more applications of this type in Section 4.5.

FIGURE 4.21

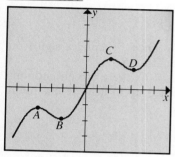

EXAMPLE 7 A long rectangular sheet of metal, 12 inches wide, is to be made into a rain gutter by turning up two sides so that they are perpendicular to the sheet. How many inches should be turned up to give the gutter its greatest capacity?

FIGURE 4.22

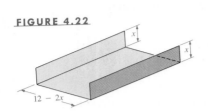

SOLUTION The gutter is illustrated in Figure 4.22, where we have let x denote the number of inches turned up on each side. The width of the base of the gutter is $12 - 2x$ inches. The capacity will be greatest when the area of the rectangle with sides of lengths x and $12 - 2x$ has its greatest value. Letting $f(x)$ denote this area,

$$f(x) = x(12 - 2x) = 12x - 2x^2.$$

Note that $0 < x < 6$, because if $x = 0$ or $x = 6$, no gutter is formed (the area of the rectangle would be $f(0) = 0 = f(6)$).

Differentiating,

$$f'(x) = 12 - 4x = 4(3 - x)$$

and hence the only critical number of f is 3. Since f' changes sign from positive to negative at 3, $f(3)$ is a local maximum for f.

Since the domain of f is the open interval $(0, 6)$, there are no endpoint extrema. It follows that 3 inches should be turned up to achieve maximum capacity. •

EXERCISES 4.3

Exer. 1–16: Find the local extrema of f. Describe the intervals in which f is increasing or decreasing, and sketch the graph of f.

1 $f(x) = 5 - 7x - 4x^2$ 2 $f(x) = 6x^2 - 9x + 5$

3 $f(x) = 2x^3 + x^2 - 20x + 1$

4 $f(x) = x^3 - x^2 - 40x + 8$

5 $f(x) = x^4 - 8x^2 + 1$

6 $f(x) = x^3 - 3x^2 + 3x + 7$

7 $f(x) = x^{4/3} + 4x^{1/3}$ 8 $f(x) = x^{2/3}(8 - x)$

9 $f(x) = x^2\sqrt[3]{x^2 - 4}$ 10 $f(x) = x\sqrt{4 - x^2}$

11 $f(x) = x^{2/3}(x - 7)^2 + 2$ 12 $f(x) = 4x^3 - 3x^4$

13 $f(x) = x^3 + (3/x)$

14 $f(x) = 8 - \sqrt[3]{x^2 - 2x + 1}$

15 $f(x) = 10x^3(x - 1)^2$ 16 $f(x) = (x^2 - 10x)^4$

Exer. 17–22: Find the local extrema of f on the interval $[0, 2\pi]$, and determine where f is increasing or decreasing on $[0, 2\pi]$. Sketch the graph of f corresponding to $0 \le x \le 2\pi$.

17 $f(x) = \cos x + \sin x$ 18 $f(x) = \cos x - \sin x$

19 $f(x) = \frac{1}{2}x - \sin x$ 20 $f(x) = x + 2 \cos x$

21 $f(x) = 2 \cos x + \sin 2x$ 22 $f(x) = 2 \cos x + \cos 2x$

Exer. 23–28: Find the local extrema of f.

23 $f(x) = \sqrt[3]{x^3 - 9x}$ 24 $f(x) = x^2/\sqrt{x + 7}$

25 $f(x) = (x - 2)^3(x + 1)^4$ 26 $f(x) = x^2(x - 5)^4$

27 $f(x) = \dfrac{2x - 5}{x + 3}$ 28 $f(x) = \dfrac{x^2 + 3}{x - 1}$

Exer. 29–32: Use the First Derivative Test to find the local extrema of f on the given interval.

29 $f(x) = \sec \frac{1}{2}x$; $[-\pi/2, \pi/2]$

30 $f(x) = \cot^2 x + 2 \cot x$; $[\pi/6, 5\pi/6]$

31 $f(x) = 2 \tan x - \tan^2 x$; $[-\pi/3, \pi/3]$

32 $f(x) = \tan x - 2 \sec x$; $[-\pi/4, \pi/4]$

Exer. 33–34: Shown in the figure is a computer-generated graph of the equation for $-2\pi \le x \le 2\pi$. Determine the x-coordinates of the points that correspond to relative extrema.

33 $y = \frac{1}{2}x + \cos x$

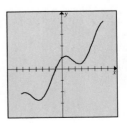

34 $y = (\sqrt{3}/2)x - \sin x$

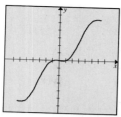

35–38 For the functions defined in Exercises 1–4, find the absolute maximum and minimum values on each of the following intervals:
(a) $[-1, 1]$ (b) $[-4, 2]$ (c) $[0, 5]$

Exer. 39–40: Sketch the graph of a differentiable function f that satisfies the given conditions.

39 $f'(-5) = 0$; $f'(0) = 0$; $f'(5) = 0$; $f'(x) > 0$ if $|x| > 5$; $f'(x) < 0$ if $0 < |x| < 5$.

40 $f'(a) = 0$ for $a = 1, 2, 3, 4, 5$, and $f'(x) > 0$ for all other values of x.

41 A herd of 100 deer is relocated to a small island. The herd increases rapidly, but eventually the food resources of the island dwindle and the population declines. Suppose that the number $N(t)$ of deer present after t years is given by $N(t) = -t^4 + 21t^2 + 100$.
(a) When does the population cease to grow? What is the maximum population level? When does the population become extinct?
(b) Sketch the graph of N for $t \geq 0$.

42 If an object of weight W pounds is pulled along a horizontal plane by a force applied to a rope that is attached to the object, and if the rope makes an angle θ with the horizontal, then the magnitude of the force is given by $F = \mu W/(\mu \sin \theta + \cos \theta)$ (see Exercise 35 of Section 3.5). If $\mu = \sqrt{3}/3$, find the angle that will minimize F.

43 A battery having fixed voltage V and fixed internal resistance r is connected to a circuit that has variable resistance R (see figure). By Ohm's law, the current I in the circuit is $I = V/(R + r)$. If the power output P is given by $P = I^2R$, show that the maximum power occurs if $R = r$.

EXERCISE 43

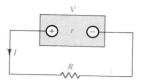

44 The power output P of an automobile battery is given by $P = VI - I^2r$ for voltage V, current I, and internal resistance r of the battery. What current corresponds to the maximum power?

45 The frame for a shipping crate is to be constructed from 24 feet of 2-by-2 lumber. If the crate is to have square ends as shown in the figure, find the dimensions that will result in the maximum outer volume. (*Hint:* First express y as a function of x, and then express V as a function of x.)

EXERCISE 45

46 One thousand feet of chain link fence will be used to construct six cages for a zoo exhibit as shown in the figure. Find the dimensions that maximize the enclosed area A. (*Hint:* First express y as a function of x, and then express A as a function of x.)

EXERCISE 46

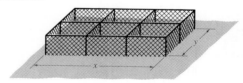

4.4 CONCAVITY AND THE SECOND DERIVATIVE TEST

The concept of *concavity* is useful for describing the graph of a differentiable function f. If $f'(c)$ exists, then the graph of f has a tangent line l with slope $f'(c)$ at the point $P(c, f(c))$. Figure 4.23 (on page 182) illustrates three possible situations that may occur if $f'(c) > 0$. Similar situations occur if $f'(c) < 0$ or $f'(c) = 0$. For the graph shown in Figure 4.23(i), we say that on the interval (a, b), the graph is *above* the tangent line through P. In (ii), the graph is *below* the tangent line. In (iii), for every open interval (a, b) containing c, the tangent line is neither above nor below the graph but instead *crosses* the graph at that point.

FIGURE 4.23

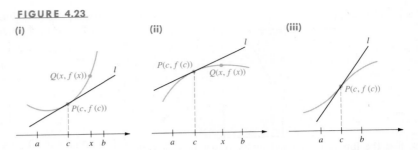

The following terminology is used to describe the graphs illustrated in Figure 4.23.

DEFINITION (4.15)

> Let f be a function that is differentiable at a number c.
>
> (i) The graph of f is **concave upward** at the point $P(c, f(c))$ if there exists an open interval (a, b) containing c such that on (a, b) the graph of f is above the tangent line through P.
>
> (ii) The graph of f is **concave downward** at $P(c, f(c))$ if there exists an open interval (a, b) containing c such that on (a, b) the graph of f is below the tangent line through P.

Consider a function f whose graph is concave upward at the point $P(c, f(c))$. As in Figure 4.23(i), there is an open interval (a, b) containing c such that on (a, b) the graph of f is above the tangent line through P. If f is differentiable on (a, b), then the graph of f has a tangent line at every point $Q(x, f(x))$ for $a < x < b$. If we consider several of these tangent lines (see Figure 4.24), it appears that as x increases, the slope $f'(x)$ of the tangent line also increases. Thus in Figure 4.24(i), a larger positive slope is obtained as P moves to the right. In (ii), the slope of the tangent line also increases, becoming less negative as P moves to the right.

FIGURE 4.24
Upward concavity

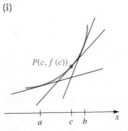

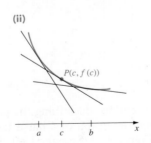

FIGURE 4.25
Downward concavity

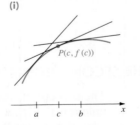

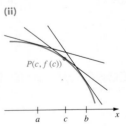

Similarly, if the graph of f is concave downward at $P(c, f(c))$, then the graphs sketched in Figure 4.25 may occur. Here the slope $f'(x)$ of the tangent line *decreases* as P moves to the right.

Conversely, we might expect that if the derivative $f'(x)$ increases as x increases through c, then the graph is concave upward at $P(c, f(c))$, and if

$f'(x)$ decreases, then the graph is concave downward. According to Theorem (4.13), if the derivative of a function is positive on an interval, then the function is increasing. Consequently, if the values of the *second derivative* f'' are positive on an interval, then the *first* derivative f' is increasing. Similarly, if f'' is negative, then f' is decreasing. This suggests that the sign of f'' can be used to test concavity, as stated in the next theorem. A proof is given at the end of this section.

TEST FOR CONCAVITY (4.16)

> Suppose that a function f is differentiable on an open interval containing c, and that $f''(c)$ exists.
>
> (i) If $f''(c) > 0$, the graph is concave upward at $P(c, f(c))$.
> (ii) If $f''(c) < 0$, the graph is concave downward at $P(c, f(c))$.

If the second derivative $f''(x)$ changes sign as x increases through a number k, then by (4.16) the concavity changes from upward to downward, or from downward to upward. The point $(k, f(k))$ is called a *point of inflection* according to the next definition.

DEFINITION (4.17)

> A point $P(k, f(k))$ on the graph of a function f is a **point of inflection** if f'' exists an open interval (a, b) containing k, and f'' changes sign at k.

The sketch in Figure 4.26 displays typical points of inflection on a graph. A graph is said to be **concave upward** (or **downward**) **on an interval** if it is concave upward (or downward) at every number in the interval. Intervals on which the graph in Figure 4.26 is concave upward or concave downward are abbreviated CU or CD, respectively. Observe that a corner may, or may not, be a point of inflection.

FIGURE 4.26

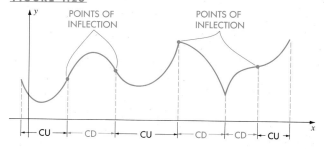

EXAMPLE 1 If $f(x) = x^3 + x^2 - 5x - 5$, determine intervals on which the graph of f is concave upward and intervals on which the graph is concave downward.

SOLUTION The function f was considered in Examples 1 and 2 of the preceding section. Since $f'(x) = 3x^2 + 2x - 5$,

$$f''(x) = 6x + 2 = 2(3x + 1).$$

FIGURE 4.27

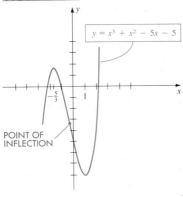

$y = x^3 + x^2 - 5x - 5$

POINT OF
INFLECTION

Hence $f''(x) < 0$ if $3x + 1 < 0$, that is, if $x < -\frac{1}{3}$. By the Test for Concavity (4.16), the graph is concave downward on the infinite interval $(-\infty, -\frac{1}{3})$. Similarly, $f''(x) > 0$ if $x > -\frac{1}{3}$ and the graph is concave upward on $(-\frac{1}{3}, \infty)$. By Definition (4.17) the point $P(-\frac{1}{3}, -\frac{88}{27})$ at which f'' changes sign (and the concavity changes from downward to upward) is a point of inflection. The graph of f obtained previously is sketched again in Figure 4.27 to show the point of inflection. •

If $P(c, f(c))$ is a point of inflection on the graph of f and if f'' is continuous on an open interval containing c, then necessarily $f''(c) = 0$. This is a consequence of the Intermediate Value Theorem (2.33) applied to the function f''. Thus, to locate points of inflection for a function whose second derivative is continuous, we begin by finding all numbers x such that $f''(x) = 0$. Each of these numbers is then tested to determine if it is an x-coordinate of a point of inflection. Before giving an example, let us state the following test for local maxima and minima.

THE SECOND DERIVATIVE TEST (4.18)

Suppose that a function f is differentiable on an open interval containing c and that $f'(c) = 0$.

(i) If $f''(c) < 0$, then f has a local maximum at c.

(ii) If $f''(c) > 0$, then f has a local minimum at c.

FIGURE 4.28
Local maximum $f(c)$

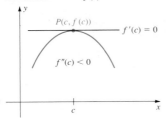

PROOF If $f'(c) = 0$, then the tangent line to the graph at $P(c, f(c))$ is horizontal. If, in addition, $f''(c) < 0$, then the graph is concave downward at c and hence there is an interval (a, b) containing c such that the graph lies below the tangent line at P. It follows that $f(c)$ is a local maximum for f, as illustrated in Figure 4.28. This proves (i). A similar proof may be given for (ii), which is illustrated in Figure 4.29. • •

If $f''(c) = 0$, the Second Derivative Test is not applicable. In such cases the First Derivative Test should be employed (see Example 3).

FIGURE 4.29
Local minimum $f(c)$

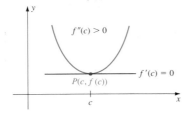

EXAMPLE 2 If $f(x) = 12 + 2x^2 - x^4$, use the Second Derivative Test to find the local maxima and minima of f. Discuss concavity, find the points of inflection, and sketch the graph of f.

SOLUTION We begin by finding the first and second derivatives and factoring them as follows:

$$f'(x) = 4x - 4x^3 = 4x(1 - x^2)$$
$$f''(x) = 4 - 12x^2 = 4(1 - 3x^2)$$

The expression for $f'(x)$ is used to find the critical numbers 0, 1, and -1. The values of f'' at these numbers are

$$f''(0) = 4 > 0, \quad f''(1) = -8 < 0, \quad \text{and} \quad f''(-1) = -8 < 0.$$

Hence, by the Second Derivative Test, the function has a local minimum at 0 and local maxima at 1 and -1. The corresponding function values are $f(0) = 12$ and $f(1) = 13 = f(-1)$. The following table summarizes our discussion.

Critical number c	$f''(c)$	Sign of $f''(c)$	Conclusion
-1	-8	$-$	Local max: $f(-1) = 13$
0	4	$+$	Local min: $f(0) = 12$
1	-8	$-$	Local max: $f(1) = 13$

To locate the possible points of inflection we solve the equation $f''(x) = 0$, that is, $4(1 - 3x^2) = 0$. Evidently, the solutions are $-\sqrt{3}/3$ and $\sqrt{3}/3$. This suggests an examination of the sign of $f''(x)$ in each of the intervals $(-\infty, -\sqrt{3}/3)$, $(-\sqrt{3}/3, \sqrt{3}/3)$, and $(\sqrt{3}/3, \infty)$. Since f'' is continuous and has no zeros on each interval, we may use test values to determine the sign of $f''(x)$. Let us arrange our work in tabular form as follows. The last row is a consequence of (4.18).

FIGURE 4.30

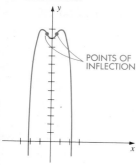

POINTS OF INFLECTION

Interval	$(-\infty, -\sqrt{3}/3)$	$(-\sqrt{3}/3, \sqrt{3}/3)$	$(\sqrt{3}/3, \infty)$
k	-1	0	1
Test value $f''(k)$	-8	4	-8
Sign of $f''(x)$	$-$	$+$	$-$
Concavity	downward	upward	downward

Since $f''(x)$ changes sign at $-\sqrt{3}/3$ and $\sqrt{3}/3$, the corresponding points $(\pm\sqrt{3}/3, 113/9)$ on the graph are points of inflection. These are the points at which the concavity changes. As shown in the table, the graph is concave upward on the open interval $(-\sqrt{3}/3, \sqrt{3}/3)$ and concave downward outside of $[-\sqrt{3}/3, \sqrt{3}/3]$. The graph is sketched in Figure 4.30. •

EXAMPLE 3 If $f(x) = x^5 - 5x^3$, use the Second Derivative Test to find the local extrema of f. Discuss concavity, find the points of inflection, and sketch the graph of f.

SOLUTION Differentiating,

$$f'(x) = 5x^4 - 15x^2 = 5x^2(x^2 - 3)$$
$$f''(x) = 20x^3 - 30x = 10x(2x^2 - 3).$$

Solving the equation $f'(x) = 0$ gives us the critical numbers 0, $-\sqrt{3}$, and $\sqrt{3}$. As in Example 2, we obtain the following table (check each entry).

Critical number c	$f''(c)$	Sign of $f''(c)$	Conclusion
$-\sqrt{3}$	$-30\sqrt{3}$	$-$	Local max: $f(-\sqrt{3}) = 6\sqrt{3}$
0	0	none	No conclusion
$\sqrt{3}$	$30\sqrt{3}$	$+$	Local min: $f(\sqrt{3}) = -6\sqrt{3}$

Since $f''(0) = 0$, the Second Derivative Test is not applicable at 0 and hence we apply the First Derivative Test. We can show, using test values, that if $-\sqrt{3} < x < 0$, then $f'(x) < 0$, and if $0 < x < \sqrt{3}$, then $f'(x) < 0$. Since $f'(x)$ does not change sign, no extremum occurs.

To find possible points of inflection we consider the equation $f''(x) = 0$, that is, $10x(2x^2 - 3) = 0$. The solutions of this equation, in order of magnitude, are $-\sqrt{6}/2$, 0, and $\sqrt{6}/2$. As in Example 2, we construct a table:

Interval	$(-\infty, -\sqrt{6}/2)$	$(-\sqrt{6}/2, 0)$	$(0, \sqrt{6}/2)$	$(\sqrt{6}/2, \infty)$
k	-2	-1	1	2
Test value $f''(k)$	-100	10	-10	100
Sign of $f''(x)$	$-$	$+$	$-$	$+$
Concavity	downward	upward	downward	upward

FIGURE 4.31

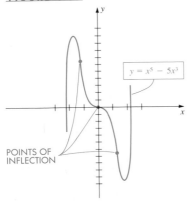

$y = x^5 - 5x^3$

POINTS OF
INFLECTION

Since the sign of $f''(x)$ changes as x increases through $-\sqrt{6}/2$, 0, and $\sqrt{6}/2$, the points $(0, 0)$, $(-\sqrt{6}/2, 21\sqrt{6}/8)$, and $(\sqrt{6}/2, -21\sqrt{6}/8)$ are points of inflection. The graph is sketched in Figure 4.31, with different scales on the x- and y-axes. •

In all of the preceding examples f'' is continuous. It is also possible for $(c, f(c))$ to be a point of inflection if either $f'(c)$ or $f''(c)$ does not exist, as illustrated in the next example.

EXAMPLE 4 If $f(x) = 1 - x^{1/3}$, find the local extrema, discuss concavity, find the points of inflection, and sketch the graph of f.

SOLUTION Differentiating,

$$f'(x) = -\frac{1}{3} x^{-2/3} = -\frac{1}{3x^{2/3}}$$

$$f''(x) = \frac{2}{9} x^{-5/3} = \frac{2}{9x^{5/3}}.$$

The first derivative does not exist at $x = 0$, and 0 is the only critical number for f. Since $f''(0)$ is undefined, the Second Derivative Test is not applicable. However, if $x \neq 0$, then $x^{2/3} > 0$ and $f'(x) = -1/(3x^{2/3}) < 0$, which means that f is decreasing throughout its domain. Consequently, $f(0)$ is not a local extremum.

The fact that f'' is undefined at $x = 0$ suggests that the point $(0, 1)$ on the graph of f is a point of inflection. Let us apply Definition (4.17) with $k = 0$. If $x < 0$, then $x^{5/3} < 0$. Hence

$$f''(x) = \frac{2}{9x^{5/3}} < 0 \quad \text{if } x < 0$$

which implies that the graph of f is concave downward on the interval $(-\infty, 0)$. If $x > 0$, then $x^{5/3} > 0$. Thus

$$f''(x) = \frac{2}{9x^{5/3}} > 0 \quad \text{if } x > 0$$

FIGURE 4.32

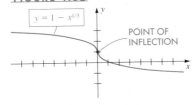

$y = 1 - x^{1/3}$

POINT OF
INFLECTION

which means that the graph is concave upward on the interval $(0, \infty)$. Thus, by (4.19), the point $(0, 1)$ is a point of inflection. Using this information and plotting several points gives us the sketch in Figure 4.32. •

EXAMPLE 5 If $f(x) = 2 \sin x + \cos 2x$, find the local extrema and sketch the graph of f on the interval $[0, 2\pi]$.

SOLUTION Differentiating,

$$f'(x) = 2 \cos x - 2 \sin 2x$$
$$f''(x) = -2 \sin x - 4 \cos 2x.$$

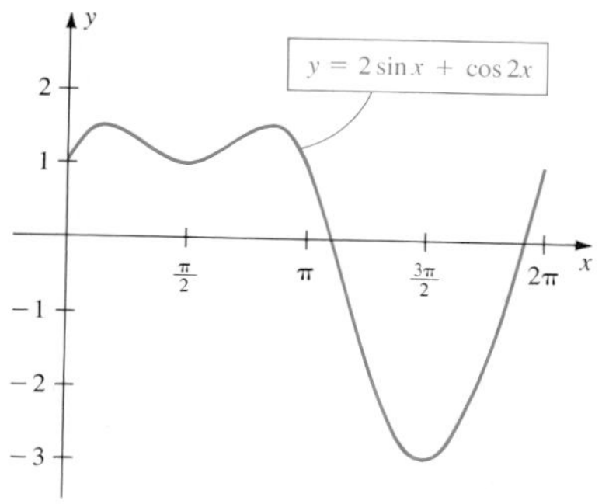

In Example 5 of Section 4.1 we found that the critical numbers of f in the interval $[0, 2\pi]$ are $\pi/6$, $5\pi/6$, $\pi/2$, and $3\pi/2$. Substituting these critical numbers for x in $f''(x)$, we obtain

$$f''(\pi/6) = -3, \quad f''(5\pi/6) = -3, \quad f''(\pi/2) = 2, \quad f''(3\pi/2) = 6.$$

Applying the Second Derivative Test we see that there are local maxima at $\pi/6$ and $5\pi/6$, and local minima at $\pi/2$ and $3\pi/2$. Thus we have

Local max: $f(\pi/6) = 3/2$ and $f(5\pi/6) = 3/2$

Local min: $f(\pi/2) = 1$ and $f(3\pi/2) = -3$

Using this information and plotting several more points gives us the sketch in Figure 4.33. •

FIGURE 4.33

$y = 2 \sin x + \cos 2x$

In Section 3.3 we discussed problems in economics that involve the cost, average cost, revenue, and profit functions—C, c, R, and P. Their derivatives C', c', R', and P' (called the marginal cost, marginal average cost, marginal revenue, and marginal profit functions) can be used to investigate maximum and minimum values, as illustrated in the following examples.

EXAMPLE 6 A furniture company estimates that the weekly cost (in dollars) of manufacturing x hand-finished reproductions of a colonial desk is given by $C(x) = x^3 - 3x^2 - 80x + 500$. Each desk produced is sold for \$2800. What weekly production rate will maximize the profit? What is the largest possible profit per week?

SOLUTION Since the income obtained from selling x desks is $2800x$, the revenue function R is given by $R(x) = 2800x$. The profit function P is the difference between the revenue function R and the cost function C, that is,

$$P(x) = R(x) - C(x) = 2800x - (x^3 - 3x^2 - 80x + 500)$$

or

$$P(x) = -x^3 + 3x^2 + 2880x - 500.$$

To find the maximum profit we differentiate, obtaining

$$P'(x) = -3x^2 + 6x + 2880 = -3(x^2 - 2x - 960).$$

The critical numbers of P are the solutions of

$$x^2 - 2x - 960 = 0 \quad \text{or} \quad (x - 32)(x + 30) = 0,$$

that is, $x = 32$ or $x = -30$. Since the negative solution is extraneous, it suffices to check $x = 32$.

The second derivative of the profit function P is

$$P''(x) = -6x + 6$$

Consequently,

$$P''(32) = -6(32) + 6 = -176 < 0.$$

Thus, by the Second Derivative Test, a maximum profit occurs if 32 desks per week are manufactured and sold. The maximum weekly profit is

$$P(32) = -(32)^3 + 3(32)^2 + 2880(32) - 500 = \$61{,}964 \quad •$$

At the end of Section 3.3 we introduced the notion of a *demand function* p such that $p(x)$ is the price per unit when there is a demand for x units of a commodity. The revenue $R(x)$ is, therefore, $R(x) = xp(x)$. If we let $S = p(x)$, then S is the selling price per unit associated with a demand of x units. Since a decrease in S would ordinarily be associated with an increase in x, a demand function p is usually decreasing, that is $p'(x) < 0$ for every x. Demand functions are sometimes defined implicitly by an equation involving S and x, as in the next example.

EXAMPLE 7 The demand for x units of a product is related to a selling price of S dollars per unit by means of the equation $2x + S^2 - 12{,}000 = 0$. Find the demand function, the marginal demand function, the revenue function, and the marginal revenue function. Find the number of units and the price per unit that yield the maximum revenue. What is the maximum revenue?

SOLUTION Since $S^2 = 12{,}000 - 2x$ and S is positive, we see that the demand function p is given by

$$S = p(x) = \sqrt{12{,}000 - 2x}.$$

The domain of p consists of every x such that $12{,}000 - 2x > 0$, or equivalently, $2x < 12{,}000$. Thus $0 \le x < 6000$. The graph of p is sketched in Figure 4.34. In theory, there are no sales if the selling price is $\sqrt{12{,}000}$, or approximately 109.54, and when the selling price is close to 0 the demand is close to 6000.

The marginal demand function p' is given by

$$p'(x) = \frac{-1}{\sqrt{12{,}000 - 2x}}.$$

The negative sign indicates that a decrease in price is associated with an increase in demand.

The revenue function R is given by

$$R(x) = xp(x) = x\sqrt{12{,}000 - 2x}.$$

Differentiating and simplifying gives us the marginal revenue function R' such that

$$R'(x) = \frac{12{,}000 - 3x}{\sqrt{12{,}000 - 2x}}.$$

A critical number for the revenue function R is $x = 12{,}000/3 = 4000$. Since $R'(x)$ is positive if $0 \le x < 4000$ and negative if $4000 < x < 6000$, the maximum revenue occurs when 4000 units are produced and sold. This corresponds to a selling price of

$$p(4000) = \sqrt{12{,}000 - 2(4000)} \approx \$63.25$$

The maximum revenue, obtained from selling 4000 units at this price is

$$4000(63.25) = \$253{,}000 \quad \bullet$$

PROOF OF TEST FOR CONCAVITY (4.16) Applying Definition (3.1') to the function f',

$$f''(c) = \lim_{x \to c} \frac{f'(x) - f'(c)}{x - c}.$$

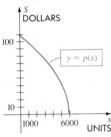

FIGURE 4.34

If $f''(c) > 0$, then by Theorem (2.12) there is an open interval (a, b) containing c such that

$$\frac{f'(x) - f'(c)}{x - c} > 0$$

for every x in (a, b) different from c; that is, $f'(x) - f'(c)$ and $x - c$ both have the same sign. We next show that this implies upward concavity.

For every x in the interval (a, b), consider the point (x, y) on the tangent line through $P(c, f(c))$ and the point $Q(x, f(x))$ on the graph of f (see Figure 4.35). If we let $g(x) = f(x) - y$, then we can establish upward concavity by showing that $g(x)$ is positive for every x such that $x \neq c$.

Since an equation of the tangent line at P is $y - f(c) = f'(c)(x - c)$, it follows that the y-coordinate of the point (x, y) on the tangent line is given by $y = f(c) + f'(c)(x - c)$. Consequently,

$$g(x) = f(x) - y = f(x) - f(c) - f'(c)(x - c).$$

Applying the Mean Value Theorem (4.12) to f and the interval $[x, c]$, we see that there exists a number w in the open interval (x, c) such that

$$f(x) - f(c) = f'(w)(x - c).$$

If we substitute for $f(x) - f(c)$ in the formula for $g(x)$, we obtain

$$g(x) = [f(x) - f(c)] - f'(c)(x - c)$$
$$= f'(w)(x - c) - f'(c)(x - c).$$

This can be factored as follows:

$$g(x) = [f'(w) - f'(c)](x - c)$$

Since w is in the interval (a, b) we know, from the first paragraph of the proof, that $f'(w) - f'(c)$ and $w - c$ have the same sign. Moreover, since w is between x and c, $w - c$ and $x - c$ have the same sign. Thus $f'(w) - f'(x)$ and $x - c$ have the same sign for every x in (a, b) such that $x \neq c$. It follows from the factored form of $g(x)$ that $g(x)$ is positive if $x \neq c$, which is what we wished to prove. Part (ii) may be established in similar fashion. • •

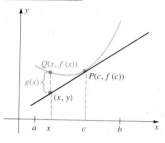

FIGURE 4.35

EXERCISES 4.4

Exer. 1–18: Use the Second Derivative Test (whenever applicable) to find the local extrema of f. Discuss concavity, find x-coordinates of points of inflection, and sketch the graph of f.

1 $f(x) = x^3 - 2x^2 + x + 1$

2 $f(x) = x^3 + 10x^2 + 25x - 50$

3 $f(x) = 3x^4 - 4x^3 + 6$

4 $f(x) = 8x^2 - 2x^4$

5 $f(x) = 2x^6 - 6x^4$

6 $f(x) = 3x^5 - 5x^3$

7 $f(x) = (x^2 - 1)^2$

8 $f(x) = x - (16/x)$

9 $f(x) = \sqrt[5]{x} - 1$

10 $f(x) = \dfrac{x + 4}{\sqrt{x}}$

11 $f(x) = x^2 - (27/x^2)$

12 $f(x) = x^{2/3}(1 - x)$

13 $f(x) = \dfrac{x}{x^2 + 1}$

14 $f(x) = \dfrac{x^2}{x^2 + 1}$

15 $f(x) = \sqrt[3]{x^2}(3x + 10)$

16 $f(x) = x^4 - 4x^3 + 10$

17 $f(x) = 8x^{1/3} + x^{4/3}$

18 $f(x) = x\sqrt{4 - x^2}$

Exer. 19–24: Use the Second Derivative Test to find the local extrema of f on the interval $[0, 2\pi]$. (These exercises appeared in the Section 4.3 (Exercises 17–22), where the method of solution involved the First Derivative Test.)

19 $f(x) = \cos x + \sin x$

20 $f(x) = \cos x - \sin x$

21 $f(x) = \tfrac{1}{2}x - \sin x$

22 $f(x) = x + 2 \cos x$

23 $f(x) = 2 \cos x + \sin 2x$

24 $f(x) = 2 \cos x + \cos 2x$

Exer. 25–28: Use the Second Derivative Test to find the local extrema of f on the given interval. (See Exercises 29–32 of Section 4.3.)

25 $f(x) = \sec \frac{1}{2}x; \ [-\pi/2, \pi/2]$

26 $f(x) = \cot^2 x + 2 \cot x; \ [-\pi/6, 5\pi/6]$

27 $f(x) = 2 \tan x - \tan^2 x; \ [-\pi/3, \pi/3]$

28 $f(x) = \tan x - 2 \sec x; \ [-\pi/4, \pi/4]$

29–30 Rework Exercises 33 and 34 of Section 4.3, using the Second Derivative Test to verify the local extrema.

Exer. 31–38: Sketch the graph of a continuous function f that satisfies all of the stated conditions.

31 $f(0) = 1; \ f(2) = 3; \ f'(0) = f'(2) = 0;$
$f'(x) < 0$ if $|x - 1| > 1; \ f'(x) > 0$ if $|x - 1| < 1;$
$f''(x) > 0$ if $x < 1; \ f''(x) < 0$ if $x > 1.$

32 $f(0) = 4; \ f(2) = 2; \ f(5) = 6; \ f'(0) = f'(2) = 0;$
$f'(x) > 0$ if $|x - 1| > 1; \ f'(x) < 0$ if $|x - 1| < 1;$
$f''(x) < 0$ if $x < 1$ or if $|x - 4| < 1;$
$f''(x) > 0$ if $|x - 2| < 1$ or if $x > 5.$

33 $f(0) = 2; \ f(2) = f(-2) = 1; \ f'(0) = 0;$
$f'(x) > 0$ if $x < 0; \ f'(x) < 0$ if $x > 0;$
$f''(x) < 0$ if $|x| < 2; \ f''(x) > 0$ if $|x| > 2.$

34 $f(1) = 4; \ f'(x) > 0$ if $x < 1; \ f'(x) < 0$ if $x > 1;$
$f''(x) > 0$ for every $x \neq 1.$

35 $f(-2) = f(6) = -2; \ f(0) = f(4) = 0; \ f(2) = f(8) = 3;$
f' is undefined at 2 and 6; $f'(0) = 1; \ f'(x) > 0$ throughout $(-\infty, 2)$ and $(6, \infty); \ f'(x) < 0$ if $|x - 4| < 2; \ f''(x) < 0$ throughout $(-\infty, 0), (4, 6),$ and $(6, \infty); \ f''(x) > 0$ throughout $(0, 2)$ and $(2, 4).$

36 $f(0) = 2; \ f(2) = 1; \ f(4) = f(10) = 0; \ f(6) = -4;$
$f'(2) = f'(6) = 0; \ f'(x) < 0$ throughout $(-\infty, 2), (2, 4), (4, 6),$ and $(10, \infty); \ f'(x) > 0$ throughout $(6, 10); \ f'(4)$ and $f'(10)$ do not exist; $f''(x) > 0$ throughout $(-\infty, 2), (4, 10),$ and $(10, \infty); \ f''(x) < 0$ throughout $(2, 4).$

37 If n is an odd integer, then $f(n) = 1$ and $f'(n) = 0;$ if n is an even integer, then $f(n) = 0$ and $f'(n)$ does not exist; if n is any integer then
(a) $f'(x) > 0$ whenever $2n < x < 2n + 1;$
(b) $f'(x) < 0$ whenever $2n - 1 < x < 2n;$
(c) $f''(x) < 0$ whenever $2n < x < 2n + 2.$

38 $f(x) = x$ if $x = -1, 2, 4,$ or $8; \ f'(x) = 0$ if $x = -1, 4, 6,$ or $8; \ f'(x) < 0$ throughout $(-\infty, -1), (4, 6),$ and $(8, \infty); \ f'(x) > 0$ throughout $(-1, 4)$ and $(6, 8); \ f''(x) > 0$ throughout $(-\infty, 0), (2, 3),$ and $(5, 7); \ f''(x) < 0$ throughout $(0, 2), (3, 5),$ and $(7, \infty).$

39 Prove that the graph of a quadratic function has no point of inflection. State conditions for which the graph is always (a) concave upward; (b) concave downward. Illustrate with sketches.

40 Prove that the graph of a polynomial function of degree 3 has exactly one point of inflection. Illustrate with sketches.

Exer. 41–42: For the given demand and cost functions, find (a) the marginal demand function; (b) the revenue function; (c) the profit function; (d) the marginal profit function; (e) the maximum profit; and (f) the marginal cost when the demand is 10 units.

41 $p(x) = 50 - (x/10); \ C(x) = 10 + 2x$

42 $p(x) = 80 - \sqrt{x - 1}; \ C(x) = 75x + 2\sqrt{x - 1}$

43 A travel agency estimates that to sell x "package deal" vacations, the price per vacation should be $1800 - 2x$ dollars for $1 \leq x \leq 100.$ If the cost to the agency for x vacations is $1000 + x + 0.01x^2$ dollars, find (a) the revenue function; (b) the profit function; (c) the number of vacations that will maximize the profit; and (d) the maximum profit.

44 A manufacturer determines that x units of a product will be sold if the selling price is $400 - 0.05x$ dollars for each unit. If the production cost for x units is $500 + 10x,$ find (a) the revenue function; (b) the profit function; (c) the number of units that maximize the profit; and (d) the price per unit when the marginal revenue is 300.

45 A kitchen specialty company determines that the cost of manufacturing and packaging x pepper mills per day is $500 + 0.02x + 0.001x^2$ dollars. If each item is sold for $8.00, what rate of production will maximize the profit? What is the maximum daily profit?

46 A company that conducts bus tours found that when the price was $9.00 per person, the average number of customers was 1000 per week. When the company reduced the price to $7.00 per person, the average number of customers increased to 1500 per week. Assuming that the demand function is linear, what price should be charged to obtain the greatest weekly revenue?

4.5 APPLICATIONS OF EXTREMA

Methods for finding extrema of functions can be applied to certain practical problems. These problems may either be described orally or stated in written words. To solve such problems we must convert their statements into formulas, functions, or equations. Since the types of applications are unlimited, it is difficult to state specific rules for finding solutions. However, we can develop a general strategy for attacking such problems. The following guidelines are often helpful.

GUIDELINES FOR SOLVING (4.19)
APPLIED PROBLEMS
INVOLVING EXTREMA

1 Read the problem carefully several times and think about the given facts, together with the unknown quantities that are to be found.

2 If possible, sketch a picture or diagram and label it appropriately, introducing variables for unknown quantities. Words such as *what, find, how much, how far,* or *when* should alert you to the unknown quantities.

3 Write down the known facts together with any relationships involving the variables.

4 Determine which variable is to be maximized or minimized and express this variable as a function of *one* of the other variables.

5 Find the critical numbers of the function obtained in Guideline 4 and test each of them for maxima or minima.

6 Check to see if extrema occur at the endpoints of the domain of the function obtained in Guideline 4.

7 Don't become discouraged if you are unable to solve a given problem. It takes a great deal of effort and practice to become proficient in solving applied problems. Keep trying!

The use of Guidelines (4.19) is illustrated in the following examples.

EXAMPLE 1 An open box with a rectangular base is to be constructed from a rectangular piece of cardboard 16 inches wide and 21 inches long by cutting out a square from each corner and then bending up the resulting sides. Find the size of the corner square that will produce a box having the largest possible volume.

FIGURE 4.36

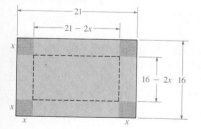

SOLUTION Applying Guideline 2, let us begin by drawing a picture of the cardboard as in Figure 4.36, where we have introduced a variable x to denote the length of the side of the square to be cut out from each corner. Note that $0 \le x \le 8$. Using Guideline 3, we specify the known facts (the size of the rectangle) by labeling the figure appropriately.

Next (see Guideline 4), we determine that the quantity to be maximized is the volume V of the box to be constructed by folding along the dashed lines (see Figure 4.37). Continuing with Guideline 4, let us express V as a function of the variable x. Referring to Figure 4.37,

$$V = x(16 - 2x)(21 - 2x) = 2(168x - 37x^2 + 2x^3).$$

This equation expresses V as a function of x, and we proceed to find the critical numbers and test them for maxima or minima (see Guideline 5). Differentiating with respect to x,

$$D_x V = 2(168 - 74x + 6x^2)$$
$$= 4(3x^2 - 37x + 84)$$
$$= 4(3x - 28)(x - 3).$$

Thus the possible critical numbers are $\frac{28}{3}$ and 3. Since $\frac{28}{3}$ is outside the domain of x, the only critical number is 3.

FIGURE 4.37

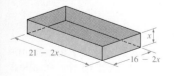

Let us employ the Second Derivative Test for extrema. The second derivative of V is

$$D_x^2 V = 2(-74 + 12x) = 4(6x - 37).$$

Substituting the critical number 3 for x, we obtain

$$D_x^2 V = 4(18 - 37) = -76 < 0$$

and hence V has a local maximum at $x = 3$.

Finally, we check for endpoint extrema (see Guideline 6). Since $0 \le x \le 8$ and since $V = 0$ if either $x = 0$ or $x = 8$, the maximum value of V does not occur at an endpoint of the domain of x. Consequently, a three-inch square should be cut from each corner of the cardboard in order to maximize the volume of the resulting box. •

In the remaining examples we shall not always point out the guidelines employed, but you should be able to determine specific guidelines by studying the solutions.

EXAMPLE 2 A circular cylindrical container, open at the top, is to have a capacity of 24π in.^{3}. If the cost of the material used for the bottom of the container is three times that used for the curved part and if there is no waste of material, find the dimensions that will minimize the manufacturing cost.

FIGURE 4.38

SOLUTION Let us begin by drawing a picture of the container as in Figure 4.38, letting r denote the radius of the base (in inches) and h the altitude (in inches). Since the volume is 24π in.^{3},

$$\pi r^2 h = 24\pi.$$

This gives us the relationship between r and h:

$$h = \frac{24}{r^2}$$

Our goal is to minimize the cost C of the material used to manufacture the container.

If a denotes the cost per square inch of the material to be used for the curved part, then the material to be used for the bottom costs $3a$ per square inch. The cost of the material for the curved part is therefore $a(2\pi rh)$, and the cost for the base is $3a(\pi r^2)$. The total cost C for material is

$$C = 3a(\pi r^2) + a(2\pi rh) = a\pi(3r^2 + 2rh)$$

or, since $h = 24/r^2$,

$$C = a\pi\left(3r^2 + \frac{48}{r}\right).$$

Since a is fixed, this expresses C as a function of one variable r as recommended in Guideline 4. To find the value of r that will lead to the smallest value for C, we find the critical numbers. Differentiating the last equation with respect to r,

$$D_r C = a\pi\left(6r - \frac{48}{r^2}\right) = 6a\pi\left(\frac{r^3 - 8}{r^2}\right).$$

Since $D_r C = 0$ if $r = 2$, we see that 2 is the only critical number. (The number 0 is not a critical number, since C is undefined if $r = 0$.) Since $D_r C < 0$ if $r < 2$ and $D_r C > 0$ if $r > 2$, it follows from the First Derivative Test that C has its minimum value if the radius of the cylinder is 2 inches. The corresponding value for the altitude (obtained from $h = 24/r^2$) is $\frac{24}{4}$, or 6 inches.

Since the domain of the variable r is the infinite interval $(0, \infty)$, there can be no endpoint extrema. •

EXAMPLE 3 Find the maximum volume of a right circular cylinder that can be inscribed in a cone of altitude 12 cm and base radius 4 cm, if the axes of the cylinder and cone coincide.

SOLUTION The problem is illustrated in Figure 4.39. Figure 4.40 represents a cross section through the axis of the cone and cylinder. The volume V of the cylinder is $V = \pi r^2 h$. To express V in terms of one variable (see Guideline 4), we must find a relationship between r and h. Referring to Figure 4.40 and using similar triangles we see that

$$\frac{h}{4-r} = \frac{12}{4} = 3 \quad \text{or} \quad h = 3(4-r).$$

Consequently,

$$V = \pi r^2 h = \pi r^2 [3(4-r)] = 3\pi r^2 (4-r).$$

If either $r = 0$ or $r = 4$ we see that $V = 0$, and hence the maximum volume is not an endpoint extremum. It is sufficient, therefore, to search for local maxima. Since $V = 3\pi(4r^2 - r^3)$,

$$D_r V = 3\pi(8r - 3r^2) = 3\pi r(8 - 3r).$$

Thus the critical numbers for V are $r = 0$ and $r = \frac{8}{3}$. Let us apply the Second Derivative Test for $r = \frac{8}{3}$. Differentiating $D_r V$ gives us

$$D_r^2 V = 3\pi(8 - 6r).$$

Substituting $\frac{8}{3}$ for r,

$$D_r^2 V = 3\pi[8 - 6(\tfrac{8}{3})] = 3\pi(8 - 16) = -24\pi < 0$$

which means that a maximum occurs at $r = \frac{8}{3}$. The corresponding value for $h = 3(4 - r)$ is

$$h = 3(4 - \tfrac{8}{3}) = 3(\tfrac{4}{3}) = 4.$$

Hence the maximum volume of the inscribed cylinder is

$$V = \pi \left(\frac{8}{3}\right)^2 (4) = \pi \left(\frac{64}{9}\right)(4) = \frac{256\pi}{9} \approx 89.4 \text{ cm}^3 \quad •$$

EXAMPLE 4 A North–South highway intersects an East–West highway at a point P. An automobile crosses P at 10:00 A.M., traveling east at a constant speed of 20 mi/hr. At that same instant another automobile is 2 miles north of P, traveling south at 50 mi/hr. Find the time at which they are closest to each other and approximate the minimum distance between the automobiles.

SOLUTION Typical positions of the automobiles are illustrated in Figure 4.41. If t denotes the number of hours after 10:00 A.M., then the slower automobile is $20t$ miles east of P. The faster automobile is $50t$ miles south of

FIGURE 4.39

FIGURE 4.40

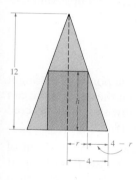

FIGURE 4.41

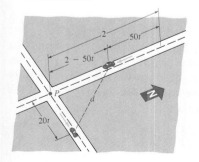

its position at 10:00 A.M. and hence its distance from P is $2 - 50t$. By the Pythagorean Theorem, the distance d between the automobiles is

$$d = \sqrt{(2 - 50t)^2 + (20t)^2}$$
$$= \sqrt{4 - 200t + 2500t^2 + 400t^2}$$
$$= \sqrt{4 - 200t + 2900t^2}.$$

Evidently d has its smallest value when the expression under the radical is minimal. Thus we may simplify our work by letting

$$f(t) = 4 - 200t + 2900t^2$$

and finding the value of t for which f has a minimum. Since

$$f'(t) = -200 + 5800t$$

the only critical number for f is

$$t = \frac{200}{5800} = \frac{1}{29}.$$

Moreover, since $f''(t) = 5800$, the second derivative is always positive and, therefore, f has a local minimum at $t = \frac{1}{29}$, and $f(\frac{1}{29}) \approx 0.55$. Since the domain of t is $[0, \infty)$ and since $f(0) = 4$, there is no endpoint extremum. Consequently the automobiles will be closest at $\frac{1}{29}$ hours (or approximately 2.07 minutes) after 10:00 A.M. The minimal distance is

$$\sqrt{f(\tfrac{1}{29})} \approx \sqrt{0.55} \approx 0.74 \text{ miles} \quad \bullet$$

EXAMPLE 5 A man in a rowboat 2 miles from the nearest point on a straight shoreline wishes to reach his house 6 miles further down the shore. If he can row at a rate of 3 mi/hr and walk at a rate of 5 mi/hr, how should he proceed in order to arrive at his house in the shortest amount of time?

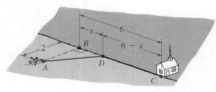

FIGURE 4.42

SOLUTION Figure 4.42 illustrates the problem: A denotes the position of the boat, B the nearest point on shore, and C the man's house; D is the point at which the boat will reach shore, and x denotes the distance between B and D. Thus x is restricted to the interval $[0, 6]$.

By the Pythagorean Theorem, the distance between A and D is $\sqrt{x^2 + 4}$. Using the formula *time = distance/rate*, the time it takes the man to row from A to D is $\sqrt{x^2 + 4}/3$, and the time it takes the man to walk from D to C is $(6 - x)/5$. Hence the total time T for the trip is

$$T = \frac{\sqrt{x^2 + 4}}{3} + \frac{6 - x}{5}$$

or, equivalently, $\qquad T = \tfrac{1}{3}(x^2 + 4)^{1/2} + \tfrac{6}{5} - \tfrac{1}{5}x.$

We wish to find the minimum value for T. Note that $x = 0$ corresponds to the extreme situation in which the man rows directly to B and then walks the entire distance from B to his house. If $x = 6$, then the man rows directly from A to his house. These numbers may be considered as endpoints of the domain of T. If $x = 0$, then from the formula for T,

$$T = \frac{\sqrt{4}}{3} + \frac{6}{5} - 0 = \frac{28}{15} \approx 1.87$$

which is approximately 1 hour 52 minutes. If $x = 6$, then

$$T = \frac{\sqrt{40}}{3} + \frac{6}{5} - \frac{6}{5} = \frac{2\sqrt{10}}{3} \approx 2.11$$

or approximately 2 hours 7 minutes.

Differentiating the general formula for T we see that

$$D_x T = \frac{1}{3} \cdot \frac{1}{2} (x^2 + 4)^{-1/2} (2x) - \frac{1}{5}$$

$$= \frac{x}{3(x^2 + 4)^{1/2}} - \frac{1}{5}.$$

To find the critical numbers we let $D_x T = 0$. This leads to the following equations:

$$\frac{x}{3(x^2 + 4)^{1/2}} = \frac{1}{5}$$

$$5x = 3(x^2 + 4)^{1/2}$$

$$25x^2 = 9(x^2 + 4)$$

$$16x^2 = 36$$

$$x^2 = \frac{36}{16}$$

$$x = \frac{6}{4} = \frac{3}{2}$$

Thus $\frac{3}{2}$ is the only critical number. We next employ the Second Derivative Test. Applying the Quotient Rule to the formula for $D_x T$,

$$D_x^2 T = \frac{3(x^2 + 4)^{1/2} - x \cdot 3(\frac{1}{2})(x^2 + 4)^{-1/2}(2x)}{9(x^2 + 4)}$$

$$= \frac{3(x^2 + 4) - 3x^2}{9(x^2 + 4)^{3/2}} = \frac{4}{3(x^2 + 4)^{3/2}}$$

which is positive if $x = \frac{3}{2}$. Consequently, T has a local minimum at $x = \frac{3}{2}$. The time T that corresponds to $x = \frac{3}{2}$ is

$$T = \frac{1}{3}(\frac{9}{4} + 4)^{1/2} + \frac{6}{5} - \frac{3}{10} = \frac{26}{15}$$

or, equivalently, 1 hour 44 minutes.

We have already examined the values of T at the endpoints of the domain, obtaining approximately 1 hour 52 minutes and over 2 hours, respectively. Hence the minimum value of T occurs at $x = \frac{3}{2}$ and, therefore, the boat should land between B and C, $1\frac{1}{2}$ miles from B. For a similar problem, but one in which the endpoints of the domain lead to minimum time, see Exercise 6. •

EXAMPLE 6 A wire 60 inches long is to be cut into two pieces. One of the pieces will be bent into the shape of a circle and the other into the shape of an equilateral triangle. Where should the wire be cut so that the sum of the areas of the circle and triangle is maximized? minimized?

SOLUTION If x denotes the length of one of the cut pieces of wire, then the length of the other piece is $60 - x$. Let the piece of length x be bent to form a circle of radius r so that $2\pi r = x$, or $r = x/2\pi$ (see Figure 4.43). If the remaining piece is bent into an equilateral triangle of side s, then $3s = 60 - x$, or $s = (60 - x)/3$. The area of the circle is

$$\pi r^2 = \pi \left(\frac{x}{2\pi}\right)^2 = \left(\frac{1}{4\pi}\right) x^2$$

and the area of the equilateral triangle is

$$\frac{\sqrt{3}}{4} s^2 = \frac{\sqrt{3}}{4} \left(\frac{60 - x}{3}\right)^2.$$

The sum A of the areas is

$$A = \left(\frac{1}{4\pi}\right) x^2 + \left(\frac{\sqrt{3}}{36}\right)(60 - x)^2.$$

We now find the critical numbers. Differentiating,

$$D_x A = \left(\frac{1}{2\pi}\right) x - \left(\frac{\sqrt{3}}{18}\right)(60 - x)$$

$$= \left(\frac{1}{2\pi} + \frac{\sqrt{3}}{18}\right) x - \frac{10\sqrt{3}}{3}.$$

Thus $D_x A = 0$ if and only if

$$x = \frac{10\sqrt{3}/3}{(1/2\pi) + (\sqrt{3}/18)} \approx 22.61.$$

The second derivative

$$D_x^2 A = \frac{1}{2\pi} + \frac{\sqrt{3}}{18}$$

is always positive and hence the indicated critical number will yield a minimum value for A. This minimum value is approximated by

$$A \approx \frac{1}{4\pi} (22.61)^2 + \frac{\sqrt{3}}{36} (60 - 22.61)^2 \approx 107.94 \text{ in.}^2$$

Since there are no other critical numbers, the maximum value of A must be an endpoint extremum. If $x = 0$, then all the wire is used to form a triangle and

$$A = \frac{\sqrt{3}}{36} (60)^2 \approx 173.21 \text{ in.}^2$$

If $x = 60$, then all the wire is used for the circle, and

$$A = \frac{1}{4\pi} (60)^2 \approx 268.48 \text{ in.}^2$$

Thus the maximum value of A occurs if the wire is not cut and the entire length of wire is bent into the shape of a circle. •

FIGURE 4.44

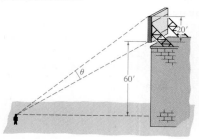

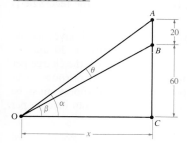

FIGURE 4.45

EXAMPLE 7 A billboard 20 feet tall is located on top of a building, with its lower edge 60 feet above the level of a viewer's eye, as shown in Figure 4.44. How far from a point directly below the sign should a viewer stand to maximize the angle θ between the lines of sight of the top and bottom of the billboard? (This angle should result in the best view of the billboard).

SOLUTION The problem is illustrated in Figure 4.45, using right triangles AOC and BOC having common side OC of (variable) length x. We see that

$$\tan \alpha = \frac{80}{x} \quad \text{and} \quad \tan \beta = \frac{60}{x}.$$

The angle $\theta = \alpha - \beta$ is a function of x and

$$\tan \theta = \tan (\alpha - \beta) = \frac{\tan \alpha - \tan \beta}{1 + \tan \alpha \tan \beta}.$$

Substituting for $\tan \alpha$ and $\tan \beta$ and simplifying, we obtain

$$\tan \theta = \frac{\dfrac{80}{x} - \dfrac{60}{x}}{1 + \left(\dfrac{80}{x}\right)\left(\dfrac{60}{x}\right)} = \frac{20x}{x^2 + 4800}.$$

The extrema of θ occur if $D_x \theta = 0$. Differentiating implicitly and using the Quotient Rule,

$$\sec^2 \theta \, D_x \theta = \frac{(x^2 + 4800)(20) - 20x(2x)}{(x^2 + 4800)^2} = \frac{96000 - 20x^2}{(x^2 + 4800)^2}.$$

Since $\sec^2 \theta > 0$, it follows that $D_x \theta = 0$ if and only if

$$96000 - 20x^2 = 0 \quad \text{or} \quad x^2 = 4800.$$

Thus the only critical number of θ is

$$x = \sqrt{4800} = 40\sqrt{3}.$$

We may verify that the sign of $D_x \theta$ changes from positive to negative at $\sqrt{4800}$, and hence a maximum value of θ occurs if $x = 40\sqrt{3}$ feet $= 69.3$ feet. •

EXERCISES 4.5

1 If a box with a square base and open top is to have a volume of 4 ft³, find the dimensions that require the least material (neglect the thickness of the material and waste in construction).

2 Rework Exercise 1 if the box has a closed top.

3 A metal cylindrical container with an open top is to hold one cubic foot. If there is no waste in construction, find the dimensions that require the least amount of material.

4 If the circular base of the container in Exercise 3 is cut from a square sheet and the remaining metal is discarded,

find the dimensions that require the least amount of material.

5 A veterinarian has 100 feet of fencing to construct six dog kennels by first building a fence around a rectangular region, and then subdividing that region into six smaller rectangles by placing five fences parallel to one of the sides. What dimensions of the region will maximize the total area?

6 Refer to Example 5 of this section. If the man is in a motorboat that can travel at an average rate of 15 mi/hr, how should he proceed in order to arrive at his house in the least time?

7 At 1:00 P.M. ship A is 30 miles due south of ship B and is sailing north at a rate of 15 mi/hr. If ship B is sailing west at a rate of 10 mi/hr, find the time at which the distance d between the ships is minimal (see figure).

EXERCISE 7

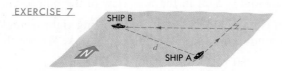

8 A window has the shape of a rectangle surmounted by a semicircle. If the perimeter of the window is 15 feet, find the dimensions that will allow the maximum amount of light to enter.

9 A fence 8 feet tall stands on level ground and runs parallel to a tall building (see figure). If the fence is 1 foot from the building, find the length of the shortest ladder that will extend from the ground over the fence to the wall of the building. (*Hint:* Use similar triangles.)

EXERCISE 9

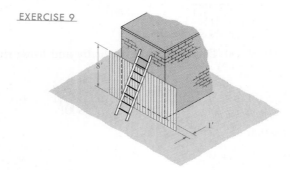

10 A page of a book is to have an area of 90 in^2, with 1-inch margins at the bottom and sides and a $\frac{1}{2}$-inch margin at the top. Find the dimensions of the page that will allow the largest printed area.

11 A builder intends to construct a storage shed having a volume of 900 ft^3, a flat roof, and a rectangular base whose width is three-fourths the length. The cost per square foot of the materials is $4 for the floor, $6 for the sides, and $3 for the roof. What dimensions will minimize the cost?

12 A water cup in the shape of a right circular cone is to be constructed by removing a circular sector from a circular sheet of paper of radius a and then joining the two straight edges of the remaining paper (see figure). Find the volume of the largest cup that can be constructed.

EXERCISE 12

13 A farmer has 500 feet of fencing to enclose a rectangular field by using a barn as part of one side of the field (see

figure). Prove that the area of the field is greatest when the rectangle is a square.

EXERCISE 13

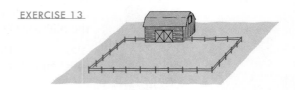

14 Refer to Exercise 13. Suppose the farmer wants the area of the rectangular field to be A square feet. Prove that the least amount of fencing is required when the rectangle is a square.

15 A hotel that charges $80 per day for a room gives special rates to organizations that reserve between 30 and 60 rooms. If more than 30 rooms are reserved, the charge per room is decreased by $1 times the number of rooms over 30. Under these conditions, how many rooms must be rented if the hotel is to receive the maximum income per day?

16 Refer to Exercise 15. Suppose that for each room rented it costs the hotel $6 per day for cleaning and maintenance. In this case how many rooms must be rented to obtain the greatest net income?

17 A steel storage tank for propane gas is to be constructed in the shape of a right circular cylinder with a hemisphere at each end (see figure). The cost per square foot of constructing the end pieces is twice that of the cylindrical piece. If the desired capacity is 10π ft^3, what dimensions will minimize the cost of construction?

EXERCISE 17

18 A pipeline for transporting oil will connect two points A and B that are 3 miles apart and on opposite banks of a straight river one mile wide (see figure). Part of the pipeline will run under water from A to a point C on the opposite bank, and then above ground from C to B. If the cost per mile of running the pipeline under water is four times the cost per mile of running it above ground, find the location of C that will minimize the cost (ignore the slope of the river bed).

EXERCISE 18

19 Find the dimensions of the rectangle of maximum area that can be inscribed in a semicircle of radius a, if two vertices lie on the diameter (see figure).

EXERCISE 19

20 Find the dimensions of the rectangle of maximum area that can be inscribed in an equilateral triangle of side a, if two vertices of the rectangle lie on one of the sides of the triangle.

21 Of all possible right circular cones which can be inscribed in a sphere of radius a, find the volume of the one that has maximum volume.

22 Find the dimensions of the right circular cylinder of maximum volume that can be inscribed in a sphere of radius a.

23 Find the point on the graph of $y = x^2 + 1$ that is closest to the point (3, 1).

24 Find the point on the graph of $y = x^3$ that is closest to the point (4, 0).

25 The strength of a rectangular beam is directly proportional to the product of its width and the square of the depth of a cross section. Find the dimensions of the strongest beam that can be cut from a cylindrical log of radius a (see figure).

EXERCISE 25

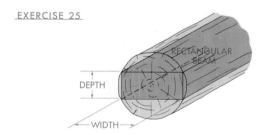

26 The illumination from a light source is directly proportional to the strength of the source and inversely proportional to the square of the distance from the source. If two light sources of strengths S_1 and S_2 are d units apart, at what point on the line segment joining the two sources is the illumination minimal?

27 A wholesaler sells running shoes at $20 per pair if less than 50 pairs are ordered. If 50 or more pairs are ordered (up to 600), the price per pair is reduced by 2 cents times the number ordered. What size order will produce the maximum amount of money for the wholesaler?

28 A paper cup is to be constructed in the shape of a right circular cone. If the volume desired is 36π in.^{3}, find the

dimensions that require the least amount of paper (neglect any waste that may occur in the construction).

29 A wire 36 cm long is to be cut into two pieces. One of the pieces will be bent into the shape of an equilateral triangle and the other into the shape of a rectangle whose length is twice its width. Where should the wire be cut if the combined area of the triangle and rectangle is (a) minimized? (b) maximized?

30 An isosceles triangle has base b and equal sides of length a. Find the dimensions of the rectangle of maximum area that can be inscribed in the triangle if one side of the rectangle lies on the base of the triangle.

31 A window has the shape of a rectangle surmounted by an equilateral triangle. If the perimeter of the window is 12 feet, find the dimensions of the rectangle that will produce the largest area for the window.

32 Two vertical poles of lengths 6 feet and 8 feet stand on level ground, with their bases 10 feet apart. Approximate the minimal length of cable that can reach from the top of one pole to some point on the ground between the poles, and then to the top of the other pole.

33 Prove that the rectangle of largest area having a given perimeter p is a square.

34 A right circular cylinder is generated by rotating a rectangle of perimeter p about one of its sides. What dimensions of the rectangle will generate the cylinder of maximum volume?

35 The owner of an apple orchard estimates that if 24 trees are planted per acre, then each mature tree will yield 600 apples per year. For each additional tree planted per acre the number of apples produced by each tree decreases by 12 per year. How many trees should be planted per acre to obtain the most apples per year?

36 A real estate company owns 180 efficiency apartments that are fully occupied when the rent is $300 per month. The company estimates that for each $10 increase in rent, five apartments will become unoccupied. What rent should be charged in order to obtain the largest gross income?

37 A package can be sent by parcel post only if the sum of its length and girth (the perimeter of the base) is not more than 108 inches. Find the dimensions of the box of maximum volume that can be sent if the base of the box is a square.

38 A North–South highway A and an East–West highway B intersect at a point P. At 10:00 A.M. an automobile crosses P traveling north on highway A at a speed of 50 mi/hr. At that same instant, an airplane flying east at a speed of 200 mi/hr and an altitude of 26,400 feet is directly above the point on highway B that is 100 miles west of P. If the automobile and airplane maintain the same speed and direction, at what time will they be closest to one another?

39 Two factories A and B that are 4 miles apart emit particles in smoke that pollute the area between the factories. Suppose that the number of particles emitted from each factory is directly proportional to the amount of smoke and inversely proportional to the cube of the distance from the factory. If factory A emits twice as much smoke as factory B, at what point between A and B is the pollution minimal?

40 An oil field contains 8 wells that produce a total of 1600 barrels of oil per day. For each additional well that is drilled, the average production per well decreases by 10 barrels per day. How many additional wells should be drilled to obtain the maximum amount of oil per day?

41 A canvas tent is to be constructed having the shape of a pyramid with a square base. A steel pole, placed in the center of the tent, will form the support (see figure). If S ft^2 of canvas is available for the four sides and x is the length of the base, show that

(a) the volume V of the tent is $V = \frac{1}{6}x\sqrt{S^2 - x^4}$.

(b) V has its maximum value when x equals the length of the pole.

EXERCISE 41

42 A boat must travel 100 miles upstream against a 10 mi/hr current. When the boat's velocity is v mi/hr, the number of gallons of gasoline consumed each hour is directly proportional to v^2.

(a) If a constant velocity of v mi/hr is maintained, show that the total number y of gallons of gasoline consumed is given by $y = 100kv^2/(v - 10)$ for $v > 10$ and for some positive constant K.

(b) Find the speed that minimizes the number of gallons of gasoline consumed during the trip.

43 Cars are crossing a bridge that is one mile long. Each car is 12 feet long and is required to stay a distance of at least d feet from the car in front of it (see figure).

(a) Show that the largest number of cars that can be on the bridge at one time is $[\![5280/(12 + d)]\!]$ ($[\![\ ]\!]$ denotes the greatest integer function).

(b) If the velocity of each car is v mi/hr, show that the maximum traffic flow rate F (in cars/hour) is $F = 5280v/(12 + d)$.

(c) The stopping distance (in feet) of a car traveling v mi/hr is approximately $0.05v^2$. If $d = 0.025v^2$, find the speed that maximizes the traffic flow across the bridge.

EXERCISE 43

44 Prove that the shortest distance from a point (x_1, y_1) to the graph of a differentiable function f is measured along a normal line to the graph, that is, a line perpendicular to the tangent line.

45 A railroad route is to be constructed from town A to town C, branching out from a point B toward C at an angle of θ degrees (see figure). Because of the mountains between A and C, the branching point B must be at least 20 miles east of A. If the construction costs are $50,000 per mile between A and B and $100,000 per mile between B and C, find the branching angle θ that minimizes the total construction cost.

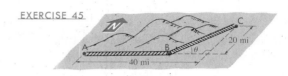
EXERCISE 45

46 A long rectangular sheet of metal, 12 inches wide, is to be made into a rain gutter by turning up two sides at angles of 120° to the sheet. How many inches should be turned up to give the gutter its greatest capacity?

47 Refer to Exercise 12. Find the central angle of the sector that will maximize the volume of the cup.

48 A square picture having sides 2 feet long is hung on a wall such that the base is 6 feet above the floor. If a person whose eye level is 5 feet above the floor looks at the picture, and if θ is the angle between the line of sight and the top and bottom of the picture, find the distance from the wall at which θ has its maximum value.

49 A rectangle made of elastic material will be made into a cylinder by joining edges AD and BC (see figure). To support the structure, a wire of fixed length L is placed along the diagonal of the rectangle. Find the angle θ that will result in the cylinder of maximum volume.

EXERCISE 49

50 When an individual is walking, the magnitude F of the vertical force of one foot on the ground (see figure) can be approximated by $F = A(\cos bt - a \cos 3bt)$ for time t (in seconds), with $A > 0$, $b > 0$, and $0 < a < 1$.

(a) Show that $F = 0$ when $t = -\pi/(2b)$ and $t = \pi/(2b)$. (The time $t = -\pi/(2b)$ corresponds to the instant that the foot first touches the ground and the weight of the body is being supported by the other foot.)

(b) Show that the maximum force occurs when $t = 0$ or when $\sin^2 bt = (9a - 1)/(12a)$.

(c) If $a = \frac{1}{3}$, express the maximum force in terms of A. If $0 < a \le \frac{1}{9}$, express the maximum force in terms of A.

EXERCISE 50

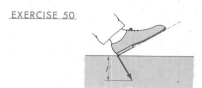

51 Two corridors 3 feet and 4 feet wide, respectively, meet at a right angle. Find the length of the longest nonbendable rod that can be carried horizontally around the corner, as shown in the figure (neglect the thickness of the rod).

EXERCISE 51

52 Light travels from one point to another by the path that requires the least time. Suppose that light has velocity v_1 in air and v_2 in water with $v_1 > v_2$. If light travels from a point P in air to a point Q in water (see figure), show that the path requiring the least amount of time

EXERCISE 52

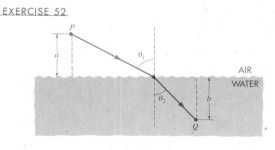

occurs if

$$\frac{\sin \theta_1}{\sin \theta_2} = \frac{v_1}{v_2}.$$

(This is an example of *Snell's law of refraction*.)

53 A circular cylinder of fixed radius R is surmounted by a cone (see figure). The ends of the cylinder are open and the total volume is to be a specified constant V.

(a) Show that the total surface area S is given by

$$S = \frac{2V}{R} + \pi R^2 \left(\csc \theta - \frac{2}{3} \cot \theta \right)$$

(b) Show that S is minimized when $\theta \approx 48.2°$.

EXERCISE 53

54 In the classic honeycomb-structure problem, a hexagonal prism of fixed radius (and side) R is surmounted by adding three identical rhombuses that meet in a common vertex (see figure). The bottom of the prism is open and the total volume is to be a specified constant V. A more elaborate geometric argument than that in Exercise 53 establishes that the total surface area S is given by

$$S = \frac{4}{3}\sqrt{3}\,\frac{V}{R} - \frac{3}{2}R^2 \cot \theta + \frac{3\sqrt{3}}{2}R^2 \csc \theta.$$

Show that S is minimized when $\theta \approx 54.7°$. (Remarkably, bees do construct their honeycombs so that the amount of wax S is minimized.)

EXERCISE 54

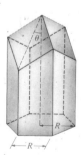

4.6 LIMITS INVOLVING INFINITY

In this section we shall consider functions whose values "stabilize" as $|x|$ becomes very large. We shall also consider functions that have no absolute minimum or absolute maximum values, but instead, have the property that

$|f(x)|$ may take on arbitrarily large values. When applied to the derivative f', our discussion will provide information about vertical tangent lines to the graph of f.

Let us begin by considering the graph of $f(x) = 2 + (1/x)$, sketched in Figure 4.46, and concentrate on values of $f(x)$ if x is large and positive. For example,

$$f(100) = 2.01$$

$$f(1000) = 2.001$$

$$f(10,000) = 2.0001$$

$$f(100,000) = 2.00001$$

We can make $f(x)$ as close to 2 as desired or, equivalently, we can make $|f(x) - 2|$ arbitrarily small by choosing x sufficiently large. This behavior of $f(x)$ is expressed by

$$\lim_{x \to \infty} \left(2 + \frac{1}{x} \right) = 2$$

which may be read: *the limit of $2 + (1/x)$ as x approaches ∞ is 2.*

It is important to remember that ∞ (infinity) is not a real number, and hence should never be substituted for the variable x. Thus, the terminology x *approaches* ∞ does not mean that the variable x gets very close to some real number. Intuitively, we think of x as increasing without bound, or being assigned arbitrarily large values. Later, in Definition (4.25), we shall employ the symbol ∞ in a different context.

Let us next define $\lim_{x \to \infty} f(x) = L$ for an arbitrary function f. As in Chapter 2, we shall give both an informal and a formal definition. We shall assume that f is defined on an infinite interval (c, ∞) for some real number c.

INFORMAL DEFINITION (4.20)

$$\lim_{x \to \infty} f(x) = L$$

means that $f(x)$ can be made arbitrarily close to L by choosing x sufficiently large.

DEFINITION (4.21)

$$\lim_{x \to \infty} f(x) = L$$

means that for every $\varepsilon > 0$, there corresponds a positive number N, such that

$$|f(x) - L| < \varepsilon \quad \text{whenever} \quad x > N$$

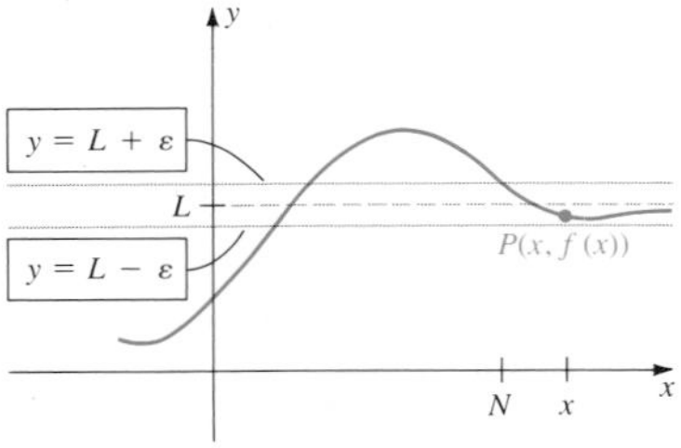

If $\lim_{x \to \infty} f(x) = L$, we say that **the limit of $f(x)$ as x approaches ∞ is L.**

Let us give a graphical interpretation of Definition (4.21). Suppose $\lim_{x \to \infty} f(x) = L$, and consider any horizontal lines $y = L \pm \varepsilon$ (see Figure 4.47). According to Definition (4.21), if x is larger than some positive number N, then every point $P(x, f(x))$ on the graph lies between these horizontal lines. Roughly speaking, the graph of f gets closer and closer to the line $y = L$ as x gets larger and larger. The line $y = L$ is a **horizontal asymptote** for the graph

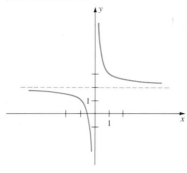

FIGURE 4.46

FIGURE 4.47
$\lim_{x \to \infty} f(x) = L$

of f. For example, the line $y = 2$ in Figure 4.46 is a horizontal asymptote for the graph of $f(x) = 2 + (1/x)$.

In Figure 4.47, the graph of f approaches the asymptote $y = L$ from below, that is, with $f(x) < L$. We could also have the graph approach $y = L$ from above, that is, with $f(x) > L$, or other variations, such as $f(x)$ alternately greater than or less than L as $x \to \infty$.

The next definition covers the situation in which $|x|$ is large and x is *negative*. We shall assume that f is defined on an infinite interval $(-\infty, c)$ for some real number c.

DEFINITION (4.22)

$$\lim_{x \to -\infty} f(x) = L$$

means that for every $\varepsilon > 0$, there corresponds a negative number N, such that

$$|f(x) - L| < \varepsilon \quad \text{whenever} \quad x < N$$

FIGURE 4.48
$\lim_{x \to -\infty} f(x) = L$

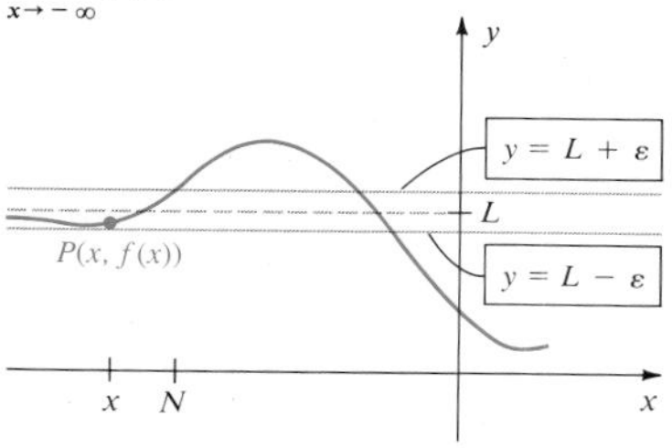

If $\lim_{x \to -\infty} f(x) = L$, we say that the **limit of $f(x)$ as x approaches $-\infty$ is L.** Definition (4.22) is illustrated in Figure 4.48. If we consider any horizontal lines $y = L \pm \varepsilon$, then every point $P(x, f(x))$ on the graph lies between these lines if x is *less* than some negative number N. As before, we refer to the line $y = L$ as a horizontal asymptote for the graph of f.

Limit theorems, analogous to those in Chapter 2, may be established for limits involving infinity. In particular, Theorem (2.15) concerning limits of sums, products, and quotients is true for the cases $x \to \infty$ or $x \to -\infty$. Similarly, Theorem (2.21) on the limit of $\sqrt[n]{f(x)}$ holds if $x \to \infty$ or $x \to -\infty$. Finally, it is trivial to show that

$$\lim_{x \to \infty} c = c \quad \text{and} \quad \lim_{x \to -\infty} c = c.$$

The next theorem is important for calculating specific limits.

THEOREM (4.23)

If k is a positive rational number and c is any real number, then

$$\lim_{x \to \infty} \frac{c}{x^k} = 0 \quad \text{and} \quad \lim_{x \to -\infty} \frac{c}{x^k} = 0$$

provided x^k is always defined.

PROOF To use (4.21) to prove that $\lim_{x \to \infty}(c/x^k) = 0$, we must show that for every $\varepsilon > 0$, there corresponds a positive number N such that

$$\left| \frac{c}{x^k} - 0 \right| < \varepsilon \quad \text{whenever} \quad x > N.$$

If $c = 0$, any $N > 0$ will suffice. If $c \neq 0$, the following four inequalities are equivalent for $x > 0$:

$$\left| \frac{c}{x^k} - 0 \right| < \varepsilon, \qquad \frac{|x|^k}{|c|} > \frac{1}{\varepsilon}, \qquad |x|^k > \frac{|c|}{\varepsilon}, \qquad x > \left(\frac{|c|}{\varepsilon} \right)^{1/k}$$

The last inequality gives us a clue to a choice for N. Letting $N = (|c|/\varepsilon)^{1/k}$, we see that whenever $x > N$ the fourth, and hence the first, inequality is true, which is what we wished to show. The second part of the theorem may be proved in similar fashion. • •

If f is a rational function, then limits as $x \to \infty$ or $x \to -\infty$ may be found by first dividing numerator and denominator of $f(x)$ by a suitable power of x and then applying Theorem (4.23). Specifically, suppose $g(x)$ and $h(x)$ are polynomials and we wish to investigate $f(x) = g(x)/h(x)$. If the degree of $g(x)$ is less than or equal to the degree of $h(x)$ and k is the degree of $h(x)$, then we divide numerator and denominator by x^k. If the degree of $g(x)$ is greater than the degree of $h(x)$, we can show that $f(x)$ has no limit as $x \to \infty$ or as $x \to -\infty$.

EXAMPLE 1 Investigate $\displaystyle\lim_{x \to -\infty} \frac{2x^2 - 5}{3x^2 + x + 2}$.

SOLUTION Since we are interested only in negative values of x, we may assume that $x \neq 0$. The degree of the numerator and denominator are both 2, and hence, by the rule stated in the preceding paragraph, we divide numerator and denominator by x^2 and then employ limit theorems. Thus

$$
\lim_{x \to -\infty} \frac{2x^2 - 5}{3x^2 + x + 2} = \lim_{x \to -\infty} \frac{2 - (5/x^2)}{3 + (1/x) + (2/x^2)}
$$

$$
= \frac{\displaystyle\lim_{x \to -\infty} \; [2 - (5/x^2)]}{\displaystyle\lim_{x \to -\infty} \; [3 + (1/x) + (2/x^2)]}
$$

$$
= \frac{\displaystyle\lim_{x \to -\infty} 2 - \lim_{x \to -\infty} (5/x^2)}{\displaystyle\lim_{x \to -\infty} 3 + \lim_{x \to -\infty} (1/x) + \lim_{x \to -\infty} (2/x^2)}
$$

$$
= \frac{2 - 0}{3 + 0 + 0} = \frac{2}{3}.
$$

It follows that the line $y = \frac{2}{3}$ is a horizontal asymptote for the graph of f. •

EXAMPLE 2 If $f(x) = 4x/(x^2 + 9)$, determine the horizontal asymptotes, find the local maxima and minima and the points of inflection, discuss concavity, and sketch the graph of f.

SOLUTION Since the degree of the numerator (one) is less than the degree of the denominator (two), we divide numerator and denominator by x^2 and use limit theorems as follows:

$$
\lim_{x \to \infty} f(x) = \lim_{x \to \infty} \frac{4x}{x^2 + 9} = \lim_{x \to \infty} \frac{(4/x)}{1 + (9/x^2)}
$$

$$
= \frac{\displaystyle\lim_{x \to \infty} (4/x)}{\displaystyle\lim_{x \to \infty} 1 + \lim_{x \to \infty} (9/x^2)} = \frac{0}{1 + 0} = \frac{0}{1} = 0
$$

Thus, $y = 0$ (the x-axis) is a horizontal asymptote for the graph of f. Similarly, $\lim_{x \to -\infty} f(x) = 0$.

We may verify that

$$f'(x) = \frac{4(9 - x^2)}{(x^2 + 9)^2} \quad \text{and} \quad f''(x) = \frac{8x(x^2 - 27)}{(x^2 + 9)^3}.$$

The critical numbers of f are the solutions of $f'(x) = 0$, and hence are ± 3. Using either the First or Second Derivative Test, we can show that $f(-3)$ is a local minimum and $f(3)$ is a local maximum.

Using $f''(x)$, we see that the points of inflection have the x-coordinates 0 and $\pm\sqrt{27} = \pm 3\sqrt{3}$. The sign of $f''(x)$ tells us that the graph is concave downward on the intervals $(-\infty, -3\sqrt{3})$ and $(0, 3\sqrt{3})$ and concave upward on $(-3\sqrt{3}, 0)$ and $(3\sqrt{3}, \infty)$. The graph is sketched in Figure 4.49, with different scales on the axes. •

FIGURE 4.49

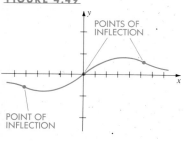

POINTS OF INFLECTION

POINT OF INFLECTION

EXAMPLE 3 Investigate $\lim\limits_{x \to \infty} \dfrac{\sqrt{9x^2 + 2}}{3 - 4x}$.

SOLUTION If x is large and positive, the numerator $\sqrt{9x^2 + 2}$ may be approximated by $\sqrt{9x^2}$, or $3x$. This suggests that we divide numerator and denominator by x. Since $x = \sqrt{x^2}$ for positive values of x, we have

$$\frac{\sqrt{9x^2 + 2}}{x} = \frac{\sqrt{9x^2 + 2}}{\sqrt{x^2}} = \sqrt{\frac{9x^2 + 2}{x^2}} = \sqrt{9 + \frac{2}{x^2}}$$

and

$$\lim_{x \to \infty} \frac{\sqrt{9x^2 + 2}}{3 - 4x} = \lim_{x \to \infty} \frac{\sqrt{9 + (2/x^2)}}{(3/x) - 4} = \frac{\lim\limits_{x \to \infty} \sqrt{9 + (2/x^2)}}{\lim\limits_{x \to \infty} [(3/x) - 4]}$$

$$= \frac{\sqrt{\lim\limits_{x \to \infty} [9 + (2/x^2)]}}{\lim\limits_{x \to \infty} (3/x) - \lim\limits_{x \to \infty} 4} = \frac{\sqrt{\lim\limits_{x \to \infty} 9 + \lim\limits_{x \to \infty} (2/x^2)}}{0 - 4}$$

$$= \frac{\sqrt{9 + 0}}{-4} = -\frac{3}{4}. \quad •$$

Let us next consider $f(x) = 1/(x - 3)^2$. The graph of f is sketched in Figure 4.50. If x is close to 3 (but $x \neq 3$) the denominator $(x - 3)^2$ is close to zero, and hence $f(x)$ is very large. For example,

FIGURE 4.50

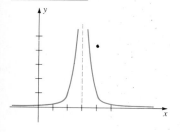

$$f(2.9) = 100 \qquad\qquad f(3.1) = 100$$
$$f(2.99) = 10,000 \qquad\qquad f(3.01) = 10,000$$
$$f(2.999) = 1,000,000 \qquad f(3.001) = 1,000,000$$

Evidently, we can make $f(x)$ as large as desired by choosing x sufficiently close to 3. This is symbolized by writing

$$\lim_{x \to 3} \frac{1}{(x - 3)^2} = \infty.$$

The following is an informal statement of this type of behavior for an arbitrary function f.

INFORMAL DEFINITION (4.24)

$$\lim_{x \to a} f(x) = \infty$$

means that $f(x)$ can be made as large as desired by choosing x sufficiently close to a.

A formal definition, which can be used in proofs of theorems, may be stated as follows.

DEFINITION (4.25)

Let f be defined on an open interval containing a, except possibly at a itself. The statement **the limit of $f(x)$ as x approaches a is ∞,** written

$$\lim_{x \to a} f(x) = \infty,$$

means that for every positive number M, there corresponds a $\delta > 0$ such that

$$\text{if} \quad 0 < |x - a| < \delta, \quad \text{then} \quad f(x) > M$$

FIGURE 4.51
$\lim_{x \to a} f(x) = \infty$

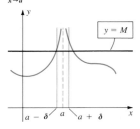

If $\lim_{x \to a} f(x) = \infty$, we sometimes say that $f(x)$ *increases without bound as x approaches a*. If, in Figure 4.51, we consider any horizontal line $y = M$, then when x is in a suitable interval $(a - \delta, a + \delta)$, but $x \neq a$, the points on the graph of f lie *above* the horizontal line.

Remember that the symbol ∞ does not represent a real number. In particular, $\lim_{x \to a} f(x) = \infty$ *does not mean that the limit exists* as x approaches a. The limit notation is merely used to denote the variation of $f(x)$ that we have described.

The notion of $f(x)$ increasing without bound as x approaches a from the right or left is indicated by writing

$$\lim_{x \to a^+} f(x) = \infty \quad \text{or} \quad \lim_{x \to a^-} f(x) = \infty,$$

respectively. Only minor modifications of Definition (4.25) are needed to define these concepts. Thus, for $x \to a^+$, we assume that $f(x)$ exists on some open interval (a, c), and the inequality $0 < |x - a| < \delta$ in Definition (4.25) is changed to $a < x < a + \delta$; that is, x may only take on values *larger* than a. Similar statements may be made for $x \to a^-$. Graphs illustrating these ideas are sketched in Figure 4.52.

FIGURE 4.52

(i) $\lim_{x \to a^+} f(x) = \infty$ (ii) $\lim_{x \to a^-} f(x) = \infty$

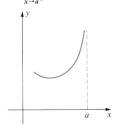

The following is analogous to Definition (4.25) if $|f(x)|$ is very large, and $f(x)$ is negative as x approaches a.

DEFINITION (4.26)

Let f be defined on an open interval containing a, except possibly at a itself. The statement **the limit of $f(x)$ as x approaches a is $-\infty$,** written

$$\lim_{x \to a} f(x) = -\infty$$

means that for every negative number M there corresponds a $\delta > 0$ such that

$$\text{if} \quad 0 < |x - a| < \delta, \quad \text{then} \quad f(x) < M$$

If $\lim_{x \to a} f(x) = -\infty$, we sometimes say that $f(x)$ *decreases without bound as x approaches a.*

Graphical illustrations of Definition (4.26), together with the cases $x \to a^+$ and $x \to a^-$, are sketched in Figure 4.53.

The line $x = a$ in Figures 4.51–4.53 is a **vertical asymptote** for the graph of f. Note that f has an *infinite discontinuity* at $x = a$ (see Section 2.6).

FIGURE 4.53

(i) $\lim\limits_{x \to a} f(x) = -\infty$ (ii) $\lim\limits_{x \to a^+} f(x) = -\infty$ (iii) $\lim\limits_{x \to a^-} f(x) = -\infty$

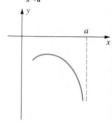

The limits of the functions whose graphs are sketched in Figure 4.53 do not exist at $x = a$, even though we use a limit notation to specify the behavior of f. In Example 3 of Section 2.3 we proved that $\lim_{x \to 0} (1/x)$ does not exist. Using the present notation, this nonexistence is denoted by

$$\lim_{x \to 0^-} \frac{1}{x} = -\infty \quad \text{and} \quad \lim_{x \to 0^+} \frac{1}{x} = \infty.$$

The next theorem, stated without proof, is useful when investigating limits of rational functions.

THEOREM (4.27)

(i) If n is an even positive integer, then

$$\lim_{x \to a} \frac{1}{(x - a)^n} = \infty$$

(ii) If n is an odd positive integer, then

$$\lim_{x \to a^+} \frac{1}{(x - a)^n} = \infty \quad \text{and} \quad \lim_{x \to a^-} \frac{1}{(x - a)^n} = -\infty$$

If n is a positive integer and $f(x) = 1/(x - a)^n$, then by Theorem (4.27), the graph of f has the line $x = a$ as a vertical asymptote. Note also that since $\lim_{x \to \infty} 1/(x - a)^n = 0$, the x-axis is a horizontal asymptote.

EXAMPLE 4 If $f(x) = \dfrac{1}{(x - 2)^3}$, discuss $\lim\limits_{x \to 2^+} f(x)$, $\lim\limits_{x \to 2^-} f(x)$, and sketch the graph of f.

SOLUTION If x is very close to 2, but larger than 2, then $x - 2$ is a small positive number and hence $1/(x - 2)^3$ is a large positive number. Thus it is evident that

$$\lim_{x \to 2^+} \frac{1}{(x - 2)^3} = \infty.$$

This fact also follows from Theorem (4.27) (ii) with $a = 2$ and $n = 3$.

If x is close to 2 but *less* than 2, then $x - 2$ is close to 0 and *negative*. Hence

$$\lim_{x \to 2^-} \frac{1}{(x - 2)^3} = -\infty.$$

This is also a consequence of (4.27) (ii). Thus, the line $x = 2$ is a vertical asymptote.

We can show that

$$\lim_{x \to \infty} \frac{1}{(x - 2)^3} = 0 \quad \text{and} \quad \lim_{x \to -\infty} \frac{1}{(x - 2)^3} = 0.$$

Hence $y = 0$ (the x-axis) is a horizontal asymptote.

The graph of f is sketched in Figure 4.54. Note that f has no absolute maximum or absolute minimum. •

We next state several properties of sums, products, and quotients of functions that become infinite. Analogous results hold for the cases $x \to a^+$ and $x \to a^-$.

FIGURE 4.54

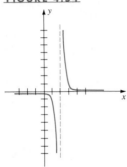

THEOREM (4.28)

If $\lim_{x \to a} f(x) = \infty$ and $\lim_{x \to a} g(x) = c$ for some number c, then:

(i) $\lim\limits_{x \to a} [g(x) + f(x)] = \infty$

(ii) If $c > 0$, then $\lim\limits_{x \to a} [g(x)f(x)] = \infty$ and $\lim\limits_{x \to a} \dfrac{f(x)}{g(x)} = \infty$

(iii) If $c < 0$, then $\lim\limits_{x \to a} [g(x)f(x)] = -\infty$ and $\lim\limits_{x \to a} \dfrac{f(x)}{g(x)} = -\infty$

(iv) $\lim\limits_{x \to a} \dfrac{g(x)}{f(x)} = 0$

The conclusions of Theorem (4.28) are intuitively evident. In (i), for example, if $g(x)$ is close to c and if $f(x)$ increases without bound as x approaches a, then $g(x) + f(x)$ can be made very large by choosing x sufficiently close to a.

In (iii), if $\lim_{x \to a} g(x) = c < 0$, then $g(x)$ is negative when x is close to a. Consequently, if $f(x)$ is large and positive, then when x is close to a the product $g(x)f(x)$ is *negative* and $|g(x)f(x)|$ is large. This suggests that $\lim_{x \to a} g(x)f(x) = -\infty$.

We shall not give a formal proof of Theorem (4.28). A similar theorem can be stated for $\lim_{x \to a} f(x) = -\infty$. Theorems involving more than two functions can also be proved.

Graphs of rational functions frequently have vertical and horizontal asymptotes. For illustrations see Figures 4.46, 4.49, 4.50, and 4.54. Another is given in the following example.

EXAMPLE 5 If $f(x) = 2x^2/(9 - x^2)$, find the vertical and horizontal asymptotes and sketch the graph of f.

SOLUTION We begin by factoring the denominator, obtaining

$$f(x) = \frac{2x^2}{(3 - x)(3 + x)}.$$

The denominator is zero at $x = 3$ and $x = -3$, and hence the corresponding lines are candidates for vertical asymptotes. Since

$$\lim_{x \to 3^-} \frac{1}{3 - x} = \infty \quad \text{and} \quad \lim_{x \to 3^-} \frac{2x^2}{3 + x} = 3$$

we see from Theorem (4.28) (ii),

$$\lim_{x \to 3^-} f(x) = \lim_{x \to 3^-} \left(\frac{1}{3 - x}\right)\left(\frac{2x^2}{3 + x}\right) = \infty.$$

Moreover, since

$$\lim_{x \to 3^+} \frac{1}{3 - x} = -\infty \quad \text{and} \quad \lim_{x \to 3^+} \frac{2x^2}{3 + x} = 3 > 0,$$

it follows that

$$\lim_{x \to 3^+} f(x) = \lim_{x \to 3^+} \left(\frac{1}{3 - x}\right)\left(\frac{2x^2}{3 + x}\right) = -\infty.$$

FIGURE 4.55

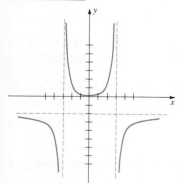

This variation of $f(x)$ near the vertical asymptote $x = 3$ is illustrated in Figure 4.55.

Since

$$\lim_{x \to -3^-} \frac{1}{3 + x} = -\infty \quad \text{and} \quad \lim_{x \to -3^-} \frac{2x^2}{3 - x} = 3 > 0,$$

we conclude that

$$\lim_{x \to -3^-} f(x) = \lim_{x \to -3^-} \left(\frac{1}{3 + x}\right)\left(\frac{2x^2}{3 - x}\right) = -\infty.$$

Similarly,

$$\lim_{x \to -3^+} \frac{1}{3 + x} = \infty \quad \text{implies that} \quad \lim_{x \to -3^+} f(x) = \infty.$$

This behavior of $f(x)$ near $x = -3$ is illustrated in Figure 4.55.

To find the horizontal asymptotes we consider

$$\lim_{x \to \infty} f(x) = \lim_{x \to \infty} \frac{2x^2}{9 - x^2} = \lim_{x \to \infty} \frac{2}{(9/x^2) - 1} = -2.$$

The same limit is obtained if $x \to -\infty$. The line $y = -2$ is a horizontal asymptote. Using this information and plotting several points gives us the sketch in Figure 4.55.

If more detailed information about the graph is desired, we could use the first derivative to show that f is decreasing on the intervals $(-\infty, -3)$ and $(-3, 0]$, and increasing on $[0, 3)$ and $(3, \infty)$, with a minimum occurring at $x = 0$ as shown in the figure. The second derivative could be used to investigate concavity. •

If $f(x) = g(x)/h(x)$ for polynomials $g(x)$ and $h(x)$ and *if the degree of $g(x)$ is 1 greater than the degree of $h(x)$*, then the graph of f has an **oblique asymptote** $y = ax + b$; that is, the graph approaches this line as x increases or decreases without bound. To establish this fact we may use long division to express $f(x)$ in the form

$$f(x) = \frac{g(x)}{h(x)} = (ax + b) + \frac{r(x)}{h(x)}$$

with either $r(x) = 0$ or the degree of $r(x)$ less than the degree of $h(x)$. It follows that $\lim_{x \to \infty} r(x)/h(x) = 0$ and $\lim_{x \to -\infty} r(x)/h(x) = 0$. Consequently, $f(x)$ gets closer and closer to $ax + b$ as x approaches either ∞ or $-\infty$. The next example illustrates this procedure for finding oblique asymptotes.

EXAMPLE 6 Find all the asymptotes and sketch the graph of f if

$$f(x) = \frac{x^2 - 9}{2x - 4}$$

SOLUTION A vertical asymptote occurs if $2x - 4 = 0$, that is, if $x = 2$. Moreover, since the degree of the numerator $x^2 - 9$ is 1 greater than the degree of the denominator $2x - 4$, the graph has an oblique asymptote. Using long division,

$$\frac{x^2 - 9}{2x - 4} = (\tfrac{1}{2}x + 1) - \frac{5}{2x - 4}.$$

By the discussion preceding this example, the line $y = \frac{1}{2}x + 1$ is an oblique asymptote. This line and the vertical asymptote $x = 2$ are sketched (with dashes) in Figure 4.56.

The x-intercepts of the graph are the solutions of the equation $x^2 - 9 = 0$, and hence are 3 and -3. The y-intercept is $f(0) = \frac{9}{4}$. The corresponding points are plotted in Figure 4.56. It is now easy to show that the graph has the shape indicated in Figure 4.57. •

If we regard a tangent line l as a limiting position of a secant line (see Figure 2.4), then it is possible for l to be vertical. In such cases, as the secant line approaches l its slope increases or decreases without bound. With this in mind we formulate the next definition.

FIGURE 4.56

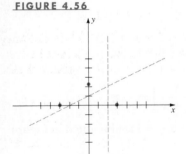

FIGURE 4.57

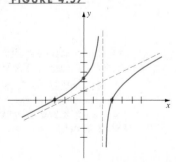

DEFINITION (4.29)

The graph of a function f has a **vertical tangent line** at the point $P(a, f(a))$ if f is continuous at a and

$$\lim_{x \to a} |f'(x)| = \infty$$

EXAMPLE 7 If $f(x) = (x-8)^{1/3} + 1$ and $g(x) = (x-8)^{2/3} + 1$, show that the graphs of f and g have vertical tangent lines at $P(8, 1)$. Sketch the graphs, showing the tangent lines at P.

SOLUTION Both f and g are continuous at 8. Differentiating, we obtain

$$f'(x) = \frac{1}{3}(x-8)^{-2/3} = \frac{1}{3(x-8)^{2/3}}$$

$$g'(x) = \frac{2}{3}(x-8)^{-1/3} = \frac{2}{3(x-8)^{1/3}}.$$

Evidently,

$$\lim_{x \to 8} |f'(x)| = \infty \quad \text{and} \quad \lim_{x \to 8} |g'(x)| = \infty.$$

FIGURE 4.58

(i)

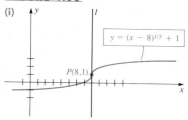

(ii)

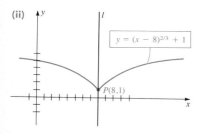

By Definition (4.29), each graph has a vertical tangent line at $P(8, 1)$.

The graphs are sketched in Figure 4.58. Note that for the graph of f in (i) the slope of the tangent line increases without bound if x approaches 8 from either the left or the right. However, for the graph of g in (ii) the slope decreases without bound if $x \to 8^-$, or increases without bound if $x \to 8^+$.

The number 8 is a critical number for each function. It follows from the First Derivative Test that $f(8) = 1$ is not an extremum for f, and $g(8) = 1$ is an absolute minimum for g. •

Definition (4.29) may be modified to include vertical tangent lines at an endpoint of the domain of a function. Thus, if f is continuous on $[a, b]$, but is undefined outside of this interval, then the graph has a vertical tangent line at $P(a, f(a))$ or $Q(b, f(b))$ if

$$\lim_{x \to a^+} |f'(x)| = \infty \quad \text{or} \quad \lim_{x \to b^-} |f'(x)| = \infty,$$

FIGURE 4.59

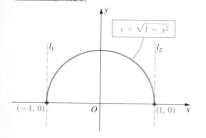

respectively. For example, if $f(x) = \sqrt{1 - x^2}$, then the graph of f is the upper half of the unit circle with center at the origin. There are vertical tangent lines l_1 and l_2 at the points $(-1, 0)$ and $(1, 0)$, respectively (see Figure 4.59).

EXERCISES 4.6

Exer. 1–8: Find the limit.

1 $\displaystyle\lim_{x \to \infty} \frac{5x^2 - 3x + 1}{2x^2 + 4x - 7}$

2 $\displaystyle\lim_{x \to \infty} \frac{3x^3 - x + 1}{6x^3 + 2x^2 - 7}$

3 $\displaystyle\lim_{x \to -\infty} \frac{4 - 7x}{2 + 3x}$

4 $\displaystyle\lim_{x \to -\infty} \frac{(3x + 4)(x - 1)}{(2x + 7)(x + 2)}$

5 $\displaystyle\lim_{x \to \infty} \sqrt[3]{\frac{8 + x^2}{x(x + 1)}}$

6 $\displaystyle\lim_{x \to -\infty} \frac{4x - 3}{\sqrt{x^2 + 1}}$

7 $\displaystyle\lim_{x \to \infty} \frac{\sqrt{4x + 1}}{10 - 3x}$

8 $\displaystyle\lim_{x \to \infty} \frac{2x^2 - x + 3}{x^3 + 1}$

Exer. 9–18: Find $\lim_{x \to a^+} f(x)$ and $\lim_{x \to a^-} f(x)$. **Identify the vertical and horizontal asymptotes, find the local maxima and minima, and sketch the graph of** f.

9 $f(x) = \dfrac{5}{x - 4}$, $a = 4$

10 $f(x) = \dfrac{5}{4 - x}$, $a = 4$

11 $f(x) = \dfrac{8}{(2x + 5)^3}$, $a = -\dfrac{5}{2}$

12 $f(x) = \dfrac{-4}{7x + 3}$, $a = -\dfrac{3}{7}$

13 $f(x) = \dfrac{3x}{(x + 8)^2}$, $a = -8$

14 $f(x) = \dfrac{3x^2}{(2x - 9)^2}$, $a = \dfrac{9}{2}$

15 $f(x) = \dfrac{2x^2}{x^2 - x - 2}$, $a = -1$, $a = 2$

16 $f(x) = \dfrac{4x}{x^2 - 4x + 3}$, $a = 1$, $a = 3$

17 $f(x) = \dfrac{1}{x(x - 3)^2}$, $a = 0$, $a = 3$

18 $f(x) = \dfrac{x^2}{x + 1}$, $a = -1$

Exer. 19–28: Determine the vertical and horizontal asymptotes, and sketch the graph of f.

19 $f(x) = \dfrac{1}{x^2 - 4}$

20 $f(x) = \dfrac{5x}{4 - x^2}$

21 $f(x) = \dfrac{4x^2}{x^2 + 1}$

22 $f(x) = \dfrac{3x}{x^2 + 1}$

23 $f(x) = \dfrac{1}{x^3 + x^2 - 6x}$

24 $f(x) = \dfrac{x^2 - x}{16 - x^2}$

25 $f(x) = \dfrac{x^2 + 3x + 2}{x^2 + 2x - 3}$

26 $f(x) = \dfrac{x^2 - 5x}{x^2 - 25}$

27 $f(x) = \dfrac{x + 4}{x^2 - 16}$

28 $f(x) = \dfrac{\sqrt[3]{16 - x^2}}{4 - x}$

Exer. 29–34: Find the vertical and oblique asymptotes, and sketch the graph of f.

29 $f(x) = \dfrac{x^2 - x - 6}{x + 1}$

30 $f(x) = \dfrac{2x^2 - x - 3}{x - 2}$

31 $f(x) = \dfrac{8 - x^3}{2x^2}$

32 $f(x) = \dfrac{x^3 + 1}{x^2 - 9}$

33 $f(x) = \dfrac{x^4 - 4}{x^3 - 1}$

34 $f(x) = \dfrac{1 - x^4}{2x^3 - 8x}$

Exer. 35–38: Find the points on the graph of f at which the tangent line is vertical.

35 $f(x) = x(x + 2)^{3/5}$

36 $f(x) = \sqrt{x + 2}$

37 $f(x) = \sqrt{16 - 9x^2} + 3$

38 $f(x) = \sqrt[3]{x} - 5$

39 Salt water of concentration 0.1 pounds of salt per gallon flows into a large tank that initially contains 50 gallons of pure water.

(a) If the flow rate of salt water into the tank is 5 gallons per minute, what is the volume $V(t)$ of water and the quantity $Q(t)$ of salt in the tank at time t?

(b) Show that the salt concentration $c(t)$ at time t is given by $c(t) = t/(10t + 100)$.

(c) Discuss the behavior of $c(t)$ as $t \to \infty$.

40 An important problem is fishery science is predicting the next year's adult breeding population R (the recruits) from the number S that are presently spawning. For some species (such as North Sea herring), the relationship between R and S takes the form $R = aS/(S + b)$. What is the interpretation of the constant a? Conclude that for large values of S, recruitment is more or less constant.

41 Coulomb's law in electricity asserts that the force of attraction F between two charged particles is directly proportional to the product of the charges and inversely proportional to the square of the distance between the particles. Suppose a particle of charge $+1$ is placed on a coordinate line between two particles of charge -1 (see figure).

(a) Show that the net force acting on the particle of charge $+1$ is given by

$$F(x) = -\dfrac{k}{x^2} + \dfrac{k}{(x - 2)^2}$$

for some $k > 0$.

(b) Let $k = 1$ and sketch the graph of F for $0 < x < 2$.

EXERCISE 41

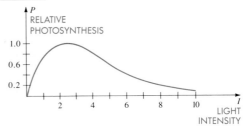

42 Biomathematicians have proposed many different functions for describing the effect of light on the rate at which photosynthesis can take place. If the function is to be realistic, then it must exhibit the *photoinhibition effect;* that is, the rate of production P of photosynthesis must decrease to 0 as the light intensity I reaches high levels (see figure). Which of the following formulas for P, with a and b constant, may be used? Give reasons for your answer.

(a) $P = \dfrac{aI}{b + I}$

(b) $P = \dfrac{aI}{b + I^2}$

EXERCISE 42

In Chapter 3 we studied problems that were stated in the form: *Given a function g, find the derivative g'*. We shall now consider the converse problem—*Given the derivative g', find the function g*. An equivalent way of stating this converse problem is:

Given a function f, find a function F such that F' = f.

As a simple illustration, suppose $f(x) = 8x^3$. In this case it is easy to find a function F such that $F'(x) = f(x)$. We know that differentiating a power of x *reduces* the exponent by 1 and, therefore, to obtain F we must *increase* the given exponent by 1. Thus $F(x) = ax^4$ for some number a. Differentiating, we obtain $F'(x) = 4ax^3$ and, in order for this to equal $f(x)$, a must equal 2. Consequently, the function F such that $F(x) = 2x^4$ has the property $F' = f$. According to the next definition, F is an *antiderivative of f*.

DEFINITION (4.30)

A function F is an **antiderivative** of a function f if $F' = f$.

The process of finding an antiderivative is called **antidifferentiation. We** shall use the phrase $F(x)$ *is an antiderivative of f(x)* synonymously with F *is an antiderivative of f*. The domain of an antiderivative will not usually be specified. It will follow from the Fundamental Theorem of Calculus (5.21) that a suitable domain is any closed interval $[a, b]$ on which f is continuous.

Results about derivatives may be used to obtain formulas for antiderivatives, as in the proof of the following.

THEOREM (4.31)

If $F(x)$ and $G(x)$ are antiderivatives of $f(x)$ and $g(x)$, respectively, then

(i) $F(x) + G(x)$ is antiderivative of $f(x) + g(x)$.
(ii) $kF(x)$ is an antiderivative of $kf(x)$ for every real number k.

PROOF By hypothesis, $D_x[F(x)] = f(x)$ and $D_x[G(x)] = g(x)$. Hence

$$D_x[F(x) + G(x)] = D_x[F(x)] + D_x[G(x)] = f(x) + g(x).$$

This proves (i). To prove (ii) we merely note that

$$D_x[kF(x)] = kD_x[F(x)] = kf(x) \quad \bullet \bullet$$

Theorem (4.31) (i) can be extended to any finite sum of functions. This fact may be stated as follows: *An antiderivative of a sum is the sum of the antiderivatives*. As usual, when working with several functions we assume that the domain is restricted to the intersection of the domains of the individual functions. A similar result is true for differences.

Antiderivatives are never unique. Since the derivative of a constant is zero, if $F(x)$ is an antiderivative of $f(x)$, so is $F(x) + C$ for every number C. For example, if $f(x) = 8x^3$, then several antiderivatives are $2x^4 + 7$, $2x^4 - 3$, and $2x^4 + \frac{3}{5}$. The next theorem indicates that functions of this type are the *only* possible antiderivatives of $f(x)$.

THEOREM (4.32)

> If F_1 and F_2 are differentiable functions and $F_1'(x) = F_2'(x)$ for every x in a closed interval $[a, b]$, then $F_2(x) = F_1(x) + C$ for some number C and every x in $[a, b]$.

PROOF If we define the function g by

$$g(x) = F_2(x) - F_1(x),$$

then

$$g'(x) = F_2'(x) - F_1'(x) = 0$$

for every x in $[a, b]$. If x is any number such that $a < x \le b$, then applying the Mean Value Theorem (4.12) to the function g and the closed interval $[a, x]$, there exists a number z in the open interval (a, x) such that

$$g(x) - g(a) = g'(z)(x - a) = 0 \cdot (x - a) = 0.$$

Hence $g(x) = g(a)$ for every x in $[a, b]$. Substitution in the first equation stated in the proof gives us

$$g(a) = F_2(x) - F_1(x).$$

Adding $F_1(x)$ to both sides, we obtain the desired conclusion, with $C = g(a)$.
• •

FIGURE 4.60

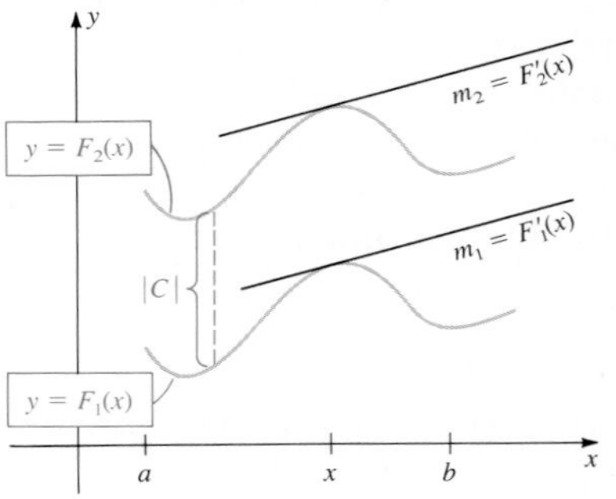

The graphical significance of Theorem (4.32) is illustrated in Figure 4.60. The figure indicates that if the slopes $m_1 = F_1'(x)$ and $m_2 = F_2'(x)$ of the tangent lines to the graphs of F_1 and F_2 are the same for every x in $[a, b]$, then one graph can be obtained from the other by a vertical shift of $|C|$ units. We shall use Theorem (4.32) to prove the following.

THEOREM (4.33)

> If $f'(x) = 0$ for every x in $[a, b]$, then f is a constant function on $[a, b]$.

PROOF The conclusion is geometrically evident, for if f is continuous and if the slope of the tangent line is zero at every point on the graph, we would expect the graph to be a horizontal line (the graph of a constant function). To prove this fact, let F_2 denote f, and let F_1 be defined by $F_1(x) = 0$ for every x. Since $F_1'(x) = 0$ and $F_2'(x) = f'(x) = 0$, we see that $F_1'(x) = F_2'(x)$ for every x in $[a, b]$. Applying Theorem (4.32), there is a number C such that $F_2(x) = F_1(x) + C$; that is, $f(x) = 0 + C$ for every x in $[a, b]$. Thus, f is a constant function. • •

If F_1 and F_2 are antiderivatives of the same function f, then $F_1' = f = F_2'$ and hence, by Theorem (4.32), $F_2(x) = F_1(x) + C$ for some C. In other words, if $F(x)$ is an antiderivative of $f(x)$, then every other antiderivative has the form $F(x) + C$, for an arbitrary constant C, that is, an unspecified real number. We shall refer to $F(x) + C$ as the **most general antiderivative** of $f(x)$.

The following rule will be used many times throughout the remainder of the text.

POWER RULE FOR **(4.34)**
ANTIDIFFERENTIATION

Let r be a rational number such that $r \neq -1$.

The most general antiderivative of x^r is $\dfrac{x^{r+1}}{r+1} + C$

for an arbitrary constant C.

PROOF Let $f(x) = x^r$ and $F(x) = x^{r+1}/(r+1)$. We can show, using the Power Rule for *derivatives*, that $F'(x) = f(x)$. Hence $F(x) + C$ is the most general antiderivative of $f(x)$. • •

It follows from the Power Rule (4.34) and Theorem (4.31) (ii) that the most general antiderivative of ax^r is $ax^{r+1}/(r+1)$ for every real number a and $r \neq -1$. Using Theorem (4.31) (i), we may then find the most general antiderivative of any *sum* of terms of the form ax^r by finding an antiderivative of each term. When finding antiderivatives of sums it is unnecessary to introduce an arbitrary constant for each term, since the constants can be added together to produce one (arbitrary) constant C (see Example 1(d)).

EXAMPLE 1 Find the most general antiderivative of $f(x)$:

(a) $f(x) = 4x^5$ (b) $f(x) = \sqrt[3]{x^2}$

(c) $f(x) = 1/x^2$ (d) $f(x) = 8x^3 - 3x + 7$

SOLUTION The antiderivatives are listed in the following table. Note that when finding an antiderivative of x^r, we *increase* the exponent r by 1, obtaining $r + 1$, and then divide by $r + 1$.

$f(x)$	Most general antiderivative of $f(x)$
$4x^5$	$4\left(\dfrac{x^6}{6}\right) + C = \dfrac{2}{3}x^6 + C$
$\sqrt[3]{x^2} = x^{2/3}$	$\dfrac{x^{5/3}}{5/3} + C = \dfrac{3}{5}x^{5/3} + C$
$\dfrac{1}{x^2} = x^{-2}$	$\dfrac{x^{-1}}{(-1)} + C = -\dfrac{1}{x} + C$
$8x^3 - 3x + 7$	$8\left(\dfrac{x^4}{4}\right) - 3\left(\dfrac{x^2}{2}\right) + 7x + C = 2x^4 - \dfrac{3}{2}x + 7x + C$

•

To avoid algebraic errors, we should check solutions of problems involving antidifferentiation by differentiating the final antiderivative. In each case the given expression should be obtained. To illustrate, in Example 1(d),

$$D_x\left(2x^4 - \tfrac{3}{2}x^2 + 7x + C\right) = D_x(2x^4) - D_x\left(\tfrac{3}{2}x^2\right) + D_x(7x) + D_x(C)$$
$$= 8x^3 - 3x + 7 + 0 = f(x).$$

The next theorem provides formulas for antiderivatives of the sine and cosine functions. The remaining four trigonometric functions will be considered in Chapter 8.

In each statement, $F(x)$ is the most general antiderivative of $f(x)$ and k is any nonzero real number:

(i) If $f(x) = \sin kx$, then $F(x) = -\dfrac{1}{k} \cos kx + C$.

(ii) If $f(x) = \cos kx$, then $F(x) = \dfrac{1}{k} \sin kx + C$.

PROOF It is sufficient to show that $F'(x) = f(x)$. Thus, for part (i) we have

$$F'(x) = D_x \left[-\frac{1}{k} \cos kx + C \right]$$

$$= -\frac{1}{k} D_x (\cos kx) + 0$$

$$= -\frac{1}{k} (-\sin kx)(k) = \sin kx.$$

The proof of (ii) is similar. • •

EXAMPLE 2 Find the most general antiderivative of $f(x)$:

(a) $f(x) = \sin 5x - \cos x$ (b) $f(x) = \sqrt{x} + 4 \cos 3x$.

SOLUTION

(a) Using (4.31) and (4.35), we obtain

$$F(x) = -\tfrac{1}{5} \cos 5x - \sin x + C$$

for an arbitrary constant C.

(b) By the Power Rule (4.34):

An antiderivative of $\sqrt{x} = x^{1/2}$ is $\dfrac{x^{3/2}}{3/2} = \tfrac{2}{3} x^{3/2}$.

By (4.31) (ii) and (4.35) (ii):

An antiderivative of $4 \cos 3x$ is $4 [\tfrac{1}{3} \sin 3x] = \tfrac{4}{3} \sin 3x$.

Thus, the most general antiderivative of $f(x) = \sqrt{x} + 4 \cos 3x$ is

$$F(x) = \tfrac{2}{3} x^{3/2} + \tfrac{4}{3} \sin 3x + C$$

for an arbitrary constant C. •

A **differential equation** involves derivatives of an unknown function. A function f is a **solution** of a differential equation if it satisfies the equation; that is, if substitution of f for the unknown function produces a true statement. To **solve** a differential equation means to find all solutions. Sometimes, in addition to the differential equation, we may know a certain value of f, called an **initial condition,** as illustrated in the next example.

EXAMPLE 3 Solve the differential equation $f'(x) = 6x^2 + x - 5$ with initial condition $f(0) = 2$.

SOLUTION From our discussion of antiderivatives,

$$f(x) = 2x^3 + \tfrac{1}{2}x^2 - 5x + C$$

for some number C. Letting $x = 0$ and using the given initial condition, we obtain

$$f(0) = 0 + 0 - 0 + C = 2$$

and hence $C = 2$. Consequently, the solution f of the differential equation with the given initial condition is

$$f(x) = 2x^3 + \tfrac{1}{2}x^2 - 5x + 2. \quad \bullet$$

The term *differential equation* stems from the fact that differentials may be used in place of derivatives. Thus, the differential equation in Example 3 could be written

$$\frac{dy}{dx} = 6x^2 + x - 5 \quad \text{or} \quad dy = (6x^2 + x - 5)\, dx$$

for $y = f(x)$.

EXAMPLE 4 Solve the differential equation

$$f''(x) = 5 \cos 3x + 2 \sin 4x$$

subject to the initial conditions $f(0) = 1$ and $f'(0) = 3$.

SOLUTION Since $f'(x)$ is an antiderivative of $f''(x)$, we have, by (4.34) and (4.35),

$$f'(x) = 5[\tfrac{1}{3} \sin 3x] + 2[-\tfrac{1}{4} \cos 4x] + C$$
$$= \tfrac{5}{3} \sin 3x - \tfrac{1}{2} \cos 4x + C$$

for a real number C. Letting $x = 0$ and using the initial condition $f'(0) = 3$ gives us

$$3 = \tfrac{5}{3}(0) - \tfrac{1}{2}(1) + C \quad \text{or} \quad C = 3 + \tfrac{1}{2} = \tfrac{7}{2}.$$

Consequently,

$$f'(x) = \tfrac{5}{3} \sin 3x - \tfrac{1}{2} \cos 4x + \tfrac{7}{2}.$$

We now find an antiderivative of $f'(x)$. Thus

$$f(x) = \tfrac{5}{3}[-\tfrac{1}{3} \cos 3x] - \tfrac{1}{2}[\tfrac{1}{4} \sin 4x] + \tfrac{7}{2}x + D$$
$$= -\tfrac{5}{9} \cos 3x - \tfrac{1}{8} \sin 4x + \tfrac{7}{2}x + D$$

for some number D. Letting $x = 0$ and using the condition $f(0) = 1$, we obtain

$$1 = -\tfrac{5}{9}(1) - \tfrac{1}{8}(0) + \tfrac{7}{2}(0) + D \quad \text{or} \quad D = 1 + \tfrac{5}{9} = \tfrac{14}{9}.$$

Hence a solution of the differential equation is

$$f(x) = -\tfrac{5}{9} \cos 3x - \tfrac{1}{8} \sin 4x + \tfrac{7}{2}x + \tfrac{14}{9} \quad \bullet$$

If a point P is moving on a coordinate line, then its position function s is an antiderivative of its velocity function v, that is, $s'(t) = v(t)$. Similarly, since $v'(t) = a(t)$, the velocity function is an antiderivative of the acceleration function. If we know the velocity or acceleration function, then given sufficient initial conditions, we can determine the position function.

EXAMPLE 5 A motorboat moves away from a dock along a straight line, with an acceleration at time t given by $a(t) = 12t - 4$ ft/sec^2. If, at time $t = 0$, the boat had a velocity of 8 ft/sec and was 15 feet from the dock, find its distance $s(t)$ from the dock at the end of t seconds.

SOLUTION Since $v'(t) = 12t - 4$ we obtain, by antidifferentiation,

$$v(t) = 6t^2 - 4t + C$$

for some number C. Letting $t = 0$ and using the fact that $v(0) = 8$ gives us $8 = 0 - 0 + C$ and hence, $C = 8$. Thus

$$v(t) = 6t^2 - 4t + 8$$

or, equivalently,

$$s'(t) = 6t^2 - 4t + 8.$$

The most general antiderivative of $s'(t)$ is

$$s(t) = 2t^3 - 2t^2 + 8t + D$$

for some number D. Letting $t = 0$ and using the fact that $s(0) = 15$ leads to $15 = 0 - 0 + 0 + D$. Consequently, $D = 15$ and

$$s(t) = 2t^3 - 2t^2 + 8t + 15 \quad \bullet$$

An object on or near the surface of the earth is acted upon by a force—gravity—that produces a constant acceleration, denoted by g. The approximation to g that is employed for most problems is 32 ft/sec^2, or 980 cm/sec^2. The use of this important physical constant is illustrated in the following example.

EXAMPLE 6 A stone is thrown vertically upward from a position 144 feet above the ground with an initial velocity of 96 ft/sec. If air resistance is neglected, find its distance above the ground after t seconds. For what length of time does the stone rise? When, and with what velocity, does it strike the ground?

SOLUTION The motion of the stone may be represented by a point moving on a vertical coordinate line l with origin at ground level and positive direction upward (see Figure 4.61). The distance above the ground at time t is $s(t)$, and the initial conditions are $s(0) = 144$ and $v(0) = 96$. Since the velocity is decreasing, $v'(t) < 0$; that is, the acceleration is negative. Hence, by the remarks preceding this example,

$$a(t) = -32.$$

Since v is an antiderivative of a,

$$v(t) = -32t + C$$

for some number C. Substituting 0 for t and using the fact that $v(0) = 96$ gives us $96 = 0 + C = C$ and, consequently,

$$v(t) = -32t + 96.$$

Since $s'(t) = v(t)$ we obtain, by antidifferentiation,

$$s(t) = -16t^2 + 96t + D$$

for some number D. Letting $t = 0$ and using the fact that $s(0) = 144$ leads to $144 = 0 + 0 + D = D$. It follows that the distance from the ground to the

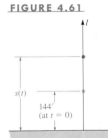

FIGURE 4.61

l

$s(t)$

144'
(at $t = 0$)

stone at time t is given by

$$s(t) = -16t^2 + 96t + 144.$$

The stone will rise until $v(t) = 0$, that is, until $-32t + 96 = 0$, or $t = 3$. The stone will strike the ground when $s(t) = 0$, or $-16t^2 + 96t + 144 = 0$. Equivalently, $t^2 - 6t - 9 = 0$. Applying the quadratic formula, $t = 3 \pm 3\sqrt{2}$. The solution $3 - 3\sqrt{2}$ is extraneous, since t is nonnegative. Hence the stone strikes the ground after $3 + 3\sqrt{2}$ sec. The velocity at that time is

$$v(3 + 3\sqrt{2}) = -32(3 + 3\sqrt{2}) + 96$$
$$= -96\sqrt{2} \approx -135.8 \text{ ft/sec} \quad \bullet$$

In economic applications, if a marginal function is known, then we may sometimes use antidifferentiation to find the function, as illustrated in the next example.

EXAMPLE 7 A manufacturer finds that the marginal cost (in dollars) associated with the production of x units of a photocopier component is given by $30 - 0.02x$. If the cost of producing one unit is $35, find the cost function and the cost of producing 100 units.

SOLUTION If C is the cost function, then the marginal cost is the rate of change of C with respect to x, that is

$$C'(x) = 30 - 0.02x.$$

Antidifferentiation gives us

$$C(x) = 30x - 0.01x^2 + K$$

for some real number K. Letting $x = 1$ and using $C(1) = 35$ we obtain

$$35 = 30 - 0.01 + K$$

and hence $K = 5.01$. Consequently,

$$C(x) = 30x - 0.01x^2 + 5.01.$$

In particular, the cost of producing 100 units is

$$C(100) = 3000 - 100 + 5.01 = \$2,905.01 \quad \bullet$$

EXERCISES 4.7

Exer. 1–28: Find the most general antiderivative of the function.

1 $f(x) = 9x^2 - 4x + 3$

2 $f(x) = 4x^2 - 8x + 1$

3 $f(x) = 2x^3 - x^2 + 3x - 7$

4 $f(x) = 10x^4 - 6x^3 + 5$

5 $f(x) = \dfrac{1}{x^3} - \dfrac{3}{x^2}$

6 $f(x) = \dfrac{4}{x^7} - \dfrac{7}{x^4} + x$

7 $f(x) = 3\sqrt{x} + \dfrac{1}{\sqrt{x}}$

8 $f(x) = \sqrt{x^3} - \frac{1}{2}x^{-2} + 5$

9 $f(x) = \dfrac{6}{\sqrt[3]{x}} - \dfrac{\sqrt[3]{x}}{6} + 7$

10 $f(x) = 3x^5 - x^{5/3}$

11 $f(x) = 2x^{5/4} + 6x^{1/4} + 3x^{-4}$

12 $f(x) = \left(x - \dfrac{1}{x}\right)^2$

13 $f(x) = 3 \sin 4x$

14 $f(x) = 4 \cos \frac{1}{2}x$

15 $f(x) = 2x^2 - 3x + 7 \sin 5x$

16 $f(x) = (4 + 3x^2 \cos 4x)/x^2$

17 $f(x) = 2 \cos 3x - 3 \sin 2x$

18 $f(x) = (\sin x + \cos x)^2$ (*Hint:* $\sin 2\theta = 2 \sin \theta \cos \theta$)

19 $f(x) = \dfrac{\sin 4x}{\cos 2x}$ (*Hint:* $\sin 2\theta = 2 \sin \theta \cos \theta$)

20 $f(x) = \cos 3x \tan 3x$

21 $f(x) = (3x - 1)^2$

22 $f(x) = (2x - 5)(3x + 1)$

23 $f(x) = \dfrac{8x - 5}{\sqrt[3]{x}}$

24 $f(x) = \dfrac{2x^2 - x + 3}{\sqrt{x}}$

25 $f(x) = \sqrt[5]{32x^4}$

26 $f(x) = \sqrt[3]{64x^5}$

27 $f(x) = \dfrac{x^3 - 1}{x - 1}$

28 $f(x) = \dfrac{x^3 + 3x^2 - 9x - 2}{x - 2}$

Exer. 29–34: Solve the differential equation subject to the given conditions.

29 $f'(x) = 12x^2 - 6x + 1$, $f(1) = 5$

30 $f'(x) = 9x^2 + x - 8$, $f(-1) = 1$

31 $f''(x) = 4x - 1$, $f'(2) = -2$, $f(1) = 3$

32 $f'''(x) = 6x$, $f''(0) = 2$, $f'(0) = -1$, $f(0) = 4$

33 $f'''(t) = \dfrac{\pi^3}{2} \sin \dfrac{\pi}{4} t$, $f(0) = 32$, $f'(0) = 0$, $f''(0) = 4 - 2\pi^2$

34 $f''(x) = 4 \sin 2x + 16 \cos 4x$, $f(0) = 6$, $f'(0) = 1$

35 A point moves on a coordinate line with $a(t) = 2 - 6t$. If the initial conditions are $v(0) = -5$ and $s(0) = 4$, find $s(t)$.

36 Rework Exercise 35 if $a(t) = 3t^2$, $v(0) = 20$, and $s(0) = 5$.

37 A charged particle is moving on a coordinate line in a magnetic field such that its velocity (in cm/sec) at time t is given by $v(t) = \frac{1}{2} \sin (3t - \frac{1}{4}\pi)$. Show that the motion is simple harmonic (see page 134).

38 The acceleration of a particle that is moving on a coordinate line is given by $a(t) = k \cos (\omega t + \varphi)$ for constants $k, \omega,$ and φ and time t (in seconds). Show that the motion is simple harmonic (see page 134).

39 A projectile is fired vertically upward from ground level with a velocity of 1600 ft/sec. If air resistance is neglected, find its distance $s(t)$ above ground at time t. What is its maximum height?

40 An object is dropped from a height of 1000 feet. Neglecting air resistance, find the distance it falls in t seconds. What is its velocity at the end of 3 seconds? When will it strike the ground?

41 A stone is thrown directly downward from a height of 96 feet with an initial velocity of 16 ft/sec. Find (a) its distance above the ground after t seconds; (b) when it will strike the ground; and (c) the velocity at which it strikes the ground.

42 A gravitational constant for objects near the surface of the moon is 5.3 ft/sec^2.

(a) If an astronaut on the moon throws a stone directly upward with an initial velocity of 60 ft/sec, find its maximum altitude.

(b) If, after returning to Earth, the astronaut throws the same stone directly upward with the same initial velocity, find the maximum altitude.

43 If a projectile is fired vertically upward from a height of s_0 feet above the ground with a velocity of v_0 ft/sec, prove that if air resistance is neglected, its distance $s(t)$ above the ground after t seconds is given by $s(t) = -\frac{1}{2}gt^2 + v_0 t + s_0$ for a gravitational constant g.

44 A ball rolls down an inclined plane with an acceleration of 2 ft/sec^2. If the ball is given no initial velocity how far will it roll in t seconds? What initial velocity must be given for the ball to roll 100 feet in 5 seconds?

45 If an automobile starts from rest, what constant acceleration will enable it to travel 500 feet in 10 seconds?

46 If a car is traveling at a speed of 60 mi/hr, what constant (negative) acceleration will enable it to stop in 9 seconds?

47 If C and F denote Celsius and Fahrenheit temperature readings, then the rate of change of F with respect to C is given by $dF/dC = \frac{9}{5}$. If $F = 32$ when $C = 0$, use antidifferentiation to obtain a general formula for F in terms of C.

48 The rate of change of the temperature T of a solution (in °C) is given by $dT/dt = \frac{1}{4}t + 10$ for time t (in minutes). If $T = 5$ °C at $t = 0$, find a formula for T at time t.

49 The volume V of a balloon is changing with respect to time t at a rate given by $dV/dt = 3\sqrt{t} + \frac{1}{4}t$ ft^3/sec. If, at $t = 4$, the volume is 20 ft^3, express V as a function of t.

50 Suppose the slope of the tangent line at any point P on the graph of an equation equals the square of the x-coordinate of P. Find the equation if the graph contains (a) the origin; (b) the point $(3, 6)$; (c) the point $(-1, 1)$. Sketch the graph in each case.

51 A small country has natural gas reserves of 100 billion ft^3. If $A(t)$ denotes the total amount of natural gas consumed after t years, then dA/dt is the *rate of consumption*. If the rate of consumption is predicted to be $5 + 0.01t$ billion ft^3/year, in approximately how many years will the country's natural gas reserves be depleted?

52 Refer to Exercise 51. Based on U.S. Department of Energy statistics, the rate of consumption of gasoline in the U.S. (in billions of gallons per year) is approximated by $dA/dt = 2.74 - 0.11t - 0.01t^2$ with $t = 0$ corresponding to the year 1980. Estimate the number of gallons of gasoline consumed in the U.S. between 1980 and 1984.

53 A reservoir supplies water to a community. In summer the demand A for water (in ft^3/day) changes according to the formula $dA/dt = 4000 + 2000 \sin (\frac{1}{90}\pi t)$ for time t (in days) and $t = 0$ corresponding to the beginning of summer. Estimate the total water consumption during 90 days of summer.

54 The pumping action of the heart consists of the systolic phase, in which blood rushes from the left ventricle into the aorta, and the diastolic phase, during which the heart muscle relaxes. The graph shown in the figure is often used to model one complete cycle of the process. For a particular individual, the systolic phase lasts $\frac{1}{4}$ second and has a maximum flow rate dV/dt of 8 liters/minute for the volume V of blood in the heart and time t.

(a) Show that $dV/dt = 8 \sin (240\pi t)$ liter/minute.

(b) Estimate the total amount of blood pumped into the aorta during a systolic phase.

EXERCISE 54

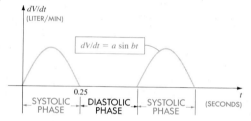

55 The rhythmic process of breathing consists of alternating periods of inhaling and exhaling. One complete cycle nor-

mally takes place every 5 seconds. If V denotes the volume of air in the lungs at time t, then dV/dt is the flow rate.

(a) If the maximum flow rate is 0.6 liters/sec, find a formula $dV/dt = a \sin bt$ that fits the given information.

(b) Using the function in part (a), estimate the amount of air inhaled during one cycle.

56 Many animal populations fluctuate over ten-year cycles. Suppose that the rate of growth of a rabbit population is given by $dN/dt = 1000 \cos (\frac{1}{5}\pi t)$ rabbits/year, with N denoting the number in the population at time t (in years) and $t = 0$ corresponding to the beginning of a cycle. If the population after 5 years is estimated to be 3000 rabbits, find a formula for N at time t. Estimate the maximum population.

57 A sportswear manufacturer determines that the marginal cost of producing x warm-up suits is given in dollars by $20 - 0.015x$. If the cost of producing one suit is $25, find the cost function and the cost of producing 50 suits.

58 If the marginal cost function of a product is given by $2/x^{1/3}$ and if the cost of producing 8 units is $20, find the cost function and the cost of producing 64 units.

59 If the marginal revenue function of a product is given by $x^2 - 6x + 15$, find the revenue function and the marginal demand function.

60 Rework Exercise 59 if the marginal revenue function is given by $x + (4/\sqrt{x})$.

4.8 REVIEW

Define or discuss each of the following.

1 Increasing or decreasing function
2 Absolute maximum or absolute minimum of a function
3 Local maximum or local minimum of a function
4 Critical numbers of a function
5 Endpoint extrema of a function
6 Rolle's Theorem
7 The Mean Value Theorem
8 The First Derivative Test
9 Upward or downward concavity

10 Tests for concavity
11 Point of inflection
12 The Second Derivative Test
13 $\lim_{x \to \infty} f(x) = L$; $\lim_{x \to -\infty} f(x) = L$
14 $\lim_{x \to a} f(x) = \infty$; $\lim_{x \to a} f(x) = -\infty$
15 Vertical and horizontal asymptotes
16 Antiderivative of a function
17 Power Rule for Antidifferentiation
18 Differential equation

EXERCISES 4.8

1 If $f(x) = x^3 + x^2 + x + 1$, find a number c that satisfies the conclusion of the Mean Value Theorem on the interval $[0, 4]$.

2 The posted speed limit on a 125-mile toll highway is 65 mi/hr. As an automobile enters the toll road, the driver

is issued a ticket that is printed with the exact time. If the driver completes the trip in 1 hour 40 minutes or less, a speeding citation is issued when the toll is paid. Use the Mean Value Theorem to explain why this citation is justified.

Exer. 3–5: Use the First Derivative Test to find the local extrema of f. Describe the intervals on which f is increasing or decreasing, and sketch the graph of f.

3 $f(x) = -4x^3 + 9x^2 + 12x$ **4** $f(x) = 1/(1 + x^2)$

5 $f(x) = (4 - x)x^{1/3}$

Exer. 6–8: Use the Second Derivative Test to find the local extrema of f. Discuss concavity, find x-coordinates of points of inflection, and sketch the graph of f.

6 $f(x) = -x^3 + 4x^2 - 3x$ **7** $f(x) = 1/(1 + x^2)$

8 $f(x) = 40x^3 - x^6$

9 If $f(x) = 2 \sin x - \cos 2x$, find the local extrema, and sketch the graph of f for $0 \le x \le 4\pi$.

10 If $f(x) = 2 \sin x - \cos 2x$, find equations of the tangent and normal lines to the graph of f at the point $(\pi/6, \frac{1}{2})$.

11 A man wishes to put a fence around a rectangular field, and then subdivide this field into three smaller rectangular plots by placing two fences parallel to one of the sides. If he can afford only 1000 yards of fencing, what dimensions will give him the maximum area?

12 An open rectangular storage shelter, consisting of two vertical sides, 4 feet wide, and a flat roof, is to be attached to an existing structure as illustrated in the figure. The flat roof is made of tin and costs $5 per ft². The two sides are made of plywood costing $2 per ft². If $400 is available for construction, what dimensions will maximize the volume of the shelter?

EXERCISE 12

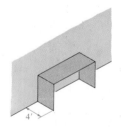

13 A V-shaped water gutter is to be constructed from two rectangular sheets of metal 10 inches wide. Find the angle between the sheets that will maximize the carrying capacity of the gutter.

14 Find the altitude of the right circular cylinder of maximum curved surface area that can be inscribed in a sphere of radius a.

15 The interior of a half-mile race track consists of a rectangle with semicircles at two opposite ends. Find the dimensions that will maximize the area of the rectangle.

16 A cable television firm presently serves 5000 households and charges $20 per month. A marketing survey indicates that each decrease of $1 in monthly charge will result in 500 new customers. Find the monthly charge that will result in maximum monthly revenue.

17 A wire 5 feet long is to be cut into two pieces. One piece is to be bent into the shape of a circle and the other into the shape of a square. Where should the wire be cut so that the sum of the areas of the circle and square is (a) a maximum; (b) a minimum?

18 An electronics company estimates that the cost of producing x calculators per day is $C(x) = 500 + 6x + 0.02x^2$. If each calculator is sold for $18, find (a) the revenue function; (b) the profit function; (c) the daily production that will maximize the profit; (d) the maximum daily profit.

Exer. 19–26: Find the limit, if it exists.

19 $\lim\limits_{x \to -\infty} \dfrac{(2x - 5)(3x + 1)}{(x + 7)(4x - 9)}$ **20** $\lim\limits_{x \to \infty} \dfrac{2x + 11}{\sqrt{x + 1}}$

21 $\lim\limits_{x \to -\infty} \dfrac{6 - 7x}{(3 + 2x)^4}$ **22** $\lim\limits_{x \to -3} \sqrt[3]{\dfrac{x + 3}{x^3 + 27}}$

23 $\lim\limits_{x \to 2/3^+} \dfrac{x^2}{4 - 9x^2}$ **24** $\lim\limits_{x \to 3/5^-} \dfrac{1}{5x - 3}$

25 $\lim\limits_{x \to 0^+} \left(\sqrt{x} - \dfrac{1}{\sqrt{x}} \right)$ **26** $\lim\limits_{x \to 1} \dfrac{x - 1}{\sqrt{(x - 1)^2}}$

Exer. 27–30: Find the horizontal, vertical, and oblique asymptotes, and sketch the graph of f.

27 $f(x) = \dfrac{3x^2}{9x^2 - 25}$ **28** $f(x) = \dfrac{x^2}{(x - 1)^2}$

29 $f(x) = \dfrac{x^2 + 2x - 8}{x + 3}$ **30** $f(x) = \dfrac{x^4 - 16}{x^3}$

31 In biochemistry, the general threshold-response curve is given by $R = kS^n/(S^n + a^n)$, where R is the chemical response that corresponds to a concentration S of a substance for positive constants k, n and a. An example is the rate R at which the liver removes alcohol from the bloodstream when the concentration of alcohol is S. Show that R is an increasing function of S and that $R = k$ is a horizontal asymptote for the curve.

32 A small office building is to contain 500 ft² of floor space. The simple floor plans are shown in the figure. If the walls cost $100 per running foot and if the wall space above the doors is disregarded,

(a) show that the cost $C(x)$ of the walls is
$$C(x) = 100[3x - 6 + (1000/x)].$$

EXERCISE 34

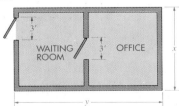

(b) find the vertical and oblique asymptotes, and sketch the graph of $C(x)$ for $x > 0$.

(c) find the design that minimizes the cost.

Exer. 33–38: Find the most general antiderivative of f.

33 $f(x) = \dfrac{8x^2 - 4x + 5}{x^4}$

34 $f(x) = 3x^5 + 2x^3 - x$

35 $f(x) = 100$

36 $f(x) = x^{3/5}(2x - \sqrt{x})$

37 $f(x) = (2x + 1)^3$

38 $f(t) = 3 \cos 2\pi t - 5 \sin 4\pi t$

39 Solve the differential equation $f''(x) = x^{1/3} - 5$ if $f'(1) = 2$ and $f(1) = -8$.

40 A stone is thrown directly downward from a height of 900 feet with an initial velocity of 30 ft/sec. Find its distance above ground after t seconds. What is its velocity after 5 seconds? When will it strike the ground?

CHAPTER 5

THE DEFINITE INTEGRAL

Calculus consists of two main parts, *differential calculus* and *integral calculus*. Differential calculus is based upon the derivative. In this chapter we define the concept that is the basis for integral calculus: the *definite integral*. One of the most important results we shall discuss is the *Fundamental Theorem of Calculus*. This theorem demonstrates that differential and integral calculus are very closely related.

5.1 AREA

In our development of the definite integral we shall employ sums of many numbers. To express such sums compactly, it is convenient to use **summation notation.** To illustrate, given a collection of numbers $\{a_1, a_2, \ldots, a_n\}$, the symbol $\sum_{k=1}^{n} a_k$ represents their sum, that is,

(5.1)
$$\sum_{k=1}^{n} a_k = a_1 + a_2 + a_3 + \cdots + a_n$$

The Greek capital letter Σ (sigma) indicates a sum, and a_k represents the kth term. The letter k is the **index of summation,** or the **summation variable,** and assumes successive integral values. The integers 1 and n indicate the extreme values of the summation variable.

EXAMPLE 1 Find $\displaystyle\sum_{k=1}^{4} k^2(k-3)$.

SOLUTION In this case, $a_k = k^2(k-3)$. To find the sum we substitute 1, 2, 3, and 4 for k and add the resulting terms. Thus,

$$\sum_{k=1}^{4} k^2(k-3) = 1^2(1-3) + 2^2(2-3) + 3^2(3-3) + 4^2(4-3)$$

$$= (-2) + (-4) + 0 + 16 = 10 \qquad \bullet$$

Letters other than k can be used for the summation variable. To illustrate,

$$\sum_{k=1}^{4} k^2(k-3) = \sum_{i=1}^{4} i^2(i-3) = \sum_{j=1}^{4} j^2(j-3) = 10.$$

If $a_k = c$ for every k, then

$$\sum_{k=1}^{2} a_k = a_1 + a_2 = c + c = 2c = \sum_{k=1}^{2} c$$

$$\sum_{k=1}^{3} a_k = a_1 + a_2 + a_3 = c + c + c = 3c = \sum_{k=1}^{3} c.$$

In general, the following result is true for every positive integer n.

THEOREM (5.2)

$$\sum_{k=1}^{n} c = nc$$

The domain of the summation variable does not have to begin at 1. For example,

$$\sum_{k=4}^{8} a_k = a_4 + a_5 + a_6 + a_7 + a_8.$$

EXAMPLE 2 Find $\displaystyle\sum_{k=0}^{3} \frac{2^k}{(k+1)}$.

SOLUTION $\displaystyle\sum_{k=0}^{3} \frac{2^k}{(k+1)} = \frac{2^0}{(0+1)} + \frac{2^1}{(1+1)} + \frac{2^2}{(2+1)} + \frac{2^3}{(3+1)}$

$$= 1 + 1 + \frac{4}{3} + 2 = \frac{16}{3}$$

THEOREM (5.3)

If n is any positive integer and $\{a_1, a_2, \ldots, a_n\}$ and $\{b_1, b_2, \ldots, b_n\}$ are sets of real numbers, then

(i) $\displaystyle\sum_{k=1}^{n} (a_k + b_k) = \sum_{k=1}^{n} a_k + \sum_{k=1}^{n} b_k$.

(ii) $\displaystyle\sum_{k=1}^{n} ca_k = c\left(\sum_{k=1}^{n} a_k\right)$ for every real number c.

(iii) $\displaystyle\sum_{k=1}^{n} (a_k - b_k) = \sum_{k=1}^{n} a_k - \sum_{k=1}^{n} b_k$.

PROOF To prove (i) we begin with

$$\sum_{k=1}^{n} (a_k + b_k) = (a_1 + b_1) + (a_2 + b_2) + (a_3 + b_3) + \cdots + (a_n + b_n).$$

Rearranging terms on the right,

$$\sum_{k=1}^{n} (a_k + b_k) = (a_1 + a_2 + a_3 + \cdots + a_n) + (b_1 + b_2 + b_3 + \cdots + b_n)$$

$$= \sum_{k=1}^{n} a_k + \sum_{k=1}^{n} b_k.$$

For (ii),

$$\sum_{k=1}^{n} (ca_k) = ca_1 + ca_2 + ca_3 + \cdots + ca_n$$

$$= c(a_1 + a_2 + a_3 + \cdots + a_n) = c\left(\sum_{k=1}^{n} a_k\right).$$

The proof of (iii) is left as an exercise. • •

The definition of the definite integral is closely related to the areas of certain regions in a coordinate plane. We can easily calculate the area if the region is bounded by lines. For example, the area of a rectangle is the product of its length and width. The area of a triangle is one-half the product of an altitude and the corresponding base. The area of any polygon can be found by subdividing it into triangles.

To find areas of more complicated regions, whose boundaries involve graphs of functions, we must introduce a limiting process and then use methods of calculus. In particular, let us consider a region R in a coordinate plane, bounded by the vertical lines $x = a$ and $x = b$, by the x-axis, and by the graph of a function f that is continuous and nonnegative on the closed interval $[a, b]$. A region of this type is illustrated in Figure 5.1. Since $f(x) \geq 0$ for every x in $[a, b]$, no part of the graph lies below the x-axis. For convenience, we shall refer to R as **the region under the graph of f from a to b.** Our objective is to define the area of R.

FIGURE 5.1

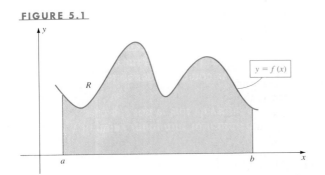

FIGURE 5.2

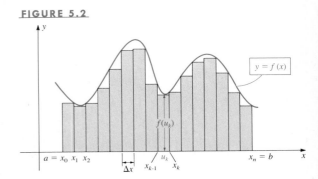

If n is any positive integer, let us begin by dividing the interval $[a, b]$ into n subintervals, all having the same length $(b - a)/n$. We can do this by choosing numbers $x_0, x_1, x_2, \ldots, x_n$ with $a = x_0$, $b = x_n$, and

$$x_k - x_{k-1} = \frac{b - a}{n}$$

for $k = 1, 2, \ldots, n$. If the length $(b - a)/n$ of each subinterval is denoted by Δx, then for each k,

$$\Delta x = x_k - x_{k-1} \quad \text{and} \quad x_k = x_{k-1} + \Delta x,$$

as illustrated in Figure 5.2.

Note that

$$x_0 = a, \quad x_1 = a + \Delta x, \quad x_2 = a + 2\,\Delta x, \quad \ldots,$$

$$x_k = a + k\,\Delta x, \quad \ldots, \quad x_n = a + n\,\Delta x = b.$$

Since f is continuous on each subinterval $[x_{k-1}, x_k]$, f takes on a minimum value at some number u_k in $[x_{k-1}, x_k]$ (see Theorem (4.3)). For each k, let us construct a rectangle of width $\Delta x = x_k - x_{k-1}$ and height equal to the minimum distance $f(u_k)$ from the x-axis to the graph of f, as shown in Figure 5.2. The area of the kth rectangle is $f(u_k)\,\Delta x$. The boundary of the region formed by the totality of these rectangles is the **inscribed rectangular polygon** associated with the subdivision of $[a, b]$ into n subintervals. The area of this inscribed polygon is the sum of the areas of the n rectangles, that is,

$$f(u_1)\,\Delta x + f(u_2)\,\Delta x + \cdots + f(u_n)\,\Delta x.$$

Using summation notation,

$$\text{Area of inscribed rectangular polygon} = \sum_{k=1}^{n} f(u_k)\,\Delta x$$

for the minimum value for $f(u_k)$ of f on $[x_{k-1}, x_k]$.

Referring to Figure 5.3, it appears that if n is very large or, equivalently, if Δx is very small, then the sum of the rectangular areas should be close to the area of the region R. From an intuitive viewpoint, if there exists a number A such that the sum $\sum_{k=1}^{n} f(u_k)\,\Delta x$ gets closer and closer to A as Δx gets closer and closer to 0 (but $\Delta x \neq 0$), then we should call A the **area** of R and write

$$A = \lim_{\Delta x \to 0} \sum_{k=1}^{n} f(u_k)\,\Delta x.$$

FIGURE 5.3

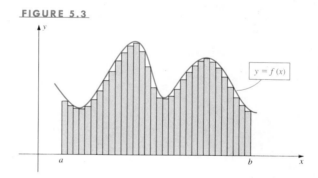

The meaning of this *limit of sums* is not the same as that for limit of a function, introduced in Chapter 2. To eliminate the phrase *closer and closer* and arrive at a satisfactory definition of A, let us take a slightly different point of view. If A denotes the area of R, then the difference

$$A - \sum_{k=1}^{n} f(u_k)\,\Delta x$$

is the area of the *unshaded* portion in Figures 5.2 or 5.3 that lies under the graph of f and over the inscribed rectangular polygon. This number may be regarded as the error involved in using the area of the inscribed rectangular polygon to approximate A. It appears that we should be able to make this error as small as desired by choosing the width Δx of the rectangles sufficiently small. This is the motivation for the following definition of the area A of R. The notation is the same as that used in the preceding discussion.

DEFINITION (5.4)

Let f be continuous and nonnegative on $[a, b]$. Let A be a real number and let $f(u_k)$ be the minimum value of f on $[x_{k-1}, x_k]$. The statement

$$A = \lim_{\Delta x \to 0} \sum_{k=1}^{n} f(u_k)\, \Delta x$$

means that for every $\varepsilon > 0$ there corresponds a $\delta > 0$ such that if $0 < \Delta x < \delta$, then

$$A - \sum_{k=1}^{n} f(u_k)\, \Delta x < \varepsilon$$

If A is the indicated limit and we let $\varepsilon = 10^{-9}$, then Definition (5.4) states that by using rectangles of sufficiently small width Δx, the difference between A and the area of the inscribed polygon is less than one-billionth of a square unit. Similarly, if $\varepsilon = 10^{-12}$ we can make this difference less than one-trillionth of a square unit. In general, the difference can be made less than *any* preassigned ε.

If f is continuous on $[a, b]$, then (as shown in more advanced texts) a number A satisfying Definition (5.4) actually exists. We shall call A **the area under the graph of f from a to b.**

The area A may also be obtained by means of **circumscribed rectangular polygons** of the type illustrated in Figure 5.4. In this case we select the number v_k in each interval $[x_{k-1}, x_k]$ such that $f(v_k)$ is the *maximum* value of f on $[x_{k-1}, x_k]$. Figure 5.5 illustrates a case with n relatively large (and Δx correspondingly small).

FIGURE 5.4

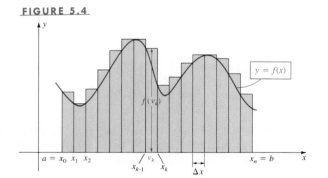

FIGURE 5.5

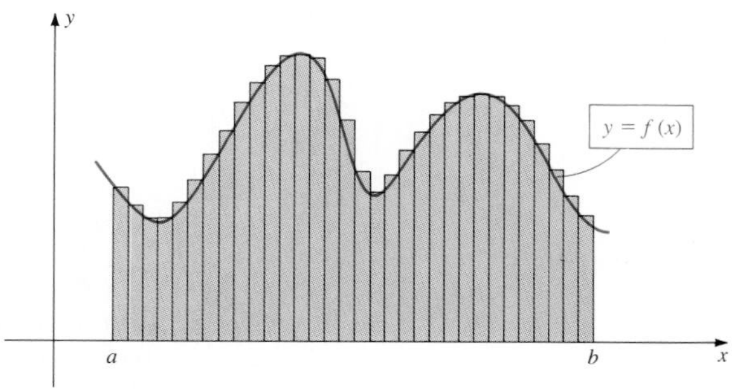

Using summation notation,

$$\text{Area of circumscribed rectangular polygon} = \sum_{k=1}^{n} f(v_k)\, \Delta x.$$

The limit of these sums as $\Delta x \to 0$ is defined as in (5.4). The only change is that we use

$$\sum_{k=1}^{n} f(v_k)\, \Delta x - A < \varepsilon,$$

since we want this difference to be nonnegative. It can be proved that the same number A is obtained using either inscribed or circumscribed rectangles.

The following formulas will be useful in some examples of Definition (5.4). They may be proved by mathematical induction (see Appendix I).

(5.5)

(i) $\displaystyle\sum_{k=1}^{n} k = 1 + 2 + \cdots + n = \frac{n(n+1)}{2}$

(ii) $\displaystyle\sum_{k=1}^{n} k^2 = 1^2 + 2^2 + \cdots + n^2 = \frac{n(n+1)(2n+1)}{6}$

(iii) $\displaystyle\sum_{k=1}^{n} k^3 = 1^3 + 2^3 + \cdots + n^3 = \left[\frac{n(n+1)}{2}\right]^2$

The next two examples provide specific illustrations of how summation properties may be used with Definition (5.4) to find the areas of certain regions in a coordinate plane.

EXAMPLE 3 If $f(x) = 16 - x^2$, find the area of the region under the graph of f from 0 to 3.

FIGURE 5.6

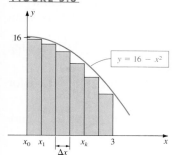

$y = 16 - x^2$

SOLUTION The region is illustrated in Figure 5.6 (with different scales on the x- and y-axes). If the interval $[0, 3]$ is divided into n equal subintervals, then the length Δx of each subinterval is $3/n$. Employing the notation used in Figure 5.2, with $a = 0$ and $b = 3$,

$$x_0 = 0, \quad x_1 = \Delta x, \quad x_2 = 2\,\Delta x, \quad \ldots, \quad x_k = k\,\Delta x, \quad \ldots, \quad x_n = n\,\Delta x = 3.$$

Since $\Delta x = 3/n$,

$$x_k = k\,\Delta x = k\frac{3}{n} = \frac{3k}{n}.$$

Since f is decreasing on $[0, 3]$, the number u_k in $[x_{k-1}, x_k]$ at which f takes on its minimum value is always the right-hand endpoint x_k of the subinterval, that is, $u_k = x_k = 3k/n$. Thus,

$$f(u_k) = f\left(\frac{3k}{n}\right) = 16 - \left(\frac{3k}{n}\right)^2 = 16 - \frac{9k^2}{n^2}$$

and the summation in Definition (5.4) is

$$\sum_{k=1}^{n} f(u_k)\,\Delta x = \sum_{k=1}^{n} \left(16 - \frac{9k^2}{n^2}\right)\left(\frac{3}{n}\right)$$

$$= \sum_{k=1}^{n} \left(\frac{48}{n} - \frac{27k^2}{n^3}\right).$$

Using Theorem (5.3) and (5.2), the last sum may be written as follows:

$$\sum_{k=1}^{n} \frac{48}{n} - \sum_{k=1}^{n} \frac{27k^2}{n^3} = \left(\frac{48}{n}\right)n - \frac{27}{n^3} \sum_{k=1}^{n} k^2$$

Next, applying (5.5) (ii) we obtain

$$\sum_{k=1}^{n} f(u_k)\,\Delta x = 48 - \frac{27}{n^3}\left[\frac{n(n+1)(2n+1)}{6}\right]$$

$$= 48 - \frac{9}{2n^3}\left[2n^3 + 3n^2 + n\right].$$

To find the area of the region, we let Δx approach 0. Since $\Delta x = 3/n$, we can accomplish this by letting n increase without bound. Although our discussion of limits involving infinity in Section 4.6 was concerned with a real variable x, a similar discussion can be given for the variable n for any integer n. Assuming that this is true, and that we can replace $\Delta x \to 0$ by $n \to \infty$, we have

$$\lim_{\Delta x \to 0} \sum_{k=1}^{n} f(u_k)\,\Delta x = \lim_{n \to \infty}\left\{48 - \frac{9}{2n^3}\left[2n^3 + 3n^2 + n\right]\right\}$$

$$= \lim_{n \to \infty} 48 - \frac{9}{2}\lim_{n \to \infty}\left[\frac{2n^3 + 3n^2 + n}{n^3}\right]$$

$$= 48 - \frac{9}{2}\lim_{n \to \infty}\left[2 + \frac{3}{n} + \frac{1}{n^2}\right]$$

$$= 48 - \tfrac{9}{2}[2 + 0 + 0] = 48 - 9 = 39$$

Thus the area of the region is 39 square units. •

The area in the preceding example may also be found by using circumscribed rectangular polygons. In this case we select, in each subinterval $[x_{k-1}, x_k]$, the number $v_k = (k - 1)(3/n)$ at which f takes on its maximum value.

The next example illustrates the use of circumscribed rectangles in finding an area.

EXAMPLE 4 If $f(x) = x^3$, find the area under the graph of f from 0 to b, for any $b > 0$.

SOLUTION Subdividing the interval $[0, b]$ into n equal parts (see Figure 5.7), we obtain a circumscribed rectangular polygon such that

$$\Delta x = \frac{b}{n} \quad \text{and} \quad x_k = k\,\Delta x.$$

Since f is an increasing function, the maximum value $f(v_k)$ in the interval $[x_{k-1}, x_k]$ occurs at the right-hand endpoint, that is,

$$v_k = x_k = k\,\Delta x = k\frac{b}{n} = \frac{bk}{n}.$$

The sum of the areas of the circumscribed rectangles is

$$\sum_{k=1}^{n} f(v_k)\,\Delta x = \sum_{k=1}^{n}\left(\frac{bk}{n}\right)^3\left(\frac{b}{n}\right) = \sum_{k=1}^{n}\frac{b^4}{n^4}k^3$$

$$= \frac{b^4}{n^4}\sum_{k=1}^{n}k^3 = \frac{b^4}{n^4}\left[\frac{n(n+1)}{2}\right]^2$$

The final step follows from (5.5) (iii). Thus,

$$\sum_{k=1}^{n} f(v_k)\,\Delta x = \frac{b^4}{4}\left(\frac{n^4 + 2n^3 + n^2}{n^4}\right) = \frac{b^4}{4}\left(1 + \frac{2}{n} + \frac{1}{n^2}\right).$$

If we let Δx approach 0, then n increases without bound and the expression in parentheses approaches 1. It follows that the area under the graph is

$$\lim_{\Delta x \to 0}\sum_{k=1}^{n} f(v_k)\,\Delta x = \frac{b^4}{4} \quad •$$

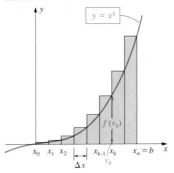

FIGURE 5.7

Exer. 1–10: Determine the sum.

1 $\sum_{k=1}^{5} (3k - 10)$

2 $\sum_{k=1}^{6} (9 - 2k)$

3 $\sum_{j=1}^{4} (j^2 + 1)$

4 $\sum_{n=1}^{10} [1 + (-1)^n]$

5 $\sum_{k=0}^{5} k(k - 1)$

6 $\sum_{k=0}^{4} (k - 2)(k - 3)$

7 $\sum_{i=1}^{8} 2^i$

8 $\sum_{s=1}^{6} (5/s)$

9 $\sum_{i=1}^{50} 10$

10 $\sum_{k=1}^{1000} 2$

11 Prove Theorem (5.3) (iii).

12 Extend Theorem (5.3) (i) to $\sum_{k=1}^{n} (a_k + b_k + c_k)$.

13 Find $\sum_{k=1}^{n} (k^2 + 3k + 5)$. (*Hint:* Write the sum as $\sum_{k=1}^{n} k^2 + 3 \sum_{k=1}^{n} k + \sum_{k=1}^{n} 5$ and employ (5.5) and (5.2).)

14 Find $\sum_{k=1}^{n} (3k^2 - 2k + 1)$.

15 Find $\sum_{k=1}^{n} (2k - 3)^2$.

16 Find $\sum_{k=1}^{n} (k^3 + 2k^2 - k + 4)$.

Exer. 17–26: Find the area under the graph of f from a to b using (a) inscribed rectangles; (b) circumscribed rectangles.

In each case sketch the graph and typical rectangles, labeling the drawing as in Figures 5.6 and 5.7.

17 $f(x) = 2x + 3$; $a = 0$, $b = 4$

18 $f(x) = 8 - 3x$; $a = 0$, $b = 2$

19 $f(x) = x^2$; $a = 0$, $b = 5$

20 $f(x) = x^2 + 2$; $a = 1$, $b = 3$

21 $f(x) = 3x^2 + 5$; $a = 1$, $b = 4$

22 $f(x) = 7$; $a = -2$, $b = 6$

23 $f(x) = 9 - x^2$; $a = 0$, $b = 3$

24 $f(x) = 4x^2 + 3x + 2$; $a = 1$, $b = 5$

25 $f(x) = x^3 + 1$; $a = 1$, $b = 2$

26 $f(x) = 4x + x^3$; $a = 0$, $b = 2$

27 Use Definition (5.4) to prove that the area of a right triangle of altitude h and corresponding base b is $\frac{1}{2}bh$. (*Hint:* Consider the area under the graph of $f(x) = (h/b)x$ from 0 to b.)

28 If $f(x) = px^2 + qx + r$ and $f(x) \geq 0$ for every x, prove that the area under the graph of f from 0 to b is
$$p(b^3/3) + q(b^2/2) + rb.$$

29 Use Mathematical induction to prove (5.5).

30 Prove (5.5) (i) by writing
$$S = 1 + \quad 2 \quad + \cdots + n$$
$$S = n + (n - 1) + \cdots + 1$$
and then adding corresponding sides of these equations.

5.2 DEFINITION OF DEFINITE INTEGRAL

In Section 5.1 we defined the area under the graph of a function f from a to b as a limit of the form

$$\lim_{\Delta x \to 0} \sum_{k=1}^{n} f(w_k)\, \Delta x.$$

In our discussion we restricted f and Δx as follows:

1 The function f is continuous on the closed interval $[a, b]$.

2 $f(x)$ is nonnegative for every x in $[a, b]$.

3 All the subintervals $[x_{k-1}, x_k]$ have the same length Δx.

4 The number w_k is chosen such that $f(w_k)$ is always the minimum (or maximum) value of f on $[x_{k-1}, x_k]$.

There are many applications involving this type of limit in which one or more of these conditions is not satisfied. Thus it is desirable to allow the following changes in **1–4**:

1' The function f may be discontinuous at some numbers in $[a, b]$.

2' $f(x)$ may be negative for some x in $[a, b]$.

3′ The lengths of the subintervals $[x_{k-1}, x_k]$ may be different.

4′ The number w_k may be *any* number in $[x_{k-1}, x_k]$.

We shall begin by introducing some new terminology and notation. A **partition** P of a closed interval $[a, b]$ is any decomposition of $[a, b]$ into subintervals of the form

$$[x_0, x_1], [x_1, x_2], [x_2, x_3], \ldots, [x_{n-1}, x_n]$$

for a positive integer n and numbers x_k such that

$$a = x_0 < x_1 < x_2 < x_3 < \cdots < x_{n-1} < x_n = b.$$

The length of the kth subinterval $[x_{k-1}, x_k]$ will be denoted by Δx_k, that is,

$$\Delta x_k = x_k - x_{k-1}.$$

A typical partition of $[a, b]$ is illustrated in Figure 5.8. The largest of the numbers $\Delta x_1, \Delta x_2, \ldots, \Delta x_n$ is the **norm** of the partition P and is denoted by $\|P\|$.

FIGURE 5.8
A partition of $[a, b]$

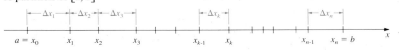

The following concept, named after the mathematician G.F.B. Riemann (1826–1866), is fundamental for the definition of the definite integral.

DEFINITION (5.6)

> Let f be a function that is defined on a closed interval $[a, b]$ and let P be a partition of $[a, b]$. A **Riemann sum** of f (or $f(x)$) for P is any expression R_P of the form
>
> $$R_P = \sum_{k=1}^{n} f(w_k)\, \Delta x_k$$
>
> for some number w_k in $[x_{k-1}, x_k]$ and $k = 1, 2, \ldots, n$.

FIGURE 5.9

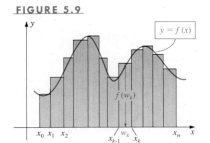

In Definition (5.6), $f(w_k)$ is not necessarily a maximum or minimum value of f on $[x_{k-1}, x_k]$. If we construct a rectangle of length $f(w_k)$ and width Δx_k as illustrated in Figure 5.9, the rectangle may be neither inscribed nor circumscribed. Moreover, since $f(x)$ may be negative for some x, some terms of the Riemann sum R_P may be negative. Consequently, R_P does not always represent a sum of areas of rectangles.

We may interpret the Riemann sum R_P in (5.6) geometrically, as follows. For each subinterval $[x_{k-1}, x_k]$, construct a horizontal line segment through the point $(w_k, f(w_k))$, thereby obtaining a collection of rectangles. If $f(w_k)$ is positive, the rectangle lies above the x-axis as illustrated by the gray rectangles in Figure 5.10, and the product $f(w_k)\, \Delta x_k$ is the area of this rectangle. If $f(w_k)$ is negative, then the rectangle lies below the x-axis as illustrated by the colored rectangles in Figure 5.10. In this case the product $f(w_k)\, \Delta x_k$ is the *negative* of the area of a rectangle. It follows that R_P is the sum of the areas of the rectangles that lie above the x-axis and the *negatives* of the areas of the rectangles that lie below the x-axis.

FIGURE 5.10

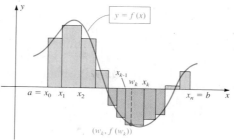

FIGURE 5.11

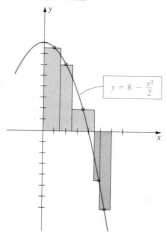

$y = 8 - \dfrac{x^2}{2}$

EXAMPLE Let $f(x) = 8 - \frac{1}{2}x^2$, and let P be the partition of $[0, 6]$ into the five subintervals determined by $x_0 = 0$, $x_1 = 1.5$, $x_2 = 2.5$, $x_3 = 4.5$, $x_4 = 5$, and $x_5 = 6$. Find the norm of the partition and the Riemann sum R_P if $w_1 = 1$, $w_2 = 2$, $w_3 = 3.5$, $w_4 = 5$, and $w_5 = 5.5$.

SOLUTION The graph of f is sketched in Figure 5.11. Also shown in the figure are the points that correspond to w_k and the rectangles of lengths $|f(w_k)|$ for $k = 1, 2, 3, 4$, and 5. Thus,

$$\Delta x_1 = 1.5, \quad \Delta x_2 = 1, \quad \Delta x_3 = 2, \quad \Delta x_4 = 0.5, \quad \Delta x_5 = 1.$$

The norm $\|P\|$ of the partition is Δx_3, or 2.

By Definition (5.6),

$$
\begin{aligned}
R_P &= f(w_1)\,\Delta x_1 + f(w_2)\,\Delta x_2 + f(w_3)\,\Delta x_3 + f(w_4)\,\Delta x_4 + f(w_5)\,\Delta x_5 \\
&= f(1)(1.5) + f(2)(1) + f(3.5)(2) + f(5)(0.5) + f(5.5)(1) \\
&= (7.5)(1.5) + (6)(1) + (1.875)(2) + (-4.5)(0.5) + (-7.125)(1) \\
&= 11.625
\end{aligned}
$$

We shall not always specify the number n of subintervals in a partition P of $[a, b]$. A Riemann sum (5.6) will then be written

$$R_P = \sum_k f(w_k)\,\Delta x_k$$

and we will assume that terms of the form $f(w_k)\,\Delta x_k$ are to be summed over all subintervals $[x_{k-1}, x_k]$ of the partition P.

In a manner similar to Definition (5.4), we may define

$$\lim_{\|P\| \to 0} \sum_k f(w_k)\,\Delta x_k = I$$

for a real number I. Intuitively, this limit means that if the norm $\|P\|$ of the partition P is close to 0, then *every* Riemann sum for P is close to I.

DEFINITION (5.7)

> Let f be a function that is defined on a closed interval $[a, b]$, and let I be a real number. The statement
>
> $$\lim_{\|P\| \to 0} \sum_k f(w_k)\,\Delta x_k = I$$
>
> means that for every $\varepsilon > 0$ there exists a $\delta > 0$ such that if P is a partition of $[a, b]$ with $\|P\| < \delta$, then
>
> $$\left| \sum_k f(w_k)\,\Delta x_k - I \right| < \varepsilon$$
>
> for any choice of numbers w_k in the subintervals $[x_{k-1}, x_k]$ of P. The number I is a **limit of (Riemann) sums.**

Note that for every $\delta > 0$ there are infinitely many partitions P with $\|P\| < \delta$. Moreover, for each such partition P there are infinitely many ways of choosing the number w_k in $[x_{k-1}, x_k]$. Consequently, an infinite number of different Riemann sums may be associated with *each* partition P. However, if the limit I exists, then for any $\varepsilon > 0$, every Riemann sum is within ε units of I, provided a small enough norm is chosen. Although Definition (5.7) differs from the definition of limit of a function, we may use a proof similar to that given for the Uniqueness Theorem in Appendix II to show that if the limit I exists, then it is unique.

We next define the definite integral as a limit of a sum.

DEFINITION (5.8)

Let f be a function that is defined on a closed interval $[a, b]$. The **definite integral of f from a to b,** denoted by $\int_a^b f(x)\, dx$, is

$$\int_a^b f(x)\, dx = \lim_{\|P\| \to 0} \sum_k f(w_k)\, \Delta x_k$$

provided the limit exists.

If the definite integral of f from a to b exists, then f is **integrable** on $[a, b]$, and we say that the integral $\int_a^b f(x)\, dx$ **exists.** The process of finding the number represented by the limit is called **evaluating the integral.**

The symbol $\int$ in Definition (5.8) is an **integral sign.** It may be thought of as an elongated letter S (the first letter of the word *sum*) and is used to indicate the connection between definite integrals and Riemann sums. The numbers a and b are the **limits of integration,** a being the **lower limit** and b the **upper limit.** In this context *limit* refers to the smallest or largest number in the interval $[a, b]$ and has no connection with any definitions of limits given earlier in the text. The expression $f(x)$, which appears to the right of the integral sign (or *behind* the integral sign), is the **integrand.** The differential symbol dx that follows $f(x)$ may be associated with the increment Δx_k of a Riemann sum of f. This association will be useful in later applications.

Letters other than x may be used in the notation for the definite integral. If f is integrable on $[a, b]$, then

$$\int_a^b f(x)\, dx = \int_a^b f(s)\, ds = \int_a^b f(t)\, dt$$

and so on. For this reason the letter x in Definition (5.8) is called a **dummy variable.**

Whenever an interval $[a, b]$ is employed we assume that $a < b$. Consequently, Definition (5.8) does not take into account the cases in which the lower limit of integration is greater than or equal to the upper limit. The definition may be extended to include the case where the lower limit is greater than the upper limit, as follows.

DEFINITION (5.9)

If $c > d$, then $\int_c^d f(x)\, dx = -\int_d^c f(x)\, dx$.

Definition (5.9) may be phrased as follows: *Interchanging the limits of integration changes the sign of the integral.* One reason for the form of Definition (5.9) will become apparent after we have considered the Fundamental Theorem of Calculus (5.22).

The case in which the lower and upper limits of integration are equal is covered by the next definition.

DEFINITION (5.10)

If $f(a)$ exists, then $\int_a^a f(x)\,dx = 0$.

If f is integrable, then the limit in Definition (5.8) exists for every choice of w_k in $[x_{k-1}, x_k]$. This allows us to specialize w_k if we wish to do so. For example, we could always choose w_k as the smallest number x_{k-1} in the subinterval, or as the largest number x_k, or as the midpoint of the subinterval, or as the number that always produces the minimum or maximum value in $[x_{k-1}, x_k]$, and so on. Moreover, since the limit is independent of the partition P of $[a, b]$ (provided that $\|P\|$ is sufficiently small) we may specialize the partitions to the case in which all the subintervals $[x_{k-1}, x_k]$ have the same length Δx. A partition of this type is a **regular partition.**

If a regular partition of $[a, b]$ contains n subintervals, then $\Delta x = (b - a)/n$. In this case the symbol $\|P\| \to 0$ is equivalent to $\Delta x \to 0$ or $n \to \infty$, and Definition (5.8) takes on the form

$$\int_a^b f(x)\,dx = \lim_{n \to \infty} \sum_{k=1}^{n} f(w_k)\,\Delta x.$$

The following theorem is a first application of these special Riemann sums. Many other applications will be discussed in Chapter 6.

THEOREM (5.11)

If f is integrable and $f(x) \geq 0$ for every x in $[a, b]$, then the area A of the region under the graph of f from a to b is

$$A = \int_a^b f(x)\,dx.$$

PROOF The area A is the limit of the sums $\sum_k f(u_k)\,\Delta x$ for the minimum value $f(u_k)$ of f on $[x_{k-1}, x_k]$ (see Definition (5.4)). Since these are Riemann sums, the conclusion follows from Definition (5.8). • •

Theorem (5.11) is illustrated in Figure 5.12. It is important to keep in mind that area is merely our first application of the definite integral. *There are many instances where $\int_a^b f(x)\,dx$ does not represent the area of a region.* In fact, if $f(x) < 0$ for some x in $[a, b]$, then this definite integral may be negative or zero.

The next theorem states that functions that are continuous on closed intervals are integrable. This fact will play a crucial role in the proof of the Fundamental Theorem of Calculus in Section 5.4.

FIGURE 5.12

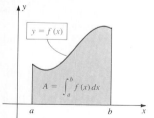

THEOREM (5.12)

If a function f is continuous on $[a, b]$, then f is integrable on $[a, b]$.

FIGURE 5.13
Nonintegrable discontinuous function

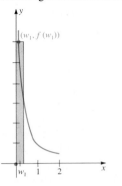

A proof of Theorem (5.12) may be found in texts on advanced calculus.

Definite integrals of discontinuous functions may or may not exist, depending on the types of discontinuities. In particular, *functions that have infinite discontinuities on a closed interval are not integrable on that interval.* To illustrate, consider the piecewise-defined function f with domain $[0, 2]$ such that

$$f(x) = 0 \text{ if } x = 0 \quad \text{and} \quad f(x) = \frac{1}{x} \text{ if } 0 < x \le 2.$$

The graph of f is sketched in Figure 5.13. Note that $\lim_{x \to 0^+} f(x) = \infty$. If M is any (large) positive number, then in the first subinterval $[x_0, x_1]$ of any partition P of $[a, b]$, we can find a number w_1 such that $f(w_1) > M/\Delta x_1$ or, equivalently, $f(w_1) \Delta x_1 > M$. It follows that there are Riemann sums $\sum_k f(w_k) \Delta x_k$ that are arbitrarily large, and hence the limit in Definition (5.8) cannot exist. Thus f is not integrable. A similar argument can be given for *any* function that has an infinite discontinuity in $[a, b]$. Consequently, *if a function f is integrable on $[a, b]$, then it is bounded on $[a, b]$*; that is, there is a real number M such that $|f(x)| \le M$ for every x in $[a, b]$.

As an illustration of a discontinuous function that *is* integrable, consider the piecewise-defined function f with domain $[0, 2]$ such that

$$f(x) = 4 \text{ if } x = 0 \quad \text{and} \quad f(x) = x^3 \text{ if } 0 < x \le 2.$$

The graph of f is sketched in Figure 5.14. Note that f has a jump discontinuity at $x = 0$. From Example 4 of Section 5.1, the area under the graph of $y = x^3$ from 0 to 2 is $2^4/4 = 2$. Thus, by Theorem (5.11), $\int_0^2 x^3 \, dx = 2$. We can also show that $\int_0^2 f(x) \, dx = 2$. Hence, f is integrable.

We have shown that a function that is discontinuous on a closed interval may, or may not, be integrable. However, by Theorem (5.12), functions that are *continuous* on a closed interval are *always* integrable.

FIGURE 5.14
Integrable discontinuous function

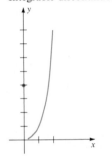

EXERCISES 5.2

Exer. 1–4: The numbers $\{x_0, x_1, x_2, \ldots, x_n\}$ determine a partition P of the interval $[a, b]$. Find $\Delta x_1, \Delta x_2, \ldots, \Delta x_n$ and the norm $\|P\|$ of the partition.

1 $[0, 5]$; $\{0, 1.1, 2.6, 3.7, 4.1, 5\}$

2 $[2, 6]$; $\{2, 3, 3.7, 4, 5.2, 6\}$

3 $[-3, 1]$; $\{-3, -2.7, -1, 0.4, 0.9, 1\}$

4 $[1, 4]$; $\{1, 1.6, 2, 3.5, 4\}$

Exer, 5–6: Find the Riemann sum R_P of f for the regular partition P of $[1, 5]$ into the four equal subintervals determined by $x_0 = 1$, $x_1 = 2$, $x_2 = 3$, $x_3 = 4$, $x_4 = 5$, and
(a) w_k is the right-hand endpoint x_k of $[x_{k-1}, x_k]$.
(b) w_k is the left-hand endpoint x_{k-1} of $[x_{k-1}, x_k]$.
(c) w_k is the midpoint of $[x_{k-1}, x_k]$.

5 $f(x) = 2x + 3$

6 $f(x) = 3 - 4x$

7 If $f(x) = 8 - \frac{1}{2}x^2$, find the Riemann sum R_P of f for the regular partition P of $[0, 6]$ into the six equal subintervals determined by $x_0 = 0$, $x_1 = 1$, $x_2 = 2$, $x_3 = 3$, $x_4 = 4$, $x_5 = 5$, $x_6 = 6$, and w_k is the midpoint of the interval $[x_{k-1}, x_k]$.

8 If $f(x) = 8 - \frac{1}{2}x^2$, find the Riemann sum R_P of f for the partition P of $[0, 6]$ into the four subintervals determined by $x_0 = 0$, $x_1 = 1.5$, $x_2 = 3$, $x_3 = 4.5$, $x_4 = 6$, and $w_1 = 1$, $w_2 = 2$, $w_3 = 4$, and $w_4 = 5$.

9 Suppose $f(x) = x^3$ and P is the partition of $[-2, 4]$ into the four subintervals determined by $x_0 = -2$, $x_1 = 0$, $x_2 = 1$, $x_3 = 3$, and $x_4 = 4$. Find the Riemann sum R_P if $w_1 = -1$, $w_2 = 1$, $w_3 = 2$, and $w_4 = 4$.

10 Suppose $f(x) = \sqrt{x}$ and P is the partition of $[1, 16]$ into the five subintervals determined by $x_0 = 1$, $x_1 = 3$, $x_2 = 5$, $x_3 = 7$, $x_4 = 9$, and $x_5 = 16$. Find the Riemann sum R_P if $w_1 = 1$, $w_2 = 4$, $w_3 = 5$, $w_4 = 9$, and $w_5 = 9$.

Exer. 11–14: Use Definition (5.8) to express each limit as a definite integral on the interval $[a, b]$.

11 $\lim\limits_{\|P\| \to 0} \sum\limits_{k=1}^{n} (3w_k^2 - 2w_k + 5) \Delta x_k; \quad [-1, 2]$

12 $\lim\limits_{\|P\| \to 0} \sum\limits_{k=1}^{n} \pi(w_k^2 - 4) \Delta x_k; \quad [2, 3]$

13 $\lim\limits_{\|P\| \to 0} \sum\limits_{k=1}^{n} 2\pi w_k(1 + w_k^3) \Delta x_k; \quad [0, 4]$

14 $\lim\limits_{\|P\| \to 0} \sum\limits_{k=1}^{n} (\sqrt[3]{w_k} + 4w_k) \Delta x_k; \quad [-4, -3]$

15 If $\int_1^4 \sqrt{x}\, dx = \frac{14}{3}$, find $\int_4^1 \sqrt{x}\, dx$.

16 Find $\int_3^3 x^2\, dx$.

17 If $\int_1^2 (5x^4 - 1)\, dx = 30$, find $\int_1^2 (5r^4 - 1)\, dr$.

18 If $\int_{-1}^{8} \sqrt[3]{s}\, ds = \frac{45}{4}$, find $\int_8^{-1} \sqrt[3]{t}\, dt$.

Exer. 19–22: Find the value of the definite integral by regarding it as the area under the graph of a function f.

19 $\int_{-3}^{2} (2x + 6)\, dx$

20 $\int_{-1}^{2} (7 - 3x)\, dx$

21 $\int_0^3 \sqrt{9 - x^2}\, dx$

22 $\int_{-a}^{a} \sqrt{a^2 - x^2}\, dx, \ a > 0$

23 Find $\int_0^b x^3\, dx$. (*Hint:* See Example 4 of Section 5.1.)

24 Let c be an arbitrary real number and suppose $f(x) = c$ for every x. If P is any partition of $[a, b]$, show that every Riemann sum R_P of f equals $c(b - a)$. Use this fact to prove that $\int_a^b c\, dx = c(b - a)$. Interpret this geometrically if $c > 0$.

25 Give an example of a function that is continuous on the interval $(0, 1)$ such that $\int_0^1 f(x)\, dx$ does not exist. Why doesn't this contradict Theorem (5.12)?

26 Give an example of a function that is not continuous on $[0, 1]$ such that $\int_0^1 f(x)\, dx$ exists.

PROPERTIES OF THE DEFINITE INTEGRAL

This section contains some fundamental properties of the definite integral. Most of the proofs are rather technical and have been placed in Appendix II.

THEOREM (5.13)

$$\int_a^b c\, dx = c(b - a); \quad c \text{ a constant}$$

PROOF Let f be the constant function defined by $f(x) = c$ for every x in $[a, b]$. If P is a partition of $[a, b]$, then for every Riemann sum of f,

$$\sum_k f(w_k) \Delta x_k = \sum_k c\, \Delta x_k = c \sum_k \Delta x_k = c(b - a).$$

(The last equality is true because the sum $\sum_k \Delta x_k$ is the length of the interval $[a, b]$). Consequently,

$$\left| \sum_k f(w_k) \Delta x_k - c(b - a) \right| = \left| c(b - a) - c(b - a) \right| = 0,$$

which is less than any positive number ε *regardless* of the size of $\|P\|$. Thus, by Definition (5.7), with $I = c(b - a)$,

$$\lim_{\|P\| \to 0} \sum_k f(w_k) \Delta x_k = \lim_{\|P\| \to 0} \sum_k c\, \Delta x_k = c(b - a).$$

By Definition (5.8), this means that

$$\int_a^b f(x)\, dx = \int_a^b c\, dx = c(b - a) \quad \bullet \ \bullet$$

FIGURE 5.15

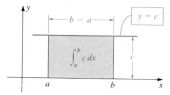

Theorem (5.13) is in agreement with the discussion of area in Section 5.1, for if $c > 0$, then as illustrated in Figure 5.15, the graph of f is the horizontal line $y = c$, and the region under the graph from a to b is a rectangle with sides of length c and $b - a$. Hence the area $\int_a^b f(x)\, dx$ of the rectangle is $c(b - a)$.

EXAMPLE 1 Evaluate $\int_{-2}^{3} 7\, dx$.

SOLUTION   Using Theorem (5.13),

$$\int_{-2}^{3} 7\, dx = 7[3 - (-2)] = 7(5) = 35 \quad \bullet$$

If $c = 1$ in Theorem (5.13), we shall abbreviate the integrand as follows:

$$\int_a^b dx = b - a.$$

If a function f is integrable on $[a, b]$ and c is a real number, then by Theorem (5.3) (ii), a Riemann sum of the function cf may be written

$$\sum_k cf(w_k)\, \Delta x_k = c \sum_k f(w_k)\, \Delta x_k.$$

We can prove that the limit of the sums on the left is equal to c times the limit of the sums on the right. Restating this fact in terms of definite integrals gives us the next theorem. A proof may be found in Appendix II.

THEOREM (5.14)

> If f is integrable on $[a, b]$ and c is any real number, then cf is integrable on $[a, b]$ and
>
> $$\int_a^b cf(x)\, dx = c \int_a^b f(x)\, dx$$

Theorem (5.14) is sometimes stated as follows: *A constant factor in the integrand may be taken outside the integral sign.* It is *not* permissible to take expressions involving variables outside the integral sign in this manner.

If two functions f and g are defined on $[a, b]$, then by Theorem (5.3) (i), a Riemann sum of $f + g$ may be written

$$\sum_k [f(w_k) + g(w_k)]\, \Delta x_k = \sum_k f(w_k)\, \Delta x_k + \sum_k g(w_k)\, \Delta x_k.$$

We can show that if f and g are integrable, then the limit of the sums on the left may be found by adding the limits of the two sums on the right. This fact is stated in integral form in (i) of the next theorem. A proof of (i) may be found in Appendix II. The analogous result for differences is stated in (ii) of the theorem. Its proof is left as an exercise.

THEOREM (5.15)

> If f and g are integrable on $[a, b]$, then $f + g$ and $f - g$ are integrable on $[a, b]$ and
>
> (i) $\int_a^b [f(x) + g(x)]\, dx = \int_a^b f(x)\, dx + \int_a^b g(x)\, dx.$
>
> (ii) $\int_a^b [f(x) - g(x)]\, dx = \int_a^b f(x)\, dx - \int_a^b g(x)\, dx.$

Theorem (5.15) (i) may be extended to any finite number of functions. Thus, if $f_1, f_2, \ldots, f_n$ are integrable on $[a, b]$, then so is their sum and

$$\int_a^b [f_1(x) + f_2(x) + \cdots + f_n(x)] \, dx$$

$$= \int_a^b f_1(x) \, dx + \int_a^b f_2(x) \, dx + \cdots + \int_a^b f_n(x) \, dx.$$

EXAMPLE 2 It will follow from the results in Section 5.5 that

$$\int_0^2 x^3 \, dx = 4 \quad \text{and} \quad \int_0^2 x \, dx = 2.$$

Use these facts to evaluate $\int_0^2 (5x^3 - 3x + 6) \, dx$.

SOLUTION We may proceed as follows:

$$\int_0^2 (5x^3 - 3x + 6) \, dx = \int_0^2 5x^3 \, dx - \int_0^2 3x \, dx + \int_0^2 6 \, dx$$

$$= 5 \int_0^2 x^3 \, dx - 3 \int_0^2 x \, dx + 6(2 - 0)$$

$$= 5(4) - 3(2) + 12 = 26 \qquad \bullet$$

If f is continuous on $[a, b]$ and $f(x) \geq 0$ for every x in $[a, b]$, then by Theorem (5.11) the integral $\int_a^b f(x) \, dx$ is the area under the graph of f from a to b. Similarly, if $a < c < b$, then the integrals $\int_a^c f(x) \, dx$ and $\int_c^b f(x) \, dx$ are the areas under the graph of f from a to c and from c to b, respectively, as illustrated in Figure 5.16. Since the area from a to b is the sum of the two smaller areas, we have

$$\int_a^b f(x) \, dx = \int_a^c f(x) \, dx + \int_c^b f(x) \, dx.$$

FIGURE 5.16

The next theorem shows that the preceding equality is true under a more general hypothesis. The proof is given in Appendix II.

THEOREM (5.16)

If $a < c < b$, and if f is integrable on both $[a, c]$ and $[c, b]$, then f is integrable on $[a, b]$ and

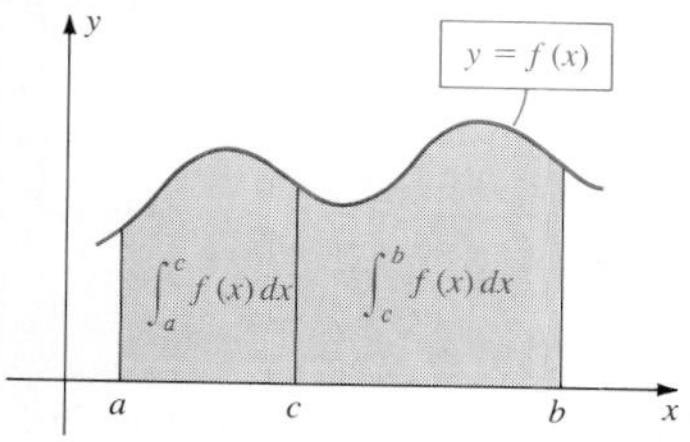

$$\int_a^b f(x) \, dx = \int_a^c f(x) \, dx + \int_c^b f(x) \, dx$$

The following result is a generalization of Theorem (5.16) to the case where c is not necessarily between a and b.

THEOREM (5.17)

If f is integrable on a closed interval and if a, b, and c are any three numbers in the interval, then

$$\int_a^b f(x) \, dx = \int_a^c f(x) \, dx + \int_c^b f(x) \, dx$$

PROOF If a, b, and c are all different, then there are six possible ways of arranging these three numbers. The theorem should be verified for each of these cases and also for the cases in which two, or all three, of the numbers are equal. We shall verify one case and leave the remaining parts as exercises. Thus, suppose $c < a < b$. By Theorem (5.16),

$$\int_c^b f(x)\, dx = \int_c^a f(x)\, dx + \int_a^b f(x)\, dx$$

which, in turn, may be written

$$\int_a^b f(x)\, dx = -\int_c^a f(x)\, dx + \int_c^b f(x)\, dx.$$

The conclusion of the theorem now follows from the fact that interchanging the limits of integration changes the sign of the integral (see Definition (5.9)). • •

If f and g are continuous on $[a, b]$ and $f(x) \ge g(x) \ge 0$ for every x in $[a, b]$, then the area under the graph of f from a to b is greater than or equal to the area under the graph of g from a to b. The corollary to the next theorem is a generalization of this fact to arbitrary integrable functions. The proof of the theorem is given in Appendix II.

THEOREM (5.18)

If f is integrable on $[a, b]$ and if $f(x) \ge 0$ for every x in $[a, b]$, then

$$\int_a^b f(x)\, dx \ge 0$$

COROLLARY (5.19)

If f and g are integrable on $[a, b]$ and $f(x) \ge g(x)$ for every x in $[a, b]$, then

$$\int_a^b f(x)\, dx \ge \int_a^b g(x)\, dx$$

PROOF By Theorem (5.15), $f - g$ is integrable. Moreover, $f(x) - g(x) \ge 0$ for every x in $[a, b]$. Hence, by Theorem (5.18),

$$\int_a^b [f(x) - g(x)]\, dx \ge 0.$$

Applying Theorem (5.15) (ii) leads to the desired conclusion. • •

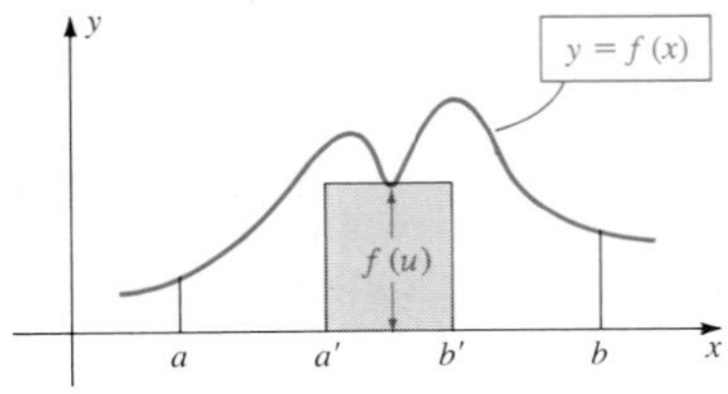

Suppose, in Theorem (5.18), that f is continuous and that, in addition to the condition $f(x) \ge 0$, we have $f(c) > 0$ for some c in $[a, b]$. In this case $\lim_{x \to c} f(x) > 0$ and by Theorem (2.12), there is a subinterval $[a', b']$ of $[a, b]$ throughout which $f(x)$ is positive. If $f(u)$ is the minimum value of f on $[a', b']$ (see Figure 5.17), then the area under the graph of f from a to b is at least as large as the area $f(u)(b' - a')$ of the pictured rectangle. Consequently, $\int_a^b f(x)\, dx > 0$. It now follows, as in the proof of Corollary (5.19), that if f and g are continuous on $[a, b]$, if $f(x) \ge g(x)$ throughout $[a, b]$, and if

FIGURE 5.17

$f(x) > g(x)$ for some x in $[a, b]$, then $\int_a^b f(x)\,dx > \int_a^b g(x)\,dx$. This fact will be used in the proof of the next theorem.

THE MEAN VALUE THEOREM (5.20)
FOR DEFINITE INTEGRALS

> If f is continuous on a closed interval $[a, b]$, then there is a number z in the open interval (a, b) such that
>
> $$\int_a^b f(x)\,dx = f(z)(b - a)$$

PROOF If f is a constant function, then $f(x) = c$ for some number c, and by Theorem (5.13),

$$\int_a^b f(x)\,dx = \int_a^b c\,dx = c(b - a) = f(z)(b - a)$$

for *any* number z in (a, b).

Next assume that f is not a constant function and suppose that m and M are the minimum and maximum values of f, respectively, on $[a, b]$. Let $f(u) = m$ and $f(v) = M$ for some u and v in $[a, b]$. This is illustrated in Figure 5.18 for the case where $f(x)$ is positive throughout $[a, b]$. Since f is not a constant function, $m < f(x) < M$ for some x in $[a, b]$ and hence by the remark immediately preceding this theorem,

$$\int_a^b m\,dx < \int_a^b f(x)\,dx < \int_a^b M\,dx.$$

Applying Theorem (5.13),

$$m(b - a) < \int_a^b f(x)\,dx < M(b - a).$$

Dividing by $b - a$ and recalling that $m = f(u)$ and $M = f(v)$ gives us

$$f(u) < \frac{1}{b - a}\int_a^b f(x)\,dx < f(v).$$

Since $[1/(b - a)]\int_a^b f(x)\,dx$ is a number between $f(u)$ and $f(v)$, it follows from the Intermediate Value Theorem (2.33) that there is a number z, with $u < z < v$, such that

$$f(z) = \frac{1}{b - a}\int_a^b f(x)\,dx.$$

Multiplying both sides by $b - a$ gives us the conclusion of the theorem. • •

The number z of Theorem (5.20) is not necessarily unique; however, the theorem guarantees that at *least* one number z will produce the desired result.

The Mean Value Theorem has an interesting geometric interpretation if $f(x) \geq 0$ on $[a, b]$. In this case $\int_a^b f(x)\,dx$ is the area under the graph of f from a to b. If, as in Figure 5.19, a horizontal line is drawn through the point $P(z, f(z))$, then the area of the rectangular region bounded by this line, the x-axis, and the lines $x = a$ and $x = b$ is $f(z)(b - a)$, which, according to Theorem (5.20), is the same as the area under the graph of f from a to b.

FIGURE 5.18

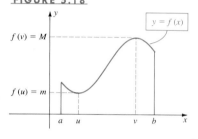

FIGURE 5.19

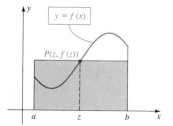

EXAMPLE 3 It will follow from the results of Section 5.4 that $\int_0^3 x^2 \, dx = 9$. Find a number z that satisfies the conclusion of the Mean Value Theorem (5.20) for this definite integral.

SOLUTION The graph of $f(x) = x^2$ for $0 \le x \le 3$ is sketched in Figure 5.20. By the Mean Value Theorem, there is a number z between 0 and 3 such that

$$\int_0^3 x^2 \, dx = f(z)(3 - 0) = z^2(3).$$

This implies that

$$9 = 3z^2 \quad \text{and} \quad z^2 = 3.$$

Thus $z = \sqrt{3}$ satisfies the conclusion of the theorem.

If we consider the horizontal line through $P(\sqrt{3}, 3)$, then the area of the rectangle bounded by this line, the x-axis, and the lines $x = 0$ and $x = 3$ is equal to the area under the graph of f from $x = 0$ to $x = 3$ (see Figure 5.20).

FIGURE 5.20

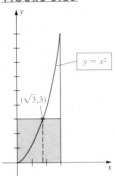

$y = x^2$

$(\sqrt{3}, 3)$

In statistics the term **arithmetic mean** is used for the **average** of a set of numbers. Thus, the arithmetic mean of two numbers a and b is $(a + b)/2$, the arithmetic mean of three numbers a, b, and c is $(a + b + c)/3$, and so on. To see the relationship between arithmetic means and the word *mean* used in Mean Value Theorem, let us rewrite the conclusion of (5.20) as

$$f(z) = \frac{1}{b - a} \int_a^b f(x) \, dx$$

and express the definite integral as a limit of sums. If we specialize Definition (5.8) by using a regular partition P with n subintervals, then

$$f(z) = \frac{1}{b - a} \lim_{n \to \infty} \sum_{k=1}^n f(w_k) \, \Delta x = \lim_{n \to \infty} \sum_{k=1}^n \left[f(w_k) \frac{\Delta x}{b - a} \right]$$

for any number w_k in the kth subinterval of P and $\Delta x = (b - a)/n$. Since $\Delta x/(b - a) = 1/n$, we obtain

$$f(z) = \lim_{n \to \infty} \sum_{k=1}^n \left[f(w_k) \frac{\Delta x}{b - a} \right] = \lim_{n \to \infty} \sum_{k=1}^n \left[f(w_k) \frac{1}{n} \right]$$

or

$$f(z) = \lim_{n \to \infty} \left[\frac{f(w_1) + f(w_2) + \cdots + f(w_n)}{n} \right].$$

This shows that we may regard the number $f(z)$ in the Mean Value Theorem (5.20) as a limit of the arithmetic means (averages) of the function values $f(w_1), f(w_2), \ldots, f(w_n)$ as n increases without bound. This is the motivation for the next definition.

DEFINITION (5.21)

Let f be continuous on $[a, b]$. The **average value** f_{av} of f on $[a, b]$ is

$$f_{av} = \frac{1}{b - a} \int_a^b f(x) \, dx$$

As an illustration of Definition (5.21), let us consider $f(x) = x^2$ on $[0, 3]$. Note that on this interval the function values range from $f(0) = 0$ to $f(3) = 9$. From Example 3, we see that $f_{av} = f(\sqrt{3}) = 3$.

EXERCISES 5.3

Exer. 1–6: Evaluate the definite integral.

1 $\int_{-2}^{4} 5\, dx$

2 $\int_{1}^{10} \sqrt{2}\, dx$

3 $\int_{6}^{2} 3\, dx$

4 $\int_{4}^{-3} dx$

5 $\int_{-1}^{1} dx$

6 $\int_{2}^{2} 100\, dx$

Exer. 7–10: It will follow from the results in Section 5.4 that

$$\int_{1}^{4} x^2\, dx = 21 \quad \text{and} \quad \int_{1}^{4} x\, dx = \tfrac{15}{2}.$$

Use these facts to evaluate the integral.

7 $\int_{1}^{4} (3x^2 + 5)\, dx$

8 $\int_{1}^{4} (6x - 1)\, dx$

9 $\int_{1}^{4} (2 - 9x - 4x^2)\, dx$

10 $\int_{1}^{4} (3x + 2)^2\, dx$

Exer. 11–12: Verify the inequality without evaluating the integral.

11 $\int_{1}^{2} (3x^2 + 4)\, dx \geq \int_{1}^{2} (2x^2 + 5)\, dx$

12 $\int_{2}^{4} (5x^2 - 4\sqrt{x} + 2)\, dx > 0$

Exer. 13–16: Express the sum or difference as a single integral of the form $\int_{a}^{b} f(x)\, dx$.

13 $\int_{5}^{1} f(x)\, dx + \int_{-3}^{5} f(x)\, dx$

14 $\int_{4}^{1} f(x)\, dx + \int_{6}^{4} f(x)\, dx$

15 $\int_{c}^{c+h} f(x)\, dx - \int_{c}^{h} f(x)\, dx$

16 $\int_{-2}^{6} f(x)\, dx - \int_{-2}^{2} f(x)\, dx$

Exer. 17–22: Each definite integral $\int_{a}^{b} f(x)\, dx$ may be verified using the results in Section 5.5. Find (a) numbers that satisfy the conclusion of the Mean Value Theorem (5.20), and (b) the average value of f on $[a, b]$.

17 $\int_{0}^{3} 3x^2\, dx = 27$

18 $\int_{-1}^{3} (3x^2 - 2x + 3)\, dx = 32$

19 $\int_{-1}^{8} 3\sqrt{x + 1}\, dx = 54$

20 $\int_{-2}^{-1} 8x^{-3}\, dx = -3$

21 $\int_{1}^{2} (4x^3 - 1)\, dx = 14$

22 $\int_{1}^{4} (2 + 3\sqrt{x})\, dx = 20$

23 Prove Theorem (5.15) (ii).
   (*Hint:* Let $f(x) - g(x) = f(x) + [-g(x)]$.)

24 If f and g are integrable on $[a, b]$ and if p and q are any real numbers, prove that

$$\int_{a}^{b} [pf(x) + qg(x)]\, dx = p \int_{a}^{b} f(x)\, dx + q \int_{a}^{b} g(x)\, dx.$$

25 If f is continuous on $[a, b]$, prove that

$$\left| \int_{a}^{b} f(x)\, dx \right| \leq \int_{a}^{b} |f(x)|\, dx.$$

26 Complete the proof of Theorem (5.17) by considering all other orderings of the numbers a, b, and c.

5.4   THE FUNDAMENTAL THEOREM OF CALCULUS

This section contains one of the most important theorems in this text. In addition to the theorem's usefulness in evaluating definite integrals, it also exhibits the relationship between differentiation and integration. This theorem, aptly called *The Fundamental Theorem of Calculus*, was discovered independently by Sir Isaac Newton (1642–1727) in England and by Gottfried Wilhelm Leibniz (1646–1716) in Germany. It is primarily because of this discovery that both men are credited with the invention of calculus.

To avoid confusion in the following discussion, we use t as the independent variable and denote the definite integral of f from a to b by $\int_{a}^{b} f(t)\, dt$. If f is continuous on $[a, b]$ and $a \leq x \leq b$, then f is continuous on $[a, x]$ and hence, by Theorem (5.12), f is integrable on $[a, x]$. Consequently, the formula

$$G(x) = \int_{a}^{x} f(t)\, dt$$

determines a function G with domain $[a, b]$, since for each x in $[a, b]$ there corresponds a unique number $G(x)$.

To obtain a geometric interpretation of $G(x)$, suppose that $f(t) \geq 0$ for every t in $[a, b]$. In this case we see from Theorem (5.11) that $G(x)$ is the area of the region under the graph of f from a to x (see Figure 5.21).

FIGURE 5.21

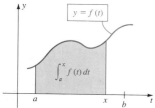

FIGURE 5.22

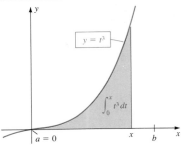

As a specific illustration, consider $f(t) = t^3$ with $a = 0$ and $b > 0$ (see Figure 5.22). In Example 4 of Section 5.1 we proved that the area under the graph of f from 0 to b is $\frac{1}{4}b^4$. Hence the area from 0 to x is

$$G(x) = \int_0^x t^3 \, dt = \frac{1}{4} x^4.$$

This gives us an explicit form for the function G if $f(t) = t^3$. Note that in this illustration,

$$G'(x) = D_x \left(\tfrac{1}{4}x^4 \right) = x^3 = f(x).$$

Thus, by Definition (4.30), G is an antiderivative of f. This result is not an accident. Part I of the next theorem brings out the remarkable fact that if f is *any* continuous function, then $G(x) = \int_a^x f(t) \, dt$ is an antiderivative of $f(x)$. Part II of the theorem shows how *any* antiderivative may be used to find the value of $\int_a^b f(x) \, dx$.

THE FUNDAMENTAL THEOREM (5.22)
OF CALCULUS

Suppose f is continuous on a closed interval $[a, b]$.

Part I   If the function G is defined by

$$G(x) = \int_a^x f(t) \, dt$$

for every x in $[a, b]$, then G is an antiderivative of f on $[a, b]$.

Part II   If F is any antiderivative of f on $[a, b]$, then

$$\int_a^b f(x) \, dx = F(b) - F(a)$$

PROOF   To establish Part I we must show that if x is in $[a, b]$, then $G'(x) = f(x)$, that is,

$$\lim_{h \to 0} \frac{G(x + h) - G(x)}{h} = f(x).$$

Before giving a formal proof, let us consider some geometric aspects of this limit. If $f(x) \geq 0$ throughout $[a, b]$, then $G(x)$ is the area under the graph of f from a to x, as illustrated in Figure 5.23. If $h > 0$, then the difference $G(x + h) - G(x)$ is the area under the graph of f from x to $x + h$, and the number h is the length of the interval $[x, x + h]$. We will show that

$$\frac{G(x + h) - G(x)}{h} = f(z)$$

for some number z between x and $x + h$. Apparently, if $h \to 0$, then $z \to x$ and $f(z) \to f(x)$, which is what we wish to prove.

Let us now give a rigorous proof that $G'(x) = f(x)$. If x and $x + h$ are in $[a, b]$ then using the definition of G, together with Definition (5.9) and Theorem (5.16),

$$G(x + h) - G(x) = \int_a^{x+h} f(t) \, dt - \int_a^x f(t) \, dt$$

$$= \int_a^{x+h} f(t) \, dt + \int_x^a f(t) \, dt$$

$$= \int_x^{x+h} f(t) \, dt.$$

FIGURE 5.23

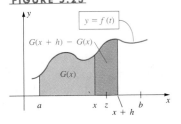

Consequently, if $h \neq 0$,

$$\frac{G(x + h) - G(x)}{h} = \frac{1}{h} \int_x^{x+h} f(t)\, dt.$$

If $h > 0$, then by the Mean Value Theorem (5.20), there is a number z in the open interval $(x, x + h)$ such that

$$\int_x^{x+h} f(t)\, dt = f(z)h$$

and, therefore,

$$\frac{G(x + h) - G(x)}{h} = f(z).$$

Since $x < z < x + h$, it follows from the continuity of f that

$$\lim_{h \to 0^+} f(z) = \lim_{z \to x^+} f(z) = f(x)$$

and hence

$$\lim_{h \to 0^+} \frac{G(x + h) - G(x)}{h} = \lim_{h \to 0^+} f(z) = f(x).$$

If $h < 0$, then we may prove in similar fashion that

$$\lim_{h \to 0^-} \frac{G(x + h) - G(x)}{h} = f(x).$$

The two preceding one-sided limits imply that

$$G'(x) = \lim_{h \to 0} \frac{G(x + h) - G(x)}{h} = f(x).$$

This completes the proof of Part I.

To prove Part II, let F be any antiderivative of f and let G be the special antiderivative defined in Part I. It follows from Theorem (4.32) that F and G differ by a constant; that is, there is a number C such that

$$G(x) - F(x) = C$$

for every x in $[a, b]$. Hence, from the definition of G,

$$\int_a^x f(t)\, dt - F(x) = C$$

for every x in $[a, b]$. If we let $x = a$ and use the fact that $\int_a^a f(t)\, dt = 0$, we obtain $0 - F(a) = C$. Consequently,

$$\int_a^x f(t)\, dt - F(x) = -F(a).$$

Since this is an identity for every x in $[a, b]$, we may substitute b for x, obtaining

$$\int_a^b f(t)\, dt - F(b) = -F(a).$$

Adding $F(b)$ to both sides of this equation and replacing the variable t by x gives us the conclusion of Part II. • •

We denote the difference $F(b) - F(a)$ either by the symbol $F(x)]_a^b$ or by $[F(x)]_a^b$. Part II of the Fundamental Theorem may then be expressed as follows.

COROLLARY (5.23)

If f is continuous on $[a, b]$ and F is any antiderivative of f, then

$$\int_a^b f(x)\, dx = F(x)\Big]_a^b = F(b) - F(a)$$

The formula in Corollary (5.23) is also valid if $a \geq b$. If $a > b$, then by Definition (5.9),

$$\int_a^b f(x)\, dx = -\int_b^a f(x)\, dx$$
$$= -[F(a) - F(b)]$$
$$= F(b) - F(a).$$

If $a = b$, then by Definition (5.10),

$$\int_a^a f(x)\, dx = 0 = F(a) - F(a).$$

Corollary (5.23) allows us to evaluate definite integrals very easily if an antiderivative of the integrand can be found. For example, since an antiderivative of x^3 is $\frac{1}{4}x^4$, we have

$$\int_0^b x^3\, dx = \frac{1}{4}x^4\Big]_0^b = \frac{1}{4}b^4 - \frac{1}{4}(0)^4 = \frac{1}{4}b^4.$$

Those who doubt the importance of the Fundamental Theorem should compare this simple computation with the limit of a sum calculation discussed in Example 4 of Section 5.1.

The following example is another illustration of Corollary (5.23).

EXAMPLE 1 Evaluate $\int_{-2}^3 (6x^2 - 5)\, dx$.

SOLUTION An antiderivative of $6x^2 - 5$ is given by $F(x) = 2x^3 - 5x$. Applying Corollary (5.23),

$$\int_{-2}^3 (6x^2 - 5)\, dx = \left[2x^3 - 5x\right]_{-2}^3$$
$$= [2(3)^3 - 5(3)] - [2(-2)^3 - 5(-2)]$$
$$= [54 - 15] - [-16 + 10] = 45 \qquad \bullet$$

Note that if $F(x) + C$ is used in place of $F(x)$ in Corollary (5.23), the same result is obtained, since

$$\left[F(x) + C\right]_a^b = \{F(b) + C\} - \{F(a) + C\}$$
$$= F(b) - F(a) = \left[F(x)\right]_a^b$$

This is in keeping with the statement that *any* antiderivative F may be employed in the Fundamental Theorem. Also note that for any number k,

$$\left[kF(x)\right]_a^b = kF(b) - kF(a) = k[F(b) - F(a)] = k\left[F(x)\right]_a^b,$$

that is, a constant factor can be "taken out" of the bracket when this notation is used. This result is analogous to Theorem (5.14) on definite integrals.

If r is a rational number and $r \neq -1$, then

$$\int_a^b x^r \, dx = \frac{x^{r+1}}{r+1}\Bigg]_a^b = \frac{1}{r+1}(b^{r+1} - a^{r+1})$$

PROOF By the Power Rule for Derivatives, $x^{r+1}/(r+1)$ is an antiderivative of x^r whenever these expressions are defined. Applying Corollary (5.23) gives us the conclusion. • •

We may use the Power Rule (5.24) together with Theorem (5.14) to obtain

$$\int_a^b k \, x^r \, dx = k \int_a^b x^r \, dx = k\left[\frac{x^{r+1}}{r+1}\right]_a^b$$

provided $r \neq -1$. For sums of terms of the form kx^r we apply the Power Rule (5.24) to each term as illustrated in the next two examples.

EXAMPLE 2 Evaluate $\int_{-1}^2 (x^3 + 1)^2 \, dx$.

SOLUTION We first square the integrand and then apply the Power Rule to each term as follows:

$$\int_{-1}^2 (x^3 + 1)^2 \, dx = \int_{-1}^2 (x^6 + 2x^3 + 1) \, dx$$

$$= \left[\frac{x^7}{7} + 2\left(\frac{x^4}{4}\right) + x\right]_{-1}^2$$

$$= \left[\frac{2^7}{7} + 2\left(\frac{2^4}{4}\right) + 2\right] - \left[\frac{(-1)^7}{7} + 2\frac{(-1)^4}{4} + (-1)\right]$$

$$= \frac{405}{14}$$

In the preceding example it is important to note that

$$\int_{-1}^2 (x^3 + 1)^2 \, dx \neq \frac{(x^3 + 1)^3}{3}\Bigg]_{-1}^2$$

EXAMPLE 3 Evaluate $\int_1^4 \left(5x - 2\sqrt{x} + \frac{32}{x^3}\right) dx$.

SOLUTION We begin by changing the form of the integrand so that the Power Rule may be applied to each term. Thus

$$\int_1^4 (5x - 2x^{1/2} + 32x^{-3}) \, dx = \left[5\left(\frac{x^2}{2}\right) - 2\left(\frac{x^{3/2}}{3/2}\right) + 32\left(\frac{x^{-2}}{-2}\right)\right]_1^4$$

$$= \left[\frac{5}{2}x^2 - \frac{4}{3}x^{3/2} - \frac{16}{x^2}\right]_1^4$$

$$= \left[\frac{5}{2}(4)^2 - \frac{4}{3}(4)^{3/2} - \frac{16}{4^2}\right] - \left[\frac{5}{2} - \frac{4}{3} - 16\right]$$

$$= \frac{259}{6}$$

EXAMPLE 4 Evaluate $\int_{-2}^{3} |x| \, dx$.

SOLUTION By Definition (1.2), $|x| = -x$ if $x < 0$ and $|x| = x$ if $x \geq 0$. This suggests that we use Theorem (5.16) to express the integral as a sum of two definite integrals as follows:

$$\int_{-2}^{3} |x| \, dx = \int_{-2}^{0} |x| \, dx + \int_{0}^{3} |x| \, dx$$

$$= \int_{-2}^{0} (-x) \, dx + \int_{0}^{3} x \, dx$$

$$= -\left[\frac{x^2}{2}\right]_{-2}^{0} + \left[\frac{x^2}{2}\right]_{0}^{3}$$

$$= -\left[0 - \frac{4}{2}\right] + \left[\frac{9}{2} - 0\right]$$

$$= 2 + \frac{9}{2} = \frac{13}{2} \qquad \bullet$$

Using the antiderivative formulas in Theorem (4.35) together with Corollary (5.23) leads to the following definite integral formulas for the sine and cosine functions, provided $k \neq 0$.

THEOREM (5.25)

$$\int_{a}^{b} \sin kx \, dx = -\frac{1}{k} \cos kx \Big]_{a}^{b}$$

$$\int_{a}^{b} \cos kx \, dx = \frac{1}{k} \sin kx \Big]_{a}^{b}$$

More general formulas for evaluating integrals involving the sine and cosine functions are stated in Theorem (5.35).

EXAMPLE 5 Find the area A of the region under the graph of $y = \sin 2x$ from $x = 0$ to $x = \pi/4$.

SOLUTION The graph of the equation and the region are shown in Figure 5.24. By Theorems (5.11) and (5.25),

$$A = \int_{0}^{\pi/4} \sin 2x \, dx = -\frac{1}{2} \cos 2x \Big]_{0}^{\pi/4}$$

$$= -\frac{1}{2}\left[\cos 2\left(\frac{\pi}{4}\right) - \cos 2(0)\right]$$

$$= -\frac{1}{2}\left[\cos \frac{\pi}{2} - \cos 0\right] = -\frac{1}{2}[0 - 1] = \frac{1}{2} \qquad \bullet$$

FIGURE 5.24

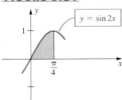

The technique of defining a function by means of a definite integral, as in Part I of the Fundamental Theorem of Calculus (5.22), will have a very important application in Chapter 7 (see Definition (7.4)). Recall, from (5.22), that if f is continuous on $[a, b]$ and $G(x) = \int_{a}^{x} f(t) \, dt$ for $a \leq x \leq b$, then

G is an antiderivative of f, that is, $D_x G(x) = f(x)$. This may be stated in integral form as follows:

$$D_x \int_a^x f(t)\, dt = f(x).$$

The preceding formula is generalized in the next theorem.

THEOREM (5.26)

> Let f be continuous on $[a, b]$. If $a \le c \le b$, then for every x in $[a, b]$,
>
> $$D_x \int_c^x f(t)\, dt = f(x)$$

PROOF If F is an antiderivative of f, then

$$D_x \int_c^x f(t)\, dt = D_x \left[F(x) - F(c) \right]$$
$$= D_x F(x) - D_x F(c)$$
$$= f(x) - 0 = f(x) \quad \bullet \ \bullet$$

EXAMPLE 6 If $G(x) = \int_1^x \dfrac{1}{t}\, dt$ and $x > 0$, find $G'(x)$.

SOLUTION We shall apply Theorem (5.26) with $c = 1$ and $f(x) = 1/x$. If we choose a and b such that $0 < a \le 1 \le b$, then f is continuous on $[a, b]$. Hence by Theorem (5.26), for every x in $[a, b]$,

$$G'(x) = D_x \int_1^x \frac{1}{t}\, dt = \frac{1}{x} \, \cdot$$

In (5.21) we defined the *average value* f_{av} of a function f on $[a, b]$ as follows:

$$f_{av} = \frac{1}{b - a} \int_a^b f(x)\, dx$$

The next example illustrates why this terminology is appropriate.

EXAMPLE 7 Suppose a point P moving on a coordinate line has a continuous velocity function v. Show that the average value of v on $[a, b]$ equals the average velocity during the time interval $[a, b]$.

SOLUTION By Definition (5.21),

$$v_{av} = \frac{1}{b - a} \int_a^b v(t)\, dt.$$

If s is the position function of P, then by Definition (3.18), $s'(t) = v(t)$; that is, $s(t)$ is an antiderivative of $v(t)$. Hence, by the Fundamental Theorem of Calculus,

$$\int_a^b v(t)\, dt = \int_a^b s'(t)\, dt = s(t) \Big]_a^b = s(b) - s(a).$$

Substituting in the formula for v_{av} gives us

$$v_{\text{av}} = \frac{s(b) - s(a)}{b - a},$$

which is the average velocity of P on $[a, b]$ (see (2.3)). •

Results similar to that in Example 7 occur in discussions of average acceleration, average marginal cost, average marginal revenue, and many other applications of the derivative (see Exercises 49–54).

EXERCISES 5.4

Exer. 1–32: Evaluate the definite integral.

1 $\int_1^4 (x^2 - 4x - 3)\, dx$

2 $\int_{-2}^3 (5 + x - 6x^2)\, dx$

3 $\int_{-2}^3 (8z^3 + 3z - 1)\, dz$

4 $\int_0^2 (w^4 - 2w^3)\, dw$

5 $\int_7^{12} dx$

6 $\int_{-6}^{-1} 8\, dx$

7 $\int_1^2 [5/(8x^6)]\, dx$

8 $\int_1^4 \sqrt{16x^5}\, dx$

9 $\int_4^9 \frac{t - 3}{\sqrt{t}}\, dt$

10 $\int_{-1}^{-2} \frac{2s - 7}{s^3}\, ds$

11 $\int_{-8}^8 (\sqrt[3]{s^2} + 2)\, ds$

12 $\int_1^0 t^2(\sqrt[3]{t} - \sqrt{t})\, dt$

13 $\int_0^1 (2x - 3)(5x + 1)\, dx$

14 $\int_{-1}^1 (x^2 + 1)^2\, dx$

15 $\int_{-1}^0 (2w + 3)^2\, dw$

16 $\int_5^5 \sqrt[3]{x^2 + \sqrt{x^5 + 1}}\, dx$

17 $\int_3^2 \frac{x^2 - 1}{x - 1}\, dx$

18 $\int_0^{-1} \frac{x^3 + 8}{x + 2}\, dx$

19 $\int_1^1 (4x^2 - 5)^{100}\, dx$

20 $\int_1^2 (4u^{-5} + 6u^{-4})\, du$

21 $\int_1^3 \frac{2x^3 - 4x^2 + 5}{x^2}\, dx$

22 $\int_{-2}^{-1} \left(r - \frac{1}{r}\right)^2 dr$

23 $\int_{-3}^6 |x - 4|\, dx$

24 $\int_{-1}^1 (x + 1)(x + 2)(x + 3)\, dx$

25 $\int_0^4 \sqrt{3t}(\sqrt{t} + \sqrt{3})\, dt$

26 $\int_{-1}^5 |2x - 3|\, dx$

27 $\int_{\pi/2}^\pi \cos(\tfrac{1}{3}x)\, dx$

28 $\int_0^{\pi/2} 3\sin(\tfrac{1}{2}x)\, dx$

29 $\int_{\pi/4}^{\pi/3} (4\sin 2\theta + 6\cos 3\theta)\, d\theta$

30 $\int_{\pi/6}^{\pi/4} (1 - \cos 4t)\, dt$

31 $\int_{-\pi/6}^{\pi/6} (x + \sin 5x)\, dx$

32 $\int_{-\pi/4}^0 (\sin x + \cos x)^2\, dx$

Exer. 33–34: Verify the identity by first using the Fundamental Theorem of Calculus (5.22) and then differentiating. (see Theorem (5.26)).

33 $D_x \int_0^x (t^3 - 4\sqrt{t} + 5)\, dt = x^3 - 4\sqrt{x} + 5$ if $x \geq 0$

34 $D_x \int_0^x (5t + 3)^2\, dt = (5x + 3)^2$

35 Find $D_x \int_0^x \frac{1}{t + 1}\, dt$

36 Find $D_x \int_0^x \frac{1}{\sqrt{1 - t^2}}\, dt$ if $|x| < 1$.

37 If $f(x) = x^2 + 1$, find the area of the region under the graph of f from -1 to 2.

38 If $f(x) = x^3$, find the area of the region under the graph of f from 1 to 3.

Exer. 39–42: Verify the formula by differentiating.

39 $\int_a^b \frac{x}{\sqrt{(x^2 + 1)^3}}\, dx = \frac{-1}{\sqrt{x^2 + 1}}\Big]_a^b$

40 $\int_a^b \frac{x^3}{\sqrt{x^2 + 1}}\, dx = \left[\frac{1}{3}\sqrt{(x^2 + 1)^3} - \sqrt{x^2 + 1}\right]_a^b$

41 $\int_a^b \frac{x}{(x^2 + c^2)^{n+1}}\, dx = \frac{-1}{2n(x^2 + c^2)^n}\Big]_a^b$, $n \neq 0$

42 $\int_a^b \frac{x^2}{(x^3 + c^3)^{n+1}}\, dx = \frac{-1}{3n(x^3 + c^3)^n}\Big]_a^b$, $n \neq 0$

Exer. 43–46: Find the number that satisfies the conclusion of the Mean Value Theorem (5.20) for the definite integral.

43 $\int_0^4 (\sqrt{x} + 1)\, dx$

44 $\int_{-1}^1 (2x + 1)^2\, dx$

45 $\int_{-1}^2 (3x^3 + 2)\, dx$

46 $\int_1^9 (3/x^2)\, dx$

Exer. 47–48: Find the average value of the function f on the interval $[a, b]$.

47 $f(x) = x^2 + 3x - 1$, $[-1, 2]$

48 $f(x) = \sin x$, $[0, 2\pi]$

49 Suppose a point P moves on a coordinate line with a continuous acceleration function a. The *average acceleration* on a time interval $[t_1, t_2]$ is $[v(t_2) - v(t_1)]/(t_2 - t_1)$ for the velocity function v. Show that the average acceleration equals the average value of a on $[t_1, t_2]$.

50 If a function f has a continuous derivative on $[a, b]$, show that the average rate of change of $f(x)$ with respect to x on $[a, b]$ (see Definition (3.19)) equals the average value of f' on $[a, b]$.

51 Let C and c denote the cost and average cost functions of a product. Show that the average value of the marginal cost function on $[0, n]$ equals the average cost $c(n)$.

52 If R is the revenue function for a product, then the average revenue for producing n units is $R(n)/n$. Show that this average revenue equals the average value of the marginal revenue function on $[0, n]$.

53 A ball is dropped from a height of s_0 feet. Use Definition (5.21) (and neglect air resistance) to show that the average velocity for the ball's journey to the ground is $4\sqrt{s_0}$ ft/sec. (*Hint:* From physics, the distance the ball falls in t seconds is $16t^2$ feet.)

54 A meteorologist determines that the temperature T (in °F) on a cold winter day is given by $T = 0.05t(t - 12)(t - 24)$ for time t (in hours) with $t = 0$ corresponding to midnight. Find the average temperature between 6 A.M. and 12 noon.

55 If g is differentiable and f is continuous for every x, prove that $D_x \int_a^{g(x)} f(t)\, dt = f(g(x))g'(x)$. (*Hint:* Use Part I of Theorem (5.22) and the Chain Rule.)

56 Extend the formula in Exercise 55 to
$$D_x \int_{k(x)}^{g(x)} f(t)\, dt = f(g(x))g'(x) - f(k(x))k'(x).$$

Exer. 57–60: Use Exercises 55 and 56 to find the derivative.

57 $D_x \int_2^{x^4} \dfrac{t}{\sqrt{t^3 + 2}}\, dt$

58 $D_x \int_0^{x^2} \sqrt[3]{t^4 + 1}\, dt$

59 $D_x \int_{3x}^{x^3} (t^3 + 1)^{10}\, dt$

60 $D_x \int_{1/x}^{\sqrt{x}} \sqrt{t^4 + t^2 + 4}\, dt$

5.5 INDEFINITE INTEGRALS AND CHANGE OF VARIABLES

Because of the connection between antiderivatives and definite integrals provided by the Fundamental Theorem of Calculus, we generally use integral signs to denote antiderivatives. In order to distinguish antiderivatives from definite integrals, no limits of integration are attached to the integral sign, as in the following definition.

DEFINITION (5.27)

The **indefinite integral** $\int f(x)\, dx$ of f (or of $f(x)$) is defined by
$$\int f(x)\, dx = F(x) + C$$
for an antiderivative F of f and an arbitrary constant C.

Note that the indefinite integral is merely another way of specifying the most general antiderivative of f (see page 214). Instead of using the term *antidifferentiation* for the process of finding F when f is given, we now use the phrase **indefinite integration**. The arbitrary constant C is the **constant of integration**, $f(x)$ is the **integrand**, and x is the **variable of integration**. We often refer to the process of finding $F(x) + C$ in Definition (5.27) as **evaluating the indefinite integral**. The domain of F will not usually be stated explicitly. We always assume that a suitable interval over which f is integrable has been chosen. In particular, a closed interval on which f is continuous could be used. As with definite integrals, the symbol employed for the variable of integration is insignificant since, for example, $\int f(t)\, dt$, $\int f(u)\, du$, and so on, specify the same function F as $\int f(x)\, dx$.

Since the indefinite integral of f is an antiderivative, the Fundamental Theorem of Calculus relates definite and indefinite integrals, as follows.

THEOREM (5.28)

$$\int_a^b f(x)\, dx = \left[\int f(x)\, dx \right]_a^b$$

Thus, if we know the indefinite integral of a function f, then we can evaluate definite integrals of f. In later chapters we will develop methods for finding indefinite integrals of many different functions. At the present we can state several useful rules. The following *power rule for indefinite integrals* may be proved by differentiating the expression on the right-hand side of the equation and showing that the integrand is obtained.

POWER RULE FOR (5.29)
INDEFINITE INTEGRALS

If r is a rational number and $r \neq -1$, then

$$\int x^r \, dx = \frac{x^{r+1}}{r+1} + C$$

EXAMPLE 1 Evaluate: (a) $\int x^2 \, dx$ (b) $\int \left(8t^3 - 6\sqrt{t} + \frac{1}{t^2} \right) dt$

SOLUTION We only need to find an antiderivative for each integrand and then add the arbitrary constant C. Using the Power Rule (5.29),

(a) $\int x^2 \, dx = \frac{x^3}{3} + C = \frac{1}{3} x^3 + C$

(b) $\int (8t^3 - 6t^{1/2} + t^{-2}) \, dt = 8 \cdot \frac{t^4}{4} - 6 \cdot \frac{t^{3/2}}{3/2} + \frac{t^{-1}}{-1} + C$

$$= 2t^4 - 4t^{3/2} - \frac{1}{t} + C \quad \bullet$$

The next theorem indicates what happens if an indefinite integration is followed by a differentiation, or vice versa.

THEOREM (5.30)

(i) $D_x \int f(x) \, dx = f(x)$

(ii) $\int D_x [f(x)] \, dx = f(x) + C$

PROOF To prove (i) we may use Definition (5.27) as follows:

$$D_x \int f(x) \, dx = D_x [F(x) + C] = F'(x) = f(x).$$

Formula (ii) is true because $f(x)$ is an antiderivative of $D_x [f(x)]$. $\bullet\bullet$

EXAMPLE 2 Verify Theorem (5.30) for the special case $f(x) = x^2$.

SOLUTION

(i) If we first integrate, and then differentiate,

$$D_x \int x^2 \, dx = D_x \left(\frac{x^3}{3} + C \right) = x^2$$

(ii) If we first differentiate, and then integrate,

$$\int D_x (x^2) \, dx = \int 2x \, dx = x^2 + C \quad \bullet$$

The following theorem is analogous to Theorems (5.14) and (5.15) for definite integrals. The proof follows from Theorems (3.11) and (3.12), and the fact that any two antiderivatives differ by an additive constant.

$$\text{(i)} \int cf(x)\, dx = c \int f(x)\, dx; \quad c \text{ a constant}$$

$$\text{(ii)} \int [f(x) + g(x)]\, dx = \int f(x)\, dx + \int g(x)\, dx$$

Every rule for differentiation can be transformed into a corresponding rule for indefinite integration. For example,

$$D_x \sqrt{x^2 + 5} = \frac{x}{\sqrt{x^2 + 5}} \quad \text{implies that} \quad \int \frac{x}{\sqrt{x^2 + 5}}\, dx = \sqrt{x^2 + 5} + C.$$

We may use the Chain Rule in similar fashion to obtain a rule for indefinite integration. Thus, suppose that F is an antiderivative of f, and that g is a differentiable function such that $g(x)$ is in the domain of F for every x in some closed interval $[a, b]$. We may then consider the composite function defined by $F(g(x))$ for every x in $[a, b]$. Applying the Chain Rule (3.27) and the fact that $F' = f$,

$$D_x\, F(g(x)) = F'(g(x))g'(x) = f(g(x))g'(x).$$

This, in turn, gives us the integration formula

$$\int f(g(x))g'(x)\, dx = F(g(x)) + C \quad \text{with } F' = f.$$

There is a simple way to remember this formula. If we let $u = g(x)$ and formally replace $g'(x)\, dx$ by the differential du, we obtain

$$\int f(u)\, du = F(u) + C \quad \text{with } F' = f,$$

which has the same form as Definition (5.27). This indicates that $g'(x)\, dx$ may be regarded as the product of $g'(x)$ and dx. This relationship is one of the main reasons for using the symbol dx in integral notation. Since the variable x has been replaced by a new variable u, finding indefinite integrals in this way is referred to as a **change of variable** or as **the method of substitution**. We may summarize our discussion as follows, provided f and g have the properties described previously.

Given $\int f(g(x))g'(x)\, dx$, let $u = g(x)$ and $du = g'(x)\, dx$. If F is an antiderivative of f, then

$$\int f(g(x))g'(x)\, dx = \int f(u)\, du = F(u) + C = F(g(x)) + C$$

As a first application, let us extend Theorem (5.29) to powers of functions. If $f(x) = x^r$ in (5.32), then $f(g(x)) = [g(x)]^r$. If we let

$$F(x) = \frac{x^{r+1}}{r+1}, \quad \text{then} \quad F(g(x)) = \frac{[g(x)]^{r+1}}{r+1}$$

and the conclusion of Theorem (5.32) takes on the form stated in (5.33).

POWER RULE FOR FUNCTIONS (5.33)

$$\int [g(x)]^r g'(x)\, dx = \frac{[g(x)]^{r+1}}{r+1} + C; \ r \neq -1$$

We may also verify (5.33) by differentiating the expression on the right-hand side of the equation. A convenient way of expressing the Power Rule is as follows.

POWER RULE (5.34)
(ALTERNATIVE FORM)

Suppose $u = g(x)$ for a differentiable function g. If r is a rational number and $r \neq -1$, then

$$\int u^r\, du = \frac{u^{r+1}}{r+1} + C$$

EXAMPLE 3 Find $\int (2x^3 + 1)^7 x^2\, dx$.

SOLUTION If an integrand involves some expression raised to a power, as in $(2x^3 + 1)^7$, we often substitute a variable for the expression. Thus we let

$$u = 2x^3 + 1, \qquad du = 6x^2\, dx.$$

Note that once u is decided upon, du (the differential of u) is determined by differentiation. To change the form of the integral to that in (5.33), it is necessary to introduce the factor 6 into the integrand. Doing this, and compensating by multiplying the integral by $\frac{1}{6}$ (this is legitimate by Theorem (5.31) (i)), we obtain

$$\int (2x^3 + 1)^7 x^2\, dx = \frac{1}{6} \int (2x^3 + 1)^7 6x^2\, dx.$$

Making the indicated substitution and then using Theorem (5.34),

$$\int (2x^3 + 1)^7 x^2\, dx = \frac{1}{6} \int u^7\, du = \frac{1}{6}\left[\frac{u^8}{8} + K \right]$$

for a constant K. It is now necessary to return to the original variable x. Since $u = 2x^3 + 1$, the last equality gives us

$$\int (2x^3 + 1)^7 x^2\, dx = \frac{1}{6}\left[\frac{(2x^3 + 1)^8}{8} + K \right]$$

$$= \frac{1}{48}(2x^3 + 1)^8 + \frac{1}{6} K.$$

We can also write this result as

$$\int (2x^3 + 1)^7 x^2\, dx = \frac{1}{48}(2x^3 + 1)^8 + C.$$

for an arbitrary constant C. •

In Example 3 the relationship between K and C is $C = \frac{1}{6}K$; however, there is no practical advantage in remembering this fact. In the future we shall manipulate constants of integration in various ways without making explicit mention of existing relationships. Moreover, in Example 3 it is permissible to integrate as follows:

$$\frac{1}{6} \int u^7 \, du = \frac{1}{6} \left[\frac{u^8}{8} \right] + C$$

A substitution in an indefinite integral can be made in different ways. To illustrate, another method for solving Example 3 is to let

$$u = 2x^3 + 1, \qquad du = 6x^2 \, dx, \qquad \frac{1}{6} du = x^2 \, dx.$$

We then substitute $\frac{1}{6} du$ for $x^2 \, dx$ as follows:

$$\int (2x^3 + 1)^7 x^2 \, dx = \int u^7 \frac{1}{6} \, du = \frac{1}{6} \int u^7 \, du$$

$$= \frac{1}{48} u^8 + C = \frac{1}{48} (2x^3 + 1)^8 + C.$$

The change of variable technique for indefinite integrals is a powerful tool, provided we recognize that the integrand is of the form $f(g(x))g'(x)$, or $f(g(x))kg'(x)$, for a real number k. The ability to recognize this form is directly proportional to the number of exercises worked.

EXAMPLE 4 Find $\int x \sqrt[3]{7 - 6x^2} \, dx$.

SOLUTION Note that the integrand contains the term $x \, dx$. If the factor x were missing, the problem would be more complicated. For integrands that involve a radical we often substitute a variable for the expression under the radical sign. Thus we let

$$u = 7 - 6x^2, \qquad du = -12x \, dx.$$

We next introduce the factor -12 in the integrand, and then compensate by multiplying the integral by $-\frac{1}{12}$ as follows:

$$\int x \sqrt[3]{7 - 6x^2} \, dx = -\frac{1}{12} \int \sqrt[3]{7 - 6x^2} (-12)x \, dx$$

$$= -\frac{1}{12} \int \sqrt[3]{u} \, du = -\frac{1}{12} \int u^{1/3} \, du$$

$$= -\frac{1}{12} \frac{u^{4/3}}{(4/3)} + C = -\frac{1}{16} u^{4/3} + C$$

$$= -\frac{1}{16} (7 - 6x^2)^{4/3} + C.$$

As in the remarks following Example 2, we could have written

$$u = 7 - 6x^2, \qquad du = -12x \, dx, \qquad -\frac{1}{12} du = x \, dx$$

and substituted directly for $x\,dx$. Thus

$$\int \sqrt[3]{7 - 6x^2}\, x\, dx = \int \sqrt[3]{u}\left(-\frac{1}{12}\right) du = -\frac{1}{12}\int \sqrt[3]{u}\, du.$$

The remainder of the solution now proceeds exactly as before. •

Letting f in Theorem (5.32) denote either the sine or cosine function gives us the following theorem. Indefinite integrals of the other trigonometric functions will be considered later in the text.

THEOREM (5.35)

> If $u = g(x)$, and g is differentiable, then
> $$\int \sin u\, du = -\cos u + C$$
> $$\int \cos u\, du = \sin u + C$$

The preceding theorem generalizes the antiderivative formulas for $\sin kx$ and $\cos kx$ given in Theorem (4.35). Those formulas may now be discarded in favor of the technique used in the next example.

EXAMPLE 5 Evaluate $\int \sin 5x\, dx$.

SOLUTION We let

$$u = 5x, \qquad du = 5\, dx$$

and proceed as follows:

$$\int \sin 5x\, dx = \frac{1}{5}\int (\sin 5x)5\, dx$$

$$= \frac{1}{5}\int \sin u\, du$$

$$= -\frac{1}{5}\cos u + C = -\frac{1}{5}\cos 5x + C \quad •$$

EXAMPLE 6 Evaluate $\displaystyle\int \frac{\cos \sqrt{x}}{\sqrt{x}}\, dx$.

SOLUTION Making the substitutions

$$u = \sqrt{x} = x^{1/2}, \qquad du = \frac{1}{2}x^{-1/2}\, dx = \frac{1}{2\sqrt{x}}\, dx,$$

the integral may be evaluated as follows:

$$\int \frac{\cos \sqrt{x}}{\sqrt{x}}\, dx = 2\int \cos \sqrt{x}\left(\frac{1}{2\sqrt{x}}\right) dx$$

$$= 2\int \cos u\, du = 2\sin u + C$$

$$= 2\sin \sqrt{x} + C \qquad •$$

EXAMPLE 7 Evaluate $\int \cos z \, (1 + \sin z)^3 \, dz$.

SOLUTION We make the substitution

$$u = 1 + \sin z, \qquad du = \cos z \, dz$$

and proceed as follows:

$$\int \cos z \, (1 + \sin z)^3 \, dz = \int (1 + \sin z)^3 \cos z \, dz$$

$$= \int u^3 \, du$$

$$= \frac{u^4}{4} + C = \frac{1}{4} u^4 + C$$

$$= \tfrac{1}{4}(1 + \sin z)^4 + C \qquad \bullet$$

The method of substitution may also be used to evaluate a definite integral. We could use Theorem (5.32) to find an indefinite integral (that is, an antiderivative) and then apply the Fundamental Theorem of Calculus. Another method, which is sometimes shorter, is to change the limits of integration. Using (5.32) together with the Fundamental Theorem gives us the following formula, with $F' = f$:

$$\int_a^b f(g(x))g'(x) \, dx = F(g(x)) \Big]_a^b.$$

The number on the right may be written

$$F(g(b)) - F(g(a)) = F(u)\Big]_{g(a)}^{g(b)} = \int_{g(a)}^{g(b)} f(u) \, du.$$

This gives us the following result, provided f and g' are integrable.

THEOREM (5.36)

> If $u = g(x)$, then $\int_a^b f(g(x))g'(x) \, dx = \int_{g(a)}^{g(b)} f(u) \, du.$

Theorem (5.36) states that after making the substitution $u = g(x)$, we may use the values of g that correspond to $x = a$ and $x = b$, respectively, as the limits of the integral involving u. It is then unnecessary to return to the variable x after integrating. This technique is illustrated in the next example.

EXAMPLE 8 Evaluate $\int_2^{10} \dfrac{3}{\sqrt{5x - 1}} \, dx$.

SOLUTION Let us begin by writing the integral as

$$3 \int_2^{10} \frac{1}{\sqrt{5x - 1}} \, dx.$$

The form of the integrand suggests the following substitution:

$$u = 5x - 1, \qquad du = 5 \, dx$$

To change the limits of integration we note from $u = 5x - 1$ that

$$\text{if } x = 2, \quad \text{then} \quad u = 9,$$

and $\text{if } x = 10, \quad \text{then} \quad u = 49.$

Applying Theorem (5.36),

$$3 \int_2^{10} \frac{1}{\sqrt{5x - 1}} \, dx = \frac{3}{5} \int_2^{10} \frac{1}{\sqrt{5x - 1}} \, 5 \, dx$$

$$= \frac{3}{5} \int_9^{49} \frac{1}{\sqrt{u}} \, du = \frac{3}{5} \int_9^{49} u^{-1/2} \, du$$

$$= \left(\frac{3}{5}\right) \frac{u^{1/2}}{1/2} \Big]_9^{49} = \frac{6}{5} [49^{1/2} - 9^{1/2}] = \frac{24}{5} \quad \bullet$$

The following example illustrates a useful technique for evaluating certain definite integrals.

EXAMPLE 9 Let f be continuous on $[-a, a]$. If f is an even function, show that

$$\int_{-a}^{a} f(x) \, dx = 2 \int_0^a f(x) \, dx$$

SOLUTION If f is an even function, then the graph of f is symmetric with respect to the y-axis (see Theorem (1.23)). As a special case, if $f(x) \geq 0$ for every x in $[0, a]$, we have a situation similar to that in Figure 5.25, and hence the area under the graph of f from $x = -a$ to $x = a$ is twice that from $x = 0$ to $x = a$. This gives us the desired integration formula.

To show that the formula is true if $f(x) < 0$ for some x, we may proceed as follows. Using, successively, Theorem (5.16), Definition (5.9), and Theorem (5.14),

$$\int_{-a}^{a} f(x) \, dx = \int_{-a}^{0} f(x) \, dx + \int_0^a f(x) \, dx$$

$$= -\int_0^{-a} f(x) \, dx + \int_0^a f(x) \, dx$$

$$= \int_0^{-a} f(x)(-dx) + \int_0^a f(x) \, dx$$

Since f is even, $f(-x) = f(x)$, and the last equality may be written

$$\int_{-a}^{a} f(x) \, dx = \int_0^{-a} f(-x)(-dx) + \int_0^a f(x) \, dx.$$

If, in the first integral on the right, we substitute $u = -x$, $du = -dx$, and observe that $u = a$ when $x = -a$, we obtain

$$\int_{-a}^{a} f(x) \, dx = \int_0^a f(u) \, du + \int_0^a f(x) \, dx.$$

The last two integrals on the right are equal, since the variables are *dummy* variables, and, therefore,

$$\int_{-a}^{a} f(x) \, dx = 2 \int_0^a f(x) \, dx \quad \bullet$$

See Exercise 53 for a corresponding result if f is an *odd* function.

FIGURE 5.25

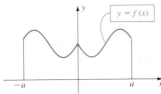

Exer. 1–33: Evaluate the integral.

1 $\int (3x + 1)^4 \, dx$

2 $\int (2x^2 - 3)^5 x \, dx$

3 $\int t^2 \sqrt{t^3 - 1} \, dt$

4 $\int z \sqrt{9 - z^2} \, dz$

5 $\int \dfrac{x - 2}{(x^2 - 4x + 3)^3} \, dx$

6 $\int \dfrac{x^2 + x}{(4 - 3x^2 - 2x^3)^4} \, dx$

7 $\int \dfrac{s}{\sqrt[3]{1 - 2s^2}} \, ds$

8 $\int \sqrt[5]{t^4 - t^2}\,(10t^3 - 5t) \, dt$

9 $\int \dfrac{(\sqrt{u} + 3)^4}{\sqrt{u}} \, du$

10 $\int \left(1 + \dfrac{1}{u}\right)^{-3} \left(\dfrac{1}{u^2}\right) du$

11 $\int_1^4 \sqrt{5 - x} \, dx$

12 $\int_1^5 \sqrt[3]{2x - 1} \, dx$

13 $\int \cos (4x - 3) \, dx$

14 $\int \sin (1 + 6x) \, dx$

15 $\int x \sin (x^2) \, dx$

16 $\int \dfrac{\cos \sqrt[3]{x}}{\sqrt[3]{x^2}} \, dx$

17 $\int \cos 3x \sqrt[3]{\sin 3x} \, dx$

18 $\int \dfrac{\sin 2x}{\sqrt{1 - \cos 2x}} \, dx$

19 $\int_{-1}^1 (t^2 - 1)^3 t \, dt$

20 $\int_{-2}^0 \dfrac{v^2}{(v^3 - 2)^2} \, dv$

21 $\int_0^1 \dfrac{1}{(3 - 2v)^2} \, dv$

22 $\int_0^4 \dfrac{x}{\sqrt{x^2 + 9}} \, dx$

23 $\int (x^2 + 1)^3 \, dx$

24 $\int (3 - x^3)^2 x \, dx$

25 $\int 5 \sqrt{8x + 5} \, dx$

26 $\int \dfrac{6}{\sqrt{4 - 5t}} \, dt$

27 $\int_1^4 \dfrac{1}{\sqrt{x}(\sqrt{x} + 1)^3} \, dx$

28 $\int (3 - x^4)^3 x^3 \, dx$

29 $\int \sin x \,(1 + \sqrt{\cos x})^2 \, dx$

30 $\int \sin^3 x \cos x \, dx$

31 $\int \dfrac{\sin x}{\cos^2 x} \, dx$

32 $\int \dfrac{\cos x}{(1 - \sin x)^2} \, dx$

33 $\int (2 + 5 \cos x)^3 \sin x \, dx$

34 Show, by evaluating in three different ways, that

$$\int \sin x \cos x \, dx = \tfrac{1}{2} \sin^2 x + C$$
$$= -\tfrac{1}{2} \cos^2 x + D = -\tfrac{1}{4} \cos 2x + E.$$

How can all three answers be correct?

Exer. 35–38: Evaluate the integral by (a) the method of substitution and (b) expanding the integrand. In what way do the constants of integration differ?

35 $\int (x + 4)^2 \, dx$

36 $\int (x^2 + 4)^2 x \, dx$

37 $\int \dfrac{(\sqrt{x} + 3)^2}{\sqrt{x}} \, dx$

38 $\int \left(1 + \dfrac{1}{x}\right)^2 \dfrac{1}{x^2} \, dx$

Exer. 39–40: Verify the formula by first integrating, and then differentiating.

39 $D_x \int x^3 \sqrt{x^4 + 5} \, dx = x^3 \sqrt{x^4 + 5}$

40 $D_x \int (3x + 2)^7 \, dx = (3x + 2)^7$

41 Find $D_x \int \dfrac{1}{\sqrt{x^3 + x + 5}} \, dx.$

42 Find $\int D_x \dfrac{x}{\sqrt[3]{x + 1}} \, dx.$

43 Find $\int_0^3 D_x \sqrt{x^2 + 16} \, dx.$

44 Find $D_x \int_0^1 x \sqrt{x^2 + 4} \, dx.$

45 If $f(x) = \sqrt{x + 1}$, find the area of the region under the graph of f from 0 to 3.

46 If $f(x) = x/(x^2 + 1)^2$, find the area of the region under the graph of f from 1 to 2.

Exer. 47–50: Find (a) a number that satisfies the conclusion of the Mean Value Theorem (5.20) for the integral $\int_a^b f(x)\,dx$, and (b) the average value of f on $[a, b]$.

47 $\int_0^4 \dfrac{x}{\sqrt{x^2 + 9}} \, dx$

48 $\int_{-2}^0 \sqrt[3]{x + 1} \, dx$

49 $\int_0^5 \sqrt{x + 4} \, dx$

50 $\int_{-3}^2 \sqrt{6 - x} \, dx$

51 The vertical distribution of velocity of the water in a river may frequently be approximated by the formula $v = c(D - y)^{1/6}$ for velocity v (in m/sec) at a depth of y meters below the water surface, depth D of the river, and a positive constant c.

(a) Find a formula for the average velocity v_{av} in terms of D and c.

(b) Show that $v_{av} = \tfrac{6}{7} v_0$ if v_0 is the velocity at the surface.

52 In the electrical circuit shown in the figure, the alternating current I can be represented by $I = I_M \sin \omega t$ for time t and maximum current I_M. The rate P at which heat is being produced in the resistor of R ohms is given by $P = I^2 R$. Compute the *average rate* of production of heat over one complete cycle (from $t = 0$ to $t = 2\pi/\omega$). (*Hint:* Use the trigonometric identity $\sin^2 u = (1 - \cos 2u)/2$.)

EXERCISE 52

53 Let f be continuous on $[-a, a]$. If f is an odd function show that $\int_{-a}^a f(x) \, dx = 0$. Interpret this result graphically.

54 Use Exercise 53 to show that

(a) $\int_{-\pi}^{\pi} \sin^5 x \, dx = 0$

(b) $\int_{-2}^{2} (2x^5 - 3x^3 + 4x)^{1/3} \, dx = 0$

Exer. 55–58: Verify the formula by differentiating.

55 $\int \dfrac{\sqrt{x^2 - a^2}}{x^4} \, dx = \dfrac{\sqrt{(x^2 - a^2)^3}}{3a^2 x^3} + C$

56 $\int \dfrac{1}{x^2 \sqrt{x^2 - a^2}} \, dx = \dfrac{\sqrt{x^2 - a^2}}{a^2 x} + C$

57 $\int \sin^2 x \, dx = \tfrac{1}{2}x - \tfrac{1}{4} \sin 2x + C$

58 $\int \cos^3 x \, dx = \tfrac{1}{3}(2 + \cos^2 x) \sin x + C$

5.6 NUMERICAL INTEGRATION

To evaluate a definite integral $\int_a^b f(x)\, dx$ by means of the Fundamental Theorem of Calculus, we must find an antiderivative of f. If an antiderivative cannot be found, then we may use numerical methods to approximate the integral to any desired degree of accuracy. For example, if the norm of a partition of $[a, b]$ is small, then by Definition (5.8) the definite integral can be approximated by any Riemann sum of f. In particular, if we use a regular partition with $\Delta x = (b - a)/n$, then

$$\int_a^b f(x)\, dx \approx \sum_{k=1}^{n} f(w_k) \, \Delta x$$

for any number w_k in the kth subinterval $[x_{k-1}, x_k]$ of the partition. Of course, the accuracy of the approximation depends upon the nature of f and the magnitude of Δx. It may be necessary to make Δx very small in order to obtain the desired degree of accuracy. This, in turn, means that n is large, and hence the preceding sum contains many terms. Figure 5.2 illustrates the case in which $f(w_k)$ is the minimum value of f on $[x_{k-1}, x_k]$. In this case the error involved in the approximation is numerically the same as the area of the unshaded region that lies under the graph of f and over the inscribed rectangles.

If we let $w_k = x_{k-1}$, that is, if f is evaluated at the left-hand endpoint of each subinterval $[x_{k-1}, x_k]$, then

$$\int_a^b f(x)\, dx \approx \sum_{k=1}^{n} f(x_{k-1}) \, \Delta x.$$

If we let $w_k = x_k$, that is, if f is evaluated at the right-hand endpoint of $[x_{k-1}, x_k]$, then

$$\int_a^b f(x)\, dx \approx \sum_{k=1}^{n} f(x_k) \, \Delta x.$$

Another, and usually more accurate, approximation can be obtained by using the average of the last two approximations, that is,

$$\frac{1}{2}\left[\sum_{k=1}^{n} f(x_{k-1}) \, \Delta x + \sum_{k=1}^{n} f(x_k) \, \Delta x \right].$$

With the exception of $f(x_0)$ and $f(x_n)$, each function value $f(x_k)$ appears twice, and hence we may write the last expression as

$$\frac{\Delta x}{2}\left[f(x_0) + \sum_{k=1}^{n-1} 2f(x_k) + f(x_n) \right].$$

Since $\Delta x = (b - a)/n$, this gives us the following rule.

If f is continuous on $[a, b]$ and if a regular partition of $[a, b]$ is determined by $a = x_0, x_1, \ldots, x_n = b$, then

$$\int_a^b f(x)\, dx \approx \frac{b - a}{2n} \left[f(x_0) + 2f(x_1) + 2f(x_2) + \cdots \right.$$
$$\left. + 2f(x_{n-1}) + f(x_n) \right].$$

FIGURE 5.26

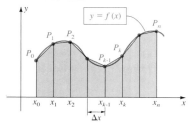

The term *trapezoidal* derives from the case in which $f(x)$ is nonnegative on $[a, b]$. As illustrated in Figure 5.26, if P_k is the point with x-coordinate x_k on the graph of $y = f(x)$, then for each $k = 1, 2, \ldots, n$, the points on the x-axis with x-coordinates x_{k-1} and x_k, together with P_{k-1} and P_k, are vertices of a trapezoid having area

$$\frac{\Delta x}{2} \left[f(x_{k-1}) + f(x_k) \right].$$

The sum of the areas of these trapezoids is the same as the sum in Rule (5.37). Hence, in geometric terms, the Trapezoidal Rule gives us an approximation to the area under the graph of f from a to b by means of trapezoids instead of the rectangles associated with Riemann sums.

The next result provides information about the maximum error that can occur if the Trapezoidal Rule is used to approximate a definite integral. The proof is omitted.

ERROR ESTIMATE FOR (5.38)
THE TRAPEZOIDAL RULE

If M is a positive real number such that $|f''(x)| \le M$ for every x in $[a, b]$, then the error involved in using the Trapezoidal Rule (5.37) is not greater than $M(b - a)^3/(12n^2)$.

EXAMPLE 1 Use the Trapezoidal Rule with $n = 10$ to approximate $\int_1^2 (1/x)\, dx$. Estimate a maximum error in the approximation.

SOLUTION It is convenient to arrange our work as follows. Each $f(x_k)$ was obtained with a calculator and is accurate to nine decimal places. The column labeled m contains the coefficient of $f(x_k)$ in the Trapezoidal Rule (5.37). Thus $m = 1$ for $f(x_0)$ or $f(x_n)$, and $m = 2$ for the remaining $f(x_h)$.

k	x_k	$f(x_k)$	m	$mf(x_k)$
0	1.0	1.000000000	1	1.000000000
1	1.1	0.909090909	2	1.818181818
2	1.2	0.833333333	2	1.666666666
3	1.3	0.769230769	2	1.538461538
4	1.4	0.714285714	2	1.428571428
5	1.5	0.666666667	2	1.333333334
6	1.6	0.625000000	2	1.250000000
7	1.7	0.588235294	2	1.176470588
8	1.8	0.555555556	2	1.111111112
9	1.9	0.526315790	2	1.052631580
10	2.0	0.500000000	1	0.500000000

The sum of the numbers in the last column is 13.875428064.

Since

$$\frac{b-a}{2n} = \frac{2-1}{20} = \frac{1}{20},$$

it follows from (5.37) that

$$\int_1^2 \frac{1}{x}\,dx \approx \frac{1}{20}(13.875428064) \approx 0.693771403$$

The error in the approximation may be estimated by means of (5.38). Since $f(x) = 1/x$, we have $f'(x) = -1/x^2$ and $f''(x) = 2/x^3$. The maximum value of $f''(x)$ on the interval $[1, 2]$ occurs at $x = 1$, and hence

$$|f''(x)| \le \frac{2}{(1)^3} = 2.$$

Applying (5.38) with $M = 2$, we see that the maximum error is not greater than

$$\frac{2(2-1)^3}{12(10)^2} = \frac{1}{600} < 0.002 \quad \bullet$$

In Chapter 7 we will see that the integral in Example 1 equals the natural logarithm of 2, denoted by ln 2. Using a calculator it is possible to verify that, to nine decimal places, ln 2 is approximated by 0.693147181. To obtain this approximation by means of the Trapezoidal Rule, it is necessary to use a very large value of n.

The following rule is often more accurate than the Trapezoidal Rule.

SIMPSON'S RULE (5.39)

> Suppose f is continuous on $[a, b]$ and n is an even integer. If a regular partition is determined by $a = x_0, x_1, \ldots, x_n = b$, then
>
> $$\int_a^b f(x)\,dx \approx \frac{b-a}{3n}\big[f(x_0) + 4f(x_1) + 2f(x_2) + 4f(x_3) + \cdots$$
> $$+ 2f(x_{n-2}) + 4f(x_{n-1}) + f(x_n)\big]$$

The idea behind the proof of Simpson's Rule is that instead of using trapezoids to approximate the graph of f, we use portions of graphs of equations of the form $y = cx^2 + dx + e$; that is, portions of parabolas or lines. If $P_0(x_0, y_0)$, $P_1(x_1, y_1)$, and $P_2(x_2, y_2)$ are points on the parabola, such that $x_0 < x_1 < x_2$, then substituting the coordinates of P_0, P_1, and P_2, respectively, into the equation gives us three equations that may be solved for c, d, and e. As a special case, suppose h, y_0, y_1, and y_2 are positive, and consider the points $P_0(-h, y_0)$, $P_1(0, y_1)$, and $P_2(h, y_2)$, as illustrated in Figure 5.27. The area A under the graph of the equation from $-h$ to h is

FIGURE 5.27

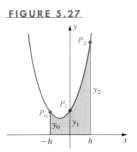

$$A = \int_{-h}^h (cx^2 + dx + e)\,dx = \frac{cx^3}{3} + \frac{dx^2}{2} + ex\bigg]_{-h}^h = \frac{h}{3}(2ch^2 + 6e).$$

Since the coordinates of $P_0(-h, y_0)$, $P_1(0, y_1)$, and $P_2(h, y_2)$ are solutions of $y = cx^2 + dx + e$, we have, by substitution,

$$y_0 = ch^2 - dh + e$$

$$y_1 = e$$

$$y_2 = ch^2 + dh + e.$$

Thus

$$y_0 + 4y_1 + y_2 = 2ch^2 + 6e$$

and

$$A = \frac{h}{3}(y_0 + 4y_1 + y_2).$$

If the points P_0, P_1, and P_2 are translated horizontally, as illustrated in Figure 5.28, then the area under the graph remains the same. Consequently, the preceding formula for A is true for *any* points P_0, P_1, and P_2, provided $x_1 - x_0 = x_2 - x_1$.

If $f(x) \geq 0$ on $[a, b]$, then Simpson's Rule is obtained by regarding the definite integral as the area under the graph of f from a to b. Thus, suppose n is an even integer and $h = (b - a)/n$. We divide $[a, b]$ into n subintervals, each of length h, by choosing numbers $a = x_0, x_1, \ldots, x_n = b$. Let $P_k(x_k, y_k)$ be the point on the graph of f with x-coordinate x_k, as illustrated in Figure 5.29.

FIGURE 5.28

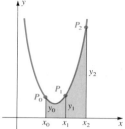

FIGURE 5.29

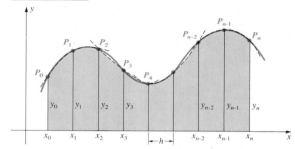

If the arc through P_0, P_1, and P_2 is approximated by the graph of an equation $y = cx^2 + dx + e$, then, as we have seen, the area under the graph of f from x_0 to x_2 is approximated by

$$\frac{h}{3}(y_0 + 4y_1 + y_2).$$

If we use the arc through P_2, P_3, and P_4, then the area under the graph of f from x_2 to x_4 is approximately

$$\frac{h}{3}(y_2 + 4y_3 + y_4).$$

We continue in this manner until we reach the last triple of points P_{n-2}, P_{n-1}, P_n and the corresponding approximation to the area under the graph, that is,

$$\frac{h}{3}(y_{n-2} + 4y_{n-1} + y_n).$$

Summing these approximations gives us

$$\int_a^b f(x)\, dx \approx \frac{h}{3}(y_0 + 4y_1 + 2y_2 + 4y_3 + \cdots + 2y_{n-2} + 4y_{n-1} + y_n)$$

which is the same as the sum in (5.39). If f is negative for some x in $[a, b]$, then negatives of areas may be used to establish Simpson's Rule.

The following result is analogous to (5.38).

ERROR ESTIMATE FOR (5.40)
SIMPSON'S RULE

> If M is a positive real number such that $\left|f^{(4)}(x)\right| \le M$ for every x in $[a, b]$, then the error involved in using Simpson's Rule (5.39) is not greater than $M(b - a)^5/(180n^4)$.

EXAMPLE 2 Use Simpson's Rule with $n = 10$ to approximate $\int_1^2 (1/x)\, dx$. Estimate the error in the approximation.

SOLUTION This is the same integral considered in Example 1. We arrange our work as follows. The column labeled m contains the coefficient of $f(x_k)$ in Simpson's Rule (5.39).

k	x_k	$f(x_k)$	m	$mf(x_k)$
0	1.0	1.000000000	1	1.000000000
1	1.1	0.909090909	4	3.636363636
2	1.2	0.833333333	2	1.666666666
3	1.3	0.769230769	4	3.076923078
4	1.4	0.714285714	2	1.428571428
5	1.5	0.666666667	4	2.666666668
6	1.6	0.625000000	2	1.250000000
7	1.7	0.588235294	4	2.352941176
8	1.8	0.555555556	2	1.111111112
9	1.9	0.526315790	4	2.105263160
10	2.0	0.500000000	1	0.500000000

The sum of the numbers in the last column is 20.794506924. Since

$$\frac{b - a}{3n} = \frac{2 - 1}{30},$$

it follows from (5.39) that

$$\int_1^2 \frac{1}{x}\, dx \approx \left(\frac{1}{30}\right)(20.794506924) \approx 0.693150231$$

We shall use (5.40) to estimate the error in the approximation. If $f(x) = 1/x$, we can verify that $f^{(4)}(x) = 24/x^5$. The maximum value of $f^{(4)}(x)$ on the interval $[1, 2]$ occurs at $x = 1$, and hence

$$\left|f^{(4)}(x)\right| \le \frac{24}{(1)^5} = 24.$$

Applying (5.40) with $M = 24$, we see that the maximum error in the approximation is not greater than

$$\frac{24(2 - 1)^5}{180(10)^4} = \frac{2}{150000} < 0.00002$$

Note that this estimated error is much less than that obtained using the Trapezoidal Rule in Example 1. •

At the beginning of this section we noted that a definite integral may be approximated by a Riemann sum $\sum_k f(w_k)\, \Delta x$. We then considered the special case where w_k was always a left-hand (or right-hand) endpoint of the kth subinterval $[x_{k-1}, x_k]$ of a regular partition of $[a, b]$. Another technique is to choose w_k as the *midpoint* of $[x_{k-1}, x_k]$. This leads to the next rule.

THE MIDPOINT RULE (5.41)

> If f is continuous on $[a, b]$ and if a regular partition of $[a, b]$ is determined by $a = x_0, x_1, \ldots, x_n = b$, then
> $$\int_a^b f(x)\, dx \approx \frac{b-a}{n} \left[f(\bar{x}_1) + f(\bar{x}_2) + \cdots + f(\bar{x}_n) \right]$$
> with $\bar{x}_k = (x_{k-1} + x_k)/2$ the midpoint of $[x_{k-1}, x_k]$.

The proof of (5.41) merely consists of noting that $\Delta x = (b - a)/n$ for each k. It can be shown that if M is a positive real number and $|f''(x)| \le M$ for every x in $[a, b]$, then the error involved in using the Midpoint Rule is not greater than $M(b - a)^3/(24n^2)$.

EXAMPLE 3 Use the Midpoint Rule with $n = 10$ to approximate $\int_1^2 (1/x)\, dx$.

SOLUTION The integral is the same as that considered in the preceding examples. Since the partition is determined by $1, 1.1, 1.2, \ldots, 1.9, 2$, the midpoints of the subintervals are $1.05, 1.15, \ldots, 1.95$, and the Midpoint Rule gives us

$$\int_1^2 \frac{1}{x}\, dx \approx 0.1 \left[\frac{1}{1.05} + \frac{1}{1.15} + \frac{1}{1.25} + \frac{1}{1.35} + \frac{1}{1.45} + \frac{1}{1.55} + \frac{1}{1.65} \right.$$
$$\left. + \frac{1}{1.75} + \frac{1}{1.85} + \frac{1}{1.95} \right]$$

We can verify that this gives us the approximation 0.693. •

An important aspect of numerical integration is that it can be used to approximate the definite integral of a function that is described by means of a table or graph. To illustrate, suppose it is found experimentally that two physical variables x and y are related as shown in the following table.

x	1.0	1.5	2.0	2.5	3.0	3.5	4.0
y	3.1	4.0	4.2	3.8	2.9	2.8	2.7

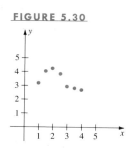

FIGURE 5.30

The points (x, y) are plotted in Figure 5.30. If we regard y as a function of x, say $y = f(x)$ with f continuous, then the definite integral $\int_1^4 f(x)\, dx$ may represent a physical quantity. In the present illustration, *the integral may be approximated without knowing an explicit form for $f(x)$*. In particular, if the Trapezoidal Rule (5.37) is used, with $n = 6$ and $(b - a)/(2n) = (4 - 1)/12 = 0.25$, then

$$\int_1^4 f(x)\, dx \approx 0.25[3.1 + 2(4.0) + 2(4.2) + 2(3.8) + 2(2.9) + 2(2.8) + 2.7]$$
$$\approx 10.3$$

Since the number of subdivisions is even, we could also approximate the integral by means of Simpson's Rule.

Exer. 1–8: Use (a) the Trapezoidal Rule and (b) Simpson's Rule to approximate the definite integral for the stated value of n. Use approximations to four decimal places for $f(x_k)$ and round off answers to two decimal places. (It is advisable to use a calculator for these exercises.)

1 $\int_1^4 \frac{1}{x} \, dx, \quad n = 6$

2 $\int_0^3 \frac{1}{1 + x} \, dx, \quad n = 8$

3 $\int_0^1 \frac{1}{\sqrt{1 + x^2}} \, dx, \quad n = 4$

4 $\int_2^3 \sqrt{1 + x^3} \, dx, \quad n = 4$

5 $\int_0^2 \frac{1}{4 + x^2} \, dx, \quad n = 10$

6 $\int_0^{0.6} \frac{1}{\sqrt{4 - x^2}} \, dx, \quad n = 6$

7 $\int_0^\pi \sqrt{\sin x} \, dx, \quad n = 6$

8 $\int_0^\pi \sin \sqrt{x} \, dx, \quad n = 4$

9–12 Rework Exercises 1–4 using the Midpoint Rule.

13 Use the Trapezoidal Rule with $(b - a)/n = 0.1$ to show that

$$\int_1^{2.7} \frac{1}{x} \, dx < 1 < \int_1^{2.8} \frac{1}{x} \, dx.$$

14 Find upper bounds for the errors in parts (a) and (b) of Exercise 1.

Exer. 15–16: Suppose the table of values for physical variables x and y was obtained experimentally. Assuming that $y = f(x)$ with f continuous, approximate $\int_2^4 f(x) \, dx$ by means of (a) the Trapezoidal Rule; (b) Simpson's Rule.

15

x	y
2.00	4.12
2.25	3.76
2.50	3.21
2.75	3.58
3.00	3.94
3.25	4.15
3.50	4.69
3.75	5.44
4.00	7.52

16

x	y
2.0	12.1
2.2	11.4
2.4	9.7
2.6	8.4
2.8	6.3
3.0	6.2
3.2	5.8
3.4	5.4
3.6	5.1
3.8	5.9
4.0	5.6

17 The graph in the figure was recorded by an instrument used to measure a physical quantity. Estimate y-coordinates of points on the graph and approximate the area of the shaded region by using (with $n = 6$) (a) the Trapezoidal Rule; (b) Simpson's Rule; (c) the Midpoint Rule.

EXERCISE 17

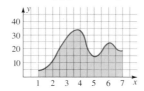

18 A man-made lake has the shape illustrated in the figure, with adjacent measurements 20 feet apart. Use the Trapezoidal Rule to estimate the surface area of the lake.

EXERCISE 18

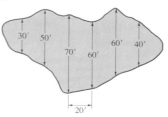

19 An important aspect of water management is the production of reliable data on *streamflow*, the number of cubic meters per second passing through a cross section of a stream or river. A first step in this computation is the determination of the average velocity $\bar{v}_x$ at a distance x meters from the river bank (see figure). If D is the depth of the stream at a point x meters from the bank, and $v(y)$ is the velocity (in m/sec) at a depth y meters, then

$$\bar{v}_x = \frac{1}{D} \int_0^D v(y) \, dy$$

(see Definition (5.21)). The *Six-Point Method* takes velocity readings at the surface; at depths $0.2D$, $0.4D$, $0.6D$, $0.8D$; and near the river bottom, and then uses the Trapezoidal Rule to estimate $\bar{v}_x$. Given the data in the following table, estimate $\bar{v}_x$.

y (m)	0	0.2D	0.4D	0.6D	0.8D	D
$v(y)$ (m/sec)	0.28	0.23	0.19	0.17	0.13	0.02

EXERCISE 19

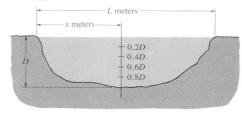

20 Refer to Exercise 19. The streamflow F (in m³/sec) can be computed using the formula

$$F = \int_0^L \bar{v}_x \, D(x) \, dx$$

for the depth $D(x)$ of the stream at a distance x meters from the bank and the length L of the cross section. Given the data in the following table, use Simpson's Rule to estimate F.

x (m)	0	3	6	9	12	15	18	21	24
$D(x)$ (m)	0	0.51	0.73	1.61	2.11	2.02	1.53	0.64	0
$\bar{v}_x$ (m/sec)	0	0.09	0.18	0.21	0.36	0.32	0.19	0.11	0

Exer. 21–22: Find the least integer n such that the error involved in approximating the definite integral is less than the indicated value if (a) the Trapezoidal Rule is used; (b) Simpson's Rule is used.

21 $\int_{1/2}^{1} (1/x) \, dx$; 0.0001

22 $\int_{0}^{3} (1 + x)^{-1} \, dx$; 0.001

23 If $f(x)$ is a polynomial of degree less than 4, prove that Simpson's Rule gives the exact value of $\int_{a}^{b} f(x) \, dx$.

24 If f is continuous and has nonnegative values, and if $f''(x) > 0$ throughout $[a, b]$, prove that $\int_{a}^{b} f(x) \, dx$ is less than the number given by the Trapezoid Rule.

5.7 REVIEW

Define or discuss each of the following.

1 The area under the graph of a nonnegative continuous function f from a to b
2 Partition of $[a, b]$
3 Norm of a partition
4 Riemann sum
5 $\lim\limits_{||P|| \to 0} \sum\limits_{k} f(w_k) \, \Delta x_k = I$
6 The definite integral of f from a to b
7 Upper and lower limits of integration
8 Integrand
9 Properties of the definite integral

10 The Mean Value Theorem for Definite Integrals
11 Average value of a function
12 The Fundamental Theorem of Calculus
13 Indefinite integral
14 Constant of integration
15 Power Rule for integration
16 Change of variables
17 Trapezoidal Rule
18 Simpson's Rule
19 Midpoint Rule

EXERCISES 5.7

Exer. 1–3: Determine the sum.

1 $\sum\limits_{k=1}^{5} (k^2 + 3)$

2 $\sum\limits_{k=1}^{50} 8$

3 $\sum\limits_{k=0}^{4} (-1/2)^{k-2}$

4 Suppose $f(x) = 1 - x^2$ is defined on the interval $[-2, 3]$, and let P be the regular partition of $[-2, 3]$ into five equal subintervals. Find the Riemann sum R_P if f is evaluated at the midpoint of each subinterval.

5 Work Exercise 4 if f is evaluated at the right-hand endpoint of each subinterval.

6 Given $\int_{1}^{4} (x^2 + 2x - 5) \, dx$, find (a) numbers that satisfy the conclusion of the Mean Value Theorem (5.20), and (b) the average value of $x^2 + 2x - 5$ on $[1, 4]$.

Exer. 7–36: Evaluate the integral.

7 $\int_{0}^{1} \sqrt[3]{8x^7} \, dx$

8 $\int \sqrt[3]{5t + 1} \, dt$

9 $\int_{0}^{1} \dfrac{z^2}{(1 + z^3)^2} \, dz$

10 $\int (x^2 + 4)^2 \, dx$

11 $\int (1 - 2x^2)^3 x \, dx$

12 $\int \dfrac{(1 + \sqrt{x})^2}{\sqrt[3]{x}} \, dx$

13 $\int_{1}^{2} \dfrac{w + 1}{\sqrt{w^2 + 2w}} \, dw$

14 $\int_{1}^{2} \dfrac{s^2 + 2}{s^2} \, ds$

15 $\int \dfrac{1}{\sqrt{x}(1 + \sqrt{x})^2} \, dx$

16 $\int_{1}^{2} \dfrac{x^2 - x - 6}{x + 2} \, dx$

17 $\int (3 - 2x - 5x^3) \, dx$

18 $\int (y + y^{-1})^2 \, dy$

19 $\int_{0}^{2} x^2 \sqrt{x^3 + 1} \, dx$

20 $\int_{1}^{1} 3x^2 \sqrt{x^3 + x} \, dx$

21 $\int (4t + 1)(4t^2 + 2t - 7)^2 \, dt$

22 $\int \dfrac{\sqrt[4]{1 - v^{-1}}}{v^2} \, dv$

23 $\int (2x^{-3} - 3x^{-2}) \, dx$

24 $\int_{1}^{9} \sqrt{2r + 7} \, dr$

25 $\int \sin (3 - 5x) \, dx$

26 $\int x^2 \cos (2x^3) \, dx$

27 $\int \cos 3x \sin^4 3x \, dx$

28 $\int \dfrac{\sin (1/x)}{x^2} \, dx$

29 $\int_{0}^{\pi/2} \cos x \sqrt{3 + 5 \sin x} \, dx$

30 $\int \dfrac{\cos 3x}{\sin^3 3x} \, dx$

31 $\int_{0}^{\pi/4} \sin 2x \cos^2 2x \, dx$

32 $\int_{\pi/6}^{\pi/4} (\sec x + \tan x)(1 - \sin x) \, dx$

33 $\int D_y \sqrt[5]{y^4 + 2y^2 + 1} \, dy$

34 $\int_{0}^{\pi/2} D_x (x \sin^3 x) \, dx$

35 $D_x \int_0^1 (x^3 + x^2 - 7)^5 \, dx$

36 $D_z \int_0^z (x^2 + 1)^{10} \, dx$

37 Evaluate $\int_0^{10} \sqrt{1 + x^4} \, dx$ by using (a) the Trapezoidal Rule with $n = 5$ and (b) Simpson's Rule with $n = 8$. Use approximations to four decimal places for $f(x_k)$ and round off answers to two decimal places.

38 If $f(x) = x^4 \sqrt{x^5 + 4}$, find the area of the region under the graph of f from 0 to 2.

39 Find the area of the region under the graph of $y = \sin\left(\frac{1}{4}x\right)$ from $x = 0$ to $x = \pi$.

40 Use a definite integral to prove that the area of a right triangle of altitude a and base b is $\frac{1}{2}ab$. (*Hint:* Take the vertices at the points $(0, 0)$, $(b, 0)$, and (b, a).)

41 To monitor the thermal pollution of a river, a biologist takes hourly temperature readings (in °F) from 9 A.M. to 5 P.M. The results are shown in the following table.

Time of day	9	10	11	12	1	2	3	4	5
Temperature	75.3	77.0	83.1	84.8	86.5	86.4	81.1	78.6	75.1

Use Simpson's Rule and Definition (5.21) to estimate the average water temperature between 9 A.M. and 5 P.M.

42 If P is the profit function for a product, show that the average profit $P(n)/n$ of manufacturing n units equals the average value of the marginal profit function on $[0, n]$.

The definite integral is useful in solving a variety of problems. In this chapter we shall discuss area, volume, lengths of curves, surfaces of revolution, work, liquid force, center of mass, and applications from many fields, including physics, engineering, biology, and economics.

APPLICATIONS OF THE DEFINITE INTEGRAL

6.1 AREA

FIGURE 6.1

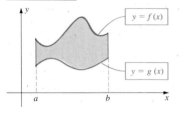

FIGURE 6.2

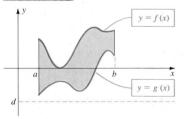

FIGURE 6.3

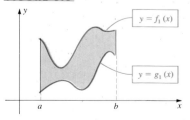

If a function f is continuous and nonnegative on a closed interval $[a, b]$, then by Theorem (5.11) the area under the graph of f from a to b is given by $\int_a^b f(x)\,dx$. If g is another such function, and if $f(x) \geq g(x)$ for every x in $[a, b]$, then the area A of the region bounded by the graphs of f, g, $x = a$, and $x = b$ (see Figure 6.1) can be found by subtracting the area under the graph of g from the area under the graph of f, that is,

$$A = \int_a^b f(x)\,dx - \int_a^b g(x)\,dx = \int_a^b [f(x) - g(x)]\,dx.$$

This formula for A is also true if f or g is negative for some x in $[a, b]$, as illustrated in Figure 6.2. To verify this, let us choose a negative number d less than the minimum value of g on $[a, b]$ and consider the two functions f_1 and g_1 defined by

$$f_1(x) = f(x) - d, \qquad g_1(x) = g(x) - d$$

for every x in $[a, b]$. Since d is negative, we find function values of f_1 and g_1 by adding the positive number $-d$ to values of f and g, respectively. Correspondingly, we raise the graphs of f and g a distance $|d|$, giving us a region having the same shape and area as the original region, but lying entirely above the x-axis (see Figure 6.3).

If A is the area of the region in Figure 6.3, then

$$A = \int_a^b [f_1(x) - g_1(x)]\,dx$$

$$= \int_a^b \{[f(x) - d] - [g(x) - d]\}\,dx$$

$$= \int_a^b [f(x) - g(x)]\,dx.$$

The preceding discussion may be summarized as follows.

THEOREM (6.1)

If f and g are continuous and $f(x) \geq g(x)$ for every x in $[a, b]$, then the area A of the region bounded by the graphs of f, g, $x = a$, and $x = b$ is

$$A = \int_a^b [f(x) - g(x)] \, dx$$

The formula for A in Theorem (6.1) may be interpreted as a limit of sums. If we let $h(x) = f(x) - g(x)$, and if w is in $[a, b]$, then $h(w)$ is the vertical distance between the graphs of f and g at $x = w$. As in the discussion of Riemann sums in Chapter 5, let P denote a partition of $[a, b]$ determined by the numbers $a = x_0, x_1, \ldots, x_n = b$. For each k, let $\Delta x_k = x_k - x_{k-1}$, and let w_k be an arbitrary number in the kth subinterval $[x_{k-1}, x_k]$ of P. By the definition of h,

$$h(w_k) \, \Delta x_k = [f(w_k) - g(w_k)] \, \Delta x_k,$$

which is the area of a rectangle of length $f(w_k) - g(w_k)$ and width Δx_k (see Figure 6.4).

FIGURE 6.4

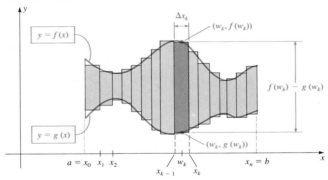

The Riemann sum

$$\sum_k h(w_k) \, \Delta x_k = \sum_k [f(w_k) - g(w_k)] \, \Delta x_k$$

is the sum of the areas of all rectangles pictured in Figure 6.4 and may be regarded as an approximation to the area of the region bounded by the graphs of f and g from $x = a$ to $x = b$. By the definition of the definite integral,

$$\lim_{\|P\| \to 0} \sum_k h(w_k) \, \Delta x_k = \int_a^b h(x) \, dx.$$

Using the definition of h gives us

(6.2) $\qquad A = \lim_{\|P\| \to 0} \sum_k [f(w_k) - g(w_k)] \, \Delta x_k = \int_a^b [f(x) - g(x)] \, dx$

When we use (6.2) to find areas, we initially think of approximating the region by means of rectangles, as in Figure 6.4. After writing a formula for the area of a typical rectangle, we sum all such rectangles and then find the limit of these sums, as illustrated in the following examples.

EXAMPLE 1 Find the area of the region bounded by the graphs of the equations $y = x^2$ and $y = \sqrt{x}$.

SOLUTION We shall use the Riemann sum approach. The region and a typical rectangle are sketched in Figure 6.5. The points $(0, 0)$ and $(1, 1)$ at which the graphs intersect can be found by solving the equations $y = x^2$ and $y = \sqrt{x}$ simultaneously. The length of a typical rectangle is $\sqrt{w_k} - w_k^2$ and its area is $(\sqrt{w_k} - w_k^2)\, \Delta x_k$. Using (6.2) with $a = 0$ and $b = 1$,

$$A = \lim_{\|P\| \to 0} \sum_k (\sqrt{w_k} - w_k^2)\, \Delta x_k$$

$$= \int_0^1 (\sqrt{x} - x^2)\, dx = \int_0^1 (x^{1/2} - x^2)\, dx$$

$$= \left[\frac{x^{3/2}}{3/2} - \frac{x^3}{3} \right]_0^1 = \frac{2}{3} - \frac{1}{3} = \frac{1}{3}.$$

The area can also be found using Theorem (6.1), with $f(x) = \sqrt{x}$ and $g(x) = x^2$. •

FIGURE 6.5

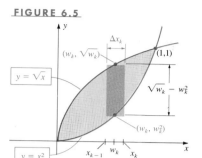

As illustrated in Example 1, to find areas by using limits of sums, we always begin by sketching the region with at least one typical rectangle, and labeling the drawing appropriately. The reason for stressing sums, instead of merely substituting in the formula of Theorem (6.1), is that similar limiting processes will be employed later for calculating many other mathematical and physical quantities. Treating areas as limits of sums will make it easier to understand future applications. At the same time it will help solidify the meaning of the definite integral.

EXAMPLE 2 Find the area of the region bounded by the graphs of $y + x^2 = 6$ and $y + 2x - 3 = 0$.

SOLUTION The region and a typical rectangle are sketched in Figure 6.6. The points of intersection $(-1, 5)$ and $(3, -3)$ of the two graphs may be found by solving the two given equations simultaneously. To use Theorem (6.1) or (6.2) it is necessary to solve each equation for y in terms of x, obtaining

$$y = 6 - x^2 \quad \text{and} \quad y = 3 - 2x.$$

The functions f and g are then given by

$$f(x) = 6 - x^2 \quad \text{and} \quad g(x) = 3 - 2x.$$

As shown in Figure 6.6, the length of a typical rectangle is

$$(6 - w_k^2) - (3 - 2w_k)$$

for some number w_k in the kth subinterval of a partition P of $[-1, 3]$. The area of this rectangle is

$$[(6 - w_k^2) - (3 - 2w_k)]\, \Delta x_k.$$

FIGURE 6.6

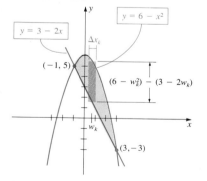

Applying (6.2),

$$A = \lim_{\|P\| \to 0} \sum_k \left[(6 - w_k^2) - (3 - 2w_k)\right] \Delta x_k$$

$$= \int_{-1}^{3} \left[(6 - x^2) - (3 - 2x)\right] dx$$

$$= \int_{-1}^{3} (3 - x^2 + 2x) \, dx$$

$$= \left[3x - \frac{x^3}{3} + x^2 \right]_{-1}^{3}$$

$$= \left[9 - \frac{27}{3} + 9 \right] - \left[-3 - \left(-\frac{1}{3}\right) + 1 \right] = \frac{32}{3} \quad \bullet$$

When we thoroughly understand the limit of sums method illustrated in Examples 1 and 2, we can bypass the subscript part of the solution and proceed immediately to "setting up" the integral. To illustrate, in Example 2 we could regard dx as the width of a typical rectangle. The length of the rectangle is represented by the distance

$$(6 - x^2) - (3 - 2x)$$

between the upper and lower boundaries of the region, for an arbitrary number x in $[-1, 3]$. Thus the area of a rectangle may be represented, in *nonsubscript* notation, by

$$\left[(6 - x^2) - (3 - 2x)\right] dx.$$

Placing the integral sign $\int_{-1}^{3}$ in front of this expression may be regarded as summing all such terms and simultaneously letting the widths of the rectangles approach 0.

Since one of our objectives is to emphasize the definition of the definite integral (5.8), we will continue to use limits of sums in many of the remaining examples in this chapter. The symbol w_k employed in solutions always denotes a number in the kth subinterval of a partition. Those who wish to use the nonsubscript approach may (at the discretion of the instructor) proceed immediately to the definite integral that follows a stated limited of sums.

The following example illustrates that it is sometimes necessary to subdivide a region and use Theorem (6.1) or (6.2) more than once in order to find the area.

EXAMPLE 3 Find the area of the region R that is bounded by the graphs of $y - x = 6$, $y - x^3 = 0$, and $2y + x = 0$.

FIGURE 6.7

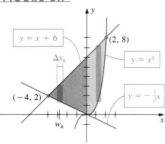

SOLUTION The graphs and region are sketched in Figure 6.7, where each equation has been solved for y in terms of x so that appropriate functions may be introduced. Typical rectangles are shown extending from the lower boundary to the upper boundary of R. Since the lower boundary consists of portions of two different graphs, the area cannot be found by using only one definite integral. However, if R is divided into two subregions R_1 and R_2, as shown in Figure 6.8, then we can determine the area of each and add them together.

For region R_1 the upper and lower boundaries are the lines $y = x + 6$ and $y = -x/2$, respectively, and hence the length of a typical rectangle is

$$(w_k + 6) - (-w_k/2).$$

FIGURE 6.8

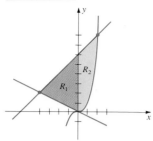

Applying (6.2), the area A_1 of R_1 is

$$A_1 = \lim_{||P|| \to 0} \sum_k \left[(w_k + 6) - \left(-\frac{w_k}{2} \right) \right] \Delta x_k$$

$$= \int_{-4}^{0} \left[(x + 6) - \left(-\frac{x}{2} \right) \right] dx$$

$$= \int_{-4}^{0} \left[\frac{3}{2} x + 6 \right] dx = \left[\frac{3}{2} \left(\frac{x^2}{2} \right) + 6x \right]_{-4}^{0}$$

$$= 0 - (12 - 24) = 12.$$

Region R_2 has the same upper boundary $y = x + 6$ as R_1; however, the lower boundary is given by $y = x^3$. In this case the length of a typical rectangle is

$$(w_k + 6) - w_k^3$$

and the area A_2 of R_2 is

$$A_2 = \lim_{||P|| \to 0} \sum_k \left[(w_k + 6) - w_k^3 \right] \Delta x_k$$

$$= \int_{0}^{2} \left[(x + 6) - x^3 \right] dx$$

$$= \left[\frac{x^2}{2} + 6x - \frac{x^4}{4} \right]_{0}^{2}$$

$$= (2 + 12 - 4) - 0 = 10.$$

The area A of the entire region R is

$$A = A_1 + A_2 = 12 + 10 = 22 \quad \bullet$$

FIGURE 6.9

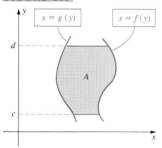

Let us next consider the area A of the region bounded by the graphs of $y = c$ and $y = d$, and of two equations of the form $x = f(y)$ and $x = g(y)$, for continuous functions f and g with $f(y) \geq g(y)$ for every y in $[c, d]$ (see Figure 6.9). As in our earlier discussion, but with the roles of x and y interchanged, we see that

$$A = \int_{c}^{d} [f(y) - g(y)] \, dy.$$

We now regard y as the independent variable and refer to this as **integration with respect to y.** The method used in Theorem (6.1) is **integration with respect to x.** As in Figure 6.10, if an equation is solved for x in terms of y such that $x = f(y)$, and if a value w is assigned to y, then $f(w)$ is the x-coordinate of the corresponding point on the graph.

FIGURE 6.10

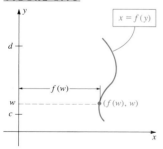

Limits of sums can also be used for the region shown in Figure 6.11. In this case we select points on the y-axis with y-coordinates $y_0, y_1, \ldots, y_n$, with $y_0 = c$ and $y_n = d$, thereby obtaining a partition of the interval $[c, d]$ into subintervals of width $\Delta y_k = y_k - y_{k-1}$. For each k we choose a number w_k in $[y_{k-1}, y_k]$ and consider horizontal rectangles that have areas $[f(w_k) - g(w_k)] \Delta y_k$, as illustrated in Figure 6.11. This leads to

$$A = \lim_{||P|| \to 0} \sum_k [f(w_k) - g(w_k)] \Delta y_k = \int_{c}^{d} [f(y) - g(y)] \, dy.$$

The last equality follows from the definition of the definite integral.

FIGURE 6.11

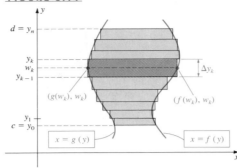

EXAMPLE 4 Find the area of the region bounded by the graphs of the equations $2y^2 = x + 4$ and $x = y^2$.

FIGURE 6.12

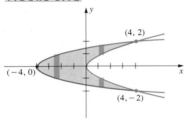

FIGURE 6.13

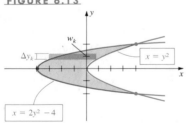

SOLUTION Sketches of the region are shown in Figures 6.12 and 6.13. Figure 6.12 illustrates the situation that occurs if we use vertical rectangles (integration with respect to x), and Figure 6.13 is the case if we use horizontal rectangles (integration with respect to y). Referring to Figure 6.12, we see that several definite integrals are required to find the area. However, in Figure 6.13 we can use integration with respect to y to find the area with only one integration. Letting $f(y) = y^2$, $g(y) = 2y^2 - 4$, and referring to Figure 6.13, the length $f(w_k) - g(w_k)$ of a *horizontal* rectangle is

$$w_k^2 - (2w_k^2 - 4).$$

Since the width is Δy_k, the area of the rectangle is

$$[w_k^2 - (2w_k^2 - 4)] \, \Delta y_k.$$

Hence, the area A is

$$A = \lim_{\|P\| \to 0} \sum_k [w_k^2 - (2w_k^2 - 4)] \, \Delta y_k$$

$$= \int_{-2}^{2} [y^2 - (2y^2 - 4)] \, dy$$

$$= \int_{-2}^{2} (4 - y^2) \, dy$$

$$= \left[4y - \frac{y^3}{3} \right]_{-2}^{2} = \left[8 - \frac{8}{3} \right] - \left[-8 - \left(-\frac{8}{3} \right) \right] = \frac{32}{3}.$$

Actually, the integration could have been simplified further. For example, since the x-axis bisects the region, it is sufficient to find the area of that part of the region that lies above the x-axis and double it, obtaining

$$A = 2 \int_{0}^{2} [y^2 - (2y^2 - 4)] \, dy.$$

(See the result about even functions in Example 9 of Section 5.5.) •

Throughout this section we have assumed that the graphs of the functions (or equations) do not cross one another in the interval under discussion. For example, in Theorem (6.1) we demanded that $f(x) \geq g(x)$ for every x in $[a, b]$.

FIGURE 6.14

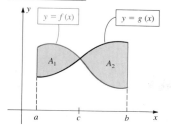

If the graphs of f and g cross at one point $P(c, d)$, with $a < c < b$, and we wish to find the area bounded by the graphs from $x = a$ to $x = b$, then the methods developed in this section may still be used; however, *two* integrations are required, one corresponding to the interval $[a, c]$ and the other to $[c, b]$. This is illustrated in Figure 6.14, with $f(x) \geq g(x)$ on $[a, c]$ and $g(x) \geq f(x)$ on $[c, b]$. The area A is given by

$$A = A_1 + A_2 = \int_a^c [f(x) - g(x)]\, dx + \int_c^b [g(x) - f(x)]\, dx.$$

If the graphs cross *several* times, then several integrals may be necessary. Problems where graphs cross one or more times appear in Exercises 33–36.

In scientific investigations, a physical quantity is often interpreted as an area. One illustration of this occurs in the *theory of elasticity*. To test the strength of a material, an investigator records values of strain that correspond to different loads (stresses). The sketch in Figure 6.15 is a typical stress-strain diagram for a sample of an elastic material, such as vulcanized rubber. (Note that stress values are assigned in the vertical direction.) Referring to the figure, we see that as the load applied to the material (the stress) increases, the strain (indicated by the arrows on the colored graph) increases until the material is stretched to five times its original length. As the load decreases, the elastic material returns to its original length; however, the same graph is not retraced. Instead, the graph shown in black is obtained. This phenomenon is called *elastic hysteresis*. (A similar occurrence takes place in the study of magnetic materials, where it is called *magnetic hysteresis*.) The two curves in the figure make up a *hysteresis loop* for the material. The area of the region enclosed by this loop is numerically equal to the energy dissipated within the elastic (or magnetic) material during the test. In the case of vulcanized rubber, the larger the area, the better the material is for absorbing vibrations.

FIGURE 6.15

Stress-strain diagram for an elastic material

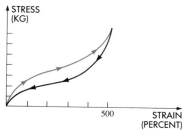

EXERCISES 6.1

Exer. 1–20: Sketch the region bounded by the graphs of the equations, show a typical vertical or horizontal rectangle, and find the area of the region.

1 $y = 1/x^2$, $y = -x^2$, $x = 1$, $x = 2$

2 $y = \sqrt{x}$, $y = -x$, $x = 1$, $x = 4$

3 $y^2 = -x$, $x - y = 4$, $y = -1$, $y = 2$

4 $x = y^2$, $y - x = 2$, $y = -2$, $y = 3$

5 $y = x^2 + 1$, $y = 5$

6 $y = 4 - x^2$, $y = -4$

7 $y = x^2$, $y = 4x$

8 $y = x^3$, $y = x^2$

9 $y = 1 - x^2$, $y = x - 1$

10 $x + y = 3$, $y + x^2 = 3$

11 $y^2 = 4 + x$, $y^2 + x = 2$

12 $x = y^2$, $x - y - 2 = 0$

13 $y = x$, $y = 3x$, $x + y = 4$

14 $x - y + 1 = 0$, $7x - y - 17 = 0$, $2x + y + 2 = 0$

15 $y = x^3 - x$, $y = 0$

16 $y = x^3 - x^2 - 6x$, $y = 0$

17 $x = 4y - y^3$, $x = 0$

18 $x = y^{2/3}$, $x = y^2$

19 $y = x\sqrt{4 - x^2}$, $y = 0$

20 $y = x\sqrt{x^2 - 9}$, $y = 0$, $x = 5$

Exer. 21–22: Find the area of the region between the graphs of the equations from $x = 0$ to $x = \pi$.

21 $y = \sin 4x$, $y = 1 + \cos \frac{1}{3}x$

22 $y = 4 + \cos 2x$, $y = 3 \sin \frac{1}{2}x$

Exer. 23–24: Express the area A of the region R bounded by the graphs of the equation(s) as a limit of sums. Find A without integrating.

23 $(x - 4)^2 + y^2 = 9$

24 $2x + 3y = 6$, $x = 0$, $y = 0$

Exer. 25–32: The expression represents the area A of a region as a limit of sums for a function f on the interval $[a, b]$. Describe the region and find A.

25 $\lim\limits_{\|P\|\to 0} \sum\limits_k (4w_k + 1)\,\Delta x_k;\ [0, 1]$

26 $\lim\limits_{\|P\|\to 0} \sum\limits_k (w_k - w_k^3)\,\Delta x_k;\ [0, 1]$

27 $\lim\limits_{\|P\|\to 0} \sum\limits_k (4 - w_k^2)\,\Delta y_k;\ [0, 1]$

28 $\lim\limits_{\|P\|\to 0} \sum\limits_k \sqrt{3w_k + 1}\,\Delta y_k;\ [0, 1]$

29 $\lim\limits_{\|P\|\to 0} \sum\limits_k [w_k/(w_k^2 + 1^2)]\,\Delta x_k;\ [2, 5]$

30 $\lim\limits_{\|P\|\to 0} \sum\limits_k (5w_k/\sqrt{9 + w_k^2})\,\Delta x_k;\ [1, 4]$

31 $\lim\limits_{\|P\|\to 0} \sum\limits_k [(5 + \sqrt{w_k})/\sqrt{w_k}]\,\Delta y_k;\ [1, 4]$

32 $\lim\limits_{\|P\|\to 0} \sum\limits_k w_k^{-2}\,\Delta y_k;\ [-5, -2]$

Exer. 33–36: Find the area of the region that lies between the graphs of f and g when x is restricted to the interval $[a, b]$. (**Hint:** See Figure 6.14.)

33 $f(x) = 6 - 3x^2,\ g(x) = 3x;\ [0, 2]$

34 $f(x) = x^2 - 4,\ g(x) = x + 2;\ [1, 4]$

35 $f(x) = \sin x,\ g(x) = \cos x;\ [0, 2\pi]$

36 $f(x) = x^2,\ g(x) = x^3,\ [-1, 2]$

37 The shape of a particular stress-strain diagram is shown in the figure (see the last paragraph of this section). Estimate y-coordinates and approximate the area of the region enclosed by the hysteresis loop using, with $n = 6$, (a) the Trapezoidal Rule; (b) Simpson's Rule.

EXERCISE 37

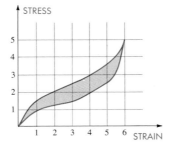

38 Suppose the function values of f and g in the table were obtained empirically. Assuming that f and g are continuous, approximate the area between their graphs from $x = 1$ to $x = 5$ using, with $n = 8$, (a) the Trapezoidal Rule; (b) Simpson's Rule.

x	1	1.5	2	2.5	3	3.5	4	4.5	5
$f(x)$	3.5	2.5	3	4	3.5	2.5	2	2	3
$g(x)$	1.5	2	2	1.5	1	0.5	1	1.5	1

39 A typical graph of rate of growth for boys is shown in the figure. The rate of growth R (in cm/yr) is given as a function of the age t in years.

(a) What does the area under the graph from $t = 10$ to $t = 15$ represent?

(b) Use the Trapezoidal Rule with $n = 6$ and the following table to estimate the area in part (a).

t (yr)	10	11	12	13	14	15
R (cm/yr)	5.3	5.2	4.9	6.5	9.3	7.0

EXERCISE 39

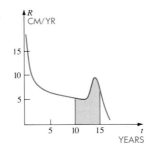

40 The *depth profile* of a species of zooplankton is shown in the figure. The density $\rho(x)$ (in number/m^3) is shown as a function of the ocean depth x (in meters) where the sample was taken. The integral

$$\int_0^{80} \rho(x)\,dx$$

represents the total number of zooplankton in the *water column*, a column formed by a square meter of ocean surface projected down to the ocean floor. Estimate this integral using Simpson's Rule with $n = 8$.

EXERCISE 40

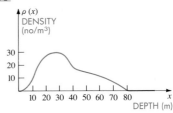

6.2 SOLIDS OF REVOLUTION

The volume of an object plays an important role in many problems in the physical sciences, such as finding centers of mass and moments of inertia. (These concepts will be discussed later in the text.) Because it is usually very

FIGURE 6.16

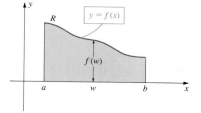

FIGURE 6.17

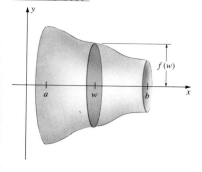

difficult to determine the volume of an irregularly shaped object, we shall begin with objects that have simple shapes. Included in this category are the solids of revolution discussed here and in the next section.

If a region in a plane is revolved about a line l in the plane, the resulting solid is a **solid of revolution,** and the solid is **generated** by the region. The line l is an **axis of revolution.** If the region R bounded by the graph of a continuous nonnegative function f, the x-axis, and the vertical lines $x = a$ and $x = b$ (see Figure 6.16) is revolved about the x-axis, a solid of the type shown in Figure 6.17 is generated. For example, if f is a constant function, say $f(x) = k$, then the region R is rectangular and the solid generated is a right circular cylinder. If the graph of f is a semicircle with endpoints of a diameter at the points $(a, 0)$ and $(b, 0)$ with $b > a$, then the solid of revolution is a sphere with diameter $b - a$. If the given region is a right triangle with base on the x-axis and two vertices at the points $(a, 0)$ and $(b, 0)$ with the right angle at one of these points, then a right circular cone is generated.

If a plane perpendicular to the x-axis intersects the solid shown in Figure 6.17, a circular cross section is obtained. If, as indicated in the figure, the plane passes through the point on the x-axis with x-coordinate w, then the radius of the circle is $f(w)$ and hence its area is $\pi[f(w)]^2$. We shall arrive at a definition for the volume of such a solid of revolution by using Riemann sums in a manner similar to that used for areas in the previous section.

Suppose f is continuous and nonnegative on $[a, b]$. Consider a Riemann sum $\sum_k f(w_k)\,\Delta x_k$ for any number w_k in the kth subinterval $[x_{k-1}, x_k]$ of a partition P of $[a, b]$. This gives us a sum of areas of rectangles of the types shown in Figure 6.18. The solid of revolution generated by the polygon formed by these rectangles has the appearance shown in Figure 6.19.

FIGURE 6.18

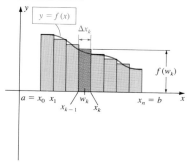

FIGURE 6.19

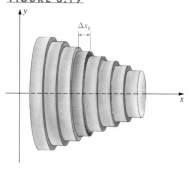

Observe that the kth rectangle generates a circular disk (that is, a "flat" right circular cylinder) of base radius $f(w_k)$ and altitude, or "thickness," $\Delta x_k = x_k - x_{k-1}$. The volume of this disk is the area of the base times the altitude, that is, $\pi[f(w_k)]^2\,\Delta x_k$. The sum of the volumes of all such disks is the volume of the solid shown in Figure 6.19 and is given by

$$\sum_k \pi[f(w_k)]^2\,\Delta x_k.$$

The sum may be regarded as a Riemann sum for $\pi[f(x)]^2$. It appears that if $\|P\|$ is close to zero, then the sum is close to the volume of the solid. Hence we define the volume of the solid of revolution as the limit of these sums.

DEFINITION (6.3)

Let f be continuous on $[a, b]$, and let R be the region bounded by the graph of f, the x-axis, and the vertical lines $x = a$ and $x = b$. The **volume** V of the solid of revolution generated by revolving R about the x-axis is

$$V = \lim_{\|P\| \to 0} \sum_k \pi[f(w_k)]^2 \, \Delta x_k = \int_a^b \pi[f(x)]^2 \, dx$$

The fact that the limit of the sums in Definition (6.3) equals $\int_a^b \pi[f(x)]^2 \, dx$ follows from the definition of the definite integral. Hereafter, when considering limits of Riemann sums, the meanings of symbols such as $\|P\|$, w_k, and Δx_k will not be explicitly stated. Moreover, we shall not ordinarily specify the units of measure for volume. If the linear measurement is inches, the volume is in in.^{3}. If x is in cm, then V is in cm^3, and so on.

The requirement that $f(x)$ be nonnegative was omitted in Definition (6.3). If f is negative for some x, as illustrated in Figure 6.20, and if the region bounded by the graphs of f, $x = a$, $x = b$, and the x-axis is revolved about the x-axis, a solid of the type shown in Figure 6.21 is obtained. This solid is the same as that generated by revolving the region under the graph of $y = |f(x)|$ from a to b about the x-axis. Since $|f(x)|^2 = [f(x)]^2$, the limit in Definition (6.3) gives us the volume.

FIGURE 6.20

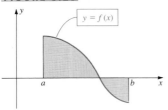

EXAMPLE 1 If $f(x) = x^2 + 1$, find the volume of the solid generated by revolving the region under the graph of f from $x = -1$ to $x = 1$ about the x-axis.

SOLUTION The solid is illustrated in Figure 6.22 together with a typical rectangle and the disk that it generates. Since the radius of the disk is $w_k^2 + 1$, its volume is

$$\pi(w_k^2 + 1)^2 \, \Delta x_k$$

and, as in Definition (6.3),

$$V = \lim_{\|P\| \to 0} \sum_k \pi(w_k^2 + 1)^2 \, \Delta x_k$$

$$= \int_{-1}^1 \pi(x^2 + 1)^2 \, dx = \pi \int_{-1}^1 (x^4 + 2x^2 + 1) \, dx$$

$$= \pi \left[\frac{x^5}{5} + 2\left(\frac{x^3}{3}\right) + x \right]_{-1}^1$$

$$= \pi \left[\left(\frac{1}{5} + \frac{2}{3} + 1 \right) - \left(-\frac{1}{5} - \frac{2}{3} - 1 \right) \right] = \frac{56}{15} \pi \approx 11.7 \quad \bullet$$

FIGURE 6.21

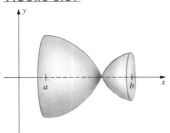

If we had used the symmetry of the graph in Example 1, then the volume could have been calculated by integrating from 0 to 1 and doubling the result.

In order to apply the preceding methods to solids obtained by revolving regions about the y-axis, we must integrate with respect to y. Consider a region bounded by the horizontal lines $y = c$ and $y = d$, by the y-axis, and by the graph of $x = g(y)$ for a function g that is continuous and nonnegative on $[c, d]$. If this region is revolved about the y-axis, the volume V of

FIGURE 6.22

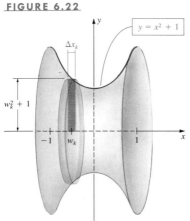

the resulting solid may be found by interchanging the roles of x and y in Definition (6.3). Thus, let P be the partition of the interval $[c, d]$ determined by the numbers $c = y_0, y_1, \ldots, y_n = d$. Let $\Delta y_k = y_k - y_{k-1}$, let w_k be any number in the kth subinterval, and consider the rectangles of length $g(w_k)$ and width Δy_k as illustrated in Figure 6.23. The solid generated by revolving these rectangles about the y-axis is illustrated in Figure 6.24.

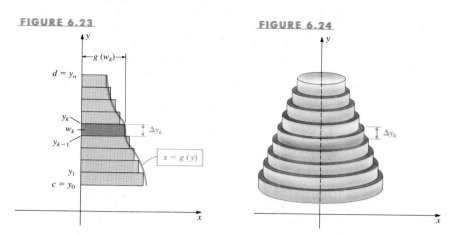

FIGURE 6.23

FIGURE 6.24

The volume of the disk generated by the kth rectangle is $\pi[g(w_k)]^2 \Delta y_k$. A limit of sums then gives us

(6.4)
$$V = \lim_{\|P\| \to 0} \sum_k \pi[g(w_k)]^2 \Delta y_k = \int_c^d \pi[g(y)]^2 \, dy$$

EXAMPLE 2 The region bounded by the y-axis and the graphs of $y = x^3$, $y = 1$, and $y = 8$ is revolved about the y-axis. Find the volume of the resulting solid.

SOLUTION The solid is sketched in Figure 6.25 together with a disk generated by a typical horizontal rectangle. Since we plan to integrate with respect to y, we solve the equation $y = x^3$ for x in terms of y, obtaining $x = y^{1/3}$. If we let $x = g(y) = y^{1/3}$, then the radius of the disk is $g(w_k) = w_k^{1/3}$ and its volume is $\pi(w_k^{1/3})^2 \Delta y_k$. Applying (6.4) with $g(y) = y^{1/3}$,

$$V = \lim_{\|P\| \to 0} \sum_k \pi(w_k^{1/3})^2 \Delta y_k$$

$$= \int_1^8 \pi(y^{1/3})^2 \, dy = \pi \int_1^8 y^{2/3} \, dy = \pi \left[\frac{y^{5/3}}{5/3} \right]_1^8$$

$$= \frac{3}{5} \pi \left[y^{5/3} \right]_1^8 = \frac{3}{5} \pi [8^{5/3} - 1] = \frac{93}{5} \pi \approx 58.4 \quad \bullet$$

Let us next consider a region bounded by the vertical lines $x = a$ and $x = b$ and by the graphs of two continuous functions f and g with $f(x) \geq g(x) \geq 0$ for every x in $[a, b]$, as illustrated in Figure 6.26(i). If this

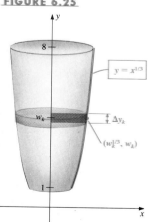

FIGURE 6.25

FIGURE 6.26

(i)

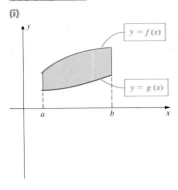

(ii)

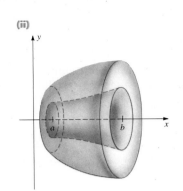

(iii)

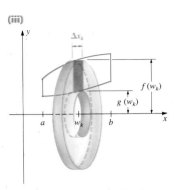

region is revolved about the x-axis, the solid illustrated in (ii) of the figure is obtained. Note that if $g(x) > 0$ for every x in $[a, b]$, then there is a hole through the solid.

The volume V may be found by subtracting the volume of the solid generated by the smaller region from the volume of the solid generated by the larger region. Using Definition (6.3) gives us

$$V = \int_a^b \pi [f(x)]^2 \, dx - \int_a^b \pi [g(x)]^2 \, dx = \int_a^b \pi \{[f(x)]^2 - [g(x)]^2\} \, dx.$$

The last integral has an interesting interpretation as a limit of sums. As illustrated in Figure 6.26(iii), a rectangle extending from the graph of g to the graph of f, through the points with x-coordinate w_k, generates a washer-shaped solid whose volume is

$$\pi [f(w_k)]^2 \, \Delta x_k - \pi [g(w_k)]^2 \, \Delta x_k = \pi \{[f(w_k)]^2 - [g(w_k)]^2\} \, \Delta x_k.$$

Summing the volumes of all such washers and considering the limit gives us the desired integral formula. When working problems of this type it is often convenient to use the following general formula.

FIGURE 6.27

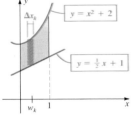

FIGURE 6.28

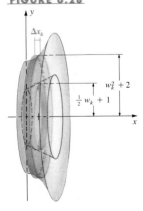

(6.5)

$$\begin{array}{c} \text{Volume of} \\ \text{a washer} \end{array} = \pi [(\text{outer radius})^2 - (\text{inner radius})^2] \cdot (\text{thickness})$$

In integration problems the thickness will be given by either Δx_k or Δy_k. To find the volume we consider a limit of sums of volumes of these washers.

EXAMPLE 3 The region bounded by the graphs of the equations $x^2 = y - 2$ and $2y - x - 2 = 0$, and by the vertical lines $x = 0$ and $x = 1$ is revolved about the x-axis. Find the volume of the resulting solid.

SOLUTION The region and a typical rectangle are sketched in Figure 6.27. Since we wish to integrate with respect to x, we solve the first two equations for y in terms of x, obtaining $y = x^2 + 2$ and $y = \frac{1}{2}x + 1$. The washer generated by the rectangle is illustrated in Figure 6.28.

Since the outer radius of the washer is $w_k^2 + 2$ and the inner radius is $\frac{1}{2}w_k + 1$, its volume by (6.5) is

$$\pi[(w_k^2 + 2)^2 - (\tfrac{1}{2}w_k + 1)^2]\,\Delta x_k.$$

The limit of sums of such volumes gives us

$$V = \int_0^1 \pi\left[(x^2 + 2)^2 - (\tfrac{1}{2}x + 1)^2\right] dx$$

$$= \pi \int_0^1 (x^4 + \tfrac{15}{4}x^2 - x + 3)\,dx$$

$$= \pi\left[\frac{x^5}{5} + \frac{15}{4}\left(\frac{x^3}{3}\right) - \frac{x^2}{2} + 3x\right]_0^1 = \frac{79\pi}{20} \approx 12.4 \quad \bullet$$

EXAMPLE 4 Find the volume of the solid generated by revolving the region described in Example 3 about the line $y = 3$.

SOLUTION The region and a typical rectangle are resketched in Figure 6.29 together with the axis of revolution $y = 3$. The washer generated by the rectangle is illustrated in Figure 6.30. Note that the radii of this washer are as follows:

$$\text{Inner radius} = 3 - (w_k^2 + 2) = 1 - w_k^2$$

$$\text{Outer radius} = 3 - (\tfrac{1}{2}w_k + 1) = 2 - \tfrac{1}{2}w_k.$$

Using (6.5), the volume of the washer is

$$\pi\left[(2 - \tfrac{1}{2}w_k)^2 - (1 - w_k^2)^2\right]\Delta x_k.$$

The limit of sums of such terms gives us

$$V = \int_0^1 \pi\left[(2 - \tfrac{1}{2}x)^2 - (1 - x^2)^2\right] dx$$

$$= \pi \int_0^1 \left[(4 - 2x + \tfrac{1}{4}x^2) - (1 - 2x^2 + x^4)\right] dx$$

$$= \pi \int_0^1 (3 - 2x + \tfrac{9}{4}x^2 - x^4)\,dx$$

$$= \pi\left[3x - x^2 + \tfrac{9}{4}\left(\frac{x^3}{3}\right) - \frac{x^5}{5}\right]_0^1$$

$$= \pi\left[3 - 1 + \tfrac{3}{4} - \tfrac{1}{5}\right] = \frac{51}{20}\pi \approx 8.01 \qquad \bullet$$

By interchanging the roles of x and y we can apply the preceding methods to solids generated by revolving regions about the y-axis, as in the next example.

EXAMPLE 5 The region in the first quadrant bounded by the graphs of $y = \tfrac{1}{8}x^3$ and $y = 2x$ is revolved about the y-axis. Find the volume of the resulting solid.

SOLUTION The region and a typical rectangle are shown in Figure 6.31. Since we wish to integrate with respect to y, we solve the given equations

FIGURE 6.29

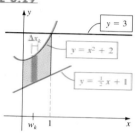

FIGURE 6.30

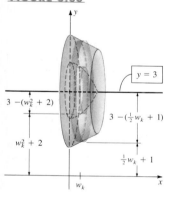

FIGURE 6.31

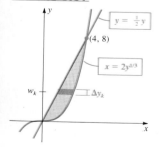

FIGURE 6.32

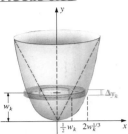

for x in terms of y, obtaining

$$x = \tfrac{1}{2}y \quad \text{and} \quad x = 2y^{1/3}.$$

As shown in Figure 6.32, the inner and outer radii of the washer generated by the rectangle are $\tfrac{1}{2}w_k$ and $2w_k^{1/3}$, respectively. Since the thickness is Δy_k, it follows from (6.5) that the volume of the washer is

$$\pi\left[(2w_k^{1/3})^2 - (\tfrac{1}{2}w_k)^2\right]\Delta y_k = \pi\left[4w_k^{2/3} - \tfrac{1}{4}w_k^2\right]\Delta y_k.$$

The limit of sums of such terms gives us

$$V = \int_0^8 \pi\left[4y^{2/3} - \tfrac{1}{4}y^2\right]dy = \pi\left[\tfrac{12}{5}y^{5/3} - \tfrac{1}{12}y^3\right]_0^8$$
$$= \pi\left[\tfrac{12}{5}(8^{5/3}) - \tfrac{1}{12}(8^3)\right] = \tfrac{512}{15}\pi \approx 107.2 \qquad \bullet$$

EXERCISES 6.2

Exer. 1–12: Sketch the region R bounded by the graphs of the equations, and find the volume of the solid generated by revolving R about the indicated axis. Show a typical rectangle, together with the disk or washer it generates. (*Remark:* Although the solids described are mathematical entities, many objects encountered in everyday life have similar shapes.)

1 $y = 1/x$, $x = 1$, $x = 3$, $y = 0$; about the x-axis.

2 $y = \sqrt{x}$, $y = 0$, $x = 4$; about the x-axis.

3 $y = x^2$, $y = 2$; about the y-axis.

4 $y = 1/x$, $x = 0$, $y = 1$, $y = 3$; about the y-axis.

5 $y = x^2 - 4x$, $y = 0$; about the x-axis.

6 $y = x^3$, $x = -2$, $y = 0$; about the x-axis.

7 $y^2 = x$, $2y = x$; about the y-axis.

8 $y = 2x$, $y = 4x^2$; about the y-axis.

9 $y = x^2$, $y = 4 - x^2$; about the x-axis.

10 $x = y^3$, $x^2 + y = 0$; about the x-axis.

11 $x = y^2$, $y - x + 2 = 0$; about the y-axis.

12 $x + y = 1$, $y = x + 1$, $x = 2$; about the y-axis.

13 The region bounded by the x-axis and the graph of $y = \sin 2x$ from $x = 0$ to $x = \pi$ is revolved about the x-axis. Find the volume of the resulting solid. (*Hint:* $\sin^2\theta = \tfrac{1}{2}(1 - \cos 2\theta)$.)

14 The region bounded by the x-axis and the graph of $y = 1 + \cos 3x$ from $x = 0$ to $x = 2\pi$ is revolved about the x-axis. Find the volume of the resulting solid. (*Hint:* $\cos^2\theta = \tfrac{1}{2}(1 + \cos 2\theta)$.)

15 The region between the graphs of $y = \sin x$ and $y = \cos x$ from $x = 0$ to $x = \pi/4$ is revolved about the x-axis. Find

the volume of the resulting solid. (*Hint:* Use a trigonometric identity to change the form of the integrand.)

16 The region bounded by the graphs of $y = \cos\tfrac{1}{2}x$, $y = 2$, $x = 0$, and $x = 4\pi$ is revolved about the line $y = 2$. Find the volume of the resulting solid.

17 Find the volume of the solid generated by revolving the region bounded by the graphs of $y = x^2$ and $y = 4$ (a) about the line $y = 4$; (b) about the line $y = 5$; (c) about the line $x = 2$.

18 Find the volume of the solid generated by revolving the region bounded by the graphs of $y = \sqrt{x}$, $y = 0$, and $x = 4$ (a) about the line $x = 4$; (b) about the line $x = 6$; (c) about the line $y = 2$.

Exer. 19–24: Sketch the region R bounded by the graphs of the equations in (a) and then set up (but do not evaluate) integrals needed to find the volume of the solid obtained by revolving R about the line given in (b). As usual, show typical rectangles and the corresponding disks or washers.

19 (a) $y = x^3$, $y = 4x$
    (b) $y = 8$

20 (a) $y = x^3$, $y = 4x$
    (b) $x = 4$

21 (a) $x + y = 3$, $y + x^2 = 3$
    (b) $x = 2$

22 (a) $y = 1 - x^2$, $x - y = 1$
    (b) $y = 3$

23 (a) $x^2 + y^2 = 1$
    (b) $x = 5$

24 (a) $y = x^{2/3}$, $y = x^2$
    (b) $y = -1$

Exer. 25–28: Use a definite integral to derive a formula for the volume of the indicated solid.

25 A right circular cone of altitude h and radius of base r.

26 A sphere of radius r.

27 A frustum of a right circular cone of altitude h, lower base radius R, and upper base radius r.

28 The volume of a spherical segment of height h and radius of sphere r.

Exer. 29–30: The expression represents the volume V of a solid of revolution as a limit of sums for a function f defined on the interval $[0, 1]$. Describe the solid and find V.

29 $\displaystyle \lim_{\|P\| \to 0} \sum_k \pi(w_k^4 - w_k^6) \, \Delta x_k$

30 $\displaystyle \lim_{\|P\| \to 0} \sum_k \pi(w_k - w_k^8) \, \Delta y_k$

31 If the region shown in the figure is revolved about the x-axis, use the Trapezoidal Rule with $n = 6$ to approximate the volume of the resulting solid.

EXERCISE 31

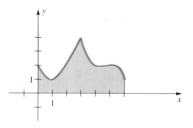

32 If the region shown in the figure is revolved about the x-axis, use Simpson's Rule with $n = 8$ to approximate the volume of the resulting solid.

EXERCISE 32

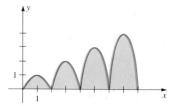

6.3 VOLUMES USING CYLINDRICAL SHELLS

The methods we introduced in Section 6.2 for finding volumes of solids of revolution are, in certain cases, difficult to apply. Another method employs hollow circular cylinders, that is, thin cylindrical shells of the type illustrated in Figure 6.33.

FIGURE 6.33

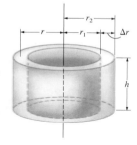

The volume of a shell having outer radius r_2, inner radius r_1, and altitude h is $\pi r_2^2 h - \pi r_1^2 h$. This expression may also be written

$$\pi(r_2^2 - r_1^2)h = \pi(r_2 + r_1)(r_2 - r_1)h = 2\pi\left(\frac{r_2 + r_1}{2}\right)h(r_2 - r_1).$$

If we let $r = (r_2 + r_1)/2$ (the **average radius** of the shell) and let $\Delta r = r_2 - r_1$ (the **thickness** of the shell), then the volume of the shell is $2\pi r h \, \Delta r$. Thus,

(6.6)

> Volume of a shell $= 2\pi$(average radius)(altitude)(thickness)

Suppose a function f is continuous and nonnegative on $[a, b]$ for $0 \le a < b$. Let R be the region bounded by the graph of f, the x-axis, and the vertical lines $x = a$ and $x = b$, as illustrated in Figure 6.34(i). The solid generated by revolving R about the y-axis is illustrated in (ii) of the figure.

FIGURE 6.34

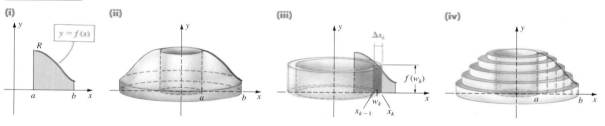

Note that if $a > 0$, there is a hole through the solid. Let P be a partition of $[a, b]$ and consider the rectangle with base corresponding to the interval $[x_{k-1}, x_k]$ and altitude $f(w_k)$, with w_k the midpoint of $[x_{k-1}, x_k]$. If this rectangle is revolved about the y-axis, then as illustrated in Figure 6.34(iii), we obtain a cylindrical shell with average radius w_k, altitude $f(w_k)$, and thickness $\Delta x_k = x_k - x_{k-1}$. By (6.6) its volume is

$$2\pi w_k f(w_k) \, \Delta x_k.$$

Finding the volume for each subinterval in the partition and summing gives us

$$\sum_k 2\pi w_k f(w_k) \, \Delta x_k.$$

This sum represents the volume of a solid of the type illustrated in Figure 6.34(iv). The smaller the norm $\|P\|$ of the partition, the better the sum approximates the volume V of the solid generated by R (shown in (ii) of the figure). This is the motivation for the following definition.

DEFINITION (6.7)

Let f be continuous and nonnegative on $[a, b]$ for $0 \le a < b$. The **volume** V of the solid of revolution generated by revolving the region bounded by the graphs of f, $x = a$, $x = b$, and the x-axis about the y-axis is

$$V = \lim_{\|P\| \to 0} \sum_k 2\pi w_k f(w_k) \, \Delta x_k = \int_a^b 2\pi x f(x) \, dx$$

The last equality in Definition (6.7) follows from the definition of the definite integral (5.8). We can prove that if the methods of Section 6.2 are also applicable, then both methods lead to the same answer (see Exercise 52, Section 9.1).

EXAMPLE 1 The region bounded by the graph of $y = 2x - x^2$ and the x-axis is revolved about the y-axis. Find the volume of the resulting solid.

FIGURE 6.35

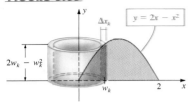

SOLUTION The graph of $y = 2x - x^2$ (a parabola) is sketched in Figure 6.35. Also shown is the region to be revolved and a shell generated by a typical rectangle, with w_k the midpoint of the kth subinterval $[x_{k-1}, x_k]$ of a partition of $[0, 2]$. The average radius of the shell is w_k, the altitude is $2w_k - w_k^2$, and the thickness is Δx_k. Using (6.6), the volume of the shell is

$$2\pi w_k (2w_k - w_k^2) \, \Delta x_k.$$

Thus, by Definition (6.7), the volume V of the solid is

$$V = \lim_{\|P\| \to 0} \sum_k 2\pi w_k (2w_k - w_k^2) \, \Delta x_k$$

$$= \int_0^2 2\pi x (2x - x^2) \, dx = 2\pi \int_0^2 (2x^2 - x^3) \, dx$$

$$= 2\pi \left[2 \left(\frac{x^3}{3} \right) - \frac{x^4}{4} \right]_0^2 = \frac{8\pi}{3} \approx 8.4$$

FIGURE 6.36

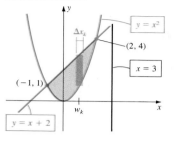

FIGURE 6.37

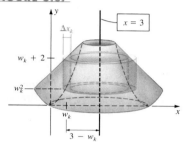

FIGURE 6.38

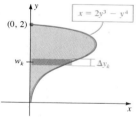

FIGURE 6.39

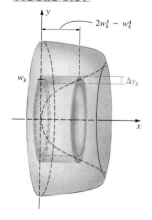

The volume V can also be found using washers; however, the calculations would be more involved, since the equation $y = 2x - x^2$ would have to be solved for x in terms of y. •

EXAMPLE 2 The region bounded by the graphs of $y = x^2$ and $y = x + 2$ is revolved about the line $x = 3$. Express the volume of the resulting solid as a definite integral.

SOLUTION The region and a typical rectangle are shown in Figure 6.36, where w_k is the midpoint of the kth subinterval $[x_{k-1}, x_k]$. The cylindrical shell generated by the rectangle is illustrated in Figure 6.37. This shell has the following dimensions:

$$\text{altitude} = (w_k + 2) - w_k^2$$
$$\text{average radius} = 3 - w_k$$
$$\text{thickness} = \Delta x_k.$$

Hence, by (6.6) the volume of the shell is

$$2\pi(3 - w_k)[(w_k + 2) - w_k^2]\,\Delta x_k.$$

The limit of sums of such terms gives us

$$V = \int_{-1}^{2} 2\pi(3 - x)(x + 2 - x^2)\,dx \quad •$$

The definite integral in the preceding example could be evaluated by multiplying the factors in the integrand and then integrating each term. Since we have already considered such integrations, it would not be very instructive to carry out all of these details. For convenience we shall refer to the process of expressing V in terms of an integral as *setting up the integral for V*.

By interchanging the roles of x and y we may sometimes use shells and integrate with respect to y, as illustrated in the next example.

EXAMPLE 3 The region in the first quadrant bounded by the graph of the equation $x = 2y^3 - y^4$ and the y-axis is revolved about the x-axis. Set up the integral for the volume of the resulting solid.

SOLUTION The region is sketched in Figure 6.38 together with a typical (horizontal) rectangle. The cylindrical shell generated by the rectangle is illustrated in Figure 6.39. The shell has the following dimensions:

$$\text{altitude} = 2w_k^3 - w_k^4$$
$$\text{average radius} = w_k$$
$$\text{thickness} = \Delta y_k.$$

Hence, by (6.6) its volume is

$$2\pi w_k(2w_k^3 - w_k^4)\,\Delta y_k.$$

The limit of sums of such terms gives us

$$V = \int_{0}^{2} 2\pi y(2y^3 - y^4)\,dy \quad •$$

It is worth noting that in the preceding example we were forced to use shells and to integrate with respect to y, since use of washers and integration with respect to x would require that we solve the equation $x = 2y^3 - y^4$ for y in terms of x, a rather formidable task.

EXERCISES 6.3

Exer. 1–10: Sketch the region R bounded by the graphs of the equations, and use the methods of this section to find the volume of the solid generated by revolving R about the indicated axis. Show a typical rectangle together with the cylindrical shell it generates.

1 $y = \sqrt{x}$, $x = 4$, $y = 0$; about the y-axis.

2 $y = 1/x$, $x = 1$, $x = 2$, $y = 0$; about the y-axis.

3 $y = x^2$, $y^2 = 8x$; about the y-axis.

4 $y = x^2 - 5x$, $y = 0$; about the y-axis.

5 $2x - y - 12 = 0$, $x - 2y - 3 = 0$, $x = 4$; about the y-axis.

6 $y = x^3 + 1$, $x + 2y = 2$, $x = 1$; about the y-axis.

7 $x^2 = 4y$, $y = 4$; about the x-axis.

8 $y^3 = x$, $y = 3$, $x = 0$; about the x-axis.

9 $y = 2x$, $y = 6$, $x = 0$; about the x-axis.

10 $2y = x$, $y = 4$, $x = 1$; about the x-axis.

11 The region bounded by the x-axis and the graph of $y = \cos(x^2)$ from $x = 0$ to $x = \sqrt{\pi/2}$ is revolved about the y-axis. Find the volume of the resulting solid.

12 The region bounded by the x-axis and the graph of $y = x \sin(x^3)$ from $x = 0$ to $x = 1$ is revolved about the y-axis. Find the volume of the resulting solid.

13 Find the volume of the solid generated by revolving the region bounded by the graphs of $y = x^2 + 1$, $x = 0$, $x = 2$, and $y = 0$ (a) about the line $x = 3$; (b) about the line $x = -1$.

14 Find the volume of the solid generated by revolving the region bounded by the graphs of $y = 4 - x^2$ and $y = 0$ (a) about the line $x = 2$; (b) about the line $x = -3$.

15–26 Use the methods of this section to solve Exercises 17–28 of Section 6.2.

Exer. 27–28: The expression represents the volume V of a solid of revolution as a limit of sums for a function f on the interval $[0, 1]$. Describe the solid, and find V.

27 $\displaystyle \lim_{\|P\| \to 0} \sum_k 2\pi(w_k^2 - w_k^3) \, \Delta x_k$

28 $\displaystyle \lim_{\|P\| \to 0} \sum_k 2\pi(w_k^5 + w_k^{3/2}) \, \Delta x_k$

29 Refer to Exercise 31 of Section 6.2. If the region is revolved about the y-axis, use the Trapezoidal Rule with $n = 6$ to approximate the volume of the resulting solid.

30 Refer to Exercise 32 of Section 6.2. If the region is revolved about the y-axis, use Simpson's Rule with $n = 8$ to approximate the volume of the resulting solid.

6.4 VOLUMES BY SLICING

If a plane intersects a solid, then the region common to the plane and the solid is a **cross section** of the solid. In Section 6.2 we used circular and washer-shaped cross sections. We shall now consider solids that have the following property: for every x in a closed interval $[a, b]$ on a coordinate line l, the plane perpendicular to l at x intersects the solid in a cross section whose area is given by $A(x)$ for a continuous function A on $[a, b]$. Figures 6.40 and 6.41 illustrate solids of the type we wish to discuss.

The solid is called a **cylinder** if, as illustrated in Figure 6.40, a line parallel to l that traces the boundary of the cross section corresponding to a also traces the boundary of the cross section corresponding to x for every x in $[a, b]$. If we are only interested in that part of the solid bounded by planes at a and b, then the cross sections determined by these planes are the **bases** of the cylinder. The distance between the bases is the **altitude** of the cylinder. By definition, the volume of the cylinder is the area of a base multiplied by

FIGURE 6.40

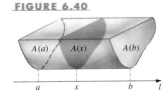

FIGURE 6.41

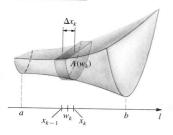

the altitude. As a special case, a right circular cylinder of base radius r and altitude h has volume $\pi r^2 h$.

The solid illustrated in Figure 6.41 is not a cylinder, since cross sections by planes perpendicular to l are not all the same. To find the volume we begin with a partition P of $[a, b]$ by choosing $a = x_0, x_1, x_2, \ldots, x_n = b$. The planes perpendicular to l at the points with these coordinates slice the solid into smaller pieces. Let $\Delta x_k = x_k - x_{k-1}$ and choose any number w_k in $[x_{k-1}, x_k]$. If Δx_k is small, then as shown in Figure 6.41, the volume of a typical slice can be approximated by the volume of a cylinder of base area $A(w_k)$ and altitude Δx_k, that is, by $A(w_k) \, \Delta x_k$. The volume V of the solid is approximated by the Riemann sum

$$V \approx \sum_{k=1}^{n} A(w_k) \, \Delta x_k.$$

Since the approximation improves as $\| P \|$ gets smaller, we define the volume V of the solid as the limit of these sums.

VOLUME BY SLICING FORMULA (6.8)

$$V = \lim_{\| P \| \to 0} \sum_k A(w_k) \, \Delta x_k = \int_a^b A(x) \, dx$$

EXAMPLE 1 Find the volume of a right pyramid that has altitude h and square base of side a.

FIGURE 6.42

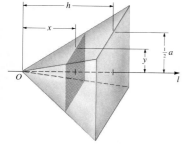

SOLUTION As shown in Figure 6.42, we introduce a coordinate line l along the axis of the pyramid, with origin O at the vertex. Cross sections by planes perpendicular to l are squares. If $0 \le x \le h$, then the cross-sectional area $A(x)$ is

$$A(x) = (2y)^2 = 4y^2$$

for the distance y indicated in the figure. Using similar triangles,

$$\frac{y}{x} = \frac{\frac{1}{2}a}{h} \quad \text{or} \quad y = \frac{ax}{2h}$$

and hence

$$A(x) = 4y^2 = \frac{4a^2 x^2}{4h^2} = \frac{a^2}{h^2} x^2.$$

Applying (6.8),

$$V = \int_0^h A(x) \, dx = \int_0^h \left(\frac{a^2}{h^2}\right) x^2 \, dx = \left(\frac{a^2}{h^2}\right)\left[\frac{x^3}{3}\right]_0^h = \frac{a^2 h}{3} \quad \bullet$$

FIGURE 6.43

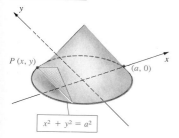

EXAMPLE 2 A solid has as its base the circular region in the xy-plane bounded by the graph of $x^2 + y^2 = a^2$ with $a > 0$. Find the volume of the solid if every cross section by a plane perpendicular to the x-axis is an equilateral triangle with one side in the base.

SOLUTION A typical cross section by a plane x units from the origin is illustrated in Figure 6.43. If the point $P(x, y)$ is on the circle and $y > 0$, then the lengths of the sides of the equilateral triangle are $2y$ and the altitude is $\sqrt{3}y$. Hence the area $A(x)$ of the pictured triangle is

$$A(x) = \tfrac{1}{2}(2y)(\sqrt{3}y) = \sqrt{3}y^2 = \sqrt{3}(a^2 - x^2).$$

Applying (6.8) gives us

$$V = \int_{-a}^{a} A(x)\,dx = \int_{-a}^{a} \sqrt{3}(a^2 - x^2)\,dx = \sqrt{3}\left[a^2 x - \frac{x^3}{3}\right]_{-a}^{a} = \frac{4\sqrt{3}a^3}{3} \, .$$

EXERCISES 6.4

1 A solid has as its base the circular region in the xy-plane bounded by the graph of $x^2 + y^2 = a^2$ with $a > 0$. Find the volume of the solid if every cross section by a plane perpendicular to the x-axis is a square.

2 Rework Exercise 1 if every cross section is an isosceles triangle with base on the xy-plane and altitude equal to the length of the base.

3 A solid has as its base the region in the xy-plane bounded by the graphs of $y = 4$ and $y = x^2$. Find the volume of the solid if every cross section by a plane perpendicular to the x-axis is an isosceles right triangle with hypotenuse on the xy-plane.

4 Rework Exercise 3 if each cross section is a square.

5 Find the volume of a pyramid of the type illustrated in Figure 6.42 if the altitude is h and the base is a rectangle of dimensions a and $2a$.

6 A solid has as its base the region in the xy-plane bounded by the graphs of $y = x$ and $y^2 = x$. Find the volume of the solid if every cross section by a plane perpendicular to the x-axis is a semicircle with diameter in the xy-plane.

7 A solid has as its base the region in the xy-plane bounded by the graphs of $y^2 = 4x$ and $x = 4$. If every cross section by a plane perpendicular to the y-axis is a semicircle, find the volume of the solid.

8 A solid has as its base the region in the xy-plane bounded by the graphs of $x^2 = 16y$ and $y = 2$. Every cross section by a plane perpendicular to the y-axis is a rectangle whose height is twice that of the side in the xy-plane. Find the volume of the solid.

9 A log having the shape of a right circular cylinder of radius a is lying on its side. A wedge is removed from the log by making a vertical cut and another cut at an angle of $45°$, both cuts intersecting at the center of the log (see figure). Find the volume of the wedge.

EXERCISE 9

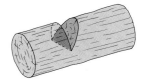

10 The axes of two right circular cylinders of radius a intersect at right angles. Find the volume of the solid bounded by the cylinders.

11 The base of a solid is the circular region in the xy-plane bounded by the graph of $x^2 + y^2 = a^2$ with $a > 0$. Find

the volume of the solid if every cross section by a plane perpendicular to the x-axis is an isosceles triangle of constant altitude b. (*Hint:* Interpret $\int_{-a}^{a} \sqrt{a^2 - x^2}\,dx$ as an area.)

12 Cross sections of a horn-shaped solid by planes perpendicular to its axis are circles. If a cross section that is s inches from the smaller end of the solid has diameter $6 + \frac{1}{36}s^2$ inches, and if the length of the solid is 2 feet, find its volume.

13 A tetrahedron has three mutually perpendicular faces and three mutually perpendicular edges of lengths 2, 3, and 4 cm, respectively. Find its volume.

14 A hole of diameter d is bored through a spherical solid of radius r such that the axis of the hole coincides with a diameter of the sphere. Find the volume of the solid that remains.

15 The base of a solid is an isosceles right triangle whose equal sides have length a. Find the volume if cross sections that are perpendicular to the base and to one of the equal sides are semicircular.

16 Rework Exercise 15 if the cross sections are regular hexagons with one side in the base.

17 A *prismatoid* is a solid whose cross-sectional areas by planes parallel to, and a distance x from, a fixed plane can be expressed as $ax^2 + bx + c$ for real numbers a, b, and c. Prove that the volume V of a prismatoid is

$$V = \frac{h}{6}(B_1 + B_2 + 4B)$$

for areas B_1 and B_2 of the bases, cross-sectional area B parallel to and halfway between the bases, and distance h between the bases. Show that the frustum of a right circular cone of base radii r_1 and r_2 and altitude h is a prismatoid, and find its volume.

18 *Cavalieri's theorem* states that if two solids have equal altitudes and if all cross sections by planes parallel to their bases and at the same distances from their bases have equal areas, then the solids have the same volume (see figure). Prove Cavalieri's theorem.

EXERCISE 18

19 Show that the disk and washer methods discussed in Section 6.2 are special cases of (6.8).

20 A circular swimming pool has diameter 26 feet. The depth of the water changes slowly from 3 feet at a point A on one side of the pool to 9 feet at a point B diametrically opposite A (see figure). Depth readings (in feet) taken along the diameter AB are given in the following table, where x is the distance (in feet) from A.

x	0	4	8	12	16	20	24	28
Water depth	3	3.5	4	5	6.5	8	8.5	9

Use the Trapezoidal Rule, with $n = 7$, to estimate the volume of water in the pool. Approximate the number of gallons of water contained in the pool (1 gal ≈ 0.134 ft³).

EXERCISE 20

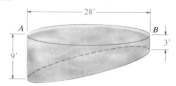

6.5 ARC LENGTH AND SURFACES OF REVOLUTION

To solve certain applied problems we must consider *lengths* of graphs. For example, we may wish to determine the distance a rocket travels during a specified interval of time or find the length of a twisted piece of wire. If the wire is flexible, we could simply straighten it and find the linear length with a ruler (or by means of the Distance Formula). However, if the wire is not flexible, then we must use other methods. As we shall see, the key to defining the length of a graph is to divide the graph into many small pieces and then approximate each piece by means of a line segment. Next, we consider a limit of sums of lengths of such line segments. This leads to a definite integral. To guarantee that the integral exists, we must place restrictions on the function, as indicated in the following discussion.

A function f is **smooth** on an interval if it has a derivative f' that is continuous throughout the interval. Roughly speaking, this means that a small change in x produces a small change in the slope $f'(x)$ of the tangent line to the graph of f. Thus, the graph has no sharp corners. We shall define the **length of arc** between the two points A and B on the graph of a smooth function.

If f is smooth on a closed interval $[a, b]$, the points $A(a, f(a))$ and $B(b, f(b))$ will be called the **endpoints** of the graph of f. Let P be the partition of $[a, b]$ determined by $a = x_0, x_1, x_2, \ldots, x_n = b$, and let Q_k denote the point with coordinates $(x_k, f(x_k))$. This gives us $n + 1$ points $Q_0, Q_1, Q_2, \ldots, Q_n$ on the graph of f. If, as illustrated in Figure 6.44, we connect each Q_{k-1} to Q_k by a line segment of length $d(Q_{k-1}, Q_k)$, then the length L_P of the resulting broken line is

FIGURE 6.44

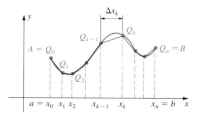

$$L_P = \sum_{k=1}^{n} d(Q_{k-1}, Q_k).$$

If the norm $\| P \|$ of the partition is small, then Q_{k-1} is close to Q_k for each k, and L_P is an approximation to the length of arc between A and B. This gives us a clue to a definition of arc length. Specifically, we shall consider the limit of the sums L_P.

By the Distance Formula,

$$d(Q_{k-1}, Q_k) = \sqrt{(x_k - x_{k-1})^2 + [f(x_k) - f(x_{k-1})]^2}.$$

Applying the Mean Value Theorem (4.12),

$$f(x_k) - f(x_{k-1}) = f'(w_k)(x_k - x_{k-1})$$

for some w_k in the open interval (x_{k-1}, x_k). Substituting this into the preceding formula and letting $\Delta x_k = x_k - x_{k-1}$, we obtain

$$d(Q_{k-1}, Q_k) = \sqrt{(\Delta x_k)^2 + [f'(w_k) \, \Delta x_k]^2}$$
$$= \sqrt{1 + [f'(w_k)]^2} \, \Delta x_k.$$

Consequently,

$$L_P = \sum_{k=1}^{n} \sqrt{1 + [f'(w_k)]^2} \, \Delta x_k.$$

Observe that L_P is a Riemann sum for $g(x) = \sqrt{1 + [f'(x)]^2}$. Moreover, g is continuous on $[a, b]$, since f' is continuous. As we have mentioned, if the norm $\|P\|$ is small, then the length L_P of the broken line approximates the length of the graph of f from A to B. Since this approximation should improve as $\|P\|$ decreases, we define the *length* (also called the *arc length*) of the graph of f from A to B as the limit of sums L_P. Since $g = \sqrt{1 + (f')^2}$ is a continuous function, the limit exists and equals the definite integral $\int_a^b \sqrt{1 + [f'(x)]^2} \, dx$. This arc length will be denoted by L_a^b.

DEFINITION (6.9)

> Let the function f be smooth on a closed interval $[a, b]$. The **arc length of the graph** of f from $A(a, f(a))$ to $B(b, f(b))$ is
>
> $$L_a^b = \int_a^b \sqrt{1 + [f'(x)]^2} \, dx$$

Definition (6.9) will be extended to more general graphs in Chapter 13. If a function f is defined implicitly by an equation in x and y, then we shall also refer to the *arc length of the graph of the equation*.

EXAMPLE 1 If $f(x) = 3x^{2/3} - 10$, find the arc length of the graph of f from the point $A(8, 2)$ to $B(27, 17)$.

SOLUTION The graph of f is sketched in Figure 6.45. Since

$$f'(x) = 2x^{-1/3} = \frac{2}{x^{1/3}},$$

we have, by Definition (6.9),

$$L_8^{27} = \int_8^{27} \sqrt{1 + \left(\frac{2}{x^{1/3}}\right)^2} \, dx = \int_8^{27} \sqrt{1 + \frac{4}{x^{2/3}}} \, dx$$

$$= \int_8^{27} \sqrt{\frac{x^{2/3} + 4}{x^{2/3}}} \, dx.$$

$$= \int_8^{27} \sqrt{x^{2/3} + 4} \, \frac{1}{x^{1/3}} \, dx.$$

To evaluate this integral we make the substitution

$$u = x^{2/3} + 4, \qquad du = \frac{2}{3} x^{-1/3} \, dx = \frac{2}{3} \frac{1}{x^{1/3}} \, dx.$$

FIGURE 6.45

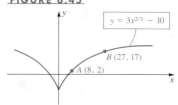

The integral can be expressed in a suitable form for integration by introducing the factor $\frac{2}{3}$ in the integrand and compensating for this by multiplying the integral by $\frac{3}{2}$. Thus

$$L_8^{27} = \frac{3}{2} \int_8^{27} \sqrt{x^{2/3} + 4} \left(\frac{2}{3} \frac{1}{x^{1/3}} \right) dx.$$

If $x = 8$, then $u = (8)^{2/3} + 4 = 8$. If $x = 27$, then $u = (27)^{2/3} + 4 = 13$. Making the substitution and changing the limits of integration,

$$L_8^{27} = \frac{3}{2} \int_8^{13} \sqrt{u}\, du = u^{3/2} \Big]_8^{13} = 13^{3/2} - 8^{3/2} \approx 24.2 \quad \bullet$$

Interchanging the roles of x and y in Definition (6.9) gives us a formula for integration with respect to y. Specifically, let the function g be defined by $x = g(y)$ with g smooth on the interval $[c, d]$. The arc length of the graph of g from the point $(g(c), c)$ to the point $(g(d), d)$ is

(6.10)

$$L_c^d = \int_c^d \sqrt{1 + [g'(y)]^2}\, dy$$

The integrands $\sqrt{1 + [f'(x)]^2}$ and $\sqrt{1 + [g'(y)]^2}$ in formulas (6.9) and (6.10) often result in expressions that have no obvious antiderivatives. In such cases, numerical integration may be used to find arc length, as illustrated in the next example.

EXAMPLE 2 Set up an integral for finding the arc length of the graph of $y^3 - y - x = 0$ from $A(0, -1)$ to $B(6, 2)$. Approximate the integral to one decimal place by using Simpson's Rule (5.39) with $n = 6$.

SOLUTION Since the equation is not of the form $y = f(x)$, Definition (6.9) cannot be applied directly. However, if we write $x = y^3 - y$, then we can employ (6.10) with $g(y) = y^3 - y$. The graph of the equation is sketched in Figure 6.46. Using (6.10) with $c = -1$ and $d = 2$ gives us

FIGURE 6.46

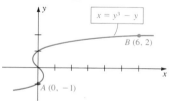

$$L_{-1}^2 = \int_{-1}^2 \sqrt{1 + (3y^2 - 1)^2}\, dy$$

$$= \int_{-1}^2 \sqrt{9y^4 - 6y^2 + 2}\, dy.$$

To use Simpson's Rule, we let $f(y) = \sqrt{9y^4 - 6y^2 + 2}$ and arrange our work as we did in Section 5.6, obtaining the following table.

k	y_k	$f(y_k)$	m	$mf(y_k)$
0	-1.0	2.2361	1	2.2361
1	-0.5	1.0308	4	4.1232
2	0.0	1.4142	2	2.8284
3	0.5	1.0308	4	4.1232
4	1.0	2.2361	2	4.4722
5	1.5	5.8363	4	23.3452
6	2.0	11.0454	1	11.0454

The sum of the numbers in the last column is 52.1737.

By Simpson's Rule with $a = -1$, $b = 2$, and $n = 6$,

$$\int_{-1}^{2} \sqrt{9y^4 - 6y^2 + 2}\, dy \approx \tfrac{1}{6}[52.1737] \approx 8.7 \quad \bullet$$

If a graph can be decomposed into a finite number of parts, each of which is the graph of a smooth function, then the arc length of the graph is defined as the sum of the arc lengths of the individual graphs. A function of this type is said to be **piecewise smooth** on its domain.

To avoid any misunderstanding in the following discussion, the variable of integration will be denoted by t. In this case the arc length formula given in Definition (6.9) becomes

$$L_a^b = \int_a^b \sqrt{1 + [f'(t)]^2}\, dt.$$

If f is smooth on $[a, b]$, then f is smooth on $[a, x]$ for every x in $[a, b]$, and the length of the graph from the point $A(a, f(a))$ to the point $Q(x, f(x))$ is

$$L_a^x = \int_a^x \sqrt{1 + [f'(t)]^2}\, dt.$$

If we change the notation and use the symbol $s(x)$ in place of L_a^x, then s may be regarded as a function with domain $[a, b]$, since to each x in $[a, b]$ there corresponds a unique number $s(x)$. As shown in Figure 6.47, $s(x)$ is the length of arc of the graph of f from $A(a, f(a))$ to $Q(x, f(x))$. We shall call s the *arc length function* for the graph of f, as in the next definition.

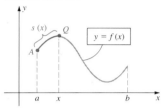
FIGURE 6.47

DEFINITION (6.11)   Let the function f be smooth on $[a, b]$. The **arc length function** s for the graph of f on $[a, b]$ is defined by

$$s(x) = \int_a^x \sqrt{1 + [f'(t)]^2}\, dt$$

for $a \le x \le b$.

If s is an arc length function, the differential $ds = s'(x)\, dx$ is called the **differential of arc length.** The next theorem specifies formulas for finding ds.

THEOREM (6.12)   Let f be smooth on $[a, b]$, and let s be the arc length function for the graph of $y = f(x)$ on $[a, b]$. If dx and dy are the differentials of x and y, then

(i) $ds = \sqrt{1 + [f'(x)]^2}\, dx$ (ii) $(ds)^2 = (dx)^2 + (dy)^2$

PROOF   By Definition (6.11) and Theorem (5.26),

$$s'(x) = D_x[s(x)] = D_x\left[\int_a^x \sqrt{1 + [f'(t)]^2}\, dt\right] = \sqrt{1 + [f'(x)]^2}.$$

Applying Definition (3.23),

$$ds = s'(x)\, dx = \sqrt{1 + [f'(x)]^2}\, dx$$

This proves (i).

To prove (ii) we square both sides of (i), obtaining

$$(ds)^2 = \{1 + [f'(x)]^2\}(dx)^2$$
$$= (dx)^2 + [f'(x)\,dx]^2$$
$$= (dx)^2 + (dy)^2$$

The last equality follows from Definition (3.23). • •

Theorem (6.12) (ii) has an interesting and useful geometric interpretation. Consider $y = f(x)$ and give x an increment Δx. Let Δy denote the change in y, and Δs the change in arc length corresponding to Δx. These increments are illustrated in Figure 6.48. Note that dy is the amount that the tangent line rises or falls if the independent variable changes from x to $x + \Delta x$ (see also Figure 3.16). Since $(ds)^2 = (dx)^2 + (dy)^2$, ds may be regarded as the length of the hypotenuse of a right triangle with sides $|dx|$ and $|dy|$ as illustrated in Figure 6.48. This figure also indicates that if Δx is small, then ds may be used to approximate the increment Δs of arc length.

FIGURE 6.48
$(ds)^2 = (dx)^2 + (dy)^2$

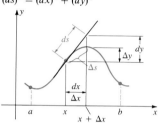

EXAMPLE 3 Use differentials to approximate the arc length of the graph of $y = x^3 + 2x$ from $A(1, 3)$ to $B(1.2, 4.128)$.

__SOLUTION__ If we let $f(x) = x^3 + 2x$, then by Theorem (6.12) (i),

$$ds = \sqrt{1 + (3x^2 + 2)^2}\,dx.$$

An approximation may be obtained by letting $x = 1$ and $dx = 0.2$. Thus

$$ds = \sqrt{1 + 5^2}\,(0.2) = \sqrt{26}\,(0.2) \approx 1.02 \quad •$$

Let f be a function that is nonnegative throughout a closed interval $[a, b]$. If the graph of f is revolved about the x-axis, a **surface of revolution** is generated (see Figure 6.49). We wish to find a formula for the area of this surface. For example, if $f(x) = \sqrt{r^2 - x^2}$ for a positive constant r, then the graph of f on $[-r, r]$ is the upper half of the circle $x^2 + y^2 = r^2$, and a revolution about the x-axis produces a sphere of radius r having surface area $4\pi r^2$.

If the graph of f is the line segment shown in Figure 6.50 then the surface generated is a frustum of a cone having base radii r_1 and r_2 and slant height s. We can show that the surface area is

$$\pi(r_1 + r_2)s = 2\pi\left(\frac{r_1 + r_2}{2}\right)s.$$

It is convenient to remember this formula as follows:

FIGURE 6.49

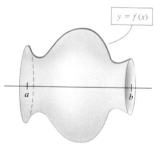

$y = f(x)$

FIGURE 6.50

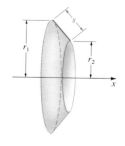

(6.13) $\quad$ Surface area of a frustum of a cone $= 2\pi$ (average radius)(slant height)

We will use this fact in the following discussion.

Let f be a smooth function that is nonnegative on $[a, b]$, and consider the surface generated by revolving the graph of f about the x-axis (see Figure 6.49). We wish to find a formula for the area S of this surface. To do so, let P be a partition of $[a, b]$ determined by $a = x_0, x_1, \ldots, x_n = b$ and, for each

FIGURE 6.51

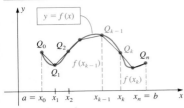

k, let Q_k denote the point $(x_k, f(x_k))$ on the graph of f (see Figure 6.51). If the norm $\| P \|$ is close to zero, then the broken line l_P obtained by connecting Q_{k-1} to Q_k for each k is an approximation to the graph of f, and hence the area of the surface generated by l_P should approximate S. The line segment $Q_{k-1}Q_k$ generates a frustum of a cone having base radii $f(x_{k-1})$ and $f(x_k)$ and slant height $d(Q_{k-1}, Q_k)$. By (6.13) its surface area is

$$2\pi \frac{f(x_{k-1}) + f(x_k)}{2} d(Q_{k-1}, Q_k).$$

Summing terms of this form from $k = 1$ to $k = n$ gives us the area S_P of the surface generated by the broken line l_P. If we use the expression for $d(Q_{k-1}, Q_k)$ on page 290, then

$$S_P = \sum_{k=1}^{n} 2\pi \frac{f(x_{k-1}) + f(x_k)}{2} \sqrt{1 + [f'(w_k)]^2}\, \Delta x_k$$

for $x_{k-1} < w_k < x_k$. We define the area S of the surface of revolution as $S = \lim_{\| P \| \to 0} S_P$. From the form of S_P, it is reasonable to expect that the limit is

$$\int_a^b 2\pi \frac{f(x) + f(x)}{2} \sqrt{1 + [f'(x)]^2}\, dx \quad \text{or} \quad \int_a^b 2\pi f(x) \sqrt{1 + [f'(x)]^2}\, dx.$$

The proof of this fact requires results from advanced calculus and is omitted. The following definition summarizes our discussion.

DEFINITION (6.14)

> If f is smooth and nonnegative on $[a, b]$, then the **area** S of the surface generated by revolving the graph of f about the x-axis is
>
> $$S = \int_a^b 2\pi f(x) \sqrt{1 + [f'(x)]^2}\, dx$$

If f is negative for some x in $[a, b]$, then the following extension of Definition (6.14) can be used to find the surface area S:

$$S = \int_a^b 2\pi |f(x)| \sqrt{1 + [f'(x)]^2}\, dx$$

We can obtain an easy way to remember the formula for S in Definition (6.14). As in Figure 6.52, let (x, y) denote an arbitrary point on the graph of f and, as in Theorem (6.12) (i), consider the differential of arc length

$$ds = \sqrt{1 + [f'(x)]^2}\, dx.$$

FIGURE 6.52

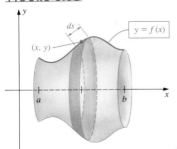

Next, *regard ds as the slant height* of the frustum of a cone with average radius $y = f(x)$ (see Figure 6.50). Applying (6.13), the surface area of this frustum is $2\pi f(x)\, ds = 2\pi y\, ds$. Integrating from $x = a$ to $x = b$ may be thought of as simultaneously summing terms $2\pi f(x)\, ds$ and considering the limit of these sums, obtaining

$$S = \int_a^b 2\pi f(x)\, ds = \int_a^b 2\pi y\, ds.$$

If we interchange the roles of x and y in the preceding discussion, then an analogous formula can be obtained for integration with respect to y. Thus, if

FIGURE 6.53

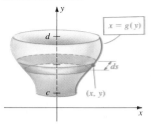

$x = g(y)$ and g is smooth and nonnegative on $[c, d]$, then the area S of the surface generated by revolving the graph of g about the y-axis (see Figure 6.53), is

$$S = \int_c^d 2\pi g(y)\sqrt{1 + [g'(y)]^2}\, dy.$$

EXAMPLE 4 The arc of the parabola $y^2 = x$ from $(1, 1)$ to $(4, 2)$ is revolved about the x-axis. Find the area of the resulting surface.

SOLUTION The surface is illustrated in Figure 6.54. Using Definition (6.14) with $y = f(x) = x^{1/2}$,

$$S = \int_1^4 2\pi x^{1/2}\sqrt{1 + \left(\frac{1}{2x^{1/2}}\right)^2}\, dx$$

$$= \int_1^4 2\pi x^{1/2}\sqrt{\frac{4x + 1}{4x}}\, dx = \pi \int_1^4 \sqrt{4x + 1}\, dx$$

$$= \frac{\pi}{6}(4x + 1)^{3/2}\Big]_1^4 = \frac{\pi}{6}(17^{3/2} - 5^{3/2}) \approx 30.85 \text{ square units} \quad \bullet$$

FIGURE 6.54

EXERCISES 6.5

Exer. 1–4: Use Definition (6.9) to find the arc length of the graph of the equation from A to B.

1 $8x^2 = 27y^3$; $A(1, \frac{2}{3})$, $B(8, \frac{8}{3})$

2 $(y + 1)^2 = (x - 4)^3$; $A(5, 0)$, $B(8, 7)$

3 $y = 5 - \sqrt{x^3}$; $A(1, 4)$, $B(4, -3)$

4 $y = 6\sqrt[3]{x^2} + 1$; $A(-1, 7)$, $B(-8, 25)$

5–6 Solve Exercises 1 and 2 by means of (6.10).

Exer. 7–10: Find the arc length of the graph of the equation from A to B.

7 $y = \dfrac{x^3}{12} + \dfrac{1}{x}$; $A(1, \frac{13}{12})$, $B(2, \frac{7}{6})$

8 $y + \dfrac{1}{4x} + \dfrac{x^3}{3} = 0$; $A(2, \frac{67}{24})$, $B(3, \frac{109}{12})$

9 $30xy^3 - y^8 = 15$; $A(\frac{8}{15}, 1)$, $B(\frac{271}{240}, 2)$

10 $x = \dfrac{y^4}{16} + \dfrac{1}{2y^2}$; $A(\frac{9}{8}, -2)$, $B(\frac{9}{16}, -1)$

Exer. 11–12: Set up (but do not evaluate) the integral for finding the arc length of the graph of the equation from A to B.

11 $2y^3 - 7y + 2x - 8 = 0$; $A(3, 2)$, $B(4, 0)$

12 $11x - 4x^3 - 7y + 7 = 0$; $A(1, 2)$, $B(0, 1)$

13 Find the arc length of the graph of $x^{2/3} + y^{2/3} = 1$. (*Hint:* Consider symmetry with respect to the line $y = x$.)

14 Find the arc length of the graph of $y = (3x^8 + 5)/(30x^3)$ from the point $(1, \frac{4}{5})$ to the point $(2, \frac{773}{240})$.

15 If f is defined by $f(x) = \sqrt[3]{x^2}$, what is the arc length function s? If x increases from 1 to 1.1, find Δs and ds.

16 Rework Exercise 15 if $f(x) = \sqrt{x^3}$.

17 Use differentials to approximate the arc length of the graph of $y = x^2$ from $A(2, 4)$ to $B(2.1, 4.41)$. Illustrate this approximation graphically and compare it with $d(A, B)$.

18 Use differentials to approximate the arc length of the graph of $y + x^3 = 0$ from $A(1, -1)$ to $B(1.1, -1.331)$. Illustrate this approximation graphically and compare it with $d(A, B)$.

19 Use differentials to approximate the arc length of the graph of $y = \cos x$ between the points with x-coordinates $\pi/6$ and $31\pi/180$.

Exer. 20–22: Use Simpson's Rule with $n = 4$ to approximate the arc length of the graph of the equation from A to B.

20 $y = x^2 + x + 3$; $A(-2, 5)$, $B(2, 9)$

21 $y = x^3$; $A(0, 0)$, $B(2, 8)$

22 $y = \sin x$; $A(0, 0)$, $B(\pi/2, 1)$

Exer. 23–26: The graph of the equation from A to B is revolved about the x-axis. Find the area of the resulting surface.

23 $4x = y^2$; $A(0, 0)$, $B(1, 2)$

24 $y = x^3$; $A(1, 1)$, $B(2, 8)$

25 $8y = 2x^4 + x^{-2}$; $A(1, \frac{3}{8})$, $B(2, \frac{129}{32})$

26 $y = 2\sqrt{x + 1}$; $A(0, 2)$, $B(3, 4)$

Exer. 27–28: The graph of the equation from A to B is revolved about the y-axis. Find the area of the resulting surface.

27 $y = 2\sqrt[3]{x}$; $A(1, 2)$, $B(8, 4)$

28 $x = 4\sqrt{y}$, $A(4, 1)$, $B(12, 9)$

29 The smaller arc of the circle $x^2 + y^2 = 25$ between the points $(-3, 4)$ and $(3, 4)$ is revolved about the y-axis. Find the area of the resulting surface.

30 If the arc described in Exercise 29 is revolved about the x-axis, find the area of the resulting surface.

31 Use Definition (6.14) to prove that the surface area of a right circular cone of altitude a and base radius b is $\pi b \sqrt{a^2 + b^2}$.

32 Prove that the area of the surface of a sphere of radius a between two parallel planes depends only on the distance between the planes.

33 If the graph in Figure 6.51 is revolved about the y-axis, show that the area of the resulting surface is given by $\int_a^b 2\pi x \sqrt{1 + [f'(x)]^2}\, dx$. (*Hint:* What is the average radius of the frustum?)

34 Use Exercise 33 to find the area of the surface generated by revolving the graph of $y = 3\sqrt[3]{x}$ from $A(1, 3)$ to $B(8, 6)$ about the y-axis.

6.6 WORK

The concept of **force** may be thought of as a push or pull on an object. For example, a force is needed to push or pull furniture across a floor, to lift an object off the ground, or to move a charged particle through an electromagnetic field.

Forces are often measured in pounds. If an object weighs ten pounds, then by definition the force required to lift it (or hold it off the ground) is ten pounds. A force of this type is a **constant force,** since its magnitude does not change while it is applied to the object.

The concept of *work* is used when a force moves an object from one place to another. The following definition covers the simplest case, where the object moves along a line in the same direction as the applied force.

DEFINITION (6.15)

> If a constant force F is applied to an object, moving it a distance d in the direction of the force, then the **work** W done on the object is
> $$W = Fd$$

In the British system, if F is measured in pounds and d in feet, the unit for W is the foot-pound (ft-lb). If F is in pounds and d in inches, then the unit of work is the inch-pound (in.-lb). In the metric system, two different units of force are used, depending on whether the cgs (centimeter-gram-second) system or the mks (meter-kilogram-second) system is employed. In the cgs system, a **dyne** is defined as the force that, when applied to a mass of 1 gram, induces an acceleration of 1 cm/sec². If F is expressed in dynes and d in centimeters, then the unit of work is the dyne-centimeter (dyn-cm), or **erg.** In the mks system, 1 **Newton** is the magnitude of the force required to impart an acceleration of 1 m/sec² to a mass of 1 kg. If F is given in Newtons and d in meters, then the unit of work is the Newton-meter (N-m), or **joule** (J). It can be shown that 1 joule $= 10^7$ ergs ≈ 0.74 ft-lb.

EXAMPLE 1 Find the work done in pushing an automobile along a level road from a point A to another point B, 20 feet from A, while exerting a constant force of 90 pounds.

FIGURE 6.55

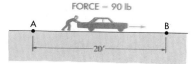

FORCE = 90 lb

SOLUTION The problem is illustrated in Figure 6.55. Since the constant force is $F = 90$ lb and the distance the automobile moves is $d = 20$ ft, it follows from Definition (6.15) that the work done is

$$W = (90)(20) = 1800 \text{ ft-lb.} \quad \bullet$$

We must sometimes determine F before applying Definition (6.15). In the next example we use *Newton's second law of motion* $F = ma$ for the mass m and acceleration a of the object.

EXAMPLE 2 Approximate the work done if an object of mass 15 kg is lifted vertically, a distance of 4 meters.

SOLUTION The required force F is given by $F = ma$. If the mass m is measured in kg and the acceleration a in m/sec^2, then F is in Newtons. In this example, a is the acceleration due to gravity, which is, in the mks system, approximately 9.81 m/sec^2. Hence

$$F = ma \approx (15)(9.81) = 147.15 \text{ N.}$$

Applying Definition (6.15) and the mks system,

$$W = Fd \approx (147.15)(4) = 588.6 \text{ N-m} = 588.6 \text{ J} \quad \bullet$$

Anyone who has pushed an automobile (or some other object) is aware that the force applied usually varies from one point to another. Thus, if an automobile is stalled, a larger force may be required to get it moving than to keep it in motion. The force could also vary because of friction, since, for example, part of the road may be smooth and another part rough. Forces that are not constant are sometimes referred to as *variable forces*.

If a variable force is applied to an object, moving the object a certain distance in the direction of the force, then methods of calculus are needed to find the work done. In this section we shall assume that the object moves along a line l. The case of motion along a nonlinear path will be considered in Chapter 18. Let us introduce a coordinate system on l and assume that the object moves from the point A with coordinate a to the point B with coordinate b for $b > a$. To find the work done, it is essential to know the force at the point on l with coordinate x, for every x in the interval $[a, b]$. This force will be denoted by $f(x)$. We will assume that the function f is continuous on $[a, b]$.

As usual, let P be the partition of $[a, b]$ determined by $a = x_0, x_1, \ldots,$ $x_n = b$ and let $\Delta x_k = x_k - x_{k-1}$ (see Figure 6.56). If w_k is a number in $[x_{k-1}, x_k]$, then the force at the point Q_k with coordinate w_k is $f(w_k)$. If Δx_k is small, then since f is continuous, the values of f change very little in the interval $[x_{k-1}, x_k]$. Roughly speaking, the function f is *almost constant* on $[x_{k-1}, x_k]$. Thus, the work W_k done as the object moves through the kth subinterval may be approximated by means of Definition (6.15), that is,

$$W_k \approx f(w_k) \Delta x_k.$$

FIGURE 6.56

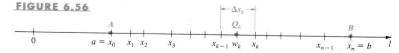

It seems evident that the smaller we choose Δx_k, the better $f(w_k) \Delta x_k$ approximates the work done in the interval $[x_{k-1}, x_k]$. If we also assume that work is additive, in the sense that the work W done as the object moves from A to B can be found by adding the work done over each subinterval, then

$$W \approx \sum_{k=1}^{n} f(w_k)\,\Delta x_k.$$

Since we expect this approximation to improve as the norms of the partitions approach zero, we define W as the limit of these sums. This limit leads to a definite integral.

DEFINITION (6.16)

Let the force at the point with coordinate x on a coordinate line l be $f(x)$ with f continuous on $[a, b]$. The **work** W done in moving an object from the point with coordinate a to the point with coordinate b is

$$W = \lim_{\|P\| \to 0} \sum_{k} f(w_k)\,\Delta x_k = \int_{a}^{b} f(x)\,dx$$

The formula in Definition (6.16) can be used to find the work done in stretching or compressing a spring. To solve problems of this type it is necessary to use the following law from physics.

HOOKE'S LAW (6.17)

The force $f(x)$ required to stretch a spring x units beyond its natural length is

$$f(x) = kx$$

for a constant k, called the **spring constant.**

The formula in (6.17) is also used to find the work done in compressing a spring x units from its natural length.

EXAMPLE 3 A force of 9 pounds is required to stretch a spring from its natural length of 6 inches to a length of 8 inches. Find the work done in stretching the spring **(a)** from its natural length to a length of 10 inches; **(b)** from a length of 7 inches to a length of 9 inches.

SOLUTION

(a) Let us introduce a coordinate line l as shown in Figure 6.57, with one end of the spring attached to some point to the left of the origin and the end to be pulled located at the origin. According to Hooke's law (6.17), the force $f(x)$ required to stretch a spring x units beyond its natural length is $f(x) = kx$ for some constant k. Using the given data, $f(2) = 9$. Substituting in $f(x) = kx$ we obtain $9 = k \cdot 2$, and hence the spring constant is $k = \frac{9}{2}$. Consequently, for this spring, Hooke's law has the form

$$f(x) = \tfrac{9}{2}x.$$

By Definition (6.16), the work done in stretching the spring 4 inches is

$$W = \int_{0}^{4} \frac{9}{2}x\,dx = \frac{9}{2}\frac{x^2}{2}\Bigg]_{0}^{4} = 36 \text{ in.-lb}$$

FIGURE 6.57

NATURAL LENGTH

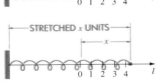

STRETCHED x UNITS

(b) We use the same function f, but change the interval to $[1, 3]$, obtaining

$$W = \int_1^3 \frac{9}{2} x \, dx = \frac{9}{2} \frac{x^2}{2} \Big]_1^3 = \frac{81}{4} - \frac{9}{4} = 18 \text{ in.-lb} \quad \bullet$$

EXAMPLE 4 A right circular conical tank of altitude 20 feet and radius of base 5 feet has its vertex at ground level and axis vertical. If the tank is full of water weighing 62.5 lb/ft³, find the work done in pumping all the water over the top of the tank.

FIGURE 6.58

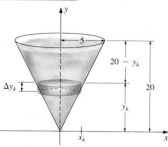

SOLUTION We begin by introducing a coordinate system as shown in Figure 6.58. The cone intersects the xy-plane along the line of slope 4 through the origin. An equation for this line is $y = 4x$.

Let P denote the partition of the interval $[0, 20]$ determined by the numbers $0 = y_0, y_1, y_2, \ldots, y_n = 20$, let $\Delta y_k = y_k - y_{k-1}$, and let x_k be the x-coordinate of the point on $y = 4x$ with y-coordinate y_k. If the cone is subdivided by means of planes perpendicular to the y-axis at each y_k, then we may think of the water as being sliced into n very thin parts. As illustrated in Figure 6.58, the volume of the kth slice may be approximated by the volume $\pi x_k^2 \, \Delta y_k$ of a circular disk or, since $x_k = y_k/4$, by $\pi(y_k/4)^2 \, \Delta y_k$. Thus

$$\text{Volume of } k\text{th slice} \approx \pi \left(\frac{y_k^2}{16} \right) \Delta y_k.$$

Since the water weighs 62.5 lb/ft³, we have

$$\text{Weight of } k\text{th slice} \approx 62.5\pi \left(\frac{y_k^2}{16} \right) \Delta y_k.$$

By (6.15) the work done in lifting the disk in the figure to the top of the tank is the product of the distance $20 - y_k$ and the weight. This gives us the following approximation:

$$\text{Work done in lifting } k\text{th slice} \approx (20 - y_k)62.5\pi \left(\frac{y_k^2}{16} \right) \Delta y_k.$$

Since the last number is an approximation to the work done in lifting the kth slice to the top, the work done in emptying the entire tank is approximately

$$\sum_{k=1}^{n} (20 - y_k)62.5\pi \left(\frac{y_k^2}{16} \right) \Delta y_k.$$

The actual work W is obtained by finding the limit of these sums as the norms approach zero. This gives us

$$W = \int_0^{20} (20 - y)62.5\pi \left(\frac{y^2}{16} \right) dy$$

$$= \frac{62.5\pi}{16} \int_0^{20} (20y^2 - y^3) \, dy$$

$$= \frac{62.5\pi}{16} \left[20\left(\frac{y^3}{3} \right) - \frac{y^4}{4} \right]_0^{20}$$

$$= \frac{62.5\pi}{16} \left(\frac{40,000}{3} \right) \approx 163,625 \text{ ft-lb} \quad \bullet$$

EXAMPLE 5 The pressure p (lb/in.^{2}) and volume v (in.^{3}) of an enclosed expanding gas are related by the formula $pv^m = c$ for constants m and c. If the gas expands from $v = a$ to $v = b$, show that the work done (in.-lb) is given by

$$W = \int_a^b p \, dv$$

SOLUTION Since the work done is independent of the shape of the container, we may assume that the gas is enclosed in a right circular cylinder of radius r and the expansion takes place against a piston head as illustrated in Figure 6.59. Let P denote a partition of the *volume interval* $[a, b]$ determined by $a = v_0, v_1, \ldots, v_n = b$, and let $\Delta v_k = v_k - v_{k-1}$. We shall regard the expansion as taking place in volume increments $\Delta v_1, \Delta v_2, \ldots, \Delta v_n$, and let $s_1, s_2, \ldots, s_n$ be the corresponding distances that the piston head moves (see Figure 6.59). It follows that for each k,

$$\Delta v_k = \pi r^2 s_k \quad \text{or} \quad s_k = \left(\frac{1}{\pi r^2}\right) \Delta v_k.$$

If p_k represents a value of the pressure p corresponding to the kth volume increment Δv_k, then the force against the piston head is the product of p_k and the area of the piston head, that is, $p_k \pi r^2$. Hence the work done in this increment is

$$(p_k \pi r^2)s_k = (p_k \pi r^2)\left(\frac{1}{\pi r^2}\right) \Delta v_k = p_k \, \Delta v_k$$

and, therefore,

$$W \approx \sum_k p_k \, \Delta v_k.$$

Since this approximation improves as the Δv_k approach zero, we conclude that

$$W = \lim_{\|P\| \to 0} \sum_k p_k \, \Delta v_k = \int_a^b p \, dv \quad \bullet$$

FIGURE 6.59

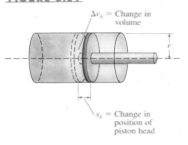

Δv_k = Change in volume

s_k = Change in position of piston head

EXERCISES 6.6

1 A spring of natural length 10 inches stretches 1.5 inches under a weight of 8 pounds. Find the work done in stretching the spring (a) from its natural length to a length of 14 inches; (b) from a length of 11 inches to a length of 13 inches.

2 A force of 25 N is required to compress a spring of natural length 0.80 m to a length of 0.75 m. Find the work done in compressing the spring from its natural length to a length of 0.70 m.

3 If a spring is 12 cm long, compare the work done in stretching it from 12 cm to 13 cm with that done in stretching it from 13 cm to 14 cm.

4 It requires 60 ergs of work to stretch a certain spring from a length of 6 cm to 7 cm, and another 120 ergs of work to stretch it from 7 cm to 8 cm. Find the spring constant and the natural length of the spring.

5 A fishtank has a rectangular base of width 2 feet and length 4 feet, and rectangular sides of height 3 feet. If the tank is filled with water, what work is required to pump all the water over the top of the tank?

6 Generalize Example 4 of this section to the case of a conical tank of altitude h and radius of base a which is filled with a liquid of density (weight per unit volume) ρ.

7 A freight elevator of mass 1500 kg is supported by a cable 4 m long, with mass of 7 kg per linear meter. Approximate the work required to lift the elevator 3 m by winding the cable onto a winch.

8 A 170-pound man climbs a vertical telephone pole 15 feet high. Find the work done if he reaches the top in (a) 10 seconds; (b) 5 seconds.

9 A vertical cylindrical tank of diameter 3 feet and height 6 feet is full of water. Find the work required to pump all the water (a) over the top of the tank; (b) through a pipe that rises to a height 4 feet above the top of the tank.

10 Answer parts (a) and (b) of Exercise 9 if the tank is only half full of water.

11 The ends of a water trough 8 feet long are equilateral triangles having sides of width 2 feet. If the trough is full of water, find the work required to pump all of it over the top.

12 A cistern has the shape of the bottom half of a sphere of radius 5 feet. If the cistern is full of water, find the work required to pump all the water to a point 4 feet above the top of the cistern.

13 The force (in dynes) with which two electrons repel one another is inversely proportional to the square of the distance (in cm) between them.

(a) If one electron is held fixed at the point $(5, 0)$, find the work done in moving a second electron along the x-axis from the origin to the point $(3, 0)$.

(b) If two electrons are held fixed at the points $(5, 0)$ and $(-5, 0)$, respectively, find the work done in moving a third electron from the origin to $(3, 0)$.

14 A uniform cable 12 m long, with mass 30 kg, hangs vertically from a pulley system at the top of a building (see figure). If a steel beam of mass 250 kg is attached to the end of the cable, what work is required to pull it to the top?

EXERCISE 14

12 m

15 A bucket containing water is lifted vertically at a constant rate of 1.5 ft/sec by means of a rope of negligible weight. As it rises, water leaks out at the rate of 0.25 lb/sec. If the bucket weighs 4 pounds when empty, and if it contained 20 pounds of water at the instant that the lifting began, determine the work done in raising the bucket 12 feet.

16 In Exercise 15, find the work required to raise the bucket until half the water has leaked out.

17 The volume and pressure of a certain gas vary in accordance with the law $pv^{1.2} = 115$, where the units of measurement are inches and pounds. Find the work done if the gas expands from 32 to 40 in.^{3}. (*Hint:* See Example 5.)

18 The pressure and volume of a quantity of enclosed steam are related by the formula $pv^{1.14} = c$ for a constant c. If the initial pressure and volume are p_0 and v_0, respectively, find a formula for the work done if the steam expands to twice its volume. (*Hint:* See Example 5.)

19 Newton's law of gravitation states that the force F of attraction between two particles having masses m_1 and m_2 is given by $F = gm_1m_2/s^2$ for a constant g and distance s between the particles. If the mass m_1 of the earth is regarded as concentrated at the center of the earth, and a rocket of mass m_2 is on the surface (a distance 4,000 miles from the center), find a general formula for the work done in firing the rocket vertically upward to an altitude h (see figure).

EXERCISE 19

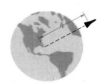

20 In the study of electricity, the formula $F = kq/r^2$, for a constant k, is used to find the force (in dyn) with which a positive charge Q of strength q units repels a unit positive charge located r cm from Q. Find the work done in moving a unit charge from a point d cm from Q to a point $\frac{1}{2}d$ cm from Q.

Exer. 21–22: Suppose the table was obtained experimentally for a force $f(x)$ acting at the point with coordinate x on a coordinate line l. Use the Trapezoidal Rule to approximate the work done on the interval $[a, b]$, where a and b are the smallest and largest values of x, respectively.

21

x ft	$f(x)$ lb
0	7.4
0.5	8.1
1.0	8.4
1.5	7.8
2.0	6.3
2.5	7.1
3.0	5.9
3.5	6.8
4.0	7.0
4.5	8.0
5.0	9.2

22

x m	$f(x)$ N
1	125
2	120
3	130
4	146
5	165
6	157
7	150
8	143
9	140

23 If the force function is constant, show that Definition (6.16) reduces to Definition (6.15).

In physics the *pressure p* at a depth *h* in a fluid is defined as the weight of the fluid contained in a column having a cross-sectional area of one square unit and altitude *h*. Pressure may also be regarded as the force per unit area exerted by the fluid, as in the following definition.

DEFINITION (6.18)

> If the **density** (the weight per unit volume) of a fluid is denoted by ρ, then the **pressure** p at a depth h is
>
> $$p = \rho h$$

FIGURE 6.60

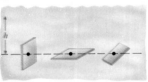

EXAMPLE 1 If the density of the water in a lake is 62.5 lb/ft³, find the pressure at a depth of **(a)** 2 feet; **(b)** 6 feet.

SOLUTION Using Definition (6.18) with $\rho = 62.5$, we obtain

(a) $p = \rho h = (62.5)(2) = 125$ lb/ft²

(b) $p = \rho h = (62.5)(6) = 375$ lb/ft² •

Pascal's principle in physics states that the pressure at any depth *h* in a fluid is the same in all directions. Thus, if a flat plate is submerged in a fluid, then the pressure at a point on the plate that is *h* units below the surface is ρh, regardless of whether the plate is submerged vertically, horizontally, or obliquely (see Figure 6.60, where the pressure at either *A*, *B*, or *C* is ρh).

If a rectangular tank, such as a fish aquarium, is filled with water, then the total force exerted by the water on the (horizontal) base can be found by multiplying the pressure at the bottom of the tank by the area of the base. For example, if, as in Figure 6.61, the depth of water is 2 feet and the area of the base is 12 ft², then from Example 1(a), the pressure at the bottom is 125 lb/ft² and the total force acting on the base is $(125)(12) = 1500$ pounds. This corresponds to 12 columns of water, each having cross-sectional area 1 ft² and each weighing 125 pounds.

FIGURE 6.61

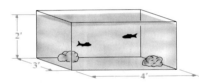

It is more complicated to find the force exerted on one of the sides of the aquarium, since the pressure is not constant there, but increases as the depth increases. Instead of investigating this particular problem, let us consider the following, more general, situation.

Suppose a flat plate is submerged in a fluid of density ρ such that the face of the plate is perpendicular to the surface of the fluid. Let us introduce a coordinate system as shown in Figure 6.62, where the width of the plate extends over the interval $[c, d]$ on the *y*-axis. Assume that for each *y* in $[c, d]$, the corresponding depth of the fluid and the length of the plate are given by $h(y)$ and $L(y)$, respectively, where *h* and *L* are continuous functions on $[c, d]$.

FIGURE 6.62

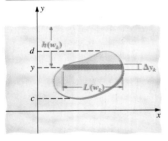

Let *P* be a partition of $[c, d]$, let w_k be any number in the *k*th subinterval $[y_{k-1}, y_k]$ and, as in Figure 6.63, consider the horizontal rectangle of length $L(w_k)$ and width $\Delta y_k = y_k - y_{k-1}$. If the norm $\| P \|$ is close to zero, then Δy_k is very small, and all points on the rectangle are roughly the same distance $h(w_k)$ from the surface of the fluid. Thus, by Definition (6.18), the pressure at every point within the rectangle is approximately $\rho h(w_k)$. Since the area of the rectangle is $L(w_k)\,\Delta y_k$, the total force on the rectangle is approximately

FIGURE 6.63

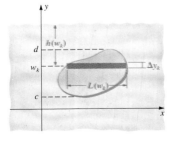

$\rho h(w_k) \cdot L(w_k) \, \Delta y_k$, and hence the force on the plate is approximately

$$\sum_k \rho h(w_k) \cdot L(w_k) \, \Delta y_k.$$

Since this approximation should improve as $\| P \|$ approaches zero, we arrive at the following formula.

DEFINITION (6.19)

> The force F exerted by a fluid of constant density ρ on a submerged region of the type illustrated in Figure 6.64 is
> $$F = \lim_{\|P\| \to 0} \sum_k \rho h(w_k) \cdot L(w_k) \, \Delta y_k = \int_c^d \rho h(y) \cdot L(y) \, dy$$

If a more complicated region is divided into subregions of the type used in (6.19), we apply the definition to each subregion and add the resulting forces. The coordinate system may be introduced in various ways. In Example 3 we shall choose the x-axis along the surface of the liquid and the positive direction of the y-axis downward.

It is often convenient to use the following formula to approximate the **force on a thin horizontal rectangle:**

(6.20)

> Force $\approx$ (density)(depth)(area of rectangle)

After using (6.20), we consider a limit of sums (or integrate) to obtain the total force on the region.

EXAMPLE 2 The ends of a water trough 8 feet long have the shape of isosceles trapezoids of lower base 4 feet, upper base 6 feet, and altitude 4 feet. Find the total force on one end if the trough is full of water.

SOLUTION Figure 6.64 illustrates one end of the trough superimposed on a rectangular coordinate system and labeled as in the preceding discussion. Using the Point-Slope Form (1.15), we can show that an equation of the line through the points (2, 0) and (3, 4) is $y = 4x - 8$, or equivalently, $x = \frac{1}{4}(y + 8)$. Referring to Figure 6.64, the depth $h(y)$ and the length $L(y)$ associated with a typical rectangle are

$$h(y) = 4 - y$$

and $$L(y) = 2x = 2 \cdot \tfrac{1}{4}(y + 8) = \tfrac{1}{2}(y + 8)$$

Using Definition (6.19), with $\rho = 62.5$, the total force is

$$F = \int_0^4 (62.5)(4 - y) \cdot \tfrac{1}{2}(y + 8) \, dy$$

$$= 31.25 \int_0^4 (32 - 4y - y^2) \, dy$$

$$= 31.25 \left[32y - 2y^2 - \frac{y^3}{3} \right]_0^4$$

$$= 31.25 \left[128 - 32 - \frac{64}{3} \right] = \frac{7000}{3} \approx 2333 \text{ lb} \quad \bullet$$

FIGURE 6.64

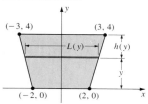

As indicated in the preceding example, the length of the water trough is irrelevant when considering the force on one end. The same is true for the oil tank in the next example.

EXAMPLE 3 A cylindrical oil storage tank 6 feet in diameter and 10 feet long is lying on its side. If the tank is half full of oil weighing 58 lb/ft^3, find the force exerted by the oil on one end of the tank.

SOLUTION Let us introduce a coordinate system so that the end of the tank is a circle of radius 3 feet with center at the origin. The equation of the circle is $x^2 + y^2 = 9$. If we choose the positive direction of the y-axis *downward*, then as shown in Figure 6.65, y represents the depth at a point in a typical horizontal rectangle. The width $L(y)$ of the tank is given by

$$L(y) = 2x = 2\sqrt{9 - y^2}.$$

Using Definition (6.19) with $\rho = 58$, we obtain

$$F = \int_0^3 58y(2\sqrt{9 - y^2})\, dy.$$

The integral may be evaluated by making the substitution

$$u = 9 - y^2, \qquad du = -2y\, dy.$$

Note that if $y = 0$, then $u = 9$, and if $y = 3$, then $u = 0$. This leads to

$$F = 58 \int_0^3 \sqrt{9 - y^2}\, 2y\, dy$$

$$= -58 \int_0^3 \sqrt{9 - y^2}(-2y\, dy)$$

$$= -58 \int_9^0 u^{1/2}\, du = -58 \frac{u^{3/2}}{3/2}\Big]_9^0$$

$$= -\frac{116}{3}\left[u^{3/2}\right]_9^0 = -\frac{116}{3}[0 - 9^{3/2}] = 1044\ \text{lb} \quad \bullet$$

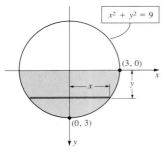

FIGURE 6.65

$x^2 + y^2 = 9$

$(3, 0)$

$(0, 3)$

EXERCISES 6.7

1 A glass aquarium tank is 3 feet long and has square ends of width 1 foot. If the tank is filled with water, find the force exerted by the water on (a) one end; (b) one side.

2 If one of the square ends of the tank in Exercise 1 is divided into two parts by means of a diagonal, find the force exerted on each part.

3 The ends of a water trough 6 feet long have the shape of isosceles triangles with equal sides of length 2 feet and the third side of length $2\sqrt{3}$ feet at the top of the trough. Find the force exerted by the water on one end of the trough if the trough is (a) full of water; (b) half full of water.

4 Answer (a) and (b) of Exercise 3 if the (vertical) altitude of the triangle is h feet, with $0 < h < 2$.

5 A cylindrical oil storage tank 4 feet in diameter and 5 feet long is lying on its side. If the tank is half full of oil weighing

60 lb/ft^3, find the force exerted by the oil on one end of the tank.

6 A rectangular gate in a dam is 5 feet long and 3 feet high. If the gate is vertical, with the top of the gate parallel to the surface of the water and 6 feet below it, find the force of the water against the gate.

7 A rectangular swimming pool is 20 feet wide and 40 feet long. The depth of the water in the pool varies uniformly from 3 feet at one end to 9 feet at the other end. Find the total force exerted by the water on the bottom of the pool.

8 Find the force exerted by the water on the side of the swimming pool described in Exercise 7.

9 A plate having the shape of an isosceles trapezoid with upper base 4 feet long and lower base 8 feet long is submerged vertically in water such that the bases are parallel

to the surface. If the distances from the surface of the water to the lower and upper bases are 10 feet and 6 feet, respectively, find the force exerted by the water on one side of the plate.

10 A circular plate of radius 2 feet is submerged vertically in water. If the distance from the surface of the water to the center of the plate is 6 feet, find the force exerted by the water on one side of the plate.

11 The ends of a water trough have the shape of the region bounded by the graphs of $y = x^2$ and $y = 4$, with x and y measured in feet. If the trough is full of water, find the force on one end.

12 A flat plate has the shape of the region bounded by the graphs of $y = x^4$ and $y = 1$, with x and y measured in feet. The plate is submerged vertically in water with the straight part of its boundary parallel to (and closest to) the surface. If the distance from the surface of the water to the straight part of the boundary is 4 feet, find the force exerted by the water on one side of the plate.

13 A rectangular plate 3 feet wide and 6 feet long is submerged vertically in oil with its short side parallel to, and 2 feet

below, the surface. If the oil weighs 50 lb/ft³, find the total force exerted on one side of the plate.

14 If the plate in Exercise 13 is divided into two parts by means of a diagonal, find the force exerted on each part.

15 A flat, irregularly shaped plate is submerged vertically in water (see figure). Measurements of its width, taken at successive depths at intervals of 0.5 feet, are compiled in the following table.

Water depth (ft)	1	1.5	2	2.5	3	3.5	4
Width of plate (ft)	0	2	3	5.5	4.5	3.5	0

Estimate the force of the water on one side of the plate by using (with $n = 6$) (a) the Trapezoidal Rule; (b) Simpson's Rule.

EXERCISE 15

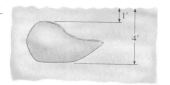

6.8 MOMENTS AND CENTER OF MASS OF A LAMINA

In this section we shall consider methods for finding the moments and center of mass of a *lamina*, that is, a thin flat plate. To motivate this work, let us first consider particles, or **point masses.**

Let l be a coordinate line and P a point on l with coordinate x. If a particle of mass m is located at P, then the **moment** (more precisely, the **first moment**) of the particle with respect to the origin O is defined as the product mx. Suppose l is horizontal with positive direction to the right, and let us imagine that l is free to rotate about O as if a fulcrum were positioned as shown in Figure 6.66. If an object has its mass m_1 concentrated at a point with positive coordinate x_1, then the moment $m_1 x_1$ is positive, and l would rotate in a clockwise direction. If an object of mass m_2 is at a point with negative coordinate x_2, then its moment $m_2 x_2$ is negative, and l would rotate in a counterclockwise direction. The set consisting of both objects is said to be in *equilibrium* if $m_1 x_1 = m_2 |x_2|$. Since $x_2 < 0$, this is equivalent to $m_1 x_1 = -m_2 x_2$, or

$$m_1 x_1 + m_2 x_2 = 0;$$

that is, *the sum of the moments with respect to the origin is zero.* This situation is similar to a seesaw that balances at the point O if two persons having masses of m_1 and m_2, respectively, are located as indicated in Figure 6.66.

If the particles are not in equilibrium, then as illustrated in Figure 6.67, there is a "balance" point P with coordinate $\bar{x}$, in the sense that

$$m_1(x_1 - \bar{x}) = m_2 |x_2 - \bar{x}| = -m_2(x_2 - \bar{x})$$

(Note that $|x_2 - \bar{x}| = -(x_2 - \bar{x})$, since $x_2 < \bar{x}$.)

FIGURE 6.66

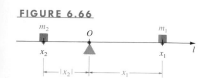

FIGURE 6.67

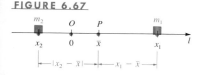

To locate P we solve for $\bar{x}$ as follows:

$$m_1(x_1 - \bar{x}) + m_2(x_2 - \bar{x}) = 0$$

$$m_1 x_1 + m_2 x_2 - (m_1 + m_2)\bar{x} = 0$$

$$(m_1 + m_2)\bar{x} = m_1 x_1 + m_2 x_2$$

$$\bar{x} = \frac{m_1 x_1 + m_2 x_2}{m_1 + m_2}.$$

Thus, to find $\bar{x}$ we divide the sum of the moments with respect to the origin by the total mass $m = m_1 + m_2$. The point with coordinate $\bar{x}$ is called the *center of mass* (or *center of gravity*) of the two particles. The next definition is an extension of this concept to more than two particles.

DEFINITION (6.21)

> Let l be a coordinate line and let S denote a collection of n particles of masses $m_1, m_2, \ldots, m_n$, located at points with coordinates $x_1, x_2, \ldots, x_n$, respectively. Let $m = \sum_{k=1}^{n} m_k$ be the total mass.
>
> (i) The **moment of S with respect to the origin** is $\sum_{k=1}^{n} m_k x_k$.
>
> (ii) The **center of mass** (or **center of gravity**) of S is the point with coordinate $\bar{x}$, such that
>
> $$\bar{x} = \frac{\sum_{k=1}^{n} m_k x_k}{m} \quad \text{or} \quad m\bar{x} = \sum_{k=1}^{n} m_k x_k$$

The number $m\bar{x}$ in Definition (6.21) may be regarded as the moment with respect to the origin of a particle of mass m located at the point with coordinate $\bar{x}$. The formula in (6.21)(ii) then states that $\bar{x}$ gives the position at which the total mass m could be concentrated without changing the moment with respect to the origin. The point with coordinate $\bar{x}$ is the balance point (in the sense of our seesaw illustration). If $\bar{x} = 0$, then by Definition (6.21) $\sum_{k=1}^{n} m_k x_k = 0$ and the collection is said to be in **equilibrium.** In this event the origin is the center of mass.

EXAMPLE 1 Three particles of masses 40, 60, and 100 grams are located at points with coordinates $-2, 3,$ and 7, respectively, on a coordinate line l. Find the center of mass.

SOLUTION If we denote the three masses by $m_1, m_2,$ and m_3, then we have the situation illustrated in Figure 6.68, with $x_1 = -2$, $x_2 = 3$, and $x_3 = 7$. Applying Definition (6.21), the coordinate $\bar{x}$ of the center of mass is

$$\bar{x} = \frac{40(-2) + 60(3) + 100(7)}{40 + 60 + 100}$$

$$= \frac{-80 + 180 + 700}{200} = \frac{800}{200} = 4 \bullet$$

FIGURE 6.68

FIGURE 6.69

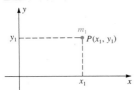

If a particle of mass m_1 is located at a point $P(x_1, y_1)$ in a coordinate plane, as illustrated in Figure 6.69, then the moments M_x and M_y of the particle with respect to the x-axis and y-axis, respectively, are defined by

$$M_x = m_1 y_1 \quad \text{and} \quad M_y = m_1 x_1.$$

Note that if x_1 and y_1 are both positive, then to find M_x we multiply the mass of the particle by its distance y_1 from the x-axis, and M_y is found by multiplying the mass by the distance x_1 from the y-axis. If either x_1 or y_1 is negative, then M_x or M_y is also negative.

As in the next definition, to find M_x and M_y for a collection of particles, we add the moments of the individual particles.

DEFINITION (6.22)

Let S be a collection of n particles of masses $m_1, m_2, \ldots, m_n$, located at points $P_1(x_1, y_1), P_2(x_2, y_2), \ldots, P_n(x_n, y_n)$, respectively, in a coordinate plane; and let $m = \sum_{k=1}^{n} m_k$ be the total mass.

(i) The moments M_x and M_y of S with respect to the x-axis and y-axis, respectively, are

$$M_x = \sum_{k=1}^{n} m_k y_k \quad \text{and} \quad M_y = \sum_{k=1}^{n} m_k x_k$$

(ii) The **center of mass** (or **center of gravity**) of S is the point $P(\bar{x}, \bar{y})$ such that

$$m\bar{x} = M_y \quad \text{and} \quad m\bar{y} = M_x$$

The numbers $m\bar{x}$ and $m\bar{y}$ in Definition (6.22) are the moments with respect to the y-axis and the x-axis of a particle of mass m located at the point $P(\bar{x}, \bar{y})$. Thus, the equations in Definition (6.22) (ii) imply that the center of mass is the point at which the total mass can be concentrated without changing the moments M_x and M_y. If we regard the n particles as fastened to the center of mass P by weightless rods, as spokes of a wheel are attached to the center of the wheel, then the collection S would balance if supported by a cord attached to P, as illustrated in Figure 6.70. The appearance would be similar to a mobile having all its objects in the same horizontal plane.

EXAMPLE 2 Particles of masses 4, 8, 3, and 2 kg are located at the points $P_1(-2, 3), P_2(2, -6), P_3(7, -3)$, and $P_4(5, 1)$, respectively. Find the moments M_x and M_y and the coordinates of the center of mass of the system.

SOLUTION The particles are illustrated in Figure 6.71, where we have also anticipated the position of $(\bar{x}, \bar{y})$. Applying Definition (6.22),

$$M_x = (4)(3) + (8)(-6) + (3)(-3) + (2)(1) = -43$$
$$M_y = (4)(-2) + (8)(2) + (3)(7) + (2)(5) = 39.$$

Since $m = 4 + 8 + 3 + 2 = 17$, it follows from (6.22) (ii) that

$$\bar{x} = \frac{M_y}{m} = \frac{39}{17} \approx 2.3, \qquad \bar{y} = \frac{M_x}{m} = -\frac{43}{17} \approx -2.5 \quad \bullet$$

Let us next consider a thin sheet of material (called a **lamina**), which is **homogeneous;** that is, it has a constant density. We wish to define the center of mass P such that if the tip of a sharp pencil is placed at P, as illustrated in Figure 6.72, the lamina would balance in a level position. If a lamina has the shape of a rectangle, then it is evident that the center of mass is the center of the rectangle, that is, the point of intersection of the diagonals. If the rectangle lies in an xy-plane, we shall assume that the mass of the rectangle

FIGURE 6.70

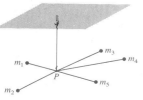

FIGURE 6.71

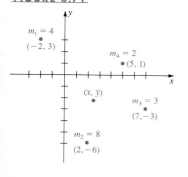

FIGURE 6.72

(more precisely, of the rectangular lamina) can be concentrated at the center of mass without changing its moments with respect to the x- or y-axes.

For simplicity let us begin by considering a lamina that has the shape of a region of the type illustrated in Figure 6.73, with f continuous on the closed interval $[a, b]$. Partition $[a, b]$ by choosing $a = x_0 < x_1 < \cdots < x_n = b$. For each k, let w_k be the midpoint of the subinterval $[x_{k-1}, x_k]$, that is, $w_k = (x_{k-1} + x_k)/2$. The Riemann sum $\sum_{k=1}^{n} f(w_k) \Delta x_k$ may be regarded as the sum of areas of rectangles of the type shown in Figure 6.73.

If the **area density** (the mass per unit area) is denoted by δ, then the mass corresponding to the kth rectangle is $\delta f(w_k) \Delta x_k$, and consequently the mass of the rectangular polygon associated with the partition is the sum $\sum_k \delta f(w_k) \Delta x_k$. As the norms of the partitions approach zero, the area of the rectangular polygon approaches the area of the face of the lamina, and the sum should approach the mass of the lamina. Thus, we define the mass m of the lamina by

$$m = \lim_{\|P\| \to 0} \sum_k \delta f(w_k) \Delta x_k = \delta \int_a^b f(x)\, dx.$$

The center of mass of the kth rectangular lamina illustrated in Figure 6.73 is located at the point $C_k(w_k, \frac{1}{2}f(w_k))$. If we assume that the mass is concentrated at C_k, then its moment with respect to the x-axis can be found by multiplying the distance $\frac{1}{2}f(w_k)$ from the x-axis to C_k by the mass $\delta f(w_k) \Delta x_k$. Using the additive property of moments, the moment of the rectangular polygon associated with the partition is $\sum_k \frac{1}{2}f(w_k) \cdot \delta f(w_k) \Delta x_k$. The moment M_x of the lamina is defined as the limit of these sums, that is,

$$M_x = \lim_{\|P\| \to 0} \sum_k \tfrac{1}{2}f(w_k) \cdot \delta f(w_k) \Delta x_k = \delta \int_a^b \tfrac{1}{2}f(x) \cdot f(x)\, dx.$$

Similarly, using the distance w_k from the y-axis to the center of mass of the kth rectangle, we arrive at the definition of the moment M_y of the lamina with respect to the y-axis. Specifically,

$$M_y = \lim_{\|P\| \to 0} \sum_k w_k \cdot \delta f(w_k) \Delta x_k = \delta \int_a^b x \cdot f(x)\, dx.$$

As with particles (see Definition 6.22), the coordinates $\bar{x}$ and $\bar{y}$ of the center of mass of the lamina are defined by $m\bar{x} = M_y$ and $m\bar{y} = M_x$.

The following summarizes this discussion.

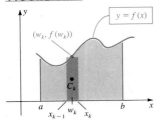

FIGURE 6.73

DEFINITION (6.23)

Let the function f be continuous and nonnegative on $[a, b]$. If a homogeneous lamina of area density ρ has the shape of the region under the graph of f from a to b, then

(i) the **mass** of the lamina is $m = \delta \int_a^b f(x)\, dx$.

(ii) the **moments M_x and M_y** of the lamina are

$$M_y = \delta \int_a^b \tfrac{1}{2}f(x) \cdot f(x)\, dx \quad \text{and} \quad M_y = \delta \int_a^b x \cdot f(x)\, dx$$

(iii) the **center of mass** (or **center of gravity**) of the lamina is the point $P(\bar{x}, \bar{y})$ such that

$$m\bar{x} = M_y \quad \text{and} \quad m\bar{y} = M_x$$

Substituting the integral forms from (i) and (ii) into (iii) of Definition (6.23), and solving for $\bar{x}$ and $\bar{y}$ gives us

$$\bar{x} = \frac{M_y}{m} = \frac{\delta \int_a^b x \cdot f(x)\, dx}{\delta \int_a^b f(x)\, dx}, \qquad \bar{y} = \frac{M_x}{m} = \frac{\delta \int_a^b \frac{1}{2} f(x) \cdot f(x)\, dx}{\delta \int_a^b f(x)\, dx}.$$

Since the constant δ in these formulas may be canceled, we see that the coordinates of the center of mass of a homogeneous lamina are independent of the density δ; that is, the coordinates depend only on the shape of the lamina and not on the density. For this reason the point $(\bar{x}, \bar{y})$ is sometimes referred to as the center of mass of a *region* in the plane, or as the **centroid** of the region. We can obtain formulas for moments of centroids by letting $\delta = 1$ in Definition (6.23).

EXAMPLE 3 Find the coordinates of the centroid of the region bounded by the graphs of $y = x^2 + 1$, $x = 0$, $x = 1$, and $y = 0$.

SOLUTION The region is sketched in Figure 6.74. Using Definition (6.23) (i) and (ii), with $\delta = 1$,

$$m = \int_0^1 (x^2 + 1)\, dx = \left[\tfrac{1}{3}x^3 + x \right]_0^1 = \tfrac{4}{3}$$

$$M_x = \int_0^1 \tfrac{1}{2}(x^2 + 1) \cdot (x^2 + 1)\, dx = \tfrac{1}{2} \int_0^1 (x^4 + 2x^2 + 1)\, dx$$

$$= \tfrac{1}{2} \left[\tfrac{1}{5}x^5 + \tfrac{2}{3}x^3 + x \right]_0^1 = \tfrac{14}{15}$$

$$M_y = \int_0^1 x(x^2 + 1)\, dx = \int_0^1 (x^3 + x)\, dx$$

$$= \left[\tfrac{1}{4}x^4 + \tfrac{1}{2}x^2 \right]_0^1 = \tfrac{3}{4}$$

Hence, by Definition (6.23) (iii),

$$\bar{x} = \frac{M_y}{m} = \frac{3/4}{4/3} = \frac{9}{16}; \qquad \bar{y} = \frac{M_x}{m} = \frac{14/15}{4/3} = \frac{7}{10} \, \bullet$$

Formulas similar to those in Definition (6.23) may be obtained for more complicated regions. Thus, consider a lamina of constant density δ having the shape illustrated in Figure 6.75, for continuous functions f and g such that $f(x) \geq g(x)$ for every x in $[a, b]$. Partitioning $[a, b]$, choosing w_k as before, and then applying the Midpoint Formula, we see that the center of mass of the kth rectangular lamina pictured in Figure 6.75 is the point

$$C_k(w_k, \tfrac{1}{2}[f(w_k) + g(w_k)]).$$

Using an argument similar to that given previously, the moment of the kth rectangle with respect to the x-axis is the distance from the x-axis to C_k multiplied by the mass, that is,

$$\tfrac{1}{2}[f(w_k) + g(w_k)] \cdot \delta[f(w_k) - g(w_k)]\, \Delta x_k.$$

FIGURE 6.74

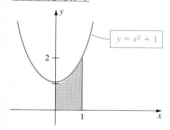

FIGURE 6.75

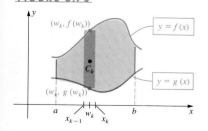

Summing and considering the limit as the norms of the partitions approach zero gives us

$$M_x = \delta \int_a^b \tfrac{1}{2}[f(x) + g(x)] \cdot [f(x) - g(x)]\, dx$$

$$= \delta \int_a^b \tfrac{1}{2}\{[f(x)]^2 - [g(x)]^2\}\, dx.$$

Similarly,

$$M_y = \delta \int_a^b x[f(x) - g(x)]\, dx.$$

The formulas in (6.23) (iii) may then be used to find $\bar{x}$ and $\bar{y}$.

EXAMPLE 4 Find the coordinates of the centroid of the region bounded by the graphs of $y + x^2 = 6$ and $y + 2x - 3 = 0$.

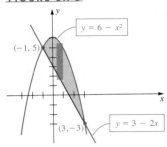

FIGURE 6.76

SOLUTION The region is the same as that considered in Example 2 of Section 6.1, where we found that its area is $\tfrac{32}{3}$. The region is resketched in Figure 6.76. If we let $f(x) = 6 - x^2$ and $g(x) = 3 - 2x$, then as in the preceding discussion, with $\delta = 1$,

$$M_x = \int_{-1}^{3} \tfrac{1}{2}[(6 - x^2) + (3 - 2x)][(6 - x^2) - (3 - 2x)]\, dx$$

$$= \tfrac{1}{2} \int_{-1}^{3} [(6 - x^2)^2 - (3 - 2x)^2]\, dx$$

$$= \tfrac{1}{2} \int_{-1}^{3} (x^4 - 16x^2 + 12x + 27)\, dx.$$

We may verify that $M_x = \tfrac{416}{15}$. Hence,

$$\bar{y} = \frac{M_x}{m} = \frac{416/15}{32/3} = \frac{13}{5}.$$

Similarly,

$$M_y = \int_{-1}^{3} x[(6 - x^2) - (3 - 2x)]\, dx$$

$$= \int_{-1}^{3} x(3 - x^2 + 2x)\, dx$$

$$= \int_{-1}^{3} (3x - x^3 + 2x^2)\, dx$$

from which it follows that $M_y = \tfrac{32}{3}$. Consequently,

$$\bar{x} = \frac{M_y}{m} = \frac{32/3}{32/3} = 1 \quad \bullet$$

Formulas for moments may also be obtained for other types of regions and for integration with respect to y; however, it is advisable to remember the *technique* of finding moments of rectangular laminas—multiplying a distance from an axis by a mass—instead of memorizing formulas that cover all possible cases.

If a homogeneous lamina has the shape of a region that has an axis of symmetry, then the center of mass must lie on that axis. This fact is used in the next example.

FIGURE 6.77

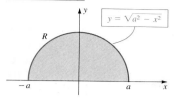

EXAMPLE 5 Find the centroid of the semicircular region R of radius a shown in Figure 6.77.

SOLUTION By symmetry, the centroid is on the y-axis, that is, $\bar{x} = 0$. Hence we need find only $\bar{y}$. Using Definition (6.23) (ii) with $\delta = 1$ and $f(x) = \sqrt{a^2 - x^2}$,

$$M_x = \int_{-a}^{a} \tfrac{1}{2}\sqrt{a^2 - x^2} \cdot \sqrt{a^2 - x^2}\, dx = \tfrac{1}{2}\int_{-a}^{a}(a^2 - x^2)\, dx$$

$$= \tfrac{1}{2}\left[a^2 x - \tfrac{1}{3}x^3\right]_{-a}^{a} = \tfrac{1}{2}\left[(a^3 - \tfrac{1}{3}a^3) - (-a^3 + \tfrac{1}{3}a^3)\right]$$

$$= \tfrac{1}{2}\left[\tfrac{4}{3}a^3\right] = \tfrac{2}{3}a^3.$$

Hence

$$\bar{y} = \frac{M_x}{m} = \frac{\tfrac{2}{3}a^3}{\tfrac{1}{2}\pi a^2} = \frac{4a}{3\pi} \approx 0.4a.$$

Thus the centroid is the point $(0, \tfrac{4}{3}a^3/\pi)$. •

EXERCISES 6.8

1 Particles of masses 2, 7, and 5 kg are located at the points $A(4, -1)$, $B(-2, 0)$, and $C(-8, -5)$, respectively. Find the moments M_x and M_y and the coordinates of the center of mass of this system.

2 Particles of masses 10, 3, 4, 1, and 8 g are located at the points $A(-5, -2)$, $B(3, 7)$, $C(0, -3)$, $D(-8, -3)$, and $O(0, 0)$. Find the moments M_x and M_y and the coordinates of the center of mass of this system.

Exer. 3–12: Sketch the region bounded by the graphs of the equations, and find the centroid of the region.

3 $y = x^3$, $y = 0$, $x = 1$

4 $y = \sqrt{x}$, $y = 0$, $x = 9$

5 $y = 4 - x^2$, $y = 0$

6 $2x + 3y = 6$, $y = 0$, $x = 0$

7 $y^2 = x$, $2y = x$

8 $y = x^2$, $y = x^3$

9 $y = 1 - x^2$, $y = x - 1$

10 $y = x^2$, $x + y = 2$

11 $x = y^2$, $x - y = 2$

12 $x = 9 - y^2$, $x + y = 3$

13 Find the centroid of the region in the first quadrant bounded by the circle $x^2 + y^2 = a^2$ and the coordinate axes.

14 Let R be the region in the first quadrant bounded by part of the parabola $y^2 = cx$ with $c > 0$, the x-axis, and the vertical line through the point (a, b) on the parab-

ola, as shown in the figure. Prove that the centroid of R is $(\tfrac{3}{5}a, \tfrac{3}{8}b)$.

EXERCISE 14

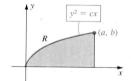

15 A region has the shape of a square of side $2a$ surmounted by a semicircle of radius a. Find the centroid. (*Hint:* Use Example 5 and the fact that moments are additive.)

16 Let $0 < a < b$ and let the points P, Q, R, and S have coordinates $(-b, 0)$, $(-a, 0)$, $(a, 0)$, and $(b, 0)$, respectively. Find the centroid of the region bounded by the graphs of $y = \sqrt{b^2 - x^2}$, $y = \sqrt{a^2 - x^2}$, and the line segments PQ and RS. (*Hint:* Use Example 5.)

17 Prove that the centroid of a triangle coincides with the intersection of the medians. (*Hint:* Take the vertices at the points $(0, 0)$, (a, b), and $(0, c)$, with a, b, and c positive.)

18 A region has the shape of a square of side a surmounted by an equilateral triangle of side a. Find the centroid. (*Hint:* Use Exercise 17 and the fact that moments are additive.)

6.9 OTHER APPLICATIONS

Definite integrals may be employed in most fields that use mathematics as a tool. In this section we shall consider several applications that illustrate the versatility of this important concept.

It should be evident from our work in this chapter that if a quantity can be approximated by a sum of many terms, then it is a candidate for representation as a definite integral. The main requirement is that as the number of terms increases, the sums approach a limit. Similarly, any quantity that can be interpreted as an area of a region in a plane may be investigated by means of a definite integral. (See, for example, the discussion of hysteresis at the end of Section 6.1.) Conversely, definite integrals allow us to represent physical quantities as areas. In the following illustrations, a quantity is *numerically equal* to an area of a region, that is, we *disregard units of measurement*, such as cm, ft-lb, and so on.

Suppose $v(t)$ is the velocity, at time t, of an object that is moving on a coordinate line. If s is the position function, then $s'(t) = v(t)$, and

$$\int_a^b v(t)\, dt = \int_a^b s'(t)\, dt = s(t)\Big]_a^b = s(b) - s(a).$$

FIGURE 6.78

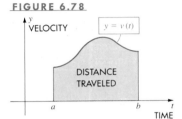

If $v(t) > 0$ throughout the time interval $[a, b]$, this tells us that the area under the graph of the function v from a to b represents the distance the object travels, as illustrated in Figure 6.78. This observation is useful to an engineer or physicist, who may not have an explicit form for $v(t)$, but merely a graph (or table) indicating the velocity at various times. The distance traveled may then be estimated by approximating the area under the graph.

If $v(t) < 0$ at certain times in $[a, b]$, the graph of v may resemble that in Figure 6.79. The figure indicates that the object moved in the negative direction from $t = c$ to $t = d$. The distance it traveled during that time is given by $\int_c^d |v(t)|\, dt$. It follows that $\int_a^b |v(t)|\, dt$ is the *total* distance traveled in $[a, b]$, whether $v(t)$ is positive or negative.

FIGURE 6.79

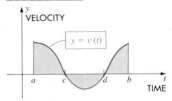

EXAMPLE 1 As an object moves along a straight path, its velocity $v(t)$ (in ft/sec) at time t is recorded each second for six seconds. The results are given in the table.

t	0	1	2	3	4	5	6
$v(t)$	1	3	4	6	5	5	3

Approximate the distance traveled by the object.

FIGURE 6.80

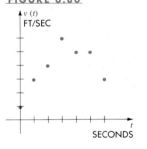

SOLUTION The points $(t, v(t))$ are plotted in Figure 6.80. If we assume that v is a continuous function, then as in the preceding discussion, the distance traveled during the time interval $[0, 6]$ is $\int_0^6 v(t)\, dt$. Let us approximate this definite integral by means of Simpson's Rule with $n = 6$. Thus, by (5.39),

$$\int_0^6 v(t)\, dt \approx \frac{6 - 0}{3 \cdot 6} \big[v(0) + 4v(1) + 2v(2) + 4v(3) + 2v(4) + 4v(5) + v(6) \big]$$

$$= \tfrac{1}{3}[1 + 4 \cdot 3 + 2 \cdot 4 + 4 \cdot 6 + 2 \cdot 5 + 4 \cdot 5 + 3] = 26 \text{ ft} \qquad \bullet$$

FIGURE 6.81

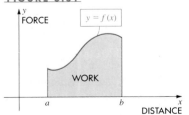

In (6.16) we defined the work W done by a variable force $f(x)$ as its point of application moves along a coordinate line from $x = a$ to $x = b$ by $W = \int_a^b f(x)\, dx$. Suppose $f(x) \geq 0$ throughout $[a, b]$. If we sketch the graph of f, then, as illustrated in Figure 6.81, the work W is numerically equal to the area under the graph from a to b.

FIGURE 6.82

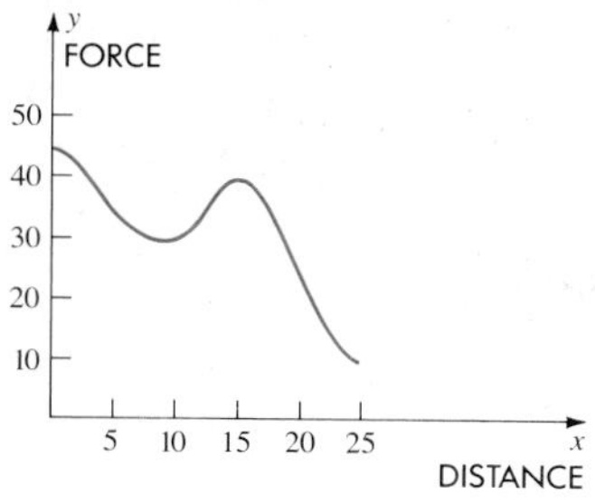

EXAMPLE 2 An engineer obtains the graph in Figure 6.82, which shows the force (in pounds) acting on a small cart as it moves 25 feet along horizontal ground. Estimate the work done.

SOLUTION Assuming the force is a continuous function f for $0 \le x \le 25$, the work done is

$$W = \int_0^{25} f(x)\, dx.$$

We do not have an explicit form for $f(x)$; however, we may estimate function values from the graph and approximate W by means of numerical integration.

Let us apply the Trapezoidal Rule with $a = 0$, $b = 25$, and $n = 5$. Referring to the graph to estimate function values gives us the following table.

k	x_k	$f(x_k)$	m	$mf(x_k)$
0	0	45	1	45
1	5	35	2	70
2	10	30	2	60
3	15	40	2	80
4	20	25	2	50
5	25	10	1	10

The sum of the numbers in the last column is 315. Since

$$(b - a)/2n = (25 - 0)/10 = 2.5$$

it follows from (5.37) that

$$W = \int_0^{25} f(x)\, dx \approx 2.5(315) \approx 788 \text{ ft-lb.}$$

For greater accuracy we could use a larger value of n or Simpson's Rule. •

Suppose that the amount of a physical entity, such as oil, water, electric power, money supply, bacteria count, or blood flow, is increasing or decreasing in some manner, and that $R(t)$ is the rate at which it is changing at time t. If $Q(t)$ is the amount of the entity present at time t, and if Q is differentiable, then we know from Section 3.3 that $Q'(t) = R(t)$. If $R(t) > 0$ (or $R(t) < 0$) in a time interval $[a, b]$, then the amount that the entity increases (or decreases) between $t = a$ and $t = b$ is

$$Q(b) - Q(a) = \int_a^b Q'(t)\, dt = \int_a^b R(t)\, dt.$$

This number may be represented as the area of the region in a ty-plane bounded by the graphs of R, $t = a$, $t = b$, and $y = 0$.

EXAMPLE 3 Starting at 9:00 A.M., oil is pumped into a storage tank at a rate of $(150t^{1/2} + 25)$ gal/hr, for time t (in hours) after 9:00 A.M. How many gallons will have been pumped into the tank at 1:00 P.M.?

SOLUTION Letting $R(t) = 150t^{1/2} + 25$ in the preceding discussion, we obtain

$$\int_0^4 (150t^{1/2} + 25)\, dt = 100t^{3/2} + 25t \Big]_0^4$$

$$= 100(4)^{3/2} + 25(4)$$

$$= 100(8) + 100 = 900 \text{ gal} \quad •$$

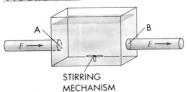

FIGURE 6.83

STIRRING
MECHANISM

Definite integrals can be applied to dye-dilution, or tracer, methods used in physiological tests and elsewhere. One example involves the measurement of cardiac output, that is, the rate at which blood flows through the aorta. A simple model for tracer experiments is sketched in Figure 6.83, where a liquid (or gas) flows into a tank at A and exits at B, with a constant flow rate F (in liters/sec). Suppose that at time $t = 0$, Q_0 grams of tracer (or dye) are introduced into the tank at A and that a stirring mechanism thoroughly mixes the solution at all times. The concentration $c(t)$ (in grams/liter) of tracer at time t is monitored at B. Thus the amount of tracer passing B at time t is $F \cdot c(t)$ g/sec.

If the amount of tracer in the tank at time t is $Q(t)$ for a differentiable function Q, then the rate of change $Q'(t)$ of Q is given by

$$Q'(t) = -F \cdot c(t)$$

(the negative sign indicates that Q is decreasing).

If T is a time at which all the tracer has left the tank, then $Q(T) = 0$ and, by the Fundamental Theorem of Calculus,

$$\int_0^T Q'(t)\, dt = Q(t) \Big]_0^T = Q(T) - Q(0)$$

$$= 0 - Q_0 = -Q_0.$$

We may also write

$$\int_0^T Q'(t)\, dt = \int_0^T \left[-F \cdot c(t) \right] dt = -F \int_0^T c(t)\, dt.$$

This gives us the following formula.

(6.24)

$$Q_0 = F \int_0^T c(t)\, dt$$

Usually an explicit form for $c(t)$ will not be known, but instead, a table of function values will be given. By employing numerical integration, we may find an approximation to the flow rate F (see Exercises 7 and 8).

Let us next consider another aspect of the flow of liquids. If a liquid flows through a cylindrical tube and if the velocity is a constant v_0, then the volume of liquid passing a fixed point per unit time is given by $v_0 A$ for the area A of a cross section of the tube (see Figure 6.84).

A more complicated formula is required to study the flow of blood in an arteriole. In this case the flow is in layers, as illustrated in Figure 6.85. In the layer closest to the wall of the arteriole, the blood tends to stick to the wall, and its velocity may be considered zero. The velocity increases as the layers approach the center of the arteriole.

For computational purposes, we may regard the blood flow as consisting of thin cylindrical shells that slide along, with the outer shell fixed and the velocity of the shells increasing as the radii of the shells decrease (see Figure 6.85). If the velocity in each shell is considered constant, then from the theory of liquids in motion, the velocity $v(r)$ in a shell having average radius r is

$$v(r) = \frac{P}{4vl} (R^2 - r^2)$$

FIGURE 6.84

FIGURE 6.85

for the radius R of the arteriole (in cm), the length l of the arteriole (in cm), the pressure difference P between the two ends of the arteriole (in dyn/cm^2), and the viscosity v of the blood (in dyn-sec/cm^2). Note that the formula gives zero velocity if $r = R$ and maximum velocity $PR^2/(4vl)$ as r approaches 0. If the radius of the kth shell is r_k and the thickness of the shell is Δr_k, then by (6.6) the volume of blood in this shell is

$$2\pi r_k v(r_k)\, \Delta r_k = \frac{2\pi r_k P}{4vl}\, (R^2 - r_k^2)\, \Delta r_k.$$

If there are n shells, then the total flow in the arteriole per unit time may be approximated by

$$\sum_{k=1}^{n} \frac{2\pi r_k P}{4vl}\, (R^2 - r_k^2)\, \Delta r_k.$$

To estimate the total flow F, that is, the volume of blood per unit time, we consider the limit of these sums as n increases without bound. This leads to the following definite integral:

$$F = \int_0^R \frac{2\pi r P}{4vl}\, (R^2 - r^2)\, dr$$

$$= \frac{2\pi P}{4vl} \int_0^R (R^2 r - r^3)\, dr$$

$$= \frac{\pi P}{2vl} \left[\frac{1}{2} R^2 r^2 - \frac{1}{4} r^4 \right]_0^R$$

Substituting the limits of integration gives us

$$F = \frac{\pi P R^4}{8vl}\ \text{cm}^3.$$

This formula for F is not exact, since the thickness of the shells cannot be made arbitrarily small. Indeed, the lower limit is the width of a red blood cell, or approximately 2×10^{-4} cm. However, we may assume that the formula gives a reasonable estimate. It is interesting to observe that a small change in the radius of an arteriole produces a large change in the flow, since F is directly proportional to the fourth power of R. A small change in pressure difference has a lesser effect, since P appears to the first power.

In many types of employment, a worker must perform the same assignment repeatedly. For example, a bicycle shop employee may be asked to assemble new bicycles. As more and more bicycles are assembled, the time required for each assembly should decrease until a certain minimum assembly time is reached. Another example of this process of learning by repetition is that of a data processor who must keyboard information from written forms to a computer system. The time required to process each entry should decrease as the number of entries increases. As a final illustration, the time required for a person to trace a path through a maze should improve with practice.

Let us consider a general situation in which a certain task is to be repeated many times. Suppose experience has shown that the time required to perform the task for the kth time can be approximated by $f(k)$ for a continuous decreasing function f on a suitable interval. The total time required to perform

FIGURE 6.86

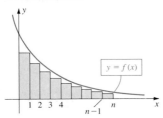

the task n times is given by the sum

$$\sum_{k=1}^{n} f(k) = f(1) + f(2) + \cdots + f(n).$$

If we consider the graph of f, then as illustrated in Figure 6.86, the preceding sum equals the area of the pictured inscribed rectangular polygon and, therefore, may be approximated by the definite integral $\int_0^n f(x)\, dx$. Evidently, the approximation will be close to the actual sum if f decreases slowly on $[0, n]$. If f changes rapidly per unit change in x, then an integral should not be used as an approximation.

EXAMPLE 4 A company that conducts polls by telephone interviews finds that the time required by an employee to complete one interview depends on the number of interviews that the employee has completed previously. Suppose it is estimated that for a certain survey, the number of minutes required to complete the kth interview is given by $f(k) = 6(1 + k)^{-1/5}$. Use a definite integral to approximate the time required for an employee to complete 100 interviews; 200 interviews. If an interviewer receives \$4.80 per hour, estimate how much more expensive it is to have two employees each conduct 100 interviews than it is to have one employee conduct 200 interviews.

SOLUTION As in the preceding discussion, the time required for 100 interviews is approximately

$$\int_0^{100} 6(1 + x)^{-1/5}\, dx = 6 \cdot \tfrac{5}{4}(1 + x)^{4/5}\Big]_0^{100}$$

$$= \tfrac{15}{2}\big[(101)^{4/5} - 1\big]$$

$$\approx 293.5 \text{ min.}$$

The time required for 200 interviews is approximately

$$\int_0^{200} 6(1 + x)^{-1/5}\, dx = \tfrac{15}{2}\big[(201)^{4/5} - 1\big]$$

$$\approx 514.4 \text{ min.}$$

Since an interviewer receives \$0.08 per minute, the cost for one employee to conduct 200 interviews is roughly (\$0.08)(514.4), or \$41.15. If two employees each conduct 100 interviews the cost is about 2(\$0.08)(293.5), or \$46.96, which is \$5.81 more than the cost of one employee. Note, however, that the time saved in using two people is approximately 221 minutes.

A computer may be used to show that

$$\sum_{k=1}^{100} 6(1 + k)^{-1/5} \approx 291.75$$

and

$$\sum_{k=1}^{200} 6(1 + k)^{-1/5} \approx 512.57$$

Hence the results obtained by integration (the area under the graph of f) are roughly 2 minutes more than the value of the corresponding sum (the area of the inscribed rectangular polygon). •

In economics, the process that a corporation uses to increase its accumulated wealth is called **capital formation**. If the amount K of capital at time

t can be approximated by $K = f(t)$ for a differentiable function f, then the rate of change of K with respect to t is called the **net investment flow.** Hence, if I denotes the investment flow, then

$$I = \frac{dK}{dt} = f'(t).$$

Conversely, if I is given by $g(t)$ for a function g that is continuous on an interval $[a, b]$, then the increase in capital over this time interval is

$$\int_a^b g(t)\, dt = f(b) - f(a).$$

EXAMPLE 5 Suppose a corporation wishes to have its net investment flow approximated by $g(t) = t^{1/3}$ for t in years and $g(t)$ in millions of dollars per year. If $t = 0$ corresponds to the present time, estimate the amount of capital formation over the next eight years.

SOLUTION From the preceding discussion, the increase in capital over the next eight years is

$$\int_0^8 g(t)\, dt = \int_0^8 t^{1/3}\, dt = \tfrac{3}{4} t^{4/3} \Big]_0^8 = 12.$$

Consequently, the amount of capital formation is $12,000,000. •

We have given only a few illustrations of the use of definite integrals. The interested reader may find many more in books on the physical and biological sciences, economics, and business, and even such areas as political science and sociology.

EXERCISES 6.9

1 The velocity (in mi/hr) of an automobile as it traveled along a freeway over a 12-minute interval is indicated in the figure. Approximate the distance traveled.

EXERCISE 1

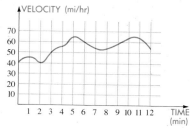

2 The acceleration (in ft/sec²) of an automobile over a period of 8 seconds is indicated in the figure. Approximate the net change in velocity in this time period.

EXERCISE 2

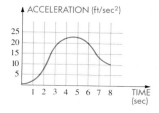

3 The following table was obtained by recording the force $f(x)$ (in dynes) acting on a particle as it moved 6 cm along a coordinate line from $x = 1$ to $x = 7$. Estimate the work done using (a) the Trapezoidal Rule with $n = 6$; (b) Simpson's Rule with $n = 6$.

x	1	2	3	4	5	6	7
$f(x)$	20	23	25	22	26	30	28

4 A bicyclist pedals directly up a hill, recording the velocity $v(t)$ (in ft/sec) at the end of every two seconds. Referring to the results recorded in following table, use numerical integration to approximate the distance traveled.

t	0	2	4	6	8	10
$v(t)$	24	22	16	10	2	0

5 A motorboat uses gasoline at the rate of $t\sqrt{9 - t^2}$ gal/hr. If the motor is started at $t = 0$, how much gasoline is used in two hours?

6 The population of a city has increased since 1985 at a rate of $1.5 + 0.3\sqrt{t} + 0.006t^2$ thousand people per year, where t is the number of years after 1985. Assuming that this rate continues and that the population was 50,000 in 1985, estimate the population in 1994.

7 Refer to (6.24). To estimate cardiac output F (the number of liters of blood per minute that the heart pumps through the aorta), a 5-mg dose of the tracer indocyanine-green is injected in a pulmonary artery and dye concentration measurements are taken every minute from a peripheral artery near the aorta. The results are given in the following table. Use Simpson's Rule with $n = 12$ to estimate the cardiac output.

t (min)	$c(t)$ (mg/liter)
0	0
1	0
2	0.15
3	0.48
4	0.86
5	0.72
6	0.48
7	0.26
8	0.15
9	0.09
10	0.05
11	0.01
12	0

8 Refer to (6.24). Suppose 1200 kg of sodium dichromate are mixed into a river at point A and sodium dichromate samples are taken every 30 seconds at a point B downstream. The results are given in the following table. Use the Trapezoidal Rule with $n = 12$ to estimate the river flow rate F.

t (sec)	$c(t)$ (mg/liter or g/m³)
0	0
30	2.14
60	3.89
90	5.81
120	8.95
150	7.31
180	6.15
210	4.89
240	2.98
270	1.42
300	0.89
330	0.29
360	0

9 A simple thermocouple, in which heat is transformed into electrical energy, is shown in the figure. To determine the total charge Q (in coulombs) transferred to the copper wire, current readings (in amperes) are recorded every $\frac{1}{2}$ second, and the results are shown in the following table.

t (sec)	0	0.5	1.0	1.5	2.0	2.5	3.0
I (amp)	0	0.2	0.6	0.7	0.8	0.5	0.2

Use the fact that $I = dQ/dt$ and the Trapezoidal Rule with $n = 6$ to estimate the total charge transferred to the copper wire during the first three seconds.

EXERCISE 9

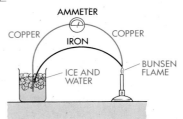

10 If, in Example 5, the rate of investment is approximated by $g(t) = 2t(3t + 1)$, with $g(t)$ in thousands of dollars, find the amount of capital formation over the intervals $[0, 5]$ and $[5, 10]$.

11 A data processor keyboards registration data for college students from written forms to electronic files. The number of minutes required to process the kth registration is estimated to be approximately $f(k) = 6(1 + k)^{-1/3}$. Use a definite integral to estimate the time required for (a) one person to keyboard 600 registrations; (b) two people to keyboard 300 each.

12 The number of minutes needed for a person to trace a path through a certain maze without error is estimated to be $f(k) = 5k^{-1/2}$, where k is the number of trials previously completed. Approximately how much time is required to complete 10 trials?

13 A manufacturer estimates that the time required for a worker to assemble a certain item depends on the number of this item the worker has previously assembled. If the time (in minutes) required to assemble the kth item is given by $f(k) = 20(k + 1)^{-0.4} + 3$, approximate the time required to assemble (a) 1 item; (b) 4 items; (c) 8 items; (d) 16 items.

14 Use a definite integral to approximate the sum
$$\sum_{k=1}^{100} k(k^2 + 1)^{-1/4}.$$

Define or discuss each of the following.

1 The area between the graphs of two continuous functions
2 Solid of revolution
3 Methods of finding volumes of solids of revolution
4 Volumes by slicing
5 Smooth function
6 Arc length of a graph
7 The arc length function
8 Surface of revolution
9 Work
10 Force exerted by a liquid

EXERCISES 6.10

Exer. 1–2: Sketch the region bounded by the graphs of the equations, and find the area (a) by integrating with respect to x; (b) integrating with respect to y.

1 $y = -x^2$, $y = x^2 - 8$

2 $y^2 = 4 - x$, $x + 2y - 1 = 0$

Exer. 3–4: Find the area of the region bounded by the graphs of the equations.

3 $x = y^2$, $x + y = 1$

4 $y + x^3 = 0$, $y = \sqrt{x}$, $3y + 7x - 10 = 0$

5 Find the area of the region between the graphs of the equations $y = \cos \frac{1}{2}x$ and $y = \sin x$ from $x = \frac{1}{3}\pi$ to $x = \pi$.

6 The region bounded by the graph of $y = \sqrt{1 + \cos 2x}$ and the x-axis, from $x = 0$ to $x = \pi/2$, is revolved about the x-axis. Find the volume of the resulting solid.

Exer. 7–10: Sketch the region R bounded by the graphs of the equations, and find the volume of the solid generated by revolving R about the indicated axis.

7 $y = \sqrt{4x + 1}$, $y = 0$, $x = 0$, $x = 2$; about the x-axis

8 $y = x^4$, $y = 0$, $x = 1$; about the y-axis

9 $y = x^3 + 1$, $x = 0$, $y = 2$; about the y-axis

10 $y = \sqrt[3]{x}$, $y = \sqrt{x}$; about the x-axis

11 Find the volume of the solid generated by revolving the region bounded by the graphs of the equations $y = 4x^2$ and $4x + y - 8 = 0$ (a) about the x-axis; (b) about the line $x = 1$; (c) about the line $y = 16$.

12 Find the volume of the solid generated by revolving the region bounded by the graphs of $y = x^3$, $x = 2$, and $y = 0$ (a) about the x-axis; (b) about the y-axis; (c) about the line $x = 2$; (d) about the line $x = 3$; (e) about the line $y = 8$; (f) about the line $y = -1$.

13 Find the arc length of the graph of $(x + 3)^2 = 8(y - 1)^3$ from $A(-2, \frac{3}{2})$ to $B(5, 3)$.

14 A solid has for its base the region in the xy-plane bounded by the graphs of $y^2 = 4x$ and $x = 4$. Find the volume of the solid if every cross section by a plane perpendicular to the x-axis is an isosceles right triangle with one of the equal sides on the base of the solid.

15 An above-ground swimming pool has the shape of a right circular cylinder of diameter 12 feet and height 5 feet. If the depth of the water in the pool is 4 feet, find the work required to empty the pool by pumping the water out over the top.

16 As a bucket is raised a distance of 30 feet from the bottom of a well, water leaks out at a uniform rate. Find the work done if the bucket originally contains 24 pounds of water and one-third leaks out. Assume that the weight of the empty bucket is 4 pounds and neglect the weight of the rope.

17 A square plate of side 4 feet is submerged vertically in water such that one of the diagonals is parallel to the surface. If the distance from the surface to the center of the plate is 6 feet, find the force exerted by the water on one side of the plate.

18 Use differentials to approximate the arc length of the graph of $y = 2 \sin \frac{1}{3}x$ between the points with x-coordinates π and $91\pi/90$.

Exer. 19–20: Sketch the region bounded by the graphs of the equations, and find the centroid of the region.

19 $y = 1 + x^3$, $x + y + 1 = 0$, $x = 1$

20 $y = 1 + x^2$, $y - x = 0$, $x = -1$, $x = 2$

21 The graph of the equation $12y = 4x^3 + (3/x)$ from $A(1, \frac{7}{12})$ to $B(2, \frac{67}{24})$ is revolved about the x-axis. Find the area of the resulting surface.

22 The shape of a reflector in a searchlight is obtained by revolving a parabola about its axis. If, as shown in the figure, the reflector is 4 feet across at the opening, and 1 foot deep, find its surface area.

EXERCISE 22

23 The velocity $v(t)$ of a rocket that is traveling directly upward is given in the following table. Use numerical integration to approximate the distance the rocket travels from $t = 0$ to $t = 5$.

t (sec)	0	1	2	3	4	5
$v(t)$ (ft/sec)	100	120	150	190	240	300

24 The electrician at a hotel suspects that the meter showing the total consumption Q in kilowatt hours (kwh) of electricity is not functioning properly. To check the accuracy, the consumption rate R is measured directly every 10 minutes, obtaining the results in the following table.

t (min)	0	10	20	30	40	50	60
R (kwh/min)	1.31	1.43	1.45	1.39	1.36	1.47	1.29

(a) Use Simpson's Rule to estimate the total consumption during this one-hour period.

(b) If the meter read 48792 kwh at the beginning of the experiment and 48887 kwh at the end, what should the electrician conclude?

Exer. 25–28: Suppose $\lim_{\|P\| \to 0} \sum_k \pi w_k^4 \, \Delta x_k$ represents the limit of sums for a function f on the interval $[0, 1]$.

25 Find the value of the limit.

26 Interpret the limit as the area of a region in the xy-plane.

27 Interpret the limit as the volume of a solid of revolution.

28 Interpret the limit as the work done by a force.

In this chapter we define logarithmic and exponential functions and investigate some of their properties. Since logarithmic and exponential functions are inverses of one another, we shall begin the chapter by discussing inverse functions.

LOGARITHMIC AND EXPONENTIAL FUNCTIONS

7.1 INVERSE FUNCTIONS

FIGURE 7.1

(i) $y = f(x)$ (ii) $x = g(y)$

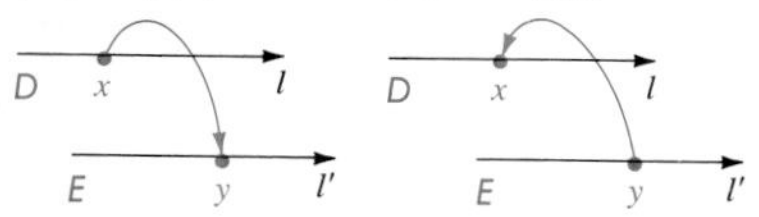

Suppose f is a one-to-one function with domain D and range E. Thus, for each number y in E, there is *exactly one* number x in D such that $f(x) = y$, as illustrated by the arrow in Figure 7.1(i). Since x is *unique*, we may define a function g from E to D such that

$$g(y) = x.$$

As in Figure 7.1(ii), g *reverses the correspondence* given by f. Since

$$g(y) = x \quad \text{and} \quad f(x) = y$$

for x in D and y in E, we see, by substitution, that

$$g(f(x)) = x \quad \text{and} \quad f(g(y)) = y.$$

We call g the *inverse function* of f and denote it by f^{-1}, as in the following definition.

DEFINITION (7.1)

Let f be a one-to-one function with domain D and range E. A function f^{-1} with domain E and range D is the **inverse function of f** if

$$f^{-1}(f(x)) = x \quad \text{for every } x \text{ in } D$$

and

$$f(f^{-1}(y)) = y \quad \text{for every } y \text{ in } E.$$

The $^{-1}$ used in the notation f^{-1} should not be mistaken for an exponent; that is, $f^{-1}(y)$ *does not mean* $1/[f(y)]$. The reciprocal $1/[f(y)]$ is denoted by $[f(y)]^{-1}$.

Remember that to define the inverse of a function f, *it is absolutely essential that f be one-to-one*. The most common examples of one-to-one functions are those that are increasing or decreasing on their domains, for in this case, if $a \neq b$ in the domain, then $f(a) \neq f(b)$ in the range.

A one-to-one function f has only one inverse function f^{-1} and, moreover, f^{-1} is one-to-one (see Exercises 20 and 21). It follows from Definition (7.1) that f is the inverse function of f^{-1}. We say that f and f^{-1} are *inverse functions of one another*. The following relationship (illustrated in Figure 7.1, with $g = f^{-1}$) always holds between a one-to-one function f and its inverse function f^{-1}.

(7.2)

$$y = f(x) \quad \text{if and only if} \quad x = f^{-1}(y).$$

Since we often let x denote an arbitrary number in the domain of a function, for the inverse function f^{-1} we may wish to consider $f^{-1}(x)$ *for x in the domain E of f^{-1}*. In this event, the two formulas in Definition (7.1) are written

$$f^{-1}(f(x)) = x \quad \text{for every } x \text{ in } D$$

and

$$f(f^{-1}(x)) = x \quad \text{for every } x \text{ in } E.$$

Figure 7.1 contains a hint for finding the inverse of a one-to-one function in certain cases: If possible, we *solve the equation $y = f(x)$ for x in terms of y*, obtaining an equation of the form $x = g(y)$. If the two conditions $g(f(x)) = x$ and $f(g(x)) = x$ are true for every x in the domains of f and g, respectively, then g is the required inverse function f^{-1}. The following guidelines summarize this procedure; in guideline 2, in anticipation of finding f^{-1}, we write $x = f^{-1}(y)$ instead of $x = g(y)$.

GUIDELINES FOR FINDING (7.3)
f^{-1} IN SIMPLE CASES

1 Verify that f is a one-to-one function (or that f is increasing or decreasing) throughout its domain.

2 Solve the equation $y = f(x)$ for x in terms of y, obtaining an equation of the form $x = f^{-1}(y)$.

3 Verify the two conditions

$$f^{-1}(f(x)) = x \quad \text{and} \quad f(f^{-1}(x)) = x$$

for every x in the domains of f and f^{-1}, respectively.

The success of this method depends on the nature of the equation $y = f(x)$, since we must be able to solve for x in terms of y. That is what we mean by *simple cases* in the heading of the guidelines.

EXAMPLE 1 If $f(x) = 3x - 5$, find the inverse function of f.

SOLUTION We shall follow the Guidelines (7.3). First, we note that the graph of the linear function f is a line of slope 3, and hence f is increasing throughout $\mathbb{R}$. Thus, the inverse function f^{-1} exists. Moreover, since the domain and range of f is $\mathbb{R}$, the same is true for f^{-1}.

As in guideline 2, we solve

$$y = 3x - 5$$

for x in terms of y, obtaining

$$x = \frac{y + 5}{3}.$$

We now formally let

$$f^{-1}(y) = \frac{y + 5}{3}.$$

Since the symbol used for the variable is immaterial, we may also write

$$f^{-1}(x) = \frac{x + 5}{3}.$$

Finally, we verify that the two conditions

$$f^{-1}(f(x)) = x \quad \text{and} \quad f(f^{-1}(x)) = x$$

are satisfied. Thus,

$$f^{-1}(f(x)) = f^{-1}(3x - 5) = \frac{(3x - 5) + 5}{3} = x$$

Also,

$$f(f^{-1}(x)) = f\left(\frac{x + 5}{3}\right) = 3\left(\frac{x + 5}{3}\right) - 5 = x$$

This proves that the inverse function of f is given by

$$f^{-1}(x) = \frac{x + 5}{3}. \quad \bullet$$

EXAMPLE 2 Find the inverse function of f if $f(x) = x^2 - 3$ and $x \geq 0$.

SOLUTION The graph of f is sketched in Figure 7.2. The domain D is $[0, \infty)$, and the range E is $[-3, \infty)$. Since f is increasing on D, it has an inverse function f^{-1} that has domain E and range D.

As in guideline 2, we consider the equation

$$y = x^2 - 3$$

and solve for x, obtaining

$$x = \pm\sqrt{y + 3}.$$

Since x is nonnegative, we reject $x = -\sqrt{y + 3}$ and let

$$f^{-1}(y) = \sqrt{y + 3} \quad \text{or, equivalently,} \quad f^{-1}(x) = \sqrt{x + 3}.$$

Finally, we verify that $f^{-1}(f(x)) = x$ for x in $D = [0, \infty)$, and that $f(f^{-1}(x)) = x$ for x in $E = [-3, \infty)$. Thus

$$f^{-1}(f(x)) = f^{-1}(x^2 - 3) = \sqrt{(x^2 - 3) + 3} = \sqrt{x^2} = x \quad \text{if } x \geq 0.$$

and

$$f(f^{-1}(x)) = f(\sqrt{x + 3}) = (\sqrt{x + 3})^2 - 3 = (x + 3) - 3 = x \quad \text{if } x \geq -3.$$

FIGURE 7.2

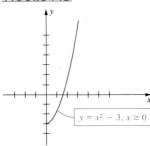

$y = x^2 - 3, x \geq 0$

This proves that the inverse function is given by

$$f^{-1}(x) = \sqrt{x+3} \quad \text{for } x \geq -3 \quad \bullet$$

There is an interesting relationship between the graphs of a function f and its inverse function f^{-1}. We first note that $b = f(a)$ means the same thing as $a = f^{-1}(b)$. These equations imply that the point (a, b) is on the graph of f if and only if the point (b, a) is on the graph of f^{-1}.

As an illustration, in Example 2 we found that the functions f and f^{-1}, given by

$$f(x) = x^2 - 3 \quad \text{and} \quad f^{-1}(x) = \sqrt{x+3},$$

FIGURE 7.3

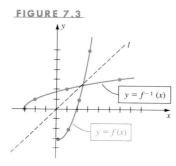

are inverse functions of one another, provided that x is suitably restricted. Some points on the graph of f are $(0, -3)$, $(1, -2)$, $(2, 1)$, and $(3, 6)$. Corresponding points on the graph of f^{-1} are $(-3, 0)$, $(-2, 1)$, $(1, 2)$, and $(6, 3)$. The graphs of f and f^{-1} are sketched on the same coordinate axes in Figure 7.3. If the page is folded along the line l that bisects quadrants I and III (as indicated by the dashes in the figure), then the graphs of f and f^{-1} coincide. Note that an equation for l is $y = x$. The two graphs are *reflections* of one another through the line l (or are *symmetric* with respect to l). This is true for the graph of any function f and its inverse function f^{-1} (see Exercise 22).

EXERCISES 7.1

Exer. 1–4: Prove that f and g are inverse functions of one another, and sketch the graphs of f and g on the same coordinate plane.

1 $f(x) = 7x + 5$; $g(x) = \frac{1}{7}(x - 5)$

2 $f(x) = x^2 - 1$, $x \geq 0$; $g(x) = \sqrt{x+1}$, $x \geq -1$

3 $f(x) = \sqrt{2x - 4}$, $x \geq 2$; $g(x) = \frac{1}{2}(x^2 + 4)$, $x \geq 0$

4 $f(x) = x^3 + 1$; $g(x) = \sqrt[3]{x - 1}$

Exer. 5–18: Find the inverse function of f.

5 $f(x) = 4x - 3$

6 $f(x) = 9 - 7x$

7 $f(x) = \dfrac{1}{2x + 5}$, $x > -\frac{5}{2}$

8 $f(x) = \dfrac{1}{3x - 1}$, $x > \frac{1}{3}$

9 $f(x) = 9 - x^2$, $x \geq 0$

10 $f(x) = 4x^2 + 1$, $x \geq 0$

11 $f(x) = 5x^3 - 2$

12 $f(x) = 7 - 2x^3$

13 $f(x) = \sqrt{3x - 5}$, $x \geq \frac{5}{3}$

14 $f(x) = \sqrt{4 - x^2}$, $0 \leq x \leq 2$

15 $f(x) = \sqrt[3]{x} + 8$

16 $f(x) = (x^3 + 1)^5$

17 $f(x) = x$

18 $f(x) = -x$

19 (a) Prove that the linear function defined by $f(x) = ax + b$ for $a \neq 0$ has an inverse function, and find $f^{-1}(x)$.

(b) Does a constant function have an inverse? Explain.

20 If f is a one-to-one function with domain D and range E, prove that f^{-1} is a one-to-one function with domain E and range D.

21 Prove that a one-to-one function has only one inverse function.

22 Establish the fact that the graph of f^{-1} is the reflection of the graph of f through the line $y = x$ by verifying (a)–(c):

(a) If $P(a, b)$ is on the graph of f, then $Q(b, a)$ is on the graph of f^{-1}.

(b) The midpoint of line segment PQ is on the line $y = x$.

(c) The line PQ is perpendicular to the line $y = x$.

Exer. 23–26: The graph of a one-to-one function is given. Use the reflection property to sketch the graph of f^{-1}. Determine the domains and ranges of f and f^{-1}.

23

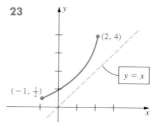

24

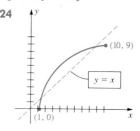

25

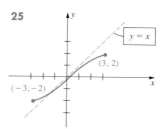

26

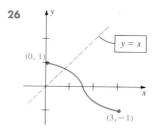

If a is a positive real number, it is easy to define a^r for every rational number r; however, the extension to irrational exponents requires advanced mathematical concepts. For example, to define a^π, we could express π as a nonterminating decimal 3.14159 . . . , and consider the successive approximations

$$a^3, \ a^{3.1}, \ a^{3.14}, \ a^{3.141}, \ a^{3.1415}, \ a^{3.14159}, \dots$$

Since we expect successive powers to get closer to a^π, it would seem natural to define

$$a^\pi = \lim_{r \to \pi} a^r$$

with r restricted to rational numbers. Unfortunately, major problems are associated with this type of definition. In addition to showing that the limit exists and is unique, we would also have to prove that the laws of exponents $a^u a^v = a^{u+v}$, $(a^u)^v = a^{uv}$, and so on, are true for all real numbers u and v, a formidable task. Due to the difficulty of such a proof, it is customary, in precalculus, to *assume* that if $a > 0$, then a^u exists for every real number u, and that the laws of exponents are true for real exponents. Logarithms with base a are introduced as follows:

$$u = \log_a v \quad \text{if and only if} \quad a^u = v.$$

Properties of logarithms may then be obtained using the (assumed) laws of exponents. Although this development is suitable in elementary algebra, it is inappropriate in advanced courses, where the standards of mathematical rigor are higher.

   In this chapter we shall state definitions of $\log_a x$ for every positive real number x and of a^x for every real number x. Our approach will be first to use a definite integral to introduce the *natural logarithmic function*, denoted by ln. We will then use the natural logarithmic function to define the *natural exponential function*, exp. Finally, we will define a^x and $\log_a x$. This approach will allow us to establish results on continuity, derivatives, and integrals in a relatively simple manner, and to prove the laws of exponents that were assumed in precalculus mathematics.

   Let f be a function that is continuous on a closed interval $[a, b]$. As in the proof of Part I of the Fundamental Theorem of Calculus (5.22), we can define a function F by

$$F(x) = \int_a^x f(t) \, dt$$

FIGURE 7.4

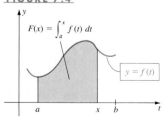

for every number x in $[a, b]$. If $f(t) \geq 0$ throughout $[a, b]$, then $F(x)$ is the area under the graph of f from a to x, as illustrated in Figure 7.4. For the special case $f(t) = t^n$, with n a rational number and $n \neq -1$, we can find an explicit form for F. Thus, by the Power Rule (5.24),

$$F(x) = \int_a^x t^n \, dt = \frac{1}{n+1} t^{n+1} \Big]_a^x$$

$$= \frac{1}{n+1} (x^{n+1} - a^{n+1})$$

provided t^n is defined throughout $[a, x]$. We must exclude $f(t) = 1/t = t^{-1}$, since $1/(n+1)$ is undefined if $n = -1$. Up to now we have not determined

an antiderivative of $f(x) = 1/x$. We shall remedy this situation by introducing a function whose derivative is $1/x$.

DEFINITION (7.4)

> The **natural logarithmic function,** denoted by **ln,** is defined by
>
> $$\ln x = \int_1^x \frac{1}{t}\, dt$$
>
> for every $x > 0$.

The notation ln is read *ell-en*. An interpretation of ln x as an area is
illustrated in Figure 7.5 for the case $x > 1$.

The expression ln x is called the **natural logarithm of x.** The restriction
$x > 0$ is necessary, because $\int_1^x (1/t)\, dt$ does not exist if $x \le 0$. The reason for
the term *logarithmic* in Definition (7.4) will become clear after we have proved
that ln satisfies the laws of logarithms studied in precalculus courses (see
Theorem (7.7)).

It follows from Theorem (5.26) that

$$D_x \int_1^x \frac{1}{t}\, dt = \frac{1}{x}$$

for every $x > 0$ (see Example 6 of Section 5.4). This gives us the next theorem.

FIGURE 7.5

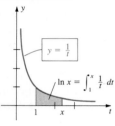

$$y = \frac{1}{t}$$

$$\ln x = \int_1^x \frac{1}{t}\, dt$$

THEOREM (7.5)

> $$D_x (\ln x) = \frac{1}{x}$$

Thus, ln x *is an antiderivative of* $1/x$. Since ln x is differentiable, and its
derivative $1/x$ is positive for every $x > 0$, it follows from Theorems (3.5) and
(4.13) that *the natural logarithmic function is continuous and increasing
throughout its domain*. Also note that

$$D_x^2 (\ln x) = D_x \left(\frac{1}{x}\right) = -\frac{1}{x^2},$$

which is negative for every $x > 0$. Hence by (4.16), the graph of the natural
logarithmic function is concave downward at every point $P(c, \ln c)$ on the
graph.

Let us sketch the graph of $y = \ln x$. The y-coordinates of points on the
graph are given by the integral in Definition (7.4). In particular, if $x = 1$, then
by Definition (5.10),

$$\ln 1 = \int_1^1 \frac{1}{t}\, dt = 0.$$

Thus the graph of $y = \ln x$ has x-intercept 1. Since ln is an increasing function, it follows that if $x > 1$, then the point (x, y) on the graph lies above
the x-axis, and if $0 < x < 1$, then (x, y) lies below the x-axis. To estimate
y if $x \ne 1$ we may apply either the Trapezoidal Rule or Simpson's Rule. If
$x = 2$, then by Example 2 in Section 5.6,

$$\ln 2 = \int_1^2 \frac{1}{t}\, dt \approx 0.693$$

We will show in Theorem (7.7) that if $a > 0$, then $\ln a^r = r \ln a$ for every rational number r. Using this result, we obtain

$$\ln 4 = \ln 2^2 = 2 \ln 2 \approx 2(0.693) \approx 1.386$$

$$\ln 8 = \ln 2^3 = 3 \ln 2 \approx 2.079$$

$$\ln \tfrac{1}{2} = \ln 2^{-1} = -\ln 2 \approx -0.693$$

$$\ln \tfrac{1}{4} = \ln 2^{-2} = -2 \ln 2 \approx -1.386$$

$$\ln \tfrac{1}{8} = \ln 2^{-3} = -3 \ln 2 \approx -2.079$$

Table C in Appendix III provides a list of natural logarithms of many other numbers, correct to three decimal places. Calculators may also be used to estimate values of ln.

Plotting the points that correspond to the y-coordinates we have calculated and using the fact that ln is continuous and increasing gives us the sketch in Figure 7.6. (Note the downward concavity at every point on the graph.)

At the end of this section we will prove that

$$\lim_{x \to \infty} \ln x = \infty \quad \text{and} \quad \lim_{x \to 0^+} \ln x = -\infty.$$

Thus $\ln x$ increases without bound as $x \to \infty$, and the y-axis is a vertical asymptote for the graph (see Figure 7.6).

The next result generalizes Theorem (7.5).

FIGURE 7.6

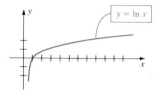
$y = \ln x$

THEOREM (7.6)

If $u = g(x)$, with g differentiable and $g(x) > 0$, then

$$D_x \ln u = \frac{1}{u} D_x u$$

PROOF If we let $y = \ln u$ and $u = g(x)$, then by the Chain Rule,

$$D_x \ln u = \frac{dy}{dx} = \frac{dy}{du}\frac{du}{dx} = \frac{1}{u} D_x u \quad \bullet \ \bullet$$

Note that if $u = x$, then Theorem (7.6) reduces to (7.5).

EXAMPLE 1

(a) Find $f'(x)$ if $f(x) = \ln (x^2 + 6)$.

(b) Find y' if $y = \ln \sqrt{x + 1}$ and $x > -1$.

SOLUTION

(a) Letting $u = x^2 + 6$ in Theorem (7.6),

$$f'(x) = D_x \ln (x^2 + 6) = \frac{1}{x^2 + 6} D_x (x^2 + 6) = \frac{2x}{x^2 + 6}$$

(b) Letting $u = \sqrt{x + 1}$ in Theorem (7.6) gives us

$$y' = D_x \ln \sqrt{x + 1}$$

$$= \frac{1}{\sqrt{x + 1}} D_x \sqrt{x + 1} = \frac{1}{\sqrt{x + 1}} \cdot \frac{1}{2}(x + 1)^{-1/2}$$

$$= \frac{1}{\sqrt{x + 1}} \cdot \frac{1}{2}\frac{1}{\sqrt{x + 1}} = \frac{1}{2(x + 1)} \quad \bullet$$

The next theorem states that natural logarithms satisfy the same laws of logarithms studied in precalculus mathematics courses.

THEOREM (7.7)

> If $p > 0$ and $q > 0$, then
>
> (i) $\ln pq = \ln p + \ln q$.
>
> (ii) $\ln \left(\dfrac{p}{q}\right) = \ln p - \ln q$.
>
> (iii) $\ln p^r = r \ln p$ for any rational number r.

PROOF

(i) If $p > 0$, then using Theorem (7.6) with $u = px$,

$$D_x \ln (px) = \frac{1}{px} D_x (px) = \frac{1}{px} p = \frac{1}{x}.$$

Since $\ln px$ has the same derivative as $\ln x$ for every $x > 0$, it follows from Theorem (4.32) that these expressions differ by a constant; that is,

$$\ln px = \ln x + C$$

for some real number C. Substituting 1 for x, we obtain

$$\ln p = \ln 1 + C.$$

Since $\ln 1 = 0$, we see that $C = \ln p$, and hence

$$\ln px = \ln x + \ln p.$$

Substituting q for x in the last equation gives us

$$\ln pq = \ln q + \ln p,$$

which is what we wished to prove.

(ii) Using the formula $\ln p + \ln q = \ln pq$ with $p = 1/q$, we see that

$$\ln \frac{1}{q} + \ln q = \ln \left(\frac{1}{q} \cdot q\right) = \ln 1 = 0$$

and hence $\ln \dfrac{1}{q} = -\ln q.$

Consequently,

$$\ln \frac{p}{q} = \ln \left(p \cdot \frac{1}{q}\right) = \ln p + \ln \frac{1}{q} = \ln p - \ln q.$$

(iii) If r is any rational number and $x > 0$, then by Theorem (7.6) with $u = x^r$,

$$D_x (\ln x^r) = \frac{1}{x^r} D_x (x^r) = \frac{1}{x^r} rx^{r-1} = r\left(\frac{1}{x}\right).$$

However, by Theorems (3.11) and (7.5), we may also write

$$D_x (r \ln x) = r D_x (\ln x) = r \left(\frac{1}{x} \right).$$

Consequently, $\qquad\qquad D_x (\ln x^r) = D_x (r \ln x).$

Since $\ln x^r$ and $r \ln x$ have the same derivative, they differ by a constant (see Theorem (4.32)). Thus,

$$\ln x^r = r \ln x + C$$

for some C. If we let $x = 1$ in the last formula we obtain

$$\ln 1 = r \ln 1 + C.$$

Since $\ln 1 = 0$, this implies that $C = 0$ and, therefore,

$$\ln x^r = r \ln x.$$

In Section 7.5 we shall extend this law to irrational exponents. • •

The next example illustrates the fact that it is sometimes convenient to apply Theorem (7.7) to a function before differentiating.

EXAMPLE 2 Find $f'(x)$ if $f(x) = \ln [\sqrt{6x - 1}(4x + 5)^3]$ and $x > \frac{1}{6}$.

SOLUTION We first write $\sqrt{6x - 1} = (6x - 1)^{1/2}$ and then use (i) and (iii) of Theorem (7.7), obtaining

$$\begin{aligned}
f(x) &= \ln [(6x - 1)^{1/2}(4x + 5)^3] \\
&= \ln (6x - 1)^{1/2} + \ln (4x + 5)^3 \\
&= \tfrac{1}{2} \ln (6x - 1) + 3 \ln (4x + 5).
\end{aligned}$$

Using Theorem (7.6),

$$\begin{aligned}
f'(x) &= \frac{1}{2} \cdot \frac{1}{6x - 1} (6) + 3 \cdot \frac{1}{4x + 5} (4) \\[2mm]
&= \frac{3}{6x - 1} + \frac{12}{4x + 5} \\[2mm]
&= \frac{84x + 3}{(6x - 1)(4x + 5)}
\end{aligned}$$

In the following examples and exercises, if a function is defined in terms of the natural logarithmic function, its domain will not usually be stated explicitly. Instead we will assume that x is restricted to values for which the given expression has meaning.

EXAMPLE 3 Find y' if $y = \ln \sqrt[3]{\dfrac{x^2 - 1}{x^2 + 1}}$.

SOLUTION We first use Theorem (7.7) to change the form of y as follows:

$$\begin{aligned}
y &= \ln \left(\frac{x^2 - 1}{x^2 + 1} \right)^{1/3} = \frac{1}{3} \ln \left(\frac{x^2 - 1}{x^2 + 1} \right) \\[2mm]
&= \tfrac{1}{3} [\ln (x^2 - 1) - \ln (x^2 + 1)]
\end{aligned}$$

Next we apply Theorem (7.6), obtaining

$$y' = \frac{1}{3}\left[\frac{1}{x^2-1}(2x) - \frac{1}{x^2+1}(2x)\right]$$

$$= \frac{2x}{3}\left[\frac{1}{x^2-1} - \frac{1}{x^2+1}\right]$$

$$= \frac{2x}{3}\left[\frac{2}{(x^2-1)(x^2+1)}\right] = \frac{4x}{3(x^4-1)} \quad \bullet$$

An application of natural logarithms to growth processes is given in the next example. Many additional practical problems involving ln appear in other examples and exercises of this chapter.

EXAMPLE 4 The *Count Model* is an empirically based formula that can be used to predict the height of a preschooler. If h denotes height (in cm) and x is age (in years), then

$$h = 70.228 + 5.104x + 9.222 \ln x$$

for $\frac{1}{4} \le x \le 6$.

(a) Predict the height and rate of growth of a typical two-year-old child.

(b) At what age is the rate of growth largest?

SOLUTION

(a) If $x = 2$, then the height is

$$h = 70.228 + 5.104(2) + 9.22 \ln 2$$
$$\approx 70.228 + 10.208 + 9.222(0.693) \approx 86.8 \text{ cm.}$$

By Definition (3.19), the rate of change of h with respect to x is

$$\frac{dh}{dx} = 5.104 + 9.222\left(\frac{1}{x}\right).$$

If $x = 2$, then

$$\frac{dh}{dx} = 5.104 + 9.222\left(\frac{1}{2}\right) = 9.715$$

Hence, the rate of growth, at age 2, is approximate 9.7 cm/year.

(b) To determine the maximum value of the rate of growth dh/dx, we first find the critical numbers of dh/dx. Differentiating dh/dx gives us

$$\frac{d}{dx}\left(\frac{dh}{dx}\right) = \frac{d^2h}{dx^2} = 9.222\left(-\frac{1}{x^2}\right) = -\frac{9.222}{x^2},$$

which exists throughout the interval $[\frac{1}{4}, 6]$ and is never zero. Hence dh/dx has no critical numbers on this interval. Since d^2h/dx^2 is always negative, it follows from Theorem (4.13) that dh/dx is decreasing throughout $[\frac{1}{4}, 6]$. Consequently, dh/dx has an endpoint maximum at $x = \frac{1}{4}$; that is, the rate of growth is largest at age 3 months. •

We shall conclude this section by investigating $\ln x$ as $x \to \infty$ and as $x \to 0^+$. If $x > 1$ we may interpret the integral $\int_1^x (1/t)\, dt = \ln x$ as the area

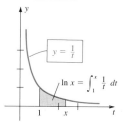

FIGURE 7.7

$$y = \frac{1}{t}$$

$$\ln x = \int_1^x \frac{1}{t}\, dt$$

FIGURE 7.8

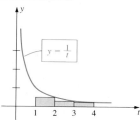

$$y = \frac{1}{t}$$

of the region shown in Figure 7.7. The sum of the areas of the three rectangles shown in Figure 7.8 is

$$\frac{1}{2} + \frac{1}{3} + \frac{1}{4} = \frac{13}{12}.$$

Since the area under the graph of $y = 1/t$ from $t = 1$ to $t = 4$ is $\ln 4$, we see that

$$\ln 4 > \frac{13}{12} > 1.$$

It follows that if M is any positive rational number, then

$$M \ln 4 > M \quad \text{or} \quad \ln 4^M > M.$$

If $x > 4^M$, then since ln is an increasing function,

$$\ln x > \ln 4^M > M.$$

This proves that $\ln x$ can be made as large as desired by choosing x sufficiently large, that is,

$$\lim_{x \to \infty} \ln x = \infty.$$

To investigate the case $x \to 0^+$ we first note that

$$\ln \frac{1}{x} = \ln 1 - \ln x = 0 - \ln x = -\ln x.$$

Hence

$$\lim_{x \to 0^+} \ln x = \lim_{x \to 0^+} \left(-\ln \frac{1}{x} \right).$$

As x approaches zero through positive values, $1/x$ increases without bound and, therefore, so does $\ln (1/x)$. Consequently, $-\ln (1/x)$ *decreases* without bound, that is,

$$\lim_{x \to 0^+} \ln x = -\infty.$$

EXERCISES 7.2

Exer. 1–34: Find $f'(x)$ if $f(x)$ equals the given expression.

1 $\ln (9x + 4)$

2 $\ln (x^4 + 1)$

3 $\ln (2 - 3x)^5$

4 $\ln (5x^2 + 1)^3$

5 $\ln \sqrt{7 - 2x^3}$

6 $\ln \sqrt[3]{6x + 7}$

7 $\ln (3x^2 - 2x + 1)$

8 $\ln (4x^3 - x^2 + 2)$

9 $\ln \sqrt[3]{4x^2 + 7x}$

10 $\ln |x|$ (*Hint:* $|x| = \sqrt{x^2}$)

11 $x \ln x$

12 $x^2/(\ln x)$

13 $\ln \sqrt{x} + \sqrt{\ln x}$

14 $\ln x^3 + (\ln x)^3$

15 $\dfrac{1}{\ln x} + \ln \left(\dfrac{1}{x} \right)$

16 $\ln \sqrt{\dfrac{4 + x^2}{4 - x^2}}$

17 $\ln (5x - 7)^4 (2x + 3)^3$

18 $\ln \sqrt[3]{4x - 5}(3x + 8)^2$

19 $\ln \dfrac{\sqrt{x^2 + 1}}{(9x - 4)^2}$

20 $\ln \dfrac{x^2(2x - 1)^3}{(x + 5)^2}$

21 $\ln \sqrt{\dfrac{x^2 - 1}{x^2 + 1}}$

22 $\ln (\ln x)$

23 $\ln (x + \sqrt{x^2 - 1})$

24 $\ln (x + \sqrt{x^2 + 1})$

25 $\ln \cos 2x$

26 $\cos (\ln 2x)$

27 $(\sin x)(\ln \sin x)$

28 $(\sin 5x)/(\ln x)$

29 $\ln (\csc x + \cot x)$

30 $\ln (\sec x + \tan x)$

31 $\ln \tan^3 3x$

32 $\ln \cot (x^2)$

33 $\ln \ln \sec 2x$

34 $\ln \csc^2 4x$

Exer. 35–40: Use implicit differentiation to find y'.

35 $3y - x^2 + \ln xy = 2$

36 $y^2 + \ln (x/y) - 4x + 3 = 0$

37 $x \ln y - y \ln x = 1$

38 $y^3 + x^2 \ln y = 5x + 3$

39 $x \sin y = 1 + y \ln x$

40 $y = \ln (\sin x + \cos y)$

41 Find an equation of the tangent line to the graph of $y = x^2 + \ln (2x - 5)$ at the point on the graph with x-coordinate 3.

42 Find an equation of the tangent line to the graph of $y = x + \ln x$ that is perpendicular to the line $2x + 6y = 5$.

43 Find the points on the graph of $y = x^2 + 4 \ln x$ at which the tangent line is parallel to the line $y - 6x + 3 = 0$.

44 Find an equation of the tangent line to the graph of $x^3 - x \ln y + y^3 = 2x + 5$ at the point $(2, 1)$.

45 Shown in the figure is a computer-generated graph of $y = 5 \ln x - \frac{1}{2}x$ for $x > 0$. Find the coordinates of the highest point and show that the graph is concave downward for $x > 0$.

EXERCISE 45

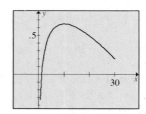

46 Shown in the figure is a computer-generated graph of $y = \ln (x^2 + 1)$. Find the coordinates of all points of inflection on the graph.

EXERCISE 46

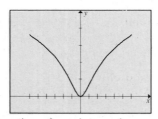

47 The position function of a point moving on a coordinate line is given by $s(t) = t^2 - 4 \ln (t + 1)$ for $0 \le t \le 4$. Find the velocity and acceleration at time t, and describe the motion of the point during the time interval $[0, 4]$.

48 The analysis of a particular type of chemical reaction involves the equation

$$t = \frac{1}{c(a - b)} \ln \frac{b(a - x)}{a(b - x)}$$

for the concentration x of a substance at time t and constants a, b, and c. Prove that $dx/dt = c(a - x)(b - x)$.

49 An approximation to the age T (in years) of a female blue whale can be obtained from a length measurement L (in feet) using the formula $T = -2.57 \ln [(87 - L)/63]$. A blue whale has been spotted by a research vessel, and her length is estimated to be 80 feet. If the error in estimating L can be as large as 2 feet, use differentials to approximate the maximum error in T.

50 The *Ehrenberg relation* $\ln W = \ln 2.4 + 1.84h$ is an empirically based formula relating the height h (in cm) to the weight W (in kg) for children aged 5 through 13. The formula, with minor changes in the constants, has been verified in many different countries. Find the relationship between the rates of change dW/dt and dh/dt, for time t (in years).

51 A rocket of mass m_1 is filled with fuel of mass m_2, which will be burned at a constant rate of b kg/sec. If the fuel is expelled from the rocket at a constant velocity, the distance $s(t)$ (in meters) that the rocket travels after t seconds is

$$s(t) = ct + \frac{c}{b} (m_1 + m_2 - bt) \ln \left(\frac{m_1 + m_2 - bt}{m_1 + m_2} \right)$$

for some constant $c > 0$.

(a) Find the initial velocity and initial acceleration of the rocket.

(b) Burnout occurs when $t = m_2/b$. Find the velocity and acceleration at burnout.

52 When a large number N of butterflies or moths has been collected from a certain locale, the expected number $S(N)$ of different species in the collected sample can be estimated from the formula $S(N) = a \ln [(a + N)/a]$ for some $a > 0$.

(a) Show that $S(0) = 0$ and $S(1) = \ln [1 + (1/a)]^a$.

(b) Sketch the graph of S by analyzing S' and S'' for $N > 0$.

53 Use Table C or a calculator to help sketch the graph of $y = \ln x$ (see Figure 7.6). Find the slope of the tangent line to the graph at the points with x-coordinates 1, 5, 10, 100, and 1000. Describe the slope as the x-coordinate a of the point of tangency increases without bound. What is true if a approaches 0?

54 Sketch the graphs of (a) $y = \ln |x|$; (b) $y = |\ln x|$.

55 Describe the difference between the graphs of $y = \ln (x^2)$ and $y = 2 \ln x$.

56 If $0 < a < b$, show that the natural logarithmic function satisfies the hypotheses of the Mean Value Theorem (4.12) on $[a, b]$, and find a general formula for the number c in the conclusion of (4.12).

57 If $f(x) = \ln x$, find a formula for the nth derivative $f^{(n)}(x)$, for any positive integer n.

58 If $0 < x < 1$, show that $\ln x$ is the negative of the area under the graph of $y = 1/t$ from $t = x$ to $t = 1$ (see Figure 7.5).

59 Prove that for every $x > 0$, $(x - 1)/x \le \ln x \le x - 1$.

60 Use Exercise 59 to prove that $\lim\limits_{x \to 0} \dfrac{\ln (x + 1)}{x} = 1$.

In Section 7.2 we saw that

$$\lim_{x \to \infty} \ln x = \infty \quad \text{and} \quad \lim_{x \to 0^+} \ln x = -\infty.$$

These facts are used in the proof of the following result.

THEOREM (7.8)

> To every real number x there corresponds a unique positive real number y such that $\ln y = x$.

PROOF First note that if $x = 0$, then $\ln 1 = 0$. Moreover, since $\ln$ is an increasing function, 1 is the only value of y such that $\ln y = 0$.

If x is positive, then we may choose a number b such that

$$\ln 1 < x < \ln b.$$

Since $\ln$ is continuous, it takes on every value between $\ln 1$ and $\ln b$ (see the Intermediate Value Theorem (2.33)). Thus there is a number y between 1 and b such that $\ln y = x$. Again, since $\ln$ is an increasing function, there is only one such number.

Finally, if x is negative, then there is a number $b > 0$ such that

$$\ln b < x < \ln 1$$

and as before, there is precisely one number y between b and 1 such that $\ln y = x$. • •

It follows from Theorem (7.8) that the range of $\ln$ is $\mathbb{R}$. Since $\ln$ is an increasing function, it is one-to-one and therefore has an inverse function, to which we give the following special name.

DEFINITION (7.9)

> The **natural exponential function,** denoted by **exp,** is the inverse of the natural logarithmic function.

FIGURE 7.9

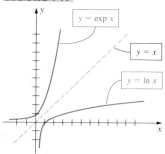

$y = \exp x$

$y = x$

$y = \ln x$

The reason for the term *exponential* in this definition will become clear shortly. Since exp is the inverse of ln, its domain is $\mathbb{R}$ and its range is the set of nonnegative real numbers. Moreover, as in (7.2),

$$y = \exp x \quad \text{if and only if} \quad x = \ln y$$

where x is any real number and $y > 0$. By Definition (7.1) we may also write

$$\ln (\exp x) = x \quad \text{and} \quad \exp (\ln y) = y.$$

As we pointed out in Section 7.1, if two functions are inverses of one another, then their graphs are reflections through the line $y = x$. Hence the graph of $y = \exp x$ can be obtained by reflecting the graph of $y = \ln x$ through this line, as illustrated in Figure 7.9. Evidently,

$$\lim_{x \to \infty} \exp x = \infty \quad \text{and} \quad \lim_{x \to -\infty} \exp x = 0.$$

By Theorem (7.8) there exists a unique positive real number whose natural logarithm is 1. This number is denoted by e. The great Swiss mathematician Leonhard Euler (1707–1783) was among the first to study its properties extensively.

The letter e denotes the positive real number such that $\ln e = 1$.

Several values of ln were calculated in Section 7.2. We can show, by means of the Trapezoidal Rule, that

$$\int_1^{2.7} \frac{1}{t}\, dt < 1 < \int_1^{2.8} \frac{1}{t}\, dt$$

(see Exercise 13 in Section 5.6). Consequently, by Definition (7.4),

$$\ln 2.7 < \ln e < \ln 2.8$$

which implies that

$$2.7 < e < 2.8$$

Later, in Theorem (7.25) we will show that e may be expressed as the following limit:

$$e = \lim_{h \to 0} (1 + h)^{1/h}$$

This formula can be used to approximate e to any degree of accuracy. In particular, to five decimal places (see page 351),

$$e \approx 2.71828$$

It can be shown that e is an irrational number.

If r is any rational number, then using Theorem (7.7) (iii),

$$\ln e^r = r \ln e = r(1) = r.$$

This equation may be used to motivate a definition of e^x for every real number x. Specifically, we shall *define* e^x as the real number y such that $\ln y = x$. The following equivalent statement is a convenient way to remember this fact.

DEFINITION OF e^x **(7.11)**   If x is any real number, then
$$e^x = y \quad \text{if and only if} \quad \ln y = x.$$

Recall, from Definition (7.9), that

$$\exp x = y \quad \text{if and only if} \quad \ln y = x.$$

Comparing this relationship with Definition (7.11), we see that

$$e^x = \exp x \qquad \text{for every } x.$$

This is the reason for calling exp an *exponential* function and referring to it as the **exponential function with base e**. The fact that $\ln (\exp x) = x$ for every x, and $\exp (\ln x) = x$ for every $x > 0$ may now be written as follows:

THEOREM **(7.12)**   (i) $\ln e^x = x$ for every x.
(ii) $e^{\ln x} = x$ for every $x > 0$.

The next theorem states that the usual laws of exponents hold for powers of the number e.

THEOREM (7.13)

If p and q are real numbers and r is a rational number, then

(i) $e^p e^q = e^{p+q}$ (ii) $\dfrac{e^p}{e^q} = e^{p-q}$ (iii) $(e^p)^r = e^{pr}$

PROOF Using Theorems (7.7) and (7.12),

$$\ln e^p e^q = \ln e^p + \ln e^q = p + q = \ln e^{p+q}.$$

Since the natural logarithmic function is one-to-one, it follows that

$$e^p e^q = e^{p+q}.$$

This proves (i). The proofs for (ii) and (iii) are similar. We will show in Section 7.5 that (iii) is also true if r is irrational. • •

It is not difficult to show that if n is a positive integer, then (as in elementary algebra) $e^n = e \cdot e \cdot \cdots \cdot e$ with n factors, all equal to e, on the right. Thus, from Theorem (7.12),

$$e^0 = e^{\ln 1} = 1 \quad \text{and} \quad e^1 = e^{\ln e} = e.$$

Next, from Theorem (7.13),

$$e^2 = e^{1+1} = e^1 e^1 = ee, \qquad e^3 = e^{2+1} = e^2 e^1 = (ee)e,$$

and so on. Negative exponents also have the usual properties, that is,

$$e^{-1} = e^{0-1} = \frac{e^0}{e^1} = \frac{1}{e} \quad \text{and} \quad e^{-n} = \frac{1}{e^n}$$

for every positive integer n. Rational powers of e may also be interpreted as they are in elementary algebra.

The graph of $y = e^x$ is the same as that of $y = \exp x$ illustrated in Figure 7.9. Hereafter, *we shall use e^x instead of* $\exp x$ *to denote values of the natural exponential function.*

In precalculus mathematics, graphs of equations such as $y = 2^x$ and $y = 3^x$ are sketched by *assuming* that these exponential expressions are defined for every real number x and increase as x increases. Using this intuitive point of view, a rough sketch of the graph of $y = e^x$ can be obtained by sketching $y = (2.7)^x$.

We will show in Section 7.7 that the inverse function of a differentiable function is differentiable. Anticipating this result, let us find the derivative of the natural exponential function implicitly. Thus, if

$$y = e^x, \quad \text{then} \quad \ln y = x.$$

Differentiating the last equation implicitly gives us

$$D_x (\ln y) = D_x (x) \quad \text{or} \quad \frac{1}{y} D_x y = 1.$$

Multiplying both sides of the last equation by y, we obtain

$$D_x y = y.$$

Since $y = e^x$, we have the following formula, which is true for every real number x.

THEOREM (7.14)

$$D_x e^x = e^x$$

Note that Theorem (7.14) states that *the natural exponential function is its own derivative*.

EXAMPLE 1 Find $f'(x)$ if $f(x) = x^2 e^x$.

SOLUTION By the Product Rule,

$$f'(x) = x^2 D_x e^x + e^x D_x x^2$$
$$= x^2 e^x + e^x (2x) = (x + 2)xe^x \quad \bullet$$

The next result is a generalization of Theorem (7.14).

THEOREM (7.15)

If $u = g(x)$ and g is differentiable, then
$$D_x e^u = e^u D_x u$$

PROOF Letting $y = e^u$ with $u = g(x)$, and using the Chain Rule,

$$\frac{dy}{dx} = \frac{dy}{du} \frac{du}{dx} = e^u D_x u \quad \bullet \bullet$$

If $u = x$ then Theorem (7.15) reduces to (7.14).

EXAMPLE 2 Find y' if $y = e^{\sqrt{x^2 + 1}}$.

SOLUTION By Theorem (7.15),

$$D_x e^{\sqrt{x^2 + 1}} = e^{\sqrt{x^2 + 1}} D_x \sqrt{x^2 + 1} = e^{\sqrt{x^2 + 1}} D_x (x^2 + 1)^{1/2}$$

$$= e^{\sqrt{x^2 + 1}} (\tfrac{1}{2})(x^2 + 1)^{-1/2}(2x) = e^{\sqrt{x^2 + 1}} \cdot \frac{x}{\sqrt{x^2 + 1}}$$

$$= \frac{xe^{\sqrt{x^2 + 1}}}{\sqrt{x^2 + 1}} \qquad\qquad \bullet$$

EXAMPLE 3 The function f defined by $f(x) = e^{-x^2/2}$ occurs in the branch of mathematics called *probability*. Find the local extrema of f, discuss concavity, find the points of inflection, and sketch the graph of f.

SOLUTION By Theorem (7.15),

$$f'(x) = e^{-x^2/2} D_x \left(-\frac{x^2}{2} \right) = e^{-x^2/2} \left(-\frac{2x}{2} \right) = -xe^{-x^2/2}.$$

Since $e^{-x^2/2}$ is always positive, the only critical number of f is 0. If $x < 0$, then $f'(x) > 0$, and if $x > 0$, then $f'(x) < 0$. It follows from the First Derivative Test that f has a local maximum at 0. The maximum value is $f(0) = e^{-0} = 1$.

Applying the Product Rule to $f'(x)$,

$$f''(x) = -xe^{-x^2/2}(-2x/2) - e^{-x^2/2}$$
$$= e^{-x^2/2}(x^2 - 1)$$

and hence the second derivative is zero at -1 and 1. If $-1 < x < 1$, then $f''(x) < 0$ and, by (4.16), the graph of f is concave downward in the open interval $(-1, 1)$. If $x < -1$ or $x > 1$, then $f''(x) > 0$ and, therefore, the graph is concave upward throughout the infinite intervals $(-\infty, -1)$ and $(1, \infty)$. Consequently, $P(-1, e^{-1/2})$ and $Q(1, e^{-1/2})$ are points of inflection. Writing

$$f(x) = \frac{1}{e^{x^2/2}},$$

it is evident that as x increases numerically, $f(x)$ approaches 0. We can prove that $\lim_{x \to \infty} f(x) = 0$ and $\lim_{x \to -\infty} f(x) = 0$; that is, the x-axis is a horizontal asymptote. The graph of f is sketched in Figure 7.10. •

FIGURE 7.10

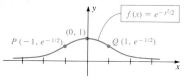

Exponential functions play an important role in the field of *radiotherapy*, the treatment of tumors by radiation. The fraction of cells in a tumor that survives a treatment, or *surviving fraction*, depends not only on the energy and nature of the radiation, but also on the depth, size, and characteristics of the tumor itself. The exposure to radiation may be thought of as a number of potentially damaging events, where only one *hit* is required to kill a tumor cell. Suppose that each cell has exactly one *target* that must be hit. If k denotes the average target size of a tumor cell, and if x is the number of damaging events (the *dose*), then the surviving fraction $f(x)$ is given by

$$f(x) = e^{-kx}.$$

This is called the *one target–one hit surviving fraction*.

Suppose next that each cell has n targets and that hitting any one of the targets results in the death of a cell. In this case, the *n target–one hit surviving fraction* is given by

$$f(x) = 1 - (1 - e^{-kx})^n.$$

In the next example we examine the case $n = 2$.

EXAMPLE 4 If each cell of a tumor has two targets, then the two target–one hit surviving fraction is given by

$$f(x) = 1 - (1 - e^{-kx})^2$$

for the average size k of a cell. Analyze the graph of f to determine what effect increasing the dosage x has on decreasing the surviving fraction of tumor cells.

SOLUTION First note that if $x = 0$, then $f(0) = 1$; that is, if there is no dose, then all cells survive. Differentiating, we obtain

$$f'(x) = 0 - 2(1 - e^{-kx}) D_x (1 - e^{-kx})$$
$$= -2(1 - e^{-kx})(ke^{-kx})$$
$$= -2ke^{-kx}(1 - e^{-kx}).$$

Since $f'(x) < 0$ and $f'(0) = 0$, the function f is decreasing and the graph has a horizontal tangent line at the point $(0, 1)$. We may verify that the second derivative is

$$f''(x) = 2k^2e^{-kx}(1 - 2e^{-kx}).$$

Evidently, $f''(x) = 0$ if $1 - 2e^{-kx} = 0$; that is, if $e^{-kx} = \frac{1}{2}$, or equivalently, $-kx = \ln\frac{1}{2} = -\ln 2$. This gives us

$$x = \frac{1}{k}\ln 2.$$

We can show that if $0 \le x < (1/k)\ln 2$, then $f''(x) < 0$, and hence the graph is concave downward. If $x > (1/k)\ln 2$, we have $f''(x) > 0$, and the graph is concave upward. This implies that there is a point of inflection with x-coordinate $(1/k)\ln 2$. The y-coordinate of this point is

$$f\left(\frac{1}{k}\ln 2\right) = 1 - (1 - e^{-\ln 2})^2$$

$$= 1 - \left(1 - \frac{1}{2}\right)^2 = \frac{3}{4}.$$

The graph is sketched in Figure 7.11 for the case $k = 1$. The "shoulder" on the curve near the point $(0, 1)$ represents the threshold nature of the treatment; that is, a small dose results in very little tumor elimination. Note that if x is large, then an increase in dosage has little effect on the surviving fraction. To determine the ideal dose that should be administered to a given patient, specialists in radiation therapy must also take into account the number of healthy cells that are killed during a treatment. •

FIGURE 7.11

Surviving fraction of tumor cells after a radiation treatment

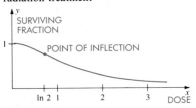

EXERCISES 7.3

Exer. 1–30: Find $f'(x)$ if $f(x)$ equals the given expression.

1 e^{-5x}

2 e^{3x}

3 e^{3x^2}

4 e^{1-x^3}

5 $\sqrt{1 + e^{2x}}$

6 $1/(e^x + 1)$

7 $e^{\sqrt{x+1}}$

8 xe^{-x}

9 x^2e^{-2x}

10 $\sqrt{e^{2x} + 2x}$

11 $e^x/(x^2 + 1)$

12 x/e^{x^2}

13 $(e^{4x} - 5)^3$

14 $(e^{3x} - e^{-3x})^4$

15 $e^{1/x} + (1/e^x)$

16 $e^{\sqrt{x}} + \sqrt{e^x}$

17 $\dfrac{e^x - e^{-x}}{e^x + e^{-x}}$

18 $e^{x\ln x}$

19 $e^{-2x}\ln x$

20 $\ln e^x$

21 $\sin e^{5x}$

22 $e^{\sin 5x}$

23 $\ln\cos e^{-x}$

24 $e^{-3x}\cos 3x$

25 $e^{3x}\tan\sqrt{x}$

26 $\sec e^{-2x}$

27 $\sec^2(e^{-4x})$

28 $e^{-x}\tan^2 x$

29 $xe^{\cot x}$

30 $\ln(\csc e^{3x})$

Exer. 31–34: Use implicit differentiation to find y'.

31 $e^{xy} - x^3 + 3y^2 = 11$

32 $xe^y + 2x - \ln(y + 1) = 3$

33 $e^x\cot y = xe^{2y}$

34 $e^x\cos y = xe^y$

35 Find an equation of the tangent line to the graph of $y = (x - 1)e^x + 3\ln x + 2$ at the point $P(1, 2)$.

36 Find an equation of the tangent line to the graph of $y = x - e^{-x}$ that is parallel to the line $6x - 2y = 7$.

37 Shown in the figure is a computer-generated graph of $y = e^{-x}\ln x$ for $x > 0$.

  (a) Show that a critical number must satisfy the equation $x\ln x = 1$.

  (b) Explain why $x\ln x = 1$ has a unique solution, and approximate this solution to two decimal places using Newton's Method.

EXERCISE 37

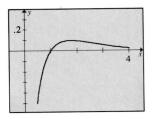

38 Shown in the figure is a computer-generated graph of $y = (1 - \ln x)^2$.

(a) Determine the points corresponding to relative extrema.

(b) Investigate the concavity of the graph and find the coordinates of all points of inflection.

EXERCISE 38

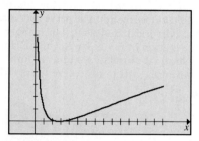

Exer. 39–44: Find the local extrema of f. Determine where f is increasing or decreasing, discuss concavity, find the points of inflection, and sketch the graph of f.

39 $f(x) = xe^x$

40 $f(x) = x^2 e^{-2x}$

41 $f(x) = e^{-2x}$

42 $f(x) = xe^{-x}$

43 $f(x) = x \ln x$

44 $f(x) = e^{-x} \sin x$ for $0 \le x \le 4\pi$

45 A radioactive substance decays according to the formula $q(t) = q_0 e^{-ct}$ for the initial amount q_0 of the substance, a positive constant c, and the amount $q(t)$ remaining after t units of time. Show that the rate at which the substance decays is proportional to $q(t)$.

46 The current $I(t)$ at time t in an electrical circuit is given by $I(t) = I_0 e^{-Rt/L}$ for the resistance R, inductance L, and the current I_0 at time $t = 0$. Show that the rate of change of the current at any time t is proportional to $I(t)$.

47 If a drug is injected into the blood stream, then its concentration t minutes later is given by

$$C(t) = \frac{k}{a - b} (e^{-bt} - e^{-at})$$

for positive constants a, b and k.

(a) At what time does the maximum concentration occur?

(b) What can be said about the concentration after a long period of time? (Justify your answer.)

48 If a beam of light that has intensity k is projected vertically downward into water, then its intensity $I(x)$ at a depth of x meters is $I(x) = ke^{-1.4x}$.

(a) At what rate is the intensity changing with respect to depth at 1 meter? 5 meters? 10 meters?

(b) At what depth is the intensity one-half its value at the surface? One-tenth its value?

49 The *Jenss Model* is generally regarded as the most accurate formula for predicting the height of preschoolers. If h is the height (in cm) and x is the age (in years), then

$$h = 79.041 + 6.39x - e^{3.261 - 0.993x}$$

for $\frac{1}{4} \le x \le 6$ (see Example 4 of Section 7.2).

(a) Approximate the height and the rate of growth of a typical one-year-old child.

(b) At what age is the rate of growth the largest? the smallest?

50 For a population of female African elephants, the weight $W(t)$ (in kg) at age t (in years) may be approximated by a *von Bertanlanffy growth function* W such that

$$W(t) = 2600(1 - 0.51e^{-0.075t})^3.$$

(a) Approximate the weight and the rate of growth of a newborn.

(b) Assuming that an adult female weights 1800 kg, estimate her age and her rate of growth at present.

(c) Find and interpret $\lim_{t \to \infty} W(t)$.

(d) Show that the rate of growth is largest between the ages of 5 years and 6 years.

51 Gamma distributions are important in traffic control studies and probability theory, and are determined by $f(x) = cx^n e^{-ax}$ for $x > 0$, a positive integer n, a positive constant a, and $c = a^{n+1}/n!$. Shown in the figure are graphs corresponding to $a = 1$ for $n = 2$, 3, and 4.

(a) Show that f has exactly one local maximum.

(b) If $n = 4$, determine where $f(x)$ is increasing most rapidly.

EXERCISE 51

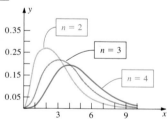

52 The relative number of gas molecules in a container that travel at a velocity of v cm/sec can be computed using the *Maxwell-Boltzmann speed distribution* $F(v) = cv^2 e^{-mv^2/(2kT)}$ for temperature T (in °K), mass m of a molecule, and positive constants c and k. Show that the absolute maximum of F occurs when $v = \sqrt{2kT/m}$.

53 An *urban density model* is a formula that relates the population density (in number/mi²) to the distance r (in miles) from the center of the city. The formula $D = ae^{-br + cr^2}$, for positive constants a, b, and c, has been found to be appropriate for certain cities. Determine the shape of the graph for $r \ge 0$.

54 The effect of light on the rate of photosynthesis can be described by

$$f(x) = x^a e^{(a/b)(1 - x^b)} \quad \text{for } x > 0.$$

(a) Show that f has an absolute maximum at $x = 1$.

(b) Conclude that if $x_0 > 0$ and $y_0 > 0$, then $g(x) = y_0 f(x/x_0)$ has an absolute maximum $g(x_0) = y_0$.

55 The rate R at which a tumor grows is related to its size x by the equation $R = rx \ln (K/x)$ for positive constants r and K. Show that the tumor is growing most rapidly when $x = e^{-1}K$.

56 If p denotes the selling price (in dollars) of a commodity and x is the corresponding demand (in number sold per day), then the relationship between p and x may be given by $p = p_0 e^{-ax}$ for positive constants p_0 and a. Suppose $p = 300e^{-0.02x}$. Find the selling price that will maximize daily revenues (see page 115).

57 In statistics the *normal distribution function* f is defined by

$$f(x) = \frac{1}{\sigma \sqrt{2\pi}} e^{(-1/2)[(x - \mu)/\sigma]^2}$$

for constants μ and σ with $\sigma > 0$. (The constant μ is the *mean* and σ is the *standard deviation*.) Find the local extrema of f and determine where f is increasing or decreasing. Discuss concavity, find points of inflection, find $\lim_{x \to \infty} f(x)$ and $\lim_{x \to -\infty} f(x)$, and sketch the graph of f (see Example 3).

58 The integral $\int_a^b e^{-x^2} dx$ has applications in statistics. Use the Trapezoidal Rule with $n = 10$ and a calculator to approximate this integral if $a = 0$ and $b = 1$.

59 Nerve impulses in the human body travel along nerve fibers that consist of an *axon*, which transports the impulse, and an insulating coating surrounding the axon called the *myelin sheath* (see figure). The nerve fiber is similar to an insulated cylindrical cable, for which the velocity v of an impulse is given by $v = -k(r/R)^2 \ln (r/R)$ for the radius r of the cable and the insulation radius R. Find the value of r/R that maximizes v. (In most nerve fibers $r/R \approx 0.6$.)

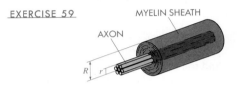

EXERCISE 59   MYELIN SHEATH   AXON

60 Sketch the graph of the *three target–one hit surviving fraction* (with $k = 1$) given by $f(x) = 1 - (1 - e^{-x})^3$ (see Example 4).

DIFFERENTIATION AND INTEGRATION

The formula for the derivative of $\ln u$, which we established in Section 7.2, is generalized in the next theorem.

THEOREM (7.16)

> If $u = g(x)$, with g differentiable and $g(x) \neq 0$, then
>
> $$D_x \ln |u| = \frac{1}{u} D_x u$$

PROOF   If $x < 0$, then $\ln |x| = \ln (-x)$ and by Theorem (7.6),

$$D_x \ln (-x) = \frac{1}{(-x)} D_x (-x) = \frac{1}{(-x)} (-1) = \frac{1}{x}.$$

Consequently,

$$D_x \ln |x| = \frac{1}{x} \quad \text{for every } x \neq 0.$$

To complete the proof we let $y = \ln |u|$, $u = g(x)$, and apply the Chain Rule as follows:

$$D_x \ln |u| = \frac{dy}{dx} = \frac{dy}{du} \frac{du}{dx} = \frac{1}{u} D_x u \quad \bullet \bullet$$

Observe that if $u = g(x) > 0$, then Theorem (7.16) is the same as (7.6).

EXAMPLE 1 Find $f'(x)$ if $f(x) = \ln |4 + 5x - 2x^3|$.

SOLUTION Using Theorem (7.16) with $u = 4 + 5x - 2x^3$,

$$f'(x) = D_x \ln |4 + 5x - 2x^3|$$

$$= \frac{1}{4 + 5x - 2x^3} (5 - 6x^2) = \frac{5 - 6x^2}{4 + 5x - 2x^3} \quad \bullet$$

We may use differentiation formulas for ln to obtain rules for integration. In particular, since

$$D_x \ln |g(x)| = \frac{1}{g(x)} g'(x),$$

the following integration formula holds.

THEOREM (7.17)

> If $u = g(x) \neq 0$, and g is differentiable, then
>
> $$\int \frac{1}{u} \, du = \ln |u| + C$$

Of course, if $u > 0$, then the absolute value sign may be deleted. A special case of Theorem (7.17) is

$$\int \frac{1}{x} \, dx = \ln |x| + C.$$

EXAMPLE 2 Evaluate $\int \frac{x}{3x^2 - 5} \, dx$.

SOLUTION Let us make the substitution

$$u = 3x^2 - 5, \qquad du = 6x \, dx.$$

Introducing a factor 6 in the integrand and using Theorem (7.17),

$$\int \frac{x}{3x^2 - 5} \, dx = \frac{1}{6} \int \frac{1}{3x^2 - 5} 6x \, dx = \frac{1}{6} \int \frac{1}{u} \, du$$

$$= \frac{1}{6} \ln |u| + C = \frac{1}{6} \ln |3x^2 - 5| + C.$$

Another technique is to replace the expression $x \, dx$ in the integral by $\frac{1}{6} \, du$ and then integrate. $\bullet$

EXAMPLE 3 Evaluate $\int_2^4 \frac{1}{9 - 2x} \, dx$.

SOLUTION Since $1/(9 - 2x)$ is continuous on $[2, 4]$, the definite integral exists. One method of evaluation consists of using an indefinite integral to find an antiderivative of $1/(9 - 2x)$. We let

$$u = 9 - 2x, \qquad du = -2 \, dx$$

and proceed as follows:

$$\int \frac{1}{9 - 2x}\, dx = -\frac{1}{2} \int \frac{1}{9 - 2x}\, (-2)\, dx$$

$$= -\frac{1}{2} \int \frac{1}{u}\, du = -\frac{1}{2} \ln |u| + C$$

$$= -\frac{1}{2} \ln |9 - 2x| + C$$

Applying the Fundamental Theorem of Calculus,

$$\int_2^4 \frac{1}{9 - 2x}\, dx = -\frac{1}{2} \left[\ln |9 - 2x| \right]_2^4$$

$$= -\frac{1}{2} [\ln 1 - \ln 5] = \frac{1}{2} \ln 5.$$

Another method is to use the same substitution in the *definite* integral and change the limits of integration. Since $u = 9 - 2x$, we see that

$$\text{if } x = 2, \quad \text{then} \quad u = 5;$$
$$\text{if } x = 4, \quad \text{then} \quad u = 1.$$

Consequently,

$$\int_2^4 \frac{1}{9 - 2x}\, dx = -\frac{1}{2} \int_2^4 \frac{1}{9 - 2x}\, (-2)\, dx$$

$$= -\frac{1}{2} \int_5^1 \frac{1}{u}\, du = -\frac{1}{2} \left[\ln |u| \right]_5^1$$

$$= -\frac{1}{2} [\ln 1 - \ln 5] = \frac{1}{2} \ln 5 \quad \bullet$$

EXAMPLE 4 Evaluate $\displaystyle\int \frac{\sqrt{\ln x}}{x}\, dx$.

<u>SOLUTION</u> If we make the substitution

$$u = \ln x, \qquad du = \frac{1}{x}\, dx,$$

then

$$\int \frac{\sqrt{\ln x}}{x}\, dx = \int \sqrt{\ln x} \cdot \frac{1}{x}\, dx = \int u^{1/2}\, du = \frac{u^{3/2}}{3/2} + C$$

$$= \frac{2}{3} (\ln x)^{3/2} + C \qquad \bullet$$

Since $D_x\, e^{g(x)} = e^{g(x)} g'(x)$, we obtain the following integration formula for the natural exponential function.

THEOREM (7.18)

If $u = g(x)$, and g is differentiable, then
$$\int e^u \, du = e^u + C$$

As a special case of Theorem (7.18), if $u = x$, then
$$\int e^x \, dx = e^x + C.$$

EXAMPLE 5 Evaluate $\int \dfrac{e^{3/x}}{x^2} \, dx$.

SOLUTION Let us make the substitution
$$u = \frac{3}{x}, \qquad du = -\frac{3}{x^2} \, dx.$$

The integrand may be written in the form (7.18) by introducing the factor -3. Doing this and compensating by multiplying the integral by $-\frac{1}{3}$, we obtain

$$\int \frac{e^{3/x}}{x^2} \, dx = -\frac{1}{3} \int e^{3/x} \left(-\frac{3}{x^2} \right) dx$$

$$= -\frac{1}{3} \int e^u \, du$$

$$= -\frac{1}{3} e^u + C$$

$$= -\frac{1}{3} e^{3/x} + C \qquad \bullet$$

EXAMPLE 6 Evaluate $\int_1^2 \dfrac{e^{3/x}}{x^2} \, dx$.

SOLUTION We found an antiderivative in Example 5. Applying the Fundamental Theorem of Calculus,

$$\int_1^2 \frac{e^{3/x}}{x^2} \, dx = -\frac{1}{3} \left[e^{3/x} \right]_1^2$$

$$= -\frac{1}{3} \left(e^{3/2} - e^3 \right).$$

This example can also be solved using the method of substitution. As in Example 5, we let $u = 3/x$, $du = (-3/x^2) \, dx$, and note that if $x = 1$, then $u = 3$, and if $x = 2$, then $u = \frac{3}{2}$. Consequently,

$$\int_1^2 \frac{e^{3/x}}{x^2} \, dx = -\frac{1}{3} \int_1^2 e^{3/x} \left(-\frac{3}{x^2} \right) dx$$

$$= -\frac{1}{3} \int_3^{3/2} e^u \, du$$

$$= -\frac{1}{3} \left[e^u \right]_3^{3/2} = -\frac{1}{3} \left(e^{3/2} - e^3 \right) \qquad \bullet$$

FIGURE 7.12

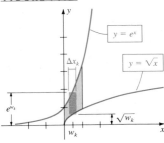

EXAMPLE 7 Find the area of the region bounded by the graphs of the equations $y = e^x$, $y = \sqrt{x}$, $x = 0$, and $x = 1$.

SOLUTION The region and a typical rectangle of the type considered in Chapter 6 are shown in Figure 7.12. The area A is

$$A = \lim_{\|P\| \to 0} \sum_k (e^{w_k} - \sqrt{w_k}) \, \Delta x_k$$

$$= \int_0^1 (e^x - x^{1/2}) \, dx$$

$$= \left[e^x - \tfrac{2}{3}x^{3/2} \right]_0^1$$

$$= (e - \tfrac{2}{3}) - (e^0 - 0) = e - \tfrac{5}{3}.$$

If an approximation is desired, then to the nearest hundredth,

$$A \approx 2.72 - 1.67 = 1.05 \quad \bullet$$

Given $y = f(x)$, we may sometimes find $D_x y$ by **logarithmic differentiation**. This method is especially useful if $f(x)$ involves complicated products, quotients, or powers. The process may be outlined as follows, provided $f(x) > 0$.

Steps in Logarithmic Differentiation

1	$y = f(x)$	(Given)
2	$\ln y = \ln f(x)$	(Take natural logarithms and simplify)
3	$D_x [\ln y] = D_x [\ln f(x)]$	(Differentiate implicitly)
4	$\dfrac{1}{y} D_x y = D_x [\ln f(x)]$	(By Theorem (7.6))
5	$D_x y = f(x) D_x [\ln f(x)]$	(Multiply by $y = f(x)$)

Of course, to complete the solution we must differentiate $\ln f(x)$ at some stage after Step 3. If $f(x) < 0$ for some x, then Step 2 is invalid, since $\ln f(x)$ is undefined. However, in this event we could replace Step 1 by $|y| = |f(x)|$ and take natural logarithms, obtaining $\ln |y| = \ln |f(x)|$. If we now differentiate implicitly and use Theorem (7.16), we again arrive at Step 4. Thus negative values of $f(x)$ do not change the outcome, and we are not concerned if $f(x)$ is positive or negative. The method should not be used to find $f'(a)$ if $f(a) = 0$, since $\ln 0$ is undefined.

EXAMPLE 8 Use logarithmic differentiation to find $D_x y$ if $y = \dfrac{(5x - 4)^3}{\sqrt{2x + 1}}$.

SOLUTION As in Step 2 of the outline, we begin by taking the natural logarithm of each side and simplifying, obtaining

$$\ln y = 3 \ln (5x - 4) - \left(\frac{1}{2} \right) \ln (2x + 1).$$

We next differentiate both sides with respect to x:

$$\frac{1}{y} D_x y = 3 \frac{1}{5x - 4} (5) - \left(\frac{1}{2}\right) \frac{1}{2x + 1} (2)$$

$$= \frac{25x + 19}{(5x - 4)(2x + 1)}.$$

Finally, multiplying both sides of the last equation by y, that is, by $(5x - 4)^3/\sqrt{2x + 1}$,

$$D_x y = \frac{25x + 19}{(5x - 4)(2x + 1)} \cdot \frac{(5x - 4)^3}{\sqrt{2x + 1}}$$

$$= \frac{(25x + 19)(5x - 4)^2}{(2x + 1)^{3/2}}.$$

We may check this result by applying the Quotient Rule to y. •

EXERCISES 7.4

Exer. 1–28: Evaluate the integral.

1 $\int \dfrac{x}{x^2 + 1} \, dx$

2 $\int \dfrac{1}{8x + 3} \, dx$

3 $\int \dfrac{1}{7 - 5x} \, dx$

4 $\int \dfrac{x^3}{x^4 - 5} \, dx$

5 $\int \dfrac{x - 2}{x^2 - 4x + 9} \, dx$

6 $\int \dfrac{(2 + \ln x)^3}{x} \, dx$

7 $\int \dfrac{x^2}{x^3 + 1} \, dx$

8 $\int_1^2 \dfrac{3x}{x^2 + 4} \, dx$

9 $\int_{-2}^1 \dfrac{1}{2x + 7} \, dx$

10 $\int_{-1}^0 \dfrac{1}{4 - 5x} \, dx$

11 $\int (x + e^{5x}) \, dx$

12 $\int (1 + e^{-3x}) \, dx$

13 $\int \dfrac{\ln x}{x} \, dx$

14 $\int \dfrac{1}{x (\ln x)^2} \, dx$

15 $\int_1^3 e^{-4x} \, dx$

16 $\int \dfrac{e^{\sqrt{x}}}{\sqrt{x}} \, dx$

17 $\int e^{\cot x} \csc^2 x \, dx$

18 $\int e^{\cos x} \sin x \, dx$

19 $\int \dfrac{\cos x}{\sin x} \, dx$

(*Hint:* Let $u = \sin x$)

20 $\int e^{-\tan x} \sec^2 x \, dx$

21 $\int \dfrac{\sec^2 x}{1 + \tan x} \, dx$

22 $\int \dfrac{3 \sin x}{1 + 2 \cos x} \, dx$

23 $\int e^x \cos e^x \, dx$

24 $\int \dfrac{\cos x - \sin x}{\cos x + \sin x} \, dx$

25 $\int \dfrac{(e^x + 1)^2}{e^x} \, dx$

26 $\int \dfrac{e^x}{(e^x + 1)^2} \, dx$

27 $\int \dfrac{e^x - e^{-x}}{e^x + e^{-x}} \, dx$

28 $\int \dfrac{e^x}{e^x + 1} \, dx$

Exer. 29–32: Find the area of the region bounded by the graphs of the given equations.

29 $y = e^{2x}, \ y = 0, \ x = 0, \ x = \ln 3$

30 $xy = 1, \ y = 0, \ x = 1, \ x = e$

31 $y = e^{-x}, \ xy = 1, \ x = 1, \ x = 2$

32 $y = e^{-2x}, \ y = -e^x, \ x = 0, \ x = 2$

33 The region bounded by the graphs of $y = e^{-x^2}$, $y = 0$, $x = 0$, and $x = 1$ is revolved about the y-axis. Find the volume of the resulting solid.

34 The region bounded by the graphs of $y = 1/\sqrt{x}$, $y = 0$, $x = 1$, and $x = 4$ is revolved about the x-axis. Find the volume of the resulting solid.

Exer. 35–40: Find dy/dx by logarithmic differentiation.

35 $y = (5x + 2)^3 (6x + 1)^2$

36 $y = \sqrt{4x + 7} (x - 5)^3$

37 $y = (x^2 + 3)^5/\sqrt{x + 1}$

38 $y = (x + 1)^2 (x + 2)^3 (x + 3)^4$

39 $y = \sqrt{(3x^2 + 2)\sqrt{6x - 7}}$

40 $y = (x^2 + 3)^{2/3} (3x - 4)^4/\sqrt{x}$

Exer. 41–42: Find $f'(x)$.

41 $f(x) = \ln |3 - 2x|$

42 $f(x) = \ln |4 - 5x^3|^2$

43 The specific heat c of a metal such as silver is constant at temperatures T above 200 °K. If the temperature of the metal increases from T_1 to T_2, the area under the curve $y = c/T$ from T_1 to T_2 is called the *change in entropy* ΔS, a measurement of the increased molecular disorder of the system. Express ΔS in terms of T_1 and T_2.

44 If the marginal cost function for a type of hang glider is $C'(x) = 150/(0.1x + 2)$, estimate the cost of producing the tenth through the twentieth hang gliders.

45 In a circuit containing a 12-volt battery, a resistor, and a capacitor, the current $I(t)$ at time t is predicted to be $I(t) = 10e^{-4t}$ amperes. If $Q(t)$ is the charge (in coulombs) on the capacitor, then $I = dQ/dt$.

(a) If $Q(0) = 0$, find $Q(t)$.

(b) Find the charge on the capacitor after a long period of time.

46 A country that presently has coal reserves of 50 million tons used 6.5 million tons last year. Based on population projections, the rate of consumption R (in million tons/year) is expected to increase according to the formula $R = 6.5e^{0.02t}$. If the country only uses its own resources, estimate how many years the coal reserves will last.

47 A very small spherical particle (on the order of 5 microns in diameter) is projected into still air with an initial velocity of v_0 m/sec, but its velocity decreases because of drag forces. Its velocity after t seconds is given by $v(t) = v_0 e^{-t/k}$ for some positive constant k.

(a) Express the distance that the particle travels as a function of t.

(b) The *stopping distance* is the distance traveled by the particle before it comes to rest. Express the stopping distance in terms of v_0 and k.

48 The pressure p and volume v of an expanding gas are related by the equation $pv = k$ for a constant k. Show that the work done if the gas expands from v_0 to v_1 is $k \ln (v_1/v_0)$. (*Hint:* See Example 5 of Section 6.6.)

Exer. 49–50: A nonnegative function f defined on a closed interval $[a, b]$ is called a *probability density function* if $\int_a^b f(x)\, dx = 1$. Determine c so that the resulting function is a probability density function.

49 $f(x) = \dfrac{cx}{x^2 + 4}$ for $0 \le x \le 3$

50 $f(x) = cxe^{-x^2}$ for $0 < x \le 10$

7.5 GENERAL EXPONENTIAL AND LOGARITHMIC FUNCTIONS

Throughout this section a will denote a positive real number. Let us begin by defining a^x for every real number x. If the exponent is a *rational* number r, then applying Theorem (7.12) (ii) and Theorem (7.7) (iii),

$$a^r = e^{\ln a^r} = e^{r \ln a}.$$

This formula is the motivation for the following definition of a^x.

DEFINITION OF a^x (7.19)

$$a^x = e^{x \ln a}$$

for every $a > 0$ and every real number x.

If $f(x) = a^x$, then f is the **exponential function with base a.** Since e^x is positive for every x, so is a^x. To approximate values of a^x, we may use a calculator or refer to tables of logarithmic and exponential functions.

It is now possible to prove that the law of logarithms stated in Theorem (7.7) (iii) is also true for irrational exponents. Thus, if u is any *real* number, then by Definition (7.19) and Theorem (7.12) (i),

$$\ln a^u = \ln e^{u \ln a} = u \ln a.$$

The next theorem states that properties of rational exponents from elementary algebra are also true for real exponents.

LAWS OF EXPONENTS (7.20)

Suppose $a > 0$ and $b > 0$. If u and v are any real numbers, then

$$a^u a^v = a^{u+v} \qquad (a^u)^v = a^{uv} \qquad (ab)^u = a^u b^u$$

$$\frac{a^u}{a^v} = a^{u-v} \qquad \left(\frac{a}{b}\right)^u = \frac{a^u}{b^u}$$

PROOF To show that $a^u a^v = a^{u+v}$ we use Definition (7.19) and Theorem (7.13) (i) as follows:

$$a^u a^v = e^{u \ln a} e^{v \ln a}$$
$$= e^{u \ln a + v \ln a}$$
$$= e^{(u+v) \ln a}$$
$$= a^{u+v}$$

To prove that $(a^u)^v = a^{uv}$ we first use Definition (7.19) with a^u in place of a and $v = x$ to write

$$(a^u)^v = e^{v \ln a^u}.$$

Using the fact that $\ln a^u = u \ln a$, and then applying Definition (7.19), we obtain

$$(a^u)^v = e^{vu \ln a} = a^{vu} = a^{uv}.$$

The proofs of the remaining laws are left as exercises. • •

THEOREM (7.21)

(i) $D_x (a^x) = a^x \ln a$

(ii) $D_x (a^u) = (a^u \ln a) D_x u \quad$ for $u = g(x)$.

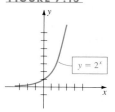

FIGURE 7.13

PROOF Applying Definition (7.19) and Theorem (7.15),

$$D_x (a^x) = D_x (e^{x \ln a}) = e^{x \ln a} D_x (x \ln a) = e^{x \ln a}(\ln a).$$

Again using (7.19), we obtain

$$D_x (a^x) = a^x \ln a.$$

This proves (i). Formula (ii) follows from the Chain Rule. • •

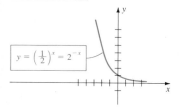

FIGURE 7.14

Note that if $a = e$, then Theorem (7.21) (i) reduces to (7.14), since $\ln e = 1$.

If $a > 1$, then $\ln a > 0$ and, therefore, $D_x (a^x) > 0$. Hence a^x is increasing on the interval $(-\infty, \infty)$ if $a > 1$.

If $0 < a < 1$, then $\ln a < 0$ and $D_x (a^x) < 0$. Thus a^x is decreasing for every x if $0 < a < 1$.

The graphs of $y = 2^x$ and $y = (\frac{1}{2})^x = 2^{-x}$, in Figures 7.13 and 7.14, may be sketched by plotting points. The graph of $y = a^x$ has the general shape illustrated in Figure 7.13 or 7.14 if $a > 1$ or $0 < a < 1$, respectively.

EXAMPLE 1 Find y' if $y = 3^{\sqrt{x}}$.

SOLUTION Using Theorem (7.21) (ii) with $a = 3$ and $u = \sqrt{x}$,

$$y' = (3^{\sqrt{x}} \ln 3)\left(\frac{1}{2} x^{-1/2}\right) = \frac{3^{\sqrt{x}} \ln 3}{2\sqrt{x}} \quad •$$

If $u = g(x)$ we must be certain to distinguish between expressions of the form a^u and u^a. To differentiate a^u we use (7.21); however, for u^a the Power Rule must be employed, as illustrated in the next example.

EXAMPLE 2 Find y' if $y = (x^2 + 1)^{10} + 10^{x^2 + 1}$.

SOLUTION Using the Power Rule for Functions and Theorem (7.21),

$$y' = 10(x^2 + 1)^9(2x) + (10^{x^2 + 1} \ln 10)(2x)$$
$$= 20x[(x^2 + 1)^9 + 10^{x^2} \ln 10] \qquad \bullet$$

The integration formula in (i) of the next theorem may be verified by showing that the integrand is the derivative of the expression on the right side of the equation. Formula (ii) follows from Theorem (7.21) (ii).

THEOREM (7.22)

(i) $\int a^x \, dx = \left(\dfrac{1}{\ln a}\right) a^x + C$

(ii) $\int a^u \, du = \left(\dfrac{1}{\ln a}\right) a^u + C \quad$ for $u = g(x)$.

EXAMPLE 3 Evaluate $\int 3^{x^2} x \, dx$.

SOLUTION We make the substitution

$$u = x^2, \qquad du = 2x \, dx$$

and proceed as follows:

$$\int 3^{x^2} x \, dx = \frac{1}{2} \int 3^{x^2}(2x) \, dx = \frac{1}{2} \int 3^u \, du$$

$$= \frac{1}{2}\left(\frac{1}{\ln 3}\right) 3^u + C = \left(\frac{1}{2 \ln 3}\right) 3^{x^2} + C \quad \bullet$$

EXAMPLE 4 The amount of light that penetrates ocean depths is studied by oceanographers. The *Beer-Lambert law* asserts that the amount of light (in calories/cm^2/sec) that reaches depth x (in meters) is given by $I(x) = I_0 a^x$ for some positive constant a.

(a) What is the significance of I_0?

(b) Show that the rate of change dI/dx at depth x is directly proportional to $I(x)$.

(c) If $a = 0.4$ and $I_0 = 10$, approximate the average amount of light between the ocean surface and a depth of 5 meters.

(d) Show that $I(x) = I_0 e^{kx}$ for some constant k.

SOLUTION

(a) If $x = 0$, then $I(0) = I_0 a^0 = I_0$. Hence I_0 is the light intensity at the surface of the ocean.

(b) Differentiating $I(x)$,

$$I'(x) = I_0(a^x \ln a) = (\ln a)(I_0 a^x) = (\ln a)I(x).$$

Hence the rate of change dI/dx at depth x is directly proportional to $I(x)$, and the constant of proportionality is $\ln a$.

(c) If $I(x) = 10(0.4)^x$, then by Definition (5.21), the average value of I on the interval $[0, 5]$ is

$$I_{av} = \frac{1}{5-0} \int_0^5 10(0.4)^x \, dx = 2 \int_0^5 (0.4)^x \, dx$$

$$= 2 \left[\frac{1}{\ln (0.4)} (0.4)^x \right]_0^5 = \frac{2}{\ln (0.4)} [(0.4)^5 - (0.4)^0]$$

$$= \frac{-1.9795}{\ln (0.4)} \approx 2.16$$

(d) Using Definition (7.19),

$$I(x) = I_0 a^x = I_0 e^{x \ln a} = I_0 e^{kx}$$

with $k = \ln a$. •

If $a \neq 1$ and $f(x) = a^x$, then f is a one-to-one function. Its inverse function is denoted by $\log_a$ and is called the *logarithmic function with base a*. Another way of stating this relationship is as follows.

DEFINITION OF $\log_a x$ **(7.23)**

$$y = \log_a x \quad \text{if and only if} \quad x = a^y.$$

The expression $\log_a x$ is called **the logarithm of x with base a.** Using this terminology, natural logarithms are logarithms with base e, that is,

$$\ln x = \log_e x.$$

The proof that laws of logarithms similar to Theorem (7.7) are true for logarithms with base a is left as an exercise.

To obtain the relationship between $\log_a$ and $\ln$, consider $y = \log_a x$ or, equivalently, $x = a^y$. Taking the natural logarithm of both sides of the last equation gives us $\ln x = y \ln a$, or $y = (\ln x)/(\ln a)$. This proves that

$$\log_a x = \frac{\ln x}{\ln a}.$$

Differentiating both sides of the last equation leads to (i) of the next theorem. Using the Chain Rule and generalizing to absolute values as in Theorem (7.16) gives us (ii).

THEOREM **(7.24)**

(i) $D_x \log_a x = \dfrac{1}{x \ln a}$

(ii) $D_x \log_a |u| = \dfrac{1}{u \ln a} D_x u \quad$ for $u = g(x) \neq 0.$

Logarithms with base 10 are useful for certain applications (see Exercises 40–43 and 45). We refer to such logarithms as **common logarithms** and use the symbol **log x** as an abbreviation for $\log_{10} x$. This notation is used in the next example.

EXAMPLE 5 Find $f'(x)$ if $f(x) = \log \sqrt[3]{(2x + 5)^2}$.

SOLUTION Using the law $\log u^r = r \log u$ (see Exercise 50 (iii)) together with $(2x + 5)^2 = |2x + 5|^2$, we obtain

$$f(x) = \tfrac{1}{3} \log (2x + 5)^2$$
$$= \tfrac{1}{3} \log |2x + 5|^2 = \tfrac{2}{3} \log |2x + 5|.$$

Next, employing Theorem (7.24) (ii) with $a = 10$,

$$f'(x) = \frac{2}{3} \frac{1}{(2x + 5) \ln 10} \quad (2) = \frac{4}{3(2x + 5) \ln 10} \quad \bullet$$

Since we have defined irrational exponents, we may now consider the **general power function** f given by $f(x) = x^c$ for any real number c. If c is irrational, then by definition, the domain of f is the set of positive real numbers. Using Definition (7.19), and Theorems (7.15) and (7.4),

$$D_x (x^c) = D_x (e^{c \ln x}) = e^{c \ln x} D_x (c \ln x)$$
$$= e^{c \ln x} \left(\frac{c}{x} \right) = x^c \left(\frac{c}{x} \right) = cx^{c-1}.$$

This proves that the Power Rule is true for irrational as well as rational exponents. The Power Rule for Functions may also be extended to irrational exponents.

EXAMPLE 6 Find dy/dx if (a) $y = x^{\sqrt{2}}$, (b) $y = (1 + e^{2x})^\pi$.

SOLUTION (a) $\dfrac{dy}{dx} = \sqrt{2} x^{\sqrt{2} - 1}$

(b) $\dfrac{dy}{dx} = \pi(1 + e^{2x})^{\pi - 1} D_x (1 + e^{2x})$

$$= \pi(1 + e^{2x})^{\pi - 1}(2e^{2x})$$
$$= 2\pi e^{2x}(1 + e^{2x})^{\pi - 1} \quad \bullet$$

EXAMPLE 7 Find $D_x y$ if $y = x^x$ and $x > 0$.

SOLUTION Since the exponent in x^x is a variable, the Power Rule may not be used. Similarly, Theorem (7.21) is not applicable, since the base a is not a fixed real number. However, by Definition (7.19), $x^x = e^{x \ln x}$ for every $x > 0$, and hence

$$D_x (x^x) = D_x (e^{x \ln x})$$
$$= e^{x \ln x} D_x (x \ln x)$$
$$= e^{x \ln x} \left[x \left(\frac{1}{x} \right) + (1) \ln x \right]$$
$$= x^x (1 + \ln x).$$

Another way of solving this problem is to use the method of logarithmic differentiation introduced in the preceding section. In this case we take the natural logarithm of both sides of the equation $y = x^x$, and then differentiate implicitly as follows:

$$\ln y = \ln x^x = x \ln x$$

$$D_x (\ln y) = D_x (x \ln x)$$

$$\frac{1}{y} D_x y = 1 + \ln x$$

$$D_x y = y(1 + \ln x) = x^x(1 + \ln x) \quad \bullet$$

We shall conclude this section by expressing the number e as a limit.

THEOREM (7.25)

(i) $\lim\limits_{h \to 0} (1 + h)^{1/h} = e$ (ii) $\lim\limits_{n \to \infty} \left(1 + \dfrac{1}{n}\right)^n = e$

PROOF Applying the derivative formula (3.4) to $f(x) = \ln x$ and using laws of logarithms gives us

$$f'(x) = \lim_{h \to 0} \frac{\ln (x + h) - \ln x}{h} = \lim_{h \to 0} \frac{1}{h} \ln \frac{x + h}{x}$$

$$= \lim_{h \to 0} \frac{1}{h} \ln \left(1 + \frac{h}{x}\right) = \lim_{h \to 0} \ln \left(1 + \frac{h}{x}\right)^{1/h}.$$

Since $f'(x) = 1/x$ we have, for $x = 1$,

$$1 = \lim_{h \to 0} \ln (1 + h)^{1/h}.$$

We next observe, from Theorem (7.12), that

$$(1 + h)^{1/h} = e^{\ln (1 + h)^{1/h}}.$$

Since the natural exponential function is continuous at 1, it follows from Theorem (2.32) that

$$\lim_{h \to 0} (1 + h)^{1/h} = \lim_{h \to 0} \left[e^{\ln (1 + h)^{1/h}}\right]$$

$$= e^{\left[\lim\limits_{h \to 0} \ln (1 + h)^{1/h}\right]} = e^1 = e.$$

This establishes part (i) of the theorem. The limit in part (ii) may be obtained by introducing the change of variable $n = 1/h$. $\quad \bullet \ \bullet$

The formulas in Theorem (7.25) are sometimes used to *define* the number e. You may find it instructive to calculate $(1 + h)^{1/h}$ for numerically small values of h. Some approximate values are given in the following table.

h	$(1 + h)^{1/h}$	h	$(1 + h)^{1/h}$
0.01	2.704814	−0.01	2.731999
0.001	2.716924	−0.001	2.719642
0.0001	2.718146	−0.0001	2.718418
0.00001	2.718268	−0.00001	2.718295
0.000001	2.718280	−0.000001	2.718283

To five decimal places, $e \approx 2.71828$.

Exer. 1–24: Find $f'(x)$ if $f(x)$ is the given expression.

1 7^x 2 5^{-x}

3 8^{x^2+1} 4 $9^{\sqrt{x}}$

5 $\log(x^4 + 3x^2 + 1)$ 6 $\log_3 |6x - 7|$

7 5^{3x-4} 8 3^{2-x^2}

9 $(x^2 + 1)10^{1/x}$ 10 $(10^x + 10^{-x})^{10}$

11 $\log(3x^2 + 2)^5$ 12 $\log\sqrt{x^2 + 1}$

13 $\log_5 \left| \dfrac{6x + 4}{2x - 3} \right|$ 14 $\log \left| \dfrac{1 - x^2}{2 - 5x^3} \right|$

15 $\log \ln x$ 16 $\ln \log x$

17 $x^e + e^x$ 18 $x^\pi \pi^x$

19 $(x + 1)^x$ 20 x^{x^2+4}

21 $2^{\sin^2 x}$ 22 $4^{\sec 3x}$

23 $x^{\tan x}$ 24 $(\cos 2x)^x$

Exer. 25–34: Evaluate the integral.

25 $\displaystyle\int 10^{3x}\, dx$ 26 $\displaystyle\int 5^{-5x}\, dx$

27 $\displaystyle\int x(3^{-x^2})\, dx$ 28 $\displaystyle\int \frac{(2^x + 1)^2}{2^x}\, dx$

29 $\displaystyle\int \frac{2^x}{2^x + 1}\, dx$ 30 $\displaystyle\int \frac{3^x}{\sqrt{3^x + 4}}\, dx$

31 $\displaystyle\int_1^2 5^{-2x}\, dx$ 32 $\displaystyle\int_{-1}^1 2^{3x-1}\, dx$

33 $\displaystyle\int \frac{1}{x \log x}\, dx$ 34 $\displaystyle\int \frac{10^{\sqrt{x}}}{\sqrt{x}}\, dx$

35 Find the area of the region bounded by the graphs of $y = 2^x$, $x + y = 1$, and $x = 1$.

36 The region under the graph of $y = 3^{-x}$ from $x = 1$ to $x = 2$ is revolved about the x-axis. Find the volume of the resulting solid.

37 An economist predicts that the buying power $B(t)$ of a dollar t years from now will decrease according to the formula $B(t) = (0.95)^t$.

  (a) At approximately what rate will the buying power be decreasing two years from now?

  (b) Estimate the *average buying power* of the dollar over the next two years.

38 When a person takes a 100-mg tablet of an asthma drug orally, the rate R at which the drug enters the bloodstream is predicted to be $R = 5(0.95)^t$ mg/min. If the bloodstream does not contain any trace of the drug when the tablet is taken, determine the number of minutes needed for 50 mg to enter the bloodstream.

39 One thousand trout, each one year old, are introduced into a large pond. The number still alive after t years is predicted to be $N(t) = 1000(0.9)^t$.

  (a) Approximate the death rate dN/dt at times $t = 1$ and $t = 5$. At what rate is the population decreasing when $N = 500$?

  (b) The weight $W(t)$ (in pounds) of an individual trout is expected to increase according to the formula $W(t) = 0.2 + 1.5t$. After approximately how many years is the total number of pounds of trout in the pond a maximum?

40 The vapor pressure P (in psi), a measure of the volatility of a liquid, is related to its temperature T (in °F) by the *Antoine equation* $\log P = a + [b/(c + T)]$ for constants a, b, and c. Vapor pressure increases rapidly with an increase in temperature. Find conditions on a, b, and c that guarantee that P is an increasing function of T.

41 Chemists use a number denoted by pH to describe quantitatively the acidity or basicity of solutions. By definition, $\text{pH} = -\log[H^+]$, where $[H^+]$ is the hydrogen ion concentration in moles per liter. For a certain brand of vinegar it is estimated (with a maximum percentage error of 0.5%) that $[H^+] \approx 6.3 \times 10^{-3}$. Calculate the pH and use differentials to estimate the maximum percentage error in the calculation.

42 Using the Richter scale, the magnitude R of an earthquake of intensity I may be found by means of the formula $R = \log(I/I_0)$ for a certain minimum intensity I_0. Suppose the intensity of an earthquake is estimated to be 100 times that of I_0. If the maximum percentage error in the estimate is 1%, use differentials to approximate the maximum percentage error in the calculated value of R.

43 Let $R(x)$ be the reaction of a subject to a stimulus of strength x. For example, if the stimulus x is "saltiness" (in grams of salt/liter), $R(x)$ may be the subject's estimate of how salty the solution tasted on a scale from 0 to 10. A function that has been proposed to relate R to x is given by the *Weber-Fechner* formula $R = a \log(x/x_0)$ for a positive constant a.

  (a) Show that $R = 0$ for the *treshold stimulus* $x = x_0$.

  (b) The derivative $S = dR/dx$ is the *sensitivity* at stimulus level x and measures the ability to detect small changes in stimulus level. Show that S is inversely proportional to x, and compare $S(x)$ to $S(2x)$.

44 If a principal of P dollars is invested in a savings account for t years, and the yearly interest rate r (expressed as a decimal) is compounded n times per year, then the amount A in the account after t years is given by the *compound interest formula* $A = P[1 + (r/n)]^{nt}$.

  (a) Let $h = r/n$ and show that $\ln A = \ln P + rt \ln(1 + h)^{1/h}$.

(b) Let $n \to \infty$ and use the expression in part (a) to establish the formula $A = Pe^{rt}$ for interest *compounded continuously*.

45 The loudness of sound, as experienced by the human ear, is based upon intensity level. A formula used for finding the intensity level α that corresponds to a sound intensity I is $\alpha = 10 \log (I/I_0)$ decibels, with a special value I_0 of I agreed to be the weakest sound that can be detected by the ear under certain conditions. Find the rate of change of α with respect to I if

(a) I is 10 times as great as I_0.

(b) I is 1,000 times as great as I_0.

(c) I is 10,000 times as great as I_0. (This is the intensity level of the average voice.)

46 Establish Theorem (7.25) (ii) by using the limit in part (i) and the change of variable $n = 1/h$.

47 Prove that $\lim_{n \to \infty} [1 + (x/n)]^n = e^x$ by letting $h = x/n$ and using Theorem 7.25 (i).

48 The approximation $(1 - p)^n \approx e^{-np}$ is used in probability theory when p is small and np is moderate in size.

(a) Let $A = (1 - p)^n$ and show that $\ln A = np \ln (1 - p)^{1/p}$.

(b) Use Exercise 47 to conclude that $\ln A \approx -np$ if p is close to 0.

49 Prove that if $a > 0$ and $b > 0$, then for all real numbers u and v:

(i) $(ab)^u = a^u b^u$

(ii) $a^u/a^v = a^{u-v}$

(iii) $(a^u)^v = a^{uv}$

50 Prove the Laws of Logarithms:

(i) $\log_a (uv) = \log_a u + \log_a v$

(ii) $\log_a \dfrac{u}{v} = \log_a u - \log_a v$

(iii) $\log_a u^r = r \log_a u$

7.6 LAWS OF GROWTH AND DECAY

Suppose that a physical quantity varies with time and that the magnitude of the quantity at time t is given by $q(t)$ for a differentiable function q such that $q(t) > 0$ for every t. The derivative $q'(t)$ is the rate of change of $q(t)$ with respect to time. In many applications, this rate of change is directly proportional to the magnitude of the quantity at time t, and the relationship can be expressed by means of the differential equation

$$q'(t) = cq(t)$$

for a constant c. The number of bacteria in certain cultures behaves in this way. If the number of bacteria $q(t)$ is small, then the rate of increase $q'(t)$ is small; however, as the number of bacteria increases, the *rate of increase* also increases. The decay of a radioactive substance obeys a similar law, since as the amount of matter decreases, the rate of decay—that is, the amount of radiation—also decreases. As a final illustration, suppose an electrical condenser is allowed to discharge. If the charge on the condenser is large at the outset, the rate of discharge is also large, but as the charge weakens, the condenser discharges less rapidly.

In applied problems, the equation $q'(t) = cq(t)$ is often expressed in terms of differentials. Thus, if $y = q(t)$ we may write

$$\frac{dy}{dt} = cy \quad \text{or} \quad dy = cy \, dt.$$

Dividing both sides of the last equation by y, we obtain

$$\frac{1}{y} dy = c \, dt.$$

Since it is possible to **separate the variables** y and t—in the sense that they can be placed on opposite sides of the equal sign—the differential equation

$dy/dt = cy$ is a **separable differential equation.** We will study such equations in more detail later in the text, and will show that solutions can be found by integrating both sides of the "separated" equation $(1/y)\, dy = c\, dt$. Thus,

$$\int \frac{1}{y}\, dy = \int c\, dt$$

and, assuming $y > 0$,

$$\ln y = ct + d$$

We have combined the two constants of integration into one constant d. It follows that

$$y = e^{ct+d} = e^d e^{ct}.$$

If y_0 denotes the initial value of y (that is, the value corresponding to $t = 0$), then letting $t = 0$ in the last equation gives us

$$y_0 = e^d e^0 = e^d$$

and hence the solution $y = e^d e^{ct}$ may be written

$$y = y_0 e^{ct}.$$

We have proved the following.

THEOREM (7.26)

> Let y be a differentiable function of t such that $y > 0$ for every t, and let y_0 be the value of y at $t = 0$. If $dy/dt = cy$, for some constant c, then $y = y_0 e^{ct}$.

The preceding theorem states that *if the rate of change of $y = q(t)$ with respect to t is directly proportional to y, then y may be expressed in terms of an exponential function.* If y increases with t, the formula $y = y_0 e^{ct}$ is a **law of growth** and if y decreases, it is a **law of decay.**

EXAMPLE 1 The number of bacteria in a culture increases from 600 to 1800 in two hours. Assuming that the rate of increase is directly proportional to the number of bacteria present, find a formula for the number of bacteria at time t. What is the number of bacteria at the end of four hours?

SOLUTION Let $y = q(t)$ denote the number of bacteria after t hours. Thus $y_0 = q(0) = 600$ and $q(2) = 1800$. By hypothesis,

$$\frac{dy}{dt} = cy.$$

Following exactly the same steps used in the proof of Theorem (7.26), we obtain

$$y = y_0 e^{ct} = 600 e^{ct}.$$

Since $y = 1800$ when $t = 2$,

$$1800 = 600 e^{2c} \quad \text{or} \quad e^{2c} = 3.$$

Consequently,

$$2c = \ln 3 \quad \text{or} \quad c = \tfrac{1}{2} \ln 3.$$

and the formula for y is

$$y = 600e^{[(1/2) \ln 3]t}.$$

If we use the fact that $e^{\ln 3} = 3$ (see Definition (7.11)), then this law of growth can be expressed in terms of an exponential function with base 3 as follows:

$$y = 600(e^{\ln 3})^{t/2} = 600(3)^{t/2}.$$

In particular, at the end of four hours the number of bacteria is

$$600(3)^{4/2} = 600(9) = 5400 \quad \bullet$$

EXAMPLE 2 Radium decays exponentially and has a half-life of approximately 1600 years; that is, given any quantity, one-half of it will disintegrate in 1600 years. Find a formula for the amount y remaining from 50 mg of pure radium after t years. When will the remaining amount be 20 mg?

SOLUTION If we let $y = q(t)$, then

$$y_0 = q(0) = 50 \quad \text{and} \quad q(1600) = \tfrac{1}{2}(50) = 25.$$

Since $dy/dt = cy$ for some c, it follows from Theorem (7.26) that

$$y = 50e^{ct}.$$

Since $y = 25$ when $t = 1600$,

$$25 = 50e^{1600c} \quad \text{or} \quad e^{1600c} = \tfrac{1}{2}.$$

Hence

$$1600c = \ln \tfrac{1}{2} = \ln 1 - \ln 2 = -\ln 2$$

and

$$c = -\frac{\ln 2}{1600}.$$

Consequently,

$$y = 50e^{-(\ln 2/1600)t}.$$

As in Example 1, we may write this formula in terms of a different base as follows:

$$y = 50(e^{\ln 2})^{-t/1600}$$

or

$$y = 50(2)^{-t/1600}.$$

To find the value of t at which $y = 20$, we solve the equation

$$20 = 50(2)^{-t/1600} \quad \text{or} \quad 2^{t/1600} = \tfrac{5}{2}.$$

Taking the natural logarithm of both sides,

$$\frac{t}{1600} \ln 2 = \ln \frac{5}{2} \quad \text{or} \quad t = \frac{1600 \ln \frac{5}{2}}{\ln 2}.$$

If an approximation is desired, then using a calculator or Table C,

$$t \approx 1600(0.916)/(0.693) \approx 2115 \text{ years} \quad \bullet$$

EXAMPLE 3 Newton's law of cooling states that the rate at which an object cools is directly proportional to the difference in temperature between the object and the surrounding medium. If an object cools from 125 °F to 100 °F in half an hour when surrounded by air at a temperature of 75 °F, find its temperature at the end of the next half hour.

SOLUTION If y denotes the temperature of the object after t hours of cooling, then by Newton's law of cooling,

$$\frac{dy}{dt} = c(y - 75)$$

for a constant c. Separating variables,

$$\frac{1}{y - 75} \, dy = c \, dt.$$

Integrating both sides gives us

$$\ln(y - 75) = ct + b$$

for a constant b. This is equivalent to

$$y - 75 = e^{ct+b} = e^b e^{ct}.$$

If we let $k = e^b$, then the last formula may be written

$$y = ke^{ct} + 75.$$

Since $y = 125$ when $t = 0$, we see that

$$125 = ke^0 + 75 = k + 75 \quad \text{or} \quad k = 50.$$

Hence $\qquad\qquad\qquad\qquad y = 50e^{ct} + 75.$

Since $y = 100$ when $t = \frac{1}{2}$ hour,

$$100 = 50e^{c/2} + 75 \quad \text{or} \quad e^{c/2} = \tfrac{25}{50} = \tfrac{1}{2}.$$

This implies that

$$c = 2 \ln \tfrac{1}{2} = \ln \tfrac{1}{4}.$$

Consequently, the temperature y after t hours is given by

$$y = 50e^{t \ln(1/4)} + 75.$$

In particular, if $t = 1$,

$$y = 50e^{\ln(1/4)} + 75$$
$$= 50(\tfrac{1}{4}) + 75 = 87.5 \text{ °F} \quad \bullet$$

In many cases, growth circumstances are more stable than that described in Theorem (7.26). Typical situations involve populations, sales of products, and values of assets. In biology, a function G is sometimes used as follows to estimate the size of a quantity at time t:

$$G(t) = ke^{(-Ae^{-Bt})}$$

for positive constants k, A, and B. The function G is called a **Gompertz growth function.** It is always positive and increasing, but has a limit as t increases without bound. The graph of G is called a **Gompertz growth curve.**

EXAMPLE 4 Discuss and sketch the graph of the Gompertz growth function G.

SOLUTION We first observe that the y-intercept is $G(0) = ke^{-A}$ and that $G(t) > 0$ for every t. Differentiating, we obtain

$$G'(t) = ke^{(-Ae^{-Bt})} D_t (-Ae^{-Bt})$$
$$= ABke^{(-Bt-Ae^{-Bt})}$$

and

$$G''(t) = ABke^{(-Bt-Ae^{-Bt})} D_t (-Bt - Ae^{-Bt})$$
$$= ABk(-B + ABe^{-Bt})e^{-Bt-Ae^{-Bt}}.$$

Since $G'(t) > 0$ for every t, the function G is increasing on $[0, \infty)$. The second derivative $G''(t)$ is zero if

$$-B + ABe^{-Bt} = 0 \quad \text{or} \quad e^{Bt} = A.$$

Solving the last equation for t gives us $t = (1/B) \ln A$, which is a critical number for the function G'. We leave it as an exercise to show that at this time the rate of growth G' has a maximum value Bk/e. We can also show that

$$\lim_{t \to \infty} G'(t) = 0 \quad \text{and} \quad \lim_{t \to \infty} G(t) = k.$$

Hence, as t increases without bound, the rate of growth approaches 0 and the graph of G has a horizontal asymptote $y = k$. A typical graph is sketched in Figure 7.15. •

FIGURE 7.15

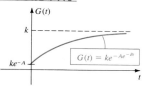

EXERCISES 7.6

1 The number of bacteria in a culture increases from 5,000 to 15,000 in 10 hours. Assuming that the rate of increase is proportional to the number of bacteria present, find a formula for the number of bacteria in the culture at any time t. Estimate the number at the end of 20 hours. When will the number be 50,000?

2 The polonium isotope ^{210}Po has a half-life of approximately 140 days. If a sample weighs 20 mg initially, how much remains after t days? Approximately how much will be left after two weeks?

3 If the temperature is constant, then the rate of change of barometric pressure p with respect to altitude h is proportional to p. If $p = 30$ inches at sea level and $p = 29$ inches when $h = 1000$ feet, find the pressure at an altitude of 5000 feet.

4 The population of a city is increasing at the rate of 5 percent per year. If the present population is 500,000, what will the population be in 10 years?

5 Agronomists use the assumption that $\frac{1}{4}$ acre of land is required to provide food for one person and estimate that there are 10 billion acres of tillable land in the world. Hence, a maximum population of 40 billion people can be sustained if no other food source is available. The world population at the beginning of 1980 was approximately 4.5 billion. Assuming that the population increases at a rate of 2 percent per year, when will the maximum population be reached?

6 A metal plate that has been heated cools from 180 °F to 150 °F in 20 minutes when surrounded by air at a temperature of 60 °F. Use Newton's law of cooling (see Example 3) to approximate its temperature at the end of one hour of cooling. When will the temperature be 100 °F?

7 An outdoor thermometer registers a temperature of 40 °F. Five minutes after it is brought into a room where the temperature is 70 °F, the thermometer registers 60 °F. When will it register 65 °F?

8 The rate at which salt dissolves in water is directly proportional to the amount that remains undissolved. If 10 pounds of salt is placed in a container of water and 4 pounds dissolves in 20 minutes, how long will it take for two more pounds to dissolve?

9 According to Kirchhoff's first law for electrical circuits, $V = RI + L(dI/dt)$ where the constants V, R, and L denote the electromotive force, the resistance, and the inductance, respectively, and I denotes the current at time t (see figure). If the electromotive force is terminated at time $t = 0$ and

if the current is I_0 at the instant of removal, prove that $I = I_0 e^{-Rt/L}$.

EXERCISE 9

10 A physicist finds that an unknown radioactive substance registers 2000 counts per minute on a Geiger counter. Ten days later the substance registers 1500 counts per minute. Approximate its half-life.

11 The air pressure P (in atmospheres) at an elevation of z meters above sea level is a solution of the differential equation $dP/dz = -9.81\rho(z)$ for the density $\rho(z)$ of air at elevation z. Assuming that air obeys the ideal gas law, this differential equation can be rewritten as $dP/dz = -0.0342P/T$ for temperature T (in °K) at elevation z. If $T = 288 - 0.01z$ and if the pressure is 1 atm at sea level, express P as a function of z.

12 During the first month of growth for crops such as maize, cotton, and soybeans, the rate of growth (in grams/day) is proportional to the present weight W. For a species of cotton, $dW/dt = 0.21W$. Predict the weight of a plant at the end of the month ($t = 30$) if the plant weighs 70 mg at the beginning of the month.

13 Radioactive strontium-90, ^{90}Sr, with a half-life of 29 years, can cause bone cancer in humans. The substance is carried by acid rain, soaks into the ground, and is passed through the food chain. The radioactivity level in a particular field is estimated to be 2.5 times the safe level S. For approximately how many years will this field be contaminated?

14 The radioactive tracer ^{51}Cr, with a half-life of 27.8 days, can be used in medical testing to locate the position of a placenta in a pregnant woman. Often the tracer must be ordered from a medical supply lab. If 35 units are needed for a test, and delivery from the lab requires two days, estimate the minimum number of units that should be ordered.

15 Veterinarians use sodium pentobarbital to anesthetize animals. Suppose that to anesthetize a dog, 30 mg is required for each kilogram of body weight. If sodium pentobarbital is eliminated exponentially from the bloodstream, and half is eliminated in four hours, approximate the single dose that will anesthetize a dog that weighs 20 kg for 45 minutes.

16 In the study of lung physiology the following differential equation is used to describe the transport of a substance across a capillary wall:

$$\frac{dh}{dt} = -\frac{V}{Q}\left(\frac{h}{k+h}\right)$$

for the hormone concentration h in the bloodstream, time t, maximum transport rate V, volume Q of the capillary, and a constant k that measures the affinity between the hor-

mones and enzymes that assist with the transport process. Find the general solution of the differential equation.

17 A space probe is shot upward from the earth. Neglecting air resistance, a differential equation for the velocity after burnout is $v(dv/dy) = -ky^{-2}$ for the distance y from the center of the earth and a positive constant k. If y_0 is the distance from the center of the earth at burnout and v_0 is the corresponding velocity, express v as a function of y. (*Hint:* Separate the variables y and v.)

18 At high temperatures, nitrogen dioxide NO_2 decomposes into NO and O_2. If $y(t)$ is the concentration of NO_2 (in moles per liter), then, at 600 °K, $y(t)$ changes according to the *second-order reaction law* $dy/dt = -0.05y^2$ for time t in seconds. Express y in terms of t and the initial concentration y_0.

19 The technique of carbon-14 dating is used to determine the age of archeological or geological specimens. This method is based on the fact that the unstable isotope carbon-14 (^{14}C) is present in the CO_2 in the atmosphere. Plants take in carbon from the atmosphere; when they die the ^{14}C that has accumulated begins to decay, with a half-life of approximately 5700 years. By measuring the amount of ^{14}C that remains in a specimen, it is possible to approximate when the organism died. Suppose that a bone fossil contains 20 percent as much ^{14}C as an equal amount of carbon in present-day bone. Approximate the age of the bone.

20 Refer to Exercise 19. The hydrogen isotope $^3_1$H, which has a half-life of 12.3 years, is produced in the atmosphere by cosmic rays and it brought to earth by rain. If the wood siding of an old house contains 10 percent as much $^3_1$H as the siding on a similar new house, approximate the age of the old house.

21 In the *law of logistic growth*, it is assumed that at time t, the rate of growth $f'(t)$ of a quantity $f(t)$ is given by $f'(t) = Af(t)[B - f(t)]$ for constants A and B. If $f(0) = C$, show that

$$f(t) = \frac{BC}{C + (B - C)e^{-ABt}}.$$

22 Certain learning processes may be described by the graph of $f(x) = a + b(1 - e^{-cx})$ for positive constants a, b, and c. Suppose a manufacturer estimates that a new employee can produce five items the first day on the job. As the employee becomes more proficient, her daily production increases until a certain maximum production is reached. Suppose that on the nth day on the job, the number $f(n)$ of items produced is approximated by the formula $f(n) = 3 + 20(1 - e^{-0.1n})$.

(a) Estimate the number of items produced on the fifth day; the ninth day; the twenty-fourth day; the thirtieth day.

(b) Sketch the graph of f from $n = 0$ to $n = 30$. (Graphs of this type are called *learning curves*, and are used frequently in education and psychology.)

(c) What happens as n increases without bound?

23 A spherical cell has volume V and surface area S. A simple model for cell growth before mitosis assumes that the rate of growth dV/dt is proportional to the surface area of the cell. Show that $dV/dt = kV^{2/3}$ for some $k > 0$, and express V as a function of t.

24 In Theorem (7.26), we assumed that the rate of change of a quantity $q(t)$ at time t is directly proportional to $q(t)$.

Show that if the rate of change is proportional to $[q(t)]^2$, then there is a time t_1 at which $\lim_{t \to t_1^+} q(t) = \infty$.

25 In Example 4, verify that (a) Bk/e is a maximum value for G'; (b) $\lim_{t \to \infty} G'(t) = 0$; (c) $\lim_{t \to \infty} G(t) = k$.

26 Sketch the graph of G in Example 4 if $k = 10$, $A = \frac{1}{2}$, and $B = 1$.

7.7 DERIVATIVES OF INVERSE FUNCTIONS

In Section 7.1 we defined the inverse function f^{-1} of a function f. In this section we prove several theorems about inverse functions and obtain a formula for finding their derivatives.

THEOREM (7.27)

> If a function f is continuous and increasing on an interval $[a, b]$, then f has an inverse function f^{-1} that is continuous and increasing on the interval $[f(a), f(b)]$.

PROOF If f is increasing, then f is one-to-one and hence f^{-1} exists. To prove that f^{-1} is increasing, we must show that if $w_1 < w_2$ in $[f(a), f(b)]$, then $f^{-1}(w_1) < f^{-1}(w_2)$ in $[a, b]$. Let us give an indirect proof of this fact. Thus, *suppose* $f^{-1}(w_2) \le f^{-1}(w_1)$. Since f is increasing, it follows that $f(f^{-1}(w_2)) \le f(f^{-1}(w_1))$ and hence $w_2 \le w_1$, which is a contradiction. Consequently, $f^{-1}(w_1) < f^{-1}(w_2)$.

We must also prove that f^{-1} is continuous on $[f(a), f(b)]$. Recall that $y = f(x)$ if and only if $x = f^{-1}(y)$. In particular, if y_0 is in an open interval $(f(a), f(b))$, let x_0 denote the number in the interval (a, b) such that $y_0 = f(x_0)$ or, equivalently, $x_0 = f^{-1}(y_0)$. We wish to show that

(∗)
$$\lim_{y \to y_0} f^{-1}(y) = f^{-1}(y_0) = x_0.$$

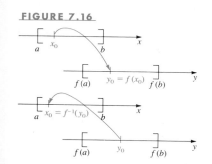

FIGURE 7.16

A geometric representation of f and its inverse f^{-1} similar to that given in Figure 7.1 is shown in Figure 7.16. The domain $[a, b]$ of f is represented by points on an x-axis and the domain $[f(a), f(b)]$ of f^{-1} by points on a y-axis. Arrows are drawn from one axis to the other to represent function values. To prove (∗), consider any interval $(x_0 - \varepsilon, x_0 + \varepsilon)$ for $\varepsilon > 0$. It is sufficient to find an interval $(y_0 - \delta, y_0 + \delta)$, of the type sketched in Figure 7.17, such that whenever y is in $(y_0 - \delta, y_0 + \delta)$, then $f^{-1}(y)$ is in $(x_0 - \varepsilon, x_0 + \varepsilon)$. We may assume that $x_0 - \varepsilon$ and $x_0 + \varepsilon$ are in $[a, b]$. As in Figure 7.18, let $\delta_1 = y_0 - f(x_0 - \varepsilon)$ and $\delta_2 = f(x_0 + \varepsilon) - y_0$. Since f determines a one-to-one correspondence between the numbers in the intervals $(x_0 - \varepsilon, x_0 + \varepsilon)$ and $(y_0 - \delta_1, y_0 + \delta_2)$, the function values of f^{-1} that correspond to numbers in $(y_0 - \delta_1, y_0 + \delta_2)$ must lie in $(x_0 - \varepsilon, x_0 + \varepsilon)$. Let δ denote the smaller of δ_1 and δ_2. It follows that if y is in $(y_0 - \delta, y_0 + \delta)$, then $f^{-1}(y)$ is in $(x_0 - \varepsilon, x_0 + \varepsilon)$, which is what we wished to prove. The continuity at the endpoints $f(a)$ and $f(b)$ of the domain of f^{-1} is proved in a similar manner using one-sided limits. • •

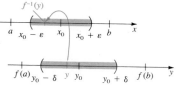

FIGURE 7.17

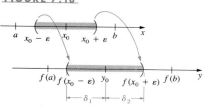

FIGURE 7.18

The proof of the next result is similar to that of Theorem (7.27).

THEOREM (7.28) If a function f is continuous and decreasing on an interval $[a, b]$, then f has an inverse function that is continuous and decreasing on the interval $[f(b), f(a)]$.

EXAMPLE 1 Verify Theorem (7.27) if $f(x) = x^2 - 3$ on the interval $[0, 6]$.

SOLUTION From Example 2 of Section 7.1, f^{-1} is given by

$$f^{-1}(x) = \sqrt{x + 3}, \quad \text{for } x \geq -3.$$

In the present example the domain of f^{-1} is $[f(0), f(6)]$, that is, $[-3, 33]$.

The graphs of f and f^{-1}, first illustrated in Figure 7.3, are resketched in Figure 7.19.

Since f is a polynomial function, it is continuous on $[0, 6]$. Moreover, since $f'(x) = 2x$, $f'(x) > 0$ for every x in $(0, 6]$ and, therefore, f is increasing on $[0, 6]$.

Since $\lim_{x \to a} \sqrt{x + 3} = \sqrt{a + 3}$ for every a in $(-3, 33]$, f^{-1} is continuous on this interval. It is also continuous at -3. Finally, since the derivative

$$D_x f^{-1}(x) = \frac{1}{2\sqrt{x + 3}}$$

is positive on $(-3, 33]$, it follows that f^{-1} is increasing on $[-3, 33]$. •

The next theorem provides a method for finding the derivative of an inverse function.

FIGURE 7.19

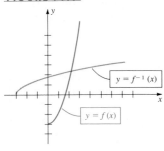

THEOREM (7.29) If a differentiable function f has an inverse function g, and if $f'(g(c)) \neq 0$, then g is differentiable at c and

$$g'(c) = \frac{1}{f'(g(c))}$$

PROOF By Definition (3.1'),

$$g'(c) = \lim_{x \to c} \frac{g(x) - g(c)}{x - c}.$$

Let us introduce a new variable z such that $z = g(x)$ and let $a = g(c)$. Since f and g are inverse functions of one another,

$$g(x) = z \quad \text{if and only if} \quad f(z) = x,$$

and $\qquad g(c) = a \quad \text{if and only if} \quad f(a) = c.$

The statement $x \to c$ in the limit formula for $g'(c)$ may be replaced by $z \to a$, since if $x \to c$, then $g(x) \to g(c)$, that is, $z \to a$. However, if $z \to a$, then it follows that $f(z) \to f(a)$. Thus we may write

$$g'(c) = \lim_{z \to a} \frac{z - a}{f(z) - f(a)} = \lim_{z \to a} \frac{1}{\left[\dfrac{f(z) - f(a)}{z - a}\right]}$$

$$= \frac{1}{f'(a)} = \frac{1}{f'(g(c))}. \qquad \bullet \bullet$$

It is convenient to restate Theorem (7.29) as follows.

THEOREM (7.30)

> If g is the inverse function of a differentiable function f, and if $f'(g(x)) \neq 0$, then
> $$g'(x) = \frac{1}{f'(g(x))}$$

EXAMPLE 2 Use Theorem (7.30) and the fact that the natural exponential function is the inverse of the natural logarithmic function to prove that $D_x e^x = e^x$.

SOLUTION If $f(x) = \ln x$ and $g(x) = e^x$, then

$$f'(x) = \frac{1}{x} \quad \text{and} \quad f'(g(x)) = f'(e^x) = \frac{1}{e^x}.$$

Hence, from Theorem (7.30),

$$g'(x) = \frac{1}{(1/e^x)} = e^x \quad \bullet$$

EXAMPLE 3 If $f(x) = x^3 + 2x - 1$, prove that f has an inverse function g, and find the slope of the tangent line to the graph of g at the point $P(2, 1)$.

SOLUTION Note that f is continuous and $f'(x) = 3x^2 + 2 > 0$ for every x. Thus, by Theorem (7.27), f has an inverse function g. Since $f(1) = 2$, it follows that $g(2) = 1$ and consequently the point $P(2, 1)$ is on the graph of g. Using Theorem (7.30), the slope of the tangent line at $P(2, 1)$ is

$$g'(2) = \frac{1}{f'(g(2))} = \frac{1}{f'(1)} = \frac{1}{5} \quad \bullet$$

An easy way to remember Theorem (7.30) is to let $y = f(x)$. If g is the inverse function of f, then $g(y) = g(f(x)) = x$. From (7.30),

$$g'(y) = \frac{1}{f'(g(y))} = \frac{1}{f'(x)}$$

or, using differential notation,

$$\frac{dx}{dy} = \frac{1}{\left(\dfrac{dy}{dx}\right)}.$$

This shows that, in a sense, the derivative of the inverse function g is the reciprocal of the derivative of f. A disadvantage of using the last two formulas is that neither is stated in terms of the independent variable for the inverse function. To illustrate, suppose in Example 3 we let $y = x^3 + 2x - 1$ and $x = g(y)$. Then

$$\frac{dx}{dy} = \frac{1}{dy/dx} = \frac{1}{3x^2 + 2},$$

that is,

$$g'(y) = \frac{1}{3x^2 + 2} = \frac{1}{3(g(y))^2 + 2}.$$

This may also be written in the form

$$g'(x) = \frac{1}{3(g(x))^2 + 2}.$$

Consequently, to find $g'(x)$ it is necessary to know $g(x)$, just as in Theorem (7.30).

Exer. 1–10: Prove that the function f, defined on the given interval, has an inverse function f^{-1}, and state its domain. Find $f^{-1}(x)$. Find $D_x f^{-1}(x)$ directly and also by means of Theorem (7.30).

1 $f(x) = \sqrt{2x + 3}$, $[1, 11]$

2 $f(x) = \sqrt[3]{5x + 2}$, $[0, 5]$

3 $f(x) = 4 - x^2$, $[0, 7]$

4 $f(x) = x^2 - 4x + 5$, $[-1, 1]$

5 $f(x) = 1/x$, $(0, \infty)$

6 $f(x) = \sqrt{9 - x^2}$, $[0, 3]$

7 $f(x) = e^{-x^2}$, $[0, \infty]$

8 $f(x) = \ln(3 - 2x)$, $(-\infty, \frac{3}{2})$

9 $f(x) = e^x - e^{-x}$, $(-\infty, \infty)$

10 $f(x) = e^x + e^{-x}$, $[0, \infty)$

Exer. 11–14: Prove that f has an inverse function f^{-1} on every closed interval $[a, b]$, and find the slope of the tangent line to the graph of f^{-1} at P.

11 $f(x) = x^5 + 3x^3 + 2x - 1$, $P(5, 1)$

12 $f(x) = 2 - x - x^3$, $P(-8, 2)$

13 $f(x) = (e^{2x} - 1)/(e^{2x} + 1)$, $P(0, 0)$

14 $f(x) = e^{2 - 3x}$, $P(1/e, 1)$

Exer. 15–18: Prove that if the domain of f is $\mathbb{R}$, then f has no inverse function. Also prove that if the domain is suitably restricted, then f^{-1} exists.

15 $f(x) = x^4 + 3x^2 + 7$ 16 $f(x) = |x - 2|$

17 $f(x) = 10^{-x^2}$ 18 $f(x) = \ln(x^2 + 1)$

19 Complete the proof of Theorem (7.27) by showing that f^{-1} is continuous at $f(a)$ and $f(b)$.

20 Prove Theorem (7.28).

7.8 REVIEW

Define or discuss each of the following.

1 Inverse function

2 The natural logarithmic function

3 Laws of logarithms

4 The natural exponential function

5 The number e

6 Derivative formulas for $\ln u$ and e^u

7 Logarithmic differentiation

8 a^x for $a > 0$

9 $\log_a$

10 Derivative formulas for $\log_a u$ and a^u

11 General power function

12 Laws of growth and decay

13 Derivatives of inverse functions

Exer. 1–30: Find $f'(x)$ if $f(x)$ is the given expression.

1 $(1 - 2x) \ln|1 - 2x|$ 2 $\sqrt{\ln \sqrt{x}}$

3 $\ln \dfrac{(3x + 2)^4 \sqrt{6x - 5}}{8x - 7}$ 4 $\log \left| \dfrac{2 - 9x}{1 - x^2} \right|$

5 $\dfrac{1}{\ln(2x^2 + 3)}$ 6 $\dfrac{\ln x}{e^{2x} + 1}$

7 $e^{\ln(x^2 + 1)}$ 8 $\ln(e^{4x} + 9)$

9 $10^x \log x$ 10 $\ln \sqrt[4]{x/(3x + 5)}$

11 $x^{\ln x}$ 12 $4^{\sqrt{2x + 3}}$

13 $x^2 e^{1 - x^2}$ 14 $\sqrt{e^{3x} + e^{-3x}}$

15 $2^{-1/x}/(x^3 + 4)$ 16 $5^{3x} + (3x)^5$

17 $(1 + \sqrt{x})^e$ 18 $\ln e^{\sqrt{x}}$

19 $10^{\ln x}$

20 $(x^2 + 1)^{2x}$

21 $(\ln x)^{\ln x}$

22 $7^{\ln |x|}$

23 $\ln |\tan x - \sec x|$

24 $\ln \csc \sqrt{x}$

25 $\csc e^{-2x} \cot e^{-2x}$

26 $x^2 e^{-\cot 2x}$

27 $\ln \cos^4 4x$

28 $3^{\sin 3x}$

29 $(\sin x)^{\cos x}$

30 $\dfrac{1}{\sin^2 e^{-3x}}$

Exer. 31–32: Find y'.

31 $1 + xy = e^{xy}$

32 $\ln (x + y) + x^2 - 2y^3 = 1$

Exer. 33–56: Evaluate the integral.

33 $\displaystyle\int \dfrac{(1 + e^x)^2}{e^{2x}}\, dx$

34 $\displaystyle\int_0^1 e^{-3x + 2}\, dx$

35 $\displaystyle\int_1^4 \dfrac{1}{\sqrt{x}\, e^{\sqrt{x}}}\, dx$

36 $\displaystyle\int \dfrac{1}{x \ln x}\, dx$

37 $\displaystyle\int \dfrac{1}{x - x \ln x}\, dx$

38 $\displaystyle\int_1^2 \dfrac{x^2 + 1}{x^3 + 3x}\, dx$

39 $\displaystyle\int \dfrac{x^2}{3x + 2}\, dx$

40 $\displaystyle\int \dfrac{e^{1/x}}{x^2}\, dx$

41 $\displaystyle\int_0^1 x 4^{x^2}\, dx$

42 $\displaystyle\int \dfrac{(e^{2x} + e^{3x})^2}{e^{5x}}\, dx$

43 $\displaystyle\int \dfrac{1}{x\sqrt{\log x}}\, dx$

44 $\displaystyle\int \dfrac{x}{x^4 + 2x^2 + 1}\, dx$

45 $\displaystyle\int \dfrac{x^2 + 1}{x + 1}\, dx$

46 $\displaystyle\int \dfrac{2e^x}{1 + e^x}\, dx$

47 $\displaystyle\int 5^x e^x\, dx$

48 $\displaystyle\int x 10^{x^2}\, dx$

49 $\displaystyle\int x^e\, dx$

50 $\displaystyle\int 5^x \sqrt{1 + 5^x}\, dx$

51 $\displaystyle\int e^{-x} \sin e^{-x}\, dx$

52 $\displaystyle\int \tan x\, e^{\sec x} \sec x\, dx$

53 $\displaystyle\int \dfrac{\csc^2 x}{1 + \cot x}\, dx$

54 $\displaystyle\int \dfrac{\cos x + \sin x}{\sin x - \cos x}\, dx$

55 $\displaystyle\int \dfrac{\cos 2x}{1 - 2 \sin 2x}\, dx$

56 $\displaystyle\int 3^x(3 + \sin 3^x)\, dx$

57 A particle moves on a coordinate line with an acceleration at time t of $e^{t/2}$ cm/sec^2. At $t = 0$ the particle is at the origin and its velocity is 6 cm/sec. How far does it travel during the time interval $[0, 4]$?

58 Find the local extrema of $f(x) = x^2 \ln x$ for $x > 0$. Discuss concavity, find the points of inflection, and sketch the graph of f.

59 Find an equation of the tangent line to the graph of the equation $y = xe^{1/x^3} + \ln |2 - x^2|$ at the point $P(1, e)$.

60 Find the area of the region bounded by the graphs of the equations $y = e^{2x}$, $y = x/(x^2 + 1)$, $x = 0$, and $x = 1$.

61 The region bounded by the graphs of $y = e^{4x}$, $x = -2$, $x = -3$, and $y = 0$ is revolved about the x-axis. Find the volume of the resulting solid.

62 The 1980 population estimate for India was 651 million, and the population has been increasing at a rate of about 2% per year. If t denotes the time (in years) after 1980, find a formula for $N(t)$, the population (in millions) at time t. Assuming that this rapid growth rate continues, estimate the population and the rate of population growth in the year 2000.

63 A radioactive substance has a half-life of five days. How long will it take for an amount A to disintegrate to the time at which only 1 percent of A remains?

64 The carbon-14 dating equation $T = -8310 \ln x$ is used to predict the age T (in years) of a fossil in terms of the percentage $100x$ of carbon still present in the specimen (see Exercise 19, Section 7.6).

(a) If $x = 0.04$, estimate the age of the fossil to the nearest 1000 years.

(b) If the error in estimating x in part (a) could be as large as 0.005, use differentials to approximate the maximum error in T.

65 The rate at which sugar decomposes in water is proportional to the amount that remains undecomposed. Suppose that 10 pounds of sugar is placed in a container of water at 1:00 P.M., and one-half is dissolved at 4:00 P.M.

(a) How long will it take two more pounds to dissolve?

(b) How much of the 10 pounds will be dissolved at 8:00 P.M.?

66 According to Newton's law of cooling, the rate at which an object cools is directly proportional to the difference in temperature between the object and its surrounding medium. If $f(t)$ denotes the temperature at time t, show that $f(t) = T + [f(0) - T]e^{-kt}$ for temperature T of the surrounding medium and a positive constant k.

67 The bacterium $E.\ coli$ undergoes cell division approximately every 20 minutes. Starting with 100,000 cells, determine the number of cells after two hours.

68 The differential equation $p\, dv + cv\, dp = 0$ describes the adiabatic change of state of air for pressure p, volume v, and a constant c. Solve for p as a function of v.

69 If $f(x) = 2x^3 - 8x + 5$ on $[-1, 1]$, prove that f has an inverse function g, and find $g'(5)$.

70 Suppose $f(x) = e^{2x} + 2e^x + 1$ for $x \geq 0$.

(a) Prove that f has an inverse function f^{-1} and state its domain.

(b) Find $f^{-1}(x)$ and $D_x f^{-1}(x)$.

(c) Find the slope of the tangent line to the graph of f at the point $(0, 4)$ and the slope of the tangent line to the graph of f^{-1} at $(4, 0)$.

CHAPTER 8

OTHER TRANSCENDENTAL FUNCTIONS

We have obtained formulas for derivatives of the trigonometric functions and integrals of the sine and cosine functions in preceding chapters. We now consider integrals of the remaining four trigonometric functions and of other important trigonometric expressions. The last two sections of the chapter contain discussions of hyperbolic and inverse hyperbolic functions.

8.1 INTEGRALS OF TRIGONOMETRIC FUNCTIONS

Formulas for indefinite integrals of important trigonometric expressions are listed in the next theorem, with $u = g(x)$ and g differentiable.

THEOREM (8.1)

$$\int \sin u \, du = -\cos u + C$$

$$\int \cos u \, du = \sin u + C$$

$$\int \tan u \, du = \ln |\sec u| + C$$

$$\int \cot u \, du = \ln |\sin u| + C$$

$$\int \sec u \, du = \ln |\sec u + \tan u| + C$$

$$\int \csc u \, du = \ln |\csc u - \cot u| + C$$

$$\int \sec^2 u \, du = \tan u + C$$

$$\int \csc^2 u \, du = -\cot u + C$$

$$\int \sec u \tan u \, du = \sec u + C$$

$$\int \csc u \cot u \, du = -\csc u + C$$

PROOF We shall verify several of the formulas and leave the remainder as exercises. It is sufficient to consider the case $u = x$, since the formulas for $u = g(x)$ then follow from Theorem (5.32).

The indefinite integrals of sin x and cos x were obtained earlier. To find $\int \tan x\, dx$ we first use a trigonometric identity to express tan x in terms of sin x and cos x as follows:

$$\int \tan x\, dx = \int \frac{\sin x}{\cos x}\, dx.$$

The form of the integrand on the right suggests that we make the substitution

$$v = \cos x, \qquad dv = -\sin x\, dx.$$

This gives us

$$\int \tan x\, dx = \int \frac{1}{\cos x} \sin x\, dx = -\int \frac{1}{v}\, dv$$

If cos $x \neq 0$, then by Theorem (7.16),

$$\int \tan x\, dx = -\ln |v| + C = -\ln |\cos x| + C.$$

Since $-\ln |\cos x| = \ln (1/|\cos x|) = \ln |\sec x|$, the last formula can be written

$$\int \tan x\, dx = \ln |\sec x| + C.$$

A formula for $\int \cot x\, dx$ may be obtained in similar fashion by first writing $\cot x = (\cos x)/(\sin x)$.

To find a formula for $\int \sec x\, dx$ we begin as follows:

$$\int \sec x\, dx = \int \sec x \frac{\sec x + \tan x}{\sec x + \tan x}\, dx$$

$$= \int \frac{\sec^2 x + \sec x \tan x}{\sec x + \tan x}\, dx.$$

Using the substitution

$$v = \sec x + \tan x, \qquad dv = (\sec x \tan x + \sec^2 x)\, dx$$

gives us

$$\int \sec x\, dx = \int \frac{1}{v}\, dv$$

$$= \ln |v| + C$$

$$= \ln |\sec x + \tan x| + C.$$

To verify a formula such as $\int \sec^2 x\, dx = \tan x + C$, it is sufficient to note that $D_x \tan x = \sec^2 x$. • •

In the following examples we shall use many of the formulas in Theorem (8.1). In each example, we let $u = g(x)$ for the expression $g(x)$ to the right of a trigonometric function symbol cot, sec, $\csc^2$, and so on. We then obtain $du = g'(x)\, dx$ from Definition (3.23).

EXAMPLE 1 Evaluate $\int x \cot x^2\, dx$.

SOLUTION To obtain the form $\int \cot u\, du$ we make the substitution

$$u = x^2, \qquad du = 2x\, dx.$$

We next introduce the factor 2 in the integrand as follows:

$$\int x \cot x^2 \, dx = \frac{1}{2} \int (\cot x^2) 2x \, dx$$

Since $u = x^2$ and $du = 2x \, dx$,

$$\int x \cot x^2 \, dx = \frac{1}{2} \int \cot u \, du = \frac{1}{2} \ln |\sin u| + C$$

$$= \frac{1}{2} \ln |\sin x^2| + C \qquad \bullet$$

EXAMPLE 2 Evaluate $\int \sec x \, (\sec x + \tan x) \, dx$.

SOLUTION Changing the integrand, and then using two of the formulas in Theorem (8.1),

$$\int \sec x \, (\sec x + \tan x) \, dx = \int (\sec^2 x + \sec x \tan x) \, dx$$

$$= \int \sec^2 x \, dx + \int \sec x \tan x \, dx$$

$$= \tan x + \sec x + C \qquad \bullet$$

As usual, we combined the two constants of integration obtained from the two indefinite integrals in Example 2 into one constant C.

EXAMPLE 3 Evaluate $\int \dfrac{\csc^2 \sqrt{x}}{\sqrt{x}} \, dx$.

SOLUTION We first write

$$\int \frac{\csc^2 \sqrt{x}}{\sqrt{x}} \, dx = \int (\csc^2 \sqrt{x}) \frac{1}{\sqrt{x}} \, dx.$$

Next we substitute

$$u = \sqrt{x}, \qquad du = \frac{1}{2\sqrt{x}} \, dx.$$

and obtain the form $\int \csc^2 u \, du$ by introducing the factor $\frac{1}{2}$ as follows:

$$\int \frac{\csc^2 \sqrt{x}}{\sqrt{x}} \, dx = 2 \int (\csc^2 \sqrt{x}) \frac{1}{2\sqrt{x}} \, dx$$

$$= 2 \int \csc^2 u \, du = -2 \cot u + C$$

$$= -2 \cot \sqrt{x} + C \qquad \bullet$$

EXAMPLE 4 Find the area A of the region under the graph of $y = \tan (x/2)$ from $x = 0$ to $x = \pi/2$.

SOLUTION The region is sketched in Figure 8.1. As in Section 6.1,

$$A = \int_0^{\pi/2} y \, dx = \int_0^{\pi/2} \tan \frac{x}{2} \, dx.$$

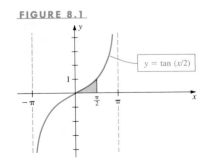

FIGURE 8.1

$y = \tan (x/2)$

We now make the substitution

$$u = \frac{x}{2}, \qquad du = \frac{1}{2} dx$$

and note that $u = 0$ if $x = 0$ and $u = \pi/4$ if $x = \pi/2$. Thus,

$$\int_0^{\pi/2} \tan \frac{x}{2} \, dx = 2 \int_0^{\pi/2} \tan \frac{x}{2} \cdot \frac{1}{2} \, dx$$

$$= 2 \int_0^{\pi/4} \tan u \, du = 2 \ln \sec u \Big]_0^{\pi/4}$$

In this case we may drop the absolute value sign given in Theorem (8.1), since $\sec u$ is positive if u is between 0 and $\pi/4$. Since $\ln \sec (\pi/4) = \ln \sqrt{2} = \frac{1}{2} \ln 2$, and $\ln \sec 0 = \ln 1 = 0$, it follows that

$$\int_0^{\pi/2} \tan \frac{x}{2} \, dx = 2 \cdot \frac{1}{2} \ln 2 = \ln 2 \approx 0.69 \quad \bullet$$

EXAMPLE 5 Evaluate $\int e^{2x} \sec e^{2x} \, dx$.

SOLUTION We let

$$u = e^{2x}, \qquad du = 2e^{2x} \, dx$$

and proceed as follows:

$$\int e^{2x} \sec e^{2x} \, dx = \frac{1}{2} \int (\sec e^{2x}) 2e^{2x} \, dx$$

$$= \frac{1}{2} \int \sec u \, du$$

$$= \frac{1}{2} \ln |\sec u + \tan u| + C$$

$$= \frac{1}{2} \ln |\sec e^{2x} + \tan e^{2x}| + C \quad \bullet$$

EXAMPLE 6 Evaluate $\int (\csc x - 1)^2 \, dx$.

SOLUTION

$$\int (\csc x - 1)^2 \, dx = \int (\csc^2 x - 2 \csc x + 1) \, dx$$

$$= \int \csc^2 x \, dx - 2 \int \csc x \, dx + \int dx$$

$$= -\cot x - 2 \ln |\csc x - \cot x| + x + C \quad \bullet$$

We will discuss additional methods for integrating trigonometric expressions in Chapter 9.

EXERCISES 8.1

Exer. 1–28: Evaluate the integral.

1 $\int \csc 4x \, dx$

2 $\int \sec^2 5x \, dx$

3 $\int \tan 3x \sec 3x \, dx$

4 $\int x^2 \cot x^3 \csc x^3 \, dx$

5 $\int (\tan 3x + \sec 3x) \, dx$

6 $\int \frac{1}{\sec 2x} \, dx$

7 $\int \frac{1}{\cos 2x} \, dx$

8 $\int \frac{\cot \sqrt[3]{x}}{\sqrt[3]{x^2}} \, dx$

9 $\int x \csc^2 (x^2 + 1)\, dx$

10 $\int (x + \csc 8x)\, dx$

11 $\int \cot 6x \sin 6x\, dx$

12 $\int \sin 2x \tan 2x\, dx$

13 $\int_0^{\pi/4} \tan x \sec^2 x\, dx$

14 $\int \csc^2 x \cot x\, dx$

15 $\int \dfrac{\tan^2 2x}{\sec 2x}\, dx$

16 $\int_{\pi/6}^{\pi/2} \dfrac{\cos^2 x}{\sin x}\, dx$

17 $\int_{\pi/6}^{\pi} \sin x \cos x\, dx$

18 $\int \dfrac{\cos^2 x}{\csc x}\, dx$

19 $\int \dfrac{1 - \sin x}{x + \cos x}\, dx$

20 $\int \dfrac{e^x}{\cos e^x}\, dx$

21 $\int_{\pi/4}^{\pi/3} \dfrac{1 + \sin x}{\cos^2 x}\, dx$

22 $\int_0^{\pi/4} (1 + \sec x)^2\, dx$

23 $\int e^x(1 + \tan e^x)\, dx$

24 $\int (\csc^2 x)2^{\cot x}\, dx$

25 $\int \dfrac{e^{\cos x}}{\csc x}\, dx$

26 $\int \dfrac{\tan e^{-3x}}{e^{3x}}\, dx$

27 $\int \dfrac{\sec^2 x}{2 \tan x + 1}\, dx$

28 $\int \dfrac{\sec x \tan x}{1 + 3 \sec x}\, dx$

29 Find the area of the region under the graph of $y = \cot \frac{1}{2}x$ from $x = \pi/3$ to $x = \pi/2$.

30 Find the area of the region under the graph of $y = 2 \tan x$ from $x = 0$ to $x = \pi/4$.

31 Find the area of the region bounded by the graphs of $y = \sec x$, $y = x$, $x = -\pi/4$, and $x = \pi/4$.

32 Find the area of the region bounded by the graphs of $y = \sin x$, $y = \csc x$, $x = \pi/4$, and $x = \pi/2$.

33 The region bounded by the graphs of the equations $y = \sec x$, $x = -\pi/3$, $x = \pi/3$, and $y = 0$ is revolved about the x-axis. Find the volume of the resulting solid.

34 The region between the graphs of $y = \tan (x^2)$ and the x-axis, from $x = 0$ to $x = \sqrt{\pi}/2$, is revolved about the y-axis. Find the volume of the resulting solid.

35 Verify the formula $\int \csc x\, dx = \ln |\csc x - \cot x| + C$

36 Verify the formula $\int \cot x\, dx = \ln |\sin x| + C$

Exer. 37–38: Verify the formula by evaluating the integral in two different ways. How can the answers be reconciled?

37 $\int \tan x \sec^2 x\, dx = \frac{1}{2} \tan^2 x + C = \frac{1}{2} \sec^2 x + D$

38 $\int \cot x \csc^2 x\, dx = -\frac{1}{2} \cot^2 x + C = -\frac{1}{2} \csc^2 x + D$

Exer. 39–40: Use (a) the Trapezoidal Rule and (b) Simpson's Rule to approximate the definite integral for the stated value of n. Use approximations to four decimal places for $f(x_k)$ and round off answers to two decimal places.

39 $\int_0^{\pi} \sqrt{\sin x}\, dx$, $n = 6$

40 $\int_0^{\pi} \sin \sqrt{x}\, dx$, $n = 4$

Exer. 41–42: (a) Set up an integral for finding the arc length L from A to B of the graph of the equation; (b) write out the formula for approximating L by means of Simpson's Rule with $n = 4$; (c) find the approximation given by the formula in part (b).

41 $y = \sec \frac{1}{2}x$; $A(0, 1)$, $B(\pi/2, \sqrt{2})$

42 $y = \tan x$; $A(0, 0)$, $B(\pi/4, 1)$

43 In *seasonal population growth* the population $q(t)$ at time t (in years) increases during spring and summer, but decreases during the fall and winter. A differential equation that is sometimes used to describe this type of growth is $q'(t)/q(t) = k \sin 2\pi t$ with $k > 0$ and $t = 0$ corresponding to the first day of spring.
(a) Show that the population $q(t)$ is seasonal.
(b) If $q_0 = q(0)$, find an explicit formula for $q(t)$.

44 The ends of a water trough 6 feet long have the shape of the graph of $y = \sec x$ from $x = -\pi/3$ to $x = \pi/3$ for x in feet. Suppose the trough is full of water.
(a) Set up a definite integral with respect to y that equals the work required to pump all the water over the top of the trough.
(b) Change the integral in (a) to an integral with respect to x and use Simpson's Rule with $n = 4$ on the interval $[-\pi/3, \pi/3]$ to approximate the work required.

45 Given the right triangle ABC shown in the figure, let θ denote angle ABC (measured in radians), let a be the length of side BC, and let P be a point on side AC. Show that the average distance of P from B is $(a/\theta) \ln (\sec \theta + \tan \theta)$. (*Hint:* Express the length of side BP as a function f of angle PBC, and then find the average value of f.)

EXERCISE 45

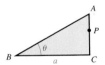

46 Refer to Exercise 45. In a 1970 survey of 244 urbanized areas, only ten were found to have shapes significantly different from a circle, square, hexagon, or a 1-by-2 rectangle. Let C denote the center of a city, P a point on the outskirts, and let A denote the area of the city in square miles. If the shape of the city is either a square (see figure) or a hexagon, use the formula in Exercise 45 to establish that the average distance from the outskirts to the center of the city is approximately $0.56\sqrt{A}$ miles.

EXERCISE 46

If f is a one-to-one function with domain D and range E, then its inverse function f^{-1} has domain E and range D. Moreover,

$$y = f^{-1}(x) \quad \text{if and only if} \quad f(y) = x$$

for every x in D and every y in E (see Section 7.1). Since the trigonometric functions are not one-to-one, they do not have inverse functions. However, by restricting their domains we may obtain functions that have the same values as the trigonometric functions (over the smaller domains), and that *do* have inverses.

Let us first consider the graph of the sine function, whose domain is $\mathbb{R}$ and range is the closed interval $[-1, 1]$ (see Figure 8.2). The sine function is not one-to-one since, for example, numbers such as $\pi/6$, $5\pi/6$, and $-7\pi/6$ lead to the same function value, $\frac{1}{2}$. If we restrict the domain to $[-\pi/2, \pi/2]$ then, as illustrated by the colored portion of the curve in Figure 8.2, we obtain an increasing function that takes on all the values of the sine function once and only once. This new function, with domain $[-\pi/2, \pi/2]$ and range $[-1, 1]$, is continuous and increasing and hence, by Theorem (7.27), has an inverse function that is continuous and increasing. The inverse function has domain $[-1, 1]$ and range $[-\pi/2, \pi/2]$. This leads to the following definition.

FIGURE 8.2

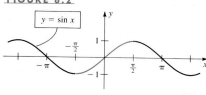
$y = \sin x$

DEFINITION (8.2)

> The **inverse sine function,** denoted $\sin^{-1}$, is defined by
> $$y = \sin^{-1} x \quad \text{if and only if} \quad \sin y = x$$
> for $-1 \le x \le 1$ and $-\pi/2 \le y \le \pi/2$.

The inverse sine function is also called the **arcsine function** and arcsin x is often used in place of $\sin^{-1} x$. The $^{-1}$ in $\sin^{-1}$ is not to be regarded as an exponent, but rather as a means of denoting this inverse function. Observe that by Definition (8.2),

$$-\frac{\pi}{2} \le \sin^{-1} x \le \frac{\pi}{2} \quad \text{or} \quad -\frac{\pi}{2} \le \arcsin x \le \frac{\pi}{2}.$$

Using the method we introduced in Section 7.1 for sketching the graph of an inverse function, we can sketch the graph of $y = \sin^{-1} x$ by reflecting the colored portion of Figure 8.2 through the line $y = x$. (We could also sketch the graph of $x = \sin y$ for $-\pi/2 \le y \le \pi/2$.) This gives us Figure 8.3. Since sin and $\sin^{-1}$ are inverse functions of one another,

FIGURE 8.3

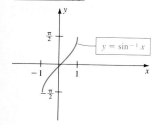
$y = \sin^{-1} x$

$$\sin^{-1}(\sin x) = x \quad \text{if} \quad -\frac{\pi}{2} \le x \le \frac{\pi}{2}$$

and

$$\sin(\sin^{-1} x) = x \quad \text{if} \quad -1 \le x \le 1.$$

These formulas may also be written

$$\arcsin(\sin x) = x \quad \text{and} \quad \sin(\arcsin x) = x.$$

EXAMPLE 1 Find $\sin^{-1}(\sqrt{2}/2)$ and $\arcsin(-\frac{1}{2})$.

SOLUTION If $y = \sin^{-1}(\sqrt{2}/2)$, then $\sin y = \sqrt{2}/2$ and hence $y = \pi/4$. Note that it is essential to choose y in the interval $[-\pi/2, \pi/2]$. A number such as $3\pi/4$ is incorrect, even though $\sin(3\pi/4) = \sqrt{2}/2$.

If $y = \arcsin(-\frac{1}{2})$, then $\sin y = -\frac{1}{2}$ and, consequently, $y = -\pi/6$. •

We may use the other five trigonometric functions to define inverse trigonometric functions. If the domain of the cosine function is restricted to the interval $[0, \pi]$ (see the colored portion of Figure 8.4), we obtain a continuous, decreasing function that has a continuous, decreasing inverse function by Theorem (7.28). This leads to the next definition.

FIGURE 8.4

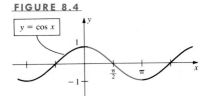

$y = \cos x$

DEFINITION (8.3)

> The **inverse cosine function,** denoted by $\cos^{-1}$, is defined by
> $$y = \cos^{-1} x \quad \text{if and only if} \quad \cos y = x$$
> for $-1 \leq x \leq 1$ and $0 \leq y \leq \pi$.

The inverse cosine function is also referred to as the **arccosine function** and the notation $\arccos x$ is used interchangeably with $\cos^{-1} x$. Note that

$$\cos^{-1}(\cos x) = \arccos(\cos x) = x \quad \text{if} \quad 0 \leq x \leq \pi$$
$$\cos(\cos^{-1} x) = \cos(\arccos x) = x \quad \text{if} \quad -1 \leq x \leq 1.$$

The graph of the inverse cosine function can be found by reflecting the colored portion of Figure 8.4 through the line $y = x$. This gives us the sketch in Figure 8.5. We could also use the equation $x = \cos y$ with $0 \leq y \leq \pi$ to finds points on the graph.

FIGURE 8.5

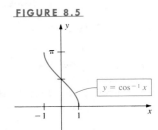

$y = \cos^{-1} x$

EXAMPLE 2 Find $\cos^{-1}(-\sqrt{3}/2)$.

SOLUTION If $y = \cos^{-1}(-\sqrt{3}/2)$, then $\cos y = -\sqrt{3}/2$. Since y must be chosen in the interval $[0, \pi]$, we see that $y = 5\pi/6$.

To use a calculator to approximate $\cos^{-1}(-\sqrt{3}/2)$, we select the radian mode and proceed as follows:

$$\text{ENTER:} \quad -\sqrt{3}/2 \approx -0.8660254$$
$$\text{PRESS } \boxed{\text{INV}}\ \boxed{\text{COS}}: \quad 2.6179939$$

The last decimal is an approximation to $5\pi/6$. •

If we restrict the domain of the tangent function to the open interval $(-\pi/2, \pi/2)$, then a one-to-one function is obtained. This leads to the following definition.

DEFINITION (8.4)

> The **inverse tangent function,** or **arctangent function,** denoted by $\tan^{-1}$ or $\arctan$, is defined by
> $$y = \tan^{-1} x = \arctan x \quad \text{if and only if} \quad \tan y = x$$
> for every real number x and $-\pi/2 < y < \pi/2$.

FIGURE 8.6

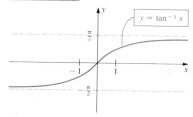

The domain of the inverse tangent function is $\mathbb{R}$ and the range is the open interval $(-\pi/2, \pi/2)$. Its graph is sketched in Figure 8.6.

EXAMPLE 3 Find $\sec\,(\arctan\frac{2}{3})$ without using a calculator or table.

SOLUTION If $y = \arctan\frac{2}{3}$, then $\tan y = \frac{2}{3}$. Using $\sec^2 y = 1 + \tan^2 y$ and the fact that $\sec y > 0$ for $0 < y < \pi/2$,

$$\sec y = \sqrt{1 + \tan^2 y} = \sqrt{1 + \left(\frac{2}{3}\right)^2}$$

$$= \sqrt{1 + \frac{4}{9}} = \sqrt{\frac{13}{9}} = \frac{\sqrt{13}}{3}.$$

Hence, $\sec\,(\arctan\frac{2}{3}) = \sec y = \sqrt{13}/3$. •

If y varies through the two intervals $[0, \pi/2)$ and $[\pi, 3\pi/2)$, then $\sec y$ takes on each of its values once and only once and an inverse function may be defined.

DEFINITION (8.5)

> The **inverse secant function,** or **arcsecant function,** denoted by $\sec^{-1}$ or arcsec, is defined by
>
> $$y = \sec^{-1} x = \text{arcsec } x \quad \text{if and only if} \quad \sec y = x$$
>
> for $|x| \geq 1$ and y in $[0, \pi/2)$ or in $[\pi, 3\pi/2)$.

FIGURE 8.7

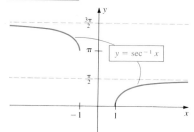

The graph of $y = \sec^{-1} x$ is sketched in Figure 8.7. We have restricted y to the intervals in Definition (8.5) instead of to the "more natural" intervals $[0, \pi/2)$ and $(\pi/2, \pi]$ because the differentiation formula for the inverse secant is simpler. We show, in the next section, that $D_x \sec^{-1} x = 1/(x\sqrt{x^2 - 1})$. Thus, the slope of the tangent line to the graph of $y = \sec^{-1} x$ is negative if $x < -1$ or positive if $x > 1$ (see Figure 8.7). For the more natural intervals, the slope is always positive, and we would have $D_x \sec^{-1} x = 1/(|x|\sqrt{x^2 - 1})$.

The following examples illustrate some manipulations with inverse trigonometric functions.

EXAMPLE 4 If $-1 \leq x \leq 1$, rewrite $\cos\,(\sin^{-1} x)$ as an algebraic expression in x.

SOLUTION Let $y = \sin^{-1} x$, so that $\sin y = x$. We wish to express $\cos y$ in terms of x. Since $-\pi/2 \leq y \leq \pi/2$, it follows that $\cos y \geq 0$, and hence

$$\cos y = \sqrt{1 - \sin^2 y} = \sqrt{1 - x^2}.$$

Consequently, $\qquad \cos\,(\sin^{-1} x) = \sqrt{1 - x^2}.$

FIGURE 8.8
$y = \sin^{-1} x$

The last identity is also evident geometrically if $0 < x < 1$. In this case $0 < y < \pi/2$, and we may regard y as the radian measure of an angle of a right triangle such that $\sin y = x$, as illustrated in Figure 8.8. (The side of length $\sqrt{1 - x^2}$ is found by the Pythagorean Theorem.) Referring to the triangle,

$$\cos\,(\sin^{-1} x) = \cos y = \frac{\sqrt{1 - x^2}}{1} = \sqrt{1 - x^2} \quad •$$

EXAMPLE 5 Verify the identity

$$\frac{1}{2} \cos^{-1} x = \tan^{-1} \sqrt{\frac{1-x}{1+x}}$$

for $|x| < 1$.

SOLUTION Let $y = \cos^{-1} x$. We wish to show that

$$\frac{y}{2} = \tan^{-1} \sqrt{\frac{1-x}{1+x}}.$$

By a half-angle formula,

$$\left| \tan \frac{y}{2} \right| = \sqrt{\frac{1 - \cos y}{1 + \cos y}}.$$

Since $y = \cos^{-1} x$ and $|x| < 1$, it follows that $0 < y < \pi$ and, hence, that $0 < (y/2) < \pi/2$. Consequently, $\tan(y/2) > 0$, and we may drop the absolute value sign, obtaining

$$\tan \frac{y}{2} = \sqrt{\frac{1 - \cos y}{1 + \cos y}}.$$

Using the fact that $\cos y = x$ gives us

$$\tan \frac{y}{2} = \sqrt{\frac{1-x}{1+x}}$$

or, equivalently,

$$\frac{y}{2} = \tan^{-1} \sqrt{\frac{1-x}{1+x}},$$

which is what we wished to show. •

EXAMPLE 6 Find the solutions of the equation

$$5 \sin^2 t + 3 \sin t - 1 = 0$$

in the interval $[-\pi/2, \pi/2]$.

SOLUTION We may regard the equation as a quadratic equation in $\sin t$. Applying the quadratic formula,

$$\sin t = \frac{-3 \pm \sqrt{9 + 20}}{10} = \frac{-3 \pm \sqrt{29}}{10}.$$

Using the definition of the inverse sine function, we obtain the solutions

$$t = \sin^{-1} \tfrac{1}{10}(-3 + \sqrt{29})$$

and

$$t = \sin^{-1} \tfrac{1}{10}(-3 - \sqrt{29}).$$

If approximations are desired, we use a calculator (in radian mode) and proceed as follows:

ENTER: $\tfrac{1}{10}(-3 + \sqrt{29}) \approx 0.2385165$

PRESS $\boxed{\text{INV}}$ $\boxed{\text{SIN}}$:   0.240838

Hence, $t \approx 0.2408$.

Similarly,

$$t = \sin^{-1} \tfrac{1}{10}(-3 - \sqrt{29}) \approx \sin^{-1}(-0.8385165) \approx -0.9946 \quad •$$

Exer. 1–18: Find the exact value without the use of a calculator or table.

1 (a) $\sin^{-1}(\sqrt{3}/2)$

  (b) $\sin^{-1}(-\sqrt{3}/2)$

2 (a) $\sin^{-1} 0$

  (b) $\arccos 0$

3 (a) $\cos^{-1}(\sqrt{2}/2)$

  (b) $\cos^{-1}(-\sqrt{2}/2)$

4 (a) $\arcsin(-1)$

  (b) $\cos^{-1}(-1)$

5 (a) $\tan^{-1}\sqrt{3}$

  (b) $\arctan(-\sqrt{3})$

6 (a) $\tan^{-1}(-1)$

  (b) $\arccos\frac{1}{2}$

7 $\sin[\cos^{-1}(\sqrt{3}/2)]$

8 $\cos(\sin^{-1} 0)$

9 $\sin(\arccos\frac{3}{5})$

10 $\tan(\tan^{-1} 10)$

11 $\arcsin(\sin\sqrt{5})$

12 $\tan^{-1}(\cos 0)$

13 $\cos(\sin^{-1}\frac{3}{5} + \tan^{-1}\frac{4}{3})$

14 $\sin(\arcsin\frac{1}{2} + \arccos 0)$

15 $\tan(\arctan\frac{3}{4} + \arccos\frac{3}{5})$

16 $\cos(2\sin^{-1}\frac{8}{17})$

17 $\sin[2\arccos(-\frac{4}{5})]$

18 $\sin(\arctan\frac{1}{2} - \arccos\frac{4}{5})$

Exer. 19–22: Rewrite the expression as an algebraic expression in x.

19 $\sin(\tan^{-1} x)$

20 $\tan(\arccos x)$

21 $\cos(\frac{1}{2}\arccos x)$

22 $\cos(2\tan^{-1} x)$

Exer. 23–30: Verify the identity.

23 $\sin^{-1} x + \cos^{-1} x = \pi/2$ (*Hint:* Let $\alpha = \sin^{-1} x$, $\beta = \cos^{-1} x$, and consider $\sin(\alpha + \beta)$.)

24 $\arctan x + \arctan(1/x) = \pi/2,\ x > 0$

25 $\arcsin\dfrac{2x}{1+x^2} = 2\arctan x,\ |x| \le 1$

26 $2\cos^{-1} x = \cos^{-1}(2x^2 - 1),\ 0 \le x \le 1$

27 $\sin^{-1}(-x) = -\sin^{-1} x$

28 $\arccos(-x) = \pi - \arccos x$

29 $\sin^{-1} x = \tan^{-1}\dfrac{x}{\sqrt{1-x^2}}$

30 $\tan^{-1} x + \tan^{-1} y = \tan^{-1}\dfrac{x+y}{1-xy}$ for $|x| < 1$ and $|y| < 1$

31 Define $\cot^{-1}$ by restricting the domain of cot to the interval $(0, \pi)$.

32 Define $\csc^{-1}$ by restricting the domain of csc to $(0, \pi/2]$ and $(\pi, 3\pi/2]$.

Exer. 33–42: Sketch the graph of the equation.

33 $y = \sin^{-1} 2x$

34 $y = \cos^{-1}\frac{1}{2}x$

35 $y = \frac{1}{2}\sin^{-1} x$

36 $y = 2\cos^{-1} x$

37 $y = 2\tan^{-1} x$

38 $y = \tan^{-1} 2x$

39 $y = \sin(\sin^{-1} x)$

40 $y = \sin(\arccos x)$

41 $y = \sin^{-1}(\sin x)$

42 $y = \arccos(\cos x)$

Exer. 43–44: Show that the equation is *not* an identity.

43 $\tan^{-1} x = 1/(\tan x)$

44 $(\arcsin x)^2 + (\arccos x)^2 = 1$

Exer. 45–50: (a) Use inverse trigonometric functions to find the solutions of the equation in the given interval and (b) use a calculator to approximate the solutions in part (a) to four decimal places.

45 $2\tan^2 t + 9\tan t + 3 = 0;\ (-\pi/2, \pi/2)$

46 $3\sin^2 t + 7\sin t + 3 = 0;\ [-\pi/2, \pi/2]$

47 $15\cos^4 x - 14\cos^2 x + 3 = 0;\ [0, \pi]$

48 $3\tan^4\theta - 19\tan^2\theta + 2 = 0;\ (-\pi/2, \pi/2)$

49 $6\sin^3\theta + 18\sin^2\theta - 5\sin\theta - 15 = 0;\ (-\pi/2, \pi/2)$

50 $6\sin 2x - 8\cos x + 9\sin x - 6 = 0;\ (-\pi/2, \pi/2)$

Exer. 51–55: Use a calculator.

51 Show that $\sin^{-1}(\sin 2) \ne 2$. Explain why 2 is not the answer. Is $\cos^{-1}(\cos 2) = 2$? Why?

52 Show that $\cos^{-1}(\cos(-1)) \ne -1$. Explain why -1 is not the answer. Is $\sin^{-1}(\sin(-1)) = -1$? Why?

53 If a number x between -1 and 1 is entered into a calculator, is it always possible, by pressing $\boxed{\text{INV}}\ \boxed{\text{SIN}}$ twice, to calculate $\sin^{-1}(\sin^{-1} x)$? If not, decide which values of x will avoid an error message.

54 For any real number x, will an error message ever result when $\boxed{\text{INV}}\ \boxed{\text{TAN}}$ is pressed three times in succession? Explain.

55 The function $\tan^{-1}$ is the only inverse trigonometric function available in many computer languages. (For example, in **BASIC** the function is denoted by ATN(X).) Express $\sin^{-1} x$, $\cos^{-1} x$, and $\cot^{-1} x$ in terms of $\tan^{-1}$ (see Exercise 29.) Are the formulas valid over the entire domain of the inverse function?

In this section we shall concentrate on the inverse sine, cosine, tangent, and secant functions. Formulas for their derivatives and for integrals that result in inverse trigonometric functions are listed in the next three theorems, with $u = g(x)$ differentiable and x restricted to values for which the indicated expressions have meaning.

THEOREM (8.6)

$$D_x \sin^{-1} u = \frac{1}{\sqrt{1 - u^2}} D_x u \qquad D_x \cos^{-1} u = -\frac{1}{\sqrt{1 - u^2}} D_x u$$

$$D_x \tan^{-1} u = \frac{1}{1 + u^2} D_x u \qquad D_x \sec^{-1} u = \frac{1}{u\sqrt{u^2 - 1}} D_x u$$

PROOF We shall consider only the special case $u = x$, since the formulas for $u = g(x)$ may then be obtained by applying the Chain Rule.

If we let $f(x) = \sin x$ and $g(x) = \sin^{-1} x$ in Theorem (7.29), then it follows that the inverse sine function g is differentiable if $|x| < 1$. We shall use implicit differentiation to find $g'(x)$. First note that the equations

$$y = \sin^{-1} x \quad \text{and} \quad \sin y = x$$

are equivalent if $-1 < x < 1$ and $-\pi/2 < y < \pi/2$. Differentiating $\sin y = x$ implicitly,

$$\cos y \, D_x y = 1$$

and hence

$$D_x \sin^{-1} x = D_x y = \frac{1}{\cos y}.$$

Since $-\pi/2 < y < \pi/2$, $\cos y$ is positive and, therefore,

$$\cos y = \sqrt{1 - \sin^2 y} = \sqrt{1 - x^2}.$$

Thus,

$$D_x \sin^{-1} x = \frac{1}{\sqrt{1 - x^2}}$$

for $|x| < 1$. The inverse sine function is not differentiable at ± 1. This fact is evident from Figure 8.3, since vertical tangent lines occur at the endpoints of the graph.

Similarly, to find the derivative of the inverse cosine function we begin with the equivalent equations

$$y = \cos^{-1} x \quad \text{and} \quad \cos y = x$$

for $|x| < 1$ and $0 < y < \pi$. Differentiating $\cos y = x$ implicitly,

$$-\sin y \, D_x y = 1$$

and, therefore,

$$D_x \cos^{-1} x = D_x y = -\frac{1}{\sin y}.$$

Since $0 < y < \pi$, $\sin y$ is positive, and $\sin y = \sqrt{1 - \cos^2 y} = \sqrt{1 - x^2}$. Hence if $|x| < 1$,

$$D_x \cos^{-1} x = -\frac{1}{\sqrt{1 - x^2}}.$$

It follows from Theorem (7.29) that the inverse tangent function is differentiable at every real number. Let us consider the equivalent equations

$$y = \tan^{-1} x \quad \text{and} \quad \tan y = x$$

for $-\pi/2 < y < \pi/2$. Differentiating $\tan y = x$ implicitly,

$$\sec^2 y \, D_x \, y = 1.$$

Consequently, $\qquad D_x \tan^{-1} x = D_x \, y = \dfrac{1}{\sec^2 y}.$

Using the fact that $\sec^2 y = 1 + \tan^2 y = 1 + x^2$ gives us

$$D_x \tan^{-1} x = \frac{1}{1 + x^2}.$$

Finally, consider the equivalent equations

$$y = \sec^{-1} x \quad \text{and} \quad \sec y = x$$

for y in either $(0, \pi/2)$ or $(\pi, 3\pi/2)$. Differentiating $\sec y = x$ implicitly,

$$\sec y \tan y \, D_x \, y = 1.$$

Since $0 < y < \pi/2$ or $\pi < y < 3\pi/2$, it follows that $\sec y \tan y \neq 0$ and, hence,

$$D_x \sec^{-1} x = D_x \, y = \frac{1}{\sec y \tan y}.$$

Using the fact that $\tan y = \sqrt{\sec^2 y - 1} = \sqrt{x^2 - 1}$,

$$D_x \sec^{-1} x = \frac{1}{x\sqrt{x^2 - 1}}$$

for $|x| > 1$. The inverse secant function is not differentiable at $x = \pm 1$. Note that the graph has vertical tangent lines at the points with these x-coordinates (see Figure 8.7). • •

EXAMPLE 1 Find dy/dx if $y = \sin^{-1} 3x - \cos^{-1} 3x$.

SOLUTION Applying Theorem (8.6) with $u = 3x$,

$$\frac{dy}{dx} = \frac{3}{\sqrt{1 - 9x^2}} - \frac{-3}{\sqrt{1 - 9x^2}} = \frac{6}{\sqrt{1 - 9x^2}} \qquad \bullet$$

EXAMPLE 2 Find $f'(x)$ if $f(x) = \tan^{-1} e^{2x}$.

SOLUTION Using (8.6) with $u = e^{2x}$,

$$f'(x) = \frac{1}{1 + (e^{2x})^2} D_x \, e^{2x} = \frac{2e^{2x}}{1 + e^{4x}} \qquad \bullet$$

EXAMPLE 3 Find y' if $y = \sec^{-1}(x^2)$.

SOLUTION Applying Theorem (8.6) with $u = x^2$,

$$y' = \frac{1}{x^2\sqrt{x^4 - 1}}(2x) = \frac{2}{x\sqrt{x^4 - 1}} \quad \bullet$$

EXAMPLE 4 A rocket is fired directly upward and burns fuel at a rate that maintains a constant acceleration of 50 ft/sec² for $0 \le t \le 5$ with time t in seconds. As illustrated in Figure 8.9, an observer 400 feet from the launching pad visually follows the flight of the rocket.

(a) Express the angle of elevation θ of the rocket as a function of t.

(b) The observer perceives the rocket to be rising fastest when $d\theta/dt$ is largest. (Of course, this is an illusion, since the velocity is steadily increasing.) Determine the height of the rocket at the moment of (perceived) maximum velocity.

SOLUTION

(a) Let $s(t)$ denote the height of the rocket at time t (see Figure 8.9). Since the acceleration is always 50, we obtain the differential equation $s''(t) = 50$. Initial conditions are $s'(0) = 0$ and $s(0) = 0$. Antidifferentiating two times (see Example 6 of Section 4.7) gives us $s(t) = 25t^2$. Referring to Figure 8.9, with $s(t) = 25t^2$,

$$\tan\theta = \frac{25t^2}{400} = \frac{t^2}{16} \quad \text{or} \quad \theta = \arctan\frac{t^2}{16}.$$

(b) By Theorem (8.6), the rate of change of θ with respect to t is

$$\frac{d\theta}{dt} = \frac{1}{1 + (t^2/16)^2}\left(\frac{2t}{16}\right) = \frac{32t}{256 + t^4}.$$

Since we wish to find the maximum value of $d\theta/dt$, we begin by finding the critical numbers of $d\theta/dt$. Using the Quotient Rule,

$$\frac{d}{dt}\left(\frac{d\theta}{dt}\right) = \frac{d^2\theta}{dt^2} = \frac{(256 + t^4)(32) - 32t(4t^3)}{(256 + t^4)^2} = \frac{32(256 - 3t^4)}{(256 + t^4)^2}.$$

Setting $d^2\theta/dt^2 = 0$ gives us the critical number $t = \sqrt[4]{256/3}$. If follows from the First (or Second) Derivative Test that $d\theta/dt$ has a maximum value at $t = \sqrt[4]{256/3} \approx 3.04$ seconds. The height of the rocket at this time is

$$s(\sqrt[4]{256/3}) = 25(\sqrt[4]{256/3})^2 = 25\sqrt{256/3} \approx 230.9 \text{ feet} \quad \bullet$$

We may use differentiation formulas for the inverse trigonometric functions to obtain the following integration formulas.

THEOREM (8.7)

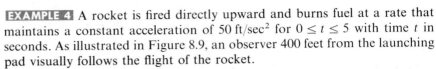

$$\int \frac{1}{\sqrt{1 - u^2}}\,du = \sin^{-1}u + C$$

$$\int \frac{1}{1 + u^2}\,du = \tan^{-1}u + C$$

$$\int \frac{1}{u\sqrt{u^2 - 1}}\,du = \sec^{-1}u + C$$

FIGURE 8.9

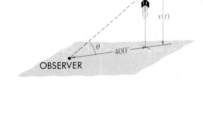

OBSERVER

EXAMPLE 5 Evaluate $\int \dfrac{e^{2x}}{\sqrt{1 - e^{4x}}} \, dx$.

SOLUTION The integral may be written as in the first formula of Theorem (8.7) by means of the substitution

$$u = e^{2x}, \qquad du = 2e^{2x} \, dx.$$

We introduce a factor 2 in the integrand and proceed as follows:

$$\int \frac{e^{2x}}{\sqrt{1 - e^{4x}}} \, dx = \tfrac{1}{2} \int \frac{1}{\sqrt{1 - (e^{2x})^2}} \, 2e^{2x} \, dx$$

$$= \tfrac{1}{2} \int \frac{1}{\sqrt{1 - u^2}} \, du$$

$$= \tfrac{1}{2} \sin^{-1} u + C$$

$$= \tfrac{1}{2} \sin^{-1} e^{2x} + C \qquad \bullet$$

The formulas in Theorem (8.7) can be extended as follows for a nonzero real number a.

THEOREM (8.8)

$$\int \frac{1}{\sqrt{a^2 - u^2}} \, du = \sin^{-1} \frac{u}{a} + C$$

$$\int \frac{1}{a^2 + u^2} \, du = \frac{1}{a} \tan^{-1} \frac{u}{a} + C$$

$$\int \frac{1}{u\sqrt{u^2 - a^2}} \, du = \frac{1}{a} \sec^{-1} \frac{u}{a} + C$$

PROOF Let us prove the second formula. If $u = g(x)$ and g is differentiable, then by Theorem (8.6),

$$D_x \left[\frac{1}{a} \tan^{-1} \frac{g(x)}{a} \right] = \frac{1}{a} \frac{1}{1 + [g(x)/a]^2} D_x \left[\frac{1}{a} g(x) \right]$$

$$= \frac{1}{a} \frac{a^2}{a^2 + [g(x)]^2} \frac{1}{a} g'(x)$$

$$= \frac{1}{a^2 + [g(x)]^2} g'(x).$$

Consequently,

$$\int \frac{1}{a^2 + [g(x)]^2} g'(x) \, dx = \frac{1}{a} \tan^{-1} \frac{g(x)}{a} + C = \frac{1}{a} \tan^{-1} \frac{u}{a} + C.$$

The remaining formulas may be verified in similar fashion. $\bullet$ $\bullet$

EXAMPLE 6 Evaluate $\int \dfrac{x^2}{5 + x^6}\, dx$.

SOLUTION The integral may be written as in the second formula of Theorem (8.8) by letting

$$u = x^3, \qquad du = 3x^2\, dx.$$

We introduce a factor 3 in the integrand and proceed as follows;

$$\int \frac{x^2}{5 + x^6}\, dx = \frac{1}{3} \int \frac{1}{5 + (x^3)^2}\, 3x^2\, dx$$

$$= \frac{1}{3} \int \frac{1}{(\sqrt{5})^2 + u^2}\, du$$

$$= \frac{1}{3} \cdot \frac{1}{\sqrt{5}} \tan^{-1} \frac{u}{\sqrt{5}} + C$$

$$= \frac{\sqrt{5}}{15} \tan^{-1} \frac{x^3}{\sqrt{5}} + C \qquad \bullet$$

EXAMPLE 7 Evaluate $\int \dfrac{1}{x\sqrt{x^4 - 9}}\, dx$.

SOLUTION The integral may be written in the proper form by means of substitution

$$u = x^2, \qquad du = 2x\, dx.$$

Thus,

$$\int \frac{1}{x\sqrt{x^4 - 9}}\, dx = \frac{1}{2} \int \frac{1}{x^2 \sqrt{(x^2)^2 - 9}}\, 2x\, dx$$

$$= \frac{1}{2} \int \frac{1}{u\sqrt{u^2 - 9}}\, du$$

$$= \frac{1}{2} \cdot \frac{1}{3} \sec^{-1} \frac{u}{3} + C$$

$$= \frac{1}{6} \sec^{-1} \frac{x^2}{3} + C \qquad \bullet$$

EXERCISES 8.3

Exer. 1–30: Find $f'(x)$ if $f(x)$ is the expression.

1 $\tan^{-1}(3x - 5)$

2 $\sin^{-1} \frac{1}{3}x$

3 $\sin^{-1} \sqrt{x}$

4 $\tan^{-1} x^2$

5 $e^{-x} \operatorname{arcsec} e^{-x}$

6 $\sqrt{\operatorname{arcsec} 3x}$

7 $x^2 \arctan x^2$

8 $\tan^{-1} \sin 2x$

9 $(1 + \cos^{-1} 3x)^3$

10 $x^2 \sec^{-1} 5x$

11 $\ln \arctan x^2$

12 $\arcsin \ln x$

13 $\dfrac{1}{\sin^{-1} x}$

14 $\arctan \dfrac{x + 1}{x - 1}$

15 $\cos(x^{-1}) + (\cos x)^{-1}$

16 $(\sin 2x)(\sin^{-1} 2x)$

17 $e^{-\tan 3x}$

18 $\dfrac{\sec \sqrt{x}}{\sqrt{x}}$

19 $(\csc 2x + \cot 2x)^4$

20 $e^{\cos x} + (\cos x)^e$

21 $\sec^{-1} \sqrt{x^2 - 1}$

22 $\left(\dfrac{1}{x} - \arcsin \dfrac{1}{x}\right)^4$

23 $(\arctan x)/(x^2 + 1)$

24 $\cos^{-1} \cos e^x$

25 $\sqrt{x} \sec^{-1} \sqrt{x}$

26 $e^{2x}/(\sin^{-1} 5x)$

27 $3^{\arcsin x^3}$

28 $x \arccos \sqrt{4x + 1}$

29 $(\tan x)^{\arctan x}$

30 $(\tan^{-1} 4x)e^{\tan^{-1} 4x}$

Exer. 31–32: Find y'.

31 $x^2 + x \sin^{-1} y = ye^x$

32 $\ln(x + y) = \tan^{-1} xy$

Exer. 33–48: Evaluate the integral.

33 $\int_0^4 \dfrac{1}{x^2 + 16}\,dx$

34 $\int_0^1 \dfrac{e^x}{1 + e^{2x}}\,dx$

35 $\int_0^{\sqrt{2}/2} \dfrac{x}{\sqrt{1 - x^4}}\,dx$

36 $\int_{2/\sqrt{3}}^2 \dfrac{1}{x\sqrt{x^2 - 1}}\,dx$

37 $\int \dfrac{\sin x}{\cos^2 x + 1}\,dx$

38 $\int \dfrac{\cos x}{\sqrt{9 - \sin^2 x}}\,dx$

39 $\int \dfrac{1}{\sqrt{x}(1 + x)}\,dx$

40 $\int \dfrac{1}{e^x\sqrt{1 - e^{-2x}}}\,dx$

41 $\int \dfrac{e^x}{\sqrt{16 - e^{2x}}}\,dx$

42 $\int \dfrac{\sec x \tan x}{1 + \sec^2 x}\,dx$

43 $\int \dfrac{1}{x\sqrt{x^6 - 4}}\,dx$

44 $\int \dfrac{x}{\sqrt{36 - x^2}}\,dx$

45 $\int \dfrac{x}{x^2 + 9}\,dx$

46 $\int \dfrac{1}{x\sqrt{x - 1}}\,dx$

47 $\int \dfrac{1}{\sqrt{e^{2x} - 25}}\,dx$

48 $\int \dfrac{e^x}{\sqrt{4 - e^x}}\,dx$

49 Find the area of the region bounded by the graphs of the equations $y = 4/\sqrt{16 - x^2}$, $x = -2$, $x = 2$, and $y = 0$.

50 If $f(x) = x^2/(1 + x^6)$, find the area of the region under the graph of f from $x = 0$ to $x = 1$.

51 The floor of a storage shed has the shape of a right triangle. The sides opposite and adjacent to an acute angle θ of the triangle are measured as 10 feet and 7 feet, respectively, with a possible error of 0.5 inch in the 10-foot measurement. Use the differential of an inverse trigonometric function to approximate the error in the calculated value of θ.

52 Use differentials to approximate the change in $\arcsin x$ if x changes from 0.25 to 0.26.

53 An airplane at a constant altitude of 5 miles and a speed of 500 mi/hr is flying in a direction away from an observer on the ground. Use inverse trigonometric functions to find the rate at which the angle of elevation is changing when the airplane flies over a point 2 miles from the observer.

54 A searchlight located $\frac{1}{8}$ mile from the nearest point P on a straight road is trained on an automobile traveling on the road at a rate of 50 mi/hr. Use inverse trigonometric functions to find the rate at which the searchlight is rotating when the car is $\frac{1}{4}$ mile from P.

55 A billboard 20 feet high is located on top of a building, with its lower edge 60 feet above the level of a viewer's eye. Use inverse trigonometric functions to find how far from a point directly below the sign a viewer should stand to maximize the angle between the lines of sight of the top and bottom of the billboard (see Example 7 of Section 4.5).

56 Given points $A(3, 1)$ and $B(6, 4)$ in a rectangular coordinate system, find the x-coordinate of the point P on the x-axis such that angle APB has its largest value.

57 Find equations for the tangent and normal lines to the graph of $y = \sin^{-1}(x - 1)$ at the point $(\frac{3}{2}, \pi/6)$.

58 Find the points on the graph of $y = \tan^{-1} 2x$ at which the tangent line is parallel to the line $13y - 2x + 5 = 0$.

59 Find the intervals in which the graph of $y = \tan^{-1} x$ is (a) concave upward; (b) concave downward.

60 The velocity, at time t, of a point moving on a coordinate line is $(1 + t^2)^{-1}$ ft/sec. If the point is at the origin at $t = 0$, find its position at the instant that the acceleration and velocity have the same absolute value.

61 The region bounded by the graphs of the equations $y = e^x$, $y = 1/\sqrt{x^2 + 1}$, and $x = 1$ is revolved about the x-axis. Find the volume of the resulting solid.

62 Use differentials to approximate the arc length of the graph of $y = \tan^{-1} x$ from $A(0, 0)$ to $B(0.1, \tan^{-1} 0.1)$.

63 A missile is fired vertically from a point that is 5 miles from a tracking station and at the same elevation. For the first 20 seconds of flight its angle of elevation changes at a constant rate of 2° per second. Use inverse trigonometric functions to find the velocity of the missile when the angle of elevation is 30° (see Exercise 38, Section 3.9).

64 Blood flowing through a blood vessel causes a loss of energy due to friction. According to *Poiseuille's Law*, this energy loss E is given by $E = kl/r^4$ for a blood vessel of radius r and length l, and a constant k. Suppose a blood vessel of radius r_2 and length l_2 branches off, at an angle θ, from a blood vessel of radius r_1 and length l_1, as illustrated in the figure, where the colored arrows indicate the direction of blood flow. The energy loss is then the sum of the individual energy losses; that is,

$$E = (kl_1/r_1^4) + (kl_2/r_2^4).$$

Express l_1 and l_2 in terms of a, b, and θ, and find the angle that minimizes the energy loss.

EXERCISE 64

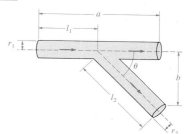

65 Prove that $\int_0^1 [4/(1 + x^2)] \, dx = \pi$ and then apply Simpson's Rule with $n = 10$ to approximate π.

66 A triangle with base b and altitude a is to be cut from a rectangle of dimensions a by b as shown in the figure. How should the rectangle be cut to maximize angle θ?

8.4 THE HYPERBOLIC FUNCTIONS

The exponential expressions

$$\frac{e^x - e^{-x}}{2} \quad \text{and} \quad \frac{e^x + e^{-x}}{2}$$

occur in advanced applications of calculus. Their properties are, in many ways, similar to those of $\sin x$ and $\cos x$. Later in our discussion, we will see why they are called the *hyperbolic sine* and the *hyperbolic cosine* of x.

DEFINITION (8.9)

> The **hyperbolic sine function**, denoted by **sinh**, and the **hyperbolic cosine function**, denoted **cosh**, are defined by
>
> $$\sinh x = \frac{e^x - e^{-x}}{2} \quad \text{and} \quad \cosh x = \frac{e^x + e^{-x}}{2}$$
>
> for every real number x.

FIGURE 8.10

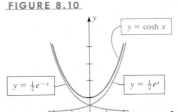

FIGURE 8.11

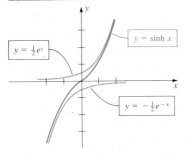

FIGURE 8.12

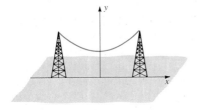

The graph of $y = \cosh x$ may be found by the method of **addition of ordinates.** To use this technique, we first sketch the graphs of $y = \frac{1}{2}e^x$ and $y = \frac{1}{2}e^{-x}$ on one coordinate system, as shown in gray in Figure 8.10. We then add the y-coordinates of points on these graphs to obtain the graph of $y = \cosh x$. Note that the range of cosh is $[1, \infty)$.

We may find the graph of $y = \sinh x$ by adding y-coordinates of the graphs of $y = \frac{1}{2}e^x$ and $y = -\frac{1}{2}e^{-x}$, as shown in Figure 8.11.

The hyperbolic cosine function can be used to describe the shape of a uniform flexible cable, or chain, whose ends are supported from the same height. As illustrated in Figure 8.12, telephone or power lines may be strung between poles in this manner. The shape of the cable appears to be a parabola, but is actually a **catenary** (after the Latin word for *chain*). If we introduce a coordinate system, as in Figure 8.12, we can show that an equation corresponding to the shape of the cable is $y = a \cosh (x/a)$ for some real number a.

The hyperbolic cosine function also occurs in the analysis of motion in a resisting medium. If an object is dropped from a given height, and if air resistance is neglected, then the distance y that it falls in t seconds is $y = \frac{1}{2}gt^2$, for a gravitational constant g. However, air resistance cannot always be neglected. As the velocity of the object increases, air resistance may significantly affect its motion. For example, if the air resistance is directly proportional to the square of the velocity, then the distance y that the object falls in t seconds is given by

$$y = A \ln (\cosh Bt)$$

for constants A and B (see Exercise 56).

Many identities similar to those for trigonometric functions hold for the hyperbolic sine and cosine functions. For example, if $\cosh^2 x$ and $\sinh^2 x$ denote $(\cosh x)^2$ and $(\sinh x)^2$, respectively, we have the following identity.

THEOREM (8.10)

$$\cosh^2 x - \sinh^2 x = 1$$

PROOF By Definition (8.9),

$$\cosh^2 x - \sinh^2 x = \left(\frac{e^x + e^{-x}}{2}\right)^2 - \left(\frac{e^x - e^{-x}}{2}\right)^2$$

$$= \frac{e^{2x} + 2 + e^{-2x}}{4} - \frac{e^{2x} - 2 + e^{-2x}}{4}$$

$$= \frac{e^{2x} + 2 + e^{-2x} - e^{2x} + 2 - e^{-2x}}{4}$$

$$= \frac{4}{4} = 1. \qquad \bullet\ \bullet$$

Theorem (8.10) is analogous to the identity $\cos^2 x + \sin^2 x = 1$. Other hyperbolic identities are given in the exercises. To verify an identity, it is sufficient to express the hyperbolic functions in terms of exponential functions and show that one side of the equation can be transformed into the other. The hyperbolic identities are similar to (but not always the same as) certain trigonometric identities—differences usually involve signs of terms.

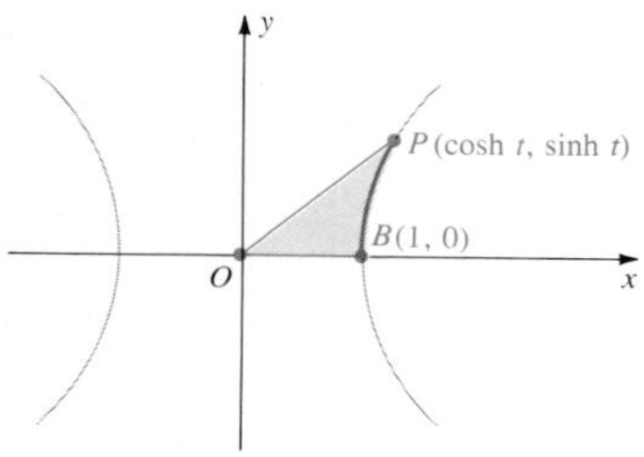

An interesting geometric relationship exists for the points $(\cos t, \sin t)$ and $(\cosh t, \sinh t)$ for a real number t. Let us consider the graphs of $x^2 + y^2 = 1$ and $x^2 - y^2 = 1$, sketched in Figures 8.13 and 8.14. The graph in Figure 8.13 is the unit circle with center at the origin. The graph in Figure 8.14 is a *hyperbola*. (Hyperbolas and their properties will be discussed in detail in Chapter 12.) Note first that since $\cos^2 t + \sin^2 t = 1$, the point $P(\cos t, \sin t)$ is on the circle $x^2 + y^2 = 1$. Next, by Theorem (8.10), $\cosh^2 t - \sinh^2 t = 1$, and hence the point $P(\cosh t, \sinh t)$ is on the hyperbola $x^2 - y^2 = 1$. These are the reasons for referring to cos and sin as *circular* functions and to cosh and sinh as *hyperbolic* functions.

The graphs in Figures 8.13 and 8.14 are related in another way. If $0 < t < \pi/2$, then t is the radian measure of angle POB, shown in Figure 8.13. By Theorem (2.25), the area A of the shaded circular sector is $A = \frac{1}{2}(1)^2 t = \frac{1}{2}t$, and hence, $t = 2A$. Similarly, if $P(\cosh t, \sinh t)$ is the point in Figure 8.14, then $t = 2A$ for the area A of the shaded hyperbolic sector (see Exercise 53).

The impressive analogies between the trigonometric and hyperbolic sine and cosine functions can be extended to define hyperbolic functions that correspond to the four remaining trigonometric functions. The **hyperbolic tangent, hyperbolic cotangent, hyperbolic secant,** and **hyperbolic cosecant**

FIGURE 8.13
$x^2 + y^2 = 1$

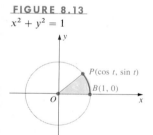

FIGURE 8.14
$x^2 - y^2 = 1$

functions, denoted by tanh, coth, sech, and csch, respectively, are defined as follows.

DEFINITION (8.11)

$$\tanh x = \frac{\sinh x}{\cosh x} = \frac{e^x - e^{-x}}{e^x + e^{-x}}$$

$$\coth x = \frac{\cosh x}{\sinh x} = \frac{e^x + e^{-x}}{e^x - e^{-x}}, \quad x \neq 0$$

$$\operatorname{sech} x = \frac{1}{\cosh x} = \frac{2}{e^x + e^{-x}}$$

$$\operatorname{csch} x = \frac{1}{\sinh x} = \frac{2}{e^x - e^{-x}}, \quad x \neq 0$$

FIGURE 8.15

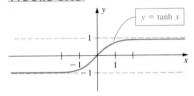

$y = \tanh x$

The graph of $y = \tanh x$ is sketched in Figure 8.15. The lines $y = 1$ and $y = -1$ are horizontal asymptotes. We leave the verification of this, and the sketching of the graphs of the remaining hyperbolic functions, as exercises.

If we divide both sides of (8.10) by $\cosh^2 x$ and use the definitions of $\tanh x$ and $\operatorname{sech} x$, we obtain the first formula of the next theorem. The second formula may be obtained by dividing both sides of (8.10) by $\sinh^2 x$.

THEOREM (8.12)

$$1 - \tanh^2 x = \operatorname{sech}^2 x \qquad \coth^2 x - 1 = \operatorname{csch}^2 x$$

Note the similarities and differences between (8.12) and the analogous trigonometric identities.

Derivative formulas for the hyperbolic functions are listed in the next theorem, for $u = g(x)$ and g differentiable.

THEOREM (8.13)

$$D_x \sinh u = \cosh u \, D_x u$$

$$D_x \cosh u = \sinh u \, D_x u$$

$$D_x \tanh u = \operatorname{sech}^2 u \, D_x u$$

$$D_x \coth u = -\operatorname{csch}^2 u \, D_x u$$

$$D_x \operatorname{sech} u = -\operatorname{sech} u \tanh u \, D_x u$$

$$D_x \operatorname{csch} u = -\operatorname{csch} u \coth u \, D_x u$$

PROOF As usual, we consider only the case $u = x$. Since $D_x e^x = e^x$ and $D_x e^{-x} = -e^{-x}$,

$$D_x \sinh x = D_x \left(\frac{e^x - e^{-x}}{2} \right) = \frac{e^x + e^{-x}}{2} = \cosh x$$

$$D_x \cosh x = D_x \left(\frac{e^x + e^{-x}}{2} \right) = \frac{e^x - e^{-x}}{2} = \sinh x.$$

To differentiate $\tanh x$ we apply the Quotient Rule as follows:

$$D_x \tanh x = D_x \frac{\sinh x}{\cosh x}$$

$$= \frac{\cosh x \, D_x \sinh x - \sinh x \, D_x \cosh x}{\cosh^2 x}$$

$$= \frac{\cosh^2 x - \sinh^2 x}{\cosh^2 x}$$

$$= \frac{1}{\cosh^2 x} = \operatorname{sech}^2 x$$

The proofs of the remaining formulas are left to the reader. • •

EXAMPLE 1 Find $f'(x)$ if $f(x) = \cosh(x^2 + 1)$.

SOLUTION Applying Theorem (8.13), with $u = x^2 + 1$,

$$f'(x) = \sinh(x^2 + 1) \cdot D_x(x^2 + 1)$$
$$= 2x \sinh(x^2 + 1) \qquad \bullet$$

The integration formulas that correspond to the derivative formulas in Theorem (8.13) are as follows.

THEOREM (8.14)

$$\int \sinh u \, du = \cosh u + C$$

$$\int \cosh u \, du = \sinh u + C$$

$$\int \operatorname{sech}^2 u \, du = \tanh u + C$$

$$\int \operatorname{csch}^2 u \, du = -\coth u + C$$

$$\int \operatorname{sech} u \tanh u \, du = -\operatorname{sech} u + C$$

$$\int \operatorname{csch} u \coth u \, du = -\operatorname{csch} u + C$$

EXAMPLE 2 Evaluate $\int x^2 \sinh x^3 \, dx$.

SOLUTION If we let $u = x^3$, then $du = 3x^2 \, dx$, and

$$\int x^2 \sinh x^3 \, dx = \tfrac{1}{3} \int (\sinh x^3) \, 3x^2 \, dx$$

$$= \tfrac{1}{3} \int \sinh u \, du = \tfrac{1}{3} \cosh u + C$$

$$= \tfrac{1}{3} \cosh x^3 + C \qquad \bullet$$

Exer. 1–14: Verify the identity.

1 $\cosh x + \sinh x = e^x$

2 $\cosh x - \sinh x = e^{-x}$

3 $\sinh(-x) = -\sinh x$

4 $\cosh(-x) = \cosh x$

5 $\sinh(x + y) = \sinh x \cosh y + \cosh x \sinh y$

6 $\cosh(x + y) = \cosh x \cosh y + \sinh x \sinh y$

7 $\sinh 2x = 2 \sinh x \cosh x$

8 $\cosh 2x = \cosh^2 x + \sinh^2 x$

9 $\tanh(x + y) = \dfrac{\tanh x + \tanh y}{1 + \tanh x \tanh y}$

10 $\tanh 2x = \dfrac{2 \tanh x}{1 + \tanh^2 x}$

11 $\cosh \dfrac{x}{2} = \sqrt{\dfrac{1 + \cosh x}{2}}$

12 $\tanh \dfrac{x}{2} = \dfrac{\sinh x}{1 + \cosh x}$

13 $\sinh x + \sinh y = 2 \sinh \frac{1}{2}(x + y) \cosh \frac{1}{2}(x - y)$

14 $(\cosh x + \sinh x)^n = \cosh nx + \sinh nx$ for every positive integer n (*Hint:* Use Exercise 1.)

Exer. 15–30: Find $f'(x)$ if $f(x)$ equals the expression.

15 $\sinh 5x$

16 $\cosh \sqrt{4x^2 + 3}$

17 $\sqrt{x} \tanh \sqrt{x}$

18 $x \operatorname{csch} e^{4x}$

19 $\dfrac{\operatorname{sech} x^2}{x^2 + 1}$

20 $\dfrac{\coth x}{\cot x}$

21 $\cosh x^3$

22 $\sinh(x^2 + 1)$

23 $\cosh^3 x$

24 $\sinh^2 3x$

25 $\ln \sinh 2x$

26 $\arctan \tanh x$

27 $e^{3x} \operatorname{sech} x$

28 $\sqrt{\operatorname{sech} 5x}$

29 $\dfrac{1}{\tanh x + 1}$

30 $\dfrac{1 + \cosh x}{1 - \cosh x}$

31 Find y' if $\sinh xy = ye^x$.

32 Find y' if $x^2 \tanh y = \ln y$.

Exer. 33–44: Evaluate the integral.

33 $\displaystyle\int \dfrac{\sinh \sqrt{x}}{\sqrt{x}} \, dx$

34 $\displaystyle\int \dfrac{\cosh \ln x}{x} \, dx$

35 $\displaystyle\int \coth x \, dx$

36 $\displaystyle\int \dfrac{1}{\cosh^2 3x} \, dx$

37 $\displaystyle\int \sinh x \cosh x \, dx$

38 $\displaystyle\int \operatorname{sech}^2 x \tanh x \, dx$

39 $\displaystyle\int \tanh 3x \operatorname{sech} 3x \, dx$

40 $\displaystyle\int \sinh x \sqrt{\cosh x} \, dx$

41 $\displaystyle\int \tanh^2 3x \operatorname{sech}^2 3x \, dx$

42 $\displaystyle\int \tanh x \, dx$

43 $\displaystyle\int \dfrac{\operatorname{sech}^2 x}{1 - 2 \tanh x} \, dx$

44 $\displaystyle\int \dfrac{e^{\sinh x}}{\operatorname{sech} x} \, dx$

45 Find the area of the region bounded by the graphs of $y = \sinh 3x$, $y = 0$, and $x = 1$.

46 Find the arc length of the graph of $y = \cosh x$ from $x = 0$ to $x = 1$.

47 Find the points on the graph of $y = \sinh x$ at which the tangent line has slope 2.

48 The region bounded by the graphs of $y = \cosh x$, $x = -1$, $x = 1$, and $y = 0$ is revolved about the x-axis. Find the volume of the resulting solid.

49 Verify the graph of $y = \tanh x$ in Figure 8.15.

Exer. 50–52: Sketch the graph of the equation.

50 $y = \coth x$

51 $y = \operatorname{sech} x$

52 $y = \operatorname{csch} x$

53 If A is the shaded region in Figure 8.14, show that $t = 2A$.

54 Sketch the graph of $x^2 - y^2 = 1$ and show that as t varies, the point $P(\cosh t, \sinh t)$ traces the part of the graph in quadrants I and IV.

55 The Gateway Arch in St. Louis has the shape of an inverted catenary. Rising 630 feet at its center and stretching 630 feet across its base, the arch's shape can be described by $y = -127.7 \cosh(x/127.7) + 757.7$ for $-315 \le x \le 315$.

(a) Approximate the total open area under the arch.

(b) Approximate the total length of the arch.

EXERCISE 55

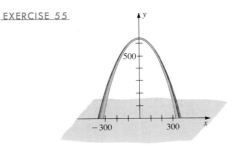

56 If a steel ball is released into water, the distance $s(t)$ it travels after t seconds is given by $s(t) = A \ln(\cosh Bt)$ for positive constants A and B.

(a) Find the velocity function v and verify that $v(0) = 0$. What is $\lim_{t \to \infty} v(t)$?

(b) Show that if $A = m/k$ and $B^2 = gk/m$, then v satisfies the differential equation

$$m \frac{dv}{dt} + kv^2 = mg$$

for mass m, gravitational constant g, and $k > 0$.

57 If a wave of length L is traveling across water of depth D (see figure), the velocity, or *celerity*, of the wave is related

to L and D by the formula

$$v^2 = \frac{gL}{2\pi} \tanh \frac{2\pi D}{L}$$

for a gravitational constant g.

(a) Find $\lim_{D \to \infty} v^2$ to establish that $v \approx \sqrt{gL/2\pi}$ in deep water.

(b) If x is small and f is a continuous function, then by the Mean Value Theorem (5.22), $f(x) - f(0) \approx f'(0)x$. Use this fact to show that $v \approx \sqrt{gD}$ if D/L is small. Conclude

that wave velocity is independent of wave length in shallow water.

58 A soap bubble formed by two parallel concentric rings is shown in the figure. If the rings are not too far apart, it can be shown that the function f whose graph generates this surface of revolution is a solution of the differential equation $y\, y'' = 1 + (y')^2$. If A and B are positive constants, show that $y = A \cosh Bx$ with $A > 0$ is a solution if and only if $AB = 1$. Conclude that the graph is a catenary.

EXERCISE 57

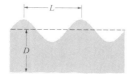

EXERCISE 58

8.5 THE INVERSE HYPERBOLIC FUNCTIONS

The hyperbolic sine function is continuous and increasing for every x and hence, by Theorem (7.27), has a continuous, increasing inverse function, denoted by $\sinh^{-1}$. Since $\sinh x$ is defined in terms of e^x, we might expect that $\sinh^{-1}$ can be expressed in terms of the inverse, ln, of the natural exponential function. The first formula of the next theorem shows that this is the case.

THEOREM (8.15)

$$\sinh^{-1} x = \ln\left(x + \sqrt{x^2 + 1}\right)$$

$$\cosh^{-1} x = \ln\left(x + \sqrt{x^2 - 1}\right), \quad x \geq 1$$

$$\tanh^{-1} x = \frac{1}{2} \ln \frac{1 + x}{1 - x}, \quad |x| < 1$$

$$\operatorname{sech}^{-1} x = \ln \frac{1 + \sqrt{1 - x^2}}{x}, \quad 0 < x \leq 1$$

PROOF To prove the formula for $\sinh^{-1} x$ we begin by noting that

$$y = \sinh^{-1} x \quad \text{if and only if} \quad x = \sinh y.$$

The equation $x = \sinh y$ can be used to find an explicit form for $\sinh^{-1} x$. Thus, if

$$x = \sinh y = \frac{e^y - e^{-y}}{2}$$

then

$$e^y - 2x - e^{-y} = 0.$$

Multiplying both sides by e^y, we obtain

$$e^{2y} - 2xe^y - 1 = 0$$

Applying the quadratic formula,

$$e^y = \frac{2x \pm \sqrt{4x^2 + 4}}{2} \quad \text{or} \quad e^y = x \pm \sqrt{x^2 + 1}.$$

Since e^y is never negative, we must discard the minus sign. Doing this, and then taking the natural logarithm of both sides of the equation gives us

$$y = \ln{(x + \sqrt{x^2 + 1})},$$

that is,

$$\sinh^{-1}{x} = \ln{(x + \sqrt{x^2 + 1})}.$$

The formulas for the remaining inverse hyperbolic functions are obtained in similar fashion. As with trigonometric functions, some inverse functions exist only if the domain is restricted. For example, if the domain of cosh is restricted to the set of nonnegative real numbers, then the resulting function is continuous and increasing, and its inverse function $\cosh^{-1}$ is defined by

$$y = \cosh^{-1}{x} \quad \text{if and only if} \quad \cosh{y} = x, \quad y \geq 0.$$

Employing the process used for $\sinh^{-1}{x}$ leads us to the logarithmic formula for $\cosh^{-1}{x}$.

Similarly, we write

$$y = \tanh^{-1}{x} \quad \text{if and only if} \quad \tanh{y} = x \quad \text{for } |x| < 1.$$

Using Definition (8.11), $\tanh{y} = x$ may be written

$$\frac{e^y - e^{-y}}{e^y + e^{-y}} = x.$$

Solving for y gives us the logarithmic form for $\tanh^{-1}{x}$.

Finally, if we restrict the domain of sech to nonnegative numbers, the result is a one-to-one function and we define

$$y = \text{sech}^{-1}{x} \quad \text{if and only if} \quad \text{sech}\,y = x, \quad y \geq 0.$$

Again introducing the exponential form leads to the fourth formula in the statement of the theorem. • •

In the next theorem $u = g(x)$, provided g is differentiable and x is suitably restricted.

THEOREM (8.16)

$$D_x \sinh^{-1}{u} = \frac{1}{\sqrt{u^2 + 1}} D_x u$$

$$D_x \cosh^{-1}{u} = \frac{1}{\sqrt{u^2 - 1}} D_x u, \quad u > 1$$

$$D_x \tanh^{-1}{u} = \frac{1}{1 - u^2} D_x u, \quad |u| < 1$$

$$D_x \text{sech}^{-1}{u} = \frac{-1}{u\sqrt{1 - u^2}} D_x u, \quad 0 < u < 1$$

PROOF Using Theorem (8.15),

$$D_x \sinh^{-1} x = D_x \ln (x + \sqrt{x^2 + 1})$$

$$= \frac{1}{x + \sqrt{x^2 + 1}} \left(1 + \frac{x}{\sqrt{x^2 + 1}} \right)$$

$$= \frac{\sqrt{x^2 + 1} + x}{(x + \sqrt{x^2 + 1})\sqrt{x^2 + 1}}$$

$$= \frac{1}{\sqrt{x^2 + 1}}.$$

This formula can be extended to $D_x \sinh^{-1} u$ by applying the Chain Rule in the usual way. Proofs of the remaining formulas are left to the reader.

● ●

EXAMPLE 1 Find y' if $y = \sinh^{-1} (\tan x)$.

SOLUTION Using Theorem (8.16) with $u = \tan x$,

$$y' = \frac{1}{\sqrt{\tan^2 x + 1}} D_x \tan x = \frac{1}{\sqrt{\sec^2 x}} \sec^2 x$$

$$= \frac{1}{|\sec x|} |\sec x|^2 = |\sec x|$$ ●

The following theorem may be verified by differentiating the right-hand side of each formula, with $u = g(x)$.

THEOREM (8.17)

$$\int \frac{1}{\sqrt{a^2 + u^2}} \, du = \sinh^{-1} \frac{u}{a} + C, \quad a > 0$$

$$\int \frac{1}{\sqrt{u^2 - a^2}} \, du = \cosh^{-1} \frac{u}{a} + C, \quad u > a > 0$$

$$\int \frac{1}{a^2 - u^2} \, du = \frac{1}{a} \tanh^{-1} \frac{u}{a} + C, \quad a > 0, \quad |u| < a$$

$$\int \frac{1}{u\sqrt{a^2 - u^2}} \, du = -\frac{1}{a} \operatorname{sech}^{-1} \frac{|u|}{a} + C, \quad a > 0, \quad 0 < |u| < a$$

If we use Theorem (8.15), then each of the integration formulas in the preceding theorem can be expressed in terms of the natural logarithmic function. To illustrate,

$$\int \frac{1}{\sqrt{a^2 + u^2}} \, du = \sinh^{-1} \frac{u}{a} + C$$

$$= \ln \left(\frac{u}{a} + \sqrt{\left(\frac{u}{a}\right)^2 + 1} \right) + C.$$

In Exercise 29 we are asked to show that if $a > 0$ this can also be written as

$$\int \frac{1}{\sqrt{a^2 + u^2}} \, du = \ln (u + \sqrt{a^2 + u^2}) + D$$

for an arbitrary constant D. In Section 9.3 we shall discuss another method for evaluating the integrals in Theorem (8.17).

EXAMPLE 2 Evaluate $\int \dfrac{1}{\sqrt{25 + 9x^2}}\, dx$.

SOLUTION We may express the integral as in the first formula of Theorem (8.17) by letting $u = 3x$ and $du = 3\, dx$. Thus,

$$\int \frac{1}{\sqrt{25 + 9x^2}}\, dx = \frac{1}{3} \int \frac{1}{\sqrt{5^2 + (3x)^2}}\, 3\, dx$$

$$= \frac{1}{3} \sinh^{-1} \frac{3x}{5} + C \qquad \bullet$$

EXAMPLE 3 Evaluate $\int \dfrac{e^x}{16 - e^{2x}}\, dx$.

SOLUTION Letting $u = e^x$, $du = e^x\, dx$, and applying Theorem (8.17),

$$\int \frac{e^x}{16 - e^{2x}}\, dx = \int \frac{1}{4^2 - (e^x)^2}\, e^x\, dx$$

$$= \int \frac{1}{4^2 - u^2}\, du$$

$$= \frac{1}{4} \tanh^{-1} \frac{u}{4} + C$$

$$= \frac{1}{4} \tanh^{-1} \frac{e^x}{4} + C$$

for $e^x < 4$. $\bullet$

EXERCISES 8.5

1 Derive the formula for $\cosh^{-1} x$ in Theorem (8.15).

2 Derive the formula for $\tanh^{-1} x$ in Theorem (8.15).

Exer. 3–6: Verify the formula.

3 $D_x \cosh^{-1} u = \dfrac{1}{\sqrt{u^2 - 1}}\, D_x u, \quad u > 1$

4 $D_x \tanh^{-1} u = \dfrac{1}{1 - u^2}\, D_x u, \quad |u| < 1$

5 $\displaystyle\int \frac{1}{\sqrt{u^2 - a^2}}\, du = \cosh^{-1} \frac{u}{a} + C, \quad u > a > 0$

6 $\displaystyle\int \frac{1}{a^2 - u^2}\, du = \frac{1}{a} \tanh^{-1} \frac{u}{a} + C, \quad a > 0,\ |u| < a$

Exer. 7–10: Sketch the graph of the equation.

7 $y = \sinh^{-1} x$ **8** $y = \cosh^{-1} x$

9 $y = \tanh^{-1} x$ **10** $y = \text{sech}^{-1} x$

Exer. 11–20: Find $f'(x)$ if $f(x)$ equals the expression.

11 $\sinh^{-1} 5x$ **12** $\sinh^{-1} e^x$

13 $\cosh^{-1} \sqrt{x}$ **14** $\sqrt{\cosh^{-1} x}$

15 $\tanh^{-1} (x^2 - 1)$ **16** $\tanh^{-1} \sin 3x$

17 $x \sinh^{-1} (1/x)$ **18** $1/(\sinh^{-1} x^2)$

19 $\ln \cosh^{-1} 4x$ **20** $\cosh^{-1} \ln 4x$

Exer. 21–28: Evaluate the integral.

21 $\displaystyle\int \frac{1}{\sqrt{81 + 16x^2}}\, dx$ **22** $\displaystyle\int \frac{1}{\sqrt{16x^2 - 9}}\, dx$

23 $\displaystyle\int \frac{1}{49 - 4x^2}\, dx$ **24** $\displaystyle\int \frac{\sin x}{\sqrt{1 + \cos^2 x}}\, dx$

25 $\displaystyle\int \frac{e^x}{\sqrt{e^{2x} - 16}}\, dx$ **26** $\displaystyle\int \frac{2}{5 - 3x^2}\, dx$

27 $\displaystyle\int \frac{1}{x\sqrt{9-x^4}}\,dx$

28 $\displaystyle\int \frac{1}{\sqrt{5-e^{2x}}}\,dx$

29 Prove that if $a > 0$, then

$$\int \frac{1}{\sqrt{a^2+u^2}}\,du = \ln\left(u + \sqrt{a^2+u^2}\right) + C.$$

30 Prove that if $u > a > 0$, then

$$\int \frac{1}{\sqrt{u^2-a^2}}\,du = \ln\left(u + \sqrt{u^2-a^2}\right) + C.$$

31 A particle moves along the line $x = 1$ in a coordinate plane at a velocity that is proportional to its distance from the origin. Express the y-coordinate of the particle as a function of time t (in seconds) if the initial position of the particle is $(1, 0)$ and the initial velocity is 3 ft/sec.

32 A rectangular coordinate system is shown in the figure to illustrate the problem of a dog seeking its master. The dog, initially at the point $(1, 0)$, sees his master at the point $(0, 0)$. The master proceeds up the y-axis at a constant speed, and the dog runs directly toward him at all times. If the speed of the dog is twice that of the master, the solution $y = f(x)$ of the differential equation that describes the path of the dog can be given by

$$2xy'' = \sqrt{1+(y')^2}.$$

Solve this equation by first letting $z = dy/dx$ and solving $2xz' = \sqrt{1+z^2}$, obtaining $z = \frac{1}{2}[\sqrt{x} - (1/\sqrt{x})]$. Finally, solve $y' = \frac{1}{2}[\sqrt{x} - (1/\sqrt{x})]$.

EXERCISE 32

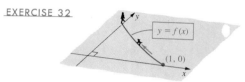

8.6 REVIEW

Define or discuss each of the following.

1 Integration formulas for trigonometric expressions
2 The inverse trigonometric functions
3 Differentiation and integration formulas involving inverse trigonometric functions
4 Hyperbolic functions
5 Inverse hyperbolic functions

EXERCISES 8.6

Exer. 1–32: Find $f'(x)$ if $f(x)$ is the expression.

1 $\arctan\sqrt{x-1}$

2 $\tan^{-1}(\ln 3x)$

3 $x^2 \operatorname{arcsec} x^2$

4 $\cot\dfrac{1}{x} + \dfrac{1}{\cot x}$

5 $\dfrac{1}{\cos^{-1}x}$

6 $\dfrac{\ln \sinh x}{x}$

7 $\cosh e^{-5x}$

8 $(\cos x)^{\cot x}$

9 $\tanh^{-1}(\tanh \sqrt[3]{x})$

10 $\sec(\sec x)$

11 $2^{\arctan 2x}$

12 $(1 + \operatorname{arcsec} 2x)^{\sqrt{2}}$

13 $\sin^3 e^{-2x}$

14 $\ln \sin(\pi/3)$

15 $e^{-x^2}\cot x^2$

16 $\sec 5x \tan 5x$

17 $\ln \tan^{-1} x^2$

18 $\dfrac{1-x^2}{\arccos x}$

19 $\sin^{-1}\sqrt{1-x^2}$

20 $\sqrt{\sin^{-1}(1-x^2)}$

21 $\tan(\sin 3x)$

22 $\tan^{-1}\sqrt{\tan 2x}$

23 $(\tan x + \tan^{-1} x)^4$

24 $e^{4x}\sec^{-1}e^{4x}$

25 $\tan^{-1}(\tan^{-1} x)$

26 $e^{x\cosh x}$

27 $e^{-x}\sinh e^{-x}$

28 $\ln \tanh(5x+1)$

29 $\dfrac{\sinh x}{\cosh x - \sinh x}$

30 $\dfrac{1}{x}\tanh\dfrac{1}{x}$

31 $\sinh^{-1} x^2$

32 $\cosh^{-1}\tan x$

Exer. 33–64: Evaluate the integral.

33 $\displaystyle\int \cos(5-3x)\,dx$

34 $\displaystyle\int \csc\frac{x}{2}\cot\frac{x}{2}\,dx$

35 $\displaystyle\int \frac{\sec^2\sqrt{x}}{\sqrt{x}}\,dx$

36 $\displaystyle\int \frac{\sec(1/x)}{x^2}\,dx$

37 $\displaystyle\int (\cot 9x + \csc 9x)\,dx$

38 $\displaystyle\int x\csc^2(3x^2+4)\,dx$

39 $\displaystyle\int e^x \tan e^x\,dx$

40 $\displaystyle\int \cot 2x \csc 2x\,dx$

41 $\displaystyle\int (\csc 3x + 1)^2\,dx$

42 $\displaystyle\int \frac{\sin x + 1}{\cos x}\,dx$

43 $\displaystyle\int \frac{\sin 4x}{\tan 4x}\,dx$

44 $\displaystyle\int x\cos x^2\,dx$

45 $\displaystyle\int \frac{x}{4+9x^2}\,dx$

46 $\displaystyle\int \frac{1}{4+9x^2}\,dx$

47 $\displaystyle\int \frac{e^{2x}}{\sqrt{1 - e^{2x}}}\,dx$

48 $\displaystyle\int \frac{e^x}{\sqrt{1 - e^{2x}}}\,dx$

49 $\displaystyle\int \frac{x}{\operatorname{sech} x^2}\,dx$

50 $\displaystyle\int \frac{1}{x\sqrt{x^4 - 1}}\,dx$

51 $\displaystyle\int_{-1/2}^{1/2} \frac{1}{\sqrt{1 - x^2}}\,dx$

52 $\displaystyle\int_0^{\pi/2} \frac{\cos x}{1 + \sin^2 x}\,dx$

53 $\displaystyle\int \sec^2 x\,(1 + \tan x)^2\,dx$

54 $\displaystyle\int (1 + \cos^2 x)\sin x\,dx$

55 $\displaystyle\int \frac{\csc^2 x}{2 + \cot x}\,dx$

56 $\displaystyle\int (\sin x)e^{\cos x + 1}\,dx$

57 $\displaystyle\int \frac{\sinh\,(\ln x)}{x}\,dx$

58 $\displaystyle\int \operatorname{sech}^2(1 - 2x)\,dx$

59 $\displaystyle\int \frac{1}{\sqrt{9 - 4x^2}}\,dx$

60 $\displaystyle\int \frac{x}{\sqrt{9 - 4x^2}}\,dx$

61 $\displaystyle\int \frac{1}{x\sqrt{9 - 4x^2}}\,dx$

62 $\displaystyle\int \frac{1}{x\sqrt{4x^2 - 9}}\,dx$

63 $\displaystyle\int \frac{x}{\sqrt{25x^2 + 36}}\,dx$

64 $\displaystyle\int \frac{1}{\sqrt{25x^2 + 36}}\,dx$

65 Find the points on the graph of $y = \sin^{-1} 3x$ at which the tangent line is parallel to the line through $A(2, -3)$ and $B(4, 7)$.

66 Find an equation of the tangent line to the graph of $y = \pi \tan\,(y/x) + x^2 - 16$ at the point $(4, \pi)$.

67 Find the local extrema of $f(x) = 8\sec x + \csc x$ on the interval $(0, \pi/2)$, and describe where $f(x)$ is increasing or decreasing on that interval.

68 Find the points of inflection and discuss the concavity of the graph of $y = x\sin^{-1} x$.

69 The region between the graphs of $y = \tan x$ and the x-axis, from $x = 0$ to $x = \pi/4$, is revolved about the x-axis. Find the volume of the resulting solid. (*Hint:* $\tan^2 x = \sec^2 x - 1$.)

70 Find the area of the region bounded by the graphs of $y = x/(1 + x^4)$, $x = 1$, and $y = 0$.

71 Damped oscillations are oscillations of decreasing magnitude that occur when frictional forces are considered (see Section 19.5). Shown in the figure is a computer-generated graph of the damped oscillation given by the equation $y = e^{-x/2}\sin 2x$.

EXERCISE 71

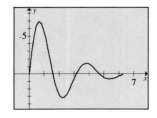

(a) Find the x-coordinates of the relative extrema for $0 \le x \le 2\pi$.

(b) Approximate the x-coordinates in part (a) to two decimal places.

72 Find the arc length of the graph of $y = \ln \tanh \frac{1}{2}x$ from $x = 1$ to $x = 2$.

73 A person on level ground observes the vertical ascent of a balloon that is released at a point $\frac{1}{2}$ km from the person. If the balloon rises at a constant rate of 2 m/sec, find the rate at which the angle of elevation of the observer's line of sight is changing at the instant the balloon is 100 meters above the level of the observer's eyes.

74 A square picture having sides 2 feet long is hung on a wall such that the base is 6 feet above the floor. A person whose eye level is 5 feet above the floor approaches the picture at a rate of 2 ft/sec. If θ is the angle between the line of sight and the top and bottom of the picture, find (a) the rate at which θ is changing when the person is 8 feet from the wall, and (b) the distance from the wall at which θ has its maximum value.

75 A stuntman jumps from a hot-air balloon that is hovering at a constant altitude, 100 feet above a lake. A television camera on shore, 200 feet from a point directly below the balloon, follows his descent (see figure). At what rate is the angle of elevation θ of the camera changing 2 seconds after the stuntman jumps? (Neglect the height of the camera.)

EXERCISE 75

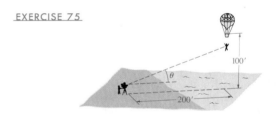

76 A man on a small island I, which is k miles from the closest point A on a straight shoreline, wishes to reach a camp that is d miles downshore from A (see figure). He plans to swim to some point P on shore and then walk the rest of the way. Suppose that he burns c_1 calories per mile while swimming and c_2 calories per mile while walking, and $c_1 > c_2$.

(a) Find a formula for the total number c of calories burned in completing the trip.

(b) For what angle AIP does c have a minimum value?

EXERCISE 76

In this chapter we shall discuss techniques that can be used to evaluate many different types of integrals. The formulas for integrals we have obtained previously are listed below. They will be used extensively throughout the chapter.

FORMULAS FOR INTEGRALS

$$\int u^n \, du = \frac{1}{n+1} u^{n+1} + C, \quad n \neq -1$$

$$\int \frac{1}{u} \, du = \ln |u| + C$$

$$\int e^u \, du = e^u + C$$

$$\int a^u \, du = \frac{1}{\ln a} a^u + C$$

$$\int \sin u \, du = -\cos u + C$$

$$\int \cos u \, du = \sin u + C$$

$$\int \tan u \, du = \ln |\sec u| + C$$

$$\int \cot u \, du = \ln |\sin u| + C$$

$$\int \sec u \, du = \ln |\sec u + \tan u| + C$$

$$\int \csc u \, du = \ln |\csc u - \cot u| + C$$

$$\int \sec u \tan u \, du = \sec u + C$$

$$\int \csc u \cot u \, du = -\csc u + C$$

$$\int \sec^2 u \, du = \tan u + C$$

$$\int \csc^2 u \, du = -\cot u + C$$

$$\int \frac{1}{\sqrt{a^2 - u^2}} \, du = \sin^{-1} \frac{u}{a} + C$$

$$\int \frac{1}{\sqrt{a^2 + u^2}} \, du = \sinh^{-1} \frac{u}{a} + C$$

$$\int \frac{1}{a^2 + u^2} \, du = \frac{1}{a} \tan^{-1} \frac{u}{a} + C$$

$$\int \frac{1}{a^2 - u^2} \, du = \frac{1}{a} \tanh^{-1} \frac{u}{a} + C$$

$$\int \frac{1}{u\sqrt{u^2 - a^2}} \, du = \frac{1}{a} \sec^{-1} \frac{u}{a} + C$$

$$\int \frac{1}{u\sqrt{a^2 - u^2}} \, du = -\frac{1}{a} \text{sech}^{-1} \frac{|u|}{a} + C$$

The following result is useful for simplifying certain types of integrals.

INTEGRATION BY PARTS FORMULA (9.1)

> If $u = f(x)$ and $v = g(x)$ and if f' and g' are continuous, then
> $$\int u \, dv = uv - \int v \, du$$

PROOF By the Product Rule,

$$D_x \left[f(x)g(x) \right] = f(x)g'(x) + g(x)f'(x)$$

or, equivalently,

$$f(x)g'(x) = D_x \left[f(x)g(x) \right] - g(x)f'(x).$$

Integrating both sides of the previous equation gives us

$$\int f(x)g'(x) \, dx = \int D_x \left[f(x)g(x) \right] dx - \int g(x)f'(x) \, dx.$$

By Theorem (5.30) (ii) the first integral on the right side equals $f(x)g(x) + C$. Since we obtain another constant of integration from the second integral, we may omit C in the formula, that is,

$$\int f(x)g'(x) \, dx = f(x)g(x) - \int g(x)f'(x) \, dx.$$

Since $du = f'(x) \, dx$ and $dv = g'(x) \, dx$, we may write the preceding formula as in (9.1). • •

When applying Formula (9.1) to an integral, we begin by letting one part of the integrand correspond to dv. The expression we choose for dv must include the differential dx. After selecting dv, we designate the remaining part of the integrand by u and then find du. Since this process involves splitting the integrand into two parts, the use of (9.1) is referred to as **integrating by parts**. A proper choice for dv is crucial. We usually choose the most complicated part of the integrand that can be readily integrated. The following examples illustrate this method of integration.

EXAMPLE 1 Evaluate $\int xe^{2x} \, dx$.

SOLUTION There are four possible choices for dv, namely dx, $x \, dx$, $e^{2x} \, dx$, or $xe^{2x} \, dx$. If we let $dv = e^{2x} \, dx$, then the remaining part of the integrand is u, that is, $u = x$. To find v we integrate dv, obtaining $v = \frac{1}{2}e^{2x}$. Note that a constant of integration is not added at this stage of the solution. (In Exercise 51 we are asked to prove that if a constant *is* added to v, the same result is obtained.) Since $u = x$, we see that $du = dx$. For ease of reference let us display these expressions as follows:

$$dv = e^{2x} \, dx \qquad u = x$$
$$v = \tfrac{1}{2}e^{2x} \qquad du = dx$$

Substituting these expressions in Formula (9.1), that is, *integrating by parts*, we obtain

$$\int xe^{2x} \, dx = x(\tfrac{1}{2}e^{2x}) - \int \tfrac{1}{2}e^{2x} \, dx.$$

We may find the integral on the right side by means of Theorem (7.22). This gives us

$$\int xe^{2x}\,dx = \tfrac{1}{2}xe^{2x} - \tfrac{1}{4}e^{2x} + C \quad \bullet$$

It takes considerable practice to become proficient in making a suitable choice for dv. To illustrate, if we had chosen $dv = x\,dx$ in Example 1, then it would have been necessary to let $u = e^{2x}$, giving us

$$dv = x\,dx \qquad u = e^{2x}$$
$$v = \tfrac{1}{2}x^2 \qquad du = 2e^{2x}\,dx.$$

Integrating by parts, we obtain

$$\int xe^{2x}\,dx = \tfrac{1}{2}x^2e^{2x} - \int x^2e^{2x}\,dx.$$

Since the exponent associated with x has increased, the integral on the right is more complicated than the given integral. This indicates that we have made an incorrect choice for dv.

EXAMPLE 2 Evaluate $\int x\sec^2 x\,dx$.

SOLUTION Since $\sec^2 x$ can be integrated readily, we let $dv = \sec^2 x\,dx$. The remaining part of the integrand is x and hence we must let $u = x$. Thus

$$dv = \sec^2 x\,dx \qquad u = x$$
$$v = \tan x \qquad du = dx$$

and integration by parts gives us

$$\int x\sec^2 x\,dx = x\tan x - \int \tan x\,dx$$
$$= x\tan x - \ln|\sec x| + C \quad \bullet$$

If, in Example 2, we had chosen $dv = x\,dx$ and $u = \sec^2 x$, then the integration by parts formula (9.1) would have led to a more complicated integral. (Verify this fact.)

In the next example we use integration by parts to find an antiderivative of the natural logarithmic function.

EXAMPLE 3 Evaluate $\int \ln x\,dx$.

SOLUTION Let

$$dv = dx \qquad u = \ln x$$
$$v = x \qquad du = \frac{1}{x}\,dx$$

and integrate by parts:

$$\int \ln x\,dx = (\ln x)x - \int (x)\frac{1}{x}\,dx$$
$$= x\ln x - \int dx$$
$$= x\ln x - x + C \quad \bullet$$

Sometimes it is necessary to use integration by parts more than once in the same problem. This is illustrated in the next example.

EXAMPLE 4 Evaluate $\int x^2 e^{2x} \, dx$.

SOLUTION Let

$$dv = e^{2x} \, dx \qquad u = x^2$$

$$v = \tfrac{1}{2} e^{2x} \qquad du = 2x \, dx$$

and integrate by parts:

$$\int x^2 e^{2x} \, dx = x^2(\tfrac{1}{2} e^{2x}) - \int (\tfrac{1}{2} e^{2x}) 2x \, dx$$

$$= \tfrac{1}{2} x^2 e^{2x} - \int x e^{2x} \, dx$$

To evaluate the integral on the right side of the last equation we must again integrate by parts. Proceeding exactly as in Example 1 leads to

$$\int x^2 e^{2x} \, dx = \tfrac{1}{2} x^2 e^{2x} - \tfrac{1}{2} x e^{2x} + \tfrac{1}{4} e^{2x} + C \quad \bullet$$

The following example illustrates another device for evaluating an integral by means of two applications of the integration by parts formula.

EXAMPLE 5 Evaluate $\int e^x \cos x \, dx$.

SOLUTION Let

$$dv = \cos x \, dx \qquad u = e^x$$

$$v = \sin x \qquad du = e^x \, dx$$

and integrate by parts:

(a)
$$\int e^x \cos x \, dx = e^x \sin x - \int (\sin x) e^x \, dx$$

$$= e^x \sin x - \int e^x \sin x \, dx$$

We next apply integration by parts to the integral on the right side of equation (a). Letting

$$dv = \sin x \, dx \qquad u = e^x$$

$$v = -\cos x \qquad du = e^x \, dx$$

and integrating by parts leads to

(b)
$$\int e^x \sin x \, dx = e^x(-\cos x) - \int (-\cos x) e^x \, dx$$

$$= -e^x \cos x + \int e^x \cos x \, dx.$$

If we now use equation (b) to substitute on the right side of equation (a) we obtain

$$\int e^x \cos x \, dx = e^x \sin x - \left[-e^x \cos x + \int e^x \cos x \, dx \right]$$

or
$$\int e^x \cos x \, dx = e^x \sin x + e^x \cos x - \int e^x \cos x \, dx.$$

Adding $\int e^x \cos x \, dx$ to both sides gives us

$$2 \int e^x \cos x \, dx = e^x (\sin x + \cos x).$$

Finally, dividing both sides by 2 and adding the constant of integration, we have

$$\int e^x \cos x \, dx = \tfrac{1}{2} e^x (\sin x + \cos x) + C \quad \bullet$$

We must choose substitutions carefully when evaluating an integral of the type given in Example 5. To illustrate, suppose that in the evaluation of the integral on the right in equation (a) of the solution we had used

$$dv = e^x \, dx \qquad u = \sin x$$
$$v = e^x \qquad du = \cos x \, dx$$

Integration by parts then leads to

$$\int e^x \sin x \, dx = (\sin x) e^x - \int e^x \cos x \, dx$$
$$= e^x \sin x - \int e^x \cos x \, dx.$$

If we now substitute in (a), we obtain

$$\int e^x \cos x \, dx = e^x \sin x - \left[e^x \sin x - \int e^x \cos x \, dx \right]$$

which reduces to

$$\int e^x \cos x \, dx = \int e^x \cos x \, dx.$$

Although this is a true statement, it is not a solution to the problem. Incidentally, the integral in Example 5 *can* be evaluated by using $dv = e^x \, dx$ for *both* the first and second applications of the integration by parts formula.

EXAMPLE 6 Evaluate $\int \sec^3 x \, dx$.

SOLUTION Let

$$dv = \sec^2 x \, dx \qquad u = \sec x$$
$$v = \tan x \qquad du = \sec x \tan x \, dx$$

and integrate by parts:

$$\int \sec^3 x \, dx = \sec x \tan x - \int \sec x \tan^2 x \, dx.$$

Instead of applying another integration by parts, let us change the form of the integral on the right by using the identity $1 + \tan^2 x = \sec^2 x$. This gives us

$$\int \sec^3 x \, dx = \sec x \tan x - \int \sec x \, (\sec^2 x - 1) \, dx$$

or

$$\int \sec^3 x \, dx = \sec x \tan x - \int \sec^3 x \, dx + \int \sec x \, dx.$$

Adding $\int \sec^3 x \, dx$ to both sides of the last equation gives us

$$2 \int \sec^3 x \, dx = \sec x \tan x + \int \sec x \, dx.$$

If we now evaluate $\int \sec x \, dx$ and divide both sides of the resulting equation by 2 (and then add the constant of integration), we obtain

$$\int \sec^3 x \, dx = \tfrac{1}{2} \sec x \tan x + \tfrac{1}{2} \ln |\sec x + \tan x| + C \quad \bullet$$

Integration by parts may sometimes be employed to obtain **reduction formulas** for integrals. We can use such formulas to write an integral involving powers of an expression in terms of integrals that involve lower powers of the expression.

EXAMPLE 7 Find a reduction formula for $\int \sin^n x \, dx$.

SOLUTION Let

$$dv = \sin x \, dx \qquad u = \sin^{n-1} x$$
$$v = -\cos x \qquad du = (n-1) \sin^{n-2} x \cos x \, dx$$

and integrate by parts:

$$\int \sin^n x \, dx = -\cos x \sin^{n-1} x + (n-1) \int \sin^{n-2} x \cos^2 x \, dx.$$

Since $\cos^2 x = 1 - \sin^2 x$, we may write

$$\int \sin^n x \, dx = -\cos x \sin^{n-1} x + (n-1) \int \sin^{n-2} x \, dx - (n-1) \int \sin^n x \, dx.$$

Consequently,

$$\int \sin^n x \, dx + (n-1) \int \sin^n x \, dx = -\cos x \sin^{n-1} x + (n-1) \int \sin^{n-2} x \, dx.$$

The left side of the last equation reduces to $n \int \sin^n x \, dx$. Dividing both sides by n, we obtain

$$\int \sin^n x \, dx = -\frac{1}{n} \cos x \sin^{n-1} x + \frac{n-1}{n} \int \sin^{n-2} x \, dx \quad \bullet$$

EXAMPLE 8 Use the reduction formula in Example 7 to evaluate $\int \sin^4 x \, dx$.

SOLUTION Using the formula with $n = 4$ gives us

$$\int \sin^4 x \, dx = -\tfrac{1}{4} \cos x \sin^3 x + \tfrac{3}{4} \int \sin^2 x \, dx.$$

Applying the reduction formula, with $n = 2$, to the integral on the right,

$$\int \sin^2 x \, dx = -\tfrac{1}{2} \cos x \sin x + \tfrac{1}{2} \int dx$$
$$= -\tfrac{1}{2} \cos x \sin x + \tfrac{1}{2} x + C.$$

Consequently,

$$\int \sin^4 x \, dx = -\tfrac{1}{4} \cos x \sin^3 x - \tfrac{3}{8} \cos x \sin x + \tfrac{3}{8} x + D$$

with $D = \tfrac{3}{4} C$. $\bullet$

It should be evident that by repeated applications of the formula in Example 7 we can find $\int \sin^n x \, dx$ for any positive integer n, because these reductions end with either $\int \sin x \, dx$ or $\int dx$, and each of these can be evaluated easily.

Exer. 1–38: Evaluate the integral.

1 $\int xe^{-x}\,dx$

2 $\int x \sin x\,dx$

3 $\int x^2 e^{3x}\,dx$

4 $\int x^2 \sin 4x\,dx$

5 $\int x \cos 5x\,dx$

6 $\int xe^{-2x}\,dx$

7 $\int x \sec x \tan x\,dx$

8 $\int x \csc^2 3x\,dx$

9 $\int x^2 \cos x\,dx$

10 $\int x^3 e^{-x}\,dx$

11 $\int \tan^{-1} x\,dx$

12 $\int \sin^{-1} x\,dx$

13 $\int \sqrt{x}\,\ln x\,dx$

14 $\int x^2 \ln x\,dx$

15 $\int x \csc^2 x\,dx$

16 $\int x \tan^{-1} x\,dx$

17 $\int e^{-x} \sin x\,dx$

18 $\int e^{3x} \cos 2x\,dx$

19 $\int \sin x \ln \cos x\,dx$

20 $\int_0^1 x^3 e^{-x^2}\,dx$

21 $\int \csc^3 x\,dx$

22 $\int \sec^5 x\,dx$

23 $\int_0^1 \dfrac{x^3}{\sqrt{x^2+1}}\,dx$

24 $\int \sin \ln x\,dx$

25 $\int_0^{\pi/2} x \sin 2x\,dx$

26 $\int_{\pi/6}^{\pi/4} x \sec^2 x\,dx$

27 $\int x(2x+3)^{99}\,dx$

28 $\int \dfrac{x^5}{\sqrt{1-x^3}}\,dx$

29 $\int e^{4x} \sin 5x\,dx$

30 $\int x^3 \cos(x^2)\,dx$

31 $\int (\ln x)^2\,dx$

32 $\int x\,2^x\,dx$

33 $\int x^3 \sinh x\,dx$

34 $\int (x+4) \cosh 4x\,dx$

35 $\int \cos \sqrt{x}\,dx$

36 $\int \cot^{-1} 3x\,dx$

37 $\int \cos^{-1} x\,dx$

38 $\int (x+1)^{10}(x+2)\,dx$

Exer. 39–42: Use integration by parts to derive the reduction formula.

39 $\int x^m e^x\,dx = x^m e^x - m \int x^{m-1} e^x\,dx$

40 $\int x^m \sin x\,dx = -x^m \cos x + m \int x^{m-1} \cos x\,dx$

41 $\int (\ln x)^m\,dx = x\,(\ln x)^m - m \int (\ln x)^{m-1}\,dx$

42 $\int \sec^m x\,dx = \dfrac{\sec^{m-2} x \tan x}{m-1} + \dfrac{m-2}{m-1} \int \sec^{m-2} x\,dx$
   for $m \ne 1$.

43 Use Exercise 39 to evaluate $\int x^5 e^x\,dx$.

44 Use Exercise 41 to evaluate $\int (\ln x)^4\,dx$.

45 If $f(x) = \sin \sqrt{x}$, find the area of the region under the graph of f from $x = 0$ to $x = \pi^2$.

46 The region between the graph of $y = x\sqrt{\sin x}$ and the x-axis from $x = 0$ to $x = \pi/2$ is revolved about the x-axis. Find the volume of the resulting solid.

47 The region bounded by the graphs of $y = \ln x$, $y = 0$, and $x = e$ is revolved about the y-axis. Find the volume of the resulting solid.

48 Suppose the force $f(x)$ acting at the point with coordinate x on a coordinate line l is given by $f(x) = x^5 \sqrt{x^3 + 1}$. Find the work done in moving an object from $x = 0$ to $x = 1$.

49 The ends of a water trough have the shape of the region between the graph of $y = \sin x$ and the x-axis from $x = \pi$ to $x = 2\pi$. If the trough is full of water, find the total force on one end.

50 The velocity (at time t) of a point moving along a coordinate line is t/e^{2t} ft/sec. If the point is at the origin at $t = 0$, find its position at time t.

51 When applying the Integration by Parts Formula (9.1), show that if, after choosing dv, we use $v + C$ in place of v, the same result is obtained.

52 In Chapter 6 the argument given for finding volumes by means of cylindrical shells (see Definition (6.7)) was incomplete because we did not show that the same result is obtained if the disk method is also applicable. Use integration by parts to prove that if f is differentiable and either $f'(x) > 0$ on $[a, b]$ or $f'(x) < 0$ on $[a, b]$, and if V is the volume of the solid obtained by revolving the region bounded by the graphs of f, $x = a$, and $x = b$ about the x-axis, then the same value of V is obtained using either the disk method or the shell method. (*Hint:* Let g be the inverse function of f, and use integration by parts on $\int_a^b \pi[f(x)]^2\,dx$.)

53 Discuss the following use of Formula (9.1).
   Given $\int (1/x)\,dx$, let $dv = dx$ and $u = 1/x$ so that $v = x$ and $du = (-1/x^2)\,dx$. Hence

$$\int \frac{1}{x}\,dx = \left(\frac{1}{x}\right)x - \int x\left(-\frac{1}{x^2}\right)dx$$

or

$$\int \frac{1}{x}\,dx = 1 + \int \frac{1}{x}\,dx.$$

Consequently, $0 = 1$.

54 If $u = f(x)$ and $v = g(x)$, prove that the analogue of Formula (9.1) for definite integrals is

$$\int_a^b u\,dv = uv\Big]_a^b - \int_a^b v\,du$$

for values a and b of x.

In Example 7 of Section 9.1 we obtained a reduction formula for $\int \sin^n x \, dx$. Integrals of this type may also be found without using integration by parts. If n is an odd positive integer, we begin by writing

$$\int \sin^n x \, dx = \int \sin^{n-1} x \sin x \, dx.$$

Since the integer $n - 1$ is even, we may then use the trigonometric identity $\sin^2 x = 1 - \cos^2 x$ to obtain a form that is easy to integrate, as illustrated in the following example.

EXAMPLE 1 Evaluate $\int \sin^5 x \, dx$.

SOLUTION As in the preceding discussion,

$$\int \sin^5 x \, dx = \int \sin^4 x \sin x \, dx$$

$$= \int (\sin^2 x)^2 \sin x \, dx$$

$$= \int (1 - \cos^2 x)^2 \sin x \, dx$$

$$= \int (1 - 2\cos^2 x + \cos^4 x) \sin x \, dx.$$

If we substitute

$$u = \cos x, \qquad du = -\sin x \, dx$$

we obtain

$$\int \sin^5 x \, dx = -\int (1 - 2\cos^2 x + \cos^4 x)(-\sin x) \, dx$$

$$= -\int (1 - 2u^2 + u^4) \, du$$

$$= -u + \tfrac{2}{3} u^3 - \tfrac{1}{5} u^5 + C$$

$$= -\cos x + \tfrac{2}{3} \cos^3 x - \tfrac{1}{5} \cos^5 x + C \qquad •$$

Similarly, for odd powers of $\cos x$ we write

$$\int \cos^n x \, dx = \int \cos^{n-1} x \cos x \, dx$$

and use the fact that $\cos^2 x = 1 - \sin^2 x$ to obtain an integrable form.

If the integrand is $\sin^n x$ or $\cos^n x$ and n is *even*, then the half-angle formulas

$$\sin^2 x = \frac{1 - \cos 2x}{2} \quad \text{or} \quad \cos^2 x = \frac{1 + \cos 2x}{2}$$

may be used to simplify the integrand.

EXAMPLE 2 Evaluate $\int \cos^2 x \, dx$.

SOLUTION Using a half-angle formula,

$$\int \cos^2 x \, dx = \tfrac{1}{2} \int (1 + \cos 2x) \, dx$$

$$= \tfrac{1}{2} x + \tfrac{1}{4} \sin 2x + C \qquad •$$

EXAMPLE 3 Evaluate $\int \sin^4 x \, dx$.

SOLUTION
$$\int \sin^4 x \, dx = \int (\sin^2 x)^2 \, dx$$
$$= \int \left(\frac{1 - \cos 2x}{2} \right)^2 dx$$
$$= \tfrac{1}{4} \int (1 - 2 \cos 2x + \cos^2 2x) \, dx.$$

We apply a half-angle formula again and write
$$\cos^2 2x = \tfrac{1}{2}(1 + \cos 4x) = \tfrac{1}{2} + \tfrac{1}{2} \cos 4x.$$

Substituting in the last integral and simplifying gives us
$$\int \sin^4 x \, dx = \tfrac{1}{4} \int (\tfrac{3}{2} - 2 \cos 2x + \tfrac{1}{2} \cos 4x) \, dx$$
$$= \tfrac{3}{8}x - \tfrac{1}{4} \sin 2x + \tfrac{1}{32} \sin 4x + C \quad \bullet$$

Integrals involving only products of $\sin x$ and $\cos x$ may be evaluated using the following guidelines.

GUIDELINES FOR EVALUATING (9.2)
INTEGRALS OF THE FORM
$\int \sin^m x \cos^n x \, dx$

1 If both m and n are even integers, use half-angle formulas for $\sin^2 x$ and $\cos^2 x$ to reduce the exponents by one-half.

2 If n is an odd integer, write the integral as
$$\int \sin^m x \cos^n x \, dx = \int \sin^m x \cos^{n-1} x \cos x \, dx$$
and express $\cos^{n-1} x$ in terms of $\sin x$ by using the trigonometric identity $\cos^2 x = 1 - \sin^2 x$. Use the substitution $u = \sin x$ to evaluate the resulting integral.

3 If m is an odd integer, write the integral as
$$\int \sin^m x \cos^n x \, dx = \int \sin^{m-1} x \cos^n x \sin x \, dx$$
and express $\sin^{m-1} x$ in terms of $\cos x$ by using the trigonometric identity $\sin^2 x = 1 - \cos^2 x$. Use the substitution $u = \cos x$ to evaluate the resulting integral.

EXAMPLE 4 Evaluate $\int \cos^3 x \sin^4 x \, dx$.

SOLUTION Using Guideline 2 of (9.2),
$$\int \cos^3 x \sin^4 x \, dx = \int \cos^2 x \sin^4 x \cos x \, dx$$
$$= \int (1 - \sin^2 x) \sin^4 x \cos x \, dx.$$

If we let $u = \sin x$, then $du = \cos x \, dx$, and the integral may be written
$$\int \cos^3 x \sin^4 x \, dx = \int (1 - u^2)u^4 \, du = \int (u^4 - u^6) \, du$$
$$= \tfrac{1}{5}u^5 - \tfrac{1}{7}u^7 + C$$
$$= \tfrac{1}{5} \sin^5 x - \tfrac{1}{7} \sin^7 x + C \quad \bullet$$

1 If n is an even integer, write the integral as

$$\int \tan^m x \sec^n x \, dx = \int \tan^m x \sec^{n-2} x \sec^2 x \, dx$$

and express $\sec^{n-2} x$ in terms of $\tan x$ by using the trigonometric identity $\sec^2 x = 1 + \tan^2 x$. Use the substitution $u = \tan x$ to evaluate the resulting integral.

2 If m is an odd integer, write the integral as

$$\int \tan^m x \sec^n x \, dx = \int \tan^{m-1} x \sec^{n-1} x \sec x \tan x \, dx.$$

Since $m - 1$ is even, $\tan^{m-1} x$ may be expressed in terms of $\sec x$ by means of the identity $\tan^2 x = \sec^2 x - 1$. Use the substitution $u = \sec x$ to evaluate the resulting integral.

3 If n is odd and m is even, then another method, such as integration by parts, should be used.

EXAMPLE 5 Evaluate $\int \tan^2 x \sec^4 x \, dx$.

SOLUTION Using Guideline 1 of (9.3),

$$\int \tan^2 x \sec^4 x \, dx = \int \tan^2 x \sec^2 x \sec^2 x \, dx$$

$$= \int \tan^2 x \, (\tan^2 x + 1) \sec^2 x \, dx.$$

If we let $u = \tan x$, then $du = \sec^2 x \, dx$, and

$$\int \tan^2 x \sec^4 x \, dx = \int u^2(u^2 + 1) \, du$$

$$= \int (u^4 + u^2) \, du$$

$$= \tfrac{1}{5}u^5 + \tfrac{1}{3}u^3 + C$$

$$= \tfrac{1}{5} \tan^5 x + \tfrac{1}{3} \tan^3 x + C \quad \bullet$$

EXAMPLE 6 Evaluate $\int \tan^3 x \sec^5 x \, dx$.

SOLUTION Using Guideline 2 of (9.3),

$$\int \tan^3 x \sec^5 x \, dx = \int \tan^2 x \sec^4 x \, (\sec x \tan x) \, dx$$

$$= \int (\sec^2 x - 1) \sec^4 x \, (\sec x \tan x) \, dx.$$

Substituting $u = \sec x$ and $du = \sec x \tan x \, dx$, we obtain

$$\int \tan^3 x \sec^5 x \, dx = \int (u^2 - 1)u^4 \, du$$

$$= \int (u^6 - u^4) \, du$$

$$= \tfrac{1}{7}u^7 - \tfrac{1}{5}u^5 + C$$

$$= \tfrac{1}{7} \sec^7 x - \tfrac{1}{5} \sec^5 x + C \quad \bullet$$

Integrals of the form $\int \cot^m x \csc^n x \, dx$ may be evaluated in similar fashion. Finally, integrals of the form $\int \sin mx \cos nx \, dx$ may be evaluated by means of the Product Formulas (1.38), as illustrated in the next example.

EXAMPLE 7 Evaluate $\int \cos 5x \cos 3x \, dx$.

SOLUTION Using the Product Formula (1.38) for $\cos u \cos v$, we obtain

$$\cos 5x \cos 3x = \tfrac{1}{2}(\cos 8x + \cos 2x).$$

Consequently,

$$\int \cos 5x \cos 3x \, dx = \tfrac{1}{2}\int(\cos 8x + \cos 2x) \, dx$$

$$= \tfrac{1}{16} \sin 8x + \tfrac{1}{4} \sin 2x + C \quad \bullet$$

EXERCISES 9.2

Exer. 1–30: Evaluate the integral.

1 $\int \cos^3 x \, dx$

2 $\int \sin^2 2x \, dx$

3 $\int \sin^2 x \cos^2 x \, dx$

4 $\int \cos^7 x \, dx$

5 $\int \sin^3 x \cos^2 x \, dx$

6 $\int \sin^5 x \cos^3 x \, dx$

7 $\int \sin^6 x \, dx$

8 $\int \sin^4 x \cos^2 x \, dx$

9 $\int \tan^3 x \sec^4 x \, dx$

10 $\int \sec^6 x \, dx$

11 $\int \tan^3 x \sec^3 x \, dx$

12 $\int \tan^5 x \sec x \, dx$

13 $\int \tan^6 x \, dx$

14 $\int \cot^4 x \, dx$

15 $\int \sqrt{\sin x} \cos^3 x \, dx$

16 $\int \dfrac{\cos^3 x}{\sqrt{\sin x}} \, dx$

17 $\int (\tan x + \cot x)^2 \, dx$

18 $\int \cot^3 x \csc^3 x \, dx$

19 $\int_0^{\pi/4} \sin^3 x \, dx$

20 $\int_0^1 \tan^2 (\tfrac{1}{4}\pi x) \, dx$

21 $\int \sin 5x \sin 3x \, dx$

22 $\int_0^{\pi/4} \cos x \cos 5x \, dx$

23 $\int_0^{\pi/2} \sin 3x \cos 2x \, dx$

24 $\int \sin 4x \cos 3x \, dx$

25 $\int \csc^4 x \cot^4 x \, dx$

26 $\int (1 + \sqrt{\cos x})^2 \sin x \, dx$

27 $\int \dfrac{\cos x}{2 - \sin x} \, dx$

28 $\int \dfrac{\tan^2 x - 1}{\sec^2 x} \, dx$

29 $\int \dfrac{\sec^2 x}{(1 + \tan x)^2} \, dx$

30 $\int \dfrac{\sec x}{\cot^5 x} \, dx$

31 The region bounded by the x-axis and the graph of $y = \cos^2 x$ from $x = 0$ to $x = 2\pi$ is revolved about the x-axis. Find the volume of the resulting solid.

32 The region between the graphs of $y = \tan^2 x$ and $y = 0$ from $x = 0$ to $x = \pi/4$ is revolved about the x-axis. Find the volume of the resulting solid.

33 The velocity (at time t) of a point moving on a coordinate line is $\cos^2 \pi t$ ft/sec. How far does the point travel in 5 seconds?

34 The acceleration (at time t) of a point moving along a coordinate line is $\sin^2 t \cos t$ ft/sec^2. At $t = 0$ the point is at the origin and its velocity is 10 ft/sec. Find its position at time t.

35 (a) Prove that if m and n are positive integers,

$$\int \sin mx \sin nx \, dx$$

$$= \begin{cases} \dfrac{\sin (m - n)x}{2(m - n)} - \dfrac{\sin (m + n)x}{2(m + n)} + C & \text{if } m \neq n \\[2ex] \dfrac{x}{2} - \dfrac{\sin 2mx}{4m} + C & \text{if } m = n \end{cases}$$

(b) Obtain formulas similar to that in part (a) for

$$\int \sin mx \cos nx \, dx$$

and

$$\int \cos mx \cos nx \, dx.$$

36 (a) Use part (a) of Exercise 35 to prove that

$$\int_{-\pi}^{\pi} \sin mx \sin nx \, dx = \begin{cases} 0 & \text{if } m \neq n \\ \pi & \text{if } m = n \end{cases}$$

(b) Find:

(i) $\int_{-\pi}^{\pi} \sin mx \cos nx \, dx$

(ii) $\int_{-\pi}^{\pi} \cos mx \cos nx \, dx$

If an integrand contains one of the expressions $\sqrt{a^2 - x^2}$, $\sqrt{a^2 + x^2}$, or $\sqrt{x^2 - a^2}$ for $a > 0$, we can eliminate the radical by using the trigonometric substitutions listed below.

TRIGONOMETRIC SUBSTITUTIONS (9.4)

Expression in integrand	Trigonometric substitution
$\sqrt{a^2 - x^2}$	$x = a \sin \theta$
$\sqrt{a^2 + x^2}$	$x = a \tan \theta$
$\sqrt{x^2 - a^2}$	$x = a \sec \theta$

When making a trigonometric substitution we shall assume that θ is in the range of the corresponding inverse trigonometric function. Thus, for the substitution $x = a \sin \theta$, we have $-\pi/2 \leq \theta \leq \pi/2$. In this case, $\cos \theta \geq 0$ and

$$\sqrt{a^2 - x^2} = \sqrt{a^2 - a^2 \sin^2 \theta}$$
$$= \sqrt{a^2(1 - \sin^2 \theta)}$$
$$= \sqrt{a^2 \cos^2 \theta}$$
$$= a \cos \theta.$$

If $\sqrt{a^2 - x^2}$ occurs in a denominator, we add the restriction $|x| \neq a$, or equivalently, $-\pi/2 < \theta < \pi/2$.

EXAMPLE 1 Evaluate $\int \dfrac{1}{x^2 \sqrt{16 - x^2}}\, dx$.

SOLUTION The integrand contains $\sqrt{16 - x^2}$, which is of the form $\sqrt{a^2 - x^2}$ with $a = 4$. Hence, by (9.4) we let

$$x = 4 \sin \theta \quad \text{for } -\pi/2 < \theta < \pi/2.$$

It follows that

$$\sqrt{16 - x^2} = \sqrt{16 - 16 \sin^2 \theta} = 4\sqrt{1 - \sin^2 \theta} = 4\sqrt{\cos^2 \theta} = 4 \cos \theta.$$

Since $x = 4 \sin \theta$, we have $dx = 4 \cos \theta\, d\theta$. Substituting in the given integral,

$$\int \frac{1}{x^2 \sqrt{16 - x^2}}\, dx = \int \frac{1}{(16 \sin^2 \theta) 4 \cos \theta}\, 4 \cos \theta\, d\theta$$

$$= \frac{1}{16} \int \frac{1}{\sin^2 \theta}\, d\theta$$

$$= \frac{1}{16} \int \csc^2 \theta\, d\theta$$

$$= -\frac{1}{16} \cot \theta + C.$$

It is now necessary to return to the original variable of integration, x. Since $\theta = \arcsin(x/4)$, we could write $-\frac{1}{16}\cot\theta$ as $-\frac{1}{16}\cot\arcsin(x/4)$. However, since the given integrand contains $\sqrt{16 - x^2}$, it is desirable to have the evaluated form also contain this radical. There is a simple geometric method for doing this. If $0 < \theta < \pi/2$ and $\sin\theta = x/4$, we may interpret θ as an acute angle of a right triangle having opposite side and hypotenuse of lengths x and 4, respectively (see Figure 9.1). By the Pythagorean Theorem, the length of the adjacent side is $\sqrt{16 - x^2}$. Referring to the triangle,

$$\cot\theta = \frac{\sqrt{16 - x^2}}{x}.$$

FIGURE 9.1

$$\frac{x}{4} = \sin\theta$$

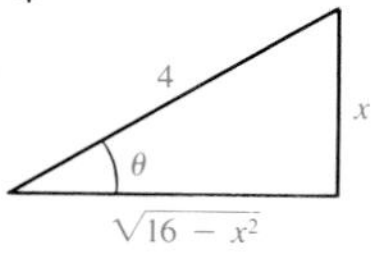

It can be shown that the last formula is also true if $-\pi/2 < \theta < 0$. Thus, Figure 9.1 may be used if θ is either positive or negative.

Substituting $\sqrt{16 - x^2}/x$ for $\cot\theta$ in our integral evaluation gives us

$$\int \frac{1}{x^2\sqrt{16 - x^2}}\, dx = -\frac{1}{16} \cdot \frac{\sqrt{16 - x^2}}{x} + C$$

$$= -\frac{\sqrt{16 - x^2}}{16x} + C \qquad \bullet$$

If an integrand contains $\sqrt{a^2 + x^2}$ for $a > 0$, then by (9.4) the substitution $x = a\tan\theta$ will eliminate the radical sign. When using this substitution we assume that θ is in the range of the inverse tangent function; that is, $-\pi/2 < \theta < \pi/2$. In this case, $\sec\theta > 0$ and

$$\sqrt{a^2 + x^2} = \sqrt{a^2 + a^2\tan^2\theta}$$

$$= \sqrt{a^2(1 + \tan^2\theta)}$$

$$= \sqrt{a^2\sec^2\theta}$$

$$= a\sec\theta.$$

After making this substitution and evaluating the resulting trigonometric integral, it is necessary to return to the variable x. The preceding formulas show that

FIGURE 9.2

$$\frac{x}{a} = \tan\theta$$

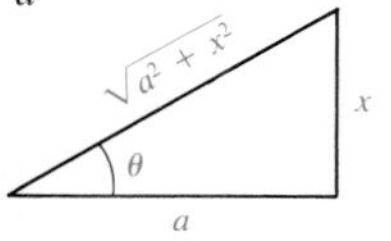

$$\tan\theta = \frac{x}{a} \quad \text{and} \quad \sec\theta = \frac{\sqrt{a^2 + x^2}}{a}.$$

As in the solution of Example 1, the trigonometric functions of θ can be found by referring to the triangle in Figure 9.2 if θ is positive or negative.

EXAMPLE 2 Evaluate $\displaystyle\int \frac{1}{\sqrt{4 + x^2}}\, dx$.

SOLUTION The integrand contains $\sqrt{4 + x^2}$, which is of the form $\sqrt{a^2 + x^2}$ with $a = 2$. Hence, by (9.4) we substitute as follows:

$$x = 2\tan\theta, \qquad dx = 2\sec^2\theta\, d\theta$$

Consequently,

$$\sqrt{4 + x^2} = \sqrt{4 + 4\tan^2\theta} = 2\sqrt{1 + \tan^2\theta} = 2\sqrt{\sec^2\theta} = 2\sec\theta$$

and

$$\int \frac{1}{\sqrt{4+x^2}}\, dx = \int \frac{1}{2 \sec \theta}\, 2 \sec^2 \theta\, d\theta$$

$$= \int \sec \theta\, d\theta$$

$$= \ln |\sec \theta + \tan \theta| + C.$$

FIGURE 9.3

$$\frac{x}{2} = \tan \theta$$

Since $\tan \theta = x/2$ we see from the triangle in Figure 9.3 that

$$\sec \theta = \frac{\sqrt{4+x^2}}{2}$$

and hence

$$\int \frac{1}{\sqrt{4+x^2}}\, dx = \ln \left| \frac{\sqrt{4+x^2}}{2} + \frac{x}{2} \right| + C.$$

The expression on the right may be written

$$\ln \left| \frac{\sqrt{4+x^2}+x}{2} \right| + C = \ln \left| \sqrt{4+x^2}+x \right| - \ln 2 + C.$$

Since $\sqrt{4+x^2}+x > 0$ for every x, the absolute value sign is unnecessary. If we also let $D = -\ln 2 + C$, then

$$\int \frac{1}{\sqrt{4+x^2}}\, dx = \ln \left(\sqrt{4+x^2}+x \right) + D \quad \bullet$$

If an integrand contains $\sqrt{x^2-a^2}$, then by (9.4) we substitute $x = a \sec \theta$ for θ chosen in the range of the inverse secant function; that is, either $0 \le \theta < \pi/2$ or $\pi \le \theta < 3\pi/2$. In this case, $\tan \theta \ge 0$ and

$$\sqrt{x^2-a^2} = \sqrt{a^2 \sec^2 \theta - a^2}$$

$$= \sqrt{a^2 (\sec^2 \theta - 1)}$$

$$= \sqrt{a^2 \tan^2 \theta}$$

$$= a \tan \theta.$$

FIGURE 9.4

$$\frac{x}{a} = \sec \theta$$

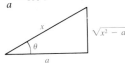

Since $\qquad \sec \theta = \dfrac{x}{a} \quad$ and $\quad \tan \theta = \dfrac{\sqrt{x^2-a^2}}{a}$

it follows that we may refer to the triangle in Figure 9.4 when changing from the variable θ to the variable x.

EXAMPLE 3 Evaluate $\displaystyle\int \frac{\sqrt{x^2-9}}{x}\, dx$.

SOLUTION   The integrand contains $\sqrt{x^2-9}$, which is of the form $\sqrt{x^2-a^2}$ with $a = 3$. Referring to (9.4), we substitute as follows:

$$x = 3 \sec \theta, \qquad dx = 3 \sec \theta \tan \theta\, d\theta$$

Consequently,

$$\sqrt{x^2-9} = \sqrt{9 \sec^2 \theta - 9} = 3\sqrt{\sec^2 \theta - 1} = 3\sqrt{\tan^2 \theta} = 3 \tan \theta$$

and, therefore,

$$\int \frac{\sqrt{x^2 - 9}}{x}\, dx = \int \frac{3 \tan \theta}{3 \sec \theta}\, 3 \sec \theta \tan \theta\, d\theta$$

$$= 3 \int \tan^2 \theta\, d\theta$$

$$= 3 \int (\sec^2 \theta - 1)\, d\theta = 3 \int \sec^2 \theta\, d\theta - 3 \int d\theta$$

$$= 3 \tan \theta - 3\theta + C.$$

Since $\sec \theta = x/3$, we have $\theta = \sec^{-1}(x/3)$. Referring to the triangle in Figure 9.5, we also see that $\tan \theta = \sqrt{x^2 - 9}/3$. Hence

$$\int \frac{\sqrt{x^2 - 9}}{x}\, dx = 3 \frac{\sqrt{x^2 - 9}}{3} - 3 \sec^{-1}\left(\frac{x}{3}\right) + C$$

$$= \sqrt{x^2 - 9} - 3 \sec^{-1}\left(\frac{x}{3}\right) + C \quad \bullet$$

FIGURE 9.5

$$\frac{x}{3} = \sec \theta$$

Hyperbolic functions may also be used to simplify certain integrations. For example, $\cosh^2 u = 1 + \sinh^2 u$ and hence, if an integrand contains the expression $\sqrt{a^2 + x^2}$, the substitution $x = a \sinh u$ leads to

$$\sqrt{a^2 + x^2} = \sqrt{a^2 + a^2 \sinh^2 u}$$

$$= \sqrt{a^2(1 + \sinh^2 u)}$$

$$= \sqrt{a^2 \cosh^2 u}$$

$$= a \cosh u.$$

EXAMPLE 4 Evaluate $\int \dfrac{1}{\sqrt{4 + x^2}}\, dx$ by using hyperbolic functions.

SOLUTION The integral is the same as that considered in Example 2. Let us make the hyperbolic substitution

$$x = 2 \sinh u, \qquad dx = 2 \cosh u\, du.$$

It follows that

$$\sqrt{4 + x^2} = \sqrt{4 + 4 \sinh^2 u} = \sqrt{4 \cosh^2 u} = 2 \cosh u$$

and hence

$$\int \frac{1}{\sqrt{4 + x^2}}\, dx = \int \frac{1}{2 \cosh u}\, 2 \cosh u\, du$$

$$= \int du = u + C$$

$$= \sinh^{-1}(x/2) + C.$$

The fact that this is equivalent to the solution in Example 2 follows from Theorem (8.15). $\bullet$

Although we now have additional integration techniques available, it is a good idea to keep earlier methods in mind. For example, the integral $\int (x/\sqrt{9 + x^2})\, dx$ could be evaluated by means of the trigonometric substitution $x = 3 \tan \theta$. However, it is simpler to use the algebraic substitution $u = 9 + x^2$ and $du = 2x\, dx$, for in this event the integral takes on the form

$\frac{1}{2} \int u^{-1/2} \, du$, which is readily integrated by means of the Power Rule. The following exercises include integrals that can be evaluated using simpler techniques than trigonometric substitutions.

EXERCISES 9.3

Exer. 1–22: Evaluate the integral.

1 $\displaystyle \int \frac{x^2}{\sqrt{4 - x^2}} \, dx$

2 $\displaystyle \int \frac{\sqrt{4 - x^2}}{x^2} \, dx$

3 $\displaystyle \int \frac{1}{x\sqrt{9 + x^2}} \, dx$

4 $\displaystyle \int \frac{1}{x^2 \sqrt{x^2 + 9}} \, dx$

5 $\displaystyle \int \frac{1}{x^2 \sqrt{x^2 - 25}} \, dx$

6 $\displaystyle \int \frac{1}{x^3 \sqrt{x^2 - 25}} \, dx$

7 $\displaystyle \int \frac{x}{\sqrt{4 - x^2}} \, dx$

8 $\displaystyle \int \frac{x}{x^2 + 9} \, dx$

9 $\displaystyle \int \frac{1}{(x^2 - 1)^{3/2}} \, dx$

10 $\displaystyle \int \frac{1}{\sqrt{4x^2 - 25}} \, dx$

11 $\displaystyle \int \frac{1}{(36 + x^2)^2} \, dx$

12 $\displaystyle \int \frac{1}{(16 - x^2)^{5/2}} \, dx$

13 $\displaystyle \int \sqrt{9 - 4x^2} \, dx$

14 $\displaystyle \int \frac{\sqrt{x^2 + 1}}{x} \, dx$

15 $\displaystyle \int \frac{x}{(16 - x^2)^2} \, dx$

16 $\displaystyle \int x\sqrt{x^2 - 9} \, dx$

17 $\displaystyle \int \frac{x^3}{\sqrt{9x^2 + 49}} \, dx$

18 $\displaystyle \int \frac{1}{x\sqrt{25x^2 + 16}} \, dx$

19 $\displaystyle \int \frac{1}{x^4 \sqrt{x^2 - 3}} \, dx$

20 $\displaystyle \int \frac{x^2}{(1 - 9x^2)^{3/2}} \, dx$

21 $\displaystyle \int \frac{(4 + x^2)^2}{x^3} \, dx$

22 $\displaystyle \int \frac{3x - 5}{\sqrt{1 - x^2}} \, dx$

23–28 Use trigonometric substitutions to obtain the formulas in Theorems (8.7) and (8.8).

29 The region bounded by the graphs of $y = x(x^2 + 25)^{-1/2}$, $y = 0$, and $x = 5$ is revolved about the y-axis. Find the volume of the resulting solid.

30 Find the area of the region bounded by the graph of $y = x^3(10 - x^2)^{-1/2}$, the x-axis, and the line $x = 1$.

31 Find the arc length of the graph of $y = \frac{1}{2}x^2$ from $A(0, 0)$ to $B(2, 2)$.

32 The region bounded by the graphs of $y = 1/(x^2 + 1)$, $y = 0$, $x = 0$, and $x = 2$ is revolved about the x-axis. Find the volume of the resulting solid.

33 Find the area of the region bounded by the graphs of the equations $y = \sqrt{x^2 + 4}$, $y = 0$, $x = 0$, and $x = 3$.

34 Find the area of the region enclosed by the graph of $4x^2 + y^2 = 16$.

35 Suppose that $y = f(x)$ and $x \, dy - \sqrt{x^2 - 16} \, dx = 0$. If $f(4) = 0$, find $f(x)$.

36 Suppose two variables x and y are related such that $\sqrt{1 - x^2} \, dy = x^3 \, dx$. If $y = 0$ when $x = 0$, express y as a function of x.

Exer. 37–40: Evaluate the integral by means of a hyperbolic substitution.

37 $\displaystyle \int \frac{1}{x^2 \sqrt{25 + x^2}} \, dx$

38 $\displaystyle \int \frac{x^2}{(x^2 + 9)^{3/2}} \, dx$

39 $\displaystyle \int \frac{1}{x^2 \sqrt{1 - x^2}} \, dx$ (*Hint:* Let $x = \tanh u$ and use (8.12).)

40 $\displaystyle \int \frac{1}{16 - x^2} \, dx$

Exer. 41–46: Use a trigonometric substitution to establish the indicated formula in the Table of Integrals (see Appendix IV).

41 Formula 21

42 Formula 27

43 Formula 31

44 Formula 36

45 Formula 41

46 Formula 44

9.4 INTEGRALS OF RATIONAL FUNCTIONS

Recall that if q is a rational function, then $q(x) = f(x)/g(x)$ for polynomials $f(x)$ and $g(x)$. In this section we shall state rules for evaluating $\int q(x) \, dx$.

Let us consider the specific example $q(x) = 2/(x^2 - 1)$. It is easy to verify that

$$\frac{2}{x^2 - 1} = \frac{1}{x - 1} + \frac{-1}{x + 1}.$$

The expression on the right side of the equation is called the *partial fraction decomposition* of $2/(x^2 - 1)$. To find $\int [2/(x^2 - 1)]\, dx$ we merely integrate each of the fractions that make up the decomposition, obtaining

$$\int \frac{2}{x^2 - 1}\, dx = \int \frac{1}{x - 1}\, dx + \int \frac{-1}{x + 1}\, dx$$

$$= \ln |x - 1| - \ln |x + 1| + C$$

$$= \ln \left| \frac{x - 1}{x + 1} \right| + C.$$

It is theoretically possible to write *any* rational expression $f(x)/g(x)$ as a sum of rational expressions whose denominators involve powers of polynomials of degree not greater than two. Specifically, if $f(x)$ and $g(x)$ are polynomials *and the degree of $f(x)$ is less than the degree of $g(x)$*, then we can prove that

$$\frac{f(x)}{g(x)} = F_1 + F_2 + \cdots + F_r$$

such that each term F_k of the sum has one of the forms

$$\frac{A}{(px + q)^m} \quad \text{or} \quad \frac{Ax + B}{(ax^2 + bx + c)^n}$$

for some real numbers A and B, nonnegative integers m and n, and where $ax^2 + bx + c$ is **irreducible,** in the sense that this quadratic polynomial has no real zeros, that is, $b^2 - 4ac < 0$. In this case, $ax^2 + bx + c$ cannot be expressed as a product of two first-degree polynomials.

The sum $F_1 + F_2 + \cdots + F_r$ is called the **partial fraction decomposition** of $f(x)/g(x)$, and each F_k is called a **partial fraction.** We shall not prove this algebraic result but will, instead, state guidelines for obtaining the decomposition.

The guidelines for finding the partial fraction decomposition of $f(x)/g(x)$ should be used only if $f(x)$ has lower degree than $g(x)$. If this is not the case, then we may use long division to arrive at the proper form. For example, given

$$\frac{x^3 - 6x^2 + 5x - 3}{x^2 - 1}$$

we obtain, by long division,

$$\frac{x^3 - 6x^2 + 5x - 3}{x^2 - 1} = x - 6 + \frac{6x - 9}{x^2 - 1}.$$

We then find the partial fraction decomposition for $(6x - 9)/(x^2 - 1)$.

1 If the degree of $f(x)$ is not lower than the degree $g(x)$, use long division to obtain the proper form.

2 Express $g(x)$ as a product of linear factors $px + q$ or irreducible quadratic factors $ax^2 + bx + c$, and collect repeated factors so that $g(x)$ is a product of *different* factors of the form $(px + q)^m$ or $(ax^2 + bx + c)^n$ for nonnegative integers m and n.

3 Apply the following rules.

RULE (a) For each factor of the form $(px + q)^m$ with $m \geq 1$, the partial fraction decomposition contains a sum of m partial fractions of the form

$$\frac{A_1}{px + q} + \frac{A_2}{(px + q)^2} + \cdots + \frac{A_m}{(px + q)^m}$$

such that each numerator A_k is a real number.

RULE (b) For each factor of the form $(ax^2 + bx + c)^n$ with $n \geq 1$ and $ax^2 + bx + c$ irreducible, the partial fraction decomposition contains a sum of n partial fractions of the form

$$\frac{A_1 x + B_1}{ax^2 + bx + c} + \frac{A_2 x + B_2}{(ax^2 + bx + c)^2} + \cdots + \frac{A_n x + B_n}{(ax^2 + bx + c)^n}$$

such that each A_k and B_k is a real number.

EXAMPLE 1 Evaluate $\displaystyle\int \frac{4x^2 + 13x - 9}{x^3 + 2x^2 - 3x} \, dx$.

SOLUTION The denominator of the integrand has the factored form $x(x + 3)(x - 1)$. Each factor has the form stated in Rule (a) of (9.5), with $m = 1$. Thus, for the factor x there corresponds a partial fraction of the form A/x. Similarly, for the factors $x + 3$ and $x - 1$ there correspond partial fractions $B/(x + 3)$ and $C/(x - 1)$, respectively. Thus the partial fraction decomposition has the form

$$\frac{4x^2 + 13x - 9}{x(x + 3)(x - 1)} = \frac{A}{x} + \frac{B}{x + 3} + \frac{C}{x - 1}.$$

Multiplying by the lowest common denominator gives us

(∗) $4x^2 + 13x - 9 = A(x + 3)(x - 1) + Bx(x - 1) + Cx(x + 3),$

where we have used the symbol (∗) for later reference. In a case such as this, in which the factors are all linear and nonrepeated, the values for A, B, and C can be found by substituting values for x that make the various factors zero. If we let $x = 0$ in (∗), then

$$-9 = -3A \quad \text{or} \quad A = 3.$$

Letting $x = 1$ in (∗) gives us

$$8 = 4C \quad \text{or} \quad C = 2.$$

Finally, if $x = -3$, then

$$-12 = 12B \quad \text{or} \quad B = -1.$$

The partial fraction decomposition is, therefore,

$$\frac{4x^2 + 13x - 9}{x(x + 3)(x - 1)} = \frac{3}{x} + \frac{-1}{x + 3} + \frac{2}{x - 1}.$$

Integrating, and letting K denote the sum of the constants of integration,

$$\int \frac{4x^2 + 13x - 9}{x(x + 3)(x - 1)}\, dx = \int \frac{3}{x}\, dx + \int \frac{-1}{x + 3}\, dx + \int \frac{2}{x - 1}\, dx$$

$$= 3 \ln |x| - \ln |x + 3| + 2 \ln |x - 1| + K$$

$$= \ln |x^3| - \ln |x + 3| + \ln |x - 1|^2 + K$$

$$= \ln \left| \frac{x^3(x - 1)^2}{x + 3} \right| + K$$

Another technique for finding A, B, and C is to compare coefficients of x. If we expand the right-hand side of $(*)$ and collect like powers of x, then we obtain

$$4x^2 + 13x - 9 = (A + B + C)x^2 + (2A - B + 3C)x - 3A.$$

We now use the fact that if two polynomials are equal, then coefficients of like powers are the same. Thus

$$A + B + C = 4$$
$$2A - B + 3C = 13$$
$$-3A = -9$$

We may show that the solution of this system of equations is $A = 3$, $B = -1$, and $C = 2$. •

EXAMPLE 2 Evaluate $\displaystyle\int \frac{3x^3 - 18x^2 + 29x - 4}{(x + 1)(x - 2)^3}\, dx.$

__SOLUTION__ By Rule (a) of (9.5), there is a partial fraction of the form $A/(x + 1)$ corresponding to the factor $x + 1$ in the denominator of the integrand. For the factor $(x - 2)^3$ we apply Rule (a) (with $m = 3$), obtaining a sum of three partial fractions $B/(x - 2)$, $C/(x - 2)^2$, and $D/(x - 2)^3$. Consequently, the partial fraction decomposition has the form

$$\frac{3x^3 - 18x^2 + 29x - 4}{(x + 1)(x - 2)^3} = \frac{A}{x + 1} + \frac{B}{x - 2} + \frac{C}{(x - 2)^2} + \frac{D}{(x - 2)^3}.$$

Multiplying both sides by $(x + 1)(x - 2)^3$ gives us

$(*)$ $3x^3 - 18x^2 + 29x - 4$
$$= A(x - 2)^3 + B(x + 1)(x - 2)^2 + C(x + 1)(x - 2) + D(x + 1).$$

Two of the unknown constants may be determined easily. If we let $x = 2$ in $(*)$, then

$$24 - 72 + 58 - 4 = 3D, \quad 6 = 3D, \quad \text{and} \quad D = 2.$$

Similarly, letting $x = -1$ in $(*)$,

$$-3 - 18 - 29 - 4 = -27A, \quad -54 = -27A, \quad \text{and} \quad A = 2.$$

The remaining constants may be found by comparing coefficients. If we expand the right-hand side of (∗) and collect like powers of x, we see that the coefficient of x^3 is $A + B$. This must equal the coefficient of x^3 on the left, that is,

$$A + B = 3.$$

Since $A = 2$, it follows that $B = 3 - A = 3 - 2 = 1$. Finally, we compare the constant terms in (∗) by letting $x = 0$. This gives us

$$-4 = -8A + 4B - 2C + D.$$

Substituting the values we have found for A, B, and D leads to

$$-4 = -16 + 4 - 2C + 2$$

which has the solution $C = -3$. The partial fraction decomposition is, therefore,

$$\frac{3x^3 - 18x^2 + 29x - 4}{(x + 1)(x - 2)^3} = \frac{2}{x + 1} + \frac{1}{x - 2} + \frac{-3}{(x - 2)^2} + \frac{2}{(x - 2)^3}.$$

To find the given integral we integrate each of the partial fractions on the right side of the last equation, obtaining

$$2 \ln |x + 1| + \ln |x - 2| + \frac{3}{x - 2} - \frac{1}{(x - 2)^2} + K$$

with K the sum of the four constants of integration. This may be written in the form

$$\ln \left[(x + 1)^2 |x - 2| \right] + \frac{3x - 7}{(x - 2)^2} + K \quad \bullet$$

EXAMPLE 3 Evaluate $\displaystyle\int \frac{x^2 - x - 21}{2x^3 - x^2 + 8x - 4} \, dx$.

SOLUTION The denominator may be factored by grouping as follows:

$$2x^3 - x^2 + 8x - 4 = x^2(2x - 1) + 4(2x - 1) = (x^2 + 4)(2x - 1).$$

Applying Rule (b) of (9.5) to the irreducible quadratic factor $x^2 + 4$, we see that one of the partial fractions has the form $(Ax + B)/(x^2 + 4)$. By Rule (a), there is also a partial fraction $C/(2x - 1)$ corresponding to the factor $2x - 1$. Consequently,

$$\frac{x^2 - x - 21}{2x^3 - x^2 + 8x - 4} = \frac{Ax + B}{x^2 + 4} + \frac{C}{2x - 1}.$$

As in previous examples, this leads to

(∗) $$x^2 - x - 21 = (Ax + B)(2x - 1) + C(x^2 + 4).$$

Substituting $x = \frac{1}{2}$, we obtain $\frac{1}{4} - \frac{1}{2} - 21 = \frac{17}{4}C$, which has the solution $C = -5$. The remaining constants may be found by comparing coefficients. Rearranging the right side of (∗) gives us

$$x^2 - x - 21 = (2A + C)x^2 + (-A + 2B)x - B + 4C.$$

Comparing the coefficients of x^2 we see that $2A + C = 1$. Since $C = -5$, it follows that $2A = 6$, or $A = 3$. Similarly, comparing the constant terms,

$-B + 4C = -21$ and hence $-B - 20 = -21$, or $B = 1$. Thus the partial fraction decomposition of the integrand is

$$\frac{x^2 - x - 21}{2x^3 - x^2 + 8x - 4} = \frac{3x + 1}{x^2 + 4} + \frac{-5}{2x - 1}$$

$$= \frac{3x}{x^2 + 4} + \frac{1}{x^2 + 4} - \frac{5}{2x - 1}.$$

The given integral may now be found by integrating the right side of the last equation. This gives us

$$\frac{3}{2} \ln (x^2 + 4) + \frac{1}{2} \tan^{-1} \frac{x}{2} - \frac{5}{2} \ln |2x - 1| + K \bullet$$

EXAMPLE 4 Evaluate $\displaystyle\int \frac{5x^3 - 3x^2 + 7x - 3}{(x^2 + 1)^2} \, dx.$

SOLUTION   Applying Rule (b) of (9.5), with $n = 2$,

$$\frac{5x^3 - 3x^2 + 7x - 3}{(x^2 + 1)^2} = \frac{Ax + B}{x^2 + 1} + \frac{Cx + D}{(x^2 + 1)^2}$$

and, therefore,

$$5x^3 - 3x^2 + 7x - 3 = (Ax + B)(x^2 + 1) + Cx + D$$

or $\quad 5x^3 - 3x^2 + 7x - 3 = Ax^3 + Bx^2 + (A + C)x + (B + D).$

Comparing the coefficients of x^3 and x^2, we obtain $A = 5$ and $B = -3$. From the coefficients of x we see that $A + C = 7$, or $C = 7 - A = 7 - 5 = 2$. Finally, the constant terms give us $B + D = -3$, or $D = -3 - B = -3 - (-3) = 0$. Therefore,

$$\frac{5x^3 - 3x^2 + 7x - 3}{(x^2 + 1)^2} = \frac{5x - 3}{x^2 + 1} + \frac{2x}{(x^2 + 1)^2}$$

$$= \frac{5x}{x^2 + 1} - \frac{3}{x^2 + 1} + \frac{2x}{(x^2 + 1)^2}.$$

Integrating, we obtain

$$\int \frac{5x^3 - 3x^2 + 7x - 3}{(x^2 + 1)^2} \, dx = \frac{5}{2} \ln (x^2 + 1) - 3 \tan^{-1} x - \frac{1}{x^2 + 1} + K \bullet$$

EXERCISES 9.4

Exer. 1–32: Evaluate the integral.

1 $\displaystyle\int \frac{5x - 12}{x(x - 4)} \, dx$

2 $\displaystyle\int \frac{x + 34}{(x - 6)(x + 2)} \, dx$

3 $\displaystyle\int \frac{37 - 11x}{(x + 1)(x - 2)(x - 3)} \, dx$

4 $\displaystyle\int \frac{4x^2 + 54x + 134}{(x - 1)(x + 5)(x + 3)} \, dx$

5 $\displaystyle\int \frac{6x - 11}{(x - 1)^2} \, dx$

6 $\displaystyle\int \frac{-19x^2 + 50x - 25}{x^2(3x - 5)} \, dx$

7 $\displaystyle\int \frac{x + 16}{x^2 + 2x - 8} \, dx$

8 $\displaystyle\int \frac{11x + 2}{2x^2 - 5x - 3}\, dx$

9 $\displaystyle\int \frac{5x^2 - 10x - 8}{x^3 - 4x}\, dx$

10 $\displaystyle\int \frac{4x^2 - 5x - 15}{x^3 - 4x^2 - 5x}\, dx$

11 $\displaystyle\int \frac{2x^2 - 25x - 33}{(x + 1)^2(x - 5)}\, dx$

12 $\displaystyle\int \frac{2x^2 - 12x + 4}{x^3 - 4x^2}\, dx$

13 $\displaystyle\int \frac{9x^4 + 17x^3 + 3x^2 - 8x + 3}{x^5 + 3x^4}\, dx$

14 $\displaystyle\int \frac{5x^2 + 30x + 43}{(x + 3)^3}\, dx$

15 $\displaystyle\int \frac{x^3 + 3x^2 + 3x + 63}{(x^2 - 9)^2}\, dx$

16 $\displaystyle\int \frac{1}{(x - 7)^5}\, dx$

17 $\displaystyle\int \frac{5x^2 + 11x + 17}{x^3 - 5x^2 + 4x + 20}\, dx$

18 $\displaystyle\int \frac{4x^3 - 3x^2 + 6x - 27}{x^4 + 9x^2}\, dx$

19 $\displaystyle\int \frac{x^2 + 3x + 1}{x^4 + 5x^2 + 4}\, dx$

20 $\displaystyle\int \frac{4x}{(x^2 + 1)^3}\, dx$

21 $\displaystyle\int \frac{2x^3 + 10x}{(x^2 + 1)^2}\, dx$

22 $\displaystyle\int \frac{x^4 + 2x^2 + 4x + 1}{(x^2 + 1)^3}\, dx$

23 $\displaystyle\int \frac{x^3 + 3x - 2}{x^2 - x}\, dx$

24 $\displaystyle\int \frac{x^4 + 2x^2 + 3}{x^3 - 4x}\, dx$

25 $\displaystyle\int \frac{x^6 - x^3 + 1}{x^4 + 9x^2}\, dx$

26 $\displaystyle\int \frac{x^5}{(x^2 + 4)^2}\, dx$

27 $\displaystyle\int \frac{2x^3 - 5x^2 + 46x + 98}{(x^2 + x - 12)^2}\, dx$

28 $\displaystyle\int \frac{-2x^4 - 3x^3 - 3x^2 + 3x + 1}{x^2(x + 1)^3}\, dx$

29 $\displaystyle\int \frac{4x^3 + 2x^2 - 5x - 18}{(x - 4)(x + 1)^3}\, dx$

30 $\displaystyle\int \frac{10x^2 + 9x + 1}{2x^3 + 3x^2 + x}\, dx$

31 $\displaystyle\int \frac{2x^4 - 2x^3 + 6x^2 - 5x + 1}{x^3 - x^2 + x - 1}\, dx$

32 $\displaystyle\int \frac{x^5 - x^4 - 2x^3 + 4x^2 - 15x + 5}{(x^2 + 1)^2(x^2 + 4)}\, dx$

Exer. 33–36: Use partial fractions to evaluate the integral (see formulas 19, 49, 50, and 52 of the Table of Integrals in Appendix IV).

33 $\displaystyle\int \frac{1}{a^2 - u^2}\, du$

34 $\displaystyle\int \frac{1}{u(a + bu)}\, du$

35 $\displaystyle\int \frac{1}{u^2(a + bu)}\, du$

36 $\displaystyle\int \frac{1}{u(a + bu)^2}\, du$

37 If $f(x) = x/(x^2 - 2x - 3)$, find the area of the region under the graph of f from $x = 0$ to $x = 2$.

38 The region bounded by the graphs of $y = 1/(x - 1)(4 - x)$, $y = 0$, $x = 2$, and $x = 3$ is revolved about the y-axis. Find the volume of the resulting solid.

39 If the region described in Exercise 38 is revolved about the x-axis, find the volume of the resulting solid.

40 The velocity (at time t) of a point moving along a coordinate line is $(t + 3)/(t^3 + t)$ ft/sec. How far does the point travel during the time interval $[1, 2]$?

41 As an alternative to partial fractions, show that an integral of the form

$$\int \frac{1}{ax^2 + bx}\, dx$$

may be evaluated by writing it as

$$\int \frac{(1/x^2)}{a + (b/x)}\, dx$$

and using the substitution $u = a + (b/x)$.

42 Generalize Exercise 41 to integrals of the form

$$\int \frac{1}{ax^n + bx}\, dx.$$

43 Suppose $g(x) = (x - c_1)(x - c_2) \cdots (x - c_n)$ for a positive integer n and distinct real numbers $c_1, c_2, \ldots, c_n$. If $f(x)$ is a polynomial of degree less than n, show that

$$\frac{f(x)}{g(x)} = \frac{A_1}{x - c_1} + \frac{A_2}{x - c_2} + \cdots + \frac{A_n}{x - c_n}$$

with $A_k = f(c_k)/g'(c_k)$ for $k = 1, 2, \ldots, n$. This is a method for finding the partial fraction decomposition if the denominator can be factored into distinct linear factors.)

44 Use Exercise 43 to find the partial fraction decomposition of

$$\frac{2x^4 - x^3 - 3x^2 + 5x + 7}{x^5 - 5x^3 + 4x}.$$

Partial fraction decompositions may lead to integrands containing an irreducible quadratic expression $ax^2 + bx + c$. If $b \neq 0$ it is sometimes necessary to complete the square as follows:

$$ax^2 + bx + c = a\left(x^2 + \frac{b}{a}x\right) + c$$

$$= a\left(x + \frac{b}{2a}\right)^2 + c - \frac{b^2}{4a}.$$

The substitution $u = x + (b/2a)$ may then lead to an integrable form.

EXAMPLE 1 Evaluate $\displaystyle\int \frac{2x - 1}{x^2 - 6x + 13}\, dx$.

SOLUTION Note that the quadratic expression $x^2 - 6x + 13$ is irreducible, since $b^2 - 4ac = 36 - 52 = -16 < 0$. We complete the square as follows:

$$x^2 - 6x + 13 = (x^2 - 6x \quad) + 13$$
$$= (x^2 - 6x + 9) + 13 - 9 = (x - 3)^2 + 4.$$

If we make the substitution

$$u = x - 3, \qquad x = u + 3, \qquad dx = du$$

then

$$\int \frac{2x - 1}{x^2 - 6x + 13}\, dx = \int \frac{2x - 1}{(x - 3)^2 + 4}\, dx$$

$$= \int \frac{2(u + 3) - 1}{u^2 + 4}\, du$$

$$= \int \frac{2u + 5}{u^2 + 4}\, du$$

$$= \int \frac{2u}{u^2 + 4}\, du + 5 \int \frac{1}{u^2 + 4}\, du$$

$$= \ln(u^2 + 4) + \frac{5}{2} \tan^{-1} \frac{u}{2} + C$$

$$= \ln(x^2 - 6x + 13) + \frac{5}{2} \tan^{-1} \frac{x - 3}{2} + C \quad \bullet$$

We may also employ the technique of completing the square if a quadratic expression appears under a radical sign.

EXAMPLE 2 Evaluate $\displaystyle\int \frac{1}{\sqrt{8 + 2x - x^2}}\, dx$.

SOLUTION We may complete the square for the quadratic expression $8 + 2x - x^2$ as follows:

$$8 + 2x - x^2 = 8 - (x^2 - 2x) = 8 + 1 - (x^2 - 2x + 1)$$
$$= 9 - (x - 1)^2$$

Letting

$$u = x - 1, \qquad du = dx,$$

we obtain

$$\int \frac{1}{\sqrt{8 + 2x - x^2}} \, dx = \int \frac{1}{\sqrt{9 - (x - 1)^2}} \, dx$$

$$= \int \frac{1}{\sqrt{9 - u^2}} \, du$$

$$= \sin^{-1} \frac{u}{3} + C$$

$$= \sin^{-1} \frac{x - 1}{3} + C \qquad \bullet$$

In the next example we shall make a trigonometric substitution after completing the square.

EXAMPLE 3 Evaluate $\int \dfrac{1}{\sqrt{x^2 + 8x + 25}} \, dx$.

SOLUTION We complete the square for the quadratic expression as follows:

$$x^2 + 8x + 25 = (x^2 + 8x \qquad) + 25$$
$$= (x^2 + 8x + 16) + 25 - 16$$
$$= (x + 4)^2 + 9.$$

Hence

$$\int \frac{1}{\sqrt{x^2 + 8x + 25}} \, dx = \int \frac{1}{\sqrt{(x + 4)^2 + 9}} \, dx.$$

If we next make the trigonometric substitution

$$x + 4 = 3 \tan \theta,$$

then

$$dx = 3 \sec^2 \theta \, d\theta$$

and

$$\sqrt{(x + 4)^2 + 9} = \sqrt{9 \tan^2 \theta + 9} = 3\sqrt{\tan^2 \theta + 1} = 3 \sec \theta.$$

Hence

$$\int \frac{1}{\sqrt{x^2 + 8x + 25}} \, dx = \int \frac{1}{3 \sec \theta} \, 3 \sec^2 \theta \, d\theta$$

$$= \int \sec \theta \, d\theta$$

$$= \ln \left| \sec \theta + \tan \theta \right| + C.$$

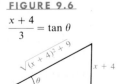

FIGURE 9.6

$$\frac{x + 4}{3} = \tan \theta$$

To return to the variable x, we use the triangle in Figure 9.6. This gives us

$$\int \frac{1}{\sqrt{x^2 + 8x + 25}} \, dx = \ln \left| \frac{\sqrt{x^2 + 8x + 25}}{3} + \frac{x + 4}{3} \right| + C$$

$$= \ln \left| \sqrt{x^2 + 8x + 25} + x + 4 \right| + K$$

with $K = C - \ln 3$.

Another way to evaluate the integral after completing the square is to use Theorem (8.17). This leads to

$$\int \frac{1}{\sqrt{(x + 4)^2 + 9}} \, dx = \sinh^{-1} \frac{x + 4}{3} + C \qquad \bullet$$

Exer. 1–18: Evaluate the integral.

1 $\int \dfrac{1}{x^2 - 4x + 8} \, dx$

2 $\int \dfrac{1}{\sqrt{7 + 6x - x^2}} \, dx$

3 $\int \dfrac{1}{\sqrt{4x - x^2}} \, dx$

4 $\int \dfrac{1}{x^2 - 2x + 2} \, dx$

5 $\int \dfrac{2x + 3}{\sqrt{9 - 8x - x^2}} \, dx$

6 $\int \dfrac{x + 5}{9x^2 + 6x + 17} \, dx$

7 $\int \dfrac{1}{x^3 - 1} \, dx$

8 $\int \dfrac{x^3}{x^3 - 1} \, dx$

9 $\int \dfrac{1}{(x^2 + 4x + 5)^2} \, dx$

10 $\int \dfrac{1}{x^4 - 4x^3 + 13x^2} \, dx$

11 $\int \dfrac{1}{(x^2 + 6x + 13)^{3/2}} \, dx$

12 $\int \sqrt{x(6 - x)} \, dx$

13 $\int \dfrac{1}{2x^2 - 3x + 9} \, dx$

14 $\int \dfrac{2x}{(x^2 + 2x + 5)^2} \, dx$

15 $\int_2^3 \dfrac{x^2 - 4x + 6}{x^2 - 4x + 5} \, dx$

16 $\int \dfrac{x}{2x^2 + 3x - 4} \, dx$

17 $\int \dfrac{e^x}{e^{2x} + 3e^x + 2} \, dx$

18 $\int_0^1 \dfrac{x - 1}{x^2 + x + 1} \, dx$

19 Find the area of the region bounded by the graphs of $y = (x^3 + 1)^{-1}$, $y = 0$, $x = 0$, and $x = 1$.

20 The region bounded by the graph of $y = 1/(x^2 + 2x + 10)$, the coordinate axes, and the line $x = 2$ is revolved about the x-axis. Find the volume of the resulting solid.

21 If the region described in Exercise 20 is revolved about the y-axis, find the volume of the resulting solid.

22 The velocity (at time t) of a point moving along a coordinate line is $(75 + 10t - t^2)^{-1/2}$ ft/sec. How far does the point travel during the time interval $[0, 5]$?

9.6 MISCELLANEOUS SUBSTITUTIONS

In this section we shall consider substitutions that are useful for evaluating certain types of integrals. The first example illustrates that if an integral contains an expression of the form $\sqrt[n]{f(x)}$, then one of the substitutions $u = \sqrt[n]{f(x)}$ or $u = f(x)$ may simplify the evaluation.

EXAMPLE 1 Evaluate $\int \dfrac{x^3}{\sqrt[3]{x^2 + 4}} \, dx$.

SOLUTION 1 The substitution $u = \sqrt[3]{x^2 + 4}$ leads to the following equivalent equations:

$$u = \sqrt[3]{x^2 + 4}, \qquad u^3 = x^2 + 4, \qquad x^2 = u^3 - 4.$$

Taking the differential of each side of the last equation, we obtain

$$2x \, dx = 3u^2 \, du \quad \text{or} \quad x \, dx = \tfrac{3}{2}u^2 \, du.$$

We now substitute in the given integral as follows:

$$\int \dfrac{x^3}{\sqrt[3]{x^2 + 4}} \, dx = \int \dfrac{x^2}{\sqrt[3]{x^2 + 4}} \cdot x \, dx$$

$$= \int \dfrac{u^3 - 4}{u} \cdot \dfrac{3}{2} u^2 \, du = \dfrac{3}{2} \int (u^4 - 4u) \, du$$

$$= \tfrac{3}{2}(\tfrac{1}{5}u^5 - 2u^2) + C = \tfrac{3}{10}u^2(u^3 - 10) + C$$

$$= \tfrac{3}{10}(x^2 + 4)^{2/3}(x^2 - 6) + C$$

SOLUTION 2 If we substitute u for the expression *underneath* the radical, then

$$u = x^2 + 4 \qquad x^2 = u - 4$$

$$2x \, dx = du \qquad x \, dx = \tfrac{1}{2} \, du.$$

In this case we may write

$$\int \frac{x^3}{\sqrt[3]{x^2 + 4}} \, dx = \int \frac{x^2}{\sqrt[3]{x^2 + 4}} \cdot x \, dx$$

$$= \int \frac{u - 4}{u^{1/3}} \cdot \frac{1}{2} \, du = \frac{1}{2} \int (u^{2/3} - 4u^{-1/3}) \, du$$

$$= \frac{1}{2} [\frac{3}{5}u^{5/3} - 6u^{2/3}] + C = \frac{3}{10}u^{2/3}[u - 10] + C$$

$$= \frac{3}{10}(x^2 + 4)^{2/3}(x^2 - 6) + C \qquad \bullet$$

EXAMPLE 2 Evaluate $\int \dfrac{1}{\sqrt{x} + \sqrt[3]{x}} \, dx$.

SOLUTION If we let $u = \sqrt[6]{x}$, then

$$x = u^6, \qquad \sqrt{x} = u^3, \qquad \sqrt[3]{x} = u^2, \qquad dx = 6u^5 \, du$$

and

$$\int \frac{1}{\sqrt{x} + \sqrt[3]{x}} \, dx = \int \frac{1}{u^3 + u^2} 6u^5 \, du = 6 \int \frac{u^3}{u + 1} \, du.$$

By long division,

$$\frac{u^3}{u + 1} = u^2 - u + 1 - \frac{1}{u + 1}.$$

Consequently,

$$\int \frac{1}{\sqrt{x} + \sqrt[3]{x}} \, dx = 6 \int \left[u^2 - u + 1 - \frac{1}{u + 1} \right] du$$

$$= 6(\tfrac{1}{3}u^3 - \tfrac{1}{2}u^2 + u - \ln|u + 1|) + C$$

$$= 2\sqrt{x} - 3\sqrt[3]{x} + 6\sqrt[6]{x} - 6 \ln(\sqrt[6]{x} + 1) + C \qquad \bullet$$

If an integrand is a rational expression in $\sin x$ and $\cos x$, then the substitution $u = \tan(x/2)$ for $-\pi < x < \pi$ will transform the integrand into a rational (algebraic) expression in u. To prove this, first note that

$$\cos \frac{x}{2} = \frac{1}{\sec(x/2)} = \frac{1}{\sqrt{1 + \tan^2(x/2)}} = \frac{1}{\sqrt{1 + u^2}},$$

$$\sin \frac{x}{2} = \tan \frac{x}{2} \cos \frac{x}{2} = u \frac{1}{\sqrt{1 + u^2}}.$$

Consequently,

$$\sin x = 2 \sin \frac{x}{2} \cos \frac{x}{2} = \frac{2u}{1 + u^2},$$

$$\cos x = 1 - 2 \sin^2 \frac{x}{2} = 1 - \frac{2u^2}{1 + u^2} = \frac{1 - u^2}{1 + u^2}.$$

Moreover, since $x/2 = \tan^{-1} u$, we have $x = 2 \tan^{-1} u$ and, therefore,

$$dx = \frac{2}{1 + u^2} \, du.$$

The following theorem summarizes this discussion.

THEOREM (9.6)

If an integrand is a rational expression in $\sin x$ and $\cos x$, the following substitutions will produce a rational expression in u:

$$\sin x = \frac{2u}{1 + u^2}, \qquad \cos x = \frac{1 - u^2}{1 + u^2}, \qquad dx = \frac{2}{1 + u^2}\, du$$

with $u = \tan(x/2)$.

EXAMPLE 3 Evaluate $\displaystyle\int \frac{1}{4 \sin x - 3 \cos x}\, dx$.

SOLUTION Applying Theorem (9.6) and simplifying the integrand,

$$\int \frac{1}{4 \sin x - 3 \cos x}\, dx = \int \frac{1}{4\left(\dfrac{2u}{1 + u^2}\right) - 3\left(\dfrac{1 - u^2}{1 + u^2}\right)} \cdot \frac{2}{1 + u^2}\, du$$

$$= \int \frac{2}{8u - 3(1 - u^2)}\, du$$

$$= 2 \int \frac{1}{3u^2 + 8u - 3}\, du$$

Using partial fractions,

$$\frac{1}{3u^2 + 8u - 3} = \frac{1}{10}\left(\frac{3}{3u - 1} - \frac{1}{u + 3}\right)$$

and hence

$$\int \frac{1}{4 \sin x - 3 \cos x}\, dx = \frac{1}{5} \int \left(\frac{3}{3u - 1} - \frac{1}{u + 3}\right) du$$

$$= \frac{1}{5}\left(\ln |3u - 1| - \ln |u + 3|\right) + C$$

$$= \frac{1}{5} \ln \left|\frac{3u - 1}{u + 3}\right| + C$$

$$= \frac{1}{5} \ln \left|\frac{3 \tan(x/2) - 1}{\tan(x/2) + 3}\right| + C \qquad \bullet$$

Other substitutions are sometimes useful; however, it is impossible to state rules that apply to all situations.

EXERCISES 9.6

Exer. 1–28: Evaluate the integral.

1 $\displaystyle\int x \sqrt[3]{x + 9}\, dx$

2 $\displaystyle\int x^2 \sqrt{2x + 1}\, dx$

3 $\displaystyle\int \frac{x}{\sqrt[5]{3x + 2}}\, dx$

4 $\displaystyle\int \frac{5x}{(x + 3)^{2/3}}\, dx$

5 $\displaystyle\int_4^9 \frac{1}{\sqrt{x} + 4}\, dx$

6 $\displaystyle\int_0^{25} \frac{1}{\sqrt{4 + \sqrt{x}}}\, dx$

7 $\displaystyle\int \frac{\sqrt{x}}{1 + \sqrt[3]{x}}\, dx$

8 $\displaystyle\int \frac{1}{\sqrt[4]{x} + \sqrt[3]{x}}\, dx$

9 $\int \dfrac{1}{(x+1)\sqrt{x-2}}\,dx$

10 $\int_0^4 \dfrac{2x+3}{\sqrt{1+2x}}\,dx$

11 $\int \dfrac{x+1}{(x+4)^{1/3}}\,dx$

12 $\int \dfrac{x^{1/3}+1}{x^{1/3}-1}\,dx$

13 $\int e^{3x}\sqrt{1+e^x}\,dx$

14 $\int \dfrac{e^{2x}}{\sqrt[3]{1+e^x}}\,dx$

15 $\int \dfrac{e^{2x}}{e^x+4}\,dx$

16 $\int \dfrac{\sin 2x}{\sqrt{1+\sin x}}\,dx$

17 $\int \sin\sqrt{x+4}\,dx$

18 $\int \sqrt{x}\,e^{\sqrt{x}}\,dx$

19 $\int_2^3 \dfrac{x}{(x-1)^6}\,dx$

20 $\int \dfrac{x^2}{(3x+4)^{10}}\,dx$

(*Hint:* Let $u = x - 1$.)

(*Hint:* Let $u = 3x + 4$.)

21 $\int \dfrac{1}{2+\sin x}\,dx$

22 $\int \dfrac{1}{3+2\cos x}\,dx$

23 $\int \dfrac{1}{1+\sin x + \cos x}\,dx$

24 $\int \dfrac{1}{\tan x + \sin x}\,dx$

25 $\int \dfrac{\sec x}{4-3\tan x}\,dx$

26 $\int \dfrac{1}{\sin x - \sqrt{3}\cos x}\,dx$

27 $\int \dfrac{\sin 2x}{\sin^2 x - 2\sin x - 8}\,dx$

28 $\int \dfrac{\sin x}{5\cos x + \cos^2 x}\,dx$

(*Hint:* Let $u = \sin x$.)

(*Hint:* Let $u = \cos x$.)

Exer. 29–30: Use an appropriate substitution to establish the identity for all real numbers n and m.

29 $\int_0^{\pi/2} \sin^n x\,dx = \int_0^{\pi/2} \cos^n x\,dx$

30 $\int_0^1 x^m(1-x)^n\,dx = \int_0^1 x^n(1-x)^m\,dx$

31 Use an appropriate substitution to show that

$$\int_x^1 \frac{1}{1+t^2}\,dt = \int_1^{1/x} \frac{1}{1+t^2}\,dt \quad \text{for } x > 0$$

and then use this result to establish the identity

$$\arctan x + \arctan\frac{1}{x} = \frac{\pi}{2}$$

(see Exercise 24 of Section 8.2).

32 Show that

$$\int \frac{1}{(x+a)^{3/2} + (x-a)^{3/2}}\,dx$$

may be transformed into the integral of a rational function by means of the substitution $x = \frac{1}{2}a(t^2 + t^{-2})$.

Exer. 33–34: Use Theorem (9.6) to verify the formula.

33 $\int \sec x\,dx = \ln\left|\dfrac{1+\tan\frac{1}{2}x}{1-\tan\frac{1}{2}x}\right| + C$

34 $\int \csc x\,dx = \dfrac{1}{2}\ln\left(\dfrac{1-\cos x}{1+\cos x}\right) + C$

9.7 TABLES OF INTEGRALS

Mathematicians and scientists who use integrals in their work sometimes refer to tables of integrals. Many of the formulas contained in these tables may be obtained by methods we have studied. In general, tables of integrals should be used only after gaining experience with standard methods of integration. For complicated integrals it is often necessary to make substitutions, or to use partial fractions, integration by parts, or other techniques to obtain integrands for which the table is applicable.

The following examples illustrate the use of several formulas stated in the brief Table of Integrals in Appendix IV. To guard against errors introduced when using the table, we should always check answers by differentiation.

EXAMPLE 1 Evaluate $\int x^3 \cos x\,dx$.

SOLUTION We first use reduction Formula 85 in the Table of Integrals with $n = 3$ and $u = x$, obtaining

$$\int x^3 \cos x\,dx = x^3 \sin x - 3\int x^2 \sin x\,dx.$$

Next we apply Formula 84 with $n = 2$, and then Formula 83, obtaining

$$\int x^2 \sin x \, dx = -x^2 \cos x + 2 \int x \cos x \, dx$$

$$= -x^2 \cos x + 2[\cos x + x \sin x] + C.$$

Substitution in the first expression gives us

$$\int x^3 \cos x \, dx = x^3 \sin x + 3x^2 \cos x - 6 \cos x - 6x \sin x + C \quad \bullet$$

EXAMPLE 2 Evaluate $\displaystyle\int \frac{1}{x^2 \sqrt{3 + 5x^2}} \, dx$ for $x > 0$.

SOLUTION   The integrand suggests that we use that part of the table dealing with the form $\sqrt{a^2 + u^2}$. Specifically, Formula 28 states that

$$\int \frac{du}{u^2 \sqrt{a^2 + u^2}} = -\frac{\sqrt{a^2 + u^2}}{a^2 u} + C$$

(for compactness, the differential du is placed in the numerator instead of to the right of the integrand). To use this formula we must adjust the given integral so that it matches *exactly* with the formula. If we let

$$a^2 = 3 \quad \text{and} \quad u^2 = 5x^2$$

then the expression underneath the radical is taken care of; however, we also need

(i) u^2 to the left of the radical.

(ii) du in the numerator.

We can obtain (i) by writing the integral as

$$5 \int \frac{1}{5x^2 \sqrt{3 + 5x^2}} \, dx.$$

For (ii) we note that

$$u = \sqrt{5} x \quad \text{and} \quad du = \sqrt{5} \, dx$$

and write the preceding integral as

$$5 \cdot \frac{1}{\sqrt{5}} \int \frac{1}{5x^2 \sqrt{3 + 5x^2}} \sqrt{5} \, dx.$$

The last integral matches exactly with that in Formula 28 and hence

$$\int \frac{1}{x^2 \sqrt{3 + 5x^2}} \, dx = \sqrt{5} \left[-\frac{\sqrt{3 + 5x^2}}{3(\sqrt{5}x)} \right] + C$$

$$= -\frac{\sqrt{3 + 5x^2}}{3x} + C \quad \bullet$$

As illustrated in the next example, it may be necessary to make a substitution of some type before a table can be used to help evaluate an integral.

EXAMPLE 3 Evaluate $\int \dfrac{\sin 2x}{\sqrt{3 - 5\cos x}}\,dx$.

SOLUTION Let us begin by rewriting the integral:

$$\int \frac{\sin 2x}{\sqrt{3 - 5\cos x}}\,dx = \int \frac{2\sin x \cos x}{\sqrt{3 - 5\cos x}}\,dx$$

Since no formulas in the table have this form, we consider making the substitution $u = \cos x$. In this case $du = -\sin x\,dx$ and the integral may be written

$$2\int \frac{\sin x \cos x}{\sqrt{3 - 5\cos x}}\,dx = -2\int \frac{\cos x}{\sqrt{3 - 5\cos x}}\,(-\sin x)\,dx$$

$$= -2\int \frac{u}{\sqrt{3 - 5u}}\,du.$$

Referring to the Table of Integrals we see that Formula 55 is

$$\int \frac{u\,du}{\sqrt{a + bu}} = \frac{2}{3b^2}(bu - 2a)\sqrt{a + bu}.$$

Using this result with $a = 3$ and $b = -5$ gives us

$$-2\int \frac{u}{\sqrt{3 - 5u}}\,du = -2\left(\frac{2}{75}\right)(-5u - 6)\sqrt{3 - 5u} + C.$$

Finally, since $u = \cos x$, we obtain

$$\int \frac{\sin 2x}{\sqrt{3 - 5\cos x}}\,dx = \frac{4}{75}(5\cos x + 6)\sqrt{3 - 5\cos x} + C \quad \bullet$$

EXERCISES 9.7

Exer. 1–30: Use the Table of Integrals in Appendix IV to evaluate the integral.

1 $\int \dfrac{\sqrt{4 + 9x^2}}{x}\,dx$

2 $\int \dfrac{1}{x\sqrt{2 + 3x^2}}\,dx$

3 $\int (16 - x^2)^{3/2}\,dx$

4 $\int x^2\sqrt{4x^2 - 16}\,dx$

5 $\int x\sqrt{2 - 3x}\,dx$

6 $\int x^2\sqrt{5 + 2x}\,dx$

7 $\int \sin^6 3x\,dx$

8 $\int x\cos^5(x^2)\,dx$

9 $\int \csc^4 x\,dx$

10 $\int \sin 5x \cos 3x\,dx$

11 $\int x\sin^{-1} x\,dx$

12 $\int x^2 \tan^{-1} x\,dx$

13 $\int e^{-3x}\sin 2x\,dx$

14 $\int x^5 \ln x\,dx$

15 $\int \dfrac{\sqrt{5x - 9x^2}}{x}\,dx$

16 $\int \dfrac{1}{x\sqrt{3x - 2x^2}}\,dx$

17 $\int \dfrac{x}{5x^4 - 3}\,dx$

18 $\int \cos x\sqrt{\sin^2 x - 4}\,dx$

19 $\int e^{2x}\cos^{-1} e^x\,dx$

20 $\int \sin^2 x \cos^3 x\,dx$

21 $\int x^3\sqrt{2 + x}\,dx$

22 $\int \dfrac{7x^3}{\sqrt{2 - x}}\,dx$

23 $\int \dfrac{\sin 2x}{4 + 9\sin x}\,dx$

24 $\int \dfrac{\tan x}{\sqrt{4 + 3\sec x}}\,dx$

25 $\int \dfrac{\sqrt{9 + 2x}}{x}\,dx$

26 $\int \sqrt{8x^3 - 3x^2}\,dx$

27 $\int \dfrac{1}{x(4 + \sqrt[3]{x})}\,dx$

28 $\int \dfrac{1}{2x^{3/2} + 5x^2}\,dx$

29 $\int \sqrt{16 - \sec^2 x}\,\tan x\,dx$

30 $\int \dfrac{\cot x}{\sqrt{4 - \csc^2 x}}\,dx$

9.8 REVIEW

Discuss each of the following.

1 Integration by parts
2 Trigonometric substitutions

3 Integrals of rational functions
4 Integrals involving quadratic expressions

Exer. 1–100: Evaluate the integral.

1 $\int x \sin^{-1} x \, dx$

2 $\int \sec^3 (3x) \, dx$

3 $\int_0^1 \ln (1 + x) \, dx$

4 $\int_0^1 e^{\sqrt{x}} \, dx$

5 $\int \cos^3 2x \sin^2 2x \, dx$

6 $\int \cos^4 x \, dx$

7 $\int \tan x \sec^5 x \, dx$

8 $\int \tan x \sec^6 x \, dx$

9 $\int \frac{1}{(x^2 + 25)^{3/2}} \, dx$

10 $\int \frac{1}{x^2 \sqrt{16 - x^2}} \, dx$

11 $\int \frac{\sqrt{4 - x^2}}{x} \, dx$

12 $\int \frac{x}{(x^2 + 1)^2} \, dx$

13 $\int \frac{x^3 + 1}{x(x - 1)^3} \, dx$

14 $\int \frac{1}{x + x^3} \, dx$

15 $\int \frac{x^3 - 20x^2 - 63x - 198}{x^4 - 81} \, dx$

16 $\int \frac{x - 1}{(x + 2)^5} \, dx$

17 $\int \frac{x}{\sqrt{4 + 4x - x^2}} \, dx$

18 $\int \frac{x}{x^2 + 6x + 13} \, dx$

19 $\int \frac{\sqrt[3]{x + 8}}{x} \, dx$

20 $\int \frac{\sin x}{2 \cos x + 3} \, dx$

21 $\int e^{2x} \sin 3x \, dx$

22 $\int \cos (\ln x) \, dx$

23 $\int \sin^3 x \cos^3 x \, dx$

24 $\int \cot^2 3x \, dx$

25 $\int \frac{x}{\sqrt{4 - x^2}} \, dx$

26 $\int \frac{1}{x \sqrt{9x^2 + 4}} \, dx$

27 $\int \frac{x^5 - x^3 + 1}{x^3 + 2x^2} \, dx$

28 $\int \frac{x^3}{x^3 - 3x^2 + 9x - 27} \, dx$

29 $\int \frac{1}{x^{3/2} + x^{1/2}} \, dx$

30 $\int \frac{2x + 1}{(x + 5)^{100}} \, dx$

31 $\int e^x \sec e^x \, dx$

32 $\int x \tan x^2 \, dx$

33 $\int x^2 \sin 5x \, dx$

34 $\int \sin 2x \cos x \, dx$

35 $\int \sin^3 x \cos^{1/2} x \, dx$

36 $\int \sin 3x \cot 3x \, dx$

37 $\int e^x \sqrt{1 + e^x} \, dx$

38 $\int x(4x^2 + 25)^{-1/2} \, dx$

39 $\int \frac{x^2}{\sqrt{4x^2 + 25}} \, dx$

40 $\int \frac{3x + 2}{x^2 + 8x + 25} \, dx$

41 $\int \sec^2 x \tan^2 x \, dx$

42 $\int \sin^2 x \cos^5 x \, dx$

43 $\int x \cot x \csc x \, dx$

44 $\int (1 + \csc 2x)^2 \, dx$

45 $\int x^2(8 - x^3)^{1/3} \, dx$

46 $\int x (\ln x)^2 \, dx$

47 $\int \sqrt{x} \sin \sqrt{x} \, dx$

48 $\int x \sqrt{5 - 3x} \, dx$

49 $\int \frac{e^{3x}}{1 + e^x} \, dx$

50 $\int \frac{e^{2x}}{4 + e^{4x}} \, dx$

51 $\int \frac{x^2 - 4x + 3}{\sqrt{x}} \, dx$

52 $\int \frac{\cos^3 x}{\sqrt{1 + \sin x}} \, dx$

53 $\int \frac{x^3}{\sqrt{16 - x^2}} \, dx$

54 $\int \frac{x}{25 - 9x^2} \, dx$

55 $\int \frac{1 - 2x}{x^2 + 12x + 35} \, dx$

56 $\int \frac{7}{x^2 - 6x + 18} \, dx$

57 $\int \tan^{-1} 5x \, dx$

58 $\int \sin^4 3x \, dx$

59 $\int \frac{e^{\tan x}}{\cos^2 x} \, dx$

60 $\int \frac{x}{\csc 5x^2} \, dx$

61 $\int \frac{1}{\sqrt{7 + 5x^2}} \, dx$

62 $\int \frac{2x + 3}{x^2 + 4} \, dx$

63 $\int \cot^6 x \, dx$

64 $\int \cot^5 x \csc x \, dx$

65 $\int x^3 \sqrt{x^2 - 25} \, dx$

66 $\int (\sin x) 10^{\cos x} \, dx$

67 $\int (x^2 - \text{sech}^2 4x) \, dx$

68 $\int x \cosh x \, dx$

69 $\int x^2 e^{-4x} \, dx$

70 $\int x^5 \sqrt{x^3 + 1} \, dx$

71 $\int \frac{3}{\sqrt{11 - 10x - x^2}} \, dx$

72 $\int \frac{12x^3 + 7x}{x^4} \, dx$

73 $\int \tan 7x \cos 7x \, dx$

74 $\int e^{1 + \ln 5x} \, dx$

75 $\int \frac{4x^2 - 12x - 10}{(x - 2)(x^2 - 4x + 3)} \, dx$

76 $\int \frac{1}{x^4 \sqrt{16 - x^2}} \, dx$

77 $\displaystyle\int (x^3 + 1) \cos x \, dx$

78 $\displaystyle\int (x - 3)^2 (x + 1) \, dx$

89 $\displaystyle\int \tan^3 x \sec x \, dx$

90 $\displaystyle\int \frac{x}{\sqrt{4 + 9x^2}} \, dx$

79 $\displaystyle\int \frac{\sqrt{9 - 4x^2}}{x^2} \, dx$

80 $\displaystyle\int \frac{4x^3 - 15x^2 - 6x + 81}{x^4 - 18x^2 + 81} \, dx$

91 $\displaystyle\int \frac{2x^3 + 4x^2 + 10x + 13}{x^4 + 9x^2 + 20} \, dx$

92 $\displaystyle\int \frac{\sin x}{(1 + \cos x)^3} \, dx$

81 $\displaystyle\int (5 - \cot 3x)^2 \, dx$

82 $\displaystyle\int x(x^2 + 5)^{3/4} \, dx$

93 $\displaystyle\int \frac{(x^2 - 2)^2}{x} \, dx$

94 $\displaystyle\int \cot^2 x \csc x \, dx$

83 $\displaystyle\int \frac{1}{x(\sqrt{x} + \sqrt[4]{x})} \, dx$

84 $\displaystyle\int \frac{x}{\cos^2 4x} \, dx$

95 $\displaystyle\int x^{3/2} \ln x \, dx$

96 $\displaystyle\int \frac{x}{\sqrt[3]{x} - 1} \, dx$

85 $\displaystyle\int \frac{\sin x}{\sqrt{1 + \cos x}} \, dx$

86 $\displaystyle\int \frac{4x^2 - 6x + 4}{(x^2 + 4)(x - 2)} \, dx$

97 $\displaystyle\int \frac{x^2}{\sqrt[3]{2x + 3}} \, dx$

98 $\displaystyle\int \frac{1 - \sin x}{\cot x} \, dx$

87 $\displaystyle\int \frac{x^2}{(25 + x^2)^2} \, dx$

88 $\displaystyle\int \sin^4 x \cos^3 x \, dx$

99 $\displaystyle\int x^3 e^{(x^2)} \, dx$

100 $\displaystyle\int (x + 2)^2 (x + 1)^{10} \, dx$

In this chapter we introduce techniques that are useful in the investigation of certain limits. We shall also study definite integrals that have discontinuous integrands or infinite limits of integration. The final section contains a method for approximating functions by means of polynomials. These topics have many mathematical and physical applications; however, we will discuss our most important use for them in the next chapter, when we study *infinite series*.

INDETERMINATE FORMS, IMPROPER INTEGRALS, AND TAYLOR'S FORMULA

10.1 THE INDETERMINATE FORMS 0/0 AND ∞/∞

Problems involving limits often take on the form $\lim_{x \to c} f(x)/g(x)$, where both f and g have the limit 0 as x approaches c. We say that $f(x)/g(x)$ has the **indeterminate form 0/0** at $x = c$. Perhaps the most important examples of the indeterminate form 0/0 occur when using the derivative formula

$$f'(c) = \lim_{x \to c} \frac{f(x) - f(c)}{x - c}.$$

If the limits of $f(x)$ and $g(x)$ are ∞ or $-\infty$ as x approaches c, we say that $f(x)/g(x)$ has the **indeterminate form ∞/∞** at $x = c$.

In Chapter 2 we considered

$$\lim_{x \to 2} \frac{2x^2 - 5x + 2}{5x^2 - 7x - 6}.$$

The quotient has the indeterminate form 0/0; however, the limit may be found as follows:

$$\lim_{x \to 2} \frac{2x^2 - 5x + 2}{5x^2 - 7x - 6} = \lim_{x \to 2} \frac{(2x - 1)(x - 2)}{(5x + 3)(x - 2)} = \lim_{x \to 2} \frac{2x - 1}{5x + 3} = \frac{3}{13}.$$

Other indeterminate forms require more complicated techniques. For example, in Chapter 2 we used a geometric argument to show that

$$\lim_{x \to 0} \frac{\sin x}{x} = 1.$$

In this section we shall prove *L'Hôpital's Rule* and apply it to indeterminate forms. The proof makes use of the following formula, which bears the name of the famous French mathematician A. Cauchy (1789–1857).

If the functions f and g are continuous on a closed interval $[a, b]$, differentiable on the open interval (a, b), and if $g'(x) \neq 0$ for every x in (a, b), then there is a number w in (a, b) such that

$$\frac{f(b) - f(a)}{g(b) - g(a)} = \frac{f'(w)}{g'(w)}$$

PROOF We first note that $g(b) - g(a) \neq 0$, for otherwise $g(a) = g(b)$, and by Rolle's Theorem (4.10) there is a number c in (a, b) such that $g'(c) = 0$, contrary to our assumption about g'.

Let us introduce a new function h as follows:

$$h(x) = [f(b) - f(a)]g(x) - [g(b) - g(a)]f(x)$$

for every x in $[a, b]$. It follows that h is continuous on $[a, b]$, differentiable on (a, b), and that $h(a) = h(b)$. By Rolle's Theorem, there is a number w in (a, b) such that $h'(w) = 0$, that is,

$$[f(b) - f(a)]g'(w) - [g(b) - g(a)]f'(w) = 0.$$

This is equivalent to Cauchy's Formula. • •

If we let $g(x) = x$ in (10.1) we obtain

$$\frac{f(b) - f(a)}{b - a} = \frac{f'(w)}{1}$$

or, equivalently,

$$f(b) - f(a) = f'(w)(b - a).$$

This shows that Cauchy's Formula is a generalization of the Mean Value Theorem (4.12).

The next result is the main theorem on indeterminate forms.

L'HÔPITAL'S RULE* (10.2)

Suppose the functions f and g are differentiable on an open interval (a, b) containing c, except possibly at c itself. If $g'(x) \neq 0$ for $x \neq c$, and if $f(x)/g(x)$ has the indeterminate form $0/0$ or ∞/∞ at $x = c$, then

$$\lim_{x \to c} \frac{f(x)}{g(x)} = \lim_{x \to c} \frac{f'(x)}{g'(x)}$$

provided $\lim_{x \to c} [f'(x)/g'(x)]$ exists or $\lim_{x \to c} [f'(x)/g'(x)] = \infty$.

*G. L'Hôpital (1661–1704) was a French nobleman who published the first calculus book. The rule appeared in that book; however, it was actually discovered by his teacher, the Swiss mathematician Johann Bernoulli (1667–1748), who communicated the result to L'Hôpital in 1694.

PROOF Suppose $f(x)/g(x)$ has the indeterminate form $0/0$ at $x = c$ and $\lim_{x \to c} [f'(x)/g'(x)] = L$ for some number L. We wish to prove that $\lim_{x \to c} [f(x)/g(x)] = L$. Let us introduce two functions F and G such that

$$F(x) = f(x) \text{ if } x \neq c \quad \text{and} \quad F(c) = 0,$$

$$G(x) = g(x) \text{ if } x \neq c \quad \text{and} \quad G(c) = 0.$$

Since

$$\lim_{x \to c} F(x) = \lim_{x \to c} f(x) = 0 = F(c),$$

the function F is continuous at c and hence is continuous *throughout* the interval (a, b). Similarly, G is continuous on (a, b). Moreover, at every $x \neq c$ we have $F'(x) = f'(x)$ and $G'(x) = g'(x)$. It follows from Cauchy's Formula, applied either to the interval $[c, x]$ or to $[x, c]$, that there is a number w between c and x such that

$$\frac{F(x) - F(c)}{G(x) - G(c)} = \frac{F'(w)}{G'(w)} = \frac{f'(w)}{g'(w)}.$$

Using the fact that $F(x) = f(x)$, $G(x) = g(x)$, and $F(c) = G(c) = 0$ gives us

$$\frac{f(x)}{g(x)} = \frac{f'(w)}{g'(w)}.$$

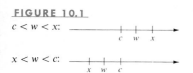

Since w is always between c and x (see Figure 10.1), it follows that

$$\lim_{x \to c} \frac{f(x)}{g(x)} = \lim_{x \to c} \frac{f'(w)}{g'(w)} = \lim_{w \to c} \frac{f'(w)}{g'(w)} = L$$

which is what we wished to prove.

A similar argument may be given if $\lim_{x \to c} [f'(x)/g'(x)] = \infty$. The proof for the indeterminate form ∞/∞ is more difficult and may be found in texts on advanced calculus. • •

L'Hôpital's Rule is sometimes used incorrectly—the Quotient Rule is applied to $f(x)/g(x)$. Note that (10.2) states that the derivatives of $f(x)$ and $g(x)$ are taken *separately*, after which the limit of the quotient $f'(x)/g'(x)$ is investigated.

EXAMPLE 1 Find $\lim\limits_{x \to 0} \dfrac{\cos x + 2x - 1}{3x}$.

SOLUTION The quotient has the indeterminate form $0/0$ at $x = 0$. By L'Hôpital's Rule (10.2),

$$\lim_{x \to 0} \frac{\cos x + 2x - 1}{3x} = \lim_{x \to 0} \frac{-\sin x + 2}{3} = \frac{2}{3}. \quad \bullet$$

To be completely rigorous in Example 1 we should have determined if $\lim_{x \to 0} (-\sin x + 2)/3$ exists *before* equating it to the given expression; however, to simplify solutions we usually assume that the limit exists, and proceed as indicated.

Sometimes it is necessary to employ L'Hôpital's Rule several times in the same problem, as illustrated in the next example.

EXAMPLE 2 Find $\lim\limits_{x \to 0} \dfrac{e^x + e^{-x} - 2}{1 - \cos 2x}$.

SOLUTION The quotient has the indeterminate form $0/0$. By L'Hôpital's Rule,

$$\lim_{x \to 0} \frac{e^x + e^{-x} - 2}{1 - \cos 2x} = \lim_{x \to 0} \frac{e^x - e^{-x}}{2 \sin 2x}$$

provided the second limit exists. Since the last quotient has the indeterminate form $0/0$, we apply L'Hôpital's Rule a second time, obtaining

$$\lim_{x \to 0} \frac{e^x - e^{-x}}{2 \sin 2x} = \lim_{x \to 0} \frac{e^x + e^{-x}}{4 \cos 2x} = \frac{2}{4} = \frac{1}{2}.$$

It follows that the given limit exists and equals $\frac{1}{2}$. •

L'Hôpital's Rule is also valid for one-sided limits, as illustrated in the following example.

EXAMPLE 3 Find $\displaystyle\lim_{x \to (\pi/2)^-} \frac{4 \tan x}{1 + \sec x}$.

SOLUTION The indeterminate form is ∞/∞. By L'Hôpital's Rule,

$$\lim_{x \to (\pi/2)^-} \frac{4 \tan x}{1 + \sec x} = \lim_{x \to (\pi/2)^-} \frac{4 \sec^2 x}{\sec x \tan x} = \lim_{x \to (\pi/2)^-} \frac{4 \sec x}{\tan x}.$$

The last quotient again has the indeterminate form ∞/∞ at $x = \pi/2$; however, additional applications of L'Hôpital's Rule always produce the form ∞/∞ (check this fact). In this case the limit may be found by using trigonometric identities to change the quotient as follows:

$$\frac{4 \sec x}{\tan x} = \frac{4/(\cos x)}{(\sin x)/(\cos x)} = \frac{4}{\sin x}.$$

Consequently,

$$\lim_{x \to (\pi/2)^-} \frac{4 \tan x}{1 + \sec x} = \lim_{x \to (\pi/2)^-} \frac{4}{\sin x} = \frac{4}{1} = 4 \quad •$$

Another form of L'Hôpital's Rule can be proved for $x \to \infty$ or $x \to -\infty$. Let us give a partial proof of this fact. Suppose

$$\lim_{x \to \infty} f(x) = \lim_{x \to \infty} g(x) = 0.$$

If we let $u = 1/x$ and apply L'Hôpital's Rule,

$$\lim_{x \to \infty} \frac{f(x)}{g(x)} = \lim_{u \to 0^+} \frac{f(1/u)}{g(1/u)} = \lim_{u \to 0^+} \frac{D_u\, f(1/u)}{D_u\, g(1/u)}.$$

By the Chain Rule,

$$D_u\, f(1/u) = f'(1/u)(-1/u^2) \quad \text{and} \quad D_u\, g(1/u) = g'(1/u)(-1/u^2).$$

Substituting in the last limit and simplifying, we obtain

$$\lim_{x \to \infty} \frac{f(x)}{g(x)} = \lim_{u \to 0^+} \frac{f'(1/u)}{g'(1/u)} = \lim_{x \to \infty} \frac{f'(x)}{g'(x)}.$$

We shall also refer to this as L'Hôpital's Rule. The next two examples illustrate the application of the rule to the form ∞/∞.

EXAMPLE 4 Find $\displaystyle\lim_{x \to \infty} \frac{\ln x}{\sqrt{x}}$.

SOLUTION The indeterminate form is ∞/∞. By L'Hôpital's Rule,

$$\lim_{x \to \infty} \frac{\ln x}{\sqrt{x}} = \lim_{x \to \infty} \frac{1/x}{1/(2\sqrt{x})}.$$

The last expression has the indeterminate form 0/0. However, further applications of L'Hôpital's Rule would again lead to 0/0 (check this fact). If, instead, we simplify the expression algebraically, we can find the limit as follows:

$$\lim_{x \to \infty} \frac{1/x}{1/(2\sqrt{x})} = \lim_{x \to \infty} \frac{2\sqrt{x}}{x} = \lim_{x \to \infty} \frac{2}{\sqrt{x}} = 0 \quad \bullet$$

EXAMPLE 5 Find $\displaystyle \lim_{x \to \infty} \frac{e^{3x}}{x^2}$, if it exists.

SOLUTION The indeterminate form is ∞/∞. In this case we must apply L'Hôpital's Rule twice, as follows:

$$\lim_{x \to \infty} \frac{e^{3x}}{x^2} = \lim_{x \to \infty} \frac{3e^{3x}}{2x} = \lim_{x \to \infty} \frac{9e^{3x}}{2} = \infty$$

Thus the given expression has no limit, but increases without bound as x approaches ∞. $\bullet$

It is extremely important to verify that a given quotient has the indeterminate form 0/0 or ∞/∞ before using L'Hôpital's Rule. If we apply the rule to a form that is not indeterminate, we may obtain an incorrect conclusion, as illustrated in the next example.

EXAMPLE 6 Find $\displaystyle \lim_{x \to 0} \frac{e^x + e^{-x}}{x^2}$, if it exists.

SOLUTION The quotient does *not* have either of the indeterminate forms 0/0 or ∞/∞ at $x = 0$. To investigate the limit we write

$$\lim_{x \to 0} \frac{e^x + e^{-x}}{x^2} = \lim_{x \to 0} (e^x + e^{-x})\left(\frac{1}{x^2}\right).$$

Since

$$\lim_{x \to 0} (e^x + e^{-x}) = 2 \quad \text{and} \quad \lim_{x \to 0} \frac{1}{x^2} = \infty,$$

Theorem (4.27) (ii) implies that

$$\lim_{x \to 0} \frac{e^x + e^{-x}}{x^2} = \infty.$$

If we had overlooked the fact that the quotient does not have either of the indeterminate forms 0/0 or ∞/∞ at $x = 0$, and had (incorrectly) applied L'Hôpital's Rule, we would have obtained

$$\lim_{x \to 0} \frac{e^x + e^{-x}}{x^2} = \lim_{x \to 0} \frac{e^x - e^{-x}}{2x}.$$

Since the last quotient has the indeterminate form 0/0 we may apply L'Hôpital's Rule, obtaining

$$\lim_{x \to 0} \frac{e^x - e^{-x}}{2x} = \lim_{x \to 0} \frac{e^x + e^{-x}}{2} = \frac{1+1}{2} = 1.$$

This gives us the (wrong) conclusion that the given limit exists and equals 1. $\bullet$

The next example illustrates an application of an indeterminate form in the analysis of an electrical circuit.

FIGURE 10.2

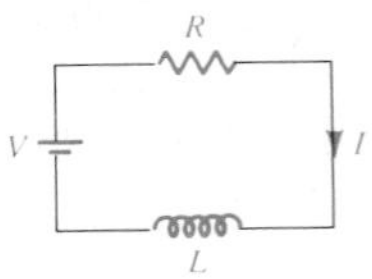

EXAMPLE 7 The current I at time t in an electrical circuit is given by $I = V(1 - e^{-Rt/L})/R$ for the electromotive force V, resistance R, and inductance L (see Figure 10.2). If L is the only independent variable, find $\lim_{L \to 0^+} I$. If R is the only independent variable, find $\lim_{R \to 0^+} I$.

SOLUTION If we consider V, R, and t as constants and L as a variable, then the expression for I is not indeterminate at $L = 0$. Using standard limit theorems we obtain

$$\lim_{L \to 0^+} I = \lim_{L \to 0^+} \frac{V}{R}(1 - e^{-Rt/L})$$

$$= \frac{V}{R}\left(1 - \lim_{L \to 0^+} e^{-Rt/L}\right)$$

$$= \frac{V}{R}(1 - 0) = \frac{V}{R}.$$

Thus, under these conditions, the current is given by Ohm's law $I = V/R$.

If V, L, and t are constant, and if R is a variable, then I takes on the indeterminate form $0/0$ at $R = 0$. Applying L'Hôpital's Rule,

$$\lim_{R \to 0^+} I = V \lim_{R \to 0^+} \frac{(1 - e^{-Rt/L})}{R}$$

$$= V \lim_{R \to 0^+} \frac{0 - e^{-Rt/L}(-t/L)}{1}$$

$$= V[0 - (1)(-t/L)] = \frac{V}{L}t.$$

This may be interpreted as follows. As $R \to 0^+$, the current I is directly proportional to the time t, with the constant of proportionality V/L. Thus, at $t = 1$ the current is V/L, at $t = 2$ it is $(V/L)(2)$, at $t = 3$ it is $(V/L)(3)$, and so on. •

EXERCISES 10.1

Exer. 1–52: Find the limit, if it exists.

1 $\lim\limits_{x \to 0} \dfrac{\sin x}{2x}$

2 $\lim\limits_{x \to 0} \dfrac{5x}{\tan x}$

3 $\lim\limits_{x \to 5} \dfrac{\sqrt{x - 1} - 2}{x^2 - 25}$

4 $\lim\limits_{x \to 4} \dfrac{x - 4}{\sqrt[3]{x + 4} - 2}$

5 $\lim\limits_{x \to 2} \dfrac{2x^2 - 5x + 2}{5x^2 - 7x - 6}$

6 $\lim\limits_{x \to -3} \dfrac{x^2 + 2x - 3}{2x^2 + 3x - 9}$

7 $\lim\limits_{x \to 1} \dfrac{x^3 - 3x + 2}{x^2 - 2x - 1}$

8 $\lim\limits_{x \to 2} \dfrac{x^2 - 5x + 6}{2x^2 - x - 7}$

9 $\lim\limits_{x \to 0} \dfrac{\sin x - x}{\tan x - x}$

10 $\lim\limits_{x \to 0} \dfrac{\sin x}{x - \tan x}$

11 $\lim\limits_{x \to 0} \dfrac{x + 1 - e^x}{x^2}$

12 $\lim\limits_{x \to 0^+} \dfrac{x + 1 - e^x}{x^3}$

13 $\lim\limits_{x \to 0} \dfrac{x - \sin x}{x^3}$

14 $\lim\limits_{x \to \pi/2} \dfrac{1 - \sin x}{\cos x}$

15 $\lim\limits_{x \to \pi/2} \dfrac{1 + \sin x}{\cos^2 x}$

16 $\lim\limits_{x \to 0^+} \dfrac{\cos x}{x}$

17 $\displaystyle\lim_{x\to(\pi/2)^-}\frac{2+\sec x}{3\tan x}$

18 $\displaystyle\lim_{x\to 0^+}\frac{\ln x}{\cot x}$

19 $\displaystyle\lim_{x\to\infty}\frac{x^2}{\ln x}$

20 $\displaystyle\lim_{x\to\infty}\frac{\ln x}{x^2}$

21 $\displaystyle\lim_{x\to 0^+}\frac{\ln\sin x}{\ln\sin 2x}$

22 $\displaystyle\lim_{x\to 0}\frac{2x}{\tan^{-1}x}$

23 $\displaystyle\lim_{x\to 0}\frac{e^x-e^{-x}-2\sin x}{x\sin x}$

24 $\displaystyle\lim_{x\to 2}\frac{\ln(x-1)}{x-2}$

25 $\displaystyle\lim_{x\to 0}\frac{x\cos x+e^{-x}}{x^2}$

26 $\displaystyle\lim_{x\to 0}\frac{2e^x-3x-e^{-x}}{x^2}$

27 $\displaystyle\lim_{x\to\infty}\frac{2x^2+3x+1}{5x^2+x+4}$

28 $\displaystyle\lim_{x\to\infty}\frac{x^3+x+1}{3x^3+4}$

29 $\displaystyle\lim_{x\to\infty}\frac{x\ln x}{x+\ln x}$

30 $\displaystyle\lim_{x\to\infty}\frac{e^{3x}}{\ln x}$

31 $\displaystyle\lim_{x\to\infty}\frac{x^n}{e^x},\ n>0$

32 $\displaystyle\lim_{x\to\infty}\frac{e^x}{x^n},\ n>0$

33 $\displaystyle\lim_{x\to 2^+}\frac{\ln(x-1)}{(x-2)^2}$

34 $\displaystyle\lim_{x\to 0}\frac{\sin^2 x+2\cos x-2}{\cos^2 x-x\sin x-1}$

35 $\displaystyle\lim_{x\to 0}\frac{\sin^{-1}2x}{\sin^{-1}x}$

36 $\displaystyle\lim_{x\to\infty}\frac{\ln(\ln x)}{\ln x}$

37 $\displaystyle\lim_{x\to 0}\frac{\tan x-\sin x}{x^3\tan x}$

38 $\displaystyle\lim_{x\to 1}\frac{2x^3-5x^2+6x-3}{x^3-2x^2+x-1}$

39 $\displaystyle\lim_{x\to-\infty}\frac{3-3^x}{5-5^x}$

40 $\displaystyle\lim_{x\to 0}\frac{2-e^x-e^{-x}}{1-\cos^2 x}$

41 $\displaystyle\lim_{x\to 1}\frac{x^4-x^3-3x^2+5x-2}{x^4-5x^3+9x^2-7x+2}$

42 $\displaystyle\lim_{x\to 1}\frac{x^4+x^3-3x^2-x+2}{x^4-5x^3+9x^2-7x+2}$

43 $\displaystyle\lim_{x\to 0}\frac{x-\tan^{-1}x}{x\sin x}$

44 $\displaystyle\lim_{x\to\infty}\frac{e^{-x}}{1+e^{-x}}$

45 $\displaystyle\lim_{x\to\infty}\frac{x^{3/2}+5x-4}{x\ln x}$

46 $\displaystyle\lim_{x\to 0}\frac{x\sin^{-1}x}{x-\sin x}$

47 $\displaystyle\lim_{x\to\infty}\frac{\sqrt{x^2+1}}{\tan^{-1}x}$

48 $\displaystyle\lim_{x\to\pi/2}\frac{\tan x}{\cot 2x}$

49 $\displaystyle\lim_{x\to\infty}\frac{2e^{3x}+\ln x}{e^{3x}+x^2}$

50 $\displaystyle\lim_{x\to 0}\frac{e^{-1/x}}{x}$

51 $\displaystyle\lim_{x\to\infty}\frac{x-\cos x}{x}$

52 $\displaystyle\lim_{x\to\infty}\frac{x+\cosh x}{x^2+1}$

53 An object of mass m is released from a hot-air balloon. If the force of resistance due to air is directly proportional to the velocity $v(t)$ of the object at time t, then
$$v(t)=(mg/k)(1-e^{-(k/m)t})$$
for some constant k and a gravitational constant g (see Section 19.2, Example 4). Prove that $\lim_{k\to 0^+}v(t)=-gt$.

54 A ball of mass m is released into water (see Exercise 56 of Section 8.4). If the force of resistance due to water is directly proportional to the square of the velocity, then the distance $s(t)$ that the ball travels in time t is given by
$$s(t)=(m/k)\ln\cosh(\sqrt{gk/m}\,t)$$
for some constant k and a gravitational constant g. Show that $\lim_{k\to 0^+}s(t)=\frac{1}{2}gt^2$.

55 Refer to Definition (3.32) for simple harmonic motion. The following is an example of the phenomenon of resonance. A weight of mass m is attached to a spring suspended from a support. The weight is set in motion by moving the support up and down according to the equation $D=A\cos\omega t$, for positive constants A and ω and time t. If frictional forces are neglected, then the displacement s of the weight from its initial position at time t is given by
$$s=\frac{A\omega^2}{\omega_0^2-\omega^2}(\cos\omega t-\cos\omega_0 t)$$
with $\omega_0=\sqrt{k/m}$ for some constant k and $\omega\neq\omega_0$. Find $\lim_{\omega\to\omega_0}s$, and show that the resulting oscillations increase in magnitude.

56 The logistic model for population growth predicts the size $y(t)$ of a population at time t by means of the formula $y(t)=K/(1+ce^{-rt})$, for positive constants r and K and $c=[K-y(0)]/y(0)$. Ecologists call K the *carrying capacity*, and interpret it as the maximum number of individuals that the environment can sustain. Find $\lim_{t\to\infty}y(t)$ and $\lim_{K\to\infty}y(t)$, and discuss the graphical significance of these limits.

57 The *sine integral* $Si(x)=\int_0^x[(\sin u)/u]\,du$ is a special function in applied mathematics. Find
(a) $\displaystyle\lim_{x\to 0}\frac{Si(x)}{x}$
(b) $\displaystyle\lim_{x\to 0}\frac{Si(x)-x}{x^3}$

58 The *Fresnel cosine integral* $C(x)=\int_0^x\cos u^2\,du$ is used in the analysis of the diffraction of light. Find
(a) $\displaystyle\lim_{x\to 0}\frac{C(x)}{x}$
(b) $\displaystyle\lim_{x\to 0}\frac{C(x)-x}{x^5}$

59 Let $x>0$. If $n\neq-1$ we know that $\int_1^x t^n\,dt=[t^{n+1}/(n+1)]_1^x$. Show that
$$\lim_{n\to-1}\int_1^x t^n\,dt=\int_1^x t^{-1}\,dt.$$

60 Find $\lim_{x\to\infty}f(x)/g(x)$ if
$$f(x)=\int_0^x e^{t^2}\,dt\quad\text{and}\quad g(x)=e^{x^2}.$$

If $\lim_{x \to c} f(x) = 0$ and either $\lim_{x \to c} g(x) = \infty$ or $\lim_{x \to c} g(x) = -\infty$, then $f(x)g(x)$ is said to have the **indeterminate form $0 \cdot \infty$** at $x = c$. The same terminology is used for one-sided limits, or if $x \to \infty$ or $x \to -\infty$. This form may be changed to one of the indeterminate forms $0/0$ or ∞/∞ by writing

$$f(x)g(x) = \frac{f(x)}{1/g(x)} \quad \text{or} \quad f(x)g(x) = \frac{g(x)}{1/f(x)}.$$

EXAMPLE 1 Find $\lim_{x \to 0^+} x^2 \ln x$.

SOLUTION The indeterminate form is $0 \cdot \infty$. We first write

$$x^2 \ln x = \frac{\ln x}{1/x^2}$$

and then apply L'Hôpital's Rule to the resulting indeterminate form ∞/∞. Thus

$$\lim_{x \to 0^+} x^2 \ln x = \lim_{x \to 0^+} \frac{\ln x}{1/x^2} = \lim_{x \to 0^+} \frac{1/x}{-2/x^3}.$$

The last quotient has the indeterminate form ∞/∞; however, further applications of L'Hôpital's Rule would again lead to ∞/∞. In this case we simplify the quotient algebraically and find the limit as follows:

$$\lim_{x \to 0^+} \frac{1/x}{-2/x^3} = \lim_{x \to 0^+} \frac{x^3}{-2x} = \lim_{x \to 0^+} \frac{x^2}{-2} = 0 \quad \bullet$$

EXAMPLE 2 Find $\lim_{x \to \pi/2} (2x - \pi) \sec x$.

SOLUTION The indeterminate form is $0 \cdot \infty$. Hence we begin by writing

$$(2x - \pi) \sec x = \frac{2x - \pi}{1/(\sec x)} = \frac{2x - \pi}{\cos x}.$$

Since the last expression has the indeterminate form $0/0$ at $x = \pi/2$, L'Hôpital's Rule may be applied as follows:

$$\lim_{x \to \pi/2} \frac{2x - \pi}{\cos x} = \lim_{x \to \pi/2} \frac{2}{-\sin x} = \frac{2}{-1} = -2 \quad \bullet$$

Indeterminate forms denoted by 0^0, ∞^0, and 1^∞ result from expressions such as $f(x)^{g(x)}$. One method for investigating these forms is to write

$$y = f(x)^{g(x)}$$

and take the natural logarithm of both sides, obtaining

$$\ln y = \ln f(x)^{g(x)} = g(x) \ln f(x).$$

If the indeterminate form for y is 0^0 or ∞^0, then the indeterminate form for $\ln y$ is $0 \cdot \infty$, which may be handled using earlier methods. Similarly, if y has the form 1^∞, then the indeterminate form for $\ln y$ is $\infty \cdot 0$. It follows that

$$\text{if} \quad \lim_{x \to c} \ln y = L, \quad \text{then} \quad \lim_{x \to c} y = \lim_{x \to c} e^{\ln y} = e^L,$$

that is,

$$\lim_{x \to c} f(x)^{g(x)} = e^L.$$

This procedure may be summarized as follows.

GUIDELINES FOR INVESTIGATING (10.3)
$\lim_{x \to c} f(x)^{g(x)}$ **IF THE INDETERMINATE FORM IS** 0^0, 1^∞, **OR** ∞^0

1 Let $y = f(x)^{g(x)}$.

2 Take logarithms: $\ln y = \ln f(x)^{g(x)} = g(x) \ln f(x)$.

3 Find $\lim_{x \to c} \ln y$, if it exists.

4 If $\lim_{x \to c} \ln y = L$, then $\lim_{x \to c} y = e^L$.

A common error is to stop after showing $\lim_{x \to c} \ln y = L$ and conclude that the given expression has the limit L. Remember that *we wish to find the limit of y*. Thus if $\ln y$ has the limit L, then y has the limit e^L. The guidelines may also be used if $x \to \infty$, or $x \to -\infty$, or for one-sided limits.

EXAMPLE 3 Find $\lim_{x \to 0} (1 + 3x)^{1/(2x)}$.

SOLUTION The indeterminate form is 1^∞. Following Guidelines (10.3), we write

1
$$y = (1 + 3x)^{1/(2x)}.$$

2
$$\ln y = \frac{1}{2x} \ln (1 + 3x) = \frac{\ln (1 + 3x)}{2x}.$$

The last expression has the indeterminate form $0/0$ at $x = 0$. By L'Hôpital's Rule,

3
$$\lim_{x \to 0} \ln y = \lim_{x \to 0} \frac{\ln (1 + 3x)}{2x} = \lim_{x \to 0} \frac{3/(1 + 3x)}{2} = \frac{3}{2}.$$

Consequently,

4
$$\lim_{x \to 0} (1 + 3x)^{1/(2x)} = \lim_{x \to 0} y = e^{3/2} \quad \bullet$$

If $\lim_{x \to c} f(x) = \lim_{x \to c} g(x) = \infty$, then $f(x) - g(x)$ *has the indeterminate form* $\infty - \infty$ *at* $x = c$. In this case the expression should be changed so that one of the forms we have discussed is obtained.

EXAMPLE 4 Find $\lim_{x \to 0} \left(\frac{1}{e^x - 1} - \frac{1}{x} \right)$.

SOLUTION The form is $\infty - \infty$; however, if the difference is written as a single fraction, then

$$\lim_{x \to 0} \left(\frac{1}{e^x - 1} - \frac{1}{x} \right) = \lim_{x \to 0} \frac{x - e^x + 1}{xe^x - x}.$$

This gives us the indeterminate form $0/0$. It is necessary to apply L'Hôpital's Rule twice, since the first application leads to the indeterminate form $0/0$. Thus

$$\lim_{x \to 0} \frac{x - e^x + 1}{xe^x - x} = \lim_{x \to 0} \frac{1 - e^x}{xe^x + e^x - 1}$$

$$= \lim_{x \to 0} \frac{-e^x}{xe^x + 2e^x} = -\frac{1}{2} \quad \bullet$$

EXAMPLE 5 Refer to Exercise 59 of Section 7.3. The velocity v of an electrical impulse in a cable is given by

$$v = -k \left(\frac{r}{R} \right)^2 \ln \left(\frac{r}{R} \right)$$

for the radius R of the insulated cable, the radius r of the cable, and a positive constant k (see Figure 10.3). Find

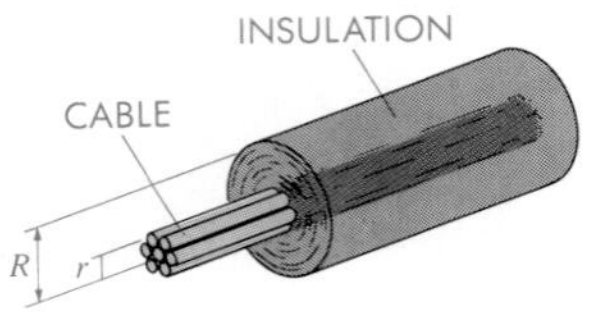

(a) $\lim\limits_{R \to r^+} v$ (b) $\lim\limits_{r \to 0^+} v$.

FIGURE 10.3

INSULATION

CABLE

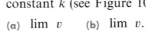

$R \quad r$

SOLUTION

(a) The limit notation implies that r is fixed and R is a variable. In this case the expression for v is not indeterminate, and

$$\lim_{R \to r^+} v = -k \lim_{R \to r^+} \left(\frac{r}{R} \right)^2 \ln \left(\frac{r}{R} \right) = -k(1)^2 \ln 1 = -k(0) = 0.$$

(b) If R is fixed and r is a variable, then the expression for v has the indeterminate form $0 \cdot \infty$ at $r = 0$, and we first change the form of the expression algebraically, as follows:

$$\lim_{r \to 0^+} v = -k \lim_{r \to 0^+} \frac{\ln (r/R)}{(r/R)^{-2}} = -kR^2 \lim_{r \to 0^+} \frac{\ln r - \ln R}{r^{-2}}$$

Since the last quotient has the indeterminate form ∞/∞ at $r = 0$, we may apply L'Hôpital's Rule, obtaining

$$\lim_{r \to 0^+} v = -kR^2 \lim_{r \to 0^+} \frac{(1/r) - 0}{-2r^{-3}}$$

$$= -kR^2 \lim_{r \to 0^+} \left(\frac{r^2}{-2} \right) = -kR^2(0) = 0 \quad \bullet$$

EXERCISES 10.2

Exer. 1–42: Find the limit, if it exists.

1 $\lim\limits_{x \to 0^+} x \ln x$

2 $\lim\limits_{x \to (\pi/2)^-} \tan x \ln \sin x$

3 $\lim\limits_{x \to \infty} (x^2 - 1)e^{-x^2}$

4 $\lim\limits_{x \to \infty} x(e^{1/x} - 1)$

5 $\lim\limits_{x \to 0} e^{-x} \sin x$

6 $\lim\limits_{x \to -\infty} x \tan^{-1} x$

7 $\lim\limits_{x \to 0^+} \sin x \ln \sin x$

8 $\lim\limits_{x \to \infty} x \left(\frac{\pi}{2} - \tan^{-1} x \right)$

9 $\lim\limits_{x \to \infty} x \sin \dfrac{1}{x}$

10 $\lim\limits_{x \to \infty} e^{-x} \ln x$

11 $\lim\limits_{x \to 0} x \sec^2 x$

12 $\lim\limits_{x \to 0} (\cos x)^{x+1}$

13 $\lim\limits_{x \to \infty} \left(1 + \dfrac{1}{x} \right)^{5x}$

14 $\lim\limits_{x \to 0^+} (e^x + 3x)^{1/2}$

15 $\lim\limits_{x \to 0^+} (e^x - 1)^x$

16 $\lim\limits_{x \to 0^+} x^x$

17 $\lim\limits_{x \to \infty} x^{1/x}$

18 $\lim\limits_{x \to (\pi/2)^-} (\tan x)^{\cos x}$

19 $\lim\limits_{x \to (\pi/2)^-} (\tan x)^x$

20 $\lim\limits_{x \to 2^+} (x - 2)^x$

21 $\lim\limits_{x \to 0^+} (2x + 1)^{\cot x}$

22 $\lim\limits_{x \to 0^+} (1 + 3x)^{\csc x}$

23 $\lim\limits_{x \to \infty} \left(\dfrac{x^2}{x - 1} - \dfrac{x^2}{x + 1} \right)$

24 $\lim\limits_{x \to 1} \left(\dfrac{1}{x - 1} - \dfrac{1}{\ln x} \right)$

25 $\lim\limits_{x \to 0} \left(\dfrac{1}{x} - \dfrac{1}{\sin x} \right)$

26 $\lim\limits_{x \to (\pi/2)^-} (\sec x - \tan x)$

27 $\lim\limits_{x \to 1^-} (1 - x)^{\ln x}$

28 $\lim\limits_{x \to \infty} (1 + e^x)^{e^{-x}}$

29 $\lim\limits_{x \to 0} \left(\dfrac{1}{\sqrt{x^2 + 1}} - \dfrac{1}{x} \right)$

30 $\lim\limits_{x \to 0} (\cot^2 x - \csc^2 x)$

31 $\lim\limits_{x \to 0} \cot 2x \tan^{-1} x$

32 $\lim\limits_{x \to \infty} x^3 \, 2^{-x}$

33 $\lim\limits_{x \to 0} (\cot^2 x - e^{-x})$

34 $\lim\limits_{x \to \infty} (\sqrt{x^2 + 4} - \tan^{-1} x)$

35 $\displaystyle\lim_{x\to(\pi/2)^-}(1+\cos x)^{\tan x}$

36 $\displaystyle\lim_{x\to 0}(1+ax)^{b/x}$

37 $\displaystyle\lim_{x\to -3}\left(\frac{4}{x^2+2x-3}-\frac{4}{x+3}\right)$

38 $\displaystyle\lim_{x\to\infty}(\sqrt{x^4+5x^2+3}-x^2)$

39 $\displaystyle\lim_{x\to 0}(x+\cos 2x)^{\csc 3x}$

40 $\displaystyle\lim_{x\to\pi/2}\sec x\cos 3x$

41 $\displaystyle\lim_{x\to\infty}(\sinh x-x)$

42 $\displaystyle\lim_{x\to\infty}\big[\ln(4x+3)-\ln(3x+4)\big]$

Exer. 43–44: Sketch the graph of f for $x>0$. Find the local extrema and discuss the behavior of $f(x)$ near $x=0$. Find horizontal asymptotes, if they exist.

43 $f(x)=x^{1/x}$ **44** $f(x)=x^x$

45 The *geometric mean* of two positive real numbers a and b is defined as $\sqrt{ab}$. Use L'Hôpital's Rule to prove that
$$\sqrt{ab}=\lim_{x\to\infty}\left(\frac{a^{1/x}+b^{1/x}}{2}\right)^x.$$

46 If a sum of money P is invested at an interest rate of $100r$ percent per year, compounded m times per year, then the principal at the end of t years is given by $P(1+rm^{-1})^{mt}$. If we regard m as a real number and let m increase without bound, then the interest is said to be *compounded continuously*. Use L'Hôpital's Rule to show that in this case the principal after t years is Pe^{rt}.

47 Refer to Exercise 45 of Section 8.1. Given the right triangle ABC (see figure), the average distance k of a point P on side AB from vertex C is $k=(a/\theta)\ln(\sec\theta+\tan\theta)$.
(a) Use geometric intuition to find $\lim_{\theta\to 0}k$.
(b) Verify your answer in part (a) by using L'Hôpital's Rule.

EXERCISE 47

48 Refer to Exercise 53 of Section 10.1. In the velocity formula
$$v(t)=(mg/k)(1-e^{-(k/m)t})$$
m represents the mass of the falling object. Find $\lim_{m\to\infty}v(t)$ and conclude that $v(t)$ is approximately proportional to time t if the mass is very large.

10.3 INTEGRALS WITH INFINITE LIMITS OF INTEGRATION

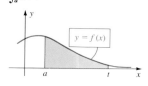

FIGURE 10.4
$\int_a^t f(x)\,dx$

Suppose a function f is continuous and nonnegative on an infinite interval $[a,\infty)$ and $\lim_{x\to\infty}f(x)=0$. If $t>a$, then the area $A(t)$ under the graph of f from a to t, as illustrated in Figure 10.4, is
$$A(t)=\int_a^t f(x)\,dx.$$

If $\lim_{t\to\infty}A(t)$ exists, then the limit may be interpreted as the area of the region that lies under the graph of f, over the x-axis, and to the right of $x=a$, as illustrated in Figure 10.5. The symbol $\int_a^\infty f(x)\,dx$ is used to denote this number.

Part (i) of the next definition generalizes the preceding remarks to the case where $f(x)$ may be negative in $[a,\infty)$.

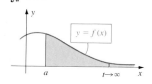

FIGURE 10.5
$\int_a^\infty f(x)\,dx$

DEFINITION (10.4)

(i) If f is continuous on $[a,\infty)$, then
$$\int_a^\infty f(x)\,dx=\lim_{t\to\infty}\int_a^t f(x)\,dx$$
provided the limit exists.

(ii) If f is continuous on $(-\infty,a]$, then
$$\int_{-\infty}^a f(x)\,dx=\lim_{t\to-\infty}\int_t^a f(x)\,dx$$
provided the limit exists.

FIGURE 10.6

$\int_{-\infty}^{a} f(x)\, dx$

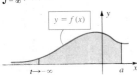

If $f(x) \geq 0$ for every x, then the limit in Definition (10.4)(ii) may be regarded as the area under the graph of f, over the x-axis, and to the *left* of $x = a$ (see Figure 10.6).

The expressions in Definition (10.4) are **improper integrals.** They differ from definite integrals because one of the limits of integration is not a real number. An improper integral is said to **converge** if the limit exists, and the limit is the **value** of the improper integral. If the limit does not exist, the improper integral **diverges.**

There are many applications of Definition (10.4). In Example 4 we shall use an improper integral to calculate the work required to project an object from the surface of the earth to a point outside of the earth's gravitational field. Another, very important application occurs in the investigation of infinite series (see Theorem (11.22)).

An improper integral may have *two* infinite limits of integration, as in the following definition.

DEFINITION (10.5)

Let f be continuous for every x. If a is any real number, then

$$\int_{-\infty}^{\infty} f(x)\, dx = \int_{-\infty}^{a} f(x)\, dx + \int_{a}^{\infty} f(x)\, dx$$

provided both of the improper integrals on the right converge.

If one of the integrals in (10.5) diverges, then $\int_{-\infty}^{\infty} f(x)\, dx$ is said to **diverge.** We can show that (10.5) is independent of the real number a. We can also show that $\int_{-\infty}^{\infty} f(x)\, dx$ is not necessarily the same as $\lim_{t \to \infty} \int_{-t}^{t} f(x)\, dx$.

EXAMPLE 1 Does the integral converge or diverge? If it converges, find its value.

(a) $\int_{2}^{\infty} \dfrac{1}{(x-1)^2}\, dx$ (b) $\int_{2}^{\infty} \dfrac{1}{x-1}\, dx$

SOLUTION

(a) By Definition (10.4)(i),

$$\int_{2}^{\infty} \frac{1}{(x-1)^2}\, dx = \lim_{t \to \infty} \int_{2}^{t} \frac{1}{(x-1)^2}\, dx = \lim_{t \to \infty} \left[\frac{-1}{x-1} \right]_{2}^{t}$$

$$= \lim_{t \to \infty} \left(\frac{-1}{t-1} + \frac{1}{2-1} \right) = 0 + 1 = 1.$$

Thus the integral converges and has the value 1.

(b) By Definition (10.4)(i),

$$\int_{2}^{\infty} \frac{1}{x-1}\, dx = \lim_{t \to \infty} \int_{2}^{t} \frac{1}{x-1}\, dx$$

$$= \lim_{t \to \infty} \left[\ln(x-1) \right]_{2}^{t}$$

$$= \lim_{t \to \infty} \left[\ln(t-1) - \ln(2-1) \right]$$

$$= \lim_{t \to \infty} \ln(t-1) = \infty.$$

Consequently, the improper integral diverges. •

FIGURE 10.7

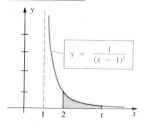

FIGURE 10.8

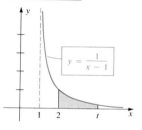

FIGURE 10.9

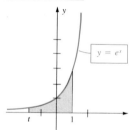

The graphs of the two functions given by the integrands in Example 1 are sketched in Figures 10.7 and 10.8. Note that although the graphs have the same general shape for $x \geq 2$, we may assign an area to the region under the graph shown in Figure 10.7, but not for that in Figure 10.8.

The graph in Figure 10.8 has an interesting property. If the region under the graph of $y = 1/(x - 1)$ from 2 to t is revolved about the x-axis, then the volume of the resulting solid is

$$\pi \int_2^t \frac{1}{(x - 1)^2} \, dx.$$

The improper integral

$$\pi \int_2^\infty \frac{1}{(x - 1)^2} \, dx$$

may be regarded as the volume of the *unbounded* solid obtained by revolving, about the x-axis, the region under the graph of $y = 1/(x - 1)$ for $x \geq 2$. By (a) of Example 1, the value of this improper integral is $\pi \cdot 1$, or π. This gives us the curious fact that although we cannot assign an area to the region, the volume of the solid of revolution generated by the region is finite. (A similar situation is described in Exercise 35.)

EXAMPLE 2 Assign an area to the region that lies under the graph of $y = e^x$, over the x-axis, and to the left of $x = 1$.

SOLUTION The region bounded by the graphs of $y = e^x$, $y = 0$, $x = 1$, and $x = t$, for $t < 1$, is sketched in Figure 10.9. The area of the *unbounded* region to the left of $x = 1$ is

$$\int_{-\infty}^1 e^x \, dx = \lim_{t \to -\infty} \int_t^1 e^x \, dx = \lim_{t \to -\infty} \left[e^x \right]_t^1$$

$$= \lim_{t \to -\infty} (e - e^t) = e - 0 = e \quad \bullet$$

EXAMPLE 3 Evaluate $\int_{-\infty}^\infty \frac{1}{1 + x^2} \, dx$. Sketch the graph of $f(x) = \frac{1}{1 + x^2}$ and interpret the integral as an area.

SOLUTION Using Definition (10.5) with $a = 0$,

$$\int_{-\infty}^\infty \frac{1}{1 + x^2} \, dx = \int_{-\infty}^0 \frac{1}{1 + x^2} \, dx + \int_0^\infty \frac{1}{1 + x^2} \, dx.$$

Next, applying Definition (10.4) (i),

$$\int_0^\infty \frac{1}{1 + x^2} \, dx = \lim_{t \to \infty} \int_0^t \frac{1}{1 + x^2} \, dx = \lim_{t \to \infty} \left[\arctan x \right]_0^t$$

$$= \lim_{t \to \infty} (\arctan t - \arctan 0) = \pi/2 - 0 = \pi/2.$$

Similarly, we may show, by using (10.4) (ii), that

$$\int_{-\infty}^0 \frac{1}{1 + x^2} \, dx = \frac{\pi}{2}.$$

Consequently, the given improper integral converges and has the value $(\pi/2) + (\pi/2) = \pi$.

FIGURE 10.10

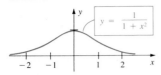

FIGURE 10.11

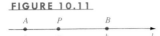

FIGURE 10.12

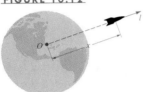

The graph of $y = 1/(1 + x^2)$ is sketched in Figure 10.10. As in our previous discussion, the unbounded region that lies under the graph and above the x-axis may be assigned an area of π square units. •

Let us conclude this section with a physical application of an improper integral. If a and b are the coordinates of two points A and B on a coordinate line l (see Figure 10.11), and if $f(x)$ is the force acting at the point P with coordinate x, then by Definition (6.16), the work done as P moves from A to B is given by

$$W = \int_a^b f(x)\, dx.$$

In similar fashion, the improper integral $\int_a^\infty f(x)\, dx$ may be used to define the work done as P moves indefinitely to the right (in applications, we use the terminology *P moves to infinity*). For example, if $f(x)$ is the force of attraction between a particle fixed at point A and a (movable) particle at P, and if $c > a$, then $\int_c^\infty f(x)\, dx$ represents the work required to move P from the point with coordinate c to infinity.

EXAMPLE 4 Refer to Exercise 19 of Section 6.6. Let l be a coordinate line with origin O at the center of the earth, as shown in Figure 10.12. The gravitational force exerted at a point on l that is a distance x from 0, is given by $f(x) = k/x^2$, for some constant k. Using 4000 miles for the radius of the earth, find the work required to project an object, weighing 100 pounds, along l, from the surface to a point outside of the earth's gravitational field.

SOLUTION Theoretically, there is *always* a gravitational force $f(x)$ acting on the object; however, we may think of projecting the object from the surface to infinity. From the preceding discussion we wish to find

$$W = \int_{4000}^\infty f(x)\, dx.$$

By definition, $f(x)$ is the weight of an object that is a distance x from O and hence,

$$100 = F(4000) = \frac{k}{(4000)^2}$$

or

$$k = 100(4000)^2 = 10^2 \cdot 16 \cdot 10^6 = 16 \cdot 10^8.$$

Thus

$$W = \int_{4000}^\infty 16 \cdot 10^8 \frac{1}{x^2}\, dx = 16 \cdot 10^8 \lim_{t \to \infty} \int_{4000}^t \frac{1}{x^2}\, dx$$

$$= 16 \cdot 10^8 \lim_{t \to \infty} \left[-\frac{1}{x} \right]_{4000}^t = 16 \cdot 10^8 \lim_{t \to \infty} \left[-\frac{1}{t} + \frac{1}{4000} \right]$$

$$= \frac{16 \cdot 10^8}{4000} = 4 \cdot 10^5 \text{ mi-lb.}$$

In terms of foot-pounds,

$$W = 5280 \cdot 4 \cdot 10^5 \approx (2.1)10^9 \text{ ft-lb}$$

or approximately two trillion ft-lb. •

Exer. 1–24: Does the integral converge or diverge? If it converges, find its value.

1 $\int_1^\infty \dfrac{1}{x^{4/3}}\,dx$

2 $\int_{-\infty}^0 \dfrac{1}{(x-1)^3}\,dx$

3 $\int_1^\infty \dfrac{1}{x^{3/4}}\,dx$

4 $\int_0^\infty \dfrac{x}{1+x^2}\,dx$

5 $\int_{-\infty}^2 \dfrac{1}{5-2x}\,dx$

6 $\int_{-\infty}^\infty \dfrac{x}{x^4+9}\,dx$

7 $\int_0^\infty e^{-2x}\,dx$

8 $\int_{-\infty}^0 e^x\,dx$

9 $\int_{-\infty}^{-1} \dfrac{1}{x^3}\,dx$

10 $\int_0^\infty \dfrac{1}{\sqrt[3]{x+1}}\,dx$

11 $\int_{-\infty}^0 \dfrac{1}{(x-8)^{2/3}}\,dx$

12 $\int_1^\infty \dfrac{x}{(1+x^2)^2}\,dx$

13 $\int_0^\infty \dfrac{\cos x}{1+\sin^2 x}\,dx$

14 $\int_{-\infty}^2 \dfrac{1}{x^2+4}\,dx$

15 $\int_{-\infty}^\infty xe^{-x^2}\,dx$

16 $\int_{-\infty}^\infty \cos^2 x\,dx$

17 $\int_1^\infty \dfrac{\ln x}{x}\,dx$

18 $\int_3^\infty \dfrac{1}{x^2-1}\,dx$

19 $\int_0^\infty \cos x\,dx$

20 $\int_{-\infty}^{\pi/2} \sin 2x\,dx$

21 $\int_{-\infty}^\infty \operatorname{sech} x\,dx$

22 $\int_0^\infty xe^{-x}\,dx$

23 $\int_{-\infty}^0 \dfrac{1}{x^2-3x+2}\,dx$

24 $\int_4^\infty \dfrac{x+18}{x^2+x-12}\,dx$

Exer. 25–28: If f and g are continuous functions such that $0 \le f(x) \le g(x)$ for every x in $[a, \infty)$, then the following comparison tests for improper integrals are true:
(i) If $\int_a^\infty g(x)\,dx$ converges, then $\int_a^\infty f(x)\,dx$ converges.
(ii) If $\int_a^\infty f(x)\,dx$ diverges, then $\int_a^\infty g(x)\,dx$ diverges.
Use these tests to determine if the improper integral converges or diverges.

25 $\int_1^\infty \dfrac{1}{1+x^4}\,dx$

26 $\int_2^\infty \dfrac{1}{\sqrt[3]{x^2-1}}\,dx$

27 $\int_2^\infty \dfrac{1}{\ln x}\,dx$

28 $\int_1^\infty e^{-x^2}\,dx$

Exer. 29–32: Assign, if possible, a value to (a) the area of the region R, and (b) the volume of the solid obtained by revolving R about the x-axis.

29 $R = \{(x, y): x \ge 1,\ 0 \le y \le 1/x\}$

30 $R = \{(x, y): x \ge 1,\ 0 \le y \le 1/\sqrt{x}\}$

31 $R = \{(x, y): x \ge 4,\ 0 \le y \le x^{-3/2}\}$

32 $R = \{(x, y): x \ge 8,\ 0 \le y \le x^{-2/3}\}$

33 The unbounded region to the right of the y-axis and between the graphs of $y = e^{-x^2}$ and $y = 0$ is revolved about the y-axis. Show that a volume can be assigned to the resulting unbounded solid. What is the volume?

34 The graph of $y = e^{-x}$ for $x \ge 0$ is revolved about the x-axis. Show that an area can be assigned to the resulting unbounded surface. What is the area?

35 The solid of revolution known as *Gabriel's horn* is generated by rotating the region under the graph of $y = 1/x$ for $x \ge 1$ about the x-axis (see figure).
(a) Show that Gabriel's horn has a finite volume of π cubic units.
(b) Is a finite volume obtained if the graph is rotated about the y-axis?
(c) Show that the surface area of Gabriel's horn is given by $\int_1^\infty 2\pi(1/x)\sqrt{x^4+1}\,dx$. Use a comparison test (see Exercises 25–28) with $f(x) = 2\pi/x$ to establish that this integral diverges. Thus we cannot assign an area to the surface even though the volume of the horn is finite.

EXERCISE 35

$y = 1/x,\ x \ge 1$

36 A spacecraft carries a fuel supply of mass m. As a conservation measure, the captain decides to burn the fuel at a rate of $R(t) = mke^{-kt}$ g/sec, for some positive constant k.
(a) What does the improper integral $\int_0^\infty R(t)\,dt$ represent?
(b) When will the spacecraft run out of fuel?

37 The force (in dynes) with which two electrons repel one another is inversely proportional to the square of the distance (in cm) between them. If, in Figure 10.11, one electron is fixed at A, find the work done if another electron is repelled along l from a point B, which is one cm from A, to infinity.

38 An electric dipole consists of opposite charges separated by a small distance d. Suppose that charges of $+q$ and $-q$ units are located on a coordinate line l at $\frac{1}{2}d$ and $-\frac{1}{2}d$, respectively (see figure). By Coulomb's law, the net force acting on a unit charge of -1 unit at $x > \frac{1}{2}d$ is given by

$$f(x) = \dfrac{-kq}{(x - \frac{1}{2}d)^2} + \dfrac{kq}{(x + \frac{1}{2}d)^2}$$

for some positive constant k. If $a > \frac{1}{2}d$, find the work done in moving the unit charge along l from a to infinity.

EXERCISE 38

39 The reliability $R(t)$ of a product is the probability that it will not require repair for at least t years. To design a warranty guarantee, a manufacturer must know the average time of service before first repair of a product. This is given by the improper integral $\int_0^\infty (-t)R'(t)\,dt$.

(a) For many high-quality products, $R(t)$ has the form e^{-kt} for some positive constant k. Find an expression in terms of k for the average time of service before repair.

(b) Is it possible to manufacture a product for which $R(t) = 1/(t + 1)$?

40 A sum of money is deposited into an account that pays interest at 8% per year, compounded continuously (see Exercise 46 of Section 10.2). Starting T years from now, money will be withdrawn at the *capital flow rate* of $f(t)$ dollars per year, continuing indefinitely. To generate future income at this rate, the minimum amount A that must be deposited, or the *present value of the capital flow*, is given by the improper integral $A = \int_T^\infty f(t)e^{-0.08t}\,dt$. Find A if the income desired 20 years from now is

(a) 12,000 dollars per year;

(b) $12{,}000e^{0.04t}$ dollars per year.

41 (a) Use integration by parts to establish the formula

$$\int_0^\infty x^2 e^{-ax^2}\,dx = \frac{1}{2a^{3/2}} \int_0^\infty e^{-u^2}\,du.$$

The value of this integral is $\sqrt{\pi}/2$ (see Exercise 26 of Section 17.7).

(b) The relative number of gas molecules in a container that travel at a speed of v cm/sec can be found by using the *Maxwell-Boltzmann speed distribution F*:

$$F(v) = cv^2 e^{-mv^2/(2kT)}$$

for temperature T (in °K), mass m of a molecule, and positive constants c and k. The constant c must be selected so that $\int_0^\infty F(v)\,dv = 1$. Use part (a) to express c in terms of k, T, and m.

42 The *Fourier transform* is useful for solving certain differential equations. The *Fourier cosine transform* of a function f is defined by

$$F_C[f(x)] = \int_0^\infty f(x) \cos sx\,dx$$

for every real number s for which the improper integral converges. Find $F_C[e^{-ax}]$ for $a > 0$.

Exer. 43–48: In the theory of differential equations, if f is a function, then the *Laplace transform L* of $f(x)$ is defined by

$$L[f(x)] = \int_0^\infty e^{-sx}f(x)\,dx$$

for every real number s for which the improper integral converges. Find $L[f(x)]$ if $f(x)$ is the expression.

43 1   **44** x **45** $\cos x$

46 $\sin x$ **47** e^{ax} **48** $\sin ax$

49 The *gamma function* Γ is defined by $\Gamma(n) = \int_0^\infty x^{n-1}e^{-x}\,dx$ for every positive real number n.

(a) Find $\Gamma(1)$, $\Gamma(2)$, and $\Gamma(3)$.

(b) Prove that $\Gamma(n + 1) = n\Gamma(n)$. (*Hint:* Integrate by parts.)

(c) Use mathematical induction to prove that if n is any positive integer, then $\Gamma(n + 1) = n!$. (This shows that factorials are special values of the gamma function.)

50 Refer to Exercise 49. Functions given by $f(x) = cx^k e^{-ax}$ with $x > 0$ are called *gamma distributions* and play an important role in probability theory. The constant c must be selected so that $\int_0^\infty f(x)\,dx = 1$. Express c in terms of the positive constants k and a, and the gamma function Γ.

10.4 INTEGRALS WITH DISCONTINUOUS INTEGRANDS

FIGURE 10.13
$\int_a^t f(x)\,dx$

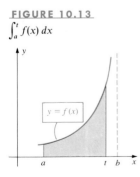

$y = f(x)$

If a function f is continuous on a closed interval $[a, b]$, then by Theorem (5.12) the definite integral $\int_a^b f(x)\,dx$ exists. If f has an infinite discontinuity at some number in the interval it may still be possible to assign a value to the integral. Suppose, for example, that f is continuous and nonnegative on the half-open interval $[a, b)$ and $\lim_{x \to b^-} f(x) = \infty$. If $a < t < b$, then the area $A(t)$ under the graph of f from a to t (see Figure 10.13) is

$$A(t) = \int_a^t f(x)\,dx$$

If $\lim_{t \to b^-} A(t)$ exists, then the limit may be interpreted as the area of the unbounded region that lies under the graph of f, over the x-axis, and between $x = a$ and $x = b$. We shall denote this number by $\int_a^b f(x)\,dx$.

For the situation illustrated in Figure 10.14, $\lim_{x \to a^+} f(x) = \infty$, and we define $\int_a^b f(x)\, dx$ as the limit of $\int_t^b f(x)\, dx$ as $t \to a^+$.

FIGURE 10.14

$\int_t^b f(x)\, dx$

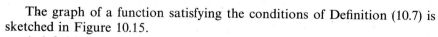

These remarks are the motivation for the following definition.

DEFINITION (10.6)

(i) If f is continuous on $[a, b)$ and discontinuous at b, then

$$\int_a^b f(x)\, dx = \lim_{t \to b^-} \int_a^t f(x)\, dx$$

provided the limit exists.

(ii) If f is continuous on $(a, b]$ and discontinuous at a, then

$$\int_a^b f(x)\, dx = \lim_{t \to a^+} \int_t^b f(x)\, dx$$

provided the limit exists.

As in the preceding section, the integrals defined in (10.6) are referred to as *improper integrals* and they *converge* if the limits exists. The limits are called the *values* of the improper integrals. If the limits do not exist, the improper integrals *diverge*.

Another type of improper integral is defined as follows.

FIGURE 10.15

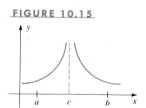

DEFINITION (10.7)

If f has a discontinuity at a number c in the open interval (a, b) but is continuous elsewhere on $[a, b]$, then

$$\int_a^b f(x)\, dx = \int_a^c f(x)\, dx + \int_c^b f(x)\, dx$$

provided *both* of the improper integrals on the right converge. If both converge, then the value of the improper integral $\int_a^b f(x)\, dx$ is the sum of the two values.

FIGURE 10.16

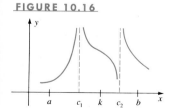

The graph of a function satisfying the conditions of Definition (10.7) is sketched in Figure 10.15.

A definition similar to (10.7) is used if f has any finite number of discontinuities in (a, b). For example, suppose f has discontinuities at c_1 and c_2, with $c_1 < c_2$, but is continuous elsewhere in $[a, b]$. One possibility is illustrated in Figure 10.16. In this case we choose a number k between c_1 and c_2 and express $\int_a^b f(x)\, dx$ as a sum of four improper integrals over the intervals $[a, c_1]$, $[c_1, k]$, $[k, c_2]$, and $[c_2, b]$, respectively. By definition, $\int_a^b f(x)\, dx$

converges if and only if each of the four improper integrals in the sum converges. We can show that this definition is independent of the number k.

Finally, if f is continuous on (a, b) but has infinite discontinuities at a and b, then we again define $\int_a^b f(x)\, dx$ by means of (10.7).

EXAMPLE 1 Evaluate $\int_0^3 \dfrac{1}{\sqrt{3-x}}\, dx$.

SOLUTION Since the integrand has an infinite discontinuity at $x = 3$, we apply Definition (10.6) (i) as follows:

$$\int_0^3 \frac{1}{\sqrt{3-x}}\, dx = \lim_{t \to 3^-} \int_0^t \frac{1}{\sqrt{3-x}}\, dx$$

$$= \lim_{t \to 3^-} \left[-2\sqrt{3-x} \right]_0^t$$

$$= \lim_{t \to 3^-} \left[-2\sqrt{3-t} + 2\sqrt{3} \right]$$

$$= 0 + 2\sqrt{3} = 2\sqrt{3} \qquad \bullet$$

EXAMPLE 2 Does the improper integral $\int_0^1 \dfrac{1}{x}\, dx$ converge or diverge?

SOLUTION The integrand is undefined at $x = 0$. Applying (10.6) (ii),

$$\int_0^1 \frac{1}{x}\, dx = \lim_{t \to 0^+} \int_t^1 \frac{1}{x}\, dx = \lim_{t \to 0^+} \left[\ln x \right]_t^1$$

$$= \lim_{t \to 0^+} \left[0 - \ln t \right] = \infty.$$

Since the limit does not exist, the improper integral diverges. $\bullet$

EXAMPLE 3 Does the improper integral $\int_0^4 \dfrac{1}{(x-3)^2}\, dx$ converge or diverge?

SOLUTION The integrand is undefined at $x = 3$. Since this number is in the interval $(0, 4)$, we use Definition (10.7) with $c = 3$:

$$\int_0^4 \frac{1}{(x-3)^2}\, dx = \int_0^3 \frac{1}{(x-3)^2}\, dx + \int_3^4 \frac{1}{(x-3)^2}\, dx$$

For the integral on the left to converge, *both* integrals on the right must converge. Applying Definition (10.6) (i) to the first integral,

$$\int_0^3 \frac{1}{(x-3)^2}\, dx = \lim_{t \to 3^-} \int_0^t \frac{1}{(x-3)^2}\, dx$$

$$= \lim_{t \to 3^-} \left[\frac{-1}{x-3} \right]_0^t$$

$$= \lim_{t \to 3^-} \left(\frac{-1}{t-3} - \frac{1}{3} \right) = \infty.$$

Thus the given improper integral diverges. $\bullet$

It is important to note that the Fundamental Theorem of Calculus cannot be applied to the integral in Example 3, since the function given by the integrand is not continuous on $[0, 4]$. If we had (incorrectly) applied the Fundamental Theorem, we would have obtained

$$\left[\frac{-1}{x - 3}\right]_0^4 = -1 - \frac{1}{3} = -\frac{4}{3}.$$

This result is obviously incorrect, since the integrand is never negative.

EXAMPLE 4 Evaluate $\int_{-2}^{7} \frac{1}{(x + 1)^{2/3}}\, dx$.

SOLUTION The integrand is undefined at $x = -1$, which is in the interval $(-2, 7)$. Hence we apply Definition (10.7) with $c = -1$:

$$\int_{-2}^{7} \frac{1}{(x + 1)^{2/3}}\, dx = \int_{-2}^{-1} \frac{1}{(x + 1)^{2/3}}\, dx + \int_{-1}^{7} \frac{1}{(x + 1)^{2/3}}\, dx$$

We next investigate each of the integrals on the right-hand side of this equation. Using (10.6) (i) with $b = -1$ gives us

$$\begin{aligned}
\int_{-2}^{-1} \frac{1}{(x + 1)^{2/3}}\, dx &= \lim_{t \to -1^-} \int_{-2}^{t} \frac{1}{(x + 1)^{2/3}}\, dx \\
&= \lim_{t \to -1^-} \left[3(x + 1)^{1/3}\right]_{-2}^{t} \\
&= \lim_{t \to -1^-} \left[3(t + 1)^{1/3} - 3(-1)^{1/3}\right] \\
&= 0 + 3 = 3.
\end{aligned}$$

Similarly, using (10.6) (ii) with $a = -1$,

$$\begin{aligned}
\int_{-1}^{7} \frac{1}{(x + 1)^{2/3}}\, dx &= \lim_{t \to -1^+} \int_{t}^{7} \frac{1}{(x + 1)^{2/3}}\, dx \\
&= \lim_{t \to -1^+} \left[3(x + 1)^{1/3}\right]_{t}^{7} \\
&= \lim_{t \to -1^+} \left[3(8)^{1/3} - 3(t + 1)^{1/3}\right] \\
&= 6 - 0 = 6.
\end{aligned}$$

Since both integrals converge, the given integral converges and has the value $3 + 6 = 9$. •

An improper integral may have both a discontinuity in the integrand and an infinite limit of integration. Integrals of this type may be investigated by expressing them as sums of improper integrals, each of which has *one* of the forms previously defined. As an illustration, since the integrand of $\int_0^{\infty} (1/\sqrt{x})\, dx$ is discontinuous at $x = 0$, we choose any number greater than 0—say 1—and write

$$\int_0^{\infty} \frac{1}{\sqrt{x}}\, dx = \int_0^{1} \frac{1}{\sqrt{x}}\, dx + \int_1^{\infty} \frac{1}{\sqrt{x}}\, dx.$$

We can show that the first integral on the right side of the equation converges and the second diverges. Hence (by definition) the given integral diverges.

FIGURE 10.17

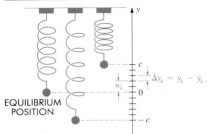

EQUILIBRIUM
POSITION

Improper integrals of the types considered in this section arise in physical applications. Figure 10.17 is a schematic drawing of a spring with an attached weight that is oscillating between points with coordinates $-c$ and c on a coordinate line y (the y-axis has been positioned at the right for clarity). The **period** T is the time required for one complete oscillation, that is, *twice* the time required for the weight to cover the interval $[-c, c]$. The next example illustrates how an improper integral results when obtaining a formula for T.

EXAMPLE 5 Let $v(y)$ denote the velocity of the weight in Figure 10.17 when it is at the point with coordinate y in $[-c, c]$. Show that the period T is given by

$$T = 2 \int_{-c}^{c} \frac{1}{v(y)} \, dy$$

SOLUTION Let us partition $[-c, c]$ in the usual way and let $\Delta y_k = y_k - y_{k-1}$ denote the distance the weight travels during the time interval Δt_k. If w_k is any number in the subinterval $[y_{k-1}, y_k]$, then $v(w_k)$ is the velocity of the weight when it is at the point with coordinate w_k. If the norm of the partition is small and if we assume v is a continuous function, then the distance Δy_k may be approximated by the product $v(w_k) \Delta t_k$, that is,

$$\Delta y_k \approx v(w_k) \Delta t_k.$$

Hence the time required for the weight to cover the distance Δy_k may be approximated by

$$\Delta t_k \approx \frac{1}{v(w_k)} \Delta y_k$$

and, therefore,

$$T = 2 \sum_k \Delta t_k \approx 2 \sum_k \frac{1}{v(w_k)} \Delta y_k.$$

Considering the limit of the sums on the right, and using the definition of definite integral,

$$T = 2 \int_{-c}^{c} \frac{1}{v(y)} \, dy.$$

Note that $v(c) = 0$ and $v(-c) = 0$, and hence the integral is improper. •

EXERCISES 10.4

Exer. 1–30: Does the integral converge or diverge? If it converges, find its value.

1 $\int_0^8 \frac{1}{\sqrt[3]{x}} \, dx$

2 $\int_0^9 \frac{1}{\sqrt{x}} \, dx$

3 $\int_{-3}^1 \frac{1}{x^2} \, dx$

4 $\int_{-2}^{-1} \frac{1}{(x+2)^{5/4}} \, dx$

5 $\int_0^{\pi/2} \sec^2 x \, dx$

6 $\int_0^1 \frac{e^{\sqrt{x}}}{\sqrt{x}} \, dx$

7 $\int_0^4 \frac{1}{(4-x)^{3/2}} \, dx$

8 $\int_0^{-1} \frac{1}{\sqrt[3]{x+1}} \, dx$

9 $\int_0^4 \frac{1}{(4-x)^{2/3}} \, dx$

10 $\int_1^2 \frac{x}{x^2-1} \, dx$

11 $\int_{-2}^2 \frac{1}{(x+1)^3} \, dx$

12 $\int_{-1}^1 x^{-4/3} \, dx$

13 $\int_{-2}^0 \frac{1}{\sqrt{4-x^2}} \, dx$

14 $\int_{-2}^0 \frac{x}{\sqrt{4-x^2}} \, dx$

15 $\int_{-1}^{2} \frac{1}{x} \, dx$

16 $\int_{0}^{4} \frac{1}{x^2 - x - 2} \, dx$

17 $\int_{0}^{1} x \ln x \, dx$

18 $\int_{0}^{\pi/2} \tan^2 x \, dx$

19 $\int_{0}^{\pi/2} \tan x \, dx$

20 $\int_{0}^{\pi/2} \frac{1}{1 - \cos x} \, dx$

21 $\int_{2}^{4} \frac{x - 2}{x^2 - 5x + 4} \, dx$

22 $\int_{1/e}^{e} \frac{1}{x (\ln x)^2} \, dx$

23 $\int_{-1}^{2} \frac{1}{x^2} \cos \frac{1}{x} \, dx$

24 $\int_{0}^{\pi} \sec x \, dx$

25 $\int_{0}^{\pi} \frac{\cos x}{\sqrt{1 - \sin x}} \, dx$

26 $\int_{0}^{9} \frac{x}{\sqrt[3]{x - 1}} \, dx$

27 $\int_{0}^{4} \frac{1}{x^2 - 4x + 3} \, dx$

28 $\int_{-1}^{3} \frac{x}{\sqrt[3]{x^2 - 1}} \, dx$

29 $\int_{0}^{\infty} \frac{1}{(x - 4)^2} \, dx$

30 $\int_{-\infty}^{0} \frac{1}{x + 2} \, dx$

Exer. 31–34: Suppose f and g are continuous such that $0 \le f(x) \le g(x)$ for every x in $(a, b]$. If f and g are discontinuous at $x = a$, then the following *comparison tests* can be proved:

(i) If $\int_{a}^{b} g(x) \, dx$ converges, then $\int_{a}^{b} f(x) \, dx$ converges.

(ii) If $\int_{a}^{b} f(x) \, dx$ diverges, then $\int_{a}^{b} g(x) \, dx$ diverges.

Analogous tests may be stated for continuity on $[a, b)$ with a discontinuity at $x = b$. Use these tests to determine if the improper integral converges or diverges.

31 $\int_{0}^{\pi} \frac{\sin x}{\sqrt{x}} \, dx$

32 $\int_{0}^{\pi/4} \frac{\sec x}{x^3} \, dx$

33 $\int_{0}^{2} \frac{\cosh x}{(x - 2)^2} \, dx$

34 $\int_{0}^{1} \frac{e^{-x}}{x^{2/3}} \, dx$

Exer. 35–36: Find all values of n for which the integral converges.

35 $\int_{0}^{1} x^n \, dx$

36 $\int_{0}^{1} x^n \ln x \, dx$

Exer. 37–40: Assign, if possible, a value to (a) the area of the region R, and (b) the volume of the solid obtained by revolving R about the x-axis.

37 $R = \{(x, y): 0 \le x \le 1, \; 0 \le y \le 1/\sqrt{x}\}$

38 $R = \{(x, y): 0 \le x \le 1, \; 0 \le y \le 1/\sqrt[3]{x}\}$

39 $R = \{(x, y): -4 \le x \le 4, \; 0 \le y \le 1/(x + 4)\}$

40 $R = \{(x, y): 1 \le x \le 2, \; 0 \le y \le 1/(x - 1)\}$

41 Refer to Example 5. If the weight in Figure 10.17 has mass m, and if the spring obeys Hooke's law (with spring constant $k > 0$), then, in the absence of frictional forces, the velocity v of the weight is a solution of the differential equation

$$mv \frac{dv}{dy} + ky = 0.$$

(a) Use separation of variables (see Section 7.6) to show that $v^2 = (k/m)(c^2 - y^2)$.
   (*Hint:* Recall that $v(c) = v(-c) = 0$.)

(b) Find the period T of the oscillation.

42 A simple pendulum consists of a bob of mass m attached to a string of length L (see figure). If we assume that the string is weightless and that no other frictional forces are present, then the angular velocity $v = d\theta/dt$ is a solution of the differential equation

$$v \frac{dv}{d\theta} + \frac{g}{L} \sin \theta = 0$$

for a gravitational constant g.

(a) If $v = 0$ at $\theta = \pm\theta_0$, use separation of variables to show that

$$v^2 = (2g/L)(\cos \theta - \cos \theta_0).$$

(b) The period T of the pendulum is twice the amount of time needed for θ to change from $-\theta_0$ to θ_0. Show that T is given by the improper integral

$$T = 2\sqrt{2L/g} \int_{0}^{\theta_0} \frac{1}{\sqrt{\cos \theta - \cos \theta_0}} \, d\theta.$$

EXERCISE 42

43 When a dose of y_0 mg of a drug is injected directly into the bloodstream, the average length of time T that a molecule remains in the bloodstream is given by the formula $T = (1/y_0) \int_{0}^{y_0} t \, dy$ for the time t at which exactly y mg is still present.

(a) If $y = y_0 e^{-kt}$ for some positive constant k, explain why the integral for T is improper.

(b) If τ is the half-life of the drug in the bloodstream, show that $T = \tau/\ln 2$.

44 In fishery science, the collection of fish that results from one annual reproduction is referred to as a *cohort*. The number N of fish still alive after t years is usually given by an exponential function. For North Sea haddock with initial size of a cohort N_0, $N = N_0 e^{-0.2t}$. The average life expectancy T (in years) of a fish in a cohort is given by $T = (1/N_0) \int_{0}^{N_0} t \, dN$ for the time t when precisely N fish are still alive.

(a) Estimate the value of T for North Sea haddock.

(b) Is it possible to have a species such that $N = N_0/(1 + kN_0 t)$ for some positive constant k? If so, compute T for such a species.

Recall that f is a polynomial function of degree n if

$$f(x) = a_0 + a_1 x + a_2 x^2 + \cdots + a_n x^n$$

where each a_k is a real number, $a_n \neq 0$, and the exponents are nonnegative integers. Polynomial functions are the simplest functions to use for calculations—their values can be found by employing only additions and multiplications of real numbers. More complicated operations are needed to calculate values of logarithmic, exponential, or trigonometric functions; however, sometimes it is possible to *approximate* values by using polynomials. For example, since $\lim_{x \to 0} (\sin x)/x = 1$, it follows that if x is close to 0, then $\sin x \approx x$; that is, $\sin x$ *may be approximated by the polynomial* x (provided x is close to 0).

As a second illustration, let f be the natural exponential function, that is, $f(x) = e^x$ for every x. Suppose we are interested in calculating values of f when x is close to 0. Since $f'(x) = e^x$, the slope of the tangent line at the point $(0, 1)$ on the graph of f is $f'(0) = e^0 = 1$. Hence an equation of the tangent line is

$$y - 1 = 1(x - 0) \quad \text{or} \quad y = 1 + x.$$

FIGURE 10.18

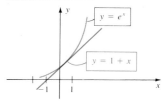

Referring to Figure 10.18, we can see that if x is very close to 0, then the point $(x, 1 + x)$ on the tangent line is close to the point (x, e^x) on the graph of f, and hence we may write $e^x \approx 1 + x$. This formula allows us to approximate e^x by means of a polynomial of degree 1. Since the approximation is obviously poor unless x is very close to 0, let us seek a second-degree polynomial

$$g(x) = a + bx + cx^2$$

such that $e^x \approx g(x)$. The first and second derivatives of $g(x)$ are

$$g'(x) = b + 2cx \quad \text{and} \quad g''(x) = 2c.$$

If we want the graph of g (a parabola) to have (i) the same y-intercept, (ii) the same tangent line, and (iii) the same concavity as the graph of f at the point $(0, 1)$, then we must have

(i) $g(0) = f(0)$,   (ii) $g'(0) = f'(0)$,   (iii) $g''(0) = f''(0)$.

Since all derivatives of e^x equal e^x, and $e^0 = 1$, these three equations imply that

$$a = 1, \quad b = 1, \quad \text{and} \quad 2c = 1$$

and hence

$$e^x \approx g(x) = 1 + x + \tfrac{1}{2}x^2.$$

The graphs of f and g are sketched in Figure 10.19. Comparing the graphs with Figure 10.18, we see that if x is close to 0, then e^x is closer to $1 + x + \tfrac{1}{2}x^2$ than to $1 + x$.

FIGURE 10.19

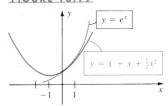

If we wish to approximate $f(x) = e^x$ by means of a *third*-degree polynomial $h(x)$, it is natural to require that $h(0)$, $h'(0)$, $h''(0)$, and $h'''(0)$ be the same as $f(0)$, $f'(0)$, $f''(0)$, and $f'''(0)$, respectively. The graphs of f and h then have the same tangent line and concavity at $(0, 1)$ and, in addition, their *rates of change of concavity* with respect to x (that is, the third derivatives) are equal. Using the same technique employed previously gives us

$$e^x \approx h(x) = 1 + x + \frac{1}{2}x^2 + \frac{1}{3!}x^3.$$

To give some idea of the accuracy of this approximation, we have displayed several values of e^x and $h(x)$, approximated to the nearest hundredth, in the following table.

x	-1.5	-1.0	-0.5	0	0.5	1.0	1.5
e^x	0.22	0.37	0.61	1	1.65	2.72	4.48
$h(x)$	0.06	0.33	0.60	1	1.65	2.67	4.19

FIGURE 10.20

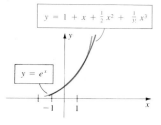

Observe that the error in the approximation increases as $|x|$ increases. The graphs of f and h are sketched in Figure 10.20. Notice the improvement in the approximation near $x = 0$.

If we continued in this manner and determined a polynomial of degree n, whose first n derivatives coincide with those of f at $x = 0$, we would arrive at

$$e^x \approx 1 + x + \frac{1}{2}x^2 + \frac{1}{3!}x^3 + \cdots + \frac{1}{n!}x^n.$$

It will follow from our later work that this remarkably simple formula can be used to approximate e^x to any desired degree of accuracy.

We now ask two questions:

1. Does there exist a *general* formula that may be used to obtain polynomial approximations for any exponential function, logarithmic function, trigonometric function, inverse trigonometric function, and other transcendental or algebraic functions?

2. Can the previous discussion be generalized to the case where x is close to an arbitrary number $c \neq 0$?

The answer to both questions is yes, provided the function has a sufficient number of derivatives. To see why this is true, suppose f is a function that has many derivatives, and consider a number c in the domain of f. Let us proceed as we did for the special case of e^x discussed previously, but with c in place of 0. First we note that the only polynomial of degree 0 that coincides with f at c is the constant $f(c)$. To approximate $f(x)$ near c by a polynomial of degree 1, we choose the polynomial whose graph is the tangent line to the graph of f at $(c, f(c))$. The equation of this tangent line is

$$y - f(c) = f'(c)(x - c) \quad \text{or} \quad y = f(c) + f'(c)(x - c).$$

Thus the desired first-degree polynomial is

$$f(c) + f'(c)(x - c).$$

A better approximation should be obtained by using a polynomial whose graph has the same *concavity* and tangent line as that of f at $(c, f(c))$. We can verify that these conditions are fulfilled by the second-degree polynomial

$$f(c) + f'(c)(x - c) + \frac{f''(c)}{2}(x - c)^2.$$

If we next consider the third-degree polynomial

$$g(x) = f(c) + f'(c)(x - c) + \frac{f''(c)}{2!}(x - c)^2 + \frac{f'''(c)}{3!}(x - c)^3$$

then it is easy to show that in addition to $g'(c) = f'(c)$ and $g''(c) = f''(c)$, we also have $g'''(c) = f'''(c)$. It appears that if this pattern is continued, we should get better approximations to $f(x)$ when x is near c. This leads to the first $n + 1$ terms on the right side of Formula (10.9). The form of the last term is a consequence of the next result.

THEOREM (10.8)

Let f be a function and n a positive integer such that the derivative $f^{(n+1)}(x)$ exists for every x in an interval I. If c and d are distinct numbers in I, then there is a number z between c and d such that

$$f(d) = f(c) + \frac{f'(c)}{1!}(d - c) + \frac{f''(c)}{2!}(d - c)^2 + \cdots$$

$$+ \frac{f^{(n)}(c)}{n!}(d - c)^n + \frac{f^{(n+1)}(z)}{(n + 1)!}(d - c)^{n+1}$$

PROOF If n is a positive integer, let us define R_n as follows:

$$R_n = f(d) - \left[f(c) + \frac{f'(c)}{1!}(d - c) + \frac{f''(c)}{2!}(d - c)^2 + \cdots + \frac{f^{(n)}(c)}{n!}(d - c)^n \right].$$

This equation may be rewritten as

$$f(d) = f(c) + \frac{f'(c)}{1!}(d - c) + \frac{f''(c)}{2!}(d - c)^2 + \cdots + \frac{f^{(n)}(c)}{n!}(d - c)^n + R_n.$$

All we need to show is that for a suitable z, R_n is the same as the last term of the formula given in the statement of the theorem.

Let g be the function defined by

$$g(x) = f(d) - f(x) - \frac{f'(x)}{1!}(d - x) - \frac{f''(x)}{2!}(d - x)^2 - \cdots$$

$$- \frac{f^{(n)}(x)}{n!}(d - x)^n - R_n \frac{(d - x)^{n+1}}{(d - c)^{n+1}}$$

for every x in I. If we differentiate each side of the equation, then many terms on the right cancel. We leave it as an exercise to show that

$$g'(x) = -\frac{f^{(n+1)}(x)}{n!}(d - x)^n + R_n(n + 1)\frac{(d - x)^n}{(d - c)^{n+1}}.$$

It is easy to see that $g(d) = 0$. Moreover, substituting c for x in the expression for $g(x)$ and making use of the second equation of the proof gives us $g(c) = 0$. According to Rolle's Theorem, there is a number z between c and d such that $g'(z) = 0$. Evaluating $g'(x)$ at z and solving for R_n, we see that

$$R_n = \frac{f^{(n+1)}(z)}{(n + 1)!}(d - c)^{n+1}$$

which is what we wished to prove. • •

Replacement of d by x in Theorem (10.8) leads to the following result, which bears the name of the English mathematician Brook Taylor (1685–1731).

<table>
<tr><td style="text-align:right; vertical-align:top;">TAYLOR'S FORMULA WITH (10.9)
THE REMAINDER</td><td>Let f have $n + 1$ derivatives throughout an interval containing c. If x is any number in the interval, then there is a number z between c and x such that

$$f(x) = f(c) + \frac{f'(c)}{1!}(x - c) + \frac{f''(c)}{2!}(x - c)^2 + \cdots$$

$$+ \frac{f^{(n)}(c)}{n!}(x - c)^n + \frac{f^{(n+1)}(z)}{(n + 1)!}(x - c)^{n+1}$$
</td></tr>
</table>

For convenience we denote the sum of the first $n + 1$ terms in Taylor's Formula by $P_n(x)$, and give it a special name, as in (i) of the next definition.

<table>
<tr><td style="text-align:right; vertical-align:top;">DEFINITION (10.10)</td><td>(i) The **nth-degree Taylor Polynomial $P_n(x)$** of f at c is

$$P_n(x) = f(x) + \frac{f'(c)}{1!}(x - c) + \frac{f''(c)}{2!}(x - c)^2 + \cdots + \frac{f^{(n)}(c)}{n!}(x - c)^n$$

(ii) The **Taylor Remainder $R_n(x)$** of f at c is

$$R_n(x) = \frac{f^{(n+1)}(z)}{(n + 1)!}(x - c)^{n+1}$$

for a number z between c and x given by (10.9).
</td></tr>
</table>

We may now state the following result.

<table>
<tr><td style="text-align:right; vertical-align:top;">THEOREM (10.11)</td><td>Let f have $n + 1$ derivatives throughout an interval containing c. If x is any number in the interval, then

(i) $f(x) = P_n(x) + R_n(x)$.

(ii) $f(x) \approx P_n(x)$ if $x \approx c$.

The error involved in this approximation is $|R_n(x)|$.
</td></tr>
</table>

PROOF Part (i) is merely a restatement of Formula (10.9) using the notation of (10.10). Part (ii) follows from the fact that $|f(x) - P_n(x)| = |R_n(x)|$. • •

In the next two examples, Taylor Polynomials are used to approximate function values. To discuss the accuracy of an approximation, we must agree on what is meant by 1-decimal-place accuracy, 2-decimal-place accuracy, and so on. Let us adopt the following convention. If E is the error in an approximation, then the approximation will be considered accurate to k decimal places if $|E| < 0.5 \times 10^{-k}$. For example, we have

$$1\text{-decimal-place accuracy if } |E| < 0.5 \times 10^{-1} = 0.05$$
$$2\text{-decimal-place accuracy if } |E| < 0.5 \times 10^{-2} = 0.005$$
$$3\text{-decimal-place accuracy if } |E| < 0.5 \times 10^{-3} = 0.0005.$$

If we are interested in k-decimal-place accuracy in the approximation of a sum, we often approximate each term of the sum to $k + 1$ decimal places and then round off the final result to k decimal places. In certain cases this may fail to produce the required degree of accuracy; however, it is customary to proceed in this way for elementary approximations. More precise techniques may be found in texts on *numerical analysis*.

EXAMPLE 1 Suppose $f(x) = \ln x$.

(a) Find Taylor's Formula with the Remainder for $n = 3$ and $c = 1$.

(b) Use part (a) to approximate $\ln (1.1)$, and estimate the accuracy of this approximation.

SOLUTION

(a) If $n = 3$ in Taylor's Formula (10.9), then we need the first four derivatives of f. It is convenient to arrange our work as follows:

$$\begin{aligned}
f(x) &= \ln x & f(1) &= 0 \\
f'(x) &= x^{-1} & f'(1) &= 1 \\
f''(x) &= -x^{-2} & f''(1) &= -1 \\
f'''(x) &= 2x^{-3} & f'''(1) &= 2 \\
f^{(4)}(x) &= -6x^{-4} & f^{(4)}(z) &= -6z^{-4}
\end{aligned}$$

By (10.9),

$$\ln x = 0 + \frac{1}{1!}(x-1) - \frac{1}{2!}(x-1)^2 + \frac{2}{3!}(x-1)^3 - \frac{6z^{-4}}{4!}(x-1)^4$$

for z between 1 and x. This simplifies to

$$\ln x = (x-1) - \frac{1}{2}(x-1)^2 + \frac{1}{3}(x-1)^3 - \frac{1}{4z^4}(x-1)^4$$

(b) Substituting 1.1 for x in the formula of part (a) gives us

$$\ln (1.1) = 0.1 - \frac{1}{2}(0.1)^2 + \frac{1}{3}(0.1)^3 - \frac{1}{4z^4}(0.1)^4$$

for $1 < z < 1.1$. Summing the first three terms we obtain $\ln (1.1) \approx 0.0953$. If $z > 1$, then $1/z < 1$ and therefore $1/z^4 < 1$. Consequently,

$$|R_3(1.1)| = \left| \frac{(0.1)^4}{4z^4} \right| < \left| \frac{0.0001}{4} \right| = 0.000025.$$

Since $0.000025 < 0.00005$, it follows from Theorem (10.11) that the approximation $\ln (1.1) \approx 0.0953$ is accurate to four decimal places. •

If we wish to approximate a function value $f(x)$ for some x, it is desirable to choose the number c in (10.9) such that the remainder $R_n(x)$ is very close to 0 when n is relatively small (say, $n = 3$ or $n = 4$). This will be true if we choose c close to x. In addition, we should choose c so that values of the first $n + 1$ derivatives of f at c are easy to calculate. This was done in Example 1, where to approximate $\ln x$ for $x = 1.1$ we selected $c = 1$. The next example provides another illustration of a suitable choice of c.

EXAMPLE 2 Use a Taylor Polynomial to approximate $\cos 61°$, and estimate the accuracy of the approximation.

SOLUTION We wish to approximate $f(x) = \cos x$ if $x = 61°$. Let us begin by observing that $61°$ is close to $60°$, or $\pi/3$ radians, and that it is easy to calculate values of trigonometric functions at $\pi/3$. This suggests that we choose $c = \pi/3$ in (10.9). The choice of n will depend on the accuracy we wish to attain. Let us try $n = 2$. In this case, the first three derivatives of f are required and we arrange our work as follows:

$$f(x) = \cos x \qquad f(\pi/3) = 1/2$$
$$f'(x) = -\sin x \qquad f'(\pi/3) = -\sqrt{3}/2$$
$$f''(x) = -\cos x \qquad f''(\pi/3) = -1/2$$
$$f'''(x) = \sin x \qquad f'''(z) = \sin z$$

By Definition (10.10) (i), the second-degree Taylor Polynomial of f at $\pi/3$ is

$$P_2(x) = \frac{1}{2} - \frac{\sqrt{3}/2}{1!}\left(x - \frac{\pi}{3}\right) - \frac{1/2}{2!}\left(x - \frac{\pi}{3}\right)^2.$$

Since x represents a real number, $61°$ must be converted to radian measure before substitution on the right side. Writing

$$61° = 60° + 1° = \frac{\pi}{3} + \frac{\pi}{180}$$

and substituting in $P_2(x)$, we obtain

$$P_2\left(\frac{\pi}{3} + \frac{\pi}{180}\right) = \frac{1}{2} - \left(\frac{\sqrt{3}}{2}\right)\left(\frac{\pi}{180}\right) - \frac{1}{4}\left(\frac{\pi}{180}\right)^2 \approx 0.48481.$$

Thus, by Theorem (10.11),

$$\cos 61° \approx 0.48481.$$

To estimate the accuracy of this approximation, we consider

$$|R_2(x)| = \left|\frac{\sin z}{3!}\left(x - \frac{\pi}{3}\right)^3\right|.$$

Substituting $x = (\pi/3) + (\pi/180)$ and using the fact that $|\sin z| \le 1$,

$$\left|R_2\left(\frac{\pi}{3} + \frac{\pi}{180}\right)\right| = \left|\frac{\sin z}{3!}\left(\frac{\pi}{180}\right)^3\right| \le \left|\frac{1}{3!}\left(\frac{\pi}{180}\right)^3\right| \le 0.000001.$$

By Theorem (10.11), the approximation $\cos 61° \approx 0.48481$ is accurate to five decimal places. If we desire greater accuracy, we must find a value of n such that the maximum value of $|R_n[(\pi/3) + (\pi/180)]|$ is within the desired range.

If we let $c = 0$ in Formula (10.9), we obtain the following formula, named after the Scottish mathematician Colin Maclaurin (1698–1746).

MACLAURIN'S FORMULA (10.12)

> Let f have $n + 1$ derivatives throughout an interval containing 0. If x is any number in the interval, then there is a number z between 0 and x such that
>
> $$f(x) = f(0) + \frac{f'(0)}{1!}x + \frac{f''(0)}{2!}x^2 + \cdots + \frac{f^{(n)}(0)}{n!}x^n + \frac{f^{(n+1)}(z)}{(n+1)!}x^{n+1}$$

EXAMPLE 3

(a) Find Maclaurin's Formula for $f(x) = e^x$ if n is any positive integer.

(b) Use part (a) with $n = 9$ to approximate e, and estimate the error involved.

SOLUTION

(a) For every positive integer k we see that $f^{(k)}(x) = e^x$, and consequently $f^{(k)}(0) = e^0 = 1$. Substituting in (10.12),

$$e^x = 1 + x + \frac{x^2}{2!} + \cdots + \frac{x^n}{n!} + \frac{e^z}{(n+1)!}x^{n+1}$$

for z between 0 and x.

(b) If we use the formula of part (a) with $n = 9$ and $x = 1$, then

$$e \approx 1 + 1 + \frac{1}{2!} + \frac{1}{3!} + \cdots + \frac{1}{9!} \approx 2.71828153.$$

To estimate the error we consider

$$R_9(x) = \frac{e^z}{10!}x^{10}.$$

If $x = 1$, then $0 < z < 1$. Using results about e^x from Chapter 7, $e^z < e^1 < 3$, and

$$|R_9(1)| = \left| \frac{e^z}{10!}(1) \right| < \frac{3}{10!} < 0.000001.$$

Hence the approximation $e \approx 2.71828$ is accurate to five decimal places. •

The formula obtained in Example 3 may be used to approximate values of the natural exponential function. We will discuss another important application in the next chapter in conjunction with representations of functions by means of infinite series.

EXAMPLE 4 Find Maclaurin's Formula for $f(x) = \sin x$ with $n = 8$, and estimate the error involved if it is used to approximate $\sin (0.1)$.

SOLUTION We need the first nine derivatives of $f(x)$. Let us begin as follows:

$$
\begin{aligned}
f(x) &= \sin x & f(0) &= 0 \\
f'(x) &= \cos x & f'(0) &= 1 \\
f''(x) &= -\sin x & f''(0) &= 0 \\
f'''(x) &= -\cos x & f'''(0) &= -1
\end{aligned}
$$

Since $f^{(4)}(x) = \sin x$, the remaining derivatives follow the same pattern as the first three, and we arrive at

$$f^{(9)}(x) = \cos x \qquad f^{(9)}(z) = \cos z.$$

Substituting in (10.12) and noting that the constant term and the coefficients of x^2, x^4, x^6, and x^8 are 0, we obtain

$$\sin x = \frac{1}{1!} x + \frac{(-1)}{3!} x^3 + \frac{1}{5!} x^5 + \frac{(-1)}{7!} x^7 + \frac{\cos z}{9!} x^9$$

which may be written

$$\sin x = x - \frac{x^3}{3!} + \frac{x^5}{5!} - \frac{x^7}{7!} + \frac{\cos z}{9!} x^9.$$

If we use this formula to approximate $\sin(0.1)$, then the error would be less than $|R_8(0.1)|$. Since $|\cos z| \le 1$,

$$|R_8(0.1)| = \left| \frac{\cos z}{9!} (0.1)^9 \right| \le \frac{(0.1)^9}{9!} < 2.7 \times 10^{-15} < 0.5 \times 10^{-14}.$$

Thus the approximation would be correct to 14 decimal places. •

EXERCISES 10.5

Exer. 1–12: Find Taylor's Formula with the Remainder for the given value of c and n.

1 $f(x) = \sin x$, $c = \pi/2$, $n = 3$

2 $f(x) = \cos x$, $c = \pi/4$, $n = 3$

3 $f(x) = \sqrt{x}$, $c = 4$, $n = 3$

4 $f(x) = e^{-x}$, $c = 1$, $n = 3$

5 $f(x) = \tan x$, $c = \pi/4$, $n = 4$

6 $f(x) = 1/(x - 1)^2$, $c = 2$, $n = 5$

7 $f(x) = 1/x$, $c = -2$, $n = 5$

8 $f(x) = \sqrt[3]{x}$, $c = -8$, $n = 3$

9 $f(x) = \tan^{-1} x$, $c = 1$, $n = 2$

10 $f(x) = \ln \sin x$, $c = \pi/6$, $n = 3$

11 $f(x) = xe^x$, $c = -1$, $n = 4$

12 $f(x) = \log x$, $c = 10$, $n = 2$

Exer. 13–24: Find Maclaurin's Formula for the given values of n.

13 $f(x) = \ln(x + 1)$, $n = 4$

14 $f(x) = \sin x$, $n = 7$

15 $f(x) = \cos x$, $n = 8$

16 $f(x) = \tan^{-1} x$, $n = 3$

17 $f(x) = e^{2x}$, $n = 5$

18 $f(x) = \sec x$, $n = 3$

19 $f(x) = 1/(x - 1)^2$, $n = 5$

20 $f(x) = \sqrt{4 - x}$, $n = 3$

21 $f(x) = \arcsin x$, $n = 2$

22 $f(x) = e^{-x^2}$, $n = 3$

23 $f(x) = 2x^4 - 5x^3 + x^2 - 3x + 7$, $n = 4$ and $n = 5$

24 $f(x) = \cosh x$, $n = 4$ and $n = 5$

Exer. 25–28: Approximate the number to four decimal places by using the indicated exercise and the fact that $\pi/180 \approx 0.0175$. Prove that your answer is correct by showing that $|R_n(x)| < 0.5 \times 10^{-4}$.

25 $\sin 89°$ (Exercise 1)

26 $\cos 47°$ (Exercise 2)

27 $\sqrt{4.03}$ (Exercise 3)

28 $e^{-1.02}$ (Exercise 4)

Exer. 29–34: Approximate the number by using the indicated exercise, and estimate the error in the approximation by means of $R_n(x)$.

29 $-1/2.2$ (Exercise 7)

30 $\sqrt[3]{-8.5}$ (Exercise 8)

31 $\ln(1.25)$ (Exercise 13)

32 $\sin 0.1$ (Exercise 14)

33 $\cos 30°$ (Exercise 15)

34 $\log 10.01$ (Exercise 12)

Exer. 35–40: Use Maclaurin's Formula to establish the approximation formula and state, in terms of decimal places, the accuracy of the approximation if $|x| \le 0.1$.

35 $\cos x \approx 1 - \dfrac{x^2}{2}$

36 $\sqrt[3]{1 + x} \approx 1 + \dfrac{1}{3} x$

37 $e^x \approx 1 + x + \dfrac{x^2}{2}$

38 $\sin x \approx x - \dfrac{x^3}{6}$

39 $\ln(1 + x) \approx x - \dfrac{x^2}{2} + \dfrac{x^3}{3}$

40 $\cosh x \approx 1 + \dfrac{x^2}{2}$

41 In the proof of Theorem (10.8) verify the formula for $g'(x)$.

42 If $f(x)$ is a polynomial of degree n and a is a real number, prove that $f(x) = P_n(x)$ for $P_n(x)$ given by Definition (10.10) (i).

10.6 REVIEW

Define or discuss each of the following.

1 Indeterminate forms
2 Cauchy's Formula
3 L'Hôpital's Rule

4 Improper integrals
5 Taylor's Formula
6 Maclaurin's Formula

EXERCISES 10.6

Exer. 1–16: Find the limit, if it exists.

1 $\displaystyle \lim_{x\to 0} \frac{\ln(2-x)}{1+e^{2x}}$

2 $\displaystyle \lim_{x\to 0} \frac{\sin 2x - \tan 2x}{x^2}$

3 $\displaystyle \lim_{x\to\infty} \frac{x^2 + 2x + 3}{\ln(x+1)}$

4 $\displaystyle \lim_{x\to 0} \frac{\tan^{-1} x}{\sin^{-1} x}$

5 $\displaystyle \lim_{x\to 0} \frac{e^{2x} - e^{-2x} - 4x}{x^3}$

6 $\displaystyle \lim_{x\to(\pi/2)^-} \frac{\tan x}{\sec x}$

7 $\displaystyle \lim_{x\to\infty} \frac{x^e}{e^x}$

8 $\displaystyle \lim_{x\to(\pi/2)^-} \cos x \ln \cos x$

9 $\displaystyle \lim_{x\to\infty} (1 - 2e^{1/x})x$

10 $\displaystyle \lim_{x\to 0} \tan^{-1} x \csc x$

11 $\displaystyle \lim_{x\to 0} (1 + 8x^2)^{1/x^2}$

12 $\displaystyle \lim_{x\to 1} (\ln x)^{x-1}$

13 $\displaystyle \lim_{x\to\infty} (e^x + 1)^{1/x}$

14 $\displaystyle \lim_{x\to 0} \left(\frac{1}{\tan x} - \frac{1}{x} \right)$

15 $\displaystyle \lim_{x\to\infty} \frac{\sqrt{x^2 + 1}}{x}$

16 $\displaystyle \lim_{x\to\infty} \frac{3^x + 2x}{x^3 + 1}$

Exer. 17–28: Does the integral converge or diverge? If it converges, find its value.

17 $\displaystyle \int_4^\infty \frac{1}{\sqrt{x}}\, dx$

18 $\displaystyle \int_4^\infty \frac{1}{x\sqrt{x}}\, dx$

19 $\displaystyle \int_{-\infty}^0 \frac{1}{x+2}\, dx$

20 $\displaystyle \int_0^\infty \sin x\, dx$

21 $\displaystyle \int_{-8}^1 \frac{1}{\sqrt[3]{x}}\, dx$

22 $\displaystyle \int_{-4}^0 \frac{1}{x+4}\, dx$

23 $\displaystyle \int_0^2 \frac{x}{(x^2 - 1)^2}\, dx$

24 $\displaystyle \int_1^2 \frac{1}{x\sqrt{x^2 - 1}}\, dx$

25 $\displaystyle \int_{-\infty}^\infty \frac{1}{e^x + e^{-x}}\, dx$

26 $\displaystyle \int_{-\infty}^0 xe^x\, dx$

27 $\displaystyle \int_0^1 \frac{\ln x}{x}\, dx$

28 $\displaystyle \int_0^{\pi/2} \csc x\, dx$

29 Find $\lim_{x\to\infty} f(x)/g(x)$ if $f(x) = \int_1^x (\sin t)^{2/3}\, dt$ and $g(x) = x^2$.

30 Guass' error integral $\mathrm{erf}\,(x) = (2/\sqrt{\pi}) \int_0^x e^{-u^2}\, du$ is used in probability theory. It has the special property $\lim_{x\to\infty} \mathrm{erf}\,(x) = 1$. Find $\lim_{x\to\infty} e^{(x^2)}[1 - \mathrm{erf}\,(x)]$.

31 Find Taylor's Formula with the Remainder:
   (a) $f(x) = \ln \cos x,\ c = \pi/6,\ n = 3$
   (b) $f(x) = \sqrt{x-1},\ c = 2,\ n = 4$

32 Find Maclaurin's Formula:
   (a) $f(x) = e^{-x^3},\ n = 3$
   (b) $f(x) = 1/(1-x),\ n = 6$

33 Use Taylor's Formula to approximate $\sin^2 43°$ to four decimal places. (*Hint:* $\sin^2 x = \frac{1}{2}(1 - \cos 2x)$.)

34 Show that the approximation $\sin x \approx x - \frac{1}{6}x^3 + \frac{1}{120}x^5$ is accurate to four decimal places for $0 \le x \le \pi/4$ (and therefore for angles between $0°$ and $45°$).

Infinite series are useful in advanced courses in mathematics, physics, and engineering because they may be employed to represent functions in a new way. In this chapter we shall discuss some of the fundamental results associated with this important mathematical concept.

11.1 INFINITE SEQUENCES

An arbitrary *infinite sequence* is often denoted by

(11.1) $$a_1, \ a_2, \ a_3, \ \ldots, \ a_n, \ \ldots$$

and may be regarded as a collection of real numbers that is in one-to-one correspondence with the positive integers. For convenience we sometimes refer to infinite sequences merely as *sequences*. Each real number a_k is called a **term** of the sequence. The sequence is *ordered*, in the sense that there is a **first term** a_1, a **second term** a_2, and, if n denotes an arbitrary positive integer, an **nth** term a_n.

Infinite sequences are introduced in precalculus mathematics. For example, the sequence

$$0.6, \ 0.66, \ 0.666, \ 0.6666, \ 0.66666, \ \ldots$$

may be used to represent the rational number $\frac{2}{3}$. In this case, the nth term gets closer and closer to $\frac{2}{3}$ as n increases.

One important use for infinite sequences is in the definition of infinite series in Section 11.2. This definition will enable us to express a rational number, such as $\frac{2}{3}$, as an *infinite sum* (or *infinite series*). For $\frac{2}{3}$, the infinite sum is

$$0.6 + 0.06 + 0.006 + 0.0006 + \cdots$$

Note that the *sequence* for $\frac{2}{3}$ may be obtained by adding more and more terms of this sum; that is,

$$0.6 = 0.6, \qquad 0.66 = 0.6 + 0.06, \qquad 0.666 = 0.6 + 0.06 + 0.006,$$

and so on.

Our study of infinite series will take us far beyond this elementary illustration. Later in the chapter we shall represent e^x, $\sin x$, $\tan^{-1} x$, and many other complicated expressions as infinite series. This use of series has far-reaching consequences, not only for basic calculations, but also in advanced mathematics and applications.

Thus far our discussion has been at an intuitive level. To make our work more precise, we may regard an infinite sequence as a function. Recall that a function f is a correspondence that associates with each number x in the domain, a unique number $f(x)$ in the range, as illustrated by the two coordinate lines in Figure 11.1. If we restrict the domain to the positive integers $1, 2, 3, \ldots$, then we obtain an infinite sequence.

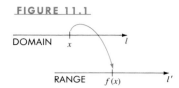

FIGURE 11.1

DOMAIN x l

RANGE $f(x)$ l'

DEFINITION (11.2)

> An **infinite sequence** is a function whose domain is the set of positive integers.

In this text the range of an infinite sequence will be a set of real numbers.

If f is an infinite sequence, then to each positive integer k there corresponds a real number $f(k)$. The numbers in the range of f may be denoted by

$$f(1), \ f(2), \ f(3), \ \ldots, \ f(n), \ \ldots$$

where the three dots at the end indicate that the sequence does not terminate. Thus (11.1) and (11.2) are essentially the same. If we wish to obtain the subscript form (11.1) from (11.2), we let $a_n = f(n)$ for each positive integer n. Conversely, given (11.1), we may obtain a function f by letting $f(n) = a_n$ for each positive integer n.

We sometimes denote a sequence with nth term a_n by $\{a_n\}$. For example, the sequence $\{2^n\}$ has nth term $a_n = 2^n$. According to Definition (11.2), the sequence $\{2^n\}$ is the function f such that $f(n) = 2^n$ for every positive integer n.

From the definition of equality of functions, a sequence $\{a_n\}$ is **equal** to a sequence $\{b_n\}$ if and only if $a_k = b_k$ for every positive integer k.

EXAMPLE 1 List the first four terms and the tenth term of each sequence:

(a) $\left\{ \dfrac{n}{n+1} \right\}$

(b) $\{2 + (0.1)^n\}$

(c) $\left\{ (-1)^{n+1} \dfrac{n^2}{3n-1} \right\}$

(d) $\{4\}$

SOLUTION To find the first four terms we substitute, successively, $n = 1, 2, 3$, and 4 in the formula for a_n. The tenth term is found by substituting 10 for n. Doing this and simplifying gives us the following:

Sequence	nth term	First four terms	Tenth term
(a) $\left\{ \dfrac{n}{n+1} \right\}$	$\dfrac{n}{n+1}$	$\dfrac{1}{2}, \dfrac{2}{3}, \dfrac{3}{4}, \dfrac{4}{5}$	$\dfrac{10}{11}$
(b) $\{2 + (0.1)^n\}$	$2 + (0.1)^n$	$2.1,\ 2.01,\ 2.001,\ 2.0001$	2.0000000001
(c) $\left\{ (-1)^{n+1} \dfrac{n^2}{3n-1} \right\}$	$(-1)^{n+1} \dfrac{n^2}{3n-1}$	$\dfrac{1}{2}, -\dfrac{4}{5}, \dfrac{9}{8}, -\dfrac{16}{11}$	$\dfrac{100}{29}$
(d) $\{4\}$	4	$4, 4, 4, 4$	4

For some sequences we state the first term a_1, together with a rule for obtaining any term a_{k+1} from the preceding term a_k whenever $k \geq 1$. We call this a **recursive definition,** and the sequence is said to be defined **recursively.**

EXAMPLE 2 Find the first four terms and the nth term of the infinite sequence defined recursively as follows:

$$a_1 = 3 \quad \text{and} \quad a_{k+1} = 2a_k \quad \text{for } k \geq 1$$

SOLUTION The sequence is defined recursively, since the first term is given and, moreover, whenever a term a_k is known, then the next term a_{k+1} can be found. Thus,

$$a_1 = 3$$
$$a_2 = 2a_1 = 2 \cdot 3 = 6$$
$$a_3 = 2a_2 = 2 \cdot 2 \cdot 3 = 2^2 \cdot 3 = 12$$
$$a_4 = 2a_3 = 2 \cdot 2 \cdot 2 \cdot 3 = 2^3 \cdot 3 = 24$$

We have written the terms as products to gain insight into the nature of the nth term. Continuing, we obtain $a_5 = 2^4 \cdot 3$ and $a_6 = 2^5 \cdot 3$; it appears that $a_n = 2^{n-1} \cdot 3$. We can use mathematical induction to prove that this guess is correct. •

An infinite sequence $\{a_n\}$ may have the property that as n increases, a_n gets very close to some real number L; that is, $|a_n - L| \approx 0$ if n is large. As an illustration, suppose

$$a_n = 2 + (-\tfrac{1}{2})^n.$$

The first few terms of the sequence $\{a_n\}$ are

$$2 - \tfrac{1}{2}, \ 2 + \tfrac{1}{4}, \ 2 - \tfrac{1}{8}, \ 2 + \tfrac{1}{16}, \ 2 - \tfrac{1}{32}, \ \ldots$$

and it appears that the terms get closer to 2 as n increases. As a matter of fact, for every positive integer n,

$$|a_n - 2| = |2 + (-\tfrac{1}{2})^n - 2| = |(-\tfrac{1}{2})^n| = (\tfrac{1}{2})^n = 1/2^n$$

and the number $1/2^n$, and hence $|a_n - 2|$, *can be made arbitrarily close to zero by choosing n sufficiently large.* According to the next definition, the sequence *has the limit* 2, or *converges* to 2, and we write

$$\lim_{n \to \infty} [2 + (-\tfrac{1}{2})^n] = 2.$$

This limit is almost identical to the definition of $\lim_{x \to \infty} f(x) = L$ given in Chapter 4. The only difference is that if $f(n) = a_n$, then the domain of f is the set of positive integers, and not an infinite interval of real numbers. As in Definition (4.21), but using a_n instead of $f(x)$, we state the following definition.

A sequence $\{a_n\}$ **has the limit L,** or **converges to L,** denoted by

$$\lim_{n \to \infty} a_n = L,$$

if for every $\varepsilon > 0$ there exists a positive number N such that

$$|a_n - L| < \varepsilon \quad \text{whenever } n > N.$$

If such a number L does not exist, the sequence **has no limit,** or **diverges.**

FIGURE 11.2

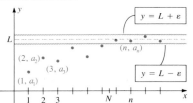

A geometric interpretation similar to that shown for the limit of a function in Figure 4.47 can be given for the limit of a sequence. The only difference is that the coordinate x of the point shown on the x-axis is always a positive integer. In Figure 11.2 we have plotted some points (k, a_k) for a specific case in which $\lim_{n \to \infty} a_n = L$. Note that for any $\varepsilon > 0$, the points (n, a_n) lie between the lines $y = L \pm \varepsilon$, provided n is sufficiently large. Of course, the approach to L may vary from that illustrated in the figure (see, for example, Figures 11.3 and 11.6).

If we can make a_n as large as desired by choosing n sufficiently large, then the sequence $\{a_n\}$ diverges, but we still use the limit notation, and write $\lim_{n \to \infty} a_n = \infty$. A more precise way of specifying this follows.

DEFINITION (11.4)

The notation

$$\lim_{n \to \infty} a_n = \infty$$

means that for every positive real number P, there exists a number N such that $a_n > P$ whenever $n > N$.

Remember that $\lim_{n \to \infty} a_n = \infty$ does *not* mean that the limit exists, but rather that the numbers a_n increase without bound as n increases. Similarly, $\lim_{n \to \infty} a_n = -\infty$ means that a_n decreases without bound as n increases.

The next theorem is important because it allows us to use results of Chapter 4 to investigate convergence or divergence of sequences. The proof follows from Definitions (11.3) and (4.21).

THEOREM (11.5)

Let $\{a_n\}$ be an infinite sequence, let $f(n) = a_n$, and suppose that $f(x)$ exists for every real number $x \geq 1$.

(i) If $\lim_{x \to \infty} f(x) = L$, then $\lim_{n \to \infty} f(n) = L$.

(ii) If $\lim_{x \to \infty} f(x) = \infty$ (or $-\infty$), then $\lim_{n \to \infty} f(n) = \infty$ (or $-\infty$).

The following example illustrates the use of Theorem (11.5).

EXAMPLE 3 If $a_n = 1 + (1/n)$, determine if $\{a_n\}$ converges or diverges.

SOLUTION We let $f(n) = 1 + (1/n)$ and consider

$$f(x) = 1 + \frac{1}{x} \quad \text{for every real number } x \geq 1.$$

From our work in Section 4.6,

$$\lim_{x \to \infty} f(x) = \lim_{x \to \infty} \left(1 + \frac{1}{x}\right) = \lim_{x \to \infty} 1 + \lim_{x \to \infty} \frac{1}{x} = 1 + 0 = 1.$$

Hence, by Theorem (11.5),

$$\lim_{n \to \infty} \left(1 + \frac{1}{n}\right) = 1.$$

Thus the sequence $\{a_n\}$ converges to 1. •

The difference between $\lim_{x \to \infty} [1 + (1/x)] = 1$ and $\lim_{n \to \infty} [1 + (1/n)] = 1$ is illustrated in Figure 11.3. Note that for $1 + (1/x)$, the function f is continuous if $x \geq 1$, and the graph has a horizontal asymptote $y = 1$. For $1 + (1/n)$ we consider the points whose x-coordinates are positive integers.

FIGURE 11.3

(i) $\lim_{x \to \infty} \left(1 + \dfrac{1}{x}\right) = 1$ (ii) $\lim_{n \to \infty} \left(1 + \dfrac{1}{n}\right) = 1$

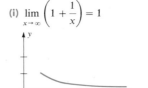 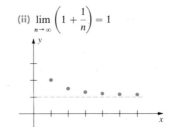

FIGURE 11.4

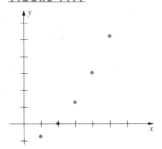

EXAMPLE 4 Determine if each sequence converges or diverges:

(a) $\left\{\frac{1}{4}n^2 - 1\right\}$ (b) $\left\{(-1)^{n-1}\right\}$

SOLUTION

(a) If we let $f(x) = \frac{1}{4}x^2 - 1$, then $f(x)$ exists for every $x \geq 1$ and

$$\lim_{x \to \infty} \left(\tfrac{1}{4}x^2 - 1\right) = \infty.$$

Hence by Theorem (11.5),

$$\lim_{n \to \infty} \left(\tfrac{1}{4}n^2 - 1\right) = \infty.$$

FIGURE 11.5

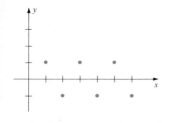

Thus the sequence has no limit. Figure 11.4 illustrates the manner in which the sequence diverges.

(b) Letting $n = 1, 2, 3, \ldots$ we see that the terms of $(-1)^{n-1}$ oscillate between 1 and -1 as follows:

$$1, \quad -1, \quad 1, \quad -1, \quad 1, \quad -1, \quad \ldots$$

This is illustrated graphically in Figure 11.5. Thus, $\lim_{n \to \infty}(-1)^{n-1}$ does not exist, and the sequence diverges. •

The next example shows how we may use L'Hôpital's Rule to find limits of certain sequences.

EXAMPLE 5 Determine if the sequence $\{5n/e^{2n}\}$ converges or diverges.

SOLUTION Let $f(x) = 5x/e^{2x}$ for every real number x. Since f takes on the indeterminate form ∞/∞ as $x \to \infty$, we may use L'Hôpital's Rule, obtaining

$$\lim_{x \to \infty} \frac{5x}{e^{2x}} = \lim_{x \to \infty} \frac{5}{2e^{2x}} = 0.$$

Hence, by Theorem (11.5), $\lim_{n \to \infty}(5n/e^{2n}) = 0$. Thus, the sequence converges to 0. •

The proof of the next theorem illustrates the use of Definition (11.3).

THEOREM (11.6)

(i) $\lim_{n \to \infty} r^n = 0$ if $|r| < 1$

(ii) $\lim_{n \to \infty} |r^n| = \infty$ if $|r| > 1$

PROOF If $r = 0$, it follows trivially that the limit is 0. Let us assume that $0 < |r| < 1$. To prove (i) by means of Definition (11.3) we must show that for every $\varepsilon > 0$ there exists a positive number N such that

$$\text{if} \quad n > N, \quad \text{then} \quad |r^n - 0| < \varepsilon.$$

The inequality $|r^n - 0| < \varepsilon$ is equivalent to each inequality in the following list:

$$|r|^n < \varepsilon, \quad \ln |r|^n < \ln \varepsilon, \quad n \ln |r| < \ln \varepsilon, \quad n > \frac{\ln \varepsilon}{\ln |r|}$$

The final inequality sign is reversed because $\ln |r|$ is negative if $0 < |r| < 1$. The last inequality in the list provides a clue to the choice of N. Specifically, if $\varepsilon < 1$, then $\ln \varepsilon < 0$ and we let $N = \ln \varepsilon/\ln |r| > 0$. In this event, if $n > N$, then the last inequality in the list is true, and hence so is the first, which is what we wished to prove. If $\varepsilon \geq 1$, then $\ln \varepsilon \geq 0$ and hence $\ln \varepsilon/\ln |r| \leq 0$. In this case if N is *any* positive number, then whenever $n > N$ the last inequality in the list is again true.

To prove (ii), let $|r| > 1$ and consider any positive real number P. The following inequalities are equivalent:

$$|r|^n > P, \quad \ln |r|^n > \ln P, \quad n \ln |r| > \ln P, \quad n > \frac{\ln P}{\ln |r|}.$$

If we choose $N = \ln P/\ln |r|$, then whenever $n > N$, the last inequality is true, and hence so is the first; that is, $|r|^n > P$. By Definition (11.4), this means that $\lim_{n \to \infty} |r|^n = \infty$. • •

EXAMPLE 6 List the first four terms of each sequence, and determine if it converges or diverges:

(a) $\{(-\frac{2}{3})^n\}$ (b) $\{(1.01)^n\}$

(a) The first four terms of $\{(-\frac{2}{3})^n\}$ are

$$-\tfrac{2}{3}, \quad \tfrac{4}{9}, \quad -\tfrac{8}{27}, \quad \tfrac{16}{81}.$$

According to Theorem (11.6) (i),

$$\lim_{n \to \infty} (-\tfrac{2}{3})^n = 0$$

(b) The first four terms of $\{(1.01)^n\}$ are

$$1.01, \quad 1.0201, \quad 1.030301, \quad 1.04060401$$

According to Theorem (11.6) (ii),

$$\lim_{n \to \infty} (1.01)^n = \infty \quad \bullet$$

Limit theorems that are analogous to those used in Section 4.6 for sums, differences, products, and quotients of functions can be established for sequences. For example, if $\{a_n\}$ and $\{b_n\}$ are convergent sequences, then

$$\lim_{n \to \infty} (a_n + b_n) = \lim_{n \to \infty} a_n + \lim_{n \to \infty} b_n,$$

$$\lim_{n \to \infty} (a_n b_n) = \left(\lim_{n \to \infty} a_n \right)\left(\lim_{n \to \infty} b_n \right),$$

and so on.

If $a_n = c$ for every n, so that the infinite sequence is $c, c, \ldots, c, \ldots$, then

$$\lim_{n \to \infty} c = c.$$

Similarly, if c is a real number and k is a positive rational number, then, as in the proof of Theorem (4.23),

$$\lim_{n \to \infty} \frac{c}{n^k} = 0.$$

EXAMPLE 7 Find the limit of the sequence $\left\{ \dfrac{2n^2}{5n^2 - 3} \right\}$.

SOLUTION We wish to find $\lim_{n \to \infty} a_n$ for $a_n = 2n^2/(5n^2 - 3)$. Dividing numerator and denominator of a_n by n^2 and applying limit theorems,

$$\lim_{n \to \infty} \frac{2n^2}{5n^2 - 3} = \lim_{n \to \infty} \frac{2}{5 - (3/n^2)} = \frac{\lim\limits_{n \to \infty} 2}{\lim\limits_{n \to \infty} [5 - (3/n^2)]}$$

$$= \frac{2}{\lim\limits_{n \to \infty} 5 - \lim\limits_{n \to \infty} (3/n^2)} = \frac{2}{5 - 0} = \frac{2}{5}.$$

Hence the sequence has the limit $\frac{2}{5}$. We can also prove this by applying L'Hôpital's Rule to $2x^2/(5x^2 - 3)$. $\bullet$

The next theorem, which is similar to Theorem (2.22), states that if the terms of an infinite sequence are always sandwiched between corresponding terms of two sequences that have the same limit L, then the given sequence also has the limit L. The proof is given in Appendix II.

| THE SANDWICH THEOREM (11.7) FOR INFINITE SEQUENCES | If $\{a_n\}$, $\{b_n\}$, and $\{c_n\}$ are infinite sequences such that $a_n \le b_n \le c_n$ for every n, and if $$\lim_{n \to \infty} a_n = L = \lim_{n \to \infty} c_n$$ then $\lim_{n \to \infty} b_n = L$. |

EXAMPLE 8 Find the limit of the sequence $\left\{ \dfrac{\cos^2 n}{3^n} \right\}$.

SOLUTION Since $0 < \cos^2 n < 1$ for every positive integer n,

$$0 < \frac{\cos^2 n}{3^n} < \frac{1}{3^n}.$$

Applying Theorem (11.6) with $r = \frac{1}{3}$,

$$\lim_{n \to \infty} \frac{1}{3^n} = \lim_{n \to \infty} \left(\frac{1}{3}\right)^n = 0.$$

Moreover $\lim_{n \to \infty} 0 = 0$. It follows from the Sandwich Theorem (11.7), with $a_n = 0$, $b_n = (\cos^2 n)/3^n$, and $c_n = (\frac{1}{3})^n$, that

$$\lim_{n \to \infty} \frac{\cos^2 n}{3^n} = 0.$$

Hence the limit of the sequence is 0. •

The next theorem follows directly from Definition (11.3).

| THEOREM (11.8) | Let $\{a_n\}$ be a sequence. If $\lim_{n \to \infty} |a_n| = 0$, then $\lim_{n \to \infty} a_n = 0$. |

EXAMPLE 9 Suppose the nth term of a sequence is $a_n = (-1)^{n+1}(1/n)$. Prove that $\lim_{n \to \infty} a_n = 0$.

SOLUTION The terms of the sequence are alternately positive and negative. For example, the first five terms are

$$1, \quad -\frac{1}{2}, \quad \frac{1}{3}, \quad -\frac{1}{4}, \quad \frac{1}{5}.$$

Since

$$\lim_{n \to \infty} |a_n| = \lim_{n \to \infty} \frac{1}{n} = 0,$$

it follows from Theorem (11.8) that $\lim_{n \to \infty} a_n = 0$. •

A sequence is **monotonic** if successive terms are nondecreasing, that is,

$$a_1 \le a_2 \le \cdots \le a_n \le \cdots$$

or nonincreasing, that is

$$a_1 \ge a_2 \ge \cdots \ge a_n \ge \cdots$$

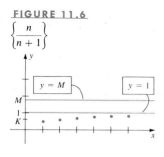

FIGURE 11.6
$$\left\{\frac{n}{n+1}\right\}$$

A sequence is **bounded** if there is a positive real number M such that $|a_k| \le M$ for every k. To illustrate, the sequence

$$\frac{1}{2}, \quad \frac{2}{3}, \quad \frac{3}{4}, \quad \frac{4}{5}, \quad \cdots \quad, \quad \frac{n}{n+1}, \quad \cdots$$

is both monotonic (the terms are increasing) and bounded (since $k/(k+1) < 1$ for every k). The sequence is illustrated in Figure 11.6. Note that any number $M \ge 1$ is a bound for the sequence; however, if $K < 1$, then K is not a bound, since $K < k/(k+1)$ when k is sufficiently large.

The next theorem is fundamental for later developments.

THEOREM (11.9) A bounded, monotonic, infinite sequence has a limit.

To prove Theorem (11.9), it is necessary to use an important property of real numbers. Let us first state several definitions. If S is a nonempty set of real numbers, then a real number u is called an **upper bound** of S if $x \le u$ for every x in S. A number v is a **least upper bound** of S if v is an upper bound and no number less than v is an upper bound of S. The least upper bound is, therefore, the smallest real number that is greater than or equal to every number in S. To illustrate, if S is the open interval (a, b), then any number greater than b is an upper bound of S; however, the least upper bound of S is unique and equals b.

The following statement is an axiom for the real number system.

THE COMPLETENESS PROPERTY (11.10) If a nonempty set S of real numbers has an upper bound, then S has a least upper bound.

FIGURE 11.7

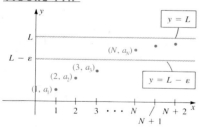

PROOF OF THEOREM (11.9) Let $\{a_n\}$ be a bounded monotonic sequence with nondecreasing terms. Thus

$$a_1 \le a_2 \le \cdots \le a_n \le \cdots$$

and there is a number M such that $a_k \le M$ for every positive integer k. Thus M is an upper bound for the set S of all numbers in the sequence, and it follows from the Completeness Property that S has a least upper bound L such that $L \le M$ (see Figure 11.7).

If $\varepsilon > 0$, then $L - \varepsilon$ is not an upper bound of S and hence at least one term of $\{a_n\}$ is greater than $L - \varepsilon$; that is,

$$L - \varepsilon < a_N \quad \text{for some positive integer } N$$

as shown in Figure 11.7. Since the terms of $\{a_n\}$ are nondecreasing, we have

$$a_N \le a_{N+1} \le a_{N+2} \le \cdots$$

and, therefore,

$$L - \varepsilon < a_n \quad \text{for every } n \ge N.$$

It follows that if $n > N$, then

$$0 \le L - a_n < \varepsilon \quad \text{or} \quad |L - a_n| < \varepsilon.$$

By Definition (11.3) this means that

$$\lim_{n \to \infty} a_n = L \leq M,$$

that is, $\{a_n\}$ has a limit.

We may obtain the proof for a sequence $\{a_n\}$ of nonincreasing terms by a similar proof or by considering the sequence $\{-a_n\}$. • •

As an illustration of Theorem (11.9), the monotonic sequence $\{n/(n + 1)\}$ pictured in Figure 11.6 has the least upper bound (and limit) 1.

APPLICATION:
S → I → S EPIDEMICS

The symbol S → I → S is an abbreviation for Susceptible → Infected → Susceptible and signifies that an infected person who becomes cured is not immune to the disease, but may contract it again. Examples of such diseases are gonorrhea and strep throat.

Infinite sequences may be applied to the investigation of the time course of an S → I → S epidemic.* Suppose that physicians issue daily reports indicating the number of persons who have become infected with the disease and those who have been cured. We shall label the reporting days as $1, 2, \ldots, n, \ldots$, and let N denote the total population. In addition, let

I_n = number of persons who have the disease on day n

F_n = number of newly infected persons on day n

C_n = number of persons cured on day n.

It follows that for every $n \geq 1$,

$$I_{n+1} = I_n + F_{n+1} - C_{n+1}.$$

Suppose health officials decide that the number of new cases on a given day is directly proportional to the product of the number ill and the number not infected on the previous day. (This is known as the *Law of Mass Action*, and is typical of a population of students on a college campus.) Moreover, suppose that the number cured each day is directly proportional to the number ill the previous day. Hence,

$$F_{n+1} = aI_n(N - I_n) \quad \text{and} \quad C_{n+1} = bI_n$$

for constants a and b. Substituting in the preceding formula for I_{n+1},

$$I_{n+1} = I_n + aI_n(N - I_n) - bI_n.$$

In the early stages of an epidemic I_n will be very small compared to N, and from the point of view of public health, it is better to *overestimate* the number ill than it is to underestimate and be unprepared for the spread of the disease. With this in mind, we may realistically investigate the early dynamics of the epidemic by examining the equation

$$I_{n+1} = I_n + aNI_n - bI_n = (1 + aN - b)I_n$$

for the initial number I_1 of infected individuals. If we let $r = 1 + aN - b$, then $I_{n+1} = rI_n$ and, therefore,

$$I_2 = rI_1, \quad I_3 = rI_2 = r^2I_1, \quad I_4 = rI_3 = r^3I_1, \quad \ldots, \quad I_n = r^{n-1}I_1, \quad \ldots$$

This gives us the following sequence of infected individuals:

$$I_1, \; rI_1, \; r^2I_1, \; r^3I_1, \; \ldots, \; r^{n-1}I_1, \; \ldots$$

The number $r = 1 + aN - b$ (which can be approximated from early data) is of critical import. If $r > 1$, then by Theorem (11.6), $\lim_{n \to \infty} I_n = \infty$ and an epidemic is in progress. In this case, when n is large, I_n is no longer small compared to N, and the formula for I_{n+1} becomes invalid. If $r < 1$, then $\lim_{n \to \infty} I_n = 0$ and health officials need not be concerned. The case $r = 1$ results in the constant sequence $I_1, I_1, \ldots, I_1, \ldots$

Exer. 1–16: The expression is the nth term a_n of an infinite sequence $\{a_n\}$. Find the first four terms and $\lim_{n \to \infty} a_n$, if it exists.

1. $\dfrac{n}{3n + 2}$

2. $\dfrac{6n - 5}{5n + 1}$

3. $\dfrac{7 - 4n^2}{3 + 2n^2}$

4. $\dfrac{4}{8 - 7n}$

5. -5

6. $\sqrt{2}$

7. $\dfrac{(2n - 1)(3n + 1)}{n^3 + 1}$

8. $8n + 1$

9. $\dfrac{2}{\sqrt{n^2 + 9}}$

10. $\dfrac{100n}{n^{3/2} + 4}$

11. $(-1)^{n+1} \dfrac{3n}{n^2 + 4n + 5}$

12. $(-1)^{n+1} \dfrac{\sqrt{n}}{n + 1}$

13. $1 + (0.1)^n$

14. $1 - (1/2^n)$

15. $1 + (-1)^{n+1}$

16. $(n + 1)/\sqrt{n}$

Exer. 17–42: Determine if the sequence converges or diverges. If it converges, find the limit.

17. $\{6(-5/6)^n\}$

18. $\{8 - (7/8)^n\}$

19. $\{\arctan n\}$

20. $\{(\tan^{-1} n)/n\}$

21. $\{1000 - n\}$

22. $\{(1.0001)^n/1000\}$

23. $\left\{(-1)^n \dfrac{\ln n}{n}\right\}$

24. $\left\{\dfrac{n^2}{\ln(n + 1)}\right\}$

25. $\left\{\dfrac{4n^4 + 1}{2n^2 - 1}\right\}$

26. $\left\{\dfrac{\cos n}{n}\right\}$

27. $\{e^n/n^4\}$

28. $\{e^{-n} \ln n\}$

29. $\left\{\left(1 + \dfrac{1}{n}\right)^n\right\}$

30. $\{(-1)^n n^3 3^{-n}\}$

31. $\{2^{-n} \sin n\}$

32. $\left\{\dfrac{4n^3 + 5n + 1}{2n^3 - n^2 + 5}\right\}$

33. $\left\{\dfrac{n^2}{2n - 1} - \dfrac{n^2}{2n + 1}\right\}$

34. $\left\{n \sin \dfrac{1}{n}\right\}$

35. $\{\cos n\pi\}$

36. $\{4 + \sin \tfrac{1}{2}\pi n\}$

37. $\{n^{1/n}\}$

38. $\{n^2/2^n\}$

39. $\left\{\dfrac{n^{-10}}{\sec n}\right\}$

40. $\left\{(-1)^n \dfrac{n^2}{1 + n^2}\right\}$

41. $\{\sqrt{n + 1} - \sqrt{n}\}$

42. $\{\sqrt{n^2 + n} - n\}$

Exer. 43–44: Refer to the Application: $S \to I \to S$ Epidemics.

43. A stable population of 35,000 birds lives on three islands. Each year 10% of the population on island A migrates to island B, 20% of the population on island B migrates to island C, and 5% of the population on island C migrates to island A. Let A_n, B_n, and C_n denote the number of birds in year n on islands A, B, and C, respectively, before migration takes place.
   (a) Show that $A_{n+1} = 0.9A_n + 0.05C_n$, $B_{n+1} = 0.1A_n + 0.80B_n$, and $C_{n+1} = 0.95C_n + 0.20B_n$.
   (b) Assuming that $\lim_{n \to \infty} A_n$, $\lim_{n \to \infty} B_n$, and $\lim_{n \to \infty} C_n$ exist, approximate the number of birds on each island after many years.

44. A bobcat population is classified by age as kittens (less than one year old) and adults (at least one year old). All adult females, including those born the prior year, have a litter each June, with an average litter size of three kittens. The survival rate of kittens is 50%, while that of adults is $66\frac{2}{3}\%$ per year. Let K_n be the number of newborn kittens in June of the nth year, let A_n be the number of adults, and assume that the ratio of males to females is always 1.
   (a) Show that $K_{n+1} = \frac{3}{2}A_{n+1}$ and $A_{n+1} = \frac{2}{3}A_n + \frac{1}{2}K_n$.
   (b) Conclude that $A_{n+1} = \frac{17}{12}A_n$ and $K_{n+1} = \frac{17}{12}K_n$, and that $A_n = (\frac{17}{12})^{n-1}A_1$ and $K_n = (\frac{17}{12})^{n-1}K_1$. What can you conclude about the population?

Exer. 45–47: Use a calculator.

45. Terms of the sequence defined recursively by $a_1 = 5$ and $a_{k+1} = \sqrt{a_k}$ may be generated by entering 5 in a calculator and pressing the square root key repeatedly.
   (a) Describe what happens to the terms of the sequence as k increases.
   (b) Show that $a_n = 5^{1/2^n}$, and find $\lim_{n \to \infty} a_n$.

46. A sequence is generated by entering a number N in a calculator and pressing the reciprocal key $\boxed{1/\text{x}}$ repeatedly. Under what conditions does the sequence have a limit?

47. Terms of the sequence defined recursively by $a_1 = 1$ and $a_{k+1} = \cos a_k$ may be generated by entering 1 in a calculator (in radian mode) and pressing the cosine key repeatedly.
   (a) Describe what happens to terms of the sequence as k increases.
   (b) Assuming that $\lim_{n \to \infty} a_n = L$, prove that $\cos L = L$. (Hint: $\lim_{n \to \infty} a_{n+1} = L$.)

48. A sequence $\{x_n\}$ is defined recursively by $x_{k+1} = x_k - \tan x_k$.
   (a) If $x_1 = 3$, approximate the first five terms of the sequence. Can you guess $\lim_{n \to \infty} x_n$?
   (b) If $x_1 = 6$, approximate the first five terms of the sequence. Can you guess $\lim_{n \to \infty} x_n$?
   (c) Assuming that $\lim_{n \to \infty} x_n = L$, prove that $L = \pi n$ for some integer n.

49. Approximations to $\sqrt{N}$ may be generated from the sequence defined recursively by
$$x_1 = \frac{N}{2}, \quad x_{k+1} = \frac{1}{2}\left(x_k + \frac{N}{x_k}\right).$$
   (a) Approximate x_2, x_3, x_4, x_5, x_6 if $N = 10$.
   (b) Assuming that $\lim_{n \to \infty} x_n = L$, prove that $L = \sqrt{N}$.

50 The famous *Fibonacci sequence* is defined recursively by $a_{k+1} = a_k + a_{k-1}$ with $a_1 = a_2 = 1$.

(a) Find the first ten terms of the sequence.

(b) The terms of the sequence $r_k = a_{k+1}/a_k$ give progres-

sively better approximations to τ, the *golden ratio*. Approximate the first ten terms of this sequence.

(c) Assuming that $\lim_{n \to \infty} r_n = \tau$, prove that $\tau = \frac{1}{2}(1 + \sqrt{5})$.

11.2 CONVERGENT OR DIVERGENT INFINITE SERIES

In Section 11.1 we stated that it is possible to express the rational number $\frac{2}{3}$ as the *infinite sum* (or *infinite series*)

$$0.6 + 0.06 + 0.006 + 0.0006 + 0.00006 + \cdots$$

Since only finite sums may be added algebraically, we must *define* what is meant by an infinite sum of this type.

Let us begin by introducing some terminology that will be used throughout the remainder of the chapter.

DEFINITION (11.11)

Let $\{a_n\}$ be an infinite sequence. The expression

$$a_1 + a_2 + \cdots + a_n + \cdots$$

is called an **infinite series,** or simply a **series.**

In summation notation, the series in Definition (11.11) will be denoted by either

$$\sum_{n=1}^{\infty} a_n \quad \text{or} \quad \sum a_n.$$

The summation variable in the last sum is understood to be n. Each number a_k is a **term** of the series, and a_n is the **nth term.**

DEFINITION (11.12)

(i) The **kth partial sum** S_k of the infinite series $\sum a_n$ is

$$S_k = a_1 + a_2 + \cdots + a_k$$

(ii) The **sequence of partial sums** associated with the infinite series $\sum a_n$ is

$$S_1, S_2, S_3, \ldots, S_n, \ldots$$

From Definition (11.12) (i),

$$S_1 = a_1$$
$$S_2 = a_1 + a_2$$
$$S_3 = a_1 + a_2 + a_3$$
$$S_4 = a_1 + a_2 + a_3 + a_4.$$

If we calculate S_5, S_6, S_7, and so on, we add more and more terms of the series. Thus S_{1000} is the sum of the first one thousand terms of $\sum a_n$. If the sequence $\{S_n\}$ has a limit S (that is, it converges to S), we call S the sum of the *infinite* series $\sum a_n$, as in the next definition.

DEFINITION (11.13)

> An infinite series $\sum a_n$ is **convergent** (or **converges**) if its sequence of partial sums $\{S_n\}$ converges; that is, if
>
> $$\lim_{n \to \infty} S_n = S \quad \text{for some real number } S.$$
>
> The limit S is the **sum** of the series $\sum a_n$, and we write
>
> $$S = a_1 + a_2 + \cdots + a_n + \cdots$$
>
> The series $\sum a_n$ is **divergent** (or **diverges**) if $\{S_n\}$ diverges. A divergent infinite series has no sum.

EXAMPLE 1 Prove that the infinite series

$$\frac{1}{1 \cdot 2} + \frac{1}{2 \cdot 3} + \frac{1}{3 \cdot 4} + \cdots + \frac{1}{n(n + 1)} + \cdots$$

converges, and find its sum.

SOLUTION The partial fraction decomposition of the nth term a_n is

$$a_n = \frac{1}{n(n + 1)} = \frac{1}{n} - \frac{1}{n + 1}.$$

Consequently, the nth partial sum of the series may be written

$$S_n = a_1 + a_2 + a_3 + \cdots + a_n$$

$$= \left(1 - \frac{1}{2}\right) + \left(\frac{1}{2} - \frac{1}{3}\right) + \left(\frac{1}{3} - \frac{1}{4}\right) + \cdots + \left(\frac{1}{n} - \frac{1}{n + 1}\right)$$

$$= 1 - \frac{1}{n + 1} = \frac{n}{n + 1}.$$

Since

$$\lim_{n \to \infty} S_n = \lim_{n \to \infty} \frac{n}{n + 1} = 1,$$

the series converges and has the sum 1. •

The series $\sum 1/[n(n + 1)]$ of Example 1 is called a **telescoping series,** since writing S_n as shown in the solution causes the terms to collapse to $1 - [1/(n + 1)]$.

EXAMPLE 2 Show that the infinite series $\sum_{n=1}^{\infty} (-1)^{n-1}$ diverges.

SOLUTION The series may be written

$$1 + (-1) + 1 + (-1) + \cdots + (-1)^{n-1} + \cdots$$

Note that $S_k = 1$ if k is odd, and $S_k = 0$ if k is even. Since the sequence of partial sums $\{S_n\}$ oscillates between 1 and 0, it follows that $\lim_{n \to \infty} S_n$ does not exist. Hence the infinite series diverges. •

EXAMPLE 3 Prove that the following series is divergent:

$$1 + \frac{1}{2} + \frac{1}{3} + \frac{1}{4} + \cdots + \frac{1}{n} + \cdots$$

SOLUTION Let us group the terms of the series as follows:

$$1 + \tfrac{1}{2} + (\tfrac{1}{3} + \tfrac{1}{4}) + (\tfrac{1}{5} + \tfrac{1}{6} + \tfrac{1}{7} + \tfrac{1}{8}) + (\tfrac{1}{9} + \cdots + \tfrac{1}{16}) + (\tfrac{1}{17} + \cdots + \tfrac{1}{32}) + \cdots$$

Note that each group contains twice the number of terms as the preceding group. Since the sum of the terms in each parenthesis is greater than $\frac{1}{2}$, we obtain the following inequalities:

$$S_2 = 1 + \tfrac{1}{2} > 2(\tfrac{1}{2})$$

$$S_4 > 1 + \tfrac{1}{2} + \tfrac{1}{2} > 3(\tfrac{1}{2})$$

$$S_8 > 1 + \tfrac{1}{2} + \tfrac{1}{2} + \tfrac{1}{2} > 4(\tfrac{1}{2})$$

Similarly, $S_{16} > 5(\tfrac{1}{2})$ and $S_{32} > 6(\tfrac{1}{2})$. We can show, by mathematical induction, that

$$S_{2^k} > (k + 1)(\tfrac{1}{2}) \quad \text{for every positive integer } k.$$

It follows that S_n can be made as large as desired by taking n sufficiently large, that is, $\lim_{n \to \infty} S_n = \infty$. Since $\{S_n\}$ diverges, the infinite series diverges.

•

The series in Example 3 will be useful in later developments. It is given the following special name.

DEFINITION (11.14)

> The **harmonic series** is the divergent infinite series
> $$1 + \frac{1}{2} + \frac{1}{3} + \cdots + \frac{1}{n} + \cdots$$

In the next section we will give another proof of the divergence of the harmonic series.

Certain types of infinite series occur frequently in solutions of applied problems. One of the most important is the **geometric series**

$$a + ar + ar^2 + \cdots + ar^{n-1} + \cdots$$

for real numbers a and r, with $a \neq 0$.

THEOREM (11.15)

> Let $a \neq 0$. The geometric series
> $$a + ar + ar^2 + \cdots + ar^{n-1} + \cdots$$
> (i) converges and has the sum $\dfrac{a}{1-r}$ if $|r| < 1$.
> (ii) diverges if $|r| \geq 1$.

PROOF If $r = 1$, then $S_n = a + a + \cdots + a = na$ and the series diverges, since $\lim_{n \to \infty} S_n$ does not exist.

If $r = -1$, then $S_k = a$ if k is odd, and $S_k = 0$ if k is even. Since the sequence of partial sums oscillates between a and 0, the series diverges.

If $r \neq 1$, consider

$$S_n = a + ar + ar^2 + \cdots + ar^{n-1}$$

and $\qquad rS_n = ar + ar^2 + ar^3 + \cdots + ar^n.$

Subtracting corresponding sides of these equations, we obtain

$$(1 - r)S_n = a - ar^n.$$

Dividing both sides by $1 - r$ gives us

$$S_n = \frac{a}{1 - r} - \frac{ar^n}{1 - r}.$$

Consequently,

$$\lim_{n \to \infty} S_n = \lim_{n \to \infty} \left(\frac{a}{1 - r} - \frac{ar^n}{1 - r} \right)$$

$$= \lim_{n \to \infty} \frac{a}{1 - r} - \lim_{n \to \infty} \frac{ar^n}{1 - r}$$

$$= \frac{a}{1 - r} - \frac{a}{1 - r} \lim_{n \to \infty} r^n.$$

If $|r| < 1$, then $\lim_{r \to \infty} r^n = 0$ (see Theorem (11.6)) and hence

$$\lim_{n \to \infty} S_n = \frac{a}{1 - r}.$$

If $|r| > 1$, then $\lim_{n \to \infty} r^n$ does not exist (see Theorem (11.6)) and hence $\lim_{n \to \infty} S_n$ does not exist. In this case the series diverges. • •

EXAMPLE 4 Prove that the following infinite series converges, and find its sum:

$$0.6 + 0.06 + 0.006 + \cdots + \frac{6}{10^n} + \cdots$$

SOLUTION This is the series considered at the beginning of this section. It is geometric with $a = 0.6$ and $r = 0.1$. By Theorem (11.15) (i), the series converges and has the sum

$$S = \frac{0.6}{1 - 0.1} = \frac{0.6}{0.9} = \frac{2}{3}.$$

Thus

$$\frac{2}{3} = 0.6 + 0.06 + 0.006 + \cdots + \frac{6}{10^n} + \cdots$$

This justifies the nonterminating decimal notation $\frac{2}{3} = 0.66666\ldots$. •

EXAMPLE 5 Prove that the following series converges, and find its sum:

$$2 + \frac{2}{3} + \frac{2}{3^2} + \cdots + \frac{2}{3^{n-1}} + \cdots$$

SOLUTION The series converges, since it is geometric with $r = \frac{1}{3} < 1$. By Theorem (11.15) (i), the sum is

$$S = \frac{2}{1 - \frac{1}{3}} = \frac{2}{\frac{2}{3}} = 3 \quad \bullet$$

THEOREM (11.16) | If an infinite series $\sum a_n$ is convergent, then $\lim_{n \to \infty} a_n = 0$.

PROOF The nth term a_n of the infinite series can be expressed as

$$a_n = S_n - S_{n-1}.$$

If $\lim_{n \to \infty} S_n = S$, then also $\lim_{n \to \infty} S_{n-1} = S$ and

$$\lim_{n \to \infty} a_n = \lim_{n \to \infty} (S_n - S_{n-1}) = \lim_{n \to \infty} S_n - \lim_{n \to \infty} S_{n-1} = 0 \quad \bullet \bullet$$

The preceding theorem states that *if* a series converges, *then* its nth term a_n has the limit 0 as $n \to \infty$. The converse is false, that is, *if* $\lim_{n \to \infty} a_n = 0$ *it does not necessarily follow* that the series $\sum a_n$ is convergent. The harmonic series (11.14) is an illustration of a divergent series $\sum a_n$ for which $\lim_{n \to \infty} a_n = 0$. Consequently, to establish convergence of an infinite series *it is not enough* to prove that $\lim_{n \to \infty} a_n = 0$, since that may be true for divergent as well as for convergent series.

The next result about *divergence* is an immediate corollary of Theorem (11.16).

nTH-TERM TEST (11.17)
FOR DIVERGENCE

> If $\lim_{n \to \infty} a_n \neq 0$, then the infinite series $\sum a_n$ is divergent.

EXAMPLE 6 Apply the nth-Term Test (11.17) to each series and comment on the result:

(a) $\displaystyle\sum_{n=1}^{\infty} \frac{n}{2n+1}$ (b) $\displaystyle\sum_{n=1}^{\infty} \frac{1}{n^2}$ (c) $\displaystyle\sum_{n=1}^{\infty} \frac{1}{\sqrt{n}}$ (d) $\displaystyle\sum_{n=1}^{\infty} \frac{e^n}{n}$

SOLUTION Taking the limit of the nth term a_n, we obtain:

(a) $\displaystyle\lim_{n \to \infty} \frac{n}{2n+1} = \frac{1}{2} \neq 0$ (The series diverges by Test (11.17).)

(b) $\displaystyle\lim_{n \to \infty} \frac{1}{n^2} = 0$ (Test (11.17) is not applicable. The series may converge or diverge.)

(c) $\displaystyle\lim_{n \to \infty} \frac{1}{\sqrt{n}} = 0$ (Test (11.17) is not applicable. The series may converge or diverge.)

(d) $\displaystyle\lim_{n \to \infty} \frac{e^n}{n} = \infty$ (The series diverges by Test (11.17).)

We shall see in the next section that the series in part (b) converges, and that in (c) diverges. •

The next theorem states that if corresponding terms of two infinite series are identical after a certain term, then both series converge or both series diverge.

THEOREM (11.18)

> If $\sum a_n$ and $\sum b_n$ are infinite series such that $a_i = b_i$ for every $i > k$, where k is a positive integer, then both series converge or both series diverge.

PROOF By hypothesis we may write

$$\sum a_n = a_1 + a_2 + \cdots + a_k + a_{k+1} + \cdots + a_n + \cdots$$
$$\sum b_n = b_1 + b_2 + \cdots + b_k + a_{k+1} + \cdots + a_n + \cdots$$

Let S_n and T_n denote the nth partial sums of $\sum a_n$ and $\sum b_n$, respectively. It follows that if $n \geq k$, then

$$S_n - S_k = T_n - T_k$$

or

$$S_n = T_n + (S_k - T_k).$$

Consequently,

$$\lim_{n \to \infty} S_n = \lim_{n \to \infty} T_n + (S_k - T_k)$$

and hence either both of the limits exist or both do not exist. This gives us the desired conclusion. Evidently, if both series converge, then their sums differ by $S_k - T_k$. • •

Theorem (11.18) implies that changing a finite number of terms of an infinite series has no effect on its convergence or divergence (although it does change the sum of a convergent series). In particular, if we replace the first k terms of $\sum a_n$ by 0, convergence is unaffected. It follows that the series

$$a_{k+1} + a_{k+2} + \cdots + a_n + \cdots$$

converges or diverges if $\sum a_n$ converges or diverges, respectively. The series $a_{k+1} + a_{k+2} + \cdots$ is obtained from $\sum a_n$ by **deleting the first k terms.**

EXAMPLE 7 Show that the following series converges:

$$\frac{1}{3 \cdot 4} + \frac{1}{4 \cdot 5} + \cdots + \frac{1}{(n+2)(n+3)} + \cdots$$

SOLUTION The series converges, since it can be obtained by deleting the first two terms of the convergent telescoping series of Example 1. •

The proof of the next theorem follows directly from Definition (11.13) and is left as an exercise.

THEOREM (11.19)

> If $\sum a_n$ and $\sum b_n$ are convergent series with sums A and B, respectively, then
>
> (i) $\sum (a_n + b_n)$ converges and has sum $A + B$.
> (ii) if c is a real number, $\sum ca_n$ converges and has sum cA.
> (iii) $\sum (a_n - b_n)$ converges and has sum $A - B$.

It is also easy to show that if $\sum a_n$ diverges, then so does $\sum ca_n$ for every $c \neq 0$.

EXAMPLE 8 Prove that the following series converges, and find its sum:

$$\sum_{n=1}^{\infty} \left[\frac{7}{n(n+1)} + \frac{2}{3^{n-1}} \right].$$

SOLUTION The telescoping series $\sum 1/[n(n+1)]$ was considered in Example 1, where we found that it converges and has the sum 1. Using Theorem (11.19) (ii) with $c = 7$ and $a_n = 1/[n(n+1)]$, we see that $\sum 7/[n(n+1)]$ converges and has the sum $7(1) = 7$.

The geometric series $\sum 2/3^{n-1}$ converges and has the sum 3 (see Example 5). Hence by Theorem (11.19) (i), the given series converges and has the sum $7 + 3 = 10$. •

THEOREM (11.20)

If $\sum a_n$ is a convergent series and $\sum b_n$ is divergent, then $\sum (a_n + b_n)$ is divergent.

PROOF We shall give an indirect proof. Thus, suppose $\sum (a_n + b_n)$ is convergent. Applying Theorem (11.19) (iii), $\sum [(a_n + b_n) - a_n] = \sum b_n$ is convergent, a contradiction. Hence our supposition is false, that is, $\sum (a_n + b_n)$ is divergent. • •

EXAMPLE 9 Determine the convergence or divergence of the series

$$\sum_{n=1}^{\infty} \left(\frac{1}{5^n} + \frac{1}{n} \right).$$

SOLUTION Since $\sum (1/5^n)$ is a convergent geometric series and $\sum (1/n)$ is the divergent harmonic series, the given series diverges by Theorem (11.20).

•

APPLICATION: S → I → S EPIDEMICS

Let us return to the discussion of the epidemic given at the end of Section 11.1. Suppose that instead of I_n (the number ill on day n) we are interested in the *total number* S_n of individuals who have been ill at some time between the first and nth days. As in our earlier discussion, let us overestimate S_n by approximating the number F_{n+1} of new cases on day $n + 1$ by aNI_n. Thus

$$S_n = I_1 + F_2 + F_3 + F_4 + \cdots + F_n$$
$$= I_1 + aNI_1 + aNI_2 + aNI_3 + \cdots + aNI_{n-1}.$$

Recalling that $I_n = r^{n-1}I_1$, with $r = 1 + aN - b$, we obtain

$$S_n = I_1 + aNI_1 + aNrI_1 + aNr^2I_1 + \cdots + aNr^{n-2}I_1$$
$$= I_1 + aNI_1(1 + r + r^2 + \cdots + r^{n-2}).$$

As in the proof of Theorem (11.15), this may be written

$$S_n = I_1 + aNI_1 \left(\frac{1}{1-r} - \frac{r^{n-1}}{1-r} \right).$$

If $r < 1$, then

$$\lim_{n \to \infty} S_n = I_1 + aNI_1 \left(\frac{1}{1-r} \right)$$
$$= I_1 \left(1 + \frac{aN}{1-r} \right)$$
$$= I_1 \left(1 + \frac{aN}{b-aN} \right)$$
$$= I_1 \left(\frac{b}{b-aN} \right).$$

If a and b are approximated from early data, this result enables health officials to determine an upper bound for the total number of individuals who will be ill at some stage of the epidemic.

Exer. 1–20: Determine if the infinite series converges or diverges. If it converges, find its sum.

1 $3 + \dfrac{3}{4} + \cdots + \dfrac{3}{4^{n-1}} + \cdots$

2 $3 + \dfrac{3}{(-4)} + \cdots + \dfrac{3}{(-4)^{n-1}} + \cdots$

3 $1 + \dfrac{(-1)}{\sqrt{5}} + \cdots + \left(\dfrac{-1}{\sqrt{5}}\right)^{n-1} + \cdots$

4 $1 + \left(\dfrac{e}{3}\right) + \cdots + \left(\dfrac{e}{3}\right)^{n-1} + \cdots$

5 $0.37 + 0.0037 + \cdots + \dfrac{37}{(100)^n} + \cdots$

6 $0.628 + 0.000628 + \cdots + \dfrac{628}{(1000)^n} + \cdots$

7 $\displaystyle\sum_{n=1}^{\infty} 2^{-n}3^{n-1}$

8 $\displaystyle\sum_{n=1}^{\infty} (-5)^{n-1}4^{-n}$

9 $\displaystyle\sum_{n=1}^{\infty} (-1)^{n-1}$

10 $\displaystyle\sum (\sqrt{2})^{n-1}$

11 $\dfrac{1}{4 \cdot 5} + \dfrac{1}{5 \cdot 6} + \cdots + \dfrac{1}{(n+3)(n+4)} + \cdots$

12 $\dfrac{-1}{1 \cdot 2} + \dfrac{-1}{2 \cdot 3} + \cdots + \dfrac{-1}{n(n+1)} + \cdots$

13 $\dfrac{5}{1 \cdot 2} + \dfrac{5}{2 \cdot 3} + \cdots + \dfrac{5}{n(n+1)} + \cdots$

14 $\dfrac{1}{4} + \dfrac{1}{5} + \cdots + \dfrac{1}{n+3} + \cdots$

15 $3 + \dfrac{3}{2} + \cdots + \dfrac{3}{n} + \cdots$

16 $\dfrac{1}{2} + \dfrac{2}{3} + \cdots + \dfrac{n}{n+1} + \cdots$

17 $\displaystyle\sum_{n=1}^{\infty} \dfrac{3n}{5n-1}$

18 $\displaystyle\sum_{n=1}^{\infty} \dfrac{1}{1 + (0.3)^n}$

19 $\displaystyle\sum_{n=1}^{\infty} \left(\dfrac{1}{8^n} + \dfrac{1}{n(n+1)}\right)$

20 $\displaystyle\sum_{n=1}^{\infty} \left(\dfrac{1}{3^n} - \dfrac{1}{4^n}\right)$

Exer. 21–28: Use the examples and theorems discussed in this section to determine if the series converges or diverges.

21 $\displaystyle\sum_{n=1}^{\infty} \dfrac{1}{\sqrt[n]{e}}$

22 $\displaystyle\sum_{n=1}^{\infty} \dfrac{n}{\ln(n+1)}$

23 $\displaystyle\sum_{n=1}^{\infty} \left(\dfrac{5}{n+2} - \dfrac{5}{n+3}\right)$

24 $\displaystyle\sum_{n=1}^{\infty} \ln\left(\dfrac{2n}{7n-5}\right)$

25 $\displaystyle\sum_{n=1}^{\infty} \left[\left(\dfrac{3}{2}\right)^n + \left(\dfrac{2}{3}\right)^n\right]$

26 $\displaystyle\sum_{n=1}^{\infty} \left[\dfrac{1}{n(n+1)} - \dfrac{4}{n}\right]$

27 $\displaystyle\sum_{n=1}^{\infty} n \sin\dfrac{1}{n}$

28 $\displaystyle\sum_{n=1}^{\infty} (2^{-n} - 2^{-3n})$

Exer. 29–32: Find a formula for S_n and prove that the series converges or diverges using $\lim_{n \to \infty} S_n$ (see Example 1).

29 $\displaystyle\sum_{n=1}^{\infty} \dfrac{1}{4n^2 - 1}$

30 $\displaystyle\sum_{n=1}^{\infty} \dfrac{-1}{9n^2 + 3n - 2}$

31 $\displaystyle\sum_{n=1}^{\infty} \ln\dfrac{n}{n+1}$

32 $\displaystyle\sum_{n=1}^{\infty} \dfrac{1}{\sqrt{n+1} + \sqrt{n}}$ (*Hint:* Rationalize the denominator.)

33 Prove or disprove: If $\sum a_n$ and $\sum b_n$ both diverge, then $\sum (a_n + b_n)$ diverges.

34 What is wrong with the following "proof" that the divergent geometric series $\sum_{n=1}^{\infty} (-1)^{n+1}$ has the sum 0 (see Example 2)?

$$\sum_{n=1}^{\infty} (-1)^{n+1} = [1 + (-1)] + [1 + (-1)]$$
$$+ [1 + (-1)] + \cdots$$
$$= 0 + 0 + 0 + \cdots = 0$$

Exer. 35–38: The bar indicates that the digits underneath repeat indefinitely. Express the repeating decimal as an infinite series, and find the rational number it represents.

35 $0.\overline{23}$

36 $5.\overline{146}$

37 $3.2\overline{394}$

38 2.71828

39 A rubber ball is dropped from a height of 10 meters. If it rebounds approximately one-half the distance after each fall, use an infinite geometric series to approximate the total distance the ball travels before coming to rest.

40 The bob of a pendulum swings through an arc 24 cm long on its first swing. If each successive swing is approximately five-sixths the length of the preceding swing, use an infinite geometric series to approximate the total distance it travels before coming to rest.

41 If a dosage of Q units of a certain drug is administered to an individual, then the amount remaining in the bloodstream at the end of t minutes is given by Qe^{-ct} for a positive constant c. Suppose this same dosage is given at successive T-minute intervals.
   (a) Show that the amount $A(k)$ of the drug in the bloodstream immediately after the kth dose is given by $A(k) = \sum_{n=0}^{k-1} Qe^{-ncT}$.
   (b) Find an upper bound for the amount of the drug in the bloodstream after any number of doses.
   (c) Find the smallest time between doses that will ensure that $A(k)$ does not exceed a certain level M.

42 Suppose that each dollar introduced into the economy recirculates as follows: 85% of the original dollar is spent, then 85% of the remaining $0.85 is spent, and so on. Find the economic impact (the total amount spent) if $1,000,000 is introduced into the economy.

43 In a pest eradication program, N sterilized male flies are released into the general population each day, and 90% of these flies will survive a given day.

(a) Show that the number of sterilized flies in the population after n days is $N + (0.9)N + \cdots + (0.9)^{n-1}N$.

(b) If the *long-range* goal of the program is to keep 20,000 sterilized males in the population, how many flies should be released each day?

44 A certain drug has a half-life in the bloodstream of about 2 hours. Doses of K mg will be administered every four hours, with K still to be determined.

(a) Show that the number of milligrams of drug in the bloodstream after the nth dose has been administered is $K + \frac{1}{4}K + \cdots + (\frac{1}{4})^{n-1}K$, and that this sum is approximately $\frac{4}{3}K$ for large values of n.

(b) If more than 500 mg of drug in the bloodstream is considered to be a dangerous level, find the largest possible dose that can be given repeatedly over a long period of time.

(c) Refer to Exercise 41. If the dose K is 50 mg, how frequently can the drug be safely administered?

45 Shown in the first figure is a nested sequence of squares $S_1, S_2, \ldots, S_k, \ldots$. Let a_k, A_k, and P_k denote the side, area, and perimeter, respectively, of the square S_k. The square S_{k+1} is constructed from S_k by selecting four points on S_k that are a distance of $\frac{1}{4}a_k$ from the vertices and connecting them (see second figure).

EXERCISE 45

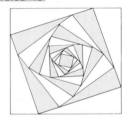

(a) Find the relationship between a_{k+1} and a_k, and determine a_n, A_n, and P_n.

(b) Calculate $\sum_{n=1}^{\infty} P_n$ and $\sum_{n=1}^{\infty} A_n$.

46 Shown in the figure is a nested sequence consisting of alternating circles and squares. Each circle is inscribed in the preceding square and each square (excluding the first) is inscribed in the preceding circle. Let S_n denote the area of the nth square, and C_n the area of the nth circle.

(a) Find the relationships between S_n and C_n, and between C_n and S_{n+1}.

(b) What portion of the original square has been shaded in the figure?

EXERCISE 46

Exer. 47–50: Use Theorem (11.15) to verify each statement.

47 $1 - x + x^2 - x^3 + \cdots + (-1)^{n+1}x^n + \cdots = \dfrac{1}{1+x}$,

if $-1 < x < 1$

48 $1 + x^2 + x^4 + \cdots + x^{2n} + \cdots = \dfrac{1}{1-x^2}$,

if $-1 < x < 1$

49 $\dfrac{1}{2} + \dfrac{(x-3)}{4} + \dfrac{(x-3)^2}{8} + \cdots + \dfrac{(x-3)^n}{2^{n+1}} + \cdots = \dfrac{1}{5-x}$,

if $1 < x < 5$

50 $3 + (x-1) + \dfrac{(x-1)^2}{3} + \cdots + \dfrac{(x-1)^n}{3^{n-1}} + \cdots = \dfrac{9}{4-x}$,

if $-2 < x < 4$

11.3 POSITIVE-TERM SERIES

It is difficult to apply the definition of convergence or divergence (11.13) to an infinite series $\sum a_n$, since in most cases it is impossible to find a simple formula for S_n. We can, however, develop techniques for using the nth term a_n to test a series for convergence or divergence. When applying these tests we shall be concerned not with the *sum* of the series but with whether the series converges or diverges.

In this section, we shall consider only **positive-term series,** that is, series such that $a_n > 0$ for every n. Although this approach may appear to be very specialized, positive-term series are the foundation for all of our future work with infinite series. As we shall see later, the convergence or divergence of an *arbitrary* infinite series can often be determined from that of a related positive-term series.

The next theorem shows that to establish convergence or divergence of a positive-term series, it is sufficient to determine if the sequence of partial sums $\{S_n\}$ is bounded.

THEOREM (11.21)

> If $\sum a_n$ is a positive-term series and if there exists a number M such that $S_n < M$ for every n, then the series converges and has a sum $S \le M$. If no such M exists the series diverges.

PROOF If $\{S_n\}$ is the sequence of partial sums of the positive-term series $\sum a_n$, then

$$S_1 < S_2 < \cdots < S_n < \cdots$$

and therefore $\{S_n\}$ is monotonic. If there exists a number M such that $S_n < M$ for every n, then $\{S_n\}$ is bounded monotonic. As in the proof of Theorem (11.9),

$$\lim_{n \to \infty} S_n = S \le M$$

for some S, and hence the series converges. If no such M exists, then $\lim_{n \to \infty} S_n = \infty$ and the series diverges. • •

Suppose a function f is defined for every real number $x \ge 1$. We may then consider the infinite series

$$\sum_{n=1}^{\infty} f(n) = f(1) + f(2) + \cdots + f(n) + \cdots$$

For example, if $f(x) = 1/x^2$, then

$$\sum_{n=1}^{\infty} f(n) = \sum_{n=1}^{\infty} \frac{1}{n^2} = \frac{1}{1^2} + \frac{1}{2^2} + \cdots + \frac{1}{n^2} + \cdots$$

A series of this type may be tested for convergence or divergence by means of an improper integral, as indicated in the following result.

THE INTEGRAL TEST (11.22)

> If a function f is positive-valued, continuous, and decreasing for $x \ge 1$, then the infinite series
>
> $$f(1) + f(2) + \cdots + f(n) + \cdots$$
>
> (i) converges if $\int_1^{\infty} f(x)\,dx$ converges.
>
> (ii) diverges if $\int_1^{\infty} f(x)\,dx$ diverges.

FIGURE 11.8

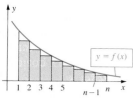

FIGURE 11.9

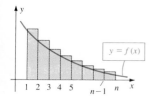

PROOF If n is a positive integer greater than 1, then the area of the inscribed rectangular polygon illustrated in Figure 11.8 is

$$\sum_{k=2}^{n} f(k) = f(2) + f(3) + \cdots + f(n).$$

Similarly, the area of the circumscribed rectangular polygon illustrated in Figure 11.9 is

$$\sum_{k=1}^{n-1} f(k) = f(1) + f(2) + \cdots + f(n-1).$$

Since $\int_1^n f(x)\,dx$ is the area under the graph of f from 1 to n,

$$\sum_{k=2}^{n} f(k) \le \int_1^n f(x)\,dx \le \sum_{k=1}^{n-1} f(k).$$

If S_n denotes the nth partial sum of the series $f(1) + f(2) + \cdots + f(n) + \cdots$, then this inequality may be written

$$S_n - f(1) \le \int_1^n f(x)\,dx \le S_{n-1}.$$

The preceding inequality implies that if the integral $\int_1^\infty f(x)\,dx$ converges and equals $K > 0$, then

$$S_n - f(1) \le K \quad \text{or} \quad S_n \le K + f(1)$$

for every positive integer n. Hence, by Theorem (11.21) the series $\sum f(n)$ converges.

If the improper integral diverges, then

$$\lim_{n \to \infty} \int_1^n f(x)\,dx = \infty$$

and since $\int_1^n f(x)\,dx \le S_{n-1}$, we also have $\lim_{n \to \infty} S_{n-1} = \infty$, that is, the series $\sum f(n)$ diverges. • •

EXAMPLE 1 Use the Integral Test (11.22) to prove that the harmonic series

$$1 + \frac{1}{2} + \frac{1}{3} + \cdots + \frac{1}{n} + \cdots$$

diverges (see Example 3 of Section 11.2).

SOLUTION If we let $f(x) = 1/x$, then f is positive-valued, continuous, and decreasing for $x \ge 1$ and, therefore, the Integral Test may be applied. Since

$$\int_1^\infty \frac{1}{x}\,dx = \lim_{t \to \infty} \int_1^t \frac{1}{x}\,dx = \lim_{t \to \infty} \Big[\ln x\Big]_1^t$$

$$= \lim_{t \to \infty} \big[\ln t - \ln 1\big] = \infty,$$

the series diverges by (11.22) (ii). •

EXAMPLE 2 Determine if the infinite series $\sum n e^{-n^2}$ converges or diverges.

SOLUTION If we let $f(x) = x e^{-x^2}$, then the given series is the same as $\sum f(n)$. If $x \ge 1$, f is positive-valued and continuous. The first derivative may be used to determine if f is decreasing. Since

$$f'(x) = e^{-x^2} - 2x^2 e^{-x^2} = e^{-x^2}(1 - 2x^2) < 0,$$

f is decreasing on $[1, \infty)$. We may, therefore, apply the Integral Test as follows:

$$\int_1^\infty x e^{-x^2}\,dx = \lim_{t \to \infty} \int_1^t x e^{-x^2}\,dx = \lim_{t \to \infty} \Big(-\tfrac{1}{2}\Big)e^{-x^2}\Big]_1^t$$

$$= (-\tfrac{1}{2}) \lim_{t \to \infty} \left[\frac{1}{e^{t^2}} - \frac{1}{e}\right] = \frac{1}{2e}.$$

Hence the series converges by (11.22) (i). •

An integral test may also be used if the function f satisfies the conditions of (11.22) for every $x \geq m$ for some positive integer m. In this case we merely replace the integral in (11.22) by $\int_m^\infty f(x)\,dx$. This corresponds to deleting the first $m - 1$ terms of the series.

If $f(x) = 1/x^p$ for $p > 0$, then the series $\sum f(n)$ has the form

$$1 + \frac{1}{2^p} + \frac{1}{3^p} + \cdots + \frac{1}{n^p} + \cdots$$

and is called the **p-series,** or **hyperharmonic series.** This series will be useful when we apply comparison tests later in this section. The following theorem provides information about convergence or divergence.

THEOREM (11.23)

The p-series

$$\sum_{n=1}^{\infty} \frac{1}{n^p} = 1 + \frac{1}{2^p} + \frac{1}{3^p} + \cdots + \frac{1}{n^p} + \cdots$$

converges if $p > 1$ or diverges if $p \leq 1$.

PROOF First we note that the special case $p = 1$ is the divergent harmonic series. Next, suppose that p is any positive real number different from 1. If we let $f(x) = 1/x^p = x^{-p}$, then f is positive-valued and continuous for $x \geq 1$. Moreover, for these values of x we see that $f'(x) = -px^{-p-1} < 0$, and hence f is decreasing. Thus f satisfies the conditions stated in the Integral Test (11.22) and we consider

$$\int_1^\infty \frac{1}{x^p}\,dx = \lim_{t \to \infty} \int_1^t x^{-p}\,dx = \lim_{t \to \infty} \frac{x^{1-p}}{1-p}\Big]_1^t$$

$$= \frac{1}{1-p} \lim_{t \to \infty} [t^{1-p} - 1].$$

If $p > 1$, then $p - 1 > 0$ and the last expression may be written

$$\frac{1}{1-p} \lim_{t \to \infty} \left[\frac{1}{t^{p-1}} - 1 \right] = \frac{1}{1-p}(0-1) = \frac{1}{p-1}.$$

Thus by (11.22) (i), the p-series converges if $p > 1$.

If $0 < p < 1$, then $1 - p > 0$ and

$$\frac{1}{1-p} \lim_{t \to \infty} [t^{1-p} - 1] = \infty.$$

Hence by (11.22) (ii), the p-series diverges.

If $p \leq 0$, then $\lim_{n \to \infty} (1/n^p) \neq 0$, and by (11.17) the series diverges. • •

EXAMPLE 3 Determine if each series converges or diverges.

(a) $1 + \dfrac{1}{2^2} + \dfrac{1}{3^2} + \cdots + \dfrac{1}{n^2} + \cdots$

(b) $5 + \dfrac{5}{\sqrt{2}} + \dfrac{5}{\sqrt{3}} + \cdots + \dfrac{5}{\sqrt{n}} + \cdots$

SOLUTION

(a) The series $\sum 1/n^2$ converges, since it is the p-series with $p = 2 > 1$.

(b) The series $\sum 1/\sqrt{n}$ diverges, since it is the p-series with $p = \frac{1}{2} < 1$.
Consequently $\sum 5/\sqrt{n}$ diverges (see remark following Theorem (11.19)). •

The next theorem allows us to use known convergent (divergent) series to establish the convergence (divergence) of other series.

BASIC COMPARISON TEST (11.24)

Suppose $\sum a_n$ and $\sum b_n$ are positive-term series.

(i) If $\sum b_n$ converges and $a_n \le b_n$ for every positive integer n, then $\sum a_n$ converges.
(ii) If $\sum b_n$ diverges and $a_n \ge b_n$ for every positive integer n, then $\sum a_n$ diverges.

PROOF Let S_n and T_n denote the nth partial sums of $\sum a_n$ and $\sum b_n$, respectively. Suppose $\sum b_n$ converges and has the sum T. If $a_n \le b_n$ for every n, then $S_n \le T_n < T$ and hence, by Theorem (11.21), $\sum a_n$ converges. This proves part (i).

To prove (ii), suppose $\sum b_n$ diverges and $a_n \ge b_n$ for every n. Then $S_n \ge T_n$ and, since T_n increases without bound as n becomes infinite, so does S_n. Consequently, $\sum a_n$ diverges. • •

Since convergence or divergence of a series is not affected by deleting a finite number of terms, the conditions $a_n \ge b_n$ or $a_n \le b_n$ of (11.24) are only required from the kth term on, for some fixed positive integer k.

A series $\sum d_n$ is said to **dominate** a series $\sum c_n$ if $c_n \le d_n$ for every positive integer n. Using this terminology, (11.24) (i) states that a positive-term series that is dominated by a convergent series is also convergent. Part (ii) states that a series that dominates a divergent positive-term series also diverges.

EXAMPLE 4 Determine if each series converges or diverges.

(a) $\displaystyle\sum_{n=1}^{\infty} \frac{1}{2 + 5^n}$　　(b) $\displaystyle\sum_{n=2}^{\infty} \frac{3}{\sqrt{n - 1}}$

SOLUTION

(a) For every $n \ge 1$,

$$\frac{1}{2 + 5^n} < \frac{1}{5^n} = \left(\frac{1}{5}\right)^n.$$

Since $\sum (1/5)^n$ is a convergent geometric series, the given series converges by (11.24) (i).

(b) The p-series $\sum 1/\sqrt{n}$ diverges and hence so does the series obtained by disregarding the first term $1/\sqrt{1}$. Since

$$\frac{3}{\sqrt{n - 1}} > \frac{1}{\sqrt{n}} \quad \text{if } n \ge 2,$$

it follows from (11.24) (ii) that the given series diverges. •

The following comparison test is often easier to apply than (11.24).

> If $\sum a_n$ and $\sum b_n$ are positive-term series and if
>
> $$\lim_{n \to \infty} \frac{a_n}{b_n} = k > 0$$
>
> then either both series converge or both diverge.

PROOF If $\lim_{n \to \infty}(a_n/b_n) = k > 0$, then a_n/b_n is close to k if n is large. Hence there exists a number N such that

$$\frac{k}{2} < \frac{a_n}{b_n} < \frac{3k}{2} \quad \text{whenever } n > N$$

FIGURE 11.10

$(n > N)$

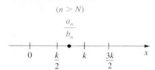

(see Figure 11.10). This is equivalent to

$$\frac{k}{2} b_n < a_n < \frac{3k}{2} b_n \quad \text{whenever } n > N.$$

If the series $\sum a_n$ converges, then $\sum (k/2)b_n$ also converges, since it is dominated by $\sum a_n$. Applying (11.19) (ii),

$$\sum b_n = \sum \left(\frac{2}{k}\right)\left(\frac{k}{2}\right) b_n$$

converges.

Conversely, if $\sum b_n$ converges, then so does $\sum a_n$, since it is dominated by the convergent series $\sum (3k/2)b_n$. We have proved that $\sum a_n$ converges if and only if $\sum b_n$ converges. Consequently, $\sum a_n$ diverges if and only if $\sum b_n$ diverges. $\bullet$ $\bullet$

Two other comparison tests are stated in Exercises 51 and 52.

EXAMPLE 5 Determine if each series converges or diverges.

(a) $\displaystyle\sum_{n=1}^{\infty} \frac{1}{\sqrt[3]{n^2 + 1}}$ (b) $\displaystyle\sum_{n=1}^{\infty} \frac{3n^2 + 5n}{2^n(n^2 + 1)}$

SOLUTION

(a) The form of the nth term, $a_n = 1/\sqrt[3]{n^2 + 1}$ suggests that we use, for comparison, the series that has nth term $b_n = 1/\sqrt[3]{n^2}$. The series $\sum b_n$ is the divergent p-series with $p = \frac{2}{3} < 1$. Note that we cannot use b_n with the Basic Comparison Test (11.24), since $a_n < b_n$ instead of $a_n \geq b_n$. However, applying the Limit Comparison Test (11.25),

$$\lim_{n \to \infty} \frac{a_n}{b_n} = \lim_{n \to \infty} \frac{\sqrt[3]{n^2}}{\sqrt[3]{n^2 + 1}} = \lim_{n \to \infty} \sqrt[3]{\frac{n^2}{n^2 + 1}} = 1 > 0,$$

and, since $\sum b_n$ diverges, so does $\sum a_n$.

(b) When we seek a suitable series $\sum b_n$ to be used for comparison with $\sum a_n$, with a_n a fraction, a good procedure is to disregard all terms in the numerator and denominator except those that have the most effect on the magnitude.

Let us consider

$$a_n = \frac{3n^2 + 5n}{2^n n^2 + 2^n}.$$

Since the term $3n^2$ has the most effect on the magnitude of the numerator, and since $2^n n^2$ has the most effect on the denominator, we are led to

$$b_n = \frac{3n^2}{2^n n^2} = \frac{3}{2^n}.$$

Applying the Limit Comparison Test (11.25),

$$\lim_{n \to \infty} \frac{a_n}{b_n} = \lim_{n \to \infty} \frac{3n^2 + 5n}{2^n(n^2 + 1)} \cdot \frac{2^n}{3} = \lim_{n \to \infty} \frac{n^2 + \frac{5}{3}n}{n^2 + 1} = 1 > 0.$$

Since, by Theorem (11.15), $\sum b_n$ is a convergent geometric series (with $r = \frac{1}{2}$ and $a = 3$), the series $\sum a_n$ is also convergent. •

EXAMPLE 6 Let $a_n = \dfrac{8n + \sqrt{n}}{5 + n^2 + n^{7/2}}$. Determine if $\sum a_n$ converges or diverges.

SOLUTION To find a suitable comparison series $\sum b_n$, we disregard all but the highest powers of n in the numerator and denominator, obtaining

$$b_n = \frac{8n}{n^{7/2}} = \frac{8}{n^{5/2}}.$$

Applying the Limit Comparison Test (11.25),

$$\lim_{n \to \infty} \frac{a_n}{b_n} = \lim_{n \to \infty} \frac{8n + n^{1/2}}{5 + n^2 + n^{7/2}} \cdot \frac{n^{5/2}}{8}$$

$$= \lim_{n \to \infty} \frac{8n^{7/2} + n^3}{40 + 8n^2 + 8n^{7/2}} = 1 > 0.$$

Since $\sum (1/n^{5/2})$ is a convergent p-series with $p = \frac{5}{2} > 1$, it follows from (11.19) (ii) that $\sum b_n$ is convergent. Hence, by (11.25), $\sum a_n$ is also convergent. •

We shall conclude this section with several general remarks about positive-term series. Suppose $\sum a_n$ is a positive-term series and the terms are grouped in some manner. For example, we could have

$$(a_1 + a_2) + a_3 + (a_4 + a_5 + a_6 + a_7) + \cdots$$

If we denote the last series by $\sum b_n$, so that

$$b_1 = a_1 + a_2, \quad b_2 = a_3, \quad b_3 = a_4 + a_5 + a_6 + a_7, \quad \ldots$$

then any partial sum of the series $\sum b_n$ is also a partial sum of $\sum a_n$. It follows that if $\sum a_n$ converges, then $\sum b_n$ converges and has the same sum. A similar argument may be used for any grouping of the terms of $\sum a_n$. Thus, *if a positive-term series converges, then the series obtained by grouping the terms in any manner also converges and has the same sum.* We cannot make a similar statement about arbitrary divergent series. For example, the terms of the divergent series $\sum (-1)^n$ may be grouped to produce a convergent series (see Exercise 34 of Section 11.2).

Next, suppose that a convergent positive-term series $\sum a_n$ has the sum S and that a new series $\sum b_n$ is formed by rearranging the terms in some way. For example, $\sum b_n$ could be the series

$$a_2 + a_8 + a_1 + a_5 + a_7 + a_3 + \cdots$$

If T_n is the nth partial sum of $\sum b_n$, then it is a sum of terms of $\sum a_n$. If m is the largest subscript associated with the terms a_k in T_n, then $T_n \le S_m < S$. Consequently, $T_n < S$ for every n. Applying Theorem (11.21), $\sum b_n$ converges and has a sum $T \le S$. The preceding proof is independent of the particular rearrangement of terms. We may also regard the series $\sum a_n$ as having been obtained by rearranging the terms of $\sum b_n$ and hence, by the same argument, $S \le T$. We have proved that *if the terms of a convergent positive-term series $\sum a_n$ are rearranged in any manner, then the resulting series converges and has the same sum.*

EXERCISES 11.3

Exer. 1–16: Use the Integral Test to determine if the series converges or diverges.

1. $\displaystyle\sum_{n=1}^{\infty} \frac{1}{(3+2n)^2}$

2. $\displaystyle\sum_{n=1}^{\infty} \frac{1}{(4+n)^{3/2}}$

3. $\displaystyle\sum_{n=1}^{\infty} \frac{1}{4n+7}$

4. $\displaystyle\sum_{n=2}^{\infty} \frac{1}{n(\ln n)^2}$

5. $\displaystyle\sum_{n=1}^{\infty} \frac{\ln n}{n}$

6. $\displaystyle\sum_{n=1}^{\infty} \frac{n}{n^2+1}$

7. $\displaystyle\sum_{n=1}^{\infty} \frac{1}{\sqrt[3]{2n+1}}$

8. $\displaystyle\sum_{n=1}^{\infty} \frac{1}{1+16n^2}$

9. $\displaystyle\sum_{n=1}^{\infty} \frac{\arctan n}{1+n^2}$

10. $\displaystyle\sum_{n=1}^{\infty} ne^{-n}$

11. $\displaystyle\sum_{n=3}^{\infty} \frac{1}{n(2n-5)}$

12. $\displaystyle\sum_{n=1}^{\infty} \frac{1}{n(n+1)(n+2)}$

13. $\displaystyle\sum_{n=1}^{\infty} n2^{-n^2}$

14. $\displaystyle\sum_{n=1}^{\infty} \frac{1}{\sqrt{n+9}}$

15. $\displaystyle\sum_{n=2}^{\infty} \frac{1}{n\sqrt[3]{\ln n}}$

16. $\displaystyle\sum_{n=2}^{\infty} \frac{1}{n\sqrt{n^2-1}}$

Exer. 17–38: Use comparison tests to determine if the series converges or diverges.

17. $\displaystyle\sum_{n=1}^{\infty} \frac{1}{n^4+n^2+1}$

18. $\displaystyle\sum_{n=1}^{\infty} \frac{\sqrt{n}}{n^2+1}$

19. $\displaystyle\sum_{n=1}^{\infty} \frac{1}{n3^n}$

20. $\displaystyle\sum_{n=1}^{\infty} \frac{n^2}{n^3+1}$

21. $\displaystyle\sum_{n=1}^{\infty} \frac{2n+n^2}{n^3+1}$

22. $\displaystyle\sum_{n=1}^{\infty} \frac{2}{3+\sqrt{n}}$

23. $\displaystyle\sum_{n=1}^{\infty} \frac{\sqrt{n}}{n+4}$

24. $\displaystyle\sum_{n=1}^{\infty} \frac{n^5+4n^3+1}{2n^8+n^4+2}$

25. $\displaystyle\sum_{n=2}^{\infty} \frac{1}{\sqrt{4n^3-5n}}$

26. $\displaystyle\sum_{n=4}^{\infty} \frac{3n}{2n^2-7}$

27. $\displaystyle\sum_{n=1}^{\infty} \frac{8n^2-7}{e^n(n+1)^2}$

28. $\displaystyle\sum_{n=1}^{\infty} \frac{\sin^2 n}{2^n}$

29. $\displaystyle\sum_{n=1}^{\infty} \frac{1+2^n}{1+3^n}$

30. $\displaystyle\sum_{n=1}^{\infty} \frac{\ln n}{n^4}$ (*Hint:* $\ln n < n$)

31. $\displaystyle\sum_{n=1}^{\infty} \frac{2+\cos n}{n^2}$

32. $\displaystyle\sum_{n=1}^{\infty} \frac{\arctan n}{n^2}$

33. $\displaystyle\sum_{n=1}^{\infty} \frac{(2n+1)^3}{(n^3+1)^2}$

34. $\displaystyle\sum_{n=1}^{\infty} \frac{1}{\sqrt{n(n+1)(n+2)}}$

35. $\displaystyle\sum_{n=1}^{\infty} \frac{1}{\sqrt[3]{5n^2+1}}$

36. $\displaystyle\sum_{n=1}^{\infty} \frac{3n+5}{n\cdot 2^n}$

37. $\displaystyle\sum_{n=1}^{\infty} \frac{1}{n^n}$

38. $\displaystyle\sum_{n=1}^{\infty} \frac{1}{n!}$

Exer. 39–45: Determine if the series converges or diverges.

39. $\displaystyle\sum_{n=1}^{\infty} \frac{n+\ln n}{n^3+n+1}$

40. $\displaystyle\sum_{n=1}^{\infty} \frac{n+\ln n}{n^2+1}$

41. $\displaystyle\sum_{n=1}^{\infty} \sin\frac{1}{n^2}$

42. $\displaystyle\sum_{n=1}^{\infty} \tan\frac{1}{n}$

43. $\displaystyle\sum_{n=1}^{\infty} \frac{\ln n}{n^3}$

44. $\displaystyle\sum_{n=1}^{\infty} \ln\left(1+\frac{1}{2^n}\right)$

45. $\displaystyle\sum_{n=1}^{\infty} \frac{n^2+2^n}{n+3^n}$

46. $\displaystyle\sum_{n=1}^{\infty} \frac{\sin n + 2^n}{n+5^n}$

Exer. 47–48: Find every real number k for which the series converges.

47. $\displaystyle\sum_{n=2}^{\infty} \frac{1}{n^k \ln n}$

48. $\displaystyle\sum_{n=2}^{\infty} \frac{1}{n(\ln n)^k}$

49. (a) Use the proof of the Integral Test (11.22) to show that, for every positive integer n,

$$\ln(n+1) < 1 + \frac{1}{2} + \frac{1}{3} + \cdots + \frac{1}{n} < 1 + \ln n$$

(b) Estimate the number of terms of the harmonic series that should be added so that $S_n > 100$.

50 Consider the hypothetical problem illustrated in the figure, where starting with a ball of radius 1 foot, additional balls are stacked vertically, such that if r_k is the radius of the kth ball, then $r_{n+1} = r_n \sqrt{n/(n+1)}$ for each positive integer n.

(a) Show that the height of the stack can be made arbitrarily large.

(b) If the balls are made of a material that weighs 1 lb/ft³, show that the total weight of the stack is always less than 4π pounds.

EXERCISE 50

51 Suppose $\sum a_n$ and $\sum b_n$ are positive-term series. Prove that if $\lim_{n \to \infty} (a_n/b_n) = 0$ and $\sum b_n$ converges, then $\sum a_n$ converges. (This is not necessarily true for series that contain negative terms.)

52 Prove that if $\lim_{n \to \infty} (a_n/b_n) = \infty$ and $\sum b_n$ diverges, then $\sum a_n$ diverges.

53 Let $\sum a_n$ be a convergent, positive-term series. Let $f(n) = a_n$, and suppose f is continuous and decreasing for $x \geq N$ for some integer N. Prove that the error in approximating the sum of the given series by $\sum_{n=1}^{N} a_n$ is less than $\int_N^\infty f(x)\, dx$.

Exer. 54–56: Use Exercise 53 to determine the smallest number of terms that can be added to approximate the sum of the series with an error less than E.

54 $\sum_{n=1}^{\infty} \dfrac{1}{n^2}$, $E = 0.001$ **55** $\sum_{n=1}^{\infty} \dfrac{1}{n^3}$, $E = 0.01$

56 $\sum_{n=2}^{\infty} \dfrac{1}{n (\ln n)^2}$, $E = 0.05$

57 Prove that if a positive-term series $\sum a_n$ converges, then $\sum (1/a_n)$ diverges.

58 Prove that if a positive-terms series $\sum a_n$ converges, then $\sum \sqrt{a_n a_{n+1}}$ converges.
(*Hint:* First show that $\sqrt{a_n a_{n+1}} \leq (a_n + a_{n+1})/2$.)

11.4 THE RATIO AND ROOT TESTS

To apply the Integral Test to a positive-term series $\sum a_n$ with $a_n = f(n)$, the terms must be decreasing and we must be able to integrate $f(x)$. This often rules out series that involve factorials and other complicated expressions. We shall now introduce two tests that can be used to help determine convergence or divergence when other tests are not applicable. Unfortunately, as indicated by part (iii), the tests are inconclusive for some series.

THE RATIO TEST (11.26)

Let $\sum a_n$ be a positive-term series, and suppose

$$\lim_{n \to \infty} \frac{a_{n+1}}{a_n} = L.$$

(i) If $L < 1$, the series is convergent.

(ii) If $L > 1$ or $\lim_{n \to \infty} \dfrac{a_{n+1}}{a_n} = \infty$, the series is divergent.

(iii) If $L = 1$, apply a different test; the series may be convergent or divergent.

PROOF

(i) Suppose $\lim_{n \to \infty} (a_{n+1}/a_n) = L < 1$. Let r be any number such that $0 \leq L < r < 1$. Since a_{n+1}/a_n is close to L if n is large, there exists an integer N such that whenever $n \geq N$

$$\frac{a_{n+1}}{a_n} < r \quad \text{or} \quad a_{n+1} < a_n r$$

Substituting $N, N + 1, N + 2, \ldots$ for n, we obtain

$$a_{N+1} < a_N r$$
$$a_{N+2} < a_{N+1} r < a_N r^2$$
$$a_{N+3} < a_{N+2} r < a_N r^3$$

and, in general,

$$a_{N+m} < a_N r^m \quad \text{whenever } m > 0.$$

It follows from the Basic Comparison Test (11.24) that the series

$$a_{N+1} + a_{N+2} + \cdots + a_{N+m} + \cdots$$

converges, since its terms are less than the corresponding terms of the convergent geometric series

$$a_N r + a_N r^2 + \cdots + a_N r^n + \cdots$$

Since convergence or divergence is unaffected by discarding a finite number of terms, the series $\sum_{n=1}^{\infty} a_n$ also converges.

(ii) Suppose $\lim_{n \to \infty} (a_{n+1}/a_n) = L > 1$. If r is a real number such that $L > r > 1$, then there exists an integer N such that

$$\frac{a_{n+1}}{a_n} > r > 1 \quad \text{whenever } n \geq N.$$

Consequently, $a_{n+1} > a_n$ if $n \geq N$. Thus $\lim_{n \to \infty} a_n \neq 0$ and by the nth-Term Test for Divergence (11.17), the series $\sum a_n$ diverges.

The proof for $\lim_{n \to \infty} (a_{n+1}/a_n) = \infty$ is similar and is left as an exercise.

(iii) The Ratio Test is inconclusive if

$$\lim_{n \to \infty} \frac{a_{n+1}}{a_n} = 1,$$

for it is easy to verify that the limit is 1 for both the convergent series $\sum (1/n^2)$ and the divergent series $\sum (1/n)$. Consequently, *if the limit is 1, then a different test must be employed.* • •

EXAMPLE 1 Determine if each series is convergent or divergent.

(a) $\displaystyle\sum_{n=1}^{\infty} \frac{3^n}{n!}$ (b) $\displaystyle\sum_{n=1}^{\infty} \frac{3^n}{n^2}$

SOLUTION

(a) Applying the Ratio Test,

$$\lim_{n \to \infty} \frac{a_{n+1}}{a_n} = \lim_{n \to \infty} \left(a_{n+1} \cdot \frac{1}{a_n} \right)$$

$$= \lim_{n \to \infty} \frac{3^{n+1}}{(n+1)!} \cdot \frac{n!}{3^n}$$

$$= \lim_{n \to \infty} \frac{3}{n+1} = 0.$$

Since $0 < 1$, the series is convergent.

(b) Since

$$\lim_{n \to \infty} \frac{a_{n+1}}{a_n} = \lim_{n \to \infty} \frac{3^{n+1}}{(n+1)^2} \cdot \frac{n^2}{3^n}$$

$$= \lim_{n \to \infty} \frac{3n^2}{n^2 + 2n + 1} = 3 > 1$$

the series diverges. •

EXAMPLE 2 Determine the convergence or divergence of $\displaystyle\sum_{n=1}^{\infty} \frac{n^n}{n!}$.

SOLUTION   Applying the Ratio Test,

$$\lim_{n \to \infty} \frac{a_{n+1}}{a_n} = \lim_{n \to \infty} \frac{(n+1)^{n+1}}{(n+1)!} \cdot \frac{n!}{n^n}$$

$$= \lim_{n \to \infty} \frac{(n+1)^{n+1}}{(n+1)} \cdot \frac{1}{n^n}$$

$$= \lim_{n \to \infty} \frac{(n+1)^n}{n^n} = \lim_{n \to \infty} \left(\frac{n+1}{n}\right)^n$$

$$= \lim_{n \to \infty} \left(1 + \frac{1}{n}\right)^n = e$$

The last equality is a consequence of Theorem (7.25). Since $e > 1$, the series diverges. •

The following test is useful if a_n contains only powers of n.

THE ROOT TEST (11.27)

Let $\sum a_n$ be a positive-term series, and suppose

$$\lim_{n \to \infty} \sqrt[n]{a_n} = L.$$

(i) If $L < 1$, the series is convergent.

(ii) If $L > 1$ or $\lim_{n \to \infty} \sqrt[n]{a_n} = \infty$, the series is divergent.

(iii) If $L = 1$, apply a different test; the series may be convergent or divergent.

PROOF   The proof is similar to that used for the Ratio Test. If $L < 1$, as in (i), let us consider any number r such that $L < r < 1$. By the definition of limit, there exists a positive integer N such that if $n \geq N$, then

$$\sqrt[n]{a_n} < r \quad \text{or} \quad a_n < r^n.$$

Since $0 < r < 1$, $\sum_{n=N}^{\infty} r^n$ is a convergent geometric series, and hence by the Basic Comparison Test (11.24), $\sum_{n=N}^{\infty} a_n$ converges. Consequently, $\sum_{n=1}^{\infty} a_n$ converges. This proves (i). The remainder of the proof is left as an exercise. • •

EXAMPLE 3 Determine the convergence or divergence of $\sum\limits_{n=1}^{\infty} \dfrac{2^{3n+1}}{n^n}$.

SOLUTION Applying the Root Test,

$$\lim_{n \to \infty} \sqrt[n]{\frac{2^{3n+1}}{n^n}} = \lim_{n \to \infty} \left(\frac{2^{3n+1}}{n^n}\right)^{1/n}$$

$$= \lim_{n \to \infty} \frac{2^{3+(1/n)}}{n} = 0.$$

Since $0 < 1$, the series converges. We could have applied the Ratio Test; however, the process of evaluating the limit is more complicated. •

EXERCISES 11.4

Exer. 1–26: Determine if the series is convergent or divergent.

1. $\sum\limits_{n=1}^{\infty} \dfrac{3n+1}{2^n}$

2. $\sum\limits_{n=1}^{\infty} \dfrac{3^n}{n^2+4}$

3. $\sum\limits_{n=1}^{\infty} \dfrac{5^n}{n(3^{n+1})}$

4. $\sum\limits_{n=1}^{\infty} \dfrac{2^{n-1}}{5^n(n+1)}$

5. $\sum\limits_{n=1}^{\infty} \dfrac{100^n}{n!}$

6. $\sum\limits_{n=1}^{\infty} \dfrac{n^{10}+10}{n!}$

7. $\sum\limits_{n=1}^{\infty} \dfrac{n!}{e^n}$

8. $\sum\limits_{n=1}^{\infty} \dfrac{\sqrt{n}}{3n+4}$

9. $\sum\limits_{n=1}^{\infty} \dfrac{\sqrt{n}}{n^2+1}$

10. $\sum\limits_{n=1}^{\infty} \dfrac{n+1}{n^3+1}$

11. $\sum\limits_{n=2}^{\infty} \dfrac{1}{n(\ln n)^2}$

12. $\sum\limits_{n=1}^{\infty} \dfrac{n!}{(n+1)^5}$

13. $\sum\limits_{n=1}^{\infty} \dfrac{2}{n^3+e^n}$

14. $\sum\limits_{n=1}^{\infty} \dfrac{n3^{2n}}{5^{n-1}}$

15. $\sum\limits_{n=1}^{\infty} \dfrac{\arctan n}{n^2}$

16. $\sum\limits_{n=1}^{\infty} \dfrac{n!}{n^n}$

17. $\sum\limits_{n=1}^{\infty} \dfrac{n^n}{10^n}$

18. $\sum\limits_{n=1}^{\infty} \dfrac{10+2^n}{n!}$

19. $\sum\limits_{n=1}^{\infty} \dfrac{(n!)^2}{(2n)!}$

20. $\sum\limits_{n=1}^{\infty} \dfrac{(2n)^n}{(5n+3n^{-1})^n}$

21. $\sum\limits_{n=1}^{\infty} \dfrac{\ln n}{(1.01)^n}$

22. $\sum\limits_{n=1}^{\infty} 3^{1/n}$

23. $\sum\limits_{n=1}^{\infty} n \tan \dfrac{1}{n}$

24. $\sum\limits_{n=2}^{\infty} \dfrac{1}{(\ln n)^n}$

25. $1 + \dfrac{1 \cdot 3}{2!} + \dfrac{1 \cdot 3 \cdot 5}{3!} + \cdots + \dfrac{1 \cdot 3 \cdot 5 \cdots (2n-1)}{n!} + \cdots$

26. $\dfrac{1}{2} + \dfrac{1 \cdot 4}{2 \cdot 4} + \dfrac{1 \cdot 4 \cdot 7}{2 \cdot 4 \cdot 6} + \cdots + \dfrac{1 \cdot 4 \cdot 7 \cdots (3n-2)}{2 \cdot 4 \cdot 6 \cdots (2n)} + \cdots$

27. Complete the proof of the Ratio Test (11.26) by showing that if $\lim_{n \to \infty} (a_{n+1}/a_n) = \infty$, then $\sum a_n$ is divergent.

28. Complete the proof of the Root Test (11.27).

11.5 ALTERNATING SERIES AND ABSOLUTE CONVERGENCE

The tests for convergence that we have discussed thus far can only be applied to positive-term series. We shall next consider infinite series that contain both positive and negative terms. One of the simplest, and most useful, series of this type is an **alternating series,** for which the terms are alternately positive and negative. It is customary to express an alternating series in one of the forms

$$a_1 - a_2 + a_3 - a_4 + \cdots + (-1)^{n-1}a_n + \cdots$$

or

$$-a_1 + a_2 - a_3 + a_4 - \cdots + (-1)^n a_n + \cdots$$

with $a_k > 0$ for every k. The next theorem provides the main test for convergence of these series. For convenience we shall consider $\sum_{n=1}^{\infty} (-1)^{n-1}a_n$. A similar proof holds for $\sum_{n=1}^{\infty} (-1)^n a_n$.

> If $a_k \geq a_{k+1} > 0$ for every k and $\lim_{n \to \infty} a_n = 0$, then the alternating series $\sum_{n=1}^{\infty} (-1)^{n-1} a_n$ is convergent.

PROOF Let us first consider the partial sums

$$S_2, \ S_4, \ S_6, \ \ldots, \ S_{2n}, \ \ldots$$

that contain an even number of terms of the series. Since

$$S_{2n} = (a_1 - a_2) + (a_3 - a_4) + \cdots + (a_{2n-1} - a_{2n})$$

and $a_k - a_{k+1} \geq 0$ for every k, we see that

$$0 \leq S_2 \leq S_4 \leq \cdots \leq S_{2n} \leq \cdots,$$

that is, $\{S_{2n}\}$ is a monotonic sequence. This fact is also evident from Figure 11.11, where we have used a coordinate line l to represent the following four partial sums of the series:

$$S_1 = a_1, \quad S_2 = a_1 - a_2, \quad S_3 = a_1 - a_2 + a_3, \quad S_4 = a_1 - a_2 + a_3 - a_4.$$

You may find it instructive to locate the points on l that correspond to S_5 and S_6.

Referring to Figure 11.11, we see that $S_{2n} \leq a_1$ for every positive integer n. This may also be proved algebraically by observing that

$$S_{2n} = a_1 - (a_2 - a_3) - (a_4 - a_5) - \cdots - (a_{2n-2} - a_{2n-1}) - a_{2n} \leq a_1.$$

Thus $\{S_{2n}\}$ is a *bounded* monotonic sequence. As in the proof of Theorem (11.9),

$$\lim_{n \to \infty} S_{2n} = S \leq a_1$$

for some number S. If we consider a partial sum S_{2n+1} having an *odd* number of terms of the series, then $S_{2n+1} = S_{2n} + a_{2n+1}$ and, since $\lim_{n \to \infty} a_{2n+1} = 0$,

$$\lim_{n \to \infty} S_{2n+1} = \lim_{n \to \infty} S_{2n} = S.$$

It follows that

$$\lim_{n \to \infty} S_n = S \leq a_1,$$

that is, the series converges. • •

FIGURE 11.11

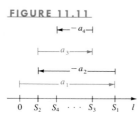

EXAMPLE 1 Determine if each alternating series converges or diverges.

(a) $\sum_{n=1}^{\infty} (-1)^{n-1} \dfrac{2n}{4n^2 - 3}$ (b) $\sum_{n=1}^{\infty} (-1)^{n-1} \dfrac{2n}{4n - 3}$

SOLUTION

(a) Let $a_n = f(n) = \dfrac{2n}{4n^2 - 3}$.

To apply the Alternating Series Test we must show that:

  (i) $a_k \geq a_{k+1}$ for every positive integer k, and

  (ii) $\lim_{n \to \infty} a_n = 0$.

One method of proving (i) is to show that $f(x) = 2x/(4x^2 - 3)$ is decreasing for $x \geq 1$. By the Quotient Rule,

$$f'(x) = \frac{(4x^2 - 3)(2) - (2x)(8x)}{(4x^2 - 3)^2}$$

$$= \frac{-8x^2 - 6}{(4x^2 - 3)^2} < 0.$$

It follows that f is decreasing and, therefore, $a_k \geq a_{k+1}$ for every positive integer k.

We can obtain the same result directly by proving that $a_k - a_{k+1} \geq 0$. Specifically, if $a_n = 2n/(4n^2 - 3)$, then

$$a_k - a_{k+1} = \frac{2k}{4k^2 - 3} - \frac{2(k+1)}{4(k+1)^2 - 3}$$

$$= \frac{8k^2 + 8k + 6}{(4k^2 - 3)(4k^2 + 8k + 1)} \geq 0$$

for every positive integer k. Another technique for proving that $a_k \geq a_{k+1}$ is to show that $a_{k+1}/a_k \leq 1$.

To prove (ii), we see that

$$\lim_{n \to \infty} a_n = \lim_{n \to \infty} \frac{2n}{4n^2 - 3} = 0.$$

Thus the alternating series converges.

(b) We can show that $a_k \geq a_{k+1}$ for every k; however,

$$\lim_{n \to \infty} \frac{2n}{4n - 3} = \frac{1}{2} \neq 0$$

and hence the series diverges by the nth-Term Test for Divergence (11.17). •

If an infinite series converges, then the nth partial sum S_n can be used to approximate the sum S of the series. In most cases it is difficult to determine the accuracy of the approximation. However, for an *alternating* series, the next theorem provides a simple way of estimating the error that is involved.

THEOREM (11.29) | If $\sum_{n=1}^{\infty} (-1)^{n-1} a_n$ is an alternating series such that $a_k > a_{k+1} > 0$ for every positive integer k and if $\lim_{n \to \infty} a_n = 0$, then the error E involved in approximating the sum S of the series by the nth partial sum S_n is less than a_{n+1}.

PROOF Note that the alternating series $\sum (-1)^{n-1} a_n$ satisfies the conditions of the Alternating Series Test (11.28) and hence has a sum S. The series obtained by deleting the first n terms, namely

$$(-1)^n a_{n+1} + (-1)^{n+1} a_{n+2} + (-1)^{n+2} a_{n+3} + \cdots$$

also satisfies the conditions of (11.28) and, therefore, has a sum R_n. Thus

$$S - S_n = R_n = (-1)^n (a_{n+1} - a_{n+2} + a_{n+3} - \cdots)$$

and

$$|R_n| = a_{n+1} - a_{n+2} + a_{n+3} - \cdots$$

Employing the same argument used in the proof of the Alternating Series Test (but replacing $\le$ by $<$) we see that $|R_n| < a_{n+1}$. Consequently,

$$E = |S - S_n| = |R_n| < a_{n+1}$$

which is what we wished to prove. • •

EXAMPLE 2 Prove that the series

$$1 - \frac{1}{3!} + \frac{1}{5!} - \cdots + (-1)^{n-1} \frac{1}{(2n-1)!} + \cdots$$

is convergent, and approximate its sum S to five decimal places.

SOLUTION Evidently, $a_n = 1/(2n-1)!$ has the limit 0 as $n \to \infty$, and $a_k > a_{k+1}$ for every positive integer k. Hence the series converges by the Alternating Series Test. If we use S_4 to approximate the sum S of the series, then

$$S \approx 1 - \frac{1}{3!} + \frac{1}{5!} - \frac{1}{7!}$$

$$= 1 - \frac{1}{6} + \frac{1}{120} - \frac{1}{5040} \approx 0.841468$$

By Theorem (11.29), the error involved in the approximation is less than

$$a_5 = \frac{1}{9!} < 0.000005.$$

Thus the approximation 0.84147 is accurate to five decimal places. In (11.43) we will see that the sum of the series is sin 1 and hence sin $1 \approx 0.84147$. •

The following concept is useful when investigating a series that contains both positive and negative terms but is not alternating. It will allow us to use tests for positive-term series to establish convergence for other types of series (see Theorem 11.31).

DEFINITION (11.30) | An infinite series $\sum a_n$ is **absolutely convergent** if the series

$$\sum |a_n| = |a_1| + |a_2| + \cdots + |a_n| + \cdots$$

is convergent.

Note that if $\sum a_n$ is a positive-term series, then $|a_n| = a_n$, and in this case absolute convergence is the same as convergence.

EXAMPLE 3 Prove that the following alternating series is absolutely convergent:

$$1 - \frac{1}{2^2} + \frac{1}{3^2} - \frac{1}{4^2} + \cdots + (-1)^n \frac{1}{n^2} + \cdots$$

SOLUTION Taking the absolute value of each term gives us

$$1 + \frac{1}{2^2} + \frac{1}{3^2} + \frac{1}{4^2} + \cdots + \frac{1}{n^2} + \cdots$$

which is a convergent p-series. Hence by Definition (11.30), the alternating series is absolutely convergent. •

The next theorem tells us that absolute convergence implies convergence.

THEOREM (11.31)   If an infinite series $\sum a_n$ is absolutely convergent, then $\sum a_n$ is convergent.

PROOF If we let $b_n = a_n + |a_n|$ and use the property $-|a_n| \le a_n \le |a_n|$, then

$$0 \le a_n + |a_n| \le 2|a_n| \quad \text{or} \quad 0 \le b_n \le 2|a_n|.$$

If $\sum a_n$ is absolutely convergent, then $\sum |a_n|$ is convergent, and hence by Theorem (11.19) (ii), $\sum 2|a_n|$ is convergent. If we apply the Basic Comparison Test (11.24), it follows that $\sum b_n$ is convergent. By (11.19) (iii), $\sum (b_n - |a_n|)$ is convergent. Since $b_n - |a_n| = a_n$, this completes the proof. • •

EXAMPLE 4 Determine if the following series is convergent or divergent:

$$\sin 1 + \frac{\sin 2}{2^2} + \frac{\sin 3}{3^2} + \cdots + \frac{\sin n}{n^2} + \cdots$$

SOLUTION The series contains both positive and negative terms, but it is not an alternating series, since, for example, the first three terms are positive and the next three are negative. The series of absolute values is

$$\sum_{n=1}^{\infty} \left| \frac{\sin n}{n^2} \right| = \sum_{n=1}^{\infty} \frac{|\sin n|}{n^2}.$$

Since

$$\frac{|\sin n|}{n^2} \le \frac{1}{n^2},$$

the series of absolute values $\sum |(\sin n)/n^2|$ is dominated by the convergent p-series $\sum (1/n^2)$ and hence is convergent. Thus the given series is absolutely convergent and, therefore, is convergent by Theorem (11.31). •

Series that are convergent, but are not absolutely convergent, are given a special name, as indicated in the next definition.

DEFINITION (11.32)   An infinite series $\sum a_n$ is **conditionally convergent** if $\sum a_n$ is convergent and $\sum |a_n|$ is divergent.

EXAMPLE 5 Show that the following series is conditionally convergent:

$$1 - \frac{1}{2} + \frac{1}{3} - \frac{1}{4} + \cdots + (-1)^{n-1} \frac{1}{n} + \cdots$$

SOLUTION The series is convergent by the Alternating Series Test. Taking the absolute value of each term, we obtain

$$1 + \frac{1}{2} + \frac{1}{3} + \frac{1}{4} + \cdots + \frac{1}{n} + \cdots$$

which is the divergent harmonic series. Hence by Definition (11.32), the alternating series is conditionally convergent. •

We see from the preceding discussion that an arbitrary infinite series may be classified in exactly *one* of the following ways: (i) absolutely convergent, (ii) conditionally convergent, (iii) divergent. Of course, for positive-term series we need only determine convergence or divergence.

The following form of the Ratio Test may be used to investigate absolute convergence.

<table>
<tr><td align="right">RATIO TEST FOR **(11.33)**
ABSOLUTE CONVERGENCE</td><td>

Let $\sum a_n$ be a series of nonzero terms, and suppose

$$\lim_{n \to \infty} \left| \frac{a_{n+1}}{a_n} \right| = L.$$

(i) If $L < 1$, the series is absolutely convergent.

(ii) If $L > 1$ or $\lim\limits_{n \to \infty} \left| \dfrac{a_{n+1}}{a_n} \right| = \infty$, the series is divergent.

(iii) If $L = 1$, apply a different test; the series may be absolutely convergent, conditionally convergent, or divergent.

</td></tr>
</table>

The proof is similar to that of (11.26). Note that for positive-term series the two ratio tests are identical.

We can also state a Root Test for absolute convergence. The statement is the same as that of (11.27) except that we replace $\sqrt[n]{a_n}$ with $\sqrt[n]{|a_n|}$.

EXAMPLE 6 Determine if the following series is absolutely convergent, conditionally convergent, or divergent:

$$\sum_{n=1}^{\infty} (-1)^n \frac{n^2 + 4}{2^n}$$

SOLUTION Using the Ratio Test (11.33),

$$\lim_{n \to \infty} \left| \frac{a_{n+1}}{a_n} \right| = \lim_{n \to \infty} \left| \frac{(n+1)^2 + 4}{2^{n+1}} \cdot \frac{2^n}{n^2 + 4} \right|$$

$$= \lim_{n \to \infty} \frac{1}{2} \left(\frac{n^2 + 2n + 5}{n^2 + 4} \right) = \frac{1}{2} (1) = \frac{1}{2} < 1.$$

Hence, by (11.33) (i), the series is absolutely convergent. •

We can prove that if a series $\sum a_n$ is absolutely convergent and if the terms are rearranged in any manner, then the resulting series converges and has the same sum as the given series. This is not true for conditionally convergent series. If $\sum a_n$ is conditionally convergent, then by suitably rearranging terms, we can obtain either a divergent series or a series that converges and has any desired sum S.*

We now have a variety of tests that can be used to investigate an infinite series for convergence or divergence. Considerable skill is needed to determine which test is best suited for a particular series. This skill can be obtained by working many exercises involving different types of series. The following summary may be helpful in deciding which test to apply; however, some infinite series cannot be investigated by any of these tests. In those cases it may be necessary to use results from advanced mathematics courses.

* See, for example, R.C. Buck, *Advanced Calculus*, Third Edition (New York: McGraw-Hill, 1978), pp. 238–239.

Test	Series	Convergence or divergence	Comments
nth-Term	$\sum a_n$	Diverges if $\lim_{n \to \infty} a_n \neq 0$.	Inconclusive if $\lim_{n \to \infty} a_n = 0$.
Geometric Series	$\sum_{n=1}^{\infty} ar^{n-1}$	(i) Converges to $a/(1-r)$ if $\lvert r \rvert < 1$. (ii) Diverges if $\lvert r \rvert \geq 1$.	Useful for comparison tests.
p-Series	$\sum_{n=1}^{\infty} \frac{1}{n^p}$	(i) Converges if $p > 1$. (ii) Diverges if $p \leq 1$.	Useful for comparison tests.
Integral	$\sum_{n=1}^{\infty} a_n$ $a_n = f(n)$	(i) Converges if $\int_1^{\infty} f(x)\,dx$ converges. (ii) Diverges if $\int_1^{\infty} f(x)\,dx$ diverges.	The function f must be continuous, positive, decreasing, and readily integrable.
Comparison	$\sum a_n, \sum b_n$ $a_n > 0, b_n > 0$	(i) $\sum a_n$ converges if $\sum b_n$ converges and $a_n \leq b_n$ for every n. (ii) $\sum a_n$ diverges if $\sum b_n$ diverges and $a_n \geq b_n$ for every n. (iii) If $\lim_{n \to \infty} (a_n/b_n) = k > 0$, then both series converge or both diverge.	The comparison series $\sum b_n$ is often geometric or a p-series.
Ratio	$\sum a_n$	If $\lim_{n \to \infty} \lvert a_{n+1}/a_n \rvert = L$ (or ∞), the series (i) converges (absolutely) if $L < 1$. (ii) diverges if $L > 1$ (or ∞).	Inconclusive if $L = 1$. Useful if a_n involves factorials or nth powers. If $a_n > 0$ for every n, the absolute value signs may be disregarded.
Root	$\sum a_n$	If $\lim_{n \to \infty} \sqrt[n]{\lvert a_n \rvert} = L$ (or ∞), the series (i) converges (absolutely) if $L < 1$. (ii) diverges if $L > 1$ (or ∞).	Inconclusive if $L = 1$. Useful if a_n involves only nth powers. If $a_n > 0$ for every n, the absolute value signs may be disregarded.
Alternating Series	$\sum (-1)^n a_n$ $a_n > 0$	Converges if $a_k \geq a_{k+1}$ for every k and $\lim_{n \to \infty} a_n = 0$.	Applicable only to an alternating series.
$\sum \lvert a_n \rvert$	$\sum a_n$	If $\sum \lvert a_n \rvert$ is convergent, then $\sum a_n$ is convergent.	Useful for series that contain both positive and negative terms.

Exer. 1–24: Determine if the series is absolutely convergent, conditionally convergent, or divergent.

1 $\displaystyle\sum_{n=1}^{\infty} (-1)^{n-1} \frac{1}{\sqrt{2n+1}}$

2 $\displaystyle\sum_{n=1}^{\infty} (-1)^{n-1} \frac{1}{n^{2/3}}$

3 $\displaystyle\sum_{n=1}^{\infty} (-1)^{n-1} \frac{1}{\ln(n+1)}$

4 $\displaystyle\sum_{n=1}^{\infty} (-1)^{n-1} \frac{n}{n^2+4}$

5 $\displaystyle\sum_{n=2}^{\infty} (-1)^{n} \frac{n}{\ln n}$

6 $\displaystyle\sum_{n=1}^{\infty} (-1)^{n} \frac{\ln n}{n}$

7 $\displaystyle\sum_{n=1}^{\infty} (-1)^{n} \frac{5}{n^3+1}$

8 $\displaystyle\sum_{n=1}^{\infty} (-1)^{n} e^{-n}$

9 $\displaystyle\sum_{n=1}^{\infty} \frac{(-10)^{n}}{n!}$

10 $\displaystyle\sum_{n=1}^{\infty} \frac{n!}{(-5)^{n}}$

11 $\displaystyle\sum_{n=1}^{\infty} (-1)^{n} \frac{n^2+3}{(2n-5)^2}$

12 $\displaystyle\sum_{n=1}^{\infty} \frac{\sin\sqrt{n}}{\sqrt{n^3+4}}$

13 $\displaystyle\sum_{n=1}^{\infty} (-1)^{n-1} \frac{\sqrt[3]{n}}{n+1}$

14 $\displaystyle\sum_{n=1}^{\infty} (-1)^{n} \frac{(n+1)^2}{n^5+1}$

15 $\displaystyle\sum_{n=1}^{\infty} \frac{\cos\frac{1}{6}n\pi}{n^2}$

16 $\displaystyle\sum_{n=1}^{\infty} (-1)^{n} \frac{\ln n}{(1.5)^{n}}$

17 $\displaystyle\sum_{n=1}^{\infty} (-1)^{n} n \sin\frac{1}{n}$

18 $\displaystyle\sum_{n=1}^{\infty} (-1)^{n} \frac{\arctan n}{n^2}$

19 $\displaystyle\sum_{n=2}^{\infty} (-1)^{n} \frac{1}{n\sqrt{\ln n}}$

20 $\displaystyle\sum_{n=1}^{\infty} (-1)^{n} \frac{2^{1/n}}{n!}$

21 $\displaystyle\sum_{n=1}^{\infty} \frac{n^{n}}{(-5)^{n}}$

22 $\displaystyle\sum_{n=1}^{\infty} \frac{(n^2+1)^{n}}{(-n)^{n}}$

23 $\displaystyle\sum_{n=1}^{\infty} (-1)^{n} \frac{1+4^{n}}{1+3^{n}}$

24 $\displaystyle\sum_{n=1}^{\infty} (-1)^{n} \frac{n^4}{e^{n}}$

Exer. 25–30: Approximate the sum of each series to three decimal places.

25 $\displaystyle\sum_{n=0}^{\infty} (-1)^{n} \frac{1}{n!}$

26 $\displaystyle\sum_{n=0}^{\infty} (-1)^{n+1} \frac{1}{(2n)!}$

27 $\displaystyle\sum_{n=1}^{\infty} (-1)^{n-1} \frac{1}{n^3}$

28 $\displaystyle\sum_{n=1}^{\infty} (-1)^{n-1} \frac{1}{n^5}$

29 $\displaystyle\sum_{n=1}^{\infty} (-1)^{n-1} \frac{n+1}{5^{n}}$

30 $\displaystyle\sum_{n=1}^{\infty} (-1)^{n} \frac{1}{n}\left(\frac{1}{2}\right)^{n}$

Exer. 31–34: Use Theorem (11.29) to find a positive integer n such that S_n approximates the sum of the series to four decimal places.

31 $\displaystyle\sum_{n=1}^{\infty} (-1)^{n} \frac{1}{n^2}$

32 $\displaystyle\sum_{n=1}^{\infty} (-1)^{n} \frac{1}{\sqrt{n}}$

33 $\displaystyle\sum_{n=1}^{\infty} (-1)^{n} \frac{1}{n^{n}}$

34 $\displaystyle\sum_{n=1}^{\infty} (-1)^{n} \frac{1}{n^3+1}$

Exer. 35–36: Show that the alternating series converges for every positive integer k.

35 $\displaystyle\sum_{n=1}^{\infty} (-1)^{n} \frac{(\ln n)^{k}}{n}$

36 $\displaystyle\sum_{n=1}^{\infty} (-1)^{n} \frac{1}{\sqrt[k]{n}}$

37 If $\sum a_n$ and $\sum b_n$ are both convergent series, is $\sum a_n b_n$ convergent? Explain.

38 If $\sum a_n$ and $\sum b_n$ are both divergent series, is $\sum a_n b_n$ divergent? Explain.

11.6 POWER SERIES

In Section 11.1 we stated that expressions such as e^x, $\sin x$, and $\tan^{-1} x$ can be represented by infinite series. These series have terms that contain the variable x. (In previous sections the terms of a series consisted only of constants.) As a simple illustration, if x is a variable, then from our work with geometric series (see Theorem (11.15)),

$$1 + x + x^2 + \cdots + x^n + \cdots = \frac{1}{1-x} \quad \text{if } |x| < 1.$$

If we let $f(x) = 1/(1-x)$ with $|x| < 1$, then

$$f(x) = 1 + x + x^2 + \cdots + x^n + \cdots$$

We say that $f(x)$ is *represented* by this infinite power series. If we wish to find function values, we can do so by finding the sum of a series. For example,

$$f(\tfrac{1}{2}) = 1 + \tfrac{1}{2} + (\tfrac{1}{2})^2 + \cdots + (\tfrac{1}{2})^n + \cdots = \frac{1}{1-\tfrac{1}{2}} = 2.$$

This new method of representing a function enables us to investigate properties of a function by using an infinite series. We shall explore this idea further later in the chapter. In this section we introduce some basic concepts for series with nonconstant terms. The next definition may be considered as a generalization of the notion of a polynomial in x to infinite series.

DEFINITION (11.34)

Let x be a variable. A **power series in x** is a series of the form

$$\sum_{n=0}^{\infty} a_n x^n = a_0 + a_1 x + a_2 x^2 + \cdots + a_n x^n + \cdots$$

such that each a_k is a real number.

If a number is substituted for x in Definition (11.34), we obtain a series of constant terms that may be tested for convergence or divergence. To simplify the nth term of the power series, we assume that $x^0 = 1$ even if $x = 0$. The main objective of this section is to determine all values of x for which a power series converges. Evidently, every power series in x converges if $x = 0$. To find other values of x that produce convergent series we often employ the Ratio Test for Absolute Convergence (11.33), as illustrated in the following examples.

EXAMPLE 1 Find all values of x for which the following power series is absolutely convergent:

$$1 + \frac{1}{5}x + \frac{2}{5^2}x^2 + \cdots + \frac{n}{5^n}x^n + \cdots$$

SOLUTION If we let

$$u_n = \frac{n}{5^n}x^n = \frac{nx^n}{5^n},$$

then

$$\lim_{n \to \infty} \left| \frac{u_{n+1}}{u_n} \right| = \lim_{n \to \infty} \left| \frac{(n+1)x^{n+1}}{5^{n+1}} \cdot \frac{5^n}{nx^n} \right|$$

$$= \lim_{n \to \infty} \left| \frac{(n+1)x}{5n} \right| = \lim_{n \to \infty} \left(\frac{n+1}{5n} \right) |x| = \tfrac{1}{5}|x|.$$

By the Ratio Test (11.33) with $L = \frac{1}{5}|x|$, the series is absolutely convergent if the following equivalent inequalities are true:

$$\tfrac{1}{5}|x| < 1, \quad |x| < 5, \quad -5 < x < 5.$$

The series diverges if $\frac{1}{5}|x| > 1$, that is, if $x > 5$ or $x < -5$.

If $\frac{1}{5}|x| = 1$, the Ratio Test is inconclusive, and hence the numbers 5 and -5 require special consideration. Substituting 5 for x in the power series, we obtain

$$1 + 1 + 2 + 3 + \cdots + n + \cdots$$

which is divergent. If we let $x = -5$ we obtain

$$1 - 1 + 2 - 3 + \cdots + n(-1)^n + \cdots$$

which is also divergent. Consequently, the power series is absolutely convergent for every x in the open interval $(-5, 5)$ and diverges elsewhere. •

EXAMPLE 2 Find all values of x for which the following power series is absolutely convergent:

$$1 + \frac{1}{1!}x + \frac{1}{2!}x^2 + \cdots + \frac{1}{n!}x^n + \cdots$$

SOLUTION We shall employ the same technique used in Example 1. If we let

$$u_n = \frac{1}{n!}x^n = \frac{x^n}{n!}$$

then

$$\lim_{n\to\infty}\left|\frac{u_{n+1}}{u_n}\right| = \lim_{n\to\infty}\left|\frac{x^{n+1}}{(n+1)!}\cdot\frac{n!}{x^n}\right|$$

$$= \lim_{n\to\infty}\left|\frac{x}{n+1}\right| = \lim_{n\to\infty}\frac{1}{n+1}|x| = 0.$$

Since the limit 0 is less than 1 for every value of x, it follows from the Ratio Test (11.33) that the power series is absolutely convergent for every real number. •

EXAMPLE 3 Find all values of x for which $\sum n!x^n$ is convergent.

SOLUTION Let $u_n = n!x^n$. If $x \neq 0$, then

$$\lim_{n\to\infty}\left|\frac{u_{n+1}}{u_n}\right| = \lim_{n\to\infty}\left|\frac{(n+1)!x^{n+1}}{n!x^n}\right|$$

$$= \lim_{n\to\infty}|(n+1)x| = \lim_{n\to\infty}(n+1)|x| = \infty$$

and, by the Ratio Test (11.33), the series diverges. Hence the power series is convergent only if $x = 0$. •

Theorem (11.36) will show that the solutions of the preceding examples are typical, in the sense that if a power series converges for nonzero values of x, then either it is absolutely convergent for every real number or it is absolutely convergent throughout some open interval $(-r, r)$ and diverges outside of the closed interval $[-r, r]$. The proof of this fact depends on the next theorem.

THEOREM (11.35)

(i) If a power series $\sum a_n x^n$ converges for a nonzero number c, then it is absolutely convergent whenever $|x| < |c|$.

(ii) If a power series $\sum a_n x^n$ diverges for a nonzero number d, then it diverges whenever $|x| > |d|$.

PROOF If $\sum a_n c^n$ converges and $c \neq 0$, then by Theorem (11.16), $\lim_{n\to\infty} a_n c^n = 0$. Employing Definition (11.3) with $\varepsilon = 1$, there is a positive integer N such that

$$|a_n c^n| < 1 \quad \text{whenever } n \geq N.$$

Consequently,

$$|a_n x^n| = \left|\frac{a_n c^n x^n}{c^n}\right| = |a_n c^n|\left|\frac{x}{c}\right|^n < \left|\frac{x}{c}\right|^n$$

provided $n \geq N$. If $|x| < |c|$, then $|x/c| < 1$ and $\sum |x/c|^n$ is a convergent geometric series. Hence, by the Basic Comparison Test (11.24), the series obtained by deleting the first N terms of $\sum |a_n x^n|$ is convergent. It follows that the series $\sum |a_n x^n|$ is also convergent, which proves (i).

To prove (ii), suppose the series diverges for $x = d \neq 0$. If the series converges for some number c_1 with $|c_1| > |d|$, then by (i) it converges whenever $|x| < |c_1|$. In particular, the series converges for $x = d$, contrary to our supposition. Hence the series diverges whenever $|x| > |d|$. • •

We may now prove the following.

THEOREM (11.36)

If $\sum a_n x^n$ is a power series, then precisely one of the following is true:

(i) The series converges only if $x = 0$.

(ii) The series is absolutely convergent for every x.

(iii) There is a positive number r such that the series is absolutely convergent if $|x| < r$ and divergent if $|x| > r$.

PROOF If neither (i) nor (ii) is true, then there exist nonzero numbers c and d such that the series converges if $x = c$ and diverges if $x = d$. Let S denote the set of all real numbers for which the series is absolutely convergent. By Theorem (11.35), the series diverges if $|x| > |d|$ and hence every number in S is less than $|d|$. By the Completeness Property (11.10), S has a least upper bound r. It follows that the series is absolutely convergent if $|x| < r$ and diverges if $|x| > r$. • •

If (iii) of Theorem (11.36) occurs, then the power series $\sum a_n x^n$ is absolutely convergent throughout the open interval $(-r, r)$ and diverges outside of 'he closed interval $[-r, r]$, as illustrated in Figure 11.12. The number r is cali·d the **radius of convergence** of the series. Either convergence or divergence may occur at $-r$ or r, depending on the nature of the series.

The totality of numbers for which a power series converges is called its **interval of convergence**. If the radius of convergence r is positive, then the interval of convergence is one of the following:

$$(-r, r), \quad (-r, r], \quad [-r, r), \quad \text{or} \quad [-r, r].$$

FIGURE 11.12
$\sum a_n x^n$ with radius of convergence r

DIVERGENT   ABSOLUTLEY CONVERGENT   DIVERGENT

$-r \quad 0 \quad r$

If (i) or (ii) of Theorem (11.36) occurs, then the radius of convergence is denoted by 0 or ∞, respectively. In Example 1 of this section, the interval of convergence is $(-5, 5)$ and the radius of convergence is 5. In Example 2 the interval of convergence is $(-\infty, \infty)$ and we write $r = \infty$. In Example 3, $r = 0$. The next example illustrates the case of a half-open interval of convergence.

EXAMPLE 4 Find the interval of convergence of the power series $\sum_{n=1}^{\infty} \frac{1}{\sqrt{n}} x^n$.

SOLUTION Note that in this example the coefficient of x^0 is 0 and the summation begins with $n = 1$. We let $u_n = x^n / \sqrt{n}$ and consider

$$\lim_{n \to \infty} \left| \frac{u_{n+1}}{u_n} \right| = \lim_{n \to \infty} \left| \frac{x^{n+1}}{\sqrt{n+1}} \cdot \frac{\sqrt{n}}{x^n} \right| = \lim_{n \to \infty} \left| \frac{\sqrt{n}}{\sqrt{n+1}} x \right|$$

$$= \lim_{n \to \infty} \sqrt{\frac{n}{n+1}} |x| = (1)|x| = |x|.$$

It follows from the Ratio Test (11.33) that the power series is absolutely convergent if $|x| < 1$, that is, if $-1 < x < 1$. The series diverges if $|x| > 1$, that is, if $x > 1$ or $x < -1$. The numbers 1 and -1 must be checked by direct substitution in the power series. If we let $x = 1$, we obtain

$$\sum_{n=1}^{\infty} \frac{1}{\sqrt{n}} (1)^n = 1 + \frac{1}{\sqrt{2}} + \frac{1}{\sqrt{3}} + \cdots + \frac{1}{\sqrt{n}} + \cdots$$

which is a divergent p-series. If we substitute $x = -1$ the result is

$$\sum_{n=1}^{\infty} \frac{1}{\sqrt{n}} (-1)^n = -1 + \frac{1}{\sqrt{2}} - \frac{1}{\sqrt{3}} + \cdots + \frac{(-1)^n}{\sqrt{n}} + \cdots$$

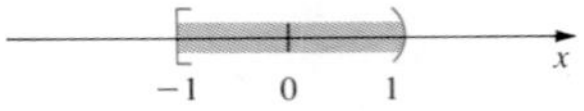

which converges by the Alternating Series Test. Hence the power series converges if $-1 \le x < 1$. The interval of convergence $[-1, 1)$ is sketched in Figure 11.13. •

FIGURE 11.13
Interval of convergence of $\sum \dfrac{1}{\sqrt{n}} x^n$

DEFINITION (11.37)

Let c be a real number and x a variable. A **power series in $x - c$** is a series of the form

$$\sum_{n=0}^{\infty} a_n(x - c)^n = a_0 + a_1(x - c) + a_2(x - c)^2 + \cdots + a_n(x - c)^n + \cdots$$

such that each a_k is a real number.

To simplify the nth term in (11.37), we assume that $(x - c)^0 = 1$ even if $x = c$. As in the proof of Theorem (11.36), but with x replaced by $x - c$, we can show that precisely one of the following is true:

(i) The series converges only if $x - c = 0$, that is, if $x = c$.

(ii) The series is absolutely convergent for every x.

(iii) There is a positive number r such that the series is absolutely convergent if $|x - c| < r$ and divergent if $|x - c| > r$.

If (iii) occurs, then the series $\sum a_n(x - c)^n$ is absolutely convergent if

$$-r < x - c < r \quad \text{or} \quad c - r < x < c + r,$$

FIGURE 11.14
$\sum a_n(x - c)^n$ with radius of convergence r

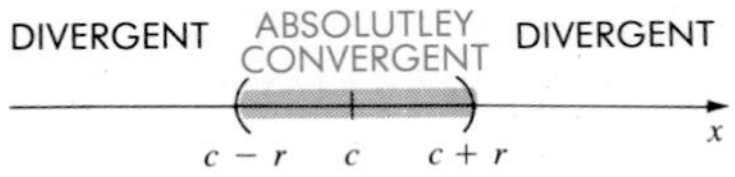

DIVERGENT   ABSOLUTLEY CONVERGENT   DIVERGENT

that is, if x is the interval $(c - r, c + r)$ as illustrated in Figure 11.14. The endpoints of the interval must be checked separately. As before, the totality of numbers for which the series converges is called the **interval of convergence,** and r is the **radius of convergence.**

EXAMPLE 5 Find the interval of convergence of the series

$$1 - \frac{1}{2}(x - 3) + \frac{1}{3}(x - 3)^2 + \cdots + (-1)^n \frac{1}{n + 1}(x - 3)^n + \cdots$$

SOLUTION If we let

$$u_n = (-1)^n \frac{(x - 3)^n}{n + 1}$$

then

$$\lim_{n \to \infty} \left| \frac{u_{n+1}}{u_n} \right| = \lim_{n \to \infty} \left| \frac{(x-3)^{n+1}}{n+2} \cdot \frac{n+1}{(x-3)^n} \right|$$

$$= \lim_{n \to \infty} \left| \frac{n+1}{n+2}(x-3) \right|$$

$$= \lim_{n \to \infty} \left(\frac{n+1}{n+2} \right) |x-3|$$

$$= (1)|x-3| = |x-3|.$$

By the Ratio Test, the series is absolute convergent if $|x-3| < 1$, that is, if

$$-1 < x - 3 < 1 \quad \text{or} \quad 2 < x < 4.$$

The series diverges if $x < 2$ or $x > 4$. The numbers 2 and 4 must both be checked. If $x = 4$, the resulting series is

$$1 - \frac{1}{2} + \frac{1}{3} - \cdots + (-1)^n \frac{1}{n+1} + \cdots$$

which converges by the Alternating Series Test. For $x = 2$, the series becomes

$$1 + \frac{1}{2} + \frac{1}{3} + \cdots + \frac{1}{n+1} + \cdots$$

which is the divergent harmonic series. Hence the interval of convergence is $(2, 4]$, as illustrated in Figure 11.15. •

FIGURE 11.15
Interval of convergence of
$\sum (-1)^n \dfrac{1}{n+1}(x-3)^n$

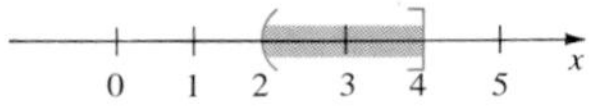

EXERCISES 11.6

Exer. 1–26: Find the interval of convergence of the power series.

1 $\displaystyle\sum_{n=0}^{\infty} \frac{1}{n+4} x^n$

2 $\displaystyle\sum_{n=0}^{\infty} \frac{1}{n^2+4} x^n$

2 $\displaystyle\sum_{n=0}^{\infty} \frac{n^2}{2^n} x^n$

4 $\displaystyle\sum_{n=1}^{\infty} \frac{(-3)^n}{n} x^{n+1}$

5 $\displaystyle\sum_{n=1}^{\infty} (-1)^{n-1} \frac{1}{\sqrt{n}} x^n$

6 $\displaystyle\sum_{n=1}^{\infty} \frac{1}{\ln(n+1)} x^n$

7 $\displaystyle\sum_{n=2}^{\infty} \frac{n}{n^2+1} x^n$

8 $\displaystyle\sum_{n=1}^{\infty} \frac{1}{4^n \sqrt{n}} x^n$

9 $\displaystyle\sum_{n=2}^{\infty} \frac{\ln n}{n^3} x^n$

10 $\displaystyle\sum_{n=0}^{\infty} \frac{10^{n+1}}{3^{2n}} x^n$

11 $\displaystyle\sum_{n=0}^{\infty} \frac{n+1}{10^n} (x-4)^n$

12 $\displaystyle\sum_{n=1}^{\infty} \frac{1}{n(n+1)} (x-2)^n$

13 $\displaystyle\sum_{n=0}^{\infty} \frac{n!}{100^n} x^n$

14 $\displaystyle\sum_{n=0}^{\infty} \frac{(3n)!}{(2n)!} x^n$

15 $\displaystyle\sum_{n=0}^{\infty} \frac{1}{(-4)^n} x^{2n+1}$

16 $\displaystyle\sum_{n=1}^{\infty} (-1)^{n-1} \frac{1}{\sqrt[3]{n3^n}} x^n$

17 $\displaystyle\sum_{n=0}^{\infty} \frac{2^n}{(2n)!} x^{2n}$

18 $\displaystyle\sum_{n=0}^{\infty} \frac{10^n}{n!} x^n$

19 $\displaystyle\sum_{n=0}^{\infty} \frac{3^{2n}}{n+1} (x-2)^n$

20 $\displaystyle\sum_{n=1}^{\infty} \frac{1}{n5^n} (x-5)^n$

21 $\displaystyle\sum_{n=0}^{\infty} \frac{n^2}{2^{3n}} (x+4)^n$

22 $\displaystyle\sum_{n=0}^{\infty} \frac{1}{2n+1} (x+3)^n$

23 $\displaystyle\sum_{n=1}^{\infty} \frac{\ln n}{e^n} (x-e)^n$

24 $\displaystyle\sum_{n=0}^{\infty} \frac{n}{3^{2n-1}} (x-1)^{2n}$

25 $\displaystyle\sum_{n=1}^{\infty} (-1)^n \frac{1}{n6^n} (2x-1)^n$

26 $\displaystyle\sum_{n=0}^{\infty} \frac{1}{\sqrt{3n+4}} (3x+4)^n$

Exer. 27–30: Find the radius of convergence of the power series.

27 $\displaystyle\sum_{n=1}^{\infty} (-1)^n \frac{1 \cdot 3 \cdot 5 \cdots \cdot (2n-1)}{3 \cdot 6 \cdot 9 \cdots \cdot (3n)} x^n$

28 $\displaystyle\sum_{n=1}^{\infty} \frac{2 \cdot 4 \cdot 6 \cdots \cdot (2n)}{4 \cdot 7 \cdot 10 \cdots \cdot (3n+1)} x^n$

29 $\displaystyle\sum_{n=1}^{\infty} \frac{n^n}{n!} x^n$

30 $\displaystyle\sum_{n=0}^{\infty} \frac{(n+1)!}{10^n} (x-5)^n$

Exer. 31–32: Find the radius of convergence of the power series for positive integers c and d.

31 $\displaystyle\sum_{n=0}^{\infty} \frac{(n+c)!}{n!(n+d)!} x^n$

32 $\displaystyle\sum_{n=0}^{\infty} \frac{(cn)!}{(n!)^c} x^n$

33 *Bessel functions* are useful in the analysis of problems that involve oscillations. For α a positive integer, the Bessel function $J_\alpha(x)$ *of the first kind of order* α is defined by the power series

$$J_\alpha(x) = \sum_{n=0}^{\infty} \frac{(-1)^n}{n!(n+\alpha)!} \left(\frac{x}{2}\right)^{2n+\alpha}$$

Show that this power series is convergent for every real number.

34 Refer to Exercise 33. The sixth-degree polynomial $1 - [x^2/2^2] + [x^4/(2^2 4^2)] - [x^6/(2^2 4^2 6^2)]$ is sometimes used to approximate the Bessel function $J_0(x)$ of the first kind of order zero for $0 \le x \le 1$. Show that the error E involved in this approximation is less than 0.00001.

35 If $\lim_{n \to \infty} |a_{n+1}/a_n| = k$, and $k \ne 0$, prove that the radius of convergence of $\sum a_n x^n$ is $1/k$.

36 If $\lim_{n \to \infty} \sqrt[n]{|a_n|} = k$, and $k \ne 0$, prove that the radius of convergence of $\sum a_n x^n$ is $1/k$.

37 If $\sum a_n x^n$ has radius of convergence r, prove that $\sum a_n x^{2n}$ has radius of convergence $\sqrt{r}$.

38 If $\sum a_n$ is absolutely convergent, prove that $\sum a_n x^n$ is absolutely convergent for every x in the interval $[-1, 1]$.

39 If the interval of convergence of $\sum a_n x^n$ is $(-r, r]$, prove that the series is conditionally convergent at r.

40 If $\sum a_n x^n$ is absolutely convergent at one endpoint of its interval of convergence, prove that it is also absolutely convergent at the other endpoint.

11.7 POWER SERIES REPRESENTATIONS OF FUNCTIONS

A power series $\sum a_n x^n$ may be used to define a function f whose domain is the interval of convergence of the series. Specifically, for each x in this interval we let $f(x)$ equal the sum of the series, that is,

$$f(x) = a_0 + a_1 x + a_2 x^2 + \cdots + a_n x^n + \cdots$$

If a function f is defined in this way, we say that $\sum a_n x^n$ is a **power series representation for $f(x)$** (or *of* $f(x)$). We also use the phrase f **is represented by the power series.**

A power series representation $f(x) = \sum a_n x^n$ enables us to find function values in a new way. Thus, if c is in the interval of convergence, then $f(c)$ can be found (or approximated) by finding (or approximating) the sum of the series

$$a_0 + a_1 c + a_2 c^2 + \cdots + a_n c^n + \cdots$$

EXAMPLE 1 Find a function f that is represented by the power series

$$1 - x + x^2 - x^3 + \cdots + (-1)^n x^n + \cdots$$

SOLUTION If $|x| < 1$, then by Theorem (11.15), the given geometric series converges and has the sum $1/[1 - (-x)] = 1/(1 + x)$. Hence we may write

$$\frac{1}{1+x} = 1 - x + x^2 - x^3 + \cdots + (-1)^n x^n + \cdots$$

This is a power series representation for $f(x) = 1/(1 + x)$ on the interval $(-1, 1)$. •

A function f defined by a power series has properties similar to those of a polynomial. In particular, f has a derivative whose power series representation may be found by differentiating each term of the given series. Similarly, definite integrals of f may be obtained by integrating each term of the series. These facts are consequences of the next theorem, which is stated without proof.

THEOREM (11.38)

Suppose a power series $\sum a_n x^n$ has a nonzero radius of convergence r, and let the function f be defined by

$$f(x) = \sum_{n=0}^{\infty} a_n x^n = a_0 + a_1 x + a_2 x^2 + \cdots + a_n x^n + \cdots$$

for every x in the interval of convergence. If $-r < x < r$, then

(i) $f'(x) = \sum_{n=0}^{\infty} D_x (a_n x^n) = \sum_{n=1}^{\infty} n a_n x^{n-1}$

$$= a_1 + 2a_2 x + 3a_3 x^2 + \cdots + n a_n x^{n-1} + \cdots$$

(ii) $\int_0^x f(t)\, dt = \sum_{n=0}^{\infty} \int_0^x (a_n t^n)\, dt = \sum_{n=0}^{\infty} \frac{a_n}{n+1} x^{n+1}$

$$= a_0 x + \frac{1}{2} a_1 x^2 + \frac{1}{3} a_2 x^3 + \cdots + \frac{1}{n+1} a_n x^{n+1} + \cdots$$

The series obtained by differentiation in (i) or integration in (ii) of Theorem (11.38) has the same radius of convergence as $\sum a_n x^n$. As a corollary of (i), *a function that is represented by a power series in an interval* $(-r, r)$ *is continuous throughout* $(-r, r)$ (see Theorem (3.5)). Similar results are true for functions represented by power series of the form $\sum a_n (x - c)^n$.

EXAMPLE 2 Find a power series representation for $\dfrac{1}{(1+x)^2}$.

SOLUTION We saw in Example 1 that

$$\frac{1}{1+x} = 1 - x + x^2 - x^3 + \cdots + (-1)^n x^n + \cdots \quad \text{if } |x| < 1.$$

If we differentiate each term of this series, then by Theorem (11.38) (i),

$$-\frac{1}{(1+x)^2} = -1 + 2x - 3x^2 - \cdots + (-1)^n n x^{n-1} + \cdots$$

Consequently, if $|x| < 1$, then

$$\frac{1}{(1+x)^2} = 1 - 2x + 3x^2 + \cdots + (-1)^{n+1} n x^{n-1} + \cdots \quad \bullet$$

EXAMPLE 3 Find a power series representation for $\ln(1 + x)$ if $|x| < 1$.

SOLUTION If $|x| < 1$, then

$$\ln(1+x) = \int_0^x \frac{1}{1+t}\, dt$$

$$= \int_0^x \left[1 - t + t^2 - \cdots + (-1)^n t^n + \cdots \right] dt.$$

The last equality follows from Example 1. By Theorem (11.38) (ii), we may integrate each term of the series as follows:

$$\ln(1+x) = \int_0^x 1\, dt - \int_0^x t\, dt + \int_0^x t^2\, dt - \cdots + (-1)^n \int_0^x t^n\, dt + \cdots$$

$$= t\Big]_0^x - \frac{t^2}{2}\Big]_0^x + \frac{t^3}{3}\Big]_0^x - \cdots + (-1)^n \frac{t^{n+1}}{n+1}\Big]_0^x + \cdots$$

Hence

$$\ln(1+x) = x - \frac{x^2}{2} + \frac{x^3}{3} - \cdots + (-1)^n \frac{x^{n+1}}{n+1} + \cdots$$

if $|x| < 1.$ •

EXAMPLE 4 Use the results of Example 3 to calculate ln (1.1) to five decimal places.

SOLUTION In Example 3 we found a series representation for $\ln(1+x)$ if $|x| < 1$. Substituting 0.1 for x in that series gives us

$$\ln(1.1) = 0.1 - \frac{(0.1)^2}{2} + \frac{(0.1)^3}{3} - \frac{(0.1)^4}{4} + \frac{(0.1)^5}{5} - \cdots$$

$$= 0.1 - 0.005 + 0.000333 - 0.000025 + 0.000002 - \cdots$$

If we sum the first four terms on the right and round off to five decimal places, then

$$\ln(1.1) \approx 0.09531$$

By Theorem (11.29), the error E is less than the absolute value 0.000002 of the fifth term and, therefore, the number 0.09531 is accurate to five decimal places. •

EXAMPLE 5 Find a power series representation for arctan x.

SOLUTION We first observe that

$$\arctan x = \int_0^x \frac{1}{1+t^2}\, dt.$$

Next we note that if $|t| < 1$, then by Theorem (11.15) with $a = 1$ and $r = -t^2$,

$$\frac{1}{1+t^2} = 1 - t^2 + t^4 - \cdots + (-1)^n t^{2n} + \cdots$$

By Theorem (11.38) (ii) we may integrate each term of the series to obtain

$$\arctan x = x - \frac{x^3}{3} + \frac{x^5}{5} - \cdots + (-1)^n \frac{x^{2n+1}}{2n+1} + \cdots$$

provided $|x| < 1$. It can be proved that this series representation is also valid if $|x| = 1.$ •

EXAMPLE 6 Prove that e^x has the power series representation

$$e^x = 1 + x + \frac{x^2}{2!} + \frac{x^3}{3!} + \cdots + \frac{x^n}{n!} + \cdots$$

SOLUTION We considered the power series in Example 2 of the preceding section, and found that it is absolutely convergent for every real number x. If we let f denote the function represented by the series, then

$$f(x) = \sum_{n=0}^{\infty} \frac{x^n}{n!}.$$

Applying Theorem (11.38) (i),

$$f'(x) = \sum_{n=1}^{\infty} \frac{nx^{n-1}}{n!} = \sum_{n=1}^{\infty} \frac{x^{n-1}}{(n-1)!}$$

$$= 1 + x + \frac{x^2}{2!} + \frac{x^3}{3!} + \cdots + \frac{x^n}{n!} + \cdots$$

that is,
$$f'(x) = f(x) \quad \text{for every } x.$$

If, in Theorem (7.26), we let $y = f(t)$, $t = x$, and $c = 1$, we obtain

$$f(x) = f(0)e^x.$$

However,

$$f(0) = 1 + 0 + \frac{0^2}{2!} + \cdots + \frac{0^n}{n!} + \cdots = 1$$

and hence
$$f(x) = e^x$$

which is what we wished to prove. •

Note that Example 6 allows us to express the number e as the sum of a convergent positive term series, namely,

$$e = 1 + 1 + \frac{1}{2!} + \frac{1}{3!} + \cdots + \frac{1}{n!} + \cdots$$

EXAMPLE 7 Approximate $\int_0^{0.1} e^{-x^2} \, dx$.

SOLUTION Letting $x = -t^2$ in the series of Example 6 gives us

$$e^{-t^2} = 1 - t^2 + \frac{t^4}{2!} - \cdots + \frac{(-1)^n t^{2n}}{n!} + \cdots$$

for every t. Applying Theorem (11.38) (ii),

$$\int_0^{0.1} e^{-x^2} \, dx = \int_0^{0.1} e^{-t^2} \, dt$$

$$= t \Big]_0^{0.1} - \frac{t^3}{3} \Big]_0^{0.1} + \frac{t^5}{10} \Big]_0^{0.1} - \cdots$$

$$= 0.1 - \frac{(0.1)^3}{3} + \frac{(0.1)^5}{10} - \cdots$$

If we use the first two terms to approximate the sum of this convergent alternating series, then by Theorem (11.29), the error is less than the third term $(0.1)^5/10 = 0.000001$. Hence

$$\int_0^{0.1} e^{-x^2} \, dx \approx 0.1 - \frac{0.001}{3} \approx 0.99667$$

which is accurate to five decimal places. •

Thus far, the methods we have used to obtain power series representations of functions are *indirect*, in the sense that we started with known series and then differentiated or integrated. In the next section we shall discuss a *direct* method that can be used to find power series representations for a large variety of functions.

Exer. 1–10: Find a power series representation for the expression, and specify the interval of convergence. (*Hint:* Use (11.15), (11.38), or long division as necessary.)

1 $\dfrac{1}{1-x}$ 2 $\dfrac{1}{1+x^2}$ 3 $\dfrac{1}{(1-x)^2}$ 4 $\dfrac{1}{1-4x}$

5 $\dfrac{x^2}{1-x^2}$ 6 $\dfrac{x}{1-x^4}$ 7 $\dfrac{x}{2-3x}$ 8 $\dfrac{x^3}{4-x^3}$

9 $\dfrac{x^2+1}{x-1}$ 10 $\dfrac{3}{2x+5}$

11 Prove that

$$\ln(1-x) = \sum_{n=1}^{\infty} \frac{-x^n}{n} \quad \text{if } |x| < 1.$$

Use this fact to approximate $\ln(1.2)$ to three decimal places.

12 Use Example 5 to prove that the sum of the series $\sum_{n=0}^{\infty} (-\frac{1}{3})^n / [\sqrt{3}(2n+1)]$ is $\pi/6$.

Exer. 13–18: Use the result obtained in Example 6 to find a power series representation for the expression.

13 e^{-x} 14 e^{2x} 15 $\cosh x$

16 $\sinh x$ 17 xe^{3x} 18 $x^2 e^{x^2}$

Exer. 19–24: Use an infinite series to approximate the integral to four decimal places.

19 $\displaystyle\int_0^{1/3} \frac{1}{1+x^6}\, dx$ 20 $\displaystyle\int_0^{1/2} \arctan x^2\, dx$

21 $\displaystyle\int_{0.1}^{0.2} \frac{\arctan x}{x}\, dx$ 22 $\displaystyle\int_0^{0.2} \frac{x^3}{1+x^5}\, dx$

23 $\displaystyle\int_0^1 e^{-x^2/10}\, dx$ 24 $\displaystyle\int_0^{0.5} e^{-x^3}\, dx$

25 Use the power series representation for $(1-x^2)^{-1}$ to find a power series representation for $2x(1-x^2)^{-2}$.

26 Use the method of Example 3 to find a power series representation for $\ln(3+2x)$.

27 Refer to Exercise 33 of Section 11.6. Use Theorem (11.38) to prove the following.
   (a) If $J_0(x)$ and $J_1(x)$ are Bessel functions of the first kind of orders 0 and 1, respectively, then $D_x\, J_0(x) = -J_1(x)$.
   (b) If $J_2(x)$ and $J_3(x)$ are Bessel functions of the first kind of orders 2 and 3, respectively, then
   $$\int x^3 J_2(x)\, dx = x^3 J_3(x) + C.$$

28 Light is absorbed by rods and cones in the retina of the eye. The number of photons absorbed by a photoreceptor during a given flash of light is governed by the *Poisson distribution*. More precisely, the probability p_n that a photoreceptor absorbs exactly n photons is given by the formula $p_n = e^{-\lambda}\lambda^n/n!$ for some $\lambda > 0$.
   (a) Show that $\sum_{n=0}^{\infty} p_n = 1$.
   (b) Sight usually occurs when two or more photons are absorbed by a photoreceptor. Show that the probability of this occuring is $1 - e^{-\lambda}(\lambda + 1)$.

Exer. 29–30: Find a power series representation for $f(x)$. If the integrand is denoted by $g(t)$, assume that the value of $g(0)$ is $\lim_{t\to 0} g(t)$.

29 $f(x) = \displaystyle\int_0^x \frac{\ln(1-t)}{t}\, dt$ 30 $f(x) = \displaystyle\int_0^x \frac{e^t-1}{t}\, dt$

11.8 TAYLOR AND MACLAURIN SERIES

Suppose a function f is represented by a power series in $x - c$, such that

$$f(x) = \sum_{n=0}^{\infty} a_n(x-c)^n$$

$$= a_0 + a_1(x-c) + a_2(x-c)^2 + a_3(x-c)^3 + a_4(x-c)^4 + \cdots$$

and the domain of f is an open interval containing c. As in the preceding section, power series representations may be found for $f'(x)$, $f''(x)$, ... by differentiating the terms of the series for $f(x)$. Thus

$$f'(x) = \sum_{n=1}^{\infty} na_n(x-c)^{n-1}$$

$$= a_1 + 2a_2(x-c) + 3a_3(x-c)^2 + 4a_4(x-c)^3 + \cdots$$

$$f''(x) = \sum_{n=2}^{\infty} n(n-1)a_n(x-c)^{n-2}$$

$$= 2a_2 + (3\cdot 2)a_3(x-c) + (4\cdot 3)a_4(x-c)^2 + \cdots$$

$$f'''(x) = \sum_{n=3}^{\infty} n(n-1)(n-2)a_n(x-c)^{n-3}$$

$$= (3 \cdot 2)a_3 + (4 \cdot 3 \cdot 2)a_4(x-c) + \cdots$$

and, for every positive integer k,

$$f^{(k)}(x) = \sum_{n=k}^{\infty} n(n-1)\cdots(n-k+1)a_n(x-c)^{n-k}.$$

Moreover, each series obtained by differentiation has the same radius of convergence as the original series. Substituting c for x in each of these series representations, we obtain

$$f(c) = a_0, \quad f'(c) = a_1, \quad f''(c) = 2a_2, \quad f'''(c) = (3 \cdot 2)a_3$$

and, for every positive integer n,

$$f^{(n)}(c) = n!a_n \quad \text{or} \quad a_n = \frac{f^{(n)}(c)}{n!}.$$

We have proved that the series for $f(x)$ has the form stated in the next theorem. It is called the *Taylor series for $f(x)$ at c*.

TAYLOR SERIES FOR $f(x)$ AT c **(11.39)**

If f is a function such that

$$f(x) = \sum_{n=0}^{\infty} a_n(x-c)^n$$

for every x in an open interval containing c, then $f^{(n)}(c)$ exists for $n = 0, 1, 2, \ldots$ and $a_n = f^{(n)}(c)/n!$. Thus

$$f(x) = f(c) + f'(c)(x-c) + \frac{f''(c)}{2!}(x-c)^2 + \cdots$$

$$+ \frac{f^{(n)}(c)}{n!}(x-c)^n + \cdots$$

If we use the convention $f^{(0)}(c) = f(c)$, the Taylor series in (11.39) may be written in the following summation form:

$$f(x) = \sum_{k=0}^{\infty} \frac{f^{(k)}(c)}{k!}(x-c)^k$$

The special case $c = 0$ is called the *Maclaurin series for $f(x)$*. Let us state this for reference as follows.

MACLAURIN SERIES FOR $f(x)$ **(11.40)**

If f is a function such that $f(x) = \sum a_n x^n$ for every x in an open interval $(-r, r)$, then

$$f(x) = f(0) + f'(0)x + \frac{f''(0)}{2!}x^2 + \cdots + \frac{f^{(n)}(0)}{n!}x^n + \cdots$$

EXAMPLE 1 From Example 6 of the preceding section, e^x has the power series representation

$$e^x = 1 + x + \frac{1}{2!} x^2 + \frac{1}{3!} x^3 + \cdots + \frac{1}{n!} x^n + \cdots$$

Verify that this is a Maclaurin series.

SOLUTION If $f(x) = e^x$, then the kth derivative of f is $f^{(k)}(x) = e^x$ and

$$f^{(k)}(0) = e^0 = 1 \quad \text{for } k = 0, 1, 2, \ldots$$

Hence the Maclaurin series (11.40) is

$$\sum_{k=0}^{\infty} \frac{f^{(k)}(0)}{k!} x^k = \sum_{k=0}^{\infty} \frac{1}{k!} x^k = 1 + x + \frac{1}{2!} x^2 + \cdots + \frac{1}{n!} x^n + \cdots$$

which is the same as the given series. •

Theorem (11.39) implies that *if* a function f is represented by a power series in $x - c$, then the power series *must* be the Taylor series. However, the theorem does not state conditions that guarantee that a power series representation actually exists. We shall now obtain such conditions. Let us begin by noting that the $(n + 1)$st partial sum of the Taylor series stated in (11.39) is the nth-degree Taylor Polynomial $P_n(x)$ of f at c (see Definition (10.10)). Moreover, by Theorem (10.11),

$$P_n(x) = f(x) - R_n(x)$$

with $\qquad\qquad R_n(x) = \dfrac{f^{(n+1)}(z)}{(n+1)!} (x - c)^{n+1}$

for some number z between c and x. In the next theorem we use $R_n(x)$ to obtain sufficient conditions for the existence of a power series representation of $f(x)$.

THEOREM (11.41)

> Let f be a function that has derivatives of all orders throughout an interval containing c, and let $R_n(x)$ be the Taylor Remainder of f at c. If
>
> $$\lim_{n \to \infty} R_n(x) = 0$$
>
> for every x in that interval, then $f(x)$ is represented by the Taylor series for $f(x)$ at c.

PROOF The Taylor polynomial $P_n(x)$ is the nth term for the sequence of partial sums of the Taylor series for $f(x)$ at c. Since $P_n(x) = f(x) - R_n(x)$,

$$\lim_{n \to \infty} P_n(x) = f(x) - \lim_{n \to \infty} R_n(x) = f(x).$$

Hence the sequence of partial sums converges to $f(x)$. This proves the theorem. • •

In Example 2 of Section 11.6 we proved that the power series $\sum x^n/n!$ is absolutely convergent for every real number. Using Theorem (11.16) gives us the following result.

THEOREM (11.42) For every real number x, $\displaystyle\lim_{n \to \infty} \frac{|x|^n}{n!} = 0$.

We shall use Theorem (11.42) in the solution of the following example.

EXAMPLE 2 Find the Maclaurin series for $\sin x$, and prove that it represents $\sin x$ for every real number x.

SOLUTION Let us arrange our work as follows:

$$f(x) = \sin x \qquad\qquad f(0) = 0$$
$$f'(x) = \cos x \qquad\qquad f'(0) = 1$$
$$f''(x) = -\sin x \qquad\qquad f''(0) = 0$$
$$f'''(x) = -\cos x \qquad\qquad f'''(0) = -1$$

Successive derivatives follow this pattern. Substitution in (11.40) gives us the Maclaurin series

$$\sin x = x - \frac{x^3}{3!} + \frac{x^5}{5!} - \frac{x^7}{7!} + \cdots + (-1)^n \frac{x^{2n+1}}{(2n+1)!} + \cdots$$

At this stage all we know is that *if* $\sin x$ is represented by a power series in x, then it is given by the preceding series. To prove that $\sin x$ *is* actually represented by this Maclaurin series, let us employ Theorem (11.41). If n is a positive integer, then either

$$\left| f^{(n+1)}(x) \right| = \left| \cos x \right| \quad \text{or} \quad \left| f^{(n+1)}(x) \right| = \left| \sin x \right|.$$

Hence $\left| f^{(n+1)}(z) \right| \le 1$ for every number z and, using the formula for $R_n(x)$ with $c = 0$,

$$\left| R_n(x) \right| = \frac{\left| f^{(n+1)}(z) \right|}{(n+1)!} \left| x \right|^{n+1} \le \frac{\left| x \right|^{n+1}}{(n+1)!}.$$

It follows from Theorem (11.42) and the Sandwich Theorem (11.7) that $\lim_{n \to \infty} \left| R_n(x) \right| = 0$. Consequently, $\lim_{n \to \infty} R_n(x) = 0$, and the Maclaurin series representation for $\sin x$ is true for every x. •

EXAMPLE 3 Find the Maclaurin series for $\cos x$.

SOLUTION We could proceed directly as in Example 2; however, let us obtain the series for $\cos x$ by differentiating the series for $\sin x$ obtained in Example 2. This gives us

$$\cos x = 1 - \frac{x^2}{2!} + \frac{x^4}{4!} - \frac{x^6}{6!} + \cdots + (-1)^n \frac{x^{2n}}{(2n)!} + \cdots$$

This series can also be obtained by integrating the terms of the series for $\sin x$.

•

The Maclaurin series for e^x was obtained in Example 6 of the preceding section by an indirect technique (see also Example 1 of this section). We shall next give a direct derivation of this important formula.

EXAMPLE 4 Find a Maclaurin series that represents e^x for every real number x.

SOLUTION If $f(x) = e^x$, then $f^{(k)}(x) = e^x$ for every positive integer k. Hence $f^{(k)}(0) = 1$ and substitution in (11.40) gives us

$$e^x = 1 + x + \frac{x^2}{2!} + \frac{x^3}{3!} + \cdots + \frac{x^n}{n!} + \cdots$$

As in the solution of Example 2, we now employ Theorem (11.41) to prove that this power series representation of e^x is true for every real number x. Using the formula for $R_n(x)$ with $c = 0$,

$$R_n(x) = \frac{f^{(n+1)}(z)}{(n+1)!} x^{n+1} = \frac{e^z}{(n+1)!} x^{n+1}$$

for a number z between 0 and x. If $0 < x$, then $e^z < e^x$, since the natural exponential function is increasing, and hence, for every positive integer n,

$$0 < R_n(x) < \frac{e^x}{(n+1)!} x^{n+1}.$$

Using Theorem (11.42),

$$\lim_{n \to \infty} \frac{e^x}{(n+1)!} x^{n+1} = e^x \lim_{n \to \infty} \frac{x^{n+1}}{(n+1)!} = 0$$

and, by the Sandwich Theorem (11.7),

$$\lim_{n \to \infty} R_n(x) = 0.$$

If $x < 0$, then also $z < 0$ and hence $e^z < e^0 = 1$. Consequently,

$$0 < |R_n(x)| < \left| \frac{x^{n+1}}{(n+1)!} \right|$$

and we again see that $R_n(x)$ has the limit 0 as n increases without bound. It follows from Theorem (11.41) that the power series representation for e^x is valid for all nonzero x. Finally, note that if $x = 0$, then the series reduces to $e^0 = 1$. •

EXAMPLE 5 Find the Taylor series for $\sin x$ in powers of $x - (\pi/6)$.

SOLUTION The derivatives of $f(x) = \sin x$ are listed in Example 2. If we evaluate them at $c = \pi/6$, we obtain

$$f\left(\frac{\pi}{6}\right) = \frac{1}{2}, \quad f'\left(\frac{\pi}{6}\right) = \frac{\sqrt{3}}{2}, \quad f''\left(\frac{\pi}{6}\right) = -\frac{1}{2}, \quad f'''\left(\frac{\pi}{6}\right) = -\frac{\sqrt{3}}{2},$$

and this pattern of four numbers repeats itself indefinitely. Substitution in (11.39) gives us

$$\sin x = \frac{1}{2} + \frac{\sqrt{3}}{2}\left(x - \frac{\pi}{6}\right) - \frac{1}{2(2!)}\left(x - \frac{\pi}{6}\right)^2 - \frac{\sqrt{3}}{2(3!)}\left(x - \frac{\pi}{6}\right)^3 + \cdots$$

The nth term u_n of this series is given by

$$u_n = \begin{cases} (-1)^{n/2} \dfrac{1}{2(n!)} \left(x - \dfrac{\pi}{6}\right)^n & \text{if } n = 0, 2, 4, 6, \dots \\[3ex] (-1)^{(n-1)/2} \dfrac{\sqrt{3}}{2(n!)} \left(x - \dfrac{\pi}{6}\right)^n & \text{if } n = 1, 3, 5, 7, \dots \end{cases}$$

The proof that the series represents $\sin x$ for every x is similar to that given in Example 2 and is, therefore, omitted. •

A function may have derivatives of all orders at some number c, but not have a Taylor series representation at that number. Thus an additional condition, such as $\lim_{n \to \infty} R_n(x) = 0$, is required to guarantee the existence of a Taylor series.

In previous examples we established some of the Maclaurin series given in the following list. Verifications of the remaining series are left as exercises.

(11.43)

MACLAURIN SERIES	INTERVAL OF CONVERGENCE
(a) $\sin x = x - \dfrac{x^3}{3!} + \dfrac{x^5}{5!} - \dfrac{x^7}{7!} + \cdots + (-1)^n \dfrac{x^{2n+1}}{(2n+1)!} + \cdots$	$(-\infty, \infty)$
(b) $\cos x = 1 - \dfrac{x^2}{2!} + \dfrac{x^4}{4!} - \dfrac{x^6}{6!} + \cdots + (-1)^n \dfrac{x^{2n}}{(2n)!} + \cdots$	$(-\infty, \infty)$
(c) $e^x = 1 + x + \dfrac{x^2}{2!} + \dfrac{x^3}{3!} + \cdots + \dfrac{x^n}{n!} + \cdots$	$(-\infty, \infty)$
(d) $\ln(1+x) = x - \dfrac{x^2}{2} + \dfrac{x^3}{3} - \dfrac{x^4}{4} + \cdots + (-1)^n \dfrac{x^{n+1}}{n+1} + \cdots$	$(-1, 1]$
(e) $\tan^{-1} x = x - \dfrac{x^3}{3} + \dfrac{x^5}{5} - \dfrac{x^7}{7} + \cdots + (-1)^n \dfrac{x^{2n+1}}{2n+1} + \cdots$	$[-1, 1]$
(f) $\sinh x = x + \dfrac{x^3}{3!} + \dfrac{x^5}{5!} + \dfrac{x^7}{7!} + \cdots + \dfrac{x^{2n+1}}{(2n+1)!} + \cdots$	$(-\infty, \infty)$
(g) $\cosh x = 1 + \dfrac{x^2}{2!} + \dfrac{x^4}{4!} + \cdots + \dfrac{x^{2n}}{(2n)!} + \cdots$	$(-\infty, \infty)$

The series in (11.43) can be used to obtain power series representations for other functions. To illustrate, since (c) is true for every x, a power series representation for e^{-x^2} can be found by substituting $-x^2$ for x. This gives us

$$e^{-x^2} = 1 - x^2 + \frac{x^4}{2!} - \cdots + (-1)^n \frac{x^{2n}}{n!} + \cdots$$

Similarly, replacing x by $-x$ in (c) leads to

$$e^{-x} = 1 - x + \frac{x^2}{2!} - \cdots + (-1)^n \frac{x^n}{n!} + \cdots$$

Since $\cosh x = \frac{1}{2}(e^x + e^{-x})$, a power series for $\cosh x$ can be found by adding corresponding terms of the series for e^x and e^{-x}, and then multiplying by $\frac{1}{2}$. This gives us

$$\cosh x = 1 + \frac{x^2}{2!} + \frac{x^4}{4!} + \cdots + \frac{x^{2n}}{(2n)!} + \cdots$$

Compare this with the series for $\cos x$ in (11.43).

Taylor and Maclaurin series may be used to approximate function values. For example, to find $\sin (0.1)$ we could use (11.43) (a) to write

$$\sin (0.1) = 0.1 - \frac{0.001}{6} + \frac{0.00001}{120} - \cdots$$

By Theorem (11.29) the error involved in using the sum of the first two terms as an approximation is less than $0.00001/120 \approx 0.00000008$. More generally, we have the following polynomial approximation formula:

$$\sin x \approx x - \frac{x^3}{6}$$

By Theorem (11.29), the error is less than $|x^5|/5!$.

The next example illustrates the use of infinite series in approximating values of definite integrals.

EXAMPLE 6 Approximate $\int_0^1 \sin x^2 \, dx$ to four decimal places.

SOLUTION Substituting x^2 for x in (11.43) (a),

$$\sin x^2 = x^2 - \frac{x^6}{3!} + \frac{x^{10}}{5!} - \frac{x^{14}}{7!} + \cdots$$

Integrating each term of this series, we obtain

$$\int_0^1 \sin x^2 \, dx = \frac{1}{3} - \frac{1}{42} + \frac{1}{1320} - \frac{1}{75600} + \cdots$$

Summing the first three terms,

$$\int_0^1 \sin x^2 \, dx \approx 0.3103.$$

By Theorem (11.29), the error is less than $\frac{1}{75600} \approx 0.00001$. •

Note that in the preceding example we achieved accuracy to four decimal places by summing only *three* terms of the integrated series for $\sin x^2$. To obtain this degree of accuracy by means of the Trapezoidal Rule or Simpson's Rule, it would be necessary to use a large value of n for the interval $[0, 1]$. This brings out an important point. For numerical applications, in addition to analyzing a given problem, we should also strive to find the most efficient method for computing the answer.

To obtain a Taylor or Maclaurin series representation for a function f by means of (11.39) or (11.40), respectively, it is necessary to find a general formula for $f^{(n)}(x)$ and, in addition, to investigate $\lim_{n \to \infty} R_n(x)$. Because of this, our examples have been restricted to expressions such as $\sin x$, $\cos x$, and e^x. The method cannot be used if, for example, $f(x)$ equals $\tan x$ or $\sin^{-1} x$, since $f^{(n)}(x)$ becomes very complicated as n increases. Most of the exercises that follow are based on functions whose nth derivatives can be determined easily, or on series representations that we have already established. In more complicated cases we shall restrict our attention to only the first few terms of a Taylor or Maclaurin series representation.

EXERCISES 11.8

Exer. 1–6: Use (11.40) to find the Maclaurin series for the expression, and prove that the series is valid for every x by means of Theorem (11.43).

1 $\cos x$ 2 e^{-x} 3 e^{2x} 4 $\cosh x$

Exer. 5–12: Use (11.43) to find a Maclaurin series for the expression, and state the radius of convergence.

5 $x^2 e^x$

6 $x e^{-2x}$

7 $\sinh x$

8 $x^2 \sin x$

9 $x \sin 3x$

10 $\cos x^2$

11 $\cos^2 x$ (*Hint:* Use $\cos^2 x = \frac{1}{2}(1 + \cos 2x)$.)

12 $\sin^2 x$

Exer. 13–18: Find the Taylor Series for the expression at c. (Do not verify that $\lim_{n \to \infty} R_n(x) = 0$.)

13 $\sin x$; $c = \pi/4$

14 $\cos x$; $c = \pi/3$

15 $1/x$; $c = 2$

16 e^x; $c = -3$

17 10^x; $c = 0$

18 $\ln (1 + x)$; $c = 0$

19 Find a series representation for e^{2x} in powers of $x + 1$.

20 Find a series representation for $\ln x$ in powers of $x - 1$.

Exer. 21–26: Find the first four terms of the Taylor series for the expression at c.

21 $\sec x$; $c = \pi/3$

22 $\tan x$; $c = \pi/4$

23 $\sin^{-1} x$; $c = 1/2$

24 $\csc x$; $c = 2\pi/3$

25 $x e^x$; $c = -1$

26 $\operatorname{sech} x$; $c = 0$

Exer. 27–38: Use an infinite series to approximate the number to four decimal places.

27 $\sqrt{e}$

28 e^{-1}

29 $\cos 3°$

30 $\sin 1°$

31 $\tan^{-1} 0.1$

32 $\ln 1.5$

33 $\sinh 0.5$

34 $\cosh 0.1$

35 $\int_0^1 e^{-x^2} \, dx$

36 $\int_0^{1/2} x \cos x^3 \, dx$

37 $\int_0^{0.5} \cos x^2 \, dx$

38 $\int_0^{0.1} \tan^{-1} x^2 \, dx$

Exer. 39–42: Approximate the integral to four decimal places. Assume that if the integrand is $f(x)$, then $f(0) = \lim_{x \to 0} f(x)$.

39 $\int_0^1 \frac{1 - \cos x}{x^2} \, dx$

40 $\int_0^1 \frac{\sin x}{x} \, dx$

41 $\int_0^{1/2} \frac{\ln (1 + x)}{x} \, dx$

42 $\int_0^1 \frac{1 - e^{-x}}{x} \, dx$

43 Use (11.43) (d) to find the Maclaurin series for $\ln [(1 + x)/(1 - x)]$.

44 Use the first five terms of the series in Exercise 43 to calculate $\ln 2$, and compare your answer to the value obtained using a calculator or table.

45 Use (11.43) (e) with $x = 1$ to represent π as the sum of an infinite series. What accuracy is obtained by using the first five terms of the series to approximate π? Approximately how many terms of the series are required to obtain four-decimal-place accuracy for π?

46 Use the identity $\tan^{-1} \frac{1}{2} + \tan^{-1} \frac{1}{3} = \pi/4$ (see Exercise 30 of Section 8.2) to express π as the sum of two infinite series. Use the first five terms of each series to approximate π and compare the result with that obtained in Exercise 45.

47 When planning a highway across a desert, a surveyor must make compensations for the curvature of the earth when measuring differences in elevation (see figure).

(a) If s is the length of the highway and R is radius of the earth, show that the correction C is given by $C = R \left[\sec (s/R) - 1 \right]$.

(b) Use the Maclaurin series for $\sec x$ to show that C is approximately $s^2/(2R) + (5s^4)/(24R^3)$.

(c) The average radius of the earth is 3959 miles. Estimate the correction, to the nearest 0.1 foot, for a stretch of highway 5 miles long.

EXERCISE 47

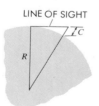

LINE OF SIGHT

48 Refer to Exercise 57 of Section 8.4. The velocity v of a water wave is related to the length L of the wave and the depth D of the water by the formula $v^2 = (gL/2\pi) \tanh (2\pi D/L)$ for a gravitational constant g.

(a) Show that $\tanh x \approx x - \frac{1}{3}x^3$ if $x \approx 0$. (The Maclaurin series for $\tanh x$ is an alternating series.)

(b) Use part (a) to obtain the approximation formula $v^2 \approx gD$ if D/L is small.

(c) The approximation in part (b) is used by ocean-ographers if $0 < D/L < \frac{1}{20}$. Show that the error $E = |v^2 - gD|$ is less than 0.002.

49 If a downward force of P pounds is applied to a cantilever column of length L at a point x units to the right of center (see figure), the column will buckle. The horizontal deflection δ can be expressed as

$$\delta = x\,\frac{1 - \cos kL}{\cos kL} \quad \text{with } k = \sqrt{P/(EI)}.$$

The constant EI is called the *flexural rigidity* of the material. Use a Maclaurin series to show that $\delta \approx PxL^2/(EI)$ if P is small.

EXERCISE 49

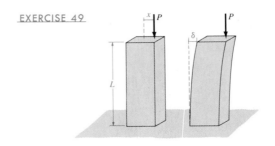

50 Show that $\cos x \approx 1 - \frac{1}{2}x^2 + \frac{1}{24}x^4 - \frac{1}{720}x^6$ is accurate to five decimal places if $0 \le x \le \pi/4$.

11.9 THE BINOMIAL SERIES

The Binomial Theorem (3.9) states that if k is a positive integer, then for all numbers a and b,

$$(a + b)^k = a^k + ka^{k-1}b + \frac{k(k - 1)}{2!}a^{k-2}b^2 + \cdots$$

$$+ \frac{k(k - 1) \cdots (k - n + 1)}{n!}a^{k-n}b^n + \cdots + b^k.$$

If we let $a = 1$ and $b = x$, then

$$(1 + x)^k = 1 + kx + \frac{k(k - 1)}{2!}x^2 + \cdots$$

$$+ \frac{k(k - 1) \cdots (k - n + 1)}{n!}x^n + \cdots + x^k.$$

If k is not a positive integer (or 0), it is useful to study the power series $\sum a_n x^n$, with $a_0 = 1$ and $a_n = k(k - 1) \cdots (k - n + 1)/n!$ for $n \ge 1$. This infinite series has the form

$$1 + kx + \frac{k(k - 1)}{2!}x^2 + \cdots + \frac{k(k - 1) \cdots (k - n + 1)}{n!}x^n + \cdots$$

and is called the **Binomial Series.** If k is a nonnegative integer, the series reduces to the finite sum given in the Binomial Theorem. Otherwise, the series does not terminate. It is left as an exercise to show that

$$\lim_{n \to \infty} \left| \frac{a_{n+1}x^{n+1}}{a_n x^n} \right| = \lim_{n \to \infty} \left| \frac{k - n}{n + 1} \right| |x| = |x|.$$

Hence, by the Ratio Test (11.33), the series is absolutely convergent if $|x| < 1$ and is divergent if $|x| > 1$. Thus the Binomial Series represents a function f such that

$$f(x) = \sum_{n=0}^{\infty} \frac{k(k - 1) \cdots (k - n + 1)}{n!}x^n \quad \text{if } |x| < 1.$$

We have already noted that if k is a nonnegative integer, then $f(x) = (1 + x)^k$. We shall now prove that the same is true for *every* real number k. Differentiating the Binomial Series gives us

$$f'(x) = k + k(k - 1)x + \cdots + \frac{nk(k - 1) \cdots (k - n + 1)}{n!} x^{n-1} + \cdots$$

and, therefore,

$$xf'(x) = kx + k(k - 1)x^2 + \cdots + \frac{nk(k - 1) \cdots (k - n + 1)}{n!} x^n + \cdots$$

If we add corresponding terms of the preceding two power series, then the coefficient of x^n is

$$\frac{(n + 1)k(k - 1) \cdots (k - n)}{(n + 1)!} + \frac{nk(k - 1) \cdots (k - n + 1)}{n!}$$

which simplifies to

$$[(k - n) + n] \frac{k(k - 1) \cdots (k - n + 1)}{n!} = ka_n.$$

Consequently,

$$f'(x) + xf'(x) = \sum_{n=0}^{\infty} ka_n x^n = kf(x)$$

and, therefore,

$$f'(x)(1 + x) - kf(x) = 0.$$

If we define the function g by $g(x) = f(x)/(1 + x)^k$, then

$$g'(x) = \frac{(1 + x)^k f'(x) - f(x)k(1 + x)^{k-1}}{(1 + x)^{2k}}$$

$$= \frac{(1 + x)f'(x) - kf(x)}{(1 + x)^{k+1}} = 0.$$

It follows from Theorem (4.33) that $g(x) = c$ for some constant c, that is,

$$\frac{f(x)}{(1 + x)^k} = c.$$

Since $f(0) = 1$, we see that $c = 1$ and hence $f(x) = (1 + x)^k$, which is what we wished to prove. The next statement summarizes this discussion.

THE BINOMIAL SERIES (11.44)

If $|x| < 1$, then for every real number k,

$$(1 + x)^k = 1 + kx + \frac{k(k - 1)}{2!} x^2 + \cdots$$

$$+ \frac{k(k - 1) \cdots (k - n + 1)}{n!} x^n + \cdots$$

EXAMPLE 1 Find a power series representation for $\sqrt[3]{1 + x}$.

SOLUTION Using (11.44) with $k = \frac{1}{3}$,

$$\sqrt[3]{1 + x} = 1 + \frac{1}{3}x + \frac{\frac{1}{3}(\frac{1}{3} - 1)}{2!}x^2 + \frac{\frac{1}{3}(\frac{1}{3} - 1)(\frac{1}{3} - 2)}{3!}x^3 + \cdots$$
$$+ \frac{\frac{1}{3}(\frac{1}{3} - 1) \cdots (\frac{1}{3} - n + 1)}{n!}x^n + \cdots$$

which may be written

$$\sqrt[3]{1 + x} = 1 + \frac{1}{3}x - \frac{2}{3^2 \cdot 2!}x^2 + \frac{1 \cdot 2 \cdot 5}{3^3 \cdot 3!}x^3 + \cdots$$
$$+ (-1)^{n+1} \frac{1 \cdot 2 \cdots (3n - 4)}{3^n \cdot n!}x^n + \cdots$$

for $|x| < 1$. The formula for the nth term of this series is valid provided $n \geq 2$. •

EXAMPLE 2 Find a power series representation for $\sqrt[3]{1 + x^4}$.

SOLUTION The power series can be obtained by substituting x^4 for x in the series of Example 1. Hence, if $|x| < 1$, then

$$\sqrt[3]{1 + x^4} = 1 + \frac{1}{3}x^4 - \frac{2}{3^2 \cdot 2!}x^8 + \cdots$$
$$+ (-1)^{n+1} \frac{1 \cdot 2 \cdots (3n - 4)}{3^n \cdot n!}x^{4n} + \cdots \quad •$$

EXAMPLE 3 Approximate $\int_0^{0.3} \sqrt[3]{1 + x^4}\, dx$.

SOLUTION Integrating the terms of the series obtained in Example 2 gives us

$$\int_0^{0.3} \sqrt[3]{1 + x^4}\, dx = 0.3 + 0.000162 - 0.000000243 + \cdots$$

Consequently, the integral may be approximated by 0.300162, which is accurate to six decimal places, since the error is less than 0.000000243. (Why?) •

The Binomial Series can be used to obtain polynomial approximation formulas for $(1 + x)^k$. To illustrate, if $|x| < 1$, then from Example 1

$$\sqrt[3]{1 + x} \approx 1 + \frac{1}{3}x,$$

where the error involved in this approximation is less than the next term, $\frac{1}{9}x^2$. •

EXERCISES 11.9

Exer. 1–12: Find a power series representation for the expression, and state the radius of convergence.

1 (a) $\sqrt{1 + x}$ (b) $\sqrt{1 - x^3}$

2 (a) $\dfrac{1}{\sqrt[3]{1 + x}}$ (b) $\dfrac{1}{\sqrt[3]{1 - x^2}}$

3 $(1 + x)^{-2/3}$ 4 $(1 + x)^{1/4}$

5 $(1 - x)^{3/5}$

6 $(1 - x)^{2/3}$

7 $(1 + x)^{-2}$

8 $(1 + x)^{-4}$

9 $(1 + x)^{-3}$

10 $x(1 + 2x)^{-2}$

11 $\sqrt[3]{8 + x}$

12 $(4 + x)^{3/2}$

13 Obtain a power series representation for $\sin^{-1} x$ by using $\sin^{-1} x = \int_0^x (1/\sqrt{1 - t^2})\, dt$. What is the radius of convergence?

14 Obtain a power series representation for $\sinh^{-1} x$ by using $\sinh^{-1} x = \int_0^x (1/\sqrt{1 + t^2})\, dt$. What is the radius of convergence?

Exer. 15–19: Approximate the integral to three decimal places.

15 $\int_0^{1/2} \sqrt{1 + x^3}\, dx$ (see Exercise 1)

16 $\int_0^{1/2} \dfrac{1}{\sqrt[3]{1 + x^2}}\, dx$ (see Exercise 2)

17 $\int_0^{0.2} (1 - x^2)^{3/5}\, dx$ (see Exercise 5)

18 $\int_0^{0.4} (1 - x^3)^{2/3}\, dx$ (see Exercise 6)

19 $\int_0^{0.3} \dfrac{1}{(1 + x^3)^2}\, dx$ (see Exercise 7)

20 Refer to Exercise 42 of Section 10.4. For a pendulum of length L, initially displaced from equilibrium to an angle of θ_0 radians, the formula for the period T is given by the improper integral

$$T = 2\sqrt{2L/g} \int_0^{\theta_0} (\cos\theta - \cos\theta_0)^{-1/2}\, d\theta.$$

The substitution $u = (\sin\frac{1}{2}\theta)/(\sin\frac{1}{2}\theta_0)$ allows us to rewrite the integral as

$$T = 2\sqrt{2L/g} \int_0^{\pi/2} (1 - k^2 \sin^2 u)^{-1/2}\, du$$

with $k = \sin\frac{1}{2}\theta_0$.

(a) Use the Binomial Series for $(1 - x)^{-1/2}$ to show that

$$T \approx 2\pi\sqrt{L/g}(1 + \tfrac{1}{4}k^2).$$

(b) Approximate T if $\theta_0 = \pi/6$.

11.10 REVIEW

Define or discuss each of the following.

1 Infinite sequence

2 Limit of an infinite sequence

3 Convergent infinite sequence

4 Divergent infinite sequence

5 Theorems on limits of sequences

6 Monotonic sequence

7 Sequence of partial sums of an infinite series

8 Convergent infinite series

9 Divergent infinite series

10 The harmonic series

11 Geometric series

12 The Integral Test

13 The p-series

14 Comparison tests

15 The Ratio Test

16 The Root Test

17 The Alternating Series Test

18 Absolute convergence

19 Conditional convergence

20 Power series

21 Radius of convergence

22 Interval of convergence

23 Power series representations of functions

24 Differentiation and integration of power series

25 Taylor series

26 Maclaurin series

27 Approximations by series

28 The Binomial Series

EXERCISES 11.10

Exer. 1–6: Determine if the sequence converges or diverges. If it converges, find the limit.

1 $\left\{ \dfrac{\ln(n^2 + 1)}{n} \right\}$

2 $\{100(0.99)^n\}$

3 $\{10^n/n^{10}\}$

4 $\left\{ \dfrac{1}{n} + (-2)^n \right\}$

5 $\left\{ \dfrac{n}{\sqrt{n + 4}} - \dfrac{n}{\sqrt{n + 9}} \right\}$

6 $\left\{ \left(1 + \dfrac{2}{n}\right)^{2n} \right\}$

Exer. 7–32: Determine if (a) a positive-term series is convergent or divergent, or (b) a series that contains negative terms is absolutely convergent, conditionally convergent, or divergent.

7 $\displaystyle\sum_{n=1}^{\infty} \dfrac{1}{\sqrt[3]{n(n + 1)(n + 2)}}$

8 $\displaystyle\sum_{n=0}^{\infty} \dfrac{(2n + 3)^2}{(n + 1)^3}$

9 $\displaystyle\sum_{n=1}^{\infty} (-2/3)^{n-1}$

10 $\displaystyle\sum_{n=0}^{\infty} \dfrac{1}{2 + (1/2)^n}$

$$11 \quad \sum_{n=1}^{\infty} \frac{3^{2n+1}}{n5^{n-1}}$$

$$12 \quad \sum_{n=1}^{\infty} \frac{1}{3^n + 2}$$

$$13 \quad \sum_{n=1}^{\infty} \frac{n!}{\ln(n+1)}$$

$$14 \quad \sum_{n=1}^{\infty} \frac{n^2 - 1}{n^2 + 1}$$

$$15 \quad \sum_{n=1}^{\infty} (n^2 + 9)(-2)^{1-n}$$

$$16 \quad \sum_{n=1}^{\infty} \frac{n + \cos n}{n^3 + 1}$$

$$17 \quad \sum_{n=1}^{\infty} \frac{e^n}{n^e}$$

$$18 \quad \sum_{n=1}^{\infty} (-1)^{n-1} \frac{n}{n^2 + 1}$$

$$19 \quad \sum_{n=1}^{\infty} (-1)^n \frac{1}{\sqrt[n]{n}}$$

$$20 \quad \sum_{n=2}^{\infty} (-1)^n \frac{(0.9)^n}{\ln n}$$

$$21 \quad \sum_{n=1}^{\infty} \frac{\sin(n5\pi/3)}{n^{5\pi/3}}$$

$$22 \quad \sum_{n=2}^{\infty} (-1)^n \frac{\sqrt[3]{n} - 1}{n^2 - 1}$$

$$23 \quad \sum_{n=1}^{\infty} (-1)^{n-1} \frac{\sqrt{n}}{n+1}$$

$$24 \quad \sum_{n=1}^{\infty} (-1)^n \frac{2n+3}{n!}$$

$$25 \quad \sum_{n=1}^{\infty} \frac{1 - \cos n}{n^2}$$

$$26 \quad \frac{2}{1!} - \frac{2 \cdot 4}{2!} + \cdots + (-1)^{n-1} \frac{2 \cdot 4 \cdots (2n)}{n!} + \cdots$$

$$27 \quad \sum_{n=1}^{\infty} \frac{(2n)^n}{n^{2n}}$$

$$28 \quad \sum_{n=1}^{\infty} \frac{3^{n-1}}{n^2 + 9}$$

$$29 \quad \sum_{n=1}^{\infty} \frac{e^{2n}}{(2n-1)!}$$

$$30 \quad \sum_{n=1}^{\infty} \left(\frac{1}{3^n} - \frac{5}{\sqrt{n}} \right)$$

$$31 \quad \sum_{n=2}^{\infty} (-1)^n \frac{\sqrt{\ln n}}{n}$$

$$32 \quad \sum_{n=1}^{\infty} \frac{\tan^{-1} n}{\sqrt{1 + n^2}}$$

Exer. 33–38: Use the Integral Test (11.22) to determine the convergence or divergence of the series.

$$33 \quad \sum_{n=1}^{\infty} \frac{1}{(3n+2)^3}$$

$$34 \quad \sum_{n=2}^{\infty} \frac{n}{\sqrt{n^2 - 1}}$$

$$35 \quad \sum_{n=1}^{\infty} n^{-2} e^{1/n}$$

$$36 \quad \sum_{n=2}^{\infty} \frac{1}{n(\ln n)^3}$$

$$37 \quad \sum_{n=1}^{\infty} \frac{10}{\sqrt[3]{n+8}}$$

$$38 \quad \sum_{n=5}^{\infty} \frac{1}{n^2 - 4n}$$

Exer. 39–40: Approximate the sum of the series to three decimal places.

$$39 \quad \sum_{n=1}^{\infty} (-1)^{n-1} \frac{1}{(2n+1)!}$$

$$40 \quad \sum_{n=1}^{\infty} (-1)^{n-1} \frac{1}{n^2(n^2 + 1)}$$

Exer. 41–44: Find the interval of convergence of the series.

$$41 \quad \sum_{n=0}^{\infty} \frac{n+1}{(-3)^n} x^n$$

$$42 \quad \sum_{n=0}^{\infty} (-1)^n \frac{4^{2n}}{\sqrt{n+1}} x^n$$

$$43 \quad \sum_{n=1}^{\infty} \frac{1}{n \cdot 2^n} (x + 10)^n$$

$$44 \quad \sum_{n=2}^{\infty} \frac{1}{n(\ln n)^2} (x - 1)^n$$

Exer. 45–46: Find the radius of convergence of the series.

$$45 \quad \sum_{n=0}^{\infty} \frac{(2n)!}{(n!)^2} x^n$$

$$46 \quad \sum_{n=0}^{\infty} \frac{1}{(n+5)!} (x + 5)^n$$

Exer. 47–52: Find the Maclaurin series for the expression, and state the radius of convergence.

$$47 \quad \begin{cases} \dfrac{1 - \cos x}{x} & \text{if } x \neq 0 \\ 0 & \text{if } x = 0 \end{cases}$$

$$48 \quad xe^{-2x}$$

$$49 \quad \sin x \cos x$$

$$50 \quad \ln(2 + x)$$

$$51 \quad (1 + x)^{2/3}$$

$$52 \quad 1/\sqrt{1 - x^2}$$

53 Find a series representation for e^{-x} in powers of $x + 2$.

54 Find a series representation for $\cos x$ in powers of $x - (\pi/2)$.

55 Find a series representation for $\sqrt{x}$ in powers of $x - 4$.

Exer. 56–60: Use an infinite series to approximate the number to three decimal places.

$$56 \quad 1/\sqrt[3]{3e}$$

$$57 \quad \int_0^1 x^2 e^{-x^2} \, dx$$

58 $\int_0^1 f(x) \, dx$ with $f(x) = (\sin x)/\sqrt{x}$ if $x \neq 0$ and $f(0) = 0$.

$$59 \quad \sqrt[5]{1.01}$$

$$60 \quad e^{-0.25}$$

Plane geometry includes the study of figures such as lines, circles, and triangles that lie in a plane. Theorems are proved by reasoning deductively from certain postulates. In *analytic* geometry, plane geometric figures are investigated by introducing a coordinate system and then using equations and formulas of various types. If the study of analytic geometry were to be summarized by means of one statement, perhaps the following would be appropriate: "Given an equation, find its graph and, conversely, given a graph, find its equation." In this chapter we shall apply coordinate methods to several basic plane figures.

12.1 CONIC SECTIONS

FIGURE 12.1

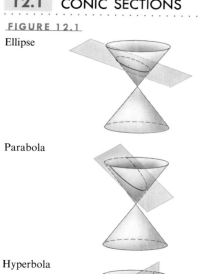

Ellipse

Parabola

Hyperbola

Each of the geometric figures discussed in this chapter can be obtained by intersecting a double-napped right circular cone with a plane, as illustrated in Figure 12.1. For this reason they are called **conic sections,** or simply **conics.** By varying the position of the plane, as shown in the figure, we obtain an **ellipse,** a **parabola,** or a **hyperbola,** respectively.

Certain positions of the plane result in **degenerate conics.** For example, if the plane intersects the cone only at the vertex, then the conic consists of one point. If the axis of the cone lies on the plane, then a pair of intersecting lines is obtained. Finally, we can begin with the case of the parabola in Figure 12.1, and move the plane parallel to its initial position until it has only one line in common with the cone.

Conic sections were studied extensively by the ancient Greeks, who discovered properties that enable us to define conics in terms of points and lines. A remarkable fact about conic sections is that although they were studied thousands of years ago, they are far from obsolete. They are important tools for present-day investigations in outer space and for the study of the behavior of atomic particles. From physics we know that if a particle moves under the influence of an *inverse square force field*, then its path may be described by means of a conic section. Examples of inverse square fields are gravitational and electromagnetic fields. Planetary orbits are elliptical. If the ellipse is very "flat," the curve resembles the path of a comet. Parabolic mirrors are sometimes used to collect solar energy. The hyperbola is useful for describing the path of an alpha particle in the electric field of the nucleus of an atom.

12.2 PARABOLAS

In this section we define a *parabola* and derive equations for parabolas that have either vertical or horizontal axes.

DEFINITION (12.1)

> A **parabola** is the set of all points in a plane equidistant from a fixed point F (the **focus**) and a fixed line l (the **directrix**) in the plane.

FIGURE 12.2

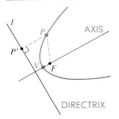

FIGURE 12.3

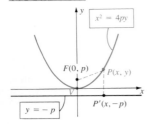

FIGURE 12.4

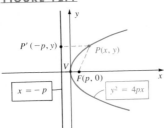

We shall assume that F is not on l, for this would result in a line rather than a parabola. If P is a point in the plane and P' is the point on l determined by a line through P that is perpendicular to l (see Figure 12.2), then by Definition (12.1), P is on the parabola if and only if $d(P, F) = d(P, P')$. The point P can lie anywhere on the curve in Figure 12.2. The line through F, perpendicular to the directrix, is the **axis** of the parabola. The point V on the axis, half-way from F to l, is the **vertex** of the parabola.

We can obtain a simple equation for a parabola by placing the y-axis along the axis of the parabola, with the origin at the vertex V, as illustrated in Figure 12.3. In this case, the focus F has coordinates $(0, p)$ for some real number $p \neq 0$, and the equation of the directrix is $y = -p$. By the Distance Formula, a point $P(x, y)$ is on the parabola if and only if

$$\sqrt{(x - 0)^2 + (y - p)^2} = \sqrt{(x - x)^2 - (y + p)^2}.$$

Squaring both sides gives us

$$(x - 0)^2 + (y - p)^2 = (y + p)^2,$$

or

$$x^2 + y^2 - 2py + p^2 = y^2 + 2py + p^2.$$

This simplifies to

$$x^2 = 4py.$$

We have shown that the coordinates of every point (x, y) on the parabola satisfy $x^2 = 4py$. Conversely, if (x, y) is a solution of this equation, then by reversing the previous steps we see that the point (x, y) is on the parabola. If $p > 0$ the parabola opens upward, as in Figure 12.3; however, if $p < 0$ the parabola opens downward. Note that the graph is symmetric with respect to the y-axis, since the solutions of the equation $x^2 = 4py$ are unchanged if $-x$ is substituted for x.

An analogous situation exists if the axis of the parabola is placed along the x-axis. If the vertex is $V(0, 0)$, the focus is $F(p, 0)$, and the directrix has equation $x = -p$ (see Figure 12.4), then using the same type of argument we obtain the equation $y^2 = 4px$. If $p > 0$ the parabola opens to the right, and if $p < 0$ it opens to the left. In this case the graph is symmetric with respect to the x-axis.

The following theorem summarizes our discussion.

THEOREM (12.2)

The graph of each of the following equations is a parabola that has its vertex at the origin and has the indicated focus and directrix.

(i) $x^2 = 4py$: focus $F(0, p)$, directrix $y = -p$

(ii) $y^2 = 4px$: focus $F(p, 0)$, directrix $x = -p$

FIGURE 12.5

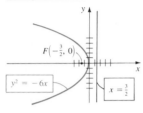

EXAMPLE 1 Find the focus and directrix of the parabola that has the equation $y^2 = -6x$, and sketch its graph.

SOLUTION The equation $y^2 = -6x$ has form (ii) of Theorem (12.2) with $4p = -6$, and hence $p = -\frac{3}{2}$. Thus the focus is $F(p, 0)$, that is, $F(-\frac{3}{2}, 0)$. The equation of the directrix is $x = -p$, or $x = \frac{3}{2}$. The graph is sketched in Figure 12.5. •

EXAMPLE 2 Find an equation of the parabola that has its vertex at the origin, opens upward, and passes through the point $P(-3, 7)$.

SOLUTION The general form of the equation is $x^2 = 4py$ (see Theorem (12.2) (i)). If $P(-3, 7)$ is on the parabola, then $(-3, 7)$ is a solution of the equation. Hence we must have $(-3)^2 = 4p(7)$, or $p = \frac{9}{28}$. Substituting for p in $x^2 = 4py$ leads to the equation $x^2 = \frac{9}{7}y$, or $7x^2 = 9y$. •

To extend the preceding discussion to the case in which the vertex of the parabola is not at the origin, we may use a **translation of axes,** as illustrated in Figure 12.6, where the x-axis and y-axis are shifted to positions, denoted by x' and y', that are parallel to the original axes. Every point P in the plane then has two different ordered pair representations: $P(x, y)$ in the xy-system and $P(x', y')$ in the $x'y'$-system. If the origin of the new $x'y'$-system has coordinates (h, k) in the xy-system, as illustrated in Figure 12.6, we see that

FIGURE 12.6

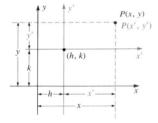

$$x = x' + h \quad \text{and} \quad y = y' + k.$$

We can show that these formulas are true for all values of h and k. Equivalent formulas are

$$x' = x - h \quad \text{and} \quad y' = y - k.$$

This gives us the following.

TRANSLATION OF AXES FORMULAS (12.3)

If (x, y) are the coordinates of a point P in an xy-coordinate system, and if (x', y') are the coordinates of P in an $x'y'$-coordinate system with origin at the point (h, k) of the xy-system, then

(i) $x = x' + h, \quad y = y' + k$

(ii) $x' = x - h, \quad y' = y - k$

The Translation of Axes Formulas enable us to go from either coordinate system to the other. Their major use is to change the form of equations of graphs. To be specific, if, in the xy-plane, a certain collection of points is the graph of an equation in x and y, then to find an equation in x' and y' that has the same graph in the $x'y'$-plane, we may substitute $x' + h$ for x and $y' + k$ for y in the given equation. Conversely, if a set of points in the $x'y'$-plane is the graph of an equation in x' and y', then to find the corresponding equation in x and y, we substitute $x - h$ for x' and $y - k$ for y'.

As a simple illustration, the equation

$$(x')^2 + (y')^2 = r^2$$

has, for its graph in the $x'y'$-plane, a circle of radius r with center at the origin. Using the Translation of Axes Formulas, an equation for this circle in the xy-plane is

$$(x - h)^2 + (y - k)^2 = r^2,$$

which is in agreement with the formula for a circle of radius r with center at $C(h, k)$ in the xy-plane (see (1.12)).

As another illustration, we know that

$$(x')^2 = 4py'$$

is an equation of a parabola with vertex at the origin of the $x'y'$-plane. Using the Translation of Axes Formulas, we see that

$$(x - h)^2 = 4p(y - k)$$

is an equation of the same parabola in the xy-plane with vertex $V(h, k)$. The focus is $F(h, k + p)$ and the directrix is $y = k - p$. Similarly, starting with $(y')^2 = 4px'$ gives us $(y - k)^2 = 4p(x - h)$. The next theorem summarizes this discussion.

THEOREM (12.4)

> The graph of each of the following equations is a parabola that has vertex $V(h, k)$ and has the indicated focus and directrix.
>
> (i) $(x - h)^2 = 4p(y - k)$: focus $F(h, k + p)$, directrix $y = k - p$
> (ii) $(y - k)^2 = 4p(x - h)$: focus $F(h + p, k)$, directrix $x = h - p$

In each case the axis of the parabola is parallel to a coordinate axis. The parabola having equation (12.4) (i) opens upward or downward, and the parabola in (ii) opens to the right or left. Typical graphs are sketched in Figure 12.7.

FIGURE 12.7

(i) $(x - h)^2 = 4p(y - k)$

(ii) $(y - k)^2 = 4p(x - h)$

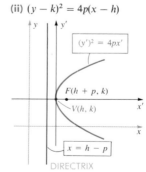

Squaring the left side of the equation in Theorem (12.4) (i) and simplifying, we obtain an equation of the form

$$y = ax^2 + bx + c$$

for real numbers a, b, and c. Conversely, given such an equation, we may complete the square in x to arrive at the form in (12.4). This technique will be illustrated in Example 3. Consequently, if $a \neq 0$, then *the graph of $y = ax^2 + bx + c$ is a parabola with a vertical axis.*

EXAMPLE 3 Discuss and sketch the graph of $y = 2x^2 - 6x + 4$.

SOLUTION The graph is a parabola with vertical axis. To obtain the proper form we begin by writing the equation as

$$y - 4 = 2x^2 - 6x = 2(x^2 - 3x).$$

Next we complete the square for the expression $x^2 - 3x$. Recall that to complete the square for *any* expression of the form $x^2 + qx$, we add the square of half the coefficient of x, that is, $(q/2)^2$. Thus, for $x^2 - 3x$ we must add $(-\frac{3}{2})^2$, or $\frac{9}{4}$. However, if we add $\frac{9}{4}$ to $x^2 - 3x$ in the equation $y - 4 = 2(x^2 - 3x)$, then because of the factor 2 outside the parentheses, this amounts to adding $\frac{9}{2}$ to the right side of the equation. Hence we must compensate by adding $\frac{9}{2}$ (rather than $\frac{9}{4}$) to the left side. This gives us

$$y - 4 + \tfrac{9}{2} = 2(x^2 - 3x + \tfrac{9}{4})$$

$$y + \tfrac{1}{2} = 2(x - \tfrac{3}{2})^2$$

or, equivalently,

$$(x - \tfrac{3}{2})^2 = \tfrac{1}{2}(y + \tfrac{1}{2}).$$

The last equation is in form (12.4) (i) with $h = \frac{3}{2}$, $k = -\frac{1}{2}$, and $4p = \frac{1}{2}$, or $p = \frac{1}{8}$. Hence the vertex of the parabola is $V(h, k)$, or $V(\frac{3}{2}, -\frac{1}{2})$. The focus is $F(h, k + p)$, or $F(\frac{3}{2}, -\frac{3}{8})$, and the directrix is

$$y = k - p = -\tfrac{1}{2} - \tfrac{1}{8} \quad \text{or} \quad y = -\tfrac{5}{8}.$$

As an aid to sketching the graph, we note that the y-intercept is 4. To find the x-intercepts we solve $2x^2 - 6x + 4 = 0$ or the equivalent equation $(2x - 2)(x - 2) = 0$, obtaining $x = 1$ and $x = 2$. The graph is sketched in Figure 12.8. •

The equation in (12.4) (ii) may be written as

$$x = ay^2 + by + c$$

for real numbers a, b, and c. Conversely, the preceding equation can be expressed in form (12.4) (ii) by completing the square in y as illustrated in the next example. Hence, if $a \neq 0$, the graph of $x = ay^2 + by + c$ is a *parabola with a horizontal axis.*

EXAMPLE 4 Discuss and sketch the graph of the equation

$$2x = y^2 + 8y + 22$$

SOLUTION The graph is a parabola with a horizontal axis. Writing

$$y^2 + 8y = 2x - 22$$

FIGURE 12.8

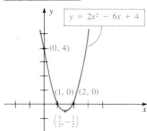

FIGURE 12.9

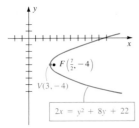

we can complete the square on the left by adding 16 to both sides. This gives us

$$y^2 + 8y + 16 = 2x - 6.$$

This equation may be written

$$(y + 4)^2 = 2(x - 3),$$

which is in form (12.4) (ii) with $h = 3$, $k = -4$, and $4p = 2$, or $p = \frac{1}{2}$. Hence the vertex is $V(3, -4)$. Since $p = \frac{1}{2} > 0$, the parabola opens to the right with focus at $F(h + p, k)$, that is, $F(\frac{7}{2}, -4)$. The equation of the directrix is $x = h - p$, or $x = \frac{5}{2}$. The parabola is sketched in Figure 12.9. •

EXAMPLE 5 Find an equation of the parabola with vertex $V(-4, 2)$ and directrix $y = 5$.

FIGURE 12.10

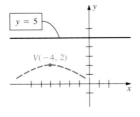

SOLUTION The vertex and directrix are shown in Figure 12.10. The dashes indicate a possible shape for the parabola. It follows that an equation of the parabola is

$$(x - h)^2 = 4p(y - k)$$

with $h = -4$, $k = 2$, and $p = -3$. This gives us

$$(x + 4)^2 = -12(y - 2).$$

This equation can be expressed in the form $y = ax^2 + bx + c$ as follows:

$$x^2 + 8x + 16 = -12y + 24$$
$$12y = -x^2 - 8x + 8$$
$$y = -\tfrac{1}{12}x^2 - \tfrac{2}{3}x + \tfrac{2}{3} \quad •$$

FIGURE 12.11
Path of a baseball

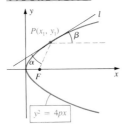

If, as shown in Figure 12.11, a baseball player throws a ball in a non-vertical direction, and if gravity is the only force acting on the ball (that is, air resistance and other outside factors are negligible), then the path of the ball (or any projected object) is parabolic. We can use this fact (which will be proved in Chapter 15) to determine where the ball (or object) will land, to find its maximum height, and so on (see Exercise 39).

The shapes of cables in certain types of suspension bridges are parabolic. However, as we pointed out in Section 8.4, for a *freely* hanging cable, the curve is a catenary, not a parabola.

An important property is associated with tangent lines to parabolas. Suppose l is the tangent line at a point $P(x_1, y_1)$ on the graph of $y^2 = 4px$, and let F be the focus. As in Figure 12.12, let α denote the angle between l and the line segment FP, and let β denote the angle between l and the horizontal half-line with endpoint P. In Exercise 34, you are asked to prove that $\alpha = \beta$. This property has many applications. For example, the shape of the mirror in a searchlight is obtained by revolving a parabola about its axis. If a light source is placed at F, then by a law of physics (*the angle of reflection equals the angle of incidence*), a beam of light will be reflected along a line parallel to the axis (see Figure 12.13(i)). The same principle is employed in the construction of mirrors for telescopes or solar ovens, where a beam of light coming toward the parabolic mirror, and parallel to the axis, will be reflected into the focus (see Figure 12.13(ii)). Antennas for radar systems, radio telescopes, and field microphones used at football games also make use of this property.

FIGURE 12.12

FIGURE 12.13

(i) Searchlight mirror

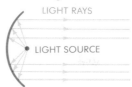

(ii) Telescope mirror

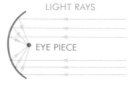

EXERCISES 12.2

Exer. 1–16: Find the vertex, focus, and directrix of the parabola, and sketch its graph.

1 $x^2 = -12y$

2 $y^2 = \frac{1}{2}x$

3 $2y^2 = -3x$

4 $x^2 = -3y$

5 $8x^2 = y$

6 $y^2 = -100x$

7 $y^2 - 12 = 12x$

8 $y = 40x - 97 - 4x^2$

9 $y = x^2 - 4x + 2$

10 $y = 8x^2 + 16x + 10$

11 $y^2 - 4y - 2x - 4 = 0$

12 $y^2 + 14y + 4x + 45 = 0$

13 $4x^2 + 40x + y + 106 = 0$

14 $y^2 - 20y + 100 = 6x$

15 $x^2 + 20y = 10$

16 $4x^2 + 4x + 4y + 1 = 0$

17 Describe a method for using a derivative to locate the vertex of the parabola $y = ax^2 + bx + c$. Use this technique to find the vertices in Exercises 8–10. Describe a similar method for $x = ay^2 + by + c$.

18 Describe how a second derivative may be used to determine whether the parabola $y = ax^2 + bx + c$ opens upward or downward. Illustrate this technique with Exercises 8–10.

Exer. 19–24: Find an equation for the parabola that satisfies the given conditions.

19 Focus $(2, 0)$, directrix $x = -2$

20 Focus $(0, -4)$, directrix $y = 4$

21 Focus $(6, 4)$, directrix $y = -2$

22 Focus $(-3, -2)$, directrix $y = 1$

23 Vertex at the origin, symmetric to the y-axis, and passing through the point $A(2, -3)$

24 Vertex $V(-3, 5)$, axis parallel to the x-axis, and passing through the point $A(5, 9)$

25 A searchlight reflector is designed so that a cross section through its axis is a parabola and the light source is at the focus. Find the focus if the reflector is 3 feet across at the opening and 1 foot deep.

26 One section of a suspension bridge has its weight uniformly distributed between twin towers that are 400 feet apart and that rise 90 feet above the horizontal roadway (see figure). A cable strung between the tops of the towers has the shape of a parabola and its center point is 10 feet above the roadway. Suppose coordinate axes are introduced as shown in the figure.

(a) Find an equation for the parabola.

(b) Set up an integral that gives the length of the cable.

(c) If nine equispaced vertical cables are used to support one parabolic cable (see figure), find the total length of these supports.

EXERCISE 26

27 Find an equation of the parabola with a vertical axis that passes through the points $A(2, 3)$, $B(-1, 6)$, and $C(1, 0)$.

28 Prove that the point on a parabola that is closest to the focus is the vertex.

29 Let R be the region bounded by the parabola $x^2 = 4y$ and the line l through the focus that is perpendicular to the axis of the parabola.

(a) Find the area of R.

(b) If R is revolved about the y-axis, find the volume of the resulting solid.

(c) If R is revolved about the x-axis, find the volume of the resulting solid.

30 Answer (a)–(c) of Exercise 29 if R is the region bounded by the graphs of $y^2 = 2x - 6$ and $x = 5$.

31 A *paraboloid of revolution* is formed by revolving a parabola about its axis. Shown in the figure on page 520 is a (finite) paraboloid of height h and base radius r.

(a) Show that the volume is $\frac{1}{2}\pi r^2 h$.

(b) The *focal length* of the paraboloid is the distance p between the vertex and focus of the parabola. Express p in terms of r and h.

EXERCISE 31

32 A parabola with equation $y^2 = 4p(x + p)$ has its focus at the origin and the line $y = 0$ as its axis. The family of all such *confocal parabolas* is illustrated in the figure. (Families of this type occur in the theory of electricity and magnetism.)

(a) Show that the family includes exactly two parabolas through any point $P(x_1, y_1)$ for $y_1 \neq 0$.

(b) Show that the two parabolas in part (a) are mutually orthogonal; that is, the tangent lines at $P(x_1, y_1)$ are perpendicular.

EXERCISE 32

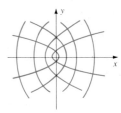

33 A vertical gate in a dam has the shape of the parabolic region of Exercise 29. If the line l lies along the surface of the water, find the force exerted on the gate by the water.

34 Prove that $\alpha = \beta$ in Figure 12.12.

35 Refer to Exercise 31. A radio telescope has the shape of a paraboloid of revolution with diameter $2a$ and focal length p.

(a) Show that the surface area S of the radio telescope is

$$S = \frac{8\pi p^2}{3}\left[\left(1 + \frac{a^2}{4p^2}\right)^{3/2} - 1\right]$$

(b) One of the largest moveable radio telescopes, located in Jodrell Bank, Cheshire, England, has a diameter of 250 feet and focal length of 50 feet. Find the total surface area S available for collecting radio waves.

36 A parabolic arch has a center height of k feet. Prove that the height of the largest rectangle that can fit under the arch (see figure) is $\frac{2}{3}k$ feet.

EXERCISE 36

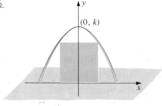

37 A cylindrical container is partially filled with a liquid such as mercury. The container is then rotated about its axis at the rate of ω radians/second. Using physics we can show that the function f whose graph generates the inside surface of the liquid is a solution of the differential equation $y' = (\omega^2/g)x$ with $g = 32$ ft/sec^2.

(a) Show that $y = \frac{1}{64}\omega^2 x^2 + f(0)$.

(b) Determine the angular velocity ω that will result in a focal length of 2 feet.

EXERCISE 37

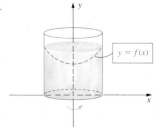

38 Suppose the tangent line l to a parabola at a point P intersects the directrix at a point Q (see figure). If F is the focus, prove that angle PFQ is a right angle.

EXERCISE 38

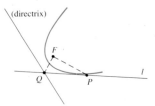

39 A boy tosses a baseball from the edge of a plateau down a hill (see figure). The ball, thrown at an angle of 45°, lands 50 feet down the hill, which is defined by the line $4y + 3x = 0$.

(a) Ignoring the height of the boy, find an equation for the parabolic path of the ball.

(b) What is the maximum height of the ball *off the ground*?

EXERCISE 39

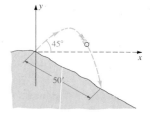

40 A *focal chord* is a line segment that passes through the focus of a parabola and has its endpoints on the parabola. Let AB be a focal chord. Prove that the tangent lines at A and B (a) are perpendicular; (b) intersect on the directrix.

An ellipse may be defined as follows. (The word *foci* is the plural of *focus*.)

DEFINITION (12.5)

> An **ellipse** is the set of all points in a plane, the sum of whose distances from two fixed points (the **foci**) in the plane is constant.

FIGURE 12.14

An ellipse can easily be constructed on paper. We begin by inserting two pushpins in the paper at points labeled F and F' and fastening the ends of a piece of string to the pins. If the string is now looped around a pencil and drawn taut at point P, as in Figure 12.14, then moving the pencil and at the same time keeping the string taut, the sum of the distances $d(F, P)$ and $d(F', P)$ is the length of the string and hence is constant. The pencil will trace out an ellipse with foci at F and F'. By varying the positions of F and F', but keeping the length of string fixed, we can change the shape of the ellipse considerably. If F and F' are far apart, so that $d(F, F')$ is almost the same as the length of the string, then the ellipse is quite flat. If $d(F, F')$ is close to zero, the ellipse is almost circular. If $F = F'$, we obtain a circle with center F.

FIGURE 12.15

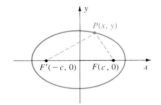

By introducing suitable coordinate systems, we may derive simple equations for ellipses. Let us choose the x-axis as the line through the two foci F and F', with the origin at the midpoint of the segment $F'F$. This point is called the **center** of the ellipse. If F has coordinates $(c, 0)$, with $c > 0$, then, as shown in Figure 12.15, F' has coordinates $(-c, 0)$, and hence the distance between F and F' is $2c$. The constant sum of the distances of P from F and F' will be denoted by $2a$. In order to get points that are not on the x-axis we must have $2a > 2c$, that is, $a > c$. By definition, $P(x, y)$ is on the ellipse if and only if

$$d(P, F) + d(P, F') = 2a$$

or, by the Distance Formula,

$$\sqrt{(x - c)^2 + (y - 0)^2} + \sqrt{(x + c)^2 + (y - 0)^2} = 2a.$$

Writing the preceding equation as

$$\sqrt{(x - c)^2 + y^2} = 2a - \sqrt{(x + c)^2 + y^2}$$

and squaring both sides, we obtain

$$x^2 - 2cx + c^2 + y^2 = 4a^2 - 4a\sqrt{(x + c)^2 + y^2} + x^2 + 2cx + c^2 + y^2,$$

which simplifies to

$$a\sqrt{(x + c)^2 + y^2} = a^2 + cx.$$

Squaring both sides gives us

$$a^2(x^2 + 2cx + c^2 + y^2) = a^4 + 2a^2cx + c^2x^2,$$

which may be written in the form

$$x^2(a^2 - c^2) + a^2y^2 = a^2(a^2 - c^2).$$

Dividing both sides by $a^2(a^2 - c^2)$ leads to

$$\frac{x^2}{a^2} + \frac{y^2}{a^2 - c^2} = 1.$$

Recalling that $a > c$ and therefore $a^2 - c^2 > 0$, we let

$$b = \sqrt{a^2 - c^2} \quad \text{or} \quad b^2 = a^2 - c^2.$$

This gives us the equation

$$\frac{x^2}{a^2} + \frac{y^2}{b^2} = 1.$$

Since $c > 0$ and $b^2 = a^2 - c^2$, it follows that $a^2 > b^2$ and hence $a > b$.

We have shown that the coordinates of every point (x, y) on the ellipse in Figure 12.16 satisfy the equation $(x^2/a^2) + (y^2/b^2) = 1$. Conversely, if (x, y) is a solution of this equation, then by reversing the preceding steps we see that the point (x, y) is on the ellipse.

The x-intercepts may be found by letting $y = 0$. Doing so gives us $x^2/a^2 = 1$, or $x^2 = a^2$, and consequently the x-intercepts are a and $-a$. The corresponding points $V(a, 0)$ and $V'(-a, 0)$ on the graph are the **vertices** of the ellipse (see Figure 12.16). The line segment $V'V$ is the **major axis.** Similarly, letting $x = 0$ in the equation of the ellipse, we obtain $y^2/b^2 = 1$, or $y^2 = b^2$. Hence the y-intercepts are b and $-b$. The segment between $M'(0, -b)$ and $M(0, b)$ is the **minor axis** of the ellipse. Note that the major axis is longer than the minor axis, since $a > b$.

Applying the Tests for Symmetry (1.11), we see that the ellipse is symmetric with respect to the x-axis, the y-axis, and the origin.

The preceding discussion may be summarized as follows.

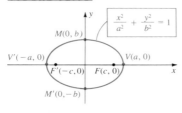

FIGURE 12.16

THEOREM (12.6)

The graph of the equation

$$\frac{x^2}{a^2} + \frac{y^2}{b^2} = 1,$$

for $a^2 > b^2$, is an ellipse with vertices $(\pm a, 0)$. The endpoints of the minor axis are $(0, \pm b)$. The foci are $(\pm c, 0)$, with $c^2 = a^2 - b^2$.

EXAMPLE 1 Discuss and sketch the graph of the equation

$$4x^2 + 18y^2 = 36.$$

SOLUTION To obtain the form in Theorem (12.6) we divide both sides of the equation by 36 and simplify. This leads to

$$\frac{x^2}{9} + \frac{y^2}{2} = 1,$$

which is in the proper form with $a^2 = 9$ and $b^2 = 2$. Thus $a = 3$, $b = \sqrt{2}$, and hence the endpoints of the major axis are $(\pm 3, 0)$ and the endpoints of the minor axis are $(0, \pm \sqrt{2})$. Since

$$c^2 = a^2 - b^2 = 9 - 2 = 7 \quad \text{or} \quad c = \sqrt{7},$$

the foci are $(\pm \sqrt{7}, 0)$. The graph is sketched in Figure 12.17. •

FIGURE 12.17

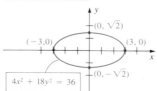

EXAMPLE 2 Find an equation of the ellipse with vertices $(\pm 4, 0)$ and foci $(\pm 2, 0)$.

SOLUTION Using the notation of Theorem (12.6), $a = 4$ and $c = 2$. Since $c^2 = a^2 - b^2$, we see that $b^2 = a^2 - c^2 = 16 - 4 = 12$. This gives us

$$\frac{x^2}{16} + \frac{y^2}{12} = 1 \ \bullet$$

It is sometimes convenient to choose the major axis of the ellipse along the y-axis. If the foci are $(0, \pm c)$, then by the same type of argument used previously, we obtain the following.

THEOREM (12.7)

The graph of the equation

$$\frac{x^2}{b^2} + \frac{y^2}{a^2} = 1,$$

for $a^2 > b^2$, is an ellipse with vertices $(0, \pm a)$. The endpoints of the minor axis are $(\pm b, 0)$. The foci are $(0, \pm c)$, with $c^2 = a^2 - b^2$.

FIGURE 12.18

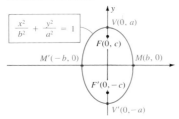

A typical graph of an ellipse with major axis on the y-axis is sketched in Figure 12.18.

The preceding discussion shows that an equation of an ellipse with center at the origin and foci on a coordinate axis can always be written in the form

$$\frac{x^2}{p} + \frac{y^2}{q} = 1 \quad \text{or} \quad qx^2 + py^2 = pq$$

with p and q positive and $p \neq q$. If $p > q$ the major axis lies on the x-axis, and if $q > p$ the major axis is on the y-axis. It is unnecessary to memorize these facts, since in any given problem the major axis can be determined by examining the x- and y-intercepts.

FIGURE 12.19

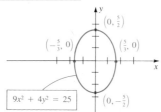

EXAMPLE 3 Sketch the graph of the equation $9x^2 + 4y^2 = 25$.

SOLUTION The graph is an ellipse with center at the origin and foci on one of the coordinate axes. To find the x-intercepts, we let $y = 0$, obtaining $9x^2 = 25$, or $x = \pm \frac{5}{3}$. Similarly, to find the y-intercepts, we let $x = 0$, obtaining $4y^2 = 25$, or $y = \pm \frac{5}{2}$. This enables us to sketch the ellipse (see Figure 12.19). Since $\frac{5}{3} < \frac{5}{2}$, the major axis is on the y-axis. $\bullet$

EXAMPLE 4 Find the area of the region bounded by an ellipse whose major and minor axes have lengths $2a$ and $2b$, respectively.

SOLUTION By (12.6), an equation for the ellipse is $(x^2/a^2) + (y^2/b^2) = 1$. Solving for y gives us

$$y = (\pm b/a)\sqrt{a^2 - x^2}.$$

The graph of the ellipse has the general shape shown in Figure 12.16 and hence, by symmetry, it is sufficient to find the area of the region in the first quadrant and multiply the result by 4. Using (5.11),

$$A = 4\frac{b}{a}\int_0^a \sqrt{a^2 - x^2}\ dx.$$

If we make the trigonometric substitution $x = a \sin \theta$, then

$$\sqrt{a^2 - x^2} = a \cos \theta \quad \text{and} \quad dx = a \cos \theta \, d\theta.$$

Moreover, the values of θ that correspond to $x = 0$ and $x = a$ are $\theta = 0$ and $\theta = \pi/2$, respectively. Consequently,

$$A = 4 \frac{b}{a} \int_0^{\pi/2} a^2 \cos^2 \theta \, d\theta = 4ab \int_0^{\pi/2} \frac{1 + \cos 2\theta}{2} \, d\theta$$

$$= 2ab \left[\theta + \tfrac{1}{2} \sin 2\theta \right]_0^{\pi/2} = 2ab[\pi/2] = \pi ab.$$

As a special case, if $b = a$ the ellipse is a circle and $A = \pi a^2$. •

By using the Translation of Axes Formulas (12.3) we can extend our work to an ellipse with center at any point $C(h, k)$ in the xy-plane. For example, since the graph of

$$\frac{(x')^2}{a^2} + \frac{(y')^2}{b^2} = 1$$

is an ellipse with center at O' in any $x'y'$-plane (see Figure 12.20), its equation relative to the xy-coordinate system is

$$\frac{(x - h)^2}{a^2} + \frac{(y - k)^2}{b^2} = 1.$$

Squaring the indicated terms in this equation and simplifying gives us an equation of the form

$$Ax^2 + Cy^2 + Dx + Ey + F = 0$$

such that the coefficients are real numbers and A and C are positive. Conversely, if we start with such an equation, then by completing squares we can obtain a form that displays the center of the ellipse and the lengths of the major and minor axes. This technique is illustrated in the next example.

FIGURE 12.20

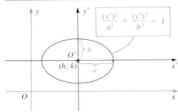

EXAMPLE 5 Discuss and sketch the graph of the equation

$$16x^2 + 9y^2 + 64x - 18y - 71 = 0.$$

SOLUTION We begin by writing the equation in the form

$$16(x^2 + 4x) + 9(y^2 - 2y) = 71.$$

Next, we complete the squares for the expressions within parentheses, obtaining

$$16(x^2 + 4x + 4) + 9(y^2 - 2y + 1) = 71 + 64 + 9.$$

Note that by adding 4 to the expression within the first parentheses we have added 64 to the left side of the equation, and hence must compensate by adding 64 to the right side. Similarly, by adding 1 to the expression within the second parentheses, we have added 9 to the left side and, consequently, we must also add 9 to the right side. The last equation may be written

$$16(x + 2)^2 + 9(y - 1)^2 = 144.$$

Dividing by 144, we obtain

$$\frac{(x+2)^2}{9} + \frac{(y-1)^2}{16} = 1,$$

FIGURE 12.21

which is of the form

$$\frac{(x')^2}{9} + \frac{(y')^2}{16} = 1$$

with $x' = x + 2$ and $y' = y - 1$. This corresponds to letting $h = -2$ and $k = 1$ in the Translation of Axes Formulas (12.3). Since the graph of the equation $(x')^2/9 + (y')^2/16 = 1$ is an ellipse with center at the origin in the $x'y'$-plane, it follows that the graph of the given equation is an ellipse with center $C(-2, 1)$ in the xy-plane and with axes parallel to the coordinate axes. The graph is sketched in Figure 12.21. •

Ellipses can be very flat or almost circular. To obtain information about the "roundness" of an ellipse we sometimes use the *eccentricity e* (not to be confused with the base of the natural logarithms). Eccentricity is defined as follows, with a, b, and c having the same meaning as in Theorems (12.6) and (12.7).

DEFINITION (12.8)

> The **eccentricity** e of an ellipse is
> $$e = \frac{c}{a} = \frac{\sqrt{a^2 - b^2}}{a}$$

Suppose, in Definition (12.8), we consider the length of the major axis $2a$ as fixed, and the length of the minor axis $2b$ as variable. Since $\sqrt{a^2 - b^2} < a$, we see that $0 < e < 1$. If $e \approx 1$, then $\sqrt{a^2 - b^2} \approx a$, and hence $b \approx 0$. In this case the ellipse is very flat. If $e \approx 0$, then $\sqrt{a^2 - b^2} \approx 0$ and $a \approx b$. In this case the ellipse is almost circular.

In Chapter 15 we will prove Kepler's three laws. Kepler's first law states that the orbit of each planet in the solar system is an ellipse with the sun at one focus. Most of these orbits are almost circular, and hence their corresponding eccentricities are close to 0. To illustrate, for Earth, $e \approx 0.017$; for Mars, $e \approx 0.093$; and for Uranus, $e \approx 0.046$. The orbits of Mercury and Pluto are less circular, with eccentricities of 0.206 and 0.249, respectively.

Many comets have elliptical orbits with the sun at a focus. In this case the eccentricity e is close to 1 and the ellipse is very flat. In the next example we use the **astronomical unit** (1 AU = 92,000,000 miles) to specify large distances.

EXAMPLE 6 Halley's comet, with a period of 76.2 years, has an elliptical orbit with eccentricity $e = 0.967$. Measuring distance in astronomical units, the closest that Halley's comet comes to the sun is 0.587 AU. Approximate the maximum distance of the comet from the sun, to the nearest 0.1 AU.

FIGURE 12.22
HALLEY'S COMET

SOLUTION Figure 12.22 illustrates the orbit of the comet, where, as usual, c is the distance from the center of the ellipse to a focus (the sun) and $2a$ is the length of the major axis.

Since $a - c$ is the minimum distance between the sun and the comet, we have (in AU),

$$a - c = 0.587 \quad \text{or} \quad a = c + 0.587.$$

Since $e = c/a = 0.967$,

$$c = 0.967a = 0.967(c + 0.587) \approx 0.967c + 0.568.$$

Thus,

$$0.033c \approx 0.568 \quad \text{and} \quad c \approx \frac{0.568}{0.033} \approx 17.2.$$

Consequently,

$$a = c + 0.587 \approx 17.2 + 0.6 = 17.8$$

and the maximum distance is

$$a + c \approx 17.8 + 17.2 = 35.0 \text{ AU} \quad \bullet$$

EXERCISES 12.3

Exer. 1–14: Sketch the graph of the ellipse, and find coordinates of the vertices and foci.

1 $\dfrac{x^2}{9} + \dfrac{y^2}{4} = 1$
2 $\dfrac{x^2}{25} + \dfrac{y^2}{16} = 1$

3 $4x^2 + y^2 = 16$
4 $y^2 + 9x^2 = 9$

5 $5x^2 + 2y^2 = 10$
6 $\frac{1}{2}x^2 + 2y^2 = 8$

7 $4x^2 + 25y^2 = 1$
8 $10y^2 + x^2 = 5$

9 $4x^2 + 9y^2 - 32x - 36y + 64 = 0$

10 $x^2 + 2y^2 + 2x - 20y + 43 = 0$

11 $9x^2 + 16y^2 + 54x - 32y - 47 = 0$

12 $4x^2 + 9y^2 + 24x + 18y + 9 = 0$

13 $25x^2 + 4y^2 - 250x - 16y + 541 = 0$

14 $4x^2 + y^2 = 2y$

Exer. 15–22: Find an equation for the ellipse satisfying the given conditions.

15 Vertices $V(\pm 8, 0)$, foci $F(\pm 5, 0)$

16 Vertices $V(0, \pm 7)$, foci $F(0, \pm 2)$

17 Vertices $V(0, \pm 5)$, length of minor axis 3

18 Foci $F(\pm 3, 0)$, length of minor axis 2

19 Vertices $V(0, \pm 6)$, passing through $(3, 2)$

20 Center at the origin, symmetric with respect to both axes, passing through the points $A(2, 3)$ and $B(6, 1)$

21 Eccentricity $\frac{3}{4}$, vertices $V(0, \pm 4)$

22 Eccentricity $\frac{1}{2}$, center at the origin, vertices on the x-axis, passing through the point $(1, 3)$

23 An arch of a bridge is semielliptical with major axis horizontal. The base of the arch is 30 feet across and the highest part of the arch is 10 feet above the horizontal roadway (see figure). Find the height of the arch 6 feet from the center of the base.

EXERCISE 23

24 Prove that an equation of the tangent line to the ellipse $(x^2/a^2) + (y^2/b^2) = 1$ at the point $P(x_1, y_1)$ is

$$\frac{xx_1}{a^2} + \frac{yy_1}{b^2} = 1.$$

25 Find an equation of the tangent line to the ellipse $5x^2 + 4y^2 = 56$ at the point $P(-2, 3)$.

26 If tangent lines to the ellipse $9x^2 + 4y^2 = 36$ intersect the y-axis at the point $(0, 6)$, find the points of tangency.

27 Find the volume of the solid obtained by revolving the region bounded by the ellipse $b^2x^2 + a^2y^2 = a^2b^2$ about (a) the x-axis; (b) the y-axis.

28 The base of a solid is a region bounded by an ellipse with major and minor axes of lengths 16 and 9, respectively. Find the volume of the solid if every cross section by a plane perpendicular to the major axis is (a) a square; (b) an equilateral triangle.

29 Find the dimensions of the rectangle of maximum area that can be inscribed in an ellipse of semiaxes a and b, if two sides of the rectangle are parallel to the major axis.

30 A cylindrical tank whose cross sections are elliptical with axes of lengths 6 feet and 4 feet, respectively, is lying on its side. If the tank is half full of water, find the force exerted by the water on one end of the tank.

31 Let l denote the tangent line at the point P on an ellipse with foci F' and F (see figure). If α is the angle between $F'P$ and l, and if β is the angle between FP and l, prove that $\alpha = \beta$. (This is analogous to the reflective property of the parabola illustrated in Figure 12.12, and is used in "whispering galleries," where a person who whispers at one focus can be heard at the other focus.)

EXERCISE 31

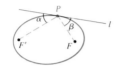

32 The base of a right elliptic cone has major and minor axes of lengths $2a$ and $2b$, respectively. Find the volume if the altitude is h.

33 Assume that the length of the major axis of the earth's orbit is 186,000,000 miles and the eccentricity is 0.017. Find the greatest and least distances from the earth to the sun, to the nearest 1000 miles.

34 (a) Show that the circumference C of an ellipse of eccentricity e is given by

$$C = 4a \int_0^{\pi/2} \sqrt{1 - e^2 \sin^2 \theta}\ d\theta.$$

(*Hint:* Use the substitution $x = a \sin \theta$ in the arc length formula (6.9). This is an *elliptic integral*, which cannot be evaluated by elementary functions.)

(b) The planet Mercury travels in an elliptical orbit with $e = 0.206$ and $a = 0.387$ AU. Use part (a) and Simpson's Rule with $n = 10$ to approximate the total length of the orbit.

35 Refer to Figure 3.20 in Section 3.6. In a spring-mass system, the total energy E (the sum of the potential and kinetic energy) is given by $E = \frac{1}{2}mv^2 + \frac{1}{2}kx^2$ for the displacement x of the mass m from equilibrium, the velocity v of the mass, and a constant k.

(a) For a fixed initial velocity and displacement, the total energy is given by $E = \frac{1}{2}mv^2 + \frac{1}{2}kx^2$ for the displacement x of the mass m from equilibrium, the velocity v of the mass, and a constant k.

(b) Find a relationship between the area A of the ellipse in part (a), the total energy E, and the frequency f of the oscillations.

36 An ellipse has a vertex at the origin and foci $F_1(p, 0)$ and $F_2(p + 2c, 0)$ (see figure). If the focus at F_1 is fixed, show that $\lim_{c \to \infty} y^2 = 4px$. Conclude that a parabola may be considered as an ellipse with "one focus at infinity."

EXERCISE 36

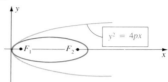

37 A common geometric figure chosen to model human limbs is the *elliptical frustrum*, in which all cross sections are elliptical and have the same eccentricity (see figure). The eccentricity of human limbs typically varies from 0.6 to values near 1. If $k = a_1/b_1 = a_2/b_2$ and L is the length of the limb, show that the volume V is given by the formula $V = (\frac{1}{3}\pi L/k)(a_1^2 + a_1 a_2 + a_2^2)$.

EXERCISE 37

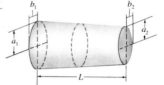

38 Prove that if a normal line to a point $P(x_1, y_1)$ on an ellipse contains the center of the ellipse, then the ellipse is a circle.

39 A line segment of length $a + b$ moves with its endpoints A and B attached to the coordinate axes, as illustrated in the figure. Prove that if $a \neq b$, then the point P traces an ellipse.

EXERCISE 39

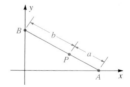

40 From a point P on the circle $x^2 + y^2 = 4$, a line segment PQ is drawn perpendicular to the diameter AB, and the midpoint M is found (see figure). Find an equation of the collection of all such midpoints M, and sketch the graph.

EXERCISE 40

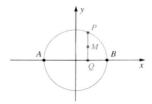

12.4 HYPERBOLAS

The definition of a hyperbola is similar to that of an ellipse. The only change is that instead of using the *sum* of distances from two fixed points, we use the *difference*.

DEFINITION (12.9)

> A **hyperbola** is the set of all points in a plane, the difference of whose distances from two fixed points (the **foci**) in the plane is a positive constant.

To find a simple equation for a hyperbola, we choose a coordinate system with foci at $F(c, 0)$ and $F'(-c, 0)$, and denote the (constant) distance by $2a$. Referring to Figure 12.23, we see that a point $P(x, y)$ is on the hyperbola if and only if either one of the following is true:

$$d(P, F') - d(P, F) = 2a$$

$$d(P, F) - d(P, F') = 2a.$$

FIGURE 12.23

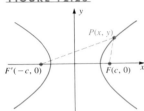

For hyperbolas (unlike ellipses), we need $a < c$ to obtain points on the hyperbola that are not on the x-axis, for if P is such a point, then from Figure 12.23 we see that

$$d(P, F) < d(F', F) + d(P, F'),$$

because the length of one side of a triangle is always less than the sum of the lengths of the other two sides. Similarly,

$$d(P, F') < d(F', F) + d(P, F).$$

Equivalent forms for the previous two inequalities are

$$d(P, F) - d(P, F') < d(F', F)$$

$$d(P, F') - d(P, F) < d(F', F).$$

Since the differences on the left both equal $2a$, and since $d(F', F) = 2c$, the last two inequalities imply that $2a < 2c$, or $a < c$.

The equations $d(P, F') - d(P, F) = 2a$ and $d(P, F) - d(P, F') = 2a$ may be replaced by the single equation

$$\left| d(P, F) - d(P, F') \right| = 2a.$$

It follows from the Distance Formula that an equation of the hyperbola is

$$\left| \sqrt{(x - c)^2 + (y - 0)^2} - \sqrt{(x + c)^2 + (y - 0)^2} \right| = 2a.$$

Employing the type of simplification procedure used to derive an equation for an ellipse, we arrive at the equivalent equation

$$\frac{x^2}{a^2} - \frac{y^2}{c^2 - a^2} = 1.$$

For convenience, we let

$$b^2 = c^2 - a^2 \qquad \text{for } b > 0$$

in the preceding equation, obtaining

$$\frac{x^2}{a^2} - \frac{y^2}{b^2} = 1.$$

We have shown that the coordinates of every point (x, y) on the hyperbola in Figure 12.23 satisfy the last equation. Conversely, if (x, y) is a solution of that equation, then by reversing the preceding steps we see that the point (x, y) is on the hyperbola.

By the Tests for Symmetry (1.11), the hyperbola is symmetric with respect to both axes and the origin. The x-intercepts are $\pm a$. The corresponding points $V(a, 0)$ and $V'(-a, 0)$ are the **vertices,** and the line segment $V'V$ is the **transverse axis** of the hyperbola. The origin is the **center** of the hyperbola. There are no y-intercepts, since the equation $-y^2/b^2 = 1$ has no solutions.

The preceding discussion may be summarized as follows.

THEOREM (12.10)

> The graph of the equation
>
> $$\frac{x^2}{a^2} - \frac{y^2}{b^2} = 1$$
>
> is a hyperbola with vertices $(\pm a, 0)$. The foci are $(\pm c, 0)$, with $c^2 = a^2 + b^2$.

If the equation $(x^2/a^2) - (y^2/b^2) = 1$ is solved for y, we obtain

$$y = \pm \frac{b}{a}\sqrt{x^2 - a^2}.$$

There are no points (x, y) on the graph if $x^2 - a^2 < 0$, that is, if $-a < x < a$. However, there *are* points $P(x, y)$ on the graph if $x \geq a$ or $x \leq -a$.

The line $y = (b/a)x$ is an asymptote for the hyperbola because the distance $p(x)$ between the point $P(x, y)$ on the hyperbola and the corresponding point $P'(x, y_1)$ on the line approaches zero as x increases without bound. To prove this we note that if $x > 0$, then

$$p(x) = \frac{b}{a}x - \frac{b}{a}\sqrt{x^2 - a^2} = \frac{b}{a}(x - \sqrt{x^2 - a^2}).$$

We can show, by rationalizing the numerator, that

$$\frac{x - \sqrt{x^2 - a^2}}{1} = \frac{a^2}{x + \sqrt{x^2 - a^2}}.$$

Hence

$$\lim_{x \to \infty} p(x) = \lim_{x \to \infty} \frac{b}{a} \frac{a^2}{x + \sqrt{x^2 - a^2}} = 0.$$

Similarly, $\lim_{x \to -\infty} p(x) = 0$. We can also show that the line $y = (-b/a)x$ is an asymptote for the hyperbola in Theorem 12.10.

The asymptotes serve as excellent guides for sketching the graph. A convenient way to sketch the asymptotes is to first plot the vertices $V(a, 0)$,

FIGURE 12.24

$$\frac{x^2}{a^2} - \frac{y^2}{b^2} = 1$$

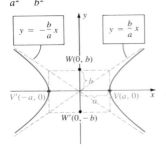

FIGURE 12.25

$9x^2 - 4y^2 = 36$

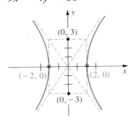

$V'(-a, 0)$ and the points $W(0, b)$, $W'(0, -b)$ (see Figure 12.24). The line segment $W'W$ of length $2b$ is called the **conjugate axis** of the hyperbola. If horizontal and vertical lines are drawn through the endpoints of the conjugate and transverse axes, respectively, then the diagonals of the resulting rectangle have slopes b/a and $-b/a$. Hence, by extending these diagonals we obtain lines with equations $y = (\pm b/a)x$. The hyperbola is then sketched as in Figure 12.24, using the asymptotes as guides. The two curves that make up the hyperbola are the **branches** of the hyperbola.

EXAMPLE 1 Discuss and sketch the graph of the equation

$$9x^2 - 4y^2 = 36.$$

SOLUTION Dividing both sides by 36, we have

$$\frac{x^2}{4} - \frac{y^2}{9} = 1,$$

which is of the form stated in Theorem (12.10) with $a^2 = 4$ and $b^2 = 9$. Hence, $a = 2$ and $b = 3$. The vertices $(\pm 2, 0)$ and the endpoints $(0, \pm 3)$ of the conjugate axis determine a rectangle whose diagonals (extended) give us the asymptotes. The graph of the equation is sketched in Figure 12.25. The equations of the asymptotes, $y = \pm\frac{3}{2}x$, can be found by referring to the graph or to the equations $y = \pm(b/a)x$. Since $c^2 = a^2 + b^2 = 4 + 19 = 13$, the foci are $(\pm\sqrt{13}, 0)$. •

The preceding example indicates that for hyperbolas it is not always true that $a < b$, as is the case for ellipses. We may have $a < b$, $a > b$, or $a = b$.

EXAMPLE 2 Find an equation, the foci, and the asymptotes of a hyperbola that has vertices $(\pm 3, 0)$ and passes through the point $P(5, 2)$.

SOLUTION Substituting $a = 3$ in $(x^2/a^2) - (y^2/b^2) = 1$, we obtain

$$\frac{x^2}{9} - \frac{y^2}{b^2} = 1.$$

If $(5, 2)$ is a solution of this equation, then

$$\frac{25}{9} - \frac{4}{b^2} = 1.$$

This gives us $b^2 = \frac{9}{4}$, and hence the desired equation is

$$\frac{x^2}{9} - \frac{4y^2}{9} = 1,$$

or equivalently, $x^2 - 4y^2 = 9$.

Since $c^2 = a^2 + b^2 = 9 + \frac{9}{4} = \frac{45}{4}$, we have $c = \sqrt{\frac{45}{4}} = \frac{3}{2}\sqrt{5}$. Thus the foci are $(\pm\frac{3}{2}\sqrt{5}, 0)$. Substituting for b and a in $y = \pm(b/a)x$ and simplifying, we obtain equations $y = \pm\frac{1}{2}x$ for the asymptotes. •

If the foci of a hyperbola are the points $(0, \pm c)$ on the y-axis, then by the same type of argument used previously, we obtain the following theorem.

THEOREM (12.11)

The graph of the equation

$$\frac{y^2}{a^2} - \frac{x^2}{b^2} = 1$$

is a hyperbola with vertices $(0, \pm a)$. The foci are $(0, \pm c)$, with $c^2 = a^2 + b^2$.

FIGURE 12.26

$\dfrac{y^2}{a^2} - \dfrac{x^2}{b^2} = 1$

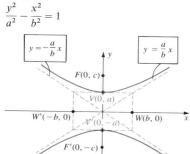

In this case the endpoints of the conjugate axis are $W(b, 0)$ and $W'(-b, 0)$. We find the asymptotes as before, by using the diagonals of the rectangle determined by the points W and W', the vertices, and lines parallel to the coordinate axes. The graph is sketched in Figure 12.26. The equations of the asymptotes are $y = \pm(a/b)x$. Note the difference between these equations and the equations $y = \pm(b/a)x$ for the asymptotes of the hyperbola considered in Theorem (12.10).

EXAMPLE 3 Discuss the graph of the equation

$$4y^2 - 2x^2 = 1.$$

SOLUTION The form in Theorem (12.11) may be obtained by writing the equation as

$$\frac{y^2}{\frac{1}{4}} - \frac{x^2}{\frac{1}{2}} = 1.$$

Thus $a^2 = \frac{1}{4}$, $b^2 = \frac{1}{2}$, and $c^2 = \frac{1}{4} + \frac{1}{2} = \frac{3}{4}$. Consequently, $a = \frac{1}{2}$, $b = \sqrt{2}/2$, and $c = \sqrt{3}/2$. The vertices are $(0, \pm\frac{1}{2})$ and the foci are $(0, \pm\sqrt{3}/2)$. The graph has the general appearance of the graph shown in Figure 12.26. •

As was the case for ellipses, we may use translations of axes to generalize our work. The following example illustrates this technique.

EXAMPLE 4 Discuss and sketch the graph of the equation

$$9x^2 - 4y^2 - 54x - 16y + 29 = 0.$$

FIGURE 12.27

$\dfrac{(x-3)^2}{4} - \dfrac{(y+2)^2}{9} = 1$

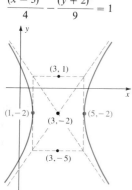

SOLUTION We arrange our work as follows:

$$9(x^2 - 6x) - 4(y^2 + 4y) = -29$$
$$9(x^2 - 6x + 9) - 4(y^2 + 4y + 4) = -29 + 81 - 16$$
$$9(x - 3)^2 - 4(y + 2)^2 = 36$$
$$\frac{(x-3)^2}{4} - \frac{(y+2)^2}{9} = 1,$$

which is of the form

$$\frac{(x')^2}{4} - \frac{(y')^2}{9} = 1$$

with $x' = x - 3$ and $y' = y + 2$. By translating the x- and y-axes to the new origin $C(3, -2)$, we obtain the sketch shown in Figure 12.27. •

The results of our work thus far indicate that the graph of every equation of the form

$$Ax^2 + Cy^2 + Dx + Ey + F = 0$$

is a conic, except for certain degenerate cases in which points, lines, or no graphs are obtained. Although we have only considered special examples, our methods are perfectly general. If A and C are equal and not zero, then the graph, when it exists, is a circle or, in exceptional cases, a point. If A and C are unequal but have the same sign, then by completing squares and properly translating axes, we obtain an equation whose graph, when it exists, is an ellipse (or a point). If A and C have opposite signs, an equation of a hyperbola is obtained, or possibly, in the degenerate case, two intersecting straight lines. Finally, if either A or C (but not both) is zero, the graph is a parabola or, in certain cases, a pair of parallel lines.

The definition of hyperbola (12.9) is based on the difference of the distances from two fixed points (the foci). This property is used in the navigational system called LORAN (for *Long Range Navigation*). This system involves two pairs of radio transmitters, such as those located at T, T' and S, S' in Figure 12.28. Suppose that signals sent out by the transmitters at T and T' reach a radio receiver in a ship located at some point P. The difference in time of arrival of the signals can be used to determine the difference of the distances of P from T and from T'. Thus P lies on one branch of a hyperbola with foci at T and T'. Repeating this process for the other pair of transmitters, we see that P also lies on one branch of a hyperbola with foci at S and S'. The intersection of these two branches determines the position of P.

FIGURE 12.28

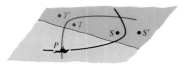

EXAMPLE 5 A LORAN station A is 200 miles directly east of another station B. A ship is sailing on a line parallel to, and 50 miles north of, the line through stations A and B. Signals are sent out from A and B at the rate of 980 ft/μsec (microsecond). If the signal from B reaches the ship 400 μsec after the signal from A, locate the position of the ship.

SOLUTION   Let us introduce a coordinate system as shown in Figure 12.29, with the stations at points A and B on the x-axis, and the ship at P, on the line $y = 50$. Since it takes 400 μsec longer for the signal to arrive from B than from A, the difference $d_1 - d_2$ in the indicated distances is

$$d_1 - d_2 = (980)(400) = 392,000 \text{ feet.}$$

Dividing by 5280 (ft/mi),

$$d_1 - d_2 = \frac{392,000}{5280} = 74.2424 \ldots \text{ miles.}$$

Recalling that in our derivation of the equation $(x^2/a^2) - (y^2/b^2) = 1$, we let $d_1 - d_2 = 2a$, it follows that in the present situation,

$$a = \frac{74.2424 \ldots}{2} = 37.1212 \ldots \quad \text{and} \quad a^2 \approx 1378.$$

Since $c = 100$,

$$b^2 = c^2 - a^2 \approx 10,000 - 1378 \approx 8622.$$

Hence an (approximate) equation for the hyperbola that has foci A and B

FIGURE 12.29

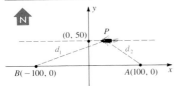

and passes through P is

$$\frac{x^2}{1378} - \frac{y^2}{8622} = 1.$$

If we let $y = 50$ (the y-coordinate of P), we obtain

$$\frac{x^2}{1378} - \frac{2500}{8622} = 1.$$

Solving for x gives us $x \approx 42.16$. Rounding off to the nearest mile, the coordinates of P are approximately $(42, 50)$. •

EXERCISES 12.4

Exer. 1–18: Sketch the graph of the hyperbola, find the coordinates of the vertices and foci and equations for the asymptotes.

1 $\dfrac{x^2}{9} - \dfrac{y^2}{4} = 1$ 2 $\dfrac{y^2}{49} - \dfrac{x^2}{16} = 1$

3 $\dfrac{y^2}{9} - \dfrac{x^2}{4} = 1$ 4 $\dfrac{x^2}{49} - \dfrac{y^2}{16} = 1$

5 $y^2 - 4x^2 = 16$ 6 $x^2 - 2y^2 = 8$

7 $x^2 - y^2 = 1$ 8 $y^2 - 16x^2 = 1$

9 $x^2 - 5y^2 = 25$ 10 $4y^2 - 4x^2 = 1$

11 $3x^2 - y^2 = -3$ 12 $16x^2 - 36y^2 = 1$

13 $25x^2 - 16y^2 + 250x + 32y + 109 = 0$

14 $y^2 - 4x^2 - 12y - 16x + 16 = 0$

15 $4y^2 - x^2 + 40y - 4x + 60 = 0$

16 $25x^2 - 9y^2 - 100x - 54y + 10 = 0$

17 $9y^2 - x^2 - 36y + 12x - 36 = 0$

18 $4x^2 - y^2 + 32x - 8y + 49 = 0$

Exer. 19–26: Find an equation for the hyperbola satisfying the given conditions.

19 Foci $F(0, \pm 4)$, vertices $V(0, \pm 1)$

20 Foci $F(\pm 8, 0)$, vertices $V(\pm 5, 0)$

21 Foci $F(\pm 5, 0)$, vertices $V(\pm 3, 0)$

22 Foci $F(0, \pm 3)$, vertices $V(0, \pm 2)$

23 Foci $F(0, \pm 5)$, length of conjugate axis 4

24 Vertices $V(\pm 4, 0)$, passing through $P(8, 2)$

25 Vertices $V(\pm 3, 0)$, equations of asymptotes $y = \pm 2x$

26 Foci $F(0, \pm 10)$, equations of asymptotes $y = \pm\frac{1}{3}x$

27 The graphs of the equations

$$\frac{x^2}{a^2} - \frac{y^2}{b^2} = 1 \quad \text{and} \quad \frac{x^2}{a^2} - \frac{y^2}{b^2} = -1$$

are called *conjugate hyperbolas*. Sketch the graphs of both

equations on the same coordinate system with $a = 2$ and $b = 5$. Describe the relationship between the two graphs.

28 Show that a point $P(x, y)$ on a branch of a hyperbola is nearest to the interior focus at the vertex V.

29 Find an equation of the tangent line to the hyperbola $2x^2 - 5y^2 = 3$ at the point $P(-2, 1)$.

30 Prove that an equation of the tangent line to the graph of the hyperbola $(x^2/a^2) - (y^2/b^2) = 1$ at the point $P(x_1, y_1)$ is

$$\frac{x_1 x}{a^2} - \frac{y_1 y}{b^2} = 1.$$

31 If tangent lines to the hyperbola $9x^2 - y^2 = 36$ intersect the y-axis at the point $(0, 6)$, find the points of tangency.

32 Find an equation of a line through $P(2, -1)$ that is tangent to the hyperbola $x^2 - 4y^2 = 16$.

33 Let R be the region bounded by the hyperbola with equation $b^2x^2 - a^2y^2 = a^2b^2$ and a vertical line through a focus. Find the volume of the solid obtained by revolving R about (a) the x-axis; (b) the y-axis.

34 Find the area of the region bounded by the hyperbola $b^2x^2 - a^2y^2 = a^2b^2$ and a line through a focus perpendicular to the transverse axis.

35 A cruise ship is traveling a course that is 100 miles from, and parallel to, a straight shoreline. The ship sends out a distress signal, which is received by two Coast Guard stations A and B, located 200 miles apart (see figure). By measuring the difference in signal reception times, officials determine that the ship is 160 miles closer to B than to A. Where is the ship?

EXERCISE 35

36 The physicist Ernest Rutherford discovered that when alpha particles are shot toward the nucleus of an atom,

they are eventually repulsed away from the nucleus along hyperbolic paths. The figure illustrates the path of a particle that starts toward the origin along the line $y = \frac{1}{2}x$ and comes within 3 units of the nucleus. Find an equation of the path.

EXERCISE 36

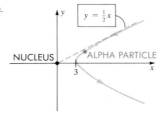

37 Let l denote the tangent line at a point P on a hyperbola with foci F and F' (see figure). If α is the angle between $F'P$ and l, and if β is the angle between FP and l, prove that $\alpha = \beta$. This is analogous to the reflective property of the ellipse (see Exercise 31 of Section 12.3). If a ray of light emanates from one focus it will be reflected back from P along a line through the other focus. This property is used in the design of telescopes of the Cassegrain type (see Exercise 38).

EXERCISE 37

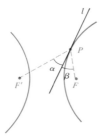

38 The Cassegrain telescope design (dating to 1672) makes use of the reflective properties of both the parabola and hyperbola (see the preceding exercise and Exercise 34 of Section 12.2). Shown in the figure is a (split) parabolic mirror with focus at F_1 and axis the line l, and a second, hyperbolic mirror with one focus also at F_1 and transverse axis along the axis of the parabola. Determine where incoming light waves parallel to the common axis will finally collect.

EXERCISE 38

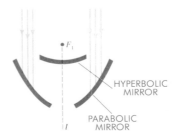

39 Some comets travel along hyperbolic paths with the sun at a focus. Shown in the figure is the graph for such a comet. The coordinate system has been chosen so that the sun is at the origin and the transverse axis is on the x-axis. From several observations of the comet's position, astronomers have determined that an equation of the path of the comet is $12x^2 + 24x - 4y^2 + 9 = 0$. Approximately how close does the comet come to the sun (in AU)? (*Hint:* See Exercise 28.)

EXERCISE 39

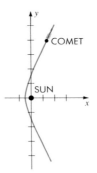

40 Let l denote the tangent line at a point P on a hyperbola (see figure). If Q and R are the points where l intersects the asymptotes, prove that P is the midpoint of the line segment QR.

EXERCISE 40

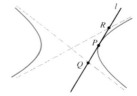

12.5 ROTATION OF AXES

We may think of the $x'y'$-coordinate system used in a translation of axes (12.3) as having been obtained by moving the origin O of the xy-system to a new point (h, k) while, at the same time, not changing the positive directions of the axes or the units of length. We shall next introduce a new coordinate system by keeping the origin O fixed and rotating the x- and y-axes about O to another position denoted by x' and y'. A transformation of this type is a **rotation of axes.**

FIGURE 12.30

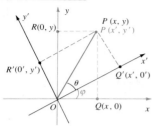

Consider the rotation of axes in Figure 12.30, and let φ denote the angle through which the positive x-axis must be rotated in order to coincide with the positive x'-axis. If (x, y) are the coordinates of a point P relative to the xy-plane, then (x', y') will denote its coordinates relative to the new $x'y'$-coordinate system.

Let the projections of P on the various axes be denoted as in Figure 12.30, and let θ denote angle POQ'. If $p = d(O, P)$, then

$$x' = p \cos \theta, \qquad y' = p \sin \theta,$$

$$x = p \cos (\theta + \varphi), \qquad y = p \sin (\theta + \varphi).$$

Applying the addition formulas for the sine and cosine, we see that

$$x = p \cos \theta \cos \varphi - p \sin \theta \sin \varphi$$

$$y = p \sin \theta \cos \varphi + p \cos \theta \sin \varphi.$$

Using the fact that $x' = p \cos \theta$ and $y' = p \sin \theta$ gives us (i) of the next theorem. The formulas in (ii) may be obtained from (i) by solving for x' and y'.

ROTATION OF AXES FORMULAS (12.12)

If the x- and y-axes are rotated about the origin O, through an angle φ, then the coordinates (x, y) and (x', y') of a point P in the two systems are related as follows:

(i) $x = x' \cos \varphi - y' \sin \varphi, \qquad y = x' \sin \varphi + y' \cos \varphi$

(ii) $x' = x \cos \varphi + y \sin \varphi, \qquad y' = -x \sin \varphi + y \cos \varphi$

FIGURE 12.31
$y = 1/x$

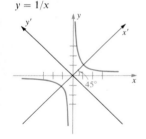

EXAMPLE 1 The graph of the equation $xy = 1$, or equivalently $y = 1/x$, is sketched in Figure 12.31. If the coordinate axes are rotated through an angle of $45°$, find an equation of the graph relative to the new $x'y'$-coordinate system.

SOLUTION Letting $\varphi = 45°$ in (i) of the Rotation of Axes Formulas (12.12),

$$x = x'\left(\frac{\sqrt{2}}{2}\right) - y'\left(\frac{\sqrt{2}}{2}\right) = \left(\frac{\sqrt{2}}{2}\right)(x' - y')$$

$$y = x'\left(\frac{\sqrt{2}}{2}\right) + y'\left(\frac{\sqrt{2}}{2}\right) = \left(\frac{\sqrt{2}}{2}\right)(x' + y').$$

Substituting for x and y in the equation $xy = 1$ gives us

$$\left(\frac{\sqrt{2}}{2}\right)(x' - y')\left(\frac{\sqrt{2}}{2}\right)(x' + y') = 1.$$

This reduces to

$$\frac{(x')^2}{2} - \frac{(y')^2}{2} = 1,$$

which is an equation of a hyperbola with vertices $(\pm\sqrt{2}, 0)$ on the x'-axis. Note that the asymptotes for the hyperbola have equations $y' = \pm x'$ in the new system. These correspond to the original x- and y-axes. •

Example 1 illustrates a method for eliminating a term of an equation that contains the product xy. This method can be used to transform any equation of the form

$$Ax^2 + Bxy + Cy^2 + Dx + Ey + F = 0,$$

with $B \neq 0$, into an equation in x' and y' that contains no $x'y'$ term. Let us prove that this may always be done. If we rotate the axes through an angle φ, then using (12.12) (i) to substitute for x and y gives us

$$A(x' \cos \varphi - y' \sin \varphi)^2$$
$$+ B(x' \cos \varphi - y' \sin \varphi)(x' \sin \varphi + y' \cos \varphi)$$
$$+ C(x' \sin \varphi + y' \cos \varphi)^2 + D(x' \cos \varphi - y' \sin \varphi)$$
$$+ E(x' \sin \varphi + y' \cos \varphi) + F = 0.$$

By performing the multiplications and rearranging terms, this equation may be written in the form

$$A'(x')^2 + B'x'y' + C'(y')^2 + D'x' + E'y' + F' = 0$$

with

(12.13)
$$A' = A \cos^2 \varphi + B \cos \varphi \sin \varphi + C \sin^2 \varphi$$
$$B' = 2(C - A) \sin \varphi \cos \varphi + B(\cos^2 \varphi - \sin^2 \varphi)$$
$$C' = A \sin^2 \varphi - B \sin \varphi \cos \varphi + C \cos^2 \varphi$$
$$D' = D \cos \varphi + E \sin \varphi$$
$$E' = -D \sin \varphi + E \cos \varphi$$
$$F' = F$$

To eliminate the $x'y'$ term, we must select φ such that $B' = 0$, that is,

$$2(C - A) \sin \varphi \cos \varphi + B(\cos^2 \varphi - \sin^2 \varphi) = 0.$$

Using double-angle formulas, this equation may be written

$$(C - A) \sin 2\varphi + B \cos 2\varphi = 0,$$

which is equivalent to
$$\cot 2\varphi = \frac{A - C}{B}.$$

This proves the next result.

THEOREM (12.14)

To eliminate the xy-term from the equation
$$Ax^2 + Bxy + Cy^2 + Dx + Ey + F = 0$$
with $B \neq 0$, choose an angle φ such that
$$\cot 2\varphi = \frac{A - C}{B}$$
and use the Rotation of Axes Formulas (12.12).

It follows that the graph of any equation in x and y of the type displayed in the preceding theorem is a conic, except for certain degenerate cases.

EXAMPLE 2 Discuss and sketch the graph of the equation
$$41x^2 - 24xy + 34y^2 - 25 = 0.$$

SOLUTION Using the notation of the preceding theorem, we have

$$A = 41, \qquad B = -24, \qquad C = 34,$$

and

$$\cot 2\varphi = \frac{41 - 34}{-24} = -\frac{7}{24}.$$

Since $\cot 2\varphi$ is negative, we may choose 2φ such that $90° < 2\varphi < 180°$, and consequently $\cos 2\varphi = -\frac{7}{25}$. (Why?) We now use the half-angle formulas to obtain

$$\sin \varphi = \sqrt{\frac{1 - \cos 2\varphi}{2}} = \sqrt{\frac{1 - (-\frac{7}{25})}{2}} = \frac{4}{5}$$

$$\cos \varphi = \sqrt{\frac{1 + \cos 2\varphi}{2}} = \sqrt{\frac{1 + (-\frac{7}{25})}{2}} = \frac{3}{5}.$$

Thus, the desired rotation formulas are

$$x = \tfrac{3}{5}x' - \tfrac{4}{5}y', \qquad y = \tfrac{4}{5}x' + \tfrac{3}{5}y'.$$

After substituting for x and y in the given equation and simplifying, we obtain

$$(x')^2 + 2(y')^2 = 1.$$

Thus the graph is an ellipse with vertices at $(\pm 1, 0)$ on the x'-axis. Since $\tan \varphi = (\sin \varphi)/(\cos \varphi) = (\tfrac{4}{5})/(\tfrac{3}{5}) = \tfrac{4}{3}$, we obtain $\varphi = \tan^{-1}(\tfrac{4}{3})$. To the nearest minute, $\varphi \approx 53°8'$. The graph is sketched in Figure 12.32. •

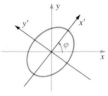

FIGURE 12.32

The next theorem indicates how to obtain information about the graph of the equation in Theorem (12.14) *before* rotating the axes.

THEOREM (12.15)

The graph of the equation

$$Ax^2 + Bxy + Cy^2 + Dx + Ey + F = 0$$

is either a conic or a degenerate conic. If the graph is a conic, then it is

(i) a parabola if $B^2 - 4AC = 0$.

(ii) an ellipse if $B^2 - 4AC < 0$.

(iii) a hyperbola if $B^2 - 4AC > 0$.

PROOF If the x and y axes are rotated through an angle φ, then using the Rotation of Axes Formulas (12.12) gives us

$$A'(x')^2 + B'x'y' + C'(y')^2 + D'x' + E'y' + F' = 0.$$

Using the relationships in (12.13) we can show that

$$B^2 - 4AC = (B')^2 - 4A'C'.$$

For a suitable rotation of axes, $B' = 0$, and we obtain

$$A'(x')^2 + C'(y')^2 + D'x' + E'y' + F' = 0.$$

Except for degenerate cases, the graph of this equation is an ellipse if $A'C' > 0$ (A' and C' have the same sign), a hyperbola if $A'C' < 0$ (A' and C' have opposite signs), or a parabola if $A'C' = 0$ (either $A' = 0$ or $C' = 0$). However, if $B' = 0$, then $B^2 - 4AC = -4A'C'$, and hence the graph is an ellipse, hyperbola, or parabola, if $B^2 - 4AC < 0$, $B^2 - 4AC > 0$, or $B^2 - 4AC = 0$, respectively. • •

The expression $B^2 - 4AC$ is called the **discriminant** of the equation in Theorem (12.15). We say that this discriminant is **invariant** under a rotation of axes, because it is unchanged by any such rotation.

EXAMPLE 3 Use Theorem (12.15) to determine if the graph of the equation

$$41x^2 - 24xy + 34y^2 - 25 = 0$$

is a parabola, an ellipse, or a hyperbola.

SOLUTION The equation was considered in Example 2, where we performed a rotation of axes. Since $A = 41$, $B = -24$, and $C = 34$, the discriminant is

$$B^2 - 4AC = 576 - 4(41)(34) = -5000 < 0.$$

Hence, by Theorem (12.15) (ii), the graph is an ellipse. •

In some cases, after eliminating the xy term, the resulting equation will contain an x' or y' term. It is then necessary to translate the axes of the $x'y'$-coordinate system to obtain the graph. Several problems of this type are included in the following exercises.

EXERCISES 12.5

Exer. 1–14: (a) Use Theorem (12.15) to determine if the graph of the equation is a parabola, ellipse, or hyperbola. (b) After a suitable rotation of axes, sketch the graph of the equation.

1 $32x^2 - 72xy + 53y^2 = 80$

2 $7x^2 - 48xy - 7y^2 = 225$

3 $11x^2 + 10\sqrt{3}xy + y^2 = 4$

4 $x^2 - xy + y^2 = 3$

5 $5x^2 - 8xy + 5y^2 = 9$

6 $11x^2 - 10\sqrt{3}xy + y^2 = 20$

7 $16x^2 - 24xy + 9y^2 - 60x - 80y + 100 = 0$

8 $x^2 + 2\sqrt{3}xy + 3y^2 + 8\sqrt{3}x - 8y + 32 = 0$

9 $5x^2 + 6\sqrt{3}xy - y^2 + 8x - 8\sqrt{3}y - 12 = 0$

10 $18x^2 - 48xy + 82y^2 + 6\sqrt{10}x + 2\sqrt{10}y - 80 = 0$

11 $x^2 + 4xy + 4y^2 + 6\sqrt{5}x - 18\sqrt{5}y + 45 = 0$

12 $15x^2 + 20xy - 4\sqrt{5}x + 8\sqrt{5}y - 100 = 0$

13 $40x^2 - 36xy + 25y^2 - 8\sqrt{13}x - 12\sqrt{13}y = 0$

14 $64x^2 - 240xy + 225y^2 + 1020x - 544y = 0$

12.6 REVIEW

Define or discuss each of the following.

1 Conic sections

2 Parabola

3 Focus, directrix, vertex, and axis of a parabola

4 Translation of axes

5 Ellipse

6 Major and minor axes of an ellipse

7 Foci and vertices of an ellipse

8 Eccentricity of an ellipse

9 Hyperbola

10 Transverse and conjugate axes of a hyperbola

11 Foci and vertices of a hyperbola

12 Asymptotes of a hyperbola

13 Reflective properties of conics

14 Rotation of axes

EXERCISES 12.6

Exer. 1–16: Find the foci and vertices, and sketch the graph of the conic that has the given equation.

1 $y^2 = 64x$

2 $y - 1 = 8(x + 2)^2$

3 $9y^2 = 144 - 16x^2$

4 $9y^2 = 144 + 16x^2$

5 $x^2 - y^2 - 4 = 0$

6 $25x^2 + 36y^2 = 1$

7 $25y = 100 - x^2$

8 $3x^2 + 4y^2 - 18x + 8y + 19 = 0$

9 $x^2 - 9y^2 + 8x + 90y - 210 = 0$

10 $x = 2y^2 + 8y + 3$

11 $4x^2 + 9y^2 + 24x - 36y + 36 = 0$

12 $4x^2 - y^2 - 40x - 8y + 88 = 0$

13 $y^2 - 8x + 8y + 32 = 0$

14 $4x^2 + y^2 - 24x + 4y + 36 = 0$

15 $x^2 - 9y^2 + 8x + 7 = 0$

16 $y^2 - 2x^2 + 6y + 8x - 3 = 0$

Exer. 17–26: Find an equation for the conic that satisfies the given conditions.

17 The hyperbola with vertices $V(0, \pm 7)$ and endpoints of conjugate axes $(\pm 3, 0)$

18 The parabola with focus $F(-4, 0)$ and directrix $x = 4$

19 The parabola with focus $F(0, -10)$ and directrix $y = 10$

20 The parabola with vertex at the origin, symmetric to the x-axis, and passing through the point $(5, -1)$

21 The ellipse with vertices $V(0, \pm 10)$ and foci $F(0, \pm 5)$

22 The hyperbola with foci $F(\pm 10, 0)$ and vertices $V(\pm 5, 0)$

23 The hyperbola with vertices $V(0, \pm 6)$ and asymptotes that have equations $y = \pm 9x$

24 The ellipse with foci $F(\pm 2, 0)$ and passing through the point $(2, \sqrt{2})$

25 The ellipse with eccentricity $\frac{2}{3}$ and endpoints of minor axis $(\pm 5, 0)$

26 The ellipse with eccentricity $\frac{3}{4}$, center at the origin, and foci $F(\pm 12, 0)$

27 A bridge is to be constructed across a river that is 200 feet wide. The arch of the bridge is to be semielliptical and must be constructed so that a ship less than 50 feet wide and 30 feet high can pass safely through (see figure). Find an equation for the arch, and calculate the height of the arch in the middle of the bridge.

EXERCISE 27

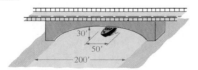

28 As illustrated in the figure, point $P(x, y)$ is the same distance from the point $(4, 0)$ as it is from the circle $x^2 + y^2 = 4$, that is, $d_1 = d_2$. Show that the collection of all such points forms a branch of a hyperbola, and sketch its graph.

EXERCISE 28

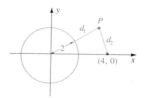

29 Find equations of the tangent and normal lines to the hyperbola $4x^2 - 9y^2 - 8x + 6y - 36 = 0$ at the point $P(-3, 2)$.

30 Tangent lines to the parabola $y = 2x^2 + 3x + 1$ pass through the point $P(2, -1)$. Find the x-coordinates of the points of tangency.

31 Prove that there is exactly one line of a given slope m which is tangent to the parabola $x^2 = 4py$, and that its equation is $y = mx - pm^2$.

32 Consider the ellipse $px^2 + qy^2 = pq$ with $p > 0$ and $q > 0$. Prove that if m is any real number, exactly two lines of slope m are tangent to the ellipse, and their equations are $y = mx \pm \sqrt{p + qm^2}$.

33 Let R be the region bounded by a parabola and the line through the focus that is perpendicular to the axis, and let p be the distance from the vertex V to the focus F (see figure). Prove that the area of R is $\frac{8}{3}p^2$.

EXERCISE 33

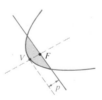

34 Let R be the region that is bounded by the parabola $9y = 5x^2 - 39x + 90$ and an asymptote of the hyperbola $9y^2 - 4x^2 = 36$.

(a) Find the area of R.

(b) Find the volume of the solid obtained by revolving R about the y-axis.

35 An ellipse having axes of lengths 8 and 4 is revolved about its major axis. Find the volume of the resulting solid.

36 A solid has, for its base, the region in the xy-plane bounded by the graph of $x^2 + y^2 = r^2$. Find the volume of the solid if every cross section by a plane perpendicular to the x-axis is half of an ellipse with one axis always of length c.

37 Find the centroid of the region in the xy-plane that is bounded by the hyperbola $b^2y^2 - a^2x^2 = a^2b^2$ and the line $y = c$ through the focus.

38 Use the discriminant to identify the graph of each equation. (Do not sketch the graph.)

(a) $2x^2 - 3xy + 4y^2 + 6x - 2y - 6 = 0$

(b) $3x^2 + 2xy - y^2 - 2x + y + 4 = 0$

(c) $x^2 - 6xy + 9y^2 + x - 3y + 5 = 0$

Exer. 39–40: After making a suitable rotation of axes, describe and sketch the graph of the equation.

39 $x^2 - 8xy + 16y^2 - 12\sqrt{17}x - 3\sqrt{17}y = 0$

40 $8x^2 + 12xy + 17y^2 - 16\sqrt{5}x - 12\sqrt{5}y = 0$

In this chapter we discuss parametric and polar equations of curves. Applications include tangent lines, areas, arc length, and surfaces of revolution.

PLANE CURVES AND POLAR COORDINATES

13.1 PLANE CURVES

The graph of an equation $y = f(x)$, for a function f, is often called a *plane curve*. However, this definition is unnecessarily restrictive, since it rules out many useful graphs. The following definition is satisfactory for most applications.

FIGURE 13.1

(i) Curve

DEFINITION (13.1)

A **plane curve** is a set C of ordered pairs of the form $(f(t), g(t))$ for functions f and g that are continuous on an interval I.

For simplicity, we shall often refer to a plane curve as a **curve.** The **graph** of the curve C in Definition (13.1) consists of all points $P(t) = (f(t), g(t))$ in a rectangular coordinate system for t in I. Each $P(t)$ is referred to as a *point on the curve*. We shall use the term *curve* interchangeably with *graph of a curve*. In some cases it is convenient to imagine that the point $P(t)$ traces the curve C as t varies through the interval I. This is especially true in applications where t represents time and $P(t)$ is the position of a moving particle at time t.

(ii) Closed curve

(iii) Simple closed curve

The graphs of several curves are sketched in Figure 13.1 for the case where I is a closed interval $[a, b]$. If, as in (i) of the figure, $P(a) \neq P(b)$, then $P(a)$ and $P(b)$ are the **endpoints** of C. The curve illustrated in (i) intersects itself, that is, two different values of t produce the same point. If $P(a) = P(b)$, as illustrated in Figure 13.1(ii), then C is a **closed curve.** If $P(a) = P(b)$ and C does not intersect itself at any other point, as illustrated in (iii), then C is a **simple closed curve.**

A convenient way to represent curves is given in the next definition.

DEFINITION (13.2)

> Let C be the curve consisting of all ordered pairs $(f(t), g(t))$ for f and g continuous on an interval I. The equations
>
> $$x = f(t), \qquad y = g(t),$$
>
> for t in I, are **parametric equations** for C with **parameter** t.

The curve C in Definition (13.2) is referred to as a **parametrized curve,** and the parametric equations are a **parametrization** for C. We often use the notation

$$x = f(t), \quad y = g(t); \qquad t \text{ in } I$$

to indicate the domain I of f and g. The continuity of f and g implies that a small change in t produces a small change in the position of the point $(f(t), g(t))$ on C. Thus, we may obtain a sketch of the graph by plotting many points and connecting them in the order of increasing t, as illustrated in Example 1. The example also shows that it is sometimes possible to eliminate the parameter and obtain an equation for C that involves only the variables x and y.

EXAMPLE 1 Sketch the graph of the curve C with parametrization

$$x = 2t, \quad y = t^2 - 1; \qquad -1 \le t \le 2.$$

SOLUTION The parametric equations can be used to tabulate coordinates of points $P(x, y)$ on C as in the following table.

t	-1	$-\frac{1}{2}$	0	$\frac{1}{2}$	1	$\frac{3}{2}$	2
x	-2	-1	0	1	2	3	4
y	0	$-\frac{3}{4}$	-1	$-\frac{3}{4}$	0	$\frac{5}{4}$	3

Plotting points and using the continuity of f and g gives us the sketch in Figure 13.2.

FIGURE 13.2
$x = 2t, \ y = t^2 - 1;$
$-1 \le t \le 2$

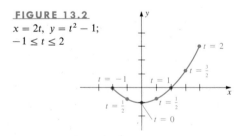

A clearer description of the graph may be obtained by eliminating the parameter. Solving the first parametric equation for t, we obtain $t = x/2$. Substituting for t in the second equation gives us

$$y = \left(\frac{x}{2}\right)^2 - 1 \quad \text{or} \quad y + 1 = \tfrac{1}{4}x^2.$$

The graph of this equation is a parabola with vertical axis and vertex at the point $(0, -1)$. Note, however, that since $x = 2t$ and $-1 \leq t \leq 2$, we have $-2 \leq x \leq 4$. Hence the curve C is that part of the parabola between the points $(-2, 0)$ and $(4, 3)$ shown in Figure 13.2. •

A curve has many different parametrizations. To illustrate, the curve C in Example 1 is given parametrically by any of the following:

$$x = 2t, \quad y = t^2 - 1; \qquad -1 \leq t \leq 2$$
$$x = t, \quad y = \tfrac{1}{4}t^2 - 1; \qquad -2 \leq t \leq 4$$
$$x = t^3, \quad y = \tfrac{1}{4}t^6 - 1; \qquad \sqrt[3]{-2} \leq t \leq \sqrt[3]{4}$$

The next example illustrates the fact that it is often useful to eliminate the parameter *before* plotting points.

EXAMPLE 2 Sketch the graph of the curve C that has the parametrization

$$x = -2 + t^2, \quad y = 1 + 2t^2; \qquad t \text{ in } \mathbb{R}.$$

SOLUTION To eliminate the parameter, we note from the first equation that $t^2 = x + 2$. Substituting for t^2 in the second equation gives us

$$y = 1 + 2(x + 2) \quad \text{or} \quad y - 1 = 2(x + 2).$$

This is an equation of the line of slope 2 through the point $(-2, 1)$, as indicated by the dashes in Figure 13.3. However, since $t^2 \geq 0$,

$$x = -2 + t^2 \geq -2 \quad \text{and} \quad y = 1 + 2t^2 \geq 1.$$

It follows that the graph of C is that part of the line to the right of the point $(-2, 1)$, corresponding to $t = 0$, as shown in Figure 13.4. This fact may also be verified by plotting several points. •

EXAMPLE 3 A particle moves in a plane such that its position $P(x, y)$ at time t is given by

$$x = a \cos t, \quad y = a \sin t; \qquad t \text{ in } \mathbb{R}$$

for a positive constant a. Describe the path of the particle.

SOLUTION If we rewrite the parametric equations as

$$\frac{x}{a} = \cos t, \qquad \frac{y}{a} = \sin t$$

and use the identity $\cos^2 t + \sin^2 t = 1$, we obtain

$$\left(\frac{x}{a}\right)^2 + \left(\frac{y}{a}\right)^2 = 1 \quad \text{or} \quad x^2 + y^2 = a^2.$$

Thus the particle moves on the circle of radius a with center at the origin (see Figure 13.5). In particular, the particle is at the point $A(a, 0)$ when $t = 0$; at $(0, a)$ when $t = \pi/2$; at $(-a, 0)$ when $t = 3\pi/2$, and so on. In general, as t increases, the particle moves around the circle in a counterclockwise direction, making one revolution every 2π units of time.

Note that in this example, the parameter t may be interpreted geometrically as the radian measure of the angle generated by the line segment OP. •

FIGURE 13.3

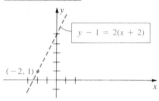

FIGURE 13.4

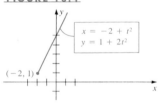

FIGURE 13.5
$x = a \cos t, \ y = a \sin t; \quad t \text{ in } \mathbb{R}$

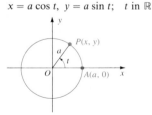

If a curve C is described by means of an equation $y = f(x)$, for a continuous function f, then an easy way to obtain parametric equations for C is to let

$$x = t, \quad y = f(t)$$

for t in the domain of f. For example, if $y = x^3$, then parametric equations are

$$x = t, \quad y = t^3; \quad t \text{ in } \mathbb{R}.$$

We can use many different substitutions for x, provided that as t varies through some interval, x takes on all values in the domain of f. Thus the graph of $y = x^3$ is also given by

$$x = t^{1/3}, \quad y = t; \quad t \text{ in } \mathbb{R}.$$

Note, however, that the parametric equations

$$x = \sin t, \quad y = \sin^3 t; \quad t \text{ in } \mathbb{R}$$

give only that part of the graph of $y = x^3$ that lies between the points $(-1, -1)$ and $(1, 1)$.

EXAMPLE 4 Find three parametrizations for the line of slope m through the point (x_1, y_1).

SOLUTION By the Point-Slope Form, an equation for the line is

$$y - y_1 = m(x - x_1).$$

If we let $x = t$, then $y - y_1 = m(t - x_1)$ and we obtain the parametric equations

$$x = t, \quad y = y_1 + m(t - x_1); \quad t \text{ in } \mathbb{R}.$$

Another pair of parametric equations for the line results if we let $x - x_1 = t$. In this case $y - y_1 = mt$, and we obtain

$$x = x_1 + t, \quad y = y_1 + mt; \quad t \text{ in } \mathbb{R}.$$

As a third illustration, if we let $x - x_1 = \tan t$, then

$$x = x_1 + \tan t, \quad y = y_1 + m \tan t; \quad -\frac{\pi}{2} < t < \frac{\pi}{2}.$$

There are many other parametrizations for the line. •

Parametric equations of the form

$$x = a \sin \omega_1 t, \quad y = b \cos \omega_2 t,$$

for constants a, b, ω_1, and ω_2 with $t \geq 0$, occur in the study of electricity. The variables x and y usually represent voltages or currents at time t. The resulting curve is often difficult to graph by plotting points or eliminating the parameter; however, using an oscilloscope and imposing voltages (or currents) on the horizontal and vertical input terminals, respectively, we can represent the graph by a figure on the screen of the oscilloscope. A figure of this type is called a **Lissajous figure.** Computers are also useful in obtaining these complicated graphs.

EXAMPLE 5

(a) Sketch the Lissajous figure corresponding to

$$x = \sin 2t, \quad y = \cos t; \qquad 0 \le t \le 2\pi.$$

(b) Find an equation in x and y for the curve.

SOLUTION

FIGURE 13.6
$x = \sin 2t,\ y = \cos t;\quad 0 \le t \le 2\pi$

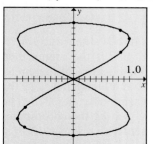

(a) The graph in Figure 13.6 was generated by a computer. To sketch the graph by plotting points, we could make a table as in Example 1. The following table gives some values of x and y for $0 \le t \le \pi$. (We could also use a calculator to approximate values of x and y.)

t	0	$\dfrac{\pi}{6}$	$\dfrac{\pi}{4}$	$\dfrac{\pi}{3}$	$\dfrac{\pi}{2}$	$\dfrac{2\pi}{3}$	$\dfrac{3\pi}{4}$	$\dfrac{5\pi}{6}$	π
x	0	$\dfrac{\sqrt{3}}{2}$	1	$\dfrac{\sqrt{3}}{2}$	0	$-\dfrac{\sqrt{3}}{2}$	-1	$-\dfrac{\sqrt{3}}{2}$	0
y	1	$\dfrac{\sqrt{3}}{2}$	$\dfrac{\sqrt{2}}{2}$	$\dfrac{1}{2}$	0	$-\dfrac{1}{2}$	$-\dfrac{\sqrt{2}}{2}$	$-\dfrac{\sqrt{3}}{2}$	-1

The points $P(x, y)$ with the indicated t-values are plotted in Figure 13.6. Note that as t increases from 0 to $\pi/2$, the point $P(x, y)$ traces the part of the curve in Quadrant I (in a generally clockwise direction); as t increases from $\pi/2$ to π, $P(x, y)$ traces the part in Quadrant III (in a counterclockwise direction); for $\pi \le t \le 3\pi/2$ we obtain the part in Quadrant IV; and $3\pi/2 \le t \le 2\pi$ gives us the part in Quadrant II.

(b) We may find an equation in x and y for the curve by employing trigonometric identities and algebraic manipulations. Writing $x = 2 \sin t \cos t$ and squaring gives us

$$x^2 = 4 \sin^2 t \cos^2 t \quad \text{or} \quad x^2 = 4(1 - \cos^2 t) \cos^2 t.$$

Since $y = \cos t$, this leads to

$$x^2 = 4(1 - y^2)y^2,$$

which is an equation for the curve.

If we wish to express y in terms of x, we could rewrite the last equation as

$$4y^4 - 4y^2 + x^2 = 0$$

and use the Quadratic Formula to solve for y^2 as follows:

$$y^2 = \frac{4 \pm \sqrt{16 - 16x^2}}{8} = \frac{1 \pm \sqrt{1 - x^2}}{2}$$

Taking square roots,

$$y = \pm \sqrt{\frac{1 \pm \sqrt{1 - x^2}}{2}}.$$

These complicated equations clearly indicate the advantage of expressing the curve in parametric form. •

A curve C is **smooth** if it has a parametrization $x = f(t)$, $y = g(t)$ on an interval I such that the derivatives f' and g' are continuous on I and are not simultaneously zero, except possibly at endpoints of I. A curve C is **piecewise smooth** if the interval I can be partitioned into closed subintervals such that C is smooth on each subinterval. The graph of a smooth curve has no corners. The curves given in Examples 1–5 are smooth. The curve described in the next example is piecewise smooth.

EXAMPLE 6 The curve traced by a fixed point P on the circumference of a circle as the circle rolls along a straight line in a plane is called a **cycloid.** Find parametric equations for a cycloid.

SOLUTION Suppose the circle has radius a and that it rolls along (and above) the x-axis in the positive direction. If one position of P is the origin, then Figure 13.7 displays part of the curve and a possible position of the circle.

Let K denote the center of the circle and T the point of tangency with the x-axis. We introduce a parameter t as the radian measure of angle TKP. The distance the circle has rolled is $d(O, T) = at$. Consequently, the coordinates of K are (at, a). If we consider an $x'y'$-coordinate system with origin at $K(at, a)$ and if $P(x', y')$ denotes the point P relative to this system, then by the Translation of Axes Formulas (12.3) with $h = at$ and $k = a$,

$$x = at + x', \quad y = a + y'.$$

If, as in Figure 13.8, θ denotes an angle in standard position on the $x'y'$-system, then $\theta = (3\pi/2) - t$. Hence

$$x' = a \cos \theta = a \cos \left[(3\pi/2) - t\right] = -a \sin t$$

$$y' = a \sin \theta = a \sin \left[(3\pi/2) - t\right] = -a \cos t,$$

and substitution in $x = at + x'$, $y = a + y'$ gives us parametric equations for the cycloid:

$$x = a(t - \sin t), \quad y = a(1 - \cos t); \qquad t \text{ in } \mathbb{R}.$$

Differentiating the parametric equations of the cycloid,

$$\frac{dx}{dt} = a(1 - \cos t), \quad \frac{dy}{dt} = a \sin t.$$

These derivatives are continuous for every t, but are simultaneously 0 if $t = 2\pi n$ for every integer n. Hence the cycloid is not smooth. Note that the points corresponding to $t = 2\pi n$ are the x-intercepts, and the cycloid has a corner at each such point (see Figure 13.7). The graph is piecewise smooth, since it is smooth on every t-interval $[2\pi n, 2\pi(n + 1)]$ for an integer n. •

If $a < 0$, then the graph of $x = a(t - \sin t)$, $y = a(1 - \cos t)$ is the inverted cycloid that results if the circle of Example 6 rolls *below* the x-axis. This curve has a number of important physical properties. In particular, suppose a thin wire passes through two fixed points A and B as illustrated in Figure 13.9, and that the shape of the wire can be changed by bending it in any manner. Suppose further that a bead is allowed to slide along the wire and the only force acting on the bead is gravity. We now ask which of all the possible paths will allow the bead to slide from A to B in the least amount of time. It is natural to believe that the desired path is the straight line segment from A to B; however, this is not the correct answer. The path that requires the least time coin-

FIGURE 13.7
Cycloid

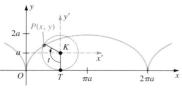

FIGURE 13.8

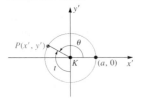

FIGURE 13.9

cides with the graph of the inverted cycloid with A at the origin. Because the velocity of the bead increases more rapidly along the cycloid than along the line through A and B, the bead reaches B more rapidly, even though the distance is greater. A proof of this result is omitted from this text.

There is another interesting property of this **curve of least descent.** Suppose that A is the origin and B is the point with x-coordinate $\pi|a|$, that is, the lowest point on the cycloid in the first arc to the right of A. If the bead is released at *any* point between A and B, the time required for it to reach B is always the same.

Variations of the cycloid occur in the investigation of applications. For example, if a motorcycle wheel rolls along a straight road, then the curve traced by a fixed point on one of the spokes is a cycloidlike curve. In this case, the curve does not have sharp corners, nor does it intersect the road (the x-axis) as does the graph of a cycloid. If the wheel of a train rolls along a railroad track, then the curve traced by a fixed point on the circumference of the wheel (which extends below the track) contains loops at regular intervals (see Exercise 44). Other cycloids are defined in Exercises 37 and 38.

EXERCISES 13.1

Exer. 1–20: (a) Sketch the graph of the curve C, and (b) find a rectangular equation of a graph that contains the points on C.

1 $x = t - 2, \; y = 2t + 3; \quad 0 \le t \le 5$

2 $x = 1 - 2t, \; y = 1 + t; \quad -1 \le t \le 4$

3 $x = t^2 + 1, \; y = t^2 - 1; \quad -2 \le t \le 2$

4 $x = t^3 + 1, \; y = t^3 - 1; \quad -2 \le t \le 2$

5 $x = 4t^2 - 5, \; y = 2t + 3; \quad t$ in $\mathbb{R}$

6 $x = t^3, \; y = t^2; \quad t$ in $\mathbb{R}$

7 $x = e^t, \; y = e^{-2t}; \quad t$ in $\mathbb{R}$

8 $x = \sqrt{t}, \; y = 3t + 4; \quad t \ge 0$

9 $x = 2 \sin t, \; y = 3 \cos t; \quad 0 \le t \le 2\pi$

10 $x = \cos t - 2, \; y = \sin t + 3; \quad 0 \le t \le 2\pi$

11 $x = \sec t, \; y = \tan t; \quad -\pi/2 < t < \pi/2$

12 $x = \cos 2t, \; y = \sin t; \quad -\pi \le t \le \pi$

13 $x = t^2, \; y = 2 \ln t; \quad t > 0$

14 $x = \cos^3 t, \; y = \sin^3 t; \quad 0 \le t \le 2\pi$

15 $x = \sin t, \; y = \csc t; \quad 0 < t \le \pi/2$

16 $x = e^t, \; y = e^{-t}; \quad t$ in $\mathbb{R}$

17 $x = \cosh t, \; y = \sinh t; \quad t$ in $\mathbb{R}$

18 $x = 3 \cosh t, \; y = 2 \sinh t; \quad t$ in $\mathbb{R}$

19 $x = t, \; y = \sqrt{t^2 - 1}; \quad |t| \ge 1$

20 $x = -2\sqrt{1 - t^2}, \; y = t; \quad |t| \le 1$

Exer. 21–24: Sketch the graph of the curve C.

21 $x = t, \; y = \sqrt{t^2 - 2t + 1}; \quad 0 \le t \le 4$

22 $x = 2t, \; y = 8t^3; \quad -1 \le t \le 1$

23 $x = (t + 1)^3, \; y = (t + 2)^2; \quad 0 \le t \le 2$

24 $x = \tan t, \; y = 1; \quad -\pi/2 < t < \pi/2$

Exer. 25–26: Curves C_1, C_2, C_3, and C_4 are given parametrically for t in $\mathbb{R}$. Sketch their graphs and discuss their similarities and differences.

25 C_1: $x = t^2, \; y = t$ C_2: $x = t^4, \; y = t^2$
   C_3: $x = \sin^2 t, \; y = \sin t$ C_4: $x = e^{2t}, \; y = -e^t$

26 C_1: $x = t, \; y = 1 - t$
   C_2: $x = 1 - t^2, \; y = t^2$
   C_3: $x = \cos^2 t, \; y = \sin^2 t$
   C_4: $x = -t + \ln t, \; y = 1 + t - \ln t$

Exer. 27–28: The parametric equations give the position of a point $P(x, y)$ at time t. Describe the motion of the point during the indicated time interval.

27 (a) $x = \cos t, \; y = \sin t; \quad 0 \le t \le \pi$
   (b) $x = \sin t, \; y = \cos t; \quad 0 \le t \le \pi$
   (c) $x = t, \; y = \sqrt{1 - t^2}; \quad -1 \le t \le 1$

28 (a) $x = t^2, \; y = 1 - t^2; \quad 0 \le t \le 1$
   (b) $x = 1 - \ln t, \; y = \ln t; \quad 1 \le t \le e$
   (c) $x = \cos^2 t, \; y = \sin^2 t; \quad 0 \le t \le 2\pi$

29 If $P_1(x_1, y_1)$ and $P_2(x_2, y_2)$ are distinct points, show that
   $$x = (x_2 - x_1)t + x_1, \quad y = (y_2 - y_1)t + y_1; \qquad t \text{ in } \mathbb{R}$$
   are parametric equations for the line l through P_1 and P_2. Find three other pairs of parametric equations for the line l. Show that l has an infinite number of parametrizations.

30 Describe the difference between the graph of the hyperbola $b^2x^2 - a^2y^2 = a^2b^2$ and the graph of
   $$x = a \cosh t, \quad y = b \sinh t; \qquad t \text{ in } \mathbb{R}.$$

31 Show that
$$x = a \cos t + h, \quad y = b \sin t + k; \qquad 0 \le t \le 2\pi$$
is a parametrization of an ellipse with center at the point (h, k) and semiaxes of lengths a and b.

32 Find parametric equations for the parabola that has
(a) vertex $V(0, 0)$; focus $F(0, p)$
(b) vertex $V(h, k)$; focus $F(h, k + p)$

33 Lissajous figures are often used in the theory of electricity to display the relationship between two voltages or currents (see Example 5).
(a) If $f(t) = a \sin \omega t$ and $g(t) = b \cos \omega t$ for time $t \ge 0$, describe the resulting Lissajous figure.
(b) Suppose $f(t) = a \sin \omega_1 t$ and $g(t) = b \sin \omega_2 t$ for positive rational numbers ω_1 and ω_2. The fraction ω_2/ω_1 may be written as m/n for positive integers m and n. Show that $f(t + p) = f(t)$ and $g(t + p) = g(t)$ for the real number $p = 2\pi n/\omega_1$. Conclude that the curve retraces itself every p units of time.

34 Shown in the figure is a computer-generated graph of the Lissajous figure given parametrically by $x = 2 \sin 3t$, $y = 3 \sin 1.5t$; $t \ge 0$.
(a) Find the period of the figure, that is, the length of the smallest t-interval that traces the curve.
(b) Find the maximum distance of a point on the curve from the origin.

EXERCISE 34

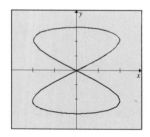

Exer. 35–36: Sketch the Lissajous figure.

35 $x = \sin 3t, \ y = \cos t; \quad t \ge 0$

36 $x = 4 \sin 2t, \ y = 2 \cos 3t; \quad t \ge 0$

37 A circle C of radius b rolls on the outside of a second circle having equation $x^2 + y^2 = a^2$, with $b < a$. Let P be a fixed point on C and let the initial position of P be $A(a, 0)$ (see figure). If the parameter t is the angle from the positive x-axis to the line segment from O to the center of C, show that parametric equations for the curve traced by P (called an *epicycloid*) are
$$x = (a + b) \cos t - b \cos \left(\frac{a + b}{b} t \right)$$
$$y = (a + b) \sin t - b \sin \left(\frac{a + b}{b} t \right)$$
for $0 \le t \le 2\pi$.

EXERCISE 37

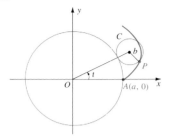

38 If the circle C of Exercise 37 rolls on the inside of the second circle (see figure) then the curve traced by P is a *hypocycloid*. Show that parametric equations for this curve are
$$x = (a - b) \cos t + b \cos \left(\frac{a - b}{b} t \right)$$
$$y = (a - b) \sin t - b \sin \left(\frac{a - b}{b} t \right)$$
for $0 \le t \le 2\pi$. If $b = \frac{1}{4} a$, show that
$$x = a \cos^3 t, \qquad y = a \sin^3 t$$
and sketch the graph of the curve.

EXERCISE 38

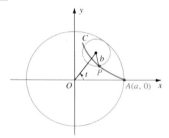

39 If $b = \frac{1}{3} a$ in Exercise 37, find parametric equations for the epicycloid and sketch the graph.

40 The radius of circle B is one-third that of circle A. How many revolutions will circle B make as it rolls around circle A until it reaches its starting point? (*Hint:* Use Exercise 39.)

41 A *nephroid* is an epicycloid obtained by letting $b = \frac{1}{2} a$ in Exercise 37. Find parametric equations for this nephroid and sketch the graph.

42 The *Folium of Descartes* (see Exercise 22 of Section 3.7) has the parametrization
$$x = \frac{3at}{1 + t^3}, \quad y = \frac{3at^2}{1 + t^3}; \qquad t \text{ in } \mathbb{R}$$
for a positive constant a.
(a) Find the t-intervals that give points in Quadrant I; in Quadrant II; in Quadrant III.
(b) Sketch the graph for $a = 1$.

43 If a string is unwound from around a circle of radius a and is kept taut in the plane of the circle, then a fixed point P on the string will trace a curve called the *involute of the* *circle*. Let the circle be chosen as in the figure. If the parameter t is the measure of the indicated angle and the initial position of P is $A(a, 0)$, show that parametric equations for the involute are

$$x = a(\cos t + t \sin t), \quad y = a(\sin t - t \cos t).$$

EXERCISE 43

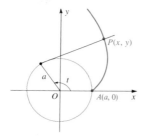

44 Generalize the cycloid of Example 6 to the case where P is any point on a fixed line through the center C of the circle. If $b = d(C, P)$, derive the parametric equations

$$x = at - b \sin t, \quad y = a - b \cos t.$$

Sketch a typical graph if $b < a$ (a *curtate cycloid*) and if $b > a$ (a *prolate cycloid*). The term *trochoid* is sometimes used for either of these curves.

45 A point B on the circle $x^2 + y^2 = 1$ is connected to a fixed point $A(2, 0)$ and the midpoint $P(x, y)$ of line segment AB is determined (see figure). If B has coordinates $(\cos \omega t, \sin \omega t)$ for $0 \le t \le 2\pi/\omega$ and a positive constant ω, find parametric equations for the curve traced by P, and sketch its graph.

EXERCISE 45

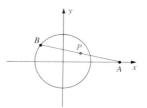

46 Given the circle $x^2 + (y - a)^2 = a^2$ and its tangent line $y = 2a$, a point P is obtained as follows. A point B is selected on the circle and chord OB is extended to meet line $y = 2a$ at A. The point $P(x, y)$ has the same x-coordinate as A and the same y-coordinate as B (see figure).

(a) If t is the radian measure of the angle made by OB and the positive x-axis, show that $x = 2a \cot t$ and $y = 2a \sin^2 t$ for $0 < t < \pi$. The curve consisting of all points $P(x, y)$ is called the *Witch of Agnesi*.

(b) Show that the area bounded by the witch is four times the area of the circle. (*Hint:* First show that $y = 8a^3/(x^2 + 4a^2)$.)

EXERCISE 46

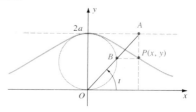

13.2 TANGENT LINES AND ARC LENGTH

The curve C with parametric equations

$$x = 2t, \quad y = t^2 - 1; \quad -1 \le t \le 2$$

can also be represented by an equation of the form $y = k(x)$ for a function k defined on a suitable interval. In Example 1 of the preceding section, we eliminated the parameter t and arrived at

$$y = k(x) = \tfrac{1}{4}x^2 - 1 \quad \text{for} \quad -2 \le x \le 4.$$

It follows that the slope of the tangent line at any point $P(x, y)$ on C is

$$k'(x) = \tfrac{1}{2}x \quad \text{or} \quad k'(x) = \tfrac{1}{2}(2t) = t.$$

We shall next derive a formula that can be used to find the slope directly from the parametric equations.

Let $x = f(t)$, $y = g(t)$ be parametric equations for a curve C and suppose that C can also be represented by an equation of the form $y = k(x)$ for a function k. We shall assume that f, g, and k are differentiable throughout their domains. If $P(f(t), g(t))$ is a point on C, then the coordinates satisfy the equation $y = k(x)$, that is,

$$g(t) = k(f(t)).$$

Applying the Chain Rule,

$$g'(t) = k'(f(t))f'(t) = k'(x)f'(t).$$

Hence the slope of the tangent line at P is

$$k'(x) = \frac{g'(t)}{f'(t)} \quad \text{provided } f'(t) \neq 0.$$

To find the points at which the tangent line is horizontal, we solve the equation $g'(t) = 0$. If $f'(t) = 0$ and $g'(t) \neq 0$ for some t, then C has a vertical tangent line at the point corresponding to t. If $f'(t)$ and $g'(t)$ are simultaneously zero, then further information is required.

The conditions imposed in the preceding discussion are fulfilled if C is smooth and $f'(t) \neq 0$ for every t in a closed interval $[a, b]$. In this case either $f'(t) > 0$ or $f'(t) < 0$ throughout $[a, b]$, and it follows from Theorem (7.27) or Theorem (7.28) that f has an inverse function f^{-1}. Since $x = f(t)$, we may write $t = f^{-1}(x)$ for x in the interval $[f(a), f(b)]$. Hence

$$y = g(t) = g(f^{-1}(x)) = k(x)$$

where k is the composite function $g \circ f^{-1}$.

An easy way to remember the formula for $k'(x)$ is to use differential notation, as in the next theorem.

THEOREM (13.3)

> If a smooth curve C is given parametrically by $x = f(t)$, $y = g(t)$, then the slope dy/dx of the tangent line to C at $P(x, y)$ is
>
> $$\frac{dy}{dx} = \frac{dy/dt}{dx/dt} \quad \text{provided} \quad \frac{dx}{dt} \neq 0$$

EXAMPLE 1 Find the slopes of the tangent and normal lines at the point $P(x, y)$ on the curve with parametrization $x = 2t$, $y = t^2 - 1$; $-1 \leq t \leq 2$.

SOLUTION The curve is the same as that considered in Example 1 of the preceding section (see Figure 13.2). Using Theorem (13.3) with $x = 2t$ and $y = t^2 - 1$, the slope of the tangent line at $P(x, y)$ is

$$\frac{dy}{dx} = \frac{dy/dt}{dx/dt} = \frac{2t}{2} = t.$$

This agrees with the discussion at the beginning of this section, where we used the form $y = k(x)$ to show that $m = \frac{1}{2}x = t$.

The slope of the normal line is the negative reciprocal $-1/t$, provided $t \neq 0$. •

EXAMPLE 2 If C has parametric equations $x = t^3 - 3t$, $y = t^2 - 5t - 1$; t in $\mathbb{R}$, find an equation of the tangent line to C at the point corresponding to $t = 2$. For what values of t is the tangent line horizontal or vertical?

SOLUTION A computer-generated graph is shown in Figure 13.10. As is usually the case for such graphs, it is difficult, if not impossible, to find the exact slope of the tangent line at any point $P(x, y)$ on the curve. However,

FIGURE 13.10
$x = t^3 - 3t, \; y = t^2 - 5t - 1; \quad t$ in $\mathbb{R}$

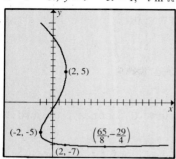

FIGURE 13.11

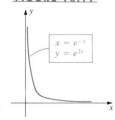

applying Theorem (13.3),

$$\frac{dy}{dx} = \frac{dy/dt}{dx/dt} = \frac{2t - 5}{3t^2 - 3}.$$

Using the parametric equations for C, the point corresponding to $t = 2$ is $(2, -7)$ (see Figure 13.10). From the formula for dy/dx, with $t = 2$, the slope m of the tangent line at this point is

$$m = \frac{2(2) - 5}{3(2^2) - 3} = -\frac{1}{9}.$$

Applying the Point-Slope Form, an equation of this tangent line is

$$y + 7 = -\tfrac{1}{9}(x - 2) \quad \text{or} \quad x + 9y + 61 = 0.$$

The tangent line is horizontal if $dy/dt = 0$, that is, if $2t - 5 = 0$ or $t = \tfrac{5}{2}$. The corresponding point on C is $(\tfrac{65}{8}, -\tfrac{29}{4})$, as shown in Figure 13.10.

The tangent line is vertical if $3t^2 - 3 = 0$. Thus, there are vertical tangent lines at the points corresponding to $t = 1$ and $t = -1$, that is, at $(2, 5)$ and $(-2, 5)$. •

If C is given by $x = f(t)$, $y = g(t)$, and if y' is a differentiable function of t, we can find d^2y/dx^2 by applying Theorem (13.3) to y' as follows:

(13.4)
$$\frac{d^2y}{dx^2} = \frac{d}{dx}(y') = \frac{dy'/dt}{dx/dt}$$

Note that $d^2y/dx^2 \neq g''(t)/f''(t)$.

EXAMPLE 3 Suppose C has parametric equations

$$x = e^{-t}, \quad y = e^{2t}; \qquad t \text{ in } \mathbb{R}.$$

(a) Sketch the graph of C.

(b) Find d^2y/dx^2 directly from the parametric equations.

(c) Define a function k that has the same graph as C, and check the answer to part (b).

(d) Discuss the concavity of C.

SOLUTION

(a) Since $x = e^{-t} = 1/e^t$, we have $e^t = 1/x$. Substitution in $y = e^{2t} = (e^t)^2$ gives us the equation

$$y = \left(\frac{1}{x}\right)^2 = \frac{1}{x^2}.$$

Remembering that $x = e^{-t} > 0$ leads to the graph in Figure 13.11.

(b) To find d^2y/dx^2 we use (13.4). Thus

$$y' = \frac{dy}{dx} = \frac{dy/dt}{dx/dt} = \frac{2e^{2t}}{-e^{-t}} = -2e^{3t},$$

and

$$\frac{d^2y}{dx^2} = \frac{dy'}{dx} = \frac{dy'/dt}{dx/dt} = \frac{-6e^{3t}}{-e^{-t}} = 6e^{4t}.$$

(c) From part (a), a function k that has the same graph as C is given by

$$k(x) = \frac{1}{x^2} = x^{-2} \qquad \text{for } x > 0.$$

Since $k'(x) = -2x^{-3}$,

$$k''(x) = 6x^{-4} = 6(e^{-t})^{-4} = 6e^{4t}$$

which is in agreement with part (b).

(d) Since $d^2y/dx^2 = 6e^{4t} > 0$ for every t, the curve C is concave upward at every point. •

If a curve is the graph of an equation $y = f(x)$ and f is smooth on an interval $[a, b]$, then we may find its arc length by evaluating the integral $\int_a^b \sqrt{1 + [f'(x)]^2}\, dx$ (see Definition (6.9)). We shall next obtain a formula for finding lengths of parametrized curves.

Suppose a smooth curve C is given parametrically by

$$x = f(t), \quad y = g(t); \qquad a \le t \le b.$$

Furthermore, suppose C does not intersect itself; that is, different values of t between a and b determine different points on C. Consider a partition P of $[a, b]$ given by $a = t_0 < t_1 < t_2 < \cdots < t_n = b$. Let $\Delta t_k = t_k - t_{k-1}$ and let $P_k = (f(t_k), g(t_k))$ be the point on C that corresponds to t_k. If $d(P_{k-1}, P_k)$ is the length of the line segment $P_{k-1}P_k$, then the length L_P of the broken line shown (in black) in Figure 13.12 is

FIGURE 13.12

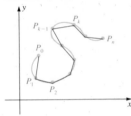

$$L_P = \sum_{k=1}^{n} d(P_{k-1}, P_k).$$

As in Section 6.5, we write

$$L = \lim_{\|P\| \to 0} L_P$$

and call L the **length of C** from P_0 to P_n if, for every $\varepsilon > 0$, there exists a $\delta > 0$ such that $|L_P - L| < \varepsilon$ for every partition P with $\|P\| < \delta$.

By the Distance Formula,

$$d(P_{k-1}, P_k) = \sqrt{[f(t_k) - f(t_{k-1})]^2 + [g(t_k) - g(t_{k-1})]^2}.$$

Applying the Mean Value Theorem (4.12), there exist numbers w_k and z_k in the open interval (t_{k-1}, t_k) such that

$$f(t_k) - f(t_{k-1}) = f'(w_k)\, \Delta t_k,$$

$$g(t_k) - g(t_{k-1}) = g'(z_k)\, \Delta t_k.$$

Substituting in the formula for $d(P_{k-1}, P_k)$ and removing the common factor $(\Delta t_k)^2$ from the radicand gives us

$$d(P_{k-1}, P_k) = \sqrt{[f'(w_k)]^2 + [g'(z_k)]^2}\; \Delta t_k.$$

Consequently

$$L = \lim_{\|P\| \to 0} L_P = \lim_{\|P\| \to 0} \sum_{k=1}^{n} \sqrt{[f'(w_k)]^2 + [g'(z_k)]^2}\; \Delta t_k,$$

provided the limit exists. If $w_k = z_k$ for every k, then the sums are Riemann sums for the function defined by $\sqrt{[f'(t)]^2 + [g'(t)]^2}$. The limit of these sums is

$$L = \int_a^b \sqrt{[f'(t)]^2 + [g'(t)]^2} \, dt.$$

The limit exists even if $w_k \neq z_k$; however, the proof requires advanced methods and is omitted. The next theorem summarizes this discussion.

THEOREM (13.5)

> If a smooth curve C is given parametrically by $x = f(t)$, $y = g(t)$; $a \leq t \leq b$, and if C does not intersect itself, except possibly at the end points of $[a, b]$, then the length L of C is
>
> $$L = \int_a^b \sqrt{[f'(t)]^2 + [g'(t)]^2} \, dt = \int_a^b \sqrt{\left(\frac{dx}{dt}\right)^2 + \left(\frac{dy}{dt}\right)^2} \, dt$$

The integral formula in Theorem (13.5) is not necessarily true if C intersects itself. For example, if C has the parametrization $x = \cos t$, $y = \sin t$; $0 \leq t \leq 4\pi$, then the graph is a unit circle with center at the origin. If t varies from 0 to 4π, the circle is traced twice, and hence intersects itself infinitely many times. If we use Theorem (13.5) with $a = 0$ and $b = 4\pi$, we obtain the incorrect value 4π for the length of C. The correct value 2π can be obtained by using the t-interval $[0, 2\pi]$. Note that in this case the curve intersects itself only at the endpoints 0 and 2π of the interval, which is allowable by the theorem.

If a curve C is described in rectangular coordinates by an equation of the form $y = k(x)$ such that k' is continuous on $[a, b]$, then parametric equations for C are

$$x = t, \quad y = k(t); \quad a \leq t \leq b.$$

In this case $\quad \dfrac{dx}{dt} = 1, \quad \dfrac{dy}{dt} = k'(t) = k'(x), \quad dt = dx$

and from Theorem (13.5),

$$L = \int_a^b \sqrt{1 + [k'(x)]^2} \, dx.$$

This is in agreement with the arc length formula given in Definition (6.9).

EXAMPLE 4 Find the length of one arch of the cycloid that has parametric equations $x = t - \sin t$, $y = 1 - \cos t$.

SOLUTION The graph has the shape illustrated in Figure 13.7. The radius a of the circle is 1. One arch is obtained if t varies from 0 to 2π. Applying Theorem (13.5),

$$L = \int_0^{2\pi} \sqrt{(1 - \cos t)^2 + (\sin t)^2} \, dt$$

$$= \int_0^{2\pi} \sqrt{1 - 2\cos^2 t + \cos^2 t + \sin^2 t} \, dt.$$

Since $\cos^2 t + \sin^2 t = 1$, the integrand reduces to

$$\sqrt{2 - 2\cos t} = \sqrt{2}\sqrt{1 - \cos t}.$$

Thus
$$L = \int_0^{2\pi} \sqrt{2}\sqrt{1 - \cos t}\ dt.$$

By a half-angle formula, $\sin^2 \frac{1}{2}t = \frac{1}{2}(1 - \cos t)$, or equivalently,
$$1 - \cos t = 2\sin^2 \tfrac{1}{2}t.$$
Hence
$$\sqrt{1 - \cos t} = \sqrt{2\sin^2 \tfrac{1}{2}t} = \sqrt{2}\,|\sin \tfrac{1}{2}t|.$$

The absolute value sign may be deleted, since if $0 \le t \le 2\pi$, then $0 \le \frac{1}{2}t \le \pi$ and hence $\sin \frac{1}{2}t \ge 0$. Consequently,
$$L = \int_0^{2\pi} \sqrt{2}\sqrt{2}\sin \tfrac{1}{2}t\ dt = 2\int_0^{2\pi} \sin \tfrac{1}{2}t\ dt$$
$$= -4\cos \tfrac{1}{2}t\Big]_0^{2\pi} = (-4)(-1) - (-4)(1) = 8 \quad \bullet$$

To remember Theorem (13.5), recall that if ds is the differential of arc length, then by Theorem (6.12),
$$(ds)^2 = (dx)^2 + (dy)^2.$$

Assuming that ds and dt are positive, we have the following.

DIFFERENTIAL OF ARC LENGTH (13.6)

$$ds = \sqrt{(dx)^2 + (dy)^2} = \sqrt{\left(\frac{dx}{dt}\right)^2 + \left(\frac{dy}{dt}\right)^2}\ dt$$

The formula for arc length in Theorem (13.5) may then be written $L = \int_{t=a}^{t=b} ds$. The limits of integration specify that the independent variable is t, and not s.

If f is a function that is smooth and nonnegative for $a \le x \le b$, then by Definition (6.14), the area S of the surface that is generated by revolving the graph of $y = f(x)$ about the y-axis (see Figure 13.13) is given by

FIGURE 13.13

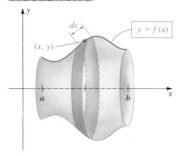

$$S = \int_{x=a}^{x=b} 2\pi y\ ds$$

with $ds = \sqrt{1 + [f'(x)]^2}\ dx$. We can regard $2\pi y\ ds$ as the surface area of a frustum of a cone of slant height ds and average radius y (see (6.13)).

If a curve C is given parametrically by $x = f(t)$, $y = g(t)$; $a \le t \le b$, and if $g(t) \ge 0$ throughout $[a, b]$, we can use an argument similar to that given in Section 6.5 to find the area of the surface generated by revolving C about the y-axis. The formula $S = \int_{t=a}^{t=b} 2\pi y\ ds$ is obtained, with ds the *parametric* differential of arc length (13.6). Let us state this for reference as follows.

THEOREM (13.7)

Let a smooth curve C be given by $x = f(t)$, $y = g(t)$; $a \le t \le b$, and suppose C does not intersect itself, except possibly at the endpoints of $[a, b]$. If $g(t) \ge 0$ throughout $[a, b]$, then the area S of the surface of revolution obtained by revolving C about the x-axis is

$$S = \int_a^b 2\pi y \sqrt{\left(\frac{dx}{dt}\right)^2 + \left(\frac{dy}{dt}\right)^2}\ dt = \int_a^b 2\pi g(t)\sqrt{[f'(t)]^2 + [g'(t)]^2}\ dt$$

FIGURE 13.14

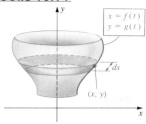

The formula for S in Theorem (13.7) can be extended to the case in which $y = g(t)$ is negative for some t in $[a, b]$ by merely replacing the variable y that precedes ds by $|y|$.

If the curve C in Theorem (13.7) is revolved about the y-axis, and if $x = f(t) \geq 0$ for $a \leq t \leq b$ (see Figure 13.14), then

$$S = \int_{t=a}^{t=b} 2\pi x \, ds = \int_a^b 2\pi x \sqrt{\left(\frac{dx}{dt}\right)^2 + \left(\frac{dy}{dt}\right)^2} \, dt.$$

In this case we may regard $2\pi x \, ds$ as the surface area of a frustum of a cone of slant height ds and average radius x.

EXAMPLE 5 Verify that the surface area of a sphere of radius a is $4\pi a^2$.

SOLUTION If C is the upper half of the circle $x^2 + y^2 = a^2$, then the spherical surface may be obtained by revolving C about the x-axis. Parametric equations for C are

$$x = a \cos t, \quad y = a \sin t; \quad 0 \leq t \leq \pi.$$

Applying Theorem (13.7) and using the identity $\sin^2 t + \cos^2 t = 1$,

$$S = \int_0^\pi 2\pi a \sin t \sqrt{a^2 \sin^2 t + a^2 \cos^2 t} \, dt = 2\pi a^2 \int_0^\pi \sin t \, dt$$

$$= -2\pi a^2 \Big[\cos t\Big]_0^\pi = -2\pi a^2[-1 - 1] = 4\pi a^2 \quad \bullet$$

EXERCISES 13.2

Exer. 1–8: Find the slope of the tangent line at the point corresponding to $t = 1$ on the curve (see Exercises 3–10 of Section 13.1).

1 $x = t^2 + 1, \; y = t^2 - 1; \quad -2 \leq t \leq 2$

2 $x = t^3 + 1, \; y = t^3 - 1; \quad -2 \leq t \leq 2$

3 $x = 4t^2 - 5, \; y = 2t + 3; \quad t \text{ in } \mathbb{R}$

4 $x = t^3, \; y = t^2; \quad t \text{ in } \mathbb{R}$

5 $x = e^t, \; y = e^{-2t}; \quad t \text{ in } \mathbb{R}$

6 $x = \sqrt{t}, \; y = 3t + 4; \quad t \geq 0$

7 $x = 2 \sin t, \; y = 3 \cos t; \quad 0 \leq t \leq 2\pi$

8 $x = \cos t - 2, \; y = \sin t + 3; \quad 0 \leq t \leq 2\pi$

Exer. 9–14: (a) Find the points on the curve at which the tangent line is either horizontal or vertical. (b) Find d^2y/dx^2.

9 $x = 4t^2, \; y = t^3 - 12t; \quad t \text{ in } \mathbb{R}$

10 $x = t^3 + 1, \; y = t^2 - 2t; \quad t \text{ in } \mathbb{R}$

11 $x = 3t^2 - 6t, \; y = \sqrt{t}; \quad t \geq 0$

12 $x = \sqrt[3]{t}, \; y = \sqrt[3]{t} - t; \quad t \text{ in } \mathbb{R}$

13 $x = \cos^3 t, \; y = \sin^3 t; \quad 0 \leq t \leq 2\pi$

14 $x = \cosh t, \; y = \sinh t; \quad t \text{ in } \mathbb{R}$

Exer. 15–16: Shown is a computer-generated Lissajous figure (see Example 5, Section 13.1). (a) Determine where the tangent lines are horizontal or vertical, and (b) find the period of the curve.

15 $x = 4 \sin 2t$
    $y = 2 \cos 3t$

16 $x = 5 \sin \frac{1}{3}t$
    $y = 4 \sin 2t$

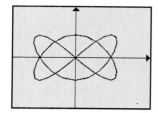
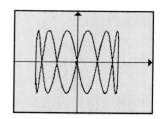

Exer. 17–22: Find the length of the curve.

17 $x = 5t^2, \; y = 2t^3; \quad 0 \leq t \leq 1$

18 $x = 3t, \; y = 2t^{3/2}; \quad 0 \leq t \leq 4$

19 $x = e^t \cos t, \; y = e^t \sin t; \quad 0 \leq t \leq \pi/2$

20 $x = \cos 2t, \; y = \sin^2 t; \quad 0 \leq t \leq \pi$

21 $x = t \cos t - \sin t, \; y = t \sin t + \cos t; \quad 0 \leq t \leq \pi/2$

22 $x = \cos^3 t, \; y = \sin^3 t; \quad 0 \leq t \leq \pi/2$

Exer. 23–28: Find the area of the surface generated by revolving the curve about the *x*-axis.

23 $x = t^2$, $y = 2t$; $\quad 0 \le t \le 4$

24 $x = 4t$, $y = t^3$; $\quad 1 \le t \le 2$

25 $x = t^2$, $y = t - \frac{1}{3}t^3$; $\quad 0 \le t \le 1$

26 $x = 4t^2 + 1$, $y = 3 - 2t$; $\quad -2 \le t \le 0$

27 $x = a(t - \sin t)$, $y = a(1 - \cos t)$; $\quad 0 \le t \le 2\pi$

28 $x = t$, $y = \frac{1}{3}t^3 + \frac{1}{4}t^{-1}$; $\quad 1 \le t \le 2$

Exer. 29–32: Find the area of the surface generated by revolving the curve about the *y*-axis.

29 $x = 4t^{1/2}$, $y = \frac{1}{2}t^2 + t^{-1}$; $\quad 1 \le t \le 4$

30 $x = 3t$, $y = t + 1$; $\quad 0 \le t \le 5$

31 $x = e^t \sin t$, $y = e^t \cos t$; $\quad 0 \le t \le \pi/2$

32 $x = 3t^2$, $y = 2t^3$; $\quad 0 \le t \le 1$

13.3 POLAR COORDINATES

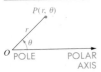

FIGURE 13.15

In a rectangular coordinate system the ordered pair (*a*, *b*) denotes the point whose *x*- and *y*-coordinates are *a* and *b*, respectively. Another method for representing points is to use *polar coordinates*. We begin with a fixed point *O* (the **origin**, or **pole**) and a directed half-line (the **polar axis**) with endpoint *O*. Next we consider any point *P* in the plane different from *O*. If, as illustrated in Figure 13.15, $r = d(O, P)$ and θ denotes the measure of any angle determined by the polar axis and *OP*, then *r* and θ are **polar coordinates** of *P* and the symbols (r, θ) or $P(r, \theta)$ are used to denote *P*. As usual, θ is considered positive if the angle is generated by a counterclockwise rotation of the polar axis and negative if the rotation is clockwise. We shall use radian measure for θ.

The polar coordinates of a point are not unique. For example $(3, \pi/4)$, $(3, 9\pi/4)$, and $(3, -7\pi/4)$ all represent the same point (see Figure 13.16). We shall also allow *r* to be negative. In this event, instead of measuring $|r|$ units along the terminal side of the angle θ, we measure along the half-line with endpoint *O* that has direction *opposite* to that of the terminal side. The points corresponding to the pairs $(-3, 5\pi/4)$ and $(-3, -3\pi/4)$ are also plotted in Figure 13.16.

FIGURE 13.16

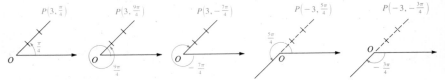

Finally, we agree that the pole *O* has polar coordinates $(0, \theta)$ for *any* θ. An assignment of ordered pairs of the form (r, θ) to points in a plane will be referred to as a **polar coordinate system** and the plane will be called an *rθ-plane.*

A **polar equation** is an equation in *r* and θ. A **solution** of a polar equation is an ordered pair (a, b) that leads to equality if *a* is substituted for *r* and *b* for θ. The **graph** of a polar equation is the set of all points (in an *rθ-plane*) that correspond to the solutions.

EXAMPLE 1 Sketch the graph of the polar equation $r = 4 \sin \theta$.

SOLUTION The following table displays some solutions of the equation. We have included a third row in the table that gives one-decimal-place approximations to *r*.

θ	0	$\pi/6$	$\pi/4$	$\pi/3$	$\pi/2$	$2\pi/3$	$3\pi/4$	$5\pi/6$	π
r	0	2	$2\sqrt{2}$	$2\sqrt{3}$	4	$2\sqrt{3}$	$2\sqrt{2}$	2	0
r (approx.)	0	2	2.8	3.4	4	3.4	2.8	2	0

In rectangular coordinates, the graph of the equation consists of sine waves of amplitude 4 and period 2π. However, if polar coordinates are used, then the points that correspond to the pairs in the table appear to lie on a circle of radius 2 and we draw the graph accordingly (see Figure 13.17). As an aid to plotting points, we have extended the polar axis in the negative direction and introduced a vertical line through the pole.

The proof that the graph of $r = 4 \sin \theta$ is a circle will be given in Example 6. Additional points obtained by letting θ vary from π to 2π lie on the same circle. For example, the solution $(-2, 7\pi/6)$ gives us the same point as $(2, \pi/6)$; the point corresponding to $(-2\sqrt{2}, 5\pi/4)$ is the same as that obtained from $(2\sqrt{2}, \pi/4)$, and so on. If we let θ increase through all real numbers, we obtain the same points again and again because of the periodicity of the sine function. •

FIGURE 13.17

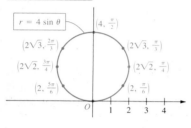

$r = 4 \sin \theta$ $(4, \frac{\pi}{2})$

$(2\sqrt{3}, \frac{2\pi}{3})$ $(2\sqrt{3}, \frac{\pi}{3})$

$(2\sqrt{2}, \frac{3\pi}{4})$ $(2\sqrt{2}, \frac{\pi}{4})$

$(2, \frac{5\pi}{6})$ $(2, \frac{\pi}{6})$

EXAMPLE 2 Sketch the graph of the polar equation $r = 2 + 2 \cos \theta$.

SOLUTION Since the cosine function decreases from 1 to -1 as θ varies from 0 to π, it follows that r decreases from 4 to 0 in this θ-interval. The following table exhibits some solutions of the equation, together with approximations to r.

θ	0	$\pi/6$	$\pi/4$	$\pi/3$	$\pi/2$	$2\pi/3$	$3\pi/4$	$5\pi/6$	π
r	4	$2 + \sqrt{3}$	$2 + \sqrt{2}$	3	2	1	$2 - \sqrt{2}$	$2 - \sqrt{3}$	0
r (approx.)	4	3.7	3.4	3	2	1	0.6	0.3	0

FIGURE 13.18
$r = 2 + 2 \cos \theta$

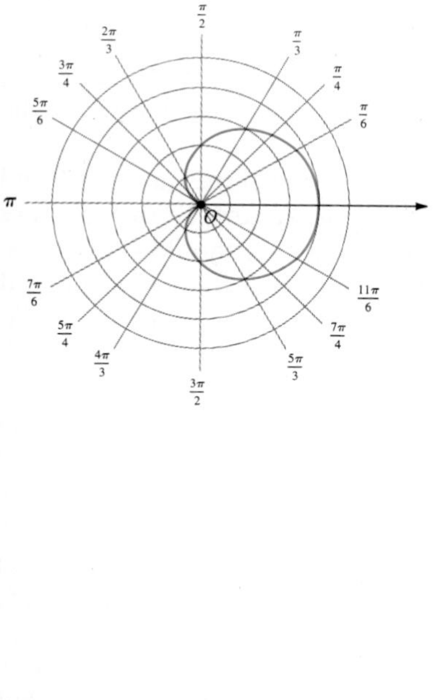

Plotting points gives us the upper half of the graph sketched in Figure 13.18, where we have used polar coordinate graph paper, which displays lines through O at various angles and equispaced concentric circles with centers at the pole.

If θ increases from π to 2π, then $\cos \theta$ increases from -1 to 1 and, consequently, r increases from 0 to 4. Plotting points for $\pi \le \theta \le 2\pi$ would give us the lower half to the graph.

The same graph may be obtained by taking other intervals of length 2π for θ. •

The heart-shaped graph in Example 2 is a **cardioid.** In general, the graph of a polar equation of any of the forms

$$r = a(1 + \cos \theta), \qquad r = a(1 + \sin \theta),$$
$$r = a(1 - \cos \theta), \qquad r = a(1 - \sin \theta),$$

for a real number a, is a cardioid.

The graph of an equation of the form

$$r = a + b \cos \theta \quad \text{or} \quad r = a + b \sin \theta,$$

for $a \neq b$, is a **limaçon**. The graph is similar in shape to a cardioid; however, there may be an added "loop" as shown in the next example.

EXAMPLE 3 Sketch the graph of $r = 2 + 4 \cos \theta$.

SOLUTION Some points corresponding to $0 \leq \theta \leq \pi$ are displayed in the following table.

θ	0	$\pi/6$	$\pi/4$	$\pi/3$	$\pi/2$	$2\pi/3$	$3\pi/4$	$5\pi/6$	π
r	6	$2 + 2\sqrt{3}$	$2 + 2\sqrt{2}$	4	2	0	$2 - 2\sqrt{2}$	$2 - 2\sqrt{3}$	-2
r (approx.)	6	5.4	4.8	4	2	0	-0.8	-1.4	-2

FIGURE 13.19

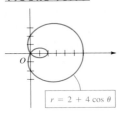

$r = 2 + 4 \cos \theta$

Note that $r = 0$ at $\theta = 2\pi/3$. The values of r are negative if $2\pi/3 < \theta \leq \pi$, and this leads to the lower half of the small loop in Figure 13.19. (Verify this fact.) Letting θ range from π to 2π gives us the upper half of the small loop and the lower half of the large loop. •

EXAMPLE 4 Sketch the graph of the equation $r = a \sin 2\theta$ for $a > 0$.

SOLUTION Instead of tabulating solutions, let us reason as follows: If θ increases from 0 to $\pi/4$, then 2θ varies from 0 to $\pi/2$ and hence $\sin 2\theta$ increases from 0 to 1. It follows that r increases from 0 to a in the θ-interval $[0, \pi/4]$. If we next let θ increase from $\pi/4$ to $\pi/2$, then 2θ changes from $\pi/2$ to π and hence $\sin 2\theta$ decreases from 1 to 0. Thus r decreases from a to 0 in the θ-interval $[\pi/4, \pi/2]$. The corresponding points on the graph constitute the first-quadrant loop illustrated in Figure 13.20. Note that the point $P(r, \theta)$ traces the loop in a *counterclockwise* direction (indicated by the arrows) as θ increases from 0 to $\pi/2$.

FIGURE 13.20

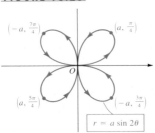

$r = a \sin 2\theta$

If $\pi/2 \leq \theta \leq \pi$, then $\pi \leq 2\theta \leq 2\pi$ and, therefore, $r = a \sin 2\theta \leq 0$. Thus, if $\pi/2 < \theta < \pi$, then r is negative and *the points $P(r, \theta)$ are in the fourth quadrant.* If θ increases from $\pi/2$ to π, then we can show, by plotting points, that $P(r, \theta)$ traces (in a counterclockwise direction) the loop shown in the *fourth* quadrant.

Similarly, for $\pi \leq \theta \leq 3\pi/2$ we get the loop in the third quadrant, and for $3\pi/2 \leq \theta \leq 2\pi$ we get the loop in the second quadrant. Both loops are traced in a counterclockwise direction as θ increases. You should verify these facts by plotting some points with, say, $a = 1$. In Figure 13.20 we have plotted only those points on the graph that correspond to the largest numerical values of r. •

The graph in Example 4 is a **four-leafed rose.** In general, an equation of the form

$$r = a \sin n\theta \quad \text{or} \quad r = a \cos n\theta,$$

for any positive integer n greater than 1 and any real number a, has a graph that consists of a number of loops attached to the origin. If n is even there are $2n$ loops, and if n is odd there are n loops (see, for example, Exercises 9, 10, and 22).

The graph of the polar equation $r = a\theta$, for any real number a, is called a **spiral of Archimedes.** The case $a = 1$ is considered in the next example.

FIGURE 13.21

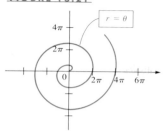

EXAMPLE 5 Sketch the graph of $r = \theta$ for $\theta \geq 0$.

SOLUTION The graph consists of all points that have polar coordinates of the form (c, c) for every real number $c \geq 0$. Thus the graph contains the points $(0, 0)$, $(\pi/2, \pi/2)$, (π, π), and so on. As θ increases, r increases at the same rate, and the spiral winds around the origin in a counterclockwise direction, intersecting the polar axis at $0, 2\pi, 4\pi, \ldots$, as illustrated in Figure 13.21.

If θ is allowed to be negative, then as θ decreases through negative values, the resulting spiral winds around the origin and is the symmetric image, with respect to the vertical axis, of the curve sketched in Figure 13.21. •

Many other interesting graphs result from polar equations. Some are included in the exercises at the end of this section. Polar coordinates are useful in applications involving circles with centers at the origin or lines that pass through the origin, since equations that have these graphs may be written in the simple forms $r = k$ or $\theta = k$ for some constant k. (Verify this fact.)

Let us next superimpose an xy-plane on an $r\theta$-plane such that the positive x-axis coincides with the polar axis. Any point P in the plane may then be assigned rectangular coordinates (x, y) or polar coordinates (r, θ). If $r > 0$ we have a situation similar to that illustrated in Figure 13.22(i). If $r < 0$ we have that shown in (ii) of the figure where, for later purposes, we have also plotted the point P' having polar coordinates $(|r|, \theta)$ and rectangular coordinates $(-x, -y)$.

The following theorem specifies relationships between (x, y) and (r, θ). In the statement of the theorem it is assumed that the positive x-axis coincides with the polar axis.

FIGURE 13.22

(i) $r > 0$

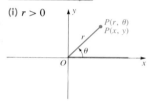

(ii) $r < 0$

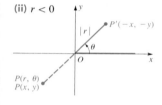

THEOREM (13.8)

The rectangular coordinates (x, y) and polar coordinates (r, θ) of a point P are related as follows:

(i) $x = r \cos \theta$, $\qquad y = r \sin \theta$

(ii) $\tan \theta = \dfrac{y}{x}$, $\qquad r^2 = x^2 + y^2$

PROOF Although we have pictured θ as an acute angle in Figure 13.22, the discussion that follows is valid for all angles. If $r > 0$, as in Figure 13.22(i), then $\cos \theta = x/r$, $\sin \theta = y/r$, and hence

$$x = r \cos \theta, \qquad y = r \sin \theta.$$

If $r < 0$, then $|r| = -r$ and from Figure 13.22(ii) we see that

$$\cos \theta = \frac{-x}{|r|} = \frac{-x}{-r} = \frac{x}{r},$$

$$\sin \theta = \frac{-y}{|r|} = \frac{-y}{-r} = \frac{y}{r}.$$

Multiplication by r produces (i) of the theorem and, therefore, these formulas hold if r is either positive or negative. If $r = 0$, then the point is the pole and we again see that the formulas in (i) are true.

The formulas in (ii) follow readily from Figure 13.22. • •

We may use Theorem (13.8) to change from one system of coordinates to the other. A more important use for (13.8) is for transforming a polar equation to an equation in x and y, and vice versa. This is illustrated in the next two examples.

EXAMPLE 6 Find an equation in x and y that has the same graph as the polar equation $r = 4 \sin \theta$.

SOLUTION The equation $r = 4 \sin \theta$ was considered in Example 1. If we multiply both sides by r, we obtain $r^2 = 4r \sin \theta$. Applying Theorem (13.8) gives us $x^2 + y^2 = 4y$, which is equivalent to $x^2 + (y - 2)^2 = 4$. Thus, the graph is a circle of radius 2 with center at $(0, 2)$ in the xy-plane. •

EXAMPLE 7 Find a polar equation of an arbitrary line.

SOLUTION We know that every line in an xy-coordinate system is the graph of a linear equation $ax + by + c = 0$. The formulas $x = r \cos \theta$ and $y = r \sin \theta$ give us the polar equation

$$ar \cos \theta + br \sin \theta + c = 0$$

or $\qquad\qquad r(a \cos \theta + b \sin \theta) + c = 0$ •

If we superimpose an xy-plane on an $r\theta$-plane, then the graph of a polar equation may be symmetric with respect to the x-axis, the y-axis, or the origin. Some typical symmetries are illustrated in Figure 13.23. This leads to the next theorem.

FIGURE 13.23
Symmetries of graphs of polar equations

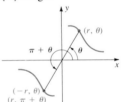

(i) x-axis (ii) y-axis (iii) Origin

TESTS FOR SYMMETRY (13.9)

(i) The graph of $r = f(\theta)$ is symmetric with respect to the x-axis if substitution of $-\theta$ for θ leads to an equivalent equation.

(ii) The graph of $r = f(\theta)$ is symmetric with respect to the y-axis if substitution of either (a) $\pi - \theta$ for θ or (b) $-r$ for r and $-\theta$ for θ leads to an equivalent equation.

(iii) The graph of $r = f(\theta)$ is symmetric with respect to the origin if substitution of either (a) $-r$ for r or (b) $\pi + \theta$ for θ leads to an equivalent equation.

To illustrate, since $\cos(-\theta) = \cos \theta$, the graph of the polar equation $r = 2 + 4 \cos \theta$ (see Figure 13.19) is symmetric with respect to the x-axis by (13.9) (i). Since $\sin(\pi - \theta) = \sin \theta$, the graph in Example 1 is symmetric with respect to the y-axis by (13.9) (ii). The graph in Example 4 is symmetric to both axes and to the origin. Other tests for symmetry may be stated; however, those listed in (13.9) are among the easiest to apply.

Unlike graphs in rectangular coordinates, it is possible for the graph of a polar equation $r = f(\theta)$ to be symmetric with respect to an axis or the origin *without* satisfying one of the tests for symmetry in (13.9). This is true because of the many different ways of specifying a point in polar coordinates.

Another difference between rectangular and polar coordinates is that the points of intersection of two graphs cannot always be found by solving the polar equations simultaneously. To illustrate, from Example 1, the graph of $r = 4 \sin \theta$ is a circle of diameter 4 with center at $(2, \pi/2)$ (see Figure 13.17). Similarly, the graph of $r = 4 \cos \theta$ is a circle of diameter 4 with center at $(2, 0)$ on the polar axis. Referring to Figure 13.24, the coordinates of the point of intersection $P(2\sqrt{2}, \pi/4)$ in Quadrant I satisfy both equations; however, the origin O, which is on each circle, *cannot* be found by solving the equations simultaneously. Thus, when searching for points of intersection of polar graphs, it is sometimes necessary to refer to the graphs themselves, in *addition* to solving the two equations simultaneously. An alternative method is to use different (equivalent) equations for the graphs.

Tangent lines to graphs of polar equations may be found by means of the next theorem.

FIGURE 13.24

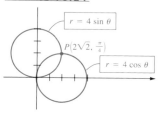

$r = 4 \sin \theta$

$P(2\sqrt{2}, \frac{\pi}{4})$

$r = 4 \cos \theta$

THEOREM (13.10)

The slope m of the tangent line to the graph of $r = f(\theta)$ at the point $P(r, \theta)$ is

$$m = \frac{\dfrac{dr}{d\theta} \sin \theta + r \cos \theta}{\dfrac{dr}{d\theta} \cos \theta - r \sin \theta}$$

PROOF If (x, y) are the rectangular coordinates of $P(r, \theta)$, then by Theorem (13.8),

$$x = r \cos \theta = f(\theta) \cos \theta$$

$$y = r \sin \theta = f(\theta) \sin \theta.$$

These may be considered as parametric equations for the graph with parameter θ. Applying Theorem (13.3), the slope of the tangent line at (x, y) is

$$\frac{dy}{dx} = \frac{dy/d\theta}{dx/d\theta} = \frac{f'(\theta) \sin \theta + f(\theta) \cos \theta}{f'(\theta) \cos \theta - f(\theta) \sin \theta}.$$

This is equivalent to the formula in the statement of the theorem. • •

Horizontal tangent lines occur if the numerator in the formula for m is 0 and the denominator is not 0. Vertical tangent lines occur if the denominator is 0 and the numerator is not 0. The case $0/0$ requires special attention.

To find the slopes of the tangent lines at the pole, we must determine the values of θ for which $r = 0$. For such values (and with $r = 0$) the formula in Theorem (13.10) reduces to $m = \tan \theta$. These remarks are illustrated in the next example.

EXAMPLE 8 For the cardioid $r = 2 + 2 \cos \theta$, find

(a) the slope of the tangent line at $\theta = \pi/6$.

(b) the points at which the tangent line is horizontal.

(c) the points at which the tangent line is vertical.

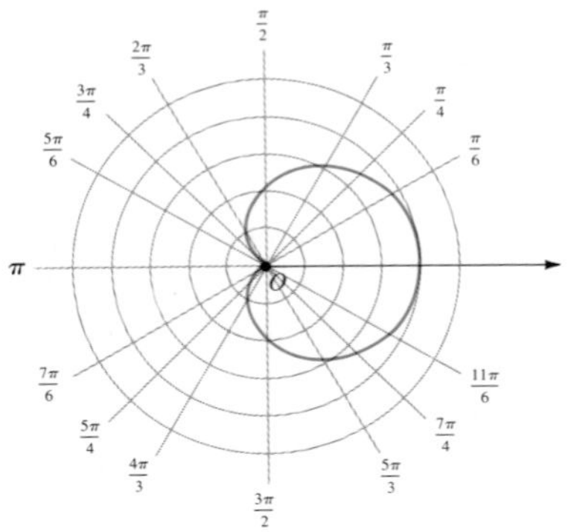

SOLUTION The graph of $r = 2 + 2 \cos \theta$ was considered in Example 2 and is resketched in Figure 13.25. Applying Theorem (13.10), the slope m of the tangent line is

$$m = \frac{(-2 \sin \theta) \sin \theta + (2 + 2 \cos \theta) \cos \theta}{(-2 \sin \theta) \cos \theta - (2 + 2 \cos \theta) \sin \theta}$$

$$= \frac{2 (\cos^2 \theta - \sin^2 \theta) + 2 \cos \theta}{-2(2 \sin \theta \cos \theta) - 2 \sin \theta}$$

$$= -\frac{\cos 2\theta + \cos \theta}{\sin 2\theta + \sin \theta}$$

(a) At $\theta = \pi/6$, that is, at the point $(2 + \sqrt{3}, \pi/6)$,

$$m = -\frac{\cos (\pi/3) + \cos (\pi/6)}{\sin (\pi/3) + \sin (\pi/6)} = -\frac{(1/2) + (\sqrt{3}/2)}{(\sqrt{3}/2) + (1/2)} = -1$$

(b) To find horizontal tangents we let

$$\cos 2\theta + \cos \theta = 0.$$

This equation may be written as

$$2 \cos^2 \theta - 1 + \cos \theta = 0,$$

or

$$(2 \cos \theta - 1)(\cos \theta + 1) = 0.$$

From $\cos \theta = \frac{1}{2}$ we obtain $\theta = \pi/3$ and $\theta = 5\pi/3$. The corresponding points are $(3, \pi/3)$ and $(3, 5\pi/3)$. Using $\cos \theta = -1$ gives us $\theta = \pi$. Since the denominator in the formula for m is 0 at $\theta = \pi$, further investigation is required. We saw in (b) that there is a horizontal tangent line at the point $(0, \pi)$ (the pole).

(c) To find vertical tangent lines we let

$$\sin 2\theta + \sin \theta = 0.$$

Equivalent equations are

$$2 \sin \theta \cos \theta + \sin \theta = 0$$

and

$$\sin \theta (2 \cos \theta + 1) = 0.$$

Letting $\sin \theta = 0$ and $\cos \theta = -\frac{1}{2}$ leads to the following values of θ: $0, \pi$, $2\pi/3$, and $4\pi/3$. We found, in part (b), that π gives us a horizontal tangent. The remaining values result in the points $(4, 0)$, $(1, 2\pi/3)$, and $(1, 4\pi/3)$, at which the graph has vertical tangent lines. •

FIGURE 13.25
$r = 2 + 2 \cos \theta$

EXERCISES 13.3

Exer. 1–24: Sketch the graph of the polar equation.

1 $r = 5$

2 $\theta = \pi/4$

3 $\theta = -\pi/6$

4 $r = -2$

5 $r = 2 \cos \theta$

6 $r = -2 \sin \theta$

7 $r = 4(1 - \sin \theta)$	8 $r = 1 + 2 \cos \theta$
9 $r = 8 \cos 3\theta$	10 $r = 2 \sin 4\theta$
11 $r^2 = 4 \cos 2\theta$ (lemniscate)	12 $r = 6 \sin^2 (\tfrac{1}{2}\theta)$
13 $r = 4 \csc \theta$	14 $r = -3 \sec \theta$
15 $r = 2 - \cos \theta$	16 $r = 2 + 2 \sec \theta$ (conchoid)
17 $r = 2^\theta, \ \theta \geq 0$ (spiral)	18 $r\theta = 1, \ \theta > 0$ (spiral)
19 $r = -6(1 + \cos \theta)$	20 $r = e^{2\theta}$ (logarithmic spiral)
21 $r = 2 + 4 \sin \theta$	22 $r = 8 \cos 5\theta$
23 $r^2 = -16 \sin 2\theta$	24 $4r = \theta$

Exer. 25–32: Find a polar equation that has the same graph as the equation in x and y.

25 $x = -3$	26 $y = 2$
27 $x^2 + y^2 = 16$	28 $x^2 = 8y$
29 $x^2 - y^2 = 16$	30 $9x^2 + 4y^2 = 36$

31 $(x^2 + y^2) \tan^{-1}(y/x) = ay, \quad a > 0$
(cochleoid, or *Oui-ja board curve*)

32 $x^3 + y^3 - 3axy = 0$ (Folium of Descartes)

Exer. 33–48: Find an equation in x and y that has the same graph as the polar equation, and use it as an aid to sketching the graph in a polar coordinate system.

33 $r \cos \theta = 5$	34 $r \sin \theta = -2$
35 $r - 6 \sin \theta = 0$	36 $r = 6 \cot \theta$
37 $r = 2$	38 $\theta = \pi/4$
39 $r = \tan \theta$	40 $r = 4 \sec \theta$

41 $r^2(4 \sin^2 \theta - 9 \cos^2 \theta) = 36$

42 $r^2 (\cos^2 \theta + 4 \sin^2 \theta) = 16$

43 $r^2 \cos 2\theta = 1$	44 $r^2 \sin 2\theta = 4$
45 $r (\sin \theta - 2 \cos \theta) = 6$	46 $r (\sin \theta + r \cos^2 \theta) = 1$
47 $r = \dfrac{1}{1 + \cos \theta}$	48 $r = \dfrac{4}{2 + \sin \theta}$

49 If $P_1(r_1, \theta_1)$ and $P_2(r_2, \theta_2)$ are points in an $r\theta$-plane, use the Law of Cosines to prove that
$$[d(P_1, P_2)]^2 = r_1^2 + r_2^2 - 2r_1 r_2 \cos (\theta_2 - \theta_1).$$

50 Prove that the graph of each polar equation (a)–(c) is a circle, and find its center and radius.
(a) $r = a \sin \theta, \quad a \neq 0$ (b) $r = b \cos \theta, \quad b \neq 0$
(c) $r = a \sin \theta + b \cos \theta, \quad ab \neq 0$

Exer. 51–60: Find the slope of the tangent line to the graph of the equation at the given value of θ. (Refer to the graphs in Exercises 5–12 and 17–18.)

51 $r = 2 \cos \theta, \quad \theta = \pi/3$	52 $r = -2 \sin \theta, \quad \theta = \pi/6$
53 $r = 4(1 - \sin \theta), \quad \theta = 0$	54 $r = 1 + 2 \cos \theta, \quad \theta = \pi/2$

55 $r = 8 \cos 3\theta, \quad \theta = \pi/4$	56 $r = 2 \sin 4\theta, \quad \theta = \pi/4$
57 $r^2 = 4 \cos 2\theta, \quad \theta = \pi/6$	58 $r = 6 \sin^2 \tfrac{1}{2}\theta, \quad \theta = 2\pi/3$
59 $r = 2^\theta, \quad \theta = \pi$	60 $r\theta = 1, \quad \theta = 2\pi$

61 If the graphs of the polar equations $r = f(\theta)$ and $r = g(\theta)$ intersect at $P(r, \theta)$, prove that the tangent lines at P are perpendicular if and only if
$$f'(\theta)g'(\theta) + f(\theta)g(\theta) = 0.$$
(The graphs are *orthogonal* at P.)

62 Use Exercise 61 to prove that the graphs of each pair of equations are orthogonal at their point of intersection.
(a) $r = a \sin \theta, \ r = a \cos \theta$ (b) $r = a\theta, \ r\theta = a$

63 If $\cos \theta \neq 0$, show that the slope of the tangent line to the graph of $r = f(\theta)$ is
$$m = \frac{(dr/d\theta) \tan \theta + r}{(dr/d\theta) - r \tan \theta}.$$

64 Let $a > 0$ and let C be the set of points in a coordinate plane, the product of whose distances from $P(a, 0)$ and $Q(-a, 0)$ is equal to a^2. Show that a polar equation for C is $r^2 = 2a^2 \cos 2\theta$. (*Hint:* Use Exercise 49.)

65 The Greek geometer Diocles (circa 180 B.C.) invented a special curve called a *cissoid*, which he used to duplicate the cube, that is, to construct a cube with twice the volume of a given cube. Using the circle $x^2 + (y - a)^2 = a^2$, a point A is selected on the tangent line $y = 2a$, and $P(r, \theta)$ is then chosen on OA so that $OP = BA$, where B is the point of intersection of OA with the circle (see figure). Show that the cissoid of Diocles has a polar equation $r = 2a \cos^2 \theta \csc \theta$.

EXERCISE 65

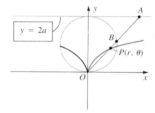

66 Show that the cissoid of Exercise 65 has a rectangular equation $x^2(2a - y) = y^3$, and sketch the graph.

67 A logarithmic spiral has a polar equation of the form $r = ae^{b\theta}$ for nonzero constants a and b (see Exercise 20). A famous "four bugs" problem illustrates such a curve. Four bugs A, B, C, and D are placed at the four corners of a square. The center of the square corresponds to the pole. The bugs begin to crawl simultaneously—bug A crawls toward B, B toward C, C toward D, and D toward A (see figure on page 563). All bugs crawl at the same rate, directly toward the next bug at all times. At any instant the position of the bugs forms the vertices of a square, which shrinks and rotates toward the center of the original square as the bugs continue to crawl. If the position of bug A has polar coordinates (r, θ), then B will have coordinates $(r, \theta + \tfrac{1}{2}\pi)$.

(a) Show that the line through A and B has slope
$$\frac{\sin \theta - \cos \theta}{\sin \theta + \cos \theta}.$$

(b) The line through A and B is tangent to the path of bug A. Use the formula in Exercise 63 to conclude that $dr/d\theta = -r$.

(c) Prove that the path of bug A is a logarithmic spiral.

EXERCISE 67

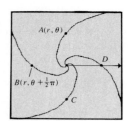

68 A line segment AB of fixed length a has endpoints A, on the y-axis, and B, on the x-axis. A perpendicular is constructed from the origin O to AB, intersecting AB at the point P as shown in the figure. If A and B may slide freely along their respective axes, show that the curve traced by P is the four-leafed rose with polar equation $r = \frac{1}{2}a \sin 2\theta$.

EXERCISE 68

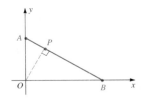

13.4 INTEGRALS IN POLAR COORDINATES

FIGURE 13.26

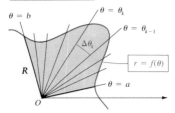

The area of a region that is bounded by graphs of polar equations can be found by using a limit of sums of areas of circular sectors. Suppose R is a region in an $r\theta$-plane bounded by the lines through O with equations $\theta = a$ and $\theta = b$, for $0 \leq a < b \leq 2\pi$, and by the graph of $r = f(\theta)$ such that f is continuous and $f(\theta) \geq 0$ on $[a, b]$. A region of this type is illustrated in Figure 13.26.

Let P denote a partition of $[a, b]$ determined by

$$a = \theta_0 < \theta_1 < \theta_2 < \cdots < \theta_n = b$$

and let $\Delta\theta_k = \theta_k - \theta_{k-1}$ for $k = 1, 2, \ldots, n$. The radial lines with equations $\theta = \theta_k$ divide R into wedge-shaped subregions. If $f(u_k)$ is the minimum value and $f(v_k)$ is the maximum value of f on $[\theta_{k-1}, \theta_k]$, then as illustrated in Figure 13.27, the area ΔA_k of the kth subregion is between the areas of the inscribed and circumscribed circular sectors having central angle $\Delta\theta_k$ and radii $f(u_k)$ and $f(v_k)$, respectively. Hence, by Theorem (2.25),

$$\tfrac{1}{2}[f(u_k)]^2 \Delta\theta_k \leq \Delta A_k \leq \tfrac{1}{2}[f(v_k)]^2 \Delta\theta_k.$$

FIGURE 13.27

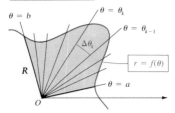

Summing from $k = 1$ to $k = n$, and using the fact that the sum of the ΔA_k is the area A of R, we obtain

$$\sum_{k=1}^{n} \tfrac{1}{2}[f(u_k)]^2 \Delta\theta_k \leq A \leq \sum_{k=1}^{n} \tfrac{1}{2}[f(v_k)]^2 \Delta\theta_k.$$

The limits of the sums, as the norm $\|P\|$ of the subdivision approaches zero, equals the integral $\int_a^b \tfrac{1}{2}[f(\theta)]^2 \, d\theta$. This gives us the following result.

THEOREM (13.11)

If f is continuous and $f(\theta) \geq 0$ on $[a, b]$ for $0 \leq a < b \leq 2\pi$, then the area A of the region bounded by the graphs of $r = f(\theta)$, $\theta = a$, and $\theta = b$ is

$$A = \int_a^b \tfrac{1}{2}[f(\theta)]^2 \, d\theta = \int_a^b \tfrac{1}{2}r^2 \, d\theta$$

FIGURE 13.28

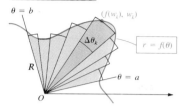

The integral in Theorem (13.11) may be interpreted as a limit of sums by writing

$$A = \int_a^b \tfrac{1}{2}[f(\theta)]^2 \, d\theta = \lim_{\|P\| \to 0} \sum_{k=1}^n \tfrac{1}{2}[f(w_k)]^2 \, \Delta\theta_k$$

for *any* number w_k in the subinterval $[\theta_{k-1}, \theta_k]$ of $[a, b]$. Figure 13.28 is a geometric illustration of a typical Riemann sum. It is convenient to think of starting with the area of a circular sector and then sweeping out the region R by letting θ vary from a to b, while at the same time each $\Delta\theta_k$ approaches zero.

EXAMPLE 1 Find the area of the region that is bounded by the cardioid $r = 2 + 2 \cos \theta$.

FIGURE 13.29

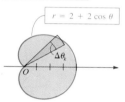

SOLUTION The graph was considered in Example 2 of Section 13.3. It is sketched again in Figure 13.29, together with a typical circular sector of the type we have discussed. Making use of symmetry, we shall find the area of the upper half of the region and double the result. The function f of Theorem (13.11) is given by $f(\theta) = 2 + 2 \cos \theta$. To find a and b we observe that the circular sectors will sweep out the upper half of the cardioid if θ varies from 0 to π. Consequently, $a = 0$, $b = \pi$, and

$$A = 2 \int_0^\pi \tfrac{1}{2}(2 + 2 \cos \theta)^2 \, d\theta$$

$$= \int_0^\pi (4 + 8 \cos \theta + 4 \cos^2 \theta) \, d\theta.$$

Using the fact that $\cos^2 \theta = \tfrac{1}{2}(1 + \cos 2\theta)$,

$$A = \int_0^\pi (6 + 8 \cos \theta + 2 \cos 2\theta) \, d\theta$$

$$= \Big[6\theta + 8 \sin \theta + \sin 2\theta \Big]_0^\pi = 6\pi.$$

We could also have found the area by using (13.11) with $a = 0$ and $b = 2\pi$.

The next result generalizes Theorem (13.11).

THEOREM (13.12)

> Let f and g be continuous functions such that $f(\theta) \ge g(\theta) \ge 0$ for every θ in $[a, b]$ and $0 \le a < b \le 2\pi$. Let R denote the region bounded by the graphs of $r = f(\theta)$, $r = g(\theta)$, $\theta = a$, and $\theta = b$. The area A of R is
>
> $$A = \tfrac{1}{2} \int_a^b \{[f(\theta)]^2 - [g(\theta)]^2\} \, d\theta$$

FIGURE 13.30

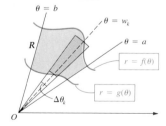

A region of the type described in Theorem (13.12) is sketched in Figure 13.30. The area A can be found by subtracting the areas of two regions of the type described in (13.11). Thus

$$A = \int_a^b \tfrac{1}{2}[f(\theta)]^2 \, d\theta - \int_a^b \tfrac{1}{2}[g(\theta)]^2 \, d\theta.$$

Writing this as one integral gives us Theorem (13.12).

The integral in (13.12) can be expressed as a limit of sums as follows:

$$A = \lim_{\|P\| \to 0} \sum_k \tfrac{1}{2}\{[f(w_k)]^2 - [g(w_k)]^2\} \, \Delta\theta_k$$

The sum may be thought of as the process of starting with the area of a region of the type illustrated by the truncated wedge in Figure 13.30 and then sweeping out R by letting θ vary from a to b.

EXAMPLE 2 Find the area A of the region that is inside the cardioid $r = 2 + 2\cos\theta$ and outside the circle $r = 3$.

SOLUTION The region is sketched in Figure 13.31, together with a typical polar element of area. Since the two graphs intersect at the points $(3, -\pi/3)$ and $(3, \pi/3)$, the region will be swept out if θ varies from $-\pi/3$ to $\pi/3$. Using Theorem (13.12),

$$A = \frac{1}{2}\int_{-\pi/3}^{\pi/3} \left[(2 + 2\cos\theta)^2 - 3^2\right] d\theta$$

$$= \frac{1}{2}\int_{-\pi/3}^{\pi/3} (8\cos\theta + 4\cos^2\theta - 5)\, d\theta.$$

As in Example 1, the integral may be evaluated by using the substitution $\cos^2\theta = \frac{1}{2}(1 + \cos 2\theta)$. We can show that $A = (9\sqrt{3}/2) - \pi \approx 4.65$. •

The next theorem provides a formula for arc length in polar coordinates.

FIGURE 13.31

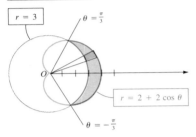

$r = 3$

$\theta = \frac{\pi}{3}$

O

$r = 2 + 2\cos\theta$

$\theta = -\frac{\pi}{3}$

THEOREM (13.13) If a curve C is the graph of a polar equation $r = f(\theta)$ and if f' is continuous on $[a, b]$, then the length L of C is

$$L = \int_a^b \sqrt{[f(\theta)]^2 + [f'(\theta)]^2}\, d\theta = \int_a^b \sqrt{r^2 + \left(\frac{dr}{d\theta}\right)^2}\, d\theta$$

PROOF As in the proof of Theorem (13.10), parametric equations for C are

$$x = f(\theta)\cos\theta, \quad y = f(\theta)\sin\theta; \quad a \leq \theta \leq b.$$

Hence

$$\frac{dx}{d\theta} = -f(\theta)\sin\theta + f'(\theta)\cos\theta$$

$$\frac{dy}{d\theta} = f(\theta)\cos\theta + f'(\theta)\sin\theta.$$

It is not difficult to show that

$$\left(\frac{dx}{d\theta}\right)^2 + \left(\frac{dy}{d\theta}\right)^2 = [f(\theta)]^2 + [f'(\theta)]^2.$$

Substitution in Theorem (13.5) with $\theta = t$ gives us the desired formula. • •

EXAMPLE 3 Find the length of the cardioid $r = 1 + \cos\theta$.

SOLUTION The cardioid is sketched in Figure 13.32. Making use of symmetry, we shall find the length of the upper half, and double the result. Applying Theorem (13.13),

$$L = 2\int_0^\pi \sqrt{(1 + \cos\theta)^2 + (-\sin\theta)}\, d\theta$$

$$= 2\int_0^\pi \sqrt{1 + 2\cos\theta + \cos^2\theta + \sin^2\theta}\, d\theta$$

$$= 2\int_0^\pi \sqrt{2 + 2\cos\theta}\, d\theta = 2\sqrt{2}\int_0^\pi \sqrt{1 + \cos\theta}\, d\theta.$$

FIGURE 13.32

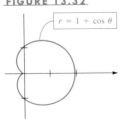

$r = 1 + \cos\theta$

The last integral may be evaluated by employing the trigonometric identity $\cos^2 \frac{1}{2}\theta = \frac{1}{2}(1 + \cos \theta)$, or equivalently, $1 + \cos \theta = 2\cos^2 \frac{1}{2}\theta$. Thus,

$$L = 2\sqrt{2} \int_0^\pi \sqrt{2 \cos^2 \tfrac{1}{2}\theta}\ d\theta$$

$$= 4 \int_0^\pi \cos \tfrac{1}{2}\theta\ d\theta$$

$$= (4 \cdot 2) \sin \tfrac{1}{2}\theta \Big]_0^\pi = 8 \qquad \bullet$$

In the solution to Example 3, it was legitimate to replace $\sqrt{\cos^2 \frac{1}{2}\theta}$ by $\cos \frac{1}{2}\theta$, because if $0 \le \theta \le \pi$, then $0 \le \frac{1}{2}\theta \le \pi/2$, and hence $\cos \frac{1}{2}\theta$ is *positive* on $[0, \pi]$. If we had *not* used symmetry, but had written L as $\int_0^{2\pi} \sqrt{r^2 + (dr/d\theta)^2}\ d\theta$, this simplification would not have been valid. Generally, when determining areas or arc lengths that involve polar coordinates, it is a good idea to use any symmetries that exist.

Let C be the graph of a polar equation $r = f(\theta)$ for $a \le \theta \le b$. We wish to obtain a formula for the area S of the surface generated by revolving C about the polar axis, as illustrated in Figure 13.33. Since parametric equations for C are

FIGURE 13.33

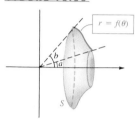

$$x = f(\theta) \cos \theta, \quad y = f(\theta) \sin \theta; \quad a \le \theta \le b,$$

we may find S by using Theorem (13.7) with $\theta = t$. This gives us the following formula, with the polar arc length differential $ds = \sqrt{[f(\theta)]^2 + [f'(\theta)]^2}\ d\theta$ obtained as in the proof of Theorem (13.13):

(13.14) $\quad S = \int_{\theta=a}^{\theta=b} 2\pi y\ ds = \int_a^b 2\pi f(\theta) \sin \theta \sqrt{[f(\theta)]^2 + [f'(\theta)]^2}\ d\theta \quad$ (about polar axis)

If C is revolved about the vertical line $\theta = \pi/2$ (the y-axis), then, as on page 555,

(13.15) $\quad S = \int_{\theta=a}^{\theta=b} 2\pi x\ ds = \int_a^b 2\pi f(\theta) \cos \theta \sqrt{[f(\theta)]^2 + [f'(\theta)]^2}\ d\theta \quad$ (about $\theta = \pi/2$)

When using either (13.14) or (13.15), *we must choose a and b such that the surface does not retrace itself when C is revolved,* as would be the case of revolving the graph of $r = \cos \theta$, with $0 \le \theta \le \pi$, about the polar axis.

EXAMPLE 4 The part of the spiral $r = e^{\theta/2}$ from $\theta = 0$ to $\theta = \pi$ is revolved about the polar axis. Find the area of the resulting surface.

FIGURE 13.34

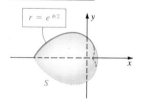

SOLUTION The surface is illustrated in Figure 13.34. From (13.14) with $f(\theta) = e^{\theta/2}$,

$$S = \int_0^\pi 2\pi e^{\theta/2} \sin \theta \sqrt{(e^{\theta/2})^2 + (\tfrac{1}{2}e^{\theta/2})^2}\ d\theta$$

$$= \int_0^\pi 2\pi e^{\theta/2} \sin \theta \sqrt{\tfrac{5}{4}e^\theta}\ d\theta = \sqrt{5}\pi \int_0^\pi e^\theta \sin \theta\ d\theta.$$

Using integration by parts or Formula 98 in the Table of Integrals (see Appendix IV),

$$S = \frac{\sqrt{5}\pi}{2} e^\theta (\sin \theta - \cos \theta)\Big]_0^\pi = \frac{\sqrt{5}\pi}{2}(e^\pi + 1) \approx 84.8 \qquad \bullet$$

Exer. 1–8: Find the area of the region bounded by the graph.

1 $r = 2 \cos \theta$

2 $r = 5 \sin \theta$

3 $r = 1 - \cos \theta$

4 $r = 6 - 6 \sin \theta$

5 $r = \sin 2\theta$

6 $r^2 = 9 \cos 2\theta$

7 $r = 4 + \sin \theta$

8 $r = 3 + 2 \cos \theta$

Exer. 9–10: Find the area of region R.

9 $R = \{(r, \theta): 0 \le \theta \le \pi/2, \ 0 \le r \le e^{\theta}\}$.

10 $R = \{(r, \theta): 0 \le \theta \le \pi, \ 0 \le r \le 2\theta\}$.

Exer. 11–14: Find the area of the region bounded by one loop of the graph of the equation.

11 $r^2 = 4 \cos 2\theta$

12 $r = 2 \cos 3\theta$

13 $r = 3 \cos 5\theta$

14 $r = \sin 6\theta$

Exer. 15–18: Find the area of the region that is outside the graph of the first equation and inside the graph of the second equation.

15 $r = 2 + 2 \cos \theta, \quad r = 3$

16 $r = 2, \quad r = 4 \cos \theta$

17 $r = 2, \quad r^2 = 8 \sin 2\theta$

18 $r = 1 - \sin \theta, \quad r = 3 \sin \theta$

Exer. 19–22: Find the area of the region that is inside the graphs of *both* equations.

19 $r = \sin \theta, \quad r = \sqrt{3} \cos \theta$

20 $r = 2(1 + \sin \theta), \quad r = 1$

21 $r = 1 + \sin \theta, \quad r = 5 \sin \theta$

22 $r^2 = 4 \cos 2\theta, \quad r = 1$

Exer. 23–26: Find the area of the region bounded by the graphs of the equations.

23 $r = 2 \sec \theta, \quad \theta = \pi/6, \quad \theta = \pi/3$

24 $r = \csc \theta \cot \theta, \quad \theta = \pi/6, \quad \theta = \pi/4$

25 $r = 4/(1 - \cos \theta), \quad \theta = \pi/4$

26 $r(1 + \sin \theta) = 2, \quad \theta = \pi/3$

27 A line crosses a circular disk of radius a at a distance b from its center with $0 < b < a$. Find the area of the smaller of the two circular segments cut from the disk by the line.

28 Let OP be the ray from the pole to the point $P(r, \theta)$ on the spiral $r = a\theta$ for $a > 0$. If the ray makes two revolutions (starting from $\theta = 0$), find the area of the region swept out in its second revolution that was not swept out in the first revolution (see figure).

EXERCISE 28

Exer. 29–34: Find the length of the curve.

29 $r = e^{-\theta}$ from $\theta = 0$ to $\theta = 2\pi$

30 $r = \theta$ from $\theta = 0$ to $\theta = 4\pi$

31 $r = \cos^2 \frac{1}{2}\theta$ from $\theta = 0$ to $\theta = \pi$

32 $r = 2^{\theta}$ from $\theta = 0$ to $\theta = \pi$

33 $r = \sin^3 \frac{1}{3}\theta$

34 $r = 2 - 2 \cos \theta$

Exer. 35–38: Find the area of the surface generated by revolving the curve about the polar axis.

35 $r = 2 + 2 \cos \theta$

36 $r^2 = 4 \cos 2\theta$

37 $r = 2a \sin \theta$

38 $r = 2a \cos \theta$

39 A *torus* is the surface generated by revolving a circle about a nonintersecting line in its plane. Use polar coordinates to find the surface area of the torus generated by revolving the circle $x^2 + y^2 = a^2$ about the line $x = b$ for $0 < a < b$.

40 The part of the spiral $r = e^{-\theta}$ from $\theta = 0$ to $\theta = \pi/2$ is revolved about the line $\theta = \pi/2$. Find the area of the resulting surface.

13.5 POLAR EQUATIONS OF CONICS

The following theorem combines the definitions of parabola, ellipse, and hyperbola into a unified description of the conic sections.

THEOREM (13.16)

Let F be a fixed point and l a fixed line in a plane. The set of all points P in the plane, such that the ratio $d(P, F)/d(P, Q)$ is a positive constant e for the distance $d(P, Q)$ from P to l, is a conic section. The conic is a parabola if $e = 1$, an ellipse if $0 < e < 1$, and a hyperbola if $e > 1$.

FIGURE 13.35

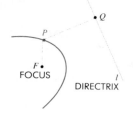

FIGURE 13.36

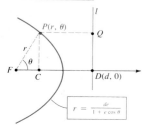

As in Definition (12.8), the constant e is the **eccentricity** of the conic. We will see that the point F is a focus of the conic. The line l is a **directrix**. A typical situation is illustrated in Figure 13.35, with possible positions of P shown by the curve.

We shall prove the theorem if $e \le 1$ and leave the case $e > 1$ as an exercise.

If $e = 1$ in Theorem (13.16), then $d(P, F) = d(P, Q)$ and, according to Definition (12.1), the resulting conic is a parabola with focus F and directrix l.

Suppose next that $0 < e < 1$. It is convenient to introduce a polar coordinate system in the plane with F as the pole and with l perpendicular to the polar axis at the point $D(d, 0)$ with $d > 0$. If $P(r, \theta)$ is a point in the plane such that $d(P, F)/d(P, Q) = e < 1$, then from Figure 13.36 we see that P lies to the left of l. Let C be the projection of P on the polar axis. Since

$$d(P, F) = r \quad \text{and} \quad d(P, Q) = \overline{FD} - \overline{FC} = d - r \cos \theta,$$

it follows that P satisfies the condition in (13.16) if and only if

$$\frac{r}{d - r \cos \theta} = e$$

or

$$r = de - er \cos \theta.$$

Solving for r gives us

$$r = \frac{de}{1 + e \cos \theta}$$

which is a polar equation of the graph. Actually, the same equation is obtained if $e = 1$; however, there is no point (r, θ) on the graph if $1 + \cos \theta = 0$.

An equation in x and y corresponding to $r = de - er \cos \theta$ is

$$\sqrt{x^2 + y^2} = de - ex.$$

Squaring both sides and rearranging terms leads to

$$(1 - e^2)x^2 + 2de^2x + y^2 = d^2e^2.$$

Completing the square and simplifying, we obtain

$$\left(x + \frac{de^2}{1 - e^2}\right)^2 + \frac{y^2}{1 - e^2} = \frac{d^2e^2}{(1 - e^2)^2}.$$

Finally, dividing both sides by $d^2e^2/(1 - e^2)^2$ gives us an equation of the form

$$\frac{(x - h)^2}{a^2} + \frac{y^2}{b^2} = 1$$

with $h = -de^2/(1 - e^2)$. Consequently, the graph is an ellipse with center at the point $(-de^2/(1 - e^2), 0)$ on the x-axis and with

$$a^2 = \frac{d^2e^2}{(1 - e^2)^2}, \qquad b^2 = \frac{d^2e^2}{1 - e^2}.$$

By Theorem (12.6),

$$c^2 = a^2 - b^2 = \frac{d^2e^4}{(1 - e^2)^2}$$

and hence $c = de^2/(1 - e^2)$. This proves that F is a focus of the ellipse. It also follows that $e = c/a$ (see Definition (12.8)). A similar proof may be given for the case $e > 1$.

We can show, conversely, that every conic that is not degenerate may be described by means of the statement in Theorem (13.16). This gives us a formulation of conic sections that is equivalent to the approach used in Chapter 12. Since (13.16) includes all three types of conics, it is sometimes regarded as a definition for the conic sections.

If we had chosen the focus F to the *right* of the directrix, as illustrated in Figure 13.37 (with $d > 0$), then the equation $r = de/(1 - e \cos \theta)$ would have resulted. (Note the minus sign in place of the plus sign.) Other sign changes occur if d is allowed to be negative.

If l is taken *parallel* to the polar axis through one of the points $(d, \pi/2)$ or $(d, 3\pi/2)$, as illustrated in Figure 13.38, then the corresponding equations would contain $\sin \theta$ instead of $\cos \theta$. The proofs of these facts are left as exercises.

FIGURE 13.37

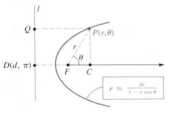

FIGURE 13.38

(i)   (ii)

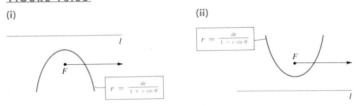

The following theorem summarizes our discussion.

THEOREM (13.17)

A polar equation having one of the four forms

$$r = \frac{de}{1 \pm e \cos \theta} \quad \text{or} \quad r = \frac{de}{1 \pm e \sin \theta}$$

is a conic section. The conic is a parabola if $e = 1$, an ellipse if $0 < e < 1$, or a hyperbola if $e > 1$.

EXAMPLE 1 Describe and sketch the graph of the equation

$$r = \frac{10}{3 + 2 \cos \theta}.$$

SOLUTION Dividing numerator and denominator of the given fraction by 3 gives us

$$r = \frac{\frac{10}{3}}{1 + \frac{2}{3} \cos \theta}$$

which has one of the forms in Theorem (13.17) with $e = \frac{2}{3}$. Thus the graph is an ellipse with focus F at the pole and major axis along the polar axis. We find the endpoints of the major axis by letting $\theta = 0$ and $\theta = \pi$. This gives us $V(2, 0)$ and $V'(10, \pi)$. Hence

$$2a = d(V', V) = 12 \quad \text{or} \quad a = 6.$$

FIGURE 13.39

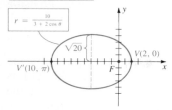

The center of the ellipse is the midpoint $(4, \pi)$ of the segment $V'V$. Using the fact that $e = c/a$, we obtain

$$c = ae = 6(\tfrac{2}{3}) = 4.$$

Hence
$$b^2 = a^2 - c^2 = 36 - 16 = 20.$$

Thus the semiminor axis has length $b = \sqrt{20}$. The graph is sketched in Figure 13.39. For reference, we have superimposed a rectangular coordinate system on the polar system. •

EXAMPLE 2 Describe and sketch the graph of the equation

$$r = \frac{10}{2 + 3 \sin \theta}.$$

SOLUTION To express the equation in one of the forms in Theorem (3.17), we divide numerator and denominator of the given fraction by 2, obtaining

$$r = \frac{5}{1 + \tfrac{3}{2} \sin \theta}.$$

Thus $e = \tfrac{3}{2}$ and the graph is a hyperbola with a focus at the pole. The expression $\sin \theta$ tells us that the transverse axis of the hyperbola is perpendicular to the polar axis. To find the vertices we let $\theta = \pi/2$ and $\theta = 3\pi/2$ in the given equation. This gives us the points $V(2, \pi/2)$, $V'(-10, 3\pi/2)$, and hence

$$2a = d(V, V') = 8 \quad \text{or} \quad a = 4.$$

The points $(5, 0)$ and $(5, \pi)$ on the graph can be used to sketch the lower branch of the hyperbola. The upper branch is obtained by symmetry, as illustrated in Figure 13.40. If we desire more accuracy or additional information, we may calculate

$$c = ae = 4(\tfrac{3}{2}) = 6$$

and
$$b^2 = c^2 - a^2 = 36 - 16 = 20.$$

Asymptotes may then be constructed in the usual way. •

FIGURE 13.40

$$r = \frac{10}{2 + 3 \sin \theta}$$

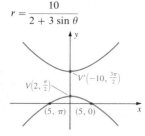

EXAMPLE 3 Sketch the graph of the equation

$$r = \frac{15}{4 - 4 \cos \theta}.$$

SOLUTION To obtain one of the forms in (13.17), we divide numerator and denominator of the given fraction by 4, obtaining

$$r = \frac{\tfrac{15}{4}}{1 - \cos \theta}.$$

Consequently, $e = 1$ and the graph is a parabola with focus at the pole. We may obtain a sketch by plotting the points that correspond to the x-and y-intercepts. These are indicated in the following table.

θ	0	$\pi/2$	π	$3\pi/2$
r	—	$\tfrac{15}{4}$	$\tfrac{15}{8}$	$\tfrac{15}{4}$

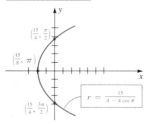

FIGURE 13.41

Note that there is no point on the graph corresponding to $\theta = 0$, since the denominator $1 - \cos \theta$ is 0 for that value. Plotting the three points and using the fact that the graph is a parabola with focus at the pole gives us the sketch in Figure 13.41. •

If we desire only a rough sketch of a conic, then the technique employed in Example 3 is recommended. To use this method we plot (if possible) points corresponding to $\theta = 0$, $\pi/2$, π, and $3\pi/2$. These points, together with the type of conic (obtained from the value of e), readily lead to the sketch.

EXERCISES 13.5

Exer. 1–10: Describe the graph of the polar equation.

1 $r = \dfrac{12}{6 + 2 \sin \theta}$

2 $r = \dfrac{12}{6 - 2 \sin \theta}$

3 $r = \dfrac{12}{2 - 6 \cos \theta}$

4 $r = \dfrac{12}{2 + 6 \cos \theta}$

5 $r = \dfrac{3}{2 + 2 \cos \theta}$

6 $r = \dfrac{3}{2 - 2 \sin \theta}$

7 $r = \dfrac{4}{\cos \theta - 2}$

8 $r = \dfrac{4 \sec \theta}{2 \sec \theta - 1}$

9 $r = \dfrac{6 \csc \theta}{2 \csc \theta + 3}$

10 $r = \csc \theta \, (\csc \theta - \cot \theta)$

11–20 Find equations in x and y for the graphs in Exercises 1–10.

Exer. 21–26: Use Theorem (13.16) to find a polar equation of the conic with focus at the pole and the given eccentricity and equation of directrix.

21 $e = \frac{1}{3}$, $r = 2 \sec \theta$

22 $e = \frac{2}{5}$, $r = 4 \csc \theta$

23 $e = 4$, $r = -3 \csc \theta$

24 $e = 3$, $r = -4 \sec \theta$

25 $e = 1$, $r \cos \theta = 5$

26 $e = 1$, $r \sin \theta = -2$

27 Find a polar equation of the parabola with focus at the pole and vertex $(4, \pi/2)$.

28 Find a polar equation of an ellipse with eccentricity $\frac{2}{3}$, a vertex at $(1, 3\pi/2)$, and a focus at the pole.

29 Prove Theorem (13.16) for the case $e > 1$.

30 (a) Use Figure 13.37 to derive the formula
$$r = de/(1 - e \cos \theta).$$
   (b) Derive the formulas $r = de/(1 \pm e \sin \theta)$ in Theorem (13.17) by using Figure 13.38.

Exer. 31–36: Express the equation in one of the forms in Theorem (13.17), and then find the eccentricity and an equation for the directrix.

31 $y^2 = 4 - 4x$

32 $x^2 = 1 - 2y$

33 $3y^2 - 16y - x^2 + 16 = 0$

34 $5x^2 + 9y^2 = 32x + 64$

35 $8x^2 + 9y^2 + 4x = 4$

36 $4x^2 - 5y^2 + 36y - 36 = 0$

Exer. 37–40: Find the slope of the tangent line to the graph of the equation at the point corresponding to the given value of θ. (Refer to the graphs in Exercise 1–4.)

37 $r = \dfrac{12}{6 + 2 \sin \theta}$, $\theta = \pi/6$

38 $r = \dfrac{12}{6 - 2 \sin \theta}$, $\theta = 0$

39 $r = \dfrac{12}{2 - 6 \cos \theta}$, $\theta = \pi/2$

40 $r = \dfrac{12}{2 + 6 \cos \theta}$, $\theta = \pi/3$

41 Kepler's first law (see Section 15.6) asserts that planets travel in elliptical orbits with the sun at one focus. To find an equation of an orbit, place the pole O at the center of the sun and the polar axis along the major axis of the ellipse (see figure).

   (a) Show that an equation of the orbit is
$$r = \frac{(1 - e^2)a}{1 - e \cos \theta}$$
   for eccentricity e and length of major axis $2a$.

   (b) The *perihelion distance* r_p and *aphelion distance* r_a are defined as the maximum and minimum distances, respectively, of a planet from the sun. Show that $r_p = a(1 + e)$ and $r_a = a(1 - e)$.

EXERCISE 41

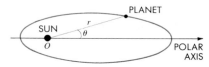

42 Refer to Exercise 41. The planet Pluto travels in an elliptical orbit of eccentricity 0.249. If the aphelion distance is 29.62 AU, find a polar equation for the orbit, and estimate the perihelion distance.

43 (a) Use the substitution $z = \tan(\theta/2)$ to show that

$$\int_{-\pi}^{\pi} \frac{1}{1 - e\cos\theta}\, d\theta = \frac{2\pi}{\sqrt{1 - e^2}}.$$

(*Hint:* See Theorem 9.6.)

(b) Use the result of part (a) to establish that the average distance of a planet from the sun is $a\sqrt{1 - e^2}$.

44 Refer to Exercises 41 and 43. Eros—with an average distance from the sun of 1.46 AU—is the smallest of the asteroids in our solar system. If Eros has an elliptical orbit of eccentricity 0.223, find a polar equation that approximates the orbit. Estimate how close Eros comes to the sun.

13.6 REVIEW

Define or discuss each of the following.

1 Plane curve
2 Closed curve
3 Simple closed curve
4 Parametric equations
5 Parametrized curve
6 Parametric form for dy/dx
7 Parametric form for d^2y/dx^2
8 Parametric form for arc length

9 Parametric form for surface area
10 Polar coordinates
11 Graphs of polar equations
12 Relationship between rectangular and polar coordinates
13 Integrals in polar coordinates
14 Polar equations of conics

EXERCISES 13.6

Exer. 1–3: Sketch the graph of the curve, find an equation in x and y of a graph that contains the points on the curve, and find the slope of the tangent line at the point corresponding to $t = 1$.

1 $x = (1/t) + 1$, $y = (2/t) - t$; $\quad 0 < t \le 4$

2 $x = \cos^2 t - 2$, $y = \sin t + 1$; $\quad 0 \le t \le 2\pi$

3 $x = \sqrt{t}$, $y = e^{-t}$; $\quad t \ge 0$

4 Let C be the curve with parametric equations $x = t^2$, $y = 2t^3 + 4t - 1$; t in $\mathbb{R}$.

(a) Find the x-coordinates of the points on C at which the tangent line passes through the origin.

(b) Find values of t that correspond to horizontal or vertical tangent lines.

(c) Find a parametric form for d^2y/dx^2.

5 Compare the graphs of the following curves for t in $\mathbb{R}$.

(a) $x = t^2$, $\quad y = t^3$ $\qquad$ (b) $x = t^4$, $\quad y = t^6$
(c) $x = e^{2t}$, $\quad y = e^{3t}$ $\qquad$ (d) $x = 1 - \sin^2 t$, $\quad y = \cos^3 t$

Exer. 6–18: Sketch the graph of the equation, and find an equation in x and y that has the same graph.

6 $r = 3 + 2\cos\theta$

7 $r = 6\cos 2\theta$

8 $r = 4 - 4\cos\theta$

9 $r^2 = -4\sin 2\theta$

10 $r = 2\sin 3\theta$

11 $r(3\cos\theta - 2\sin\theta) = 6$

12 $r = e^{-\theta}$, $\quad \theta \ge 0$

13 $r^2 = \sec 2\theta$

14 $r = 8\sec\theta$

15 $r = 4\cos^2 \tfrac{1}{2}\theta$

16 $r = 6 - r\cos\theta$

17 $r = 8/(3 + \cos\theta)$

18 $r = 8/(1 - 3\sin\theta)$

19 Find the slope of the tangent line to the graph of $r = 3/(2 + 2\cos\theta)$ at the point corresponding to $\theta = \pi/2$.

Exer. 20–21: Find a polar equation for the graph.

20 $x^2 + y^2 = 2xy$

21 $y^2 = x^2 - 2x$

22 Find a polar equation of the hyperbola that has focus at the pole, eccentricity 2, and equation of directrix $r = 6\sec\theta$.

23 Find the area of the region bounded by one loop of the graph of $r^2 = 4\sin 2\theta$.

24 Find the area of the region that is inside the graph of $r = 3 + 2\sin\theta$ and outside the graph of $r = 4$.

25 The position (x, y) of a particle at time t is given by $x = 2\sin t$, $y = \sin^2 t$. Find the distance the particle travels from $t = 0$ to $t = \pi/2$.

26 Find the length of the spiral $r = 1/\theta$ from $(1, 1)$ to $(\tfrac{1}{2}, 2)$.

27 The curve $x = 2t^2 + 1$, $y = 4t - 3$; $0 \le t \le 1$, is revolved about the y-axis. Find the area of the resulting surface.

28 The arc of the spiral $r = e^\theta$ from $(1, 0)$ to $(e, 1)$ is revolved about the line $\theta = \pi/2$. Find the area of the resulting surface.

29 Find the area of the surface generated by revolving the lemniscate $r^2 = a^2 \cos 2\theta$ about the polar axis.

30 A line segment of fixed length has endpoints A, on the y-axis, and B, on the x-axis. A fixed point P on AB is selected such that $d(A, P) = a$ and $d(B, P) = b$ (see figure). If A and B may slide freely along their respective axes, a curve C is traced by P. If t is the radian measure of angle ABO, find parametric equations for C with parameter t. What type of curve is formed?

EXERCISE 30

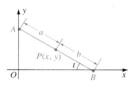

In the first two sections of this chapter we shall extend the concepts of limits, continuity, and derivatives to functions of more than one variable. We then consider *differential equations*—equations that involve derivatives or differentials. If an equation involves only derivatives of a function of one variable, then it is said to be *ordinary*. A *partial differential equation* contains partial derivatives. We will develop techniques for solving basic types of ordinary differential equations. The discussion is not intended to be a treatise on the subject, but rather to serve as an introduction to this vast and important area of mathematics.

AN INTRODUCTION TO PARTIAL DERIVATIVES AND DIFFERENTIAL EQUATIONS

14.1 FUNCTIONS OF SEVERAL VARIABLES

The functions in previous chapters involved only one independent variable. Such functions have many applications; however, in some problems *several* independent variables occur:

The area of a rectangle depends on *two* quantities, length and width.

If an object is located in space, then the temperature at a point P in the object may depend on *three* rectangular coordinates x, y, and z of P.

If the temperature of an object in space changes with time t, then *four* variables x, y, z, and t are involved.

A manufacturer may find that the cost C of producing a certain item depends on material, labor, equipment, overhead, and maintenance charges. Thus C is influenced by *five* different variables.

In this section we shall define functions of several variables and discuss some of their properties. Recall that a function f is a correspondence that associates with each element of a set D a unique element of a set E. If D is a subset of $\mathbb{R} \times \mathbb{R}$ and E is a subset of $\mathbb{R}$, then f is called a *function of two (real) variables*. Another way of stating this is as follows.

DEFINITION (14.1)

> Let D be a set of ordered pairs of real numbers. A **function f of two variables** is a correspondence that associates with each pair (x, y) in D a unique real number, denoted by $f(x, y)$. The set D is the **domain** of f. The **range** of f consists of all real numbers $f(x, y)$ for (x, y) in D.

We may represent the domain D in Definition (14.1) by points in an xy-plane and the range by points on a real line, say a w-axis, as illustrated in

FIGURE 14.1

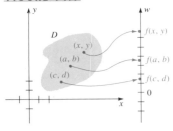

Figure 14.1. Several curved arrows are drawn from ordered pairs in D to the corresponding numbers in the range. As an application, consider a flat metal plate having the shape of D. To each point (x, y) on the plate there corresponds a temperature $f(x, y)$ that can be recorded on a thermometer, represented by the w-axis. We could also regard D as the surface of a lake and let $f(x, y)$ denote the depth of the water under the point (x, y). There are many other physical interpretations for Definition (14.1).

We often use an expression in x and y to specify $f(x, y)$, and assume that the domain is the set of all pairs (x, y) for which the given expression is meaningful. We then call f **a function of x and y.**

EXAMPLE 1 If $f(x, y) = \dfrac{xy - 5}{2\sqrt{y - x^2}}$, find the domain D of f. Sketch D, and represent the numbers $f(2, 5)$, $f(1, 2)$, and $f(-1, 2)$ on a w-axis, as in Figure 14.1.

FIGURE 14.2

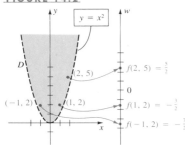

SOLUTION Referring to the denominator of $f(x, y)$, we see that the domain D is the set of all pairs (x, y) such that $y - x^2$ is positive, that is, $y > x^2$. Thus the graph of D is that part of the xy-plane that lies above the parabola $y = x^2$, as shown in Figure 14.2.

By substitution,

$$f(2, 5) = \frac{(2)(5) - 5}{2\sqrt{5 - 4}} = \frac{5}{2}.$$

Similarly, $f(1, 2) = -\frac{3}{2}$ and $f(-1, 2) = -\frac{7}{2}$. These function values are indicated on the w-axis in Figure 14.2. •

Formulas are often used to define functions of two variables. For example, the formula $V = \pi r^2 h$ expresses the volume V of a right circular cylinder as a function of the altitude h and base radius r. The symbols r and h are referred to as **independent variables,** and V is the **dependent variable.**

A function f of three (real) variables is defined as in (14.1), except that each element of the domain D is an ordered triple (x, y, z), that is, a set $\{x, y, z\}$ of three numbers in which x is considered as the first number, y as the second number, and z as the third number. In this case, with each ordered triple (x, y, z) in D there is associated a unique real number $f(x, y, z)$.

Functions of three variables are often defined by expressions. For example,

$$f(x, y, z) = \frac{xe^y + \sqrt{z^2 + 1}}{xy \sin z}$$

determines a function of x, y, and z.

Let f be a function of two variables, and consider $f(x, y)$ as (x, y) varies through the domain D of f. As a physical illustration, suppose a flat metal plate has the shape of the region D in Figure 14.1. To each point (x, y) on the plate there corresponds a temperature $f(x, y)$, which is recorded on a thermometer represented by the w-axis. As the point (x, y) moves on the plate the temperature may increase, decrease, or remain constant, and hence the point on the w-axis that corresponds to $f(x, y)$ will move in a positive direction, negative direction, or remain fixed, respectively. If the temperature

$f(x, y)$ gets closer to a fixed value L as (x, y) gets closer and closer to a fixed point (a, b), we use the following notation.

LIMIT NOTATION (14.2)

$$\lim_{(x,y)\to(a,b)} f(x, y) = L$$

This may be read: **the limit of $f(x, y)$ as (x, y) approaches (a, b) is L.**

To make (14.2) mathematically precise, we may proceed as follows. For any $\varepsilon > 0$, consider the open interval $(L - \varepsilon, L + \varepsilon)$ on the w-axis, as illustrated in Figure 14.3. If (14.2) is true, then there is a $\delta > 0$ such that for every point (x, y) inside the circle of radius δ with center (a, b), except possibly (a, b) itself, the function value $f(x, y)$ is in the interval $(L - \varepsilon, L + \varepsilon)$. This is equivalent to the following statement:

$$\text{If} \quad 0 < \sqrt{(x - a)^2 + (y - b)^2} < \delta, \quad \text{then} \quad |f(x, y) - L| < \varepsilon.$$

Thus we have the following.

FIGURE 14.3

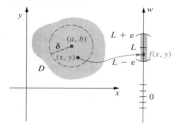

DEFINITION OF LIMIT (14.3)

Let a function f of two variables be defined throughout the interior of a circle with center (a, b), except possibly at (a, b) itself. The statement

$$\lim_{(x,y)\to(a,b)} f(x, y) = L$$

means that for every $\varepsilon > 0$ there corresponds a $\delta > 0$ such that

$$\text{if} \quad 0 < \sqrt{(x - a)^2 + (y - b)^2} < \delta, \quad \text{then} \quad |f(x, y) - L| < \varepsilon$$

We can show that *if the limit L exists, it is unique.*

If f and g are functions of two variables, then $f + g$, $f - g$, fg, and f/g are defined in the usual way, and Theorem (2.15) concerning limits of sums, products, and quotients can be extended. For example, if f and g have limits as (x, y) approaches (a, b), then

$$\lim_{(x,y)\to(a,b)} [f(x, y) + g(x, y)] = \lim_{(x,y)\to(a,b)} f(x, y) + \lim_{(x,y)\to(a,b)} g(x, y),$$

$$\lim_{(x,y)\to(a,b)} \frac{f(x, y)}{g(x, y)} = \frac{\lim_{(x,y)\to(a,b)} f(x, y)}{\lim_{(x,y)\to(a,b)} g(x, y)} \quad \text{if} \quad \lim_{(x,y)\to(a,b)} g(x, y) \neq 0,$$

$$\lim_{(x,y)\to(a,b)} \sqrt{f(x, y)} = \sqrt{\lim_{(x,y)\to(a,b)} f(x, y)} \quad \text{if} \quad \lim_{(x,y)\to(a,b)} f(x, y) > 0,$$

and so on.

A function f of two variables is a **polynomial function** if $f(x, y)$ can be expressed as a sum of terms of the form $cx^m y^n$ for a real number c and non-negative integers m and n. A **rational function** is a quotient of two polynomial functions. As for single-variable functions, limits of polynomial and rational functions in two variables may be found by substituting for x and y.

EXAMPLE 2 Find

(a) $\displaystyle\lim_{(x,y)\to(2,-3)} (x^3 - 4xy^2 + 5y - 7)$ (b) $\displaystyle\lim_{(x,y)\to(3,4)} \frac{x^2 - y^2}{\sqrt{x^2 + y^2}}$

SOLUTION

(a) Since $x^3 - 4xy^2 + 5y - 7$ is a polynomial, we may find the limit by substituting 2 for x and -3 for y. Thus

$$\lim_{(x,y)\to(2,-3)} (x^3 - 4xy^2 + 5y - 7) = 2^3 - 4(2)(-3)^2 + 5(-3) - 7$$
$$= 8 - 72 - 15 - 7 = -86$$

(b) We may proceed as follows:

$$\lim_{(x,y)\to(3,4)} \frac{x^2 - y^2}{\sqrt{x^2 + y^2}} = \frac{\displaystyle\lim_{(x,y)\to(3,4)} (x^2 - y^2)}{\displaystyle\lim_{(x,y)\to(3,4)} \sqrt{x^2 + y^2}} = \frac{9 - 16}{\sqrt{\displaystyle\lim_{(x,y)\to(3,4)} (x^2 + y^2)}}$$

$$= \frac{-7}{\sqrt{9 + 16}} = -\frac{7}{5}$$

EXAMPLE 3 Show that $\displaystyle\lim_{(x,y)\to(0,0)} \frac{x^2 - y^2}{x^2 + y^2}$ does not exist.

SOLUTION   Let

$$f(x, y) = \frac{x^2 - y^2}{x^2 + y^2}.$$

The function f is rational; however, we cannot substitute 0 for both x and y, since that would lead to a zero denominator.

If we consider any point $(x, 0)$ on the x-axis, then

$$f(x, 0) = \frac{x^2 - 0}{x^2 + 0} = 1 \quad \text{provided} \quad x \neq 0.$$

For any point $(0, y)$ on the y-axis, we have

$$f(0, y) = \frac{0 - y^2}{0 + y^2} = -1 \quad \text{provided} \quad y \neq 0.$$

FIGURE 14.4

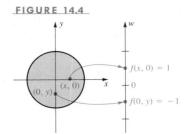

Consequently, as illustrated in Figure 14.4, *every* circle with center $(0, 0)$ contains points at which the value of f is 1 and points at which the value of f is -1. It follows that the limit does not exist, for if we take $\varepsilon = 1$ in Definition (14.3), there is no open interval $(L - \varepsilon, L + \varepsilon)$ on the w-axis containing both 1 and -1. Hence it is impossible to find a $\delta > 0$ that satisfies the conditions of the definition. •

In Chapter 2, for a function with a jump discontinuity at $x = a$, we proved that $\lim_{x\to a} f(x)$ does not exist by showing that $\lim_{x\to a^-} f(x)$ and $\lim_{x\to a^+} f(x)$ are not equal. When considering such one-sided limits we may regard the point on the x-axis with coordinate x as "approaching" the point with coordinate a either from the left or from the right, respectively. The analogous situation for functions of two variables is more complicated, since in a coordinate plane there are an infinite number of different curves, or **paths,** along which (x, y) can approach (a, b). However, if the limit in Definition (14.3) exists, then $f(x, y)$ must have the limit L, regardless of the path taken. This illustrates the following rule for investigating limits.

> If two different paths to a point $P(a, b)$ produce two different limiting values for f, then $\lim\limits_{(x, y) \to (a, b)} f(x, y)$ does not exist.

We shall next rework Example 3 using (14.4).

EXAMPLE 4 Show that $\lim\limits_{(x, y) \to (0, 0)} \dfrac{x^2 - y^2}{x^2 + y^2}$ does not exist.

SOLUTION If (x, y) approaches $(0, 0)$ along the x-axis, then the y-coordinate is always zero and the expression $(x^2 - y^2)/(x^2 + y^2)$ reduces to x^2/x^2, or 1. Hence the limiting value along this path is 1.

If (x, y) approaches $(0, 0)$ along the y-axis, then the x-coordinate is 0 and $(x^2 - y^2)/(x^2 + y^2)$ reduces to $-y^2/y^2$, or -1. Since two different values are obtained, the limit does not exist by the Two-Path Rule (14.4).

We could, of course, have chosen other paths to the origin $(0, 0)$. For example, if we let (x, y) approach $(0, 0)$ along the line $y = 2x$, then

$$\frac{x^2 - y^2}{x^2 + y^2} = \frac{x^2 - 4x^2}{x^2 + 4x^2} = \frac{-3x^2}{5x^2} = -\frac{3}{5}$$

and hence the limiting value along the line $y = 2x$ is $-\frac{3}{5}$.

As an alternative solution, we could change the expression in x and y to polar coordinates, as follows:

$$\frac{x^2 - y^2}{x^2 + y^2} = \frac{r^2 \cos^2 \theta - r^2 \sin^2 \theta}{r^2} = \cos^2 \theta - \sin^2 \theta = \cos 2\theta.$$

As $(x, y) \to (0, 0)$ along any ray $\theta = k$, the function value is always $\cos 2k$, and hence $f(x, y)$ would have this limiting value. By choosing suitable values of k, we could make $f(x, y)$ approach any value between -1 and 1. Hence, by (14.4), the limit does not exist. •

EXAMPLE 5 Show that $\lim\limits_{(x, y) \to (0, 0)} \dfrac{x^2 y}{x^4 + y^2}$ does not exist.

SOLUTION If we let (x, y) approach $(0, 0)$ along any line $y = mx$ that passes through the origin, we see that if $m \neq 0$,

$$\lim_{(x, y) \to (0, 0)} \frac{x^2 y}{x^4 + y^2} = \lim_{(x, y) \to (0, 0)} \frac{x^2(mx)}{x^4 + (mx)^2} = \lim_{(x, y) \to (0, 0)} \frac{mx^3}{x^4 + m^2 x^2}$$

$$= \lim_{(x, y) \to (0, 0)} \frac{mx}{x^2 + m^2} = \frac{0}{0 + m^2} = 0.$$

It is tempting to conclude that the limit is 0 as (x, y) approaches $(0, 0)$; however, if we let (x, y) approach $(0, 0)$ along the parabola $y = x^2$, then

$$\lim_{(x, y) \to (0, 0)} \frac{x^2 y}{x^4 + y^2} = \lim_{(x, y) \to (0, 0)} \frac{x^2(x^2)}{x^4 + (x^2)^2}$$

$$= \lim_{(x, y) \to (0, 0)} \frac{x^4}{2x^4} = \lim_{(x, y) \to (0, 0)} \frac{1}{2} = \frac{1}{2}.$$

Thus not every path through $(0, 0)$ leads to the same limiting value, and hence by the Two-Path Rule (14.4), the limit does not exist. •

FIGURE 14.5

(i) Interior point of R

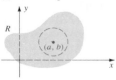

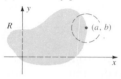

If R is a region in the xy-plane, then a point (a, b) is an **interior point** of R if there exists a disk with center (a, b) that contains only points of R. An illustration of an interior point is shown in Figure 14.5(i). A point (a, b) is a **boundary point** of R if every disk with center (a, b) contains points that are in R and points that are not in R, as illustrated in Figure 14.5(ii).

A region is **closed** if it contains all of its boundary points. A region is **open** if it contains *none* of its boundary points; that is, every point of the region is an interior point. A region that contains some, but not all, of its boundary points is neither open nor closed. These concepts are analogous to closed intervals, open intervals, and half-open intervals of real numbers.

Suppose a function f of two variables is defined for every (x, y) in a region R, except possibly at (a, b). If (a, b) is an interior point, then Definition (14.3) can be used to investigate the limit of f as (x, y) approaches (a, b). If (a, b) *is a boundary point we shall use Definition* (14.3) *with the added restriction that* (x, y) *must be both in R and inside the circle of radius* δ. Theorems on limits may then be extended to boundary points.

A function f of two variables is **continuous** at an interior point (a, b) of a region R if $f(a, b)$ exists, $f(x, y)$ has a limit as (x, y) approaches (a, b), and

$$\lim_{(x, y) \to (a, b)} f(x, y) = f(a, b).$$

This notion can be extended to a boundary point (a, b) of R by applying the preceding restrictions on limits involving boundary points. If R is in the domain D of f, then f is **continuous on R** if it is continuous at every pair (a, b) in R. If f is continuous on R, then a small change in (x, y) produces a small change in $f(x, y)$.

We may prove theorems on continuity for functions in two variables that are analogous to those for functions of one variable. In particular, polynomial functions are continuous everywhere and rational functions are continuous except at points where the denominator is zero.

The preceding discussion on limits and continuity can be extended to functions of three or more variables. For example, if f is a function of three variables, then

$$\lim_{(x, y, z) \to (a, b, c)} f(x, y, z) = L$$

means that for every $\varepsilon > 0$ there corresponds a $\delta > 0$ such that if

$$0 < \sqrt{(x - a)^2 + (y - b)^2 + (z - c)^2} < \delta,$$

then
$$|f(x, y, z) - L| < \varepsilon$$

A function f of three variables is **continuous** at (a, b, c) if $f(a, b, c)$ exists, $f(x, y, z)$ has a limit as (x, y, z) approaches (a, b, c), and

$$\lim_{(x, y, z) \to (a, b, c)} f(x, y, z) = f(a, b, c).$$

We may also consider composite functions of several variables. As a simple illustration, if f is a function of two variables x and y, and g is a function of one variable t, then we may consider $f(x, y)$ and $g(t)$. A function h of two variables may be obtained by substituting $f(x, y)$ for t; that is, by letting $h(x, y) = g(f(x, y))$, provided the range of f is in the domain of g. This is illustrated in the next example.

EXAMPLE 6 Express $g(f(x, y))$ in terms of x and y, and find the domain of the resulting composite function.

(a) $f(x, y) = xe^y$, $g(t) = 3t^2 + t + 1$

(b) $f(x, y) = y - 4x^2$, $g(t) = \sin \sqrt{t}$

SOLUTION   In each case we substitute $f(x, y)$ for t. Thus:

(a) $g(f(x, y)) = g(xe^y) = 3x^2e^{2y} + xe^y + 1$

(b) $g(f(x, y)) = g(y - 4x^2) = \sin \sqrt{y - 4x^2}$

The domain in part (a) is $\mathbb{R} \times \mathbb{R}$, and the domain in part (b) consists of every ordered pair (x, y) such that $y \geq 4x^2$. •

The following analogue of Theorem (2.32) may be proved for functions of the type illustrated in Example 6.

THEOREM (14.5)
> If a function f of two variables is continuous at (a, b), and a function g of one variable is continuous at $f(a, b)$, then the function h defined by $h(x, y) = g(f(x, y))$ is continuous at (a, b).

Theorem (14.5) allows us to establish the continuity of composite functions of several variables, as illustrated in the next example.

EXAMPLE 7 If $h(x, y) = e^{x^2 + 5xy + y^3}$, show that h is continuous at every pair (a, b).

SOLUTION   If we let $f(x, y) = x^2 + 5xy + y^3$ and $g(t) = e^t$, it follows that $h(x, y) = g(f(x, y))$. Since f is a polynomial function, it is continuous at every pair (a, b). Moreover, g is continuous at every $t = f(a, b)$. Thus, by Theorem (14.5), h is continuous at (a, b). •

EXERCISES 14.1

Exer. 1–6: Determine the domain of f and the value of f at the indicated points.

1 $f(x, y) = 2x - y^2$; $(-2, 5)$, $(5, -2)$, $(0, -2)$

2 $f(x, y) = (y + 2)/x$; $(3, 1)$, $(1, 3)$, $(2, 0)$

3 $f(u, v) = uv/(u - 2v)$; $(2, 3)$, $(-1, 4)$, $(0, 1)$

4 $f(r, s) = \sqrt{1 - r} - e^{r/s}$; $(1, 1)$, $(0, 4)$, $(-3, 3)$

5 $f(x, y, z) = \sqrt{25 - x^2 - y^2 - z^2}$; $(1, -2, 2)$, $(-3, 0, 2)$

6 $f(x, y, z) = 2 + \tan x + y \sin z$; $(\pi/4, 4, \pi/6)$, $(0, 0, 0)$

Exer. 7–16: Find the limit, if it exists.

7 $\displaystyle \lim_{(x, y) \to (0, 0)} \frac{x^2 - 2}{3 + xy}$

8 $\displaystyle \lim_{(x, y) \to (0, 0)} \frac{x^3 - x^2y + xy^2 - y^3}{x^2 + y^2}$

9 $\displaystyle \lim_{(x, y) \to (0, 0)} \frac{2x^2 - y^2}{x^2 + 2y^2}$

10 $\displaystyle \lim_{(x, y) \to (0, 0)} \frac{x^2 - 2xy + 5y^2}{3x^2 + 4y^2}$

11 $\displaystyle \lim_{(x, y) \to (0, 0)} \frac{xy^2}{x^2 + y^2}$ (*Hint:* Use polar coordinates.)

12 $\displaystyle \lim_{(x, y) \to (0, 0)} \frac{x^4 - y^4}{x^2 + y^2}$

13 $\displaystyle \lim_{(x, y) \to (1, 2)} \frac{xy - 2x - y + 2}{x^2 + y^2 - 2x - 4y + 5}$

14 $\displaystyle \lim_{(x, y) \to (0, 0)} \frac{3xy}{5x^4 + 2y^4}$

15 $\displaystyle \lim_{(x, y) \to (0, 0)} \frac{3x^3 - 2x^2y + 3y^2x - 2y^3}{x^2 + y^2}$

16 $\displaystyle \lim_{(x, y, z) \to (0, 0, 0)} \frac{xy + yz + xz}{x^2 + y^2 + z^2}$

Exer. 17–22: Discuss the continuity of the function f.

17 $f(x, y) = \ln (x + y - 1)$

18 $f(x, y) = \dfrac{xy}{x^2 - y^2}$

19 $f(x, y, z) = \dfrac{1}{x^2 + y^2 - z^2}$

20 $f(x, y, z) = \sqrt{xy} \, \tan z$

21 $f(x, y) = \sqrt{x} e^{\sqrt{1 - y^2}}$

22 $f(x, y) = \sqrt{25 - x^2 - y^2}$

Exer. 23–26: Find $h(x, y) = g(f(x, y))$ and use Theorem (14.5) to determine where h is continuous.

23 $f(x, y) = x^2 - y^2, \quad g(t) = (t^2 - 4)/t$

24 $f(x, y) = 3x + 2y - 4, \quad g(t) = \ln (t + 5)$

25 $f(x, y) = x + \tan y, \quad g(z) = z^2 + 1$

26 $f(x, y) = y \ln x, \quad g(w) = e^w$

27 (a) Define a function of four (real) variables.

(b) Extend Definition (14.3) to functions of four variables.

28 Prove that if f is a continuous function of two variables and $f(a, b) > 0$, then there is a circle C in the xy-plane with center (a, b) such that $f(x, y) > 0$ for every pair (x, y) that is in the domain of f and within C.

29 Prove, directly from Definition (14.3), that

(a) $\displaystyle \lim_{(x, y) \to (a, b)} x = a.$ (b) $\displaystyle \lim_{(x, y) \to (a, b)} y = b.$

30 If $\lim_{(x, y) \to (a, b)} f(x, y) = L$ and c is any real number, prove, directly from Definition (14.3), that

$$\lim_{(x, y) \to (a, b)} cf(x, y) = cL.$$

14.2 PARTIAL DERIVATIVES

In Chapter 3 we defined the derivative $f'(x)$ of a function of *one* variable as

$$f'(x) = \lim_{h \to 0} \frac{f(x + h) - f(x)}{h}.$$

We may interpret this formula as follows: First change the independent variable x by an amount h, then divide the corresponding change $f(x + h) - f(x)$ in f by h, and finally, let h approach 0. An analogous procedure can be applied to functions of several variables. Thus, given $f(x, y)$, first change *one* of the variables, say x, by an amount h, then divide the corresponding change $f(x + h, y) - f(x, y)$ in f by h, and finally, let h approach 0. This leads to the concept of the *partial derivative $f_x(x, y)$ of f with respect to x*, defined in (14.6). A similar process is used for the variable y.

DEFINITION (14.6)

Let f be a function of two variables. The **first partial derivatives of f with respect to x and y** are the functions f_x and f_y such that

$$f_x(x, y) = \lim_{h \to 0} \frac{f(x + h, y) - f(x, y)}{h}$$

$$f_y(x, y) = \lim_{h \to 0} \frac{f(x, y + h) - f(x, y)}{h}$$

In Definition (14.6), x and y are fixed (but arbitrary), and h is the only variable; hence we use the notation for limits of functions of one variable instead of the "$(x, y) \to (a, b)$" notation introduced in the previous section. If we let $y = b$ and define a function g of one variable by $g(x) = f(x, b)$, then $g'(x) = f_x(x, b) = f_x(x, y)$. Thus, *to find $f_x(x, y)$ we may regard y as a constant and differentiate $f(x, y)$ with respect to x.* Similarly, *to find $f_y(x, y)$ the variable x is regarded as a constant and $f(x, y)$ is differentiated with respect to y.* This technique is illustrated in Example 1.

Other common notations for partial derivatives are listed in the next box.

If $w = f(x, y)$, then

$$f_x = \frac{\partial f}{\partial x}, \qquad f_y = \frac{\partial f}{\partial y}$$

$$f_x(x, y) = \frac{\partial}{\partial x} f(x, y) = \frac{\partial w}{\partial x} = w_x$$

$$f_y(x, y) = \frac{\partial}{\partial y} f(x, y) = \frac{\partial w}{\partial y} = w_y$$

For brevity we often speak of $\partial f/\partial x$ or $\partial f/\partial y$ as the *partial of f with respect to x or to y*, respectively.

EXAMPLE 1 If $f(x, y) = x^3 y^2 - 2x^2 y + 3x$, find
(a) $f_x(x, y)$ and $f_y(x, y)$.   (b) $f_x(2, -1)$ and $f_y(2, -1)$.

SOLUTION

(a) We regard y as a constant and differentiate with respect to x:

$$f_x(x, y) = 3x^2 y^2 - 4xy + 3$$

Regarding x as a constant and differentiating with respect to y gives us

$$f_y(x, y) = 2x^3 y - 2x^2$$

(b) Substituting $x = 2$ and $y = -1$ in part (a),

$$f_x(2, -1) = 3(2)^2(-1)^2 - 4(2)(-1) + 3 = 23$$

$$f_y(2, -1) = 2(2)^3(-1) - 2(2)^2 = -24 \qquad \bullet$$

Formulas similar to those for functions of one variable are true for partial derivatives. For example, if $u = f(x, y)$ and $v = g(x, y)$, then a Product Rule and Quotient Rule for partial derivatives are

$$\frac{\partial}{\partial x}(uv) = u \frac{\partial v}{\partial x} + v \frac{\partial u}{\partial x}, \qquad \frac{\partial}{\partial x}\left(\frac{u}{v}\right) = \frac{v \dfrac{\partial u}{\partial x} - u \dfrac{\partial v}{\partial x}}{v^2}$$

or, using subscript notation,

$$(fg)_x = fg_x + gf_x \quad \text{and} \quad \left(\frac{f}{g}\right)_x = \frac{gf_x - fg_x}{g^2}.$$

A Power Rule for partial differentiation is

$$\frac{\partial}{\partial x}(u^n) = nu^{n-1} \frac{\partial u}{\partial x}$$

for any real number n. Similarly,

$$\frac{\partial}{\partial x} \cos u = -\sin u \frac{\partial u}{\partial x}, \qquad \frac{\partial}{\partial x} e^u = e^u \frac{\partial u}{\partial x},$$

and so on.

EXAMPLE 2 Find $\dfrac{\partial w}{\partial y}$ if $w = xy^2 e^{xy}$.

SOLUTION We write $w = (xy^2)(e^{xy})$ and differentiate:

$$\frac{\partial w}{\partial y} = (xy^2) \frac{\partial}{\partial y} (e^{xy}) + (e^{xy}) \frac{\partial}{\partial y} (xy^2)$$

$$= xy^2(xe^{xy}) + e^{xy}(2xy) = (xy + 2)xye^{xy} \bullet$$

FIGURE 14.6

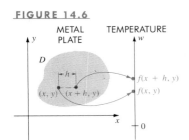

FIGURE 14.6

Let us apply Definition (14.6) to the physical illustration given at the beginning of the previous section: $f(x, y)$ is the temperature of a flat metal plate in an xy-plane at the point (x, y). Note that the points (x, y) and $(x + h, y)$ lie on a horizontal line, as indicated in Figure 14.6. If a point moves horizontally from (x, y) to $(x + h, y)$, then the difference $f(x + h, y) - f(x, y)$ is the net change in temperature, and we have

Average change in temperature: $\dfrac{f(x + h, y) - f(x, y)}{h}$

For example, if the temperature change is 2 degrees and the distance h is 4, then the average change in temperature from (x, y) to $(x + h, y)$ is $\frac{2}{4}$, or $\frac{1}{2}$. Thus, *on the average*, the temperature changes at a rate of $\frac{1}{2}$ degree per unit change in distance. Taking the limit of the average change as h approaches 0, we see that $f_x(x, y)$ is the (instantaneous) rate of change of temperature with respect to distance as the point (x, y) moves in a horizontal direction.

Similarly, if a point moves on a vertical line from (x, y) to $(x, y + h)$, as shown in Figure 14.7, we obtain

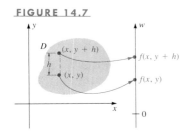

FIGURE 14.7

Average change in temperature: $\dfrac{f(x, y + h) - f(x, y)}{h}$

Letting h approach 0, we see that $f_y(x, y)$ is the (instantaneous) rate of change of temperature as the point (x, y) moves in a vertical direction.

As another application, suppose D represents the surface of a lake and $f(x, y)$ is the depth of the water under the point (x, y) on the surface. In this case $f_x(x, y)$ is the rate at which the depth changes as a point moves away from (x, y) parallel to the x-axis. Similarly, $f_y(x, y)$ is the rate of change of the depth in the direction of the y-axis.

First partial derivatives of functions of three or more variables are defined in the same manner as that used in Definition (14.6). Specifically, all variables except one are regarded as constant, and we differentiate with respect to the remaining variable. Given $f(x, y, z)$, we may find $f_x, f_y,$ and f_z (or equivalently $\partial f/\partial x, \partial f/\partial y,$ and $\partial f/\partial z$). For example,

$$f_z(x, y, z) = \lim_{h \to 0} \frac{f(x, y, z + h) - f(x, y, z)}{h}.$$

EXAMPLE 3 If $w = x^2 y^3 \sin z + e^{xz}$, find $\partial w/\partial x$, $\partial w/\partial y$, and $\partial w/\partial z$.

SOLUTION We may proceed as follows:

Regard y and z as constant: $\dfrac{\partial w}{\partial x} = 2xy^3 \sin z + ze^{xz}$

Regard x and z as constant: $\dfrac{\partial w}{\partial y} = 3x^2 y^2 \sin z$

Regard x and y as constant: $\dfrac{\partial w}{\partial z} = x^2 y^3 \cos z + xe^{xz}$ •

If f is a function of two variables x and y, then f_x and f_y are also functions of two variables and we may consider *their* first partial derivatives. These are called the **second partial derivatives** of f and are denoted as follows.

SECOND PARTIAL DERIVATIVES (14.8)

$$\frac{\partial}{\partial x} f_x = (f_x)_x = f_{xx} = \frac{\partial}{\partial x}\left(\frac{\partial f}{\partial x}\right) = \frac{\partial^2 f}{\partial x^2}$$

$$\frac{\partial}{\partial y} f_x = (f_x)_y = f_{xy} = \frac{\partial}{\partial y}\left(\frac{\partial f}{\partial x}\right) = \frac{\partial^2 f}{\partial y\, \partial x}$$

$$\frac{\partial}{\partial x} f_y = (f_y)_x = f_{yx} = \frac{\partial}{\partial x}\left(\frac{\partial f}{\partial y}\right) = \frac{\partial^2 f}{\partial x\, \partial y}$$

$$\frac{\partial}{\partial y} f_y = (f_y)_y = f_{yy} = \frac{\partial}{\partial y}\left(\frac{\partial f}{\partial y}\right) = \frac{\partial^2 f}{\partial y^2}$$

If $w = f(x, y)$ we write

$$\frac{\partial^2}{\partial x^2} f(x, y) = f_{xx}(x, y) = \frac{\partial^2 w}{\partial x^2} = w_{xx},$$

$$\frac{\partial^2}{\partial y\, \partial x} f(x, y) = f_{xy}(x, y) = \frac{\partial^2 w}{\partial y\, \partial x} = w_{xy},$$

and so on.

Note that f_{xy} means that we differentiate first with respect to x and then with respect to y, and f_{yx} means that we differentiate first with respect to y and then with respect to x. For the "∂" notation, the reverse order is used; that is, for $\partial^2 f/\partial x\, \partial y$ we differentiate first with respect to y and then with respect to x.

We shall refer to f_{xy} and f_{yx} as the **mixed second partial derivatives of f** (or simply *mixed partials of f*). The next theorem states that, under suitable conditions, these mixed partials are equal; that is, the order of differentiation is immaterial. The proof may be found in texts on advanced calculus.

THEOREM (14.9)

Let f be a function of two variables x and y. If f, f_x, f_y, f_{xy}, and f_{yx} are continuous on an open region R, then $f_{xy} = f_{yx}$ throughout R.

The hypothesis of Theorem (14.9) is satisfied for most functions encountered in calculus and its applications. Similarly, if $w = f(x, y, z)$ and f has continuous second partial derivatives, then the following equalities hold for mixed partials:

$$\frac{\partial^2 w}{\partial y \, \partial x} = \frac{\partial^2 w}{\partial x \, \partial y}; \qquad \frac{\partial^2 w}{\partial z \, \partial x} = \frac{\partial^2 w}{\partial x \, \partial z}; \qquad \frac{\partial^2 w}{\partial z \, \partial y} = \frac{\partial^2 w}{\partial y \, \partial z}.$$

EXAMPLE 4 Find the second partial derivatives of f if

$$f(x, y) = x^3 y^2 - 2x^2 y + 3x.$$

SOLUTION This function was considered in Example 1, where we obtained

$$f_x(x, y) = 3x^2 y^2 - 4xy + 3 \quad \text{and} \quad f_y(x, y) = 2x^3 y - 2x^2.$$

Hence

$$f_{xx}(x, y) = \frac{\partial}{\partial x} f_x(x, y) = \frac{\partial}{\partial x} (3x^2 y^2 - 4xy + 3) = 6xy^2 - 4y$$

$$f_{xy}(x, y) = \frac{\partial}{\partial y} f_x(x, y) = \frac{\partial}{\partial y} (3x^2 y^2 - 4xy + 3) = 6x^2 y - 4x$$

$$f_{yx}(x, y) = \frac{\partial}{\partial x} f_y(x, y) = \frac{\partial}{\partial x} (2x^3 y - 2x^2) = 6x^2 y - 4x$$

$$f_{yy}(x, y) = \frac{\partial}{\partial y} f_y(x, y) = \frac{\partial}{\partial y} (2x^3 y - 2x^2) = 2x^3 \qquad \bullet$$

Third and higher partial derivatives are defined in similar fashion. For example,

$$\frac{\partial}{\partial x} f_{xx} = f_{xxx} = \frac{\partial}{\partial x} \left(\frac{\partial^2 f}{\partial x^2} \right) = \frac{\partial^3 f}{\partial x^3},$$

$$\frac{\partial}{\partial x} f_{xy} = f_{xyx} = \frac{\partial}{\partial x} \left(\frac{\partial^2 f}{\partial y \, \partial x} \right) = \frac{\partial^3 f}{\partial x \, \partial y \, \partial x},$$

and so on. If first, second, and third partial derivatives are continuous, then the order of differentiation is immaterial, that is,

$$f_{xyx} = f_{yxx} = f_{xxy} \quad \text{and} \quad f_{yxy} = f_{xyy} = f_{yyx}.$$

Of course, letters other than x and y may be used. If f is a function of r and s, then symbols such as

$$f_r(r, s), \quad f_s(r, s), \quad f_{rs}, \quad \frac{\partial f}{\partial r}, \quad \frac{\partial^2 f}{\partial r^2}$$

are employed for partial derivatives.

Similar notations and results apply to partials of functions of more than two variables.

EXERCISES 14.2

Exer. 1–18: Find the first partial derivatives of f.

1 $f(x, y) = 2x^4 y^3 - xy^2 + 3y + 1$

2 $f(x, y) = (x^3 - y^2)^2$

3 $f(r, s) = \sqrt{r^2 + s^2}$

4 $f(s, t) = (t/s) - (s/t)$

5 $f(x, y) = xe^y + y \sin x$

6 $f(x, y) = e^x \ln xy$

7 $f(t, v) = \ln \sqrt{(t + v)/(t - v)}$

8 $f(u, w) = \arctan (u/w)$

9 $f(x, y) = x \cos (x/y)$

10 $f(x, y) = \sqrt{4x^2 - y^2} \sec x$

11 $f(r, s, t) = r^2 e^{2s} \cos t$

12 $f(x, y, t) = (x^2 - t^2)/(1 + \sin 3y)$

13 $f(x, y, z) = (y^2 + z^2)^x$

14 $f(r, s, v) = (2r + 3s)^{\cos v}$

15 $f(x, y, z) = xe^z - ye^x + ze^{-y}$

16 $f(r, s, v, p) = r^3 \tan s + \sqrt{se^{v^2}} - v \cos 2p$

17 $f(q, v, w) = \sin^{-1} \sqrt{qv} + \sin vw$

18 $f(x, y, z) = xyze^{xyz}$

Exer. 19–24: Verify that $w_{xy} = w_{yx}$.

19 $w = xy^4 - 2x^2y^3 + 4x^2 - 3y$

20 $w = x^2/(x + y)$

21 $w = x^3 e^{-2y} + y^{-2} \cos x$

22 $w = y^2 e^{x^2} + [1/(x^2 y^3)]$

23 $w = x^2 \cosh (z/y)$

24 $w = \sqrt{x^2 + y^2 + z^2}$

25 If $w = 3x^2 y^3 z + 2xy^4 z^2 - yz$, find w_{xyz}.

26 If $w = u^4 vt^2 - 3uv^2 t^3$, find w_{tut}.

27 If $u = v \sec rt$, find u_{rvr}.

28 If $v = y \ln (x^2 + z^4)$, find v_{zzy}.

29 If $w = \sin xyz$, find $\partial^3 w/\partial z\, \partial y\, \partial x$.

30 If $w = x^2/(y^2 + z^2)$, find $\partial^3 w/\partial z\, \partial y^2$.

31 If $w = r^4 s^3 t - 3s^2 e^{rt}$, verify that $w_{rrs} = w_{rsr} = w_{srr}$.

32 If $w = \tan uv + 2 \ln (u + v)$, verify that $w_{uvv} = w_{vuv} = w_{vvu}$.

Exer. 33–36: A function f of x and y is *harmonic* if $(\partial^2 f/\partial x^2) + (\partial^2 f/\partial y^2) = 0$ throughout the domain of f. Prove that the function is harmonic.

33 $f(x, y) = \ln \sqrt{x^2 + y^2}$

34 $f(x, y) = \arctan (y/x)$

35 $f(x, y) = \cos x \sinh y + \sin x \cosh y$

36 $f(x, y) = e^{-x} \cos y + e^{-y} \cos x$

37 If $w = \cos (x - y) + \ln (x + y)$, show that $(\partial^2 w/\partial x^2) - (\partial^2 w/\partial y^2) = 0$.

38 If $w = (y - 2x)^3 - \sqrt{y - 2x}$, show that $w_{xx} - 4w_{yy} = 0$.

39 If $w = e^{-c^2 t} \sin cx$, show that $w_{xx} = w_t$ for every real number c.

40 The ideal gas law may be stated as $PV = knT$ for the number of moles of gas n, volume V, temperature T, pressure P, and a constant k. Show that

$$\frac{\partial V}{\partial T} \frac{\partial T}{\partial P} \frac{\partial P}{\partial V} = -1.$$

Exer. 41–42: Show that v satisfies the *wave equation*

$$\frac{\partial^2 v}{\partial t^2} = a^2 \frac{\partial^2 v}{\partial x^2}.$$

41 $v = (\sin akt)(\sin kx)$ **42** $v = (x - at)^4 + \cos (x + at)$

Exer. 43–46: Show that the functions u and v satisfy the *Cauchy-Riemann equations* $u_x = v_y$ and $u_y = -v_x$.

43 $u(x, y) = x^2 - y^2$, $v(x, y) = 2xy$

44 $u(x, y) = y/(x^2 + y^2)$, $v(x, y) = x/(x^2 + y^2)$

45 $u(x, y) = e^x \cos y$, $v(x, y) = e^x \sin y$

46 $u(x, y) = \cos x \cosh y + \sin x \sinh y$
$v(x, y) = \cos x \cosh y - \sin x \sinh y$

47 List all possible second partial derivatives of $w = f(x, y, z)$.

48 If $w = f(x, y, z, t, v)$, define w_t as a limit.

49 A flat metal plate lies on an xy-plane such that the temperature T at (x, y) is given by $T = 10(x^2 + y^2)^2$ for T in degrees and x, y in centimeters. Find the rate of change of T with respect to distance at $(1, 2)$ in the direction of (a) the x-axis; (b) the y-axis.

50 The surface of a certain lake is represented by a region D in the xy-plane such that the depth under the point corresponding to (x, y) is given by $f(x, y) = 300 - 2x^2 - 3y^2$ for x, y, and $f(x, y)$ in feet. If a boy in the water is at the point $(4, 9)$, find the rate at which the depth changes if he swims in the direction of (a) the x-axis; (b) the y-axis.

51 When a pollutant such as nitric oxide is emitted from a smoke stack of height h meters, the long-range concentration $C(x, z)$ (in $\mu g/m^3$) of the pollutant at a point x km from the stack and at a height of z meters (see figure) can often be represented by

$$C(x, z) = \frac{a}{x^2} \left[e^{-b(z - h)^2/x^2} + e^{-b(z + h)^2/x^2} \right]$$

for positive constants a and b that depend on atmospheric conditions and the pollution emission rate. Suppose that

$$C(x, z) = \frac{200}{x^2} \left[e^{-0.02(z - 10)^2/x^2} + e^{-0.02(z + 10)^2/x^2} \right].$$

Compute and interpret $\partial C/\partial x$ and $\partial C/\partial z$ at the point $(2, 5)$.

EXERCISE 51

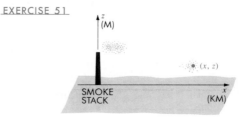

52 The analysis of certain electrical circuits involves the formula $I = V/\sqrt{R^2 + L^2\omega^2}$ for current I, voltage V, resistance R, inductance L, and a positive constant ω. Find and interpret $\partial I/\partial R$ and $\partial I/\partial L$.

53 In the study of frost penetration in highway engineering, the temperature T at time t hours and depth x feet can be approximated by
$$T = T_0 e^{-\lambda x} \sin(\omega t - \lambda x)$$
for constants T_0, ω, and λ. The period of $\sin(\omega t - \lambda x)$ is 24 hours.
(a) Find and interpret $\partial T/\partial t$ and $\partial T/\partial x$.
(b) Show that T satisfies the one-dimensional heat equation
$$\frac{\partial T}{\partial t} = k\frac{\partial^2 T}{\partial x^2} \quad \text{for some constant } k.$$

54 Show that any function given by
$$w = (\sin ax)(\cos by)\, e^{-\sqrt{a^2 + b^2}\, z}$$

satisfies the three-dimensional heat equation
$$\frac{\partial^2 w}{\partial x^2} + \frac{\partial^2 w}{\partial y^2} + \frac{\partial^2 w}{\partial z^2} = 0.$$

55 The vital capacity V of the lungs is the largest volume (in ml) that can be exhaled after a maximum inhalation of air. For a typical male, V is approximated by $V = 27.63y - 0.112xy$ for age x (in years) and height y (in cm).
(a) Compute and interpret $\partial V/\partial x$.
(b) Explain why it is difficult to interpret $\partial V/\partial y$.

56 The intensity of sunlight $I(x, t)$ (in foot-candles) at time t on a clear day and at ocean depth x can be approximated by
$$I(x, t) = I_0 e^{-kx} \sin^3(\pi t/D)$$
for the intensity at midday I_0, day length D (in hours), and $k > 0$. If $I_0 = 1000$, $D = 12$, and $k = 0.10$, compute and interpret $\partial I/\partial t$ and $\partial I/\partial x$ when $t = 6$ hours and $x = 5$ meters.

14.3 SEPARABLE DIFFERENTIAL EQUATIONS

An equation that involves x, y, y', y'', ..., and $y^{(n)}$, for a function y of x with nth derivative $y^{(n)}$ of y with respect to x, is an **ordinary differential equation of order n**. The following are examples of ordinary differential equations of specified orders:

$$\text{Order 1:} \quad y' = 2x$$

$$\text{Order 2:} \quad \frac{d^2 y}{dx^2} + x^2\left(\frac{dy}{dx}\right)^3 - 15y = 0$$

$$\text{Order 3:} \quad (y''')^4 - x^2(y'')^5 + 4xy = xe^x$$

$$\text{Order 4:} \quad \left(\frac{d^4 y}{dx^4}\right)^2 - 1 = x^3\frac{dy}{dx}$$

We considered differential equations briefly in Sections 4.7 and 7.6. Recall that a function f (or $f(x)$) is a **solution** of a differential equation if substitution of $f(x)$ for y results in an identity for every x in some interval. For example, the differential equation

$$y' = 6x^2 - 5$$

has the solution

$$f(x) = 2x^3 - 5x + C$$

for every real number C, since substitution of $f(x)$ for y leads to the identity $6x^2 - 5 = 6x^2 - 5$. We call $f(x) = 2x^3 - 5x + C$ the **general solution** of $y' = 6x^2 - 5$, since every solution is of this form. A **particular solution** is obtained by assigning C a specific value. To illustrate, if $C = 0$ we obtain the particular solution $y = 2x^3 - 5x$. Sometimes **initial conditions** are stated that determine a particular solution, as illustrated in the next example (see also Examples 3 and 4 of Section 4.7).

EXAMPLE 1

(a) Find the general solution of the differential equation $y' = 2x$, and illustrate it graphically.

(b) Find the particular solution of $y' = 2x$ that satisfies the following condition: $y = 3$ if $x = 0$.

SOLUTION

(a) If f is a solution of $y' = 2x$, then $f'(x) = 2x$. Indefinite integration gives us the general solution

$$y = f(x) = x^2 + C.$$

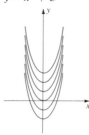

FIGURE 14.8
$y = x^2 + C$

Particular solutions may be found by assigning specific values to C. This leads to the family of parabolas $y = x^2 + C$ illustrated in Figure 14.8.

(b) If $y = 3$ when $x = 0$, then substitution in $y = x^2 + C$ gives us $3 = 0 + C$, or $C = 3$. Hence the particular solution is $y = x^2 + 3$. The graph is the parabola in Figure 14.8 with y-intercept 3. •

Differential equations of the type given in the next example occur in the analysis of vibrations. We will consider these in detail in Sections 14.4–14.7.

EXAMPLE 2 Show that the differential equation $y'' - 25y = 0$ has the solution

$$f(x) = C_1 e^{5x} + C_2 e^{-5x}$$

for all real numbers C_1 and C_2.

SOLUTION Differentiating, we obtain

$$f'(x) = 5C_1 e^{5x} - 5C_2 e^{-5x}$$

and

$$f''(x) = 25C_1 e^{5x} + 25C_2 e^{-5x}.$$

Substituting $f(x)$ for y in the differential equation $y'' - 25y = 0$ gives us

$$(25C_1 e^{5x} + 25C_2 e^{-5x}) - 25(C_1 e^{5x} + C_2 e^{-5x}) = 0$$

or

$$(25C_1 e^{5x} - 25C_1 e^{5x}) + (25C_2 e^{-5x} - 25C_2 e^{-5x}) = 0.$$

Since the left side is 0 for every x, it follows that $f(x)$ is a solution of $y'' - 25y = 0$. •

The solution $C_1 e^{5x} + C_2 e^{-5x}$ in Example 2 is the general solution of $y'' - 25y = 0$. Observe that the differential equation is of order 2 and the general solution contains two arbitrary constants (called **parameters**) C_1 and C_2. The precise definition of general solution involves the concept of **independent parameters** and is left for more advanced courses. General solutions of nth-order differential equations contain n independent parameters C_1, $C_2, \ldots, C_n$. A particular solution is obtained by assigning a specific value to every parameter. Some differential equations have solutions that are not special cases of the general solution. Such **singular solutions** will not be discussed in this text.

The solutions in Examples 1 and 2 express y explicitly in terms of x. Solutions of certain differential equations are stated implicitly, and we use implicit differentiation to check such solutions, as illustrated in the following example.

EXAMPLE 3 Show that $x^3 + x^2y - 2y^3 = C$ is an implicit solution of

$$(x^2 - 6y^2)y' + 3x^2 + 2xy = 0.$$

SOLUTION   If the first equation is satisfied by $y = f(x)$, then differentiating implicitly,

$$3x^2 + 2xy + x^2y' - 6y^2y' = 0$$

or $$(x^2 - 6y^2)y' + 3x^2 + 2xy = 0.$$

Thus $f(x)$ is a solution of the differential equation.

One of the simplest types of differential equations is

$$M(x) + N(y)y' = 0 \quad \text{or} \quad M(x) + N(y)\frac{dy}{dx} = 0$$

for continuous functions M and N. If $y = f(x)$ is a solution, then

$$M(x) + N(f(x))f'(x) = 0.$$

If $f'(x)$ is continuous, then indefinite integration leads to

$$\int M(x)\, dx + \int N(f(x))f'(x)\, dx = C$$

or $$\int M(x)\, dx + \int N(y)\, dy = C.$$

The last equation is an (implicit) solution of the differential equation. The differential equation $M(x) + N(y)y' = 0$ is said to be **separable,** since the variables x and y may be separated as we have indicated.

An easy way to remember the method of separating the variables is to change the equation

$$M(x) + N(y)\frac{dy}{dx} = 0$$

to the *differential form*

$$M(x)\, dx + N(y)\, dy = 0$$

and then integrate each term.

EXAMPLE 4 Solve the differential equation $y^4e^{2x} + \dfrac{dy}{dx} = 0.$

SOLUTION   Writing the equation in differential form and then separating the variables gives us

$$y^4e^{2x}\, dx + dy = 0$$

and, if $y \neq 0$, $$e^{2x}\, dx + \frac{1}{y^4}\, dy = 0.$$

Integrating each term, we obtain the (implicit) solution

$$\frac{1}{2}e^{2x} - \frac{1}{3y^3} = C.$$

If we multiply both sides by 6, then another form of the solution is

$$3e^{2x} - \frac{2}{y^3} = K$$

with $K = 6C$. If an explicit form is desired we may solve for y, obtaining

$$y = \left(\frac{2}{3e^{2x} - K}\right)^{1/3}$$

We have assumed that $y \neq 0$; however, $y = 0$ *is* a solution of the differential equation. In future examples we will not point out such singular solutions.

•

EXAMPLE 5 Solve the differential equation $2y + (xy + 3x)\dfrac{dy}{dx} = 0$ with $x \neq 0$.

SOLUTION   We may express the equation in differential form as

$$2y\,dx + (xy + 3x)\,dy = 0$$

or

$$2y\,dx + x(y + 3)\,dy = 0.$$

The variables may be separated by dividing both sides of the last equation by xy. This gives us

$$\frac{2}{x}\,dx + \left(\frac{y + 3}{y}\right)dy = 0$$

or

$$\frac{2}{x}\,dx + \left(1 + \frac{3}{y}\right)dy = 0.$$

Integrating each term leads to the following equivalent equations:

$$2\ln|x| + y + 3\ln|y| = c$$
$$\ln|x|^2 + \ln|y|^3 = c - y$$
$$\ln|x|^2|y|^3 = c - y$$

Changing to exponential form, we obtain

$$|x|^2|y|^3 = e^{c-y} = e^c e^{-y} = ke^{-y}$$

with $k = e^c$. We can show, by differentiation, that

$$x^2y^3 = ke^{-y} \quad \text{or} \quad x^2y^3e^y = k$$

is an implicit solution of the differential equation.   •

EXAMPLE 6 Solve the differential equation

$$(1 + y^2) + (1 + x^2)\frac{dy}{dx} = 0.$$

SOLUTION Writing the equation in differential form and then separating the variables gives us

$$(1 + y^2)\,dx + (1 + x^2)\,dy = 0$$

$$\frac{1}{1 + x^2}\,dx + \frac{1}{1 + y^2}\,dy = 0.$$

Integrating each term, we obtain the (implicit) solution

$$\tan^{-1} x + \tan^{-1} y = C.$$

To find an explicit solution we may solve for y as follows:

$$\tan^{-1} y = C - \tan^{-1} x$$

$$y = \tan(C - \tan^{-1} x)$$

The form of this solution may be changed by using the subtraction formula for the tangent function. Thus

$$y = \frac{\tan C - \tan(\tan^{-1} x)}{1 + \tan C \tan(\tan^{-1} x)}.$$

If we let $k = \tan C$ and use the fact that $\tan(\tan^{-1} x) = x$, this may be written

$$y = \frac{k - x}{1 + kx} \quad \bullet$$

An **orthogonal trajectory** of a family of curves is a curve that intersects each curve of the family orthogonally. We shall restrict our discussion to curves in a coordinate plane. To illustrate, given the family $y = 2x + b$ of all lines of slope 2, every line $y = -\frac{1}{2}x + c$ of slope $-\frac{1}{2}$ is an orthogonal trajectory. Several curves from each family are sketched in Figure 14.9. We say that two such families of curves are **mutually orthogonal.** As another example, the family $y = mx$ of all lines through the origin and the family $x^2 + y^2 = a^2$ of all concentric circles with centers at the origin are mutually orthogonal (see Figure 14.10).

Pairs of mutually orthogonal families of curves occur frequently in physical applications of mathematics. In the theory of electricity and magnetism, the lines of force associated with a given field are orthogonal trajectories of the corresponding equipotential curves. Similarly, the stream lines studied in aerodynamics and hydrodynamics are orthogonal trajectories of the *velocity-equipotential* curves. As a final illustration, in the study of thermodynamics the flow of heat across a plane surface is orthogonal to the isothermal curves.

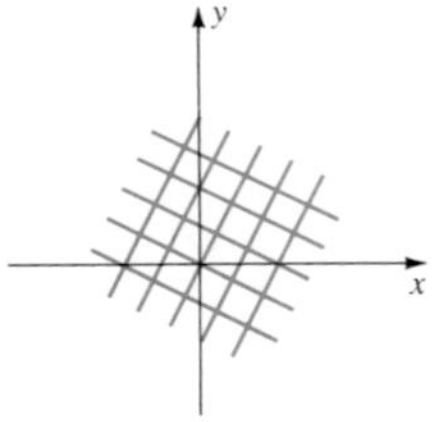

The next example illustrates a technique for finding the orthogonal trajectories of a family of curves.

EXAMPLE 7 Find the orthogonal trajectories of the family of ellipses $x^2 + 3y^2 = c$, and sketch several members of each family.

SOLUTION Differentiating the given equation implicitly, we obtain

$$2x + 6yy' = 0 \quad \text{or} \quad y' = -\frac{x}{3y}.$$

FIGURE 14.9
$y = 2x + b; \quad y = -\frac{1}{2}x + c$

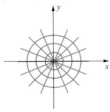

FIGURE 14.10
$y = mx; \quad x^2 + y^2 = a^2$

Hence the slope of the tangent line at any point (x, y) on each of the ellipses is $y' = -x/(3y)$. If dy/dx is the slope of the tangent line on a corresponding orthogonal trajectory, then it must equal the negative reciprocal of y'. This gives us the following differential equation for the family of orthogonal trajectories:

$$\frac{dy}{dx} = \frac{3y}{x}$$

Separating the variables,

$$\frac{dy}{y} = 3\frac{dx}{x}.$$

Integrating, and writing the constant of integration as $\ln |k|$ gives us

$$\ln |y| = 3 \ln |x| + \ln |k| = \ln |kx^3|.$$

It follows that $y = kx^3$ is an equation for the family of orthogonal trajectories. We have sketched several members of the family of ellipses (in gray) and corresponding orthogonal trajectories (in color) in Figure 14.11. •

FIGURE 14.11
$x^2 + 3y^2 = c; \; y = kx^3$

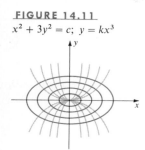

Exer. 1–4: (a) Find the general solution of the differential equation, and illustrate it graphically. (b) Find the particular solution that satisfies the condition $y = 2$ when $x = 0$.

1 $y' = 3x^2$

2 $y' = x - 1$

3 $y' = \dfrac{-x}{\sqrt{4 - x^2}}$

4 $y' = 3$

Exer. 5–10: Prove that y is a solution of the given differential equation.

5 $y'' - 3y' + 2y = 0; \quad y = C_1 e^x + C_2 e^{2x}$

6 $y' + 3y = 0; \quad y = Ce^{-3x}$

7 $2xy^3 + 3x^2y^2 \dfrac{dy}{dx} = 0; \quad y = Cx^{-2/3}$

8 $x^3 y''' + x^2 y'' - 3xy' - 3y = 0; \quad y = Cx^3$

9 $(x - 2y)\dfrac{dy}{dx} + 2x + y = 0; \quad y^2 - x^2 - xy = C$

10 $y\dfrac{dy}{dx} = x; \quad x^2 - y^2 = C$

Exer. 11–26: Solve the differential equation.

11 $\sec x \, dy - 2y \, dx = 0$

12 $x^2 \, dy - \csc 2y \, dx = 0$

13 $x \, dy - y \, dx = 0$

14 $(4 + y^2) \, dx + (9 + x^2) \, dy = 0$

15 $3y \, dx + (xy + 5x) \, dy = 0$

16 $(xy - 4x) \, dx + (x^2 y + y) \, dy = 0$

17 $y' = x - 1 + xy - y$

18 $(y + yx^2) \, dy + (x + xy^2) \, dx = 0$

19 $e^{x+2y} \, dx - e^{2x-y} \, dy = 0$

20 $\cos x \, dy - y \, dx = 0$

21 $y(1 + x^3)y' + x^2(1 + y^2) = 0$

22 $x^2 y' - yx^2 = y$

23 $x \tan y - y' \sec x = 0$

24 $xy + y'e^{(x^2)} \ln y = 0$

25 $e^y \sin x \, dx - \cos^2 x \, dy = 0$

26 $\sin y \cos x \, dx + (1 + \sin^2 x) \, dy = 0$

Exer. 27–34: Find the particular solution of the differential equation that satisfies the given condition.

27 $2y^2 y' = 3y - y'; \quad y = 1$ when $x = 3$

28 $\sqrt{x}y' - \sqrt{y} = x\sqrt{y}; \quad y = 4$ when $x = 9$

29 $x \, dy - (2x + 1)e^{-y} \, dx = 0; \quad y = 2$ when $x = 1$

30 $\sec 2y \, dx - \cos^2 x \, dy = 0; \quad y = \pi/6$ when $x = \pi/4$

31 $(xy + x) \, dx + \sqrt{4 + x^2} \, dy = 0; \quad y = 1$ when $x = 0$

32 $x \, dy - \sqrt{1 - y^2} \, dx = 0; \quad y = \frac{1}{2}$ when $x = 1$

33 $\cot x \, dy - (1 + y^2) \, dx = 0; \quad y = 1$ when $x = 0$

34 $\csc y \, dx - e^x \, dy = 0; \quad y = 0$ when $x = 0$

Exer. 35–40: Find the orthogonal trajectories of the family of curves. Describe the graphs.

35 $x^2 - y^2 = c$

36 $xy = c$

37 $y^2 = cx$

38 $y = cx^2$

39 $y^2 = cx^3$

40 $y = ce^{-x}$

The following type of differential equation occurs frequently in the study of physical phenomena.

DEFINITION (14.10)

A **first-order linear differential equation** is an equation of the form

$$y' + P(x)y = Q(x)$$

for continuous functions P and Q.

If, in Definition (14.10), $Q(x) = 0$ for every x, we obtain $y' + P(x)y = 0$, which is separable. Specifically, we may write

$$\frac{1}{y}\frac{dy}{dx} = -P(x) \quad \text{or} \quad \frac{1}{y}\,dy = -P(x)\,dx,$$

provided $y \neq 0$. Integration gives us

$$\ln|y| = -\int P(x)\,dx + \ln|C|.$$

We have expressed the constant of integration as $\ln|C|$ in order to change the form of the last equation as follows:

$$\ln|y| - \ln|C| = -\int P(x)\,dx$$

$$\ln\left|\frac{y}{C}\right| = -\int P(x)\,dx$$

$$\frac{y}{C} = e^{-\int P(x)\,dx}$$

$$ye^{\int P(x)\,dx} = C$$

We next observe that

$$D_x\left[ye^{\int P(x)\,dx}\right] = y'e^{\int P(x)\,dx} + P(x)ye^{\int P(x)\,dx}$$
$$= e^{\int P(x)\,dx}[y' + P(x)y].$$

Consequently, if we multiply both sides of $y' + P(x)y = Q(x)$ by $e^{\int P(x)\,dx}$, then the resulting equation may be written

$$D_x\left[ye^{\int P(x)\,dx}\right] = Q(x)e^{\int P(x)\,dx}.$$

Integrating both sides gives us the following implicit solution of the first-order differential equation in Definition (14.10).

$$ye^{\int P(x)\,dx} = \int Q(x)e^{\int P(x)\,dx}\,dx + K$$

for a constant K. Solving this equation for y leads to an explicit solution. The expression $e^{\int P(x)\,dx}$ is an **integrating factor** of the differential equation. We have proved the following result.

THEOREM (14.11)

The first-order linear differential equation $y' + P(x)y = Q(x)$ may be transformed into a separable differential equation by multiplying both sides by the integrating factor $e^{\int P(x)\,dx}$.

EXAMPLE 1 Solve the differential equation $\dfrac{dy}{dx} - 3x^2y = x^2$.

SOLUTION The differential equation has the form in Theorem (14.10) with $P(x) = -3x^2$ and $Q(x) = x^2$. By Theorem (14.11) an integrating factor is

$$e^{\int -3x^2\, dx} = e^{-x^3}.$$

We do not need to introduce a constant of integration, since $e^{-x^3+c} = e^c e^{-x^3}$, which differs from e^{-x^3} by a constant factor e^c. Multiplying both sides of the differential equation by the integrating factor e^{-x^3}, we obtain

$$e^{-x^3}\dfrac{dy}{dx} - 3x^2 e^{-x^3}y = x^2 e^{-x^3}$$

or $\qquad\qquad\qquad D_x\left(e^{-x^3}y\right) = x^2 e^{-x^3}.$

Integrating both sides of the last equation,

$$e^{-x^3}y = \int x^2 e^{-x^3}\, dx = -\tfrac{1}{3}e^{-x^3} + C.$$

Finally, multiplying by e^{x^3} gives us the explicit solution

$$y = -\tfrac{1}{3} + Ce^{x^3} \quad \bullet$$

EXAMPLE 2 Solve the differential equation $x^2y' + 5xy + 3x^5 = 0$ with $x \neq 0$.

SOLUTION To find an integrating factor we begin by expressing the differential equation in the "standardized" form (14.11) with the coefficient 1 for y'. Thus, dividing both sides by x^2, we obtain

$$y' + \dfrac{5}{x}y = -3x^3$$

which has the form given in (14.10) with $P(x) = 5/x$. By Theorem (14.11), an integrating factor is

$$e^{\int P(x)\, dx} = e^{5\ln|x|} = e^{\ln|x|^5} = |x|^5.$$

If $x > 0$, then $|x|^5 = x^5$ and if $x < 0$, then $|x|^5 = -x^5$. In either case, multiplying both sides of the standardized form by $|x|^5$ yields

$$x^5y' + 5x^4y = -3x^8$$

or $\qquad\qquad\qquad D_x\left(x^5y\right) = -3x^8.$

Integrating both sides of the last equation gives us the solution

$$x^5y = \int -3x^8\, dx = -\dfrac{x^9}{3} + C$$

or $\qquad\qquad\qquad y = -\dfrac{x^4}{3} + \dfrac{C}{x^5} \quad \bullet$

EXAMPLE 3 Solve the differential equation

$$y' + y\tan x = \sec x + 2x\cos x$$

SOLUTION The equation is a first-order linear differential equation (see Definition (14.10). By Theorem (14.11), an integrating factor is

$$e^{\int \tan x\, dx} = e^{\ln|\sec x|} = |\sec x|.$$

Multiplying both sides of the differential equation by $|\sec x|$ and discarding the absolute value sign gives us

$$y' \sec x + y \sec x \tan x = \sec^2 x + 2x \cos x \sec x$$

or

$$D_x (y \sec x) = \sec^2 x + 2x.$$

Integrating both sides yields the (implicit) solution

$$y \sec x = \tan x + x^2 + C.$$

Finally, multiplying both sides of the last equation by $1/\sec x = \cos x$, we obtain the explicit solution

$$y = \sin x + (x^2 + C) \cos x \quad \bullet$$

We have previously used antidifferentiation to derive laws of motion for falling bodies, assuming that air resistance could be neglected (see Example 6 in Section 4.7). This is a valid assumption for small, slowly moving objects; however, in many cases air resistance must be taken into account. This frictional force often increases as the speed of the object increases. In the following example we shall derive the law of motion for a falling body under the assumption that the resistance due to the air is directly proportional to the speed of the body.

EXAMPLE 4 An object of mass m is released from a hot-air balloon. Find the distance it falls in t seconds if the force of resistance due to the air is directly proportional to the speed of the object.

FIGURE 14.12

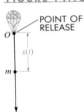

POINT OF RELEASE

O

$s(t)$

m

SOLUTION Let us introduce a vertical axis with positive direction downward and origin at the point of release, as illustrated in Figure 14.12. We wish to find the distance $s(t)$ from the origin to the object at time t. The speed of the object is $v = s'(t)$ and the magnitude of the acceleration is $a = dv/dt = s''(t)$. If g is the gravitational constant, then the object is attracted toward the earth with a force of magnitude mg. By hypothesis, the force of resistance due to the air is kv for some constant k, and this force is directed opposite to the motion. It follows that the downward force F on the object is $mg - kv$. Since Newton's second law of motion states that $F = ma = m(dv/dt)$, we arrive at the following differential equation:

$$m \frac{dv}{dt} = mg - kv$$

or, equivalently,

$$\frac{dv}{dt} + \frac{k}{m} v = g.$$

If we denote the constant k/m by c, this equation may be written

$$\frac{dv}{dt} + cv = g,$$

which is a first-order differential equation with t as the independent variable. By (14.11), an integrating factor is

$$e^{\int c \, dt} = e^{ct}.$$

Multiplying both sides of the last differential equation by e^{ct} gives us

$$e^{ct}\frac{dv}{dt} + ce^{ct}v = ge^{ct}$$

or

$$D_t(ve^{ct}) = ge^{ct}.$$

Integrating both sides,

$$ve^{ct} = \frac{g}{c}e^{ct} + K \quad \text{or} \quad v = \frac{g}{c} + Ke^{-ct}$$

for a constant K.

If we let $t = 0$, then $v = 0$ and hence

$$0 = \frac{g}{c} + K \quad \text{and} \quad K = -\frac{g}{c}.$$

Consequently,

$$v = \frac{g}{c} - \frac{g}{c}e^{-ct}.$$

Integrating both sides of this equation with respect to t and using the fact that $v = s'(t)$, we see that

$$s(t) = \frac{g}{c}t + \frac{g}{c^2}e^{-ct} + E.$$

We may find the constant E by letting $t = 0$. Since $s(0) = 0$,

$$0 = 0 + \frac{g}{c^2} + E \quad \text{or} \quad E = -\frac{g}{c^2}.$$

Thus, the distance the object falls in t seconds is

$$s(t) = \frac{g}{c}t + \frac{g}{c^2}e^{-ct} - \frac{g}{c^2}.$$

It is interesting to compare this formula for $s(t)$ with that obtained when the air resistance is neglected. In the latter case the differential equation $m(dv/dt) = mg - kv$ reduces to $dv/dt = g$, and hence $s'(t) = v = gt$. Integrating both sides leads to the much simpler formula $s(t) = \frac{1}{2}gt^2$. •

FIGURE 14.13

EXAMPLE 5 A simple electrical circuit consists of a resistance R and an inductance L connected in series, as illustrated schematically in Figure 14.13, with a constant electromotive force V. If the switch S is closed at $t = 0$, then it follows from one of Kirchhoff's rules for electrical circuits that if $t > 0$, the current I satisfies the differential equation

$$L\frac{dI}{dt} + RI = V$$

Express I as a function of t.

SOLUTION The differential equation may be written

$$\frac{dI}{dt} + \frac{R}{L}I = \frac{V}{L},$$

which is a first-order linear differential equation. Applying Theorem (14.11), we multiply both sides by the integrating factor $e^{\int (R/L)\,dt} = e^{(R/L)t}$, obtaining

$$e^{(R/L)t}\frac{dI}{dt} + \frac{R}{L}e^{(R/L)t}I = \frac{V}{L}e^{(R/L)t}$$

or

$$D_t\left[Ie^{(R/L)t}\right] = \frac{V}{L}e^{(R/L)t}.$$

Integrating with respect to t,

$$Ie^{(R/L)t} = \int \frac{V}{L}e^{(R/L)t}\,dt = \frac{V}{R}e^{(R/L)t} + C.$$

Since $I = 0$ when $t = 0$, it follows that $C = -V/R$. Substituting for C leads to

$$Ie^{(R/L)t} = \frac{V}{R}e^{(R/L)t} - \frac{V}{R}.$$

Finally, multiplying both sides by $e^{-(R/L)t}$ gives us

$$I = \frac{V}{R}\left[1 - e^{-(R/L)t}\right].$$

Observe that as t increases without bound, I approaches V/R, which is the current when no inductance is present. •

EXERCISES 14.4

Exer. 1–22: Solve the differential equation.

1 $y' + 2y = e^{2x}$

2 $y' - 3y = 2$

3 $xy' - 3y = x^5$

4 $y' + y\cot x = \csc x$

5 $xy' + y + x = e^x$

6 $xy' + (1 + x)y = 5$

7 $x^2\,dy + (2xy - e^x)\,dx = 0$

8 $x^2\,dy + (x - 3xy + 1)\,dx = 0$

9 $y' + y\cot x = 4x^2\csc x$

10 $y' + y\tan x = \sin x$

11 $(y\sin x - 2)\,dx + \cos x\,dy = 0$

12 $(x^2y - 1)\,dx + x^3\,dy = 0$

13 $(x^2\cos x + y)\,dx - x\,dy = 0$

14 $y' + y = \sin x$

15 $xy' + (2 + 3x)y = xe^{-3x}$

16 $(x + 4)y' + 5y = x^2 + 8x + 16$

17 $x^{-1}y' + 2y = 3$

18 $y' - 5y = e^{5x}$

19 $\tan x\,dy + (y - \sin x)\,dx = 0$

20 $\cos x\,dy - (y\sin x + e^{-x})\,dx = 0$

21 $y' + 3x^2y = x^2 + e^{-x^3}$

22 $y' + y\tan x = \cos^3 x$

Exer. 23–26: Find the particular solution of the differential equation that satisfies the given condition.

23 $xy' - y = x^2 + x$; $y = 2$ when $x = 1$

24 $y' + 2y = e^{-3x}$; $y = 2$ when $x = 0$

25 $xy' + y + xy = e^{-x}$; $y = 0$ when $x = 1$

26 $y' + 2xy - e^{-x^2} = x$; $y = 1$ when $x = 0$

Exer. 27–28: Solve the differential equation by (a) using an integrating factor; (b) separating the variables.

27 The differential equation

$$R\frac{dQ}{dt} + \frac{Q}{C} = V$$

describes the charge Q on a condenser of capacity C during a charging process involving a resistance R and electromotive force V. If the charge is 0 when $t = 0$, express Q as a function of t.

28 The differential equation

$$R\frac{dI}{dt} + \frac{I}{C} = \frac{dV}{dt}$$

describes an electrical circuit consisting of an electromotive force V with a resistance R and capacity C connected in series. Express I as a function of t if $I = I_0$ when $t = 0$, and V is a constant.

29 At time $t = 0$ a tank contains K pounds of salt dissolved in 80 gallons of water. Suppose that water containing $\frac{1}{3}$ pound of salt per gallon is being added to the tank at a rate of 6 gal/min, and that the well-stirred solution is being drained from the tank at the same rate. Find a formula for the amount $f(t)$ of salt in the tank at time t.

30 An object of mass m is moving on a coordinate line subject to a force given by $F(t) = e^{-t}$ for time t. The motion is resisted by a frictional force that is numerically equal to twice the speed of the object. If $v = 0$ at $t = 0$, find a formula for v at any time $t > 0$.

31 A *learning curve* is used to describe the rate at which a skill is acquired. To illustrate, suppose a manufacturer estimates that a new employee will produce A items the first day on the job, and that as the employee's proficiency increases, items will be produced more rapidly, until the employee produces a maximum of M items per day. Let $f(t)$ denote the number produced on day t for $t \geq 1$. Suppose that the rate of production $f'(t)$ is proportional to $M - f(t)$.

(a) Find a formula for $f(t)$.

(b) If $M = 30$, $f(1) = 5$, and $f(2) = 8$, estimate the number of items produced on day 20.

32 A room having dimensions 10 ft $\times$ 15 ft $\times$ 8 ft originally contains 0.001% carbon monoxide (CO). Suppose that at time $t = 0$, fumes containing 5% CO begin entering the room at a rate of 0.12 ft³/min, and that the well-circulated mixture is eliminated from the room at the same rate.

(a) Find a formula for the volume $f(t)$ of CO in the room at time t.

(b) The minimum level of CO considered to be hazardous to health is 0.015%. After approximately how many minutes will the room contain this level of CO?

33 The von Bertalanffy model for the growth of an animal assumes that there is an upper bound of y_L cm on length. If y is the length at age t years, the rate of growth is assumed to be proportional to the length yet to be achieved. Find the general solution of the resulting differential equation.

34 A cell is in a liquid containing a solute, such as potassium, of constant concentration c_0. If $c(t)$ is the concentration of the solute inside the cell at time t, then *Fick's principle* for passive diffusion across the cell membrane asserts that the rate at which the concentration changes is proportional to the *concentration gradient* $c_0 - c(t)$. Find $c(t)$ if $c(0) = 0$.

35 Many common drugs (such as forms of penicillin) are eliminated from the bloodstream at a rate that is proportional to the amount y still present.

(a) If y_0 mg is injected directly into the bloodstream, show that $y = y_0 e^{-kt}$ for some $k > 0$.

(b) If the drug is fed intravenously into the bloodstream at a rate of I mg/min, then $y' = -ky + I$. If $y = 0$ at $t = 0$, express y as a function of t, and find $\lim_{t \to \infty} y$.

(c) If the half-life of the drug in the bloodstream is 2 hours, estimate the infusion rate that will result in a long-term amount of 100 mg in the bloodstream.

36 Shown in the figure is a two-tank system in which water flows in and out of the tanks at the rate of 5 gal/min. Each tank contains 50 gallons of water.

(a) If 1 pound of dye is thoroughly mixed into the water in Tank 1, and if $x(t)$ denotes the amount of dye in the tank after t minutes, show that $x'(t) = -0.1x(t)$ and that $x(t) = e^{-0.1t}$.

(b) From part (a), the dye is entering Tank 2 at the rate of $5 \cdot \frac{1}{50}x(t) = 0.1e^{-0.1t}$ lb/min. Show that if $y(t)$ is the amount of dye in Tank 2 after t minutes, then $y'(t) = -0.1y(t) + 0.1e^{-0.1t}$. Assuming that $y(0) = 0$, find $y(t)$.

(c) What is the maximum amount of dye in Tank 2?

(d) Rework part (b) if the volume of Tank 2 is 40 gallons.

EXERCISE 36

TANK 1   TANK 2

5 gal/min →   5 gal/min →   5 gal/min

14.5 SECOND-ORDER LINEAR DIFFERENTIAL EQUATIONS

The following definition is a generalization of (14.10).

DEFINITION (14.12)

A **linear differential equation of order n** is an equation of the form
$$y^{(n)} + f_1(x)y^{(n-1)} + \cdots + f_{n-1}(x)y' + f_n(x)y = k(x)$$
for functions $f_1, f_2, \ldots, f_n$ and k of one variable that have the same domain. If $k(x) = 0$ for every x, the equation is **homogeneous.** If $k(x) \neq 0$ for some x, the equation is **nonhomogeneous.**

A thorough analysis of (14.12) may be found in textbooks on differential equations. We shall restrict our work to second-order equations in which f_1 and f_2 are constant functions. In this section we shall consider the homogeneous case. Nonhomogeneous equations will be discussed in the next section. Applications are considered in Section 14.7.

The general second-order homogeneous linear differential equation with constant coefficients has the form

$$y'' + by' + cy = 0$$

for constants b and c. Before attempting to find particular solutions, let us establish the following result.

THEOREM (14.13)

> If $y = f(x)$ and $y = g(x)$ are solutions of $y'' + by' + cy = 0$, then
> $$y = C_1 f(x) + C_2 g(x)$$
> is a solution for all real numbers C_1 and C_2.

PROOF Since $f(x)$ and $g(x)$ are solutions of $y'' + by' + cy = 0$,

$$f''(x) + bf'(x) + cf(x) = 0$$

and

$$g''(x) + bg'(x) + cg(x) = 0.$$

If we multiply the first of these equations by C_1, the second by C_2, and add, the result is

$$[C_1 f''(x) + C_2 g''(x)] + b[C_1 f'(x) + C_2 g'(x)] + c[C_1 f(x) + C_2 g(x)] = 0.$$

Thus $C_1 f(x) + C_2 g(x)$ is a solution. • •

We can show that if the solutions f and g in Theorem (14.13) have the property that $f(x) \neq Cg(x)$ for every real number C, and if $g(x)$ is not identically 0, then $y = C_1 f(x) + C_2 g(x)$ is the general solution of the differential equation $y'' + by' + cy = 0$. Thus, to determine the general solution it is sufficient to find two such functions f and g and employ (14.13).

In our search for a solution of $y'' + by' + cy = 0$, we shall use $y = e^{mx}$ as a trial solution. Since $y' = me^{mx}$ and $y'' = m^2 e^{mx}$, it follows that $y = e^{mx}$ is a solution if and only if

$$m^2 e^{mx} + bme^{mx} + ce^{mx} = 0$$

or, since $e^{mx} \neq 0$, if and only if

$$m^2 + bm + c = 0.$$

The last equation is very important in finding solutions of $y'' + by' + cy = 0$, and is given the following special name.

DEFINITION (14.14)

> The **auxiliary equation** of the differential equation $y'' + by' + cy = 0$ is $m^2 + bm + c = 0$.

Note that the auxiliary equation can be obtained from the differential equation by replacing y'' by m^2, y' by m, and y by 1. In simple cases the roots of the auxiliary equation can be found by factoring. If a factorization is not evident, then by the quadratic formula, the roots are

$$m = \frac{-b \pm \sqrt{b^2 - 4c}}{2}.$$

Thus the auxiliary equation has unequal real roots m_1 and m_2, a double real root m, or two complex conjugate roots if $b^2 - 4c$ is positive, zero, or negative, respectively. The next theorem is a consequence of the remark following the proof of Theorem (14.13).

THEOREM (14.15) If the roots m_1, m_2 of the auxiliary equation are real and unequal, then the general solution of $y'' + by' + cy = 0$ is

$$y = C_1 e^{m_1 x} + C_2 e^{m_2 x}$$

EXAMPLE 1 Solve the differential equation $y'' - 3y' - 10y = 0$.

SOLUTION The auxiliary equation is $m^2 - 3m - 10 = 0$, or equivalently, $(m - 5)(m + 2) = 0$. Since the roots $m_1 = 5$ and $m_2 = -2$ are real and unequal, it follows from Theorem (14.15) that the general solution is

$$y = C_1 e^{5x} + C_2 e^{-2x} \quad \bullet$$

THEOREM (14.16) If the auxiliary equation has a double root m, then the general solution of $y'' + by' + cy = 0$ is

$$y = C_1 e^{mx} + C_2 x e^{mx}$$

PROOF The roots of $m^2 + bm + c = 0$ are $m = (-b \pm \sqrt{b^2 - 4c})/2$. If $b^2 - 4c = 0$, we obtain $m = -b/2$ or $2m + b = 0$. Since m satisfies the auxiliary equation, $y = e^{mx}$ is a solution of the differential equation. According to the remark following the proof of Theorem (14.13), it is sufficient to show that $y = x e^{mx}$ is also a solution. Substitution of xe^{mx} for y in the equation $y'' + by' + cy = 0$ gives us

$$(2me^{mx} + m^2 xe^{mx}) + b(mxe^{mx} + e^{mx}) + cxe^{mx}$$
$$= (m^2 + bm + c)xe^{mx} + (2m + b)e^{mx}$$
$$= 0xe^{mx} + 0e^{mx} = 0,$$

which is what we wished to show. $\bullet \bullet$

EXAMPLE 2 Solve the differential equation $y'' - 6y' + 9y = 0$.

SOLUTION The auxiliary equation $m^2 - 6m + 9 = 0$, or equivalently $(m - 3)^2 = 0$, has a double root 3. Hence, by Theorem (14.16), the general solution is

$$y = C_1 e^{3x} + C_2 x e^{3x} = e^{3x}(C_1 + C_2 x) \quad \bullet$$

We may also consider second-order differential equations of the form

$$ay'' + by' + cy = 0$$

with $a \neq 1$. It is possible to obtain the form stated in Theorems (14.15) and (14.16) by dividing both sides by a; however, it is usually simpler to employ the auxiliary equation

$$am^2 + bm + c = 0$$

as illustrated in the next example.

EXAMPLE 3 Solve the differential equation $6y'' - 7y' + 2y = 0$.

SOLUTION   The auxiliary equation $6m^2 - 7m + 2 = 0$ can be factored as follows:

$$(2m - 1)(3m - 2) = 0.$$

Hence the roots are $m_1 = \frac{1}{2}$ and $m_2 = \frac{2}{3}$. By Theorem (14.15), the general solution of the given equation is

$$y = C_1 e^{x/2} + C_2 e^{2x/3} \quad \bullet$$

The final case to consider is that in which the roots of the auxiliary equation $m^2 + bm + c = 0$ of $y'' + by' + cy = 0$ are complex numbers. Recall that complex numbers may be represented by expressions of the form $a + bi$, for real numbers a and b and a symbol i that may be manipulated in the same manner as a real number, but has the additional property that $i^2 = -1$. Two complex numbers $a + bi$ and $c + di$ are said to be **equal,** and we write $a + bi = c + di$, if and only if $a = c$ and $b = d$. Operations of addition, subtraction, multiplication, and division are defined *just as though* all letters denote real numbers, with the additional stipulation that whenever i^2 occurs, it may be replaced by -1. For example, the formulas for addition and multiplication of two complex numbers $a + bi$ and $c + di$ are

$$(a + bi) + (c + di) = (a + c) + (b + d)i$$

$$(a + bi)(c + di) = (ac - bd) + (ad + bc)i.$$

We may regard the real numbers as a subset of the complex numbers by identifying the real number a with the complex number $a + 0i$. A complex number of the form $0 + bi$ is abbreviated bi.

Complex numbers are often required for solving equations of the form $f(x) = 0$ for a polynomial $f(x)$. For example, if only real numbers are allowed, then the equation $x^2 = -4$ has no solutions. However, if complex numbers are available, then the equation has a solution $2i$, since

$$(2i)^2 = 2^2 i^2 = 4(-1) = -4.$$

Similarly, $-2i$ is a solution of $x^2 = -4$.

Since $i^2 = -1$, we sometimes use the symbol $\sqrt{-1}$ in place of i and write

$$\sqrt{-13} = \sqrt{13}i, \quad 2 + \sqrt{-25} = 2 + \sqrt{25}i = 2 + 5i,$$

and so on. A quadratic equation $ax^2 + bx + c = 0$, for real numbers a, b, c and $a \neq 0$, has roots given by the quadratic formula

$$x = \frac{-b \pm \sqrt{b^2 - 4ac}}{2a}.$$

If $b^2 - 4ac < 0$, then the roots are complex numbers. To illustrate, if we apply the quadratic formula to the equation $x^2 - 4x + 13 = 0$, we obtain

$$x = \frac{4 \pm \sqrt{16 - 52}}{2} = \frac{4 \pm \sqrt{-36}}{2} = \frac{4 \pm 6i}{2} = 2 \pm 3i,$$

Thus the equation has the two complex roots $2 + 3i$ and $2 - 3i$.

The complex number $a - bi$ is the **conjugate** of the complex number $a + bi$. We see from the quadratic formula that if a quadratic equation with real coefficients has complex roots, then they are necessarily conjugates of one another.

It follows from the preceding discussion that if the auxiliary equation $m^2 + bm + c = 0$ of $y'' + by' + cy = 0$ has complex roots, then they are of the form

$$z_1 = s + ti \quad \text{and} \quad z_2 = s - ti$$

for real numbers s and t. We may anticipate, from Theorem (14.15), that the general solution of the differential equation is

$$y = C_1 e^{z_1 x} + C_2 e^{z_2 x} = C_1 e^{(s + ti)x} + C_2 e^{(s - ti)x}.$$

To define complex exponents we must extend some of the concepts of calculus to include functions whose domains include complex numbers. Since a complete development requires advanced methods, we shall merely outline the main ideas.

In Section 11.8 we discussed how certain functions can be represented by power series. We can easily extend the definitions and theorems of Chapter 11 to infinite series that involve complex numbers. Since this is true, we shall use the power series representations (11.43) to *define* e^z, $\sin z$, and $\cos z$ for every complex number z as follows.

FUNCTIONS OF $z = a + bi$ **(14.17)**

$$e^z = 1 + z + \frac{z^2}{2!} + \cdots + \frac{z^n}{n!} + \cdots$$

$$\sin z = z - \frac{z^3}{3!} + \frac{z^5}{5!} - \cdots + (-1)^n \frac{z^{2n+1}}{(2n+1)!} + \cdots$$

$$\cos z = 1 - \frac{z^2}{2!} + \frac{z^4}{4!} - \cdots + (-1)^n \frac{z^{2n}}{(2n)!} + \cdots$$

Using the first formula in (14.17) gives us

$$e^{iz} = 1 + (iz) + \frac{(iz)^2}{2!} + \frac{(iz)^3}{3!} + \frac{(iz)^4}{4!} + \frac{(iz)^5}{5!} + \cdots$$

$$= 1 + iz + i^2 \frac{z^2}{2!} + i^3 \frac{z^3}{3!} + i^4 \frac{z^4}{4!} + i^5 \frac{z^5}{5!} + \cdots$$

Since $i^2 = -1$, $i^3 = -i$, $i^4 = 1$, $i^5 = i$, and so on, we see that

$$e^{iz} = 1 + iz - \frac{z^2}{2!} - i \frac{z^3}{3!} + \frac{z^4}{4!} + i \frac{z^5}{5!} - \cdots$$

which may also be written in the form

$$e^{iz} = \left(1 - \frac{z^2}{2!} + \frac{z^4}{4!} - \cdots\right) + i\left(z - \frac{z^3}{3!} + \frac{z^5}{5!} - \cdots\right).$$

If we now use the formulas for $\cos z$ and $\sin z$ in (14.17), we obtain the following result, named after the Swiss mathematician Leonhard Euler (1707–1783).

EULER'S FORMULA (14.18)

> For every complex number z,
> $$e^{iz} = \cos z + i \sin z$$

The Laws of Exponents are true for complex numbers. In addition, formulas for derivatives may be extended to functions of a *complex* variable z. For example, $D_z e^{kz} = k e^{kz}$ for a complex number k. If the auxiliary equation of $y'' + by' + cy = 0$ has complex roots $s \pm ti$, then the general solution of this differential equation may be written

$$\begin{aligned} y &= C_1 e^{(s+ti)x} + C_2 e^{(s-ti)x} \\ &= C_1 e^{sx+txi} + C_2 e^{sx-txi} \\ &= C_1 e^{sx} e^{txi} + C_2 e^{sx} e^{-txi} \end{aligned}$$

or equivalently,

$$y = e^{sx}(C_1 e^{itx} + C_2 e^{-itx}).$$

This can be further simplified by using Euler's formula. Specifically, we see from (14.18) that

$$e^{itx} = \cos tx + i \sin tx,$$

$$e^{-itx} = \cos tx - i \sin tx$$

from which it follows that

$$\cos tx = \frac{e^{itx} + e^{-itx}}{2}, \quad \sin tx = \frac{e^{itx} - e^{-itx}}{2i}.$$

If we let $C_1 = C_2 = \frac{1}{2}$ in the preceding discussion, and then use the formula for $\cos tx$, we obtain

$$\begin{aligned} y &= \tfrac{1}{2} e^{sx}(e^{itx} + e^{-itx}) \\ &= \tfrac{1}{2} e^{sx}(2 \cos tx) = e^{sx} \cos tx. \end{aligned}$$

Thus, $y = e^{sx} \cos tx$ is a particular solution of $y'' + by' + cy = 0$. Letting $C_1 = -C_2 = i/2$ leads to the particular solution $y = e^{sx} \sin tx$. This is a partial proof of the next theorem.

THEOREM (14.19)

> If the auxiliary equation $m^2 + bm + c = 0$ has distinct complex roots $s \pm ti$, then the general solution of $y'' + by' + cy = 0$ is
> $$y = e^{sx}(C_1 \cos tx + C_2 \sin tx)$$

EXAMPLE 4 Solve the differential equation $y'' - 10y' + 41y = 0$.

SOLUTION The roots of the auxiliary equation $m^2 - 10m + 41 = 0$ are

$$m = \frac{10 \pm \sqrt{100 - 164}}{2} = \frac{10 \pm \sqrt{-64}}{2} = \frac{10 \pm 8i}{2} = 5 \pm 4i.$$

Hence by Theorem (14.19), the general solution of the differential equation is

$$y = e^{5x}(C_1 \cos 4x + C_2 \sin 4x) \quad \bullet$$

EXERCISES 14.5

Exer. 1–22: Solve the differential equation.

1 $y'' - 5y' + 6y = 0$

2 $y'' - y' - 2y = 0$

3 $y'' - 3y' = 0$

4 $y'' + 6y' + 8y = 0$

5 $y'' + 4y' + 4y = 0$

6 $y'' - 4y' + 4y = 0$

7 $y'' - 4y' + y = 0$

8 $6y'' - 7y' - 3y = 0$

9 $y'' + 2\sqrt{2}y' + 2y = 0$

10 $4y'' + 20y' + 25y = 0$

11 $8y'' + 2y' - 15y = 0$

12 $y'' + 4y' + y = 0$

13 $9y'' - 24y' + 16y = 0$

14 $4y'' - 8y' + 7y = 0$

15 $2y'' - 4y' + y = 0$

16 $2y'' + 7y' = 0$

17 $y'' - 2y' + 2y = 0$

18 $y'' - 2y' + 5y = 0$

19 $y'' - 4y' + 13y = 0$

20 $y'' + 4 = 0$

21 $\dfrac{d^2y}{dx^2} + 6\dfrac{dy}{dx} + 2y = 0$

22 $\dfrac{d^2y}{dx^2} + 2\dfrac{dy}{dx} + 6y = 0$

Exer. 23–30: Find the particular solution of the differential equation that satisfies the stated conditions.

23 $y'' - 3y' + 2y = 0$; $y = 0$ and $y' = 2$ when $x = 0$

24 $y'' - 2y' + y = 0$; $y = 1$ and $y' = 2$ when $x = 0$

25 $y'' + y = 0$; $y = 1$ and $y' = 2$ when $x = 0$

26 $y'' - y' - 6y = 0$; $y = 0$ and $y' = 1$ when $x = 0$

27 $y'' + 8y' + 16y = 0$; $y = 2$ and $y' = 1$ when $x = 0$

28 $y'' + 5y = 0$; $y = 4$ and $y' = 2$ when $x = 0$

29 $\dfrac{d^2y}{dx^2} - 2\dfrac{dy}{dx} + 5y = 0$; $y = 0$ and $\dfrac{dy}{dx} = 1$ when $x = 0$

30 $\dfrac{d^2y}{dx^2} - 6\dfrac{dy}{dx} + 13y = 0$; $y = 2$ and $\dfrac{dy}{dx} = 3$ when $x = 0$

14.6 NONHOMOGENEOUS LINEAR DIFFERENTIAL EQUATIONS

In this section we shall consider second-order nonhomogeneous linear differential equations with constant coefficients, that is, equations of the form

$$y'' + by' + cy = k(x)$$

for constants b and c and a continuous function k.

It is convenient to use the differential operator symbols D and D^2 such that if $y = f(x)$, then

$$Dy = y' = f'(x) \quad \text{and} \quad D^2 y = y'' = f''(x).$$

DEFINITION (14.20)

If $y = f(x)$ and f'' exists, then the **linear differential operator** $L = D^2 + bD + c$ is defined by

$$L(y) = (D^2 + bD + c)y = D^2 y + b\,Dy + cy = y'' + by' + cy$$

Using L, the differential equation $y'' + by' + cy = k(x)$ may be written in the compact form $L(y) = k(x)$. In Exercises 19 and 20 you are asked to verify that for every real number C

$$L(Cy) = CL(y)$$

and that if $y_1 = f_1(x)$ and $y_2 = f_2(x)$, then

$$L(y_1 \pm y_2) = L(y_1) \pm L(y_2).$$

Given the differential equation $y'' + by' + cy = k(x)$, that is, $L(y) = k(x)$, the corresponding homogeneous equation $L(y) = 0$ is the **complementary equation.**

THEOREM (14.21)

> Let $y'' + by' + cy = k(x)$ be a second-order nonhomogeneous linear differential equation. If y_p is a particular solution of $L(y) = k(x)$ and if y_c is the general solution of the complementary equation $L(y) = 0$, then the general solution of $L(y) = k(x)$ is $y = y_p + y_c$.

PROOF Since $L(y_p) = k(x)$ and $L(y_c) = 0$,

$$L(y_p + y_c) = L(y_p) + L(y_c) = k(x) + 0 = k(x),$$

which means that $y_p + y_c$ is a solution of $L(y) = k(x)$. Moreover, if $y = f(x)$ is any other solution, then

$$L(y - y_p) = L(y) - L(y_p) = k(x) - k(x) = 0.$$

Consequently, $y - y_p$ is a solution of the complementary equation. Thus $y - y_p = y_c$ for some y_c, which is what we wished to prove. • •

If we use the results of Section 14.5 to find the general solution y_c of $L(y) = 0$, then according to Theorem (14.21) all that is needed to determine the general solution of $L(y) = k(x)$ is *one* particular solution y_p.

EXAMPLE 1 Solve the differential equation $y'' - 4y = 6x - 4x^3$.

SOLUTION We see by inspection that $y_p = x^3$ is a particular solution of the given equation. The complementary equation is $y'' - 4y = 0$, which by Theorem (14.15) has the general solution

$$y_c = C_1 e^{2x} + C_2 e^{-2x}.$$

Applying Theorem (14.21), the general solution of the given nonhomogeneous equation is

$$y = C_1 e^{2x} + C_2 e^{-2x} + x^3 \quad •$$

If a particular solution of $y'' + by' + cy = k(x)$ cannot be found by inspection, then the following technique, called **variation of parameters,** may be employed. Given the differential equation $L(y) = k(x)$, let y_1 and y_2 be the expressions that appear in the general solution $y = C_1 y_1 + C_2 y_2$ of the complementary equation $L(y) = 0$. For example, we might have, as in (14.15), $y_1 = e^{m_1 x}$ and $y_2 = e^{m_2 x}$. Let us now attempt to find a particular solution of $L(y) = k(y)$ that has the form

$$y_p = u y_1 + v y_2$$

such that $u = g(x)$ and $v = h(x)$ for some functions g and h. The first and second derivatives of y_p are

$$y_p' = (u y_1' + v y_2') + (u' y_1 + v' y_2)$$

$$y_p'' = (u y_1'' + v y_2'') + (u' y_1' + v' y_2') + (u' y_1 + v' y_2)'.$$

Substituting these into $L(y_p) = y_p'' + by_p' + cy_p$ and rearranging terms, we obtain

$$L(y_p) = u(y_1'' + by_1' + cy_1) + v(y_2'' + by_2' + cy_2)$$
$$+ b(u'y_1 + v'y_2) + (u'y_1 + v'y_2)' + (u'y_1' + v'y_2').$$

Since y_1 and y_2 are solutions of $y'' + by' + cy = 0$, the first two terms on the right side are 0. Hence, to obtain $L(y_p) = k(x)$, it is sufficient to choose u and v such that

$$u'y_1 + v'y_2 = 0$$
$$u'y_1' + v'y_2' = k(x).$$

We can show that this system of equations always has a unique solution u' and v'. We may then determine u and v by integration and use the fact that $y_p = uy_1 + vy_2$ to find a particular solution of the differential equation. Our discussion may be summarized as follows.

VARIATION OF PARAMETERS (14.22)

If $y = C_1y_1 + C_2y_2$ is the general solution of the complementary equation $L(y) = 0$ of $y'' + by' + cy = k(x)$, then a particular solution of $L(y) = k(x)$ is $y_p = uy_1 + vy_2$ such that $u = g(x)$ and $v = h(x)$ satisfy the following system of equations:

$$u'y_1 + v'y_2 = 0$$
$$u'y_1' + v'y_2' = k(x)$$

EXAMPLE 2 Solve the differential equation $y'' + y = \cot x$.

SOLUTION The complementary equation is $y'' + y = 0$. Since the auxiliary equation $m^2 + 1 = 0$ has roots $\pm i$, we see from Theorem (14.19) that the general solution of the homogeneous differential equation $y'' + y = 0$ is $y = C_1 \cos x + C_2 \sin x$. Thus we seek a particular solution of the differential equation that has the form $y_p = uy_1 + vy_2$ with $y_1 = \cos x$ and $y_2 = \sin x$. The system of equations in Theorem (14.22) is, therefore,

$$u' \cos x + v' \sin x = 0$$
$$-u' \sin x + v' \cos x = \cot x.$$

Solving for u' and v' gives us

$$u' = -\cos x, \quad v' = \csc x - \sin x.$$

If we integrate each of these expressions (and drop the constants of integration), we obtain

$$u = -\sin x, \quad v = \ln|\csc x - \cot x| + \cos x.$$

Applying Theorem (14.22), a particular solution of the given equation is

$$y_p = -\sin x \cos x + \sin x \ln|\csc x - \cot x| + \sin x \cos x$$

or $y_p = \sin x \ln|\csc x - \cot x|.$

Finally, by Theorem (14.21), the general solution of $y'' + y = \cot x$ is

$$y = C_1 \cos x + C_2 \sin x + \sin x \ln|\csc x - \cot x| \quad \bullet$$

Given the differential equation

$$L(y) = y'' + by' + cy = e^{nx}$$

where e^{nx} *is not a solution of* $L(y) = 0$, it is reasonable to expect that there exists a particular solution of the form $y_p = Ae^{nx}$, since e^{nx} is the result of finding $L(Ae^{nx})$. This suggests that we use Ae^{nx} as a trial solution in the given equation and attempt to find the value of the coefficient A. This is called the **method of undetermined coefficients** and is illustrated in the next example.

EXAMPLE 3 Solve the differential equation $y'' + 2y' - 8y = e^{3x}$.

SOLUTION The auxiliary equation $m^2 + 2m - 8 = 0$ of the differential equation $y'' + 2y' - 8y = 0$ has roots 2 and -4. By Theorem (14.15), the general solution of the complementary equation is

$$y_c = C_1 e^{2x} + C_2 e^{-4x}.$$

From the preceding remarks we seek a particular solution of the form $y_p = Ae^{3x}$. Since $y_p' = 3Ae^{3x}$ and $y_p'' = 9Ae^{3x}$, substitution in the given equation leads to

$$9Ae^{3x} + 6Ae^{3x} - 8Ae^{3x} = e^{3x}.$$

Dividing both sides by e^{3x}, we obtain

$$9A + 6A - 8A = 1 \quad \text{or} \quad A = \tfrac{1}{7}.$$

Thus $y_p = \tfrac{1}{7} e^{3x}$, and by Theorem (14.21) the general solution is

$$y = C_1 e^{2x} + C_2 e^{-4x} + \tfrac{1}{7} e^{3x}. \quad \bullet$$

Three rules for arriving at trial solutions of second-order nonhomogeneous differential equations with constant coefficients are stated without proof in the next theorem. (See texts on differential equations for a more extensive treatment of this topic.)

THEOREM (14.23)

> (i) If $y'' + by' + cy = e^{nx}$ and n is not a root of the auxiliary equation $m^2 + bm + c = 0$, then there is a particular solution of the form $y_p = Ae^{nx}$.
>
> (ii) If $y'' + by' + cy = xe^{nx}$ and n is not a solution of the auxiliary equation $m^2 + bm + c = 0$, then there is a particular solution of the form $y_p = (A + Bx)e^{nx}$.
>
> (iii) If either $\qquad y'' + by' + cy = e^{sx} \sin tx$
>
> or $\qquad\qquad y'' + by' + cy = e^{sx} \cos tx$
>
> and the complex number $s + ti$ is not a solution of the auxiliary equation $m^2 + bm + c = 0$, then there is a particular solution of the form
>
> $$y_p = Ae^{sx} \cos tx + Be^{sx} \sin tx$$

Rule (i) of Theorem (14.23) was used in the solution of Example 3. Illustrations of rules (ii) and (iii) are given in the following examples.

EXAMPLE 4 Solve $y'' - 3y' - 18y = xe^{4x}$.

SOLUTION Since the auxiliary equation $m^2 - 3m - 18 = 0$ has roots 6 and -3, it follows from the preceding section that the general solution of $y'' - 3y' - 18y = 0$ is

$$y = C_1 e^{6x} + C_2 e^{-3x}.$$

Since 4 is not a root of the auxiliary equation, we see from Theorem (14.23) (ii) that there is a particular solution of the form

$$y_p = (A + Bx)e^{4x}.$$

Differentiating, we obtain

$$y_p' = (4A + 4Bx + B)e^{4x}$$
$$y_p'' = (16A + 16Bx + 8B)e^{4x}.$$

Substitution in the given differential equation produces

$$(16A + 16Bx + 8B)e^{4x} - 3(4A + 4Bx + B)e^{4x} - 18(A + Bx)e^{4x} = xe^{4x},$$

which reduces to $\qquad -14A + 5B - 14Bx = x$.

Thus y_p is a solution, provided

$$-14A + 5B = 0 \quad \text{and} \quad -14B = 1.$$

This gives us $B = -\frac{1}{14}$ and $A = -\frac{5}{196}$. Consequently,

$$y_p = (-\tfrac{5}{196} - \tfrac{1}{14}x)e^{4x} = -\tfrac{1}{196}(5 + 14x)e^{4x}.$$

Applying Theorem (14.21), the general solution is

$$y = C_1 e^{6x} + C_2 e^{-3x} - \tfrac{1}{196}(5 + 14x)e^{4x} \quad \bullet$$

EXAMPLE 5 Solve $y'' - 10y' + 41y = \sin x$.

SOLUTION We found the general solution $y_c = e^{5x}(C_1 \cos 4x + C_2 \sin 4x)$ of the complementary equation in Example 4 of the preceding section. Referring to Theorem (14.23) (iii) with $s = 0$ and $t = 1$, we seek a particular solution of the form

$$y_p = A \cos x + B \sin x.$$

Since

$$y_p' = -A \sin x + B \cos x \quad \text{and} \quad y_p'' = -A \cos x - B \sin x,$$

substitution in the given equation produces

$$-A \cos x - B \sin x + 10A \sin x - 10B \cos x$$
$$+ 41A \cos x + 41B \sin x = \sin x,$$

which can be written

$$(40A - 10B) \cos x + (10A + 40B) \sin x = \sin x.$$

Consequently, y_p is a solution, provided

$$40A - 10B = 0 \quad \text{and} \quad 10A + 40B = 1.$$

The solution of this system of equations is $A = \frac{1}{170}$ and $B = \frac{4}{170}$. Hence

$$y_p = \frac{1}{170}\cos x + \frac{4}{170}\sin x = \frac{1}{170}(\cos x + 4\sin x)$$

and the general solution is

$$y = e^{5x}(C_1 \cos 4x + C_2 \sin 4x) + \frac{1}{170}(\cos x + 4\sin x) \quad \bullet$$

EXERCISES 14.6

Exer. 1–10: Solve the differential equation by using the method of variation of parameters.

1 $y'' + y = \tan x$

2 $y'' + y = \sec x$

3 $y'' - 6y' + 9y = x^2 e^{3x}$

4 $y'' + 3y' = e^{-3x}$

5 $y'' - y = e^x \cos x$

6 $y'' - 4y' + 4y = x^{-2}e^{2x}$

7 $y'' - 9y = e^{3x}$

8 $y'' + y = \sin x$

9 $\dfrac{d^2y}{dx^2} - 3\dfrac{dy}{dx} - 4y = 2$

10 $\dfrac{d^2y}{dx^2} - \dfrac{dy}{dx} = x + 1$

Exer. 11–18: Solve the differential equation by using undetermined coefficients.

11 $y'' - 3y' + 2y = 4e^{-x}$

12 $y'' + 6y' + 9y = e^{2x}$

13 $y'' + 2y' = \cos 2x$

14 $y'' + y = \sin 5x$

15 $y'' - y = xe^{2x}$

16 $y'' + 3y' - 4y = xe^{-x}$

17 $\dfrac{d^2y}{dx^2} - 6\dfrac{dy}{dx} + 13y = e^x \cos x$

18 $\dfrac{d^2y}{dx^2} - 2\dfrac{dy}{dx} + 2y = e^{-x} \sin 2x$

Exer. 19–20: Prove the identity if L is the linear differential operator (14.20) and y, y_1, y_2 represent functions of x.

19 $L(Cy) = CL(y)$

20 $L(y_1 \pm y_2) = L(y_1) \pm L(y_2)$

14.7 VIBRATIONS

Vibrations in mechanical systems are caused by external forces. A violin string vibrates if bowed, a steel beam vibrates if struck by a hammer, and a bridge vibrates if a marching band crosses it in cadence. In this section we shall use differential equations to analyze vibrations that may occur in a spring.

According to Hooke's law, the force required to stretch a spring y units beyond its natural length is ky for some positive real number k, which is the *spring constant* (see (6.17)). The **restoring force** of the spring is $-ky$. Suppose that a weight W is attached to the spring and, at the equilibrium position, the spring is stretched a distance l_1 beyond its natural length l_0, as illustrated in Figure 14.14. If g is the gravitational constant and m is the mass of the weight, then $W = mg$, and at the equilibrium position

$$mg = kl_1 \quad \text{or} \quad mg - kl_1 = 0$$

if we assume that the mass of the spring is negligible compared to m.

Suppose the weight is pulled down and released. Consider the coordinate line shown in Figure 14.15, where y denotes the (signed) distance from the equilibrium point to the center of mass of the weight after t seconds. The force F acting on the weight when its acceleration is a is given by Newton's second law of motion $F = ma$. If we assume that the motion is **undamped**—that is, there is no external retarding force—and the weight moves in a frictionless medium, we see that

$$F = mg - k(l_1 + y) = mg - kl_1 - ky = -ky.$$

FIGURE 14.14

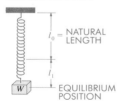

l_0 = NATURAL LENGTH

l_1

W · EQUILIBRIUM POSITION

FIGURE 14.15

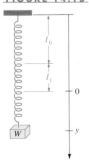

l_0

l_1

0

W

y

Since $F = ma$ and $a = d^2y/dt^2$, this implies that

$$m\frac{d^2y}{dt^2} = -ky.$$

Dividing both sides of this equation by m leads to the following differential equation.

FREE, UNDAMPED VIBRATION (14.24)

$$\frac{d^2y}{dt^2} + \frac{k}{m}y = 0$$

EXAMPLE 1 If a weight of mass m is in free, undamped vibration, show that the motion is simple harmonic. Find the displacement if the weight in Figure 14.15 is pulled down a distance l_2 and released with zero velocity.

SOLUTION If we denote k/m by ω^2, then (14.24) may be written

$$\frac{d^2y}{dt^2} + \omega^2 y = 0.$$

The solutions of the auxiliary equation $m^2 + \omega^2 = 0$ are $\pm \omega i$. Hence, from Theorem (14.19), the general solution is

$$y = C_1 \cos \omega t + C_2 \sin \omega t.$$

It follows that the weight is in simple harmonic motion (see Exercise 70 of Section 3.6).

If the weight is pulled down a distance l_2 and is released with zero velocity, then at $t = 0$

$$l_2 = C_1(1) + C_2(0) \quad \text{or} \quad C_1 = l_2.$$

Since

$$\frac{dy}{dt} = -\omega C_1 \sin \omega t + \omega C_2 \cos \omega t,$$

we also have (at $t = 0$)

$$0 = -\omega C_1(0) + \omega C_2(1) \quad \text{or} \quad C_2 = 0.$$

FIGURE 14.16

Simple harmonic motion

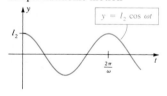

Hence the displacement y of the weight at time t is

$$y = l_2 \cos \omega t.$$

This type of motion was discussed earlier in the text (see Example 8 of Section 3.6). The *amplitude* (the maximum displacement) is l_2, and the *period* (the time for one complete vibration) is $2\pi/\omega = 2\pi\sqrt{m/k}$. A typical graph indicating this type of motion is illustrated in Figure 14.16. •

EXAMPLE 2 An 8-pound weight stretches a spring 2 feet beyond its natural length. The weight is then pulled down another $\frac{1}{2}$ foot and released with an initial upward velocity of 6 ft/sec. Find a formula for the displacement of the weight at any time t.

SOLUTION From Hooke's law, $8 = k(2)$, or $k = 4$. If y is the displacement of the weight from the equilibrium position at time t, then by (14.24)

$$\frac{d^2y}{dt^2} + \frac{4}{m}y = 0.$$

Since $W = mg$, it follows that $m = W/g = \frac{8}{32} = \frac{1}{4}$. Hence

$$\frac{d^2y}{dt^2} + 16y = 0.$$

As in the solution to Example 1, this implies that

$$y = C_1 \cos 4t + C_2 \sin 4t.$$

At $t = 0$ we have $y = \frac{1}{2}$ and therefore

$$\tfrac{1}{2} = C_1(1) + C_2(0) \quad \text{or} \quad C_1 = \tfrac{1}{2}.$$

Since

$$\frac{dy}{dt} = -4C_1 \sin 4t + 4C_2 \cos 4t$$

and $dy/dt = -6$ at $t = 0$, we obtain

$$-6 = -4C_1(0) + 4C_2(1) \quad \text{or} \quad C_2 = -\tfrac{3}{2}.$$

Hence the displacement at time t is given by

$$y = \tfrac{1}{2} \cos 4t - \tfrac{3}{2} \sin 4t \quad \bullet$$

FIGURE 14.17
Damping force

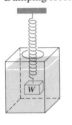

Let us next consider the motion of the spring if a *damping* (or frictional) force is present, as is the case when the weight moves through a fluid (see Figure 14.17). A shock absorber in an automobile is a good illustration of this situation. We shall assume that the direction of the damping force is opposite that of the motion and that the force is directly proportional to the velocity of the weight. Thus, the damping force is given by $-c(dy/dt)$ for some positive constant c. According to Newton's second law, the differential equation that describes the motion is

$$m\frac{d^2y}{dt^2} = -ky - c\frac{dy}{dt}.$$

Dividing by m and rearranging terms gives us the following differential equation.

FREE, DAMPED VIBRATION (14.25)

$$\frac{d^2y}{dt^2} + \frac{c}{m}\frac{dy}{dt} + \frac{k}{m}y = 0$$

EXAMPLE 3 Discuss the motion of a weight of mass m that is in free, damped vibration.

SOLUTION As in Example 1, let $k/m = \omega^2$. To simplify the roots of the auxiliary equation, we also let $c/m = 2p$. Using this notation, (14.25) may be written

$$\frac{d^2y}{dt^2} + 2p\frac{dy}{dt} + \omega^2 y = 0$$

The roots of the auxiliary equation $m^2 + 2pm + \omega^2 = 0$ are

$$\frac{-2p \pm \sqrt{4p^2 - 4\omega^2}}{2} = -p \pm \sqrt{p^2 - \omega^2}.$$

The following three possibilities for roots correspond to three possible types of motion of the weight:

$$p^2 - \omega^2 > 0, \quad p^2 - \omega^2 = 0, \quad \text{or} \quad p^2 - \omega^2 < 0.$$

Let us classify the three types of motion as follows.

CASE (i) $p^2 - \omega^2 > 0$: **Overdamped vibration**

By Theorem (19.6), the general solution of the differential equation is

$$y = e^{-pt}(C_1 e^{\sqrt{p^2 - \omega^2}\,t} + C_2 e^{-\sqrt{p^2 - \omega^2}\,t}).$$

In the case under consideration,

$$p^2 - \omega^2 = \frac{c^2}{4m^2} - \frac{k}{m} = \frac{c^2 - 4mk}{4m^2} > 0$$

so that $c^2 > 4mk$. This shows that c is usually large compared to k; that is, the damping force dominates the restoring force of the spring, and the weight returns to the equilibrium position relatively quickly. In particular, this happens if the fluid has a high viscosity, as is true for heavy oil or grease.

The manner in which y approaches 0 depends on the constants C_1 and C_2 in the general solution. If the constants are both positive, the graph has the general shape shown in Figure 14.18(i), and if they have opposite signs, the graph resembles that shown in (ii) of the figure.

CASE (ii) $p^2 - \omega^2 = 0$: **Critically damped vibration**

In this case the auxiliary equation has a double root $-p$, and by Theorem (14.16) the general solution of the differential equation is

$$y = e^{-pt}(C_1 + C_2 t).$$

The typical graph sketched in Figure 14.19 is similar to that in Figure 14.18(i); however, any decrease in the damping force leads to the oscillatory motion discussed in the following case.

CASE (iii) $p^2 - \omega^2 < 0$: **Underdamped vibration**

The roots of the auxiliary equation are complex conjugates $a \pm bi$, and by Theorem (14.19) the general solution of the differential equation is

$$y = e^{-at}(C_1 \sin bt + C_2 \cos bt).$$

In this case c is usually smaller than k, and the spring oscillates while returning to the equilibrium position, as illustrated in Figure 14.20. Underdamped vibration could occur in a worn shock absorber in an automobile. •

FIGURE 14.18

Overdamped vibration

(i)
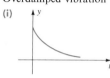

(ii)
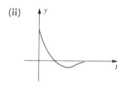

FIGURE 14.19

Critically damped vibration

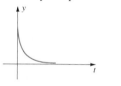

FIGURE 14.20

Underdamped vibration

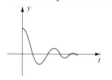

EXAMPLE 4 A 24-pound weight stretches a spring 1 foot beyond its natural length. If the weight is pushed downward from its equilibrium position with an initial velocity of 2 ft/sec and if the damping force is $-9(dy/dt)$, find a formula for the displacement y of the weight at any time t.

SOLUTION By Hooke's law, $24 = k(1)$, or $k = 24$. The mass of the weight is $m = W/g = \frac{24}{32} = \frac{3}{4}$. Using the notation introduced in Example 3,

$$2p = \frac{c}{m} = \frac{9}{\frac{3}{4}} = 12 \quad \text{and} \quad \omega^2 = \frac{k}{m} = \frac{24}{\frac{3}{4}} = 32.$$

Equation (14.25) is

$$\frac{d^2 y}{dt^2} + 12 \frac{dy}{dt} + 32y = 0.$$

We can verify that the roots of the auxiliary equation are -8 and -4. Thus the motion is overdamped, as in Case (i) of Example 3, and the general solution of the differential equation is

$$y = C_1 e^{-4t} + C_2 e^{-8t}.$$

Letting $t = 0$,

$$0 = C_1 + C_2 \quad \text{or} \quad C_2 = -C_1.$$

Since

$$\frac{dy}{dt} = -4C_1 e^{-4t} - 8C_2 e^{-8t}$$

we have, at $t = 0$,

$$2 = -4C_1 - 8C_2 = -4C_1 - 8(-C_1) = 4C_1.$$

Consequently, $C_1 = \frac{1}{2}$ and $C_2 = -C_1 = -\frac{1}{2}$. Thus the displacement y of the weight at time t is

$$y = \tfrac{1}{2}e^{-4t} - \tfrac{1}{2}e^{-8t} = \tfrac{1}{2}e^{-4t}(1 - e^{-4t}).$$

The graph is similar in appearance to that in Figure 14.18(i). In Exercise 9 you are asked to find the damping force that will produce critical damping. •

EXERCISES 14.7

1 A 5-pound weight stretches a spring 6 inches beyond its natural length. If the weight is raised 4 inches and then released with zero velocity, find a formula for the displacement of the weight at any time t.

2 A 10-pound weight stretches a spring 8 inches beyond its natural length. If the weight is pulled down another 3 inches and then released with a downward velocity of 6 in./sec, find a formula for the displacement at any time t.

3 A 10-pound weight stretches a spring 1 foot beyond its natural length. If the weight is released from its equilibrium position with a downward velocity of 2 ft/sec and if the frictional force is $-5(dy/dt)$, find a formula for the displacement of the weight.

4 A 6-pound weight is suspended from a spring whose spring constant is 48 lb/ft. Initially the weight has a downward velocity of 4 ft/sec from a position 5 inches below the equilibrium point. Find a formula for the displacement.

5 A 4-pound weight stretches a spring 3 inches beyond its natural length. If the weight is pulled down another 4 inches and released with zero velocity in a medium where the damping force is $-2(dy/dt)$, find the displacement y at any time t.

6 A spring stretches $\frac{1}{2}$ foot when an 8-pound weight is attached. If the weight is released from the equilibrium position with an upward velocity of 1 ft/sec, and if the damping force is $-4(dy/dt)$, find a formula for the displacement y at any time t.

7 Describe a spring-weight system that leads to the following differential equation and initial values:

$$\frac{1}{4}\frac{d^2y}{dt^2} + \frac{dy}{dt} + 6y = 0;$$

$$y = -2 \quad \text{and} \quad \frac{dy}{dt} = -1 \quad \text{at } t = 0.$$

8 A 4-pound weight stretches a spring 1 foot. If the weight moves in a medium whose damping force is $-c(dy/dt)$ with $c > 0$, for what values of c is the motion (a) overdamped? (b) critically damped? (c) underdamped?

9 In Example 4 of this section, find the damping force that will produce critical damping.

10 Suppose an undamped weight of mass m is attached to a spring having spring constant k, and an external periodic force given by $F \sin \alpha t$ is applied to the weight.

(a) Show that the differential equation that describes the motion of the weight is

$$\frac{d^2y}{dt^2} + \omega^2 y = \frac{F}{m} \sin \alpha t$$

with $\omega^2 = k/m$.

(b) Prove that the displacement y of the weight is given by

$$y = C_1 \cos \omega t + C_2 \sin \omega t + C \sin \alpha t$$

with $C = F/m(\omega^2 - \alpha^2)$. This type of motion is called *forced vibration*.

As shown in Chapter 11, a power series $\sum a_n x^n$ determines a function f such that

$$y = f(x) = a_0 + a_1 x + a_2 x^2 + a_3 x^3 + a_4 x^4 + \cdots$$

for every x in the interval of convergence. Moreover, series representations for the derivatives of f may be obtained by differentiating each term. Thus

$$y' = a_1 + 2a_2 x + 3a_3 x^2 + 4a_4 x^3 + \cdots = \sum_{n=1}^{\infty} na_n x^{n-1},$$

$$y'' = 2a_2 + 3 \cdot 2a_3 x + 4 \cdot 3a_4 x^2 + \cdots = \sum_{n=2}^{\infty} n(n-1)a_n x^{n-2},$$

and so on. We may use power series to solve certain differential equations. In this case, the solution is often expressed as an infinite series and is called a **series solution** of the differential equation.

EXAMPLE 1 Find a series solution of the differential equation $y' = 2xy$.

SOLUTION If the solution is given by $y = \sum a_n x^n$, then $y' = \sum na_n x^{n-1}$, and substitution in the differential equation gives us

$$\sum_{n=1}^{\infty} na_n x^{n-1} = 2x \sum_{n=0}^{\infty} a_n x^n = \sum_{n=0}^{\infty} 2a_n x^{n+1}.$$

It is convenient to change the summation on the left so that it contains the same power of x as in the summation on the right. We may accomplish this by replacing n by $n + 2$ and beginning the summation at $n = -1$. Thus

$$\sum_{n=-1}^{\infty} (n+2)a_{n+2} x^{n+1} = \sum_{n=0}^{\infty} 2a_n x^{n+1}, \quad \text{or}$$

$$a_1 + 2a_2 x + \cdots + (n+2)a_{n+2} x^{n+1} + \cdots = 2a_0 x + \cdots + 2a_n x^{n+1} + \cdots$$

Comparing coefficients, we see that $a_1 = 0$ and $(n+2)a_{n+2} = 2a_n$ if $n \geq 0$. Consequently, the coefficients are given by

$$a_1 = 0 \quad \text{and} \quad a_{n+2} = \frac{2}{n+2} a_n \quad \text{if} \quad n \geq 0.$$

In particular,

$$a_1 = 0, \quad a_2 = a_0, \quad a_3 = \tfrac{2}{3}a_1 = 0, \quad a_4 = \tfrac{1}{2}a_2 = \tfrac{1}{2}a_0, \quad a_5 = \tfrac{2}{5}a_3 = 0,$$

$$a_6 = \tfrac{1}{3}a_4 = \frac{1}{2 \cdot 3} a_0, \quad a_7 = \tfrac{2}{7}a_5 = 0, \quad a_8 = \tfrac{1}{4}a_6 = \frac{1}{2 \cdot 3 \cdot 4} a_0,$$

and so on. We can show that if n is odd, then $a_n = 0$, and $a_{2n} = (1/n!)a_0$ for every positive integer n. The series solution is therefore

$$y = \sum_{n=0}^{\infty} a_n x^n = a_0 \left(1 + x^2 + \frac{1}{2!} x^4 + \cdots + \frac{1}{n!} x^{2n} + \cdots \right). \quad \bullet$$

It follows from (14.17) that we could have written the series solution in Example 1 as $y = a_0 e^{(x^2)}$. This form may be found directly from $y' = 2xy$

by using the separation of variables technique. The objective in Example 1, however, was to illustrate series solutions and not to find the solution in the simplest manner. In many instances it is impossible to find the sum of $\sum a_n x^n$ and the solution must be left in series form.

EXAMPLE 2 Solve the differential equation $y'' - xy' - 2y = 0$.

SOLUTION Substituting

$$y = \sum_{n=0}^{\infty} a_n x^n, \quad y' = \sum_{n=1}^{\infty} n a_n x^{n-1}, \quad \text{and} \quad y'' = \sum_{n=2}^{\infty} n(n-1) a_n x^{n-2},$$

we obtain

$$\sum_{n=2}^{\infty} n(n-1) a_n x^{n-2} - x \sum_{n=1}^{\infty} n a_n x^{n-1} - 2 \sum_{n=0}^{\infty} a_n x^n = 0$$

or

$$\sum_{n=2}^{\infty} n(n-1) a_n x^{n-2} = \sum_{n=0}^{\infty} n a_n x^n + \sum_{n=0}^{\infty} 2 a_n x^n.$$

We next adjust the summation on the left so that the power x^n appears instead of x^{n-2}. This can be accomplished by replacing n by $n+2$ and starting the summation at $n = 0$. This gives us

$$\sum_{n=0}^{\infty} (n+2)(n+1) a_{n+2} x^n = \sum_{n=0}^{\infty} (n+2) a_n x^n.$$

Comparing coefficients, we see that $(n+2)(n+1) a_{n+2} = (n+2) a_n$, that is,

$$a_{n+2} = \frac{1}{n+1} a_n.$$

Letting $n = 0, 1, 2, \ldots, 7$ leads to the following form for the coefficients a_k:

$$a_2 = a_0 \qquad\qquad a_3 = \frac{1}{2} a_1$$

$$a_4 = \frac{1}{3} a_2 = \frac{1}{3} a_0 \qquad\qquad a_5 = \frac{1}{4} a_3 = \frac{1}{2 \cdot 4} a_1$$

$$a_6 = \frac{1}{5} a_4 = \frac{1}{3 \cdot 5} a_0 \qquad\qquad a_7 = \frac{1}{6} a_5 = \frac{1}{2 \cdot 4 \cdot 6} a_1$$

$$a_8 = \frac{1}{7} a_6 = \frac{1}{3 \cdot 5 \cdot 7} a_0 \qquad\qquad a_9 = \frac{1}{8} a_7 = \frac{1}{2 \cdot 4 \cdot 6 \cdot 8} a_1$$

In general,

$$a_{2n} = \frac{1}{1 \cdot 3 \cdots (2n-1)} a_0, \qquad a_{2n+1} = \frac{1}{2 \cdot 4 \cdots (2n)} a_1 = \frac{1}{2^n n!} a_1.$$

The solution $y = \sum a_n x^n$ may therefore be expressed as a sum of two infinite series:

$$y = a_0 \left[1 + \sum_{n=1}^{\infty} \frac{1}{1 \cdot 3 \cdots (2n-1)} x^{2n} \right] + a_1 \sum_{n=0}^{\infty} \frac{1}{2^n n!} x^{2n+1} \quad \bullet$$

Exer. 1–12: Find a series solution for the differential equation.

1 $y'' + y = 0$

2 $y'' - 4y = 0$

3 $y'' - 2xy = 0$

4 $y'' + 2xy' + y = 0$

5 $\dfrac{d^2y}{dx^2} - x\dfrac{dy}{dx} + 2y = 0$

6 $\dfrac{d^2y}{dx^2} + x^2y = 0$

7 $(x + 1)y' = 3y$

8 $y' = 4x^3y$

9 $y'' - y = 5x$

10 $y'' - xy = x^4$

11 $(x^2 - 1)y'' + 6xy' + 4y = -4$

12 $y'' + y = e^x$

14.9 REVIEW

Define or discuss each of the following.

1 Function of several variables

2 Limits of functions of several variables

3 Continuity of functions of several variables

4 First partial derivatives

5 Higher partial derivatives

6 Separable differential equation

7 First-order linear differential equation

8 Second-order linear differential equation

9 Series solutions of a differential equation

Exer. 1–4: Determine the domain of f.

1 $f(x, y) = \sqrt{36 - 4x^2 + 9y^2}$

2 $f(x, y) = \ln xy$

3 $f(x, y, z) = (z^2 - x^2 - y^2)^{-3/2}$

4 $f(x, y, z) = (\sec z)/(x - y)$

Exer. 5–6: Find the limit, if it exists.

5 $\displaystyle\lim_{(x, y)\to(0, 0)} \dfrac{3xy + 5}{y^2 + 4}$

6 $\displaystyle\lim_{(x, y)\to(0, 0)} \left(\dfrac{x^2 - y^2}{x^2 + y^2}\right)^2$

Exer. 7–12: Find the first partial derivatives of f.

7 $f(x, y) = x^3 \cos y - y^2 + 4x$

8 $f(r, s) = r^2e^{rs}$

9 $f(x, y, z) = (x^2 + y^2)/(y^2 + z^2)$

10 $f(u, v, t) = u \ln (v/t)$

11 $f(x, y, z, t) = x^2z\sqrt{2y + t}$

12 $f(v, w) = v^2 \cos w + w^2 \cos v$

Exer. 13–14: Find the second partial derivatives of f.

13 $f(x, y) = x^3y^2 - 3xy^3 + x^4 - 3y + 2$

14 $f(x, y, z) = x^2e^{y^2 - z^2}$

Exer. 15–48: Solve the differential equation.

15 $xe^y \, dx - \csc x \, dy = 0$

16 $x \, dy - (x + 1)y \, dx = 0$

17 $x(1 + y^2) \, dx + \sqrt{1 - x^2} \, dy = 0$

18 $y' + 4y = e^{-x}$

19 $y' + (\sec x)y = 2 \cos x$

20 $(x^2y + x^2) \, dy + y \, dx = 0$

21 $y \tan x + y' = 2 \sec x$

22 $y' + (\tan x)y = 3$

23 $y\sqrt{1 - x^2}y' = \sqrt{1 - y^2}$

24 $(2y + x^3) \, dx - x \, dy = 0$

25 $e^x \cos y \, dy - \sin^2 y \, dx = 0$

26 $(\cos x)y' + (\sin x)y = 3 \cos^2 x$

27 $y' + (2 \cos x)y = \cos x$

28 $y'' + y' - 6y = 0$

29 $y'' - 8y' + 16y = 0$

30 $y'' - 6y' + 25y = 0$

31 $\dfrac{d^2y}{dx^2} - 2\dfrac{dy}{dx} = 0$

32 $\dfrac{d^2y}{dx^2} = y + \sin x$

33 $y'' - y = e^x \sin x$

34 $y'' - y' - 6y = e^{2x}$

35 $y' + y = e^{4x}$

36 $y'' + 2y' = 0$

37 $y'' - 3y' + 2y = e^{5x}$

38 $xe^y \, dx - (x + 1)y \, dy = 0$

39 $xy' + y = (x - 2)^2$

40 $\sec^2 y \, dx = \sqrt{1 - x^2} \, dy - x \sec^2 y \, dx$

41 $\dfrac{d^2y}{dx^2} + 5\dfrac{dy}{dx} + 7y = 0$

42 $\dfrac{d^2y}{dx^2} + y = \csc x$

43 $e^{x+y} \, dx - \csc x \, dy = 0$

44 $y'' + 10y' + 25y = 0$

45 $\cot x \, dy = (y - \cos x) \, dx$ **46** $y'' + y' + y = e^x \cos x$

47 $y' + y \csc x = \tan x$ **48** $y'' - y' - 20y = xe^{-x}$

49 The bird population of an island experiences seasonal growth described by the differential equation $dy/dt = (3 \sin 2\pi t)y$ for time t in years, with $t = 0$ corresponding to the beginning of spring. Migration to and from the island is also seasonal. The rate of migration is given by $M(t) = 2000 \sin 2\pi t$ birds per year. Hence the complete differential equation for the population is

$$\frac{dy}{dt} = (3 \sin 2\pi t)y + 2000 \sin 2\pi t.$$

Solve for y if the population is 500 at $t = 0$. Determine the maximum population.

50 As a jar containing 10 moles of gas A is heated, the velocity of the gas molecules increases and a second gas B is formed. When two molecules of gas A collide, two molecules of gas B are formed. The rate dy/dt at which gas B is formed is proportional to $(10 - y)^2$, the number of pairs of molecules of gas A. Find a formula for y if $y = 2$ moles after 30 seconds.

51 In chemistry, the notation $A + B \rightarrow Y$ is used to denote the production of a substance Y due to the interaction of two substances A and B. Let a and b be the initial amounts of A and B, respectively. If, at time t, the concentration of Y is $y = f(t)$, then the concentrations of A and B are $a - f(t)$ and $b - f(t)$, respectively.. If the rate at which the production of Y takes place is given by $dy/dt = k(a - y)(b - y)$ for some positive constant k, and if $f(0) = 0$, find $f(t)$.

52 Let $y = f(t)$ be the population, at time t, of a collection, such as insects or bacteria. If the rate of growth dy/dt is proportional to y, that is, if $dy/dt = cy$ for some positive constant c, then $f(t) = f(0)e^{ct}$ (see Theorem (7.26)). In most cases the rate of growth is dependent on available resources such as food supplies, and as t becomes large, $f'(t)$ begins to decrease. An equation that is often used to describe this type of variation in population is the *law of logistic growth* $dy/dt = y(c - by)$ for some positive constants c and b.

(a) If $f(0) = k$, find $f(t)$.

(b) Find $\lim_{t \to \infty} f(t)$.

(c) Show that $f'(t)$ is increasing if $f(t) < c/(2b)$ and is decreasing if $f(t) > c/(2b)$.

(d) Sketch a typical graph of f. (A graph of this type is a *logistic curve*.)

53 Use a differential equation to describe all functions such that the tangent line at any point $P(x, y)$ on the graph is perpendicular to the line segment joining P to the origin. What is the graph of a typical solution of the differential equation?

54 The following differential equation occurs in the study of electrostatic potentials:

$$\frac{d}{d\theta}\left(\sin \theta \, \frac{dV}{d\theta}\right) = 0.$$

Find a solution $V(\theta)$ that satisfies the conditions $V(\pi/2) = 0$ and $V(\pi/4) = V_0$.

APPENDICES

I MATHEMATICAL INDUCTION

The method of proof known as **mathematical induction** may be used to show that certain statements or formulas are true for all positive integers. For example, if n is a positive integer, let P_n denote the statement

$$(xy)^n = x^n y^n$$

for real numbers x and y. Thus P_1 represents the statement $(xy)^1 = x^1 y^1$, P_2 denotes $(xy)^2 = x^2 y^2$, P_3 is $(xy)^3 = x^3 y^3$, and so on. It is easy to show that P_1, P_2, and P_3 are *true* statements. However, since the set of positive integers is infinite, it is impossible to check the validity of P_n for every positive integer n. To give a proof, the method provided by (B) is required. This method is based on the following fundamental axiom.

AXIOM OF (A)
MATHEMATICAL INDUCTION

Suppose a set S of positive integers has the following two properties:

(i) S contains the integer 1.

(ii) Whenever S contains a positive integer k, S also contains $k + 1$.

Then S contains every positive integer.

We should have little reluctance about accepting (A). If S is a set of positive integers satisfying property (ii), then whenever S contains an arbitrary positive integer k, it must also contain the next positive integer, $k + 1$. If S also satisfies property (i), then S contains 1 and hence by (ii), S contains $1 + 1$, or 2. Applying (ii) again, we see that S contains $2 + 1$, or 3. Once again, S must contain $3 + 1$, or 4. If we continue in this manner, we can argue that if n is any *specific* positive integer, then n is in S, since we can proceed a step at a time as above, eventually reaching n. Although this argument does not *prove* (A), it certainly makes it plausible.

We shall use (A) to establish the following fundamental principle.

PRINCIPLE OF **(B)**
MATHEMATICAL INDUCTION

> If with each positive integer n there is associated a statement P_n, then every statement P_n is true provided the following two conditions hold:
>
> (i) P_1 is true.
>
> (ii) Whenever k is a positive integer such that P_k is true, then P_{k+1} is also true.

PROOF Assume that (i) and (ii) of (B) hold, and let S denote the set of all positive integers n such that P_n is true. By assumption, P_1 is true and, consequently, 1 is in S. Thus S satisfies property (i) of (A). Whenever S contains a positive integer k, then by the definition of S, P_k is true and hence from assumption (ii) of (B), P_{k+1} is also true. This means that S contains $k+1$. We have shown that whenever S contains a positive integer k, then S also contains $k+1$. Consequently, property (ii) of (A) is true. Hence by (A), S contains every positive integer; that is, P_n is true for every positive integer n. • •

There are other variations of the principle of mathematical induction. One variation appears in (D). In most of our work the statement P_n will usually be given in the form of an equation involving the arbitrary positive integer n, as in our illustration $(xy)^n = x^n y^n$.

When applying (B), we always use the following two steps.

(C)

> Step (i) Prove that P_1 is true.
>
> Step (ii) Assume that P_k is true and prove that P_{k+1} is true.

Step (ii) is usually the most confusing for the beginning student. We do not *prove* that P_k is true (except for $k = 1$). Instead, we show that *if P_k is true, then the statement P_{k+1} is true.* That is all that is necessary according to (B). The assumption that P_k is true is referred to as the **induction hypothesis.**

EXAMPLE 1 Prove that for every positive integer n, the sum of the first n positive integers is $n(n + 1)/2$.

SOLUTION For any positive integer n, let P_n denote the statement

$$1 + 2 + 3 + \cdots + n = \frac{n(n + 1)}{2}.$$

By convention, when $n \le 4$ the left side is adjusted so that there are precisely n terms in the sum. The following are some special cases of P_n:

If $n = 2$, then P_2 is

$$1 + 2 = \frac{2(2 + 1)}{2} \quad \text{or} \quad 3 = 3.$$

If $n = 3$, then P_3 is

$$1 + 2 + 3 = \frac{3(3 + 1)}{2} \quad \text{or} \quad 6 = 6.$$

If $n = 5$, then P_5 is

$$1 + 2 + 3 + 4 + 5 = \frac{5(5 + 1)}{2} \quad \text{or} \quad 15 = 15.$$

We wish to show that P_n is true for every n. Although it is instructive to check P_n for several values of n as we have done, it is unnecessary to do so. We need only follow steps (i) and (ii) of (C).

(i) If we substitute $n = 1$ in P_n, then by convention the left side collapses to 1 and the right side is $\frac{1(1 + 1)}{2}$, which also equals 1. This proves that P_1 is true.

(ii) *Assume* that P_k is true. Thus the induction hypothesis is

$$1 + 2 + 3 + \cdots + k = \frac{k(k + 1)}{2}.$$

Our goal is to prove that P_{k+1} is true, that is,

$$1 + 2 + 3 + \cdots + (k + 1) = \frac{(k + 1)[(k + 1) + 1]}{2}.$$

By the induction hypothesis we already have a formula for the sum of the first k positive integers. Hence a formula for the sum of the first $k + 1$ positive integers may be found simply by adding $(k + 1)$ to both sides of the induction hypothesis. Doing so and simplifying, we obtain

$$\begin{aligned}
1 + 2 + 3 + \cdots + k + (k + 1) &= \frac{k(k + 1)}{2} + (k + 1) \\
&= \frac{k(k + 1) + 2(k + 1)}{2} \\
&= \frac{k^2 + 3k + 2}{2} \\
&= \frac{(k + 1)(k + 2)}{2} \\
&= \frac{(k + 1)[(k + 1) + 1]}{2}.
\end{aligned}$$

We have shown that P_{k+1} is true, and therefore the proof by mathematical induction is complete. •

Consider a positive integer j, and suppose that with every integer $n \geq j$ there is associated a statement P_n. For example, if $j = 6$, then the statements are numbered P_6, P_7, P_8, and so on. The principle of mathematical induction may be extended to cover this situation. Just as before, two steps are used. Specifically, to prove that the statements S_n are true for $n \geq j$, we use the following two steps.

(D)

> Step (i') Prove that S_j is true,
>
> Step (ii') *Assume* that S_k is true for $k \geq j$ and *prove* that S_{k+1} is true.

EXAMPLE 2 Let a be a nonzero real number such that $a > -1$. Prove that $(1 + a)^n > 1 + na$ for every integer $n \geq 2$.

SOLUTION For any positive integer n, let P_n denote $(1 + a)^n > 1 + na$. Note that P_1 is *false*, since $(1 + a)^1 = 1 + (1)(a)$. However, we can show that P_n is true for $n \geq 2$ by using (D) with $j = 2$.

(i') We first note that $(1 + a)^2 = 1 + 2a + a^2$. Since $a \neq 0$, we have $a^2 > 0$ and therefore $1 + 2a + a^2 > 1 + 2a$. Thus $(1 + a)^2 > 1 + 2a$ and hence P_2 is true.

(ii') Assume that P_k is true for $k \geq 2$. Thus the induction hypothesis is

$$(1 + a)^k > 1 + ka.$$

We wish to show that P_{k+1} is true, that is,

$$(1 + a)^{k+1} > 1 + (k + 1)a.$$

Since $a > -1$, we have $a + 1 > 0$, and hence multiplying both sides of the induction hypothesis by $1 + a$ will not change the inequality sign. Consequently,

$$(1 + a)^k(1 + a) > (1 + ka)(1 + a)$$

which may be rewritten as

$$(1 + a)^{k+1} > 1 + ka + a + ka^2$$

or as

$$(1 + a)^{k+1} > 1 + (k + 1)a + ka^2.$$

Since $ka^2 > 0$, we have

$$1 + (k + 1)a + ka^2 > 1 + (k + 1)a$$

and therefore

$$(1 + a)^{k+1} > 1 + (k + 1)a.$$

Thus, P_{k+1} is true and the proof is complete. •

EXERCISES

Exer. 1–18: Prove that the formula is true for every positive integer n.

1 $2 + 4 + 6 + \cdots + 2n = n(n + 1)$

2 $1 + 4 + 7 + \cdots + (3n - 2) = \dfrac{n(3n - 1)}{2}$

3 $1 + 3 + 5 + \cdots + (2n - 1) = n^2$

4 $3 + 9 + 15 + \cdots + (6n - 3) = 3n^2$

5 $2 + 7 + 12 + \cdots + (5n - 3) = \dfrac{n}{2}(5n - 1)$

6 $2 + 6 + 18 + \cdots + 2 \cdot 3^{n-1} = 3^n - 1$

7 $1 + 2 \cdot 2 + 3 \cdot 2^2 + 4 \cdot 2^3 + \cdots + n \cdot 2^{n-1}$
$$= 1 + (n - 1) \cdot 2^n$$

8 $(-1)^1 + (-1)^2 + (-1)^3 + \cdots + (-1)^n = \dfrac{(-1)^n - 1}{2}$

9 $1^2 + 2^2 + 3^2 + \cdots + n^2 = \dfrac{n(n + 1)(2n + 1)}{6}$

10 $1^3 + 2^3 + 3^3 + \cdots + n^3 = \left[\dfrac{n(n + 1)}{2} \right]^2$

11 $\dfrac{1}{1 \cdot 2} + \dfrac{1}{2 \cdot 3} + \dfrac{1}{3 \cdot 4} + \cdots + \dfrac{1}{n(n + 1)} = \dfrac{n}{n + 1}$

12 $\dfrac{1}{1 \cdot 2 \cdot 3} + \dfrac{1}{2 \cdot 3 \cdot 4} + \dfrac{1}{3 \cdot 4 \cdot 5} + \cdots + \dfrac{1}{n(n+1)(n+2)}$

$$= \dfrac{n(n+3)}{4(n+1)(n+2)}$$

13 $3 + 3^2 + 3^3 + \cdots + 3^n = \frac{3}{2}(3^n - 1)$

14 $1^3 + 3^3 + 5^3 + \cdots + (2n-1)^3 = n^2(2n^2 - 1)$

15 $n < 2^n$

16 $1 + 2n \le 3^n$

17 $1 + 2 + 3 + \cdots + n < \frac{1}{8}(2n+1)^2$

18 If $0 < a < b$, then $\left(\dfrac{a}{b}\right)^{n+1} < \left(\dfrac{a}{b}\right)^n$.

Exer. 19–26: Prove that the statement is true for every positive integer n.

19 3 is a factor of $n^3 - n + 3$.

20 2 is a factor of $n^2 + n$.

21 4 is a factor of $5^n - 1$.

22 9 is a factor of $10^{n+1} + 3 \cdot 10^n + 5$.

23 If a is any real number greater than 1, then $a^n > 1$.

24 If $a \ne 0$, then $1 + a + a^2 + \cdots + a^{n-1} = \dfrac{a^n - 1}{a - 1}$.

25 $a - b$ is a factor of $a^n - b^n$.

(*Hint:* $a^{k+1} - b^{k+1} = a^k(a - b) + (a^k - b^k)b$.)

26 $a + b$ is a factor of $a^{2n-1} + b^{2n-1}$.

27 Prove that
$$\log(a_1 a_2 \cdots a_n) = \log a_1 + \log a_2 + \cdots + \log a_n$$
for all $n \ge 2$, where each a_k is a positive real number.

28 Prove the *Generalized Distributive Law*
$$a(b_1 + b_2 + \cdots + b_n) = ab_1 + ab_2 + \cdots + ab_n$$
for all $n \ge 2$, where a and each b_k are real numbers.

29 Prove that
$$a + ar + ar^2 + \cdots + ar^{n-1} = \dfrac{a(1 - r^n)}{1 - r}$$
where n is any positive integer and a and r are real numbers with $r \ne 1$.

30 Prove that
$$a + (a + d) + (a + 2d) + \cdots + [a + (n-1)d]$$
$$= (n/2)[2a + (n-1)d]$$
where n is any positive integer and a and d are real numbers.

This appendix contains proofs for some theorems stated in the text. The numbering system corresponds to that given in previous chapters.

UNIQUENESS THEOREM FOR LIMITS

> If $f(x)$ has a limit as x approaches a, then the limit is unique.

FIGURE 1

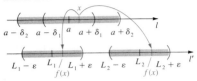

PROOF Suppose that $\lim_{x \to a} f(x) = L_1$ and $\lim_{x \to a} f(x) = L_2$ with $L_1 \neq L_2$. We may assume that $L_1 < L_2$. Choose ε such that $\varepsilon < \frac{1}{2}(L_2 - L_1)$ and consider the open intervals $(L_1 - \varepsilon, L_1 + \varepsilon)$ and $(L_2 - \varepsilon, L_2 + \varepsilon)$ on the coordinate line l' (see Figure 1). Since $\varepsilon < \frac{1}{2}(L_2 - L_1)$, these two intervals do not intersect. By Definition (2.11), there is a $\delta_1 > 0$ such that whenever x is in $(a - \delta_1, a + \delta_1)$, and $x \neq a$, then $f(x)$ is in $(L_1 - \varepsilon, L_1 + \varepsilon)$. Similarly, there is a $\delta_2 > 0$ such that whenever x is in $(a - \delta_2, a + \delta_2)$, and $x \neq a$, then $f(x)$ is in $(L_2 - \varepsilon, L_2 + \varepsilon)$. This is illustrated in Figure 1, with $\delta_1 < \delta_2$. If an x is selected that is in *both* $(a - \delta_1, a + \delta_1)$ and $(a - \delta_2, a + \delta_2)$, then $f(x)$ is in $(L_1 - \varepsilon, L_1 + \varepsilon)$ and also in $(L_2 - \varepsilon, L_2 + \varepsilon)$, contrary to the fact that these two intervals do not intersect. Hence our original supposition is false and, consequently, $L_1 = L_2$. • •

THEOREM (2.15)

> If $\lim_{x \to a} f(x) = L$ and $\lim_{x \to a} g(x) = M$, then
>
> (i) $\displaystyle\lim_{x \to a} [f(x) + g(x)] = L + M$
>
> (ii) $\displaystyle\lim_{x \to a} [f(x) \cdot g(x)] = L \cdot M$
>
> (iii) $\displaystyle\lim_{x \to a} \left[\frac{f(x)}{g(x)}\right] = \frac{L}{M}$, provided $M \neq 0$

PROOF

(i) According to Definition (2.10) we must show that for every $\varepsilon > 0$ there corresponds a $\delta > 0$ such that

(a) if $0 < |x - a| < \delta$, then $|f(x) + g(x) - (L + M)| < \varepsilon$.

We begin by writing

(b) $|f(x) + g(x) - (L + M)| = |(f(x) - L) + (g(x) - M)|$.

Employing the Triangle Inequality (1.4),

$$|(f(x) - L) + (g(x) - M)| \leq |f(x) - L| + |g(x) - M|.$$

Combining the last inequality with (b) gives us

(c) $|f(x) + g(x) - (L + M)| \leq |f(x) - L| + |g(x) - M|$.

Since $\lim_{x \to a} f(x) = L$ and $\lim_{x \to a} g(x) = M$, the numbers $|f(x) - L|$ and $|g(x) - M|$ can be made arbitrarily small by choosing x sufficiently close to a. In particular, they can be made less than $\varepsilon/2$. Thus there exist $\delta_1 > 0$ and $\delta_2 > 0$ such that

(d) if $0 < |x - a| < \delta_1$, then $|f(x) - L| < \varepsilon/2$, and

if $0 < |x - a| < \delta_2$, then $|g(x) - M| < \varepsilon/2$.

If δ denotes the *smaller* of δ_1 and δ_2, then whenever $0 < |x - a| < \delta$, the inequalities in (d) involving $f(x)$ and $g(x)$ are both true. Consequently, if $0 < |x - a| < \delta$, then from (d) and (c)

$$|f(x) + g(x) - (L + M)| < \varepsilon/2 + \varepsilon/2 = \varepsilon,$$

which is the desired statement (a).

(ii) We first show that if k is a function and

(e)
$$\text{if} \quad \lim_{x \to a} k(x) = 0, \quad \text{then} \quad \lim_{x \to a} f(x)k(x) = 0.$$

Since $\lim_{x \to a} f(x) = L$, it follows from Definition (2.10) (with $\varepsilon = 1$) that there is a $\delta_1 > 0$ such that if $0 < |x - a| < \delta_1$, then $|f(x) - L| < 1$ and hence also

$$|f(x)| = |f(x) - L + L| \leq |f(x) - L| + |L| < 1 + |L|.$$

Consequently,

(f)
$$\text{if} \quad 0 < |x - a| < \delta_1, \quad \text{then} \quad |f(x)k(x)| < (1 + |L|)|k(x)|.$$

Since $\lim_{x \to a} k(x) = 0$, for every $\varepsilon > 0$ there corresponds a $\delta_2 > 0$ such that

(g)
$$\text{if} \quad 0 < |x - a| < \delta_2, \quad \text{then} \quad |k(x) - 0| < \frac{\varepsilon}{1 + |L|}.$$

If δ denotes the smaller of δ_1 and δ_2, then whenever $0 < |x - a| < \delta$ both inequalities (f) and (g) are true and, consequently,

$$|f(x)k(x)| < (1 + |L|) \cdot \frac{\varepsilon}{1 + |L|}.$$

Therefore,

$$\text{if} \quad 0 < |x - a| < \delta, \quad \text{then} \quad |f(x)k(x) - 0| < \varepsilon,$$

which proves (e).

Next consider the identity

(h)
$$f(x)g(x) - LM = f(x)[g(x) - M] + M[f(x) - L].$$

Since $\lim_{x \to a} [g(x) - M] = 0$, it follows from (e), with $k(x) = g(x) - M$, that $\lim_{x \to a} f(x)[g(x) - M] = 0$. In addition, $\lim_{x \to a} M[f(x) - L] = 0$ and hence, from (h), $\lim_{x \to a} [f(x)g(x) - LM] = 0$. The last statement is equivalent to $\lim_{x \to a} f(x)g(x) = LM$.

(iii) It is sufficient to show that $\lim_{x \to a} 1/(g(x)) = 1/M$, for once this is done, the desired result may be obtained by applying (ii) to the product $f(x) \cdot 1/(g(x))$. Consider

(i)
$$\left| \frac{1}{g(x)} - \frac{1}{M} \right| = \left| \frac{M - g(x)}{g(x)M} \right| = \frac{1}{|M||g(x)|} |g(x) - M|.$$

Since $\lim_{x \to a} g(x) = M$, there exists a $\delta_1 > 0$ such that if $0 < |x - a| < \delta_1$, then $|g(x) - M| < |M|/2$. Consequently, for every such x,

$$|M| = |g(x) + (M - g(x))|$$
$$\leq |g(x)| + |M - g(x)|$$
$$< |g(x)| + |M|/2$$

and, therefore,

$$\frac{|M|}{2} < |g(x)| \quad \text{or} \quad \frac{1}{|g(x)|} < \frac{2}{|M|}.$$

Substitution in equation (i) leads to

(j) $$\left| \frac{1}{g(x)} - \frac{1}{M} \right| < \frac{2}{|M|^2} |g(x) - M|, \quad \text{provided } 0 < |x - a| < \delta_1.$$

Again using the fact that $\lim_{x \to a} g(x) = M$, it follows that for every $\varepsilon > 0$ there corresponds a $\delta_2 > 0$ such that

(k) $$\text{if} \quad 0 < |x - a| < \delta_2, \quad \text{then} \quad |g(x) - M| < \frac{|M|^2}{2} \varepsilon.$$

If δ denotes the smaller of δ_1 and δ_2, then both inequalities (j) and (k) are true. Thus

$$\text{if} \quad 0 < |x - a| < \delta, \quad \text{then} \quad \left| \frac{1}{g(x)} - \frac{1}{M} \right| < \varepsilon,$$

which means that $\lim_{x \to a} 1/(g(x)) = 1/M$. • •

THEOREM (2.20)

> If $a > 0$ and n is a positive integer, or if $a \leq 0$ and n is an odd positive integer, then $\lim_{x \to a} \sqrt[n]{x} = \sqrt[n]{a}$.

PROOF Suppose $a > 0$ and n is any positive integer. It must be shown that for every $\varepsilon > 0$ there corresponds a $\delta > 0$ such that

$$\text{if} \quad 0 < |x - a| < \delta, \quad \text{then} \quad |\sqrt[n]{x} - \sqrt[n]{a}| < \varepsilon$$

or, equivalently,

(a) $$\text{if} \quad -\delta < x - a < \delta, \ x \neq a, \quad \text{then} \quad -\varepsilon < \sqrt[n]{x} - \sqrt[n]{a} < \varepsilon.$$

It is sufficient to prove (a) if $\varepsilon < \sqrt[n]{a}$, for if a δ exists under this condition then the same δ can be used for any *larger* value of ε. Thus, in the remainder of the proof $\sqrt[n]{a} - \varepsilon$ is considered as a positive number less than ε. The inequalities in the following list are all equivalent:

$$-\varepsilon < \sqrt[n]{x} - \sqrt[n]{a} < \varepsilon$$
$$\sqrt[n]{a} - \varepsilon < \sqrt[n]{x} < \sqrt[n]{a} + \varepsilon$$
$$(\sqrt[n]{a} - \varepsilon)^n < x < (\sqrt[n]{a} + \varepsilon)^n$$
$$(\sqrt[n]{a} - \varepsilon)^n - a < x - a < (\sqrt[n]{a} + \varepsilon)^n - a$$
$$-[a - (\sqrt[n]{a} - \varepsilon)^n] < x - a < (\sqrt[n]{a} + \varepsilon)^n - a.$$

If δ denotes the smaller of the two positive numbers $a - (\sqrt[n]{a} - \varepsilon)^n$ and $(\sqrt[n]{a} + \varepsilon)^n - a$, then whenever $-\delta < x - a < \delta$ the last inequality in the list is true and hence so is the first. This gives us (a).

Next suppose $a < 0$ and n is an odd positive integer. In this case $-a$ and $\sqrt[n]{-a}$ are positive and, by the first part of the proof, we may write

$$\lim_{-x \to -a} \sqrt[n]{-x} = \sqrt[n]{-a}.$$

Thus for every $\varepsilon > 0$, there corresponds a $\delta > 0$ such that

$$\text{if } \ 0 < |-x - (-a)| < \delta, \quad \text{then} \quad |\sqrt[n]{-x} - \sqrt[n]{-a}| < \varepsilon$$

or, equivalently,

$$\text{if } \ 0 < |x - a| < \delta, \quad \text{then} \quad |\sqrt[n]{x} - \sqrt[n]{a}| < \varepsilon.$$

The last inequalities imply that $\lim_{x \to a} \sqrt[n]{x} = \sqrt[n]{a}$. • •

THE SANDWICH THEOREM (2.22)

If $f(x) \leq h(x) \leq g(x)$ for every x in an open interval containing a, except possibly at a, and if $\lim_{x \to a} f(x) = L = \lim_{x \to a} g(x)$, then $\lim_{x \to a} h(x) = L$.

PROOF For every $\varepsilon > 0$ there correspond $\delta_1 > 0$ and $\delta_2 > 0$ such that

(a)
$$\text{if } \ 0 < |x - a| < \delta_1, \quad \text{then} \quad |f(x) - L| < \varepsilon,$$
$$\text{if } \ 0 < |x - a| < \delta_2, \quad \text{then} \quad |g(x) - L| < \varepsilon.$$

If δ denotes the smaller of δ_1 and δ_2, then whenever $0 < |x - a| < \delta$, both inequalities in (a) that involve ε are true, that is,

$$-\varepsilon < f(x) - L < \varepsilon \quad \text{and} \quad -\varepsilon < g(x) - L < \varepsilon.$$

Consequently, if $0 < |x - a| < \delta$, then $L - \varepsilon < f(x)$ and $g(x) < L + \varepsilon$. Since $f(x) \leq h(x) \leq g(x)$, if $0 < |x - a| < \delta$, then $L - \varepsilon < h(x) < L + \varepsilon$ or, equivalently, $|h(x) - L| < \varepsilon$, which is what we wished to prove. • •

THEOREM (2.31)

If f and g are functions such that $\lim_{x \to a} g(x) = b$, and if f is continuous at b, then

$$\lim_{x \to a} f(g(x)) = f(b) = f\left(\lim_{x \to a} g(x) \right)$$

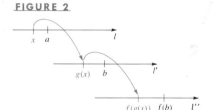

FIGURE 2

FIGURE 3

PROOF The composite function $f(g(x))$ may be represented geometrically by means of three real lines, l, l', and l'', as shown in Figure 2. To each coordinate x on l there corresponds the coordinate $g(x)$ on l' and then, in turn, $f(g(x))$ on l''. We wish to prove that $f(g(x))$ has the limit $f(b)$ as x approaches a. In terms of Definition (2.10) we must show that for every $\varepsilon > 0$ there exists a $\delta > 0$ such that

(a)
$$\text{if } \ 0 < |x - a| < \delta, \quad \text{then} \quad |f(g(x)) - f(b)| < \varepsilon.$$

Let us begin by considering the interval $(f(b) - \varepsilon, f(b) + \varepsilon)$ on l'' shown in Figure 3. Since f is continuous at b, $\lim_{z \to b} f(z) = f(b)$ and hence, as illustrated in the figure, there exists a number $\delta_1 > 0$ such that

(b)
$$\text{if } \ |z - b| < \delta_1, \quad \text{then} \quad |f(z) - f(b)| < \varepsilon.$$

In particular, if we let $z = g(x)$ in (b), it follows that

(c)
$$\text{if } \ |g(x) - b| < \delta_1, \quad \text{then} \quad |f(g(x)) - f(b)| < \varepsilon.$$

FIGURE 4

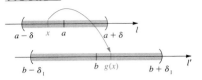

Next, turning our attention to the interval $(b - \delta_1, b + \delta_1)$ on l' and using the definition of $\lim_{x \to a} g(x) = b$, we obtain the fact illustrated in Figure 4 that there exists a $\delta > 0$ such that

(d) $\qquad\qquad$ if $\quad 0 < |x - a| < \delta, \quad$ then $\quad |g(x) - b| < \delta_1.$

Finally, combining (d) and (c), we see that

$$\text{if} \quad 0 < |x - a| < \delta, \quad \text{then} \quad |f(g(x)) - f(b)| < \varepsilon$$

which is the desired conclusion (a). • •

DEFINITIONS (3.1) AND (3.1)′

If f is defined on an open interval containing a, then
$$\lim_{x \to a} \frac{f(x) - f(a)}{x - a} = \lim_{h \to 0} \frac{f(a + h) - f(a)}{h}$$
provided either limit exists.

PROOF Suppose

$$\lim_{x \to a} \frac{f(x) - f(a)}{x - a} = L$$

for some number L. According to the definition of limit (2.10), this means that for every $\varepsilon > 0$ there exists a $\delta > 0$ such that

$$\text{if} \quad 0 < |x - a| < \delta, \quad \text{then} \quad \left| \frac{f(x) - f(a)}{x - a} - L \right| < \varepsilon.$$

If we let $h = x - a$, then $x = a + h$ and the last statement may be written

$$\text{if} \quad 0 < |h| < \delta, \quad \text{then} \quad \left| \frac{f(a + h) - f(a)}{h} - L \right| < \varepsilon$$

which means that

$$\lim_{h \to 0} \frac{f(a + h) - f(a)}{h} = L.$$

Conversely, starting with the last limit and reversing the steps leads to the first limit. • •

THE CHAIN RULE (3.27)

If $y = f(u)$, $u = g(x)$, and the derivatives dy/du and du/dx both exist, then the composite function defined by $y = f(g(x))$ has a derivative given by
$$\frac{dy}{dx} = \frac{dy}{du} \frac{du}{dx} = f'(u)g'(x) = f'(g(x))g'(x)$$

PROOF If $y = f(x)$ and $\Delta x \approx 0$, then by (3.24), the difference between the derivative $f'(x)$ and the ratio $\Delta y / \Delta x$ is numerically small. Since this difference depends on the size of Δx, we shall represent it by means of the function notation $\eta(\Delta x)$. Thus, for each $\Delta x \neq 0$,

(a) $$\eta(\Delta x) = \frac{\Delta y}{\Delta x} - f'(x).$$

It should be noted that $\eta(\Delta x)$ does *not* represent the product of η and Δx, but rather than η is *a function of* Δx, whose values are given by (a). Moreover, applying (3.24), we see that

(b) $$\lim_{\Delta x \to 0} \eta(\Delta x) = \lim_{\Delta x \to 0} \left[\frac{\Delta y}{\Delta x} - f'(x) \right] = 0.$$

The function η has been defined only for nonzero values of Δx. It is convenient to extend the definition of η to include $\Delta x = 0$ by letting $\eta(0) = 0$. It then follows from (b) that η *is continuous at* 0.

Multiplying both sides of (a) by Δx and rearranging terms gives us

(c) $$\Delta y = f'(x)\, \Delta x + \eta(\Delta x) \cdot \Delta x,$$

which is true if either $\Delta x \neq 0$ or $\Delta x = 0$. Since $f'(x)\, \Delta x = dy$, it follows from (c) that

(d) $$\Delta y - dy = \eta(\Delta x) \cdot \Delta x.$$

Let us now consider the situation stated in the hypothesis of the theorem:

$$y = f(u) \quad \text{and} \quad u = g(x).$$

If $g(x)$ is the domain of f, then we may write

$$y = f(u) = f(g(x)),$$

that is, y is a function of x. If we give x an increment Δx there corresponds an increment Δu of u and, in turn, an increment Δy of $y = f(u)$. Thus

$$\Delta u = g(x + \Delta x) - g(x)$$
$$\Delta y = f(u + \Delta u) - f(u).$$

Since dy/du exists, we may use (c) with u as the independent variable to write

(e) $$\Delta y = f'(u)\, \Delta u + \eta(\Delta u) \cdot \Delta u$$

for a function η of Δu such that, by (b),

(f) $$\lim_{\Delta u \to 0} \eta(\Delta u) = 0.$$

Moreover, η is continuous at $\Delta u = 0$ and (e) is true if $\Delta u = 0$. Dividing both sides of (e) by Δx gives us

$$\frac{\Delta y}{\Delta x} = f'(u) \frac{\Delta u}{\Delta x} + \eta(\Delta u) \cdot \frac{\Delta u}{\Delta x}.$$

If we now take the limit as Δx approaches zero and use the facts

$$\lim_{\Delta x \to 0} \frac{\Delta y}{\Delta x} = \frac{dy}{dx} \quad \text{and} \quad \lim_{\Delta x \to 0} \frac{\Delta u}{\Delta x} = \frac{du}{dx},$$

we see that

$$\frac{dy}{dx} = f'(u) \frac{du}{dx} + \lim_{\Delta x \to 0} \eta(\Delta u) \cdot \frac{du}{dx}.$$

Since $f'(u) = dy/du$, we may complete the proof by showing that the limit indicated in the last equation is 0. To accomplish this we first observe that since g is differentiable, it is continuous and hence

$$\lim_{\Delta x \to 0} [g(x + \Delta x) - g(x)] = 0$$

or, equivalently,

$$\lim_{\Delta x \to 0} \Delta u = 0.$$

In other words, Δu approaches 0 as Δx approaches 0. Using this fact, together with (f), gives us

(g)
$$\lim_{\Delta x \to 0} \eta(\Delta u) = \lim_{\Delta u \to 0} \eta(\Delta u) = 0$$

and the theorem is proved. The fact that $\lim_{\Delta x \to 0} \eta(\Delta u) = 0$ can also be established by means of an ε-δ argument using (2.10). • •

THEOREM (5.14)

If f is integrable on $[a, b]$ and c is any number, then cf is integrable on $[a, b]$ and

$$\int_a^b cf(x)\, dx = c \int_a^b f(x)\, dx.$$

PROOF If $c = 0$ the result follows from Theorem (5.13). Assume, therefore, that $c \neq 0$. Since f is integrable, $\int_a^b f(x)\, dx = I$ for some number I. If P is a partition of $[a, b]$, then each Riemann sum R_P for the function cf has the form $\sum_k cf(w_k)\, \Delta x_k$, such that for every k, w_k is in the kth subinterval $[x_{k-1}, x_k]$ of P. We wish to show that for every $\varepsilon > 0$ there corresponds a $\delta > 0$ such that whenever $\|P\| < \delta$, then

(a)
$$\left| \sum_k cf(w_k)\, \Delta x_k - cI \right| < \varepsilon$$

for every w_k in $[x_{k-1}, x_k]$. If we let $\varepsilon' = \varepsilon/|c|$, then since f is integrable there exists a $\delta > 0$ such that whenever $\|P\| < \delta$,

$$\left| \sum_k f(w_k)\, \Delta x_k - I \right| < \varepsilon' = \frac{\varepsilon}{|c|}.$$

Multiplying both sides of this inequality by $|c|$ leads to (a). Hence

$$\lim_{\|P\| \to 0} \sum_k cf(w_k)\, \Delta x_k = cI = c \int_a^b f(x)\, dx \quad • •$$

THEOREM (5.15)

If f and g are integrable on $[a, b]$, then $f + g$ is integrable on $[a, b]$ and

$$\int_a^b [f(x) + g(x)]\, dx = \int_a^b f(x)\, dx + \int_a^b g(x)\, dx$$

PROOF By hypothesis there exist real numbers I_1 and I_2 such that

$$\int_a^b f(x)\, dx = I_1 \quad \text{and} \quad \int_a^b g(x)\, dx = I_2.$$

Let P denote a partition of $[a, b]$ and let R_P denote an arbitrary Riemann sum for $f + g$ associated with P, that is,

(a)
$$R_P = \sum_k [f(w_k) + g(w_k)] \, \Delta x_k$$

such that w_k is in $[x_{k-1}, x_k]$ for every k. We wish to show that for every $\varepsilon > 0$ there corresponds a $\delta > 0$ such that whenever $\| P \| < \delta$, then $|R_P - (I_1 + I_2)| < \varepsilon$. Using Theorem (5.3) (i) we may write (a) in the form

$$R_P = \sum_k f(w_k) \, \Delta x_k + \sum_k g(w_k) \, \Delta x_k.$$

Rearranging terms and using the Triangle Inequality (1.4),

(b)
$$|R_P - (I_1 + I_2)| = \left| \left(\sum_k f(w_k) \, \Delta x_k - I_1 \right) + \left(\sum_k g(w_k) \, \Delta x_k - I_2 \right) \right|$$

$$\leq \left| \sum_k f(w_k) \, \Delta x_k - I_1 \right| + \left| \sum_k g(w_k) \, \Delta x_k - I_2 \right|.$$

By the integrability of f and g, if $\varepsilon' = \varepsilon/2$, then there exist $\delta_1 > 0$ and $\delta_2 > 0$ such that whenever $\| P \| < \delta_1$ and $\| P \| < \delta_2$,

(c)
$$\left| \sum_k f(w_k) \, \Delta x_k - I_1 \right| < \varepsilon' = \varepsilon/2,$$

and
$$\left| \sum_k g(w_k) \, \Delta x_k - I_2 \right| < \varepsilon' = \varepsilon/2$$

for every w_k in $[x_{k-1}, x_k]$. If δ denotes the smaller of δ_1 and δ_2, then whenever $\| P \| < \delta$, both inequalities in (c) are true and hence, from (b),

$$|R_P - (I_1 + I_2)| < (\varepsilon/2) + (\varepsilon/2) = \varepsilon,$$

which is what we wished to prove. • •

THEOREM (5.16)

> If $a < c < b$, and if f is integrable on both $[a, c]$ and $[c, b]$, then f is integrable on $[a, b]$ and
>
> $$\int_a^b f(x) \, dx = \int_a^c f(x) \, dx + \int_c^b f(x) \, dx$$

PROOF By hypothesis there exist real numbers I_1 and I_2 such that

(a)
$$\int_a^c f(x) \, dx = I_1 \quad \text{and} \quad \int_c^b f(x) \, dx = I_2.$$

Let us denote a partition of $[a, c]$ by P_1, of $[c, b]$ by P_2, and of $[a, b]$ by P. Arbitrary Riemann sums associated with P_1, P_2, and P will be denoted by R_{P_1}, R_{P_2}, and R_P, respectively. We must show that for every $\varepsilon > 0$ there corresponds a $\delta > 0$ such that if $\| P \| < \delta$, then $|R_P - (I_1 + I_2)| < \varepsilon$.

If we let $\varepsilon' = \varepsilon/4$, then by (a) there exist positive numbers δ_1 and δ_2 such that if $\| P_1 \| < \delta_1$ and $\| P_2 \| < \delta_2$, then

(b)
$$|R_{P_1} - I_1| < \varepsilon' = \varepsilon/4 \quad \text{and} \quad |R_{P_2} - I_2| < \varepsilon' = \varepsilon/4.$$

If δ denotes the smaller of δ_1 and δ_2, then both inequalities in (b) are true whenever $\| P \| < \delta$. Moreover, since f is integrable on $[a, c]$ and $[c, b]$, it is bounded on both intervals and hence there exists a number M such that $|f(x)| \leq M$ for every x in $[a, b]$. We shall now assume that δ has been chosen so that, in addition to the previous requirement, we also have $\delta < \varepsilon/(4M)$.

Let P be a partition of $[a, b]$ such that $\|P\| < \delta$. If the numbers that determine P are

$$a = x_0, x_1, x_2, \ldots, x_n = b,$$

then there is a unique half-open interval of the form $(x_{d-1}, x_d]$ that contains c. If $R_P = \sum_{k=1}^{n} f(w_k) \Delta x_k$ we may write

(c)
$$R_P = \sum_{k=1}^{d-1} f(w_k) \Delta x_k + f(w_d) \Delta x_d + \sum_{k=d+1}^{n} f(w_k) \Delta x_k.$$

Let P_1 denote the partition of $[a, c]$ determined by $\{a, x_1, \ldots, x_{d-1}, c\}$, let P_2 denote the partition of $[c, b]$ determined by $\{c, x_d, \ldots, x_{n-1}, b\}$, and consider the Riemann sums

$$R_{P_1} = \sum_{k=1}^{d-1} f(w_k) \Delta x_k + f(c)(c - x_{d-1})$$

(d)
$$R_{P_2} = f(c)(x_d - c) + \sum_{k=d+1}^{n} f(w_k) \Delta x_k.$$

Using the Triangle Inequality (1.4) and (b),

$$\left|(R_{P_1} + R_{P_2}) - (I_1 + I_2)\right| = \left|(R_{P_1} - I_1) + (R_{P_2} - I_2)\right|$$
$$\leq |R_{P_1} - I_1| + |R_{P_2} - I_2|$$

(e)
$$< \frac{\varepsilon}{4} + \frac{\varepsilon}{4} = \frac{\varepsilon}{2}.$$

It follows from (c) and (d) that

$$\left|R_P - (R_{P_1} + R_{P_2})\right| = |f(w_d) - f(c)| \, \Delta x_d.$$

Employing the Triangle Inequality and the choice of δ gives us

(f)
$$\left|R_P - (R_{P_1} + R_{P_2})\right| \leq \{|f(w_d)| + |f(c)|\} \, \Delta x_d$$
$$\leq (M + M)[\varepsilon/(4M)] = \varepsilon/2,$$

provided $\|P\| < \delta$. If we write

$$\left|R_P - (I_1 + I_2)\right| = \left|R_P - (R_{P_1} + R_{P_2}) + (R_{P_1} + R_{P_2}) - (I_1 + I_2)\right|$$
$$\leq \left|R_P - (R_{P_1} + R_{P_2})\right| + \left|(R_{P_1} + R_{P_2}) - (I_1 + I_2)\right|$$

then it follows from (f) and (e) that whenever $\|P\| < \delta$

$$\left|R_P - (I_1 + I_2)\right| < (\varepsilon/2) + (\varepsilon/2) = \varepsilon$$

for every Riemann sum R_P. This completes the proof. • •

THEOREM (5.18)

If f is integrable on $[a, b]$ and if $f(x) \geq 0$ for every x in $[a, b]$, then

$$\int_a^b f(x) \, dx \geq 0$$

PROOF We shall give an indirect proof. Thus let $\int_a^b f(x) \, dx = I$ and suppose that $I < 0$. Consider any partition P of $[a, b]$ and let $R_P = \sum_k f(w_k) \Delta x_k$ be an arbitrary Riemann sum associated with P. Since $f(w_k) \geq 0$ for every w_k in $[x_{k-1}, x_k]$, it follows that $R_P \geq 0$. If we let $\varepsilon = -(I/2)$, then according to Definition (5.7), whenever $\|P\|$ is sufficiently small,

$$|R_P - I| < \varepsilon = -(I/2).$$

It follows that $R_P < I - (I/2) = I/2 < 0$, a contradiction. Therefore, the supposition $I < 0$ is false and hence $I \geq 0$. • •

THEOREM (17.31)

If $x = f(u, v)$, $y = g(u, v)$ is a transformation of coordinates, then

$$\iint_R F(x, y)\, dx\, dy = \pm \iint_S F(f(u, v), g(u, v)) \frac{\partial(x, y)}{\partial(u, v)}\, du\, dv.$$

The $+$ sign or the $-$ sign is chosen if, as (u, v) traces the boundary K of S once in the positive direction, the corresponding point (x, y) traces the boundary C of R once in the positive or negative direction, respectively.

PROOF Let us begin by choosing $G(x, y)$ such that $\partial G / \partial x = F$. Applying Green's Theorem (18.19) with $G = N$ gives us

(a)
$$\iint_R F(x, y)\, dx\, dy = \iint_R \frac{\partial}{\partial x}[G(x, y)]\, dx\, dy = \oint_C G(x, y)\, dy.$$

Suppose the curve K in the uv-plane has a parametrization

$$u = \varphi(t),\ v = \psi(t); \qquad a \leq t \leq b.$$

From our assumptions on the transformation, parametric equations for the curve C in the xy-plane are

(b)
$$x = f(u, v) = f(\varphi(t), \psi(t))$$
$$y = g(u, v) = g(\varphi(t), \psi(t))$$

for $a \leq t \leq b$. We may therefore evaluate the line integral $\oint_C G(x, y)\, dy$ in (a) by formal substitutions for x and y. To simplify the notation, let

$$H(t) = G[f(\varphi(t), \psi(t)), g(\varphi(t), \psi(t))].$$

Applying a chain rule to y in (b) gives us

$$\frac{dy}{dt} = \frac{\partial y}{\partial u}\frac{du}{dt} + \frac{\partial y}{\partial v}\frac{dv}{dt} = \frac{\partial y}{\partial u}\varphi'(t) + \frac{\partial y}{\partial v}\psi'(t).$$

Consequently,

$$\oint_C G(x, y)\, dy = \oint_C H(t)\frac{dy}{dt}\, dt$$

$$= \int_a^b H(t)\left[\frac{\partial y}{\partial u}\varphi'(t) + \frac{\partial y}{\partial v}\psi'(t)\right] dt.$$

Since $du = \varphi'(t)\, dt$ and $dv = \psi'(t)\, dt$, we may regard the last line integral as a line integral around the curve K in the uv-plane. Thus

(c)
$$\oint_C G(x, y)\, dy = \pm \oint_K G\frac{\partial y}{\partial u}\, du + G\frac{\partial y}{\partial v}\, dv.$$

For simplicity we have used G as an abbreviation for $G(f(u, v), g(u, v))$. The choice of the $+$ sign or the $-$ sign is determined by letting t vary from a to b and noting if (x, y) traces C in the same or opposite direction, respectively, as (u, v) traces K.

The line integral on the right in (c) has the form

$$\oint_K M \, du + N \, dv$$

with $$M = G \frac{\partial y}{\partial u} \quad \text{and} \quad N = G \frac{\partial y}{\partial v}.$$

Applying Green's Theorem, we obtain

$$\oint_K M \, du + N \, dv = \iint_S \left(\frac{\partial N}{\partial u} - \frac{\partial M}{\partial v} \right) du \, dv$$

$$= \iint_S \left(G \frac{\partial^2 y}{\partial u \, \partial v} + \frac{\partial G}{\partial u} \frac{\partial y}{\partial v} - G \frac{\partial^2 y}{\partial v \, \partial u} - \frac{\partial G}{\partial v} \frac{\partial y}{\partial u} \right) du \, dv$$

$$= \iint_S \left[\left(\frac{\partial G}{\partial x} \frac{\partial x}{\partial u} + \frac{\partial G}{\partial y} \frac{\partial y}{\partial u} \right) \frac{\partial y}{\partial v} - \left(\frac{\partial G}{\partial x} \frac{\partial x}{\partial v} + \frac{\partial G}{\partial y} \frac{\partial y}{\partial v} \right) \frac{\partial y}{\partial u} \right] du \, dv$$

$$= \iint_S \frac{\partial G}{\partial x} \left(\frac{\partial x}{\partial u} \frac{\partial y}{\partial v} - \frac{\partial y}{\partial u} \frac{\partial x}{\partial v} \right) du \, dv.$$

Using the fact that $\partial G / \partial x = F(x, y)$, together with the definition of Jacobian (17.30) gives us

$$\oint_K M \, du + N \, dv = \iint_S F(f(u, v), g(u, v)) \frac{\partial(x, y)}{\partial(u, v)} du \, dv.$$

Combining this formula with (a) and (c) leads to the desired result. • •

TABLE A TRIGONOMETRIC FUNCTIONS

Degrees	Radians	sin	tan	cot	cos		
0	0.000	0.000	0.000		1.000	1.571	90
1	0.017	0.017	0.017	57.29	1.000	1.553	89
2	0.035	0.035	0.035	28.64	0.999	1.536	88
3	0.052	0.052	0.052	19.081	0.999	1.518	87
4	0.070	0.070	0.070	14.301	0.998	1.501	86
5	0.087	0.087	0.087	11.430	0.996	1.484	85
6	0.105	0.105	0.105	9.514	0.995	1.466	84
7	0.122	0.122	0.123	8.144	0.993	1.449	83
8	0.140	0.139	0.141	7.115	0.990	1.431	82
9	0.157	0.156	0.158	6.314	0.988	1.414	81
10	0.175	0.174	0.176	5.671	0.985	1.396	80
11	0.192	0.191	0.194	5.145	0.982	1.379	79
12	0.209	0.208	0.213	4.705	0.978	1.361	78
13	0.227	0.225	0.231	4.331	0.974	1.344	77
14	0.244	0.242	0.249	4.011	0.970	1.326	76
15	0.262	0.259	0.268	3.732	0.966	1.309	75
16	0.279	0.276	0.287	3.487	0.961	1.292	74
17	0.297	0.292	0.306	3.271	0.956	1.274	73
18	0.314	0.309	0.325	3.078	0.951	1.257	72
19	0.332	0.326	0.344	2.904	0.946	1.239	71
20	0.349	0.342	0.364	2.747	0.940	1.222	70
21	0.367	0.358	0.384	2.605	0.934	1.204	69
22	0.384	0.375	0.404	2.475	0.927	1.187	68
23	0.401	0.391	0.424	2.356	0.921	1.169	67
24	0.419	0.407	0.445	2.246	0.914	1.152	66
25	0.436	0.423	0.466	2.144	0.906	1.134	65
26	0.454	0.438	0.488	2.050	0.899	1.117	64
27	0.471	0.454	0.510	1.963	0.891	1.100	63
28	0.489	0.469	0.532	1.881	0.883	1.082	62
29	0.506	0.485	0.554	1.804	0.875	1.065	61
30	0.524	0.500	0.577	1.732	0.866	1.047	60
31	0.541	0.515	0.601	1.664	0.857	1.030	59
32	0.559	0.530	0.625	1.600	0.848	1.012	58
33	0.576	0.545	0.649	1.540	0.839	0.995	57
34	0.593	0.559	0.675	1.483	0.829	0.977	56
35	0.611	0.574	0.700	1.428	0.819	0.960	55
36	0.628	0.588	0.727	1.376	0.809	0.942	54
37	0.646	0.602	0.754	1.327	0.799	0.925	53
38	0.663	0.616	0.781	1.280	0.788	0.908	52
39	0.681	0.629	0.810	1.235	0.777	0.890	51
40	0.698	0.643	0.839	1.192	0.766	0.873	50
41	0.716	0.656	0.869	1.150	0.755	0.855	49
42	0.733	0.669	0.900	1.111	0.743	0.838	48
43	0.750	0.682	0.933	1.072	0.731	0.820	47
44	0.768	0.695	0.966	1.036	0.719	0.803	46
45	0.785	0.707	1.000	1.000	0.707	0.785	45
		cos	cot	tan	sin	Radians	Degrees

TABLE B EXPONENTIAL FUNCTIONS

x	e^x	e^{-x}	x	e^x	e^{-x}
0.00	1.0000	1.0000	2.50	12.182	0.0821
0.05	1.0513	0.9512	2.60	13.464	0.0743
0.10	1.1052	0.9048	2.70	14.880	0.0672
0.15	1.1618	0.8607	2.80	16.445	0.0608
0.20	1.2214	0.8187	2.90	18.174	0.0550
0.25	1.2840	0.7788	3.00	20.086	0.0498
0.30	1.3499	0.7408	3.10	22.198	0.0450
0.35	1.4191	0.7047	3.20	24.533	0.0408
0.40	1.4918	0.6703	3.30	27.113	0.0369
0.45	1.5683	0.6376	3.40	29.964	0.0334
0.50	1.6487	0.6065	3.50	33.115	0.0302
0.55	1.7333	0.5769	3.60	36.598	0.0273
0.60	1.8221	0.5488	3.70	40.447	0.0247
0.65	1.9155	0.5220	3.80	44.701	0.0224
0.70	2.0138	0.4966	3.90	49.402	0.0202
0.75	2.1170	0.4724	4.00	54.598	0.0183
0.80	2.2255	0.4493	4.10	60.340	0.0166
0.85	2.3396	0.4274	4.20	66.686	0.0150
0.90	2.4596	0.4066	4.30	73.700	0.0136
0.95	2.5857	0.3867	4.40	81.451	0.0123
1.00	2.7183	0.3679	4.50	90.017	0.0111
1.10	3.0042	0.3329	4.60	99.484	0.0101
1.20	3.3201	0.3012	4.70	109.95	0.0091
1.30	3.6693	0.2725	4.80	121.51	0.0082
1.40	4.0552	0.2466	4.90	134.29	0.0074
1.50	4.4817	0.2231	5.00	148.41	0.0067
1.60	4.9530	0.2019	6.00	403.43	0.0025
1.70	5.4739	0.1827	7.00	1096.6	0.0009
1.80	6.0496	0.1653	8.00	2981.0	0.0003
1.90	6.6859	0.1496	9.00	8103.1	0.0001
2.00	7.3891	0.1353	10.00	22026.0	0.00005
2.10	8.1662	0.1225			
2.20	9.0250	0.1108			
2.30	9.9742	0.1003			
2.40	11.0232	0.0907			

TABLE C NATURAL LOGARITHMS

n	0.0	0.1	0.2	0.3	0.4	0.5	0.6	0.7	0.8	0.9
0*		7.697	8.391	8.796	9.084	9.307	9.489	9.643	9.777	9.895
1	0.000	0.095	0.182	0.262	0.336	0.405	0.470	0.531	0.588	0.642
2	0.693	0.742	0.788	0.833	0.875	0.916	0.956	0.993	1.030	1.065
3	1.099	1.131	1.163	1.194	1.224	1.253	1.281	1.308	1.335	1.361
4	1.386	1.411	1.435	1.459	1.482	1.504	1.526	1.548	1.569	1.589
5	1.609	1.629	1.649	1.668	1.686	1.705	1.723	1.740	1.758	1.775
6	1.792	1.808	1.825	1.841	1.856	1.872	1.887	1.902	1.917	1.932
7	1.946	1.960	1.974	1.988	2.001	2.015	2.028	2.041	2.054	2.067
8	2.079	2.092	2.104	2.116	2.128	2.140	2.152	2.163	2.175	2.186
9	2.197	2.208	2.219	2.230	2.241	2.251	2.262	2.272	2.282	2.293
10	2.303	2.313	2.322	2.332	2.342	2.351	2.361	2.370	2.380	2.389

* Subtract 10 if $n < 1$; for example, $\ln 0.3 \approx 8.796 - 10 = -1.204$.

BASIC FORMS

1. $\displaystyle\int u\,dv = uv - \int v\,du$

2. $\displaystyle\int u^n\,du = \frac{1}{n+1}u^{n+1} + C, \quad n \neq -1$

3. $\displaystyle\int \frac{du}{u} = \ln|u| + C$

4. $\displaystyle\int e^u\,du = e^u + C$

5. $\displaystyle\int a^u\,du = \frac{1}{\ln a}a^u + C$

6. $\displaystyle\int \sin u\,du = -\cos u + C$

7. $\displaystyle\int \cos u\,du = \sin u + C$

8. $\displaystyle\int \sec^2 u\,du = \tan u + C$

9. $\displaystyle\int \csc^2 u\,du = -\cot u + C$

10. $\displaystyle\int \sec u \tan u\,du = \sec u + C$

11. $\displaystyle\int \csc u \cot u\,du = -\csc u + C$

12. $\displaystyle\int \tan u\,du = \ln|\sec u| + C$

13. $\displaystyle\int \cot u\,du = \ln|\sin u| + C$

14. $\displaystyle\int \sec u\,du = \ln|\sec u + \tan u| + C$

15. $\displaystyle\int \csc u\,du = \ln|\csc u - \cot u| + C$

16. $\displaystyle\int \frac{du}{\sqrt{a^2 - u^2}} = \sin^{-1}\frac{u}{a} + C$

17. $\displaystyle\int \frac{du}{a^2 + u^2} = \frac{1}{a}\tan^{-1}\frac{u}{a} + C$

18. $\displaystyle\int \frac{du}{u\sqrt{u^2 - a^2}} = \frac{1}{a}\sec^{-1}\frac{u}{a} + C$

19. $\displaystyle\int \frac{du}{a^2 - u^2} = \frac{1}{2a}\ln\left|\frac{u+a}{u-a}\right| + C$

20. $\displaystyle\int \frac{du}{u^2 - a^2} = \frac{1}{2a}\ln\left|\frac{u-a}{u+a}\right| + C$

FORMS INVOLVING $\sqrt{a^2 + u^2}$

21. $\displaystyle\int \sqrt{a^2 + u^2}\,du = \frac{u}{2}\sqrt{a^2 + u^2} + \frac{a^2}{2}\ln|u + \sqrt{a^2 + u^2}| + C$

22. $\displaystyle\int u^2\sqrt{a^2 + u^2}\,du = \frac{u}{8}(a^2 + 2u^2)\sqrt{a^2 + u^2} - \frac{a^4}{8}\ln|u + \sqrt{a^2 + u^2}| + C$

23. $\displaystyle\int \frac{\sqrt{a^2 + u^2}}{u}\,du = \sqrt{a^2 + u^2} - a\ln\left|\frac{a + \sqrt{a^2 + u^2}}{u}\right| + C$

24. $\displaystyle\int \frac{\sqrt{a^2 + u^2}}{u^2}\,du = -\frac{\sqrt{a^2 + u^2}}{u} + \ln|u + \sqrt{a^2 + u^2}| + C$

25. $\displaystyle\int \frac{du}{\sqrt{a^2 + u^2}} = \ln|u + \sqrt{a^2 + u^2}| + C$

26. $\displaystyle\int \frac{u^2\,du}{\sqrt{a^2 + u^2}} = \frac{u}{2}\sqrt{a^2 + u^2} - \frac{a^2}{2}\ln|u + \sqrt{a^2 + u^2}| + C$

27. $\displaystyle\int \frac{du}{u\sqrt{a^2 + u^2}} = -\frac{1}{a}\ln\left|\frac{\sqrt{a^2 + u^2} + a}{u}\right| + C$

28. $\displaystyle\int \frac{du}{u^2\sqrt{a^2 + u^2}} = -\frac{\sqrt{a^2 + u^2}}{a^2 u} + C$

29. $\displaystyle\int \frac{du}{(a^2 + u^2)^{3/2}} = \frac{u}{a^2\sqrt{a^2 + u^2}} + C$

FORMS INVOLVING $\sqrt{a^2 - u^2}$

30. $\displaystyle\int \sqrt{a^2 - u^2}\,du = \frac{u}{2}\sqrt{a^2 - u^2} + \frac{a^2}{2}\sin^{-1}\frac{u}{a} + C$

31. $\displaystyle\int u^2\sqrt{a^2 - u^2}\,du = \frac{u}{8}(2u^2 - a^2)\sqrt{a^2 - u^2} + \frac{a^4}{8}\sin^{-1}\frac{u}{a} + C$

32. $\displaystyle\int \frac{\sqrt{a^2 - u^2}}{u}\,du = \sqrt{a^2 - u^2} - a\ln\left|\frac{a + \sqrt{a^2 - u^2}}{u}\right| + C$

33. $\displaystyle\int \frac{\sqrt{a^2 - u^2}}{u^2}\,du = -\frac{1}{u}\sqrt{a^2 - u^2} - \sin^{-1}\frac{u}{a} + C$

34. $\displaystyle\int \frac{u^2\,du}{\sqrt{a^2 - u^2}} = -\frac{u}{2}\sqrt{a^2 - u^2} + \frac{a^2}{2}\sin^{-1}\frac{u}{a} + C$

35. $\displaystyle\int \frac{du}{u\sqrt{a^2 - u^2}} = -\frac{1}{a}\ln\left|\frac{a + \sqrt{a^2 - u^2}}{u}\right| + C$

36. $\displaystyle\int \frac{du}{u^2\sqrt{a^2 - u^2}} = -\frac{1}{a^2 u}\sqrt{a^2 - u^2} + C$

37. $\displaystyle\int (a^2 - u^2)^{3/2}\,du = -\frac{u}{8}(2u^2 - 5a^2)\sqrt{a^2 - u^2} + \frac{3a^4}{8}\sin^{-1}\frac{u}{a} + C$

38. $\displaystyle\int \frac{du}{(a^2 - u^2)^{3/2}} = \frac{u}{a^2\sqrt{a^2 - u^2}} + C$

FORMS INVOLVING $\sqrt{u^2 - a^2}$

39 $\displaystyle\int \sqrt{u^2 - a^2}\, du = \frac{u}{2}\sqrt{u^2 - a^2} - \frac{a^2}{2}\ln|u + \sqrt{u^2 - a^2}| + C$

40 $\displaystyle\int u^2\sqrt{u^2 - a^2}\, du = \frac{u}{8}(2u^2 - a^2)\sqrt{u^2 - a^2} - \frac{a^4}{8}\ln|u + \sqrt{u^2 - a^2}| + C$

41 $\displaystyle\int \frac{\sqrt{u^2 - a^2}}{u}\, du = \sqrt{u^2 - a^2} - a\cos^{-1}\frac{a}{u} + C$

42 $\displaystyle\int \frac{\sqrt{u^2 - a^2}}{u^2}\, du = -\frac{\sqrt{u^2 - a^2}}{u} + \ln|u + \sqrt{u^2 - a^2}| + C$

43 $\displaystyle\int \frac{du}{\sqrt{u^2 - a^2}} = \ln|u + \sqrt{u^2 - a^2}| + C$

44 $\displaystyle\int \frac{u^2\, du}{\sqrt{u^2 - a^2}} = \frac{u}{2}\sqrt{u^2 - a^2} + \frac{a^2}{2}\ln|u + \sqrt{u^2 - a^2}| + C$

45 $\displaystyle\int \frac{du}{u^2\sqrt{u^2 - a^2}} = \frac{\sqrt{u^2 - a^2}}{a^2 u} + C$

46 $\displaystyle\int \frac{du}{(u^2 - a^2)^{3/2}} = -\frac{u}{a^2\sqrt{u^2 - a^2}} + C$

FORMS INVOLVING $a + bu$

47 $\displaystyle\int \frac{u\, du}{a + bu} = \frac{1}{b^2}(a + bu - a\ln|a + bu|) + C$

48 $\displaystyle\int \frac{u^2\, du}{a + bu} = \frac{1}{2b^3}[(a + bu)^2 - 4a(a + bu) + 2a^2\ln|a + bu|] + C$

49 $\displaystyle\int \frac{du}{u(a + bu)} = \frac{1}{a}\ln\left|\frac{u}{a + bu}\right| + C$

50 $\displaystyle\int \frac{du}{u^2(a + bu)} = -\frac{1}{au} + \frac{b}{a^2}\ln\left|\frac{a + bu}{u}\right| + C$

51 $\displaystyle\int \frac{u\, du}{(a + bu)^2} = \frac{a}{b^2(a + bu)} + \frac{1}{b^2}\ln|a + bu| + C$

52 $\displaystyle\int \frac{du}{u(a + bu)^2} = \frac{1}{a(a + bu)} - \frac{1}{a^2}\ln\left|\frac{a + bu}{u}\right| + C$

53 $\displaystyle\int \frac{u^2\, du}{(a + bu)^2} = \frac{1}{b^3}\left(a + bu - \frac{a^2}{a + bu} - 2a\ln|a + bu|\right) + C$

54 $\displaystyle\int u\sqrt{a + bu}\, du = \frac{2}{15b^2}(3bu - 2a)(a + bu)^{3/2} + C$

55 $\displaystyle\int \frac{u\, du}{\sqrt{a + bu}} = \frac{2}{3b^2}(bu - 2a)\sqrt{a + bu}$

56 $\displaystyle\int \frac{u^2\, du}{\sqrt{a + bu}} = \frac{2}{15b^3}(8a^2 + 3b^2u^2 - 4abu)\sqrt{a + bu}$

57 $\displaystyle\int \frac{du}{u\sqrt{a + bu}} = \frac{1}{\sqrt{a}}\ln\left|\frac{\sqrt{a + bu} - \sqrt{a}}{\sqrt{a + bu} + \sqrt{a}}\right| + C, \quad \text{if } a > 0$

$\displaystyle\qquad\qquad\quad = \frac{2}{\sqrt{-a}}\tan^{-1}\sqrt{\frac{a + bu}{-a}} + C, \quad \text{if } a < 0$

58 $\displaystyle\int \frac{\sqrt{a + bu}}{u}\, du = 2\sqrt{a + bu} + a\int \frac{du}{u\sqrt{a + bu}}$

59 $\displaystyle\int \frac{\sqrt{a + bu}}{u^2}\, du = -\frac{\sqrt{a + bu}}{u} + \frac{b}{2}\int \frac{du}{u\sqrt{a + bu}}$

60 $\displaystyle\int u^n\sqrt{a + bu}\, du = \frac{2}{b(2n + 3)}\left[u^n(a + bu)^{3/2} - na\int u^{n-1}\sqrt{a + bu}\, du\right]$

61 $\displaystyle\int \frac{u^n\, du}{\sqrt{a + bu}} = \frac{2u^n\sqrt{a + bu}}{b(2n + 1)} - \frac{2na}{b(2n + 1)}\int \frac{u^{n-1}\, du}{\sqrt{a + bu}}$

62 $\displaystyle\int \frac{du}{u^n\sqrt{a + bu}} = -\frac{\sqrt{a + bu}}{a(n - 1)u^{n-1}} - \frac{b(2n - 3)}{2a(n - 1)}\int \frac{du}{u^{n-1}\sqrt{a + bu}}$

(Continued on page A20)

TRIGONOMETRIC FORMS

63 $\int \sin^2 u \, du = \frac{1}{2}u - \frac{1}{4}\sin 2u + C$

64 $\int \cos^2 u \, du = \frac{1}{2}u + \frac{1}{4}\sin 2u + C$

65 $\int \tan^2 u \, du = \tan u - u + C$

66 $\int \cot^2 u \, du = -\cot u - u + C$

67 $\int \sin^3 u \, du = -\frac{1}{3}(2 + \sin^2 u)\cos u + C$

68 $\int \cos^3 u \, du = \frac{1}{3}(2 + \cos^2 u)\sin u + C$

69 $\int \tan^3 u \, du = \frac{1}{2}\tan^2 u + \ln|\cos u| + C$

70 $\int \cot^3 u \, du = -\frac{1}{2}\cot^2 u - \ln|\sin u| + C$

71 $\int \sec^3 u \, du = \frac{1}{2}\sec u \tan u + \frac{1}{2}\ln|\sec u + \tan u| + C$

72 $\int \csc^3 u \, du = -\frac{1}{2}\csc u \cot u + \frac{1}{2}\ln|\csc u - \cot u| + C$

73 $\int \sin^n u \, du = -\frac{1}{n}\sin^{n-1} u \cos u + \frac{n-1}{n}\int \sin^{n-2} u \, du$

74 $\int \cos^n u \, du = \frac{1}{n}\cos^{n-1} u \sin u + \frac{n-1}{n}\int \cos^{n-2} u \, du$

75 $\int \tan^n u \, du = \frac{1}{n-1}\tan^{n-1} u - \int \tan^{n-2} u \, du$

76 $\int \cot^n u \, du = \frac{-1}{n-1}\cot^{n-1} u - \int \cot^{n-2} u \, du$

77 $\int \sec^n u \, du = \frac{1}{n-1}\tan u \sec^{n-2} u + \frac{n-2}{n-1}\int \sec^{n-2} u \, du$

78 $\int \csc^n u \, du = \frac{-1}{n-1}\cot u \csc^{n-2} u + \frac{n-2}{n-1}\int \csc^{n-2} u \, du$

79 $\int \sin au \sin bu \, du = \frac{\sin(a-b)u}{2(a-b)} - \frac{\sin(a+b)u}{2(a+b)} + C$

80 $\int \cos au \cos bu \, du = \frac{\sin(a-b)u}{2(a-b)} + \frac{\sin(a+b)u}{2(a+b)} + C$

81 $\int \sin au \cos bu \, du = -\frac{\cos(a-b)u}{2(a-b)} - \frac{\cos(a+b)u}{2(a+b)} + C$

82 $\int u \sin u \, du = \sin u - u \cos u + C$

83 $\int u \cos u \, du = \cos u + u \sin u + C$

84 $\int u^n \sin u \, du = -u^n \cos u + n \int u^{n-1} \cos u \, du$

85 $\int u^n \cos u \, du = u^n \sin u - n \int u^{n-1} \sin u \, du$

86 $\int \sin^n u \cos^m u \, du = -\frac{\sin^{n-1} u \cos^{m+1} u}{n+m} + \frac{n-1}{n+m}\int \sin^{n-2} u \cos^m u \, du$

$\qquad\qquad\qquad\quad = \frac{\sin^{n+1} u \cos^{m-1} u}{n+m} + \frac{m-1}{n+m}\int \sin^n u \cos^{m-2} u \, du$

INVERSE TRIGONOMETRIC FORMS

87 $\int \sin^{-1} u \, du = u \sin^{-1} u + \sqrt{1 - u^2} + C$

88 $\int \cos^{-1} u \, du = u \cos^{-1} u - \sqrt{1 - u^2} + C$

89 $\int \tan^{-1} u \, du = u \tan^{-1} u - \frac{1}{2}\ln(1 + u^2) + C$

90 $\int u \sin^{-1} u \, du = \frac{2u^2 - 1}{4}\sin^{-1} u + \frac{u\sqrt{1 - u^2}}{4} + C$

91 $\int u \cos^{-1} u \, du = \frac{2u^2 - 1}{4}\cos^{-1} u - \frac{u\sqrt{1 - u^2}}{4} + C$

92 $\int u \tan^{-1} u \, du = \frac{u^2 + 1}{2}\tan^{-1} u - \frac{u}{2} + C$

93 $\int u^n \sin^{-1} u \, du = \frac{1}{n+1}\left[u^{n+1}\sin^{-1} u - \int \frac{u^{n+1}\, du}{\sqrt{1 - u^2}} \right], \quad n \neq -1$

94 $\int u^n \cos^{-1} u \, du = \frac{1}{n+1}\left[u^{n+1}\cos^{-1} u + \int \frac{u^{n+1}\, du}{\sqrt{1 - u^2}} \right], \quad n \neq -1$

95 $\int u^n \tan^{-1} u \, du = \frac{1}{n+1}\left[u^{n+1}\tan^{-1} u - \int \frac{u^{n+1}\, du}{1 + u^2} \right], \quad n \neq -1$

EXPONENTIAL AND LOGARITHMIC FORMS

96 $\displaystyle\int ue^{au}\,du = \frac{1}{a^2}(au - 1)e^{au} + C$

97 $\displaystyle\int u^n e^{au}\,du = \frac{1}{a}u^n e^{au} - \frac{n}{a}\int u^{n-1}e^{au}\,du$

98 $\displaystyle\int e^{au}\sin bu\,du = \frac{e^{au}}{a^2 + b^2}(a\sin bu - b\cos bu) + C$

99 $\displaystyle\int e^{au}\cos bu\,du = \frac{e^{au}}{a^2 + b^2}(a\cos bu + b\sin bu) + C$

100 $\displaystyle\int \ln u\,du = u\ln u - u + C$

101 $\displaystyle\int u^n \ln u\,du = \frac{u^{n+1}}{(n+1)^2}[(n+1)\ln u - 1] + C$

102 $\displaystyle\int \frac{1}{u\ln u}\,du = \ln|\ln u| + C$

HYPERBOLIC FORMS

103 $\displaystyle\int \sinh u\,du = \cosh u + C$

104 $\displaystyle\int \cosh u\,du = \sinh u + C$

105 $\displaystyle\int \tanh u\,du = \ln\cosh u + C$

106 $\displaystyle\int \coth u\,du = \ln|\sinh u| + C$

107 $\displaystyle\int \text{sech }u\,du = \tan^{-1}|\sinh u| + C$

108 $\displaystyle\int \text{csch }u\,du = \ln|\tanh \tfrac{1}{2}u| + C$

109 $\displaystyle\int \text{sech}^2 u\,du = \tanh u + C$

110 $\displaystyle\int \text{csch}^2 u\,du = -\coth u + C$

111 $\displaystyle\int \text{sech }u\tanh u\,du = -\text{sech }u + C$

112 $\displaystyle\int \text{csch }u\coth u\,du = -\text{csch }u + C$

FORMS INVOLVING $\sqrt{2au - u^2}$

113 $\displaystyle\int \sqrt{2au - u^2}\,du = \frac{u - a}{2}\sqrt{2au - u^2} + \frac{a^2}{2}\cos^{-1}\left(\frac{a - u}{a}\right) + C$

114 $\displaystyle\int u\sqrt{2au - u^2}\,du = \frac{2u^2 - au - 3a^2}{6}\sqrt{2au - u^2} + \frac{a^3}{2}\cos^{-1}\left(\frac{a - u}{a}\right) + C$

115 $\displaystyle\int \frac{\sqrt{2au - u^2}}{u}\,du = \sqrt{2au - u^2} + a\cos^{-1}\left(\frac{a - u}{a}\right) + C$

116 $\displaystyle\int \frac{\sqrt{2au - u^2}}{u^2}\,du = -\frac{2\sqrt{2au - u^2}}{u} - \cos^{-1}\left(\frac{a - u}{a}\right) + C$

117 $\displaystyle\int \frac{du}{\sqrt{2au - u^2}} = \cos^{-1}\left(\frac{a - u}{a}\right) + C$

118 $\displaystyle\int \frac{u\,du}{\sqrt{2au - u^2}} = -\sqrt{2au - u^2} + a\cos^{-1}\left(\frac{a - u}{a}\right) + C$

119 $\displaystyle\int \frac{u^2\,du}{\sqrt{2au - u^2}} = -\frac{(u + 3a)}{2}\sqrt{2au - u^2} + \frac{3a^2}{2}\cos^{-1}\left(\frac{a - u}{a}\right) + C$

120 $\displaystyle\int \frac{du}{u\sqrt{2ua - u^2}} = -\frac{\sqrt{2au - u^2}}{au} + C$

ANSWERS TO ODD-NUMBERED EXERCISES

CHAPTER 1

EXERCISES 1.1, PAGE 8

1 (a) > (b) < (c) = (d) > (e) = (f) <
3 (a) 3 (b) 7 (c) 7 (d) 3 (e) $\frac{22}{7} - \pi$ (f) -1 (g) 0
(h) 9 (i) $x - 5$ (j) $b - a$
5 (a) 4 (b) 8 (c) 8 (d) 12 **7** $(\frac{17}{5}, \infty)$
9 $[-2, \infty)$ **11** $(-\infty, -3) \cup (2, \infty)$ **13** $(5, \infty)$
15 $(-\frac{4}{5}, 3]$ **17** $[-3, 1)$ **19** $(-\infty, \frac{7}{2})$ **21** $(9.7, 10.3)$
23 $[\frac{5}{3}, 3]$ **25** $(-\infty, \frac{1}{25}) \cup (\frac{3}{5}, \infty)$ **27** $(-2, \frac{1}{3})$
29 $(-\infty, -4] \cup [-\frac{1}{2}, \infty)$ **31** $(-\infty, -\frac{1}{10}) \cup (\frac{1}{10}, \infty)$
33 $[-\frac{2}{3}, \frac{7}{2})$ **35** $\frac{140}{9} \leq C \leq \frac{80}{3}$ **37** $\frac{20}{9} \leq x \leq 4$
39 $5 < p < \frac{60}{7}$

EXERCISES 1.2, PAGE 15

1 (a) 5 (b) $(4, -\frac{1}{2})$ **3** (a) $\sqrt{26}$ (b) $(-\frac{1}{2}, -\frac{9}{2})$
5 (a) 5 (b) $(-\frac{11}{2}, -2)$ **7** 35

9 **11** **13**

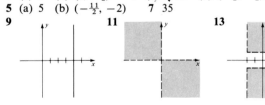

15 **17**

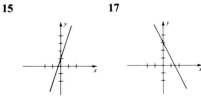

19 Symmetry:
y-axis

21 Symmetry:
y-axis

23 Symmetry:
origin

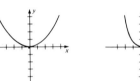

25 **27** **29**

31 **33** **35**

37 $(x - 3)^2 + (y + 2)^2 = 16$ **39** $x^2 + y^2 = 34$
41 $(x + 4)^2 + (y - 2)^2 = 4$
43 $(x - 1)^2 + (y - 2)^2 = 34$ **45** $(-2, 3), 3$
47 $(-3, 0), 3$ **49** $(\frac{1}{4}, -\frac{1}{4}), \sqrt{26}/4$

1 4   **3** Does not exist.
5 The slopes of opposite sides are equal.
7 *Hint*: Show that opposite sides are parallel and two adjacent sides are perpendicular.
9 $(-12, 0)$ **11** $x - 2y - 14 = 0$
13 $3x - 8y - 41 = 0$ **15** $x - 8y - 24 = 0$
17 (a) $x = 10$ (b) $y = -6$ **19** $5x + 2y - 29 = 0$
21 $5x - 7y + 15 = 0$
23 $m = \frac{3}{4}, b = 2$ **25** $m = -\frac{1}{2}, b = 0$

27 $m = -\frac{5}{4}, b = 5$ **29** $m = \frac{1}{3}, b = -\frac{7}{3}$

          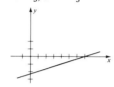

31 -3 **33** $x/(3/2) + y/(-3) = 1$
35 (a)

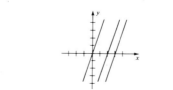

(b) Approximately
13 months

37 (a) R_0 is the resistance when $T = 0\ °C$ (b) $\frac{1}{273}$
   (c) $273\ °C$
39 (a) $L = \frac{29}{7}t + 24$ for t in months; 0.138 ft ≈ 1.66 in.
   (b) $W = \frac{20}{7}t + 3$; 0.095 ton/day or 190.476 lb/day

1 $2, -8, -3, 6\sqrt{2} - 3$
3 (a) $3a^2 - a + 2$ (b) $3a^2 + a + 2$ (c) $-3a^2 + a - 2$
   (d) $3a^2 + 6ah + 3h^2 - a - h + 2$
   (e) $3a^2 + 3h^2 - a - h + 4$ (f) $6a + 3h - 1$
5 (a) $a^2/(1 + 4a^2)$ (b) $a^2 + 4$ (c) $1/(a^4 + 4)$
   (d) $1/(a^2 + 4)^2$ (e) $1/(a + 4)$ (f) $1/\sqrt{a^2 + 4}$
7 $[\frac{5}{3}, \infty)$ **9** $[-2, 2]$
11 All real numbers except $0, 3$, and -3 **13** Yes
15 No   **17** Yes   **19** No   **21** Odd   **23** Even
25 Even   **27** Neither   **29** Neither
31 $(-\infty, \infty); (-\infty, \infty)$ **33** $(-\infty, \infty); \{-3\}$

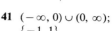 **35** $(-\infty, \infty); (-\infty, 4]$ **37** $[-2, 2]; [0, 2]$

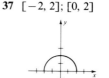

39 $(-\infty, 4) \cup (4, \infty);$ **41** $(-\infty, 0) \cup (0, \infty);$
   $(-\infty, 0) \cup (0, \infty)$ $\{-1, 1\}$

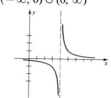   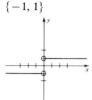

43 $(-\infty, 4]; [0, \infty)$ **45**

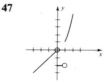   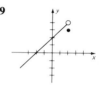

47   **49**

51 If $-1 < x < 1$, then two different points on the graph have x-coordinate x.
53 $V = 4x^3 - 100x^2 + 600x$ **55** $d = 2\sqrt{t^2 + 2500}$
57 (a) $y = \sqrt{2rh + h^2}$ (b) 1280.6 mi
59 $d = \sqrt{90,400 + x^2}$

1   **3**   **5**

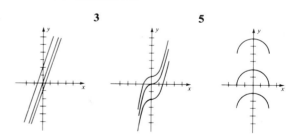

7   **9**

11 (a) (b) (c) (d)

(e) (f) (g) (h)

13 $3x^2 + 1/(2x - 3)$; $3x^2 - 1/(2x - 3)$; $3x^2/(2x - 3)$;
$3x^2(2x - 3)$
15 $2x$; $2/x$; $x^2 - (1/x^2)$; $(x^2 + 1)/(x^2 - 1)$
17 $2x^3 + x^2 + 7$; $2x^3 - x^2 - 2x + 3$;
$2x^5 + 2x^4 + 3x^3 + 4x^2 + 3x + 10$;
$(2x^3 - x + 5)/(x^2 + x + 2)$
19 $98x^2 - 112x + 37$; $-14x^2 - 31$ **21** $(x + 1)^3$; $x^3 + 1$
23 $3/(3x^2 + 2)^2 + 2$; $1/(27x^4 + 36x^2 + 14)$
25 $\sqrt{2x^2 + 7}$; $2x + 4$ **27** 5; -5 **29** $1/x^4$; $1/x^4$
31 x; x **35** $A = 36\pi t^2$ **37** $h = 5\sqrt{t^2 + 8t}$
39 $d = \sqrt{90{,}400 + (500 + 150t)^2}$

EXERCISES 1.6, PAGE 48

3 (a) II (b) III (c) IV
Exer. 5, 9: The order is sin, cos, tan, csc, sec, cot.
5 (a) $1, 0, -, 1, -, 0$
 (b) $\sqrt{2}/2, -\sqrt{2}/2, -1, \sqrt{2}, -\sqrt{2}, -1$
 (c) $0, 1, 0, -, 1, -$
 (d) $-\frac{1}{2}, \sqrt{3}/2, -\sqrt{3}/3, -2, 2\sqrt{3}/3, -\sqrt{3}$
7 $810°, -120°, 315°, 900°, 36°$
9 (a) $-\frac{4}{5}, \frac{3}{5}, -\frac{4}{3}, -\frac{5}{4}, \frac{5}{3}, -\frac{3}{4}$
 (b) $2\sqrt{13}/13, -3\sqrt{13}/13, -\frac{2}{3}, \sqrt{13}/2, -\sqrt{13}/3, -\frac{3}{2}$
 (c) $-1, 0, -, -1, -, 0$
31 (a) $\pi/2, 3\pi/2, \pi/4, 3\pi/4, 5\pi/4, 7\pi/4$
 (b) $90°, 270°, 45°, 135°, 225°, 315°$
33 (a) $0, \pi$ (b) $0°, 180°$
35 (a) $0, \pi, 2\pi/3, 4\pi/3$ (b) $0°, 180°, 120°, 240°$
37 (a) $\pi/2, 7\pi/6, 11\pi/6$ (b) $90°, 210°, 330°$
39 (a) $2\pi/3, \pi, 4\pi/3$ (b) $120°, 180°, 240°$
41 (a) $\pi/3, 5\pi/3$ (b) $60°, 300°$ **43** $\frac{84}{85}$ **45** $-\frac{36}{77}$
47 $\frac{240}{289}$ **49** $\frac{24}{7}$ **51** $\frac{1}{3}$ **53** No; $|\sin t| \le 1$
55 (a) $\sqrt{2}/2$ (b) $-\frac{1}{4}$
57 (a) (b)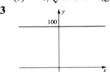

59 (a) 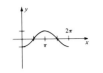 (b)

EXERCISES 1.7, PAGE 50

1 $(-\infty, -\frac{3}{5})$ **3** $[3.495, 3.505]$ **5** $(1, \frac{3}{2})$
7 $(-5, \frac{1}{3}) \cup (\frac{7}{5}, \infty)$ **9** (a) 12 (b) $(\frac{1}{2}, \frac{5}{2})$ (c) 7
11 Symmetry: y-axis **13** Symmetry: origin

15 **17**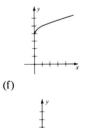

19 $(x + 4)^2 + (y + 3)^2 = 81$ **21** $(5, -7)$; 9
23 $6x - 7y + 24 = 0$ **25** $x = -4$
27 $(-\infty, 0) \cup (0, 1) \cup (1, \infty)$
29 The interval $(5, 7)$
31 (a) $1/\sqrt{2}$ (b) $\frac{1}{2}$ (c) 1 (d) $1/\sqrt[4]{2}$ (e) $1/\sqrt{1 - x}$
 (f) $-1/\sqrt{x + 1}$ (g) $1/\sqrt{x^2 + 1}$ (h) $1/(x + 1)$
33 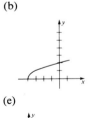 **35**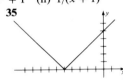

37 (a) (b) (c)

(d) (e) (f)

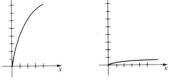

39 If $a \ne b$, then $5 - 7a \ne 5 - 7b$; that is, $f(a) \ne f(b)$.
41 $x^2 + 4 + \sqrt{2x + 5}$; $x^2 + 4 - \sqrt{2x + 5}$; $(x^2 + 4)\sqrt{2x + 5}$;
 $(x^2 + 4)/\sqrt{2x + 5}$; $2x + 9$; $\sqrt{2x^2 + 13}$
43 8 ft **45** (a) $-\sqrt{2}/2$ (b) $\frac{1}{2}$ (c) $\sqrt{3}/3$ (d) -1

49

CHAPTER 2

EXERCISES 2.1, PAGE 57

1 (a) $10a - 4$ (b) $y = 16x - 20$
3 (a) $3a^2$ (b) $y = 12x - 16$
5 (a) 3   **7** (a) $-1/a^2$
   (b) $y = 3x + 2$ (b) $x + 4y - 4 = 0$

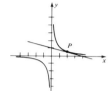

11 *Hint*: Show that the equation of the tangent line is
   $y = 3x + 2$.
13 Creature at $x = 3$; no hit
15 (a) In cm/sec: 11.8, 11.4, 11.04, 11.004   (b) 11 cm/sec
   (c) $(-\frac{3}{8}, \infty)$ (d) $(-\infty, -\frac{3}{8})$
17 In ft/sec: (a) -32 (b) $-32\sqrt{10} \approx -101$
19 In ft/sec: (a) 8   (b) 10   (c) $2\sqrt{29} \approx 10$

EXERCISES 2.2, PAGE 63

1 (a) 3   (b) 1   (c) Does not exist   (d) 2   (e) 2   (f) 2
3 (a) 1   (b) 1   (c) 1   (d) 3   (e) 3   (f) 3
5 (a) 1   (b) 0   (c) Does not exist   (d) 1   (e) 0
   (f) Does not exist
7 (a) 2   (b) 2   (c) 2   **9** (a) 2   (b) 2   (c) 2

11 (a) Does not exist
   (b) 1
   (c) Does not exist

13 (a) -1 (b) 1   (c) Does not exist   **15** -6
17 Does not exist   **19** 4   **21** $\frac{1}{9}$ **23** $\frac{17}{13}$ **25** 32
27 $2x$ **29** 12
31 (a) 2G's, the G-force at liftoff.
   (b) Left-hand limit of 8, the G-force just before second
     booster is released; right-hand limit of 1, the G-force
     just after second booster is released.
   (c) Left-hand limit of 3, the G-force just before engines are
     shut down; right-hand limit of 0, the G-force just after
     engines are shut down.

EXERCISES 2.3, PAGE 69

1 Given any ε, choose $\delta \leq \varepsilon/3$.
3 Given any ε, choose $\delta \leq \varepsilon/5$.
5 Given any ε, choose $\delta \leq \varepsilon/9$.
7, 9 Given any ε, let δ be any positive number.
11 Given any ε, choose $\delta \leq \varepsilon$.
19 Every interval $(3 - \delta, 3 + \delta)$ contains numbers for which the
   quotient equals 1, and other numbers for which the quotient
   equals -1.
21 Every interval $(-1 - \delta, -1 + \delta)$ contains numbers for
   which the quotient equals 3, and other numbers for which
   the quotient equals -3.
23 $1/x^2$ can be made as large as desired by choosing x
   sufficiently close to 0.
25 $1/(x + 5)$ can be made as large as desired by choosing x
   sufficiently close to -5.
27 There are many examples; one is $f(x) = (x^2 - 1)/(x - 1)$ if
   $x \neq 1$ and $f(1) = 3$.
29 Every interval $(a - \delta, a + \delta)$ contains numbers such that
   $f(x) = 0$ and other numbers such that $f(x) = 1$.

EXERCISES 2.4, PAGE 76

1 -13 **3** $5\sqrt{2} - 20$ **5** -2 **7** 0   **9** 15
11 -7 **13** $\frac{1}{12}$ **15** 8   **17** -23 **19** 2
21 $\frac{72}{7}$ **23** -2 **25** $-\frac{1}{8}$ **27** 0   **29** -810
31 (a) 0   (b) Does not exist   (c) Does not exist
33 (a) 0   (b) 0   (c) 0   **35** 3   **37** 1   **39** $\frac{1}{8}$
41 $(-1)^{n-1}; (-1)^n$ **43** 0; 0

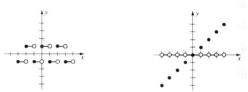

45 1; 0
51 (a) 0   (b) Temperature cannot be less than absolute zero.
53 (a) Does not exist.   (b) As $p \to f^+$, the image moves farther
   and farther to the right.

EXERCISES 2.5, PAGE 81

1 1   **3** $\frac{1}{8}$ **5** $\frac{2}{3}$ **7** 0   **9** $-\frac{3}{4}$ **11** 0
13 7   **15** 1   **17** 0   **19** 2   **21** 1   **23** 1
25 -1

EXERCISES 2.6, PAGE 88

17 $\{x : x \neq \frac{3}{2}, x \neq -1\}$ **19** $[\frac{3}{2}, \infty)$
21 $(-\infty, -1) \cup (1, \infty)$ **23** $\{x : x \neq -9\}$
25 $\{x : x \neq 0, x \neq 1\}$ **27** $[-5, -3] \cup [3, 4) \cup (4, 5]$
29 Removable discontinuity at $x = 1$
31 No discontinuities
33 Infinite discontinuity at $x = 1$
35 $\frac{5}{2}$ **37** $c = d = 8$
39 $f \circ g$ is continuous at 0; $g \circ f$ is not continuous at 0.
41 f is discontinuous on any open interval containing the
   origin.

43 No; $\lim_{x\to 3} f(x)$ does not exist.
45 Yes; all conditions of (2.28) are fulfilled.

47 Discontinuous at
50,000 $(3 + x)$
for $x = 0, 1, 2, \ldots$

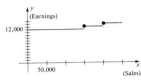

49 $c = \sqrt[3]{w - 1}$ **51** $c = \sqrt{w} - 2$
55 Show that $g \approx 9.79745$ when $\theta = 35°$ and $g \approx 9.80180$ when $\theta = 40°$.
57 (a) $T > 0°$ for $0 < t < 12$; $T < 0°$ for $12 < t < 24$
    (b) $T = 32.4$ at noon and $T = 29.75$ at 1 P.M.
59 (a) $(-\infty, 0) \cup (0, 1) \cup (3, \infty)$ (b) The interval $(1, 3)$

EXERCISES 2.7, PAGE 90

1 13   **3** $-4 - \sqrt{14}$ **5** $\frac{7}{8}$ **7** $\frac{32}{3}$
9 Does not exist   **11** 3   **13** -1 **15** $4a^3$
17 $-\frac{3}{16}$ **19** 0   **21** $\frac{2}{3}$ **23** $\frac{3}{5}$ **25** 2
27 (a) 6   (b) 4   (c) Does not exist
29 (a) $\frac{1}{11}$ (b) -1 (c) Does not exist
31 (a) 1   (b) 3   (c) Does not exist
27   **29**   **31**

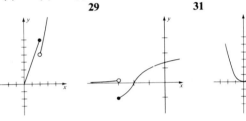

33 -6 **37** $\mathbb{R}$ **39** $[-3, -2) \cup (-2, 2) \cup (2, 3]$
41 At $4, -4$ **43** At $0, 2$ **47** $c = 1/\sqrt{w}$
49 (a) $6a - 2$ (b) $y = 16x - 27$

CHAPTER 3

EXERCISES 3.1, PAGE 98

1 (a) 0   (b) $\mathbb{R}$ (c) $y = 37$
3 (a) 9   (b) $\mathbb{R}$ (c) $y = 9x - 2$
5 (a) $8 - 10x$ (b) $\mathbb{R}$ (c) $y = -2x + 7$
7 (a) $-1/(x - 2)^2$ (b) $\{x: x \neq 2\}$ (c) $y = -x$
9 (a) $3/(2\sqrt{3x + 1})$ (b) $(-\frac{1}{3}, \infty)$ (c) $4y = 3x + 5$
11 $-7/(2\sqrt{x^3})$ **13** $6x^2 - 4$ **15** $2a$ **17** $-12/a^3$
19 $-1/(a + 5)^2$
23 $\{x: x \neq 0\}$ **25** $\{x: x \neq 0, x \neq \pm 1\}$

      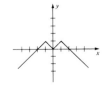

27 f is not differentiable at ± 1, ± 2

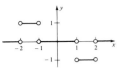

31 If $f(x) = ax + b$, then $f'(x) = a$ has degree 0. If $f(x)$ has degree 2, then $f'(x)$ has degree 1. If $f(x)$ has degree 3, then $f'(x)$ has degree 2.

EXERCISES 3.2, PAGE 106

1 $20x + 9$ **3** $-1 + 8s - 20s^3$ **5** $10x^4 + 9x^2 - 28x$
7 $18r^5 - 21r^2 + 4r$ **9** $23/(3x + 2)^2$
11 $(-27z^2 + 12z + 70)/(2 - 9z)^2$
13 $9x^2 - 4x + 4$ **15** $2t - (2/t^3)$
17 $416x^3 - 195x^2 + 64x - 20$ **19** $6v^2/(v^3 + 1)^2$
21 $-(1 + 2x + 3x^2)/(1 + x + x^2 + x^3)^2$
23 $72z^5 - 64z^3 - 18z^2 - 70z - 7$ **25** $-\frac{4}{81}s^{-5}$
27 $10(5x - 4)$ **29** $(6 - 20t - 21t^2)/[5(2 + 7t^2)^2]$ for $t \neq 0$
31 $2 - (4/x^2) - (6/x^3)$ **33** $(-3x + 2)/x^3$
35 (a) $y = 5$ (b) $5x + 2y - 10 = 0$
    (c) $4x - 5y + 13 = 0$
37 (a) $\frac{2}{3}, -2$ (b) $0, -\frac{4}{3}$ **39** $-\frac{1}{2}$
41 $y - 9 = 2(x - 5)$, point of tangency is $(1, 1)$;
    $y - 9 = 18(x - 5)$, point of tangency is $(9, 81)$.
43 (a) 1   (b) -3 (c) -4 (d) 11   (e) $-\frac{1}{25}$
45 (a) -4 (b) 1   (c) -20 (d) $-\frac{1}{4}$
49 $(8x - 1)(x^2 + 4x + 7)(3x^2) + (8x - 1)(2x + 4)(x^3 - 5) + 8(x^2 + 4x + 7)(x^3 - 5)$
51 (a) $y = -\frac{38}{6125}x^2 + x + 15$ (b) 55.3 ft
53 $y = \frac{1}{8000}x^2$; $B(800, 80)$
55 In ft/sec: (a) 4, 10, 18   (b) $6\sqrt{5} \approx 13.4$
57 (a) Second runner   (b) Runners tie   (c) First runner

EXERCISES 3.3, PAGE 115

1 (a) $\frac{1}{4}$ cm/min   (b) 36π cm^3/min   (c) 12π cm^2/min
3 In (beats/min)/sec: (a) 7   (b) 15   (c) 23
5 In cm^2/sec: (a) 3200π (b) 6400π (c) 9600π
7 At $t = \sqrt{6}$; rate of growth is positive for $0 < t < \sqrt{6}$, negative for $t > \sqrt{6}$.
Exer. 9–13: The symbols $[a, b)$, $(a, b]$, and (a, b) denote time intervals.
9 $v(t) = 6t - 12$, $a(t) = 6$; to the left in $[0, 2)$; to the right in $(2, 5]$
11 $v(t) = 3t^2 - 9$, $a(t) = 6t$; to the right in $[-3, -\sqrt{3})$; to the left in $(-\sqrt{3}, \sqrt{3})$; to the right in $(\sqrt{3}, 3]$
13 $v(t) = 8t^3 - 12t$, $a(t) = 24t^2 - 12$; to the left in $[-2, -\sqrt{3/2})$; to the right in $(-\sqrt{3/2}, 0)$; to the left in $(0, \sqrt{3/2})$; to the right in $(\sqrt{3/2}, 2]$
15 $v(t) = 144 - 32t$, $a(t) = -32$; $v(3) = 48$, $a(3) = -32$; 324 ft; at $t = 9$
17 -0.12 unit/ft
21 $-2k/d^3$ with k the constant of proportionality

23 (a) 806
(b) $c(x) = (800/x) + 0.04 + 0.0002x$;
$C'(x) = 0.04 + 0.0004x$;
$c(100) = 8.06$; $C'(100) = 0.08$

25 (a) 11,250
(b) $c(x) = (250/x) + 100 + 0.001x^2$;
$C'(x) = 100 + 0.003x^2$;
$c(100) = 112.50$; $C'(100) = 130$

27 $C'(5) = 46.00$; $C(6) - C(5) = 46.67$

EXERCISES 3.4, PAGE 120

1 $-4 \sin x$ **3** $t^2(t \cos t + 3 \sin t)$
5 $(\theta \cos \theta - \sin \theta)/\theta^2$
7 $2 \cot x - 2x \csc^2 x + 2x \tan x + x^2 \sec^2 x$
9 $(2 \sin z)/(1 + \cos z)^2$
11 $\cot x - \csc x \cot^2 x - x \csc^2 x - \csc^3 x$ **13** $-\sin x$
15 $(\sec^2 x + x^2 \sec^2 x - 2x \tan x)/(1 + x^2)^2$
17 $-\csc^2 v$ **19** $\sin \varphi + \sec \varphi \tan \varphi$
21 $y - \sqrt{2} = \sqrt{2}[x - (\pi/4)]$; $y - \sqrt{2} = -(\sqrt{2}/2)[x - (\pi/4)]$
23 $(\pi/4, \sqrt{2}), (5\pi/4, -\sqrt{2})$ **25** $(\pi/4, 2\sqrt{2})$
27 (a) $(\pi/6) + 2\pi n$ and $(5\pi/6) + 2\pi n$ for any integer n
(b) $y = x + 2$

EXERCISES 3.5, PAGE 127

1 (a) $(4x - 4) \Delta x + 2(\Delta x)^2$ (b) -0.72
3 (a) $-[(2x + \Delta x) \Delta x]/[x^2(x + \Delta x)^2]$
(b) $-\frac{189}{9801} \approx -0.019$
5 (a) $(6x + 5) \Delta x + 3 (\Delta x)^2$ (b) $(6x + 5) dx$ (c) $-3 (\Delta x)^2$
7 (a) $-\Delta x/[x(x + \Delta x)]$ (b) $(-dx)/x^2$
(c) $-(\Delta x)^2/[x^2(x + \Delta x)]$
9 (a) $-9 \Delta x$ (b) $-9 \Delta x$ (c) 0
11 (a) $4x^3 (\Delta x) + 6x^2 (\Delta x)^2 + 4x (\Delta x)^3 + (\Delta x)^4$
(b) $4x^3 \Delta x$ (c) $-6x^2 (\Delta x)^2 - 4x (\Delta x)^3 - (\Delta x)^4$
13 0.06
15 $dy = (2 \cos \theta - \sin \theta) d\theta$; $(\sqrt{3} - 0.5)(-\pi/60) \approx -0.065$
17 $\pm 1.92\pi \approx \pm 6.03$ in^2; ± 0.0075; $\pm 0.75\%$
19 30 in^3; 30.301 in^3
21 3301.661; ± 11.459; ± 0.00347; $\pm 0.347\%$
23 $1/(50\pi) \approx 0.00637$ **25** -1 **27** 0.92; 0.92236816
29 dA is the shaded region

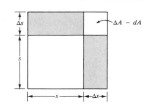

31 40% increase
33 $-2\pi/135 \approx -0.047$ **35** $dF \approx 0.27$ lb
37 $\pm \pi/9 \approx \pm 0.35$ ft   **39** $\pm 0.19°$

EXERCISES 3.6, PAGE 136

1 $3(x^2 - 3x + 8)^2(2x - 3)$ **3** $-40(8x - 7)^{-6}$
5 $-(7x^2 + 1)/(x^2 - 1)^5$
7 $5(8x^3 - 2x^2 + x - 7)^4(24x^2 - 4x + 1)$
9 $17,000(17v - 5)^{999}$ **11** $2x \cos (x^2 + 2)$

13 $-15 \cos^4 3\theta \sin 3\theta$ **15** $x \sin (1/x) + 3x^2 \cos (1/x)$
17 $8 \sec^2 (8v + 3)$
19 $4(2z + 1) \sec [(2z + 1)^2] \tan [(2z + 1)^2]$
21 $(2 - 3s^2) \csc^2 (s^3 - 2s)$
23 $-6x \sin (3x^2) - 6 \cos 3x \sin 3x$
25 $-4 \csc^2 2\varphi \cot 2\varphi$ **27** $-5z^2 \csc^2 5z + 2z \cot 5z$
29 $3 \tan^3 \theta \sec^3 \theta + 2 \tan \theta \sec^5 \theta$
31 $32x(6x - 7)^3(8x^2 + 9) + 18(8x^2 + 9)^2(6x - 7)^2$, which
reduces to $(6x - 7)^2(8x^2 + 9)(336x^2 - 224x + 162)$
33 $6[z^2 - (1/z^2)]^5[2z + (2/z^3)]$
35 $25 (\sin 5x - \cos 5x)^4(\cos 5x + \sin 5x)$
37 $-9 \cot^2 (3w + 1) \csc^2 (3w + 1)$ **39** $4/(1 - \sin 4w)$
41 $6 \tan 2x \sec^2 2x (\tan 2x - \sec 2x)$
43 (a) $y - 81 = 864(x - 2)$; $y - 81 = -\frac{1}{864}(x - 2)$
(b) $(1, 1), (\frac{1}{2}, 0), (\frac{3}{2}, 0)$
45 (a) $y - 1 = 20(x - 1)$; $y - 1 = -\frac{1}{20}(x - 1)$ (b) $(\frac{1}{2}, 0)$
47 $0, \pm 2\pi/5, \pm 2\pi/3, \pm 4\pi/5, \pm 6\pi/5, \pm 4\pi/3, \pm 8\pi/5, \pm 2\pi$
49 $5; 8; \frac{1}{8}$; the particle oscillates on a coordinate line between 5
and -5, completing one such oscillation every 8 seconds.
51 $6; 3; \frac{1}{3}$; the particle oscillates on a coordinate line between 6
and -6, completing one such oscillation every 3 seconds.
53 $79,200\pi$; $3600\sqrt{2\pi}$
55 (a) $y = 4.5 \sin [(\pi/6)(t - 10)] + 7.5$ (b) 1.178 ft/hr
57 (a) In in./sec: $0, -\pi, 0, \pi, 0$
(b) Rising during $(n, n + 1)$ when n is an odd integer; falling
during $(n, n + 1)$ when n is an even integer.
59 $dK/dt = mv(dv/dt)$ **61** $dW/dt \approx -0.819$ lb/sec
63 $-4; 15$ **65** $-\frac{2}{5}$

EXERCISES 3.7, PAGE 142

1 $f(x) = -\frac{2}{5}x^2 + 2x + \frac{4}{5}$, $\mathbb{R}$
3 $f(x) = \sqrt{16 - x^2}$, $[-4, 4]$. There are other answers.
5 $f(x) = x + \sqrt{x}$, $[0, \infty)$
7 $f(x) = 1 - 2\sqrt{x} + x$, $[0, 1]$ **9** $-8x/y$
11 $-(6x^2 + 2xy)/(x^2 + 3y^2)$ **13** $(10x - y)/(x + 8y)$
15 $1/(6 \sin 3y \cos 3y - 1) = 1/(3 \sin 6y - 1)$
17 $-(2xy^3 + 4y + 1)/(3x^2y^2 + 4x - 6)$
19 $(\cos y)/(2y + x \sin y)$ **21** $-\sqrt{2/5}$ **23** -1
25 $4x - y + 16 = 0$ **27** $y + 3 = -\frac{36}{23}(x - 2)$
29 If it did, then $[f(x)]^2 + x^2 = -1$, an impossibility.
31 (a) Infinitely many
(b) One, $f(x) = 0$, with domain $x = 0$
(c) None
33 0.09

EXERCISES 3.8, PAGE 147

1 $\frac{2}{3}x^{-1/3} + 6x^{1/2}$ **3** $8r^2(8r^3 + 27)^{-2/3}$
5 $-5v^4(v^5 - 32)^{-6/5}$ **7** $f'(x) = 1/\sqrt{2x}$
9 $15\sqrt{z} - 1/(z\sqrt[3]{z})$ **11** $(w^2 + 4w - 9)/(2w^{5/2})$
13 $(\cos \sqrt{x})/(2\sqrt{x}) + (\cos x)/(2\sqrt{\sin x})$
15 $(8 \sin \sqrt{3 - 8\theta} \cos \sqrt{3 - 8\theta})/\sqrt{3 - 8\theta}$
17 $x \sec^2 \sqrt{x^2 + 1} + (x \tan \sqrt{x^2 + 1})/\sqrt{x^2 + 1}$
19 $(2 \sec \sqrt{4x + 1} \tan \sqrt{4x + 1})/\sqrt{4x + 1}$
21 $(-3 \csc^2 3x \cot 3x)/\sqrt{4 + \csc^2 3x}$

23 $(18 - 12x)/(4x^2 + 9)^{3/2}$ 25 $2x + \sqrt{3}y - 1 = 0$
27 (4, 2)   29 $-\sqrt{y/x}$
31 $4x\sqrt{\sin y}/(4y\sqrt{\sin y} - \cos y)$
33 $dW/dt = (1.644 \times 10^{-4})L^{1.74}\,(dL/dt)$; 7.876 cm/month
35 60π cm²; ± 1.508; $\pm 0.8\%$
37 $f'(x) = \dfrac{x - 1}{|x - 1|}$; {x: x ≠ 1}
39 $f'(x) = \dfrac{2x(x^2 - 1)}{|x^2 - 1|}$; {x: x ≠ ±1}

41 $3/(2\sqrt{3z + 1})$; $-9/[4\sqrt{(3z + 1)^3}]$
43 $20(4r + 7)^4$; $320(4r + 7)^3$
45 $3\sin^2 x \cos x$; $3\sin x(2\cos^2 x - \sin^2 x)$
47 $120x^2 + 18$ 49 $-12x\sin(x^2) - 8x^3\cos(x^2)$
51 $(2xy^3 - 2x^4)/y^5 = -2x/y^5$
53 $10(y^2 - 3xy + x^2)/(2y - 3x)^3 = 40/(2y - 3x)^3$
55 $f^{(n)}(x) = (-1)^n n!/x^{n+1}$; $f^{(n)}(1) = (-1)^n n!$
57 The degree of f' is $n - 1$, of f'' is $n - 2$, …, of $f^{(n)}$ is 0. Since $f^{(n)}$ is a constant function, all higher derivatives are zero.
59 $D_x^2 y = f''(g(x))(g'(x))^2 + f'(g(x))g''(x)$

EXERCISES 3.9, PAGE 152

1 $-3\sqrt{336}/8 \approx -6.9$ ft/sec   3 $20/9\pi \approx 0.71$ ft/min
5 $\frac{64}{11}$ ft/sec; $\frac{20}{11}$ ft/sec   7 $-7442\pi \approx -23{,}368$ in³/hr
9 $\frac{10}{3}$ ft/sec   11 Increasing at a rate of 5 in³/min
13 $15\sqrt{3}/32 \approx 0.8$ ft/min   15 $5/(8\pi) \approx 0.1989$ in./yr
17 $-4\sqrt{\sqrt{3}/600} \approx -0.215$ cm/min   19 π m/sec
21 $\frac{11}{1600}$ ohm/sec   23 $13.37/112\pi \approx 0.38$ ft/min
25 64 ft/sec   27 $(6 + \sqrt{2})180/\sqrt{10 + 3\sqrt{2}} \approx 353.6$ mi/hr
29 $-27/(25\pi) \approx 0.344$ in./hr   31 $10{,}000\pi/135 \approx 232.7$ ft/sec
33 $\sqrt{3\pi}/10 \approx 0.544$ in²/min   35 70.63 mi/hr
37 $1000\pi/3$ ft/sec ≈ 714 mi/hr
39 In mi/hr: ground speed $= \frac{175}{88}(d\theta/dt)$
41 (a) $2v\dfrac{dv}{dt} = Rg\,\dfrac{1 + \cos^2\theta}{\cos^2\theta}\,\dfrac{d\theta}{dt}$ (b) $2v\dfrac{dv}{dt} = \dfrac{Rg}{L}\,\dfrac{1 + \cos^2\theta}{\cos^3\theta}\,\dfrac{dR}{dt}$

EXERCISES 3.10, PAGE 158

1 1.2599   3 0.5641   5 1.3315   7 −1.7321
9 1.4958   11 ±3.34   13 −1, 1.35
15 −1.88, 0.35, 1.53   17 2.71
19 (a) 3, 3.1425465, 3.1415926, 3.1415926, 3.1415926
   (b) They approach 2π.

EXERCISES 3.11, PAGE 159

1 $-24x/(3x^2 + 2)^2$ 3 $6x^2 - 7$ 5 $3/\sqrt{6t + 5}$
7 $\frac{1}{3}(7z^2 - 4z + 3)^{-2/3}(14z - 4)$ 9 $-144x/(3x^2 - 1)^5$
11 $-2(y^2 - y^{-2})^{-3}(2y + 2y^{-3})$ 13 $\frac{12}{5}(3x + 2)^{-1/5}$
15 $4(8s^2 - 4)^3(72s^4 - 108s^2 + 16s)/(1 - 9s^3)^5$
17 $(x^6 + 1)^4(3x + 2)^2(99x^6 + 60x^5 + 9)$

19 $(-\sin 2y)/\sqrt{1 + \cos 2y}$
21 $24x^2 \sin(4x^3)\cos(4x^3) = 12x^2 \sin 8x^3$
23 $5\sec x(\sec x + \tan x)^5$ 25 $2x(-x\csc^2 2x + \cot 2x)$
27 $(1/z^2)\csc(1/z)\cot(1/z) - \sin z$
29 $(9s - 1)^3(108s^2 - 139s + 39)$
31 $-53/[2\sqrt{(2w + 5)(7w - 9)^3}]$
33 $4\theta^3 \tan(\theta^2)\sec^2(\theta^2) + 2\theta\tan^2(\theta^2)$
35 $-x^{-2/3}(\cos\sqrt[3]{x} + \sin\sqrt[3]{x})(\cos\sqrt[3]{x} - \sin\sqrt[3]{x})^2$
37 $\csc u(1 - \cot u + \csc u)/(\cot u + 1)^2$ 39 $10\tan 5x \sec^2 5x$
41 $\theta^{-3/4}\tan^3(\sqrt[4]{\theta})\sec^2(\sqrt[4]{\theta})$
43 $(4xy^2 - 15x^2)/(12y^2 - 4x^2y)$
45 $1/[\sqrt{x}(3\sqrt{y} + 2)]$ 47 $\dfrac{\cos(x + 2y) - y^2}{2xy - 2\cos(x + 2y)}$
49 $9x - 4y - 12 = 0$; $4x + 9y = 70$
51 $(7\pi/12) + \pi n$ and $(11\pi/12) + \pi n$ for any integer n
53 $15x^2 + (2/\sqrt{x})$; $30x - (1/\sqrt{x^3})$; $30 + 3/(2\sqrt{x^5})$
55 $5(y^2 - 4xy - x^2)/(y - 2x)^3 = -40/(y - 2x)^3$
57 $f^{(n)}(x) = n!/(1 - x)^{n+1}$
59 ± 0.06; $\pm 1.5\%$ 61 -0.57
63 (a) 2 (b) −7 (c) −14 (d) 21 (e) $-\frac{10}{9}$ (f) $-\frac{19}{27}$
65 $C'(100) = 116$; $C(101) - C(100) = 116.11$
67 2%
69 $v(t) = 3(1 - t^2)/(t^2 + 1)^2$; $a(t) = 6t(t^2 - 3)/(t^2 + 1)^3$; to the left in $[-2, -1)$; to the right in $(-1, 1)$; to the left in $(1, 2]$.
71 5/6 ft³/min   73 $dp/dv = -p/v$
75 (a) $h = 60 - 50\cos(\pi/15)t$ (b) 10.4 ft/sec   77 4.493

CHAPTER 4

EXERCISES 4.1, PAGE 168

1 5; −3   3 1; −3
5 (a) Since $f'(x) = 1/(3x^{2/3})$, $f'(0)$ does not exist. If $a \neq 0$, then $f'(a) \neq 0$. Hence 0 is the only critical number of f. The number $f(0) = 0$ is not a local extremum, since $f(x) < 0$ if $x < 0$ and $f(x) > 0$ if $x > 0$.
(b) The facts that 0 is the only critical number and that there is a vertical tangent line at (0, 0) follow as in part (a). The number $f(0) = 0$ is a local minimum, since $f(x) > 0$ if $x \neq 0$.
7 There is a critical number 0, but $f(0)$ is not a local extremum, since $f(x) < f(0)$ if $x < 0$ and $f(x) > f(0)$ if $x > 0$. The function is continuous at every number a, since $\lim_{x \to a} f(x) = f(a)$. If $0 < x_1 < x_2 < 1$, then $f(x_1) < f(x_2)$ and hence there is neither a maximum nor minimum on (0, 1). This does not contradict (4.3), because the interval (0, 1) is open.
9 $\frac{3}{8}$ 11 $\frac{5}{3}$ and −2   13 2   15 4 and −4 (not 0)
17 0, $\frac{15}{7}$, and $\frac{5}{2}$ 19 None
Exer. 21-29: n denotes any integer.
21 $(2\pi/3) + 2\pi n$, $(4\pi/3) + 2\pi n$, and πn
23 $(\pi/6) + 2\pi n$, $(5\pi/6) + 2\pi n$, and $(3\pi/2) + 2\pi n$
25 $(3\pi/2) + 2\pi n$ 27 πn 29 None
31 0 and $\pm\sqrt{n\pi - 1}$ for $n = 1, 2, 3, \ldots$

33 If $f(x) = cx + d$ and $c \neq 0$, then $f'(x) = c \neq 0$. Hence there are no critical numbers. On $[a, b]$ the function has absolute extrema at a and b.

35 If $x = n$ is an integer, then $f'(n)$ does not exist. Otherwise, $f'(x) = 0$ for every $x \neq n$.

37 If $f(x) = ax^2 + bx + c$ and $a \neq 0$, then $f'(x) = 2ax + b$. Hence $-b/(2a)$ is the only critical number of f.

39 Since $f'(x) = nx^{n-1}$, the only possible critical number is $x = 0$, and $f(0) = 0$. If n is even then $f(x) > 0$ if $x \neq 0$ and hence 0 is a local minimum. If n is odd, then 0 is not an extremum, since $f(x) < 0$ if $x < 0$ and $f(x) > 0$ if $x > 0$.

EXERCISES 4.2, PAGE 173

1 f is not differentiable at the number 0 in the interval $(-1, 1)$.

3 f is not continuous on the interval $[-1, 4]$.

5 $c = 2$ **7** $c = 0$ **9** $c = 2$ **11** $c = 2$

13 Hypotheses not satisfied   **15** $c = 2$

17 $c = (2 - \sqrt{7})/3$

19 If $f(x) = cx + d$, then $f'(x) = c$ for every x. Moreover,
$$f(b) - f(a) = (cb + d) - (ca + d)$$
$$= c(b - a) = f'(x)(b - a).$$

21 If f has degree 3, then $f'(x)$ is a polynomial of degree 2. Consequently, the equation $f(b) - f(a) = f'(x)(b - a)$ has at most two solutions, x_1 and x_2. If f has degree 4, there are at most three such solutions. If f has degree n, there are at most $n - 1$ solutions.

EXERCISES 4.3, PAGE 180

1 MAX : $f(-\frac{7}{8}) = \frac{129}{16}$; increasing on $(-\infty, -\frac{7}{8}]$; decreasing on $[-\frac{7}{8}, \infty)$

3 MAX: $f(-2) = 29$; MIN: $f(\frac{5}{3}) = -\frac{548}{27}$; increasing on $(-\infty, -2]$ and $[\frac{5}{3}, \infty)$; decreasing on $[-2, \frac{5}{3}]$

5 MAX: $f(0) = 1$; MIN: $f(-2) = -15$ and $f(2) = -15$; increasing on $[-2, 0]$ and $[2, \infty)$; decreasing on $(-\infty, -2]$ and $[0, 2]$

  1   **3**   **5**

 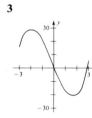 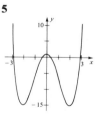

7 MIN: $f(-1) = -3$; increasing on $[-1, \infty)$; decreasing on $(-\infty, -1]$ *(see graph)*

9 MAX: $f(0) = 0$; MIN: $f(-\sqrt{3}) = f(\sqrt{3}) = -3$; increasing on $[-\sqrt{3}, 0]$ and $[\sqrt{3}, \infty)$; decreasing on $(-\infty, -\sqrt{3}]$ and $[0, \sqrt{3}]$ *(see graph)*

11 MAX: $f(\frac{7}{4}) \approx 42$; MIN: $f(0) = 2$ and $f(7) = 2$; increasing on $[0, \frac{7}{4}]$ and $[7, \infty)$; decreasing on $(-\infty, 0]$ and $[\frac{7}{4}, 7]$ *(see graph)*

  7   **9**   **11**

 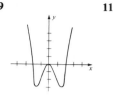

13 MAX: $f(-1) = -4$; MIN: $f(1) = 4$; increasing on $(-\infty, -1]$ and $[1, \infty)$; decreasing on $[-1, 0)$ and $(0, 1]$.

15 MAX: $f(\frac{3}{5}) \approx 0.346$; MIN: $f(1) = 0$; increasing on $(-\infty, \frac{3}{5}]$ and $[1, \infty)$; decreasing on $[\frac{3}{5}, 1]$

17 MAX: $f(\pi/4) = \sqrt{2}$; MIN: $f(5\pi/4) = -\sqrt{2}$; increasing on $[0, \pi/4]$ and $[5\pi/4, 2\pi]$; decreasing on $[\pi/4, 5\pi/4]$

  13   **15**   **17**

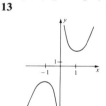  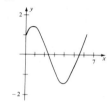

19 MAX: $f(5\pi/3) = (5\pi + 3\sqrt{3})/6$; MIN: $f(\pi/3) = (\pi - 3\sqrt{3})/6$; decreasing on $[0, \pi/3]$ and $[5\pi/3, 2\pi]$; increasing on $[\pi/3, 5\pi/3]$

21 MAX: $f(\pi/6) = 3\sqrt{3}/2$; MIN: $f(5\pi/6) = -3\sqrt{3}/2$; increasing on $[0, \pi/6]$ and $[5\pi/6, 2\pi]$; decreasing on $[\pi/6, 5\pi/6]$

  19   **21**

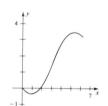 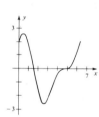

23 MAX: $f(-\sqrt{3}) = (6\sqrt{3})^{1/3}$; MIN: $f(\sqrt{3}) = -(6\sqrt{3})^{1/3}$

25 MAX: $f(-1) = 0$; MIN: $f(\frac{5}{7}) = -9^3 12^4/7^7$ **27** None

29 MIN: $f(0) = 1$ **31** MAX: $f(\pi/4) = 1$

33 $-11\pi/6, -7\pi/6, \pi/6$, and $5\pi/6$

35 (a) MAX: $f(-\frac{7}{8}) = \frac{129}{16}$; MIN: $f(1) = -6$
    (b) MAX: $f(-\frac{7}{8}) = \frac{129}{16}$; MIN: $f(-4) = -31$
    (c) MAX: $f(0) = 5$; MIN: $f(5) = -130$

37 (a) MAX: $f(-1) = 20$; MIN: $f(1) = -16$
    (b) MAX: $f(-2) = 29$; MIN: $f(-4) = -31$
    (c) MAX: $f(5) = 176$; MIN: $f(\frac{5}{3}) = -\frac{548}{27}$

39

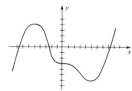

41 (a) At $t = \sqrt{10.5} \approx 3.24$ yr; 210; at $t = 5$ yr

(b)

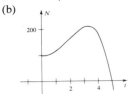

45 $x = y = 2$

EXERCISES 4.4, PAGE 189

Exer. 1–11: The notations CU and CD mean that the graph is concave upward or downward, respectively, in the interval that follows. PI denotes point(s) of inflection.

1 MAX: $f(\frac{1}{3}) = \frac{31}{27}$; MIN: $f(1) = 1$; CD on $(-\infty, \frac{2}{3})$; CU on $(\frac{2}{3}, \infty)$; x-coordinate of PI is $\frac{2}{3}$.

3 MIN: $f(1) = 5$; CU on $(-\infty, 0)$ and $(\frac{2}{3}, \infty)$; CD on $(0, \frac{2}{3})$; x-coordinates of PI are 0 and $\frac{2}{3}$.

1   **3**

5 MAX: $f(0) = 0$ (by first derivative test); MIN: $f(-\sqrt{2}) = f(\sqrt{2}) = -8$; CU on $(-\infty, -\sqrt{\frac{6}{5}})$ and $(\sqrt{\frac{6}{5}}, \infty)$; CD on $(-\sqrt{\frac{6}{5}}, \sqrt{\frac{6}{5}})$; x-coordinates of PI are $\pm\sqrt{\frac{6}{5}}$.

7 MAX: $f(0) = 1$; MIN: $f(-1) = f(1) = 0$; CU on $(-\infty, -1/\sqrt{3})$ and $(1/\sqrt{3}, \infty)$; CD on $(-1/\sqrt{3}, 1/\sqrt{3})$; x-coordinates of PI are $\pm 1/\sqrt{3}$.

5   **7**

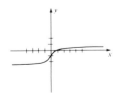

9 No local extrema; CU on $(-\infty, 0)$; CD on $(0, \infty)$; PI $(0, -1)$

11 No MAX or MIN; CU on $(-\infty, -3)$ and $(3, \infty)$; CD on $(-3, 0)$ and $(0, 3)$; x-coordinates of PI are ± 3.

9   **11**

 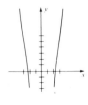

13 MIN: $f(-1) = -\frac{1}{2}$; MAX: $f(1) = \frac{1}{2}$; CU on $(-\sqrt{3}, 0)$ and $(\sqrt{3}, \infty)$; CD on $(-\infty, -\sqrt{3})$ and $(0, \sqrt{3})$; x-coordinates of PI are $0, \pm\sqrt{3}$.

15 MAX: $f(-\frac{4}{3}) \approx 7.27$; MIN: $f(0) = 0$; CD on $(-\infty, 0)$ and $(0, \frac{2}{3})$; CU on $(\frac{2}{3}, \infty)$; PI is $(\frac{2}{3}, \frac{10}{3}\sqrt[3]{12})$

17 MIN: $f(-2) \approx -7.55$; CU on $(-\infty, 0)$ and $(4, \infty)$; CD on $(0, 4)$; x-coordinates of PI are 0 and 4.

15   **17**

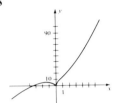 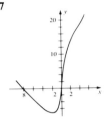

19, 21, 23: See Exercises 17, 19, 21 of Section 4.3.

25 MIN: $f(0) = 1$ **27** MAX: $f(\pi/4) = 1$

31   **33**

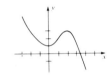 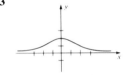

35   **37**

 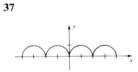

39 If $f(x) = ax^2 + bx + c$, then $f''(x) = 2a$.

(a) CU if $a > 0$ (b) CD if $a < 0$

41 (a) $-\frac{1}{10}$ (b) $50x - \frac{1}{10}x^2$ (c) $48x - \frac{1}{10}x^2 - 10$

(d) $48 - \frac{1}{5}x$ (e) 5750   (f) 2

43 (a) $1800x - 2x^2$ (b) $1799x - 2.01x^2 - 1000$ (c) 100

(d) $158,800

45 3990 units; $15,420.10

EXERCISES 4.5, PAGE 197

1 Side of base $= 2$ ft; height $= 1$ ft

3 Radius of base $=$ height $= 1/\sqrt[3]{\pi}$

5 25 ft by $\frac{50}{7}$ ft   **7** 2:23:05 P.M.   **9** $5\sqrt{5} \approx 11.2$ ft

11 Length $= 2\sqrt[3]{300} \approx 13.38$ ft;

width $= \frac{3}{2}\sqrt[3]{300} \approx 10.04$ ft;

height $= \sqrt[3]{300} \approx 6.69$ ft

15 55   **17** Radius $= \sqrt[3]{15}/2$; length of cylinder $= 2\sqrt[3]{15}$

19 Length of base $= \sqrt{2}a$; height $= a/\sqrt{2}$

21 $\frac{32}{81}\pi a^3$ **23** $(1, 2)$

25 Width $= 2a/\sqrt{3}$; depth $= 2\sqrt{2}a/\sqrt{3}$ **27** 500

29 (a) Use $36\sqrt{3}/(2 + \sqrt{3}) \approx 16.71$ cm for the rectangle
(b) Use all the wire for the rectangle

31 Width $= 12/(6 - \sqrt{3}) \approx 2.81$ ft;
height $= (18 - 6\sqrt{3})/(6 - \sqrt{3}) \approx 1.78$ ft

35 37   **37** 18 in., 18 in., 36 in.

39 $4/(1 + \sqrt[4]{\tfrac{1}{2}}) \approx 2.17$ mi from A

43 (c) 21.9 mi/hr   **45** $\theta = 60°$ **47** $2\pi(1 - \sqrt{\tfrac{2}{3}})$

49 $\tan \theta = \sqrt{2}/2; \theta \approx 35.3°$

51 $L = (4/\sin \theta) + (3/\cos \theta)$ with $\tan \theta = \sqrt[3]{\tfrac{4}{3}}$.
$L = (4^{2/3} + 3^{2/3})^{3/2} \approx 9.87$ ft

53 (b) $\cos \theta = \tfrac{2}{3}; \theta \approx 48.2°$

1 $\tfrac{5}{2}$ **3** $-\tfrac{7}{3}$ **5** 1   **7** 0

9 $\infty; -\infty; x = 4; y = 0$;
no MAX or MIN

11 $\infty; -\infty; x = -\tfrac{5}{2}$;
$y = 0$; no MAX or MIN

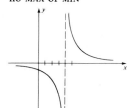
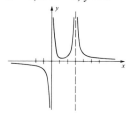

13 $-\infty - \infty; x = -8; y = 0$; MAX: $f(8) = \tfrac{3}{32}$

15 $-\infty, \infty$ for $a = -1; \infty, -\infty$ for $a = 2; x = -1, x = 2$,
$y = 2$; MAX: $f(0) = 0$; MIN: $f(-4) = \tfrac{16}{9}$

13   **15**

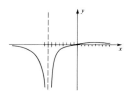
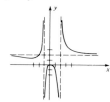

17 $\infty, -\infty$ for $a = 0; \infty, \infty$ for $a = 3; x = 0, x = 3, y = 0$;
MIN: $f(1) = \tfrac{1}{4}$

19 $x = 2, x = -2, y = 0$

17   **19**

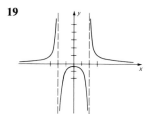

21 $y = 2$ **23** $x = -3, x = 0, x = 2, y = 0$

25 $x = -3, x = 1, y = 1$ **27** $x = 4, y = 0$

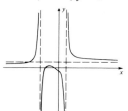
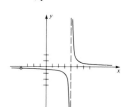

29 $x = -1; y = x - 2$ **31** $x = 0; y = -\tfrac{1}{2}x$

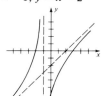
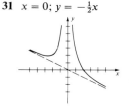

33 $x = 1; y = x$

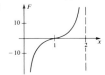

35 $(-2, 0)$ **37** $(\tfrac{4}{3}, 3), (-\tfrac{4}{3}, 3)$

39 (a) $V(t) = 50 + 5t; Q(t) = 0.5t$
(c) As $t \to \infty, c(t) \to 0.1$ lb of salt per gal.

41 (b)

1 $3x^3 - 2x^2 + 3x + C$

3 $\tfrac{1}{2}x^4 - \tfrac{1}{3}x^3 + \tfrac{3}{2}x^2 - 7x + C$

5 $-\tfrac{1}{2}x^{-2} + 3x^{-1} + C$ **7** $2x^{3/2} + 2x^{1/2} + C$

9 $9x^{2/3} - \tfrac{1}{8}x^{4/3} + 7x + C$

11 $\tfrac{8}{9}x^{9/4} + \tfrac{24}{5}x^{5/4} - x^{-3} + C$ **13** $-\tfrac{3}{4}\cos 4x + C$

15 $\tfrac{2}{3}x^3 - \tfrac{3}{2}x^2 - \tfrac{7}{5}\cos 5x + C$

17 $\tfrac{2}{3}\sin 3x + \tfrac{3}{2}\cos 2x + C$ **19** $-\cos 2x + C$

21 $3x^3 - 3x^2 + x + C$ **23** $\tfrac{24}{5}x^{5/3} - \tfrac{15}{2}x^{2/3} + C$

25 $\tfrac{10}{9}x^{9/5} + C$ **27** $\tfrac{1}{3}x^3 + \tfrac{1}{2}x^2 + x + C$ $(x \neq 1)$

29 $f(x) = 4x^3 - 3x^2 + x + 3$

31 $f(x) = \tfrac{2}{3}x^3 - \tfrac{1}{2}x^2 - 8x + \tfrac{65}{6}$

33 $f(t) = 32 \cos (\pi/4)t + 2t^2$ **35** $-t^3 + t^2 - 5t + 4$

39 $s(t) = -16t^2 + 1600t; s(50) = 40{,}000$

41 (a) $s(t) = -16t^2 - 16t + 96$ (b) $t = 2$ (c) 80 ft/sec

45 $a(t) = 10$ ft/sec^2 **47** $F = \tfrac{9}{5}C + 32$

49 $V = 2t^{3/2} + \tfrac{1}{8}t^2 + 2$ **51** After 19.6 yr

53 474,592 ft^3

55 (a) $dV/dt = 0.6 \sin (\tfrac{2}{3}\pi t)$ (b) $3/\pi \approx 0.95$ liter

57 $C(x) = 20x - (0.015/2)x^2 + 5.0075; \$986.26$

59 $R(x) = \tfrac{1}{3}x^3 - 3x^2 + 15x; p'(x) = \tfrac{2}{3}x - 3$

EXERCISES 4.8, PAGE 221

1 $(\sqrt{61} - 1)/3$

3 MAX: $f(2) = 28$; MIN: $f(-\frac{1}{2}) = -\frac{13}{4}$; decreasing on $(-\infty, -\frac{1}{2}]$ and $[2, \infty)$; increasing on $[-\frac{1}{2}, 2]$

5 MAX: $f(1) = 3$: increasing on $(-\infty, 1]$; decreasing on $[1, \infty)$

3    **5**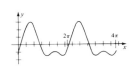

7 MAX: $f(0) = 1$; CU on $(-\infty, -1/\sqrt{3})$ and $(1/\sqrt{3}, \infty)$; CD on $(-1/\sqrt{3}, 1/\sqrt{3})$; x-coordinates of PI are $\pm 1/\sqrt{3}$

9 MAX: $f(\pi/2) = f(5\pi/2) = 3$ and $f(3\pi/2) = f(7\pi/2) = -1$; MIN: $f(7\pi/6) = f(19\pi/6) = f(11\pi/6) = f(23\pi/6) = -\frac{3}{2}$

7 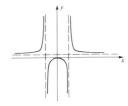   **9**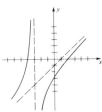

11 125 yd by 250 yd   **13** $\pi/2$

15 Radius of semicircle is $1/(8\pi)$ mi, length of rectangle is $\frac{1}{8}$ mi.

17 (a) Use all the wire for the circle.
(b) Use length $5\pi/(4 + \pi) \approx 2.2$ ft for circle; remainder for square.

19 $\frac{3}{2}$ **21** 0   **23** $-\infty$ **25** $-\infty$

27 $y = \frac{1}{3}, x = \frac{5}{3}, x = -\frac{5}{3}$ **29** $x = -3, y = x - 1$

33 $-8x^{-1} + 2x^{-2} - \frac{5}{3}x^{-3} + C$ **35** $100x + C$

37 $\frac{1}{8}(2x + 1)^4 + C$ **39** $f(x) = \frac{9}{28}x^{7/3} - \frac{5}{2}x^2 + \frac{25}{4}x - \frac{169}{14}$

CHAPTER 5

EXERCISES 5.1, PAGE 231

1 -5 **3** 34   **5** 40   **7** 510   **9** 500

13 $\frac{1}{3}(n^3 + 6n^2 + 20n)$ **15** $\frac{1}{3}(4n^3 - 12n^2 + 11n)$

17 28   **19** $\frac{125}{3}$ **21** 78   **23** 18   **25** $\frac{19}{4}$

EXERCISES 5.2, PAGE 236

1 1.1, 1.5, 1.1, 0.4, 0.9; $\|P\| = 1.5$

3 0.3, 1.7, 1.4, 0.5, 0.1; $\|P\| = 1.7$

5 (a) 40   (b) 32   (c) 36   **7** $\frac{49}{4}$ **9** 79

11 $\int_{-1}^{2} (3x^2 - 2x + 5)\, dx$ **13** $\int_{0}^{4} 2\pi x(1 + x^3)\, dx$

15 $-\frac{14}{3}$ **17** 30   **19** 25   **21** $9\pi/4$ **23** $\frac{1}{4}b^4$

25 Any unbounded function. For example, $f(x) = 1/x$, $f(x) = 1/\sqrt{1 - x}$, or $f(x) = \csc x$. There is no contradiction since the interval in (5.12) is closed.

EXERCISES 5.3, PAGE 243

1 30   **3** -12 **5** 2   **7** 78   **9** $-\frac{291}{2}$

13 $\int_{-3}^{1} f(x)\, dx$ **15** $\int_{h}^{c+h} f(x)\, dx$ **17** (a) $\sqrt{3}$ (b) 9

19 (a) 3   (b) 6   **21** (a) $\sqrt[3]{15/4}$ (b) 14

25 *Hint*: $-|f(x)| \le f(x) \le |f(x)|$

EXERCISES 5.4, PAGE 250

1 -18 **3** $\frac{265}{2}$ **5** 5   **7** $\frac{31}{256}$ **9** $\frac{20}{3}$ **11** $\frac{352}{5}$

13 $-\frac{37}{6}$ **15** $\frac{13}{3}$ **37** $-\frac{7}{2}$ **19** 0   **21** $\frac{10}{3}$

23 $\frac{53}{2}$ **25** $8\sqrt{3} + 16$ **27** $\frac{3}{2}(\sqrt{3} - 1)$ **29** $1 - \sqrt{2}$

31 0   **35** $1/(x + 1)$ **37** 6   **43** $\frac{16}{9}$ **45** $\sqrt[3]{\frac{5}{4}}$ **47** $\frac{3}{2}$

57 $12x^7/\sqrt{x^{12} + 2}$ **59** $3x^2(x^9 + 1)^{10} - 3(27x^3 + 1)^{10}$

EXERCISES 5.5, PAGE 259

1 $\frac{1}{15}(3x + 1)^5 + C$ **3** $\frac{2}{9}(t^3 - 1)^{3/2} + C$

5 $-\frac{1}{4}(x^2 - 4x + 3)^{-2} + C$ **7** $-\frac{3}{8}(1 - 2s^2)^{2/3} + C$

9 $\frac{2}{5}(\sqrt{u} + 3)^5 + C$ **11** $\frac{14}{3}$ **13** $\frac{1}{4} \sin (4x - 3) + C$

15 $-\frac{1}{2} \cos (x^2) + C$ **17** $\frac{1}{4} (\sin 3x)^{4/3} + C$ **19** 0

21 $\frac{1}{3}$ **23** $\frac{1}{7}x^7 + \frac{3}{5}x^5 + x^3 + x + C$ **25** $\frac{5}{12}(8x + 5)^{3/2} + C$

27 $\frac{5}{36}$ **29** $-\cos x - \frac{4}{3}(\cos x)^{3/2} - \frac{1}{2}\cos^2 x + C$

31 $(1/\cos x) + C = \sec x + C$ **33** $-\frac{1}{20}(2 + 5 \cos x)^4 + C$

35 (a) $\frac{1}{3}(x + 4)^3 + C_1$
(b) $\frac{1}{3}x^3 + 4x^2 + 16x + C_2$; $C_2 = C_1 + \frac{64}{3}$

37 (a) $\frac{2}{3}(\sqrt{x} + 3)^3 + C_1$
(b) $\frac{2}{3}x^{3/2} + 6x + 18x^{1/2} + C_2$; $18 + C_1 = C_2$

41 $1/\sqrt{x^3 + x + 5}$ **43** 1   **45** $\frac{14}{3}$ **47** (a) $\sqrt{3}$ (b) $\frac{1}{2}$

49 (a) $\frac{544}{225}$ (b) $\frac{38}{15}$ **51** (a) $v_{av} = \frac{6}{7}cD^{1/6}$

EXERCISES 5.6, PAGE 266

1 (a) 1.41   (b) 1.39   **3** (a) 0.88   (b) 0.88

5 (a) 0.39   (b) 0.39   **7** (a) 2.24   (b) 2.34   **9** 1.38

11 0.88   **15** (a) 8.65   (b) 8.59

17 (a) 128   (b) 132   (c) 128   **19** $\bar{v}_x \approx 0.174$ m/sec

21 (a) 41   (b) 7

EXERCISES 5.7, PAGE 267

1 70   **3** $\frac{11}{4}$ **5** -10 **7** $\frac{3}{5}$ **9** $\frac{1}{6}$

11 $-\frac{1}{16}(1 - 2x^2)^4 + C$ **13** $\sqrt{8} - \sqrt{3} \approx 1.10$

15 $-2/(1 + \sqrt{x}) + C$ **17** $3x - x^2 - \frac{5}{4}x^4 + C$ **19** $\frac{52}{9}$

21 $\frac{1}{6}(4t^2 + 2t - 7)^3 + C$ **23** $-x^{-2} + 3x^{-1} + C$

25 $\frac{1}{5} \cos (3 - 5x) + C$ **27** $\frac{1}{15} \sin^5 3x + C$

29 $\frac{2}{15}(16\sqrt{2} - 3\sqrt{3})$ **31** $\frac{1}{6}$ **33** $\sqrt[5]{y^4 + 2y^2 + 1} + C$

35 0   **37** (a) 341.36   (b) 334.42   **39** $4 - 2\sqrt{2}$

41 81.625 °F

CHAPTER 6

1 $\frac{17}{6}$ **3** $\frac{33}{2}$ **5** $\frac{32}{3}$

7 $\frac{32}{5}$ **9** $\frac{9}{2}$ **11** $8\sqrt{3}$

 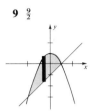

13 2   **15** $\frac{1}{2}$ **17** 8

 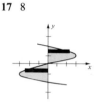

19 $\frac{16}{3}$ **21** $\pi + (3\sqrt{3}/2)$

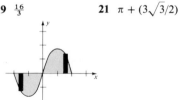

23 Let $f(x) = \sqrt{9 - (x - 4)^2}$ on $[1, 7]$. Then
$A = \lim_{\|P\| \to 0} \sum_k 2f(w_k) \Delta x_k$. Since the region is bounded
by a circle of radius 3, $A = 9\pi$.
25 The limit is the area of the region under the graph of
$y = 4x + 1$ from 0 to 1. $A = 3$
27 The limit is the area of the region to the left of the graph of
$x = 4 - y^2$ and to the right of the y-axis from $y = 0$ to
$y = 1$. $A = \frac{11}{3}$.
29 The area A of $\{(x, y): 2 \le x \le 5, 0 \le y \le x(x^2 + 1)^{-2}\}$,
$A = \frac{21}{260}$
31 The area A of $\{(x, y): 1 \le y \le 4, 0 \le x \le (5 + \sqrt{y})/\sqrt{y}\}$,
$A = 13$
33 9   **35** $4\sqrt{2}$ **37** (a) 4.25   (b) 4.50
39 (a) Change in height between $t = 10$ and $t = 15$
(b) 32.05 cm^2

1 $2\pi/3$ **3** 2π **5** $512\pi/15$

7 $64\pi/15$ **9** $64\sqrt{2}\pi/3$ **11** $72\pi/5$

13 $\pi^2/2$ **15** $\pi/2$
17 (a) $512\pi/15$ (b) $832\pi/15$ (c) $128\pi/3$
19 (a)   (b) $V = \pi \int_{-2}^{0} [(8 - 4x)^2 - (8 - x^3)^2] \, dx +$
$\pi \int_0^2 [(8 - x^3)^2 - (8 - 4x)^2] \, dx$

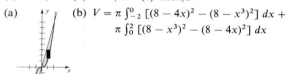

21 (a) *(see graph)*
(b) $V = \pi \int_2^3 [(y - 1)^2 - (2 - \sqrt{3 - y})^2] \, dy$
23 (a) *(see graph)*
(b) $V = \pi \int_{-1}^{1} [(5 + \sqrt{1 - y^2})^2 - (5 - \sqrt{1 - y^2})^2] \, dy$
$= 20\pi \int_{-1}^{1} \sqrt{1 - y^2} \, dy$

21   **23**

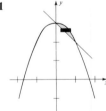 

25 $V = \frac{1}{3}\pi r^2 h$ **27** $V = \frac{1}{3}\pi h(r^2 + rR + R^2)$
29 The limit is the volume of the solid obtained by revolving
the region between $y = x^2$ and $y = x^3$, $0 \le x \le 1$, about the
x-axis. $V = 2\pi/35$.
31 $63\pi/2 \approx 99$

1 $128\pi/5$ **3** $24\pi/5$

5 $135\pi/2$ **7** $512\pi/5$

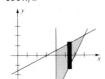

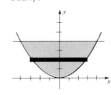

9 72π

11 π **13** (a) 16π (b) $64\pi/3$
15 (a) $512\pi/15$ (b) $832\pi/15$ (c) $128\pi/3$
17 $V = 2\pi \int_{-8}^{0} (8 - y)(\frac{1}{4}y - y^{1/3})\,dy +$
    $2\pi \int_{0}^{8} (8 - y)(y^{1/3} - \frac{1}{4}y)\,dy$
19 $V = 2\pi \int_{0}^{1} (2 - x)(x - x^2)\,dx$
21 $V = 4\pi \int_{-1}^{1} (5 - x)\sqrt{1 - x^2}\,dx$ **23** $V = \frac{1}{3}\pi r^2 h$
25 $V = \frac{1}{3}\pi h(r^2 + rR + R^2)$
27 The limit is the volume of the solid obtained by revolving
    the region between $y = x$ and $y = x^2$, $0 \le x \le 1$, about the
    y-axis. $V = \pi/6$.
29 $76\pi \approx 239$

EXERCISES 6.4, PAGE 288

1 $16a^3/3$ **3** $\frac{128}{15}$ **5** $2a^2h/3$ **7** $128\pi/15$ **9** $2a^3/3$
11 $\pi a^2 b/2$ **13** 4 **15** $\pi a^3/24$

EXERCISES 6.5, PAGE 295

1 $(4 + \frac{16}{81})^{3/2} - (1 + \frac{16}{81})^{3/2} \approx 7.29$
3 $\frac{8}{27}[10^{3/2} - (13^{3/2}/8)] \approx 7.63$ **7** $\frac{13}{12}$ **9** $\frac{353}{240}$
11 $s = \int_{0}^{2} \sqrt{\frac{53}{4} - 21y^2 + 9y^4}\,dy$ **13** 6
15 $s(x) = (x^{2/3} + \frac{4}{9})^{3/2} - (1 + \frac{4}{9})^{3/2}$;
    $\Delta s = \frac{1}{27}[(9(1.1)^{2/3} + 4)^{3/2} - 13^{3/2}] \approx 0.1196$;
    $ds = \sqrt{13}/30 \approx 0.1202$
17 $ds = \sqrt{17}(0.1) \approx 0.412$; $d(A, B) = \sqrt{0.1781} \approx 0.422$
19 $\sqrt{5\pi}/360 \approx 0.0195$ **21** 8.61
23 $\frac{8}{3}\pi(2^{3/2} - 1) \approx 15.3$ **25** $16{,}911\pi/1024 \approx 51.9$
27 $(\pi/27)(8 \cdot 37^{3/2} - 13^{3/2}) \approx 204$ **29** 10π

EXERCISES 6.6, PAGE 300

1 (a) $\frac{128}{3}$ in.-lb   (b) $\frac{64}{3}$ in.-lb   **3** $F_2 = 3F_1$
5 2250 ft-lb   **7** $44{,}660$ J
9 (a) $81(62.5)\pi/2 \approx 7952$ ft-lb
    (b) $(189)(62.5)\pi/2 \approx 18{,}555$ ft-lb
11 500 ft-lb
13 (a) $3c/10$ erg   (b) $9c/40$ erg (c a constant)
15 276 ft-lb   **17** $575(\frac{1}{2} - 1/\sqrt[5]{40}) \approx 12.55$ in.-lb
19 $W = gm_1 m_2 h/[4000(4000 + h)]$ **21** 36.85 ft-lb

EXERCISES 6.7, PAGE 304

1 (a) 31.25 lb   (b) 93.75 lb
3 (a) $62.5/\sqrt{3}$ lb   (b) $62.5\sqrt{3}/24$ lb   **5** 320 lb
7 $303{,}356.25$ lb   **9** $(592)(62.5)/3$ lb $\approx 12{,}333.3$ lb
11 $\frac{3200}{3}$ lb   **13** 4500 lb   **15** (a) 1516 lb   (b) 1614.6 lb

EXERCISES 6.8, PAGE 311

1 $M_x = -27$; $M_y = -46$; $\bar{x} = -\frac{23}{7}$; $\bar{y} = -\frac{27}{14}$
3 $(\frac{4}{5}, \frac{2}{7})$ **5** $(0, \frac{8}{5})$

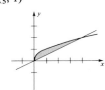   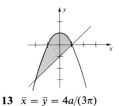

7 $(\frac{8}{5}, 1)$ **9** $(-\frac{1}{2}, -\frac{3}{5})$

11 $(\frac{8}{5}, \frac{1}{2})$ **13** $\bar{x} = \bar{y} = 4a/(3\pi)$

15 $\bar{x} = 0$, $\bar{y} = -20a/[3(8 + \pi)]$. The figure is positioned
    vertically with the origin at the center of the circle.

EXERCISES 6.9, PAGE 317

1 11 (Trapezoidal Rule)
3 (a) 150 dyn-cm   (b) 150 dyn-cm
5 $(27 - 5\sqrt{5})/3 \approx 5.27$ gal   **7** 1.56 l/min
9 1.45 coulombs
11 (a) $9[(601)^{2/3} - 1] \approx 632$ min
    (b) $2 \cdot 9[(301)^{2/3} - 1] \approx 790$ min
13 In min: (a) 18.16 (b) 66.22 (c) 115.24 (d) 197.12

EXERCISES 6.10, PAGE 319

1 $\frac{64}{3}$ **3** $\frac{5}{6}\sqrt{5}$ **5** $\frac{1}{2}$

7 10π

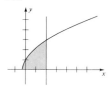

9 $3\pi/5$

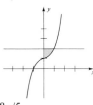

11 (a) $1152\pi/5$ (b) 54π (c) $1728\pi/5$
13 $(37^{3/2} - 10^{3/2})/27 \approx 7.16$
15 $432(62.5)\pi \approx 84{,}823$ ft-lb   **17** 6,000 lb
19 $\left(\frac{4}{15}, -\frac{1}{21}\right)$

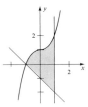

21 $515\pi/64 \approx 25.3$ **23** 900 ft (Trapezoidal Rule)   **25** $\pi/5$
27 Two possibilities exist: the solid could be obtained by
revolving the region under $y = x^2$ from $x = 0$ to $x = 1$
around the x-axis, or by revolving the region under $y = \frac{1}{2}x^3$
from $x = 0$ to $x = 1$ around the y-axis.

CHAPTER 7

EXERCISES 7.1, PAGE 324

1, 3: Show that $f(g(x)) = x = g(f(x))$.
1

3

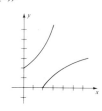

5 $f^{-1}(x) = (x + 3)/4$ **7** $f^{-1}(x) = (1 - 5x)/(2x),\ x > 0$
9 $f^{-1}(x) = \sqrt{9 - x},\ x \le 9$ **11** $f^{-1}(x) = \sqrt[3]{(x + 2)/5}$
13 $f^{-1}(x) = (x^2 + 5)/3,\ x \ge 0$ **15** $f^{-1}(x) = (x - 8)^3$
17 $f^{-1}(x) = x$
19 (a) $f^{-1}(x) = (x - b)/a$ (b) No (not one-to-one)
21 If g and h are both inverse functions of f, then $f(g(x)) =$
$x = f(h(x))$ for every x. Since f is one-to-one, this implies
that $g(x) = h(x)$ for every x, that is, $g = h$.

23

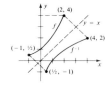

25

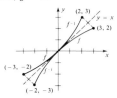

EXERCISES 7.2, PAGE 331

1 $9/(9x + 4)$ **3** $-15/(2 - 3x)$ **5** $-3x^2/(7 - 2x^3)$
7 $(6x - 2)/(3x^2 - 2x + 1)$ **9** $(8x + 7)/[3(4x^2 + 7x)]$
11 $1 + \ln x$
13 $\dfrac{1}{2x}\left(1 + \dfrac{1}{\sqrt{\ln x}}\right)$ **15** $-\dfrac{1}{x}\left(\dfrac{1}{(\ln x)^2} + 1\right)$
17 $\dfrac{20}{5x - 7} + \dfrac{6}{2x + 3}$ **19** $\dfrac{x}{x^2 + 1} - \dfrac{18}{9x - 4}$
21 $\dfrac{2x}{x^4 - 1}$ **23** $\dfrac{1}{\sqrt{x^2 - 1}}$
25 $-2 \tan 2x$ **27** $(\cos x)(1 + \ln \sin x)$ **29** $-\csc x$
31 $9 \csc 3x \sec 3x$ **33** $(2 \tan 2x)/(\ln \sec 2x)$
35 $\dfrac{(2x^2 - 1)y}{x(3y + 1)}$ **37** $\dfrac{y^2 - xy \ln y}{x^2 - xy \ln x}$ **39** $\dfrac{y - x \sin y}{x(x \cos y - \ln x)}$
41 $y = 8x - 15$ **43** $(1, 1), (2, 4 + 4 \ln 2)$
45 $(10, 5 \ln 10 - 5) \approx (10, 6.51)$
47 $v(t) = 2t - [4/(t + 1)];\ a(t) = 2 + [4/(t + 1)^2]$; to the left in
$[0, 1)$; to the right in $(1, 4]$.
49 0.73 yr
51 (a) $s'(0) = 0;\ s''(0) = bc/(m_1 + m_2)$
(b) $s'(m_2/b) = -c \ln [m_1/(m_1 + m_2)];\ s''(m_2/b) = bc/m_1$
53 $1, \frac{1}{5}, \frac{1}{10}, \frac{1}{100}, \frac{1}{1000}$; the slope approaches 0; the slope increases
without bound.
55 The graphs coincide if $x > 0$; however, the graph of
$y = \ln (x^2)$ contains points with negative x-coordinates.
57 $(-1)^{n-1}(n - 1)!x^{-n}$

EXERCISES 7.3, PAGE 338

1 $-5e^{-5x}$ **3** $6xe^{3x^2}$ **5** $e^{2x}/\sqrt{1 + e^{2x}}$
7 $e^{\sqrt{x+1}}/(2\sqrt{x + 1})$ **9** $(-2x^2 + 2x)e^{-2x}$
11 $\dfrac{e^x(x^2 + 1) - 2xe^x}{(x^2 + 1)^2}$ or $\dfrac{e^x(x - 1)^2}{(x^2 + 1)^2}$
13 $12(e^{4x} - 5)^2e^{4x}$ **15** $(-e^{1/x}/x^2) - e^{-x}$
17 $\dfrac{(e^x + e^{-x})^2 - (e^x - e^{-x})^2}{(e^x + e^{-x})^2}$ or $\dfrac{4}{(e^x + e^{-x})^2}$
19 $e^{-2x}[(1/x) - 2 \ln x]$ **21** $5e^{5x} \cos e^{5x}$ **23** $e^{-x} \tan e^{-x}$
25 $e^{3x}[\frac{1}{2}(1/\sqrt{x}) \sec^2 \sqrt{x} + 3 \tan \sqrt{x}]$
27 $-8e^{-4x} \sec^2 (e^{-4x}) \tan (e^{-4x})$
29 $e^{\cot x}(1 - x \csc^2 x)$ **31** $\dfrac{3x^2 - ye^{xy}}{xe^{xy} + 6y}$
33 $(e^x \cot y - e^{2y})/(e^x \csc^2 y + 2xe^{2y})$
35 $y = (e + 3)x - e - 1$
37 (b) $x \ln x < 1$ if $x = 1/e$, $x \ln x > 1$ if $x = e$, and $x \ln x$ is
increasing on $[1/e, e]$; 1.76
39 MIN: $f(-1) = -e^{-1}$; decreasing on $(-\infty, -1)$; increasing
on $(-1, \infty)$; CU on $(-2, \infty)$; CD on $(-\infty, -2)$; PI at
$(-2, -2e^{-2})$

41 No local extrema; decreasing on $(-\infty, \infty)$; CU on $(-\infty, \infty)$; no PI

43 MIN: $f(1/e) = -1/e$; decreasing on $(0, 1/e]$; increasing on $[1/e, \infty)$; CU on $(0, \infty)$; no PI

41   **43**

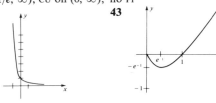

47 (a) $(\ln a - \ln b)/(a - b)$ (b) $\lim_{t \to \infty} C(t) = 0$

49 (a) 75.8 cm; 15.98 cm/yr   (b) 3 mo $(t = \frac{1}{4})$; 6 yr

51 (a) $f(n/a)$ (b) At $x = 2/a$

53

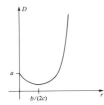

57 MAX: $1/(\sigma\sqrt{2\pi})$; increasing on $(-\infty, \mu]$; decreasing on $[\mu, \infty)$; CU on $(-\infty, \mu - \sigma)$ and $(\mu + \sigma, \infty)$; CD on $(\mu - \sigma, \mu + \sigma)$; PI at $(\mu \pm \sigma, c)$ where
$c = e^{-1/2}/(\sigma\sqrt{2\pi})$; $\lim_{x \to \infty} f(x) = 0$; $\lim_{x \to -\infty} f(x) = 0$

59 $r/R = e^{-1/2} \approx 0.607$

EXERCISES 7.4, PAGE 345

1 $\frac{1}{2}\ln(x^2 + 1) + C$ **3** $-\frac{1}{5}\ln|7 - 5x| + C$

5 $\frac{1}{2}\ln|x^2 - 4x + 9| + C$ **7** $\frac{1}{3}\ln|x^3 + 1| + C$

9 $\frac{1}{2}(\ln 9 - \ln 3)$, or $\ln\sqrt{3}$ **11** $\frac{1}{2}x^2 + \frac{1}{5}e^{5x} + C$

13 $\frac{1}{2}(\ln x)^2 + C$ **15** $-\frac{1}{4}(e^{-12} - e^{-4})$ **17** $-e^{\cot x} + C$

19 $\ln|\sin x| + C$ **21** $\ln|1 + \tan x| + C$ **23** $\sin e^x + C$

25 $e^x + 2x - e^{-x} + C$ **27** $\ln(e^x + e^{-x}) + C$ **29** 4

31 $\ln 2 + e^{-2} - e^{-1} \approx 0.46$ **33** $\pi(1 - e^{-1})$

35 $(5x + 2)^2(6x + 1)(150x + 39)$

37 $(19x^2 + 20x - 3)(x^2 + 3)^4/[2(x + 1)^{3/2}]$

39 $\left[\dfrac{3x}{3x^2 + 2} + \dfrac{3}{2(6x - 7)}\right]\sqrt{(3x^2 + 2)\sqrt{6x - 7}}$

41 $-2/(3 - 2x)$

43 $\Delta S = c \ln(T_2/T_1)$

45 (a) $Q(t) = 2.5 - 2.5e^{-4t}$ (b) 2.5 coulombs

47 (a) $s(t) = kv_0(1 - e^{-t/k})$ (b) kv_0 (Hint: Let $t \to \infty$ in (a).)

49 $2/\ln(3.25)$

EXERCISES 7.5, PAGE 352

1 $7^x \ln 7$ **3** $8^{x^2+1}(2x \ln 8)$

5 $(4x^3 + 6x)/[(\ln 10)(x^4 + 3x^2 + 1)]$ **7** $5^{3x-4}3 \ln 5$

9 $[-(x^2 + 1)10^{1/x} \ln 10]/x^2 + (2x10^{1/x})$

11 $30x/[(3x^2 + 2) \ln 10]$ **13** $\left(\dfrac{6}{6x + 4} - \dfrac{2}{2x - 3}\right)\dfrac{1}{\ln 5}$

15 $1/(x \ln x \ln 10)$ **17** $e x^{e-1} + e^x$

19 $(x + 1)^x\left(\dfrac{x}{x + 1} + \ln(x + 1)\right)$ **21** $(\ln 2)(\sin 2x)2^{\sin^2 x}$

23 $(\sec^2 x \ln x + x^{-1} \tan x)x^{\tan x}$ **25** $[1/(3 \ln 10)]10^{3x} + C$

27 $[-1/(2 \ln 3)]3^{-x^2} + C$ **29** $(1/\ln 2) \ln(2^x + 1) + C$

31 $24/(1250 \ln 5)$ **33** $(\ln 10) \ln|\log x| + C$

35 $(1/\ln 2) - \frac{1}{2}$ **37** (a) \$0.05/yr   (b) \$0.95

39 (a) In trout/yr: 95; 62; 53   (b) 9.36 yr

41 pH ≈ 2.201; $\pm 0.1\%$

43 (b) $S(x) = 2S(2x)$ (twice as sensitive)

45 (a) $1/(I_0 \ln 10)$ (b) $1/(100I_0 \ln 10)$ (c) $1/(1000I_0 \ln 10)$

EXERCISES 7.6, PAGE 357

1 $q(t) = 5000(3)^{t/10}$; 45,000; $(10 \ln 10)/\ln 3 \approx 21$ hr

3 $30(\frac{29}{30})^5 \approx 25.3$ in.

5 Approximately 109.24 yr after Jan. 1, 1980 (March 29, 2089)

7 $(5 \ln 6)/\ln 3 \approx 8.2$ min

9 Determine I such that at $t = 0$, $RI + L(dI/dt) = 0$ and $I = I_0$. Proceeding in a manner similar to the solution of Example 1, $(1/I)dI = (-R/L) dt$; $\ln I = (-R/L)t + C$ and $I = e^C e^{(-R/L)t}$. Since $I = I_0$ at $t = 0$, $I = I_0 e^{-Rt/L}$

11 $P = [(288 - 0.01z)/288]^{3.42}$ **13** $38\frac{1}{3}$ yr   **15** 683.3 mg

17 $v = [2k(y^{-1} - y_0^{-1}) + v_0^2]^{1/2}$ **19** 13,235 yr

23 $V = \frac{1}{27}(kt + c)^3$ with constants c and k and $k > 0$.

EXERCISES 7.7, PAGE 362

1 $(x^2 - 3)/2$; $[\sqrt{5}, 5]$; x

3 $\sqrt{4 - x}$; $[-45, 4]$; $-1/(2\sqrt{4 - x})$

5 $1/x$; $(0, \infty)$; $-1/x^2$

7 $\sqrt{-\ln x}$; $(0, 1]$; $-1/(2x\sqrt{-\ln x})$

9 $\ln(x + \sqrt{x^2 + 4}) - \ln 2$; $\mathbb{R}$; $1/\sqrt{x^2 + 4}$

11 f is increasing, since $f'(x) > 0$ for every x; $\frac{1}{16}$

13 f is increasing, since $f'(x) > 0$ for every x; 1

15 Since f decreases on $(-\infty, 0]$ and increases on $[0, \infty)$, there is no inverse function. If the domain is restricted to a subset of one of these intervals, then f will have an inverse function.

17 f increases on $(-\infty, 0]$ and decreases on $[0, \infty)$. If the domain is restricted to a subset of one of these intervals, then f^{-1} will exist.

EXERCISES 7.8, PAGE 362

1 $-2(1 + \ln|1 - 2x|)$

3 $12/(3x + 2) + 3/(6x - 5) - 8/(8x - 7)$

5 $-4x/(2x^2 + 3)[\ln(2x^2 + 3)]^2$ **7** $2x$

9 $(\ln 10)10^x \log x + 10^x/(x \ln 10)$ **11** $(1/x)(2 \ln x)x^{\ln x}$

13 $2x(1 - x^2)e^{1-x^2}$

15 $2^{-1/x}[(x^3 + 4) \ln 2 - 3x^4]/x^2(x^3 + 4)^2$

17 $e(1 + \sqrt{x})^{e-1}/(2\sqrt{x})$ **19** $(10^{\ln x} \ln 10)/x$

21 $(1/x)[1 + \ln(\ln x)](\ln x)^{\ln x}$ **23** $-\sec x$

25 $2e^{-2x} \csc e^{-2x}(\csc^2 e^{-2x} + \cot^2 e^{-2x})$ **27** $-16 \tan 4x$

29 $(\sin x)^{\cos x}(\cos x \cot x - \sin x \ln \sin x)$

31 $-y/x$ **33** $-\frac{1}{2}e^{-2x} - 2e^{-x} + x + C$
35 $2(e^{-1} - e^{-2})$ **37** $-\ln|1 - \ln x| + C$
39 $\frac{1}{6}x^2 - \frac{2}{9}x + \frac{4}{27}\ln|3x + 2| + C$ **41** $3/(2\ln 4)$
43 $2\ln 10\sqrt{\log x} + C$ **45** $\frac{1}{2}x^2 - x + 2\ln|x + 1| + C$
47 $(5e)^x/(1 + \ln 5) + C$ **49** $x^{e+1}/(e + 1) + C$
51 $\cos e^{-x} + C$ **53** $-\ln|1 + \cot x| + C$
55 $-\frac{1}{4}\ln|1 - 2\sin 2x| + C$ **57** $4e^2 + 12$ cm
59 $y - e = -2(1 + e)(x - 1)$ **61** $(\pi/8)(e^{-16} - e^{-24})$
63 Approximately 33.2 days
65 The amount in solution at any time t hours past 1:00 P.M. is $10(1 - 2^{-t/3})$.
   (a) 2.21 hours (approximately 6:14 P.M.)
   (b) $10(1 - 2^{-7/3}) \approx 8.016$ lb
67 6,400,000   **69** $-\frac{1}{8}$

CHAPTER 8

1 $\frac{1}{4}\ln|\csc 4x - \cot 4x| + C$ **3** $\frac{1}{3}\sec 3x + C$
5 $\frac{1}{3}(\ln|\sec 3x| + \ln|\sec 3x + \tan 3x|) + C$
7 $\frac{1}{2}\ln|\sec 2x + \tan 2x| + C$
9 $-\frac{1}{2}\cot(x^2 + 1) + C$ **11** $\frac{1}{6}\sin 6x + C$
13 $\frac{1}{2}\tan^2 x]_0^{\pi/4} = \frac{1}{2}$
15 $\frac{1}{2}(\ln|\sec 2x + \tan 2x| - \sin 2x) + C$
17 $\frac{1}{2}\sin^2 x]_{\pi/6}^{\pi} = -\frac{1}{8}$ **19** $\ln|x + \cos x| + C$
21 $\tan x + \sec x]_{\pi/4}^{\pi/3} = \sqrt{3} - \sqrt{2} + 1$
23 $e^x + \ln|\sec e^x| + C$ **25** $-e^{\cos x} + C$
27 $\frac{1}{2}\ln|2\tan x + 1| + C$ **29** $\frac{1}{2}\ln 2$
31 $\ln\left(\dfrac{\sqrt{2} + 1}{\sqrt{2} - 1}\right) = \ln(3 + 2\sqrt{2}) = 2\ln(1 + \sqrt{2})$
33 $2\pi\sqrt{3}$ **39** (a) 2.24   (b) 2.34
41 (a) $L = \frac{1}{2}\int_0^{\pi/2}\sqrt{4 + \sec^2(x/2)\tan^2(x/2)}\,dx$
   (b) If $f(x) = \frac{1}{2}\sqrt{4 + \sec^2(x/2)\tan^2(x/2)}$, then
   $L \approx (\pi/24)[f(0) + 4f(\pi/8) + 2f(\pi/4) + 4f(3\pi/8) + f(\pi/2)]$
   (c) $L \approx 1.65$
43 (b) $q(t) = q_0 e^u$ for $u = [k/(2\pi)](1 - \cos 2\pi t)$

1 (a) $\pi/3$ (b) $-\pi/3$ **3** (a) $\pi/4$ (b) $3\pi/4$
5 (a) $\pi/3$ (b) $-\pi/3$ **7** $\frac{1}{2}$ **9** $\frac{4}{5}$ **11** $\pi - \sqrt{5}$
13 0   **15** Undefined   **17** $-\frac{24}{25}$ **19** $x/\sqrt{x^2 + 1}$
21 $\sqrt{2 + 2x}/2$
31 $\cot^{-1} x = y$ if and only if $\cot y = x$ for $0 < y < \pi$
33   **35**

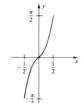

37   **39**

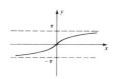

41

45 (a) $\arctan(-9 \pm \sqrt{57})/4$ (b) $-0.3478; -1.3336$
47 (a) $\arccos(\pm\sqrt{15}/5); \arccos(\pm\sqrt{3}/3)$
   (b) $0.6847; 2.4569; 0.9553; 2.1863$
49 (a) $\arcsin(\pm\sqrt{30}/6)$ (b) ± 1.1503
51 $\sin^{-1}(\sin 2)$ is between $-\pi/2$ and $\pi/2$ and 2 is not in $[-\pi/2, \pi/2]$. $\cos^{-1}(\cos 2)$ lies between 0 and π. Hence $\cos^{-1}(\cos 2) = 2$.
53 An error message will be avoided if $-\sin 1 \le x \le \sin 1$.
55 $\sin^{-1} x = \tan^{-1}(x/\sqrt{1 - x^2})$ for $-1 < x < 1$;
   $\cos^{-1} x = \tan^{-1}(\sqrt{1 - x^2}/x)$ for $0 < x < 1$;
   $\cot^{-1} x = \tan^{-1}(1/x)$ for $0 < x < \infty$

1 $3/(9x^2 - 30x + 26)$ **3** $1/(2\sqrt{x}\sqrt{1 - x})$
5 $(-e^{-x}/\sqrt{e^{-2x} - 1}) - e^{-x}\operatorname{arcsec} e^{-x}$
7 $2x^3/(1 + x^4) + 2x\arctan x^2$
9 $-[9(1 + \cos^{-1} 3x)^2]/\sqrt{1 - 9x^2}$
11 $2x/[(1 + x^4)\arctan x^2]$
13 $-1/[(\sin^{-1} x)^2\sqrt{1 - x^2}]$
15 $x^{-2}\sin(x^{-1}) + \sec x\tan x$ **17** $-3e^{-\tan 3x}\sec^2 3x$
19 $-8\csc 2x(\csc 2x + \cot 2x)^4$ **21** $x/[(x^2 - 1)\sqrt{x^2 - 2}]$
23 $(1 - 2x\arctan x)/(x^2 + 1)^2$
25 $[1/(2\sqrt{x})]\sec^{-1}\sqrt{x} + 1/(2\sqrt{x}\sqrt{x - 1})$
27 $(1 - x^6)^{-1/2}(3\ln 3)x^2 3^{\arcsin x^3}$
29 $(\tan x)^{\arctan x}[\cot x\sec^2 x\arctan x + (\ln\tan x)/(1 + x^2)]$
31 $(ye^x - \sin^{-1} y - 2x)/[(x/\sqrt{1 - y^2}) - e^x]$ **33** $\pi/16$
35 $\pi/12$ **37** $-\arctan(\cos x) + C$
39 $2\arctan\sqrt{x} + C$ **41** $\sin^{-1}(e^x/4) + C$
43 $\frac{1}{6}\sec^{-1}(x^{3/2}) + C$ **45** $\frac{1}{2}\ln(x^2 + 9) + C$
47 $\frac{1}{5}\sec^{-1}(e^x/5) + C$ **49** $4\pi/3$ **51** $\pm\frac{7}{3576} \approx \pm 0.002$
53 $-\frac{25}{1044}$ rad/sec   **55** $40\sqrt{3}$
57 $y - (\pi/6) = (2/\sqrt{3})(x - \frac{3}{2}); \; y - (\pi/6) = (-\sqrt{3}/2)(x - \frac{3}{2})$
59 (a) $(-\infty, 0)$ (b) $(0, \infty)$
61 $(\pi e^2/2) - (\pi^2/4) - (\pi/2) \approx 7.57$
63 $2\pi/27 \approx 0.233$ mi/sec

15 $5\cosh 5x$ **17** $[1/(2\sqrt{x})](\sqrt{x}\operatorname{sech}^2\sqrt{x} + \tanh\sqrt{x})$
19 $-2x\operatorname{sech} x^2[(x^2 + 1)\tanh x^2 + 1]/(x^2 + 1)^2$
21 $3x^2\sinh x^3$ **23** $3\cosh^2 x\sinh x$ **25** $2\coth 2x$
27 $-e^{3x}\operatorname{sech} x\tanh x + 3e^{3x}\operatorname{sech} x$

29 $-\mathrm{sech}^2\, x/(\tanh x + 1)^2$ **31** $\dfrac{y(e^x - \cosh xy)}{(x \cosh xy - e^x)}$

33 $2 \cosh \sqrt{x} + C$ **35** $\ln |\sinh x| + C$

37 $\frac{1}{2} \sinh^2 x + C$ (or $\frac{1}{2} \cosh^2 x + C$, or $\frac{1}{4} \sinh 2x + C$)

39 $-\frac{1}{3} \mathrm{sech}\, 3x + C$ **41** $\frac{1}{9} \tanh^3 3x + C$

43 $-\frac{1}{2} \ln |1 - 2 \tanh x| + C$ **45** $\frac{1}{3}(-1 + \cosh 3)$

47 $(\ln (2 + \sqrt{3}), \sqrt{3}), (\ln (2 - \sqrt{3}), -\sqrt{3})$

51

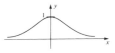

53 Let $A = \frac{1}{2} \sin t \cos t - \int_1^{\cosh t} \sqrt{x^2 - 1}\, dx$ and show that $dA/dt = \frac{1}{2}$.

55 (a) $286{,}574\ \mathrm{ft}^2$ (b) $1494\ \mathrm{ft}$

7 **9**

11 $5/\sqrt{25x^2 + 1}$ **13** $1/(2\sqrt{x}\sqrt{x - 1})$

15 $2x/(2x^2 - x^4)$ **17** $-|x|/(x\sqrt{x^2 + 1}) + \sinh^{-1}(1/x)$

19 $4/(\sqrt{16x^2 - 1} \cosh^{-1} 4x)$ **21** $\frac{1}{4} \sinh^{-1} \frac{4}{9}x + C$

23 $\frac{1}{14} \tanh^{-1} \frac{2}{7}x + C$ **25** $\cosh^{-1}(e^x/4) + C$

27 $-\frac{1}{6} \mathrm{sech}^{-1}(x^2/3) + C$ **31** $y = \sinh 3t$

1 $1/(2x\sqrt{x} - 1)$ **3** $2x \arcsec x^2 + (2x/\sqrt{x^4 - 1})$

5 $1/[\sqrt{1 - x^2} (\cos^{-1} x)^2]$ **7** $-5e^{-5x} \sinh e^{-5x}$

9 $\frac{1}{3}x^{-2/3}$ **11** $2^{\arctan 2x}(2 \ln 2)/(1 + 4x^2)$

13 $-6e^{-2x} \sin^2 e^{-2x} \cos e^{-2x}$

15 $-2xe^{-x^2}(\csc^2 x^2 + \cot x^2)$ **17** $2x/[(1 + x^4) \tan^{-1}(x^2)]$

19 $-x/\sqrt{x^2(1 - x^2)}$ **21** $3 (\sec^2 \sin 3x) \cos 3x$

23 $4(\tan x + \tan^{-1} x)^3[(\sec^2 x) + 1/(1 + x^2)]$

25 $(1 + x^2)^{-1}[1 + (\tan^{-1} x)^2]^{-1}$

27 $-e^{-x}(e^{-x} \cosh e^{-x} + \sinh e^{-x})$

29 $(\cosh x - \sinh x)^{-2}$, or e^{2x}

31 $2x/\sqrt{x^4 + 1}$ **33** $-\frac{1}{3} \sin (5 - 3x) + C$

35 $2 \tan \sqrt{x} + C$

37 $\frac{1}{9} \ln |\sin 9x| + \frac{1}{9} \ln |\csc 9x - \cot 9x| + C$

39 $-\ln |\cos e^x| + C$

41 $-\frac{1}{3} \cot 3x + \frac{2}{3} \ln |\csc 3x - \cot 3x| + x + C$

43 $\frac{1}{4} \sin 4x + C$ **45** $\frac{1}{18} \ln (4 + 9x^2) + C$

47 $-\sqrt{1 - e^{2x}} + C$ **49** $\frac{1}{2} \sinh x^2 + C$ **51** $\pi/3$

53 $\frac{1}{3}(1 + \tan x)^3 + C$ **55** $-\ln |2 + \cot x| + C$

57 $\cosh (\ln x) + C$ **59** $\frac{1}{2} \sin^{-1}(2x/3) + C$

61 $-\frac{1}{3} \mathrm{sech}^{-1} |2x/3| + C$ **63** $\frac{1}{25}\sqrt{25x^2 + 36} + C$

65 $(\frac{4}{15}, \sin^{-1}(\frac{4}{5})), (-\frac{4}{15}, \sin^{-1}(-\frac{4}{5}))$

67 MIN: $f(\tan^{-1} \frac{1}{2}) = 5\sqrt{5}$; decreasing on $(0, c)$; increasing on $(c, \pi/2)$ for $c = \tan^{-1}(\frac{1}{2})$.

69 $\pi(4 - \pi)/4$

71 (a) $x = \frac{1}{2} \tan^{-1} 4 + n(\pi/2)$ for $n = 0, 1, 2, 3$
(b) $0.66, 2.23, 3.80, 5.38$

73 $\frac{1}{260}$ rad/sec **75** -0.31 rad/sec

CHAPTER 9

1 $-(x + 1)e^{-x} + C$ **3** $e^{3x}(\frac{1}{3}x^2 - \frac{2}{9}x + \frac{2}{27}) + C$

5 $\frac{1}{25} \cos 5x + \frac{1}{5}x \sin 5x + C$

7 $x \sec x - \ln |\sec x + \tan x| + C$

9 $(x^2 - 2) \sin x + 2x \cos x + C$

11 $x \tan^{-1} x - \frac{1}{2} \ln (1 + x^2) + C$

13 $\frac{2}{9}x^{3/2}(3 \ln x - 2) + C$ **15** $-x \cot x + \ln |\sin x| + C$

17 $-\frac{1}{2}e^{-x}(\cos x + \sin x) + C$

19 $\cos x(1 - \ln \cos x) + C$

21 $-\frac{1}{2} \csc x \cot x + \frac{1}{2} \ln |\csc x - \cot x| + C$

23 $\frac{1}{3}(2 - \sqrt{2})$ **25** $\pi/4$

27 $\frac{1}{40400}(2x + 3)^{100}(200x - 3) + C$

29 $\frac{1}{41}e^{4x}(4 \sin 5x - 5 \cos 5x) + C$

31 $x (\ln x)^2 - 2x \ln x + 2x + C$

33 $x^3 \cosh x - 3x^2 \sinh x + 6x \cosh x - 6 \sinh x + C$

35 $2 \cos \sqrt{x} + 2\sqrt{x} \sin \sqrt{x} + C$

37 $x \cos^{-1} x - \sqrt{1 - x^2} + C$

43 $e^x(x^5 - 5x^4 + 20x^3 - 60x^2 + 120x - 120) + C$

45 2π **47** $(\pi/2)(e^2 + 1)$ **49** $(62.5\pi)/4 = 125\pi/8$

1 $\sin x - \frac{1}{3} \sin^3 x + C$ **3** $\frac{1}{8}x - \frac{1}{32} \sin 4x + C$

5 $\frac{1}{5} \cos^5 x - \frac{1}{3} \cos^3 x + C$

7 $\frac{1}{8}(\frac{5}{2}x - 2 \sin 2x + \frac{3}{8} \sin 4x + \frac{1}{6} \sin^3 2x) + C$

9 $\frac{1}{4} \tan^4 x + \frac{1}{6} \tan^6 x + C$ **11** $\frac{1}{5} \sec^5 x - \frac{1}{3} \sec^3 x + C$

13 $\frac{1}{5} \tan^5 x - \frac{1}{3} \tan^3 x + \tan x - x + C$

15 $\frac{2}{3} \sin^{3/2} x - \frac{2}{7} \sin^{7/2} x + C$

17 $\tan x - \cot x + C$ **19** $\frac{2}{3} - 5/(6\sqrt{2})$

21 $\frac{1}{4} \sin 2x - \frac{1}{16} \sin 8x + C$ **23** $\frac{3}{5}$

25 $-\frac{1}{5} \cot^5 x - \frac{1}{7} \cot^7 x + C$

27 $-\ln |2 - \sin x| + C$ **29** $-1/(1 + \tan x) + C$

31 $3\pi^2/4$ **33** $\frac{5}{2}$

35 (b) $\displaystyle\int \sin mx \cos nx\, dx$

$$= \begin{cases} -\dfrac{\cos (m + n)x}{2(m + n)} - \dfrac{\cos (m - n)x}{2(m - n)} + C & \text{if } m \neq n \\[2mm] -\dfrac{\cos 2mx}{4m} + C & \text{if } m = n \end{cases}$$

$\displaystyle\int \cos mx \cos nx\, dx$

$$= \begin{cases} \dfrac{\sin (m + n)x}{2(m + n)} + \dfrac{\sin (m - n)x}{2(m - n)} + C & \text{if } m \neq n \\[2mm] \dfrac{x}{2} + \dfrac{\sin 2mx}{4m} & \text{if } m = n \end{cases}$$

1 $2 \sin^{-1}(x/2) - \frac{1}{2}x\sqrt{4 - x^2} + C$

3 $\frac{1}{3} \ln \left| \dfrac{\sqrt{9 + x^2} - 3}{x} \right| + C$ **5** $\sqrt{x^2 - 25}/(25x) + C$

7 $-\sqrt{4 - x^2} + C$ **9** $-x/\sqrt{x^2 - 1} + C$

11 $\frac{1}{432}[(\tan^{-1}\frac{1}{6}x) + 6x/(36 + x^2)] + C$

13 $\frac{9}{4}(\sin^{-1}\frac{2}{3}x) + \frac{1}{2}x\sqrt{9 - 4x^2} + C$

15 $1/[2(16 - x^2)] + C$

17 $\frac{1}{243}(9x^2 + 49)^{3/2} - \frac{49}{81}(9x^2 + 49)^{1/2} + C$

19 $(3 + 2x^2)\sqrt{x^2 - 3}/(27x^3) + C$

21 $\frac{1}{2}x^2 + 8 \ln |x| - 8x^{-2} + C$

29 $25\pi[\sqrt{2} - \ln(\sqrt{2} + 1)] \approx 41.849$

31 $\sqrt{5} + \frac{1}{2} \ln(2 + \sqrt{5}) \approx 2.96$

33 $\frac{3}{2}\sqrt{13} + 2 \ln \frac{1}{2}(3 + \sqrt{13})$

35 $\sqrt{x^2 - 16} - 4 \tan^{-1}(\frac{1}{4}\sqrt{x^2 - 16})$

37 $-\sqrt{25 + x^2}/(25x) + C$ **39** $-\sqrt{1 - x^2}/x + C$

1 $3 \ln |x| + 2 \ln |x - 4| + C$, or $\ln |x|^3(x - 4)^2 + C$

3 $4 \ln |x + 1| - 5 \ln |x - 2| + \ln |x - 3| + C$, or $\ln [(x + 1)^4 |x - 3|/|x - 2|^5] + C$

5 $6 \ln |x - 1| + 5/(x - 1) + C$

7 $3 \ln |x - 2| - 2 \ln |x + 4| + C$, or $\ln \dfrac{|x - 2|^3}{(x + 4)^2} + C$

9 $2 \ln |x| - \ln |x - 2| + 4 \ln |x + 2| + C$, or $\ln [x^2(x + 2)^4/|x - 2|] + C$

11 $5 \ln |x + 1| - 1/(x + 1) - 3 \ln |x - 5| + C$, or $\ln [|x + 1|^5/|x - 5|^3] - 1/(x + 1) + C$

13 $5 \ln |x| - \dfrac{2}{x} + \dfrac{3}{2x^3} - \dfrac{1}{3x^2} + 4 \ln |x + 3| + C$

15 $\frac{1}{6} \ln |x - 3| - \dfrac{7}{2(x - 3)} + \frac{5}{6} \ln |x + 3| - \dfrac{3}{2(x + 3)} + C$

17 $3 \ln |x + 5| + \ln(x^2 + 4) + \frac{1}{2} \tan^{-1}\frac{1}{2}x + C$, or $\ln(x^2 + 4)|x + 5|^3 + \frac{1}{2} \tan^{-1}\frac{1}{2}x + C$

19 $\ln \sqrt{(x^2 + 1)/(x^2 + 4)} + \frac{1}{2} \tan^{-1}\frac{1}{2}x + C$

21 $\ln(x^2 + 1) - 4/(x^2 + 1) + C$

23 $\frac{1}{2}x^2 + x + 2 \ln |x| + 2 \ln |x - 1| + C$, or $\frac{1}{2}(x^2 + 2x) + \ln(x^2 - x)^2 + C$

25 $\frac{1}{3}x^3 - 9x - [1/(9x)] - \frac{1}{2} \ln(x^2 + 9) + \frac{728}{27} \tan^{-1}\frac{1}{3}x + C$

27 $2 \ln |x + 4| + 6(x + 4)^{-1} - 5(x - 3)^{-1} + C$

29 $2 \ln |x - 4| + 2 \ln |x + 1| - \frac{3}{2}(x + 1)^{-2} + C$

31 $\frac{3}{2} \ln(x^2 + 1) + \ln |x - 1| + x^2 + C$

37 $\frac{1}{2} \ln 3$ **39** $(\pi/27)(4 \ln 2 + 3) \approx 0.672$

1 $\frac{1}{2} \tan^{-1}\frac{1}{2}(x - 2) + C$ **3** $\sin^{-1}\frac{1}{2}(x - 2) + C$

5 $-2\sqrt{9 - 8x - x^2} - 5 \sin^{-1}\frac{1}{5}(x + 4) + C$

7 $\frac{1}{3} \ln |x - 1| - \frac{1}{6} \ln(x^2 + x + 1)$
$\qquad - (1/\sqrt{3}) \tan^{-1}[(2x + 1)/\sqrt{3}] + C$

9 $\frac{1}{2}[\tan^{-1}(x + 2) + (x + 2)/(x^2 + 4x + 5)] + C$

11 $(x + 3)/(4\sqrt{x^2 + 6x + 13}) + C$

13 $[2/(3\sqrt{7})] \tan^{-1}[(4x - 3)/(3\sqrt{7})] + C$

15 $1 + (\pi/4)$ **17** $\ln [(1 + e^x)/(2 + e^x)] + C$

19 $\frac{1}{3} \ln 2 + [2\pi/(6\sqrt{3})] \approx 0.8356$

21 $\pi \ln(1.8) + (2\pi/3)[\tan^{-1}\frac{1}{3} - (\pi/4)] \approx 0.8755$

1 $\frac{3}{7}(x + 9)^{7/3} - \frac{27}{4}(x + 9)^{4/3} + C$

3 $\frac{5}{81}(3x + 2)^{9/5} - \frac{5}{18}(3x + 2)^{4/5} + C$ **5** $2 + 8 \ln \frac{6}{7}$

7 $\frac{6}{7}x^{7/6} - \frac{6}{5}x^{5/6} + 2x^{1/2} - 6x^{1/6} + 6 \arctan x^{1/6} + C$

9 $(2/\sqrt{3}) \tan^{-1}\sqrt{(x - 2)/3} + C$

11 $\frac{3}{10}(x + 4)^{2/3}(2x - 7) + C$

13 $\frac{2}{7}(1 + e^x)^{7/2} - \frac{4}{5}(1 + e^x)^{5/2} + \frac{2}{3}(1 + e^x)^{3/2} + C$

15 $e^x - 4 \ln(e^x + 4) + C$

17 $2 \sin \sqrt{x + 4} - 2\sqrt{x + 4} \cos \sqrt{x + 4} + C$ **19** $\frac{137}{320}$

21 $(2/\sqrt{3}) \tan^{-1}[(2 \tan \frac{1}{2}x + 1)/\sqrt{3}] + C$

23 $\ln |1 + \tan \frac{1}{2}x| + C$

25 $\frac{1}{5} \ln |(2 + \tan \frac{1}{2}x)/(-1 + 2 \tan \frac{1}{2}x)| + C$

27 $\frac{4}{3} \ln |\sin x - 4| + \frac{2}{3} \ln |\sin x + 2| + C$

1 $\sqrt{4 + 9x^2} - 2 \ln |(2 + \sqrt{4 + 9x^2})/(3x)| + C$

3 $-\frac{1}{8}x(2x^2 - 80)\sqrt{16 - x^2} + 96 \sin^{-1}\frac{1}{4}x + C$

5 $-\frac{2}{135}(9x + 4)(2 - 3x)^{3/2} + C$

7 $-\frac{1}{18} \sin^5 3x \cos 3x - \frac{5}{72} \sin^3 3x \cos 3x$
$\qquad - \frac{5}{48} \sin 3x \cos 3x + \frac{5}{16}x + C$

9 $-\frac{1}{3} \cot x \csc^2 x - \frac{2}{3} \cot x + C$

11 $\frac{1}{2}x^2 \sin^{-1} x + \frac{1}{4}x\sqrt{1 - x^2} - \frac{1}{4} \sin^{-1} x + C$

13 $\frac{1}{13}e^{-3x}(-3 \sin 2x - 2 \cos 2x) + C$

15 $\sqrt{5x - 9x^2} + \frac{5}{6} \cos^{-1}\frac{1}{5}(5 - 18x) + C$

17 $[1/(4\sqrt{15})] \ln |(\sqrt{5}x^2 - \sqrt{3})/(\sqrt{5}x^2 + \sqrt{3})| + C$

19 $\frac{1}{4}(2e^{2x} - 1) \cos^{-1} e^x - \frac{1}{4}e^x\sqrt{1 - e^{2x}} + C$

21 $\frac{2}{315}(35x^3 - 60x^2 + 96x - 128)(2 + x)^{3/2} + C$

23 $\frac{2}{81}(4 + 9 \sin x - 4 \ln |4 + 9 \sin x|) + C$

25 $2\sqrt{9 + 2x} + 3 \ln |(\sqrt{9 + 2x} - 3)/(\sqrt{9 + 2x} + 3)| + C$

27 $\frac{3}{4} \ln |\sqrt[3]{x}/(4 + \sqrt[3]{x})| + C$

29 $\sqrt{16 - \sec^2 x} - 4 \ln |(4 + \sqrt{16 - \sec^2 x})/\sec x| + C$

1 $\frac{1}{2}x^2 \sin^{-1} x - \frac{1}{4} \sin^{-1} x + \frac{1}{4}x\sqrt{1 - x^2} + C$

3 $2 \ln 2 - 1$ **5** $\frac{1}{6} \sin^3 2x - \frac{1}{10} \sin^5 2x + C$

7 $\frac{1}{5} \sec^5 x + C$ **9** $x/(25\sqrt{x^2 + 25}) + C$

11 $2 \ln |(2 - \sqrt{4 - x^2})/x| + \sqrt{4 - x^2} + C$

13 $2 \ln |x - 1| - \ln |x| - x/(x - 1)^2 + C$

15 $\ln [(x + 3)^2(x^2 + 9)^2/|x - 3|^5] + \frac{1}{3} \tan^{-1}\frac{1}{3}x + C$

17 $-\sqrt{4 + 4x - x^2} + 2 \sin^{-1}[(x - 2)/\sqrt{8}] + C$

19 $3(x + 8)^{1/3} + \ln [(x + 8)^{1/3} - 2]^2$
$\qquad - \ln [(x + 8)^{2/3} + 2(x + 8)^{1/3} + 4]$
$\qquad - (6/\sqrt{3}) \tan^{-1}[((x + 8)^{1/3} + 1)/\sqrt{3}] + C$

21 $\frac{1}{13}e^{2x}(2 \sin 3x - 3 \cos 3x) + C$

23 $\frac{1}{4} \sin^4 x - \frac{1}{6} \sin^6 x + C$ **25** $-\sqrt{4 - x^2} + C$

27 $\frac{1}{3}x^3 - x^2 + 3x - 1/(2x) - \frac{1}{4} \ln |x| - \frac{23}{4} \ln |x + 2| + C$

29 $2 \tan^{-1}(x^{1/2}) + C$ **31** $\ln |\sec e^x + \tan e^x| + C$

33 $\frac{1}{125}[10x \sin 5x - (25x^2 - 2)\cos 5x] + C$

35 $\frac{2}{7}\cos^{7/2} x - \frac{2}{3}\cos^{3/2} x + C$ **37** $\frac{2}{3}(1 + e^x)^{3/2} + C$

39 $\frac{1}{16}[2x\sqrt{4x^2 + 25} - 25\ln(2x + \sqrt{4x^2 + 25})] + C$

41 $\frac{1}{3}\tan^3 x + C$ **43** $-x\csc x + \ln|\csc x - \cot x| + C$

45 $-\frac{1}{4}(8 - x^3)^{4/3} + C$

47 $-2x \cos\sqrt{x} + 4\cos\sqrt{x} + 4\sqrt{x}\sin\sqrt{x} + C$

49 $\frac{1}{2}e^{2x} - e^x + \ln(1 + e^x) + C$

51 $\frac{2}{5}x^{5/2} - \frac{8}{3}x^{3/2} + 6x^{1/2} + C$

53 $\frac{1}{3}(16 - x^2)^{3/2} - 16(16 - x^2)^{1/2} + C$

55 $\frac{11}{2}\ln|x + 5| - \frac{15}{2}\ln|x + 7| + C$

57 $x\tan^{-1} 5x - \frac{1}{10}\ln|1 + 25x^2| + C$ **59** $e^{\tan x} + C$

61 $(1/\sqrt{5})\ln|\sqrt{5}x + \sqrt{7 + 5x^2}| + C$

63 $-\frac{1}{5}\cot^5 x - \frac{1}{3}\cot^3 x - \cot x - x + C$

65 $\frac{1}{5}(x^2 - 25)^{5/2} + \frac{25}{3}(x^2 - 25)^{3/2} + C$

67 $\frac{1}{6}x^3 - \frac{1}{4}\tanh 4x + C$

69 $-\frac{1}{4}x^2 e^{-4x} - \frac{1}{8}xe^{-4x} - \frac{1}{32}e^{-4x} + C$

71 $3\sin^{-1}[(x + 5)/6] + C$ **73** $-\frac{1}{7}\cos 7x + C$

75 $18\ln|x - 2| - 9\ln|x - 1| - 5\ln|x - 3| + C$

77 $x^3 \sin x + 3x^2 \cos x - 6x \sin x - 6\cos x + \sin x + C$

79 $(-1/x)\sqrt{9 - 4x^2} - 2\sin^{-1}\frac{2}{3}x + C$

81 $24x - \frac{10}{3}\ln|\sin 3x| - \frac{1}{3}\cot 3x + C$

83 $-\ln x - (4/\sqrt[4]{x}) + 4\ln(1 + \sqrt[4]{x}) + C$

85 $-2\sqrt{1 + \cos x} + C$

87 $-x/[2(25 + x^2)] + \frac{1}{10}\tan^{-1}\frac{1}{5}x + C$

89 $\frac{1}{3}\sec^3 x - \sec x + C$

91 $\ln(x^2 + 4) - \frac{3}{2}\tan^{-1}\frac{1}{2}x + (7/\sqrt{5})\tan^{-1}(x/\sqrt{5}) + C$

93 $\frac{1}{4}x^4 - 2x^2 + 4\ln|x| + C$

95 $\frac{2}{5}x^{5/2}\ln x - \frac{4}{25}x^{5/2} + C$

97 $\frac{3}{64}(2x + 3)^{8/3} - \frac{9}{20}(2x + 3)^{5/3} + \frac{27}{16}(2x + 3)^{2/3} + C$

99 $\frac{1}{2}e^{x^2}(x^2 - 1) + C$

CHAPTER 10

EXERCISES 10.1, PAGE 428

1 $\frac{1}{2}$ **3** $\frac{1}{40}$ **5** $\frac{3}{13}$ **7** 0 **9** $-\frac{1}{2}$

11 $-\frac{1}{2}$ **13** $\frac{1}{6}$ **15** ∞ **17** $\frac{1}{3}$ **19** ∞

21 1 **23** 0 **25** Does not exist   **27** $\frac{2}{5}$

29 ∞ **31** 0 **33** ∞ **35** 2

37 Does not exist   **39** $\frac{3}{5}$ **41** -3 **43** 0

45 ∞ **47** ∞ **49** 2 **51** 1

55 $\frac{1}{2}A\omega_0 t \sin \omega_0 t$ **57** (a) 1 (b) $-\frac{1}{18}$

EXERCISES 10.2, PAGE 432

1 0 **3** 0 **5** 0 **7** 0 **9** 1 **11** 0

13 e^5 **15** 1 **17** 1 **19** Does not exist

21 e^2 **23** 2 **25** 0 **27** 1

29 Does not exist   **31** $\frac{1}{2}$ **33** Does not exist

35 e **37** Does not exist   **39** $e^{1/3}$ **41** ∞

43 $f(e) = e^{1/e}$ is a local max;

$\lim_{x \to 0^+} f(x) = 0$; $y = 1$

47 (a) a

EXERCISES 10.3, PAGE 437

Exer. 1–27: C = Converges, D = Diverges.

1 C, 3 **3** D   **5** D   **7** C, $\frac{1}{2}$ **9** C, $-\frac{1}{2}$

11 D   **13** D   **15** 0 **17** D   **19** D   **21** C, π

23 C, $\ln 2$ **25** C   **27** D

29 (a) Does not exist   (b) π

31 (a) 1 (b) $\pi/32$ **33** π **35** (b) No

37 If $F(x) = k/x^2$, then $W = k$.

39 (a) $1/k$

(b) No, the improper integral for average repair time diverges.

41 (b) $c = (4/\sqrt{\pi})\left(\dfrac{m}{2kT}\right)^{3/2}$ **43** $1/s$, $s > 0$

45 $s/(s^2 + 1)$, $s > 0$ **47** $1/(s - a)$, $s > a$ **49** (a) $1, 1, 2$

EXERCISES 10.4, PAGE 442

Exer. 1–33: C = Converges, D = Diverges.

1 C, 6 **3** D   **5** D   **7** D   **9** C, $3\sqrt[3]{4}$

11 D   **13** C, $\pi/2$ **15** D   **17** C, $-\frac{1}{4}$

19 D   **21** D   **23** D   **25** C, 0 **27** D

29 D   **31** C   **33** D   **35** $n > -1$

37 (a) 2 (b) Value cannot be assigned.

39 Values cannot be assigned to either the area or the volume.

41 (b) $T = 2\pi\sqrt{m/k}$ **43** (a) t is undefined at $y = 0$.

EXERCISES 10.5, PAGE 451

1 $\sin x = 1 - \dfrac{1}{2}\left(x - \dfrac{\pi}{2}\right)^2 + \dfrac{\sin z}{4!}\left(x - \dfrac{\pi}{2}\right)^4$,

z is between x and $\pi/2$.

3 $\sqrt{x} = 2 + \frac{1}{4}(x - 4) - \frac{1}{64}(x - 4)^2 + \frac{1}{512}(x - 4)^3 - \frac{5}{128}z^{-7/2}(x - 4)^4$, z is between x and 4.

5 $\tan x = 1 + 2\left(x - \dfrac{\pi}{4}\right) + 2\left(x - \dfrac{\pi}{4}\right)^2 + \dfrac{8}{3}\left(x - \dfrac{\pi}{4}\right)^3 + \dfrac{10}{3}\left(x - \dfrac{\pi}{4}\right)^4 + \dfrac{g(z)}{5!}\left(x - \dfrac{\pi}{4}\right)^5$,

z is between x and $\pi/4$ and

$g(z) = 16\sec^6 z + 88\sec^4 z\tan^2 z + 16\sec^2 z\tan^4 z$

7 $1/x = -\frac{1}{2} - \frac{1}{4}(x + 2) - \frac{1}{8}(x + 2)^2 - \frac{1}{16}(x + 2)^3 - \frac{1}{32}(x + 2)^4 - \frac{1}{64}(x + 2)^5 + z^{-7}(x + 2)^6$.

z is between x and -2

9 $\tan^{-1} x = \dfrac{\pi}{4} + \dfrac{1}{2}(x - 1) - \dfrac{1}{4}(x - 1)^2 + \dfrac{3z^2 - 1}{3(1 + z^2)^3}(x - 1)^3$,

z is between 1 and x

11 $xe^x = -\dfrac{1}{e} + \dfrac{1}{2e}(x - 1)^2 + \dfrac{1}{3e}(x + 1)^3 + \dfrac{1}{8e}(x + 1)^4 + \dfrac{ze^z + 5e^z}{120}(x + 1)^5$, z is between x and -1

13 $\ln(x + 1) = x - \dfrac{1}{2}x^2 + \dfrac{1}{3}x^3 - \dfrac{1}{4}x^4 + \dfrac{x^5}{5(z + 1)^5}$,

z is between 0 and x

15 $\cos x = 1 - \dfrac{x^2}{2!} + \dfrac{x^4}{4!} - \dfrac{x^6}{6!} + \dfrac{x^8}{8!} - \dfrac{\sin z}{9!}x^9$,

z is between x and 0

17 $e^{2x} = 1 + 2x + 2x^2 + \frac{4}{3}x^3 + \frac{2}{3}x^4 + \frac{4}{15}x^5 + \frac{4}{45}e^{2z}x^6$, z is between x and 0

19 $1/(x-1)^2 = 1 + 2x + 3x^2 + 4x^3 + 5x^4 + 6x^5 + 7x^6/(z-1)^8$, z is between 0 and x

21 $\arcsin x = x + \dfrac{1 + 2z^2}{6(1 - z^2)^{5/2}} x^3$, z is between 0 and x

23 $f(x) = 7 - 3x + x^2 - 5x^3 + 2x^4$ **25** 0.9998

27 2.0075 **29** 0.454545; error ≤ 0.0000005

31 0.223; error ≤ 0.0002

33 0.8660254; error $\leq (8.1)(10^{-9})$

35 Five decimal places, since $|R_3(x)| \leq 4.2 \times 10^{-6} < 0.5 \times 10^{-5}$

37 Three decimal places **39** Four decimal places

1 $\frac{1}{2}\ln 2$ **3** ∞ **5** $\frac{8}{3}$ **7** 0 **9** $-\infty$ **11** e^8

13 e **15** 1 **17** Diverges **19** Diverges **21** $-\frac{9}{2}$

23 Diverges **25** $\pi/2$ **27** Diverges **29** 0

31 (a) $\ln \cos x = \ln \dfrac{\sqrt{3}}{2} - \dfrac{1}{\sqrt{3}}\left(x - \dfrac{\pi}{6}\right) - \dfrac{2}{3}\left(x - \dfrac{\pi}{6}\right)^2$

$-\dfrac{4}{9\sqrt{3}}\left(x - \dfrac{\pi}{6}\right)^3 - \dfrac{1}{12}(\sec^4 z + 2\sec^2 z \tan^2 z)\left(x - \dfrac{\pi}{6}\right)^4$,

z is between x and $\pi/6$

(b) $\sqrt{x-1} = 1 + \frac{1}{2}(x-2) - \frac{1}{8}(x-2)^2 + \frac{1}{16}(x-2)^3$
$- \frac{5}{128}(x-2)^4 + \frac{7}{256}(z-1)^{-9/2}(x-2)^5$,

z is between x and 2

33 0.4651 (with $n = 3$, $|R_3(x)| \leq 1.6 \times 10^{-6}$)

CHAPTER 11

1 $\frac{1}{5}, \frac{1}{4}, \frac{3}{11}, \frac{2}{7}; \frac{1}{3}$ **3** $\frac{3}{5}, -\frac{9}{11}, -\frac{29}{21}, -\frac{57}{35}; -2$

5 $-5, -5, -5, -5; -5$ **7** $2, \frac{7}{3}, \frac{25}{14}, \frac{7}{5}; 0$

9 $2/\sqrt{10}, 2/\sqrt{13}, 2/\sqrt{18}, \frac{2}{5}; 0$ **11** $\frac{3}{10}, -\frac{6}{17}, \frac{9}{26}, -\frac{12}{37}; 0$

13 $1.1, 1.01, 1.001, 1.0001; 1$

15 $2, 0, 2, 0$; the limit does not exist.

Exer. 17–41: C = Converges, D = Diverges.

17 C, 0 **19** C, $\pi/2$ **21** D **23** C, 0 **25** D

27 D **29** C, e **31.** C, 0 **33** C, $\frac{1}{2}$ **35** D

37 C, 1 **39** C, 0 **41** C, 0

43 (b) $10{,}000$ on A; $5{,}000$ on B; $20{,}000$ on C

45 (a) The sequence appears to converge to 1.
(b) $\lim_{n \to \infty} a_n = 1$

47 (a) The sequence appears to converge to approximately 0.739.

49 (a) 3.5; 3.178571429; 3.162319422; 3.162277661; 3.162277660

Exer. 1–27: C = Converges, D = Diverges.

1 C, 4 **3** C, $\sqrt{5}/(\sqrt{5}+1)$ **5** C, $\frac{37}{99}$ **7** D

9 D **11** C, $\frac{1}{4}$ **13** C, 5 **15** D **17** D **19** C, $\frac{8}{7}$

21 D **23** C **25** D **27** D

29 $S_n = \frac{1}{2}[1 - 1/(2n+1)]$; converges to $\frac{1}{2}$

31 $S_n = -\ln(n+1)$; diverges

33 The result is false (find an example). **35** $\frac{23}{99}$

37 $\frac{16181}{4995}$ **39** 30 m

41 (b) $Q/(1 - e^{-ct})$ (c) $-(1/c)\ln[(M-Q)/M]$

43 (b) 2000

45 (a) $a_{k+1} = \frac{1}{4}\sqrt{10}a_k$; $a_n = (\frac{1}{4}\sqrt{10})^{n-1}a_1$; $A_n = (\frac{5}{8})^{n-1}A_1$; $P_n = (\frac{1}{4}\sqrt{10})^{n-1}P_1$

(b) $[16/(4 - \sqrt{10})]a_1$; $\frac{8}{3}a_1^2$

Exer. 1–45: C = Converges, D = Diverges.

1 C **3** D **5** D **7** D **9** C **11** C **13** C

15 D **17** C **19** C **21** D **23** D **25** C

27 C **29** C **31** C **33** C **35** D **37** C

39 C **41** C **43** C **45** C

47 Converges if $k > 1$, diverges if $k \leq 1$

49 (b) $n > e^{100} - 1 \approx 2.688 \times 10^{43}$ **55** 8

Exer. 1–25: C = Convergent, D = Divergent.

1 C **3** D **5** C **7** D **9** C **11** C **13** C

15 C **17** D **19** C **21** C **23** D **25** D

Exer. 1–23: AC = Absolutely Convergent; CC = Conditionally Convergent; D = Divergent.

1 CC **3** CC **5** D **7** AC **9** AC **11** D

13 CC **15** AC **17** D **19** CC **21** D **23** D

25 0.368 **27** 0.901 **29** 0.306 **31** 141 **33** 5

37 No. Consider $a_n = b_n = (-1)^n/\sqrt{n}$.

1 $[-1, 1)$ **3** $(-2, 2)$ **5** $(-1, 1]$ **7** $[-1, 1)$

9 $[-1, 1]$ **11** $(-6, 14)$

13 Converges only for $x = 0$ **15** $(-2, 2)$

17 $(-\infty, \infty)$ **19** $[\frac{17}{9}, \frac{19}{9})$ **21** $(-12, 4)$ **23** $(0, 2e)$

25 $(-\frac{5}{2}, \frac{7}{2}]$ **27** $\frac{3}{2}$ **29** $1/e$ **31** ∞

1 $\sum_{n=0}^{\infty} x^n$; $-1 < x < 1$ **3** $\sum_{n=1}^{\infty} nx^{n-1}$; $-1 < x < 1$

5 $\sum_{n=0}^{\infty} x^{2(n+1)}$; $-1 < x < 1$

7 $\sum_{n=0}^{\infty} (3^n/2^{n+1})x^{n+1}$; $-\frac{2}{3} < x < \frac{2}{3}$

9 $-1 - x - 2\sum_{n=2}^{\infty} x^n$; $-1 < x < 1$ **11** 0.182

13 $\sum_{n=0}^{\infty} \dfrac{(-1)^n}{n!} x^n$ **15** $\sum_{n=0}^{\infty} \dfrac{1}{(2n)!} x^{2n}$ **17** $\sum_{n=0}^{\infty} \dfrac{3^n}{n!} x^{n+1}$

19 0.3333 **21** 0.0992 **23** 0.9677 **25** $\sum_{n=1}^{\infty} 2nx^{2n-1}$

29 $-\sum_{n=1}^{\infty} \dfrac{1}{n^2} x^n$

EXERCISES 11.8, PAGE 507

1 $\displaystyle\sum_{n=0}^{\infty} \frac{(-1)^n}{(2n)!} x^{2n}$
3 $\displaystyle\sum_{n=0}^{\infty} \frac{2^n}{n!} x^n$
5 $\displaystyle\sum_{n=0}^{\infty} \frac{1}{n!} x^{n+2}$; ∞

7 $\displaystyle\sum_{n=0}^{\infty} \frac{1}{(2n+1)!} x^{2n+1}$; ∞
9 $\displaystyle\sum_{n=0}^{\infty} \frac{(-1)^n 3^{2n+1}}{(2n+1)!} x^{2n+2}$; ∞

11 $1 + \displaystyle\sum_{n=1}^{\infty} \frac{(-1)^n 2^{2n-1}}{(2n)!} x^{2n}$; ∞

13 $\displaystyle\sum_{n=0}^{\infty} \frac{(-1)^n}{\sqrt{2}(2n+1)!} \left(x - \frac{\pi}{4}\right)^{2n+1} + \sum_{n=0}^{\infty} \frac{(-1)^n}{\sqrt{2}(2n)!} \left(x - \frac{\pi}{4}\right)^{2n}$

15 $\displaystyle\sum_{n=0}^{\infty} \frac{(-1)^n}{2^{n+1}} (x-2)^n$
17 $\displaystyle\sum_{n=0}^{\infty} \frac{(\ln 10)^n}{n!} x^n$

19 $\displaystyle\sum_{n=0}^{\infty} \frac{e^{-2} 2^n}{n!} (x+1)^n$

21 $2 + 2\sqrt{3}\left(x - \frac{\pi}{3}\right) + 7\left(x - \frac{\pi}{3}\right)^2 + \frac{23\sqrt{3}}{3}\left(x - \frac{\pi}{3}\right)^3 + \cdots$

23 $\dfrac{\pi}{6} + \dfrac{2}{\sqrt{3}}\left(x - \dfrac{1}{2}\right) + \dfrac{2}{3\sqrt{3}}\left(x - \dfrac{1}{2}\right)^2 + \dfrac{8}{9\sqrt{3}}\left(x - \dfrac{1}{2}\right)^3$

25 $-\dfrac{1}{e} + \dfrac{1}{2e}(x+1)^2 + \dfrac{1}{3e}(x+1)^3 + \dfrac{1}{8e}(x+1)^4$

27 1.6487 **29** 0.9986 **31** 0.0997 **33** 0.5211
35 0.7468 **37** 0.4969 **39** 0.4864 **41** 0.4484

43 $2\left(x + \dfrac{x^3}{3} + \dfrac{x^5}{5} + \cdots + \dfrac{x^{2n-1}}{2n-1} + \cdots\right)$ if $|x| < 1$

45 $\pi = 4\left(1 - \dfrac{1}{3} + \dfrac{1}{5} - \dfrac{1}{7} + \cdots + (-1)^n \dfrac{1}{2n+1} + \cdots\right)$;
using five terms, $\pi \approx 3.34$; at least 40,000 terms are required.
47 (c) 16.7 ft

EXERCISES 11.9, PAGE 510

1 (a) $1 + \frac{1}{2}x - \frac{1}{8}x^2 + \displaystyle\sum_{n=3}^{\infty} (-1)^{n-1} \frac{1 \cdot 3 \cdot 5 \cdots (2n-3)}{2^n n!} x^n$; 1

(b) $1 - \frac{1}{2}x^3 - \frac{1}{8}x^6 - \displaystyle\sum_{n=3}^{\infty} \frac{1 \cdot 3 \cdot 5 \cdots (2n-3)}{2^n n!} x^{3n}$; 1

3 $1 - \frac{2}{3}x + \frac{5}{9}x^2 + \displaystyle\sum_{n=3}^{\infty} \frac{(-2)(-5) \cdots (1-3n)}{3^n n!} x^n$; 1

5 $1 - \frac{3}{5}x - \frac{3}{25}x^2 + \displaystyle\sum_{n=3}^{\infty} \frac{(3)(-2) \cdots (8-5n)}{5^n n!} (-x)^n$; 1

7 $1 - 2x + 3x^2 + \displaystyle\sum_{n=3}^{\infty} \frac{(-2)(-3) \cdots (-1-n)}{n!} x^n$; 1

9 $1 - 3x + 6x^2 + \displaystyle\sum_{n=3}^{\infty} (-1)^n \frac{1}{2}(n+1)(n+2)x^n$; 1

11 $2 + \frac{1}{12}x - \frac{1}{288}x^2 + 2\displaystyle\sum_{n=3}^{\infty} (-1)^{n-1} \frac{2 \cdot 5 \cdot 8 \cdots (3n-4)}{3^n 8^n n!} x^n$; 8

13 $x + \displaystyle\sum_{n=1}^{\infty} \frac{1 \cdot 3 \cdot 5 \cdots (2n-1)}{2^n n!(2n+1)} x^{2n+1}$; 1   **15** 0.508

17 0.198   **19** 0.297

EXERCISES 11.10, PAGE 511

1 Converges to 0   **3** Diverges   **5** Converges to 5

Exer. 7–37: C = Convergent, AC = Absolutely Convergent, CC = Conditionally Convergent, D = Divergent

7 D	**9** AC	**11** D	**13** D	**15** AC	**17** D
19 D	**21** AC	**23** CC	**25** AC	**27** C	**29** C
31 CC	**33** C	**35** C	**37** D	**39** 0.159	

41 $(-3, 3)$ **43** $[-12, -8)$ **45** $\frac{1}{4}$

47 $\displaystyle\sum_{n=1}^{\infty} \frac{(-1)^{n+1}}{(2n)!} x^{2n-1}$; ∞ **49** $\displaystyle\sum_{n=0}^{\infty} \frac{(-1)^n 2^{2n}}{(2n+1)!} x^{2n+1}$; ∞

51 $1 + \frac{2}{3}x + 2\displaystyle\sum_{n=2}^{\infty} (-1)^{n-1} \frac{1 \cdot 4 \cdot 7 \cdots (3n-5)}{3^n n!} x^n$; 1

53 $e^2 \displaystyle\sum_{n=0}^{\infty} \frac{(-1)^n}{n!} (x+2)^n$

55 $2 + \dfrac{x-4}{4} + \displaystyle\sum_{n=2}^{\infty} (-1)^{n-1} \frac{1 \cdot 3 \cdot 5 \cdots (2n-3)}{2^{3n-1} n!} (x-4)^n$

57 0.189   **59** 1.002

CHAPTER 12

EXERCISES 12.2, PAGE 519

1 $V(0, 0)$; $F(0, -3)$; $y = 3$
3 $V(0, 0)$; $F(-\frac{3}{8}, 0)$; $x = \frac{3}{8}$

5 $V(0, 0)$; $F(0, \frac{1}{32})$; $y = -\frac{1}{32}$
7 $V(-1, 0)$; $F(2, 0)$; $x = -4$

9 $V(2, -2)$; $F(2, -\frac{7}{4})$; $y = -\frac{9}{4}$
11 $V(-4, 2)$; $F(-\frac{7}{2}, 2)$; $x = -\frac{9}{2}$

13 $V(-5, -6)$; $F(-5, -\frac{97}{16})$; $y = -\frac{95}{16}$ **15** $V(0, \frac{1}{2})$; $F(0, -\frac{9}{2})$; $y = \frac{11}{2}$

17 Let $y' = 2ax + b = 0$ to obtain the x-coordinate $-b/(2a)$ of the vertex. Given $x = ay^2 + by + c$, let $2ay + b = 0$ to obtain the y-coordinate $-b/(2a)$ of the vertex.

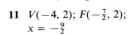

19 $y^2 = 8x$ **21** $(x - 6)^2 = 12(y - 1)$

23 $3x^2 = -4y$ **25** $\frac{9}{16}$ ft from the vertex

27 $y = 2x^2 - 3x + 1$ **29** (a) $\frac{8}{3}$ (b) 2π (c) $16\pi/5$

31 (b) $p = r^2/(4h)$ **33** $\frac{200}{3}$ lb   **35** (b) 64,968 ft²

37 $2\sqrt{2}$ rad/sec ≈ 0.45 rev/sec

39 (a) $y = -\frac{7}{160}x^2 + x$ (b) 17.5 ft

EXERCISES 12.3, PAGE 526

1 $V(\pm 3, 0)$; $F(\pm\sqrt{5}, 0)$ **3** $V(0, \pm 4)$; $F(0, \pm 2\sqrt{3})$

5 $V(0, \pm\sqrt{5})$; $F(0, \pm\sqrt{3})$ **7** $V(\pm\frac{1}{2}, 0)$; $F(\pm\sqrt{21}/10, 0)$

9 Center (4, 2), vertices (1, 2) and (7, 2); $F(4 \pm \sqrt{5}, 2)$; endpoints of minor axis (4, 4) and (4, 0)

11 Center (−3, 1), vertices (−7, 1) and (1, 1); $F(-3 \pm \sqrt{7}, 1)$; endpoints of minor axis (−3, 4) and (−3, −2)

13 Center (5, 2), vertices (5, 7) and (5, −3); $F(5, 2 \pm \sqrt{21})$; endpoints of minor axis (3, 2) and (7, 2)

    9   **11**   **13**

15 $(x^2/64) + (y^2/39) = 1$ **17** $(4x^2/9) + (y^2/25) = 1$

19 $(8x^2/81) + (y^2/36) = 1$ **21** $(x^2/7) + (y^2/16) = 1$

23 $2\sqrt{21}$ ft   **25** $y - 3 = \frac{5}{6}(x + 2)$

27 (a) $\frac{4}{3}\pi ab^2$ (b) $\frac{4}{3}\pi a^2 b$ **29** $2a/\sqrt{2}$ and $2b/\sqrt{2}$

33 94,581,000; 91,419,000   **35** (b) $A = E/fm$

EXERCISES 12.4, PAGE 533

1 $V(\pm 3, 0)$; $F(\pm\sqrt{13}, 0)$; $y = \pm 2x/3$ **3** $V(0, \pm 3)$; $F(0, \pm\sqrt{13})$; $y = \pm 3x/2$

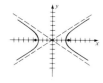

5 $V(0, \pm 4)$; $F(0, \pm 2\sqrt{5})$; $y = \pm 2x$ **7** $V(\pm 1, 0)$; $F(\pm\sqrt{2}, 0)$; $y = \pm x$

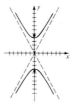

9 $V(\pm 5, 0)$; $F(\pm\sqrt{30}, 0)$; $y = \pm(\sqrt{5}/5)x$ **11** $V(0, \pm\sqrt{3})$; $F(0, \pm 2)$; $y = \pm\sqrt{3}x$

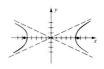

13 Center (−5, 1), vertices (−5 ± 2√5, 1); $F(-5 \pm \sqrt{205}/2, 1)$; $y - 1 = \pm\frac{5}{4}(x + 5)$

15 Center (−2, −5), vertices (−2, −2) and (−2, −8); $F(-2, -5 \pm 3\sqrt{5})$; $y + 5 = \pm\frac{1}{2}(x + 2)$

   13   **15**

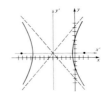
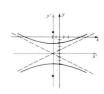

17 Center (6, 2), vertices (6, 4) and (6, 0); $F(6, 2 \pm 2\sqrt{10})$; $y - 2 = \pm\frac{1}{3}(x - 6)$

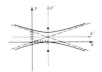

19 $15y^2 - x^2 = 15$ **21** $(x^2/9) - (y^2/16) = 1$

23 $(y^2/21) - (x^2/4) = 1$ **25** $(x^2/9) - (y^2/36) = 1$

27 Conjugate hyperbolas have the same asymptotes.

29 $y - 1 = -\frac{4}{5}(x + 2)$ **31** $(\pm 2\sqrt{2}, -6)$

33 (a) $\frac{1}{3}\pi b^2[\sqrt{a^2 + b^2}(b^2 - 2a^2) + 2a^3]/a^2$ (b) $\frac{4}{3}\pi b^4/a$

35 If a coordinate system is introduced with the x-axis along AB and the y-axis through the midpoint of AB, then the ship is at the point $(\frac{80}{3}\sqrt{34}, 100) \approx (155.5, 100)$.

39 $\frac{1}{2}$ AU

Exer. 1-13: The answer consists of (a) the value of $B^2 - 4AC$ and the type of conic; (b) an equation in x' and y', resulting from a rotation of axes, and a sketch of the graph.

1 (a) -1600, ellipse
   (b) $(x')^2 + 16(y')^2 = 16$

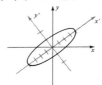

3 (a) 256, hyperbola
   (b) $4(x')^2 - (y')^2 = 1$

5 (a) -36, ellipse
   (b) $(x')^2 + 9(y')^2 = 9$

7 (a) 0, parabola
   (b) $(y')^2 = 4(x' - 1)$

9 (a) 128, hyperbola *(see graph)*
   (b) $2(x')^2 - (y')^2 - 4y' - 3 = 0$

11 (a) 0, parabola *(see graph)*
   (b) $(x')^2 - 6x' - 6y' + 9 = 0$

13 (a) -2704, ellipse *(see graph)*
   (b) $(x')^2 + 4(y')^2 - 4x' = 0$

9

11

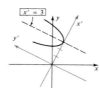

13

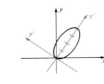

1 $F(16, 0)$;
   $V(0, 0)$

3 $F(0, \pm\sqrt{7})$;
   $V(0, \pm 4)$

5 $F(\pm 2\sqrt{2}, 0)$;
   $V(\pm 2, 0)$

7 $F(0, -\frac{9}{4})$;
   $V(0, 4)$

9 $F(-4 \pm \sqrt{10}/3, 5)$;
   vertices $(-5, 5)$ and $(-3, 5)$

11 Center $(-3, 2)$; vertices $(-6, 2)$ and $(0, 2)$; endpoints of minor axis $(-3, 0)$ and $(-3, 4)$

13 $V(2, -4)$; $F(4, -4)$

15 Center $(-4, 0)$; vertices $(-7, 0)$ and $(-1, 0)$; asymptotes $y = \pm(x + 4)/3$

11

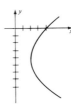

13

15

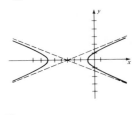

17 $9y^2 - 49x^2 = 441$ **19** $x^2 = -40y$

21 $4x^2 + 3y^2 = 300$ **23** $y^2 - 81x^2 = 36$

25 $(x^2/25) + (y^2/45) = 1$

27 $(x^2/10000) + (y^2/960) = 1$; $8\sqrt{15} \approx 30.98$ ft

29 $y - 2 = -\frac{16}{15}(x + 3)$; $y - 2 = \frac{15}{16}(x + 3)$

35 $\frac{64}{3}\pi$

37 $\bar{x} = 0$, $\bar{y} = \frac{2}{3}b^3/K$, $K = bc - a^2$ $[\ln(c + b) - \ln a]$

39 Parabola; $(y')^2 - 3x' = 0$ (after a rotation of axes)

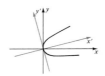

CHAPTER 13

1 $y = 2x + 7$ **3** $y = x - 2$ **5** $x = y^2 - 6y + 4$

  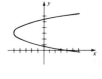

7 $y = 1/x^2$ **9** $(x^2/4) + (y^2/9) = 1$ **11** $x^2 - y^2 = 1$

13 $y = \ln x$ **15** $y = 1/x$ **17** $x^2 - y^2 = 1$

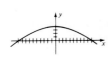

19 $y = \sqrt{x^2 - 1}$ **21** **23**

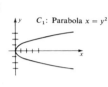 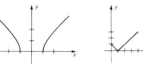 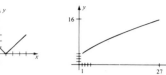

25

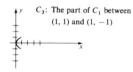 C_1: Parabola $x = y^2$ 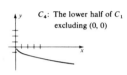 C_2: Upper half of C_1

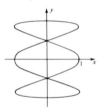 C_3: The part of C_1 between $(1, 1)$ and $(1, -1)$ 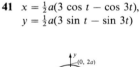 C_4: The lower half of C_1 excluding $(0, 0)$

27 (a) P moves counterclockwise along the circle $x^2 + y^2 = 1$ from $(1, 0)$ to $(-1, 0)$.
   (b) P moves clockwise along the circle $x^2 + y^2 = 1$ from $(0, 1)$ to $(0, -1)$.
   (c) P moves clockwise along the circle $x^2 + y^2 = 1$ from $(-1, 0)$ to $(1, 0)$.
29 $x = (x_2 - x_1)t^n + x_1$, $y = (y_2 - y_1)t^n + y_1$ are parametric equations for l if n is any odd positive integer.
33 (a) Ellipse with center at the origin
35

39 $x = 4b \cos t - b \cos 4t$,   **41** $x = \frac{1}{2}a(3 \cos t - \cos 3t)$,
$y = 4b \sin t - b \sin 4t$ $y = \frac{1}{2}a(3 \sin t - \sin 3t)$

45 $x = 1 + \frac{1}{2} \cos \omega t$,
$y = \frac{1}{2} \sin \omega t$

EXERCISES 13.2, PAGE 555
1 1 **3** $\frac{1}{4}$ **5** $-2e^{-3}$ **7** $-\frac{3}{2} \tan 1$
9 (a) Horizontal at $(16, -16)$ and $(16, 16)$; vertical at $(0, 0)$
   (b) $(3t^2 + 12)/(64t^3)$
11 (a) Vertical at $(0, 0)$ and $(-3, 1)$; no horizontal
   (b) $(1 - 3t)/[144t^{3/2}(t - 1)^3]$
13 (a) Horizontal at $(\pm 1, 0)$; vertical at $(0, \pm 1)$
   (b) $\frac{1}{3} \sec^4 t \csc t$
15 (a) Horizontal at $(0, \pm 2)$, $(2\sqrt{3}, \pm 2)$, $(-2\sqrt{3}, \pm 2)$
      Vertical at $(4, \pm\sqrt{2})$, $(-4, \pm\sqrt{2})$
   (b) 2π
17 $\frac{2}{27}(34^{3/2} - 125)$ **19** $\sqrt{2}(e^{\pi/2} - 1)$ **21** $\frac{1}{8}\pi^2$
23 $\frac{8}{3}\pi(17^{3/2} - 1)$ **25** $\frac{11}{9}\pi$ **27** $\frac{64}{3}\pi a^3$ **29** $\frac{536}{5}\pi$
31 $\frac{2}{5}\sqrt{2}\pi(2e^{\pi} + 1)$

EXERCISES 13.3, PAGE 562
1 **3** **5**

    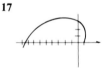

7 **9** **11**

    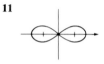

13 **15** **17**

    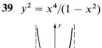

19 **21** **23**

25 $r = -3 \sec \theta$ **27** $r = 4$
29 $r^2 = 16 \sec 2\theta$ **31** $r\theta = a \sin \theta$
33 $x = 5$ **35** $x^2 + y^2 - 6y = 0$

37 $x^2 + y^2 = 4$ **39** $y^2 = x^4/(1 - x^2)$

ANSWERS TO EXERCISES 13.3 / **A45**

41 $(y^2/9) - (x^2/4) = 1$

43 $x^2 - y^2 = 1$

45 $y - 2x = 6$

47 $y^2 = 1 - 2x$

51 $\sqrt{3}/3$ **53** -1 **55** 2   **57** 0
59 $1/\ln 2 \approx 1.44$ **67** (c) Show that $r = r_0 e^{\pi/4} e^{-\theta}$

 1 π **3** $3\pi/2$ **5** $\pi/2$ **7** $33\pi/2$ **9** $(e^\pi - 1)/4$
11 2   **13** $9\pi/20$ **15** $2\pi + (9\sqrt{3}/2)$
17 $4\sqrt{3} - (4\pi/3)$ **19** $(5\pi/24) - \sqrt{3}/4$
21 $11 \sin^{-1} \frac{1}{4} + (3\pi/4) - \sqrt{15}/4$ **23** $4/\sqrt{3}$
25 $(64\sqrt{2})/3$ **27** $a^2 \arccos (b/a) - b\sqrt{a^2 - b^2}$
29 $\sqrt{2(1 - e^{-2\pi})}$ **31** 2   **33** $\frac{3}{2}\pi$ **35** $\frac{128}{5}\pi$
37 $4\pi^2 a^2$ **39** $4\pi^2 ab$

 1 Ellipse; vertices $(\frac{3}{2}, \pi/2)$ and $(3, 3\pi/2)$; foci $(0, 0)$ and $(\frac{3}{2}, 3\pi/2)$
 3 Hyperbola; vertices $(-3, 0)$ and $(\frac{3}{2}, \pi)$; foci $(0, 0)$ and
     $(-\frac{9}{2}, 0)$
 5 Parabola; $V(\frac{3}{4}, 0)$; $F(0, 0)$
 7 Ellipse; vertices $(-4, 0)$ and $(-\frac{4}{3}, \pi)$; foci $(0, 0)$ and $(-\frac{8}{3}, 0)$
 9 Hyperbola (except for the points $(\pm 3, 0)$); vertices $(\frac{6}{5}, \pi/2)$
     and $(-6, 3\pi/2)$; foci $(0, 0)$ and $(-\frac{36}{5}, 3\pi/2)$
11 $9x^2 + 8y^2 + 12y - 36 = 0$
13 $y^2 - 8x^2 - 36x - 36 = 0$ **15** $4y^2 = 9 - 12x$
17 $3x^2 + 4y^2 + 8x - 16 = 0$
19 $4x^2 - 5y^2 + 36y - 36 = 0, y \neq 0$
21 $r = 2/(3 + \cos \theta)$ **23** $r = 12/(1 - 4 \sin \theta)$
25 $r = 5/(1 + \cos \theta)$ **27** $r = 8/(1 + \sin \theta)$
31 $r = 2/(1 + \cos \theta)$; $e = 1$; $r = 2 \sec \theta$
33 $r = 4/(1 + 2 \sin \theta)$; $e = 2$; $r = 2 \csc \theta$
35 $r = 2/(3 + \cos \theta)$; $e = \frac{1}{3}$; $r = 2 \sec \theta$ **37** $-3\sqrt{3}/5$
39 3

 1 $y = 2(x - 1) - 1/(x - 1)$; 3   **3** $y = e^{-x^2}$; $-2e^{-1}$

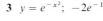

5 (a)   (b)   (c)   (d)

7 $(x^2 + y^2)^{3/2} = 6(x^2 - y^2)$ **9** $(x^2 + y^2)^2 + 8xy = 0$

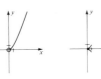

11 $3x - 2y = 6$ **13** $x^2 - y^2 = 1$

15 $x^2 + y^2 = 2(\sqrt{x^2 + y^2} + x)$
17 $8x^2 + 9y^2 + 16x - 64 = 0$

15   **17**

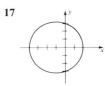

19 -1 **21** $r = 2 \cos \theta \sec 2\theta$ **23** 2
25 $\sqrt{2} + \ln (1 + \sqrt{2})$ **27** $2\pi[5\sqrt{2} + \ln (1 + \sqrt{2})]$
29 $2\pi a^2(2 - \sqrt{2})$

CHAPTER 14

 1 $\mathbb{R} \times \mathbb{R}$; $-29, 6, -4$ **3** $\{(u, v): u \neq 2v\}$; $-\frac{3}{2}, \frac{4}{9}, 0$
 5 $\{(x, y, z): x^2 + y^2 + z^2 \leq 25\}$; $4, 2\sqrt{3}$
 7 $-\frac{2}{3}$ **9** Does not exist   **11** 0   **13** Does not exist
15 0   **17** Continuous on $\{(x, y): x + y > 1\}$
19 Continuous at (x, y, z) if $x^2 + y^2 \neq z^2$
21 Continuous on $\{(x, y): x > 0, -1 < y < 1\}$
23 $(x^4 - 2x^2y^2 + y^4 - 4)/(x^2 - y^2)$; $\{(x, y): y \neq \pm x\}$
25 $x^2 + 2x \tan y + \tan^2 y + 1$;
     $\{(x, y): y \neq (\pi/2) + k\pi, k \text{ any integer}\}$
27 (a) Replace the ordered pairs in Definition (14.1) by ordered
     quadruples (x, y, z, w) for real numbers $x, y, z,$ and w.
     (b) $\lim_{(x, y, z, w) \to (a, b, c, d)} f(x, y, z, w) = L$ means that for
     every $\varepsilon > 0$ there corresponds a $\delta > 0$ such that if
     $0 < \sqrt{(x - a)^2 + (y - b)^2 + (z - c)^2 + (w - d)^2} < \delta$,
     then $|f(x, y, z, w) - L| < \varepsilon$.

1 $f_x(x, y) = 8x^3y^3 - y^2$; $f_y(x, y) = 6x^4y^2 - 2xy + 3$
3 $f_r(r, s) = r/\sqrt{r^2 + s^2}$; $f_s(r, s) = s/\sqrt{r^2 + s^2}$
5 $f_x(x, y) = e^y + y \cos x$; $f_y(x, y) = xe^y + \sin x$
7 $f_t(t, v) = -v/(t^2 - v^2)$; $f_v(t, v) = t/(t^2 - v^2)$
9 $f_x(x, y) = \cos(x/y) - (x/y) \sin(x/y)$;
   $f_y(x, y) = (x/y)^2 \sin(x/y)$
11 $f_r(r, s, t) = 2re^{2s} \cos t$; $f_s(r, s, t) = 2r^2e^{2s} \cos t$;
   $f_t(r, s, t) = -r^2e^{2s} \sin t$
13 $f_x(x, y, z) = (y^2 + z^2)^x \ln(y^2 + z^2)$;
   $f_y(x, y, z) = 2xy(y^2 + z^2)^{x-1}$;
   $f_z(x, y, z) = 2xz(y^2 + z^2)^{x-1}$
15 $f_x(x, y, z) = e^z - ye^x$; $f_y(x, y, z) = -e^x - ze^{-y}$;
   $f_z(x, y, z) = xe^z + e^{-y}$
17 $f_q(q, v, w) = v/(2\sqrt{qv}\sqrt{1 - qv})$;
   $f_v(q, v, w) = [q/(2\sqrt{qv}\sqrt{1 - qv})] + w \cos vw$;
   $f_w(q, v, w) = v \cos vw$
25 $18y^2 + 16y^3z$ 27 $t^2 (\sec rt)(\sec^2 rt + \tan^2 rt)$
29 $(1 - x^2y^2z^2) \cos xyz - 3xyz \sin xyz$
47 $w_{xx}, w_{xy}, w_{xz}, w_{yx}, w_{yy}, w_{yz}, w_{zx}, w_{zy}, w_{zz}$
49 In degrees per cm: (a) 200 (b) 400
51 $\partial C/\partial x = -36.58$ μg/m^3 per meter is the rate at which the
   concentration changes in the horizontal direction at (2, 5);
   $\partial C/\partial z = -0.229$ μg/m^3 per meter is the rate at which the
   concentration changes in the vertical direction at (2, 5).
53 (a) $\partial T/\partial t = T_0\omega e^{-\lambda x} \cos(\omega t - \lambda x)$ is the rate of change of
      temperature with respect to time at the fixed depth x;
      $\partial T/\partial x = -T_0\lambda e^{-\lambda x} [\cos(\omega t - \lambda x) + \sin(\omega t - \lambda x)]$ is
      the rate of change of temperature with respect to depth
      at a fixed time of the day.
55 (a) $\partial V/\partial x = -0.112y$ ml/yr is the rate at which lung
      capacity decreases with age for an adult male.
   (b) $\partial V/\partial y = 27.63 - 0.112x$ ml/cm is difficult to interpret
      because we usually think of adult height as fixed, or
      height y as a function of age x.

1 (a) $y = x^3 + c$ 3 (a) $y = \sqrt{4 - x^2} + c$
  (b) $y = x^3 + 2$ (b) $y = \sqrt{4 - x^2}$

11 $y = Ce^{2\sin x}$ 13 $y = Cx$
15 $x^3y^5e^y = C$, $C \neq 0$ and $y = 0$
17 $y = -1 + Ce^{x^2/2 - x}$
19 $y = -\frac{1}{3} \ln(3C + 3e^{-x})$
21 $y^2 = C(1 + x^3)^{-2/3} - 1$
23 $\cos x + x \sin x - \ln|\sin y| = C$
25 $\sec x + e^{-y} = C$ 27 $y^2 + \ln y = 3x - 8$
29 $y = \ln(2x + \ln x + e^2 - 2)$ 31 $y = 2e^{2 - \sqrt{4 + x^2}} - 1)$
33 $\tan^{-1} y - \ln|\sec x| = \pi/4$
35 $xy = k$; hyperbolas   37 $2x^2 + y^2 = k$; ellipses
39 $2x^2 + 3y^2 = k$; ellipses

1 $y = \frac{1}{4}e^{2x} + Ce^{-2x}$ 3 $y = \frac{1}{2}x^5 + Cx^3$
5 $y = (1/x)e^x - \frac{1}{2}x + (C/x)$ 7 $y = (e^x + C)/x^2$
9 $y = \frac{4}{3}x^3 \csc x + C \csc x$ 11 $y = 2 \sin x + C \cos x$
13 $y = x \sin x + Cx$ 15 $y = [\frac{1}{3}x + (C/x^2)]e^{-3x}$
17 $y = \frac{3}{2} + Ce^{-x^2}$ 19 $y = \frac{1}{2} \sin x + (C/\sin x)$
21 $y = \frac{1}{3} + (x + C)e^{-x^3}$ 23 $y = x(x + \ln x + 1)$
25 $y = e^{-x}(1 - x^{-1})$ 27 $Q = CE(1 - e^{-t/RC})$
29 $f(t) = \frac{80}{3}(1 - e^{-0.075t}) + Ke^{-0.075}$ lb
31 (a) $f(t) = M + (A - M)e^{k(1 - t)}$ (k a constant)
   (b) 28
33 $y = y_L(1 - ce^{-kt})$, $k > 0$
35 (b) $y = (I/k)(1 - e^{-kt})$ (c) 0.58 mg/min

1 $y = C_1e^{2x} + C_2e^{3x}$ 3 $y = C_1 + C_2e^{3x}$
5 $y = C_1e^{-2x} + C_2xe^{-2x}$ 7 $y = C_1e^{(2 + \sqrt{3})x} + C_2e^{(2 - \sqrt{3})x}$
9 $y = C_1e^{-\sqrt{2}x} + C_2xe^{-\sqrt{2}x}$ 11 $y = C_1e^{5x/4} + C_2e^{-3x/2}$
13 $y = C_1e^{4x/3} + C_2xe^{4x/3}$
15 $y = C_1e^{(2 + \sqrt{2})x/2} + C_2e^{(2 - \sqrt{2})x/2}$
17 $y = C_1e^x \cos x + C_2e^x \sin x$
19 $y = C_1e^{2x} \sin 3x + C_2e^{2x} \cos 3x$
21 $y = C_1e^{(-3 + \sqrt{7})x} + C_2e^{(-3 - \sqrt{7})x}$ 23 $y = 2e^{2x} - 2e^x$
25 $y = \cos x + 2 \sin x$ 27 $y = (2 + 9x)e^{-4x}$
29 $y = \frac{1}{2}e^x \sin 2x$

1 $y = C_1 \sin x + C_2 \cos x - \cos x \ln|\sec x + \tan x|$
3 $y = (C_1 + C_2x + \frac{1}{12}x^4)e^{3x}$
5 $y = C_1e^x + C_2e^{-x} + \frac{2}{5}e^x \sin x - \frac{1}{5}e^x \cos x$
7 $y = (C_1 + \frac{1}{6}x)e^{3x} + C_2e^{-3x}$ 9 $y = C_1e^{4x} + C_2e^{-x} - \frac{1}{2}$
11 $y = C_1e^x + C_2e^{2x} + \frac{2}{3}e^{-x}$
13 $y = C_1 + C_2e^{-2x} + \frac{1}{8} \sin 2x - \frac{1}{8} \cos 2x$
15 $y = C_1e^x + C_2e^{-x} + \frac{1}{9}(3x - 4)e^{2x}$
17 $y = e^{3x}(C_1 \cos 2x + C_2 \sin 2x) + \frac{1}{65}(7 \cos x - 4 \sin x)e^x$

1 $y = -\frac{1}{3} \cos 8t$ 3 $y = (\sqrt{2}/8)e^{-8t}(e^{4\sqrt{2}t} - e^{-4\sqrt{2}t})$
5 $y = \frac{1}{3}e^{-8t}(\sin 8t + \cos 8t)$
7 If m is the mass of the weight, then the spring constant is
   $24m$ and the damping force is $-4m(dy/dt)$. The motion is
   begun by releasing the weight from 2 ft above the
   equilibrium position with an initial velocity of 1 ft/sec in the
   upward direction.
9 $-6\sqrt{2}(dy/dt)$

1 $y = a_0 \sum_{n=0}^{\infty} \frac{(-1)^n}{(2n)!} x^{2n} + a_1 \sum_{n=0}^{\infty} \frac{(-1)^n}{(2n + 1)!} x^{2n + 1}$
   ($= a_0 \cos x + a_1 \sin x$)

3 $y = a_0\left[1 + \sum_{n=1}^{\infty} \frac{2^n(3n - 2)(3n - 5) \cdots 7 \cdot 4 \cdot 1}{(3n)!} x^{3n}\right]$

   $+ a_1\left[x + \sum_{n=1}^{\infty} \frac{2^n(3n - 1)(3n - 4) \cdots 8 \cdot 5 \cdot 2}{(3n + 1)!} x^{3n + 1}\right]$

5 $y = a_0(1 - x^2)$

$$+ a_1\left[x - \sum_{n=1}^{\infty} \frac{(2n-3)(2n-5)\cdots 5\cdot 3\cdot 1}{(2n+1)!}x^{2n+1}\right]$$

7 $y = a_0(x + 1)^3$

9 $y = -5x + a_0\sum_{n=0}^{\infty}\frac{x^n}{n!} + a_1\sum_{n=0}^{\infty}\frac{(-x)^n}{n!}$

$$= -5x + a_0 e^x + a_1 e^{-x}$$

11 $y = a_0\sum_{n=0}^{\infty}(n+1)x^{2n} + a_1\sum_{n=0}^{\infty}\left(\frac{2n+3}{3}\right)x^{2n+1}$

$$+ \sum_{n=1}^{\infty}(n+1)x^{2n}$$

EXERCISES 14.9, PAGE 617

1 $\{(x, y): 4x^2 - 9y^2 \le 36\}$　　**3** $\{(x, y): z^2 > x^2 + y^2\}$

5 $\frac{5}{4}$

7 $f_x(x, y) = 3x^2\cos y + 4;\ f_y(x, y) = -x^3\sin y - 2y$

9 $f_x(x, y, z) = 2x/(y^2 + z^2);$
$f_y(x, y, z) = 2y(z^2 - x^2)/(y^2 + z^2)^2;$
$f_z(x, y, z) = -(x^2 + y^2)2z/(y^2 + z^2)^2$

11 $f_x(x, y, z, t) = 2xz\sqrt{2y + t};\ f_y(x, y, z, t) = x^2 z/\sqrt{2y + t};$
$f_z(x, y, z, t) = x^2\sqrt{2y + t};\ f_t(x, y, z, t) = x^2 z/(2\sqrt{2y + t})$

13 $f_{xx}(x, y) = 6xy^2 + 12x^2,\ f_{yy}(x, y) = 2x^3 - 18xy,$
$f_{xy}(x, y) = 6x^2 y - 9y^2$

15 $\sin x - x\cos x + e^{-y} = C$　　**17** $y = \tan(\sqrt{1 - x^2} + C)$

19 $y = (2x - \cos x + C)/(\sec x + \tan x)$

21 $y = 2\sin x + C\cos x$　　**23** $\sqrt{1 - y^2} + \sin^{-1} x = C$

25 $\csc y = e^{-x} + C$　　**27** $y = Ce^{-2\sin x} + \frac{1}{2}$

29 $y = (C_1 + C_2 x)e^{4x}$　　**31** $y = C_1 + C_2 e^{2x}$

33 $y = C_1 e^x + C_2 e^{-x} - \frac{1}{5}e^x(\sin x + 2\cos x)$

35 $y = Ce^{-x} + \frac{1}{5}e^{4x}$　　**37** $y = C_1 e^x + C_2 e^{2x} + \frac{1}{12}e^{5x}$

39 $y = (x - 2)^3/(3x) + C/x$

41 $y = e^{-(5/2)x}[C_1\cos(\sqrt{3}x/2) + C_2\sin(\sqrt{3}x/2)]$

43 $e^x(\sin x - \cos x) + 2e^{-y} = C$　　**45** $y = \frac{1}{2}\cos x + C\sec x$

47 $y = (\ln|\sec x + \tan x| - x + C)/(\csc x - \cot x)$

49 $y = \frac{3500}{3}e^{-3/(2\pi)}\cos 2\pi t - \frac{2000}{3};\ 1214$

51 $f(t) = ab[e^{k(b-a)t} - 1]/[be^{k(b-a)t} - a]$

53 $y\,dy + x\,dx = 0;$ a circle with center at the origin.